# DNA Tumor Viruses

# CANCER CELLS

## COLD SPRING HARBOR LABORATORY
**1986**

# 4
# DNA Tumor Viruses

## Control of Gene Expression and Replication

Edited by
**Michael Botchan**
University of California, Berkeley

**Terri Grodzicker**
Cold Spring Harbor Laboratory

**Phillip A. Sharp**
Massachusetts Institute of Technology

**CANCER CELLS**
1 / The Transformed Phenotype
2 / Oncogenes and Viral Genes
3 / Growth Factors and Transformation
4 / DNA Tumor Viruses: Control of Gene Expression and Replication

**DNA Tumor Viruses: Control of Gene Expression and Replication**
© 1986 by Cold Spring Harbor Laboratory
Printed in the United States of America
Cover design by Emily Harste

**Library of Congress Cataloging-in-Publication Data**

DNA tumor viruses.

(Cancer cells, ISSN 0743-2194 : 4)
Includes index.
1. Oncogenic viruses—Reproduction. 2. Viruses, DNA—
Reproduction. 3. Gene expression. 4. Viral carcino-
genesis—Genetic aspects. I. Botchan, Michael.
II. Grodzicker, Terri. III. Sharp, Phillip A.
[DNLM: 1. DNA Replication. 2. DNA Tumor Viruses.
3. Gene Expression Regulation. 4. Virus Replication.
W1 CA677BG v.4 / QW 166 D6288]
QR372.06D59 1986      616.99'4071      86-50649
ISBN 0-87969-192-1

*Cover*: (*Top*) A critical step in the proposed scheme of pre-mRNA splicing. The 5′ exon (Exon 1) is joined by a 3′–5′ phosphodiester bond to Exon 2, releasing the lariat structure. Dots indicate sequence complementarity. See P.A. Sharp (this volume) for details. (*Bottom*) Electron micrograph showing R-loops formed by annealing of the r strand of the *Bam*HI-B DNA restriction fragment to the 5′ end of Ad2 mRNA for the hexon protein. (Reprinted, with permission, from T.R. Broker, 1978, *Cold Spring Harbor Symp. Quant. Biol.* **42**: 548.)

All Cold Spring Harbor Laboratory publications may be or-dered directly from Cold Spring Harbor Laboratory, Box 100, Cold Spring Harbor, New York 11724. (Phone: 1-800-843-4388). In New York (516) 367-8423.

# Conference Participants

**Acheson, Nicholas,** Department of Microbiology and Immunology, McGill University, Montreal, Canada

**Albrecht, Glenn,** LGME, University of Strasbourg, France

**Ambinder, Richard,** Department of Pharmacology, Johns Hopkins University, Baltimore, Maryland

**Anderson, Carl,** Department of Biology, Brookhaven National Laboratory, Upton, New York

**Arsenakis, Minas,** Kovler Laboratories, University of Chicago, Illinois

**Atchinson, Michael,** Fox Chase Cancer Center, Philadelphia, Pennsylvania

**Baim, Steven,** Department of Molecular Biology, Princeton University, New Jersey

**Basilico, Claudio,** Department of Pathology, New York University School of Medicine, New York

**Battula, Nara,** NCI, National Institutes of Health, Bethesda, Maryland

**Benjamin, Thomas,** Department of Pathology, Harvard Medical School, Boston, Massachusetts

**Botchen, Michael,** Department of Molecular Biology, University of California, Berkeley

**Bradac, James,** Department of Pharmacology, Johns Hopkins School of Medicine, Baltimore, Maryland

**Brady, John,** NCI, National Institutes of Health, Bethesda, Maryland

**Broker, Thomas,** Department of Biochemistry, University of Rochester School of Medicine, New York

**Brown, Maryanne,** Department of Molecular Pharmacology, Albert Einstein College of Medicine, Bronx, New York

**Campadelli-Fiume, G.,** Department of Microbiology and Virology, University of Bologna, Italy

**Cashion, Linda,** Cordon Corp., Brisbane, California

**Cassill, Aaron,** University of California, San Diego

**Chambon, Pierre,** Institute of Chemical Biology, University of Strasbourg, France

**Chang, L.-S.,** Department of Molecular Biology, Princeton University, New Jersey

**Cheng, Seng,** Integrated Genetics, Framingham, Massachusetts

**Chiang Yawen,** Cetus Corp., Emeryville, California

**Chou, Chen-Kung,** Department of Medical Research, Veterans General Hospital, Taipei, Taiwan

**Chow, Louise,** Department of Biochemistry, University of Rochester School of Medicine, New York

**Christie, Karen,** University of Tromsø, Norway

**Clancy, Suzanne,** Department of Genetics, Stanford University, California

**Cowie, Allison,** Genetics Institute, Cambridge, Massachusetts

**Cullen, Bryan,** Hoffmann-La Roche Inc., Nutley, New Jersey

**Cuzin, Francoise,** Biochemistry Center, University of Nice, France

**Dalie, Barbara,** Department of Microbiology, UMDNJ-New Jersey Medical School, Newark

**DePamphilis, Melvin,** Department of Biological Chemistry, Harvard Medical School, Boston Massachusetts

**Deyerle, Karen,** Department of Biology, University of California, San Diego

**Dubrewski, Christine,** University of Pennsylvania, Philadelphia

**Dulbecco, Renato,** Salk Institute, La Jolla, California

**Dunn, Ashley,** Ludwig Institute for Cancer Research, Royal Melbourne Hospital, Australia

**Falck-Pedersen, Eric,** Rockefeller University, New York, New York

**Fanning, Ellen,** Department of Biochemistry, Ludwig Maximillian University, Munich, Federal Republic of Germany

**Feunteun, Jean,** Institut Gustave Roussy, Villejuif, France

**Finney, David,** Department of Biological Sciences, Carnegie-Mellon University, Pittsburgh, Pennsylvania

**Frenke, Niza,** University of Chicago, Illinois

**Frey, Alan B.,** Department of Molecular Biology, Princeton University, New Jersey

**Fried, Michael,** Imperial Cancer Research Fund, London, England

**Fu, Xin-Yuan,** Department of Biological Sciences, Columbia University, New York, New York

**Gallimore, Phillip H.,** Department of Cancer Studies, University of Birmingham, England

**Gallo, Gregory,** University of Pennsylvania, Philadelphia

**Galloway, Denise,** Fred Hutchinson Cancer Research Center, Seattle, Washington

**Garramone, Anthony,** Integrated Genetics, Framingham, Massachusetts

**Gesteland, Raymond,** University of Utah, Salt Lake City

**Gilbert, James,** Cetus Corp., Emeryville, California

**Gluzman, Yakov,** Cold Spring Harbor Laboratory, New York

**Goding, C.R.,** Marie Curie Memorial Foundation Research Institute, Surrey, England

**Grodzicker, Terri,** Cold Spring Harbor Laboratory, New York

**Haigwood-Scandella, Nancy,** Chiron Corp., Emeryville, California

**Haley, Kevin,** Department of Biochemistry, Dartmouth Medical School, Hanover, New Hampshire

**Hanahan, Douglas,** Cold Spring Harbor Laboratory, New York

**Hansen, Ulla,** Dana-Farber Cancer Institute, Boston, Massachusetts

**Harlow, Edward,** Cold Spring Harbor Laboratory, New York

**Harter, Nikki,** Department of Microbiology, UMDNJ-New Jersey Medical School, Newark

**Hassell, John A.,** Department of Microbiology and Immunology, McGill University, Montreal, Canada

**Hayward, Gary,** Department of Pharmacology, Johns Hopkins School of Medicine, Baltimore, Maryland

**Herr, Winship,** Cold Spring Harbor Laboratory, New York

**Hirt, Bernhard,** Swiss Institute for Cancer Research, Lausanne

**Hoeffler, Warren,** Rockefeller University, New York, New York

**Howley, Peter,** NCI, National Institutes of Health, Bethesda, Maryland

**Huang, Pearl,** Department of Molecular Biology, Princeton University, New Jersey

**Ikeda, Joh-E.,** Department of Biochemistry, University of Rochester School of Medicine, New York

**Imperiale, Michael,** Department of Microbiology and Immunology, University of Michigan Medical School, Ann Arbor

**Ito, Yoshiaki,** NCI-Frederick Cancer Research Facility, Frederick, Maryland

**Jat, Parmjit,** Massachusetts Institute of Technology, Cambridge

**Kalyan, Narender,** Wyeth Laboratories Inc., Philadelphia, Pennsylvania

**Kaplan, Paul,** Salk Institute, San Diego, California

**Karger, Brian,** Department of Biological Sciences, Carnegie-Mellon University, Pittsburgh, Pennsylvania

**Israel, David,** Genetics Institute, Cambridge, Massachusetts

**Kedinger, Claude,** Institute of Chemical Biology, University of Strasbourg, France

**Kelly, F.,** CNRS, Institute for Scientific Research for Cancer, Villejuif, France

**Kelly, Thomas, Jr.,** Department of Molecular Biology and Genetics, Johns Hopkins University School of Medicine, Baltimore, Maryland

**Klessig, Daniel,** Department of Cellular, Viral, and Molecular Biology, University of Utah, Salt Lake City

**Kristie, Thomas,** University of Chicago, Illinois

**Kwong, Ann D.,** University of Chicago, Illinois

**Lane, David,** Department of Biochemistry, Imperial College, London, England

**Larsen, Pamela,** Department of Microbiology, Vanderbilt University School of Medicine, Nashville, Tennessee

**Lewis, James,** Fred Hutchinson Cancer Research Center, Seattle, Washington

**Li, Joachim,** Department of Molecular Biology and Genetics, Johns Hopkins University School of Medicine, Baltimore, Maryland

**Lieberman, Paul,** Department of Pharmacology, Johns Hopkins University School of Medicine, Baltimore, Maryland

**Lillie, James,** Department of Biochemistry and Molecular Biology, Harvard University, Cambridge, Massachusetts

**Lindenbaum, Jeff,** Memorial Sloan-Kettering Cancer Research Center, New York, New York

**Lindgren, Valerie,** Department of Human Genetics, Yale University, New Haven, Connecticut

**Livingston, David M.,** Dana-Farber Cancer Institute, Boston, Massachusetts

**Louis, M. John,** NCI, National Institutes of Health, Bethesda, Maryland

**Lowy, Douglas,** NCI, National Institutes of Health, Bethesda, Maryland

**Lupton, Stephen,** Department of Molecular Biology, Princeton University, New Jersey

**Manley, James,** Columbia University, New York, New York

**Manos, Michele,** Cetus Corp., Emeryville, California

**Markland, William,** Integrated Genetics, Framingham, Massachusetts

**Masanobu, Satake,** NCI-Frederick Cancer Research Facility, Frederick, Maryland

**Mathews, Michael,** Cold Spring Harbor Laboratory, New York

**May, Evelyne,** Department of Molecular Oncology, IRSC, Villejuif, France

**McConlogue, Lisa,** Cetus Corp., Emeryville, California

**McKnight, Jennifer,** University of Chicago, Illinois

**Melli, Marialuisa,** Sclavo Research Center, Siena, Italy

**Michaeli, Tamr,** Columbia University, New York, New York

**Miller, George,** Yale University School of Medicine, New Haven, Connecticut

**Molnar-Kimber, Katherine,** Wyeth Laboratories Inc., Philadelphia, Pennsylvania

**Morin, John E.,** Wyeth Laboratories Inc., Philadelphia, Pennsylvania

**Mosca, Joseph,** Oncology Center, Johns Hopkins School of Medicine, Baltimore, Maryland

**Muller, William,** McGill University, Montreal, Canada

**Murakami, Yasufumi,** Memorial Sloan-Kettering Cancer Research Center, New York, New York

**Nakata Kotoko,** NCI-Frederick Cancer Research Facility, Frederick, Maryland

**Naruto, Masanobu,** Department of Molecular Biology, University of California, Berkeley

**Natarajan, Venkatachala,** National Institutes of Health, Bethesda, Maryland

**Neufeld, David,** Hunter College, New York, New York

**Nevins, Joseph,** Rockefeller University, New York, New York

**O'Neill, Edward,** Department of Molecular Biology and Genetics, Johns Hopkins University School of Medicine, Baltimore, Maryland

**Oka, Takami,** National Institutes of Health, Bethesda, Maryland

**Ozer, Harvey,** Hunter College, New York, New York

**Parker, Ron,** Department of Biological Sciences, Columbia University, New York, New York

**Pater, Mary M.,** Department of Medicine, Memorial University of Newfoundland, St. Johns, Canada

**Pati, Uttam,** Department of Genetics, Yale University, New Haven, Connecticut

**Perkins, Karin,** Memorial Sloan-Kettering Cancer Research Center, New York, New York

**Pettersson, Ulf,** Department of Medical Genetics, Uppsala University, Sweden

**Prives, Carol,** Department of Biological Sciences, Columbia University, New York, New York

**Qasba, Pradman,** NCI, National Institutes of Health, Bethesda, Maryland

**Radna, Rachel,** Hunter College, New York, New York

**Ralston, Robert,** University of California, San Francisco

**Rawlins, Daniel,** Department of Microbiology and Immunology, Emory University School of Medicine, Atlanta, Georgia

**Reddel, Roger,** NCI, National Institutes of Health, Bethesda, Maryland

**Reddy, E.S.P.,** NCI, National Institutes of Health, Bethesda, Maryland

**Rekosh, David,** Department of Biochemistry, State University of New York, Buffalo

**Resnick, James,** Department of Molecular Biology, Princeton University, New Jersey

**Roberts, James,** Fred Hutchinson Cancer Center, Seattle, Washington

**Robinson, Robin,** Department of Microbiology, University of Texas Health Science Center, Dallas

**Roe, Grace,** University of Wisconsin, Madison

**Roizman, Bernard,** Viral Oncology Laboratory, University of Chicago, Illinois

**Rosa, Margaret,** Biogen Research Corp., Cambridge, Massachusetts

**Rosenfeld, Philip,** Department of Molecular Biology and Genetics, Johns Hopkins University School of Medicine, Baltimore, Maryland

**Ross, David,** Department of Biochemistry, New York University Medical Center, New York

**Rossini, Mara,** Sclavo Research Center, Siena, Italy

**Ruley, Earl,** Massachusetts Institute of Technology, Cambridge

**Rundell, Kathleen,** Department of Microbiology, Northwestern University, Chicago, Illinois

**Rutila, Joan,** Department of Medical Science, University of Michigan, Ann Arbor

**Sadasiv, Eileen,** Department of Animal Veterinary Sciences, University of Rhode Island, Kingston

**Sambrook, Joseph,** Cold Spring Harbor Laboratory, New York

**Samulski, Richard,** Department of Molecular Biology, Princeton University, New Jersey

**Sawadogo, Michele,** Rockefeller University, New York, New York

**Schaack, Jerome,** Department of Molecular Biology, Princeton University, New Jersey

**Schiller, John T.,** NCI, National Institutes of Health, Bethesda, Maryland

**Schlegel, Richard,** NCI, National Institutes of Health, Bethesda, Maryland

**Schmitt, Robert,** Fred Hutchinson Cancer Research Center, Seattle, Washington

**Schwarz, Elisabeth,** Department of Virus Research, German Cancer Research Center, Heidelberg, Federal Republic of Germany

**Scott, Walter A.,** Department of Biochemistry, University of Miami, Florida

**Senger, Donald,** Department of Pathology, Beth Israel Hospital, Boston, Massachusetts

**Sharp, Phillip,** Massachusetts Institute of Technology, Cambridge

**Shenk, Thomas,** Department of Molecular Biology, Princeton University, New Jersey

**Silver, Sandra,** Department of Microbiology, State University of New York, Stony Brook

**Sinha, A.M.,** American Cyanamid Co., Pearl River, New York

**Sippel, Albrecht,** Center for Molecular Biology, University of Heidelberg, Federal Republic of Germany

**Smith, C.A. Dale,** NCI, National Institutes of Health, Bethesda, Maryland

**Smith, Alan E.,** Integrated Genetics, Framingham, Massachusetts

**Solnick, David,** Department of Molecular Biophysics and Biochemistry, Yale University Medical School, New Haven, Connecticut

**Spalholz, B.A.,** NCI, National Institutes of Health, Bethesda, Maryland

**Spandidos, Demetrios,** Beatson Institute for Cancer Research, Glasgow, Scotland

**Sprangler, Rudy,** Department of Microbiology, UMDNJ-New Jersey Medical School, Newark

**Stacy, Terry,** Department of Biochemistry, Dartmouth Medical School, Hanover, New Hampshire

**Stillman, Bruce,** Cold Spring Harbor Laboratory, New York

**Stow, Nigel,** Institute of Virology, Glasgow, Scotland

**Strauss, Phyllis,** Department of Biology, Northeastern University, Boston, Massachusetts

**Sugden, William,** University of Wisconsin, Madison

**Sunstrom, Noelle A.,** Department of Microbiology and Immunology, McGill University, Montreal, Canada

**Tate, Peri,** Montreal, Canada

**Tjian, Robert,** Department of Biochemistry, University of California, Berkeley

**Vaessen, Ruud,** Sylvius Laboratories, Leiden, The Netherlands

**Vasudevachari, M.B.,** National Institutes of Health, Bethesda, Maryland

**Wang, Lotte,** Bristol-Myers, Co., Syracuse, New York

**Weinberg, David,** Department of Molecular Biology and Genetics, Johns Hopkins University School of Medicine, Baltimore, Maryland

**Westphal, Heiner,** National Institutes of Health, Bethesda, Maryland

**White, Kristin,** American Cancer Society, New York, New York

**Williams, James,** Department of Biochemical Sciences, Princeton University, New Jersey

**Winnacker, Ernest,** Department of Biochemistry, University of Munich, Federal Republic of Germany

**Winocour, Ernest,** Department of Virology, Weizmann Institute of Science, Rehovot, Israel

**Wishart, William,** Department of Molecular Biology, Princeton University, New Jersey

**Wold, Marc S.,** Department of Molecular Biology and Genetics, Johns Hopkins University School of Medicine, Baltimore, Maryland

**Wolowodiuk, Vera,** Waksman Institute of Microbiology, Piscataway, New Jersey

**Woodworth-Gutai, Mary,** Department of Cell and Tumor Biology, Roswell Park Memorial Institute, Buffalo, New York

**Wu, Kun Chi,** La Jolla Cancer Research Foundation, California

**Wurm, Florian,** Department of Molecular Biology, Massachusetts General Hospital, Boston

**Yaniv, Moshe,** Department of Molecular Biology, Pasteur Institute, Paris, France

**Yu Xiang-Ming,** University of Wisconsin, Madison

**Zajchowski, Deborah,** University of Strasbourg, France

**Ziff, Edward,** Department of Biochemistry, New York University School of Medicine, New York

*First row*: B. Hirt; J. Hassell, P. Chambon; D. Galloway
*Second row*: M. Yaniv; U. Pettersson, D. Klessig; B. Sugden
*Third row*: L. Chow; T. Grodzicker, B. Stillman, M. Botchan, Y. Gluzman; T. Broker
*Fourth row*: E. Winocour; C. Anderson, R. Gesteland; M. Fried, A. Smith

# Preface

The DNA tumor viruses have long served as model systems for the study of eukaryotic gene expression, DNA replication, and transformation. The enormous growth of this field in the last several years has led to the occurrence of specialized meetings covering the work on one or two viruses. However, the Third Cold Spring Harbor Meeting on Cancer Cells took a broader view, bringing together investigators who presented their latest findings in a variety of areas. The studies discussed at the meeting and collected in this volume encompass the molecular biology and biochemistry of SV40, polyomavirus, adenoviruses, papillomaviruses, herpes simplex virus, and Epstein-Barr virus.

To enhance this volume of *Cancer Cells*, several leading investigators were asked to write introductory articles covering the history, current developments, and possible future of research into these DNA tumor viruses. The authors of the introductions were charged to (1) provide personal, anecdotal histories of their fields; (2) update the 1980 edition of *DNA Tumor Viruses* edited by John Tooze; and (3) chart the future of research in their areas. We thank these colleagues not only for accepting this challenge, but also for the extra effort required to write the resulting insightful overviews.

A great deal of progress is reported herein in the development of in vitro systems that are now being used to dissect crucial processes such as DNA replication; RNA transcription and the interaction of proteins with promoter and enhancer elements; RNA processing, transcription termination, and poly(A) addition; and control of translation. The roles of viral *trans*-activating regulatory proteins such as SV40 T antigen, adenovirus E1A proteins, herpes virus immediate early proteins, and the bovine papillomavirus E2 protein are emphasized. These proteins affect the transcriptional activity of viral and in some cases cellular genes and, at least for SV40 and adenoviruses, are also transforming proteins.

During the Cancer Cells Meeting the new Sambrook Laboratory, which adjoins James Laboratory, was dedicated. This laboratory is named for Joe Sambrook, who over the years had directed much of the research on DNA tumor viruses at Cold Spring Harbor. The dedicatory remarks were given by Jim Watson; Mike Botchan provided remembrances of life in James Lab during Joe's tenure; and Renato Dulbecco discussed the future of molecular biology, in particular the now possible but still daunting task of sequencing the human genome. The meeting also served as a reunion for many of the scientists who had worked at Cold Spring Harbor and returned to present their latest findings. Appropriately, the occasion marked the end of Joe's productive term as Assistant Director of Research at Cold Spring Harbor; he has now taken up the chairmanship of the Biochemistry Department at the University of Texas Health Sciences Center at Dallas.

Our thanks for helping to organize and run the meeting goes, as usual, to numerous Cold Spring Harbor staff members. In particular, the Meetings Office under the direction of Gladys Kist, who has since retired, performed to its consistently high standard; the James Lab secretary, Marilyn Goodwin, was indispensable in helping the organizers coordinate the plans of all the scientists attending the meeting; Herb Parsons and his co-workers once again orchestrated the audiovisuals with care and precision; Dave Micklos and Susan Cooper organized the dedication of the Sambrook Laboratory; and LIBA's continued generosity to the Laboratory made the building of the Sambrook addition possible.

We also thank the Publications Department, in particular, Douglas Owen and Joan Ebert, for their persistence in bringing *Cancer Cells 4* to press. We are also grateful to Emily Harste for the stunning design of the cover.

**The Editors**

J.D. Watson addressing audience

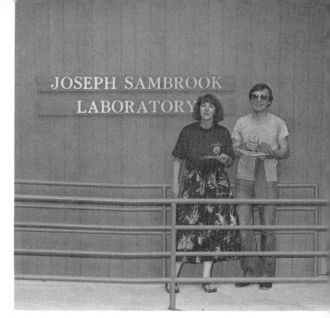

M.-J. Gething and J. Sambrook

# The Joseph Sambrook Laboratory
## Dedication Ceremony
### *September 5, 1985*

R. Dulbecco

West view

J. Sambrook

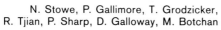

N. Stowe, P. Gallimore, T. Grodzicker,
R. Tjian, P. Sharp, D. Galloway, M. Botchan

East view

# Contents

## REGULATION OF TRANSCRIPTION

## RNA PROCESSING AND TRANSLATION

## TRANSFORMATION

## TRANSFORMING PROTEINS

## REPLICATION

# The Biology of Simian Virus 40 and Polyomavirus

## M. Fried* and C. Prives[†]

*Imperial Cancer Research Fund, Lincoln's Inn Fields, London WC2A 3PX, Great Britain; [†]Department of Biological Sciences, Columbia University, New York, New York 10027

The two small DNA tumor viruses polyomavirus and simian virus 40 (SV40) can undergo two types of interactions with susceptible cells. One type of interaction, the lytic cycle, leads to the production of new infectious virus particles and eventual cell death. The polyomavirus lytic cycle occurs in permissive mouse cells, and the SV40 lytic cycle takes place in permissive monkey cells. The other type of interaction leads to a neoplastically transformed cell that is able to continue to divide under conditions that inhibit the multiplication of normal, untransformed cells. In vitro, both polyomavirus and SV40 can transform a variety of rodent cells and, in addition, SV40 can transform primate cells. The cells transformed in vitro will induce tumors when transplanted into susceptible animals. Under specified conditions, both viruses can also induce tumor formation in a variety of rodents upon injection of large amounts of virus.

The major impetus for the extensive study of polyomavirus and SV40 was initially their oncogenic potential. It soon became clear that because these viruses are very easy to grow in tissue culture and are easily purified and manipulated and because their DNA genomes are relatively small (5 kb), the viruses are ideal for laboratory study. Thus many workers, mainly molecular biologists, began to utilize these viruses as simple model systems for the more complex mammalian genome. Soon there were a large number of groups using polyomavirus and SV40 for studies on transcription and RNA processing, DNA replication, gene regulation, DNA structure, and DNA-protein interactions as well as neoplastic transformation. Major basic findings on differential splicing, origins of DNA replication, supercoiled DNA, promoter and enhancer elements involved in gene expression, DNA-protein interactions, proteins that positively and negatively regulate gene expression as well as many other facets of the molecular biology of mammalian cells were first derived from the concerted study of these small DNA tumor viruses.

The tumorigenic capacity of polyomavirus was first identified, by Ludwik Gross, as an ability to induce tumors in newborn mice (see Gross 1983). The initial study of polyomavirus was confused by the fact that the filterable extract used by Gross also contained mouse leukemia virus. Polyomavirus was eventually separated by the selective heat inactivation of the more labile leukemia virus. Sarah Stewart and Bernice Eddy were able to easily propagate the isolated polyomavirus in mouse embryo cells in tissue culture and show that the virus could cause tumor formation in a variety of different organs—hence the name polyomavirus (for reviews, see Eddy 1960; Stewart 1960). Plaque assays were soon established to quantitate virus growth (Dulbecco and Freeman 1959; Winocour and Sachs 1959). Vogt and Dulbecco (1960) were the first to show that polyomavirus could induce transformation of rodent embryo cells in vitro. The cells transformed in vitro were indistinguishable from tumor cells induced in vivo and gave rise to tumors when inoculated into susceptible animals. In vitro transformation assays using the BHK continuous cell line were soon established (Stoker and MacPherson 1961; MacPherson and Montagnier 1964).

SV40 was first identified by Sweet and Hilleman (1960) as a noncytopathic virus present in Rhesus monkey cells that produced a strong cytopathic effect when grown on African green monkey cells. It soon became evident that many batches of polio vaccine, prepared in Rhesus monkey cells, were contaminated with infectious SV40. Although large amounts of SV40 will induce tumor formation when inoculated into newborn baby hamsters (Eddy et al. 1962; Girardi et al. 1962), and although the virus can transform a variety of cell types, including human cells, in tissue culture, after 25 years of study no adverse effects attributable to SV40 have been detected in people who received the contaminated vaccine preparations. Quantitative cell culture systems for studying the SV40 lytic interaction in monkey cells and transforming activity in rodent cells were quickly developed (e.g., Todaro and Green 1964; Black 1966). The initial rationale that SV40 could serve as a model for a human tumor virus—that it was capable of transforming human cells in vitro—encouraged a number of workers to initiate studies of this virus. The availability of continuous monkey cell lines (e.g., CV-1, BSC-1, VERO) for its lytic growth and the continuous 3T3 mouse cell line for assaying its transforming ability allowed SV40 to be studied with relative ease and was responsible for initiating research in many small as well as large laboratories.

The big boon to the early researchers working with polyomavirus and SV40 was to be able to study the viruses under reproducible and controllable conditions in vitro instead of having to use unreliable and time-consuming animal assays. Yet today it is hard for researchers who have recently entered the field to imagine that these viruses were actually grown in mammalian cells in tissue culture and not as recombinant DNA molecules in bacteria. With the great influx of workers, especially those studying SV40, the small DNA tumor virus field rapidly expanded. Initially, new findings for one virus also appeared to be true for the other virus.

**1**

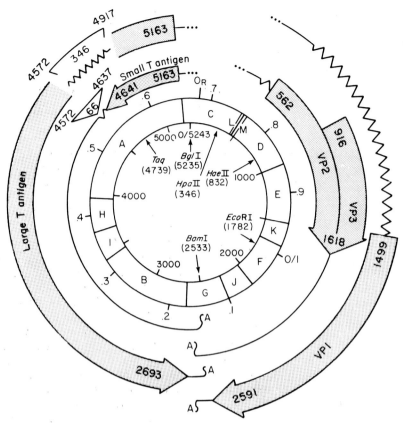

**Figure 1** Schematic representation of the SV40 genome, indicating locations of some principal features. The genome is represented as a circle with the origin of replication $(O_R)$ at the top. Numbers outside the circle refer to locations in fractional genome length relative to the single *Eco*RI cleavage site within SV40 DNA. Residue numbers and labels refer to the numbers used in the SV40 sequence in Appendix A of Tooze (1981). Sequence numbers given for codons include termination codons. Shaded arcs with arrowheads indicate coding positions of mRNAs, with the arrowheads pointing in the 5' to 3' direction. Outer boxed areas indicate continuous coding segments within the viral mRNA. (Redrawn from original material provided by A.R. Buchman, L. Burnett, and P. Berg. Reprinted, with permission, from Tooze 1981.)

Polyomavirus and SV40 were thought to have their genetic information organized in a nearly identical fashion, mainly differing in their primary sequence (e.g., see review by Fried and Griffin 1977). Indeed, SV40 and polyomavirus were the first eukaryotic genomes to be entirely sequenced (Fiers et al. 1978; Reddy et al. 1978; Deininger et al. 1979; Friedmann et al. 1979; Soeda et al. 1980). Figures 1 and 2 are maps of the SV40 and polyomavirus genomes, indicating the relative locations of the early, late, and regulatory regions. From the mid-1970s it was clear from genetic and structural analyses that the two viruses, although similar in a number of respects, had in some cases chosen quite different strategies for their gene expression. This is best exemplified by SV40 encoding one protein (large T antigen [TAg]) for use in both viral DNA replication and virally induced transformation, whereas polyomavirus encodes a large TAg as its DNA replication protein and a second protein, middle TAg, for transformation. In addition, polyomavirus and SV40 show some striking differences in their regulatory regions (i.e., the promoter, enhancer, origin of DNA replication, and TAg-binding regions). The study of other human and rodent papovaviruses has so far only shown slight variations of the polyomavirus and SV40 strategies for gene expression. The genomes of polyomavirus and SV40 have fulfilled early expectations and served as excellent models for the mammalian cell genome.

**Polyomavirus and SV40 in the wild**

Neither polyomavirus nor SV40 appear to behave as tumor viruses or cause disease in their natural host species in the wild. Wild mice trapped in some (but not all) apartment buildings in Harlem in New York City as well as on farms and in grain mills in Maryland, Indiana, Georgia, and Florida were found to be infected with polyomavirus but showed no pathological symptoms of virus infection (Rowe et al. 1961; Huebner 1963). More worrying for the researcher was that many laboratory mice colonies were also infected with polyomavirus. The original isolates of polyomavirus in tumor-bearing mice were most likely made as the result of its passenger status due to persistent infection rather than its cancer-inducing capacity. Although the presence of SV40 in its simian host has not been well studied, no definite symptoms of disease have been documented in the carrier rhesus monkeys or in other related monkey species.

It is not surprising that polyomavirus and SV40 do not

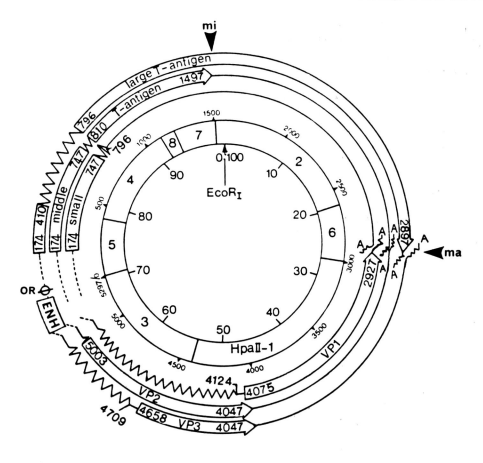

**5297 bp**

**Figure 2** The polyomavirus map of the A-2 strain (5297 bp). The early and late region mRNAs (thin lines) and their coding regions (boxed regions) are shown relative to the HpaII physical map. Map units are shown inside and nucleotide numbers outside the physical map. The nucleotide numbers at the starts and ends of the coding regions and at the splice donors and acceptors are shown. Introns are indicated by jagged lines. The positions of the origin of DNA replication (OR), the enhancer region (ENH), and the major (ma) and minor (mi) early region polyadenylation sites are shown. (Modified from Tooze 1981 and reprinted from Fried et al., this volume.)

appear to be tumorigenic in the wild, since under laboratory conditions tumors are induced only after large amounts of virus are innoculated into either immunologically immature (newborn) or immunologically deficient animals. This appears to be due to the immune rejection response that immunologically competent animals mount against the tumor-specific transplantation antigen (TSTA) induced by these viruses in infected and transformed cells. The polyomavirus TSTA was first detected by Habel (1961) and Sjogren et al. (1961), who found that animals immunized with live polyomavirus were resistant to challenge with polyoma transplantable tumor cells. A similar but distinct TSTA was also found in SV40-infected and -transformed cells (Defendi 1963; Habel and Eddy 1963; Khera et al. 1963; Koch and Sabin 1963). An immunologically related TSTA is produced by the same virus in cells of different species, but there is no cross-reaction between the polyomavirus and SV40 TSTAs (Sjogren 1965). The TSTAs appear to be on the cell surface, and tumor rejection is mediated by sensitized immune lymphocytes (Tevethia and Rapp

1966; Coggin et al. 1967; Tevethia 1980, 1983). One of the major unsolved problems in the field is to determine the nature of these TSTAs. It has been found that SV40 TSTA is associated with large TAg and the polyomavirus TSTA is associated with middle and/or large TAg. It is not clear whether the TSTAs are due entirely to variants of these virally specified proteins found on the cell surface (Chang et al. 1979; Tevethia et al. 1980; Flyer and Tevethia 1982) or are in part host-coded cell-surface proteins induced as a result of the recently recognized property of the SV40 large TAg and polyomavirus middle TAg to activate cellular genes (Ide et al. 1977; Schutzbank et al. 1982; Learned et al. 1983; Scott et al. 1983; Majello et al. 1985; Singh et al. 1985).

**Viral proteins**

The major effort toward characterizing the papovavirus proteins has focused on the structure and function of the early region proteins, the viral TAg's. The identification of the early viral proteins through examination of the DNA sequence, the analysis of spliced RNA, the in vitro translation of viral mRNA, and the utilization of cell fractionation and immunoprecipitation techniques led to the identification of two early SV40 proteins and of three early polyomavirus proteins. It has become clear that although both viruses encode rather similar (structurally and functionally) large and small TAg's, the membrane-associated middle TAg is unique to polyomavirus. The SV40 early region contains an extensive open reading frame (ORF) near the carboxyl terminus of the large TAg that could result in an alternative and rather hydrophobic carboxyterminal early region—encoded protein. However, attempts to identify such a product in infected or transformed cells have not been successful. Generation of frameshift mutants that utilize the alternative reading frame yielded a large TAg that was unable to either replicate viral DNA or transform cells (Lewis et al. 1983; Tornow and Cole 1983b). However, the existence of the ORF in both SV40 and BK primate papovaviruses is intriguing and may eventually lead to further insights. The viral small TAg's of SV40 and polyomavirus remain the most elusive in studies of their function. Both SV40 (Ellman et al. 1984) and polyomavirus (Zhu et al. 1984) small TAg's are found in nuclear and cytoplasmic cellular compartments. The SV40 small TAg induces the dissolution of actin cables (Bikel et al. 1986 and references therein) and affects other cellular structures, such as centrioles (Shyamala et al. 1982), and binds to host proteins (Yang et al. 1979; Rundell et al. 1981; Rundell 1982).

The viral large TAg's have been studied extensively, and it can probably be claimed that the SV40 large TAg is the best characterized eukaryotic nuclear regulatory protein. The initial characterization of its specific viral DNA-binding properties has been refined by using highly purified, authentic TAg and a combination of DNase footprinting and alkylation interference procedures (De-Lucia et al. 1983; Jones and Tjian 1984). The consensus sequence, pentanucleotide 5′ G(A > G)GGC 3′ is most likely the major determinant in sequence-specific DNA binding by large TAg. However, other factors, such as spacing between pentanucleotides and neighboring sequences (Ryder et al. 1985; Scheller and Prives 1985), also contribute to the binding affinity of the SV40 and polyomavirus TAg's for DNA. The role of the ATPase function of TAg (Clark et al. 1984) is not yet well understood other than that its importance is most likely relevant to the role of TAg in DNA replication and not cell transformation (Manos and Gluzman 1984). The multifunctional nature of the SV40 TAg has been established in studies using conditional (Cosman and Tevethia 1981; Pintel et al. 1981) and nonconditional (Tornow and Cole 1983a; Manos and Gluzman 1985; Rutila et al. 1986 and references therein) mutants. Because its multiplicity of functions—in viral DNA replication, autoregulation, cell immortalization and transformation, host range for replication, helping adenovirus grow in monkey cells, and activation of viral and cellular genes—suggests a rather complicated mode of action, several studies have been undertaken to identify functional domains. Experiments using naturally occurring deletion mutants (Prives et al. 1982) and deletion and point

mutants generated by in vitro mutagenesis (Paucha et al. 1986 and references therein) have localized sequences required for sequence-specific binding of the origin of DNA replication (*ori*) to the region between amino acids 139 and 220. Simmons (1986), using proteolytic fragments, has provided confirmatory evidence for the region of the SV40 TAg involved in *ori* binding. These approaches have opened the way for additional TAg domain analyses. Sequences required for ATPase function are most likely located in a region more toward the carboxyterminal portion of the SV40 large TAg relative to the *ori*-binding domain (Clark et al. 1983; Manos and Gluzman 1984). An additional property of the SV40 TAg, namely, its ability to bind adenylated nucleotides (Bradley et al. 1984), is of interest and may relate to its function in the initiation of viral DNA replication. The ability of the SV40 large TAg to bind to the host p53 protein (for review, see Klein 1982) is of considerable interest because of the possibility that this interaction is related to the SV40 large TAg transforming function. Indeed, this appears to be one property that the structurally and functionally homologous polyomavirus large TAg lacks. Since the SV40 but not the polyomavirus large TAg can transform cells, investigation of the mode by which the SV40 large TAg can bind to and stabilize p53 should be worthy of serious effort, and the role of this interaction in altering the growth properties of cells is expected to provide major insights into the differential modes by which SV40 and polyomavirus transform cells. The establishment of conditions for binding p53 to purified SV40 large TAg in vitro (McCormick et al. 1981) should provide impetus for further experiments.

Another outcome of examining the properties of naturally occurring (Lanford and Butel 1984) and genetically engineered (Kalderon et al. 1984a) mutants of the SV40 large TAg was the identification of a short, highly basic amino acid sequence whose function is apparently to relay TAg into the nucleus. That this sequence alone can impart this property to a heterologous protein was shown by an elegant experiment in which sequences encoding this short nuclear-localization sequence were joined to the gene encoding chicken pyruvate kinase, resulting in the appearance of this protein, normally a cytoplasmic constituent, in the cell nucleus (Kalderon et al. 1984b).

The polyomavirus large TAg is presently less well characterized than its counterpart in SV40. It too has the ability to bind specifically to DNA containing repeats of the consensus pentanucleotide 5′ G$^A$/$_G$GGC 3′ (Cowie and Kamen 1986) and exhibits ATPase (Gaudray et al. 1980) and nucleotide-binding (Clertant et al. 1984) activities. The polyomavirus large TAg has two discrete nuclear-localization-sequence elements, one of which is highly homologous to the SV40 sequence and the second of which has little direct sequence homology (Richardson et al. 1986). Both SV40 and polyomavirus are phosphoproteins whose sites of phosphorylation have been partly mapped (Scheidtmann et al. 1982; Hassauer et al. 1986). The relationship between protein phosphorylation and other functions of the viral large TAg's remains to be clarified.

Since the identification and initial characterization of the viral late region—encoded capsid proteins of SV40 and polyomavirus, their functions in virus assembly have not been extensively elucidated. However, considerable characterization of the polyomavirus capsid proteins has proceeded (e.g., Yuen and Consigli 1985), and a functional role for the polyomavirus middle TAg in effecting modifications of the VP1 viral capsid protein required for virus maturation has been suggested (Garcea and Benjamin 1983). In recent years the late region—encoded "agnoprotein" predicted earlier from the sequence of SV40 has been identified by using a combination of mutants that delete agnoprotein-coding sequences and immunoprecipitation using antipeptide antisera (Jay et al. 1981). This small protein is found in the perinuclear cell fraction and can bind to DNA cellulose. Its function in SV40 productive infection has not been elucidated, but roles in virus assembly (Ng et al. 1985) and late gene regulation (Hay et al. 1982) have been proposed. No counterpart in polyomavirus to the SV40 agnoprotein has yet been identified.

### Viral mRNAs

Most if not all of the stable mRNAs and proteins encoded by SV40 and polyomavirus have been identified. SV40 and polyomavirus early RNAs are both somewhat heterogeneous with respect to their 5′ ends. Early in lytic infection, the 5′ ends of early strand—specific mRNAs map downstream of the *ori* region, whereas late in infection additional early strand mRNAs with 5′ ends located further upstream are detected. The change in position of early RNA 5′ ends is not yet clearly understood but may be related to the changes in template structure as a result of the onset of DNA replication or to the increased quantities of TAg bound specifically to viral DNA in this region as infection proceeds. The stable late and early region mRNAs of SV40 and polyomavirus overlap at their 3′ ends by 88 and 50 nucleotides, respectively. This unusual structural feature may play a role in posttranscriptional regulation of early region mRNA late in infection, when there is an overwhelming abundance of late region transcripts.

The SV40 late region encodes mRNAs generated by a complex series of RNA-processing events that collectively exist as the two 19S and 16S size classes. 19S mRNAs are translated in vitro to the minor capsid proteins VP2 and VP3, as well as the major capsid protein VP1, while 16S mRNAs are translated primarily to VP1 (Prives and Shure 1979). The complexity of the various species of RNAs that comprise the 19S and 16S SV40 mRNAs has defied attempts to determine which individual RNA species are translated to which viral proteins in vivo or in vitro. Creation and characterization of viral mutants that affect utilization of the various potential 5′ and 3′ late region splice sites may eventually elucidate the relationship between late viral mRNAs and their protein products.

Polyomavirus late region RNAs comprise 19S, 18S, and 16S size classes, whose cell-free translation led to assignments of these RNAs as messengers for VP2, VP3, and VP1, respectively (Smith et al. 1976). Transcribed RNAs from both viral late regions have unusually heterogeneous 5′ ends comprising major and minor capped species (Cowie et al. 1981; Piatek et al. 1983, and references therein). The polyomavirus late strand—specific nuclear pre-mRNA has been characterized and shown to consist of extremely long transcripts, as much as several genome lengths (Acheson 1978). Late RNA from polyomavirus, but not from SV40, contains an unusual 5′ leader sequence that is reiterated and most likely generated by splicing of these giant nuclear precursors (Legon et al. 1979; Treisman 1980; Adami and Carmichael 1986).

### Processing of viral RNA

The discovery of RNA splicing in 1976 provided the answer to a paradox concerning the coding capacity of the early regions of SV40 and polyomavirus. Mutants that deleted sequences within the early coding region of SV40 that did affect the size of the large TAg had been isolated (Shenk et al. 1976). Furthermore, both viruses appeared to express too many early region polypeptides encoded by too little DNA. However, the discovery of RNA splicing and the determination of the sequences of SV40 and polyomavirus suggested the answers, which were soon confirmed by the development of S1 mapping: Namely, the early region—specific SV40 and polyomavirus pre-mRNAs are subjected to alternative splicing events, yielding different mature mRNAs (Berk and Sharp 1978; Kamen et al. 1980). SV40 early pre-mRNA has two discrete donor 5′ splice sites and one common acceptor 3′ splice site that are utilized to form the large TAg and small TAg mRNAs. The polyomavirus early region provides one of the rare eukaryotic examples of the usage of all three potential reading frames. It contains two 5′ donor sites and two 3′ acceptor splice sites. Three out of the four combinations are used to produce three functional mRNAs that are translated to the large, middle, and small TAg's. It is not clear why mRNA from the fourth possible combination (i.e., utilization of the large TAg 5′ donor and the middle TAg 3′ acceptor splice site) is not detected. The factors controlling utilization of alternative splice sites in a given pre-mRNA are of considerable interest because of the possibilities of tissue-specific and developmental regulation of protein production as a function of alternative splicing pathways. The in vitro splicing of SV40 pre-mRNA has recently been demonstrated, with the somewhat unexpected finding that only large TAg and not small TAg mRNA is produced in vitro from wild-type pre-mRNA sequences (Noble et al. 1986), even though the same pre-mRNA yields abundant quantities of small TAg pre-mRNA in vivo (Manley et al., this volume). This suggests that factors may exist in intact cells that regulate the choice of alternative splicing pathways. Evidence for variations in the requirements of different SV40 pre-mRNAs for factors involved in RNA splicing was provided by experiments involving microinjection of SV40 DNA into *Xenopus laevis* oocytes (Fradin et al. 1984).

The spliced structures of SV40 late mRNAs were largely deduced by Weissman and colleagues from analysis of mRNAs isolated from infected cells, using primer extension methodologies, and of polyomavirus late mRNAs by Kamen and colleagues, using S1 analysis. The pattern to emerge for SV40 was one of an exceedingly complex array of RNA "leader regions," some with internal splices, to two 19S and 16S "bodies." The combination of the considerable heterogeneity of the 5' ends of late RNAs with the complexity of spliced late region RNAs provides a considerable challenge to determine the relationship between different capped late pre-mRNAs, their spliced RNA products, and ultimately their respective encoded protein products.

Polyadenylation of SV40 and polyomavirus RNA has also been examined by both mutagenesis and the development of in vitro systems. The consensus sequence 5' AAUAAA 3' was shown to be required for polyadenylation of late SV40 mRNA (Fitzgerald and Shenk 1981; Wickens and Stephenson 1984). The development of systems for polyadenylation of mRNA in vitro confirmed the importance of this consensus sequence for SV40 early RNA processing (Manley et al. 1985). Additional sequences downstream of this site have been also identified as important for polyadenylation of SV40 and polyomavirus late RNA (Conway and Wickens 1985; Hart et al. 1985; Sadofsky et al. 1985).

**Transcriptional enhancers**

The discovery of gene enhancers has been one of the most exciting findings in recent years to emerge from papovavirus transcription studies. As DNA cloning and mutagenesis techniques became available, several groups embarked upon the dissection of the DNA sequence elements regulating early and late gene expression. Sequences far upstream of the polyomavirus (Tyndall et al. 1981) and SV40 (Benoist and Chambon 1981; Gruss et al. 1981; Fromm and Berg 1982) early transcriptional start sites were shown to be required for activation of early transcription. These sequences, existing as direct repeats of 72 bp in SV40, and, depending on the strain of polyomavirus, either a single-copy or repeated elements (Ruley and Fried 1983a), function as unique transcriptional activators that are distinct from previously determined eukaryotic or prokaryotic promoter elements. This is highlighted by their ability to activate transcription when placed in either orientation and at variable distances from the early region TATA box (for review, see Gluzman and Shenk 1983; Khoury and Gruss 1983). Further evidence for their unique nature derived from studies demonstrating the ability of these enhancer elements to activate transcription from several heterologous promoters (Banerji et al. 1981; de Villiers and Schaffner 1981; Moreau et al. 1981; Green et al. 1983) as well as for heterologous enhancers to activate SV40 and polyomavirus early region transcription (Conrad and Botchan 1982; de Villiers et al. 1982; Fried et al. 1983; Kriegler and Botchan 1983; Weber and Schaffner 1985).

These observations opened the way to a number of studies designed to probe the involvement of enhancers in tissue-specific gene regulation, to investigate the roles of specific DNA sequences and potential structures in enhancer function, and to search for cellular factors that may interact with enhancer elements. Considerable evidence has been provided that the SV40 enhancer can, in some cases, confer a tissue-specific or host-range response to the promoter it activates (de Villiers et al. 1982; Laimins et al. 1982; Byrne et al. 1983; Spandidos and Wilkie 1983). More recently, studies with transgenic mice have provided evidence suggesting that different, specific enhancer elements can control the response of various tissues in the intact animal to the SV40 early region—encoded TAg (Brinster et al. 1984; Hanahan 1985). The role of the polyomavirus enhancers in altering the host range of the virus for early gene expression and replication has been suggested by a number of studies. These have demonstrated that strains of polyomavirus that have adapted to grow in normally nonpermissive differentiated and undifferentiated embryonal carcinoma cells contain various alterations in their enhancer sequences (for review, see Amati 1985).

Analysis of the sequences in SV40 and polyomavirus enhancers revealed differences between the two elements. The SV40 enhancer can function with only one of the two 72-bp repeat elements. Within each element exists a crucial "core" sequence (Weiher et al. 1983) and alternating purine / pyrimidine structures. Point mutations within the core enhancer region reduce enhancer function, and revertants of these mutants contain tandem duplications of the mutant core sequences (Herr and Gluzman 1985). The function of these core sequences is not yet well understood, and it appears that other sets of reiterated sequences can function as enhancers (Swimmer and Shenk 1984). The polyomavirus enhancer region is more complex. It consists of two basic elements referred to as α and β (Mueller et al. 1984), or A and B (Herbomel et al. 1984), that each exhibit different and independent cell specificity within the polyomavirus enhancer (e.g., teratocarcinoma cells). Within the polyomavirus enhancer, homologies to the SV40 core sequence, the adenovirus E1A core sequence, and the mouse IgM enhancer region have been noted (Banerji et al. 1983). Apparently unique to polyomavirus is an additional role for the enhancer in viral DNA replication. Substitution of the polyomavirus enhancer by SV40 72-bp repeats in either orientation allows replication of polyomavirus DNA in monkey cells, and, interestingly, substitution of this region with the IgM enhancer enables polyomavirus DNA to replicate efficiently in lymphoma cells (de Villiers et al. 1984).

The mode by which enhancers activate early region gene expression is not yet clarified. It has been shown that enhancers increase the number of RNA polymerase molecules on a given template (Treisman and Maniatis 1985; Weber and Schaffner 1985) and correlate well with the presence of DNase I—hypersensitive sites on viral chromatin (Fromm and Berg 1983; Herbomel et al. 1984). Possible interaction of enhancer sequences with host factors has been inferred from in vivo competition studies (Schöler and Gruss 1984; Wildeman et al.

1984) and gel retention assays (Bohnlein and Gruss 1986). Furthermore, evidence of repression of enhancer function by heterologous viral gene products such as the adenovirus E1A proteins (Borrelli et al. 1984; Velcich and Ziff 1985) also indicates the complexity of cellular interactions with enhancer sequences.

What is the function of the enhancer sequences? By now several other similar sorts of sequence elements (with varying degrees of position and orientation independence) that activate cellular genes have been identified. Therefore, enhancers are not unique to viruses. Nevertheless, one function of enhancers may be to provide an early boost to viral gene expression, possibly by prevention of nucleosome assembly at the viral regulatory region, such that after entry into the cell and uncoating of the viral genome, early region DNA can be rapidly transcribed. Whether or not the viral enhancers directly affect late papovavirus transcription is not yet clear.

### Promoter elements

Analysis of viral gene expression was facilitated by the development of methods for transfection of viral or cloned wild-type or mutated DNA into cells, coupled with suitable transient expression assays or selection systems for identifying stable transformants. A crucial dimension was added when whole cell and nuclear extracts were developed that allowed the analysis of transcription in vitro.

Of the papovavirus transcription units, the SV40 early region has been the most extensively dissected, leading to the identification of sequences and factors involved in its regulation. The early promoter TATA box, located within 30 bp upstream of the major early strand mRNA cap site, serves as a positioning element for the start of transcription of the early-early RNAs (Benoist and Chambon 1981; Buchman et al. 1984) but not of the late-early RNAs. Additional elements, located within 115 bp upstream of the early cap sites, include the three 21-bp repeats, whose direct copies of the 5′ CCGCCC 3′ sequence were shown to be required for early region transcription. Different types of cellular factors interact with viral DNA sequences, including those that interact with the enhancer region (Sassone-Corsi et al. 1985) as well as a protein, Sp1, that interacts with the 21-bp repeat region (Dynan and Tjian 1983). Recent experiments suggest that protein-protein interaction between these and possibly additional binding proteins are required for efficient polymerase II transcription (Takahashi et al. 1986). In addition, SV40 TAg autoregulation of early region transcription was inferred from prior studies on early region–specific viral RNA levels in cells infected with tsA mutants. This was subsequently demonstrated in soluble in vitro systems transcribing SV40 early region DNA in the presence of purified SV40 TAg (Rio et al. 1980; Hansen et al. 1981). An intriguing mode of early region regulation was observed in experiments using human 293 cells, a line that contains and expresses adenovirus E1A sequences. SV40 DNA replication in these cells leads to the repression of viral early

transcription (Lebkowski et al. 1985; Lewis and Manley 1985). It is possible that replicating molecules may be unable to serve as templates for early transcription in these and other cells. Whether this is related to the function of the adenovirus E1A product in 293 cells remains to be determined.

Substantial differences exist between sequences and factors controlling polyomavirus and SV40 early region transcription. The polyomavirus early region may contain two TATA boxes, each of which direct transcription initiation in vitro (Jat et al. 1982a,b). Both the quantities and 5′ end positions of early mRNAs are modulated by the function of the large TAg (Fenton and Basilico 1982; Kamen et al. 1982). However, the unique organization of the polyomavirus enhancer region and the apparent lack of polyomavirus sequences equivalent to the 21-bp repeats (and therefore the absence of a role for Sp1 in polyoma transcription) make it likely that SV40 and polyomavirus early region transcription operate by somewhat different mechanisms. Furthermore, the mode of repression by the polyomavirus large TAg may be unique (Farmiere and Folk 1984). The application of similar approaches to fine-structure sequence analysis and further parallel comparisons of transcription in vitro and in vivo will be required before the control of polyomavirus early region transcription can be understood at the same level of detail as its SV40 counterpart.

Control of the expression of the SV40 late region is less well understood than that of the early region. Mutants deleting sequences in the viral origin, the 21-bp repeat sequences, and in the enhancers have been utilized to define the late promoter. Lacking a defined TATA box region, the late promoter resembles other "TATA-less" promoters in that the 5′ ends of late RNA are extremely heterogeneous and are spread over a region of approximately 300 bp. Attempts to identify late region promoter elements suggest that sequences located (1) within the 21-bp repeats (Brady et al. 1982; Fromm and Berg 1982; Hansen and Sharp 1983; Hartzell et al. 1984a), (2) at sites 25 bp upstream of the major late 5′ end, (3) within the 72-bp repeats (Hartzell et al. 1984b and references therein), and (4) within the vicinity of each initiation site (Piatak et al. 1983) all affect the abundance and utilization of late RNA 5′ ends. In addition, sequences within the viral replication origin were reported to be required for late transcription in X. laevis oocytes (Contreras et al. 1982), although contradictory findings in other laboratories were observed (e.g., Prives et al., this volume). The general impression generated by all these findings is one of a rather complicated system in which multiple, overlapping promoters may function to generate the complex pattern of late transcripts. Late RNA synthesis occurs to significant extents only in infected permissive cells or in nonpermissive cells microinjected with high quantities of viral DNA (Graessman et al. 1981 and references therein; Miller et al. 1982; Michaeli and Prives 1985). The reason why the late transcription unit is inactive in infected or transformed nonpermissive cells is not clear. The finding that the SV40 large TAg can function in permissive cells to activate the late promoter in a manner independent of its direct role in viral DNA replication (Brady et al. 1984; Keller and Alwine 1984) has stimulated interest in the viral A-gene product as a potential activator of other cellular genes whose expression is altered in infected or transformed cells. It is not yet clear whether this activation is mediated by its binding specifically to viral DNA (Brady and Khoury 1985), by a mechanism more akin to that proposed for the adenovirus E1A region gene products (Keller and Alwine 1985; Robbins et al. 1986), or whether both functional modes are involved. An additional mechanism by which late region expression is regulated, involving attenuation of late RNA transcription, has been proposed (Hay et al. 1982).

### Viral DNA replication

The production of infectious progeny by SV40 and polyomavirus occurs only in cells that have the ability to replicate viral DNA as free, unintegrated minichromosomes. There are three basic components involved in viral DNA replication: (1) host replication factors, including those that delineate permissive from nonpermissive cells; (2) the viral replication protein, large TAg, specified by the A gene; (3) the viral replication origin (*ori*). Host-cell permissiveness was first shown to be a dominant process as a result of studies from Dulbecco's, Koprowski's, and Basilico's laboratories in which hybrids of permissive and nonpermissive cells were shown to produce infectious virus progeny. The roles of SV40 or polyomavirus large TAg's in viral DNA replication were identified when temperature-sensitive A-gene mutants were isolated. Temperature-shift experiments using these mutants led to the conclusion that the SV40 (Tegtmeyer 1972) or polyomavirus (Franke and Eckhart 1973) A-gene product is required for initiating new rounds of viral DNA replication. There is no reported evidence for roles of large TAg in DNA chain elongation or termination. The *ori* region of SV40 was first mapped to sequences at or near 0.67 map units on the viral genome (Danna and Nathans 1972). More-detailed mapping was accomplished with the advent of techniques for in vitro mutagenesis (Bergsma et al. 1982 and references therein; DiMaio and Nathans 1982). The minimal SV40 *ori*, identified as a 65-bp segment between nucleotides 5208 and 30 on the viral genome, contains a 27-bp perfect inverted repeat adjacent to a 17-bp sequence of A/T nucleotides. It contains one of the specific binding sites (site II) for large TAg. The polyomavirus *ori* region was initially located to the region at 0.72 map units (Griffin et al. 1974) and subsequently shown to have both common and unique features when compared with SV40. This includes an inverted repeat adjacent to a 14-bp A + T−rich region. The polyomavirus core *ori* region contains only very weak polyomavirus large TAg−binding sites (Cowie and Kamen 1986), which may indicate that its interaction with TAg differs substantially from that of SV40. The minimal sequences on the early side of the *ori* region that are required for polyomavirus DNA replication have been mapped at or near nucleotide 65 (Katinka and Yaniv 1983; Dailey and

Basilico 1985). On the late side of the polyomavirus minimal *ori*, Luthman et al. (1982) and Muller et al. (1983) identified discrete segments of DNA that are also required for DNA replication, indicating that the polyomavirus *ori* region comprises multiple elements. One of the elements required in *cis* is within the polyomavirus enhancer region, and the observation that other enhancers can replace these sequences indicates that polyomavirus DNA replication may require transcriptional activation (de Villiers et al. 1984). Corresponding roles of either the 21-bp or the 72-bp transcriptional elements in SV40 DNA replication have not yet been clearly established. Studies on the in vivo processes involved in SV40 DNA replication have identified Okazaki fragments (Kaufman 1981) and have shown that nascent DNA chains contain oligoribonucleotides attached to the DNA 5′ ends. The 5′ ends of the nascent chains map to several sites within the *ori* region (Hay and DePamphilis 1982). Speculation on the role of large TAg in DNA replication may be inspired by the observations that (1) the SV40 and polyomavirus TAg's can be adenylated (Bradley et al. 1984; Clertant et al. 1984) and (2) the putative SV40 RNA primer chains all begin with rA (Hay and DePampilis 1982). Sundin and Varshavsky (1980) have characterized replication products and have observed catenated dimers as intermediates in the final stages of DNA replication in vivo.

The recent development of a reproducible system that replicates SV40 DNA in vitro (Li and Kelly 1984) has opened the way toward understanding the process of viral DNA synthesis. The success of this system was made possible because large quantities of biologically active SV40 large TAg can be purified from cells, using immunoaffinity procedures (Dixon and Nathans 1985; Simanis and Lane 1985). SV40 DNA replicates in extracts of human or monkey cells but not mouse cells in the presence of purified SV40 TAg (Li and Kelly 1984). The replication of SV40 in vitro proceeds bidirectionally from the viral origin, producing relaxed, covalently closed DNA circles that could be negatively supercoiled in the presence of additional nuclear extracts (Stillman and Gluzman 1985). Interestingly, a smaller sequence is required for the initiation of viral DNA replication in vitro than in vivo when sets of deletion mutants are compared (Stillman et al. 1985; Li et al. 1986). Since the only viral protein required in these systems is the large TAg, it is likely that they can be used as the basis for elucidating some of the features of eukaryotic cellular and viral DNA replication. Indeed, it has been observed that SV40 TAg-mediated DNA replication occurs in extracts of mouse cells only when they are supplemented with purified human or monkey polymerase α-primase (Murakami et al. 1986b). Recently, the in vitro replication of polyomavirus DNA in extracts of mouse cells has been established, using procedures similar to those developed for SV40, with the reciprocal observation that polyomavirus DNA will replicate in HeLa cell extracts only in the presence of purified mouse polymerase α−primase complex (Murakami et al. 1986a). These experiments may provide the basis for understanding the molecular basis of host-range permissiveness.

## Transformation

In cells transformed either in vivo or in vitro by polyomavirus or SV40, viral DNA containing a functional early region is invariably found integrated into cellular chromosomal DNA (Sambrook et al. 1968; Botchan et al. 1976; Ketner and Kelly 1976; Birg et al. 1979; Basilico et al. 1980; Lania et al. 1980b). No common sequence or structural features in either the viral or cellular DNA sequences at the integration sites can be detected. Integration appears to involve nonspecific, illegitimate recombination events involving 2–5 bp of homology at the recombinant join. Polyomavirus or SV40 integration causes a gross rearrangement (deletion or chromosomal translocation) of cellular DNA at the integration site (Hayday et al. 1982; Ruley et al. 1982; Stringer 1982).

The virally induced transformed phenotype does not appear to be due to integration per se (Lania et al. 1980a) or integration at specific regions in the cellular DNA that result in either the inactivation or activation of a host gene present at the integration site. Transformation results from the addition of a specific viral gene (oncogene) to the cellular genome. The determination of which of the viral genes is the oncogene was initially confusing, especially in the case of polyomavirus. Early studies showed that after virus infection a functional large TAg is required for the efficient initiation of the transformed state both in vitro and in vivo (Fried 1965; Di Mayorca et al. 1969; Eckhart 1969). On the other hand, neither an intact nor a functional large TAg–coding region was found in a number of cell lines transformed after infection with virus (Fried 1965; Basilico et al. 1980; Lania et al. 1980b,c, 1981; Hayday et al. 1982, 1983; Ruley et al. 1982; Ruley and Fried 1983b), indicating that the polyomavirus large TAg is not required for the maintenance of the transformed state. In addition, cells isolated after virus infection containing an integrated polyomavirus genome that expressed only a functional large TAg (not middle or small TAg's) displayed a normal phenotype (Lania et al. 1980a). Similar studies using SV40 A-gene mutants clearly showed that the SV40 large TAg is indeed the SV40 oncogene product required both for the initiation and maintenance of the transformed state (Brugge and Butel 1975; Kimura and Itagaki 1975; Martin and Chou 1975; Osborn and Weber 1975; Tegtmeyer 1975; Steinberg et al. 1978). That the polyomavirus and SV40 large TAg's play different roles in transformation was further reinforced by the studies of Benjamin and his colleagues, who isolated and analyzed host-range transformation-defective (hr-t) mutants of polyomavirus (Benjamin 1970, 1982). The mutations responsible for the hr-t phenotype were found to be located in a different gene than the polyomavirus A gene, which encodes the large TAg (Eckhart 1977; Fluck et al. 1977). It was eventually shown that the hr-t mutations map to the polyomavirus early region sequence within the large TAg intron unique to the middle and small TAg's coding regions (Feunteun et al. 1976; Soeda and Griffin 1978; Carmichael and Benjamin 1980; Lania et al. 1980a). The demonstration that a cDNA encoding just the middle TAg is sufficient to transform established cell lines clearly showed that the polyomavirus oncogene codes for the middle TAg (Treisman et al. 1981). Exactly how the SV40 (large TAg) and the polyomavirus (middle TAg) oncogenes function to cause transformation is unknown.

Although early studies showed that after virus infection a functional polyomavirus large TAg is required for the efficient initiation of transformation (Fried 1965; Di Mayorca et al. 1969; Eckhart 1969), later work utilizing DNA transfection indicated that a functional large TAg is not required for initiation of the transformed state (Israel et al. 1979; Chowdhury et al. 1980; Hassell et al. 1980; Novak et al. 1980). This apparent anomaly is best explained if the efficiency of transformation is dependent on the intracellular copy number of viral DNA molecules. Thus, the DNA replication function of the polyomavirus large TAg would be important to increase the copy number of viral DNA where only one or a few of the infecting viral particles enter the cell. Replication, to reach a high copy number of viral DNA, is less important after DNA transfection, since large amounts of DNA precipitate are taken up by a successfully transfected cell. Although the polyomavirus large TAg is not required for the maintenance of the middle TAg–induced transformed state in established cell lines, it is required for the maintenance of transformation in primary cells. This is due to the ability of the polyomavirus large TAg, with a detectable frequency, to "establish" or "immortalize" primary cells (Land et al. 1983; Rassoulzadegan et al. 1983). Thus, even though the middle TAg is capable of inducing a transformed phenotype in primary cells, such cells usually enter a crisis period after a specified number of cell divisions and cease dividing. In the presence of an immortalizing protein, like the polyomavirus large TAg, a proportion of the transformed cells will continue to divide through the crisis period and are capable of being expanded into cell lines. The sequence encoding the aminoterminal region of the large TAg gene is sufficient to induce establishment. Furthermore, the continued presence of a functional large TAg is required for cell division to be maintained (Rassoulzadegan et al. 1983).

Transformation does not appear to be a primary function of the polyomavirus and SV40 oncogenes in their normal hosts. Because of their small genetic content, these papovaviruses rely heavily on cellularly encoded proteins for the replication of their DNA genomes. They appear to achieve this via a virally encoded protein(s) that induces the cell to ignore existing regulatory signals and enter into its cycle, even if in a resting state, so that the viral DNA can be replicated to a high copy number, but in concert with the cellular DNA, when the S phase of the cell cycle is reached. If a complete viral life cycle is completed, infectious virus particles are produced and the cell dies. Most likely, the rare transformation event arises in a cell nonpermissive for viral DNA replication, when by chance viral DNA becomes integrated into the host chromosome and continues to generate the virally specified signal used during normal viral DNA replication. This would result in a cancer cell, which would continue to cycle and divide under conditions where normal cell division is inhibited. Both polyomavirus and SV40 induce cellular DNA synthesis and cell division in resting cell populations (Dulbecco et al. 1965; Henry et al. 1966; Kit et al. 1966, 1967; Stoker 1968; Smith et al. 1971). In case of SV40, these functions are induced by the SV40 large TAg, whereas in the case of polyomavirus the stimulation appears to be due to the polyomavirus middle TAg and not the polyomavirus large TAg (Stoker and Dulbecco 1969; Fried 1970; Graessman et al. 1981).

Whether the polyomavirus or SV40 oncogenes act directly or indirectly on their intracellular target(s) to induce a transformed phenotype is unknown. The polyomavirus middle TAg contains a hydrophobic carboxyl terminus and is thought to be located on the inner surface of the cellular membrane (Ito et al. 1977). A tyrosyl-specific protein kinase activity is found associated with the polyomavirus middle TAg (Eckhart et al. 1979; Schaffhausen and Benjamin 1979; Smith et al. 1979). This kinase activity is thought to be derived from the small amount of c-*src* protein found complexed to a small proportion of the middle TAg molecules (Courtneidge and Smith 1983, 1984). The detection of this complex has led to the theory that polyomavirus transformation is the result of the subversion of the c-*src* proto-oncogene. Under this hypothesis, the complexed c-*src* becomes activated (Bolen et al. 1984; Courtneidge 1985), possibly by differences in its phosphorylation (Courtneidge 1985; Yonemoto et al. 1985; Cartwright et al. 1986), to display its oncogenic properties. Cells containing a kinase-positive complex between c-*src* and a truncated middle TAg species have a normal phenotype (Wilson et al. 1986). This indicates either that the quality, quantity, or location of the c-*src*/middle TAg complex is important for the transformed phenotype or that transformation occurs by virtue of some other property of a functional middle TAg and the middle TAg–c-*src* complex is unimportant. Middle TAg appears to be able to activate cellular genes (Majello et al. 1985). It is possible that such activation of cellular genes under inappropriate conditions results in a transformed phenotype.

The SV40 large TAg contains a number of properties distributed between the polyomavirus middle and large TAg's. The SV40 large TAg oncogene is capable of immortalizing primary cells and transforming primary and established cells; it is also involved in viral DNA replication and can *trans*-activate cellular genes. A proportion of the SV40 large TAg is found complexed to the cellular p53 protein (Lane and Crawford 1979; Klein 1982). The role of this complex in transformation, if any, is unclear. Although the majority of large TAg is in the nucleus, a small component has been detected associated with cell membranes (Deppert et al. 1980; Santos and Butel 1982; Gooding et al. 1984). Viral mutants that have lost their nuclear location signal and whose large TAg is predominately located in the cytoplasm can still efficiently transform cell lines (Kalderon et al. 1984a; Lanford and Butel 1984) but not primary cells (Fischer-Fantuzzi and Vesco 1985; Lanford et al. 1985). This indicates either that only a small amount of

nuclear large TAg is required for transformation or that transformation is mediated by a nonnuclear form of large TAg. The DNA-binding, DNA replication, ATPase, and helper-function activities of the large TAg do not appear to be required for transformation (Pipas et al. 1983; Prives et al. 1983; Manos and Gluzman 1985; Cole et al. 1986).

The roles of the polyomavirus and SV40 small TAg's in transformation are unclear. Under certain circumstances, their presence in conjunction with the viral oncogene is important for the efficient expression of the transformed phenotype (Bouck et al. 1978; Sleigh et al. 1978; Martin et al. 1979; Rassoulzedegan et al. 1982; Asselin et al. 1983; Cuzin et al. 1984; Bikel et al. 1986).

The manner in which the papovaviruses induce transformation still remains the major problem to be solved in the study of these viruses. The main interest is in the intracellular targets involved in oncogenesis. The polyomavirus and SV40 oncogenes share a number of properties that may be involved in their ability to induce a transformed phenotype. The polyomavirus middle TAg and the SV40 large TAg both form complexes with cellular proteins; are found, at least in part, associated with cell membranes; and are capable of activating specific cellular genes. Research within the next few years will hopefully reveal how these proteins are involved in neoplastic transformation.

## Acknowledgments

We are indebted to Nancy Hogg, Chris Norbury, and Jim Manley for their critical reading, and Georgina Briody and Amelia Rugland for typing this manuscript.

## References

Acheson, N.A. 1978. Polyoma giant RNAs contain tandem repeats of the nucleotide sequence of the entire genome. *Proc. Natl. Acad. Sci.* **75:** 4754.

Adami, G.R. and G.G. Carmichael. 1986. Polyoma late leader region serves as an essential spacer function for viability and late gene expression. *J. Virol.* **58:** 417.

Amati, P. 1985. Polyoma regulatory region: A potential probe for mouse cell differentiation. *Cell* **43:** 562.

Asselin, C., C. Gelinas, and M. Bastia. 1983. Role of the three polyoma virus early proteins in tumorigenesis. *Mol. Cell. Biol.* **3:** 1451.

Banerji, J., L. Olson, and W. Schaffner. 1983. A lymphocyte specific cellular enhancer is located downstream of the joining region in immunoglobulin heavy chain genes. *Cell* **33:** 729.

Banerji, J., S. Rusconi, and W. Schaffner. 1981. Expression of a β-globin gene is enhanced by remote SV40 DNA sequences. *Cell* **27:** 299.

Basilico, C., D. Zouzias, G. Della-Valle, S. Gattoni, V. Colantuoni, R. Fenton, and L. Dailey. 1980. Integration and excision of polyoma virus genomes. *Cold Spring Harbor Symp. Quant. Biol.* **44:** 611.

Benjamin, T.L. 1970. Host range mutants of polyoma virus. *Proc. Natl. Acad. Sci.* **54:** 394.

———. 1982. The *hr-t* gene of polyoma virus. *Biochim. Biophys. Acta* **695:** 69.

Benoist, C. and P. Chambon. 1981. *In vivo* sequence requirements of the SV40 early promoter region. *Nature* **290:** 304.

Bergsma, D.J., D.M. Olive, S.W. Hartze, and K.N. Subramanian. 1982. Territorial limits and functional anatomy of the simian virus to replication origin. *Proc. Natl. Acad. Sci.* **79**: 381.

Berk, A.J. and P.A. Sharp. 1978. Spliced early mRNAs of simian virus 40. *Proc. Natl. Acad. Sci.* **75**: 1274.

Bikel, I., H. Mamon, E.L. Brown, J. Boltax, M. Agha, and D. Livingston. 1986. The t-unique coding domain is important to the transformation maintenance function of the simian virus 40 small t antigen. *Mol. Cell. Biol.* **6**: 1172.

Birg, F., R. Dulbecco, M. Fried, and R. Kamen. 1979. State and organization of polyoma virus DNA in transformed rat cell lines. *J. Virol.* **29**: 633.

Black, P.H. 1966. Transformation of mouse cell line 3T3 by SV40: Dose response relationship and correlation with SV40 tumor antigen production. *Virology* **28**: 760.

Bohnlein, E. and P. Gruss. 1986. Interaction of distinct nuclear proteins with sequences controlling the expression of polyoma virus early genes. *Mol. Cell. Biol.* **6**: 1401.

Bolen, J.B., C.J. Thiele, M.A. Israel, W. Yonemoto, L.A. Lipsich, and J.S. Brugge. 1984. Enhancement of cellular *src* gene product associated tyrosyl kinase activity following polyoma virus infection and transformation. *Cell* **38**: 767.

Borrelli, E., R. Hen, and P. Chambon. 1984. Adenovirus-2 EIA products repress enhancer-induced stimulation of transcription. *Nature* **312**: 608.

Botchan, M., W.C. Topp, and J. Sambrook. 1976. The arrangement of simian virus 40 sequences in the DNA of transformed cells. *Cell* **9**: 269.

Bouck, N., N. Beales, T. Shenk, P. Beard, and G. Di Mayorca. 1978. New region of simian virus 40 genome required for efficient viral transformation. *Proc. Natl. Acad. Sci.* **75**: 2473.

Bradley, M.K., J. Hudson, M. Villanueva, and D.M. Livingston. 1984. Specific *in vitro* adenylation of the simian virus 40 large T antigen. *Proc. Natl. Acad. Sci.* **81**: 6574.

Brady, J. and G. Khoury. 1985. *Trans* activation of the simian virus 40 late transcription unit by T antigen. *Mol. Cell. Biol.* **5**: 1391.

Brady, J., J.B. Bolen, N. Radonovich, N. Salzman, and G. Khoury. 1984. Stimulation of SV40 late gene expression by simian virus 40 tumor antigen. *Proc. Natl. Acad. Sci.* **81**: 2040.

Brady, J., M. Radonovich, M. Vodkin, V. Natarajan, M. Thoren, G. Das, J. Janik, and N.P. Salzman. 1982. Site specific base substitution and deletion mutants that enhance or suppress transcription of the SV40 major late RNA. *Cell* **31**: 625.

Brinster, R.L., H.Y. Chen, A. Messing, T. van Dyke, A.J. Levine, and R.D. Palmiter. 1984. Transgenic mice harboring SV40 T-antigen genes develop characteristic brain tumors. *Cell* **32**: 367.

Brugge, J.S. and J.S. Butel. 1975. Involvement of the simian virus 40 gene. A function in maintenance of transformation. *J. Virol.* **15**: 619.

Buchman, A.R., M. Fromm, and P. Berg. 1984. Complex regulation of simian virus 40 early region transcription from different overlapping promoters. *Mol. Cell. Biol.* **4**: 1900.

Byrne, B.J., M.S. Davis, Y. Yamaguchi, D.J. Bergsma, and K.N. Subramanian. 1983. Definition of the simian virus early promoter region and demonstration of a host range bias in the enhancement effect of the simian virus 40 72 base pair repeats. *Proc. Natl. Acad. Sci.* **80**: 721.

Carmichael, G.G. and T.L. Benjamin. 1980. Identification of DNA sequence changes leading to loss of transforming ability in polyoma virus. *J. Biol. Chem.* **255**: 230.

Cartwright, C.A., P.L. Kaplan, J.A. Cooper, T. Hunter, and W. Eckhart. 1986. Altered sites of tyrosine phosphorylation in pp60$^{c\text{-}src}$ associated with polyomavirus middle tumor antigen. *Mol. Cell Biol.* **6**: 1562.

Chang, C., R. Martin, D. Livingston, S. Luborsky, C. Hu, and P. Mora. 1979. Relationship between T antigen and tumor-specific transplantation antigen in simian virus 40-transformed cells. *J. Virol.* **29**: 69.

Chowdhury, K., S.E. Light, C.F. Garon, Y. Ito, and M.A. Israel.

1980. A cloned polyoma DNA fragment representing the 5′ half of the early gene region is oncogenic. *J. Virol* **36**: 566.

Clark, R., M.J. Tevethia, and R. Tjian. 1984. The ATPase activity of SV40 large T antigen. *Cancer Cells* **2**: 363.

Clark, R., K. Peden, J.M. Pipas, D. Nathans, and R. Tjian. 1983. Biochemical activities of T-antigen proteins encoded by simian virus 40 A gene deletion mutants. *Mol. Cell. Biol.* **3**: 220.

Clertant, P., P. Gaudray, E. May, and F. Cuzin. 1984. The nucleotide binding site detected by affinity labeling in the large T antigen of polyoma and SV40 viruses is distinct from their ATPase catalytic site. *J. Biol. Chem.* **259**: 15196.

Coggin, J.H., V.M. Larson, and M.R. Hilleman. 1967. Immunologic responses in hamster to homologous tumor antigens measured *in vivo* and *in vitro. Proc. Soc. Exp. Biol. Med.* **124**: 1295.

Cole, C.N., J. Tornow, R. Clark, and R. Tjian. 1986. Properties of the simian virus 40 (SV40) large T antigens encoded by SV40 mutants with deletions in gene A. *J. Virol.* **57**: 539.

Conrad, S.E. and M. Botchan. 1982. Isolation and characterization of human DNA fragments with nucleotide sequence homologies with the simian virus 40 regulatory region. *Mol. Cell. Biol.* **2**: 949.

Conway, L. and M. Wickens. 1985. A sequence downstream of AAUAAA is required for formation of SV40 late mRNA 3′-ends in frog oocytes. *Proc. Natl. Acad. Sci.* **82**: 3949.

Contreras, R., D. Gheysen, J. Knowland, A. van de Voorde, and W. Fiers. 1982. Evidence for direct involvement of DNA replication origin in synthesis of late SV40 RNA. *Nature* **300**: 500.

Cosman, D.J. and M.J. Tevethia. 1981. Characterization of a temperature-sensitive DNA-positive, nontransforming mutant of simian virus 40. *Virology* **112**: 605.

Courtneidge, S.A. 1985. Activation of the pp60$^{c\text{-}src}$ kinase by middle T antigen binding or by dephosphorylation. *EMBO J.* **4**: 1471.

Courtneidge, S.A. and A.E. Smith. 1983. Polyoma virus transforming protein associates with the product of the c-*src* cellular gene. *Nature* **303**: 435.

———. 1984. The complex of polyom avirus middle T-antigen and pp60$^{c\text{-}src}$. *EMBO J.* **3**: 585.

Cowie, A. and R. Kamen. 1986. Guanine nucleotide contacts within viral DNA sequences bound by polyoma large T antigen. *J. Virol.* **57**: 505.

Cowie, A., C. Tyndall, and R. Kamen. 1981. Sequences at the capped 5′ ends of polyomavirus late region mRNA's: An example of extreme terminal heterogeneity. *Nucleic Acids Res.* **9**: 6305.

Cuzin, F., M. Rassoulzadegan, and L. Limieux. 1984. Multigenic control of tumorigenesis: Three distinct oncogenes are required for transformation of rat embryo fibroblasts by polyoma virus. *Cancer Cells* **2**: 109.

Dailey, L. and C. Basilico. 1985. Sequences in the polyoma virus DNA regulatory region involved in viral DNA replication and early gene expression. *J. Virol.* **54**: 739.

Danna, K.J. and D. Nathans. 1972. Studies of SV40 DNA IV bidirectional replication of SV40 DNA. *Proc. Natl. Acad. Sci.* **64**: 3097.

Defendi, V. 1963. Effect of SV40 virus immunization on growth of transplantable SV40 and polyoma virus tumors in hamsters. *Proc. Soc. Exp. Biol. Med.* **113**: 12.

Deininger, P., A. Esty, P. LaPorte, and T. Friedmann. 1979. Nucleotide sequence and genetic organization of the polyoma late region: Features common to the polyoma early region and SV40. *Cell* **18**: 771.

DeLucia, A.L., B.A. Lewton, R. Tjian, and P. Tegtmeyer. 1983. Topography of simian virus 40 A protein–DNA complexes: Arrangement of pentanucleotide interaction sites at the origin of replication. *J. Virol.* **46**: 216.

Deppert, W., K. Hanke, and R. Henning. 1980. Simian virus 40 T antigen-serological demonstration on simian virus 40–transformed monolayer cells *in situ. J. Virol.* **35**: 505.

de Villiers, J. and W. Schaffner. 1981. A small segment of

polyoma virus DNA enhances the expression of a cloned β-globin gene over a distance of 1,400 base pairs. *Nucleic Acids Res.* **9:** 6251.

de Villiers, J., L. Olson, C. Tyndall, and W. Schaffner. 1982. Transcriptional "enhancers" from SV40 and polyoma virus show a cell type preference. *Nucleic Acids Res.* **10:** 7965.

de Villiers, J., W. Schaffner, C. Tyndall, S. Lupton, and R. Kamen. 1984. Polyoma virus DNA replication requires an enhancer. *Nature* **312:** 242.

DiMaio, D. and D. Nathans. 1982. Regulatory mutants of simian virus 40: Effect of mutations at a T antigen binding site on DNA replication and expression of viral genes. *J. Mol. Biol.* **156:** 531.

Di Mayorca, G., J. Callender, G. Marin, and R. Giordano. 1969. Temperature-sensitive mutants of polyoma virus. *Virology* **38:** 126.

Dixon, R.A.F. and D. Nathans. 1985. Purification of simian virus 40 large T antigen by immunoaffinity chromatography. *J. Virol.* **53:** 1001.

Dulbecco, R. and G. Freeman. 1959. Plaque production by the polyoma virus. *Virology* **8:** 396.

Dulbecco, R., L.H. Hartwell, and M. Vogt. 1965. Induction of cellular DNA synthesis by polyoma virus. *Proc. Natl. Acad. Sci.* **53:** 403.

Dynan, W.S. and R. Tjian. 1983. Isolation of transcription factors that discriminate between different promoters recognized by RNA polymerase II. *Cell* **32:** 669.

Eckhart, W. 1969. Complementation and transformation by temperature-sensitive mutants of polyoma virus. *Virology* **38:** 120.

———. 1977. Complementation between temperature-sensitive and host range nontransforming mutants of polyoma virus. *Virology* **77:** 589.

Eckhart, W., M.A. Hutchinson, and T. Hunter 1979. An activity phosphorylating tyrosine in polyoma T-antigen immunoprecipitates. *Cell* **18:** 925.

Eddy, B.E. 1960. The polyoma virus section B. *Adv. Virus Res.* **7:** 91.

Eddy, B.E., G.S. Borman, G.E. Grubbs, and R.D. Young. 1962. Identification of the oncogenic substance in rhesus monkey kidney cell cultures as simian virus 40. *Virology* **17:** 65.

Ellman, M., I. Bikel, J. Figge, T. Roberts, R. Schlossman, and D.M. Livingston. 1984. Localization of the simian virus 40 small t antigen in the nucleus and cytoplasm of monkey and mouse cells. *J. Virol.* **50:** 623.

Farmiere, W.G. and W.R. Folk. 1984. Regulation of polyomavirus transcription by large tumor antigen. *Proc. Natl. Acad. Sci.* **81:** 6919.

Fenton, R.G. and C. Basilico. 1982. Changes in the topography of early region transcription during polyoma virus lytic infection. *Proc. Natl. Acad. Sci.* **79:** 7142.

Feunteun, J., L. Sompayrac, M. Fluck, and T. Benjamin. 1976. Localization of gene functions in polyoma virus DNA. *Proc. Natl. Acad. Sci.* **73:** 4169.

Fiers, W., R. Contreras, G. Haegeman, R. Rogiers, A. van der Voorde, H. van Heuverswyn, J. van Herreseghe, G. Volckaert, and M. Ysebaert. 1978. The complete nucleotide sequence of SV40 DNA. *Nature* **273:** 113.

Fischer-Fantuzzi, L. and C. Vesco. 1985. Deletion of 43 amino acids in the NH$_2$-terminal half of the large tumor antigen of simian virus 40 results in a non-karyophilic protein capable of transforming established cells. *Proc. Natl. Acad. Sci.* **82:** 1891.

Fitzgerald, M. and T. Shenk. 1981. The sequence 5'-AAUAAA-3' forms part of the recognition site for polyadenylation of late SV40 mRNAs. *Cell* **24:** 251.

Fluck, M., R.J. Staneloni, and T.L. Benjamin. 1977. *Hr-t* and *ts-a:* Two early gene functions in polyoma virus DNA. *Virology* **77:** 610.

Flyer, D. and S. Tevethia. 1982. Biology of simian virus 40 (SV40) transplantation antigen (TrAg) VIII. Retention of SV40 T-Ag sites on purified SV40 T antigen following denaturation with sodium dodecyl sulfate. *Virology* **117:** 267.

Fradin, A., R. Jove, C. Hemenway, H.D. Keiser, J.L. Manley, and C. Prives. 1984. Splicing pathways of SV40 mRNAs differ in their requirements for snRNPs. *Cell* **37:** 927.

Francke, B. and W. Eckhart. 1973. Polyoma gene function required for viral DNA synthesis. *Virology* **55:** 127.

Fried, M. 1965. Cell-transforming ability of a temperature-sensitive mutant of polyoma virus. *Proc. Natl. Acad. Sci.* **53:** 486.

———. 1970. Characterization of a temperature-sensitive mutant of polyoma virus. *Virology* **40:** 605.

Fried, M. and B.E. Griffin. 1977. Organization of the genomes of polyoma virus and SV40. *Adv. Cancer Res.* **24:** 67.

Fried, M., M. Griffiths, B. Davies, G. Bjursell, G. Lamatia, and L. Lania. 1983. Isolation of cellular DNA sequences that allow expression of adjacent genes. *Proc. Natl. Acad. Sci.* **80:** 2117.

Friedmann, T., A. Esty, P. LaPorte, and P. Deininger. 1979. Nucleotide sequence and genetic organization of the polyoma early region: Features common to the polyoma late region and SV40. *Cell* **17:** 751.

Fromm, M. and P. Berg. 1982. Deletion mapping of DNA required for SV40 early region promoter function *in vivo. J. Mol. Appl. Genet.* **1:** 457.

———. 1983. Simian virus 40 early and late region promoter functions are enhanced by the 72 base pair repeat inserted at distant locations and inverted orientations. *Mol. Cell. Biol.* **3:** 991.

Garcea, A.L. and T.L. Benjamin. 1983. Host range transforming gene of polyoma virus plays a role in virus assembly. *Proc. Natl. Acad. Sci.* **80:** 3613.

Gaudray, P., P. Clertant, and F. Cuzin. 1980. ATP phosphohydrolase (ATPase) activity of a polyomavirus T antigen. *Eur. J. Biochem.* **109:** 553.

Girardi, A.J., B.H. Sweet, V.B. Slotnick, and M.R. Hilleman. 1962. Development of tumors in hamsters inoculated in the neo-natal period with vacuolating virus, SV40. *Proc. Soc. Exp. Biol. Med.* **109:** 649.

Gluzman, Y. and T. Shenk, eds. 1983. *Enhancers and eukaryotic gene expression.* Cold Spring Harbor Laboratory, Cold Spring Harbor, New York.

Gooding, L.R., R.W.D. Geib, K.A. Connell, and E. Harlow. 1984. Antibody and cellular detection of SV40 T-antigenic determinants on the surfaces of transformed cells. *Cancer Cells* **1:** 263.

Graessman, A., M. Graessman, and C. Mueller. 1981. Regulation of SV40 gene expression. *Adv. Cancer Res.* **35:** 111.

Green, M.R, R.H. Treisman, and T. Maniatis. 1983. Activation of globin gene transcription in transient assays by *cis* and *trans*-activating function. *Cell* **35:** 137.

Griffin, B.E., M. Fried, and A. Cowie. 1974. Polyoma DNA: A physical map. *Proc. Natl. Acad. Sci.* **71:** 2077.

Gross, L. 1983. *Oncogenic viruses.* Pergamon Press, Oxford, England.

Gruss, P., R. Dhar, and G. Khoury. 1981. Simian virus 40 tandem repeated sequences as an element of the early promoter. *Proc. Natl. Acad. Sci.* **78:** 943.

Habel, K. 1961. Resistance of polyoma virus immune animals to transplanted polyoma tumors. *Proc. Soc. Exp. Biol. Med.* **106:** 722.

Habel, K. and B.E. Eddy. 1963. Specificity of resistance to tumor challenge of polyoma and SV40 virus-immune hamsters. *Proc. Soc. Exp. Biol. Med.* **113:** 1.

Hanahan, D. 1985. Heritable formation of pancreatic β-cell tumours in transgenic mice expressing recombinant insulin/simian virus 40 oncogenes. *Nature* **315:** 115.

Hansen, U. and P. Sharp. 1983. Sequences controlling in vitro transcription of SV40 promoters. *EMBO J.* **2:** 2293.

Hansen, U., D.G. Tenen, D.M. Livingston, and P.A. Sharp. 1981. T antigen repression of SV40 early transcription from two promoters. *Cell* **27:** 603.

Hart, R.P., M.A. McDevitt, and J.R. Nevins. 1985. Poly(A) site cleavage in a HeLa nuclear extract is dependent on downstream sequences. *Cell* **43:** 677.

Hartzell, S.W., B.J. Byrne, and K.N. Subramanian. 1984a.

Mapping of the late promoter of simian virus 40. *Proc. Natl. Acad. Sci.* **81:** 23.

———. 1984b. The simian virus 40 minimal origin and the 72-base pair repeat are required simultaneously for efficient induction of late gene expression with large tumor antigen. *Proc Natl. Acad. Sci.* **81:** 6335.

Hassauer, M., K.H. Scheidtmann, and G. Walter. 1986. Mapping of phosphorylation sites in polyomavirus large T antigen. *J. Virol.* **58:** 805.

Hassell, J.A., W.C. Topp, D.B. Rifkin, and P. Moreau. 1980. Transformation of rat embryo fibroblasts by cloned polyoma virus DNA fragments containing only part of the early region. *Proc. Natl. Acad. Sci.* **77:** 3978.

Hay, R.T. and M.L. DePamphilis. 1982. Initiation of SV40 DNA replication *in vivo*: Location and structure of 5′ ends of DNA synthesized in the *ori* region. *Cell* **28:** 767.

Hay, N., H. Skolnick-David, and Y. Aloni. 1982. Attenuation in the control of SV40 gene expression. *Cell* **29:** 183.

Hayday, A.C., F. Chaudry, and M. Fried. 1983. Loss of polyoma virus infectivity as a result of a single amino acid change in a region which has extensive amino acid homology with simian virus 40 large T-antigen. *J. Virol.* **45:** 693.

Hayday, A., H.E. Ruley, and M. Fried. 1982. Structural and biological analysis of integrated polyoma virus DNA and its adjacent host sequences cloned from transformed rat cells. *J. Virol.* **44:** 67.

Henry, P., P.H. Black, M.N. Oxman, and S.M. Weissman. 1966. Stimulation of DNA synthesis in mouse cell line 3T3 by simian virus 40. *Proc. Natl. Acad. Sci.* **56:** 1170.

Herbomel, P., B. Bourachot, and M. Yaniv. 1984. Two distinct enhancers with different cell specificities coexist in the regulatory region of polyoma. *Cell* **39:** 653.

Herr, W. and Y. Gluzman. 1985. Duplications of a mutated simian virus 40 enhancer restores its activity. *Nature* **313:** 711.

Huebner, R.J. 1963. Tumor virus study systems. *Ann. N. Y. Acad. Sci.* **108:** 1129.

Ide, T., S. Whelly, and R. Baserga. 1977. Stimulation of RNA synthesis in isolated nuclei by partially purified preparations of SV40 T-antigen. *Proc. Natl. Acad. Sci.* **74:** 3189.

Israel, M.A., D.T. Simmons, S.L. Hourihan, W.P. Rowe, and M.A. Martin. 1979. Interrupting the early region of polyoma virus DNA enhances tumorigenicity. *Proc. Natl. Acad. Sci.* **76:** 3713.

Ito, Y., J.R. Brocklehurst, and R. Dulbecco. 1977. Virus-specific proteins in the plasma membrane of cells lytically infected or transformed by polyoma virus. *Proc. Natl. Acad. Sci.* **74:** 4666.

Jat, P., J.W. Roberts, A. Cowie, and R. Kamen. 1982a. Comparison of the polyoma virus early and late promoters by transcription *in vitro*. *Nucleic Acids Res.* **10:** 871.

Jat, P., V. Novak, A. Cowie, C. Tyndall, and R. Kamen. 1982b. DNA sequences required for specific and efficient initiation of transcription at the polyoma virus early promoter. *Mol. Cell. Biol.* **2:** 737.

Jay, G., S. Nomura, C.W. Anderson, and G. Khoury. 1981. Identification of the SV40 agnogene product: A DNA binding protein. *Nature* **291:** 346.

Jones, K.A. and R. Tjian. 1984. Essential contact residues within SV40 large T-antigen binding sites I and II identified by alkylation interference. *Cell* **36:** 155.

Kalderon, D., W.D. Richardson, A.F. Markham, and A.E. Smith. 1984a. Sequence requirements for nuclear localization of SV40 large T antigen. *Nature* **311:** 33.

Kalderon, D., B.L. Roberts, W.D. Richardson, and A.E. Smith. 1984b. A short amino acid sequence able to specify nuclear location. *Cell* **39:** 499.

Kamen, R., P. Jat, R. Treisman, J. Favaloro, and W.R. Folk. 1982. 5′ Termini of polyoma virus early region transcripts synthesized *in vivo* by wild-type virus and viable deletion mutants. *J. Mol. Biol.* **159:** 189.

Kamen, R., J. Favaloro, J. Parker, R. Treisman, L. Lania, M. Fried, and A. Mellor. 1980. Comparison of polyoma virus transcription in productively infected mouse cells and trans-

formed rodent cell lines. *Cold Spring Harbor Symp. Quant. Biol.* **44:** 63.

Katinka, M. and M. Yaniv. 1983 DNA replication origin of polyoma virus: Early proximal boundary. *J. Virol.* **47:** 244.

Kaufman, G. 1981. Characterization of initiator RNA from replicating SV40 DNA synthesized in isolated nuclei. *J. Mol. Biol.* **147:** 25.

Keller, J.M. and J.C. Alwine. 1984. Activation of the SV40 late promoter: Direct effects of T antigen in the absence of viral DNA replication. *Cell* **26:** 381.

———. 1985. Analysis of an activatable promoter: Sequences in the simian virus 40 late promoter required for T-antigen-mediated *trans* activation. *Mol. Cell. Biol.* **5:** 1859.

Ketner, G. and T.J. Kelly. 1976. Integrated simian virus 40 sequences in transformed cell DNA: Analysis using restriction endonucleases. *Proc. Natl. Acad. Sci.* **73:** 1102.

Khera, K.S., A. Ashkenazi, F. Rapp, and J.L. Melnick. 1963. Immunity in hamsters to cells transformed *in vitro* and *in vivo* by SV40. Tests for antigenic relationship among the papova viruses. *J. Immunol.* **91:** 604.

Khoury, G. and P. Gruss. 1983. Enhancer elements. *Cell* **33:** 313.

Kimura, G. and A. Itagaki. 1975. Initiation and maintenance of cell transformation by simian virus 40: A viral genetic property. *Proc. Natl. Acad. Sci.* **72:** 673.

Kit, S., R.A. deTorres, D.R. Dubbs, and M.C. Salvi. 1967. Induction of cellular deoxyribonucleic acid synthesis by simian virus 40. *J. Virol.* **1:** 738.

Kit, S., D.R. Dubbs, P.M. Frearson, and J.C. Melnick. 1966. Enzyme induction in SV40-infected green monkey kidney cultures. *Virology* **29:** 69.

Klein, G., ed. 1982. The transformation-associated cellular p53 protein. *Adv. Viral Oncol.* **2.**

Koch, M.A. and A.B. Sabin. 1963. Specificity of virus-induced resistance to transplantation of polyoma and SV40 tumors in adult hamsters. *Proc. Soc. Exp. Biol. Med.* **113:** 4.

Kriegler, M. and M. Botchan. 1983. Enhanced transformation by a simian virus 40 recombinant virus containing a Harvey murine sarcoma virus long terminal repeat. *Mol. Cell. Biol.* **3:** 325.

Laimins, L.A., G. Khoury, C. Gorman, B. Howard, and P. Gruss. 1982. Activation of SV40 genome by 72 base pair tandem repeats from simian virus 40 and Moloney murine sarcoma virus. *Proc. Natl. Acad. Sci.* **79:** 6453.

Land, H., L.F. Parada, and R.A. Weinberg. 1983. Tumorigenic conversion of primary embryo fibroblast requires at least two cooperating oncogenes. *Nature* **304:** 596.

Lane, D.P. and L.V. Crawford. 1979. T antigen is bound to a host protein in SV40-transformed cells. *Nature* **278:** 261.

Lanford, R.E. and J. Butel. 1984. Construction and characterization of an SV40 mutant defective in nuclear transport of T antigen. *Cell* **37:** 801.

Lanford, R.E., C. Wong, and J.S. Butel. 1985. Differential ability of a T-antigen transport-defective mutant of simian virus 40 to transform primary and established rodent cells. *Mol. Cell. Biol.* **5:** 1043.

Lania, L., A. Hayday, and M. Fried. 1981. Loss of functional large T-antigen and free viral genomes from cells transformed *in vitro* by polyoma virus after passage *in vivo* as tumor cells. *J. Virol.* **39:** 422.

Lania, L., M. Griffiths, B. Cooke, Y. Ito, and M. Fried. 1980a. Untransformed rat cells containing free and integrated DNA of a polyoma nontransforming (Hrt) mutant. *Cell* **18:** 793.

Lania, L., A. Hayday, G. Bjursell, D. Gandini-Attardi, and M. Fried. 1980b. Organization and expression of integrated polyoma virus DNA in transformed rodent cells. *Cold Spring Harbor Symp. Quant. Biol.* **44:** 597.

Lania, L., D. Gandini-Attardi, M. Griffiths, B. Cooke, D. deCicco, and M. Fried. 1980c. The polyoma virus 100k large T antigen is not required for the maintenance of transformation. *Virology* **100:** 217.

Learned, R.M., S.T. Smale, M.M. Haltiner, and R. Tjian. 1983. Regulation of human ribosomal RNA transcription. *Proc. Natl. Acad. Sci.* **80:** 3558.

Lebkowski, J.S., S. Clancy, and M.P. Calos. 1985. Simian virus 40 replication in adenovirus-transformed human cells antagonizes gene expression. *Nature* **317**: 169.

Legon, S., A. Flavell, A. Cowie, and R. Kamen. 1979. Amplification in the leader sequence of late polyoma virus mRNAs. *Cell* **16**: 373.

Lewis, E.D. and J.L. Manley. 1985. Repression of simian virus 40 early transcription by viral DNA replication in human 293 cells. *Nature* **317**: 172.

Lewis, E.D., S. Chen, A. Kumar, G. Blanck, R.E. Pollack, and J.L. Manley. 1983. Frameshift mutation in the carboxy terminal encoding region of SV40 large T antigen results in a replication and transformation defective virus. *Proc. Natl. Acad. Sci.* **80**: 7065.

Li, J.J. and T.J. Kelly. 1984. Simian virus 40 DNA replication *in vitro*. *Proc. Natl. Acad. Sci.* **81**: 6973.

Li, J.J., K.W.C. Peden, R.A.F. Dixon, and T. Kelly. 1986. Functional organization of the simian virus 40 origin of DNA replication. *Mol. Cell. Biol.* **6**: 1117.

Luthman, H., M.G. Nilsson, and G. Magnusson. 1982. Noncontiguous segments of the polyoma genome required in *cis* for DNA replication. *J. Mol. Biol.* **161**: 533.

MacPherson, I. and L.M. Montagnier. 1964. Agar suspension culture for the selective assay of cells transformed by polyoma virus. *Virology* **23**: 291.

Majello, B., G. La Mantia, A. Simeone, E. Boncinelli, and L. Lania. 1985. Activation of major histocompatibility complex class I mRNA containing an *Alu*-like repeat in polyoma virus-transformed rat cells. *Nature* **314**: 457.

Manley, J.L., H. Yu, and L. Ryner. 1985. RNA sequence containing hexanucleotide AAUAAA directs efficient mRNA polyadenylation *in vitro*. *Mol. Cell. Biol.* **5**: 373.

Manos, M.M. and Y. Gluzman. 1984. Simian virus 40 large T-antigen point mutants that are defective in viral DNA replication but competent in oncogenic transformation. *Mol. Cell. Biol.* **4**: 1125.

―――. 1985. Genetic and biochemical analysis of transformation-competent, replication-defective simian virus 40 large T antigen mutants. *J. Virol.* **53**: 120.

Martin. R.G. and J.Y. Chou. 1975. Simian virus 40 functions required for the establishment and maintenance of the transformed state. *J. Virol.* **15**: 599.

Martin, R.G., V.P. Setlow, C.A.F. Edwards, and D. Vembo. 1979. The roles of the simian virus 40 tumor antigens in transformation of Chinese hamster lung cells. *Cell* **17**: 635.

McCormick, F., R. Clark, E. Harlow, and R. Tjian. 1981. SV40 T antigen binds specifically to a cellular 53K protein *in vitro*. *Nature* **292**: 63.

Michaeli, T. and C. Prives. 1985. Regulation of simian virus 40 gene expression in *Xenopus laevis* oocytes. *Mol. Cell. Biol.* **5**: 2019.

Miller, T.J., D.L. Stephens, and J.E. Mertz. 1982. Kinetics of accumulation and processing of simian virus 40 RNA in *Xenopus laevis* oocytes injected with simian virus 40 DNA *Mol. Cell. Biol.* **2**: 1581.

Moreau, P., R. Hen, B. Wasylyk, R. Everett, M.P. Gaw, and P. Chambon. 1981. The SV40 72 base pair repeat has a striking effect on gene expression both in SV40 and other chimeric recombinants. *Nucleic Acids Res.* **9**: 6047.

Mueller, C.R., A.-M. Mes-Masson, M. Bouvier, and J.A. Hassell. 1984. Location of sequences in polyomavirus DNA that are required for early gene expression *in vivo* and *in vitro*. *Mol. Cell. Biol.* **4**: 2594.

Muller, W.J., C.R. Mueller, A.-M. Mes, and J.A. Hassell. 1983. Polyomavirus origin for DNA replication comprises multiple genetic elements. *J. Virol.* **47**: 586.

Murakami, Y., T. Eki, M. Yamada, C. Prives, and J. Hurwitz. 1986a. *In vitro* synthesis of DNA containing the polyoma virus replication origin and requiring the polyoma T antigen. *Proc. Natl. Acad. Sci.* **83**: (in press).

Murakami, Y., C.R. Wobbe, L. Weissbach, F.B. Dean, and J. Hurwitz. 1986b. Role of DNA polymerase α and DNA primase in simian virus 40 DNA replication *in vitro*. *Proc. Natl. Acad. Sci.* **83**: 2869.

Ng, S.C., J.E. Mertz, S. Sander-Will, and M. Bina. 1985. Simian virus 40 maturation in cells harboring mutants deleted in the agnogene. *J. Biol. Chem.* **260**: 1127.

Noble, J.C.S., C. Prives, and J. Manley. 1986. *In vitro* splicing of simian virus 40 early pre mRNA. *Nucleic Acids Res.* **3**: 1219.

Novak, Y., S.M. Dilworth, and B.E. Griffin. 1980. Coding capacity of a 35% fragment of the polyoma virus genome is sufficient to initiate and maintain cellular transformation. *Proc. Natl. Acad. Sci.* **77**: 3278.

Osborn, M. and K. Weber. 1975. Simain virus 40 gene A function and maintenance of transformation. *J. Virol.* **15**: 636.

Paucha, E., D. Kalderon, R.W. Harvey, and A.E. Smith. 1986. Simian virus 40 origin binding domain on large T antigen. *J. Virol.* **57**: 50.

Piatak, M., P.K. Ghosh, L.C. Norkin, and S.M. Weissman. 1983. Sequences locating the 5′ ends of the major simian virus 40 late mRNA forms. *J. Virol.* **48**: 503.

Pintel, D., N. Bouck, and G. Di Mayorca. 1981. Separation of lytic and transforming functions of the simian virus 40 A region: Two mutants which are temperature sensitive for lytic functions have opposite effects on transformation. *J. Virol.* **38**: 518.

Pipas, J.M., K.W.C. Peden, and D. Nathans. 1983. Mutational analysis of simian virus 40 T antigen: Isolation and characterization of mutants with deletions in the T-antigen gene. *Mol. Cell. Biol.* **3**: 203.

Prives, C. and H. Shure. 1979. Cell free translation of simian virus 40 16S and 19S L-strand specific mRNA classes to simian virus 40 major VP-1 and minor VP-2 and VP-3 capsid proteins. *J. Virol.* **29**: 1204.

Prives, C., L. Covey, A. Scheller, and Y. Gluzman. 1983. DNA-binding properties of simian virus 40 T-antigen mutants defective in viral DNA replication. *Mol. Cell. Biol.* **3**: 1958.

Prives, C., B. Barnet, A. Scheller, G. Khoury, and G. Jay. 1982. Discrete regions of simian virus 40 large T antigen are required for nonspecific and viral origin-specific DNA binding. *J. Virol.* **43**: 73.

Rassoulzadegan, M., A. Cowie, A. Carr, N. Glaichenhaus, R. Kamen, and F. Cuzin. 1982. The role of individual polyoma virus early proteins in oncogenic transformation. *Nature* **300**: 713.

Rassoulzadegan, M., Z. Naghashfar, A. Cowie, A. Carr, M. Grisoni, R. Kamen, and F. Cuzin. 1983. Expression of the large T protein of polyoma virus promotes the establishment in culture of "normal" rodent fibroblast cell lines. *Proc. Natl. Acad. Sci.* **80**: 4354.

Reddy, V.B., B. Thimmappaya, R. Dhar, K.N. Subramanian, B.S. Zain, J. Pan, P.K. Ghosh, M.L. Celma, and S.M. Weissman. 1978. The genome of simian virus 40. *Science* **200**: 494.

Richardson, W.D., B.L. Roberts, and A.E. Smith. 1986. Nuclear location signals in polyoma virus large-T. *Cell* **44**: 77.

Rio, D., A. Robbins, R. Myers, and R. Tjian. 1980. Regulation of simian virus 40 early transcription *in vitro* by a purified tumor antigen. *Proc. Natl. Acad. Sci.* **77**: 5706.

Robbins, D., D.C. Rio, and M.R. Botchan. 1986. *Trans* activation of the simian virus 40 enhancer. *Mol. Cell. Biol.* **6**: 1283.

Rowe, W.P., R.J. Huebner, and J.W. Hartley. 1961. B. Ecology of a mouse tumor virus. *Perspect. Virol.* **2**: 177.

Ruley, H.E. and M. Fried. 1983a. Sequence repeats in a polyoma virus DNA region important for gene expression. *J. Virol.* **47**: 233.

―――. 1983b. Clustered illegitimate recombination events in mammalian cells involving very short sequence homologies. *Nature* **304**: 181.

Ruley, H.E., L. Lania, F. Chaudry, and M. Fried. 1982. Use of a cellular polyadenylation signal by viral transcripts in polyoma virus transformed cells. *Nucleic Acids Res.* **10**: 4515.

Rundell, K. 1982. Presence in growth-arrested cells of cellular proteins that interact with simian virus 40 small t antigen. *J. Virol.* **42**: 1135.

Rundell, K., E.O. Major, and M. Lampert. 1981. Association of cellular 56,000 and 32,000 molecular weight proteins with

BK virus and polyoma virus t antigens. *J. Virol.* **37**: 1090.

Rutila, J.E., M.J. Imperiale, and W.W. Brockman. 1986. Replication and transformation functions of *in vitro*-generated simian virus 40 large T antigen mutants. *J. Virol.* **58**: 526.

Ryder, K., E. Vakalopoulou, R. Mertz, I. Mastrangelo, P. Hoagh, P. Tegtmeyer, and E. Fanning. 1985. Seventeen base pairs of region 1 encode a novel tripartite binding signal for SV40 T antigen. *Cell* **42**: 539.

Sadofsky, M., S. Connelly, J. Manley, and J.C. Alwine. 1985. Identification of a sequence element on the 3′ side of AAUAAA which is necessary for simian virus 40 late mRNA 3′-end processing. *Mol. Cell. Biol.* **5**: 2713.

Sambrook, J., H. Westphal, P.R. Srinivasan, and R. Dulbecco. 1968. The integrated state of viral DNA in SV40-transformed cells. *Proc. Natl. Acad. Sci.* **60**: 1288.

Santos, M. and J.S. Butel. 1982. Association of SV40 large tumor antigen and cellular proteins on the surface of SV40 transformed mouse cells. *Virology* **120**: 1.

Sassone-Corsi, P., A. Wildeman, and P. Chambon. 1985. A transacting factor is responsible for the simian virus 40 enhancer activity *in vitro*. *Nature* **313**: 458.

Schaffhausen, B.S. and T.L. Benjamin. 1979. Phosphorylation of polyoma T antigens. *Cell* **30**: 481.

Scheidtmann, K.H., B. Echle, and G. Walter. 1982. Simian virus 40 large T antigen is phosphorylated at multiple sites clustered in two separate regions. *J. Virol.* **44**: 116.

Scheller, A. and C. Prives. 1985. Simian virus 40 and polyoma virus large tumor antigens have different requirements for high affinity sequence-specific DNA binding. *J. Virol.* **54**: 532.

Schöler, H.R. and P. Gruss. 1984. Specific interaction between enhancer-containing molecules and cellular components. *Cell* **36**: 403.

Schutzbank, T., R. Robinson, M. Oren, and A.J. Levine. 1982. SV40 large tumor antigen can regulate some cellular transcripts in a positive fashion. *Cell* **30**: 481.

Scott, M.R.D., K.-H. Westphal, and P.W.J. Rigby. 1983. Activation of mouse genes in transformed cells. *Cell* **34**: 557.

Shenk, T.E., J. Carbon, and P. Berg. 1976. Construction and analysis of viable deletion mutants of simian virus 40. *J. Virol.* **18**: 664.

Shyamala, M., C.L. Atcheson, and H. Kasamatsu. 1982. Stimulation of host centriolar antigen in TC7 cells by simian virus 40: Requirement for RNA and protein synthesis and an intact simian virus 40 small t gene function. *J. Virol.* **43**: 721.

Simanis, V. and D.P. Lane. 1985. An immunoaffinity purification procedure for SV40 large T antigen. *Virology* **144**: 88.

Simmons, D.T. 1986. DNA-binding region of the simian virus 40 tumor antigens. *J. Virol.* **57**: 776.

Singh, K., M. Carey, S. Saragosti, and M. Botchan. 1985. Expression of enhanced levels of small RNA polymerase III transcripts encoded by the B2 repeats in simian virus 40—transformed mouse cells. *Nature* **314**: 553.

Sjogren, H.O. 1965. Transplantation methods as a tool for detection of tumor-specific antigens. *Prog. Exp. Tumor Res.* **6**: 289.

Sjogren, H.O., I. Helstrom, and G. Klein. 1961. Transplantation of polyomavirus-induced tumors in mice. *Cancer Res.* **21**: 329.

Sleigh, M.J., W.C. Topp, R. Hanich, and J.F. Sambrook. 1978. Mutants of SV40 with an altered small t protein are reduced in their ability to transform cells. *Cell* **14**: 79.

Smith, A.E., R. Smith, B.E. Griffin, and M. Fried. 1979. Protein kinase activity associated with polyoma virus middle T-antigen *in vitro*. *Cell* **18**: 915.

Smith, A.E., R. Kamen, W.F. Mangel, H. Shure, and T. Wheeler. 1976. Location of the sequences coding for capsid proteins VP-1 and VP-2 on polyoma virus DNA. *Cell* **9**: 481.

Smith, H.S., C.D. Scher, and G.J. Todaro. 1971. Induction of cell division in medium lacking serum growth factors by SV40. *Virology* **44**: 359.

Soeda, E. and B.E. Griffin. 1978. Sequences from the genome of a non-transforming mutant of polyoma virus. *Nature* **276**: 294.

Soeda, E., J.R. Arrand, N. Smolar, J. Walsh, and B.E. Griffin. 1980. Coding potential and regulatory signals of the polyoma virus genome. *Nature* **283**: 445.

Spandidos, D.A. and N.M. Wilkie. 1983. Host specificity of papilloma virus, Moloney murine sarcoma virus and simian virus 40 enhancer sequences. *EMBO J.* **2**: 1193.

Steinberg, B., R. Pollack, W. Topp, and M. Botchan. 1978. Isolation and characterization of T-antigen negative revertants from a line of transformed rat cells containing one copy of the SV40 genome. *Cell* **13**: 19.

Stewart, S.E. 1960. The polyoma virus section A. *Adv. Virus Res.* **7**: 61.

Stillman, B.W. and Y. Gluzman. 1985. Replication and supercoiling of simian virus 40 DNA in cell extracts from human cells. *Mol. Cell. Biol.* **5**: 2051.

Stillman, B., R.D. Gerard, R.A. Guggenheimer, and Y. Gluzman. 1985. T antigen and template requirements for SV40 DNA replication *in vitro*. *EMBO J.* **4**: 2933.

Stoker, M. 1968. Abortive transformation by polyoma virus. *Nature* **218**: 234.

Stoker, M. and R. Dulbecco. 1969. Abortive transformation by the *tsA* mutant of polyoma virus. *Nature* **223**: 397.

Stoker, M. and I. MacPherson. 1961. Studies on transformation of hamster cells by polyoma virus *in vitro*. *Virology* **14**: 359.

Stringer, J.R. 1982. DNA sequence homology and chromosomal deletion at a site of SV40 DNA integration. *Nature* **296**: 363.

Sundin, O. and A. Varshavsky. 1980. Terminal stages of SV40 DNA replication proceed via multiply intertwined catenated dimers. *Cell* **21**: 103.

Sweet, B.H. and M.R. Hilleman. 1960. The vacuolating virus, SV40. *Proc. Soc. Exp. Biol. Med.* **105**: 420.

Swimmer, C. and T. Shenk. 1984. A viable simian virus 40 variant that carries a newly generated sequence reiteration in place of the normal duplicated enhancer element. *Proc. Natl. Acad. Sci.* **81**: 6642.

Takahashi, K., M. Vigneron, H. Matthes, A. Wildeman, M. Zenke, and P. Chambon. 1986. Requirement for sterospecific alignments for initiation from the simian virus 40 early promoter. *Nature* **319**: 121.

Tegtmeyer, P. 1972. Simian virus 40 DNA synthesis: The viral replicon. *J Virol.* **10**: 591.

———. 1975. Function of simian virus 40 gene A in transforming infection. *J. Virol.* **15**: 613.

Tevethia, S.S. 1980. Immunology of simian virus 40. In *Viral oncology* (ed. G. Klein), p. 581. Raven Press, New York.

———. 1983. Cytolytic T lymphocyte responses to SV40. *Surv. Immunol. Rev.* **2**: 312.

Tevethia, S.S. and F. Rapp. 1966. Prevention and interruption of SV40 induced transplantation immunity with tumor cell extracts. *Proc. Soc. Exp. Biol. Med.* **123**: 612.

Tevethia, S.S., D.C. Flyer, and R. Tjian. 1980. Biology of simian virus 40 (SV40) transplantation antigen (TrAg) VI: Mechanism of induction of SV40 T antigen (D2 protein). *Virology* **107**: 13.

Todaro, G.J and H. Green. 1964. An assay for cellular transformation by SV40. *Virology* **23**: 117.

Tooze, J., ed. 1981. *Molecular biology of tumor viruses*, 2nd edition, revised: *DNA tumor viruses*. Cold Spring Harbor Laboratory, Cold Spring Harbor, New York.

Tornow, J. and C.N. Cole. 1983a. Intracistronic complementation in the simian virus 40 A gene. *Proc. Natl. Acad. Sci.* **80**: 6312.

———. 1983b. Nonviable mutants of simian virus 40 with deletions near the 3′ end of gene A define a function for large T antigen required after onset of viral DNA replication. *J. Virol.* **47**: 47.

Treisman, R. 1980. Characterization of polyoma late mRNA leader sequence by molecular cloning and DNA sequence analysis. *Nucleic Acids Res.* **8**: 4867.

Treisman, R. and T. Maniatis. 1985. Simian virus 40 enhancer increases the number of RNA polymerase II molecules on linked DNA. *Nature* **315**: 72.

Treisman, R.H., V. Novack, J. Favaloro, and R. Kamen. 1981. Transformation of rat cells by an altered polyoma virus genome expressing only the middle T protein. *Nature* **292**: 595.

Tyndall, C., G. LaMantia, C.M. Thacker, J. Favaloro, and R. Kamen. 1981. A region of the polyoma virus genome between the replication origin and late protein coding sequences is required in *cis* for both early gene expression and viral DNA replication. *Nucleic Acids Res.* **9**: 6231.

Velcich, A. and E. Ziff. 1985. Adenovirus E1a proteins repress transcription from the SV40 early promoter. *Cell* **40**: 705.

Vogt, M. and R. Dulbecco. 1960. Virus-cell interaction with a tumor-producing virus. *Proc. Natl. Acad. Sci.* **46**: 365.

Weber, F. and W. Schaffner. 1985. Simian virus 40 enhancer increases RNA polymerase density with the linked gene. *Nature* **315**: 75.

Weiher, H., M. Konig, and P. Gruss. 1983. Multiple point mutations affecting the simian virus 40 enhancer. *Science* **219**: 626.

Wickens, M. and P. Stephenson. 1984. Role of the conserved RNA 3′ end formation. *Science* **226**: 1045.

Wildeman, A.G., P. Sassone-Corsi, T. Grundstrom, M. Zenke, and P. Chambon. 1984. Stimulation of *in vitro* transcription from the SV40 early promoter by the enhancer involves a specific *trans*-acting factor. *EMBO J.* **3**: 3129.

Wilson, J.B., A. Hayday, S. Courtneidge, and M. Fried. 1986. Frameshift in polyoma early region generates two new proteins that define T-antigen functional domains. *Cell* **44**: 477.

Winocour, E. and L. Sachs. 1959. A plaque assay for the polyoma virus. *Virology* **8**: 397.

Yang, Y.-C., P. Hearing, and K. Rundell. 1979. Cellular proteins associated with simian virus 40 early gene products in newly infected cells. *J. Virol.* **32**: 147.

Yonemoto, W., M. Jarvis-Morar, J.S. Brugge, J.B. Bolen, and M.A. Israel. 1985. Novel tyrosine phosphorylation within the aminoterminal domain of pp60$^{c\text{-}src}$ molecules associated with polyoma virus middle tumor antigen. *Proc. Natl. Acad. Sci.* **82**: 4568.

Yuen, L.-L.-C. and R.A. Consigli. 1985. Identification and protein analysis of polyomavirus assembly intermediate from infected primary mouse embryo cells. *Virology* **144**: 127.

Zhu, Z., G.M. Veldman, A. Cowie, A. Carr, B. Schaffhausen, and R. Kamen. 1984. Construction and functional characterization of polyomavirus genomes that separately encode the three early proteins. *J. Virol.* **51**: 170.

# Papillomaviruses: Retrospectives and Prospectives

**T.R. Broker\* and M. Botchan**[†]

\*Biochemistry Department and Cancer Center, University of Rochester School of Medicine, Rochester, New York 14642; [†]Molecular Biology and Virus Laboratory, University of California, Berkeley, Berkeley, California 94720

## Historical overview

The papillomaviruses occupy a unique place in the development of tumor virus research. Ciuffo (1907) demonstrated that the infectious agent for human verruca vulgaris (common warts) persisted in filtered homogenates, ruling out a bacterial or protozoan etiology and raising the likelihood they were caused by the newly recognized submicroscopic entities that came to be called viruses. The cottontail rabbit (Shope) papillomavirus (CRPV) was the first oncogenic DNA virus to be isolated and characterized (Shope and Hurst 1933). Shope virus–induced benign rabbit papillomas were observed progressing to carcinomas (Rous and Beard 1935), and this experimental model system became, and remains today, a critical tool for the study of viral oncology. The notion of co-carcinogenesis involving papillomavirus infection and chemical mutagenesis was discovered and intensively exploited in the evaluation of potential tumor promoters (Rous and Kidd 1936; Rous and Friedewald 1944; Syverton 1952). X-ray irradiation also was shown to contribute to progression of papillomas to carcinomas in experimental animals and in human patients (Syverton et al. 1941; Holinger and Rabbett 1953). Cancer thus became regarded as the consequence of a multistep process. This work set much of the ensuing pattern of experimentation on mechanisms of tumorigenesis. Another landmark event was the establishment of transplantable Shope virus–induced tumors (Kidd and Rous 1940; Rogers et al. 1960), which have since been passed continuously from rabbit to rabbit, serving as a reproducible source of experimental material. Recently, permanent cell lines have been established from these tumors and shown to contain and express the viral CRPV DNA (McVay et al. 1982; Nasseri and Wettstein 1984b; Georges et al. 1985). Shope papillomavirus DNA was among the first genomes to be recovered in homogeneous preparations and physically and chemically characterized (Watson and Littlefield 1960). The first demonstration of eukaryotic cell transformation and induction of carcinomas in mammals using isolated CRPV DNA predated DNA transfection in cell cultures by many years (Ito and Evans 1961). Bovine papillomavirus (BPV) was shown to be capable of transforming both bovine (Black et al. 1963) and mouse (Thomas et al. 1964) cells, which set the stage for the genetic dissection of BPV over the past few years.

Ironically, papillomaviruses received relatively little attention in the modern era of molecular virology compared with the SV40 and polyomaviruses because no in vitro cell system could be found for their lytic propagation and genetic manipulation. Despite this handicap, several separate lines of investigation have yielded results that emphasize the scientific as well as medical importance of the papillomaviruses. First, with the advent of recombinant DNA technology to permit the clonal isolation and characterization of infectious virus types, there is clear epidemiological, clinical, and molecular evidence that links the papillomaviruses to widespread, serious human diseases, particularly to carcinomas of the genital and oral mucosa. Second, with the technical means for site-directed mutagenesis of DNA and its transfection into cell cultures, it has been possible to make progress in defining the *cis* and *trans* functions involved in BPV transformation, replication, and regulation. Third, because the viruses replicate as controlled episomes in vivo and in transformed cells in culture (Amtmann et al. 1980; Lancaster 1981; Law et al. 1981; Moar et al. 1981; Pfister et al. 1981a; Wettstein and Stevens 1982), they provide an interesting model for viral latency and cell cycle control of DNA replication. The episomal state of the viral DNA also elicited the attention of biotechnologists with the intriguing possibility of serving as stable, high-copy-number vectors for expressing foreign genes in mammalian cell cultures. These viruses provide exciting challenges for the molecular biologist who wishes to pursue research of direct clinical relevance; clearly, more-detailed definition of the structures and regulation of the viral products will have direct medical applications. Accordingly, we emphasize in this review the molecular, cellular, and clinical aspects of the papillomaviruses and attempt to show how a more-detailed understanding will interface with cancer research.

## Basic properties

The papillomaviruses have been classified as members of the papovavirus family, along with mouse polyomavirus, simian vacuolating virus 40 (SV40), and the human BK and JC viruses (Melnick 1962; Melnick et al. 1974). All have a closed circular, double-stranded DNA genome that is complexed with histones, condensed into nucleosomes, and encapsidated in an icosahedral virion. The mature particles are composed of protein capsomeres and lack a lipid-containing membrane envelope. Recent molecular and genetic studies of the member viruses reveal that this taxonomic association is incorrect and that the papillomaviruses constitute a distinct group of viruses. Briefly, the capsids of the papillomaviruses (55 nm diameter) are considerably larger than those of the SV40-polyomavirus group (45 nm). This size difference accommodates the 50% longer DNA molecules of the papillomaviruses (~7900 bp) compared with those of the SV40-polyomavirus group (~5250 bp). Most importantly, the genomes of the two groups of viruses bear virtually no similarity in the sequences of the DNA, in their genetic organization of open reading frames (ORFs), in their patterns of RNA synthesis and processing, or in their growth requirements. For instance, in SV40 and polyomavirus, RNA transcription of genes for early and late functions diverges from a single, complex control region along opposite strands of the DNA, with early and late messages converging about halfway around the molecule. In the papillomaviruses, all major ORFs are encoded and transcribed along only one of the two DNA strands (Figs. 1 and 2), and production of messages for early and late functions requires interspersed transcription and processing signals (cf. other papillomavirus papers, this volume). Furthermore, the SV40-polyomavirus group naturally infects a variety of internal organs and neural tissues and can be propagated in vitro. Most papillomaviruses have a single host and grow only in differentiated cutaneous or mucosal epithelium at specific anatomical sites. Bovine and deer papillomaviruses (DPV) will also infect dermal fibroblasts in vivo.

There are papillomaviruses of diverse kinds of vertebrates, including amphibians, reptiles, birds, and a wide range of mammals (Lancaster and Olson 1982; Pfister 1984; Pfister et al. 1986). From some species, multiple types and subtypes of papillomaviruses have been isolated. They can be a particular problem in domesticated herds and polygamous animals, probably because of the ease of transmission between individuals. Yet because of the stability of the virions to heat and dessication, communication need not involve direct contact. For instance, farm animals can acquire cutaneous infections through abrasions resulting from rubbing against contaminated fences.

The clinical importance of the papillomaviruses stems from their nearly ubiquitous affliction of the human population and the severity to which some lesions progress. Forty-two types of human papillomaviruses (HPVs) have been molecularly cloned from various kinds of lesions (Gissmann and Schwarz 1985; Pfister et al. 1986) (Table 1). In general, viruses trophic for external skin do not establish primary infections of the mucosa and vice-versa. On closer examination, there is a low level of crossover, possibly in association with coinfection by a type normally found in the particular tissue.

**Virus-host interactions**
The papillomaviruses appear to be restricted to the differentiated keratinocytes of epithelia for vegetative replication, and they remain undetectable, presumably latent, in the basal stem cells. This indicates that the stages of the viral life cycle may require specific factors only provided by the sequential differentiated states of the permissive tissue. A brief description of differentiation of the epithelium is necessary to understand what appears to be the natural progression of viral infection (see Fig. 3). The epidermis generally matures over a 10–14-day interval by progressive vertical differentiation. The proliferating cells are confined to the basal monolayer. The suprabasal daughter cells do not divide again, become committed to differentiate, and begin a succession of changes in keratin gene expression (Franke et al. 1986). The genes for sequentially higher molecular-weight species of keratins are activated and, in turn, inactivated to yield the sets of intermediate filaments characteristic of the stratum spinosum and stratum granulosum. In cutaneous epithelia, the process culminates with dense cross-linking of keratins via disulfide linkages to form the macrofibrils in the tough and relatively impermeable stratum corneum. The superficial cells are anucleate and incapable of supporting further genetic expression of associated papillomaviruses. This layer constantly desquamates (sloughs off) as it is replaced from beneath. Internal mucosal epithelium expresses somewhat different sets of keratins as it differentiates and does not develop a cornified surface. It is kept moist by association with interspersed secretory glands. Epithelial cell gene expression exhibits significant variations with respect to anatomical site and it changes throughout the life of the organism.

Infection with papillomaviruses apparently requires direct physical access to the basal cells exposed in a wound, and it is speculated that establishment of the infection may depend on activation of cell division during healing. The "transformation zones" between columnar and squamous epithelium located in the nasal mucosa, the larynx, the uterine cervix, and at the borders of healing wounds are highly susceptible to the mucosatrophic viral types. The reservoir of viral DNA in infected but clinically asymptomatic epithelia almost certainly must be the basal cell monolayer since those are the only cells in the epithelia that divide.

The papillomaviruses induce warts, which are localized lesions with many different morphological and histological manifestations (Croissant et al. 1985). Hyperplasia may result from a combination of accelerated cell division of the basal cells and delayed maturation of the superficial keratinocytes, protracting their transit time to the surface (Steinberg 1986). Some types of warts remain strictly benign, with the dividing cells confined as usual to the basal monolayer. Others can become dysplastic to varying degrees, and may exhibit nuclear atypia with continued replication and division of suprabasal cells, abnormal mitoses, and possible shifts in chromosome ploidy (Spriggs et al. 1971; Fu et al. 1981a,b; Reid et al. 1984). Some papillomavirus lesions can progress to carcinoma in situ, characterized by the presence of dividing cells through the entire thickness of the epithelium and by host chromosomal breakage and aneuploidy. Penetration of the transformed epithelial cells through the basement membrane results in invasive carcinoma, and this can lead to epithelial cell metastases to remote sites and to death of the host.

## Epidemiology and clinical manifestations

The World Health Organization estimates that the annual worldwide incidence of cervical carcinoma is about 450,000 cases, and that other malignancies of the genital and oral mucosa account for approximately 150,000 additional cases. Even with medical intervention, about 45% of the patients eventually die of the diseases; left untreated, the disease is generally fatal due to metastasis. Epidemiological evidence distinctly indicates that cervical carcinoma and other high-grade lesions of the female and male genital tracts derive from a sexually transmitted disease. For example, cervical cancer is not observed in nuns and other virgins (Rigoni-Stern 1842), whereas sexual consorts of men whose first wives had cervical cancer are at much higher risk themselves (see Gross et al. 1985; Kessler 1986). Many of the "risk factors" for these cancers are indirect parameters of life style but are indicative of a venereal etiology: onset of sexual relations at an early age, sexual promiscuity, use of oral contraceptives, lower socioeconomic status, and multiple kinds and episodes of infection (Brinton 1986; Doll 1986). Cervical dysplasias and carcinomas are more frequent among tobacco smokers than nonsmokers (Vessey 1986), and the presence of tobacco metabolites in cervical mucus suggests a direct chemical effect on the infected cells (Sasson et al. 1985). Attention focused successively on a variety of microorganisms as causative or contributory agents, including bacteria (*Treponema pallidum* (syphillis), *Neisseria gonorrhoeae*, and *Chlamydia trachomatis*), protozoans (*Trichomonas vaginalis*), and viruses (herpes simplex and cytomegalovirus), individually or in combination. None have withstood a rigorous examination and a sine qua non etiologic criterium.

Establishing the association of papillomaviruses with a variety of benign and malignant genital tract lesions was a gradual process that had parallels in the characterization of wart histopathology in experimental animal systems. Genital warts (condylomata acuminata) have been recognized since classical antiquity as a venereal disease (Rowson and Mahy 1967; Oriel 1971) and received renewed epidemiological attention as a consequence of wartime activities (Barrett and Silbar 1954). Characteristic cytological changes—notably koilocytosis, a large clear zone surrounding a distorted, pyknotic nucleus—were observed in a spectrum of low-to-moderate-grade lesions of the female genital tract (Koss and Durfee 1956; see Croissant et al. 1985). Papillomavirus particles were visualized in such condylomas (Dunn and Ogilvie 1968) and in papillomas of the oral mucosa (Frithiof and Wersall 1967). Additional cytological abnormalities were recognized as very similar to the histopathologies of papillomavirus-infected epithelia of experimental animals (Meisels et al. 1977). HPV-6 and HPV-11 DNA have been molecularly cloned from benign genital and laryngeal lesions (de Villiers et al. 1981; Gissmann et al. 1982b; Mounts et al. 1982).

Progression of genital warts to carcinomas occasionally occurs (zur Hausen 1977; Syrjänen 1986). Such carcinomas have no evidence of viral capsid proteins but became linked to the papillomaviruses through their physical and temporal association with the lower-grade condylomata and dysplasias and, ultimately, their containing papillomavirus DNA and RNA, which were revealed when appropriate types of HPV nucleic acid probes became available in the past few years (Dürst et al. 1983; Boshart et al. 1984; Beaudenon et al. 1986; Kahn et al. 1986; Kawashima et al. 1986; Lorincz et al. 1986a,b). The correlation of several different manifestations of genital carcinomas with papillomaviruses is now very convincing. Many primary carcinomas, metastatic tumors (Ostrow et al. 1982; Stremlau et al. 1985; Lancaster et al. 1986), and most cell lines derived from cervical tumors such as HeLa, Siha, and Caski contain HPV-16, HPV-18, or related types of HPV DNA, often integrated into one or more host chromosomal sites and usually actively transcribed (Lehn et al. 1984, 1985; Grussendorf-Conen et al. 1985; Pater and Pater 1985; Schwarz et al. 1985; Yee et al. 1985; DiLuca et al. 1986; Dürst et al. 1986; McCance et al. 1986; Tsunokawa et al. 1986b; H. zur Hausen, pers. comm.). Comparable molecular investigations of some carcinomas of the larynx (Brandsma et al. 1986; de Villiers et al. 1986; Kahn et al. 1986) also revealed integrated HPV-16 or HPV-30 sequences. These findings may be the most crucial evidence linking the HPVs to tumorigenesis. It is important to recognize that there are at least two distinct diseases of the mucosa, and they have partially overlapping pathological manifestations. Those associated with viruses related to HPV-6 will usually remain benign, and those with viruses related to HPV types 16 and 18 pose a significant risk of oncogenic conversion. Thus, regarding all the papillomavirus-containing lesions as a continuum of pathologies in a single disease process comparable to the rabbit (Shope) virus paradigm may be a misappraisal.

The correlation of certain papillomaviruses with high-grade genital dysplasias and cancers is qualified only by two concerns: A low percentage of lesions do not give a positive signal when screened for the presence of HPVs, but as more and more viral types have been isolated from such lesions and become available as cloned probes, fewer samples remain negative. The second issue is that the presence of HPV-16 or HPV-18 does not seem to be sufficient to assure the development of a high-grade lesion into a carcinoma, since only a few percent of all such cases progress (H. zur Hausen, pers. comm.). This emphasizes that other etiological agents or cofactors must be involved. Among those seriously regarded as possibly contributing are:

1. the presence of other viral and microbial infections, which may result in alterations of host and papillomavirus gene expression;
2. growth factor and oncogene activation;
3. host immune status;
4. local inflammatory responses to antigens and metabolites;
5. radiation exposure;

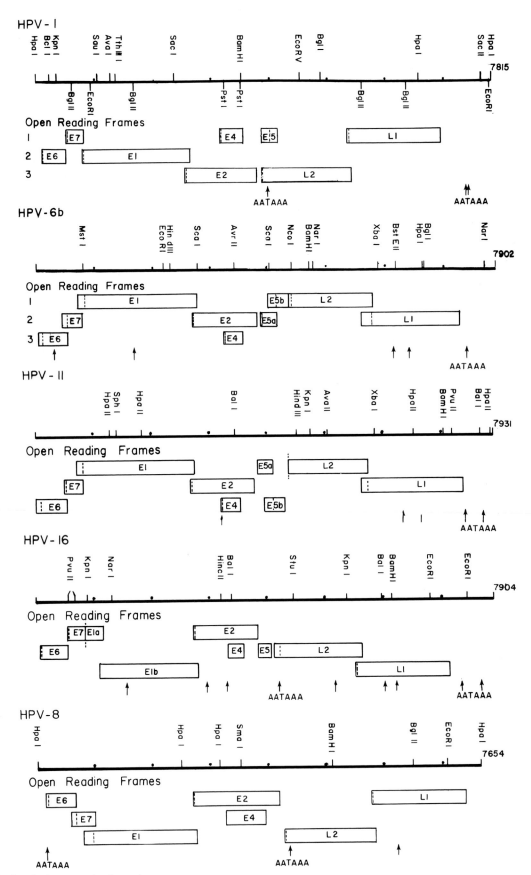

**Figure 1** *See facing page for legend.*

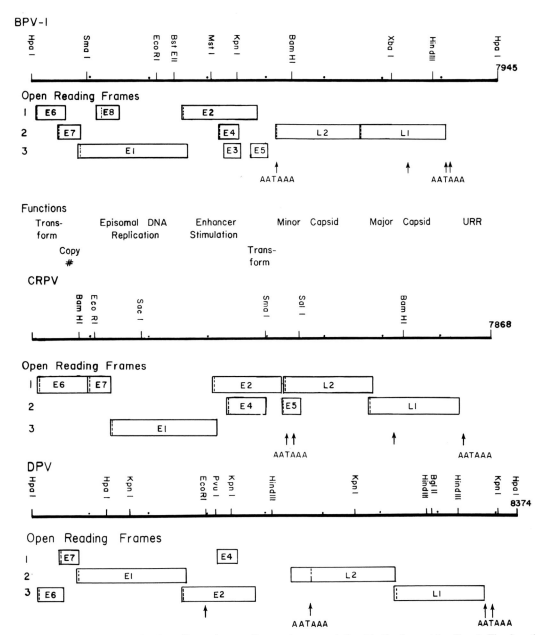

**Figure 2** Genetic organization of animal papillomaviruses. Conventions are defined in the legend for Fig. 1. The functions of the ORFs were assigned as discussed and referenced in the text. Bovine papillomavirus 1 (BPV-1; Chen et al. 1982; see also Ahola et al. 1983; Danos et al. 1983; Schwarz et al. 1983); cottontail rabbit (Shope) papillomavirus: (CRPV; Giri et al. 1985); deer papillomavirus (DPV; Groff and Lancaster 1985).

**Figure 1** (*See facing page*). Genetic organization of human papillomaviruses. The papillomavirus DNA molecule is circular, but the genome is represented according to convention as a linear molecule opened at a homologous position (nucleotide 1, often an *Hpa*I restriction site) in the upstream regulatory region (URR) immediately preceding the E6 open reading frame (ORF). Consequently, many of the transcriptional control signals for the E- and L-region messages are displaced to the right end of the maps. The positions of restriction endonuclease cleavage sites, ORFs in the three possible translation phases, and AATAAA sequences (↑), some of which specify 3' cleavage and polyadenylation of viral transcripts, are based on the complete DNA sequences (some of which have had minor revisions since original publication). The E region encodes functions involved in DNA replication, regulation of transcription, and cellular tranformation, and its transcripts are polyadenylated between the E and L regions. The L region encodes functions involved in viral morphogenesis and the transcripts are polyadenylated in the URR (see Fig. 2). The ORFs and restriction sites are based on the DNA sequences: HPV-1 (Danos et al. 1982,1983; see also Clad et al. 1982; Schwarz et al. 1983); HPV-6b (Schwarz et al. 1983); HPV-11 (Dartmann et al. 1986); HPV-16 (Seedorf et al. 1985); HPV-8 (Fuchs et al. 1986). (The sequence of HPV-33 was published by Cole and Streeck 1986.)

**Table 1** Human Papillomaviruses

| Type | Disease | Oncongenic potential |
|---|---|---|
| 1 | deep plantar and palmar myrmecia | benign |
| 2 | common warts (verrucae vulgares), some associated with anogenital condylomata | benign |
| 3, 10, 28 | juvenile flat warts (verrucae planae); associated with some types of epidermodysplasia verruciformis and genital infections and some common warts | rarely malignant |
| 4 | plantar warts and common warts | benign |
| 5, 8 | pityriasis-versicolor macules; epidermodysplasia verruciformis in patients with congenital cell-mediated immune deficiency and those undergoing immunosuppression for transplantation | 30% progress to malignancy |
| 6, 11 | ano-genital condylomata acuminata; atypical (flat) c.a. (particularly HPV-11); dysplasias and intraepithelial neoplasias, grades I and II; penile warts; juvenile and adult-onset laryngeal papillomas (particularly HPV-11) | usually benign |
| 7 | common warts of meat and animal handlers | benign |
| 9, 12, 14 15, 17 19–25, 36, 40 | epidermodysplasia verruciformis basilioma (type 20) | HPV-12, -17, and -20 lesions, at least, can progress to carcinomas |
| 13, 32 | oral focal epithelial hyperplasia (Heck's disease) | possible progression to carcinoma |
| 16, 18, 31 33, 35, 39 | high-grade dysplasias, intraepithelial neoplasias and carcinomas (CIS) of genital mucosa; Bowenoid papulosis, Bowen's disease; laryngeal, esophageal, and probably some bronchial carcinomas | high corelation with genital and oral carcinomas |
| 26 | cutaneous wart, patient with immune deficiencies | unknown |
| 27 | cutaneous wart, renal transplant recipient | unknown |
| 29 | cutaneous, intermediate warts | unknown |
| 30, 40 | larynx | carcinoma |
| 34 | nongenital Bowen's disease | carcinoma-in-situ |
| 37 | keratoacanthoma | benign |
| 38 | in a melanoma | malignant |
| 41 | multiple condylomata and cutaneous flat warts | benign |
| 42 | genital warts | benign |

References to the cloning, restriction mapping, and sequencing of the human papillomaviruses by **type. 1:** Clad et al. 1982; Danos et al. 1980, 1982, 1983; Gissmann et al. 1977; Heilman et al. 1980. **2:** Fuchs and Pfister 1984; Heilman et al. 1980; Orth et al. 1977; Orth and Favre 1985. **3:** Kremsdorf et al. 1983; Ostrow et al. 1983a. **4:** Gissmann et al. 1977; Heilman et al. 1980. **5:** Kremsdorf et al. 1982; Ostrow et al. 1983b. **6:** Boshart and zur Hausen 1986; de Villiers et al. 1981; Mounts et al. 1982; Rando et al. 1986; Schwarz et al. 1983. **7:** Oltersdorf et al. 1986; Orth et al. 1981; Ostrow et al. 1981. **8:** Fuchs et al. 1986; Kremsdorf et al. 1983; Pfister et al. 1981b. **9:** Kremsdorf et al. 1982. **10:** Green et al. 1982; Kremsdorf et al. 1983. **11:** Dartmann et al. 1986; Gissmann et al. 1982b. **12:** Kremsdorf et al. 1983. **13:** Pfister et al. 1983b. **14:** Kremsdorf et al. 1984; Tsumori et al. 1983. **15:** Kremsdorf et al. 1984. **16:** Dürst et al. 1983; Ikenberg et al. 1983; Matsukura et al. 1986; Seedorf et al. 1985. **17:** Kremsdorf et al. 1984; Yutsudo et al. 1985. **18:** Boshart et al. 1984. **19:** Gassenmaier et al. 1984; Kremsdorf et al. 1984. **20:** Gassenmaier et al. 1984; Kremsdorf et al. 1984. **21:** Kremsdorf et al. 1984. **22:** Kremsdorf et al. 1984; **23:** Kremsdorf et al. 1984. **24:** Kremsdorf et al. 1984. **25:** Gassenmaier et al. 1984; **26:** Ostrow et al. 1984. **27:** Ostrow et al. 1985. **28:** Orth and Favre 1985; Jablonska et al. 1985a. **29:** G. Orth et al., pers. comm. **30:** Kahn et al. 1986. **31:** Lorincz et al. 1986a,b. **32:** G. Orth et al., pers. comm. **33:** Beaudenon et al. 1986; Cole and Streek 1986. **34:** Kawashima et al. 1986. **35:** Lorincz et al. 1986b. **36:** Orth 1986. **37:** Scheurlen et al. 1986. **38:** Scheurlen et al. 1986. **39:** G. Orth et al., pers. comm. **40:** Kahn et al. 1986; H. zur Hausen et al., pers. comm. **41:** H. zur Hausen et al., pers. comm. **42:** G. Orth et al., pers. comm.

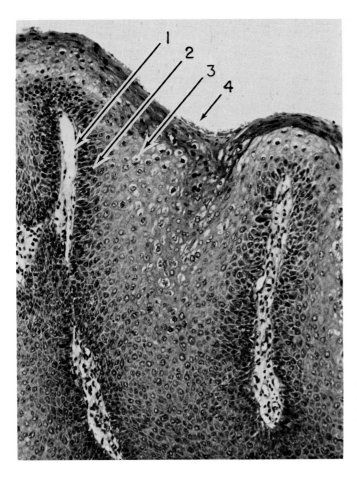

**Figure 3** Histological section of a vulvar condyloma. A 3-μm section of a formalin-fixed, paraffin-embedded biopsy of a vulvar condyloma was stained with hematoxylin and eosin. The layers of the stratified tissue are indicated: (1) underlying dermis of fibroblasts and connective tissue; (2) basal cells of epithelium; (3) spinous and granular keratinocytes, many showing koilocytosis; (4) stratum corneum. (Courtesy of M. Stoler, University of Rochester.)

6. radiomimetic drugs, tobacco products, and other carcinogens;
7. natural and contraceptive hormones;
8. mechanical irritation resulting in the recurrent presence of wound epithelium.

Fortunately, sophisticated molecular diagnostic reagents and physical tests are now making objective measurements of many of these parameters possible.

Recent molecular epidemiology has quite convincingly demonstrated a central role for additional HPV types in another group of neoplasias: those developing from the very rare, cutaneous disease called epidermodysplasia verruciformis (e.v.) (Orth et al. 1978, 1979, 1980; Green et al. 1982; Ostrow et al. 1982; Lutzner et al. 1983; Pfister et al. 1983a; Yutsudo et al. 1985). Progression of these lesions to carcinomas is observed in about one third of the e.v. patients harboring HPV types 5, 8, 12, 17, and 20, generally in areas exposed to solar radiation (Pfister 1984). Epidermodysplasias associated with the other viral types remain benign. The plethora of types causing e.v. (Table 1) are very rare, but they have a worldwide distribution (Pfister et al. 1981b; Kremsdorf et al. 1983, 1984; Tsumori et al. 1983; Gassenmaier et al. 1984; Ostrow et al. 1984). The reservoir of these viruses in the human population is unclear, but suspicions are that the viruses are fairly common but hardly ever show overt penetrance. Afflic-

tions exhibit a familial pattern and the patients have some yet to be defined, heritable cell-mediated immune deficiency (Prawer et al. 1977; Jablonska et al. 1982; Androphy et al. 1985a).

The role of the immune system in papillomavirus diseases is of more general concern and uncertainty. The ubiquity of latent HPV infections is emphasized by the frequent, often acute outbreak of wart development in pharmacologically immunosuppressed allograft recipients, cancer and AIDS patients, pregnant women, and other immunocompromised individuals (Obalek et al. 1980; Shokri-Tabibzadeh et al. 1981; Jablonska et al. 1982, 1985b; Chardonnet et al. 1985; Lutzner 1985; Gassenmaier et al. 1986). Infection with one type of virus does not confer resistance to all others, nor does it mean susceptibility to all others. A person will generally experience a limited set of simultaneous active infections. It is unknown whether there are cell-surface markers that distinguish the latently infected, transformed, or lytically productive cell, how the host keeps the viruses from entering their productive phase, or what the basis for the sudden emergence or regression of lesions at discrete sites may be.

**Molecular biology of bovine papillomavirus Type 1**

The correlation between papillomaviruses and human carcinomas has spurred a great deal of research over the past decade on the biology of the viruses. The key impediment to this work has been the inability to define a productive in vitro host cell system for any papillomavirus, since the viruses naturally require differentiating but nondividing keratinocytes that, so far, cannot be generated and amplified in tissue culture. With very few exceptions, the HPVs neither infect nor transform readily any animal cells in vivo or in culture (but see Brackmann et al. 1983; Watts et al. 1984, 1985; Burnett and Gallimore 1985; Tsunokawa et al. 1986a; Yasumoto et al. 1986; Chow et al. this volume). Furthermore, it is relatively difficult to obtain clinical specimens of sufficient size or quality for direct isolation of viral products.

The animal papillomaviruses, notably BPV-1 and Shope papillomavirus, are far more amenable to biochemical, genetic, and physiological dissection than are any of the human viruses, for several reasons:

1. Animals can be experimentally infected.
2. Such warts are particularly productive of viral DNA, RNA, proteins, and mature virions.
3. The Shope virus lesions can progress to carcinomas, and some (Vx2 and Vx7) have been carried as transplantable tumors and adapted to continuous cell cultures.
4. The bovine virus readily transforms the mouse C127 and NIH-3T3 cell lines.

The ability of the BPVs to transform somatic cells in culture has been of great utility in defining viral functions. Two aspects of viral function can be assayed: cellular transformation and plasmid replication. The assignment of functions to particular ORFs has relied on successively more refined genetic dissection through standard reverse genetic technology: introducing block deletions; engineering missense, nonsense, frameshift, or termination mutations into particular sections of individual ORFs while leaving overlapping ORFs intact; and expressing isolated ORFs from foreign promoters. Mouse cells are then transfected with such altered viral DNA, and transformation and plasmid replication are subsequently monitored. The inherent limitation of this approach is that distinctions between roles in the initiation and maintenance steps cannot be inferred from the results, for the approach analyzes only the end point. For example, the BPV-1 genome contains at least two separate transforming genes, E5 and E6 (Nakabayashi et al. 1983; Sarver et al. 1984; Schiller et al. 1984, 1986; DiMaio et al. 1985, 1986; Yang et al. 1985b). It is not certain that these genes are required for continued maintenance of transformation, but one indication comes from the retention of the E6 ORF in all human cervical carcinoma lines, whereas the E5 ORF is separated from the E-region promoters by the integrative recombination and, in some lines, is deleted. Making distinctions between initiation and maintenance also received much attention in the early discussions about cellular transformation by SV40 and polyoma and, for these viruses, both reversion data and the availability of temperature-sensitive (*ts*) mutations helped resolve those issues. For example, SV40 *tsA* mutations show that the large T antigen is continually required for initia-

tion of viral replication but only transiently required for the activation of late gene expression. Only very indirect arguments are now available for discussion of these points with respect to papillomavirus functions. Primarily one assumes that, in any hit-and-run scenario, continued expression of the gene in question would not be required and that finding such continued expression is prima facie evidence for a maintenance function. The pitfalls of this argument are self-evident. For instance, the phenomenological parameters measured are almost always complex (e.g., growth in semisolid media), and some aspects of the measured phenotypes may require continued expression and others not. Furthermore, once a gene is expressed, it may be difficult to shut it off. Another indirect approach for demonstration of a maintenance function involves gene transfer. In such experiments, total transformed cell DNA, including the resident papilloma sequences, are used as donor DNA for second-round transformation (DiMaio et al. 1985; Tsunokawa et al. 1986a). These experiments address a very special issue. They simply show that the papilloma sequences maintain their ability to start the transformation cycle all over again. The observations that the papillomavirus DNA is transferred in the second round seem to argue against induction of mutation in the host or activation of a potent cellular oncogene but does not address the more complicated issue of a programmed switching of cellular gene expression in an epigenetic sense. In this context the experiments reported by Turek et al. (1982) are of importance. These authors showed that continued interferon treatment of transformed mouse cells harboring BPV-1 plasmids leads to a high rate of reversion from the transformed state and a concommitant loss of viral plasmids. If one assumes that the target of interferon action is viral gene expression, then this experiment implies that, for both plasmid replication and transformation, continued viral expression is required.

The nature of the transformation events is not clear. Several routes are possible and testable: (1) direct alteration of host transcription via *trans*-acting regulatory proteins; (2) direct activation of host cell division via *trans*-acting DNA-binding proteins and/or *cis* integration of replication enhancers at central regulatory sites; and (3) initiation of a chain of alterations in host regulation through viral transformation proteins of undefined function.

The BPV-1 genome has received intensive scrutiny with respect to structure and function. The first surprise forthcoming from these endeavors is the complexity of the system. At least five separate complementation groups are involved either directly or indirectly with plasmid replication. The E1 ORF is associated with replication functions most directly. All mutations introduced at the 3' end of the E1 ORF lead to morphologically transformed cells that harbor the viral DNA in an integrated state. The mutations clearly define a *trans*-acting function, since the ability for episomal replication can be complemented by the presence of the wild-type E1 ORF on a separate DNA molecule. The 5' end of this same E1 ORF seems to code for a negative regulator of plasmid replication (Berg et al. 1986b; Roberts and Weintraub 1986). Interestingly, frameshift mutations at the 5′ end of the E1 ORF that destroy the negative regulatory function are still capable of normal transient replication. Presumably any frameshift mutation at the 5′ end of an ORF would abolish sense translation of any integral coding information 3′ to the frameshift. Since 5′ E1 mutants and 3′ E1 mutants complement each other, these results are most simply interpreted to mean that the E1 ORF encodes two distinct polypeptides, a possible consequence of messenger RNAs (mRNAs) with alternative 5′ ends or splicing, or of internal initiation of protein translation. Why the E1 ORF is maintained uninterrupted in all sequenced papillomaviruses (see Figs. 1 and 2; Matsukura et al. 1986) is then a puzzle. This dilemma points out the pitfalls in assuming that individual ORFs can be simply correlated to functions. A resolution may lie in a more-detailed description of each gene. For example, the two genes may share a common sequence encoding either *cis* or *trans* functions which does not allow for the evolution of frameshifting. Stenlund et al. (1985) found an RNA splice donor encoded at nucleotide position 1234 in the BPV-1 E1 ORF. This donor could mark the 3′ end of the negative regulator (within the E1 frame) and the same sequence could define a splice acceptor to dictate the 5′ end of the putative replication protein encoded by E1.

In addition to the role of the E6 ORF in transformation, the E6 ORF and the E6-E7 ORFs together play a role in maintaining high copy numbers of extrachromosomal plasmids in the nuclei of transformed cells (Lusky and Botchan 1985; Berg et al. 1986a). It is not clear if these genes manifest their effects directly upon the viral genome or work more indirectly through alterations in cellular gene expression. The latter seems more likely, for the E6 gene is capable of oncogenic transformation in isolation. When the E6 gene is transcribed from exogenous expression vectors, transformation of mouse C127 cells is readily detected (Schiller et al. 1984; Androphy et al. 1985b; Yang et al. 1985a). Interestingly, the transformation seems to require some particular cell condition, for NIH-3T3 cells are resistant to "E6-only" transformation. The role of E6 transformation in the context of the intact viral DNA is more problematic. E6 mutations in some laboratories have little or no effect on transformation efficiencies, whereas in other groups a quantitative decrease in the frequency of transformation is detected. Since the E5 ORF codes for another potent transforming protein (Schiller et al. 1984; Yang et al. 1985b; DiMaio et al. 1986; Groff and Lancaster 1986), these differences may relate to how sensitive a particular assay system is to the effects of E5. For example, if the E6 function does affect the DNA copy number, then E6 mutations may indirectly dictate the E5 levels (see Berg et al. 1986a). Moreover, the number of cell divisions required to produce a transformed colony is small, and the levels of input DNA during and shortly after transfection may not reflect the steady-state copy number. Clearly, more-direct experimentation on this transforma-

tion system is required. Specifically, it will be important to relate different transfection protocols and subsequent DNA copy number with general transcriptional activity and ultimately with the synthesis of the transformation proteins. Fortunately, the opportunity for biochemical characterization may be at hand, for the proteins encoded by the E6 and E5 ORFs have recently been identified. The BPV-1 E5 ORF encodes a small (6000 daltons) protein that is found associated with cell membranes (R. Schlegel et al., pers. comm.), and the E6 gene encodes a 15,500-dalton polypeptide that seems to be distributed both in the cytoplasm and nucleus (Androphy et al. 1985b). It is notable that some animal and human papillomaviruses do not have an equivalent to the BPV-1 E5 ORF (Figs. 1 and 2), particularly if sequence homologies are considered.

The E2 ORF encodes a *trans*-activator of E-region gene expression (Spalholz et al. 1985; Yang et al. 1985b). As expected for such a gene, its action is pleiotropic upon other viral functions, including transformation and replication (Sarver et al. 1984; DiMaio 1986; Groff and Lancaster 1986). Spalholz and colleagues initially assayed the *trans*-activator function through its effects upon an SV40 promoter–chloramphenicol acetyltransferase gene (CAT) cassette (Gorman et al. 1982). Interestingly, E2-mediated stimulation of CAT expression from the vector required certain BPV-1 sequences in *cis* to be manifest. This inducible enhancer sequence maps to the viral upstream regulatory region (URR) (also termed the long control region [LCR] or noncoding region [NCR] in the literature). An important feature of this *trans*-activation by the E2 product is that it is at once specific to a *cis*-acting signal in the URR and also is able to effect expression from other distantly related papillomaviruses (W. Phelps and P. Howley, pers. comm.; H. Hirochika, T. Broker, and L. Chow, unpubl.). This implies that a common cellular factor recognizes the viral DNA and/or that the E2 protein itself recognizes a conserved sequence. A search for such elements reveals that a repeated tract with a consensus sequence of $ACCN_6GGT$ or $ACC-A/G-N_6-C/T-GGT$ is indeed found in all of the sequenced papillomaviruses, predominantly in the URR regions (Fig. 4) (Cole and Streeck 1986; A. Stenlund, pers. comm.). In the Shope papillomavirus, Giri et al. (1985) found that the URR contains almost three complete copies of a 32-bp repetition; the large repetition itself contains an internal duplication that conforms to the consensus sequence. To summarize, the BPV-1 plasmid replication system seems to be intricately tied to the transformation system. The five genes clearly involved in replication control are the E6 and E6-E7 genes, the positive and negative factors of the E1 ORF, and the E2 *trans*-activator.

The URR is one of the two regions in the genome most likely to diverge, even between closely related papillomavirus types, the other being near the E5 ORF (Broker and Chow, this volume). It is of particular interest that duplications of tracts within the URR are observed with HPVs and may be affecting pathogenicity by

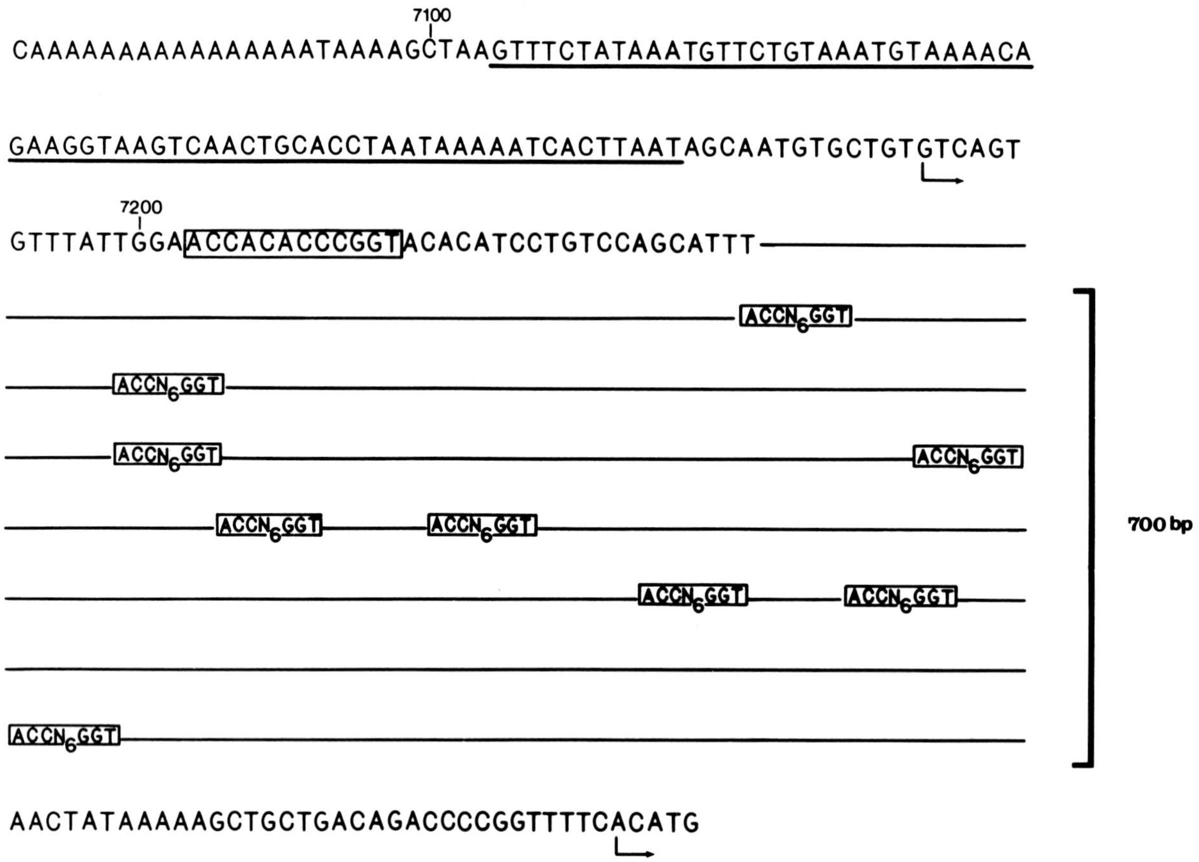
7100
CAAAAAAAAAAAAAAAAATAAAAGCTAAGTTTCTATAAATGTTCTGTAAATGTAAAACA
GAAGGTAAGTCAACTGCACCTAATAAAAATCACTTAATAGCAATGTGCTGTGTCAGT
7200
GTTTATTGGAACCACACCCGGTACACATCCTGTCCAGCATTT
ACCN6GGT
ACCN6GGT
ACCN6GGT
ACCN6GGT
ACCN6GGT
ACCN6GGT
ACCN6GGT
ACCN6GGT
ACCN6GGT
700 bp
AACTATAAAAAGCTGCTGACAGACCCCGGTTTTCACATG

**Figure 4** A prototype upstream regulatory region: BPV-1. The nucleotide sequences of the core replication origin of BPV-1 are shown at top. The region of homology between this sequence and another origin in the E1 ORF of the virus is underlined. Two RNA transcription-initiation sites, one within the origin of replication and another at nucleotide 89, are also indicated by arrows. This 900-bp region is relatively devoid of ORFs and contains the *cis*-acting transcriptional regulatory signals for both initiation sites. Nine $ACCN_6GGT$ tracts are positioned within this region. The tenth such tract also contains a consensus enhancer core element interdigitated within the repeat. (An extra tract is shown in a box at nucleotide 7770. This sequence is not in some previous publications of the BPV-1 DNA sequence because of an apparent sequencing error.)

augmenting transcriptional activity (Boshart and zur Hausen 1986; Cole and Streeck 1986; Rando et al. 1986).

The BPV-1 URR contains at least two transcriptional start sites, as measured by in vitro RNA synthesis (Fig. 4). How the *trans*-activation may directly and indirectly affect each of these promoters and other sites that modulate transcription looms as an important issue for future work. The downstream promoter with initiation at nucleotide 89 is known to be one of the major sources of mRNA in the stably transformed cell. Presumably, the E2 gene product increases activity of this promoter, although this has not been directly shown. The promoter lying within the origin of replication (Fig. 4) is actually down-regulated in transformed mouse cells. In a speculative vein, one may posit that the *trans*-activation of the promoter near the E6 ORF is linked to the down-regulation of the 5′-most promoter.

Another prominent feature of the URR is the viral origin of DNA replication. The viral DNA harbors two plasmid maintenance sequences (PMS-1 and PMS-2), which can function in replication assays performed in C127 cells. PMS-1 is found in the URR, while PMS-2 is in the middle of the E1 ORF. The sequences can work independently (Lusky and Botchan 1984), and electron microscopy of replicating forms (Waldeck et al. 1984) shows that PMS-1 is indeed a region within which bidirectional θ forms of DNA replication initiate. These two PMS elements share about 75% homology, as indicated in Figure 4. Interestingly, the origin sequences do not show significant inverted repetitions, which are often characteristic of other circular DNA viruses such as SV40, polyoma, and EBV, which also replicate bidirectionally. Recent linker-insertion analysis (Lusky and Botchan 1986) shows that, in addition to the core sequences drawn in Figure 4, the PMS-1 element requires an enhancer in *cis* to function as a replication signal. This requirement probably underlies a common requirement of replication origins, for it is found in other genomes, notably in polyoma and EBV. The nucleic acid–protein interactions that regulate BPV-1 replication remain to be elucidated.

Specific questions concern the location of the putative binding sites for both positive and negative host and viral factors and how transcriptional signals may influence plasmid replication control. Any complete picture of the virus replication cycle must address the following points:

1. the initial amplification of the viral DNA after infection; these events eventually lead to cells that harbor plasmids at about 100–200 copies per nucleus;
2. the subsequent mode of steady-state, stable replication;
3. the high rate of vegetative replication in the upper epithelial layers of productive warts.

These issues are clearly related to the more biologically relevant questions of how viral replication is modulated over the course of tissue differentiation during transit to the surface and what the identity and nature of the relevant viral and cellular factors involved may be. An intriguing question, difficult to answer at this time, concerns the relevance of the C127 model system to the in vivo process. For example, does the putative negative regulator identified as critical for plasmid maintenance in C127 cells play a similar role in controlling or stopping BPV-1 replication in nondividing basal cells? Does it help repress vegetative replication in the histologically normal suprabasal cells, and are its effects released in the productive cells of overt lesions? Answers to these questions could clearly have interesting applications to the ability to study the complete viral life cycle in a cell culture system.

Mutations in the E4 ORF, which is contained entirely within the E2 ORF in a different translation phase, affect no assayable function (D. DiMaio, pers. comm.). Surprisingly then, probing western blots of plantar wart extracts with antibodies prepared against HPV-1 E4 peptide expressed in bacteria revealed a 16,000/17,000-dalton (16/17-kD) doublet and a 10/11-kD doublet that together represent about one third of the SDS-extractable protein (Doorbar et al. 1986; see also Croissant et al. 1985). The 16/17-kD and 10/11-kD proteins were purified from warts, biochemically confirmed as the products of the E4 ORF, and used to raise additional antisera (Doorbar et al. 1986). Indirect immunofluorescent studies of sections of plantar warts revealed that the E4 proteins were primarily in the cytoplasm of superficial keratinocytes (Croissant et al. 1985; Doorbar et al. 1986). Spliced transcripts with an internal exon covering most of the E4 ORF and extending uninterrupted through the E-region polyadenylation site are the major species of viral RNAs in transformed cells or warts associated with BPV-1 (Engel et al. 1983; Stenlund et al. 1985; Yang et al. 1985a), Shope virus (Nasseri and Wettstein 1984a,b; Danos et al. 1985), and HPV-11 and HPV-6 (M. Nasseri, R. Hirochika, S. Wolinsky, T. Broker, and L. Chow, unpubl.) and in cells transfected with HPV-1 and HPV-6 (Chow and Broker 1984; Chow et al., this volume). In some of these RNA species, an AUG initiation codon present in a short 5′ leader is spliced in phase to the E4 ORF, but in other cases the AUG in the leader is in the wrong reading frame or is preceded by upstream ORFs such that translation of E4 would have to depend on internal reinitiation of protein synthesis.

The L1 ORF is the coding region for the major capsid protein, based on antibodies generated against L1 protein reacting with mature viral capsids (Pilacinski et al. 1984). The papillomaviruses are quite unusual in showing their least genetic variation in the major coat protein which, in many families of viruses, is the most divergent due to selective pressures from the host immune system. This suggests that the major host immune response to papillomavirus infection is not directed against the intact virus particle. The role of the L2 ORF is not certain, though it is likely to be a minor capsid protein (Pilacinski et al. 1984). Neither L1 nor L2 protein is produced in transformed cells in culture, so refined analyses have not been possible.

## Molecular biology of the human papillomaviruses

Among the HPVs, only HPV-1 is very productive in vivo, and the plantar and palmar warts it induces are reasonably available as sources of viral products. Obviously, experimental inoculation of humans is unethical, although in the early days of clinical research it was commonplace (Rowson and Mahy 1967). Infections of epithelial mucosa by other HPV types tend to produce few virions. Transformation of C127 or NIH-3T3 cells, although possible (Watts et al. 1984, 1985; Tsunokawa et al. 1986a; Yasumoto et al. 1986), is far from routine, and extensive experimental manipulation such as site-directed mutagenesis to define viral functions has yet to be demonstrated.

The genomes of a number of the HPV types have been sequenced (Fig. 1). Striking similarities with the animal papillomaviruses are evident in their genetic organization. Detailed comparisons among the DNA sequences of the animal and the many types of human papillomaviruses reveal homologies that range from 45% to more than 85% in the most conserved genes, E1 and L1 (Schwarz et al. 1983; Cole and Streeck 1986; Fuchs et al. 1986; Pfister et al. 1986; Broker and Chow, this volume). Antibodies against disrupted capsid (predominantly the L1 ORF product) understandably are broadly cross-reactive but also can be raised with some type-specificity (Nakai et al. 1986). The similarities have fostered a tendency to ascribe properties defined with any particular virus to the whole family of papillomaviruses. However, researchers and clinicians may be at risk for assuming too much similarity between the model systems and all the other viruses. At the minimum, one must account for differences in host range, tissue tropism, and pathogenicity. Ultimately, identification of the unique features of the various animal and human papillomaviruses will be essential to the development of vaccines, diagnosis of infectious agents, and defining distinctions among viruses with very different oncogenic potentials, thereby dictating appropriate strategies for the treatment of patients.

## DNA replication in vivo

Tandem multimers of viral episomes are seen in some high-grade lesions induced by CRPV (Wettstein and Stevens 1982), HPV-6 (Gissmann et al. 1982a; Lancaster et al. 1983) and HPV-16 (Dürst et al. 1985), and in some BPV-1–transformed mouse cell lines (Allshire and Bostock 1986). They might possibly arise as a consequence of defective segregation during termination of replication or through recombination if replication intermediates with an excessive amount of single-stranded DNA accumulate in cells that have been exposed to mutagens or in which the normal replicative proteins are deficient. Multimers contain several replication origins per episome, which can confer a survival advantage because initiation of any of them would enable the molecule to duplicate. That initiation of replication may become limiting in natural infections is suggested—but certainly not demonstrated—by the integration of the viral DNA into the host chromosomes in high-grade lesions, particularly in carcinomas. At present, it is difficult to evaluate whether the integration alters host gene expression, resulting in the unscheduled replication, abnormal mitoses, and polyploidy and aneuploidy associated with malignancies, or whether other events make DNA replication more difficult for both the virus and the host, leading to selection of transformed cells with integrated viral genes. Integration of episomes would necessarily interrupt some viral gene; intriguingly, this is reproducibly the E1 ORF and usually results in a deletion of the E2 ORF (Pater and Pater 1985; Dürst et al. 1986; Matsukura et al. 1986). Homologous recombination between the tandem repetitions could eventually reduce the insertion to an interrupted monomer. Illegitimate recombination between viral and flanking host sequences would result in deletion of some viral sequences. This may be the origin of the partially deleted monomer seen in most tumors and cervical carcinoma cell lines. Often there seems to be a reexpansion of these subgenomic viral sequences together with flanking host sequences into high-copy-number, tandem repetitions at more than one host locus (Schwarz et al. 1985; Yee et al. 1985; Dürst et al. 1986).

## Messenger RNA transcription

Studies of the transcription of the various types of papillomavirus DNAs have been hampered by the difficulty in recovering undegraded RNA from natural warts of heavily keratinized epithelia, which is extremely tough tissue in which only a small percentage of the cells support viral gene expression (Engel et al. 1983). BPV-1–transformed mouse C127 cells do make viral RNA, but it is restricted to the E region (Amtmann and Sauer 1982; Heilman et al. 1982; Stenlund et al. 1985; Yang et al. 1985a). The Shope virus–induced carcinoma cell lines also express mRNA, but again transcription is confined to the E region (Georges et al. 1984; Nasseri and Wettstein 1984a,b; Danos et al. 1985). The same limitation is usually observed among the species of HPV mRNAs recovered from genital condylomata, cervical carcinomas, and permanent cell lines derived from such carcinomas (Lehn et al. 1984, 1985; Schwarz et al. 1985, 1986). One solution to producing and recovering sufficient E- and L-region RNA for analysis of minor species has been to express HPV-1 from surrogate promoters cloned into the upstream regulatory region, with the caveat that the system is unnatural (Chow and Broker 1984; Chow et al., this volume).

The techniques of mRNA analysis have been varied and often were dictated by the extremely limited material recovered: northern transfer blot hybridization to determine the sizes and general map positions of the various species; modified S1 nuclease protection; primer extension to localize 5′ termini and splice junctions; transmission electron microscopy of RNA:DNA heteroduplexes, with acute sensitivity to minor species; and, ultimately, preparation and sequencing of cDNA libraries to establish the conjunctions of translation frames achieved in individual species by RNA splicing.

From these varied systems and approaches, a rather consistent picture of papillomavirus transcription has emerged. All transcription is along a single strand of the DNA, consistent with the asymmetric distribution of ORFs. Most RNAs are spliced, and the species form families with alternative processing patterns. The map positions of 5′ ends reveal that several different promoters are used (see Fig. 4; Danos et al. 1985; Stenlund et al. 1985; Yang et al. 1985a; L. Chow et al., pers. comm.), though little is know about their activity as a function of tissue differentiation or timing during the lytic cycle. There are two primary 3′ polyadenylation sites, at the distal ends of the E and the L regions. The L1 mRNA invariably has two 5′ leader segments from the E region and it is transcribed from promoters in the URR or the E region. Accordingly, transcription into the L region may depend in part on antitermination or inhibition of cleavage of nascent transcripts at the E-region polyadenylation site. The major messages are those that have a 3′ exon spanning the E4- through E-region poly(A) site. The accumulation of transcripts of E1 and L2 is particularly rare.

## Future Concerns and Prospects

### Experimental animal and tissue culture systems

Numerous attempts have been made to develop an in vitro culture system for epithelial cells able to carry out a normal program of differentiation and stratification comparable to tissue in vivo and suitable for the establishment of papillomavirus infections and for their experimental manipulations (Taichman et al. 1984).

Keratinocyte differentiation in culture can be modulated by the amount of vitamin A or calcium in the media (Fuchs and Green 1981; Brown et al. 1985) or by elevating the cell monolayer, or early-stage stratified culture, to the air interface to mimic the normal environment of skin (Prunieras et al. 1983; Asselineau and Prunieras 1984).

A tissue culture system would be a source of viral intermediate complexes and structures in the pathways for DNA replication, transcription and mRNA processing, and virion assembly. Neonatal foreskin appears to be a promising source of epithelium for several reasons:

1. It is readily available.
2. It has been cultured and exhibits the initial stages of differentiation of basal cells to suprabasal keratinocytes (Taichman et al. 1983).
3. It can support at least limited replication of HPV-1 (LaPorta and Taichman 1982).
4. Some foreskins are naturally infected with HPV-6 in newborns (Roman and Fife 1986).
5. Foreskin develops histologically typical condylomas when mixed with HPV-11 virions and implanted under the renal capsule of athymic mice (Kreider et al. 1985).

This nude mouse system holds great promise for dissecting the HPV genomes and developing therapeutic agents and protocols.

### DNA replication

There are many questions about replication of episomal DNA in productive warts and about integrated DNA in carcinomas. The mechanisms for modulating papillomavirus replication over the course of tissue differentiation during keratinocyte transit to the surface and the relative roles of viral and cellular factors in plasmid replication can presumably be partially addressed by the study of replication in cultured somatic cells. However, the intriguing question of how the virus replicates (in its vegetative mode) in a cell that is not dividing is more problematic. One wonders whether certain cellular enzymes in DNA replication are present in these nondividing cells or whether the virus utilizes a special mode of DNA replication in this part of its life cycle. Another question concerns the source of the commonly observed papillomavirus DNA sequence divergences, even within the same lesion, giving rise to restriction site polymorphism and small sequence duplications. Is it a consequence of error-prone replication in cells for which DNA synthesis is normally unscheduled; or is it the exceptional exposure of the epithelium to physical, chemical, and biological insults? Are the viral replication origins active when the DNA is integrated into host chromosomes and, if so, does viral replication extend out into flanking host sequences. Such in situ replication of the viral origins could clearly lead to the production of free defective forms by a mechanism analogous to the one documented for SV40 and polyomavirus. Conversely, in many situations the viral replication system may be inoperative and the DNA sequences thus carried passively by the chromosome.

### Viral gene expression in vivo

Homogenizing tissues and calculating average copy numbers of viral DNA, RNA, proteins, or virions gives a misleading picture about the nature of the infection and its natural history. It will desensitize diagnostic tests by underestimating or missing the presence of viral prod-

ucts in the subset of productive cells. Within a single lesion, there is tremendous microheterogeneity in neighboring cells, which are supposedly of clonal origin, as to viral DNA copy number and RNA transcription (Fig. 5). Why this is so is not at all evident. Since the viral transcription and replication programs are finely coordinated with cellular differentiation in a set of closely spaced microenvironments, they can only be revealed through monitoring the virus and host in situ by using appropriate diagnostic antibodies directed against viral or host differentiation-specific proteins or by using probes for viral DNA or individual mRNA species to determine the physical and temporal progression of their synthesis (Orth et al. 1971; Naghashfar et al. 1985; Crum et al. 1986; Grussendorf-Conen 1986; McDougall et al. 1986; Ostrow et al. 1986; Stoler and Broker 1986). In situ assays can also reveal the presence and possible interactions of multiple viruses in naturally and experimentally infected tissues.

### Molecular virology

Among the questions emerging from early studies of the papillomaviruses that merit additional attention are:

1. the basis for host range and tissue tropism, two of the properties that contribute to viral taxonomy and pathology and that have not been assigned to specific genetic signals or proteins;
2. the identification of the genetic regions specifying the important medical parameters, such as pathogenic potential, that should be distinguished in clinical specimens;
3. the nature of the viral proteins and induced host proteins that might be presented to the host immune surveillance system during latent and productive infections and in dysplasias and carcinomas;
4. the significance of the various RNA transcriptional promoters and the differences in their relative strengths and responses to regulatory signals;
5. the determinants of the alternative RNA splicing patterns that connect coding regions;
6. the identification of the functions of the various proteins in vivo;
7. the ability and likelihood of the viral DNA to integrate into the host chromosomes and to become involved in the cellular progression toward carcinoma;
8. the causes of the huge differences in productivity of the different viral types and the individual cells within a single lesion with respect to DNA replication, RNA transcription, protein synthesis, and virion production;
9. the basis for local outbreaks of viral productivity and host cell hyperproliferation within much wider zones of latent infection;
10. the roles of intercellular paracrines or other means of communication within an infected tissue that coordinate field effects possibly dictating the boundaries of the overt lesion (Burghardt 1986).

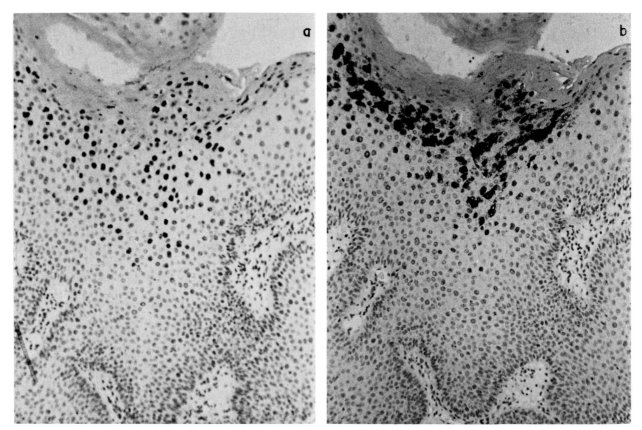

**Figure 5** In situ hybridization detection of HPV-11 DNA and RNA in a genital condyloma. [3]H-labeled, asymmetric HPV-11 RNA probes generated in vitro from an SP6 or a T7 promoter flanking whole genomic HPV-11 DNA were hybridized to adjacent sections of a vulvar condyloma processed to expose their nucleic acid sequences. The sections were radioautographed for 30 days and visualized using brightfield illumination. (*a*) HPV-11 viral DNA yields a nuclear signal; (*b*) HPV-11 mRNA yields a predominantly cytoplasmic signal, with low expression in the middle layers and high expression in the upper layers of the lesion (Stoler and Broker 1986).

## Clinical issues

The cellular mechanisms involved in penetration of an HPV-induced lesion through the basement membrane, resulting in invasive carcinomas, and the possible role of the virus in that major event are not understood. The lethality of carcinomas associated with papillomaviruses often results from metastases to remote sites, primarily to abdominal lymph nodes and lung. It will be most important to screen for the type and state of possible viral DNA in such metastases to determine if they match those found in the original carcinoma. Identity in the viral and host integration sites would prove a clonal origin with dissemination. Alterations in viral DNA copy number or translocations to additional chromosomal sites would reveal ongoing virus-host interactions that would merit continued investigation in terms of regulation of DNA replication and RNA transcription. Understanding what features of a wart are recognized by the immune system will be critical to the development of vaccines and therapeutics and to controlling penetration of the basement membrane.

There are a vast variety of clinical treatments for cutaneous or mucosal warts (Rees 1985; Leventhal and Kashima 1985), including neglect with observation, hypnosis, autologous vaccines, hypersensitization, excision, cryosurgery, laser cauterization, topical applications of salysilic acid or the antimitotic agents in the family of podophyllins, intralesional bleomycin (Bunney 1985), retinoids to alter cellular differentiation (Weiner et al. 1985), photosensitization with hematoporphyrin derivatives (Shikowitz et al. 1986), and systemic or intralesional interferon (Weck and Whisnant 1985). What is notable is the general ignorance about the mechanism and enzymology of DNA replication, thus preventing the design of selective inhibitors akin to the use of nucleoside analogs to block herpes simplex virus replication. It may be that selective intercalating dyes can take advantage of the rather unique episomal state of papillomavirus DNAs to interfere with the elongation, termination, or segregation steps of replication (just as acridines can be used to cure bacterial plasmids). Of a general concern, limited interference with viral DNA replication could be counterproductive. Such DNA, with an accumulation of single-stranded nicks and gaps or double-stranded ends, would be highly recombinogenic and more likely to integrate. Appropriate therapeutics aimed at viral DNA should result in total destruction and rapid elimination, not in protracting replication times or in simply diminishing copy numbers.

The use of radiation to treat papillomavirus lesions containing episomal viral DNA would be expected to be ineffective or counterproductive, for the following reasons: The target size for the 7900-bp viral genomes is vastly smaller than that for host chromosomal and mitochondrial DNAs, and the HPV DNAs will therefore be the last survivors of radiation (as is observed with small plasmids in bacterial maxicells); radiation damage generally stimulates the DNA repair processes necessary to mend broken ends, typically with the result of chromosomal rearrangements and/or viral integration, such as are correlated with papillomavirus-derived carcinomas. This view is supported by the extremely high rate of progression of irradiated laryngeal papillomas to carcinomas. It is notable that radiation is a standard and effective pre- or posthysterectomy treatment for cervical carcinoma patients, in whom the HPV DNA is presumably already integrated. Now that nucleic acid probes are available, it will be important to evaluate such patients by using molecular diagnostics to monitor HPV DNA in the irradiated zone and in remote locations if metastases develop.

Other critical medical concerns include the delivery of diagnosis, treatment, or preventative efforts such as possible vaccines to the places in the world where morbidity and mortality are high. In countries with well-established health care systems that have the luxury to emphasize early diagnosis and sophisticated therapy, the outcome of papillomavirus infections and of epithelial dysplasias is extremely favorable. Many of the newly developing or impoverished nations have so many health concerns such as malnutrition, childhood and maternal infections, and parasitic diseases that cancer is of less immediate priority. Indeed, it is a sobering lesson that polio, rubella, and DPT vaccines are only slowly and inconsistently applied where most needed. Accordingly, basic research and epidemiological surveys as well as clinical protocols, marketing strategies, and socio-political planning must take into account the tremendous differences in health care delivery to define and emphasize effective one-time contact strategies for prevention or for detection and therapy for those societies where the absolute numbers of the cancers are most elevated, but relative priorities diminished.

## Conclusions

Most of the critical problems of papillomavirus research can be clearly stated, and many of the approaches toward the elucidation of the molecular biology of the viruses are technically possible. An urgent need is for a better description of the normal epithelium and of the interplay of host and viral gene expression and proteins during cellular differentiation, transformation, invasion, and metastasis and throughout lytic production. These studies technically and intellectually necessitate the continuing close cooperation of basic biochemi-

cal, molecular genetic, cellular, and immunological research, clinical diagnosis and therapy, epidemiology, biotechnology, and pharmacology. Because of the increasing availability of patient specimens, progress has been rapid in the molecular isolation of a large and representative set of papillomavirus types. The unraveling of their genetic and biochemical properties will lead to the development of a new generation of molecular diagnostics, including specific antibodies and nucleic acid probes. In turn, such reagents are assisting the clinician and patient and have permitted the beginning of molecular epidemiology that has already linked the papillomaviruses to a number of diseases of the epithelium.

## Acknowledgments

T.R.B. was supported by research grants from the National Cancer Institute (CA 36200) and from the Council for Tobacco Research, USA (no. 1587). M.B. was supported by research grants from the National Cancer Institute (CA 30490) and from the American Cancer Society (MV-91). We are grateful to Dr. Louise T. Chow for critical review of this manuscript.

## References

Ahola, H., A. Stenlund, J. Moreno-Lopez, and U. Pettersson. 1983. Sequences of bovine papillomavirus type 1 DNA—Functional and evolutionary implications. *Nucleic Acids Res.* **11:** 2639.

Allshire, R.C. and C.J. Bostock. 1986. Structure of bovine papillomavirus type 1 DNA in a transformed mouse cell line. *J. Mol. Biol.* **188:** 1.

Amtmann, E. and G. Sauer. 1982. Bovine papilloma virus transcription: Polyadenylated RNA species and assessment of the direction of transcription. *J. Virol.* **43:** 59.

Amtmann, E., H. Muller, and G. Sauer. 1980. Equine connective tissue tumors contain unintegrated bovine papilloma virus DNA. *J. Virol.* **35:** 962.

Androphy, E.J., I. Dvoretzky, and D.R. Lowy. 1985a. X-linked inheritance of epidermodysplasia verruciformis. Genetic and virologic studies of a kindred. *Arch. Dermatol.* **121:** 864.

Androphy, E.J., J.T. Schiller, and D.R. Lowy. 1985b. Identification of the protein encoded by the E6 transforming gene of bovine papillomavirus. *Science* **230:** 442.

Asselineau, D. and M. Prunieras. 1984. Reconstruction of "simplified" skin: Control of fabrication. *Br. J. Dermatol.* (suppl. 27) **111:** 219.

Barrett, T.J. and J.D. Silbar. 1954. Genital warts—A venereal disease. *J. Am. Med. Assoc.* **154:** 333.

Beaudenon, S., D. Kremsdorf, O. Croissant, S. Jablonska, S. Wain-Hobson, and G. Orth. 1986. A novel type of human papillomavirus associated with genital neoplasias. *Nature* **321:** 246.

Berg, L.J., K. Singh, and M. Botchan. 1986a. Complementation of a bovine papilloma virus low-copy-number mutant: Evidence for a temporal requirement of the complementing gene. *Mol. Cell. Biol.* **6:** 859.

Berg, L.J., M. Lusky, A. Stenlund, and M. Botchan. 1986b. Repression of BPV replication is mediated by a virally encoded *trans*-acting factor. *Cell* **47:** (in press).

Black, P.H., J.W. Hartley, W.P. Rowe, and R.J. Huebner. 1963. Transformation of bovine tissue culture cells by bovine papillomavirus. *Nature* **199:** 1016.

Boshart, M. and H. zur Hausen. 1986. Human papillomaviruses in Buschke-Lowenstein tumors: Physical state of the DNA and identification of a tandem duplication in the noncoding region of a human papillomavirus 6 subtype. *J. Virol.* **58:** 963.

Boshart, M., L. Gissmann, H. Ikenberg, A. Kleinheinz, W. Scheurlen, and H. zur Hausen. 1984. A new type of papillomavirus DNA and its presence in genital cancer biopsies and in cell lines derived from cervical cancer. *EMBO J.* **3:** 1151.

Brackmann, K.H., M. Green, W.S.M. Wold, A. Rankin, P.M. Loewenstein, M.A. Cartas, P.R. Sanders, K. Olson, G. Orth, S. Jablonska, D. Kremsdorf, and M. Favre. 1983. Introduction of cloned human papillomavirus genomes into mouse cells and expression at the RNA level. *Virology* **29:** 12.

Brandsma, J.L., B.M. Steinberg, A.L. Abramson, and B. Winkler. 1986. Presence of human papillomavirus type 16 related sequences in verrucous carcinoma of the larynx. *Cancer Res.* **46:** 2185.

Brinton, L.A. 1986. Current epidemiological studies—Emerging hypotheses. *Banbury Rep.* **21:** 17.

Brown, R., R.H. Gray, and I.A. Bernstein. 1985. Retinoids alter the direction of differentiation in primary cultures of cutaneous keratinocytes. *Differentiation* **28:** 268.

Bunney, M.H. 1985. Intralesional bleomycin sulfate in treatment of recalcitrant warts. *Clin. Dermatol.* **3:** 189.

Burghardt, E.. 1986. Natural history of cervical lesions. *Banbury Rep.* **21:** 81.

Burnett, T.S. and P.H. Gallimore. 1985. Introduction of cloned human papillomavirus 1a DNA into rat fibroblasts: Integration, de novo methylation and absence of cellular morphological transformation. *J. Gen. Virol.* **66:** 1063.

Chardonnet, Y., J. Viac, M.J. Stawuet, and J. Thivolet. 1985. Cell-mediated immunity to human papillomavirus. *Clin. Dermatol.* **3:** 156.

Chen, E.Y., P.M. Howley, A.D. Levinson, and P.H. Seeburg. 1982. The primary structure and genetic organization of the bovine papillomavirus type 1 genome. *Nature* **299:** 529.

Chow, L.T. and T.R. Broker. 1984. Human papilloma virus type 1 RNA transcription and processing in COS-1 cells. *Prog. Cancer Res. Ther.* **30:** 125.

Ciuffo, G. 1907. Innesto postiveo con filtrado di verrucae volgare. *G. Ital. Mal. Venereol.* **48:** 12.

Clad, A., L. Gissmann, B. Meier, U.K. Freese, and E. Schwarz. 1982. Molecular cloning and partial nucleotide sequence of human papillomavirus type 1a DNA. *Virology* **118:** 254.

Cole, S.T. and R.E. Streeck. 1986. Genome organization and nucleotide sequence of human papillomavirus type 33, which is associated with cervical cancer. *J. Virol.* **58:** 991.

Croissant, O., F. Breitburd, and G. Orth. 1985. Specificity of cytopathic effect of cutaneous human papillomaviruses. *Clin. Dermatol.* **3:** 43.

Crum, C.P., N. Nagai, R.U. Levine, and S. Silverstein. 1986. In situ hybridization analysis of HPV 16 DNA sequences in early cervical neoplasia. *Am. J. Pathol.* **123:** 174.

Danos, O., M. Katinka, and M. Yaniv. 1980. Molecular cloning, refined physical map and heterogeneity and methylation sites of papilloma virus type 1a DNA. *Eur. J. Biochem.* **109:** 457.

———. 1982. Human papillomavirus 1a complete DNA sequence: A novel type of genome organization among papovaviridae. *EMBO J.* **1:** 231.

Danos, O., E. Georges, G. Orth, and M. Yaniv. 1985. Fine structure of the cottontail rabbit papillomavirus mRNAs expressed in the transplantable Vx2 carcinoma. *J. Virol.* **53:** 735.

Danos, O., L.W. Engel, E.Y. Chen, M. Yaniv, and P.M. Howley. 1983. A comparative analysis of the human type 1a and bovine type 1 papillomavirus genomes. *J. Virol.* **46:** 557.

Dartmann, K., E. Schwarz, L. Gissmann, and H. zur Hausen. 1986. The nucleotide sequence and genome organization of human papilloma virus type 11. *Virology* **151:** 124.

de Villiers, E.-M., L. Gissmann, and H. zur Hausen. 1981. Molecular cloning of viral DNA from human genital warts. *J. Virol.* **40:** 932.

de Villiers, E.-M., C. Neumann, J.Y. Le, H. Weidauer, and H. zur Hausen. 1986. Infection of the oral mucosa with defined types of human papillomavirus. *Med. Microbiol. Immunol.* **174:** 287.

DiLuca, D., S. Pilotti, B. Stefanon, A. Rotola, P. Monini, M. Tognon, G. DePalo, F. Rilke, and E. Cassai. 1986. Human papillomavirus type 16 DNA in genital tumours: A pathological and molecular analysis. *J. Gen. Virol.* **67:** 583.

DiMaio, D. 1986. Nonsense mutation in open reading frame E2 of bovine papillomavirus DNA. *J. Virol.* **57:** 475.

DiMaio, D., D. Guralski, and J.T. Schiller. 1986. Translation of open reading frame E5 of bovine papillomavirus is required for its transforming activity. *Proc. Natl. Acad. Sci.* **83:** 1797.

DiMaio, D., J. Metherall, K. Neary, and D. Guralski. 1985. Genetic analysis of cell transformation by bovine papillomavirus. *UCLA Symp. Mol. Cell. Biol. New Ser.* **32:** 437.

Doll, R. 1986. Implications of epidemiological evidence for future progress. *Banbury Rep.* **21:** 321.

Doorbar, J., D. Campbell, R.J.A. Grand, and P.H. Gallimore. 1986. Identification of the human papilloma virus-la E4 gene products. *EMBO J.* **5:** 355.

Dunn, A.E. and M.M. Ogilvie. 1968. Intranuclear virus particles in human genital wart tissue: Observations on the ultrastructure of the epidermal layer. *J. Ultrastruct. Res.* **22:** 282.

Dürst, M., E. Schwarz, and L. Gissmann. 1986. Integration and persistence of human papillomavirus DNA in genital tumors. *Banbury Rep.* **21:** 273.

Dürst, M., L. Gissmann, H. Ikenberg, and H. zur Hausen. 1983. A papillomavirus DNA from a cervical carcinoma and its prevalence in cancer biopsy samples from different geographic regions. *Proc. Natl. Acad. Sci.* **80:** 3812.

Dürst, M. A. Kleinheinz, M. Hotz, and L. Gissmann. 1985. The physical state of human papillomavirus type 16 DNA in benign and malignant genital tumors. *J. Gen. Virol.* **66:** 1515.

Engel, L.W., C.A. Heilman, and P.M. Howley. 1983. Transcriptional organization of the bovine papillomavirus type 1. *J. Virol.* **47:** 516.

Franke, W.W., R. Moll, T. Achtstaetter, and C. Kuhn. 1986. Cell typing of epithelia and carcinomas of the female genital tract using cytoskeletal proteins as markers. *Banbury Rep.* **21:** 121.

Frithiof, L. and J. Wersall. 1967. Virus-like particles in papillomas of the human oral cavity. *Arch. Virusforschung* **21:** 31.

Fu, Y.S., Y.W. Reagan, and R.M. Richart. 1981a. Definition of precursors. *Gynecol. Oncol.* **12:** S220.

Fu, Y.S., L.T. Temmin, Y.M. Olaizola, and J.W. Reagan. 1981b. Nuclear DNA characteristics of microinvasive squamous carcinoma of the uterine cervix. In *Progress in surgical pathology* (ed. C.M. Fenoglio and M.W. Wolff), vol. 1, p. 233. Masson, New York.

Fuchs, E. and H. Green. 1981. Regulation of terminal differentiation of cultured human keratinocytes by vitamin A. *Cell* **25:** 617.

Fuchs, P.G. and H. Pfister. 1984. Cloning and characterization of papillomavirus type 2c DNA. *Intervirology* **22:** 177.

Fuchs, P.G., T. Iftner, J. Weninger, and H. Pfister. 1986. Epidermodysplasia verruciformis-associated human papillomavirus 8: Genomic sequence and comparative analysis. *J. Virol.* **58:** 626.

Gassenmaier, A., M. Lammel, and H. Pfister. 1984. Molecular cloning and characterization of the DNAs of human papillomavirues 19, 20, and 25 from a patient with epidermodysplasia verruciformis. *J. Virol.* **52:** 1019.

Gassenmaier, A., P. Fuchs, H. Schell, and H. Pfister. 1986. Papillomavirus DNA in warts of immunosuppressed renal allograft recipients. *Arch. Dermatol. Res.* **278:** 219.

Georges, E., F. Breitburd, N. Jibard, and G. Orth. 1985. Two Shope papillomavirus-associated Vx2 carcinoma cell lines with different levels of keratinocyte differentiation and transplantability. *J. Virol.* **55:** 246.

Georges, E., O. Croissant, N. Bonneaud, and G. Orth. 1984. Physical state and transcription of the genome of the cottontail rabbit papillomavirus in the warts and in the transplantable Vx2 and Vx7 carcinomas of the domestic rabbit. *J. Virol.* **51:** 530.

Giri, I., O. Danos, and M. Yaniv. 1985. Genomic structure of the cottontail rabbit (Shope) papillomavirus. *Proc. Natl. Acad. Sci.* **82**: 1580.

Gissmann, L. and E. Schwarz. 1985. Cloning of papillomavirus DNA. In *Recombinant DNA research and virus* (ed. Y. Becker), p. 173. Martinus Nijhoff, Boston.

Gissmann, L., E.-M. de Villiers, and H. zur Hausen. 1982a. Analysis of human genital warts (condylomata acuminata) and other genital tumors for human papillomavirus type 6 DNA. *Int. J. Cancer* **29**: 143.

Gissmann, L., H. Pfister, and H. zur Hausen. 1977. Human papilloma viruses (HPV): Characterization of four different isolates. *Virology* **76**: 569.

Gissmann, L., V. Diehl, H.J. Schultz-Coulon, and H. zur Hausen. 1982b. Molecular cloning and characterization of human papilloma virus DNA derived from a laryngeal papilloma. *J. Virol.* **44**: 393.

Gorman, C.M., L.F. Moffat, and B.H. Howard. 1982. Recombinant genomes which express chloramphenicol acetyltransferase in mammalian cells. *Mol. Cell. Biol.* **2**: 1044.

Green, M., K.H. Brackmann, P.R. Sanders, P.M. Loewenstein, J.H. Freel, M. Eisinger, and S.A. Switlyk. 1982. Isolation of a human papillomavirus from a patient with epidermodysplasia verruciformis: Presence of related viral DNA genomes in human urogenital tumors. *Proc. Natl. Acad. Sci.* **79**: 4437.

Groff, D.E. and W.D. Lancaster. 1985. Molecular cloning and nucleotide sequence of deer papillomavirus. *J. Virol.* **56**: 85.

———. 1986. Genetic analysis of the 3′ early region transformation and replication functions of bovine papillomavirus type 1. *Virology.* **150**: 221.

Gross, G., D. Wagner, B. Hauser-Brauner, H. Ikenberg, and L. Gissmann. 1985. Bowenoid papulosis and carcinoma in situ of the cervix uteri in sex partners. An example of the transmissibility of HPV-16 infection. *Hautarzt* **36**: 465.

Grussendorf-Conen, E.I. 1986. In situ hybridization with papillomavirus DNA in genital lesions. *Banbury Rep.* **21**: 239.

Grussendorf-Conen, E.I., H. Ikenberg, and L. Gissmann. 1985. Demonstration of HPV-16 genomes in the nuclei of cervix carcinoma cells. *Dermatologica* **170**: 199.

Heilman, C.A., L. Engel, D.R. Lowy, and P.M. Howley. 1982. Virus-specific transcription in bovine papillomavirus-transformed mouse cells. *Virology* **119**: 22.

Heilman, C.A., M.F. Law, M.A. Israel, and P.M. Howley. 1980. Cloning of human papilloma virus genomic DNAs and analysis of homologous polynucleotide sequences. *J. Virol.* **36**: 395.

Holinger, P.H. and W.F. Rabbett. 1953. Late development of laryngeal and pharyngeal carcinoma in previously irradiated areas. *Laryngoscope* **63**: 105.

Ikenberg, H., L. Gissmann, G. Gross, E.I. Grussendorf-Conen, and H. zur Hausen. 1983. Human papillomavirus type-16-related DNA in genital Bowens' disease and in Bowenoid papulosis. *Int. J. Cancer* **32**: 563.

Ito, Y. and C.A. Evans. 1961. Induction of tumors in domestic rabbits with nucleic acid preparations from partially purified Shope papillomavirus and from extracts of papillomas of domestic and cottontail rabbits. *J. Exp. Med.* **114**: 485.

Jablonska, S., G. Orth, and M.A. Lutzner. 1982. Immunopathology of papillomavirus-induced tumors in different tissues. *Springer Semin. Immunopathol.* **5**: 33.

Jablonska, S., G. Orth, O. Croissant, and S. Obalek. 1985a. The clinical morphology, pathology and immunology of papillomavirus infections of the skin as related to the virus type. *UCLA Symp. Mol. Cell. Biol. New Ser.* **32**: 69.

Jablonska, S., G. Orth, S. Obalek, and O. Croissant. 1985b. Cutaneous warts, clinical, histologic and virologic correlations. *Clin. Dermatol.* **3**: 71.

Kahn, T., E. Schwarz, and H. zur Hausen. 1986. Molecular cloning and characterization of the DNA of a new human papillomavirus (HPV 30) from a laryngeal carcinoma. *Int. J. Cancer* **37**: 61.

Kawashima, M., S. Jablonska, M. Favre, S. Obalek, O. Croissant and G. Orth. 1986. Characterization of a new type of

human papillomavirus found in a lesion of Bowen's disease of the skin. *J. Virol.* **57**: 688.

Kessler, I.I. 1986. Cervical cancer: Social and sexual correlates. *Banbury Rep.* **21**: 55.

Kidd, J.G. and P. Rous. 1940. Cancers deriving from the virus papillomas of wild rabbits under natural conditions. *J. Exp. Med.* **71**: 469.

Koss, L.G. and G.R. Durfee. 1956. Unusual patterns of squamous epithelium of the uterine cervix and pathologic study of koilocytotic atypia. *Ann. N.Y. Acad. Sci.* **63**: 1245.

Kreider, J.W., M.K. Howett, S.A. Wolfe, G.L. Bartlett, R.J. Zaino, T.V. Sedlacek and R. Mortel. 1985. Morphological transformation in vivo of human uterine cervix with papillomavirus from condylomata acuminata. *Nature* **317**: 639.

Kremsdorf, D., S. Jablonska, M. Favre, and G. Orth. 1982. Biochemical characterization of two types of human papillomaviruses associated with epidermodysplasia verruciformis. *J. Virol.* **43**: 436.

———. 1983. Human papillomaviruses associated with epidermodysplasia verruciformis. II. Molecular cloning and biochemical characterization of human papillomavirus 3a, 8, 10, and 12 genomes. *J. Virol.* **48**: 340.

Kremsdorf, D., M. Favre, S. Jablonska, S. Obalek, A.L. Rueda, M.A. Lutzner, C. Blanchet-Bardon, P.C. Van Voorst Vader, and G. Orth. 1984. Molecular cloning and biochemical characterization of the genomes of nine newly recognized human papillomavirus types associated with epidermodysplasia verruciformis. *J. Virol.* **52**: 1013.

Lancaster, W.D. 1981. Apparent lack of integration of bovine papillomavirus DNA in virus-induced equine and bovine tumor cells and virus-transformed mouse cells. *Virology* **108**: 251.

Lancaster, W.D. and C. Olson. 1982. Animal papillomaviruses. *Microbiol. Rev.* **46**: 191.

Lancaster, W.D., R.J. Kurman, L.E. Sanz, S. Perry, and A.B. Jenson. 1983. Human papillomavirus: Detection of viral DNA sequences and evidence for molecular heterogeneity in metaplasias and dysplasias of the uterine cervix. *Intervirology* **20**: 202.

Lancaster, W.D., C. Castellano, C. Santos, G. Delgado, R.J. Kurman, and A.B. Jenson. 1986. Human papillomavirus deoxyribonucleic acid in cervical carcinoma from primary and metastatic sites. *Am. J. Obstet. Gynecol.* **154**: 115.

LaPorta, R.F. and L.B. Taichman. 1982. Human papillloma viral DNA replicates as a stable episome in cultured epidermal keratinocytes. *Proc. Natl. Acad. Sci.* **79**: 3393.

Law, M.F., D.R. Lowy, I. Dvoretzky, and P.M. Howley. 1981. Mouse cells transformed by bovine papillomavirus contain only extrachromosomal viral DNA sequences. *Proc. Natl. Acad. Sci.* **78**: 2727.

Lehn, H., T.-M. Ernst, and G. Sauer. 1984. Transcription of episomal papillomavirus DNA in human condylomata acuminata and Buschke-Lowenstein tumours. *J. Gen. Virol.* **65**: 2003.

Lehn, H., P. Krieg, and G. Sauer. 1985. Papillomavirus genomes in human cervical tumors: Analysis of their transcriptional activity. *Proc. Natl. Acad. Sci.* **82**: 5540.

Leventhal, B.G. and H.K. Kashima. 1985. Chemotherapy of papillomavirus infections. *UCLA Symp. Mol. Cell. Biol. New Ser.* **32**: 235.

Lorincz, A.T., W.D. Lancaster, and G.F. Temple. 1986a. Cloning and characterization of the DNA of a new human papilloma virus from a woman with dysplasia of the uterine cervix. *J. Virol.* **58**: 225.

Lorincz, A.T., W.D. Lancaster, R.J. Kurman, A.B. Jenson, and G.F. Temple. 1986b. Characterization of human papillomaviruses in cervical neoplasia and their detection in routine clinical screening. *Banbury Rep.* **21**: 225.

Lusky, M. and M.R. Botchan. 1984. Characterization of the bovine papilloma virus plasmid maintenance sequences. *Cell* **36**: 391.

———. 1985. Genetic analysis of bovine papillomavirus type 1 *trans*-acting replication factors. *J. Virol.* **53**: 955.

————. 1986. Transient replication of bovine papilloma virus type 1 plasmids: *cis* and *trans* requirements. *Proc. Natl. Acad. Sci.* **83**: 3609.

Lutzner, M.A. 1985. Papillomavirus lesions in immunodepression and immunosuppression. *Clin. Dermato.* **3**: 165.

Lutzner, M.A., G. Orth, V. Dutronquay, M.F. Ducasse, H. Kreis, and J. Crosnier. 1983. Detection of human papillomavirus type 5 DNA in skin cancers of an immunosuppressed renal allograft recipient. *Lancet* **II**: 422.

Matsukura, T., T. Kanda, A. Furuno, H. Yoshikawa, T. Kawana, and K. Yoshiike. 1986. Cloning of monomeric human papillomavirus type 16 DNA integrated within cell DNA from a cervical carcinoma. *J. Virol.* **58**: 979.

McCance, D.J., A. Kalache, K. Ashdown, L. Andrade, F. Menezes, P. Smith, and R. Doll. 1986. Human papillomavirus types 16 and 18 in carcinomas of the penis from Brazil. *Int. J. Cancer* **37**: 55.

McDougall, J.K., D. Myerson and A.M. Beckmann. 1986. Detection of viral DNA and RNA by in situ hybridization. *J. Histochem. Cytochem.* **34**: 33.

McVay, P., M. Fretz, F. Wettstein, J. Stevens, and Y. Ito. 1982. Integrated Shope virus DNA is present and transcribed in the transplantable rabbit tumor Vx-7. *J. Gen. Virol.* **60**: 271.

Meisels, A., R. Fortin, and M. Roy. 1977. Condylomatous lesions of the cervix. II: Cytologic, colposcopic and histopathologic study. *Acta Cytol.* **21**: 379

Melnick, J.L. 1962. Papova virus group. *Science* **135**: 1128.

Melnick, J.L., A.C. Allison, J.S. Butel, W. Eckhart, B.E. Eddy, S. Kit, A.J. Levine, J.A.R. Miles, J.S. Pagano, L. Sachs, and V. Vonka. 1974. Papovaviridae. *Intervirology* **3**: 106.

Moar, M.H., M.S. Campo, H. Laird and W.H.F. Jarrett. 1981. Persistence of non-integrated viral DNA in bovine cells transformed in vitro by bovine papillomavirus type 2. *Nature* **293**: 749.

Mounts, P., K.B. Shah, and H. Kashima. 1982. Viral etiology of juvenile and adult onset squamous papilloma of the larynx. *Proc. Natl. Acad. Sci.* **79**: 5425.

Naghashfar, Z., E. Sawada, M.J. Kutcher, J. Swancar, J. Gupta, R. Daniel, H. Kashima, J.D. Woodruff, and K. Shah. 1985. Identification of genital tract papillomaviruses HPV-6 and HPV-16 in warts of the oral cavity. *J. Med. Virol.* **17**: 313.

Nakabayashi, Y., S.K. Chattopadhyay, and D.R. Lowy. 1983. The transformation function of bovine papilloma virus DNA. *Proc. Nat. Acad. Sci.* **80**: 5832.

Nakai, Y., W.D. Lancaster, L.Y. Lim, and A.B. Jenson. 1986. Monoclonal antibodies to genus- and type-specific papillomavirus structural antigens. *Intervirology* **25**: 30.

Nasseri, M. and F. Wettstein. 1984a. Differences exist between viral transcripts in cottontail rabbit papillomavirus-induced benign and malignant tumors as well as non-virus-producing and virus-producing tumors. *J. Virol.* **51**: 706.

————. 1984b. Cottontail rabbit papillomavirus-specific transcripts in transplantable tumors with integrated DNA. *Virology* **138**: 362.

Obalek, S., W. Glinski, M. Haftek, G. Orth, and S. Jablonska. 1980. Comparative studies on cell mediated immunity in patients with different warts. *Dermatologica* **161**: 73.

Oltersdorf, T., M.S. Campo, M. Favre, K. Dartmann, and L. Gissmann. 1986. Molecular cloning and characterization of human papillomavirus type 7 DNA. *Virology* **149**: 247.

Oriel, J.D. 1971. Natural history of genital warts. *Br. J. Vener. Dis.* **47**: 1.

Orth, G. 1986. Human papillomaviruses and neoplasia of the skin. *J. Cell. Biochem.* (suppl.) **10A**: 180.

Orth, G. and M. Favre. 1985. Human papillomaviruses: Biochemical and biologic properties. *Clin. Dermatol.* **3**: 27.

Orth, G., M. Favre, and O. Croissant. 1977. Characterization of a new type of human papillomavirus that causes skin warts. *J. Virol.* **24**: 108.

Orth, G., P. Jeanteur and O. Croissant. 1971. Evidence for and localization of vegative viral DNA replication by autographic detection of RNA-DNA hybrids in sections of tumors induced by Shope papilloma virus. *Proc. Natl. Acad. Sci.* **68**: 1876.

Orth, G., S. Jablonska, M. Favre, O. Croissant, M. Jarzabek-Chorzelska, and G. Rzesa. 1978. Characterization of two types of human papillomaviruses in lesions of epidermodysplasia verruciformis. *Proc. Natl. Acad. Sci.* **75**: 1537.

Orth, G., S. Jablonska, M. Favre, O. Croissant, S. Obalek, M. Jarzabek-Chorzelska, and N. Jibard. 1981. Identification of papillomaviruses in butchers' warts. *J. Invest. Deramatol.* **76**: 97.

Orth, G., S. Jablonska, M. Jarzabek-Chorzelska, G. Rzesa, S. Obalek, M. Favre, and O. Croissant. 1979. Characteristics of the lesions and risk of malignant conversion as related to the type of the human papillomavirus involved in epidermodysplasia verruciformis. *Cancer Res.* **39**: 1074.

Orth, G., M. Favre, F. Breitburd, O. Croissant, S. Jablonska, S. Obalek, M. Jarzabek-Chorzelska, and G. Rzesa. 1980. Epidermodysplasia verruciformis: A model for the role of papillomaviruses in human cancer. *Cold Spring Harbor Conf. Cell Proliferation* **7**: 259.

Ostrow, R.S., R. Krzyzek, F. Pass, and A.J. Faras. 1981. Identification of a novel human papilloma virus in cutaneous warts of meat handlers. *Virology* **108**: 21.

Ostrow, R.S., K.R. Zachow, O. Thompson, and A.J. Faras. 1984. Molecular cloning and characterization of a unique type of human papillomavirus from an immune deficient patient. *J. Invest. Dermatol.* **82**: 362.

Ostrow, R., D. Manias, B. Clar, T. Okagaki, L. Twiggs, and A. Faras. 1986. Detection of papillomavirus-specific sequences in human genital tumors by in situ hybridization. *Banbury Rep.* **21**: 253.

Ostrow, R.S., K. Zachow, S. Watts, M. Bender, F. Pass, and A. Faras. 1983a. Characterization of two HPV-3 related papillomaviruses from common warts which are distinct clinically from flat warts or epidermodysplasia verruciformis. *J. Invest. Dermatol.* **80**: 436.

Ostrow, R., M. Bender, M. Niimura, T. Seki, M. Kawaskima, F. Pass, and A. Faras. 1982. Human papillomavirus DNA in cutaneous primary and metastasized squamous cell carcinoma from patients with epidermodysplasia verruciformis. *Proc. Natl. Acad. Sci.* **79**: 1634.

Ostrow, R.S., S. Watts, M. Bender, M. Niimura, T. Seki, M. Kawashima, F. Pass, and A.J. Faras. 1983b. Identification of three distinct papillomavirus genomes in a single patient with epidermodysplasia verruciformis. *J. Am. Acad. Dermatol.* **8**: 398.

Ostrow, R.S., K. Zachow, D. Weber, T. Okagaki, M. Fukushima, B.A. Clark, L.B. Twiggs, and A.J. Faras. 1985. Presence and possible involvement of HPV DNA in premalignant and malignant tumours. *UCLA Symp. Mol. Cell. Biol. New Ser.* **32**: 101.

Pater, M.M. and A. Pater. 1985. Human papillomavirus types 16 and 18 sequences in carcinoma cell lines of the cervix. *Virology* **145**: 313.

Pfister, H. 1984. Biology and biochemistry of papillomaviruses. *Rev. Physiol. Biochem. Pharmacol.* **99**: 111.

Pfister, H., B. Fink, and C. Thomas. 1981a. Extrachromosomal bovine papillomavirus type 1 DNA in hamster fibromas and fibrosarcomas. *Virology* **115**: 414.

Pfister, H., A. Gassenmaier, F. Nürnberger and G. Stuttgen. 1983a. Human papilloma virus 5-DNA in a carcinoma of an epidermodysplasia verruciformis patient infected with various human papilloma virus types. *Cancer Res.* **43**: 1436.

Pfister, H., F. Nürnberger, L. Gissmann, and H. zur Hausen. 1981b. Characterization of a human papillomavirus from epidermodysplasia verruciformis lesions of a patient from Upper-Volta. *Int. J. Cancer* **27**: 645.

Pfister, H., I. Hettich, U. Runne, L. Gissmann, and G.N. Chilf. 1983b. Characterization of human papillomavirus type 13 from focal epithelial hyperplasia Heck lesions. *J. Virol.* **47**: 363.

Pfister, H., J. Krubke, W. Dietrich, T. Iftner, and P.G. Fuchs. 1986. Classification of the papillomaviruses-mapping the genome. *Ciba Found. Symp.* **120**: 3.

Pilacinski, W.P., D.L. Glassman, R.A. Krzyzek, P.L. Sadowski, and A.K. Robbins. 1984. Cloning and expression in *Escherichia coli* of the bovine papillomavirus L1 and L2 open reading frames. *Biotechnology* **2**: 356.

Prawer, S.E., F. Pass, J.C. Vance, E.J. Greenberg, E.J. Yunis, and A.S. Zelickson. 1977. Depressed immune function in epidermodysplasia verruciformis. *Arch. Dermatol.* **113**: 495.

Prunieras, M., M. Regnier, and D. Woodley. 1983. Methods for cultivation of keratinocytes with an air-liquid interface. *J. Invest. Dermatol.* (suppl. 1) **81**: 28s.

Rando, R.F., D.E. Groff, J.G. Chirikjian, and W.D. Lancaster. 1986. Isolation and characterization of a novel human papillomavirus type 6 DNA from an invasive vulvar carcinoma. *J. Virol.* **57**: 353.

Rees, R.B.. 1985. The treatment of warts. *Clin. Dermatol.* **3**: 179.

Reid, R., Y.S. Fu, B.R. Herschman, C.P. Crum, L. Braun, V.D. Shah, S.J. Agronow, and C. R. Stanhope. 1984. Genital warts and cervical cancer. VI. The relationship between aneuploid and polyploid cervical lesions. *Am. J. Obstet. Gynecol.* **150**: 189.

Rigoni-Stern, D. 1842. Fatti statistici relativi alle melattie cancerose. *G. Servire Progr. Pathol. Terap.* **2**: 507.

Roberts, J. and H. Weintraub. 1986. Negative regulation of the BPV replicon. *Cell* **47**: (in press).

Rogers, S., J.G. Kidd, and P. Rous. 1960. Relationships of the Shope papilloma virus to the cancers it determines in domestic rabbits. *Acta Unio. Int. Contra Cancrum* **16**: 129.

Roman, A. and K. Fife. 1986. Human papillomavirus DNA associated with foreskins of normal newborns. *J. Infect. Dis.* **153**: 855.

Rous, P. and J.W. Beard. 1935. The progression to carcinoma of virus-induced rabbit papilloma (Shope). *J. Exp. Med.* **62**: 523.

Rous, P. and W.F. Friedewald. 1944. The effect of chemical carcinogens on virus-induced rabbit carcinomas. *J. Exp. Med.* **79**: 511.

Rous, P. and J.G. Kidd. 1936. The carcinogenic effect of a virus upon tarred skin. *Science* **83**: 468.

Rowson, K.E.K. and B.W.J. Mahy. 1967. Human papova (wart) virus. *Bacteriol. Rev.* **31**: 110.

Sarver, N., M.S. Rabson, Y.C. Yang, J.C. Byrne, and P.M. Howley. 1984. Localization and analysis of bovine papillomavirus type 1 transforming functions. *J. Virol.* **52**: 377.

Sasson, I.M., N.J. Haley, D. Hoffmann, E.L. Wynder, D. Hellberg, and S. Nilsson. 1985. Cigarette smoking and neoplasia of the uterine cervix: Smoke constituents in cervical mucus. *N. Engl. J. Med.* **312**: 315.

Scheurlen, W., L. Gissmann, G. Gross and H. zur Hausen. 1986. Molecular cloning of two new HPV types (HPV 37 and HPV 38) from a keratoacanthoma and a malignant melanoma. *Int. J. Cancer* **37**: 505.

Schiller, J.T., W.C. Vass, and D.R. Lowy. 1984. Identification of a second transforming region in bovine papillomavirus DNA. *Proc. Natl. Acad. Sci.* **81**: 7880.

Schiller J.T., W.C. Vass, K.H. Vousdan, and D.R. Lowy. 1986. The E5 open reading frame of bovine papillomavirus type 1 encodes a transforming gene. *J. Virol.* **57**: 1.

Schwarz, E., A. Schneider-Gadicke, B. Roggenbuck, W. Mayer, L. Gissmann, and H. zur Hausen. 1986. Expression of human papillomavirus DNA in cervical carcinoma cell lines. *Banbury Rep.* **21**: 281.

Schwarz, E., U.K. Freese, L. Gissmann, W. Mayer, B. Roggenbuck, A. Stremlau, and H. zur Hausen. 1985. Structure and transcription of human papillomavirus sequences in cervical carcinoma cells. *Nature* **314**: 111.

Schwarz, E., M. Dürst, C. Demankowski, O. Lattermann, R. Zech, E. Wolfsperger, S. Suhai, and H. zur Hausen. 1983. DNA sequence and genome organization of genital human papillomavirus type 6b. *EMBO J.* **2**: 2341.

Seedorf, K., G. Krammer, M. Dürst, S. Suhai, and W. Rowekamp. 1985. Human papillomavirus type 16 DNA sequence. *Virology* **145**: 181.

Shikowitz, M.J., B.M. Steinberg, and A.L. Abramson. 1986.

Hematoporphyrin derivative therapy of papillomas. Experimental study. *Arch. Otolaryngol.* **112**: 42.

Shokri-Tabibzadeh, S., L.G. Koss, J. Molnar, and S. Romney. 1981. Association of human papillomavirus with neoplastic processes in the genital tract of four women with impaired immunity. *Gynecol. Oncol.* **12**: S129.

Shope, R.E. and E.W. Hurst. 1933. Infectious papillomatosis of rabbits; with a note on the histopathology. *J. Exp. Med.* **58**: 607.

Spalholz, B.A., Y.C. Yang, and P.M. Howley. 1985. Transactivation of a bovine papilloma virus transcriptional regulatory element by the E2 gene product. *Cell* **42**: 183.

Spriggs, A.I., C.E. Bowey, and R.H. Cowdell. 1971. Chromosomes of precancerous lesions of the cervix uteri: New data and a review. *Cancer* **27**: 1239.

Steinberg, B.M. 1986. Laryngeal papillomatosis is associated with a defect in cellular differentiation. *Ciba Found. Symp.* **120**: 208.

Stenlund, A., J. Zabielski, H. Ahola, J. Moreno-Lopez, and U. Pettersson. 1985. Messenger RNAs from the transforming region of bovine papilloma virus type 1. *J. Mol. Biol.* **182**: 541.

Stoler, M.H. and T.R. Broker. 1986. In situ hybridization detection of human papilloma virus DNA and messenger RNA in genital condylomas and a cervical carcinoma. *Hum. Pathol.* **17**: (in press).

Stremlau, A., L. Gissmann, H. Ikenberg, M. Stark, P. Bannasch, and H. zur Hausen. 1985. Human papillomavirus type 16 related DNA in an anaplastic carcinoma of the lung. *Cancer* **55**: 1737.

Syrjänen, K.J. 1986. Human papillomavirus (HPV) infections of the female genital tract and their association with intraepithelial neoplasia and squamous cell carcinoma. *Pathol. Annu.* (part 1) **21**: 53.

Syverton, J.T. 1952. The pathogenesis of the rabbit papilloma-to-carcinoma sequence. *Ann. N.Y. Acad. Sci.* **54**: 1126.

Syverton, J.T., R.A. Harvey, G.P. Berry, and S.L. Warren. 1941. The Roentgen radiation of papilloma virus (Shope) I. The effect of x-rays upon papillomas on domestic rabbits. *J. Exp. Med.* **73**: 243.

Taichman, L.B., S.S. Reilly, and R.F. LaPorta. 1983. The role of keratinocyte differentiation in the expression of epitheliotropic viruses. *J. Invest. Dermatol.* **81**: 137s.

Taichman, L., F. Breitburd, O. Croissant, and G. Orth. 1984. The search for a culture system for papillomaviruses. *J. Invest. Dermatol.* **83**: 2s.

Thomas, M., M. Boiron, J. Tanzer, J.P. Levy, and J. Bernard. 1964. In vitro transformation of mice cells by bovine papilloma virus. *Nature* **202**: 709.

Tsumori, T., M. Yutsudo, Y. Nakano, T. Tanigaki, H. Kitamura, and A. Hakura. 1983. Molecular cloning of a new human papilloma virus isolated from epidermodysplasia verruciformis lesions. *J. Gen. Virol.* **64**: 967.

Tsunokawa, Y., N. Takebe, T. Kasamatsu, M. Terada, and T. Sugimura. 1986a. Transforming activity of human papillomavirus type 16 DNA sequences in a cervical cancer. *Proc. Natl. Acad. Sci.* **83**: 2200.

Tsunokawa, Y., N. Takebe, S. Nozawa, T. Kasamatsu, L. Gissmann, H. zur Hausen, M. Terada, and T. Sugimura. 1986b. Presence of human papillomavirus type-16 and type-18 DNA sequences and their expression in cervical cancers and cell lines from Japanese parents. *Int. J. Cancer* **37**: 499.

Turek, L.P., J.C. Byrne, D.R. Lowy, I. Dvoretzky, R.M. Friedman, and P.M. Howley. 1982. Interferon induces morphologic reversion with elimination of extrachromosomal viral genomes in bovine papillomavirus-transformed mouse cells. *Proc. Natl. Acad. Sci.* **79**: 7914.

Vessey, M.P.. 1986. Epidemiology of cervical cancer: Role of hormonal factors, cigarette smoking and occupation. *Banbury Rep.* **21**: 29.

Waldeck, S., F. Rösl, and H. Zentgraf. 1984. Origin of replication in episomal bovine papilloma virus type I DNA isolated from transformed cells. *EMBO J.* **3**: 2173.

Watson, J.D. and J.W. Littlefield. 1960. Some properties of DNA from Shope papilloma virus. *J. Mol. Biol.* **2**: 161.

Watts, S.L., L.T. Chow, R.S. Ostrow, A.J. Faras, and T.R. Broker. 1985. Localization of HPV-5 transforming functions. *UCLA Symp. Mol. Cell. Biol. New Ser.* **32**: 501.

Watts, S.L., W.C. Phelps, R.S. Ostrow, K.R. Zachow, and A.J. Faras. 1984. Cellular transformation by human papillomavirus DNA in vitro. *Science* **225**: 634.

Weck, P.K. and J.K. Whisnant. 1985. Clinical approaches to human papilloma virus diseases: The use of interferons. *UCLA Symp. Mol. Cell. Biol. New Ser.* **32**: 185.

Weiner, S.A., F.L. Meyskens, Jr., E.A. Surwit, D.S. Alberts, and N.S. Levine. 1985. Response of human papilloma-associated diseases to retinoids (vitamin A derivatives). *UCLA Symp. Mol. Cell. Biol. New Ser.* **32**: 249.

Wettstein, F.O. and J.G. Stevens. 1982. Variable-sized free episomes of Shope papilloma virus DNA are present in all non-virus producing neoplasms and integrated episomes are detected in some. *Proc. Natl. Acad. Sci.* **79**: 790.

Yang, Y., H. Okayama, and P.M. Howley. 1985a. Bovine papillomavirus contains multiple transforming genes. *Proc. Natl. Acad. Sci.* **82**: 1030.

Yang, Y.C., B.A. Spalholz, M.S. Rabson, and P.M. Howley. 1985b. Dissociation of transforming and *trans*-activation functions for bovine papillomavirus type 1. *Nature* **318**: 75.

Yasumoto, S., A.L. Burkhardt, J. Doniger and J.A. DiPaolo. 1986. Human papillomavirus type 16 DNA-induced malignant transformation of NIH 3T3 cells. *J. Virol.* **57**: 572.

Yee, C., I. Krishnan-Hewlett, C. Baker, R. Schlegel, and P. Howley. 1985. Presence and expression of human papillomavirus sequences in human cervical carcinoma cell lines. *Am. J. Pathol.* **119**: 361.

Yutsudo, M., T. Shimakage, and A. Hakura. 1985. Human papillomavirus type 17 DNA in skin carcinoma tissue of a patient with epidermodysplasia. *Virology* **144**: 295.

zur Hausen, H. 1977. Human papillomaviruses and their possible role in squamous cell carcinomas. *Curr. Top. Microbiol. Immunol.* **78**: 1.

# Adenovirus Gene Expression and Replication: A Historical Review

**U. Pettersson\* and R.J. Roberts[†]**

*Department of Medical Genetics, The Biomedical Center, Uppsala, Sweden; [†]Cold Spring Harbor Laboratory, Cold Spring Harbor, New York 11724

## The pioneering years

The adenoviruses were discovered in 1953 in the laboratories of Wallace Rowe (NIH)[1] and Maurice Hilleman (Merck) (Rowe et al. 1953; Hilleman and Werner 1954). This was a period in medical research when it was fashionable to search for new viruses that could be linked to different kinds of human disease. It became apparent rather early that the adenoviruses, although not an important human pathogen, had many properties that made them useful experimental systems for virologists. In the late fifties, biochemistry began to have a great impact on virology, and extensive efforts were made to study viral components by using modern biochemical methods. By 1959, using the newly developed negative-staining method for electron microscopy, it was shown that adenoviruses were icosahedral viruses, consisting of 252 capsomers (Horne et al. 1959). The adenoviruses became even more popular with the biochemical virologists when it was discovered that, during their replication, they produced vast quantities of soluble antigens, much in excess of the amounts needed for packaging. These soluble antigens were amenable to purification by the biochemical separation methods available, like DEAE chromatography, electrophoresis, and gel filtration.

A major share of adenovirus research in the sixties was devoted to the structural proteins. One early highlight of the structural work occurred in 1965, when the elegant studies of Robin Valentine and Helio Pereira (Mill Hill) first solved the adenovirus structure and showed that the particle is composed of two types of capsomers, hexons and pentons (Valentine and Pereira 1965). Very unexpected was the discovery of the antenna-like projections (now known to be the fibers) that extend from the vertices of the adenovirus icosahedron. These had apparently been overlooked in several earlier electron microscopic studies of adenovirus morphology. Erling Norrby (Stockholm), working with another serotype, Ad3, independently demonstrated the same basic structure for adenovirus particles and made the interesting observation that adenoviruses belonging to different subgroups vary with respect to the length of their fibers

(Norrby 1966). Four groups showed independently that the soluble adenovirus proteins could be fractionated, by DEAE chromatography, into three components (Klemperer and Pereira 1959; Philipson 1960; Haruna et al. 1961; Wilcox and Ginsberg 1961). They were designated hexons, pentons, and fibers by Harry Ginsberg (Philadelphia, Columbia) and fellow researchers in the adenovirus field (Ginsberg et al. 1966). It is worth remembering that the adenovirus hexon was the first animal virus protein to be crystallized (Pereira et al. 1968). Subsequent studies of the structure of the hexon protein were performed in the laboratories of Richard Franklin (Basel) and Roger Burnett (Basel, Columbia). By the late sixties, all of the major adenovirus components had been purified to homogeneity by several different groups and had been characterized in terms of their biochemical, structural, and immunological properties.

The adenoviruses differ from other DNA tumor viruses in coding for their own histone-like proteins. It was first noted in Maurice Green's laboratory (St. Louis) that the amino acid composition of the adenovirus particle was remarkably similar to that of the major capsid protein with the exception of one amino acid, arginine (Polasa and Green 1967). It was postulated that the excess arginine would be accounted for if the adenovirus particle contained a highly basic core protein. This prediction was verified independently by the groups of Lennart Philipson (Uppsala, Heidelberg), Willie Russell (Mill Hill, Dundee), and W. Graeme Laver (Canberra), who showed that the adenoviruses have a core that is composed of two major polypeptides, one of which contains 24% arginine (Laver et al. 1968; Prage et al. 1968; Russell et al. 1971). One might have expected that the adenovirus chromatin with its unique polypeptides would become an interesting model to study chromatin structure. This has not turned out to be the case since it has proved extremely difficult to study the core, even with the methods that have been applied so successfully to cellular chromatin. Despite many valiant efforts in the laboratories of Joe Weber (Quebec), Ellen Daniell (CSHL, Berkeley), Milan Nermut (Mill Hill), and Jane Flint (CSHL, MIT, Princeton), we still lack a comprehensive model for the adenovirus core.

The antigenic structure of the human adenoviruses was elucidated in a highly systematic way by Norrby and Goran Wadell (Stockholm, Umea). By applying classical techniques, such as hemagglutination, complement fixation, immunodiffusion, and virus neutralization, they

---

[1] Throughout this discussion we have indicated the affiliations of investigators in parentheses after the first mention of their names, using abbreviations of the names of the institutions where the work was done. When two affiliations are listed, the first indicates the place at which the research under discussion was performed and the second identifies the laboratory at which the individual is working at the time of this writing.

were able to show that adenoviruses were extremely complex antigenically with many cross-reacting epitopes (Norrby 1968). This has subsequently been confirmed by the use of monoclonal antibodies (Cepko et al. 1981; Russell et al. 1981; Adam et al. 1985). The molecular dissection of the capsid components was greatly facilitated by the use of sodium dodecyl sulfate (SDS)–polyacrylamide gels, originally introduced by Jake Maizel (NIH) and his coworkers. By 1966 they had shown that the adenovirus capsid was composed of a surprisingly large number of polypeptides, around a dozen (Maizel 1966). In the following years it was established that 12 unique polypeptides were present, which nowadays are designated II, III, IIIa, IV, V, VI, VII, VIII, IX, X, XI, and XII (Maizel 1971; Anderson et al. 1973; Everitt et al. 1973). The localization of the various polypeptide components in the capsid was studied by sequential degradation of the particle by using chemical and physical methods and by analyzing immature virions, empty particles, etc. On the basis of results obtained in many different laboratories, Philipson's group proposed a model for the adenovirus particle in 1973 (Everitt et al. 1973). Although the basic features of this model are still valid (Fig. 1), it should be emphasized that it is hypothetical

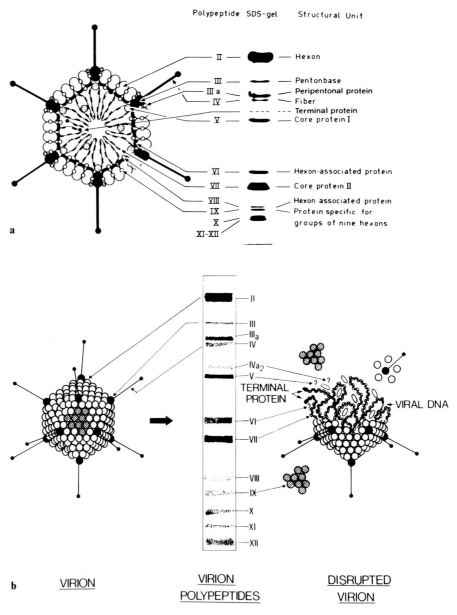

**Figure 1** Two tentative models for the location of the proteins in the Ad2 virion. The polypeptide composition of the virion proteins are shown in a stained exponential 10–16% SDS-polyacrylamide gel. A covalently linked terminal protein is also indicated but cannot be seen in a protein gel, and the corresponding polypeptide band is therefore indicated with a broken line. (*a*) A model based on protein-protein cross-linkage studies. (Modified, with permission, from Everitt et al. 1975). (*b*) A schematic model (Russell and Precious 1982). (Parts *a* and *b* are reprinted, with permission, from Philipson 1983; see this source for further details.)

and that the localization of several of the minor polypeptides is still tentative.

The biochemical studies of the adenovirus particle, which represented the highlights of adenovirus research in the sixties, have since played a less prominent role. Instead the emphasis has been on replication and gene expression. Nevertheless, some very significant findings have emerged in the protein field. Among those should be mentioned the discovery in Ray Gesteland's laboratory (CSHL, Utah) that the major core protein matures through proteolytic trimming (Anderson et al. 1973). It was subsequently demonstrated that polypeptides VI and VIII and terminal protein (pTP) follow a similar pathway (Oberg et al. 1975; Challberg and Kelly 1981). Pierre Boulanger (Lille) and his coworkers have performed very detailed physical-chemical studies of nearly all of the adenovirus capsid components. The early interaction between the virus and its receptor has been studied by many investigators, including the groups of Boulanger, Einar Everitt (Uppsala, Lund), Vivian Mautner (Glasgow) and Philipson. The discovery of two major nonstructural proteins, the 33-kilodalton (kD) and the 100-kD polypeptides are noteworthy (Russell and Skehel 1972; Anderson et al. 1973; Levinson and Levine 1977; Russell and Blair 1977). Several other proteins have been discovered subsequently and are described in later sections.

## The nucleic acid component

In the late sixties the nucleic acid component of the adenoviruses was investigated, and the groundwork for our present understanding was laid in Green's laboratory. The first physical studies of adenovirus DNA were reported from the laboratory of Alex van der Eb (Leiden) in 1966 and from Green's laboratory in 1967. The adenovirus genome was shown to be a double-stranded, nonpermuted, nonredundant DNA molecule, lacking terminal repeats and containing approximately 35,000 bp. It was subsequently demonstrated that the genome contains inverted terminal repetitions (ITRs) (Garon et al. 1972; Wolfson and Dressler 1972). Jim Rose's laboratory showed that this phenomenon was found in many other adenovirus serotypes, although the length of the ITR was quite variable (Garon et al. 1972). Green and his collaborators undertook the monumental task of characterizing the genomes of most of the adenovirus serotypes that were known at the time ($\sim$30) (Lacy and Green 1964, 1965, 1967; Pina and Green 1965). Some interesting observations were made regarding nucleotide composition and classification. The adenoviruses were originally classified by Leon Rosen (Hawaii) based on their hemagglutination properties. He found that the adenovirus family could be subdivided into three groups that formed different patterns and agglutinated erythrocytes from different animal species (Rosen et al. 1962). It was later shown that the adenovirus family could also be divided into subgroups based on their oncogenic properties (Huebner 1967), and these largely coincided with the subgroups that were identified on the basis of hemagglutination. A comprehensive study,

performed in Green's laboratory, using the best nucleic acid hybridization methods available, showed that adenovirus DNAs from members of a subgroup were quite closely related to each other but only distantly related to members of other subgroups. Analysis of a collection of human adenovirus DNAs revealed that the different subgroups each had a characteristic G + C content, ranging from 48% to 58%, and the G + C contents correlated with the oncogenic properties of the virus. These observations, which were reported almost 20 years ago, remain unexplained.

The adenovirus field experienced a dramatic transition in the early seventies as a result of the belief that adenoviruses would prove an exceptionally good model system in which to study eukaryotic gene expression. There were several reasons for this belief. First, viruses like Ad2 and Ad5 are easy to propagate in spinner cultures of HeLa cells; they are easy to purify, and large quantities of viral RNA are produced during an infection. Second, it was expected that the adenovirus genome, which has an intermediate complexity, would be replicated and transcribed by cellular polymerases. It was thus reasonable to assume that the virus would mimic the host cell in regulating its gene expression. The adenovirus genome could be regarded as a miniature chromosome, and information gathered from this system was likely to be applicable to eukaryotes in general. These expectations have to a large extent been fulfilled. Many fundamental discoveries regarding eukaryotic gene organization and expression have come as a result of adenovirus research. Adenovirus DNA replication, in contrast, has turned out to have unique properties and to be controlled largely by viral functions. The belief in the adenovirus model for eukaryotic gene expression led to prolific activity and excitement in many different laboratories around the world. These included some that had been in the field since the early days (Green in St. Louis; Ginsberg in Philadelphia; Russell in London; the Uppsala group) as well as several newcomers to the field. Notable among these were groups at Cold Spring Harbor Laboratory and Jim Darnell's laboratory at Rockefeller, which at the time were visited by two experienced adenovirologists from Sweden, Ulf Pettersson and Lennart Philipson.

## The restriction enzymes made the difference

Despite extensive efforts, progress in adenovirus molecular biology was quite slow until the early seventies. Numerous investigators were studying early and late viral mRNA with the best techniques that were available at the time (e.g., fractionation by sucrose gradient centrifugation, filter hybridization, and competition hybridization). Messenger RNA peaks with different sedimentation rates could be found early and late after infection, and a larger fraction of the genome was found to be expressed at late times as compared with early times. The results suggested that certain classes of genes were selectively active early. Given our current knowledge of the complexity of adenovirus gene expression, it is now clear that these early results and the conclusions drawn from them were hopelessly simplistic.

The idea of using nucleases to fragment the adenovirus genome was first considered in the late sixties, before the first restriction enzyme had been isolated. Walter Doerfler's group (Cologne) reported in 1971 that the adenovirus capsid contains an endonucleolytic activity, and a search among the known adenovirus capsid components for this activity suggested that it was carried by the pentons (Burlingham et al. 1971; Burlingham and Doerfler 1972). However, subsequent studies indicated that the activity was most likely of cellular origin (Reif et al. 1977). The biochemical properties of the enzyme indicated that there might be a certain specificity in the cleavage and hence that it could be used to fragment the genome into distinct components. Later studies showed that this was not feasible; the specificity was not good enough and the available separation methods were not sufficiently powerful to resolve the resulting components.

The breakthrough was provided by the discovery of the restriction enzymes. In 1971 Carel Mulder and his colleagues at Cold Spring Harbor were trying to find a means to cut the circular DNA at one unique site. Attempts with the *E. coli* PI restriction enzyme were frustrating. However, news from Herb Boyer's laboratory in San Francisco of a different restriction enzyme, *Eco*RI, prompted Mulder in collaboration with Hajo Delius, who was in charge of the electron microscopy laboratory at Cold Spring Harbor, to try this enzyme, and they succeeded in showing that *Eco*RI cleaved SV40 DNA at one unique position (Mulder and Delius 1972). Another collaboration sprang up between Mulder, Delius, Sharp, and Pettersson, who had prepared large quantities of Ad2 DNA for transcription studies, and it was quickly shown by electron microscopy that the viral DNA was cleaved into six unique fragments (Pettersson et al. 1973). It was also realized that these fragments could provide a physical basis for mapping the adenovirus chromosome.

In the early days the work was hampered by obstacles that nowadays seem trivial. The Ad2 fragments were too large to be separated in polyacrylamide gels, as had been successfully used by Dan Nathans (Johns Hopkins) for the *Hin*d fragments of SV40. However, an alternative matrix consisting of a mixture of agarose and a low concentration of polyacrylamide was able to resolve the six *Eco*RI fragments of Ad2 DNA. But then it was extremely difficult to elute the fragments from these sorts of gels. Diffusion, in combination with phenol extraction, a procedure that worked well for the much smaller SV40 DNA fragments, gave a terrible recovery of the large adenovirus fragments. A method for electrophoretic elution of the fragments had to be invented before the fragments could be successfully recovered. Subsequently, the whole technology was very much improved when Phil Sharp, Bill Sugden, and Joe Sambrook (CSHL) demonstrated that gels consisting solely of agarose were useful for the separation of large DNA fragments (Aaij and Borst 1972), and that the fragmenta-

tion pattern could be visualized after ethidium bromide staining (Sharp et al. 1973). The first restriction enzyme cleavage map of Ad2 was reported during the 1974 Cold Spring Harbor Symposium; Delius, using the elegant method of partial denaturation mapping, had succeeded in ordering the *Eco*RI fragments of Ad2 (Mulder et al. 1975). Using what are by now traditional redigestion methods, the *Hpa*I map was deduced as well (Mulder et al. 1975). Furthermore, it was shown at the time that the different adenovirus serotypes gave rise to unique fragment patterns, which could be used as a fingerprint of the virus (Mulder et al. 1974).

Rich Roberts joined Cold Spring Harbor Laboratory in the fall of 1972. He had a strong interest in nucleic acid sequencing and realized immediately the need for a collection of restriction enzymes. He had a vision of sequencing large DNA molecules by the use of enzymes that recognized tetranucleotide sequences ("the golden sixteen"). His marvelous technician Phyllis Myers (CSHL) was assigned the task of systematically testing bacteria for new restriction enzymes. The project met with immediate success and new enzymes were continuously discovered and made widely available. A useful collection of restriction enzyme cleavage maps of the Ad2 and Ad5 genomes became available in the midseventies. Many office walls still carry historic copies of the map compiled by Marc Zabeau (CSHL, EMBL)

## The genome organization

It was recognized early in the history of adenovirus research that the virus life cycle was composed of an early and a late phase. The onset of DNA replication provided the dividing line. Moreover, it was clear from early hybridization studies in Green's and Ginsberg's laboratories that different sets of genes were expressed during the two phases (Fujinaga and Green 1970; Lucas and Ginsberg 1971). With the availability of restriction fragment maps, it was natural to begin assigning the early and late genes to this map. The picture was somewhat complicated by the finding that sequences from both complementary strands were expressed early as well as late after infection (Landgraf-Leurs and Green 1973). It therefore became essential to develop methods for strand separation of the restriction fragments. One method was based on the use of the homocopolymer poly(U,G) to separate the strands of the complete viral genome (Landgraf-Leurs and Green 1971). These separated strands could then be used to effect the separation of the strands of purified restriction fragments (Tibbetts and Pettersson 1974). A more direct solution came when it was found that in many cases strand separation of individual restriction fragments occurred during gel electrophoresis under denaturing conditions (Flint and Sharp 1974). During the 1974 Cold Spring Harbor Symposium, a transcription map of the Ad2 genome was reported for the first time by the groups from Cold Spring Harbor and Uppsala. The map revealed an unexpected complexity: Four blocks of early genes (nowadays known as E1–E4) were identified that were scattered along the genome, two being transcribed from one strand and two from the complementary strand.

During the following years, individual genes were assigned to the different gene blocks. This was achieved in two principal ways. One was through mapping crossover points in recombinants resulting from mutants carrying defined genetic lesions, a novel procedure that was developed at Cold Spring Harbor by Terri Grodzicker, Joe Sambrook, Phil Sharp, Jim Williams, and their coworkers (Grodzicker et al. 1975). The other method was based on in vitro translation of mRNAs that had been selected by hybridization to restriction enzyme fragments. Much of the latter work was performed by Jim Lewis (CSHL, Seattle), Carl Anderson (CSHL, Brookhaven), and their colleagues in Gesteland's laboratory (Lewis et al. 1975). They had developed an efficient system for the in vitro translation of adenovirus mRNA as well as methods to purify the viral mRNA efficiently by hybridization to restriction enzyme fragments. In this way it was possible to assign most of the late and some of the early adenovirus polypeptides to specific locations on the viral chromosome. The late gene map was subsequently refined by Bryan Roberts (NIH, Brandeis) and his coworkers, who developed the method of hybrid arrest of translation (HAT). The early gene map has been improved considerably since then as a result of contributions from many laboratories.

The next technology to be used for transcription mapping of the adenovirus chromosome was electron microscopy. The method of R-loop mapping, which is based on the superior stability of RNA/DNA hybrids as opposed to DNA/DNA duplexes in formamide-containing solutions, was invented by Ray White and Dave Hogness at Stanford in the mid-seventies. This method was immediately used for detailed mapping of the adenovirus chromosome. Contributions were made by the groups of Louise Chow and Tom Broker (CSHL, Rochester) and Heiner Westphal (NIH). Amazingly detailed maps, which are still correct in most of their details (Fig. 2), were reported in the late seventies by Chow and Broker (Chow et al. 1977b, 1979). Sequencing, cDNA cloning, and S1 nuclease mapping have allowed a refinement of this map to the sequence level but have revealed relatively few additional details.

## Adenovirus sequences

The first RNA sequences encoded by an adenovirus were reported by Sherman Weissman's laboratory (Yale) for the VA RNAs (Ohe and Weissman 1971). By 1975, DNA sequencing was still in its most primitive stages, and the first adenovirus DNA sequence reported comprised only the first four nucleotides at the terminus of the chromosome (Steenbergh et al. 1975). However, in the next few years, DNA sequencing techniques progressed rapidly, and sequences were reported for pieces of the hexon gene (Akusjärvi and Pettersson 1978a,b) and the complete ITR (Arrand and Roberts 1979). Today the complete sequence of 35,937 nucleotides is known for Ad2 (Roberts et al. 1986). It is based on sequences from the groups of Roberts,

Pettersson, and Goran Akusjärvi (Uppsala) together with important contributions from a large number of laboratories, including those of Francois Galibert (Paris), John Sussenbach (Utrecht), and Michael Sung (Illinois).

The transforming region was first sequenced for Ad5 by Hans van Ormondt and his colleagues (Leiden), and thanks to contributions from the laboratories of van Ormondt and Sussenbach, extensive stretches of the Ad5 sequence are now available. The ITR from a large number of serotypes have been sequenced and considerable additional sequences are known for Ad7 and Ad12. With these the groups of Jeff Engler (CSHL, Alabama), Kei Fujinaga (Sapporo), and van Ormondt have made important contributions.

As a result of the Ad2 sequence, some genes that had previously escaped detection have been identified, while others still remain as unidentified open reading frames. One highlight was provided by the unexpected finding of two long open reading frames in the left third of the genome. A combination of electron microscopy together with peptide analysis and in vitro translation resulted in the discovery of a new early region, E2B, which encodes proteins that are engaged in DNA replication (Stillman et al. 1981).

It seems from the available sequence information that the genomes of all mammalian adenoviruses are likely to be organized in a very similar way. The avian adenoviruses, on the other hand, seem to be quite different, although homologies have been detected between human Ad2 and the avian CELO virus (Alestrom et al. 1982b).

## Adenovirus transcription provides many highlights

The adenoviruses have been particularly useful for studies of eukaryotic transcription and have probably had their most profound impact in this field. The early work on adenovirus transcription was performed in the laboratories of Green and Ginsberg. Viral RNA sequences that were expressed in productively infected cells and in transformed cells were studied by using filter hybridization techniques. Green's laboratory was one of the first to demonstrate that viral RNA sequences are present in the polysome fraction of transformed cells (Fujinaga and Green 1966). Much of the early work used competition hybridization, a method that unfortunately has severe limitations. No restriction enzyme maps were available at the time, and it was impossible to correlate the different hybridization patterns with the topography of the adenovirus genome. The discovery of the restriction enzymes and the subsequent development of detailed restriction enzyme maps marked a turning point in these studies.

In the late seventies, several new techniques were developed that made it possible to study viral mRNA synthesis in greater detail. Of particular importance were several new methods to map transcriptional start sites. These included promoter mapping after UV inactivation of the template. Other approaches were based on the isolation of pulse-labeled, nascent RNA chains or

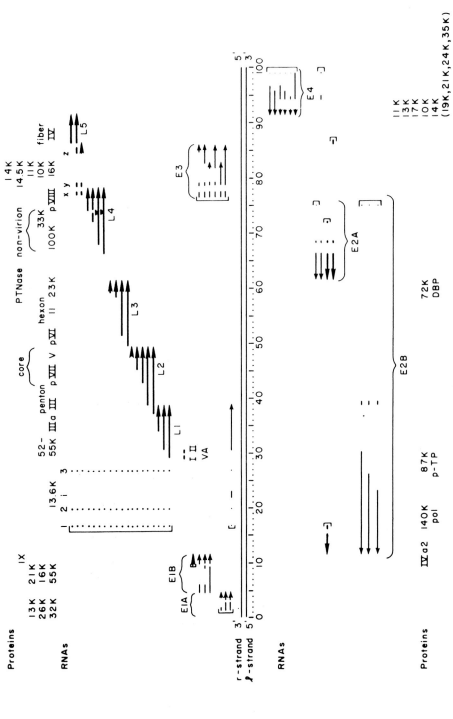

**Figure 2** The Ad2 cytoplasmic RNA transcripts, characterized by electron microscopy of RNA:DNA heteroduplexes. (Reprinted, with permission, from Broker et al. 1984).

the use of a selective inhibitor (DRB) that terminates transcription a short distance away from the initiation site. With these methods the laboratories of Darnell and Sharp succeeded in demonstrating that the four early regions were controlled by individual promoters (Berk and Sharp 1977; Evans et al. 1977). Further analysis of the E1 region showed that it in fact contained three transcription units—E1A, E1B, and that of polypeptide IX (Wilson et al. 1979).

In 1976, investigations of the capped oligonucleotides that are present at the 5' end of adenovirus mRNAs led to some unexpected and exciting results. Rich Gelinas (CSHL, Seattle), a postdoctoral fellow in Roberts's laboratory, used DBAE columns to isolate the capped $T_1$-oligonucleotides from adenovirus mRNAs. A surprising result came from this study. Late Ad2 mRNAs appeared to contain a single, unique capped undecanucleotide, suggesting that all late viral mRNAs contain an identical capped 5' end (Gelinas and Roberts 1977). This was unexpected in light of the large number of late mRNA species that were known to exist. Two interpretations seemed plausible; one was that many cap sites exist that are composed of sequences that are repeated at many locations along the viral DNA. The other possibility was that all the late adenovirus mRNAs are equipped with a common 5' leader sequence by some novel mechanism. Support for the latter alternative came from promoter-mapping studies in Darnell's laboratory, which indicated that a single major promoter was operating late after infection (Bachenheimer and Darnell 1975). More convincing were the sequence studies of Saida Zain (CSHL, Rochester), who showed that preceding the start of the coding region for fiber mRNA there was no sequence capable of coding for the capped undecanucleotide (Zain and Roberts 1978).

All these experiments were preludes to the discovery of RNA splicing in 1977. At that time, Sharp's laboratory was engaged in an electron microscopic study of late adenovirus mRNA. Sue Berget (MIT, Austin) made an unexpected observation while examining hybrids between hexon mRNA and DNA fragments encompassing its coding region. Heteroduplex molecules were obtained that contained unmatched tails of RNA at both the 5' and the 3' ends (Berget et al. 1977). A tail at the 3' end was expected, since earlier studies had shown that adenovirus mRNA is polyadenylated (Philipson et al. 1971). However, the short tail at the 5' end was entirely unexpected. Further electron microscopic studies revealed that it was composed of discontinuous sequences derived from three separate locations along the viral chromosome. Concurrent with these experiments, Gelinas and Roberts had established a collaboration with Chow and Broker, based on the earlier findings that all late viral mRNAs seemed to have a common cap structure. They used R-loop analysis to study the structure of viral mRNAs and made the same observation that had been made by the team in Sharp's laboratory. In addition, they found that the same tripartite leader sequence is attached not only to hexon mRNA but essentially to all late mRNAs (Chow et al. 1977a). These

revolutionary findings found immediate acceptance because of the wealth of biochemical evidence that supported the dramatic electron micrographs.

Thus, in 1977 the term "RNA splicing" was coined and indisputable proof was provided for the discontinuous relationship between eukaryotic genes and many of the mRNAs derived from them. In that year the Cold Spring Harbor Symposium was devoted to eukaryotic chromosomes, and the sessions dealing with transcription and mRNAs were considerably enlivened as a result of the adenovirus findings. Several scientists presented data that at first sight had seemed confusing but that could suddenly be interpreted in a very plausible fashion by invoking RNA splicing. Soon everyone was convinced that RNA splicing was not merely some weird viral phenomenon.

From then on the research moved apace. Ed Ziff (Rockefeller, NYU) and his coworkers mapped the major late promoter at the nucleotide level and demonstrated that the late mRNA cap site coincides with the transcription initiation site (Ziff and Evans 1978). The tripartite leader was sequenced in Cold Spring Harbor (Zain et al. 1979b) and Uppsala (Akusjärvi and Pettersson 1979) and the sites on the genome from which it is derived were mapped and sequenced (Zain et al. 1979a; Akusjärvi and Pettersson 1979). Berget and Sharp studied pre-mRNA in the electron microscope and were able to observe splicing intermediates, thus demonstrating that splicing takes place at the RNA level (Berget and Sharp 1979). Work in the laboratories of Heschel Raskas (St. Louis) and Darnell showed that groups of late mRNAs also shared polyadenylation sites, and it was established that five different families of late mRNA exist (L1–L5) whose members share 3' termini (McGrogan and Raskas 1978; Nevins and Darnell 1978) (Fig. 2). These were mapped at the nucleotide level (Ziff and Fraser 1978). Darnell's group also demonstrated that the mechanism of late transcription involves readthrough from the major late promoter almost all the way to the right-hand end of the genome and that the 3' termini are generated by endonucleolytic cleavage from giant precursor RNAs (Nevins and Darnell 1978; Fraser et al. 1979).

The early mRNA picture was quickly solved. Electron microscopic studies performed in the laboratories of Chow, Broker, and Westphal were particularly informative in showing that the early adenovirus mRNAs were also composed of sequence mosaics (Kitchingman et al. 1977; Chow et al. 1979; Kitchingman and Westphal 1980). To study the structure of these spliced mRNAs, Arnie Berk (MIT, UCLA) and Sharp devised an elegant and sensitive method based on the use of S1 nuclease (Berk and Sharp 1978). This proved to be a major technical breakthrough and is now often the method of choice for the analysis of mRNA structure. Finally, the molecular structure of the E1 mRNAs was solved by analysis of cDNA clones (Perricaudet et al. 1979, 1980). The cap sites and their associated promoters were localized by analyzing the capped 5' oligonucleotides from early mRNA (Baker and Ziff 1981). At the

1979 Cold Spring Harbor Symposium, which was devoted to tumor viruses, a relatively complete picture of the early and late transcription units was presented.

In the late seventies, the time was ripe for mechanistic studies of eukaryotic transcription, and once again the adenoviruses played a very significant role. Several investigators had tried, in the early seventies, to obtain faithful in vitro transcription of naked adenovirus DNA by the use of purified RNA polymerase fractions, but this approach had proved fruitless. However, the use of isolated nuclei to perform in vitro transcription turned out to be more promising (Vennstrom et al. 1978). In 1979 Jim Manley (MIT, Columbia) and his colleagues were able to demonstrate that isolated nuclei could initiate transcription (Manley et al. 1979). However, the crucial development came when soluble systems were described that could carry out transcription in vitro. Pioneering work was performed by Wu, who developed a soluble system for in vitro transcription of VA RNA genes by RNA polymerase III (Wu 1978). During the SV40/adenovirus meeting in Cambridge in 1979, Bob Roeder (St. Louis, Rockefeller) and his group reported that they had achieved successful transcription with RNA polymerase II in a soluble extract, using cloned adenovirus DNA fragments as templates. Apparently the two keys to success were to use crude extracts rather than purified polymerase preparations and to use truncated templates to assay run-off transcription (Weil et al. 1979). Simultaneously, Manley and his colleagues at MIT were developing a simpler, soluble system for RNA polymerase II, which was to become widely used (Manley et al. 1980).

After these developments, the field advanced quickly, and essentially all adenovirus promoter regions had been mapped to the nucleotide level by the early eighties. Extensive studies using in vitro–manipulated templates were performed to define the sequences that are necessary for template activity. Here the groups of Pierre Chambon (Strasbourg), Claude Kedinger (Strasbourg), Manley, and Berk played important roles.

Although the in vitro transcription studies advanced very rapidly, attempts to establish soluble systems for splicing met with little success. In vitro splicing activity was detected in isolated nuclei by Darnell's and Flint's laboratories (Blanchard et al. 1978; Yang et al. 1981), and Raskas' laboratory (Goldenberg and Raskas 1981) demonstrated in vitro splicing of the precursor for the E2A mRNAs, albeit with an extremely low efficiency. In 1983 efficient splicing was demonstrated in vitro in soluble extracts for the first time in the laboratories of Walter Keller (Heidelberg) (Hernandez and Keller 1983) and Sharp (Hardy et al. 1984). Using adenovirus RNAs as a tool in these in vitro systems, many important discoveries were quickly made concerning the mechanisms involved in RNA processing. Strong evidence was provided for the involvement of U1 snRNP in the splicing of adenovirus mRNA (Padgett et al. 1983; Kramer et al. 1984), and using the major late transcription unit, Sharp's laboratory succeeded in determining the chemical structure of the splicing intermediates (Padgett et al.

1984). Finally, in vitro systems that are able to polyadenylate the 3′ ends of precursor RNAs have been described by Sharp's laboratory (Moore and Sharp 1985).

## The regulatory circuits

Although the adenovirus growth cycle at first looked simple, consisting of an early phase and a late phase, it soon became clear that adenovirus genes are regulated at many different levels. Studies in the laboratories of Darnell, Philipson, Lewis, and Mike Mathews (CSHL) indicated that the early promoters are activated with different kinetics. Region E1A is expressed first, followed by E1B, E3, and E4 (Nevins et al. 1979; Lewis and Mathews 1980). The E2 promoter seems to be the last one to be activated. This simple observation indicates that early gene regulation is complex, and we know today that regulatory mechanisms operate at many different levels. It was discovered in the laboratories of Tom Shenk (Stony Brook, Princeton), Joe Nevins (Rockefeller), and Berk that the E1A promoter is preceded by enhancer elements, which presumably are responsible for the immediate activation of the promoter (Hearing and Shenk 1983; Hen et al. 1983; Imperiale et al. 1983; Osborne et al. 1984). The other early promoters are *trans*-activated by the E1A region, as outlined in the next section.

It was originally discovered in Green's and Raskas's laboratories that the protein synthesis inhibitor cycloheximide increases the accumulation of early adenovirus mRNA (Parsons and Green 1971; Craig and Raskas 1974). Subsequent studies in Raskas' and Darnell's laboratories showed that one important factor modulating this effect is mRNA stability, thus identifying a control mechanism for the turnover of mRNA (Eggerding and Raskas 1978; Wilson et al. 1980). The stability of viral mRNA seems to be regulated in several ways during the adenovirus growth cycle. For instance, the prominent accumulation of the 13S mRNA from region E1B is in part caused by a drastic change in the half-life of this RNA species (Wilson and Darnell 1981) but is also partly a result of differential splicing (Montell et al. 1984b). The mechanisms behind the regulation of mRNA stability are poorly understood, although it is known that the 72-kD DNA-binding protein influences mRNA stability, thereby autoregulating its own synthesis (Carter and Blanton 1978). Studies by Nevins and his coworkers suggest that several factors are involved in the regulation of mRNA stability.

The major late transcription unit illustrates additional control mechanisms for adenovirus gene expression. The major late promoter is active at a low level before the onset of DNA replication (Chow et al. 1979; Kitchingman and Westphal 1980) and appears to be a target of the *trans*-activation mediated by E1A, although this result is apparently still controversial (Leff and Chambon 1986). The transcription pattern within the major late transcription unit differs drastically at early and late times after infection (Lewis and Mathews 1980; Shaw and Ziff 1980; Akusjärvi and Persson 1981;

Nevins and Wilson 1981). Early after infection, transcription stops near the middle of the genome, and predominantly mRNAs belonging to the L1 family are made. In contrast, late after infection transcription terminates near the right-hand end of the genome, and the full complement of L1–L5 mRNAs are made.

Yet another level of regulation was discovered when it was observed that although the L1 region is expressed early after infection, the pattern of splicing is dramatically different from that observed at late times (Akusjärvi and Persson 1981; Nevins and Wilson 1981). No mRNA for polypeptide IIIa is made at early times, whereas it is made late even though the same nuclear RNA precursor is present throughout the infectious cycle. From these results the existence of a control mechanism operating at the level of RNA splicing was postulated. Finally, it has been demonstrated that replication of the viral DNA is a necessary step for expression of the full repertoire of late genes (Thomas and Mathews 1980).

## Oncogenicity adds attraction to the adenoviruses

It was discovered in 1962 that adenoviruses are oncogenic. While performing a comprehensive survey of viruses for their oncogenic potential, Trentin et al. (1962) found that Ad12 causes tumors in newborn hamsters. Shortly thereafter, it was shown that Ad18 also has this property (Huebner et al. 1962). Today it is known that many adenovirus serotypes are oncogenic in rodents, and systematic studies have shown that the adenovirus family can be subdivided into three categories based on their oncogenic properties; these are designated as highly, weakly, and nononcogenic adenoviruses (Huebner 1967).

It was first shown in 1964 that Ad12 transforms newborn hamster kidney cells (McBride and Wiener 1964). In 1967, it was shown that even nononcogenic adenoviruses can transform cells in vitro (Freeman et al. 1967), provided that the calcium concentration in the growth medium is reduced. Since then the list has gradually been expanded, and today we know that all human adenoviruses tested (sterotypes 1–31) have the capacity to transform rat cells in vitro. Nearly all of our current knowledge about the genes and proteins involved in adenovirus transformation has come from in vitro transformation studies using the nononcogenic serotypes 2 and 5.

In the late sixties, the newly developed filter hybridization procedures were used to study the state of the viral genome in the transformed cell. Doerfler reported that the viral genome becomes integrated after infection of nonpermissive hamster cells (Doerfler 1968). From results obtained using filter hybridization, it was suggested that adenovirus-transformed cell lines, like SV40-transformed cell lines, contained multiple integrated copies of the viral genome. However, in both cases these estimates turned out to be incorrect due to certain shortcomings when calibrating the filter hybridization assay. The technique of reassociation kinetics that had been cleverly applied by Malcolm Martin (NIH) to SV40-transformed cells (Gelb et al. 1971) gave a different picture when used to examine adenovirus transformants. It was shown that the adenovirus-transformed cell line 8617 contained approximately one copy of adenovirus DNA per haploid genome equivalent (Pettersson and Sambrook 1973). Subsequent analysis, employing the *Eco*RI restriction fragments of Ad2 DNA, showed surprisingly that approximately 50% of the viral genome, corresponding to the central portion, was missing in the transformed cell (Sharp et al. 1974).

Examination of additional Ad2-transformed cell lines by Phil Gallimore (Birmingham) and his coworkers showed that the deletion of sequences from the viral genome in transformed cells is the rule rather than the exception (Gallimore et al. 1974). The results clearly demonstrated that only a portion of the viral genome is required for maintenance of the transformed phenotype. This finding was not entirely unexpected since it had been shown early in the history of adenovirus research, while performing UV inactivation studies, that the target required for inhibition of transformation was substantially smaller than that required for the inhibition of infectivity. The importance of the Gallimore cell lines was that they clearly pointed to the left end, particularly the left 14%, of the Ad2 genome as containing the "transforming region." The hunt was on to find the transforming gene.

The Gallimore cell lines had all been produced by using Ad2 for the transformation. While these cell lines were being analyzed at Cold Spring Harbor, some important developments were taking place in Holland. Frank Graham (Leiden, McMaster) and Alex van der Eb were busy inventing the extremely important calcium phosphate method for introducing DNA into mammalian cells (Graham and van der Eb 1973). The method immediately opened the way to map the transforming region of the adenovirus genome. Using exonucleases that degraded the viral genome from the ends, it was shown that the transforming genes are located terminally (Graham et al. 1974). Shortly thereafter, Graham and his coworkers showed that the *Hind*III-G fragment, comprising the leftmost 7.8% (2800 bp) of the genome was sufficient to transform primary rat embryo cells in vitro (Graham et al. 1975). Thus it was established that a rather small portion of the adenovirus genome is required both for the induction and the maintenance of the transformed state. The transforming gene mapped in the early gene block that is known as E1 (Fig. 2).

Attempts to narrow the transforming region even further employed alternative restriction fragments. It was shown that the *Hpa*I-E fragment of Ad5 (0–4.5%) is able to produce "immortalized cells" but does not give cell lines with a fully transformed phenotype (Houweling et al. 1980). This showed conclusively that several functions are involved in adenovirus transformation.

With the development of recombinant DNA techniques, it became possible to study the viral sequences in adenovirus-transformed cells in greater detail. Integrated sequences were isolated in the laboratories of Sambrook, Pettersson, Sussenbach, and Doerfler by molecular cloning from cells transformed by Ad2, Ad5, and Ad12, and several virus–host cell junctions have been sequenced. The results have largely been disappointing; no clear-cut homologies between host and viral sequences have been found at the integration sites, and no regular integration pattern has emerged. The significance of the patch-like stretches of homology that have been observed in some cases (Gahlman et al. 1982) is unclear, and from our present perspective integration looks like a rather nonspecific process. Ad12-transformed cells, which have been extensively studied by Doerfler's group, seem to differ from Ad2-transformed cell lines in that they often contain multiple copies of complete adenovirus genomes (Stabel et al. 1980). The integrated viral sequences in adenovirus-transformed cells have also been very useful for studies of DNA methylation. The viral sequences show one of the better correlations between expressed genes and the absence of methylation (Doerfler 1982).

Viral mRNAs in transformed cells were studied in the early days by many groups, including those of Green, Flint, Raskas, Doerfler, Sharp, and Darnell. The main conclusions from these studies were that the mRNA sequences that are expressed in transformed cells represent a subset of sequences that are expressed early during a productive infection. Late sequences, even when present, were usually found to be silent. In all cases, mRNA sequences from the left 4.5% of the genome, early region 1A, are expressed, and, when present, mRNA sequences from the adjacent early region 1B (4.5–7.8%) are also expressed.

The proteins that are encoded by the transforming region have been studied by many different laboratories, including those of Green, Bill Wold (St. Louis), Raskas, van der Eb, Graham, Lewis, and Mathews. The conclusions from these studies are that the E1B region encodes two major and two minor polypeptides, whereas the E1A region encodes two *sets* of related polypeptides (Esche et al. 1980). One set, which has polypeptides with apparent masses of 41,000 and 53,000 daltons is encoded by the E1A 13S mRNA. The other set, which consists of 35-kD and 47-kD polypeptides, is encoded by the 12S mRNA. An initial analysis of these proteins by two-dimensional gel electrophoresis resolved the E1A polypeptides into a few additional components (Harter and Lewis 1978), while very recent studies in Ed Harlow's (CSHL) laboratory have revealed an even greater complexity, with perhaps as many as 50 components (Harlow et al. 1985). The significance of this heterogeneity is not known. The E1B region encodes two major polypeptides of 21,000 and 55,000 daltons. Recently, additional minor polypeptides have been assigned to the E1B region that apparently result from differential splicing (Anderson et al. 1984; Virtanen and Pettersson 1985). Most studies on adenovirus transformation have involved Ad2 and Ad5. However, a wealth of information is also available for Ad12 transformation, thanks to studies in the laboratories of Doerfler, van der Eb, Stanley Mak, Hiroto Shimojo, and others.

As to the function of the transforming proteins, the E1A products have been subjected to a very detailed analysis, which is described below. The E1B proteins are less well characterized. It has been reported by Branton's group that a protein kinase is induced in adenovirus-infected cells (Branton et al. 1981). However, it is not firmly established that this kinase is of viral origin, nor has it been assigned to a specific E1B polypeptide. The E1B 21-kD polypeptide has been purified to homogeneity (Persson et al. 1982). It was found preferentially in the membrane fraction, although its predicted sequence lacks the characteristics that are typical for membrane proteins. Recently mutants have been constructed that are defective in the gene for the 21-kD polypeptide. These mutants have an interesting phenotype, which manifests itself in extensive DNA degradation during the infectious cycle (Pilder et al. 1984; White et al. 1984).

### E1A: A bandwagon in adenovirus research

Since 1979, when it was discovered that E1A activates transcription from the other viral promoters, a significant portion of adenovirus research has been devoted to this region (Berk et al. 1979; Jones and Shenk 1979). The discovery by Earl Ruley (CSHL, MIT) that the products of region E1A can interact with cellular oncogenes like c-*ras* to produce a fully transformed phenotype has also stimulated interest in E1A (Ruley 1983). It became clear from results obtained by electron microscopy and S1 nuclease analysis that E1A is transcribed into three overlapping mRNAs designated 9S, 12S, and 13S (Berk and Sharp 1978; Chow et al. 1979). These seem to be processed from a common pre-mRNA by differential splicing. The relative levels of the three species differ considerably during the course of infection. This is most striking for the 9S mRNA, which is abundant late after infection but hardly detectable at early times. The protein product of the 9S mRNA has yet to be identified unambiguously.

cDNA cloning in combination with sequence analysis has made it possible to deduce that the products of the 12S and 13S mRNAs are completely overlapping, differing by an internal stretch of 46 amino acids (Perricaudet et al. 1979). Biochemical studies of the E1A proteins have progressed extremely slowly due to the low abundance of the proteins and the concomitant difficulties of purifying them from infected cells. They are extensively modified posttranslation, since they appear as multiple species that differ in size and charge (see above). Recently it has become possible to produce unmodified 12S and 13S protein products in large quantities by using prokaryotic expression vectors (Ko and Harter 1984; Ferguson et al. 1985).

It is well-documented that the E1A proteins are functionally complex, exerting both positive and negative effects on transcription. An important role for region E1A is to *trans*-activate the other viral promoters during the course of the infection. This is logical since E1A is itself preceded by multiple enhancers and hence is activated early after the entry of the viral DNA into the cell nucleus. In contrast, the other viral promoters apparently lack enhancers and need to be stimulated before they can be expressed efficiently. It has also become clear that E1A can stimulate the transcription of certain cellular genes. Thus, transcription of both preproinsulin and β-globin is stimulated when transfected into cells together with plasmids containing region E1A (Green et al. 1983; Gaynor et al. 1984; Svensson and Akusjärvi 1984a). However, the endogenous β-globin and preproinsulin genes seem to be insensitive to E1A trans-activation. In contrast, other endogenous genes, such as the heat-shock genes and the β-tubulin genes (Kao and Nevins 1983; Stein and Ziff 1984), are turned on by E1A. Studies in several different laboratories have shown that RNA polymerase III transcription is also stimulated by E1A (Berger and Folk 1985; Gaynor et al. 1985; Hoeffler and Roeder 1985). Recently, several investigators have shown that E1A can also exert a negative effect on transcription. For instance, the SV40 early promoter is down-regulated by E1A (Borrelli et al. 1984; Velcich and Ziff 1985), and it has been proposed that it controls its own synthesis by a negative feedback mechanism (Smith et al. 1985).

A very interesting type of repression was discoverd by van der Eb and his colleagues while studying possible differences between nononcogenic and highly oncogenic adenoviruses (Schrier et al. 1983). They observed that the major histocompatibility complex (MHC) class-I antigens were repressed in cells transformed by the highly oncogenic Ad12 but were not repressed in Ad5-transformed cells. Further studies based on the construction of hybrid genomes showed that repression was associated with region E1A. The conclusions drawn from this study have subsequently been challenged (Mellow et al. 1984) on the basis that the cells that are the target for Ad12 transformation might inherently lack MHC antigens. More-recent results in support of the original conclusion are discussed in Vaessen et al. (this volume).

Many models have been proposed to explain the molecular basis for trans-activation. The relative contributions of the 12S and 13S products are unclear from the present data. It could be argued that the 12S mRNA is dispensable since it contains no sequences in addition to those present in the 13S mRNA product. However, studies of a mutant that is unable to express the 12S mRNA suggest that the product of the 12S mRNA has an essential function when the virus infects growth-arrested cells (Montell et al. 1984a; Spindler et al. 1985). Early studies using mutants that were unable to express the 13S mRNA indicated that this mRNA is indeed necessary for the trans-activation effect (Montell et al. 1982). However, more recent data from mutants that lack various segments of the E1A region have blurred the picture, and different investigators seem to obtain different results depending on the experimental system used (Leff et al. 1984; Winberg and Shenk 1984). The available data suggest that trans-activation works at the level of initiation of transcription (Nevins 1981). However, with the exception of regions E2 and E4 (Gilardi and Perricaudet 1984; Imperiale and Nevins

1984), it has not been possible to identify a specific target for the stimulatory factor in the upstream region of other genes stimulated by E1A. Rather, it seems that the critical region coincides with the promoter itself, making it impossible in most cases to separate the target without inhibiting the basal level of transcription.

The mechanism of E1A trans-activation is still enigmatic. Philipson's group, as well as Mathews and Lewis, have studied adenovirus gene expression in the presence of protein synthesis inhibitors (Lewis and Mathews 1980; Persson et al. 1981). The results indicate that the strict requirement for the E1A function is alleviated if the cells are treated with protein synthesis inhibitors prior to infection. A possible interpretation of these results is that the cells produce a short-lived repressor that during a normal infection is counteracted by the E1A function. However, other experimental data do not fit this model, and further studies are essential if we are to understand trans-activation.

The other early regions have been studied in much less detail than E1A. Although one notable exception is region E2A, which encodes the famous 72-kD DNA-binding protein (see below). Recently, region E3 has attracted a great deal of attention. This region has been studied in great detail by Wold and collaborators who have found that it appears to encode a set of membrane proteins (Cladaras and Wold 1985). One of these, the 19-kD glycoprotein, has been characterized in detail (Jeng et al. 1978; Persson et al. 1979, 1980a,b). It seems to have an interesting function in that it interacts with the class-I MHC antigens, as originally discovered in a collaboration between Philipson's and Per Peterson's laboratories in Uppsala (Kvist et al. 1978; Persson et al. 1980c; Signas et al. 1982). Recent studies indicate that the 19-kD protein might prevent the expression of MHC antigens on the surface of the adenovirus-infected cell by trapping them intracellularly (Andersson et al. 1985; Burgert and Kvist 1985). This is likely to be an important mechanism by which the virus interferes with the immune surveillance system of the host.

## The adenovirus replicon has unique properties

Adenovirus replication studies got off to a late start. The first information about adenovirus DNA replication was published in the fall of 1972, a time when the basic mechanism of SV40 replication had already been worked out (Sussenbach et al. 1972). In retrospect this was quite surprising since the adenoviruses offer many advantages for mechanistic studies of DNA replication. Today the situation is quite different, and the adenovirus replicon is now one of the best-characterized mammalian replicons. A useful property of the Ad2 and Ad5 genomes is that they have a significantly higher G + C content than the genome of their host, which facilitates the purification of replicating viral DNA molecules. Moreover, the virus shuts off host DNA replication very efficiently.

During the 1972 SV40/adenovirus meeting in Cold Spring Harbor, several investigators including van der Eb, George Pearson (Oregon), and Pettersson, reported that replication intermediates of adenovirus DNA have an increased buoyant density as compared with the mature viral DNA. Although some investigators believed that this was caused by attached RNA sequences, it soon became clear that it was due to the single-stranded nature of the replication intermediates (Sussenbach et al. 1972; Pettersson 1973; Robin et al. 1973; van der Eb 1973). In the fall of 1972, Sussenbach and his colleagues published the first electron microscopic pictures of replicating adenovirus DNA and a model was proposed that has turned out to be correct in its basic aspects (Sussenbach et al. 1972). The essential point of the model is that replication of adenovirus DNA occurs by strand displacement such that the two complementary strands replicate asynchronously. Adenovirus DNA replication has many properties in common with mitochondrial DNA replication, which at the time had been elucidated in Vinograd's laboratory at the California Institute of Technology.

In the original model it was believed that replication starts preferentially at one end of the genome. Subsequent studies demonstrated that replication initiates at either end with approximately the same frequency (Lavelle et al. 1975; Tolun and Pettersson 1975; Weingartner et al. 1976). This is not unexpected since the two ends of adenovirus DNA have identical structures because of the ITRs. Later studies performed in the laboratories of Ernst Winnacker (Cologne, Munich) and Pettersson mapped the termination sites of viral DNA replication to the molecular ends (Weingartner et al. 1976). A slight revision of the original Sussenbach model was described in 1976 in a paper by Ellen Daniell (CSHL, Berkeley) (Daniell 1976). This model, which apparently was conceived by Joe Sambrook, takes into account the fact that adenovirus DNA contains ITRs. This model has been confirmed in most of its details in later studies (Lechner and Kelly 1977). From the model it is predicted that the origins of replication are located at the termini of the genome, and this has since been proven experimentally. Numerous comparative sequence studies of adenovirus termini have been carried out. Surprisingly, these revealed that the extreme ends differ and it is only the sequence between nucleotides 9 and 22 that is conserved in nearly all adenovirus termini that have been sequenced (Tolun et al. 1979; Stillman et al. 1982). Subsequent studies using in vitro DNA replication assays have confirmed the significance of this region and characterized the origin in more detail (see below).

Work on the biochemistry of adenovirus replication began in the early seventies but has recently progressed rapidly thanks to the in vitro systems for adenovirus DNA replication that have been worked out in the laboratories of Tom Kelly (Johns Hopkins), Jerry Hurwitz (Einstein, Sloan Kettering), and Bruce Stillman (CSHL). Early in the history of adenovirus research, Peter van der Vliet (Utrecht) and Arnie Levine (Princeton) discovered that adenovirus encodes its own DNA-binding protein, the so-called 72-kD (or 72K) protein (van der Vliet and Levine 1973). Thanks to the existence of the tempera-

ture-sensitive mutant, *ts*125 of Ad5, which has a lesion in the gene for the 72-kD protein, it was established that the adenovirus DNA-binding protein is essential for replication (van der Vliet et al. 1975). The protein was found to bind efficiently to single-stranded DNA, and knowing the structure of adenovirus replication intermediates, it was easy to imagine a role for this protein during replication. Subsequent studies have shown that the 72-kD protein is functionally complex in that it also binds to the termini of double-stranded DNA, is involved in the host range of the virus, and affects the half-life of certain mRNAs (Fowlkes et al. 1979; Klessig and Grodzicker 1979; Schechter et al. 1980; Babich and Nevins 1981).

The enzymology of adenovirus DNA replication has proved tricky to unravel, and the current picture is still very incomplete. From the high efficiency of adenovirus DNA replication, it was assumed that an adenovirus DNA polymerase, if it existed, would be an abundant protein in the infected cell. Several investigators searched for novel DNA polymerase activities in adenovirus-infected cells without success. Consequently, for a long time it was believed that adenovirus DNA replication utilized host polymerases only. From studies of differential inhibition, both the α and γ DNA polymerases were implicated (van der Vliet and Kwant 1978). Only after the discovery of the E2B transcription unit (Stillman et al. 1981) was a systematic search for the adenovirus-specific DNA polymerase undertaken. It was shown that the E2B region encodes the terminal protein (see below) as well as a larger protein in the 100–120-kD molecular mass range (Stillman et al. 1981). Sequence analysis of the region from 11% to 30% confirmed that two very large, open translational reading frames were present encoding a 72-kD and a 105-kD polypeptide (Alestrom et al. 1982a; Gingeras et al. 1982). Peptide analysis showed that the 72-kD reading frame encoded the terminal protein (Smart and Stillman 1982), while studies using mutants belonging to the so called N-group made it possible to demonstrate that the predicted 105-kD protein was the missing adenovirus DNA polymerase.

The linear structure of adenovirus DNA raises the important question of how DNA replication is initiated. The answer turned out to have an unusual solution. Alan Bellet (Canberra) and his colleagues discovered in 1973 that a protein was associated with the ends of adenovirus DNA (Robinson et al. 1973). This protein is known as the terminal protein, and subsequent studies showed that it is covalently attached to the ends of the DNA (Rekosh et al. 1977) via a serine residue linked to the 5′ ends of the two polynucleotide chains (Challberg et al. 1980). Thanks to the elegant studies performed in the laboratories of Kelly, Marshal, Horwitz (Einstein), Hurwitz, and Stillman, it is now clear that the terminal protein initiates DNA replication. It functions by first forming a covalent complex with dCTP, whose 3′-hydroxyl then serves as a primer for DNA replication after the complex is bound to the terminus of the viral template (Lichy et al. 1981; Pincus et al. 1981; Challberg et al. 1982; Tamanoi and Stillman 1982). This model has been confirmed in its basic aspects by the use of in vitro systems.

The development in 1979 of an in vitro replication system that was totally dependent on an exogenous template for initiation and elongation (Challberg and Kelly 1979) allowed the purification and characterization by functional assay of each individual component of the replication machinery. As purification proceeded, it became apparent that three viral proteins from cytoplasmic fractions and two host factors from nuclear fractions were sufficient to allow complete replication in vitro. The three viral factors were the 140-kD DNA polymerase, the 80-kD terminal protein, and the 72-kD DNA-binding protein. The two host factors have been purified from the nuclei of uninfected HeLa cells (Nagata et al. 1982; Lichy et al. 1983). Factor I is a 47-kD DNA-binding protein that interacts specifically with nucleotides 17–48 at the termini of Ad2 DNA (Nagata et al. 1983). Factor II is a 30-kD complex of two polypeptides (15 kD and 17 kD) that is required late during elongation and shows topoisomerase activity (Guggenheimer et al. 1983).

## Genetics leads the way to functions

Adenovirus genetics got off to a really slow start. There were several reasons for this; plaquing of the virus is inefficient and tedious, and the infectivity of the naked viral DNA is very low. In the early days, several naturally occurring mutants were isolated from virus stocks that differed from wild type in plaque morphology, temperature resistance, and so forth. Of particular interest are the so-called cyt mutants (Takemori et al. 1968) that have later been found to carry interesting mutations in the 21-kD protein encoded by E1B (Takemori et al. 1984). In the late sixties, major efforts were made to produce collections of conditional lethal adenovirus mutants, mostly based on Ad2 and Ad5. Much of the early work was performed in the laboratories of Ginsberg, Williams, Doerfler, Weber, and Boulanger. Parallel work on Ad12 mutants came from the laboratory of Hiroto Shimojo (Tokyo). In all cases the approach was rather conventional; virus stocks were mutated with various chemicals and ts mutants were selected based on their ability to grow at a reduced temperature. Using this approach, a large collection of mutants was obtained and a crude map of the viral genome was constructed based on recombinational analysis (Williams et al. 1975). The results were somewhat disappointing since most of the mutants isolated were defective in the synthesis of capsid proteins, and little information could be obtained regarding gene regulation. However, there were some notable exceptions; the ts125 mutant of Ad5, originally isolated in Ginsberg's laboratory, was clearly defective in an early gene since it was negative for viral DNA replication (Ensinger and Ginsberg 1972; van der Vliet et al. 1975). Subsequent studies showed that the defect is in the gene for the 72-kD DNA-binding protein, and this mutant has since been of monumental importance for the functional characterization of this protein. Another example is the ts1 mutant of Ad2, which was

originally isolated by Weber (Weber 1976). Studies of its phenotype at the nonpermissive temperature showed that it is defective in virion maturation due to a lesion in the gene for the 23-kD polypeptide (Liang et al. 1983) which is the viral endopeptidase. Studies of this mutant made it possible to demonstrate for the first time that a viral protein is responsible for the proteolytic trimming that is connected with virion maturation.

Mutants belonging to the so-called N-group (Galos et al. 1979) comprise another set of interesting conditional lethal mutants. These mutants, which were found to be defective in viral DNA replication, played a crucial role in the discovery of the adenovirus-specific DNA polymerase as described above. Yet another example of interesting ts mutants are the so-called hexon-transport mutants, first isolated in an avian adenovirus (Ishibashi 1970). Later, when corresponding Ad5 ts mutants were characterized in Williams's and Russell's laboratories, it became apparent that there exist mutants that are defective in hexon transport and that form their own complementation group (Williams et al. 1971; Russell et al. 1972). Recombinational studies mapped the defect outside of the hexon gene. Later studies have shown that these mutants are defective in the nonstructural 100-kD protein, which has subsequently been shown to be responsible for the correct folding of the hexon polypeptide (Cepko and Sharp 1982). It is surprising that the virus uses about 10% of its coding capacity for a protein whose primary role appears to be the folding of another protein.

One landmark in adenovirus genetics was the isolation of the 293 cell line in 1977 by Graham and coworkers (Graham et al. 1977). This cell line was obtained by transformation of human embryonic kidney cells with fragmented Ad5 DNA. Mild fragmentation of the viral DNA was necessary to abolish the infectivity of the DNA; otherwise the cells would succumb to the strong lytic effect of the virus. The conversion of the original transformants into a permanent cell line was extremely tedious and apparently required a considerable portion of luck, because no one has subsequently succeeded in isolating an equivalent cell line. Roberto Weinmann (Wistar) and coworkers have shown that 293 cells contain integrated viral sequences, derived from the left-hand end of the Ad5 genome, including an expressed E1 region (Aiello et al. 1979). It was clear that the 293 cell line can be used for the isolation of adenovirus mutants that are defective in their transforming region since they can be complemented by the E1 products expressed from the integrated viral genome. This was particularly important since no mutants had hitherto been isolated that were defective in the E1 region. Two types of mutants have been constructed based on the 293 cell line. The first were the so-called host-range mutants isolated in Williams's laboratory. These mutants are defective for growth on HeLa cells but replicate efficiently on 293 cells (Harrison et al. 1977). It was thus assumed that these mutants are defective in the E1 region and that the products from the integrated Ad5 sequences in the 293 cells compensated for the defect. Further analysis of these host-range mutants showed that they could be grouped into two complementation groups, nowadays known to be defective in regions E1A and E1B, respectively (Graham et al. 1978). The so-called *hr*1 mutant has been extremely useful for the functional dissection of the E1A region.

The second type of mutants were isolated from 293 cells after manipulation of the genome in vitro. This work was initiated in Shenk's laboratory and has yielded a number of mutants with interesting phenotypes. Nick Jones (Connecticut, ICRF) and Shenk reported in 1979 that they had isolated mutants of Ad5 that contained a single cleavage site for the restriction endonuclease *Xba*I located at coordinate 4.5 (normal Ad5 DNA contains four cleavage sites for *Xba*I) (Jones and Shenk 1979). It was then possible to manipulate the smaller of the two fragments before they were ligated together to form mutants that are defective in the E1 region. This technology has since been improved consistently and today represents one of the most valuable tools for functional studies of the adenovirus genome. One improvement includes the use of DNA-protein complexes carrying the terminal protein, rather than naked DNA, since this enhances the infectivity drastically (Sharp et al. 1976). The reconstitution of viral genomes through in vivo recombination rather than through ligation in vitro also facilitates the construction (Chinnadurai et al. 1979). Finally, Nigel Stow (CSHL, Glasgow) discovered that cloned fragments can be used for reconstruction of the viral genome, provided that the fragment is excised from the plasmid prior to transfection (Stow 1981). This procedure opened up many experimental possibilities. Particularly noteworthy is the work by Berk's laboratory that has led to the construction of many useful point mutations in the E1 region that influence single proteins (Montell et al. 1982; Osborne et al. 1982).

The methods developed for the in vitro manipulation of the adenovirus genome have allowed adenoviruses to be used as eukaryotic cloning vectors. Yasha Gluzman (CSHL), Grodzicker, and their colleagues at Cold Spring Harbor have constructed a series of clever vectors that express foreign genes when introduced into the appropriate cell lines (Solnick 1981; Thummel et al. 1981; Logan and Shenk 1984).

### VA RNA: A 20-year search for a function

It was reported from Weissman's laboratory in 1966 that adenovirus-infected cells contain an abundant class of low-molecular-weight RNA, designated VA RNA (virus-associated RNA) (Reich et al. 1966). The nucleotide sequence of VA RNA (Ohe and Weissman 1971) showed that it had a compact secondary structure but gave no clues as to its function. VA RNA was originally believed to comprise a single RNA species. However, it is now known that two unique species, VA RNA$_I$ and VA RNA$_{II}$, with very similar secondary structures are present (Mathews 1975; Soderlund et al. 1976). They map to position 30 on the adenovirus genome, and their genes are separated by a spacer of approximately 90

nucleotides (Akusjärvi et al. 1980). The VA RNAs are unique among the adenovirus-encoded RNAs in being transcribed by RNA polymerase III (Price and Penman 1972; Weinmann et al. 1974). They have typical polymerase III control regions embedded within their genes (Fowlkes and Shenk 1980; Guilfoyle and Weinmann 1981).

Although VA RNA is an extremely abundant RNA class in infected cells, it took almost two decades to assign a function to it. After the discovery of splicing in 1977, many investigators believed that VA RNA might serve its function by hybridizing to exon/intron borders, thereby aligning the exons during the splicing reaction. It was even reported that VA RNA$_I$ specifically hybridizes to a cDNA clone that contained the assembled tripartite leader (Mathews 1980). However, more-substantial experimental support for this hypothesis is lacking.

One function of VA RNA was deciphered through the construction in Shenk's laboratory of mutants that lacked either of the two VA RNA genes (Thimmappaya et al. 1982). Analysis of these mutants showed that they were defective in the synthesis of viral proteins. Since the mutants produced viral mRNA, it was concluded that the defect lay at the level of translation. The nature of the defect was worked out quite quickly. Experiments performed in Shenk's and Mathews's laboratories demonstrated that VA RNA exerts its effects during the initiation of translation (Schneider et al. 1984; Reichel et al. 1985), and later it was shown that VA RNA$_I$ interferes with the induction of the kinase that phosphorylates eIF-2 (Schneider et al. 1985; Siekierka et al. 1985).

The precise role of VA RNA during the infectious cycle has been difficult to define. It is clear that the virus is still able to replicate, albeit inefficiently, in the absence of a functional gene for VA RNA$_I$. It has been shown that VA RNA does not exclusively enhance the translation of late viral proteins (Svensson and Akusjärvi 1984b), since both early proteins and some cellular proteins are more efficiently translated in the presence of VA RNA. Recent experiments suggest that its primary role could be to counteract the effect of interferon.

### The restricted growth of human adenoviruses in monkey cells reveals sophisticated regulatory circuits

It was recognized quite early that adenoviruses have a narrow host range. For instance, it was discovered that human adenoviruses replicate poorly in most monkey cells. Alan Rabson (NIH) and his coworkers observed that this apparent host restriction could be overcome by coinfection with SV40 virus or another infectious monkey agent called "MAC" (Rabson et al. 1964). The latter viruses apparently provided some helper function in *trans*. Studies of the adenovirus-SV40 hybrids, originally isolated by Andy Lewis (NIH), gave insight into the nature of the helper function (Lewis et al. 1966). The adenovirus-SV40 hybrids were the result of an experimental accident. Because it was proscribed to propagate human viruses in human cells for the purpose of vaccine production, attempts were made to propagate human adenoviruses in primary monkey cells (Hartley et al. 1956). In some experiments it was noticed that the adenoviruses quickly adapted to growth in the particular monkey cells used. However, all hopes to produce vaccines by this route vanished when it was discovered that the cultures were contaminated with SV40. Attempts were made to remove the contaminating SV40 virus by the use of neutralizing antisera, but this was unsuccessful because the cocultivation of SV40 and human adenoviruses in monkey cells had led to recombination between the two viral genomes.

Two types of hybrid viruses were detected in the original stock: defective and nondefective hybrid viruses. In the defective stocks, recombination had resulted in the integration of a large piece of SV40 DNA. To allow the recombinant genome to be packaged, a large segment of the adenovirus genome had been deleted, including some sequences essential for growth. These hybrid viruses therefore require an adenovirus helper for growth. In contrast, the nondefective hybrids contained smaller segments of inserted SV40 DNA that replaced part of the E3 region and that is dispensible for growth in tissue culture. In these viruses, sufficient SV40 sequences are present and expressed to provide the necessary SV40 helper function. A series of nondefective hybrid viruses, designated Ad2$^+$ND1 to Ad2$^+$ND5, were isolated by Lewis and coworkers from the original stock of adenovirus-SV40 hybrids (Lewis et al. 1969). Electron microscopic heteroduplex studies, carried out by Kelly and coworkers, unmasked the structure of the hybrid genomes (Kelly and Rose 1971; Kelly et al. 1974). From the structure of the inserted piece of SV40 DNA, it could be concluded that the helper function was encoded within the carboxyterminal part of the gene for SV40 large T antigen. These nondefective viruses have been exploited by Grodzicker (Grodzicker et al. 1974, 1976) for a variety of genetic experiments, as well as to overproduce SV40 large T antigen and its derivatives (Fey et al. 1979).

Studies of the mechanism behind the monkey cell restriction have given very complex results. Early studies, performed by Steve Baum (Einstein), Lewis, and others showed that one result of the defect was a decreased synthesis of capsid proteins (Baum et al. 1968, 1972). Analysis of the viral mRNA population in adenovirus-infected monkey cells by Dan Klessig (CSHL, Utah) showed that the fiber mRNA was the most severely affected (Klessig and Anderson 1975). Electron microscopy of nuclear and cytoplasmic RNAs from infected monkey cells (Klessig and Chow 1980) has shown that splicing of late mRNA is defective and gives rise to populations of incorrectly assembled RNAs. More-detailed studies by Klessig's group have demonstrated that the defect affects several steps in mRNA production, including the initiation of transcription at the major late adenovirus promoter, mRNA splicing, and the termination of transcription.

The restricted replication of human adenoviruses in monkey cells has also been studied from another angle. Klessig's group has isolated a series of mutants of human Ad2 that replicate efficiently in monkey cells (Klessig 1977). Further characterization of these mutants has shown that the lesion is in the gene for the 72-kD DNA-binding protein encoded by region E2A (Klessig and Grodzicker 1979). This finding was unexpected since an essential function for the DNA-binding protein was already well established in connection with viral DNA replication. It thus appears that this protein and perhaps others wear many hats during the adenovirus life cycle. The restricted replication of human adenoviruses has turned out to be a very complex phenomenon, presumably reflecting the multiple levels of control that operate in an adenovirus-infected cell.

## Concluding Remarks

We are now in the fourth decade of adenovirus research and yet important problems still abound. During the fifties the biology of these viruses was the focus of attention, and their basic properties as infectious agents were described. In the sixties, biochemistry took over and the basic structural features of the virus particle were deciphered. This was followed by the frenzied activities of the seventies, during which the molecular biologists dissected the genome and its transcription and translation. The eighties have seen yet another change in emphasis; now the regulatory circuits are being investigated in detail.

Throughout these endeavors, lessons learned by studying adenoviruses have been a prelude for similar discoveries in the host. The viruses have proven themselves excellent model systems. It would be a great mistake to abandon them just yet.

## References

The reference list includes selected publications which illustrate the points that are discussed in the chapter. The authors apologize for the strong bias toward publications originating from their own laboratories or laboratories of their collaborators.

Aaij, C. and P. Borst. 1972. The gel electrophoresis of DNA. *Biochim. Biophys. Acta* **269**: 192.

Adam, E., J. Erdei, A. Lengyel, G. Berencsi, J. Fachet, and I. Nasz. 1985. Delineation of antigenic determinants of adenovirus hexons by means of monoclonal antibodies. *Intervirology* **23**: 222.

Aiello, L., R. Guilfoyle, K. Huebner, and R. Weinmann. 1979. Adenovirus 5 DNA sequences present and RNA sequences transcribed in transformed human embryo kidney cells (HEK-Ad5 or 293). *Virology* **94**: 460.

Akusjärvi, G. and H. Persson. 1981. Controls of RNA splicing and termination in the major late adenovirus transcription unit. *Nature* **292**: 420.

Akusjärvi, G. and U. Pettersson. 1978a. Nucleotide sequence at the junction between the coding region of the adenovirus 2 hexon messenger RNA and its leader sequence. *Proc. Natl. Acad. Sci.* **75**: 5822.

———. 1978b. Sequence analysis of adenovirus DNA. I. Nucleotide sequence at the carboxy-terminal end of the gene for adenovirus type 2. *Virology* **91**: 477.

———. 1979. Sequence analysis of adenovirus DNA: Complete nucleotide sequence of the spliced 5′ noncoding region of adenovirus 2 hexon messenger RNA. *Cell* **16**: 841.

Akusjärvi, G., M.B. Mathews, P. Andersson, B. Vennstrom, and U. Pettersson. 1980. Structure of genes for virus-associated RNA$_I$ and RNA$_{II}$ of adenovirus type 2. *Proc. Natl. Acad. Sci.* **77:** 2424.

Alestrom, P., G. Akusjärvi, M. Pettersson, and U. Pettersson. 1982a. DNA sequence analysis of the region encoding the terminal protein and the hypothetical N-gene product of adenovirus type 2. *J. Biol. Chem.* **257:** 13492.

Alestrom, P., A. Stenlund, P. Li, A. Bellett, and U. Pettersson. 1982b. Sequence homology between avian and human adenoviruses. *J. Virol.* **42:** 306.

Anderson, C.W., J.B. Lewis, P.R. Baum, and R.F. Gesteland. 1973. Processing of adenovirus 2–induced proteins. *J. Virol.* **12:** 241.

Anderson, C.W., R.C. Schmitt, J.E. Smart, and J.B. Lewis. 1984. Early region 1b of adenovirus serotype 2 encodes two co-terminal proteins of 495 and 155 amino acid residues. *J. Virol.* **50:** 387.

Andersson, M., S. Pääbo, T. Nilsson, and P.A. Petterson. 1985. Impaired intracellular transport of class I MHC antigens as a possible means for adenoviruses to evade immune surveillance. *Cell* **43:** 215.

Arrand, J.R. and R.J. Roberts. 1979. The nucleotide sequences at the termini of adenovirus-2 DNA. *J. Mol. Biol.* **128:** 577.

Babich, A. and J.R. Nevins. 1981. The stability of early adenovirus mRNA is controlled by the viral 72K DNA binding protein. *Cell* **26:** 371.

Bachenheimer, S. and J.E. Darnell. 1975. Adenovirus-2 mRNA is transcribed as part of a high-molecular-weight precursor RNA. *Proc. Natl. Acad. Sci.* **72:** 4445.

Baker, C.C. and E.B. Ziff. 1981. Promoters and heterogeneous 5′-termini of the messenger RNAs of adenovirus serotype 2. *J. Mol. Biol.* **149:** 189.

Baum, S., M. Horwitz, and J. Maizel, Jr. 1972. Studies on the mechanism of enhancement of human adenovirus infection in monkey cells by simian virus 40. *J. Virol.* **10:** 211.

Baum, S.G., W.H. Wiese, and P.R. Reich. 1968. Studies on the mechanism of enhancement of adenovirus 7 infection in African green monkey cells by simian virus 40: Formation of adenovirus-specific RNA. *Virology* **34:** 373.

Berger, S.L. and W.K. Folk. 1985. Differential activation of RNA polymerase III transcribed genes by the polyomavirus enhancer and the E1A gene products. *Nucleic Acids Res.* **13:** 1413.

Berget, S.M. and P.A. Sharp. 1979. Structure of late adenovirus 2 heterogeneous nuclear RNA. *J. Mol. Biol.* **129:** 547.

Berget, S.M., C. Moore, and P.A. Sharp. 1977. Spliced segments at the 5′ terminus of adenovirus 2 late mRNA. *Proc. Natl. Acad. Sci.* **74:** 3171.

Berk, A.J. and P.A. Sharp. 1977. Ultraviolet mapping of the adenovirus 2 early promoters. *Cell* **12:** 45.

———. 1978. Structure of the adenovirus 2 early mRNAs. *Cell* **14:** 695.

Berk, A.J., F. Lee, T. Harrison, J. Williams, and P.A. Sharp. 1979. Pre-early Ad5 gene product regulates synthesis of early viral mRNAs. *Cell* **17:** 935.

Blanchard, J.-M., J. Weber, W. Jelinek, and J.E. Darnell. 1978. In vitro RNA-RNA splicing in adenovirus 2 mRNA formation. *Proc. Natl. Acad. Sci.* **75:** 5344.

Borrelli, E., R. Hen, and P. Chambon. 1984. Adenovirus-2 E1A products repress enhancer-induced stimulation of transcription. *Nature* **312:** 608.

Branton, P.E., N.J. Lassam, J.F. Downey, S.-P. Yee, F.L. Graham, S. Mak, and S.T. Bayley. 1981. Protein kinase activity immunoprecipitated from adenovirus infected cells by sera from tumor bearing hamsters. *J. Virol.* **37:** 601.

Broker, T.R., C. Keller, and R.J. Roberts. 1984. Human adenovirus serotypes 2, 4, 8, 40 and 41 and adenovirus–simian virus 40 hybrids. In *Genetic maps 1984* (ed. S.J. O'Brien), p. 99. Cold Spring Harbor Laboratory, Cold Spring Harbor, New York.

Burgert, H.-G. and S. Kvist. 1985. An adenovirus type 2 glycoprotein blocks cell surface expression of human histocompatibility class 1 antigens. *Cell* **41:** 987.

Burlingham, B.T. and W. Doerfler. 1972. An endonuclease in cells infected with adenovirus and associated with adenovirions. *Virology* **48:** 1.

Burlingham, B.T., W. Doerfler, U. Pettersson, and L. Philipson. 1971. Adenovirus endonuclease: Association with the penton of adenovirus type 2. *J. Mol. Biol.* **60:** 45.

Carter, T.H. and R.A. Blanton. 1978. Autoregulation of adenovirus type 5 early gene expression. II. Effect of temperature-sensitive early mutations on virus RNA accumulation. *J. Virol.* **28:** 450.

Cepko, C.L. and P.A. Sharp. 1982. Assembly of adenovirus major capsid protein is mediated by a nonvirion protein. *Cell* **31:** 407.

Cepko, C.L., P.A. Changelian, and P.A. Sharp. 1981. Immunoprecipitation with two-dimensional pools as a hybridoma screening technique: Production and characterization of monoclonal antibodies against adenovirus 2 proteins. *Virology* **110:** 385.

Challberg, M.D. and T.J. Kelly, Jr. 1979. Adenovirus DNA replication in vitro. *Proc. Natl. Acad. Sci.* **76:** 655.

———. 1981. Processing of the adenovirus terminal protein. *J. Virol.* **38:** 272.

Challberg, M.D., S.V. Desiderio, and T.J. Kelly, Jr. 1980. Adenovirus DNA replication in vitro: Characterization of a protein covalently linked to nascent DNA strands. *Proc. Natl. Acad. Sci.* **77:** 5105.

Challberg, M.D., J.M. Ostrove, and T.J. Kelly, Jr. 1982. Initiation of adenovirus DNA replication: Detection of covalent complexes between nucleotide and the 80-kilodalton terminal protein. *J. Virol.* **41:** 265.

Chinnadurai, G., S. Chinnadurai, and J. Brusca. 1979. Physical mapping of a large-plaque mutation of adenovirus type 2. *J. Virol.* **32:** 623.

Chow, L.T., T.R. Broker, and J.B. Lewis. 1979. Complex splicing patterns of RNAs from the early regions of adenovirus-2. *J. Mol. Biol.* **134:** 265.

Chow, L.T., R.E. Gelinas, T.R. Broker, and R.J. Roberts. 1977a. An amazing sequence arrangement at the 5′ ends of adenovirus 2 messenger RNA. *Cell* **12:** 1.

Chow, L.T., J.M. Roberts, J.B. Lewis, and T.R. Broker. 1977b. A map of cytoplasmic RNA transcripts from lytic adenovirus type 2, determined by electron microscopy of RNA:DNA hybrids. *Cell* **11:** 819.

Cladaras, C. and W.S.M. Wold. 1985. DNA sequence of E3 transcription unit of adenovirus 5. *Virology* **140:** 28.

Craig, E.A. and H.J. Raskas. 1974. Effect of cycloheximide on RNA metabolism early in productive infection with adenovirus 2. *J. Virol.* **14:** 26.

Daniell, E. 1976. Genome structure of incomplete particles of adenovirus. *J. Virol.* **19:** 684.

Doerfler, W. 1968. The fate of the DNA of adenovirus type 12 in baby hamster kidney cells. *Proc. Natl. Acad. Sci.* **60:** 636.

———. 1982. Uptake, fixation and expression of foreign DNA in mammalian cells: The organization of integrated adenovirus DNA sequences. *Curr. Top. Microbiol. Immunol.* **101:** 127.

Eggerding, F. and H.J. Raskas. 1978. Effect of protein synthesis inhibitors on viral mRNAs synthesized early in adenovirus type 2 infection. *J. Virol.* **25:** 453.

Ensinger, M.J. and H.S. Ginsberg. 1972. Selection and preliminary characterization of temperature-sensitive mutants of type 5 adenovirus. *J. Virol.* **10:** 328.

Esche, H., M.B. Mathews, and J.B. Lewis. 1980. Proteins and messenger RNAs of the transforming region of wild type and mutant adenoviruses. *J. Mol. Biol.* **142:** 399.

Evans, R.M., N. Fraser, E. Ziff, J. Weber, M. Wilson, and J.E. Darnell. 1977. The initiation sites for RNA transcription in Ad2 DNA. *Cell* **12:** 733.

Everitt, E., L. Lutter, and L. Philipson. 1975. Structural proteins of adenoviruses. XII. Location and neighbor relationship among proteins of adenovirion type 2 as revealed by enzymatic iodination, immuno-precipitation and chemical cross-linking. *Virology* **67**: 197.

Everitt, E., B. Sundquist, U. Pettersson, and L. Philipson. 1973. Structural proteins of adenoviruses. X. Isolation and topography of low molecular weight antigens from the virion of adenovirus type 2. *Virology* **52**: 130.

Ferguson, B., B. Krippl, O. Andrisani, N. Jones, H. Westphal, and M. Rosenberg. 1985. E1A 13S and 12S mRNA products made in *Escherichia coli* both function as nucleus-localized transcription activators but do not directly bind DNA. *Mol. Cell. Biol.* **5**: 2653.

Fey, G., J.B. Lewis, T. Grodzicker, and A. Bothwell. 1979. Characterization of a fused protein specified by the adenovirus type 2−simian virus 40 hybrid Ad2$^+$ND1 dp2. *J. Virol.* **30**: 201.

Flint, S.J. and P.A. Sharp. 1974. Mapping of viral-specific RNA in the cytoplasm and nucleus of adenovirus 2−infected human cells. *Brookhaven Symp. Biol.* **26**: 333.

Fowlkes, D.M. and T. Shenk. 1980. Transcriptional control regions of the adenovirus VAI RNA gene. *Cell* **22**: 405.

Fowlkes, D.M., S.T. Lord, T. Linne, U. Pettersson, and L. Philipson. 1979. Interaction between the adenovirus DNA-binding protein and double-stranded DNA. *J. Mol. Biol.* **132**: 163.

Fraser, N.W., P.B. Sehgal, and J.E. Darnell, Jr. 1979. Multiple discrete sites for premature RNA chain termination late in adenovirus-2 infection: Enhancement by 5,6-dichloro-1-b-D-ribofuranosylbenzimidazole. *Proc. Natl. Acad. Sci.* **76**: 2571.

Freeman, A.E., P.H. Black, E.A. Vanderpool, P.H. Henry, J.B. Austin, and R.J. Huebner. 1967. Transformation of primary rat embryo cells by adenovirus type 2. *Proc. Natl. Acad. Sci.* **58**: 1205.

Fujinaga, K. and M. Green. 1966. The mechanism of viral carcinogenesis by DNA mammalian viruses: Viral-specific RNA in polyribosomes of adenovirus tumor and transformed cells. *Proc. Natl. Acad. Sci.* **55**: 1567.

─────. 1970. Mechanism of viral carcinogenesis by DNA mammalian viruses. VII. Viral genes transcribed in adenovirus type 2 infected and transformed cells. *Proc. Natl. Acad. Sci.* **65**: 375.

Gahlman, R., R. Leister, L. Vadimon, and W. Doerfler. 1982. Patch homologies and the integration of adenovirus DNA in mammalian cells. *EMBO J.* **1**: 1101.

Gallimore, P.H., P.A. Sharp, and J. Sambrook. 1974. Viral DNA in transformed cells. II. A study of the sequences of adenovirus 2 DNA in 9 lines of transformed rat cells using specific fragments of the viral genome. *J. Mol. Biol.* **89**: 49.

Galos, R.S., J. Williams, M.-H. Binger, and S.J. Flint. 1979. Location of additional early gene sequences in the adenoviral chromosome. *Cell* **17**: 945.

Garon, C.F., K.W. Berry, and J.A. Rose. 1972. A unique form of terminal redundancy in adenovirus DNA molecules. *Proc. Natl. Acad. Sci.* **69**: 2391.

Gaynor, R.B., L.T. Feldman, and A.J. Berk. 1985. Transcription of class III genes activated by viral immediate early proteins. *Science* **230**: 447.

Gaynor, R.B., D. Hillman, and A.J. Berk. 1984. Adenovirus early region 1A protein activates transcription of a non-viral gene introduced into mammalian cells by infection or transfection. *Proc. Natl. Acad. Sci.* **81**: 1193.

Gelb, L.D., D.E. Kohne, and M.A. Martin. 1971. Quantitation of simian virus 40 sequences in African green monkey, mouse and virus-transformed cell genomes. *J. Mol. Biol.* **57**: 129.

Gelinas, R.E. and R.J. Roberts 1977. One predominant 5'-undecanucleotide in adenovirus 2 late messenger RNAs. *Cell* **11**: 533.

Gilardi, P. and M. Perricaudet. 1984. The E4 transcriptional unit of Ad2: Far upstream sequences are required for its transactivation by E1A. *Nucleic Acids Res.* **12**: 7877.

Gingeras, T.R., D. Sciaky, R.E. Gelinas, J. Bingdong, C. Yen, M.M. Kelly, P.A. Bullock, B.L. Parsons, K.E. O'Neill, and R.J. Roberts. 1982. Nucleotide sequences from the adenovirus 2 genome. *J. Biol. Chem.* **257**: 13475.

Ginsberg, H.S., H.G. Pereira, R.C. Valentine, and W.C. Wilcox. 1966. A proposed terminology for the adenovirus antigens and virion morphological subunits. *Virology* **28**: 782.

Goldenberg, C.J. and H.J. Raskas. 1981. In vitro splicing of purified precursor RNAs specified by early region 2 of the adenovirus 2 genome. *Proc. Natl. Acad. Sci.* **78**: 5430.

Graham, F.L. and A.J. van der Eb. 1973. A new technique for the assay of infectivity of human adenovirus DNA. *Virology* **52**: 456.

Graham, F.L., T. Harrison, and J. Williams. 1978. Defective transforming capacity of adenovirus type 5 host-range mutants. *Virology* **86**: 10.

Graham, F.L., A.J. van der Eb., and H.L. Heijneker. 1974. Size and location of the transforming region in human adenovirus type 5 DNA. *Nature* **251**: 687.

Graham, F.L., J. Smiley, W.C. Russell, and R. Nairn. 1977. Characteristics of a human cell line transformed by DNA from human adenovirus 5. *J. Gen. Virol.* **36**: 59.

Graham, F.L., P.S. Abrahams, C. Mulder, H.L. Heijneker, S.O. Warnaar, F.A.J. deVries, W. Fiers, and A.J. van der Eb. 1975. Studies on in vitro transformation by DNA and DNA fragments of human adenoviruses and simian virus 40. *Cold Spring Harbor Symp. Quant. Biol.* **39**: 637

Green, M.R., R. Treisman, and T. Maniatis. 1983. Transcriptional activation of cloned human β-globin genes by viral immediate-early gene products. *Cell* **35**: 137.

Grodzicker, T., J.B. Lewis, and C.W. Anderson. 1976. Conditional lethal mutants of adenovirus type 2−simian virus 40 hybrids. II. Ad2$^+$ND1 host-range mutants that synthesize fragments of Ad2$^+$ND1 30K protein. *J. Virol.* **19**: 559.

Grodzicker, T., C. Anderson, P.A. Sharp, and J. Sambrook. 1974. Conditional lethal mutants of adenovirus 2−simian virus 40 hybrids. *J. Virol.* **13**: 1237.

Grodzicker, T., J.F. Williams, P.A. Sharp and J. Sambrook. 1975. Physical mapping of crossover events between adenovirus 5 and adenovirus 2. *Cold Spring Harbor Symp. Quant. Biol.* **39**: 439.

Guggenheimer, R.A., K. Nagata, J. Field, J. Lindinbaum, R.N. Gronostajski, M.S. Horwitz, and J. Hurwitz. 1983. In vitro synthesis of full-length adenoviral DNA. *UCLA Symp. Mol. Cell. Biol.* **10**: 395.

Guilfoyle, R. and R. Weinmann. 1981. Control region for adenovirus VA RNA transcription. *Proc. Natl. Acad. Sci.* **78**: 3378.

Hardy, S.F., P.J. Grabowski, R.A. Padgett, and P.A. Sharp. 1984. Cofactor requirements of splicing of purified messenger RNA precursors. *Nature* **308**: 375.

Harlow, E., B.R. Franza, and C. Schley. 1985. Monoclonal antibodies specific for adenovirus early region 1A proteins: Extensive heterogeneity in early region 1A products. *J. Virol.* **55**: 533.

Harrison, T., F. Graham, and J. Williams. 1977. Host range mutants of adenovirus type 5 defective for growth in HeLa cells. *Virology* **77**: 319.

Harter, M.L. and J.B. Lewis. 1978. Adenovirus type 2 early proteins synthesized in vitro and in vivo: Identification in infected cells of the 38,000 to 50,000 molecular weight protein encoded by the left end of the adenovirus type 2 genome. *J. Virol.* **26**: 736.

Hartley, J.W., R.J. Huebner, and W.P. Rowe. 1956. Serial propagation of adenovirus (APC) in monkey kidney tissue cultures. *Proc. Soc. Exp. Biol. Med.* **92**: 667.

Haruna, J., H. Yaosi, R. Kono, and I. Watanabe. 1961. Separation of adenovirus by chromatography on DEAE-cellulose. *Virology* **13**: 254.

Hearing, P. and T. Shenk. 1983. The adenovirus type 5 E1A transcriptional control region contains a duplicated enhancer element. *Cell* **33**: 695.

Hen, R., E. Borrelli, P. Sassone-Corsi, and P. Chambon. 1983. An enhancer element is located 340 base pairs upstream from the adenovirus-2 E1A capsite. *Nucleic Acids Res.* **11**: 8747.

Hernandez, N. and W. Keller. 1983. Splicing of in vitro synthesized messenger RNA precursors in HeLa cell extracts. *Cell* **35**: 89.

Hilleman, M.R. and J.R. Werner. 1954. Recovery of new agent from patients with acute respiratory illness. *Proc. Soc. Exp. Biol. Med.* **85**: 183.

Hoeffler, W.K. and R.G. Roeder. 1985. Enhancement of RNA polymerase III transcription by the E1A gene product of adenovirus. *Cell* **41**: 955.

Horne, R.W., S. Brenner, A.P. Waterson, and P. Wildy. 1959. The icosahedral form of an adenovirus. *J. Mol. Biol.* **1**: 84.

Houweling, A., P.J. van den Elsen, and A.J. van der Eb. 1980. Partial transformation of primary rat cells by the leftmost 4.5% fragment of adenovirus 5 DNA. *Virology* **105**: 537.

Huebner, R.J. 1967. Adenovirus-directed tumor and T antigens. *Perspect. Virol.* **5**: 147.

Huebner, R.J., W.P. Rowe, and W.T. Lane. 1962. Oncogenic effects in hamsters of human adenovirus types 12 and 18. *Proc. Natl. Acad. Sci.* **48**: 2051.

Imperiale, M.J. and J.R. Nevins. 1984. Adenovirus 5 E2 transcription unit: An E1A-inducible promoter with an essential element that functions independently of position or orientation. *Mol. Cell. Biol.* **4**: 875.

Imperiale, M.J., L.T. Feldman, and J.R. Nevins. 1983. Activation of gene expression by adenovirus and herpesvirus regulatory genes acting in *trans* and by a *cis*-acting adenovirus enhancer element. *Cell* **35**: 127.

Ishibashi, M. 1970. Retention of viral antigen in the cytoplasm of cells infected with temperature-sensitive mutants of an avian adenovirus. *Proc. Natl. Acad. Sci.* **65**: 304.

Jeng, T.H., W.S.M. Wold, K. Sugawara, and M. Green. 1978. Evidence for an adenovirus type 2–coded early glycoprotein. *J. Virol.* **28**: 314.

Jones, N. and T. Shenk. 1979. An adenovirus type 5 early gene function regulates expression of other early viral genes. *Proc. Natl. Acad. Sci.* **76**: 3665.

Kao, H.-T. and J.R. Nevins. 1983. Transcriptional activation and subsequent control of the human heat shock gene during adenovirus infection. *Mol. Cell. Biol.* **3**: 2058.

Kelly, T. and J. Rose. 1971. Simian virus 40 integration site in an adenovirus 7–SV40 hybrid DNA molecule. *Proc. Natl. Acad. Sci.* **68**: 1037.

Kelly, T., A.M. Lewis, A.S. Levine, and S. Siegel. 1974. Structure of two adenovirus–simian virus 40 hybrids which contain the entire SV40 genome. *J. Mol. Biol.* **89**: 113.

Kitchingman, G.R. and H. Westphal. 1980. The structure of adenovirus 2 early nuclear and cytoplasmic RNAs. *J. Mol. Biol.* **137**: 23.

Kitchingman, G.R., S.-P. Lai, and H. Westphal. 1977. Loop structures in hybrids of early RNA and the separated strands of adenovirus DNA. *Proc. Natl. Acad. Sci.* **74**: 4392.

Klemperer, H.G. and H.G. Pereira. 1959. Study of adenovirus antigens fractionated by chromatography on DEAE-cellulose. *Virology* **9**: 536.

Klessig, D.F. 1977. Isolation of a variant of human adenovirus serotype 2 that multiplies efficiently on monkey cells. *J. Virol.* **21**: 1243.

Klessig, D.F. and C.W. Anderson. 1975. Block to multiplication of adenovirus serotype 2 in monkey cells. *J. Virol.* **16**: 1650.

Klessig, D.F. and L.T. Chow. 1980. Deficient accumulation and incomplete splicing of several late viral RNAs in monkey cells infected by human adenovirus type 2. *J. Mol. Biol.* **139**: 221.

Klessig, D.F. and T. Grodzicker. 1979. Mutations that allow human Ad2 and Ad5 to express late genes in monkey cells map in the gene encoding the 72K DNA binding protein. *Cell* **17**: 957.

Ko, J.L. and M.L. Harter. 1984. Plasmid-directed synthesis of genuine adenovirus 2 early-region 1A and 1B proteins in *E. coli. Mol. Cell. Biol.* **4**: 1427.

Kramer, A., W. Keller, B. Appel, and R. Luhrmann. 1984. The 5′ terminus of the RNA moiety of U1 small nuclear ribonucleoprotein particles is required for the splicing of messenger RNA precursors. *Cell* **38**: 299.

Kvist, S., L. Ostberg, H. Persson, L. Philipson, and P.A. Pettersson. 1978. Molecular association between transplantation antigens and a cell surface antigen in an adenovirus-transformed cell line. *Proc. Natl. Acad. Sci.* **75**: 5674.

Lacy, S., Sr. and M. Green. 1964. Biochemical studies on adenovirus multiplication. VII. Homology between DNAs of tumorigenic and nontumorigenic human adenoviruses. *Proc. Natl. Acad. Sci.* **52**: 1053.

———. 1965. Adenovirus multiplication. Genetic relatedness of tumorigenic human adenovirus type 7, 12, 18. *Science* **150**: 1296.

———. 1967. The mechanism of viral carcinogenesis by DNA mammalian viruses: DNA-DNA homology relationships among the ''weakly'' oncogenic human adenoviruses. *J. Gen. Virol.* **1**: 413.

Landgraf-Leurs, M. and M. Green. 1971. Adenovirus DNA. III. Separation of the complementary strands of adenovirus types 2, 7 and 12 DNA molecules. *J. Mol. Biol.* **60**: 185.

———. 1973. DNA strand selection during the transcription of the adenovirus 2 genome in infected and transformed cells. *Biochim. Biophys. Acta* **312**: 667.

Lavelle, G., C. Patch, G. Khoury, and J. Rose. 1975. Isolation and partial characterization of single-stranded adenoviral DNA produced during synthesis of adenovirus. *J. Virol.* **16**: 775.

Laver, W.G., H.G. Pereira, W.C. Russell, and R. Valentine. 1968. Isolation of an internal component from adenovirus type 5. *J. Mol. Biol.* **37**: 379.

Lechner, R.L. and T.J. Kelly, Jr. 1977. The structure of replicating adenovirus 2 DNA molecules. *Cell* **12**: 1007.

Leff, T. and P. Chambon. 1986. Sequence-specific activation of transcription by adenovirus E1a gene products is observed in HeLa cells but not in 293 cells. *Mol. Cell. Biol.* **6**: 201.

Leff, T., T. Elkaim, C.R. Goding, P. Jalinot, P. Sassone-Corsi, M. Perricaudet, C. Kédinger, and P. Chambon. 1984. Individual products of the adenovirus 12S and 13S mRNAs stimulate viral E1A and E3 expression at the transcriptional level. *Proc. Natl. Acad. Sci.* **81**: 4381.

Levinson, A.D. and A.J. Levine. 1977. The group C adenovirus tumor antigens: Identification in infected and transformed cells and a peptide map analysis. *Cell* **11**: 871.

Lewis, A.M., Jr., S.G. Baum, K.O. Prigge, and W.P. Rowe. 1966. Occurrence of adenovirus-SV40 hybrids among monkey kidney cell–adapted strains of adenovirus. *Proc. Soc. Exp. Biol.. Med.* **122**: 214.

Lewis, A.M., Jr., M.J. Levin, W.H. Wiese, C.S. Crumpacker, and P.H. Henry. 1969. A non-defective (competent) adenovirus-SV40 hybrid isolated from the Ad2-SV40 hybrid population. *Proc. Natl. Acad. Sci.* **63**: 1128.

Lewis, J.B. and M.B. Mathews. 1980. Control of adenovirus early gene expression: A class of immediate early products. *Cell* **21**: 303.

Lewis, J.B., J.F. Atkins, C.W. Anderson, P.R. Baum, and R.F. Gesteland. 1975. Mapping of late adenovirus genes by cell free translation of RNA selected by hybridization to specific DNA fragments. *Proc. Natl. Acad. Sci.* **72**: 1344.

Liang, Y.-K., C. Akusjärvi, P. Alestrom, U. Pettersson, M. Tremblay, and J. Weber. 1983. Genetic identification of an endoproteinase encoded by the adenovirus genome. *J. Mol. Biol.* **167**: 217.

Lichy, J.H., M.S. Horwitz, and J. Hurwitz. 1981. Formation of a covalent complex between the 80,000-dalton adenovirus terminal protein and 5′-dCMP in vitro. *Proc. Natl. Acad. Sci.* **78**: 2678.

Lichy, J.H., T. Enomoto, J. Field, B.R. Friefeld, R.A. Guggenheimer, J. Ikeda, K. Nagata, M.S. Horwitz, and J. Hurwitz. 1983. Isolation of proteins involved in the replication of adenovirus DNA in vitro. *Cold Spring Harbor Symp. Quant. Biol.* **47**: 731.

Logan, J. and T. Shenk. 1984. Adenovirus tripartite leader sequence enhances translation of mRNAs late after infection. *Proc. Natl. Acad. Sci.* **81:** 3655.

Lucas, J.J. and H.S. Ginsberg. 1971. Synthesis of virus-specific ribonucleic acid in KB cells infected with type 2 adenovirus. *J. Virol.* **8:** 203.

Maizel, J.V., Jr. 1966. Acrylamide-gel electrophorograms by mechanical fractionation: Radioactive adenovirus proteins. *Science* **151:** 988.

———. 1971. Polyacrylamide gel electrophoresis of viral proteins. *Methods Virol.* **5:** 179.

Manley, J.L., P.A. Sharp, and M.L. Gefter. 1979. RNA synthesis in isolated nuclei in vitro initiation of adenovirus 2 major late mRNA precursor. *Proc. Natl. Acad. Sci.* **76:** 160.

Manley, J.L., A. Fire, A. Cano, P.A. Sharp, and M.L. Gefter. 1980. DNA-dependent transcription of adenovirus genes in a soluble whole-cell extract. *Proc. Natl. Acad. Sci.* **77:** 3855.

Mathews, M.B. 1975. Genes for VA-RNA and adenovirus 2. *Cell* **6:** 223.

———. 1980. Binding of adenovirus VA RNA to mRNA: A possible role in splicing? *Nature* **285:** 575.

McBride, W.D. and A. Wiener. 1964. In vitro transformation of hamster kidney cells by human adenovirus type 12. *Proc. Soc. Exp. Biol. Med.* **115:** 870.

McGrogan, M. and H.J. Raskas. 1978. Two regions of the adenovirus 2 genome specify families of late polysomal RNAs containing common sequences. *Proc. Natl. Acad. Sci.* **75:** 625.

Mellow, G.H., B. Fohring, J. Dougherty, P.H. Gallimore, and K. Raska. 1984. Tumorigenicity of adenovirus-transformed rat cells and expression of class I major histocompatibility antigen. *Virology* **134:** 460.

Montell, C., G. Courtois, C. Eng, and A. Berk. 1984a. Complete transformation by adenovirus 2 requires both E1A proteins. *Cell* **36:** 951.

Montell, C., E.F. Fisher, M.H. Caruthers, and A.J. Berk. 1982. Resolving the functions of overlapping viral genes by site-specific mutagenesis at a mRNA splice site. *Nature* **295:** 380.

———. 1984b. Control of adenovirus E1B mRNA synthesis by a shift in the activities of RNA splice sites. *Mol. Cell. Biol.* **4:** 966.

Moore, C.L. and P.A. Sharp. 1985. Accurate cleavage and polyadenylation of exogenous RNA substrate. *Cell* **41:** 845.

Mulder, C. and H. Delius. 1972. Specificity of the break produced by restriction endonuclease R₁ in simian virus 40 DNA, as revealed by partial denaturation mapping. *Proc. Natl. Acad. Sci.* **69:** 3215.

Mulder, C., P.A. Sharp, H. Delius, and U. Pettersson. 1974. Specific fragmentation of DNA of adenovirus serotypes 3, 5, 7, and 12, and adeno-simian virus 40 hybrid virus Ad⁺ND1 by restriction endonuclease R.EcoRI. *J. Virol.* **14:** 68.

Mulder, C., J.R. Arrand, H. Delius, W., Keller, U. Pettersson, R.J. Roberts, and P.A. Sharp. 1975. Cleavage maps of DNA from adenovirus types 2 and 5 by restriction endonucleases *EcoRI* and *HpaI*. *Cold Spring Harbor Symp. Quant. Biol.* **39:** 397.

Nagata, K., R.A. Guggenheimer, and J. Hurwitz. 1983. Specific binding of a cellular DNA replication protein to the origin of replication of adenovirus DNA. *Proc. Natl. Acad. Sci.* **80:** 6177.

Nagata, K., R.A. Guggenheimer, T. Enomoto, J.H. Lichy, and J. Hurwitz. 1982. Adenovirus DNA replicated in vitro. Identification of a host factor that stimulates the synthesis of the preterminal protein dCMP complex. *Proc. Natl. Acad. Sci.* **79:** 6438.

Nevins, J.R. 1981. Mechanism of activation of early viral transcription by the adenovirus E1A gene product. *Cell* **26:** 213.

Nevins, J.R. and J.E. Darnell. 1978. Groups of adenovirus type 2 mRNAs derived from a large primary transcript: Probable nuclear origin and possible common 3′ ends. *J. Virol.* **25:** 811.

Nevins, J.R. and M.C. Wilson. 1981. Regulation of adenovirus-2 gene expression at the level of transcriptional termination and RNA processing. *Nature* **290:** 113.

Nevins, J.R., H.S. Ginsberg, J.-M. Blanchard, M.C. Wilson, and J.E. Darnell. 1979. Regulation of the primary expression of the early adenovirus transcription units. *J. Virol.* **32:** 727.

Norrby, E. 1966. The relationship between the soluble antigens and the virion of adenovirus type 3. I. Morphological characteristics. *Virology* **28:** 236.

———. 1968. Biological significance of structural adenovirus components. *Curr. Top. Microbiol. Immunol.* **43:** 1.

Oberg, B., J. Saborio, T. Persson, E. Everitt, and L. Philipson. 1975. Identification of the in vitro translation products of adenovirus mRNA by immunoprecipitation. *J. Virol.* **15:** 199.

Ohe, K. and S.M. Weissman. 1971. The nucleotide sequence of a low molecular weight ribonucleic acid from cells infected with adenovirus 2. *J. Biochem.* **246:** 6991.

Osborn, T.F., R.B. Gaynor, and A.J. Berk. 1982. The TATA homology and the mRNA 5′ untranslated sequence are not required for expression of essential adenovirus E1A functions. *Cell* **29:** 139.

Osborne, T.F., D.N. Arvidson, E.S. Tyau, M. Dunsworth-Brown, and A.J. Berk. 1984. Transcription control region within the protein-coding portion of adenovirus E1A genes. *Mol. Cell. Biol.* **4:** 1293.

Padgett, R.A., S.M. Mount, J.A. Steitz, and P.A. Sharp. 1983. Splicing of messenger RNA precursors is inhibited by antisera to small nuclear ribonucleoprotein. *Cell* **35:** 101.

Padgett, R.A., M.M. Konarska, P.J. Grabowski, S.F. Hardy, and P.A. Sharp. 1984. Lariat RNAs as intermediates and products in the splicing of messenger RNA precursors. *Science* **225:** 898.

Parsons, J.T. and M. Green. 1971. Biochemical studies on adenovirus multiplication. VII. Resolution of early virus specific RNA species in Ad2 infected and transformed cells. *Virology* **45:** 154.

Pereira, H.G., R.C. Valentine, and W.C. Russell. 1968. Crystallization of an adenovirus protein (the hexon). *Nature* **219:** 946.

Perricaudet, M., J.M. LeMoullec, and U. Pettersson. 1980. Predicted structure of two adenovirus tumor antigens. *Proc. Natl. Acad. Sci.* **77:** 3778.

Perricaudet, M., G. Akusjärvi, A. Virtanen, and U. Pettersson. 1979. Structure of two spliced mRNAs from the transforming region of human subgroup C adenoviruses. *Nature* **281:** 694.

Persson, H., M. Jansson, and L. Philipson. 1980a. Synthesis and genomic site for an adenovirus type 2 early glycoprotein. *J. Mol. Biol.* **136:** 375.

Persson, H., H. Jornvall, and J. Zabielski. 1980b. Multiple mRNA species for the precursor to an adenovirus-encoded glycoprotein: Identification and structure of the signal sequence. *Proc. Natl. Acad. Sci.* **77:** 6349.

Persson, H., M.G. Katze, and L. Philipson. 1981. Control of adenovirus early gene expression: Accumulation of viral mRNA after infection of transformed cells. *J. Virol.* **40:** 358.

———. 1982. An adenovirus tumor antigen associated with membranes in vivo and in vitro. *J. Virol.* **42:** 905.

Persson, H., C. Signas, and L. Philipson. 1979. Purification and characterization of an early glycoprotein from adenovirus type 2–infected cells. *J. Virol.* **29:** 938.

Persson, H., S. Kvist, L. Ostberg, P.A. Peterson, and L. Philipson. 1980c. The early adenovirus glycoprotein E3-19K and its association with transplantation antigens. *Cold Spring Harbor Symp. Quant. Biol.* **44:** 409.

Pettersson, U. 1973. Some unusual properties of replicating adenovirus type 2 DNA. *J. Mol. Biol.* **81:** 521.

Pettersson, U. and J. Sambrook. 1973. The amount of viral DNA in the genome of cells transformed by adenovirus type 2. *J. Mol. Biol.* **73:** 125.

Pettersson, U., C. Mulder, H. Delius, and P.A. Sharp. 1973. Cleavage of adenovirus type 2 DNA into six unique fragments by endonuclease R.RI. *Proc. Natl. Acad. Sci.* **70:** 200.

Philipson, L. 1960. Separation on DEAE cellulose of components associated with adenovirus reproduction. *Virology* **10**: 459.

————. 1983. Structure and assembly of adenoviruses. In *The molecular biology of adenoviruses 1* (ed. W. Doerfler), p. 1. Springer-Verlag, New York.

Philipson, L., R. Wall, G. Glickman, and J.E. Darnell. 1971. Addition of polyadenylated sequences to virus specific RNA during adenovirus replication. *Proc. Natl. Acad. Sci.* **68**: 2806.

Pilder, S., J. Logan, and T. Shenk. 1984. Deletion of the gene encoding the adenovirus 5 early region 1B 21,000 molecular weight polypeptide leads to degradation of viral and host cell DNA. *J. Virol.* **52**: 664.

Pina, M. and M. Green. 1965. Biochemical studies on adenovirus multiplication. IX. Chemical and base composition analysis of 28 human adenoviruses. *Proc. Natl. Acad. Sci.* **54**: 547.

Pincus, S., W. Robertson, and D. Rekosh. 1981. Characterization of the effect of aphidicolin on adenovirus DNA replication: Evidence in support of a protein primer model of initiation. *Nucleic. Acids Res.* **9**: 4919.

Polasa, M. and M. Green. 1967. Adenovirus proteins. I. Amino acid composition of oncogenic and nononcogenic human adenoviruses. *J. Virol.* **31**: 565.

Prage, L., U. Pettersson, and L. Philipson. 1968. Internal basic proteins in adenovirus. *Virology* **36**: 508.

Price, R. and S. Penman. 1972. A distinct RNA polymerase activity, synthesizing 5.5S, 5S and 4S RNA in nuclei from adenovirus 2-infected HeLa cells. *J. Mol. Biol.* **70**: 435.

Rabson, A.S., G.T. O'Conor, I.K. Berezesky, and F.J. Paul. 1964. Enhancement of adenovirus growth in African green monkey kidney cell cultures by SV40. *Proc. Soc. Exp. Biol. Med.* **116**: 187.

Reich, P.R., B.G. Forget, and S.M. Weissman. 1966. RNA of low molecular weight in KB cells infected with adenovirus type 2. *J. Mol. Biol.* **17**:428.

Reichel, P.A., W.C. Merrick, J. Siekierka, and M.B. Mathews. 1985. Regulation of a protein synthesis initiation factor by adenovirus virus-associated RNA. *Nature* **313**: 196.

Reif, U.K., U. Winterhoff, and W. Doerfler. 1977. Characterization of the pH 4.0 endonuclease from adenovirus-type-2-infected KB cells. *Eur. J. Biochem.* **73**: 327.

Rekosh, D.M.K., W.C. Russell, A.J.D. Bellett, and A.J. Robinson. 1977. Identification of a protein covalently linked to the ends of adenovirus DNA. *Cell* **11**: 283.

Roberts, R.J., G. Akusjärvi, P. Alestrom, R.E. Gelinas, T.R. Gingeras, D. Sciaky, and U. Pettersson. 1986. A consensus sequence for the adenovirus-2 genome. In *Developments in molecular virology* (ed. W. Doerfler), p. 1. Martinus Nijhoff, Boston.

Robin, J., D. Bourgoux-Ramoizy, and P. Bourgaux. 1973. Single-stranded regions in replicating DNA of adenovirus type 2. *J. Gen. Virol.* **20**: 233.

Robinson, A.J., H.B. Younghusband, and A.J.D. Bellett. 1973. A circular DNA-protein complex from adenoviruses. *Virology* **56**: 54.

Rosen, L., J.F. Hovis, and J.A. Bell. 1962. Further observation on typing adenoviruses and a description of two possible additional serotypes. *Proc. Soc. Exp. Biol. Med.* **110**: 710.

Rowe, W.P., R.J. Huebner, L.K. Gillmore, R.H. Parrott, and T.G. Ward. 1953. Isolation of a cytopathogenic agent from human adenoids undergoing spontaneous degeneration in tissue culture. *Proc. Soc. Exp. Biol. Med.* **84**: 570.

Ruley, H.E. 1983. Adenovirus early region 1A enables viral and cellular transforming genes to transform primary cells in culture. *Nature* **304**: 602.

Russell, W.C. and G.E. Blair. 1977. Polypeptide phosphorylation in adenovirus-infected cells. *J. Gen. Virol.* **34**: 19.

Russell, W.C. and B. Precious. 1982. Nucleic acid binding properties of adenovirus structural polypeptides. *J. Gen. Virol.* **63**: 69.

Russell, W.C. and J.J. Skehel. 1972. The polypeptides of adenovirus-infected cells. *J. Gen. Virol.* **15**: 45.

Russell, W.C., K. McIntosh, and J.J. Skehel. 1971. The preparation and properties of adenovirus cores. *J. Gen. Virol.* **11**: 35.

Russell, W.C., C. Newman, and J.F. Williams. 1972. Characterization of temperature-sensitive mutants of adenovirus type 5-serology. *J. Gen. Virol.* **17**: 265.

Russell, W.C., G. Patel, B. Precious, I. Sharp, and P.S. Gardner. 1981. Monoclonal antibodies against adenovirus type 5: Preparation and preliminary characterization. *J. Gen. Virol.* **56**: 393.

Schechter, N.M., W. Davies, and C.W. Anderson. 1980. Adenovirus coded DNA binding protein. Isolation physical properties and effects of proteolytic degradation. *Biochemistry.* **19**: 2802.

Schneider, R.J., C. Weinberger, and T. Shenk. 1984. Adenovirus VAI RNA facilitates the initiation of translation in virus-infected cells. *Cell* **27**: 291.

Schneider, R.J., B. Safer, S.M. Munemitsu, C.E. Samuel, and T. Shenk. 1985. Adenovirus VAI RNA prevents phosphorylation of the eukaryotic initiation factor 2 α subunit subsequent to infection. *Proc. Natl. Acad. Sci.* **82**: 4321.

Schrier, P.I., R. Bernards, R.T.M.J. Vaessen, A. Houweling, and A.J. van der Eb. 1983. Expression of class I major histocompatibility antigens switched off by highly oncogenic adenovirus 12 in transformed rat cells. *Nature* **305**: 771.

Sharp, P.A., C. Moore, and J.L. Haverty. 1976. The infectivity of adenovirus 5 DNA-protein complex. *Virology* **75**: 442.

Sharp, P.A., U. Pettersson, and J. Sambrook. 1974. Viral DNA in transformed cells. I. A study of the sequences of adenovirus 2 DNA in a line of transformed rat cells using specific fragments of the viral genome. *J. Mol. Biol.* **86**: 709.

Sharp, P.A., B. Sugden, and J. Sambrook. 1973. Detection of two restriction endonuclease activities in *Haemophilus parainfluenzae* using analytical agarose ethidium bromide electrophoresis. *Biochemistry* **12**: 3055.

Shaw, A.R. and E.B. Ziff. 1980. Transcripts from the adenovirus-2 major late promoter yield a single family of 3′ co-terminal mRNAs during early infection and five families at late times. *Cell* **22**: 905.

Siekierka, J., T.M. Mariano, P.A. Reichel, and M.B. Mathews. 1985. Translational control by adenovirus: Lack of virus-associated RNA$_I$ during adenovirus infection results in phosphorylation of initiation factor eIF-2 and inhibition of protein synthesis. *Proc. Natl. Acad. Sci.* **82**: 1959.

Signas, C., M.G. Katze, H. Persson, and L. Philipson. 1982. An adenovirus glycoprotein binds heavy chains of class I transplantation antigens from man and mouse. *Nature* **299**: 175.

Smart, J.E. and B.W. Stillman. 1982. Adenovirus terminal protein precursor: Partial amino acid sequence and site of covalent linkage to virus DNA. *J. Biochem.* **257**: 13499.

Smith, D.H., D.M. Kegler, and E.B. Ziff. 1985. Vector expression of adenovirus type 5 E1A proteins: Evidence for E1A autoregulation. *Mol. Cell. Biol.* **5**: 2684.

Soderlund, H., U. Pettersson, B. Vennstrom, L. Philipson, and M.B. Mathews. 1976. A new species of virus-coded low molecular weight RNA from cells infected with adenovirus type 2. *Cell* **7**: 585.

Solnick, D. 1981. Construction of an adenovirus SV40 recombinant producing SV40 T-antigen from an adenovirus late promoter. *Cell* **24**: 135.

Spindler, K.R., C.Y. Eng, and A.J. Berk. 1985. An adenovirus early region 1A protein is required for maximal viral DNA replication in growth arrested human cells. *J. Virol.* **53**: 742.

Stabel, S., W. Doerfler, and R.R. Friis. 1980. Integration sites of adenovirus type 12 DNA in transformed hamster cells and hamster tumor cells. *J. Virol.* **36**: 22.

Steenbergh, P.H., J.S. Sussenbach, R.J. Roberts, and H.S. Jansz. 1975. The 3′ terminal nucleotide sequences of adenovirus types 2 and 5 DNA. *J. Virol.* **15**: 268.

Stein, R. and E.B. Ziff. 1984. HeLa cell β-tubulin gene transcription is stimulated by adenovirus 5 in parallel with viral early genes by an E1a-dependent mechanism. *Mol. Cell. Biol.* **4**: 2792.

Stillman, B.W., W.C. Topp, and J.A. Engler. 1982. Conserved sequences at the origin of adenovirus DNA replication. *J. Virol.* **44:** 530.

Stillman, B.W., J.B. Lewis, L.T. Chow, M.B. Mathews, and J.E. Smart. 1981. Identification of the gene and mRNA for the adenovirus terminal protein precursor. *Cell* **23:** 497.

Stow, N.D. 1981. Cloning of a DNA fragment from the left-hand terminus of the adenovirus type 2 genome and its use in site-directed mutagenesis. *J. Virol.* **37:** 171.

Sussenbach, J.S., P.C. van der Vliet, D.J. Ellens, and H.S. Jansz. 1972. Linear intermediates in the replication of adenovirus DNA. *Nat. New Biol.* **239:** 47.

Svensson, C. and G. Akusjärvi. 1984a. Adenovirus 2 early region 1A stimulates expression of both viral and cellular genes. *EMBO J.* **3:** 789.

———. 1984b. Adenovirus VA RNA₁: A positive regulator of mRNA translation. *Mol. Cell. Biol.* **4:** 736.

Takemori, N., J.L. Riggs, and C. Aldrich. 1968. Genetic studies with tumorigenic adenoviruses. I. Isolation of cytocidal (*cyt*) mutants of adenovirus type 12. *Virology* **36:** 575.

Takemori, N., C. Cladaras, B. Bhat, A.J. Conley, and W.S.M. Wold. 1984. *Cyt* gene of adenoviruses 2 and 5 is an oncogene for transforming function in early region E1B and encodes the E1B 19,000 molecular weight polypeptide. *J. Virol.* **52:** 793.

Tamanoi, F. and B.W. Stillman. 1982. Function of the adenovirus terminal protein in the initiation of DNA replication. *Proc. Natl. Acad. Sci.* **79:** 2221.

Thimmappaya, B., C. Weinberger, R.J. Schneider, and T. Shenk. 1982. Adenovirus VAI RNA is required for efficient translation of viral mRNAs at late times after infection. *Cell* **31:** 543.

Thomas, G.P. and M.B. Mathews. 1980. DNA replication and the early to late transition in adenovirus infection. *Cell* **22:** 523.

Thummel, C., R. Tjian, and T. Grodzicker. 1981. Expression of SV40 T antigen under control of adenovirus promoters. *Cell* **23:** 825.

Tibbetts, C. and U. Pettersson. 1974. Complementary strand specific sequences from unique fragments of adenovirus type 2 DNA for hybridization mapping experiments. *J. Mol. Biol.* **88:** 767.

Tolun, A. and U. Pettersson. 1975. Termination sites for adenovirus type 2 DNA replication. *J. Virol.* **16:** 759.

Tolun, A., P. Alestrom, and U. Pettersson. 1979. Sequence of inverted terminal repetitions from different adenoviruses: Demonstration of conserved sequences and homology between SA7 termini and SV40 DNA. *Cell* **17:** 705.

Trentin, J.J., Y. Yabe, and G. Taylor. 1962. The quest for human cancer viruses. *Science* **137:** 835.

Valentine, R.C. and H.G. Pereira. 1965. Antigens and structure of the adenovirus. *J. Mol. Biol.* **13:** 13.

van der Eb, A.J. 1973. Intermediates in type 5 adenovirus DNA replication. *Virology* **51:** 11.

van der Vliet, P.C. and M.M. Kwant. 1978. Role of DNA polymerase gamma in adenovirus DNA replication. *Nature* **276:** 532.

van der Vliet, P.C. and A.J. Levine. 1973. DNA binding proteins specific for cells infected by adenovirus. *Nat. New Biol.* **246:** 170.

van der Vliet, P.C., A.J. Levine, M.J. Ensinger, and H.S. Ginsberg. 1975. Thermolabile DNA binding proteins from cells infected with a temperature-sensitive mutant of adenovirus defective in viral DNA synthesis. *J. Virol.* **15:** 348.

Velcich, A. and E. Ziff. 1985. Adenovirus E1a proteins repress transcription from the SV40 early promoter. *Cell* **40:** 705.

Vennstrom, B., U. Pettersson, and L. Philipson. 1978. Initiation of transcription in nuclei isolated from adenovirus infected cells. *Nucleic Acids Res.* **5:** 205.

Virtanen, A. and U. Pettersson. 1985. Organization of early region 1B of human adenovirus type 2: Identification of four differentially spliced mRNAs. *J. Virol.* **54:** 383.

Weber, J. 1976. Genetic analysis of adenovirus type 2: III. Temperature sensitivity of processing of viral proteins. *J. Virol.* **17:** 462.

Weil, P.A., D.S. Luse, J. Segall, and R.G. Roeder. 1979. Selective and accurate initiation of transcription at the Ad2 major late promoter in a soluble system dependent upon purified RNA polymerase II and DNA. *Cell* **18:** 469.

Weingartner, B., E.-L. Winnacker, A. Tolun, and U. Pettersson. 1976. Two complementary strand-specific termination sites for adenovirus DNA replication. *Cell* **9:** 259.

Weinmann, R., H.J. Raskas, and R.G. Roeder. 1974. Role of DNA-dependent RNA polymerases II and III in transcription of the adenovirus genome late in productive infection. *Proc. Natl. Acad. Sci.* **71:** 3426.

White, E., T. Grodzicker, and B.W. Stillman. 1984. Mutations in the gene encoding the adenovirus early region 1B 19,000-molecular-weight tumor antigen cause the degradation of chromosomal DNA. *J. Virol.* **52:** 410.

Wilcox, W.C. and H.S. Ginsberg. 1961. Purification and immunological characterization of types 4 and 5 adenovirus soluble antigens. *Proc. Natl. Acad. Sci.* **47:** 512.

Williams, J.F., M. Gharpure, S. Ustacelebi, and S. McDonald. 1971. Isolation of temperature-sensitive mutants of adenovirus type 5. *J. Gen. Virol.* **11:** 95.

Williams, J.T., T. Grodzicker, P. Sharp, and J. Sambrook. 1975. Adenovirus recombination: Physical mapping of cross-over events. *Cell* **4:** 113.

Wilson, M.C. and J.E. Darnell, Jr. 1981. Control of messenger RNA concentration by differential cytoplasmic half-life. Adenovirus messenger RNAs from transcription units 1A and 1B. *J. Mol. Biol.* **148:** 231.

Wilson, M.C., N.W. Fraser, and J.E. Darnell, Jr. 1979. Mapping of RNA initiation sites by high doses of UV irradiation: Evidence for three independent promoters within the left 11% of the Ad2 genome. *Virology* **94:** 175.

Wilson, M.C., J.R. Nevins, J.-M. Blanchard, H.S. Ginsberg, and J.E. Darnell. 1980. Metabolism of mRNA from the transforming region of adenovirus 2. *Cold Spring Harbor Symp. Quant. Biol.* **44:** 447.

Winberg, C. and T. Shenk. 1984. Dissection of overlapping functions within the adenovirus type 5 E1A gene. *EMBO J.* **3:** 1907.

Wolfson, J. and D. Dressler. 1972. Adenovirus 2 DNA contains an inverted terminal repetition. *Proc. Natl. Acad. Sci.* **69:** 3054.

Wu, G.-J. 1978. Adenovirus DNA directed transcription of 5.5S RNA in vitro. *Proc. Natl. Acad. Sci.* **75:** 2175.

Yang, V.W., M.R. Lerner, J.A. Steitz, and S.J. Flint. 1981. A small nuclear ribonucleoprotein is required for splicing of adenoviral early RNA sequences. *Proc. Natl. Acad. Sci.* **78:** 1371.

Zain, B.S. and R.J. Roberts. 1978. Characterization and sequence analysis of a recombination site in the hybrid virus Ad2⁺ND1. *J. Mol. Biol.* **120:** 13.

Zain, B.S., T.R. Gingeras, P. Bullock, G. Wong, and R.E. Gelinas. 1979a. Determination and analysis of adenovirus-2 DNA sequences which may include signals for late messenger RNA processing. *J. Mol. Biol.* **135:** 413.

Zain, B.S., J. Sambrook, R.J. Roberts, W. Keller, M. Fried, and A.R. Dunn. 1979b. Nucleotide sequence analysis of the leader segments in a cloned copy of adenovirus 2 fiber mRNA. *Cell* **16:** 851.

Ziff, E.B. and R.M. Evans. 1978. Coincidence of the promoter and capped 5' terminus of RNA from the adenovirus 2 major late transcription unit. *Cell* **15:** 1463.

Ziff, E. and N. Fraser. 1978. Adenovirus type 2 late mRNAs structural evidence for 3'-coterminal species. *J. Virol.* **25:** 897.

# Herpesviruses:
# I. Genome Structure and Regulation

**G.S. Hayward**

The Virology Laboratories, Department of Pharmacology, Johns Hopkins University,
School of Medicine, Baltimore, Maryland 21205

# II. Latent and Oncogenic Infections by Human Herpesviruses

**B. Sugden**

McArdle Laboratory for Cancer Research, University of Wisconsin, Madison, Wisconsin 53706

## I. GENOME STRUCTURE AND REGULATION

A complete overview of all herpesvirus molecular biology today would be a massive and perhaps impossible task. Therefore, because of my specific interests and the topic of this volume, the subject matter in part I has been deliberately confined to the structure and evolution of herpesvirus genomes and to the control of gene expression. There is also a heavy bias toward those genes and sequences thought to be involved in some way in transcriptional control and DNA replication, at the expense of covering any of the fine work of recent years in areas such as glycoprotein structure, enzymology, neurovirulence, and so forth. The still mysterious molecular mechanisms of morphological or oncogenic transformation by these viruses will also not be considered in this section (but see Hayward and Reyes 1983; Spear 1983; Galloway et al., this volume). The biology of latent and oncogenic infections by human herpesviruses is dealt with in part II.

### Reflections on the Growth of Herpesvirus Molecular Biology

The genomes of the five human herpesviruses together comprise a total of more than 800,000 bp of DNA sequence and approximately 350 genes, equivalent to about 1/400th of the expected size of the gene-coding regions in the entire human genome (assuming 90% noncoding sequences). The first restriction enzyme cleavage maps of herpes simplex virus type-1 (HSV-1) DNA were completed almost exactly 10 years ago, and it was while celebrating this event that we first started to consider seriously whether the advent of DNA-cloning technology and improved DNA-sequencing procedures might make it possible to determine the complete nucleotide sequence of HSV DNA. The prospect was utterly staggering; we could not conceive how one would make sense of the data or how one could usefully handle information that would take two weeks to write on a blackboard. Of course, time has changed all that, with the biochemical wizardry of Maxam and Gilbert and

Sanger, the ability to recognize open reading frames and transcription punctuation signals, plus the advent of routine computer analysis of DNA. True, the total DNA sequence of HSV is still not yet at hand, but determination of the sequences of the 172,282-bp Epstein-Barr virus (EBV B95-8) DNA molecule (Baer et al. 1984) and the 124,884 bp of the varicella-zoster virus (VZV) genome (A. Davison, pers. comm.) has been completed, and work to reveal the sequence of the 240-kb cytomegalovirus (CMV AD169) genome is well on the way in Cambridge (B. Barrell, pers. comm.). Indeed, the entire sequence of one strand of EBV DNA fits snugly into a single 320-kilobyte 5-1/4'' floppy disk and can be compressed into a 48-page printout in just over 30 minutes.

### The early years: HSV DNA structure

In 1972, all that was known about HSV DNA molecules was their large size at 100 million daltons (Becker et al. 1968) and the overall base composition of 67% G + C content (Kieff et al. 1971). Furthermore, although the various herpesvirus subgroups displayed major biological differences and the base composition of individual herpesvirus DNAs varied greatly (e.g., VZV DNA at 46% G + C content; Ludwig et al. 1972; Goodheart and Plummer 1975), no one at that time anticipated that their genomic sizes or structural organization might differ significantly. Indeed, since VZV, CMV, and EBV were supposed to be nearly impossible to grow and work with in cell culture because of slow infection cycles, high cell-associatedness, and lack of convenient permissive host systems, HSV was expected to serve as a model for all of them.

Our initial hopes to dissect the HSV genome by agarose gel electrophoresis, utilizing the single-strand nicks that were supposed to exist in the virion DNA molecule, quickly evaporated when we discovered that the nicks were not at specific locations (Wilkie et al. 1975; G.S. Hayward and S.C. Wadsworth, unpubl.). However, subsequent progress was very rapid. Spear and Roizman (1972) and Gibson and Roizman (1972) developed ways to isolate and purify virions and intracellular capsids on sucrose density gradients, thus providing an abundant source of clean virion DNA. With low-phosphate medium for $^{32}$P labeling, gentle phenol extraction, plus the long, soft agarose gel cylinders of those days, we had all of the tools needed to proceed except for a way to dissect the genome. Although some authorities had ventured the opinion that we would probably never find a restriction enzyme giving few enough bands to be useful for mapping such a large DNA molecule, we were fortunate to have access (unknowingly) to an early *Hin*d restriction enzyme preparation in which only the more robust *Hin*dIII activity had survived. Therefore, in early 1974 we were extremely surprised and excited to discover for the first time the whole of HSV-1 DNA dissected into only 12 or so bands (8 cleavage sites). Subsequently *Eco*RI (12 sites) and *Hpa*I (19 sites) proved to be useful also, as did eventually *Bgl*II (10 sites) and *Xba*I (which amazingly gave only 4 cleavage sites).

We were now confronted with the strange patterns of major and minor bands, especially in the very early days when many of the laboratory stocks that we received also contained tandem-repeat defective DNA (Hayward et al. 1975a). At the Cold Spring Harbor tumor virus meeting in 1974, Peter Sheldrick excited everyone with his "dumbbell"-shaped structure of self-annealed HSV DNA strands seen through the electron microscope (Sheldrick and Berthelot 1975); by the end of that year, Samuel Wadsworth and Robert Jacob had found from partial denaturation maps that the L- and S-segment inverted repeats were of different sizes and sequence and of even higher G + C content than the rest of the genome (Wadsworth et al. 1975). At the same time, the four structural isomers and the L- and S-segment inversions that created submolar fragments were finally sorted out and explained (Roizman et al. 1974; Hayward et al. 1975b). We also realized that restriction enzyme cleavage allowed, for the first time, unambiguous subtyping of HSV-1 and HSV-2 and that polymorphism in the DNA "fingerprint" patterns enabled all human HSV isolates to be distinguished from one another (Hayward et al. 1975a). Next, our electron microscopy and restriction enzyme analysis showed that the aberrant, high G + C–content genomes generated in populations of defective virus passaged at high moi were composed of tandemly repeated *Hin*dIII-resistant segments from the right end of the viral genome, which led to predictions of concatemeric replicative forms (Roizman et al. 1974; Frenkel et al. 1975). The first restriction maps of HSV-1(MP) virion DNA were completed by the end of 1975 (published eventually in Morse et al. 1977), and in the following year, with Timothy Buchman, we completed the HSV-2(333) maps for *Bgl*II, *Hpa*I, *Hin*dIII, *Eco*RI, *Xba*I, and *Kpn*I, using only redigestion of isolated $^{32}$P-labeled fragments without any of the modern advantages of cloned DNA, Southern blot hybridization, and nick translation.

The HSV-1 and HSV-2 restriction maps were first presented publicly on a large posterboard kept near the front of the auditorium at the 1976 Cold Spring Harbor workshop—at least until Fred Rapp (accidentally?)

knocked it over. Of course, Bill Summers's and Neil Wilkie's groups had also been independently mapping the HSV-1 genome over this period (Wilkie 1976; Skare and Summers 1977) and everyone involved assembled together for a "DNA Nomenclature Agreement" during the 1976 meeting, when rules about the standard orientation and fragment designations for HSV-1 and HSV-2 (and hopefully all the herpesvirus genomes) were agreed upon. Although not all laboratories, nor all herpesvirus systems, have adhered to it completely, this sensible compromise solution, in which each of the major protagonists agreed either to turn some portion of their HSV maps around or to abandon some particular aspect of their fragment nomenclature (see Jones et al. 1977), has served remarkably well over the years. Similar agreements were subsequently made about HSV glycoprotein nomenclature and genetic complementation groups (Schaffer et al. 1978), but it is a pity that agreement has not been forthcoming about conformity in nomenclature for HSV proteins and the different kinetic classes of gene products. Furthermore, there is now a rapidly growing need for rational and consistent nomenclature rules for naming individual identified genes and gene functions in all herpesvirus systems.

The close of 1976 represented a clear-cut end to the early "teething" period of herpesvirus molecular biology. Other lines of research of great importance for gene regulation that had been developed by that time included the following: (1) the assignment of HSV proteins into the three major kinetic classes (α, β, and γ) by Honess and Roizman (1974, 1975); (2) the recognition that the HSV thymidine kinase enzyme (TK) could substitute for cellular TK in the powerful HAT selection procedure for cells that had stably integrated the viral gene after surviving infection with UV-inactivated virus (Munyon et al. 1971); (3) the isolation and characterization of sets of reasonably well defined HSV temperature-sensitive (*ts*) mutants (Schaffer et al. 1973; Timbury and Subak-Sharpe 1973; Schaffer 1975); (4) proof from the electron microscope studies of Lindahl et al. (1976) that, as predicted from sedimentation analysis by Nonoyama and Pagano (1972), many of the multicopy EBV genomes resident in "immortalized" B cells and in EBV-associated tumor cells (Pagano 1975) occur in the form of extrachromosomal circular plasmids or episomes.

### The descriptive phase

After 1976, an intermediate, descriptive phase of herpesvirus molecular biology continued in many parallel directions. There was a major effort to characterize, map, and clone all of the human herpesvirus genomes and several key animal herpesvirus DNAs, as well. In HSV, the TK gene was identified in Saul Silverstein's laboratory by transfection of defined DNA fragments (Wigler et al. 1977; Pellicer et al. 1978) and subsequently used to establish the powerful techniques for coselection of unlinked DNA. Genetically selected intertypic HSV-1/HSV-2 recombinant viruses produced in both Chicago and Glasgow were used to crudely map the locations of genes for many proteins (Marsden et al.

1978; Preston et al. 1978; Morse et al. 1977; Wilkie et al. 1979; Halliburton 1980). DNA transfection and marker-rescue procedures were introduced to define the approximate map locations of *ts* mutants (Knipe et al. 1978; Stow et al. 1978; Chartrand et al. 1979; Ruyechan et al. 1979; Parris et al. 1980). Similarly, RNA from different gene classes (Clements et al. 1977; Jones et al. 1977; Stringer et al. 1978; Holland et al. 1980) and eventually individual mRNA transcripts were mapped with ever-increasing precision (Clements et al. 1979; Easton and Clements 1980; Mackem and Roizman 1980). Conditional *ts* mutants in the immediate-early IE175(ICP4) regulatory protein, the DNA polymerase, and other viral proteins were also characterized and mapped (Parris et al. 1978; Watson and Clements 1978; Preston 1979a; Chartrand et al. 1980; Dixon and Schaffer 1980).

In the EBV system, the development of good producer cell lines such as EBV(B95-8) and eventually the use of the tumor promoter TPA and sodium butyrate as chemical inducers (zur Hausen et al. 1978; Saemundsen et al. 1980) eased the previous burdens of working with only submicrogram quantities of virus DNA. The development of the EBV(P3HR-1) superinfection system in Raji cells also permitted some approaches to studying the lytic cycle (Yajima and Nonoyama 1976).

In CMV, the key event was the recognition that, although most DNA preparations up to that time were composed of molecules 150 kb in size, these represented defective genomes carried over by high-moi passaging. The real infectious CMV DNA molecules were actually 50% larger, at 240 kb one of the largest known animal DNA viruses (Kilpatrick and Huang 1977; Mosmann and Hudson 1977; Geelen et al. 1978; Stinski et al. 1979). Furthermore, we found that CMV DNA could be routinely prepared in 100-μg quantities by using fibroblast-adapted strains, providing that they were free of both defectives and mycoplasma and that they had been grown in cultured diploid human fibroblasts with carefully selected batches of fetal calf serum. By early 1978, the restriction enzyme cleavage maps of the linear 220-kb SCMV(Colburn) genome had been completed, but here the best enzymes that we could find were *Sal*I (25 sites) and *Xba*I (28 sites). The following year, Bob LaFemina recognized and mapped the four isomers of the 240-kb HCMV(Towne) genome (LaFemina and Hayward 1980; revised in Thomsen and Stinski 1981). Although we were able to utilize hybridization techniques, those original HCMV maps (*Xba*I, 17 sites; *Hin*dIII, 26 sites; *Bam*HI, 33 sites) also had to be tediously constructed without the benefit of cloned DNA fragments.

During this period the Glasgow group was laboring at the old Porton Downs Laboratories to clone HSV DNA in *Escherichia coli* plasmids, and Lynn Enquist and George Vande Woude were using the containment laboratories of Building 41 at the National Institutes of Health to clone HSV DNA fragments in debilitated phage λ WES-B or Charon 4A vector/host systems (Enquist et al. 1979). Fortunately for the rest of us, and

largely through the highly appreciated efforts of Wallace Rowe and Malcolm Martin, the USA recombinant DNA guidelines for animal viral genomes were subsequently downgraded to the P2 level in late 1979. Steven McKnight was among the first to take advantage of this opportunity. He believed that the easily assayed TK gene offered a unique opportunity to study a regulated eukaryotic promoter, and he immediately cloned and sequenced the HSV-1 TK gene and started to analyze the key elements in its promoter region (McKnight 1980; McKnight and Gavis 1980). After this, most of the more active laboratories quickly established overlapping cloned sets of fragments in pBR322 from all of the genomes that interested them. This now meant that researchers who wished to study herpesvirus molecular biology no longer needed to have a "green thumb" in the virus cell culture laboratory but merely needed to be able to handle simple *E. coli* genetics. It also presaged an enormous outpouring of papers about sets of cloned DNA fragments and physical maps for the genomes of all of the human and major animal herpesviruses.

One of the first things that we noticed with cloned herpesvirus DNA was that all inserts from HSV-1 or HSV-2 that covered the minimal $ori_L$ region at 0.40–0.41, including clones of the defective repeat units themselves, rapidly underwent specific 120-bp deletions upon passaging in *E. coli* (Reyes 1982). However, it was not until 5 years later that the sequencing studies of Gray and Kaerner (1984) and Weller et al. (1985) confirmed the initial speculations about this representing an unstable palindromic structure in the replication origin itself. Other than the $ori_L$ site and some $ori_P$ clones in EBV (together with terminal fragments requiring special linker procedures), there do not appear to be any portions of herpesvirus genomes that cannot be cloned intact in pBR322-derived systems in $recA^-$ hosts. However, rapid losses of relatively large tandem repeated sequences have been a problem in bacteriophage λ systems (particularly for EBV).

## Toward correlating structure and function

Largely as a result of this massive effort to clone and map the herpesvirus genomes, together with the precise definition of mRNA transcripts and the DNA sequencing that followed, herpesvirus molecular biology has entered a much more mature (although perhaps still adolescent) stage in the mid-1980s. Several laboratories have started to ask questions about the effects of perturbations in genetic content or of promoter structure, to employ assays for specific functions, and to reconstruct regulatory and replicative events in subgenomic or in vitro systems. The major examples that come to mind are: (1) Steven McKnight's elegant linker scanner analysis of the HSV-1 TK promoter (McKnight et al. 1981, 1984; McKnight 1982), followed by the recognition of binding sites for cellular Sp1 and CCAAT transcription factors (Jones et al. 1985); (2) the development of positive and negative TK selection systems to introduce foreign or mutated genes into specific locations in HSV genomes through DNA transfection and marker-rescue procedures (Mocarski et al. 1980; Smiley 1980; Post and Roizman 1981; Lee et al. 1982; Shih et al. 1984; Tackney et al. 1984; DeLuca et al. 1985; Roizman and Jenkins 1985); (3) the discovery of the virion factor involved in HSV gene regulation (Post et al. 1981), its characterization (Batterson et al. 1983), and the subsequent, bold "shotgun" screening experiment to identify the gene involved (Campbell et al. 1984); (4) the DNA transfection and superinfection assays used to identify HSV lytic replication origins (Vlazny and Frenkel 1981; Stow 1982; Stow and McMonagle 1983; Weller et al. 1985); (5) the direct evidence for specific *trans*-activation functions for HSV IE nuclear proteins in transient gene expression assays (Everett 1984b; Gelman and Silverstein 1985; O'Hare and Hayward 1985a,b; Quinlan and Knipe 1985) or in DNA-transfected cell lines (DeLuca and Schaffer 1985; Persson et al. 1985); (6) the identification and analysis by Bill Sugden's group of the plasmid maintenance region (*ori$_P$*) of the EBV genome (Yates et al. 1984; Reisman et al. 1985), its functional requirement for the viral nuclear antigen 1 (EBNA-1) (Yates et al. 1985), and the demonstration of direct sequence-specific DNA–protein interactions at multiple EBNA-binding sites (Rawlins et al. 1985).

Hopefully, further pursuits of these and other leads toward structure-function relationships, together with the explosion of primary DNA sequence data and an awakening of interest in posttranscriptional regulatory events, should make the next period in the history of herpesvirus molecular biology just as exciting and fast-

moving for future students of the discipline as the past decade and a half has been for those of us intimately involved.

## Genome Structure and Organization

### Characteristic internal tandem and inverted repeats

The diagram in Figure 1 summarizes our current understanding of the relative sizes and locations of major tandem or inverted repeat structures in the genomes of the five human herpesviruses VZV, HSV-1, HSV-2, EBV, and CMV. The locations of DNA replication origins are also shown (where information is available). The enormous size and structural diversity of herpesvirus genomes were, and still are, very surprising features of the group. At least seven different structural forms of herpesvirus genomes are known—four of them represented among the human viruses. The DNA of the T-cell lymphotrophic virus (*Herpesvirus saimiri*) of squirrel monkeys was the first genome other than HSV to be well characterized; in this case a 100-kb segment of low G + C–content sequences (L) is bounded on both sides by a partially circularly permuted arrangement of 30 copies of an approximately 1-kb tandemly repeated G + C–rich sequence (H) (Fleckenstein et al. 1975; Bornkamm et al. 1976). After the inverted repeats and invertible L and S segments of HSV were described (Hayward et al. 1975b; Sheldrick and Berthelot 1975; Wadsworth et al. 1975; Delius and Clements 1976), examples of genomes that retain S inverted repeats but lack detectable inverted repeats around L were described in equine

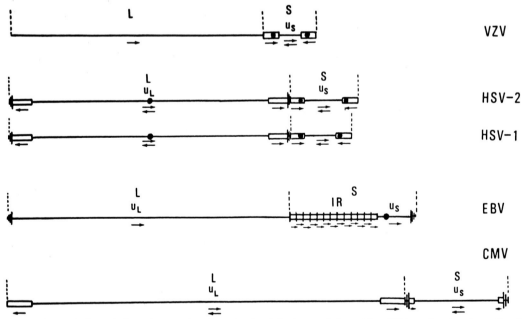

**Figure 1** Comparison of the relative sizes and major structural features of the virion DNA molecules from the five human herpesviruses: VZV (varicella-zoster virus, 125 kb); HSV-1 and HSV-2 (herpes simplex virus subtypes 1 and 2, 150 kb); EBV (Epstein-Barr virus, 172 kb); and CMV (cytomegalovirus, 240 kb). Open bars denote duplicated or tandemly repeated regions. Arrows indicate relative orientations of either the repeated regions or the unique sequences between them. (U$_L$ and U$_S$) unique-sequence portions of the L (or long) and S (or short) segment. (IR) internal 3072-bp repeats in EBV. Short horizontal lines denote the locations of extensive terminal heterogeneity. Solid circles represent locations of known lytic replication origins *ori$_S$* and *ori$_L$* in HSV and *ori$_S$* in VZV and of the *ori$_P$* plasmid maintenance region in EBV.

abortion virus (EAV; P. Sheldrick, pers. comm.) and pseudorabies virus (PRV; Stevely 1977). Subsequently, it was recognized that in these molecules the L segment is frozen in one orientation (Ben-Porat et al. 1979; Henry et al. 1981). Peter Sheldrick also described large, direct terminal repeats as the only structural feature of the relatively small (120 kb) channel catfish virus (CCV) DNA (Chousterman et al. 1979).

The structure of human and new world monkey CMV virion DNA proved somewhat surprising, displaying essentially the same pattern of L and S inverted repeats and four isomeric forms in equilibrium that had been found in HSV, although of course the CMV genomes were 50% larger (Kilpatrick and Huang 1977; LaFemina and Hayward 1980; Ebeling et al. 1983). However, the DNA of the almost equally large (220 kb) old world monkey isolate CMV(Colburn) (and subsequently that from both equine and murine CMV, as well) proved to have a linear, noninverted organization, lacking structurally defined L and S segments (LaFemina and Hayward 1980; Mercer et al. 1983). The genome of EBV was initially something of a puzzle; the first restriction maps of EBV(B95-8) DNA failed to indicate any unusual structural features (Given and Kieff 1978), but then Rymo and Forsblom (1978) followed by Given and Kieff (1979) and Hayward et al. (1980) recognized the consistent appearance of a 10 to 12 molar fragment of 3.1 kb in a number of digests and mapped all of these to an internal, tandemly repeated region (IR) toward one end of the molecule. Otherwise, EBV DNA displays a linear, noninverting structure. Finally, VZV DNA was found to fit the S-segment only inversion pattern described for EAV and PRV (Dumas et al. 1981; Straus et al. 1982; Ecker and Hyman 1982), whereas the structure of Marek's disease virus (MDV) of chickens was shown to resemble that of HSV (Cebrian et al. 1982).

To put all of this into the hypothetical perspective of an evolutionary tree, it seems that the oldest, most primitive vertebrate herpesviruses may have had simple, terminal direct repeats as exemplified by the present-day viruses of catfish (CCV) and the Lucke frog tumor virus. At some point, the nonlymphotrophic group of herpesviruses with descendants in both birds and mammals developed the L and S inverted repeat structure (HSV, MDV, plus one group of CMVs), and then some variants either lost the L-segment inversions (VZV, PRV, EAV, etc.) or both sets of inversions (the second group of CMVs and perhaps *H. tupaia* of tree shrews also). In the case of one old world simian CMV (SCMV) genome, the internal inverted repeat region appears to have been replaced with stretches of repetitive CA dinucleotides from the cellular genome (Jeang and Hayward 1983a). On the other hand, the lymphotrophic herpesviruses developed large sets of direct tandem repeats either internally (e.g., EBV-like viruses in human and old world monkeys) or externally (e.g., *H. saimiri* and *H. ateles* in new world monkeys) or a combination of both (e.g., *H. sylvilagus* of cottontail rabbits).

One or both genomic termini of all herpesviruses examined so far show some form of heterogeneity,

representing different copy numbers of tandemly repeated sequences. These range usually from between 1 and 4 additional copies of the entire 400-bp terminal repetition in the L-segment inverted repeats of HSV, through four or five 550-bp direct repeats distributed across both termini of EBV genomes, and up to 20 additional copies of only a portion of the terminal repetition in some strains of human CMV (HCMV). The entire structure of *H. saimiri* can be envisaged as an extreme example of this kind of phenomenon, where tandem repeats of the packaging/cleavage signals result in permuted termini through a packaging/envelopment mechanism that combines sequence-specific recognition with "head-full" features.

In addition to these relatively large-scale structural features, all herpesvirus genomes have multiple internal sites of heterogeneous or microvariable sequences (Hayward et al. 1975a; Locker and Frenkel 1979; Davison and Wilkie 1981). In all cases examined so far (see Table 1), these represent some form of tandemly repeated structure ranging from variable numbers of only 5- to 15-bp units up to the 102-bp *Pst*I and 125-bp *Not*I repeats in EBV (Hayward et al. 1982; Hudewentz et al. 1982; Jones and Griffin 1983). Numerous examples of these structures occur in and around the inverted repeats in HSV, for example (Rixon et al. 1984), and they are nearly all extremely rich in G + C content with a strong bias toward strings of G on one strand. They lie within coding regions in some cases (e.g., EBNA-1 and EBNA-2; Heller et al. 1982), at the 3' ends of the HSV IE175 gene (Davison and Wilkie 1981; Mocarski and Roizman 1981), and in the intron of HSV IE68 and IE12 (Watson et al. 1981). Some of them also account for the bulk of the cell-virus DNA homology described in HSV and EBV and perhaps also CMV DNA (Heller et al. 1982; Peden et al. 1982; Puga et al. 1982; Ruger et al. 1984), which appears to result primarily from accidental matches with similar, short G-rich tandem repeats that occur in relatively high abundance in mammalian genomic DNA (for review, see Hayward et al. 1984; Heller et al. 1985). In HSV, one might propose that short, repetitive elements have a role in promoting the recombination events involved in inversions of the L and S segments, but this hardly accounts for the several examples in EBV DNA. Alternatively, a role in high-level expression of immediate-early and latency genes could be suggested, although one study has implied that deletion of most of one such site in HSV has no obvious effect on viability (Voss-Hubenthal and Roizman 1985).

### Terminal repeats, inversions, and integration

A directly repeated, terminally redundant sequence (referred to as "a") was described initially in HSV by Grafstrom et al. (1975) and also by Wadsworth et al. (1975, 1976). DNA sequence analysis revealed that the "a" region is 400 bp in size in HSV-1(F) and HSV-1(17) and 250 bp in HSV-2(333) (Davison and Wilkie 1981; Mocarski and Roizman 1981). Several short, internal, tandemly repeated sequences contribute to the size variability of the "a" region in different strains of HSV-1,

**Table 1** Short, Internal, Tandemly Repeated Sequence Elements in Herpesvirus Genomes

*Class I: Major cell-virus homology site*

| | | |
|---|---|---|
| SCMV | $(CA)_{15}$, $(CA)_{22}$, $(CA)_{21}$ | *Sal*I-P at 0.83 map units |

*Class II: Variable copy numbers; correlate with heterogeneity and cell-virus homology*

| | | |
|---|---|---|
| HSV-1 | $(GGCGGAGGAGGGGGGACGCGGG)_8$ | IE68/12 intron (ICP22/47) |
| HSV-1 | $(GGGGAGGAGCGG)_8$ | terminal redundancy "a" |
| HSV-1 | $(CTGGGGCTGGGGAGGG)_{18}$ | 3' to IE175 (ICP4) |
| HSV-1 | $(GAGGGGGCGAGGGGCGG)_{22}$ | 3' to IE175 (ICP4) |
| HSV-1 | $(GGGGGTGCGTGGGAGT)_8$ | 3' to IE110 (ICP0) |
| HSV-1 | $(CCGGGGCTCCCGGGGAGA)_3$ | $U_S$ gene 10, coding |
| HSV-1 | $(TGGGTGGGTGGGGAG)_{10}$ | 3' to $U_S$ 10, 11, 12 |
| HSV-1 | $(TGGTGGTCGAGGGGGTGGAGG)_3$ | $U_S$ gene 7, coding |
| HSV-2[a] | $(CAGGGGCGGCTGGGG)_{11}$ | *Bam*HI-Y, $U_L$ |
| VZV[b] | $(CCCCGCCGATGGGGAGGGGGCGCGGTA)_6$ | |
| EBV | $(GGGGCAGGA)_{80}$ [plus rearrangements] | *Bam*HI-K, EBNA-1 coding |
| EBV | $(TGGTGGGGG)_{14}$ | *Bam*HI-Y |
| EBV | $(GGGGCT)_9$ | *Bam*HI-H, EBNA-2 coding |

*Class III: Clustered protein-binding sites*

| | | |
|---|---|---|
| SCMV | $(GGTGGACTTGGCACTGTGCCAATTCAATAT)_{23}$ | NFI binding sites, 5' to IE94 highly diverged |
| EBV | $(AGATTAGGATAGCATATGCTACCCAGATAT)_{20}$ | EBNA-1 binding sites, *ori*$_P$ weakly diverged |

*Class IV: $DS_R$ and $DS_L$, Duplicated homologs*

| | | |
|---|---|---|
| EBV | $(125 \text{ bp, GC rich})_{11}$ | *Bam*HI-H, *Not*I repeats 2.5-kb mRNA |
| EBV | $(105 \text{ bp, GC rich})_{20-25}$ | *Bgl*II-K, *Pst*-I repeats 2.5-kb mRNA |

[a]R. Jariwalla (pers. comm.).
[b]W. Ruyechan (pers. comm.).

and there is little homology between the "a" regions of HSV-1 and HSV-2. The element is characterized by a short (20-bp) direct repeat at its boundaries, and different numbers of direct copies of the "a" unit correspond to the stepwise terminal heterogeneity observed in early studies at both the internal and external termini of the L segment (Hayward et al. 1975a; Wagner and Summers 1978). A similar but larger structure ($\sim$800 bp) occurs at the joint between the L and S segments in HCMV DNA and in variable copy number at the S-segment termini (LaFemina and Hayward 1980; Tamashiro et al. 1984; Spaete and Mocarski 1985b), but the linear murine CMV genome contains only a short direct repeat of 30 bp (J. Marks and D. Spector, pers. comm.). VZV DNA apparently lacks any proper terminal redundancy, but instead (and unexpectedly) the sequence analysis of VZV termini revealed an 88-bp inverted repeat around the L segment (Davison 1984). Further, some VZV joint clones indicated a low frequency of either circular structures or inversions of the L segment among virion DNA molecules (Davison 1984; Kinchington et al. 1985). Both HSV and VZV DNAs end with a single-base 3'-C overhang on one side and a complementary unpaired 3'-G at the other end (Mocarski and Roizman 1982; Davison 1984).

Signals for cleavage and packaging of mature HSV DNA molecules from replicating forms and possibly also recombination "hot spots" were predicted to occur within the terminally redundant "a" region to account for L and S inversions. The former certainly appears to be correct, although the exact nature and location of the signals is still uncertain, and surprisingly the conserved 28-bp sequence—CCCCGGGGGGGGTGTTTTTGATGGGGGGG—found near at least one terminus in all herpesvirus genomes examined to date was not implicated (Vlazny et al. 1982; Stow et al. 1983; Spaete and Mocarski 1985b; Varmuza and Smiley 1985). Insertion of fragments that include all of the "a" sequence at a second site in L induces the formation of either additional inverted segments or deletions at high efficiency depending upon the orientation of the insert (Mocarski et al. 1980; Smiley et al. 1981; Chou and Roizman 1985) and this inversion process requires *trans*-acting viral gene products (Mocarski and Roizman 1982). However, because packaging events themselves can create inverted segments and the insertion of inverted copies of other unrelated sequences will also sometimes lead to new inversions (Pogue-Giele et al. 1985), the specificity of these events is still unclear. Despite a good deal of speculation and earlier claims to the contrary, there appears to be no biologically perferred orientation of HSV DNA (Davison and Wilkie 1983b). Viable HSV genomes that either have nonhomologous "a" regions around the L segment as a result of heterotypic recombination (Davison and Wilkie 1983a) or that lack a portion or all of the internal joint region (Poffenberger et al. 1983) have been demonstrated to display at least partially frozen L-segment orientations but tend to regenerate some form of homozygous internal joint structure upon passaging. Similarly, it was noticed in early intertypic recombinants as well as in many subsequent studies that any alteration in one copy of the S-segment inverted repeat, although creating a viable terminal heterozygote, tends to become corrected such that both copies in some daughter molecules take the larger homozygous form and others take the smaller homozygous forms (Knipe et al. 1978; Varmuza and Smiley 1984). The exact nature and mechanism of this corrective mechanism is not known, but it also applies to *ts* lesions and deletion mutants in the IE175 gene encoded by the S repeats (DeLuca and Schaffer 1985). Perhaps as a result of these kinds of evolutionary events, the exact boundaries of the U$_S$ and S-segment inverted repeat sequences differ slightly in HSV-1 and HSV-2 (Whitton and Clements 1984a). Furthermore, in VZV two additional small genes are now totally enclosed with both copies of the S-segment inverted repeat sequences in contrast to their location outside the repeats in HSV (Davison and Scott 1985). It seems likely that the size and position of the inverted repeats in herpesvirus genomes are evolutionarily quite fluid and not dependent on specific gene or sequence boundaries. A strange feature of all of the genomes with inverted repeats or internal and external tandem repeats is the tendency of these sequences to maintain at least 10–20% higher G + C content than that of their surrounding sequences (see discussions of base composition biases in Honess 1984). Even in VZV DNA, the S-segment inverted repeats have large tracts of 60–70% G + C content (Davison and Scott 1985; Ruyechan et al. 1985), and in HSV they reach as high as 85% or more (Murchie and McGeoch 1982). An elaborate secondary structure for the entire HSV S-repeat region has been proposed (Rayfield et al. 1983), and the inverted repeats themselves have been implicated in various complicated replication schemes. However, despite the fact that so many herpesvirus genomes possess or retain this characteristic structural feature, the functional significance of the internal inverted repeats is still a total mystery.

The 6 to 12 internal, direct tandem repeats (IR) of 3072 bp in size in EBV DNA were at first thought to contain a standard promoter sequence and a large coding region for transcripts proceeding rightward across the *Bam*HI-W portion of the genome and into *Bam*HI-Y and -H (Cheung and Kieff 1982; Jones and Griffin 1983). However, the story has become more complicated with the discovery of latency transcripts that contain only small portions of this region in a very complex spliced structure that itself maintains an open reading frame (Bodescot et al. 1984). These transcripts are now believed to be highly spliced leader sequences from very large primary transcripts (up to 100 kb in size) proceeding from the *Bam*HI-C region, right across the tandem repeats, and on past the coding regions for EBNA-2 and possibly EBNA-3 and into EBNA-1 (Speck and Strominger 1985; P. Farrell et al., pers. comm.). Indeed, inspection of the sequence of *Bam*HI-W shows

it to contain numerous splicing acceptor signals that occur mostly in a single reading frame. As is often the case with even much shorter internal tandem repeats in EBV DNA, the copy number of the 3072-bp repeats is not an exact integer and varies from approximately 5.6 in EBV(P3HR-1) to 11.6 in EBV(B95-8) DNA molecules (Hayward et al. 1980), seemingly in compensation for large deletions or expansions that occur elsewhere in the genome (Pritchett et al. 1975; Delius and Bornkamm 1978; Given et al. 1979). Again, some mechanism for correcting or regenerating those sequences would seem to be necessary to compensate for losses by homologous recombination, although this could no doubt also be accomplished by unequal crossing-over exchanges during intermolecular recombination events.

The ends of EBV(B95-8) DNA consist of permutations of up to five copies of a 550-bp element per molecule, distributed randomly between the two termini in linear virion DNA (Given et al. 1979; Kintner and Sugden 1979) or fused as a single tandemly repeated "joint" array in clones from intracellular plasmid DNA. Some rare EBV-positive "nonproducer" and "converted" cell lines (e.g., Namalwa and IB4 or B95-8/Ramos) that retain only one or two copies of the genome have been shown to have these sequences integrated in a "linear" fashion with the cell-virus joints occurring within the EBV terminally redundant sequences (Henderson et al. 1983; Matsuo et al. 1984). However, since cellular sequences were deleted at the site of integration, this does not appear to be the result of a specialized integration mechanism like that in retroviruses. Therefore, the significance (if any) of these forms of integrated EBV DNA for the life cycle of the virus is not clear. These low-copy-number cell lines still express EBNA-1 and maintain a highly restricted transcription pattern typical of latent EBV infections.

### Comparison of overall gene content

Just how similar or different are the genes and genetic organization of the various human herpesviruses? HSV-1 and HSV-2 were recognized early on, by means of solution hybridization studies, to share up to 50% of their nucleotide sequences (Kieff et al. 1972). Later mapping studies also showed the two DNA molecules to be very similar in physical structure, except that the U$_S$ region of HSV-2(333) is approximately 3 kb larger than that of HSV-1(MP) (perhaps representing an additional gene). Otherwise, the ability to obtain numerous, viable intertypic recombinants and the results of all gene-mapping studies suggest that the two genomes are perfectly collinear. Even some earlier confusion about the glycoprotein-C region near 0.65 map units has now been resolved in favor of collinearity. In blot hybridization experiments between total HSV-1 probes and cleaved HSV-2 DNA, the least homologous regions appear to be the high–G + C–content L- and S-segment inverted repeats (Peden et al. 1982). Mapping and sequencing data comparing the collinear HSV-1 and HSV-2 genes for TK (McKnight 1980; Wagner et al. 1981; Reyes et al. 1982a; Swain and Galloway 1983), the ribonucleotide reductase 38-kD subunit (Dutia 1983; McLaughlan and Clements 1983; Galloway and Swain 1984), glycoprotein D (Watson et al. 1982; Lasky and Dowbenko 1984), glycoprotein C (Dowbenko and Lasky 1984), and the $ori_S$ region (Murchie and McGeoch 1982; Whitton and Clements 1984b) have been available for some time, but in general DNA sequence analysis of HSV-2 DNA lags far behind that for the other human herpesviruses. Other HSV-1 genes, including those encoding glycoprotein B (Bzik et al. 1984), alkaline exonuclease (Costa et al. 1983), DNA polymerase (Gibbs et al. 1985; K.W. Knopf, pers. comm.), DNA topoisomerase, a dUTPase, and the major capsid protein genes also have been clearly identified or sequenced.

The closest known animal virus to that of HSV is bovine mammilitis virus (BMV), which shows extensive homology throughout the genome and, unlike many of the other animal herpesvirus DNAs, still retains the inverted L-segment feature (Buchman and Roizman 1978). In contrast, early high-stringency hybridization studies with total HSV DNA probes against the supposedly weakly related animal herpesvirus genomes showed a major region of sequence conservation only in the central portion of the L segment. In particular, a probe representing the DNA polymerase/major DNA-binding protein region of HSV-1 DNA (map coordinates 0.35–0.43) easily detected the homologous locus in BMV, PRV, or EAV DNA and even in *H. tupaia* DNA, whereas a TK gene probe (map coordinates 0.28–0.32) barely hybridized to an equivalent region in BMV and PRV DNA and not at all to the others. More-extensive studies have shown that the frozen orientations of the L segment of PRV, EAV, and VZV can be aligned with each other and that all correspond to the opposite orientation of the HSV L segment rather than that defined as the prototype or P-arrangement (Ben-Porat et al. 1983; Davison and Wilkie 1983c).

DNA sequence comparisons of the HSV-1 and VZV unique S segment has recently revealed the presence of unmistakably analogous genes in considerable detail (Murchie and McGeoch 1982; Davison 1983; McGeoch et al. 1985). The $U_S$ sequence of VZV DNA is very much smaller than that of HSV (5 kb vs. 13 kb) and this primarily represents the absence of 6 of the 12 genes present here in HSV-1 DNA, combined with some movement in the positions of the inverted repeats relative to the genes nearby (Davison and Scott 1985). Similar analysis of the whole VZV genome (A. Davison, pers. comm.) shows many analogous genes to those known in HSV (and EBV) and appears to confirm the relatively close relationship of the HSV and VZV groups despite the enormous shift of overall base composition (Ruyechan et al. 1985). More-complete details of this analysis are eagerly awaited. Unpublished DNA sequence analysis of the much larger 25-kb $U_S$ segment of CMV DNA has apparently revealed a family of adjacent, closely related genes, suggesting that some expansion of this region of HCMV may have occurred relatively recently by a tandem duplication mechanism (K. Weston and B. Barrell, pers. comm.).

One major benefit of having the complete sequence of EBV DNA available has been the ability to identify several EBV genes through residual amino acid homology with known HSV counterparts. This has been accomplished for the DNA polymerase (Baer et al. 1984; Gibbs et al. 1985), the ribonucleotide reductase A and B subunits (Gibson et al. 1984), and glycoprotein B (Pellett et al. 1985a). The analog of an unidentified HSV spliced gene product has also been detected through DNA homology (Costa et al. 1985a). Other identified EBV structural or metabolic genes that are not yet known to have homologous HSV counterparts include a possible unusual nucleotide kinase gene product that contains a recognizable nucleotide-binding site, a 150-kD major capsid protein, the gp350/250 membrane antigen (Biggin et al. 1984; Hummel et al. 1984) and the putative LYDMA or latency membrane antigen (Fennewald et al. 1984; Hudson et al. 1985). Some rather startling examples of alterations in the transcriptional organization pattern of equivalent genetic regions in HSV and EBV have been noted. For example, the adjacent DNA polymerase and glycoprotein B genes in EBV are both transcribed leftward, whereas in HSV they are transcribed in divergent orientations and the major DNA-binding protein that lies between them in HSV is absent from this location in EBV. In the case of the A and B subunits of ribonucleotide reductase, the two HSV transcripts overlap and both terminate at a single poly(A) site immediately after the open reading frame for the small subunit. In EBV, however, the A-subunit transcript terminates at a poly(A) site between the two genes, and the nonoverlapping B-subunit transcript terminates at a distal poly(A) site far to the right. Unfortunately, there are relatively few published sequences for HSV genes and that remains the major limitation in this kind of analysis.

Relatively recent DNA hybridization studies at low stringency with HSV DNA polymerase gene probes has led to identification of the CMV DNA polymerase locus by at least two groups (R. D'Aquilla and W. Summers; T. Kousarides and B. Barrell; both pers. comm.), and the sequence of that gene is reputed to be completed also. All four herpesvirus DNA polymerase genes apparently share extensive amino acid homology and also show limited homology at one particular locus (9 matches over a stretch of 13 amino acids) to the DNA polymerases of adenovirus and vaccinia virus (Gibbs et al. 1985).

Although all herpesviruses clearly share many common and evolutionarily conserved structural and biochemical genes, some other genes appear either to have evolved too far apart to be recognizable or to have totally different origins and functions. This may be especially true in the case of the early regulatory nuclear proteins between the three major subclasses of human herpesviruses (i.e., HSV, EBV, and CMV). Of the five known IE genes of HSV-1 and HSV-2, two map within the L- or S-segment inverted repeats (IE110 and IE175) and the others all map nearby. Equivalent genes for IE175, IE68, and IE63 have all been recognized in VZV DNA by their genome locations and by sequence comparisons; however, apparently corresponding to the lack of L-segment inverted repeats in VZV, that genome lacks an equivalent of the IE110 gene (Davison and Scott 1985). Nevertheless, no sequences homologous to any of the HSV or VZV IE genes or the VF65 gene have yet been detected in EBV DNA. Similarly, none of the EBV gene products that have attracted greatest attention for their possible roles in gene regulation are known to have homologs in HSV: for example, the latency nuclear antigens EBNA-1, -2, and -3; the two early nuclear antigens from the BamHI-M region that may be butyrate-inducible (Sample et al. 1984; Cho et al. 1985a,b); and the BamHI-Z gene product, which may have some role in disrupting latency (Countryman and Miller 1985). Furthermore, in contrast to the situation in HSV, the major spliced IE gene of HCMV maps between 0.72 and 0.74 in one orientation on the $U_L$ segment and at least 15 kb away from the nearest L-segment inverted repeat structure (Stenberg et al. 1984; Akrigg et al. 1985). A second minor IE gene also lies just downstream from IE72 in the same leftward orientation (Stenberg et al. 1985). Equivalent spliced IE genes map at the identical location in the "frozen" genome of SCMV (Jeang 1984), but again, in both cases, although the human and simian genes show detectable (and in places strong) amino acid homology, no related genes have yet been detected within the EBV or VZV DNA sequences or among the accessible HSV DNA data.

These observations further emphasize the evolutionary similarity between the two neurotrophic herpesviruses HSV and VZV, in contrast to the strikingly different biology of EBV and CMV. Admittedly, we may yet find highly spliced equivalents of the EBV latency genes in HSV or the equivalents of the HSV IE nuclear proteins in CMV, but that is beginning to seem extremely unlikely. Obviously, an extraordinary amount of intriguing information (and speculation) about herpesvirus genome evolution will become available over the next few years as we continue to compare the overall sequence and map locations of equivalent genes in HSV, VZV, EBV, and CMV.

## Defective genomes and DNA replication

Input parental herpesvirus DNA molecules are believed to circularize (or form end-to-end concatemers) immediately after entering the cell nucleus, even in the absence of de novo protein synthesis (Ben-Porat and Veach 1980; Marks and Spector 1984; Poffenberger and Roizman 1985); however, definitive evidence that circularization represents an obligatory step in the pathway toward productive replication is lacking. Certainly, all newly replicated, unpackaged intracellular DNA is present in "endless" forms, without the usual half-molar terminal fragments, in both HSV (Jacob et al. 1979) and HCMV (LaFemina and Hayward 1983), and also in CCV (Cebrian et al. 1983). Furthermore, in PRV in particular, the existence of fast-sedimenting, double-stranded concatemers has been convincingly demonstrated (Ben-Porat et al. 1976; Jean et al. 1977; Jacob et al. 1979). When the tandemly repeated nature of defective HSV DNA molecules was first recognized, we suggested that a rolling-circle type of replication mechanism may have generated these structures as well as be the normal

replication mechanism for wild-type HSV DNA (Roizman et al. 1974). However, although some electron micrographic studies have been interpreted as showing tailed-circle forms (Friedmann and Becker 1977), the occurrence of such structures is far from proven. No monomeric replicating circle (or θ) forms have been reported (except for EBV plasmids; Gussander and Adams 1984), and preliminary, unpublished evidence for interlocked catenated forms in HSV has apparently not been pursued.

Although there has been little progress in defining the structural forms of replicating HSV DNA in recent years, implications about the locations of replication origins, which also came from our early studies of packaged defective genomes, have proved to be correct. Like many other viruses, high-moi passaging of HSV leads to production and propagation of defective viruses with aberrant genomes that rapidly evolve into homogeneous populations of tandemly repeated defective DNA together with wild-type helper virus (Frenkel et al. 1975; Murray et al. 1975; Wagner et al. 1975). Two forms of tandem-repeat defectives were recognized, one from the high–G + C S-repeat regions (Frenkel et al. 1976) and another of average G + C content from close to 0.4 map units near the center of the L segment (Kaerner et al. 1979; Ciufo and Hayward 1981). The packaged defective DNA often contains up to 20 direct repeats of identical units of approximately 7–10 kb in size. Exactly integral numbers of copies totalling close to the normal genomic size of 150-kb are packaged into enveloped virions. These type-I and type-II repeats were predicted to encompass separate replication origins from the S and L segments, referred to as $ori_S$ and $ori_L$, respectively (Ciufo and Hayward 1981), and to have thus gained some form of replicative advantage over standard genomes. Similar tandem-repeat defectives have also been observed in both PRV and EAV, even on passaging through animal hosts. Biologically, these populations often show competitive interference with wild-type virus (Frenkel et al. 1975), and in some cases in HSV they overproduce either the IE175 or major DNA-binding proteins, which happen to be encoded intact within some but not all tandem repeat species (Locker et al. 1982; Knopf et al. 1983). Both types of HSV repeat units also always retain sequences from the right-hand terminus of the wild-type genome, presumably to permit correct packaging and envelopment (Ciufo and Hayward 1981; Spaete and Frenkel 1982; Gray and Kaerner 1984). Interestingly, a 20-kD late polypeptide (the product of $U_S$ gene 12) apparently binds with some specificity to the "a" sequence, which might suggest a role for this protein in either packaging or recombination/inversion events (Dalziel and Marsden 1984).

Transfection of isolated, monomeric, defective repeat units followed by superinfection and high-moi passaging led to expansion and packaging of these monomers as tandem-repeat DNA (Vlazny and Frenkel 1981). However, it was not until Stow (1982) and Stow and McMonagle (1983) demonstrated virally induced replication of transfected pBR322 sequences attached to small segments of the S repeats that formal proof of the existence of $ori_S$ was obtained. It maps entirely within the S-inverted repeats between the divergent promoters for the IE175 and IE68/12 genes and encompasses a 45-bp palindrome with a central A + T—rich feature. In comparison to $ori_S$, all fragments of HSV-1 or HSV-2 DNA containing $ori_L$ are unstable when cloned in *E. coli* plasmids even in *recA⁻* hosts, and they always give rise to 120-bp deleted forms after one or two passages. Gray and Kaerner (1984) and Weller et al. (1985) obtained sufficient quantities of the undeleted form from defective and wild-type genomes to carry out DNA sequencing and have proved that this too is a functional DNA replication origin in the presence of factors supplied in *trans* by virus infection. Remarkably, $ori_L$ and $ori_S$ share extensive DNA sequence homology, with the $ori_S$ sequence lacking a portion of one side of the perfect 144-bp palindromic structure included within $ori_L$. Intriguingly, whereas $ori_S$ lies between two divergent IE promoters, $ori_L$ lies between two divergent DE promoters: those for the major DNA-binding protein on the left and for the viral DNA polymerase gene on the right. There is no evidence at this point whether $ori_S$ or $ori_L$ function at the same or different stages during virus infection or whether they promote unidirectional or bidirectional modes of DNA replication. The highly conserved $ori_S$ region in at least one strain of HSV-2 is duplicated (Whitton and Clements 1984b) and the same applies to $ori_L$-defective DNA populations (Gray and Kaerner 1984). The presence of just the $ori_S$ sequences in transfected plasmids has been shown to be sufficient to cause competitive interference with the replication of wild-type HSV DNA molecules (Schroeder et al. 1984; Stow 1985). The VZV DNA sequence analysis indicates a sequence with partial homology to HSV $ori_S$ at an equivalent location within the inverted S repeats (Davison and Scott 1985; Stow et al., this volume), but there is apparently no origin-like sequence equivalent to $ori_L$ (A. Davison, pers. comm.). Preliminary evidence indicates the probable presence of a lytic replication origin within the L repeats of HCMV DNA (R. LaFemina, unpubl.), but no other data are available. No tandem-repeat types of defective CMV DNA have ever been reported.

In EBV, the elegant analysis by Sugden and his colleagues has pinpointed two adjacent regions in *Bam*HI-C, referred to as $ori_P$, that are necessary for plasmid maintenance functions (Reisman et al. 1985). One of these sites can potentially form an imperfect dyad symmetry structure and represents an obvious candidate for the plasmid DNA replication origin. However, there is no information at all about EBV lytic DNA replication origins. Only the P3HR-1 line is known to produce significant quantities of defective EBV genomes (Hayward and Kieff 1977). Most of these molecules represent a substantially rearranged repeat structure in which four segments representing 30% of the genome complexity have been recombined together and then tandemly repeated (Delius and Bornkamm 1978; Cho et al. 1984a). The portions of the genome represented in this structure include both termini, all of *Bam*HI-C including

$ori_P$ and the EBER genes, *Bam*HI-W and -Y, the DNA polymerase and glycoprotein B genes, and both the *Bam*HI-MS EA and *Bam*HI-Z *trans*-acting genes (Cho et al. 1984b). Conspicuously absent are the EBNA-1, -2, and -3 genes and the *Bam*HI-Q EBNA-binding site. P3HR-1 cultures can be totally cured of defective genomes by subcloning the cells (Heston et al. 1982): Apparently, the defectives survive as persistent low-grade lytic infections, and their genomes, which cannot establish a latent state, behave as independent replicons encoding functions that trigger reactivation of the lytic cycle after superinfection of Raji cells (Miller et al. 1984, 1985).

Extensive characterization of partially purified 145-kD HSV DNA polymerase, its exonuclease activities, and a tightly associated 54-kD protein has been reported (Powell and Purifoy 1977; Knopf 1979; Vaughan et al. 1985). Some properties of the alkaline exonuclease have been examined (Hoffman 1981), and the major 132-kD DNA-binding protein has been purified and characterized (Powell et al. 1981; Lee and Knipe 1983). However, there has been little progress on in vitro or reconstituted DNA replication systems for any herpesvirus subsequent to an early report by Jongeneel and Bachenheimer (1980). The papers by Stow et al., Lupton and Levine, Countryman et al., and Rawlins et al. in this volume deal in much greater depth with aspects of HSV DNA replication, EBV plasmid maintenance, EBV defective DNA functions and herpesvirus DNA-protein interactions that might be involved in replication events.

## Transcription

### HSV lytic-cycle transcription patterns

Before progressing any further, it is necessary to pay some attention to the confused state of herpesvirus gene nomenclature. The three main classes of HSV genes are referred to as α, β, and γ under the system established by Roizman and his colleagues in Chicago and as immediate-early (IE), early or delayed-early (DE), and late (L) by Subak-Sharp's group in Glasgow and several others. Although it has been claimed that the original α, β, and γ description of protein kinetic classes involved quite different criteria than those used to define, for example, IE, DE, and L genes in phage T4, there seems to be far too much historical precedent from almost every other DNA virus system studied for us not to universally embrace the immediate-early, delayed-early, and late terminology. The use of α, β, and γ was partially as a precautionary measure in case the groupings did not fit the pattern established in other virus systems. However, nothing has happened since 1974 to suggest that these three classes do not fit equally well into a standardized, descriptive temporal classification. Labels of this type will continue to convey important and useful concepts, and particularly with the newer, more mechanistic definitions that are now available, they should have meaning to all molecular virologists.

By the simplest criteria, early genes are defined as those that are synthesized before DNA replication begins, and late genes are those that are either absolutely dependent on DNA replication (γ-2 or true lates) or those that, although synthesized earlier and in the absence of DNA replication, are made in much greater abundance after the DNA copy number has been increased by replication (γ-1, β/γ, quasilate, or early/late). Subdivision of the early genes is based on whether or not they are dependent on the synthesis of other viral gene products. Thus IE (α) mRNA is synthesized even after infection in the presence of cycloheximide or anisomycin to prevent de novo protein synthesis, and their protein products are synthesized immediately after reversal of the cycloheximide block in the presence of actinomycin D. DE (β) genes are not transcribed unless and until IE protein synthesis has occurred, which has been assumed to indicate a direct involvement of one or more IE gene products in the transcription of the DE class. Synthesis or accumulation of IE mRNA proteins is usually greatly reduced at DE times, and that of DE mRNA and proteins tends to be reduced at late times.

Early work on HSV transcription revealed the continuous need for cellular RNA polymerase II rather than a completely new viral enzyme (Preston and Newton 1976). Other studies showed that limited expression of the genome occurred at early prereplicative times, followed by virtually complete representation of the genome in late postreplicative RNA, and also documented the presence of high-molecular-weight primary precursor transcripts, symmetrical transcripts, and capped and polyadenylated cytoplasmic mRNA (Wagner and Roizman 1972; Kozak and Roizman 1975). Infection in the presence of cycloheximide to block de novo protein synthesis revealed an even more limited set of transcripts (representing less than 10% of the genome) encoding the IE (or α) proteins, with the most abundant IE mRNA template being the inverted repeat regions (Jones et al. 1977; Anderson et al. 1980). R-loop mapping and RNA:DNA hybridization studies also implied an essentially random distribution rather than grouping of early and late transcription units across the genome (Clements et al. 1977; Stringer et al. 1978; Anderson et al. 1979; Holland et al. 1980). The advent of northern blotting with more-refined fragment probes and S1-mapping techniques eventually enabled precise determination of the locations and orientations of many individual mRNAs. In HSV, the TK mRNA (Smiley et al. 1980) and IE mRNAs (Clements et al. 1979; Watson et al. 1979; Easton and Clements 1980) were the first to be defined, followed by several 3′ coterminal transcripts in the 0.55–0.6 region (Draper et al. 1982; McLaughlan and Clements 1983) that were subsequently shown to be encoding the two subunits of ribonucleotide reductase. Many years of heroic studies of this type, primarily by Edward Wagner's group, have defined the sizes, map locations, orientations, and temporal class assignments of over 60 transcripts across the entire HSV-1 genome (for review, see Wagner 1984). In many cases the molecular masses of polypeptides synthesized in vitro from DNA-fragment-selected mRNAs have also been assigned and in some cases promoter and coding-region sequence data are also available. The major trends recognized from transcription mapping are as follows:

1. There is some concentration of L-class genes and transcripts to the left of the major capsid protein gene at 0.26 map units and of IE genes in the inverted repeats, but otherwise the genes from the DE and L classes are widely dispersed, and there is about equal and random representation of leftward- and rightward-oriented transcripts.
2. In general, the pattern shows individual promoters for each gene, although the sharing of 3′ cotermini for several adjacent overlapping transcripts is a common feature. The overlapping coterminal transcripts frequently belong to several different temporal classes.
3. Another apparently common finding where sufficient detail is available shows divergent transcripts that overlap for short (up to 500 bp) distances at their 5′ ends. Invariably, one of these pairs is from the DE class and the other from the L class.
4. Evidence for splicing events is extremely rare among DE or L genes, and usually where it does occur, abundant unspliced versions are also present.
5. DE-type promoters frequently contain good "TATAAA" concensus features at approximately position −30 (e.g., TATAAA, CATAAA, or TATATTA), plus upstream CCAAT boxes, G + C–rich Sp1-binding sites, and/or A + C–rich elements. Occasionally, these upstream regions are buried in the carboxyterminal coding regions of another gene (e.g., the 38-kD ribonucleotide reductase B subunit).
6. L-type promoters frequently possess much less obvious "TATAAA"-box elements (e.g., TTAAT, TAAAT) and have little else in the way of recognizable concensus promoter elements.
7. A single example is known ($U_S$ gene 10/11) of two overlapping genes using multiple reading frames in the same DNA sequence (Rixon and McGeoch 1984). One of these two genes ($U_S$ 11) also provides an example of a repetitive amino acid pattern (X-Pro-Arg)$_{24}$, a phenomenon that has been observed quite frequently also in EBV.

Attempts to develop in vitro transcription systems that are specific for HSV promoters have been only partially successful (Beck and Millette 1982), and although uninfected cell extract systems may show a strong preference for IE or E promoters (Frink et al. 1981), they will transcribe some L genes accurately in vitro (Read et al. 1984). Little is known of the extent of HSV gene expression in the latent state in neurons. The genomes occur in "endless" and apparently multicopy forms that would be compatible with a plasmid state (Rock and Fraser 1985), but, although expression of IE and other antigens has been reported in neurons, no really good in vivo or in vitro models are yet available that permit proper biochemical analysis and manipulation. Although one assumes from the EBV and other virus plasmid/episome models that limited gene expression and repression of lytic-cycle function will occur, we really have no idea whether to expect the HSV IE genes, for example, to serve also as latency genes or whether there may be a separate and possibly as yet totally unrecognized group of genes involved.

## CMV gene expression, nonpermissive hosts, and differentiation

Except for at most five or six mapped genes and transcripts for polypeptides of known molecular mass but of unknown function, little detailed transcriptional information is available yet for CMV. The lytic replication cycle in diploid human fibroblasts is, of course, expanded over several days rather than being completed within approximately 16 hr as is the case for HSV. The generalized overall pattern and temporal classes operationally fit those defined for HSV to the extent that there is very limited transcription under IE conditions with a single, very dominant major IE species and that a large group of prereplicative DE transcripts and probably a relatively small number of true L transcripts occur (DeMarchi 1981; Wathen et al. 1981). Evidence for quite extensive posttranscriptional regulation over the timing of transport of some transcripts has been presented (DeMarchi 1983).

Expression of HCMV and SCMV is strictly limited to IE polypeptides at all times after infection of some nonpermissive cell types such as rodent fibroblasts (Jeang et al. 1982; LeFemina and Hayward 1983). At DE times in these cells, additional nuclear transcripts appear from across much of the viral genome, but unlike the IE mRNA they never become polyadenylated or reach the cytoplasm (Jeang 1984). Even further restrictions to HCMV expression occur after infection of many "transformed" cell lines, including 293 cells, human teratocarcinoma stem cells, mouse L cells, and even VERO cells. In these cultures, transcription or stable accumulation of even the major IE mRNA is blocked at all times with the exception that differentiation of the teratocarcinoma cells with retinoic acid leads to a reversal of the block and a full permissive infection cycle (Gonczol et al. 1984; R.L. LaFemina and G.S. Hayward, in prep.). These blocks to IE expression in "transformed cells" do not apply to SCMV infection except in mouse L cells and do not apply to the isolated HCMV or SCMV IE genes or hybrid promoter-CAT constructs when introduced by DNA transfection procedures in either short-term assays or after the establishment of stable cell lines. The negative regulation observed in these complex interactions appears to involve either IE autoregulation or viral factors introduced with the infecting virion as well as host-cell specific factors and may also depend on the presence or absence of tandemly repeated nuclear factor I (NFI)–binding sites upstream of the IE enhancer-promoter regions. Little is known for certain about even the true host cell site for latency of CMV (believed to be monocytes), let alone about the state of the viral genomes or the extent of gene expression during latency. Some limited interactions between fresh isolates of HCMV with lymphocytes and monocytes have

been suggested, which include alterations in the differentiation phenotype of the cells, but standard fibroblast-adapted HCMV laboratory strains do not produce these effects (Rice et al. 1984).

## Induction of EBV expression in lymphocytes

The earliest analyses of viral transcripts in EBV systems were of necessity limited to comparisons of non-producer and producer B-lymphoblast cell lines. The most repressed lines (e.g., Raji and Namalwa) showed at most 5% of the genome represented in cytoplasmic or polysomal/poly(A) containing transcripts although up to 20–30% of the genome was represented in nuclear RNA (Hayward and Kieff 1976; Orellana and Kieff 1977). As expected, cell lines spontaneously producing virus (such as B95-8 and P3HR-1) had virtually the entire genome represented at some level in polysomal RNA. Surprisingly, the most abundant transcripts in either situation turned out to be two small 190-bp RNA species referred to as EBERs (Lerner et al. 1981; Rosa et al. 1981; Arrand and Rymo 1982). Like the virus-associated (VA) RNAs of adenoviruses, these species are polymerase III products and they can apparently substitute for the VA RNAs, although only very weakly (Bhat and Thimmappaya 1983). The levels of EBERs do not appear to differ appreciatively between latent and productive infection. Numerous subsequent studies have defined the detailed locations of polymerase II transcripts in both latently infected and tumor cells with steadily increasing precision, and we now recognize at least four independent, highly spliced latency transcripts that give rise to at least three known nuclear antigens (Heller et al. 1982) and the putative "LYDMA" or latency membrane protein (Fennewald et al. 1984). The possibility that all three of the EBNA genes may be under the transcriptional control of a single promoter region in *Bam*HI-C, which gives rise to huge primary transcripts (Speck and Strominger 1985), may bring the story full circle to explain the original observations of the much higher proportion of the genome represented in nuclear compared with cytoplasmic transcripts in latently infected Raji cells.

EBV early and late gene products and transcripts are now usually operationally defined as those expressed after P3HR-1 superinfection of Raji cells or after 12-O-tetradecanoylphorbol-13-acetate (TPA) and butyrate induction of B95-8 cells in the presence of phosphonoacetic acid (PAA) (prereplicative) or in the absence of PAA (postreplicative). In general, the features of the lytic cycle transcript maps in EBV resemble those in HSV, especially with regard to: (1) splicing being uncommon; (2) individual promoters for each gene; (3) the seemingly randomly assorted distribution of E and L genes (and the latency genes); (4) interspersion of leftward- and rightward oriented transcripts; and (5) frequent small clusters of 3′ coterminal transcripts. One remarkable feature of the organization of the EBV genome is the occurrence of tandemly repetitive DNA sequence elements within coding regions, and in some cases this is reflected as repetitive amino acid patterns

(Baer et al. 1984). The best known examples are the variable 700-bp (GGG,GGA,GCA) element in EBNA-1, which gives rise to the (Gly,Gly,Ala) repeats (Hennessy and Kieff 1983); a 9 × GGGGCA sequence giving rise to (Gly-Arg) repeats in EBNA-2; and a larger 7-amino-acid repeat within the central portion of the 350/250-kD envelope glycoprotein gene. After the EBV(B95-8) DNA sequence was completed, detailed analysis of a large number of EBV promoters by hybridization mapping procedures and in vitro transcription has been carried out by Farrell et al. (1983), using narrowly defined M13 hybridization probes. However, these studies failed to yield clear, definitive concensus sequences for the different classes of promoters other than the typical presence of TATAAA-box features.

## Gene Regulation

### Characteristics of HSV IE promoter structure
Because of the peculiarities of HSV genome organization, two of the five IE class genes are diploid and lie entirely within either the L repeats (IE110) or the S repeats (IE175). Two other IE genes lie across the S-segment repeat—unique boundaries and in fact share identical promoter structures within the repeats but have entirely different coding regions. Furthermore, the shared IE68/IE12 promoter region represents the opposite side of the divergent IE175-IE68/12 promoter complex, which occupies approximately 700 bp and includes the $ori_S$ signal. Only IE63 lies totally within a unique region, but it is still relatively close to the right-hand L-segment repeat—unique boundary. These four IE promoters were originally identified and sequenced by Mackem and Roizman (1980, 1981), who also presented evidence that, even after infection in the presence of anisomycin, they appear to be transcribed as a coherent unit rather than in a hierarchical pathway. Thus, there is no evidence that any of the IE gene products is dependent in any way upon prior expression of the others (at least at the level of initial accumulation of stable transcripts). Mackem and Roizman (1982c) also recognized within each of these promoters the existence of common AT-rich features that are usually embedded in extraordinarily high G + C−content sequences. In the case of the IE175-IE68/12 complex, we now recognize that 12 of these G + C−rich elements are Sp1 protein−binding sites (Jones and Tjian 1985). Interspersed among the Sp1 sites are four TAATGARAT-like elements, two oriented toward IE175 at positions −116 and −265 and two oriented toward IE68/12 at positions −310 and −365 relative to the IE175 mRNA start site. Portions of the 5′-upstream IE175 and IE68 regions that include at least one TAATGARAT box convey IE character to other genes when placed in *cis* upstream of the TK gene or other heterologous promoters (Mackem and Roizman 1982a,b; Herz and Roizman 1983; Preston et al. 1984). Available evidence implies that the Sp1-binding sites in HSV IE175 and IE68 provide much of the basal strength of these promoters, whereas the TAATGARAT elements, either alone or in combination with adjacent Sp1

sites provide the IE-specific signals for response to the VF65 virion factor (Cordingley et al. 1983; Kristie and Roizman 1984). In vitro reconstruction experiments show that the addition of the cellular Sp1 protein stimulates basal transcription of the IE175 promoter (Jones and Tjian 1985). Furthermore, in vitro transcription analysis suggests that the HSV IE68 promoter may give at least fivefold stronger basal expression with infected-cell extracts than does IE175 when both are present together as a single divergent unit in the same plasmid (P. O'Hare, unpubl.); however, similar comparative data are not available for IE110 and IE63. Further details of the identification and interaction of target signals for VF65 and Sp1 in the HSV IE175 promoter are presented in McKnight et al. (this volume).

Between the IE68/12 mRNA cap site and the initial AUG codon lies one of the sets of short G + C−rich, tandemly repeated elements, but in this case they are not Sp1 sites, and they are spliced out of the IE68 and IE12 mRNAs (Watson et al. 1981). In contrast, the IE175 mRNA is not known to be spliced, whereas that for IE110 contains several introns (F. Rixon and D. McGeoch, pers. comm.). The IE110 and IE63 upstream promoter regions again contain several TAATGARAT-like elements but relatively few Sp1 concensus sites. Part of the IE110 5′ region can also be considered to consist of three highly diverged 50- to 70-bp direct repeats (each including a TAATGARAT box). The TAAT-GARAT elements are not known to occur in any other viral or cellular promoters, although a very similar sequence is found adjacent to the NFI-binding site within the adenovirus type-2 terminal repetition. Surprisingly, possible TAATGARAT motifs are not obvious within the VZV inverted repeats in the divergent promoter-$ori_S$ region that lies between the two VZV genes that have homology to HSV IE175 and IE68 (Davison and Scott 1985). Neither are there any Sp1 concensus sites in this region in VZV DNA. However, remember that VZV can only be grown in cell culture in human diploid fibroblasts and that the control of IE gene expression is likely to profoundly influence the host range. On the other hand, the VZV L segment does have a gene equivalent to VF65 (Dalrymple et al. 1985).

### HSV IE proteins
Five gene products—IE175K (Vmw 175 or ICP4), IE110K (Vmw 110 or ICP0), IE68K (Vmw 68 or ICP22), IE63K (Vmw 63 or ICP27), and IE12K (Vmw 12 or ICP47)—are recognized as belonging to the IE regulatory class in HSV-1 and HSV-2. They are encoded by the IE175 (IE3 or α4), the IE110 (IE1 or α0), the IE68 (EI4 or α22), the IE63 (IE2 or α27), and the IE12 (IE5 or α47) genes. To reduce confusion, I use the nomenclature involving the apparent molecular masses of the gene products from HSV-1 for both the gene and protein names. However, the two largest IE gene products of HSV-2 are sometimes discriminated from their HSV-1 counterparts as IE185 and IE120. As is typical of some other early viral regulatory genes (note especially

adenovirus E1A), the calculated molecular masses of both IE175 and IE110 from DNA sequence data are apparently far below those observed in polyacrylamide gels with standard marker proteins, a discrepancy that presumably reflects the very hydrophilic and often proline-rich nature of these gene products. All of the HSV IE gene products except IE12 (Marsden et al. 1982) are nuclear, phosphorylated proteins (Wilcox et al. 1980), giving multiple forms on two-dimensional gel electrophoresis when detected with monoclonal antibodies (Ackermann et al. 1984).

The central importance of the IE175 gene product has been recognized because of the availability of several *ts* mutant viruses that fail to synthesize DE mRNA or proteins under nonpermissive conditions (Preston 1979b). Mutants with this phenotype such as *tsK* and *tsB2* were shown to form a single complementation group and to have lesions mapping within the portion of the S-segment inverted repeats that we now recognize as the coding region for IE175. The *tsK* and *tsB2* mutants synthesize all of the other IE gene products but usually fail to transport the IE175 protein to the nucleus (Cabral et al. 1980). Watson and Clements (1980) showed by performing shift-up experiments at late times after infection that the IE175 gene product was needed continuously throughout the virus life cycle and demonstrated that under these circumstances the transcription pattern reverted to IE mRNA synthesis only. The IE175 protein has been described as having DNA-binding properties (Freeman and Powell 1982), but most attempts to purify the protein have been frustrated by insolubility problems. Some biophysical properties implying an elongated shape and homodimeric structure for the partially purified protein have been described recently (Metzler and Wilcox 1985). Genetic and physical maps of many clustered mutations in IE175 have been prepared (Dixon and Shaffer 1980) and the lesion in *tsK* has been identified (Davison et al. 1984). Recently, several IE175 partial deletion mutants have also been constructed and their properties appear to match those of the *ts* mutants (DeLuca et al. 1985). One especially interesting new IE175 mutation was described recently that produces a different phenotype showing blocks to all late gene expression but allowing expression of DE genes and normal levels of DNA synthesis (DeLuca et al. 1984). Furthermore, this mutation maps in a more carboxyterminal region of the protein than the others. Jim Smiley's group has recently established a number of Ltk$^+$ cell lines with cotransfected DNA fragments that contain the intact IE175 gene (Persson et al. 1985). Some of these lines (e.g., Z4) appear to constitutively express low levels of IE175 and permit synthesis of many DE gene products from superinfecting *ts* IE175 mutants at nonpermissive temperature.

It is believed that IE175 has an essential role in transcriptional activation of DE-gene and possibly also in L-gene transcription. However, the degree of promoter specificity displayed by this protein, and questions about whether or not it interacts directly with DE promoter signals or interacts with other viral or cellular proteins are still under intensive investigation. The ability of superinfecting HSV virus to activate or enhance expression from a resident HSV TK gene integrated into cellular DNA was first recognized by Leiden et al. (1976). Activation of TK mRNA was shown to be blocked by cycloheximide and therefore to be dependent on IE gene expression (Leung et al. 1980). Many subsequent studies have confirmed this observation with other viral genes (Sandri-Goldin et al. 1983) and with hybrid TK-ovalbumin, TK-IFN, or TK-CAT targets (Post et al. 1982; Reyes et al. 1982b; Herz and Roizman 1983; Mosca et al. 1985), but questions of specificity remain (Everett 1984a, 1985). One complication of viral superinfection studies is the possibility of *trans*-activating or modifying activities among the other IE genes (e.g., IE110), and even when the activation by *tsK* infection is shown to be temperature-sensitive, one cannot be sure that it directly reflects the absence of the active IE175 protein only.

Direct evidence that IE175 itself is capable of *trans*-activation awaited the use of cell lines expressing the isolated IE175 gene (Persson et al. 1985) and cotransfection studies in transient assay systems with the isolated IE175 gene (O'Hare and Hayward 1985a). The latter studies have also produced evidence for a great deal of specificity toward HSV DE targets for positive *trans*-activation by IE175 (either alone or in the presence of IE110) and for specific negative autoregulation of IE175 on the IE175-CAT target constructs (O'Hare and Hayward 1985b). However, Feldman et al. (1982) and Imperiale et al. (1983) showed that the PRV IE gene product (or at least PRV infection) could substitute efficiently for E1A in activation or other adenovirus early genes, and Tremblay et al. (1985) have recently provided support for this notion by infecting the Z4 cell line with E1A mutants of adenovirus. However, no one has been able to activate HSV DE genes with E1A, and indeed some studies also show interference effects between both the PRV IE protein or IE175 and E1A (Tremblay et al. 1985; L. Feldman, pers. comm.). Even the question of whether or not there is a specific DE response signal in the TK and other HSV DE genes is still contentious. Several careful studies with deleted or mutated TK promoter targets have been unable to clearly discriminate between the location of response signals to viral infection and either the distal or proximal elements needed for basal expression (El Kareh et al. 1984, 1985; Eisenberg et al. 1985). Furthermore, Green et al. (1983) and Everett (1984a,b, 1985) have reported activation by HSV superinfection, or cotransfecting IE gene plasmids, of β-globin in both transient and long-term assays, implying a lack of specificity for the activation mechanism. On the other hand, Tackney et al. (1984) demonstrated that an intact globin gene inserted into the HSV genome did not behave like a viral DE gene, and O'Hare and Hayward (1984) demonstrated the absence of induction of a minimum SV40 promoter construct (A10-CAT) by virus infection under transient assay conditions in which HSV DE-CAT targets were strongly activated. Most likely, HSV provides both specific and nonspecific factors involved in *trans*-activation of gene expression, and some of these activities might be detectable only with particular specific constructs or only in circumstances involving appropriate states of the target DNA (e.g., endogenous cellular genes compared with integrated foreign DNA or plasmid forms and viral prereplicative forms compared with viral postreplication forms). One should also pay careful attention to the levels of *trans*-activation being measured; for example, a fourfold activation of one target may be considered significant by one group but nearly meaningless alongside a 40-fold activation of another target DNA under the same conditions by another group.

Until recently, none of the other HSV IE gene products were known to have essential roles in HSV infection and there were no clues about their possible functional roles. However, it is now recognized that IE63 *ts* mutants give normal DE expression and DNA synthesis but have deficiencies in late gene expression (Sacks et al. 1985). Furthermore, viruses with a deleted IE68 gene (although unaffected in human cells) appear to be unable to grow efficiently in rodent cell cultures (Sears et al. 1985). Finally, the isolated IE110 gene has been shown to be capable of *trans*-activating gene expression in transient assay systems, in some cases on its own (DeLuca and Shaffer 1985; O'Hare and Hayward 1985a,b; Quinlan and Knipe 1985) and in others only when acting in concert with IE175 (Everett 1984b; Gelman and Silverstein 1985). However, mutants with lesions in IE110 are not available as yet to evaluate the significance and role of this protein in viral infection. Further details about the action and specificity of IE175 and IE110 *trans*-activation are discussed in O'Hare et al. (this volume).

## The virion factor and other potential HSV regulatory gene products

The pre-IE virion factor involved in transcriptional activation of HSV IE promoters is also a nuclear phosphorylated protein with an apparent molecular mass of 65 kD (hence the gene name VF65, although it is also known as the Vmw65 or ICP16 protein). Fully permissive HSV infection occurs in cultured cells of almost all cell types and host species, which could potentially be a consequence of bringing in a preformed specific initiator of viral IE transcription as a component of the virion, thus becoming somewhat independent of the need for appropriate specific cellular factors to initiate the transcription process. However, until Post et al. (1981) carried out superinfection experiments in Ltk$^+$ cells receiving hybrid IE175-TK genes, there was no reason to suspect the existence of such a factor. Both double-stranded and single-stranded forms of naked HSV DNA are infectious (Graham et al. 1973; Sheldrick et al. 1973), and obviously the IE175 gene and IE175 hybrids are expressed to some extent in the absence of the virion factor in both transient DNA transfection assays (Middleton et al. 1982; Cordingley et al. 1983; O'Hare and Hayward 1985a,b) and in long-term, stably transfected cell lines (Post et al. 1981; Lang et al. 1984). Therefore,

the factor is probably not strictly essential. Some of the most compelling evidence for the virion transcription factor comes from studies with the *tsB7* or similar mutants, which were originally thought of as uncoating mutants because not even the IE mRNA or proteins are expressed under nonpermissive conditions (Knipe et al. 1981). Batterson et al. (1983) showed that after infection at nonpermissive temperature, the *tsB7* virions lose their envelopes but the capsids accumulate at nuclear pores and do not release viral DNA into the nucleus. Surprisingly, under these conditions one can still demonstrate the full activity of the virion factor on a resident IE175 hybrid promoter already integrated into the recipient cell (Herz and Roizman 1983; Mosca et al. 1985). Similar results can also be obtained with UV-irradiated input virus that is incapable of expressing its own IE genes (Batterson and Roizman 1983; Preston and Tannahill 1984). Consequently, the virion transcription factor appears to be a (presumably abundant) tegument protein released before disruption of the input capsids. Campbell et al. (1984) carried out a "shotgun"-approach experiment to screen an HSV DNA library by cotransfection for evidence of an activity capable of stimulating IE175 expression. Amazingly, although encoded by a late gene, the virion transcription factor was synthesized in sufficient quantity to give a fourfold to fivefold stimulation of the specific target when the *Bam*HI-F fragment from map coordinates 0.65 to 0.70 was employed. The map location of this activity was subsequently refined sufficiently to conclude that it corresponded to a previously well known 65-kD phosphorylated component of the virion for which monoclonal antibodies are available (McLean et al. 1982). The HSV-1 VF65 gene has been sequenced by Dalrymple et al. (1985) and by Pellett et al. (1985b). The protein is proline-rich but otherwise shows no homology to any other previously known regulatory genes. Since the VF65 protein isolated from infected cells or virions is not known to have DNA-binding properties, most speculation implies that (like the adenovirus E1A gene product) this protein may work indirectly through interaction with a cellular factor that may bind to the TAATGARAT elements.

The 128-kD major DNA-binding protein (ICP8), an abundant DE product, is also implicated as an important regulatory factor in HSV (Lee and Knipe 1983). This protein is essential for viral DNA replication and for normal levels of synthesis of late polypepetides (Conley et al. 1981). Mutations mapping in the gene lead to altered patterns of DE and L gene products (Godowski and Knipe 1983). More recently, Godowski and Knipe (1985) report that, in the absence of DNA synthesis, expression of the L-class glycoprotein-C mRNAs are greatly increased with an ICP8 *ts* mutant deficient in 128-kD expression in comparison with wild-type virus. They speculate that one function or consequence of major DNA-binding protein expression is to repress or reduce the stability of late gene transcripts from parental DNA molecules until after DNA synthesis changes the structure of the genome, amplifies the DNA template, and perhaps sequesters the 128-kD protein for other purposes. Like the 72-kD protein of adenovirus, the 128-kD protein also binds single-stranded DNA and may also have a role in early gene expression (Godowski and Knipe 1983). Incidentally, other studies by Dennis and Smiley (1984), Silver and Roizman (1985), and Costa et al. (1985b) also imply that DNA replication per se is not an absolute requirement for expression of integrated or plasmid-borne late promoters in the presence of wild-type virus infection, although their expression from a parental viral DNA molecule is normally delayed or inhibited until after DNA synthesis occurs. Chiou et al. (1985) have provided genetic evidence for a functional interaction between the major DNA-binding protein and the viral DNA polymerase. Unlike most of the other nuclear regulatory proteins described here, the 128-kD protein is not known to be phosphorylated. The protein has been extensively purified by Powell et al. (1981) and monoclonal antibodies are available (Showalter et al. 1981; McLean et al. 1982).

Another HSV early protein with interesting and unexplained properties is the 136-kD-subunit cytoplasmic phosphoprotein (ICP6), which has been identified as the A subunit of the viral ribonucleotide reductase enzyme (Frame et al. 1985). Surprisingly, the 38-kD B subunit of ribonucleotide reductase, which coprecipitates with the 136-kD subunit using monoclonal antibodies to either protein is exclusively a nuclear product by immunofluorescence analysis. However, similar studies with the 136-kD protein show a uniform distribution throughout the cytoplasm at early and late times after infection, but some is retained transiently in the nucleus at approximately 6 hr after infection. The most abundant immunoprecipitating antibodies in hyperimmune sera are against this protein, and there appears to be far too much to it to function solely as a catalytic enzyme subunit. Strangely, although the structure of its promoter is clearly of the DE type, most preparations of IE mRNA contain some of the 5.2-kb 136-kD transcript, and several studies with *tsB2* or IE175 deletion mutants indicate that its expression may be completely independent of the presence of a functional IE175 protein (Middleton et al. 1982; DeLuca et al. 1985). Both virion factor(s) and subsequent viral gene products are implicated in the multistage shut-off of host protein, mRNA, and DNA synthesis after HSV infection (T.M. Hill et al. 1983; Read and Frenkel 1983), but no specific gene products have yet been associated with these events. The early stages of infection also lead to stimulation of expression of cellular heat-shock proteins (LaThangue et al. 1984), and presumably many other complex host-virus interactions occur at both the transcriptional and posttranscriptional level. HSV also acts as complete helper for adeno-associated virus (Buller et al. 1981).

**The complete CMV IE promoter-enhancer region**
We currently recognize only a single unidirectional IE promoter complex in CMV DNA, although this may drive expression of more than one gene product (Stinski et al.

1983; Jahn et al. 1984; Wilkinson et al. 1984). Synthesis and overexpression of a single, dominant 72-kD (HCMV) or 94-kD (SCMV) protein from a cycloheximide block imposed at the time of infection was first recognized by Gibson (1981) and Jeang and Gibson (1980). This protein was also shown to be the only abundant viral species produced at any time after infection of nonpermissive rodent fibroblasts even in the absence of inhibitors (Jeang et al. 1982; LaFemina and Hayward 1983). Mapping of the major IE RNA species led to localization of a strong in vitro promoter approximately 1500 bp upstream from the main body of the IE94 mRNA and eventually to the recognition of multiple splice sites and introns at the aminoterminal end of both simian and human major IE proteins (Jeang 1984; Thomsen et al. 1984). In our hands, either the IE72 or IE94 5'-upstream promoter region fused to IFN or CAT coding regions gives enormously high basal expression of these products in transient DNA transfection or in oocyte microinjection assays (O'Hare et al., this volume). Indeed these constructs have been measured to produce 3 to 4 times more CAT enzyme than from constructs containing the complete SV40 early T-antigen promoter in VERO cells and up to 50 times more in transfected diploid human fibroblasts. Furthermore, the isolated intact IE94 gene is expressed constitutively after integration into a coselected Ltk$^+$ cell line (Jeang et al. 1984) and the IE72 upstream region behaves as a stronger enhancer than that of the SV40 72-bp repeats in a enhancer-trap assay (Boshart et al. 1985). DNA sequencing of the IE94 5'-upstream promoter region revealed the presence of multiple palindromes and interspersed repetitive elements, and we recognized at least four sets of 13–18-bp repeats within the region between −50 and −580 (Jeang 1984). These elements are totally conserved in sequence between the two primate CMV IE promoters, although their arrangement is greatly altered (Thomsen et al. 1984). A cluster of four adjacent copies of the Series II repeats may be the key elements in IE94 because the basal strength of 5'-deleted CAT constructs drops dramatically after removal of sequences between positions −260 and −140. However, both this set of repeats and the features contributing to high basal strength appear to be much more uniformly distributed across the entire 500-bp ''enhancer'' region in IE72 (Stinski and Roehr 1985). The significance of the other repeat elements is unknown, but they obviously provide complex possibilities for both positive and negative regulation by viral and cellular transcription factors. The presence and possible functional significance of additional far upstream elements, including tandemly repeated NFI-binding sites between −700 and −1280 in IE94, is described in detail by Rawlins et al. in this volume. Evidence for specific negative autoregulation of both the IE72 and IE94 promoters by their own gene products has been obtained in transient CAT assays, but the signals for this response appear to lie around the cap site within the region −70 to +30 and not in the far 5'-upstream regions (P. O'Hare et al., in prep.).

## CMV IE proteins

An early nuclear antigen (CMNA) of CMV was described by Reynolds (1978) in infected fibroblasts, using a complement-mediated immunofluorescence assay with convalescent human sera. Subsequent studies have implied that this activity corresponds to the major IE gene product of 68–76 kD synthesized in human diploid fibroblasts after infection with different strains of HCMV (Gibson 1981). Monoclonal antibodies against IE72 detect a nuclear antigen in IE stage–infected cells (Goldstein et al. 1982) and after DNA transfection with the isolated gene (R.L. LaFemina and G.S. Hayward, unpubl.). Unfortunately, there are no mutants available with defects in this gene. The major IE genes of both HCMV and SCMV have been sequenced (Jeang 1984; Stenberg et al. 1984; Akrigg et al. 1985). Both genes are highly spliced and possess highly conserved 5′-untranslated leader exons. However, the HCMV IE72 and SCMV IE94 polypeptides have diverged enormously and retain only small patches of amino acid homology. They both also possess unusual, large glutamic acid–rich domains near their carboxyl termini (50% glutamic acid over 46 adjacent amino acids in IE72 and 50% glutamic acid over 78 adjacent amino acids in IE94) and are relatively acidic and phosphorylated at multiple sites (Jeang and Gibson 1980; Jeang et al. 1982). Several minor IE gene products have been defined in HCMV(Towne) (Stinski et al. 1983; Jahn et al. 1984; Akrigg et al. 1985). These are encoded by the IE-2 gene, which maps immediately 3′ of the IE72 (or IE1) gene, and are apparently also transcribed under the control of the major IE promoter. In fact, the IE-2 mRNA transcripts contain the same initial, small 5′ exons as IE72, followed by an alternative splice into the IE-2 open reading frame instead of the preceding fourth exon of IE72 (Stenberg et al. 1985). A similar downstream open reading frame with highly conserved amino acid homology to IE-2 occurs also in SCMV(Colburn), but its mRNA transcript and protein product have yet to be detected under IE conditions. It is not clear whether expression of the IE-2 genes is independent of or subsequent to IE72/94 expression. Several major DE transcripts map within the inverted L repeats (McDonough et al. 1985), but their functions and the relationship, if any, of these genes to those of the HSV IE nuclear proteins are unknown.

In transient expression assays, we have found that cotransfection with plasmids encompassing both IE genes (i.e., the HindIII-C fragment of HCMV[Towne] or with the HindIII-H fragment of SCMV[Colburn]), gives strong trans-activation of all heterologous promoter-CAT constructs tested. Therefore, the CMV IE gene products display properties similar to those of HSV IE110. In both cases, the major IE proteins appear to be necessary for this trans-activating activity, but the adjacent minor IE gene (or genes) possibly also has a role.

Both Stinski and Roehr (1985) and Spaete and Mocarski (1985a) have presented evidence for an HCMV virion factor that gives positive trans-activation of IE72-hybrid constructs. However, the activity reported was relatively weak and we have found no evidence supporting this idea in SCMV, either in transient assays or in permanent cell lines (J. Mosca et al., unpubl.; O'Hare et al., this volume). HCMV infection can apparently complement adenovirus E1A mutants (Tevethia and Spector 1984).

## Potential regulatory gene products in EBV: EBNAs

The classical diagnostic EBNA antigen present in all EBV-containing B-lymphoblast cell lines was first described by Reedman and Klein (1973). However, the EBNA gene was not identified until Summers et al. (1982) described the establishment of a BamHI-K DNA–transfected Ltk⁺ cell line that expressed an EBV nuclear antigen detectable by anticomplement immunofluorescence with EBNA-positive human sera. Other studies confirmed these findings (Hennessy and Kieff 1983; Hearing et al. 1984) and led to the recognition of a 56-kD open reading frame encoding the apparently complete EBNA-1 polypeptide. The BamHI-K fragment reportedly contains only a 3′ 2.0-kb exon portion from a larger 3.7-kb latency transcript (Heller et al. 1982), but the intact protein is expressed directly in transient assays from transfected BamHI-K plasmids, apparently utilizing cryptic promoters in the pBR322 sequences that are considerably activated by the addition of SV40 enhancers. The apparent molecular mass of the polypeptide varies from 70 kD to 95 kD in different strains because of variable numbers of internal Gly,Gly,Ala–encoding triplet repeats. The aminoterminal portion of the protein also contains a $(Gly-Arg)_{15}$ repetitive region and the extreme carboxyl terminus is acidic. Spelsberg et al. (1982) have described a basic, loosely chromatin-associated, unmodified, non-DNA-binding form of the protein and a less basic, phosphorylated DNA-binding form that is tightly chromatin-associated. The relationship of an earlier described 48-kD "EBNA" polypeptide and its associated 53-kD cellular protein to the authentic 76-kD EBNA first reported by Strnad et al. (1981) is still unclear. Antibodies against synthetic (Gly,Gly,Ala) oligopeptides, monoclonal anti-EBNA antibodies, and rabbit monospecific sera against the E. coli–synthesized protein have all been reported. The complete protein shows both double-stranded and single-stranded DNA-binding properties and a carboxyterminal 28-kD portion displays sequence-specific DNA-binding to two multisite loci in $ori_P$ (BamHI-C) and also to a third site in BamHI-Q (Rawlins et al. 1985). The intact EBNA-1 protein has been shown to be essential for maintenance of the plasmid state in cell-culture model systems and appears to also be a positive transcription trans-activator acting specifically at the tandemly repeated binding sites in $ori_P$ (see Lupton and Levine; Rawlins et al. [both this volume] for further details).

A number of workers have used human sera lacking antibodies to EBNA-1 to describe a second latency nuclear antigen encoded by a transcript of approximately 3.1 kb (Dambaugh et al. 1984; Mueller-Lantzsch et al. 1985; Rymo et al. 1985). This region of the genome has been of great interest for a number of years because large deletions in the nontransforming strains P3HR-1 and Daudi map to this location (Bornkamm et al. 1982; Hayward et al. 1982; King et al. 1982; Rabson et al. 1982) and eliminate all or part of the EBNA-2 gene (Jeang and Hayward 1983b; Jones et al. 1984). It is also now recognized that there are two very distinct subclasses of the EBNA-2 gene, of which EBV(B95-8) represents the commoner A prototype and EBV(AG876) and EBV(Jijoye) are examples of the B prototype. The EBNA-2 proteins are usually approximately 85 kD to 90 kD in size, and Sendai-fusion transfection studies indicate a role in stimulation of host-cell DNA synthesis (Volsky et al. 1984). Tentative evidence for a third latency nuclear polypeptide of 70 kD (EBNA-3), which is encoded at least partly by the *Bam*HI-E fragment, has also been presented (Hennessy et al. 1985).

Are the EBNAs exclusively latency gene products or could they really also be the EBV equivalent of IE genes? All we know is that introduction of the intact EBV genome into monolayer cells by microinjection (Graessmann et al. 1980) and into mouse lymphocytes by envelope fusion techniques (Gross and Volsky 1984) gave EA and VCA expression, but EBNA was not detected. However, any potential IE-specific virion factor would have been lost by these techniques.

## Potential regulatory gene products in EBV: Early nuclear proteins

Sample et al. (1984) have reported that two transcripts from the *Bam*HI-M region are the only major viral mRNAs synthesized in abundance after EBV (P3HR-1) superinfection of Raji cells in the presence of cycloheximide, and they have suggested that these are therefore "immediate-early" genes. The *Bam*HI-M transcripts correspond to the two nuclear EA(D) gene products identified by Cho et al. (1985a,b), one of which (*Bam*HI-MS EA[D]) is also present in defective DNA. Interestingly, in stable DNA-transfected cell lines both of these isolated genes can be induced by treatment of the cells with sodium butyrate (Cho et al. 1985a,b). The *Bam*HI-MS EA(D) gene product, but not that of *Bam*HI-M EA(D), behaves as a strong *trans*-activator of gene expression in transient DNA transfection assays with cotransfected hybrid CAT genes as targets (P. Lieberman et al., in prep.). It displays a similar broad promoter target specificity to that of the HSV IE110 and the CMV IE72/94 gene products. Both EA(D) proteins undergo modifications involving phosphorylation and the *Bam*HI-M EA(D) gene product, which is recognized by the $R_3$ monoclonal antibody, has DNA-binding properties (G. Pearson, pers. comm.).

On the other hand, Paul Farrell's group (pers. comm.) has found predominantly transcripts from the *Bam*HI-Z region after EBV(P3HR-1) superinfection of Raji cells in the presence of anisomycin. Cloned *Bam*HI-Z DNA encodes a gene product(s) that appears to be under the control of the latency promoters in rearranged defective DNA and has been reported to stimulate EA and VCA induction after transfection of some latent cell lines (Countryman and Miller 1985). However, the wild-type

*Bam*HI-Z fragment does not do this unless fused to SV40 or long terminal repeat enhancers. This region of the genome had also been shown by Takaki et al. (1984) to activate EA(R)-like cytoplasmic antigen expression from the cotransfected *Bam*HI-HF region.

Do either the *Bam*HI-M and -MS or *Bam*HI-Z gene products really represent the EBV equivalents of IE class proteins? The *Bam*HI-Z gene under the control of a latency promoter would logically be expected to be transcribed immediately after EBV(P3HR-1) superinfection in the absence of de novo protein synthesis, but its classification and role in the wild-type genome is really the critical question. If its normal role were that of a virion factor triggering lytic infection, it would be capable of disrupting latency from defective genomes, but why then would not any nondefective virus preparation carrying the virion factor be capable of producing superinfection induction? Newly synthesized viral mRNA produced after TPA or butyrate induction would be more acceptable evidence for an authentic IE gene product. However, the only really valid definition of EBV IE genes would be those transcribed in the absence of protein synthesis after infection of primary lymphocytes or epithelial cells in the absence of a preexisting EBV latent state. Sample et al. (1984) have described preliminary experiments of this type and reported transcripts from both *Bam*HI-HF and *Bam*HI-M.

A final but important criterion for an IE gene, by analogy with SV40 large T antigen, Ad2 E1A, HSV IE175, and CMV IE68, would almost certainly be the presence of a complex upstream promoter-regulatory region with repetitive elements, palindromes, or enhancers. There are three such candidate regions known at present in EBV: first, the EBNA-1–binding sites in $ori_P$ that may act as EBNA-dependent enhancer elements (Reisman et al. 1985) and may be part of the latency transcription control region; second, an intriguing $2 \times$ 75-bp repeat lying between the convergent 3′ ends of the two early lytic transcripts encoding the *Bam*HI-M and -MS EA(D) nuclear proteins; and third, the duplicated promoter region ($DS_R$ and $DS_L$) upstream from the *Not*I repeat gene (in *Bam*HI-H of B95-8) and the *Pst*I repeat genes (absent form B95-8). These latter two species are also produced as abundant, internally repetitive RNA transcripts relatively early after TPA induction (Hudewentz et al. 1982; Freese et al. 1983; Jeang and Hayward 1983b; Jones and Griffin 1983) and they contain multiple A + T–rich and other palindromes, plus G + A–rich elements and even TAATGARAT-like sequences in their 5′-upstream promoter-regulatory regions (Jeang and Hayward 1983b; Laux et al. 1985).

## Acknowledgments

My special thanks and appreciation go to many colleagues over the years for their hard work, enthusiasm, creative ideas, encouragement and indulgence, especially Diane Hayward, Sam Wadsworth, Bob Jacob, Tim Buchman, Larry Morse, Greg Reyes, Bob LaFemina, Sandy Buchan, Keith Peden, Kuan-Teh

Jeang, Richard Ambinder, Myung-Sam Cho, Dan Rawlins, Peter O'Hare, Mabel Chiu, Michael McLane, Pamela Wright, Nancy Standish, and Dolores Ciufo. Work from this laboratory has been funded by grants CA 22130, CA 28473, and CA 37314 to G.S.H. and grant CA 30356 to S. Diane Hayward from the Department of Health and Human Services. The author is a Faculty Research Scholar of the American Cancer Society (ACS 247).

## II. LATENT AND ONCOGENIC INFECTIONS BY HUMAN HERPESVIRUSES

One hallmark of human herpesviruses is their ability to infect us latently. Work with experimental infections of mice and natural infections of people has identifed three general criteria of herpesviral latent infections: (1) The immune response of the host affects the frequency with which the escape from latency is detected; (2) a complete viral genome is present in cells; on escape, it can replicate productively; (3) the structure of the viral genome in cells differs from that found in productive infections and in virions. Although latent infections can be fully appreciated only at the organismal level, attempts to study those virus-cell relationships in culture that mimic infected cells in latently infected hosts are likely to be fruitful. B lymphoblasts immortalized by Epstein-Barr virus (EBV) appear to fulfill the three cited criteria and are readily studied in culture. Work with these cells may help us to understand in particular the control of expression of the viral genome in latently infected cells.

It has been claimed that herpesviruses are associated with human cancers. Although a prospective survey now indicates that infection with herpes simplex virus type 2 (HSV-2) is not causally associated with cervical carcinoma (Vonka et al. 1984a,b), a variety of observations link infection with EBV to endemic Burkitt's lymphoma and to rare B-cell neoplasms in immunocompromised patients. Transformation of cells in culture by herpesviruses is defined by different assays that require different viral contributions. Specific fragments of HSV-1, HSV-2, and cytomegalovirus (CMV) DNAs can transform rodent cells morphologically, but these small DNAs need not encode proteins. EBV immortalizes infected B lymphocytes. The immortalized cells contain all the viral DNA, and multiple viral functions appear to be required for the induction and maintenance of cell proliferation.

The biology of human herpesviruses is well reviewed by Tooze (1980), Roizman (1982), and Fields (1985). In this discussion I focus on latent and oncogenic infections of human herpesviruses and rely on these earlier reviews to provide an introduction to these topics. In particular, I compare and contrast observations made with HSV-1 and -2, CMV, and EBV in an attempt to develop models for some aspects of latent infections by these viruses, to assess their roles in oncogenic infections, and to describe their modes of transforming cells in culture.

### Latent Infections In Vivo

HSV-1, HSV-2, varicella-zoster virus (VZV), and EBV all produce latent infections in their human hosts; that is, after primary infection, these viruses remain sequestered somewhere in the host but are not usually detectable, or at least the pathogenesis associated with their primary infection is not detected. Occasionally the latent virus is reactivated; that is, the virus or its pathogenic effects may be detected again in the host in the apparent absence of exogenous infection. HSV-1, HSV-2, and VZV latently infect neurons. This conclusion is supported, for example, by the recovery of HSV from human trigeminal ganglia (Bastian et al. 1972), by the identification of HSV and VZV viral nucleic acids in neurons in in situ hybridization (Tenser et al. 1982; Hyman et al. 1983), and by the identification of neurons as cells from which HSV-1 is reactivated by dissociation of ganglia in culture (Kennedy et al. 1983).

In all likelihood EBV latently infects both lymphoid cells and epithelial cells. Support for the latent infection of lymphoid cells by EBV comes, first, from the observation that EBV-infected B lymphoblasts grew spontaneously from the peripheral blood of 23 out of 24 normal, seropositive donors tested (Yao et al. 1985a). Second, it has been demonstrated that most EBV-immortalized B lymphoblasts derived by infection in vitro of cells from both neonates (Wilson and Miller 1979) and adults (Sugden 1984) release small but detectable quantities of virus. These two observations taken together indicate that lymphoid cells infected in vivo by EBV may both maintain the virus in a latent state and infrequently permit its maturation and release.

Three findings are consistent with the conclusion that EBV also latently infects epithelial cells. First, EBV can infect epithelial cells derived from the ectocervix in vitro (Sixbey et al. 1983). Second, its DNA is maintained in vivo in nasopharyngeal carcinoma cells (Klein et al. 1974). Third, its DNA has been detected in the parotid gland of three normal donors by both in situ hybridization and renaturation kinetics (Wolf et al. 1984). In this last study the results of the in situ hybridization ruled out infiltrating lymphocytes as the source of the hybridization signal but could not establish that the positive cells were in the epithelia.

Two kinds of observations indicate that latent infections by herpesviruses are affected by the immune status of the host. First, reactivation is detected more frequently in immunocompromised hosts than in healthy control populations. This increased reactivation in immunocompromised hosts has been scored by an increase in herpetic lesions (Montgomerie et al. 1969), by an increased severity of herpes zoster lesions (Schimpff et al. 1972), by an increased release of EBV in the saliva (Yao et al. 1985b), and by an increased frequency of malignancies associated with EBV (Hanto et al. 1984). The second kind of observation that implicates the immune system as affecting latent infections comes from a classic study by Stevens and Cook (1974). Ganglia from mice latently infected with HSV-1 were transplanted in Millipore chambers into the peritoneal cavities of recipient mice. Those recipients that had been actively immunized with HSV-1 or passively immunized with rabbit anti-HSV-1 antibodies inhibited the detectable reactivation of HSV-1 in the transplanted ganglia relative to unimmunized recipients. These observations do not define the role the immune response plays in latent infections by herpesviruses. However, they are consistent with a model in which the immune response does not affect release of the herpesvirus from the cell in which it usually resides but does affect the subsequent amplification and spread of the virus. In a normal host, detection of reactivation would be rare; in an immunocompromised host, viral amplification and spread would be frequent, and thus reactivation would be detected frequently.

Examination of two kinds of tissues has revealed some of the features of cells usually infected by herpesviruses in latently infected animals. HSV-1 and HSV-2 have been studied in experimentally infected mice, and EBV has been analyzed in naturally infected people. As mentioned, a study with the mouse model has permitted the identification of neurons as cells from which HSV-1 is reactivated on dissociation of ganglia in culture (Kennedy et al. 1983). Work with infected mice has also demonstrated that a humoral immune response can inhibit the detection of reactivation of HSV-1 from transplanted ganglia (Stevens and Cook 1974).

Two additional, important observations have been made studying mice latently infected with HSV-1. J. Spivack and N. Fraser (pers. comm.), using nucleic acid hybridization, have enumerated the average number of HSV-1 DNA molecules per cell in the trigeminal ganglia of latently infected mice. They found that on the average there are approximately 0.2–0.4 molecules of viral DNA per cell. Neurons compose approximately 10% of the cells of the trigeminal ganglia. If all of the viral DNA is in the neurons and each of the neurons is latently infected, then each neuron will have approximately 2–4 molecules of HSV-1 DNA. If only 1% of the neurons contain all of the viral DNA present in the ganglia, then these neurons should each have 200–400 molecules on the average. This latter amount can be detected readily by in situ hybridization, whereas the former cannot. If some of the viral DNA in the trigeminal ganglia is found in nonneuronal cells that could support productive infections, then it is possible that these cells could contribute significantly to the total viral DNA content measured by Spivack and Fraser. Clearly, in situ hybridization to detect viral DNA, coupled with immunofluorescence assays to identify neurons unequivocally, is a necessary approach to reveal the distribution of HSV-1 DNA in neurons of latently infected mice.

Similar problems in identifying cells that contain EBV DNA and measuring the amount of this viral DNA per cell exist for EBV latent infections. The work of Wolf et al. (1984) indicates that EBV DNA is found in nonlymphoid cells of parotid glands isolated from normal and, therefore, presumably latently infected donors. This work could now be followed with a combination of immuno-

fluorescence plus in situ hybridization to positively identify the number and kinds of infected cells. This new information, when coupled with the nucleic acid hybridization of Wolf et al. (1984), which found between 0.5 and 5 copies of EBV DNA per cell on the average, would permit an approximation of the amount of viral DNA per positive cell in the parotid gland. Information on the number of EBV-infected lymphoblasts in latently infected donors also needs to be acquired. Their presence has been established (Yao et al. 1985a). Their number could be measured by culturing peripheral lymphocytes of normal donors in cyclosporin to eliminate T-cell-mediated killing of those cells, as Yao et al. (1985a) did, and counting the number that grow out in the presence of neutralizing antiserum. This approach will yield the number of EBV-immortalized cells in the peripheral blood of latently infected donors; the number of viral DNA molecules per immortalized cell could then be readily measured.

The structures of HSV-1 and EBV DNA in cells in latently infected hosts have also been analyzed. Rock and Fraser (1983) demonstrated by nucleic acid hybridization, using Southern transfers, that the bulk of HSV-1 DNA in the brain and trigeminal ganglia of latently infected mice is in a form distinct from that found in virions. In particular, the nucleotide sequences found at the termini of linear virion DNA are underrepresented in the viral DNA isolated from the latently infected tissues. The latent form of the viral DNA could be circular, polymeric, or integrated into the host chromosome. This finding is exciting because it indicates that the viral DNA detected in this latently infected tissue is not participating in a typical productive infection. Infections in vivo are necessarily asynchronous, and the bulk of viral DNA in any productive infection would be likely to be replicated and linear, as is found in virions. This expectation is supported by the finding of Rock and Fraser (1983) that the structure of HSV-1 in acutely infected brain tissue is identical with that found in virions.

The structure of EBV DNA in immortalized lymphoblasts established from normal donors likewise differs from that found in virions. In virions, EBV DNA is a linear molecule (Pritchett et al. 1975). In lymphoblasts, the viral DNA is maintained as a covalently closed, supercoiled molecule of full length (Kaschka-Dierich et al. 1977). The EBV-immortalized lymphoblasts studied by Kaschka-Dierich et al. (1977) were amplified by growth in vitro so that the structure of EBV DNA in infected cells from latently infected donors has not been analyzed directly. However, intracellular EBV DNA has been shown to be a covalently closed, circular molecule in fresh lymphoma and carcinoma biopsies (Kaschka-Dierich et al. 1976). It seems likely that therefore the circular structure of EBV DNA in established lymphoblasts reflects the structure of viral DNA in infected lymphoblasts present in latently infected people.

The cited studies of latent infections by HSV-1 in mice and EBV in people provide three criteria by which to characterize latent herpesviral infections: (1) The immune response of the host affects the efficiency of detection of reactivation; (2) the complete viral genome is present, so that reactivation can occur; (3) the structure of the viral DNA differs from that found in the virion.

## Latent Infections In Vitro

An understanding of latent viral infections requires consideration of the virus, the cells it infects, and the response of the host organism to that infection. However, a major contribution to the understanding of latent infections may be derived by studying cells infected in tissue culture, in which the virus—host cell relationship mimics that found in vivo in latently infected animals. This particular virus—host cell relationship will be optimistically termed a "latent infection in vitro." Many attempts have been made to develop latent infections by HSV-1 in vitro. Models using neurons are hampered because primary neurons do not divide in culture, and they exhibit a life span of only weeks in vitro. The work of Rapp and his colleagues indicates that rat and human neurons isolated from the dorsal root ganglia support productive infections by HSV-1 (Wigdahl et al. 1984a,b). These productive infections can be inhibited or retarded by treatment of the cells with inhibitors of DNA synthesis or by altering the temperature of incubation. However, there is now no evidence that this inhibition is mechanistically related to the control of latent infections in vivo. In particular, the structure of the viral DNA in these infected cells is found to be linear, as it is in virions (Wigdahl et al. 1984b), and differs from that which is characteristic of latent infections in vivo (Rock and Fraser 1983). Stevens and his colleagues (Gerdes et al. 1979) have tried to develop conditions in which HSV-1 or various temperature-sensitive (ts) mutants of HSV-1 would establish latent infections in a murine neuronal cell line. This line, the N115 clone of the C1300 neuroblastoma cell line, attains a quiescent state if serum is removed and the culture is maintained at 31°C. Under these conditions only 0.3% of the cells incorporate [$^3$H]thymidine during a 24-hr labeling period. However, both wild-type and HSV-1 ts mutants that do and do not form latent infections in vivo grow productively in this quiescent neuroblast culture.

The reason that these in vitro models fail to mimic latent infections in vivo is not known. Among the most obvious reasons are the lack of the immune response in vitro, which would act to limit the amount of infectious virus to which a neuron is exposed in vivo, and the route of entry of the virus into the neuron in vitro. In vitro the cell body of a neuron will probably be the most frequent locale for infection by HSV-1, because the processes of the neuron will be lost during its isolation. However, in vivo studies indicate that HSV-1 moves from the site of infection of the host to the ganglia by axonal transport (Cook and Stevens 1973). Furthermore, the virus must also move along axons from the ganglia to the periphery to be scored as a reactivated lesion of the skin (T.J. Hill et al. 1983). The route of entry in vivo may limit the expression of HSV-1 immediate early genes by inactivating a virion factor and thereby inhibiting the viral produc-

tive infection. Whatever the specific reasons for the failure to develop a model for latent infections by HSV-1 in vitro are, a general one may be that cultured neurons differ essentially from their normal counterparts in vivo.

The EBV-immortalized human B lymphoblast in vitro does appear to be a faithful mimic of its latently infected counterpart in vivo. It is sensitive to being killed by cytotoxic T cells educated in vivo (Svedmyr and Jondal 1975) and in vitro (Meuer et al. 1983). It contains the entire viral genome as assayed physically (Kintner and Sugden 1981a) and, more compellingly, as assayed biologically (Wilson and Miller 1979; Sugden 1984). The state of this intracellular viral genome is circular and differs from the linear molecule found in virions (Lindahl et al. 1976). Given that the EBV-immortalized human B lymphoblast is a tractable model for studying latent infection in vitro, what has its study contributed to the understanding of the control of latent infections in vivo?

If we assume that latent infections by HSV-1 and EBV are regulated similarly, then the study of EBV-immortalized cells may aid our understanding of two aspects of latent infections by HSV-1. One aspect concerns the extracellular influence on the detection of reactivation; the other concerns the viral contribution to the escape from latent infection. We know that most clones of EBV-immortalized lymphoblasts release infectious EBV. They do so inefficiently, so that the virus can only rarely be detected free in solution but can be readily detected by cocultivation with nontransformed B lymphocytes (Wilson and Miller 1979; Sugden 1984). These observations from the study of EBV can be recast in the language of the study of HSV-1. A worker with HSV-1 would conclude that a small fraction of EBV-immortalized cells regularly permit reactivation of the endogenous virus if cocultivation were used as his assay for reactivation. The conclusion may also be correct for HSV-1 in infected neurons. Note that HSV-1 cannot be detected in cell-free extracts of trigeminal ganglia but can be detected by cocultivation of fibroblasts with such ganglia. Reactivation as scored by a lesion of the skin would be a function not of the rare escape of the virus from the neuron but of whether the low amount of virus released regularly could be amplified sufficiently to be detected.

This description fits well with the observations that the treatment of mice with inhibitors of HSV-1 productive infections can prevent the establishment of a latent infection and prevent reactivation as scored by recurrent skin lesions, but it does not eliminate already established latent infections (Blyth et al. 1980; Field and DeClercq 1981). The similarity between latent infections by EBV and HSV-1 extends to the effects of these inhibitors of productive infections. These inhibitors fail to affect EBV DNA in the bulk of immortalized cells but do inhibit the maturation and subsequent release of virus from that small fraction of cells that would otherwise permit viral maturation (Summers and Klein 1976; Colby et al. 1980). These results underscore the conclusion that the mode of maintaining viral DNA in neurons and lymphoblasts of latently infected hosts differs from viral replication in productively infected cells.

An enigma that the above hypothetical description of reactivation of HSV-1 does not explain is the finding that recurrent reactivations of HSV-1 from trigeminal ganglia fail to eliminate the function of the ganglia. Productive infections by HSV-1 in cell culture generally kill the host cell. If neurons in vivo were to behave like cells in culture, then recurrent reactivations of HSV-1 might be expected to lead to the destruction of a significant fraction of the neurons in the trigeminal ganglia during a person's lifetime. However, such destruction is not observed. We must conclude either that release of HSV-1 from neurons in vivo does not necessitate cell death or that only a small fraction of the neurons of a ganglion die as a result of HSV-1 production during a human lifetime.

The study of the escape from latent infection in vitro by EBV is likely to help us understand the intracellular regulation of HSV-1 in latent infections. Countryman and Miller (1985) have demonstrated that the introduction of a fragment of EBV DNA into some EBV-immortalized cells leads to the induction of viral maturation functions. The easiest interpretation of this finding is that the added DNA contains a viral gene not normally expressed in immortalized cells. Introduction of the DNA into the cell leads to its expression, and its gene product induces maturation of the endogenous genome in *trans*. In a somewhat glib description, the control of a latent infection in vitro might then be equated with repression in the host cell of specific viral genes; the occasional expression of these genes in a rare member of a population of cells would lead to maturation of the viral genomes in that cell. It is entertaining to speculate more specifically that the neuron could limit the expression of HSV-1 genes by eliminating the virion protein that is required for the stimulation of the expression of the HSV-1 α genes in productive infections (Batterson and Roizman 1983). This elimination might take place as the unenveloped capsid migrates along the axon toward the cell body. Whatever the mode of establishing a latent infection of a neuron by HSV-1, it is clear that the regulation of the latent viral genome is likely to be distinct from that of the viral genome found in a productively infected cell.

## Oncogenesis Associated with Human Herpesviruses

A wide variety of observations have been cited to associate both HSV-2 and EBV with different human neoplasms. In particular, retrospective epidemiologic studies have linked high titers of antibodies to HSV-2 in patients with an increased risk of developing cervical carcinoma (Adam et al. 1973). HSV-2 DNA has been detected at least once in a carcinoma biopsy (Frenkel et al. 1972). Although several groups have reported the presence of HSV-2 RNA in preneoplastic lesions and carcinomas (McDougall et al. 1980, 1982; Eglin et al. 1981), a correlation between detectable levels of HSV-2 RNA in biopsies and the severity of the disease (Maitland et al. 1981) has not been established. The plethora of sometimes conflicting observations linking HSV-2 and

cervical carcinoma has been succeeded by a thorough prospective survey of 10,000 women by Vonka and his colleagues (Vonka et al. 1984a,b). Their findings "do not support the assumption that HSV-2 plays a role in cervical neoplasia" (Vonka et al. 1984a,b). In all likelihood, many of the earlier studies linking HSV-2 to cervical cancer were actually identifying venereal infections as factors that increase the risk of developing neoplasms. HSV-2 was one of the infectious agents usually detected that can now probably be viewed as a passenger virus. The causative, infectious agent for developing cervical carcinoma seems likely to be several types of human papillomaviruses (Durst et al. 1983). There is now, therefore, no persuasive evidence that associates infection by HSV-2 with cancer in people.

Infection with EBV is, however, associated with human cancers. A prospective survey of 42,000 children in the West Nile district of Uganda found that those with antibody titers to EBV viral capsid antigens 4-fold higher than the average titer of the control group had a 30-fold increased risk of developing Burkitt's lymphoma (de Thé et al. 1978). Several additional observations also indicate that EBV probably plays some role in the etiology of Burkitt's lymphoma, but that additional factors are required to produce the disease. The tumor biopsies from Uganda, where Burkitt's lymphoma is endemic, usually express one or more EBV-encoded antigens in all tumor cells (Reedman et al. 1974). As would be expected, therefore, these biopsies also contain viral DNA (zur Hausen et al. 1970), which is present in multiple copies per cell (Lindahl et al. 1974). Second, the tumor appears to be clonal when assayed from the biopsies of females who are heterozygous for isozymes of the X-linked gene, glucose-6-phosphate dehydrogenase. When the tumors were tested for the expression of these isozymes, one allele or the other was found to be expressed almost exclusively (Fialkow et al. 1970). This observed clonality, coupled with the uniform expression of one or more EBV antigens in the tumor cells, means that the virus probably infected the tumor early in its evolution, perhaps before it was a tumor. Consistent with this notion is the observation that in the Ugandan survey, infection with EBV preceded diagnosis of Burkitt's lymphoma by 7 to 54 months. This long latent period means that Burkitt's lymphoma is not simply the result of a primary infection by EBV. We can conclude that infection of children with EBV in places where Burkitt's lymphoma is endemic predisposes them to this disease but that other events must occur subsequently for the disease to arise. What those events may be is discussed below.

EBV is associated with a second human cancer, nasopharyngeal carcinoma. Retrospective seroepidemiologic surveys have found that patients with nasopharyngeal carcinoma have titers of antibodies to EBV viral capsid antigens 4-fold to 10-fold higher than those of controls (Henle et al. 1970; Huang et al. 1978). As with Burkitt's lymphoma, EBV antigens and DNA are usually present in the nasopharyngeal carcinoma tumor cell (Klein et al. 1974; Desgranges et al. 1975; Andersson-Anvret et al. 1977). This carcinoma arises in adults who were probably infected with EBV as children, and therefore it does not result from a primary infection with EBV. Clearly, EBV is associated with nasopharyngeal carcinoma; the role the virus plays in the etiology of the tumor is not yet clear.

EBV is also associated with rare neoplasms that occur in immunocompromised patients. As with Burkitt's lymphoma, the proliferating tumor cell is a B lymphoblast that expresses EBV-encoded antigens and contains multiple copies of viral DNA. These tumors arise in patients who are immunosuppressed after cardiac, renal, or bone marrow allografts (Bieber et al. 1984; Hanto et al. 1984). In patients receiving bone marrow allografts, the tumors can originate in the cells of the donor (Shubach et al. 1982; Martin et al. 1984) and grow to kill the hosts in 2 to 3 months. This rapid growth indicates that the precursors to the tumor cells, which were normal in the bone marrow donors, evolved lethal variants rapidly in the immunosuppressed recipients or required no evolution to grow fatally in these hosts. In either case the proliferative capacity of the B lymphoblasts induced by infection with EBV appears likely to play a major role in causing these tumors in immunocompromised patients.

The knowledge that EBV-immortalized B lymphoblasts derived from normal bone marrow can proliferate fatally in an immunosuppressed recipient may help us to understand the ontogeny of Burkitt's lymphoma. It is known from a variety of observations that EBV-infected, proliferating lymphoblasts are most likely to be eliminated from a normal host by a specific immune response (Svedmyr and Jondal 1975; Rocchi et al. 1977; Svedmyr et al. 1978; Tosato et al. 1979). It appears that patients with Burkitt's lymphoma may not have the immune capacity of normal individuals, but rather may be partially immunosuppressed. Burkitt's lymphoma occurs in regions of the world in which malaria is holoendemic (Morrow et al. 1976), and individuals with malaria have reduced T-cell responses to EBV-infected cells (Moss et al. 1983; Whittle et al. 1984). In addition, cell lines derived from Burkitt's lymphoma biopsies are often less susceptible to being killed by T cells educated in vitro than are cell lines established from normal B cells from the same lymphoma patients by immortalization with EBV in vitro (Rooney et al. 1985). The cause for this frequent resistance of Burkitt's lymphoma cell lines to being killed in vitro is not known. There are surface antigens that are expressed rarely in Burkitt's lymphoma cell lines but are present uniformly on cells transformed in vitro by EBV (Kintner and Sugden 1981b; Rowe et al. 1985). Whether or not these surface antigens could be targets for killing in vitro is not clear. If the resistance of the lymphoma cell lines detected in vitro reflects a parallel resistance in vivo, then it is clear that this phenotype, coupled with the proliferation induced by EBV, could aid in rendering an infected cell tumorigenic in a host with an impaired immune response.

The mechanisms that contribute to the observed antigenic differences between B cells immortalized in vitro and Burkitt's lymphoma cells and their differences in resistance to specific immunocytolysis in vitro are not known. One genetic change that Burkitt's lymphoma cells consistently display, in addition to infection with EBV, is a chromosomal translocation that juxtaposes the c-*myc* gene locus with one of several immunoglobulin gene loci (Lenoir et al. 1982; Taub et al. 1982). This provocative genetic change may directly influence the cell's resistance to cell-mediated cytotoxicity; it may decrease the cell's dependence on exogenous factors, whose concentration would otherwise limit proliferation of EBV-transformed cells; or it may affect a raft of cellular changes that are yet to be identified.

## Transformation by Human Herpesviruses

HSV-1, HSV-2, and human CMV grow productively in the cells they infect in vitro. These viruses therefore normally kill those infected cells. The early belief that HSV-2 might be causally associated with cervical carcinoma led investigators to try to transform cells with HSV-1 and HSV-2; that is, they tried to determine whether infection with these viruses could induce a heritable change in vitro that would render the cells more akin to tumor cells than to their nontransformed parents. These investigators were clearly influenced by the knowledge that infection of rodent cells in vitro with SV40 and polyomavirus can yield cells that are tumorigenic in vivo. To transform cells with these three herpesviruses, it was necessary to eliminate their capacity to grow productively and kill the infected host cell. Duff and Rapp (1971) used UV-inactivated HSV-2 to transform mouse 3T3 cells morphologically. Eventually, specific fragments of HSV-1 and HSV-2 DNA were shown to induce morphological transformation of rodent cells in culture (Comacho and Spear 1978; Reyes et al. 1980; Galloway and McDougall 1981). An unexpected outcome of all of this work was the finding that most of the viral DNA was lost from these transformed cells (Galloway and McDougall 1983). Similar experiments with CMV have shown that a fragment of viral DNA of less than 500 bp, a size that is unlikely to encode a protein, is sufficient to transform rodent cells in culture (Nelson et al. 1984). The mechanism of these transformations is still obscure and obviously different from the mechanism of transformation by papovaviruses and adenoviruses in which the continued expression of virally encoded proteins is required for maintaining the transformed state.

Two observations about HSV-1 have been made that at first appraisal might shed light on the mechanism(s) by which specific fragments of DNA from HSV-1 could transform cells in culture. Infection with inactivated HSV-1 is mutagenic at the *HGPRT* locus of human cells (Schlehofer and zur Hausen 1982). Infection of some SV40-transformed hamster cells with HSV-1 induces the amplification of the SV40 DNA in those cells, and the amplificaton is dependent upon the HSV-1 DNA polymerase (Matz et al. 1984). It is not unreasonable to posit that either of these virus-associated functions could contribute to the induction of a transformed phenotype. However, the fragments of HSV-1 and HSV-2 DNA that transform are not homologous (Reyes et al. 1980) and do not contain the gene for DNA polymerase. The transforming fragments of HSV-2 and CMV do share a structural feature: Both fragments contain DNA sequences that can be represented as stem and loop structures (Galloway et al. 1984). However, the fragments are not homologous, and, although the HSV-2 fragment is contained within an open reading of the virus, that from CMV has multiple stop codons in all reading frames (Galloway et al. 1984). It now seems unlikely that these DNA fragments could mediate transformation in culture by inducing the amplification of cellular DNA sequences, because they lack the viral DNA polymerase required for that function (Matz et al. 1984). Whether these fragments could be mutagenic for the recipient cells remains to be determined.

EBV immortalizes human B lymphocytes in vitro as it does in vivo. This immortalization is often termed "transformation." The virus both induces and maintains the proliferation of the infected cell. The evidence for the virus's being required to induce proliferation of the infected cell is experimental (Pope et al. 1966); that for maintaining proliferation is circumstantial: No cells immortalized by EBV have been identified that have lost the viral DNA and continue to proliferate. The mechanism by which EBV transforms B lymphocytes is unresolved, but two characteristics of the process have been studied. First, an indirect experiment to measure the target size for inactivation of transforming functions with a frameshift mutagen indicates that between 20% and 25% of the EBV genome is required to initiate and maintain transformation (Mark and Sugden 1982). This experiment suffers from being indirect but points to the current notion that the viral contribution by EBV to immortalization of B lymphocytes is more complex than that by the papovaviruses to transformation of rodent cells. Second, a number of studies have been performed to identify virally encoded RNAs and proteins present in transformed cells. Complex viral RNAs and several viral proteins have been identified (see below), but it is not yet known whether all or only some of these are required for maintaining transformation. Those viral products required for initiating transformation will be even more difficult to identify.

Several studies have identified viral RNAs in transformed cells that map to at least four regions of the genome and encompass approximately 40,000 bp of template DNA (Lerner et al. 1981; van Santen et al. 1981; Arrand and Rymo 1982; Bodescot et al. 1984; Hudson et al. 1985). For the RNAs that are being analyzed now, only a small fraction of the primary transcripts are found in the mature messenger RNAs. Several virally encoded proteins, or candidates for virally encoded proteins, have been identified in EBV-transformed cells (Summers et al. 1982; Hennessy et al. 1983, 1984, 1985; Hennessy and Kieff 1985; Mueller-Lantzsch et al. 1985; Rymo et al. 1985). However, functions have been associated with only one of these viral proteins.

A *cis*-acting element of EBV DNA termed $ori_P$ has been identified that is required for the replication of plasmids derived from EBV (Yates et al. 1984; Reisman et al. 1985). Circular DNAs that contain $ori_P$ will replicate as plasmids in EBV-immortalized cells; control DNAs lacking $ori_P$ do not replicate (Yates et al. 1984; Reisman et al. 1985). Because EBV DNA is usually maintained as a plasmid in transformed cells (Lindahl et al. 1976), it is likely that $ori_P$ is required for efficient viral transformation. One viral protein, EBV nuclear antigen 1 (EBNA-1), which is expressed in transformed cells (Reedman and Klein 1973), is required for $ori_P$ function (Yates et al. 1985). The EBNA-1 protein binds to the repetitive sites within $ori_P$ (Rawlins et al. 1985) that have been shown to be essential for replication of $ori_P$-bearing plasmids (Reisman et al. 1985). The replication function of EBNA-1 protein is, therefore, likely to be required for efficient transformation by EBV. A second function associated with the EBNA-1 protein may also be involved in transformation of B lymphocytes by EBV (Reisman et al. 1985; Reisman and Sugden 1986). EBNA-1 is a *trans*-acting activator of transcription from promoters linked to a *cis*-acting component within $ori_P$ (Reisman et al. 1985; Reisman and Sugden 1986). This *trans*-activation of a transcriptional enhancing component of $ori_P$ by EBNA-1 may well be important for the expression of viral genes in the transformed cell. It may also be important for affecting expression of cellular genes if there are sites in the human genome to which EBNA-1 binds as it does within $ori_P$. As yet we do not know the functions of the other EBV genes expressed in transformed cells, nor do we know if they are involved in initiating or maintaining the transformed state of cell proliferation.

## Note Added in Proof

The membrane protein encoded by EBV that is expressed in immortalized lymphoblasts has recently been shown by Wang et al. (1985) to transform an established rodent cell to anchorage-independent growth. Although the function of this protein is not yet known, it seems likely that it will be required for immortalization of the EBV-infected B lymphocyte.

## Acknowledgments

I thank Vijay Baichwal, Joyce Knutson, Joan Mecsas, Dave Reisman, Ilse Riegel, and Stan Metzenberg for reviewing this manuscript critically. I thank Jordan Spivack for communicating the results of his collaboration with Nigel Fraser prior to its publication. I was supported by grants from the U.S. Public Health Service (CA-22443 and CA-07175).

## REFERENCES

Ackermann, M., D.K. Braun, L. Pereira, and B. Roizman. 1984. Characterization of herpes simplex virus 1 alpha-proteins 0, 4, and 27 with monoclonal antibodies. *J. Virol.* **52:** 108.

Adam, E., R.H. Kaufman, J.L. Melnick, R.H Levy, and W.E. Rawls. 1973. Sero-epidemiological studies of herpesvirus type 2 and carcinoma of the cervix. IV. *Am. J. Epidemiol.* **98**: 77.

Akrigg, A., G.W.G. Wilkinson, and J.D. Oram. 1985. The structure of the major immediate early gene of human cyto-megalovirus strain AD169. *Virus Res.* **2**: 107.

Anderson, K.P., R.H. Costa, L.E. Holland, and E.K. Wagner. 1980. Characterization of herpes simplex virus type 1 RNA present in the absence of *de novo* protein synthesis. *J. Virol.* **34**: 9.

Anderson, K.P., J.R. Stringer, L.E. Holland, and E.K. Wagner. 1979. Isolation and localization of herpes simplex virus type 1 mRNA. *J. Virol.* **30**: 805.

Andersson-Anvret, M., N. Forsby, G. Klein, and W. Henle. 1977. Relationship between the Epstein-Barr virus and undifferentiated nasopharyngeal carcinoma: Correlated nucleic acid hybridization and histopathological examination. *Int. J. Cancer* **20**: 486.

Arrand, J.R. and L. Rymo. 1982. Characterization of the major Epstein-Barr virus specific RNA in Burkitt lymphoma-derived cells. *J. Virol.* **41**: 376.

Baer, R., A.T. Bankier, M.D. Biggin, P.L. Deininger, P.J. Farrell, T.J. Gibson, G. Hatfull, G.S. Hudson, S.C. Satchwell, C. Sequin, P.S. Tuffnell, and B.G. Barrell. 1984. DNA sequence and expression of the B95-8 Epstein-Barr virus genome. *Nature* **310**: 207.

Bastian, F.O., A.S. Rabson, C.L. Yee, and T.S. Tralka. 1972. *Herpesvirus hominis*: Isolation from human trigeminal ganglion. *Science* **178**: 306.

Batterson, W. and B. Roizman. 1983. Characterization of the herpes simplex virion-associated factor responsible for the induction of alpha-genes. *J. Virol.* **46**: 371.

Batterson, W., D. Furlong, and B. Roizman. 1983. Molecular genetics of herpes simplex virus. VIII. Further characterization of a temperature-sensitive mutant defective in release of viral DNA and in other stages of the viral reproductive cycle. *J. Virol.* **45**: 397.

Beck, T.W. and R.L. Millette. 1982. *In vitro* transcription of herpes simplex virus genes. *J. Biol. Chem.* **257**: 12780.

Becker, Y., H. Dym, and I. Sarov. 1968. Herpes simplex virus DNA. *Virology* **36**: 184.

Ben-Porat, T. and R.A. Veach. 1980. Origin of replication of the DNA of a herpesvirus (pseudorabies). *Proc. Natl. Acad. Sci.* **77**: 172.

Ben-Porat, T., F.J. Rixon, and M.L. Blankenship. 1979. Analysis of the structure of the genome of pseudorabies virus. *Virology* **95**: 285.

Ben-Porat, T., R.A. Veach, and S. Ihara. 1983. Localization of the regions of homology between the genomes of herpes simplex virus, type 1 and pseudorabies virus. *Virology* **127**: 194.

Ben-Porat, T., A. Kaplan, B. Stehn, and A.S. Rubenstein. 1976. Concatemeric forms of intracellular herpesvirus DNA. *Virology* **69**: 547.

Bhat, R. and B. Thimmappaya. 1983. Two small RNAs encoded by Epstein-Barr virus can functionally substitute for virus-associated RNAs in the lytic growth of adenovirus 5. *Proc. Natl. Acad. Sci.* **80**: 4789.

Bieber, C.P., R.L. Heberling, S.W. Jamieson, P.E. Oyer, M. Cleary, R. Warnke, A. Saemundsen, G. Klein, W. Henle, and E.B. Stinson. 1984. Lymphoma in cardiac transplant recipients associated with cyclosporin A, prednisone, and antithymocyte globulin (ATG). In *Immune deficiency and cancer* (ed. D.T. Purtilo), p. 309. Plenum Press, New York.

Biggin, M.D., P.J. Farrell, and B.G. Barrel. 1984. Transcript and DNA sequence of the *Bam*HI-L region of B95-8 Epstein-Barr virus. *EMBO J.* **3**: 1083.

Blyth, W.A., D.A. Harbour, and T.J. Hill. 1980. Effect of acyclovir on recurrence of herpes simplex skin lesions in mice. *J. Gen. Virol.* **48**: 417.

Bodescot, M., B. Chambraud, P. Farrell, and M. Pericaudet. 1984. Spliced RNA from the IR1-U2 region of Epstein-Barr virus: Presence of an open reading frame for a repetitive polypeptide. *EMBO J.* **3**: 1913.

Bornkamm, G.W., J. Hudenwentz, U.K. Freese, and U. Zimber.

1982. Deletion of the nontransforming Epstein-Barr virus strain P3HR-1 causes fusion of the large internal repeat to the DS$_L$ region. *J. Virol.* **43**: 952.

Bornkamm, G.W., H. Delius, B. Fleckenstein, F.-J. Werner, and C. Mulder. 1976. Structure of herpesvirus *Saimiri* genome: Arrangement of heavy and light sequences in the *M* genome. *J. Virol.* **19**: 154.

Boshart, M., F. Weber, G. Jahn, K. Dorsch-Hasler, B. Fleckenstein, and W. Schaffner. 1985. A very strong enhancer is located upstream of an immediate-early gene of human cyto-megalovirus. *Cell* **41**: 521.

Buchman, T.G. and B. Roizman. 1978. Anatomy of bovine mammillitis DNA. II. Size and arrangements of the deoxynucleotide sequences. *J. Virol.* **27**: 239.

Buller, R.M.L., J.E. Janik, E.D. Sebring, and J.A. Rose. 1981. Herpes simplex virus types 1 and 2 completely help adenovirus-associated virus replication. *J. Virol.* **40**: 241.

Bzik, D.J., B.A. Fox, N.A. DeLuca, and S. Person. 1984. Nucleotide sequence specifying the glycoprotein gene, gB of herpes simplex virus type 1. *Virology* **137**: 185.

Cabral, G.A., R.J. Courtney, P.A. Schaffer, and F. Marciano-Cabral. 1980. Ultrastructural characterization of an early, nonstructural polypeptide of herpes simplex type 1. *J. Virol.* **33**: 1192.

Camacho, A. and P.G. Spear. 1978. Transformation of hamster embryo fibroblasts by a specific fragment of the herpes simplex virus genome. *Cell* **15**: 993.

Campbell, M.E.M., J.W. Palfreyman, and C.M. Preston. 1984. Identification of herpes simplex virus DNA sequences which encode a *trans*-acting polypeptide responsible for stimulation of immediate-early transcription. *J. Mol. Biol.* **180**: 1.

Cebrian, J., D. Bucchini, and P. Sheldrick. 1983. "Endless" viral DNA in cells infected with channel catfish virus. *J. Virol.* **46**: 405.

Cebrian, J., C. Kaschka-Dierich, N. Berthelot, and P. Sheldrick. 1982. Inverted repeat nucleotide sequences in the genomes of Marek disease virus and the herpesvirus of the turkey. *Proc. Natl. Acad. Sci.* **79**: 555.

Chartrand, P., C.S. Crumpacker, P.A. Schaffer, and N.M. Wilkie. 1980. Physical and genetic analysis of the herpes simplex virus DNA polymerase locus. *Virology* **103**: 311.

Chartrand, P., N.D. Stow, M.G. Timbury, and N.M. Wilkie. 1979. Physical mapping of PAA$^r$ mutations of herpes simplex virus type 1 and type 2 by intertypic marker rescue. *J. Virol.* **31**: 265.

Cheung, A. and E. Kieff. 1982. Long internal direct repeat in Epstein-Barr virus DNA. *J. Virol.* **44**: 286.

Chiou, H.C., S.K. Weller, and D.M. Coen. 1985. Mutations in the herpes simplex virus major DNA-binding protein gene leading to altered sensitivity to DNA polymerase inhibitors. *Virology* **145**: 213.

Cho, M.-S., G.W. Bornkamm, and H. zur Hausen. 1984a. Structure of defective DNA molecules in Epstein-Barr virus preparations from P3HR-1 cells. *J. Virol.* **51**: 199.

Cho, M.-S., L. Gissmann, and S.D. Hayward. 1984b. Epstein-Barr virus (P3HR-1) defective DNA codes for components of both the early antigen and viral capsid antigen complexes. *Virology* **137**: 9.

Cho, M.-S., K.-T. Jeang, and S.D. Hayward. 1985a. Localization of the coding region of an Epstein-Barr virus early antigen and inducible expression of this 60-kilodalton nuclear protein in transfected fibroblast cells lines. *J. Virol.* **56**: 852.

Cho, M.-S., G. Milman, and S.D. Hayward. 1985b. A second Epstein-Barr virus early antigen gene in *Bam*HI fragment M encodes a 48- to 50-kilodalton nuclear protein. *J. Virol.* **56**: 860.

Chou, J. and B. Roizman. 1985. Isomerization of the herpes simplex 1 genome: Identification of the *cis*-acting and recombination sites within the domain of the sequence. *Cell* **41**: 803.

Chousterman, S., M. Lacasa, and P. Sheldrick. 1979. Physical map of the channel catfish virus genome: Location of sites for restriction endonucleases *Eco*RI, *Hind*III, *Hpa*I, and *Xba*I. *J. Virol.* **31**: 73.

Ciufo, D.M. and G.S. Hayward. 1981. Tandem repeat defective DNA from the L-segment of herpes simplex virus genome. In *Herpesvirus DNA. Developments in molecular virology* (ed. Y. Becker), vol. 1, p. 107. Martinus-Nijhoff, Netherlands.

Clements, J.B., J. McLauchlan, and D.J. McGeoch. 1979. Orientation of herpes simplex type 1 immediate-early mRNAs. *Nucleic Acids Res.* **7:** 77.

Clements, J.B., R.J. Watson, and N.M. Wilkie. 1977. Temporal regulation of herpes simplex type 1 transcription: Localization of transcripts on the viral genome. *Cell* **12:** 275.

Colby, B.M., J.E. Shaw, G.B. Elion, and J.S. Pagano. 1980. Effect of acyclovir [9-(2-hydroxyethoxymethyl)guanine] on Epstein-Barr virus DNA replication. *J. Virol.* **34:** 560.

Conley, A.J., D.M. Knipe, P.C. Jones, and B. Roizman. 1981. Molecular genetics of herpes simplex virus. VII. Characterization of a temperature-sensitive mutant produced by *in vitro* mutagenesis and defective in DNA synthesis and accumulation of gamma-polypeptides. *J. Virol.* **37:** 191.

Cook, M.L. and J.G. Stevens. 1973. Pathogenesis of herpetic neuritis and ganglionitis in mice: Evidence for intra-axonal transport of infection. *Infect. Immun.* **7:** 272.

Cordingley, M.G., M.E.M. Campbell, and C.M. Preston. 1983. Functional analysis of a herpes simplex virus type 1 promoter; identification of far-upstream regulatory sequences. *Nucleic Acids Res.* **11:** 2347.

Countryman, J. and G. Miller. 1985. Activation of expression of latent Epstein-Barr herpesvirus after gene transfer with a small cloned subfragment of heterogeneous viral DNA. *Proc. Natl. Acad. Sci.* **82:** 4085.

Costa, R.H., K.G. Draper, T.J. Kelly, and E.K. Wagner. 1985a. An unusual spliced herpes simplex type 1 transcript with sequence homology to Epstein-Barr virus DNA. *J. Virol.* **54:** 317.

Costa, R.H., K.G. Draper, G.A. Devi-Rao, R.L. Thompson, and E.K. Wagner. 1985b. Virus-induced modification of the host cell is required for expression of the bacterial chloramphenicol acetyltransferase gene controlled by a late herpes simplex virus promoter (VP5). *J. Virol.* **56:** 19.

Costa, R.H., K.G. Draper, L. Banks, K.L. Powell, G. Cohen, R. Eisenberg, and E.K. Wagner. 1983. High resolution characterization of herpes simplex type 1 transcripts encoding alkaline exonuclease and a 50,000 dalton protein tentatively identified as a capsid protein. *J. Virol.* **48:** 591.

Dalrymple, M.A., D.J. McGeoch, A.J. Davison, and C.M. Preston. 1985. DNA sequence of the herpes simplex virus type 1 gene whose product is responsible for transcriptional activation of immediate-early promoters. *Nucleic Acids Res.* **13:** 7865.

Dalziel, R.G. and H.S. Marsden. 1984. Identification of two herpes simplex virus type 1-induced proteins (21K and 22K) which interact specifically with the *a*-sequence of herpes simplex virus DNA. *J. Gen. Virol.* **65:** 1467.

Dambaugh, T., K. Hennessy, L. Chamnankit, and E. Kieff. 1984. U2 region of Epstein-Barr virus DNA may encode Epstein-Barr nuclear antigen 2. *Proc. Natl. Acad. Sci.* **81:** 7632.

Davison, A.J. 1983. DNA sequence of the $U_S$ component of the varicella-zoster virus genome. *EMBO J.* **2:** 2203.

———. 1984. Structure of the genome termini of varicella-zoster virus. *J. Gen. Virol.* **65:** 1969.

Davison, A.J. and J.E. Scott. 1985. DNA sequence of the major inverted repeat in the varicella-zoster virus genome. *J. Gen. Virol.* **66:** 207.

Davison, A.J. and N.M. Wilkie. 1981. Nucleotide sequences of the joint between the L and S segments of herpes simplex virus types 1 and 2. *J. Gen. Virol.* **55:** 315.

———. 1983a. Inversion of the two segments of the herpes simplex virus genome in intertypic recombinants. *J. Gen. Virol.* **64:** 1.

———. 1983b. Either orientation of the L segment of the herpes simplex virus genome may participate in the production of viable intertypic recombinants. *J. Gen. Virol.* **64:** 247.

———. 1983c. Location and orientation of homologous sequences in the genomes of five herpesviruses. *J. Gen. Virol.* **64:** 1927.

Davison, M.-J., V.G. Preston, and D.J. McGeoch. 1984. Determination of the sequence alteration in the DNA of the herpes simplex virus type 1 temperature-sensitive mutant *ts* K. *J. Gen. Virol.* **65:** 859.

Delius, H. and G.W. Bornkamm. 1978. Heterogeneity of Epstein-Barr virus. III. Comparison of a transforming and a non-transforming virus by partial denaturation mapping of their DNAs. *J. Virol.* **27:** 81.

Delius, H. and J.B. Clements. 1976. A partial denaturation map of herpes simplex virus type 1 DNA: Evidence for inversion of the unique DNA regions. *J. Gen. Virol.* **33:** 125.

DeLuca, N.A. and P.A. Schaffer. 1985. Activation of immediate-early, early, and late promoters by temperature-sensitive and wild-type forms of herpes simplex virus type 1 protein ICP4. *Mol. Cell. Biol.* **5:** 1997.

DeLuca, N.A., M.A. Courtney, and P.A. Schaffer. 1984. Temperature sensitive mutants in herpes simplex virus type 1 ICP4 permissive for early gene expression. *J. Virol.* **52:** 767.

DeLuca, N.A., A.M. McCarthy, and P.A. Schaffer. 1985. Isolation and characterization of deletion mutants of herpes simplex virus type 1 in the gene encoding immediate-early regulatory protein ICP4. *J. Virol.* **56:** 558.

DeMarchi, J.M. 1981. Human cytomegalovirus DNA: Restriction enzyme cleavage maps and map locations for immediate-early, early, and late RNAs. *Virology* **114:** 23.

———. 1983. Post-transcriptional control of human cytomegalovirus gene expression. *Virology* **124:** 390.

Dennis, D. and J.R. Smiley. 1984. Transactivation of a late herpes simplex virus promoter. *Mol. Cell. Biol.* **4:** 544.

Desgranges, C., H. Wolf, G. de Thé, K. Shanmugaratnam, N. Cammoun, R. Ellouz, G. Klein, K. Lennert, N. Munoz, and H. zur Hausen. 1975. Nasopharyngeal carcinoma. X. Presence of Epstein-Barr genomes in separated epithelial cells of tumors in patients from Singapore, Tunisia and Kenya. *Int. J. Cancer* **16:** 7.

de Thé, G., A. Geser, N.E. Day, P.M. Tukei, E.H. Williams, D.P. Beri, P.G. Smith, A.G. Dean, G.W. Bornkamm, P. Feorinio, and W. Henle. 1978. Epidemiological evidence for causal relationship between Epstein-Barr virus and Burkitt's lymphoma from Ugandan prospective study. *Nature* **274:** 756.

Dixon, R.A.F. and P.A. Schaffer. 1980. Fine-structure mapping and functional analysis of temperature-sensitive mutants in the gene encoding the herpes simplex virus type 1 immediate early protein VP175. *J. Virol.* **36:** 189.

Dowbenko, D.J. and L.A. Lasky. 1984. Extensive homology between the herpes simplex virus type 2 glycoprotein F gene and the herpes simplex virus type 1 glycoprotein C gene. *J. Virol.* **52:** 154.

Draper, K.G., R.J. Frink, and E.K. Wagner. 1982. Detailed characterization of an apparently unspliced beta-herpes simplex virus type 1 gene mapping in the interior of another. *J. Virol.* **43:** 1123.

Duff, R. and F. Rapp. 1971. Properties of hamster embryo fibroblasts transformed in vitro after exposure to ultraviolet-irradiated herpes simplex virus type 2. *J. Virol.* **8:** 469.

Dumas, A.M., J.L.M.C. Geelen, M.W. Weststrate, P. Wertheim, and J. van der Noordaa. 1981. *Xba*I, *Pst*I and *Bgl*II restriction enzyme maps of the two orientations of the varicella-zoster virus genome. *J. Virol.* **39:** 390.

Durst, M., L. Gissman, H. Ikenberg, and H. zur Hausen. 1983. A papillomavirus DNA from a cervical carcinoma and its prevalence in cancer biopsy samples from different geographic regions. *Proc. Natl. Acad. Sci.* **80:** 3812.

Dutia, B.M. 1983. Ribonucleotide reductase specified by herpes simplex virus has a virus-specified constituent. *J. Gen. Virol.* **64:** 513.

Easton, A.J. and J.B. Clements. 1980. Temporal regulation of herpes simplex virus type 2 transcription and characterization of virus immediate early mRNAs. *Nucleic Acids Res.* **8:** 2627.

Ebeling, A., G. Keil, B. Nowak, B. Fleckenstein, N. Berthelot, and P. Sheldrick. 1983. Genome structure and virion polypeptides of the primate herpesviruses *Herpesvirus aotus* type 1 and 3: Comparison with human cytomegalovirus. *J. Virol.* **45:** 715.

Ecker, J.R. and R.W. Hyman. 1982. Varicella zoster virus DNA exists as two isomers. *Proc. Natl. Acad. Sci.* **79:** 156.

Eglin, R.P., F. Sharp, A.B. MacLean, J.C.M. MacNab, J.B. Clements, and N.M. Wilkie. 1981. Detection of RNA complementary to herpes simplex virus DNA in human cervical squamous cell neoplasms. *Cancer Res.* **41:** 3597.

Eisenberg, S.P., D.M. Coen, and S.L. McKnight. 1985. Promoter domains required for expression of plasmid-borne copies of the herpes simplex virus thymidine kinase gene in virus-infected mouse fibroblasts and microinjected frog oocytes. *Mol. Cell. Biol.* **5:** 1940.

El Kareh, A., S. Silverstein, and J. Smiley. 1984. Control of expression of the herpes simplex virus thymidine kinase gene in biochemically transformed cells. *J. Gen. Virol.* **65:** 19.

El Kareh, A., A.J.M. Murphy, T. Fichter, A. Efstratiadis, and S. Silverstein. 1985. "Transactivation" control signals in the promoter of the herpesvirus thymidine kinase gene. *Proc. Natl. Acad. Sci.* **82:** 1002.

Enquist, L.W., M.J. Madden, P. Schiop-Stansly, and G.F. Vande Woude. 1979. Cloning of herpes simplex type 1 DNA fragments in a bacteriophage lambda vector. *Science* **203:** 541.

Everett, R.D. 1984a. A detailed analysis of an HSV-1 early promoter: Sequences involved in trans-activation by viral immediate-early gene products are not early-gene specific. *Nucleic Acids Res.* **12:** 3037.

————. 1984b. Transactivation of transcription by herpesvirus products: Requirement for two HSV-1 immediate-early polypeptides for maximum activity. *EMBO J.* **3:** 3135.

————. 1985. Activation of cellular promoters during herpesvirus infection of biochemically transformed cells. *EMBO J.* **4:** 1973.

Farrell, P.J., A. Bankier, C. Seguin, P. Deininger, and B.G. Barrell. 1983. Latent and lytic cycle promoters of Epstein-Barr virus. *EMBO J.* **2:** 1331.

Feldman, L.T., M.J. Imperiale, and J.R. Nevins. 1982. Activation of early adenovirus transcription by the herpesvirus immediate-early gene: Evidence for a common cellular control factor. *Proc. Natl. Acad. Sci.* **79:** 4952.

Fennewald, S., V. van Santen, and E. Kieff. 1984. Nucleotide sequence of an mRNA transcribed in latent growth-transforming virus infection indicates that it may encode a membrane protein. *J. Virol.* **51:** 411.

Fialkow, P.J., G. Klein, S.M. Gartler, and P. Clifford. 1970. Clonal origin for individual Burkitt tumors. *Lancet* I: 384.

Field, H.J. and E. DeClercq. 1981. Effects of oral treatment with acyclovir and bromovinyldeoxyuridine on the establishment and maintenance of latent herpes simplex infection in mice. *J. Gen. Virol.* **56:** 259.

Fields, B.N., ed. 1985. *Virology.* Raven Press, New York.

Fleckenstein, B., G. Bornkamm, and H. Ludwig. 1975. Repetitive sequences in complete and defective genomes of *Herpesvirus saimiri. J. Virol.* **15:** 398.

Frame, M.C., H.S. Marsden, and B.M. Dutia. 1985. The ribonucleotide reductase induced by herpes simplex virus type 1 involves minimally a complex of two polypeptides (136K and 38K). *J. Gen. Virol.* **66:** 1581.

Freeman, M.J. and K.L. Powell. 1982. DNA-binding properties of a herpes simplex virus immediate-early protein. *J. Virol.* **44:** 1084.

Freese, U.K., G. Laux, J. Hudewentz, E. Schwarz, and G.W. Bornkamm. 1983. Two distinct clusters of partially homologous small repeats of Epstein-Barr virus are transcribed upon induction of an abortive or lytic cycle of the virus. *J. Virol.* **48:** 731.

Frenkel, N., B. Roizman, E. Cassai, and A. Nahmias. 1972. DNA fragment of herpes simplex virus type 2 and its transcription in human cervical cancer tissue. *Proc. Natl. Acad. Sci.* **69:** 3784.

Frenkel, N., H. Locker, W. Batterson, G. Hayward, and B. Roizman. 1976. Anatomy of herpes simplex virus DNA. VI.

Defective DNA originates from the "S" component. *J. Virol.* **20:** 527.

Frenkel, N., R.J. Jacob, R.W. Honess, G.S. Hayward, H. Locker, and B. Roizman. 1975. Anatomy of herpes simplex virus DNA. III. Characterization of defective DNA molecules and biological properties of virus populations containing them. *J. Virol.* **16:** 163.

Friedmann, A. and Y. Becker. 1977. Circular and circular-linear DNA molecules of herpes simplex virus. *J. Gen. Virol.* **37:** 205.

Frink, R.J., K.G. Draper, and E.K. Wagner. 1981. Uninfected cell polymerase efficiently transcribes early but not late herpes simplex virus type 1 mRNA. *Proc. Natl. Acad. Sci.* **78:** 6139.

Galloway, D.A. and J.K. McDougall. 1981. Transformation of rodent cells by a cloned DNA fragment of herpes simplex virus type 2. *J. Virol.* **38:** 749.

————. 1983. The oncogenic potential of herpes simplex viruses: Evidence for a "hit-and-run" mechanism. *Nature* **301:** 21.

Galloway, D.A. and M.A. Swain. 1984. Organization of the left-hand end of the early simplex virus type 2 *Bgl*II-N fragment. *J. Virol.* **49:** 724.

Galloway, D.A., J.A. Nelson, and J.K. McDougall. 1984. Small fragments of herpesvirus DNA with transforming activity contain insertion sequence-like structures. *Proc. Natl. Acad. Sci.* **81:** 4736.

Geelen, J.L.M.C., C. Walig, P. Wertheim, and J. van der Noordaa. 1978. Human cytomegalovirus DNA. I. Molecular weight and infectivity. *J. Virol.* **26:** 813.

Gelman, I.H. and S. Silverstein. 1985. Identification of immediate early genes from herpes simplex virus that transactivate the virus thymidine kinase gene. *Proc. Natl. Acad. Sci.* **82:** 5265.

Gerdes, J.C., H.S. Marsden, M.L. Cook, and J.G. Stevens. 1979. Acute infection of differentiated neuroblastoma cells by latency-positive and latency-negative herpes simplex virus *ts* mutants. *Virology* **94:** 430.

Gibbs, J.S., H.C. Chiou, J.D. Hall, D.W. Mount, M. Retondo, S.K. Weller, and D.M. Coen. 1985. Sequence and mapping analyses of the herpes simplex virus DNA polymerase gene predict a C-terminal substrate binding domain. *Proc. Natl. Acad. Sci.* **82:** 7969.

Gibson, T., P. Stockwell, M. Ginsburg, and B. Barrell. 1984. Homology between two EBV early genes and HSV ribonucleotide reductase and 38K genes. *Nucleic Acids Res.* **12:** 5087.

Gibson, W. 1981. Immediate-early proteins of human cytomegalovirus strains AD169, Davis and Towne differ in electrophoretic mobility. *Virology* **112:** 350.

Gibson, W. and B. Roizman. 1972. Proteins specified by herpes simplex virus. VIII. Characterization and composition of multiple capsid forms of subtypes 1 and 2. *J. Virol.* **10:** 1044.

Given, D. and E. Kieff. 1978. DNA of Epstein-Barr virus. IV. Linkage map of restriction enzyme fragments of the B95-8 and W91 strains of Epstein-Barr virus. *J. Virol.* **28:** 524.

————. 1979. DNA of Epstein-Barr virus. VI. Mapping of the internal tandem reiteration. *J. Virol.* **31:** 315.

Given, D., D. Yee, K. Griem, and E. Kieff. 1979. DNA of Epstein-Barr virus. V. Direct repeats of the ends of Epstein-Barr virus DNA. *J. Virol.* **30:** 852.

Godowski, P.J. and D.M. Knipe. 1983. Mutations in the major DNA binding gene of herpes simplex virus type 1 result in increased levels of viral gene expression. *J. Virol.* **47:** 478.

————. 1985. Identification of a herpes simplex virus function that represses late gene expression from parental virus genomes. *J. Virol.* **55:** 357.

Goldstein, L.C., J. McDougall, R. Hackman, J.D. Meyers, E.D. Thomas, and R.C. Nowinski. 1982. Monoclonal antibodies to cytomegalovirus: Rapid identification of clinical isolates and preliminary use in diagnosis of cytomegalovirus pneumonia. *Infect. Immun.* **38:** 273.

Gonczol, E., P.W. Andrews, and S.A. Plotkin. 1984. Cytomegalovirus replicates in differentiated but not in undifferentiated human embryonal carcinoma cells. *Science* **224:** 159.

Goodheart, C.R. and G. Plummer. 1975. The densities of herpesvirus DNAs. *Prog. Med. Virol.* **19:** 324.

Graessmann, A., H. Wolf, and G.W. Bornkamm. 1980. Expression of Epstein-Barr virus genes in different cell types after microinjection of viral DNA. *Proc. Natl. Acad. Sci.* **77:** 433.

Grafstrom, R.H., J.C. Alwine, W.L. Steinhart, C.W. Hill, and R.W. Hyman. 1975. The terminal repetition of herpes simplex virus DNA. *Virology* **67:** 144.

Graham, F.L., G. Veldhuisen, and N.M. Wilkie. 1973. Infectious herpesvirus DNA. *Nat. New Biol.* **245:** 265.

Gray, C.P. and H.C. Kaerner. 1984. Sequence of the putative origin of replication in the $U_L$ region of herpes simplex virus type 1 ANG DNA. *J. Gen. Virol.* **65:** 2109.

Green, M.R., R. Treisman, and T. Maniatis. 1983. Transcriptional activation of cloned human β-globin genes by viral immediate-early gene products. *Cell* **35:** 137.

Gross, T.G. and D.J. Volsky. 1984. Infection of mouse lymphocytes by Epstein-Barr virus. II. Stimulation of cellular DNA synthesis by EBV in the absence of EBNA induction and cell transformation. *Virology* **133:** 211.

Gussander, E. and A. Adams. 1984. Electron microscopic evidence for replication of circular Epstein-Barr virus genomes in latently infected Raji cells. *J. Virol.* **52:** 549.

Halliburton, I.W. 1980. Review article. Intertypic recombinants of herpes simplex viruses. *J. Gen. Virol.* **48:** 1.

Hanto, D.W., G. Frizzera, K.J. Gajl-Peczalska, D.T. Purtilo, and R.L. Simmons. 1984. Lymphoproliferative diseases in renal allograft recipients. In *Immune deficiency and cancer* (ed. D.T. Purtilo), p. 321. Plenum Press, New York.

Hayward, G.S. and G.R. Reyes. 1983. Biochemical aspects of transformation by herpes simplex viruses. *Adv. Virol. Oncol.* **3:** 271.

Hayward, G.S., N. Frenkel, and B. Roizman. 1975a. Anatomy of herpes simplex virus DNA. I. Strain differences and heterogeneity in the locations of restriction endonuclease cleavage sites. *Proc. Natl. Acad. Sci.* **72:** 1768.

Hayward, G.S., R.J. Jacob, S.C. Wadsworth, and B. Roizman. 1975b. Anatomy of herpes simplex virus DNA. IV. Evidence for four populations of molecules that differ in the relative orientations of their long and short components. *Proc. Natl. Acad. Sci.* **72:** 4243.

Hayward, G.S., R. Ambinder, D. Ciufo, S.D. Hayward, and R.L. LaFemina. 1984. Structural organization of human herpesvirus DNA molecules. *J. Invest. Dermatol.* **83:** 293.

Hayward, S.D. and E.D. Kieff. 1976. Epstein-Barr virus-specific RNA. I. Analysis of viral RNA in cellular extracts and in the polyribosomal fraction of permissive and non-permissive lymphoblastoid cell lines. *J. Virol.* **18:** 518.

———. 1977. DNA of Epstein-Barr virus. II. Comparison of the molecular weights of restriction endonuclease fragments of the DNA of Epstein-Barr virus strains and identification of end fragments of the B95-8 strain. *J. Virol.* **23:** 421.

Hayward, S.D., S.G. Lazarowitz, and G.S. Hayward. 1982. Organization of the Epstein-Barr virus DNA molecule. II. Fine mapping of the boundaries of the internal repeat cluster of B95-8 and identification of additional small tandem repeats adjacent to the HR-1 deletion. *J. Virol.* **43:** 201.

Hayward, S.D., L. Nogee, and G.S. Hayward. 1980. Organization of repeated regions within the Epstein-Barr virus DNA molecule. *J. Virol.* **33:** 507.

Hearing, J.C., J.-C. Nicolas, and A.J. Levine. 1984. Identification of Epstein-Barr virus sequences that encode a nuclear antigen expressed in latently infected lymphocytes. *Proc. Natl. Acad. Sci.* **81:** 4373.

Heller, M., V. van Santen, and E. Kieff. 1982. Simple repeat sequence in Epstein-Barr virus DNA is transcribed in latent and productive infections. *J. Virol.* **44:** 311.

Heller, M., E. Flemington, E. Kieff, and P. Deininger. 1985. Repeat arrays in cellular DNA related to the Epstein-Barr virus IR3 repeat. *Mol. Cell. Biol.* **5:** 457.

Henderson, A., S. Ripley, M. Heller, and E. Kieff. 1983. Chromosome site for Epstein-Barr virus DNA in a Burkitt tumor cell line and in lymphocytes growth-transformed *in vitro*. *Proc. Natl. Acad. Sci.* **80:** 1987.

Henle, W., G. Henle, H.-C. Ho, P. Burtin, Y. Cachin, P. Clifford, A. De Schryver, G. de Thé, V. Diehl, and G. Klein. 1970. Antibodies to Epstein-Barr virus in nasopharyngal carcinoma, other head and neck neoplasms, and control groups. *J. Natl. Cancer Inst.* **44:** 225.

Hennessy, K. and E. Kieff. 1983. One of two Epstein-Barr virus nuclear antigens contains a glycine-alanine copolymer domain. *Proc. Natl. Acad. Sci.* **80:** 5665.

———. 1985. A second nuclear protein is encoded by Epstein-Barr virus in latent infection. *Science* **227:** 1238.

Hennessy, K., S. Fennewald, and E. Kieff. 1985. A third viral nuclear protein in lymphoblasts immortalized by Epstein-Barr virus. *Proc. Natl. Acad. Sci.* **82:** 5944.

Hennessy, K., M. Heller, V. van Santen, and E. Kieff. 1983. Simple repeat array in Epstein-Barr virus DNA encodes part of the Epstein-Barr virus nuclear antigen. *Science* **220:** 1396.

Hennessy, K., S. Fennewald, M. Hummel, T. Cole, and E. Kieff. 1984. A membrane protein encoded by Epstein-Barr virus in latent growth-transforming infection. *Proc. Natl. Acad. Sci.* **81:** 7207.

Henry, B.E., R.A. Robinson, S.A. Dauenhauer, S.S. Atherton, G.S. Hayward, and D.J. O'Callaghan. 1981. Structure of the genome of equine hepesvirus type 1. *Virology* **115:** 97.

Herz, C. and B. Roizman. 1983. The alpha-promoter regulator-ovalbumin chimeric gene resident in human cells is regulated like the authentic alpha-4 gene after infection with herpes simplex virus 1 mutants in alpha-4 gene. *Cell* **33:** 145.

Heston, L., M. Rabson, and G. Miller. 1982. New Epstein-Barr virus variants from cellular subclones of P3J-HR-1 Burkitt's lymphoma. *Nature* **295:** 160.

Hill, T.J., W.A. Blyth, and D.A. Harbour. 1983. Recurrence of herpes simplex in the mouse requires an intact nerve supply to the skin. *J. Gen. Virol.* **64:** 2763.

Hill, T.M., R.R. Sinden, and J.R. Sadler. 1983. Herpes simplex virus types 1 and 2 induce shutoff of host protein synthesis by different mechanisms in Friend erythroleukemia cells. *J. Virol.* **45:** 241.

Hoffman, P.J. 1981. Mechanism of degradation of duplex DNA by the DNase induced by herpes simplex virus. *J. Virol.* **38:** 1005.

Holland, L.E., K.P. Anderson, C.J. Shipman, and E.K. Wagner. 1980. Viral DNA synthesis is required for the efficient expression of specific herpes simplex virus type 1 mRNA species. *Virology* **101:** 10.

Honess, R.W. 1984. Herpes simplex and the "The herpes complex:" Diverse observations and a unifying hypothesis. *J. Gen. Virol.* **65:** 2077.

Honess, R.W. and B. Roizman. 1974. Regulation of herpesvirus macromolecular synthesis. I. Cascade regulation of the synthesis of three groups of viral proteins. *J. Virol.* **14:** 8.

———. 1975. Regulation of herpesvirus macromolecular synthesis: Sequential transition of polypeptide synthesis requires functional virus polypeptides. *Proc. Natl. Acad. Sci.* **72:** 1276.

Huang, D.P., H.C. Ho, W. Henle, G. Henle, D. Saw, and M. Lui. 1978. Presence of EBNA in nasopharyngeal carcinoma and control patient tissues related to EBV serology. *Int. J. Cancer* **22:** 266.

Hudewentz, J., H. Delius, U.K. Freese, U. Zimber, and G.W. Bornkamm. 1982. Two distant regions of the Epstein-Barr virus genomes with sequence homologies have the same orientation and involve small tandem repeats. *EMBO J.* **1:** 21.

Hudson, G.S., P.J. Farrell, and B.G. Barrell. 1985. Two related but differentially expressed potential membrane proteins encoded by the *Eco*RI Dhet region of Epstein-Barr virus B95-8. *J. Virol.* **53:** 528.

Hummel, M., D. Thorley-Lawson, and E. Kieff. 1984. An Epstein-Barr virus DNA fragment encodes messages for the two major envelope glycoproteins (gp 350/300 and gp 220/200). *J. Virol.* **49:** 413.

Hyman, R.W., J.R. Ecker, and R.B. Tenser. 1983. Varicella-Zoster virus RNA in human trigeminal ganglia. *Lancet* II: 814.

Imperiale, M.J., L.T. Feldman, and J.R. Nevins. 1983. Activation of gene expression by adenovirus and herpesvirus regulatory genes acting in *trans* and by a *cis*-acting adenovirus enhancer element. *Cell* **35:** 127.

Jacob, R.J., L.S. Morse, and B. Roizman. 1979. Anatomy of herpes simplex virus DNA. XII. Accumulation of head-to-tail concatemers in nuclei of infected cells and their role in the generation of the four isomeric arrangements of viral DNA. *J. Virol.* **29:** 448.

Jahn, G., E. Knust, H. Schmolla, T. Sarre, J.A. Nelson, J.K. McDougall, and B. Fleckenstein. 1984. Predominant immediate-early transcripts of human cytomegalovirus AD 169. *J. Virol.* **49:** 363.

Jean, J.-H., M.L. Blankenship, and T. Ben-Porat. 1977. Replication of herpesvirus DNA. I. Electron microscopic analysis of replicative structures. *Virology* **79:** 281.

Jeang, K.-T. 1984. "Molecular analysis of cytomegalovirus Colburn gene expression." Ph.D. thesis, Johns Hopkins University, Baltimore, Maryland.

Jeang, K.-T. and W. Gibson. 1980. A cycloheximide-enhanced protein in cytomegalovirus-infected cells. *Virology* **107:** 362.

Jeang, K.-T. and G.S. Hayward. 1983a. A cytomegalovirus DNA sequence containing tracts of tandemly repeated CA dinucleotides hybridizes to highly repetitive dispersed elements in mammalian cell genomes. *Mol. Cell. Biol.* **3:** 1389.

Jeang, K.-T. and S.D. Hayward. 1983b. Part of the template for an abundant TPA-inducible transcript containing multiple 125 bp repeats is deleted in the non-transforming HR-1 isolate of Epstein-Barr virus. *J. Virol.* **48:** 135.

Jeang, K.-T., G. Chin, and G.S. Hayward. 1982. Characterization of cytomegalovirus immediate-early genes. I. Nonpermissive rodent cells overproduce the IE94K protein from CMV (Colburn). *Virology* **121:** 393.

Jeang, K.-T., M.-S. Cho, and G.S. Hayward. 1984. Abundant constitutive expression of the immediate-early 94K protein from cytomegalovirus (Colburn) in a DNA transfected mouse cell line. *Mol. Cell. Biol.* **4:** 2214.

Jones, K.A. and R. Tjian. 1985. Sp1 binds to promoter sequences and activates herpes simplex virus "immediate-early" gene transcription *in vitro*. *Science* **317:** 179.

Jones, K.A., K.R. Yamamoto, and R. Tjian. 1985. Two distinct transcription factors bind to the HSV thymidine kinase promoter *in vitro*. *Cell* **42:** 559.

Jones, M.D. and B.E. Griffin. 1983. Clustered repeat sequences in the genome of Epstein-Barr virus. *Nucleic Acids Res.* **12:** 3919.

Jones, M.D., L. Foster, T. Sheedy, and B.E. Griffin. 1984. The EB virus genome in Daudi Burkitt's lymphoma cells has a deletion similar to that observed in a non-transforming strain (P3HR-1) of the virus. *EMBO J.* **3:** 813.

Jones, P.C., G.S. Hayward, and B. Roizman. 1977. Anatomy of herpes simplex virus DNA. VII. Alpha-RNA is homologous to noncontiguous sites in both the *L* and *S* components of viral DNA. *J. Virol.* **21:** 268.

Jongeneel, C.V. and S.L. Bachenheimer. 1980. Replication of herpes simplex virus type 1 DNA in permeabilized infected cells. *Nucleic Acids Res.* **8:** 1661.

Kaerner, H.C., I.B. Maichele, and C.H. Schroder. 1979. Origin of different classes of defective HSV-1 Angellotti DNA. *Nucleic Acids Res.* **6:** 1467.

Kaschka-Dierich, C., L. Falk, G. Bjursell, A. Adams, and T. Lindahl. 1977. Human lymphoblastoid cell lines derived from individuals without lymphoproliferative disease contain the same latent forms of Epstein-Barr virus DNA as those found in tumor cells. *Int. J. Cancer* **20:** 173.

Kaschka-Dierich, C., A. Adams, T. Lindahl, G.W. Bornkamm, G. Bjursell, G. Klein, B.C. Giovanella, and S. Singh. 1976. Intracellular forms of Epstein-Barr virus DNA in human tumor cells *in vivo*. *Nature* **260:** 302.

Kennedy, P.G.E., S.A. Al-Saadi, and G.B. Clements. 1983. Reactivation of latent herpes simplex virus from dissociated identified dorsal root ganglion cells in culture. *J. Gen. Virol.* **64:** 1629.

Kieff, E.D., S.L. Bachenheimer, and B. Roizman. 1971. Size, composition and structure of the deoxyribonucleic acid of herpes simplex virus subtypes 1 and 2. *J. Virol.* **8:** 125.

Kieff, E.D., B. Hoyer, S. Bachenheimer, and B. Roizman. 1972. Genetic relatedness of type 1 and type 2 herpes simplex viruses. *J. Virol.* **9:** 738.

Kilpatrick, B.A. and E.-S. Huang. 1977. Human cytomegalovirus genome: Partial denaturation map and organization of genome sequences. *J. Virol.* **24:** 261.

Kinchington, P.R., W.C. Reinhold, T.A. Casey, S.E. Straus, J. Hay, and W.T. Ruyechan. 1985. Inversion and circularization of the varicella-zoster virus genome. *J. Virol.* **56:** 194.

King, W., T. Dambaugh, M. Heller, J. Dowling, and E.D. Kieff. 1982. Epstein-Barr virus DNA. XII. A variable region of the Epstein-Barr virus genome is included in the P3HR-1 deletion. *J. Virol.* **43:** 979.

Kintner, C.R. and B. Sugden. 1979. The structure of the termini of the DNA of Epstein-Barr virus. *Cell* **17:** 661.

———. 1981a. Conservation and progressive methylation of Epstein-Barr viral DNA sequences in transformed cells. *J. Virol.* **38:** 305.

———. 1981b. Identification of antigenic determinants unique to the surfaces of cells transformed by Epstein-Barr virus. *Nature* **294:** 458.

Klein, G., B.C. Giovanella, T. Lindahl, P.J. Fialkow, S. Singh, and J.S. Stehlin. 1974. Direct evidence for the presence of Epstein-Barr virus DNA and nuclear antigen in malignant epithelial cells from patients with poorly differentiated carcinoma of the nasopharynx. *Proc. Natl. Acad. Sci.* **71:** 4737.

Knipe, D.M., W.T. Ruyechan, B. Roizman, and I.W. Halliburton. 1978. Molecular genetics of herpes simplex virus: Demonstration of regions of obligatory and nonobligatory identity within diploid regions of the genome by sequence replacement and insertion. *Proc. Natl. Acad. Sci.* **75:** 3896.

Knipe, D.M., W. Batterson, C. Nosal, B. Roizman, and A. Buchan. 1981. Molecular genetics of herpes simplex virus. VI. Characterization of a temperature-sensitive mutant defective in the expression of all early viral gene products. *J. Virol.* **38:** 539.

Knopf, K.W. 1979. Properties of herpes simplex virus DNA polymerase and characterization of its associated exonuclease activity. *Eur. J. Biochem.* **98:** 231.

Knopf, K.W., G. Strauss, A. Ott-Hartman, R. Schatten, and H.C. Kaerner. 1983. Herpes simplex virus defective genomes; structures of HSV-1 (ANG) defective DNA in class II and encoded polypeptides. *J. Gen. Virol.* **64:** 2455.

Kozak, M. and B. Roizman. 1975. RNA synthesis in cells infected with herpes simplex virus. IX. Evidence for accumulation of abundant symmetric transcripts in nuclei. *J. Virol.* **15:** 36.

Kristie, T.M. and B. Roizman. 1984. Separation of sequences defining basal expression from those conferring alpha-gene recognition within the regulatory domains of herpes simplex virus 1 alpha-genes. *Proc. Natl. Acad. Sci.* **81:** 4065.

LaFemina, R.L. and G.S. Hayward. 1980. Structural organization of the DNA molecules from human cytomegalovirus. *ICN-UCLA Symp. Mol. Cell. Biol.* **18:** 39.

———. 1983. Replicative forms of human cytomegalovirus DNA with joined termini are found in permissively infected human cells but not in non-permissive Balb/c-3T3 mouse cells. *J. Gen. Virol.* **64:** 373.

Lang, J.C., D.A. Spandidos, and N.M. Wilkie. 1984. Transcriptional regulation of a herpes simplex virus immediate early gene is mediated through an enhancer-type sequence. *EMBO J.* **3:** 389.

Lasky, L.A. and D.J. Dowbenko. 1984. DNA sequence analysis of the type common-glycoprotein D genes of herpes simplex virus type 1 and 2. *DNA* **3:** 23

LaThangue, N.B., K. Shriver, C. Dawson, and W.L. Chan.

1984. Herpes simplex virus infection causes the accumulation of a heat-shock protein. *EMBO J.* **3**: 267.

Laux, G., U.K. Freese, and G.W. Bornkamm. 1985. Structure and evolution of two related transcription units of Epstein-Barr virus carrying small tandem repeats. *J. Virol.* **56**: 987.

Lee, C.K. and D.M. Knipe. 1983. Thermolabile *in vivo* DNA binding activity associated with a protein encoded by mutants of herpes simplex type 1. *J. Virol.* **46**: 909.

Lee, G.T.-Y., K.L. Pogue-Geile, L. Pereira, and P.G. Spear. 1982. Expression of herpes simplex virus glycoprotein C from a DNA fragment inserted into the thymidine kinase gene of this virus. *Proc. Natl. Acad. Sci.* **79**: 6612.

Leiden, J.M., R. Buttyan, and P. Spear. 1976. Herpes simplex virus gene expression in transformed cells. I. Regulation of the viral thymidine kinase gene in transformed L cells by products of superinfecting virus. *J. Virol.* **20**: 413.

Lenoir, G.M., J.L. Preud'homme, A. Berheim, and R. Berger. 1982. Correlation between immunoglobulin light chain expression and variant translocation in Burkitt's lymphoma. *Nature* **298**: 474.

Lerner, M.R., N.C. Andrews, G. Miller, and J.A. Steitz. 1981. Two small RNAs encoded by Epstein-Barr virus and complexed with protein are precipitated by antibodies from patients with systemic lupus erythematosus. *Proc. Natl. Acad. Sci.* **78**: 805.

Leung, W.-C., K. Dimock, J.R. Smiley, and S. Bacchetti. 1980. Herpes simplex virus thymidine kinase transcripts are absent from both nucleus and cytoplasm during infection in the presence of cycloheximide. *J. Virol.* **36**: 361.

Lindahl, T., G. Klein, B.M. Reedman, B. Johansson, and S. Singh. 1974. Relationship between Epstein-Barr virus (EBV) DNA and the EBV-determined nuclear antigen (EBNA) in Burkitt lymphoma biopsies and other lymphoproliferative malignancies. *Int. J. Cancer* **13**: 764.

Lindahl, T., A. Adams, B. Bjursell, G.W. Bornkamm, C. Kaschka-Dierich, and U. Jehn. 1976. Covalently closed circular duplex DNA of EBV in a human lymphoid cell line. *J. Mol. Biol.* **102**: 511.

Locker, H. and N. Frenkel. 1979. *Bam*I, *Kpn*I and *Sal*I restriction enzyme maps of the DNAs of herpes simplex virus strains Justin and F: Occurrence of heterogeneities in defined regions of the viral DNA. *J. Virol.* **32**: 429.

Locker, H., N. Frenkel, and I. Halliburton. 1982. Structure and expression of class II defective herpes simplex virus genomes encoding infected cell polypeptide number 8. *J. Virol.* **43**: 574.

Ludwig, H., H.G. Haines, N. Biswal, and D. Benyesh-Melnick. 1972. The characterization of varicella-zoster virus DNA. *J. Gen. Virol.* **14**: 111.

Mackem, S. and B. Roizman. 1980. Regulation of herpesvirus macromolecular synthesis: Transcription-initiation sites and domains of alpha-genes. *Proc. Natl. Acad. Sci.* **77**: 7122.

———. 1981. Regulation of herpesvirus macromolecular synthesis: Temporal order of transcription of alpha genes is not dependent on the stringency of inhibition of protein synthesis. *J. Virol.* **40**: 319.

———. 1982a. Regulation of alpha-genes of herpes simplex virus: The alpha-27 gene promoter-thymidine kinase chimera is positively regulated in converted L cells. *J. Virol.* **43**: 1015.

———. 1982b. Differentiation between alpha-promoter and regulatory regions of herpes simplex virus 1: The functional domains and sequence of a movable alpha-regulator. *Proc. Natl. Acad. Sci.* **79**: 4917.

———. 1982c. Structural features of the herpes simplex virus alpha-gene 4, 0, and 27 promoter-regulatory sequences which confer alpha-regulation on chimeric thymidine kinase genes. *J. Virol.* **44**: 939.

Maitland, N.J., J.H. Kinross, A. Busuttil, S.M. Ludgate, G.E. Smart, and K.W. Jones. 1981. The detection of DNA tumor virus-specific RNA sequences in abnormal human cervical biopsies by *in situ* hybridization. *J. Gen. Virol.* **55**: 123.

Mark, W. and B. Sugden. 1982. Transformation of lymphocytes by Epstein-Barr virus requires only one-fourth of the viral genome. *Virology* **122**: 431.

Marks, J.R. and D.H. Spector. 1984. Fusion of the termini of the murine cytomegalovirus genome after infection. *J. Virol.* **52**: 24.

Marsden, H.S., J. Lang, A.J. Davison, R.G. Hope, and D.M. MacDonald. 1982. Genomic location and lack of phosphorylation of the HSV immediate-early polypeptide IE12. *J. Gen. Virol.* **62**: 17.

Marsden, H.S., N.D. Stow, V.G. Preston, M.C. Timbury, and N.M. Wilkie. 1978. Physical mapping of herpes simplex virus-induced polypeptides. *J. Virol.* **28**: 624.

Martin, P.J., H.M. Shulman, W.H. Shubach, J.A. Hansen, A. Fefer, G. Miller, and E.D. Thomas. 1984. Fatal Epstein-Barr-virus-associated proliferation of donor B cells after treatment of acute graft-versus-host disease with a murine anti-T-cell antibody. *Ann. Int. Med.* **101**: 310.

Matsuo, T., M. Heller, L. Letti, E. O'Shiro, and E. Kieff. 1984. Persistence of the entire Epstein-Barr virus genomes integrated into human lymphocyte DNA. *Science* **226**: 1322.

Matz, B., J.R. Schlehofer, and H. zur Hausen. 1984. Identification of a gene function of herpes simplex virus type 1 essential for amplification of simian virus 40 DNA sequences in transformed hamster cells. *Virology* **134**: 328.

McDonough, S.H., S.I. Staprans, and D.H. Spector. 1985. Analysis of the major transcripts encoded by the long repeat of human cytomegalovirus strain AD169. *J. Virol.* **53**: 711.

McDougall, J.K., D.A. Galloway, and C.M. Fenoglio. 1980. Cervical carcinoma: Detection of herpes simplex virus RNA in cells undergoing neoplastic change. *Int. J. Cancer* **25**: 1.

McDougall, J.K., C.P. Crum, C.M. Fenoglio, L.C. Goldstein, and D.A. Galloway. 1982. Herpesvirus-specific RNA and protein in carcinoma of the uterine cervix. *Proc. Natl. Acad. Sci.* **79**: 3853.

McGeoch, D.J., A. Dolan, S. Donald, and F.J. Rixon. 1985. Sequence determination and genetic content of the short unique region in the genome of herpes simplex virus type I. *J. Mol. Biol.* **181**: 1.

McKnight, S.L. 1980. The nucleotide sequence and transcript map of the herpes simplex virus thymidine kinase gene. *Nucleic Acids Res.* **8**: 5949.

———. 1982. Functional relationships between transcriptional control signals of the thymidine kinase gene of herpes simplex virus. *Cell* **31**: 355.

McKnight, S.L. and E.R. Gavis. 1980. Expression of the herpes thymidine kinase gene in *Xenopus laevis* oocytes: An assay for the study of deletion mutants constructed *in vitro*. *Nucleic Acids Res.* **8**: 5931.

McKnight, S.L., E.R. Gavis, R. Kingsbury, and R. Axel. 1981. Analysis of transcriptional regulatory signals of the HSV thymidine kinase gene: Identification of an upstream control region. *Cell* **25**: 385.

McKnight, S.L., R.C. Kingsbury, A. Spence, and M. Smith. 1984. The distal transcription signals of the herpesvirus tk gene share a common hexanucleotide control sequence. *Cell* **37**: 253.

McLaughlan, J. and J.B. Clements. 1983. DNA sequence homology between two co-linear loci on the HSV genome which have different transforming abilities. *EMBO J.* **2**: 1953.

McLean, C., A. Buckmaster, D. Hancock, A. Buchan, A. Fuller, and A. Minson. 1982. Monoclonal antibodies to three nonglycosylated antigens of herpes simplex virus type 2. *J. Gen. Virol.* **63**: 297.

Mercer, J.A., J.R. Marks, and D.H. Spector. 1983. Molecular cloning and restriction endonuclease mapping of the murine cytomegalovirus genome (Smith strain). *Virology* **129**: 94.

Metzler, D.W. and K.W. Wilcox. 1985. Isolation of herpes simplex virus regulatory protein ICP4 as a homodimeric complex. *J. Virol.* **55**: 329.

Meuer, S.C., J.C. Hodgdon, D.A. Cooper, R.E. Hussey, K.A. Fitzgerald, S.F. Schlossman, and E.L. Reinherz. 1983. Human cytotoxic T cell clones directed at autologous virus-transformed targets: Further evidence for linkage of genetic restriction to T4 and T8 surface glycoproteins. *J. Immunol.* **131**: 186.

Middleton, M., G.R. Reyes, D.M. Ciufo, A. Buchan, J.C.M. Macnab, and G.S. Hayward. 1982. Expression of cloned herpesvirus genes. I. Detection of nuclear antigens from herpes simplex virus type 2 inverted repeat regions in transfected mouse cells. *J. Virol.* **43**: 1091.

Miller, G., L. Heston, and J. Countryman. 1985. P3HR-1 Epstein-Barr virus with heterogeneous DNA is an independent replicon maintained by cell-to-cell spread. *J. Virol.* **54**: 45.

Miller, G., M. Rabson, and L. Heston. 1984. Epstein-Barr virus with heterogeneous DNA disrupts latency. *J. Virol.* **50**: 174.

Mocarski, E.S. and B. Roizman. 1981. Site-specific inversion sequence of the herpes simplex virus genome: Domain and structural features. *Proc. Natl. Acad. Sci.* **78**: 7047.

———. 1982. Structure and role of the herpes simplex virus DNA termini in inversion, circularization and generation of virion DNA. *Cell* **31**: 89.

Mocarski, E.S., L.E. Post, and B. Roizman. 1980. Molecular engineering of the herpes simplex virus genome: Insertion of a second L-S junction into the genome causes additional genome inversions. *Cell* **22**: 243.

Montgomerie, J.Z., D.M.O. Becroft, M.C. Croxson, P.B. Doak, and J.D.K. North. 1969. Herpes simplex virus infection after renal transplantation. *Lancet* **II**: 867.

Morrow, R.H., A. Kisuule, M.C. Pike, and P.G. Smith. 1976. Burkitt's lymphoma in the Mengo districts of Uganda: Epidemiologic features and their relationship to malaria. *J. Natl. Cancer Inst.* **56**: 479.

Morse, L.S., T.G. Buchman, B. Roizman, and P.A. Schaffer. 1977. Anatomy of herpes simplex virus DNA. IX. Apparent exclusion of some parental DNA arrangements in the generation of intertypic (HSV-1 × HSV-2) recombinants. *J. Virol.* **24**: 231.

Mosca, J.D., G.R. Reyes, P.M. Pitha, and G.S. Hayward. 1985. Differential activation of hybrid genes containing herpes simplex virus immediate-early or delayed-early promoters after superinfection of stable DNA-transfected cell lines. *J. Virol.* **56**: 867.

Mosmann, T.R. and J.B. Hudson. 1977. Some properties of the genome of murine cytomegalovirus (MCMV). *Virology* **54**: 135.

Moss, D.J., S.R. Burrows, D.J. Castelino, R.G. Kane, J.H. Pope, A.B. Rickinson, M.P. Alpers, and P.F. Heywood. 1983. A comparison of Epstein-Barr virus-specific T-cell immunity in malaria-endemic and non-endemic regions of Papua New Guinea. *Int. J. Cancer* **31**: 727.

Mueller-Lantzsch, N., G.M. Lenoir, M. Sauter, K. Takaki, J.-M. Bechet, C. Kuklik-Roos, D. Wunderlich, and G.W. Bornkamm. 1985. Identification of the coding region for a second Epstein-Barr virus nuclear antigen (EBNA2) by transfection of cloned DNA fragments. *EMBO J.* **4**: 1805.

Munyon, W., E. Kraiselburd, D. Davis, and J. Mann. 1971. Transfer of thymidine kinase to thymidine kinaseless L cells by infection with ultraviolet-irradiated herpes simplex virus. *J. Virol.* **7**: 813.

Murchie, M.-J. and D.J. McGeoch. 1982. DNA sequence analysis of an immediate-early gene region of the herpes simplex virus type 1 genome (map coordinates 0.950 to 0.978). *J. Gen. Virol.* **62**: 1.

Murray, B.K., N. Biswall, J.B. Bookout, R.E. Lanford, R.J. Courtney, and M.L. Melnick. 1975. Cyclic appearance of defective interfering particles of herpes simplex virus and the concomitant accumulation of early viral polypeptide VP175. *Intervirology* **5**: 173.

Nelson, J.A., B. Fleckenstein, G. Jahn, D.A. Galloway, and J.K. McDougall. 1984. Structure of the transforming region of human cytomegalovirus AD169. *J. Virol.* **49**: 109.

Nonoyama, N. and J.S. Pagano. 1972. Separation of Epstein-Barr virus DNA from large chromosomal DNA in non-virus-producing cells. *Nat. New Biol.* **238**: 169.

O'Hare, P. and G.S. Hayward. 1984. Expression of recombinant genes containing herpes simplex virus delayed-early and immediate-early regulatory regions and *trans*-activation by herpesvirus infection. *J. Virol.* **52**: 522.

———. 1985a. Evidence for a direct role for both the 175,000- and 110,000-molecular weight immediate-early proteins of herpes simplex virus in the transactivation of delayed-early promoters. *J. Virol.* **53**: 751.

———. 1985b. Three *trans*-acting regulatory proteins of herpes simplex virus modulate immediate-early gene expression in a pathway involving positive and negative feedback regulation. *J. Virol.* **56**: 723.

Orellana, T. and E.D. Kieff. 1977. Epstein-Barr virus specific RNA. II. Analysis of polyadenylated viral RNA in restringent abortive and productive infection. *J. Virol.* **22**: 321.

Pagano, J.S. 1975. The Epstein-Barr virus and malignancy: Molecular evidence. *Cold Spring Harbor Symp. Quant. Biol.* **39**: 797.

Parris, D.S., R.J. Courtney, and P.A. Schaffer. 1978. Temperature-sensitive mutants of herpes simplex virus type 1 defective in transcriptional and post-transcriptional functions required for viral DNA synthesis. *Virology* **90**: 177.

Parris, D.S., R.A.F. Dixon, and P.A. Schaffer. 1980. Physical mapping of herpes simplex virus type 1 ts mutants by marker rescue: Correlation of the physical and genetic maps. *Virology* **100**: 275.

Peden, K., P. Mounts, and G.S. Hayward. 1982. Homology between repetitive mammalian cell DNA sequences and human herpesvirus genomes detected by a high complexity probe hybridization procedure. *Cell* **31**: 71.

Pellett, P.E., M.D. Biggin, B. Barrell, and B. Roizman. 1985a. Epstein-Barr virus genome may encode a protein showing significant amino acid and predicted secondary structure homology with glycoprotein B of herpes simplex virus 1. *J. Virol.* **56**: 807.

Pellett, P.E., J.L.C. McKnight, F.J. Jenkins, and B. Roizman. 1985b. Nucleotide sequence and predicted amino acid sequence of a protein encoded in a small herpes simplex virus DNA fragment capable of *trans*-inducing alpha-genes. *Proc. Natl. Acad. Sci.* **82**: 5870.

Pellicer, A., M. Wigler, R. Axel, and S. Silverstein. 1978. The transfer and stable integration of the HSV thymidine kinase gene into mouse cells. *Cell* **14**: 133.

Persson, R.H., S. Bacchetti, and J.R. Smiley. 1985. Cells that constitutively express the herpes simplex virus immediate-early protein ICP4 allow efficient activation of viral delayed-early genes in *trans*. *J. Virol.* **54**: 414.

Poffenberger, K.L. and B. Roizman. 1985. A non-inverting genome on a viable herpes simplex virus 1: Presence of head-to-tail linkages in packaged genomes and requirement circularization after infection. *J. Virol.* **53**: 587.

Poffenberger, K.L., E. Tabares, and B. Roizman. 1983. Characterization of a viable, noninverting herpes simplex type 1 genome derived by insertion and deletion of sequences at the junction of components L and F. *Proc. Natl. Acad. Sci.* **80**: 2690.

Pogue-Giele, K.L., G.T.-Y. Lee, and P. Spear. 1985. Novel rearrangements of herpes simplex virus DNA sequences resulting from duplication of a sequence within the unique region of the L component. *J. Virol.* **53**: 456.

Pope, J.H., M.K. Horne, and W. Scott. 1966. Transformation of foetal human leukocytes in vitro by filtrates of a human leukaemic cell line containing herpes-like virus. *Int. J. Cancer* **3**: 857.

Post, L.E. and B. Roizman. 1981. A generalized technique for deletion of specific genes in large genomes: Alpha-gene 22 of herpes simplex virus 1 is not essential for growth. *Cell* **25**: 227.

Post, L.E., S. Mackem, and B. Roizman. 1981. Regulation of alpha-genes of herpes simplex virus: Expression of chimeric genes produced by fusion of thymidine kinase with alpha-gene promoters. *Cell* **24**: 555.

Post, L.E., B. Norrild, T. Simpson, and B. Roizman. 1982. Chicken ovalbumin gene fused to a herpes simplex virus alpha-promoter and linked to a thymidine kinase gene is regulated like a viral gene. *Mol. Cell. Biol.* **2:** 233.

Powell, K.L. and D.J.M. Purifoy. 1977. Nonstructural proteins of herpes simplex virus I. Purification of the induced DNA polymerase. *J. Virol.* **24:** 618.

Powell, K.L., E. Littler, and D.J.M. Purifoy. 1981. Nonstructural proteins of herpes simplex virus II. Major virus specific DNA binding protein. *J. Virol.* **24:** 618.

Preston, C.M. 1979a. Abnormal properties of an immediate-early polypeptide in cells infected with the herpes simplex virus type 1 mutant *ts*K. *J. Virol.* **32:** 357.

———. 1979b. Control of herpes simplex virus type 1 mRNA synthesis in cells infected with wild-type virus or the temperature-sensitive mutant *ts*K. *J. Virol.* **29:** 275.

Preston, C.M. and A.A. Newton. 1976. The effects of herpes simplex virus type 1 on cellular DNA-dependent RNA polymerase activities. *J. Gen. Virol.* **33:** 471.

Preston, C.M. and D. Tannahill. 1984. Effects of orientation and position on the activity of a herpes simplex virus immediate early gene far-upstream region. *Virology* **137:** 439.

Preston, C.M., M.G. Cordingley, and N.D. Stow. 1984. Analysis of DNA sequences which regulate the transcription of a herpes simplex virus immediate-early gene. *J. Virol.* **50:** 708.

Preston, V.G., A.J. Davison, H.S. Marsden, J.H. Timbury, J.H. Subak-Sharpe, and N.M. Wilkie. 1978. Recombinants between herpes simplex virus types 1 and 2: Analyses of genome structure and expression of immediate early polypeptides. *J. Virol.* **28:** 499.

Pritchett, R., S.D. Hayward, and E.D. Kieff. 1975. DNA of Epstein-Barr virus. I. Comparative studies of the DNA of Epstein-Barr virus from HR1 and B958 cells: Size, structure, and relatedness. *J. Virol.* **15:** 556.

Puga, A., E.M. Cantin, and A.L. Notkins. 1982. Homology between murine and human cellular DNA sequences and the terminal repetition of the S component of herpes simplex virus type 1 DNA. *Cell* **31:** 81.

Quinlan, M.P. and D.M. Knipe. 1985. Stimulation of expression of a herpes simplex virus DNA-binding protein by two viral functions. *Mol. Cell. Biol.* **5:** 957.

Rabson, M., L. Gradoville, L. Heston, and G. Miller. 1982. Non-immortalizing P3JHR-1 Epstein-Barr virus: A deletion mutant of the transforming parent, Jijoye. *J. Virol.* **44:** 834.

Rawlins, D.R., G. Milman, S.D. Hayward, and G.S. Hayward. 1985. Sequence-specific DNA binding of the Epstein-Barr virus nuclear antigen (EBNA-1) to clustered sites in the plasmid maintenance region. *Cell* **42:** 859.

Rayfield, M., G.S. Michaels, N. Muzyczka, R. Feldman, and K.I. Berns. 1983. Comparison of the DNA sequence and secondary structure of the herpes simplex virus L/S junction and the adeno-associated virus terminal repeat sequences suggests an alternative model for HSV DNA replication. *J. Theor. Biol.* **115:** 477.

Read, G.S. and N. Frenkel. 1983. Herpes simplex virus mutants defective in the virion-associated shutoff of host polypeptide synthesis and exhibiting abnormal synthesis of alpha (immediate early) viral polypeptides. *J. Virol.* **46:** 498.

Read, G.S., J.A. Sharp, and W.C. Summers. 1984. *In vitro* and *in vivo* transcription initiation sites on the TK encoding *Bam*HI-Q fragment of HSV-1 DNA. *Virology* **138:** 368.

Reedman, B.M. and G. Klein. 1973. Cellular localization of an Epstein-Barr virus (EBV)-associated complement-fixing antigen in producer and nonproducer lymphoblastoid cell lines. *Int. J. Cancer* **11:** 499.

Reedman, B.M., G. Klein, J.H. Pope, M.K. Walters, J. Hilgers, S. Singh, and B. Johansson. 1974. Epstein-Barr virus-associated complement-fixing and nuclear antigens in Burkitt lymphoma biopsies. *Int. J. Cancer* **13:** 755.

Reisman, D. and B. Sugden. 1986. *Trans*-activation of gene expression by the Epstein-Barr viral nuclear antigen (EBNA-1). *Mol. Cell. Biol.* (in press).

Reisman, D., J. Yates, and B. Sugden. 1985. A putative origin of replication of plasmids derived from Epstein-Barr virus is composed of two *cis*-acting components. *Mol. Cell. Biol.* **5:** 1822.

Reyes, G.R. 1982. "Morphological and biochemical transformation with the DNA of herpes simplex virus." Ph.D. thesis, Johns Hopkins University, Baltimore, Maryland.

Reyes, G.R., K.-T. Jeang, and G.S. Hayward. 1982a. Transfection with the isolated herpes simplex virus thymidine kinase gene. I. Minimal size of the active fragments from HSV-1 and HSV-2. *J. Gen. Virol.* **62:** 19.

Reyes, G.R., R. LaFemina, S.D. Hayward, and G.S. Hayward. 1980. Morphological transformation by DNA fragments of human herpesviruses: Evidence for two distinct transforming regions in herpes simplex virus types 1 and 2 and lack of correlation with biochemical transfer of the thymidine kinase gene. *Cold Spring Harbor Symp. Quant. Biol.* **44:** 629.

Reyes, G.R., E.R. Gavis, A. Buchan, N.B. Raj, G.S. Hayward, and P.M. Pitha. 1982b. Expression of human β-interferon cDNA under the control of a thymidine kinase promoter from herpes simplex virus. *Nature* **297:** 598.

Reynolds, D.W. 1978. Development of early nuclear antigen in cytomegalovirus infected cells in the presence of RNA and protein synthesis inhibitors. *J. Gen. Virol.* **40:** 475.

Rice, G.P.A., R.D. Schrier, and M.B.A. Oldstone. 1984. Cytomegalovirus infects human lymphocytes and monocytes: Virus expression is restricted to immediate-early gene product. *Proc. Natl. Acad. Sci.* **81:** 6134.

Rixon, F.J. and D.J. McGeoch. 1984. A 3′ co-terminal family of mRNAs from the herpes simplex virus type 1 short region: Two overlapping reading frames encode unrelated polypeptides, one of which has a highly reiterated amino acid sequence. *Nucleic Acids Res.* **12:** 2473.

Rixon, F.J., M.E. Campbell, and J.B. Clements. 1984. A tandemly reiterated DNA sequence in the long repeat region of herpes simplex virus type 1 found in close proximity to immediate-early mRNA 1. *J. Virol.* **52:** 715.

Rocchi, G., A. De Felici, G. Ragona, and A. Heinz. 1977. Quantitative evaluation of Epstein-Barr-virus-infected mononuclear peripheral blood leukocytes in infectious mononucleosis. *N. Engl. J. Med.* **296:** 132.

Rock, D.L. and N.W. Fraser. 1983. Detection of HSV-1 genome in central nervous system of latently infected mice. *Nature* **302:** 523.

———. 1985. Latent herpes simplex virus type I DNA contains two copies of the virion DNA joint region. *J. Virol.* **55:** 849.

Roizman, B., ed. 1982. *The herpesviruses*, vol. 1. Plenum Press, New York.

Roizman, B. and F.L. Jenkins. 1985. Genetic engineering of viral genomes of large DNA viruses. *Science* **229:** 1208.

Roizman, B., G. Hayward, R. Jacob, S. Wadsworth, and R.W. Honess. 1974. Human herpesvirus. I. A model for molecular organization of herpesvirus virions and their DNA. *Excerpta Med. Int. Congr. Ser.* **350(2):** 188.

Rooney, C.M., M. Rowe, L.E. Wallace, and A.B. Rickenson. 1985. Epstein-Barr virus-positive Burkitt's lymphoma cells not recognized by virus-specific T-cell surveillance. *Nature* **317:** 629.

Rosa, M.D., E. Gottlieb, M.R. Lerner, and J.A. Steitz. 1981. Striking similarities are exhibited by two small Epstein-Barr virus-encoded ribonucleic acids and the adenovirus-associated ribonucleic acids VAI and VAII. *Mol. Cell. Biol.* **1:** 785.

Rowe, M., C.M. Rooney, A.B. Rickinson, G.M. Lenoir, H. Rupani, D.J. Moss, H. Stein, and M.A. Epstein. 1985. Distinctions between endemic and sporadic forms of Epstein-Barr virus-positive Burkitt's lymphoma. *Int. J. Cancer* **35:** 435.

Ruger, R., G.W. Bornkamm, and B. Fleckenstein. 1984. Human cytomegalovirus DNA sequences with homologies to the cellular genome. *J. Gen. Virol.* **65:** 1351.

Ruyechan, W.T., L.S. Morse, D.M. Knipe, and B. Roizman. 1979. Molecular genetics of herpes simplex virus. II. Mapping of the major viral glycoproteins and of the genetic loci specifying the social behavior of infected cells. *J. Virol.* **29:** 677.

Ruyechan, W.T., T.A. Casey, W. Reinhold, A.C. Weir, M. Wellman, S.E. Strauss, and J. Hay. 1985. Distribution of (G + C)-rich regions in varicella-zoster virus DNA. *J. Gen. Virol.* **66:** 43.

Rymo, L. and S. Forsblom. 1978. Cleavage of Epstein-Barr virus DNA by restriction endonucleases *Eco*RI, *Hind*III and *Bam*HI. *Nucleic Acids Res.* **5:** 1387.

Rymo, L., G. Klein, and A. Ricksten. 1985. Expression of a second Epstein-Barr virus determined nuclear antigen in mouse cells after gene transfer with a cloned fragment of the viral genome. *Proc. Natl. Acad. Sci.* **82:** 3435.

Sacks, W.R., C.C. Green, D.P. Aschman, and P.A. Schaffer. 1985. Herpes simplex virus type 1 ICP27: An essential regulatory protein. *J. Virology* **55:** 796.

Saemundsen, A.K., B. Kallin, and G. Klein. 1980. Effect of *n*-butyrate on cellular and viral DNA synthesis in cells latently infected with Epstein-Barr virus. *Virology* **107:** 557.

Sample, J., A. Tanaka, G. Lancz, and M. Nonoyama. 1984. Identification of Epstein-Barr virus genes expressed during the early phase of virus replication and during lymphocyte immortalization. *Virology* **139:** 1.

Sandri-Goldin, R.M., A.L. Goldin, L.E. Holland, J.C. Glorioso, and M. Levine. 1983. Expression of herpes simplex virus β and alpha-genes integrated in mammalian cells and their induction by an alpha-gene product. *Mol. Cell. Biol.* **3:** 2028.

Schaffer, P.A. 1975. Temperature-sensitive mutants of herpesviruses. *Curr. Top. Microbiol. Immunol.* **70:** 51.

Schaffer, P.A., V.C. Carter, and M.C. Timbury. 1978. Collaborative complementation study of temperature-sensitive mutants of herpes simplex virus type 1 and 2. *J. Virol.* **27:** 490.

Schaffer, P.A., G.M. Aron, N. Biswal, and M. Benyesh-Melnick. 1973. Temperature sensitive mutants of herpes simplex type I: Isolation, complementation and partial characterization. *Virology* **52:** 57.

Schimpff, S., A. Serpick, B. Stoler, B. Rumack, H. Mellin, J.M. Joseph, and J. Block. 1972. Varicella-zoster infection in patients with cancer. *Ann. Intern. Med.* **76:** 241.

Schlehofer, J.R. and H. zur Hausen. 1982. Induction of mutations within the host cell genome by partially inactivated herpes simplex virus type 1. *Virology* **122:** 471.

Schroeder, C.H., B. Furst, K. Weise, and C.P. Gray. 1984. A study of interfering herpes simplex virus DNA preparations containing defective genomes of either class I or II and the identification of minimal requirements for interference. *J. Gen. Virol.* **65:** 493.

Schubach, W.H., R. Hackman, P.E. Neiman, G. Miller, and E.D. Thomas. 1982. A monoclonal immunoblastic sarcoma in donor cells bearing Epstein-Barr virus genomes following allogeneic marrow grafting for acute lymphoblastic leukemia. *Blood* **60:** 180.

Sears, A.M., I.N. Halliburton, B. Meignier, S. Silver, and B. Roizman. 1985. Herpes simplex virus 1 mutant deleted in the alpha-22 gene: Growth and gene expression in permissive and restrictive cells and establishment of latency in mice. *J. Virol.* **55:** 338.

Sheldrick, P. and N. Berthelot. 1975. Inverted repetitions in the chromosome of herpes simplex virus. *Cold Spring Harbor Symp. Quant. Biol.* **39:** 667.

Sheldrick, P., M. Laithler, D. Lando, and M.L. Ryhiner. 1973. Infectious DNA from herpes simplex virus: Infectivity of double-stranded and single-stranded molecules. *Proc. Natl. Acad. Sci.* **70:** 3621.

Shih, M.-F., M. Arsenakis, P. Tiollais, and B. Roizman. 1984. Expression of hepatitis B virus S gene by herpes simplex virus type 1 vectors carrying gamma- and beta-regulated gene chimeras. *Proc. Natl. Acad. Sci.* **81:** 5867.

Showalter, S.D., M. Zweig, and B. Hampar. 1981. Monoclonal antibodies to herpes simplex virus type 1 proteins, including the immediate-early protein ICP4. *Infect. Immun.* **34:** 684.

Silver, S. and B. Roizman. 1985. Gamma-2-thymidine kinase

chimeras are identically transcribed but regulated as gamma-2 genes in herpes simplex virus genomes and as β-genes in cell genomes. *Mol. Cell. Biol.* **5:** 518.

Sixbey, J.W., E.H. Vesterinen, J.G. Nedrud, N. Raab-Traub, L.A. Walton, and J.S. Pagano. 1983. Replication of Epstein-Barr virus in human epithelial cells infected *in vitro*. *Nature* **306:** 480.

Skare, J. and W.C. Summers. 1977. Structure and function of herpesvirus genomes. II. *Eco*RI, *Xba*I and *Hind*III endonuclease cleavage sites on HSV type 1 DNA. *Virology* **76:** 581.

Smiley, J.R. 1980. Construction *in vitro* and rescue of a thymidine kinase-deficient deletion mutation of herpes simplex virus. *Nature* **285:** 333.

Smiley, J.R., B.S. Fong, and W.-C. Leung. 1981. Construction of a double-jointed herpes simplex viral DNA molecule: Inverted repeats are required for segment inversion, and direct repeats promote deletions. *Virology* **113:** 345.

Smiley, J.R., M.J. Wagner, and W.C. Summers. 1980. Genetic and physical evidence for the polarity of transcription of the thymidine kinase gene of herpes simplex virus. *Virology* **102:** 83.

Spaete, R.R. and N. Frenkel. 1982. The herpes simplex virus amplicon: A new eukaryotic defective virus cloning-amplifying vector. *Cell* **30:** 295.

Spaete, R.R. and E.S. Mocarski. 1985a. Regulation of cytomegalovirus gene expression: Alpha- and β-promoters are *trans*-activated by viral functions in permissive human fibroblasts. *J. Virol.* **56:** 135.

——. 1985b. The *a* sequence of the cytomegalovirus genome functions as a cleavage/packaging signal for herpes simplex virus defective genomes. *J. Virol.* **54:** 817.

Spear, P.G. 1983. Transformation of cultured cells by human herpesviruses. *Int. Rev. Exp. Pathol.* **25:** 327.

Spear, P.G. and B. Roizman. 1972. Proteins specified by herpes simplex virus. V. Purification and structural proteins of the herpesvirion. *J. Virol.* **9:** 143.

Speck, S.H. and J.L. Strominger. 1985. Analysis of the transcript encoding the latent Epstein-Barr virus nuclear antigen I: A potentially polycistronic message generated by long range splicing of several exons. *Proc. Natl. Acad. Sci.* **82:** 8305.

Spelsberg, T.C., T.B. Scully, G.M. Pikler, J.A. Gilbert, and G.R. Pearson. 1982. Evidence for two classes of chromatin-associated Epstein-Barr virus-determined nuclear antigen. *J. Virol.* **43:** 555.

Stenberg, R.M., D.R. Thomsen, and M.F. Stinski. 1984. Structural analysis of the major immediate-early gene of human cytomegalovirus. *J. Virol.* **49:** 190.

Stenberg, R.M., P.R. Witte, and M.F. Stinski. 1985. Multiple spliced and unspliced transcripts from human cytomegalovirus immediate-early region 2 and evidence for a common initiation site within immediate-early region 1. *J. Virol.* **56:** 665.

Stevely, W.S. 1977. Inverted repetition in the chromosome of pseudorabies virus. *J. Virol.* **22:** 232.

Stevens, J.G. and M.L. Cook. 1974. Maintenance of latent herpetic infection: An apparent role for anti-viral IgG. *J. Immunol.* **113:** 1685.

Stinski, M.F. and T.J. Roehr. 1985. Activation of the major immediate-early genes of human cytomegalovirus by *cis*-acting elements in the promoter-regulatory sequence and by virus-specific *trans*-acting components. *J. Virol.* **55:** 431.

Stinski, M.F., E.S. Mocarski, and D.R. Thomsen. 1979. DNA of human cytomegalovirus: Size, heterogeneity and defectiveness resulting from serial undiluted passage. *J. Virol.* **31:** 231.

Stinski, M.F., D.R. Thomsen, R.M. Stenberg, and L.C. Goldstein 1983. Organization and expression of immediate-early genes of human cytomegalovirus. *J. Virol.* **46:** 1.

Stow, N.D 1982. Localization of an origin of DNA replication within the TR$_S$/IR$_S$ repeated region of the herpes simplex virus type 1 genome. *EMBO J.* **1:** 863.

——. 1985. Mutagenesis of a herpes simplex virus origin of

DNA replication and its effect on viral interference. *J. Gen. Virol.* **66:** 31.

Stow, N.D. and E.C. McMonagle. 1983. Characterization of the TR$_S$/IR$_S$ origin of DNA replication of herpes simplex virus type 1. *Virology* **130:** 427.

Stow, N.D., E.C. McMonagle, and A.V. Davison. 1983. Fragments from both termini of the herpes simplex virus type 1 genome contain signals required for the encapsidation of viral DNA. *Nucleic Acids Res.* **11:** 8205.

Stow, N.D., J.H. Subak-Sharpe, and N.M. Wilkie. 1978. Physical mapping of herpes simplex virus type 1 mutation by marker rescue. *J. Virol.* **28:** 182.

Straus, S.E., J. Owens, W.T. Ruyechan, H.E. Takiff, T.A. Casey, G.F. Vande Woude, and J. Hay. 1982. Molecular cloning and physical mapping of varicella-zoster virus DNA. *Proc. Natl. Acad. Sci.* **79:** 993.

Stringer, J.R., L.E. Holland, and E.K. Wagner. 1978. Mapping early transcripts of herpes simplex virus type 1 by electron microscopy. *J. Virol.* **27:** 56.

Strnad, B.C., T.C. Schuster, R.F. Hopkins III, R.H. Neubauer, and H. Rabin. 1981. Identification of an Epstein-Barr virus nuclear antigen by fluoroimmunoelectrophoresis and radioimmunoelectrophoresis. *J. Virol.* **38:** 996.

Sugden, B. 1984. Expression of virus-associated functions in cells transformed in vitro by Epstein-Barr virus: Epstein-Barr virus cell surface antigen and virus-release from transformed cells. In *Immune deficiency and cancer* (ed. D.T. Purtilo), p. 165. Plenum Press, New York.

Summers, W.C. and G. Klein. 1976. Inhibition of EBV DNA synthesis and late gene expression by phosphonoacetic acid. *J. Virol.* **18:** 151.

Summers, W.P., E.A. Grogan, D. Shedd, M. Robert, C.-R. Liu, and G. Miller. 1982. Stable expression in mouse cells of nuclear neoantigen after transfer of a 3.4 megadalton cloned fragment of Epstein-Barr virus DNA. *Proc. Natl. Acad. Sci.* **79:** 5688.

Svedmyr, E. and M. Jondal. 1975. Cytotoxic effector cells specific for B cell lines transformed by Epstein-Barr virus are present in patients with infectious mononucleosis. *Proc. Natl. Acad. Sci.* **72:** 1622.

Svedmyr, E., M. Jondal, W. Henle, O. Weiland, L. Rombo, and G. Klein. 1978. EBV specific killer T cells and serologic responses after onset of infectious mononucleosis. *J. Clin. Lab. Immunol.* **1:** 225.

Swain, M.A. and D.A. Galloway. 1983. Nucleotide sequence of the herpes simplex virus type 2 thymidine kinase gene. *J. Virol.* **46:** 1045.

Tackney, C., G. Cachianes, and S. Silverstein. 1984. Transduction of the Chinese hamster ovary *aprt* gene by herpes simplex virus. *J. Virol.* **52:** 606.

Takaki, K., A. Polack, and G.W. Bornkamm. 1984. Expression of a nuclear and a cytoplasmic Epstein-Barr virus early antigen after DNA transfer: Cooperation of two distant parts of the genome for expression of the cytoplasmic antigen. *Proc. Natl. Acad. Sci.* **81:** 4568.

Tamashiro, J.C., D. Filpula, T. Friedmann, and D.H. Spector. 1984. Structure of the heterogeneous L-S junction region of human cytomegalovirus strain AD169 DNA. *J. Virol.* **52:** 541.

Taub, R., I. Kirsch, C. Morton, G. Lenoir, D. Swan, S. Tronick, S. Aaronson, and P. Leder. 1982. Translocation of the *c-myc* gene into the immunoglobulin heavy chain locus in human Burkitt lymphoma and murine plasmacytoma cells. *Proc. Natl. Acad. Sci.* **79:** 7837.

Tenser, R.B., M. Dawson, S.J. Ressel, and M.E. Dunstan. 1982. Detection of herpes simplex virus mRNA in latently infected trigeminal ganglion neurons by in situ hybridization. *Ann. Neurol.* **11:** 285.

Tevethia, M.J. and D.J. Spector. 1984. Complementation of an adenovirus 5 immediate-early mutant by human cytomegalovirus. *Virology* **137:** 428.

Thomsen, D.R. and M.F. Stinski. 1981. Cloning of the human cytomegalovirus genome as endonuclease *Xba*I fragments. *Gene* **16:** 207.

Thomsen, D.R., R.M. Stenberg, W.F. Goins, and M.F. Stinski. 1984. Promoter-regulatory region of the major immediate-early gene of human cytomegalovirus. *Proc. Natl. Acad. Sci.* **81:** 659.

Timbury, M.C. and J.H. Subak-Sharpe. 1973. Genetic interactions between temperature-sensitive mutants of types 1 and 2 herpes simplex viruses. *J. Gen. Virol.* **18:** 347.

Tooze, J., ed. 1980. *Molecular biology of tumor viruses*, 2nd edition: *DNA tumor viruses*. Cold Spring Harbor Laboratory, Cold Spring Harbor, New York.

Tosato, G., I. Magrath, I. Koski, N. Dooley, and M. Blaese. 1979. Activation of suppressor T cells during Epstein-Barr-virus-induced infectious mononucleosis. *N. Engl. J. Med.* **301:** 1133.

Tremblay, M.D., S.-P. Yee, R.H. Persson, S. Bacchetti, J.R. Smiley, and P.E. Branton. 1985. Activation and inhibition of expression of the 72,000-Da early protein of adenovirus type 5 in mouse cells constitutively expressing an immediate-early protein of herpes simplex virus type 1. *Virology* **144:** 35.

van Santen, V., A. Cheung, and E Kieff. 1981. Epstein-Barr virus RNA. VII. Size and direction of transcription of virus-specific cytoplasmic RNAs in a transformed cell line. *Proc. Natl. Acad. Sci.* **78:** 1930.

Varmuza, S.L. and J.R. Smiley. 1984. Unstable heterozygosity in a diploid region of herpes simplex virus DNA. *J. Virol.* **49:** 356.

————. 1985. Signals for site-specific cleavage of HSV DNA: Maturation involves two separate cleavage events at sites distal to the recognition sequences. *Cell* **41:** 793.

Vaughan, P.J., D.J.M. Purifoy, and K.L. Powell. 1985. DNA-binding protein associated with herpes simplex virus DNA polymerase. *J. Virol.* **53:** 501.

Vlazny, D.A. and N. Frenkel. 1981. Replication of herpes simplex virus DNA: Localization of replication recognition signals within defective virus genomes. *Proc. Natl. Acad. Sci.* **78:** 742.

Vlazny, D.A., A. Kwong, and N. Frenkel. 1982. Site-specific cleavage/packaging of herpes simplex virus DNA and the selective maturation of nucleocapsids containing full-length viral DNA. *Proc. Natl. Acad. Sci.* **79:** 1423.

Volsky, D.J., T. Gross, F. Sinangil, C. Kiszynski, R. Bartzatt, T. Dambaugh, and E. Kieff. 1984. Expression of Epstein-Barr virus (EBV) DNA and cloned DNA fragments in human lymphocytes following Sendai virus envelope-mediated gene transfer. *Proc. Natl. Acad. Sci.* **81:** 5926.

Vonka, V., J. Kanka, I. Hirsch, H. Zavadova, M. Krcmar, A. Suchankova, D. Rezacova, J. Broucek, M. Press, E. Domorazkova, B. Svoboda, A. Havrankova, and J. Jelinek. 1984a. Prospective study on the relationship between cervical neoplasia and herpes simplex type-2-virus. II. Herpes simplex type-2 antibody presence in sera taken at enrollment. *Int. J. Cancer* **33:** 61.

Vonka, V., J. Kanka, J. Jelinek, I. Subrt, A. Suchanek, A. Havrankova, M. Vachal, I. Hirsch, E. Domorazkova, H. Zavadova, V. Richterova, J. Naprstkova, V. Dvorakova, and B. Svoboda. 1984b. Prospective study on the relationship between cervical neoplasia and herpes simplex types-2 virus. I. Epidemiological characteristics. *Int. J. Cancer* **33:** 49.

Voss-Hubenthal, J. and B. Roizman. 1985. Herpes simplex virus 1 reiterated sequences situated between the alpha sequence and alpha-4 gene are not essential for virus replication. *J. Virol.* **54:** 509.

Wadsworth, S.C., G.S. Hayward, and B. Roizman. 1976. Anatomy of herpes simplex virus DNA. V. Terminally repetitive sequences. *J. Virol.* **17:** 503.

Wadsworth, S.C., R.J. Jacob, and B. Roizman. 1975. Anatomy of herpes simplex virus DNA. II. Size, composition and arrangement of inverted terminal repetitions. *J. Virol.* **15:** 1487.

Wagner, E.K. 1984. Regulation of HSV transcription. *J. Invest. Dermatol.* **83:** 48.

Wagner, E.K. and B. Roizman. 1972. RNA synthesis in cells infected with herpes simplex virus. II. Evidence that a class of viral mRNA is derived from a high molecular weight precursor synthesized in the nucleus. *Proc. Natl. Acad. Sci.* **64**: 626.

Wagner, M.J. and W.C. Summers. 1978. Structure of the joint region and the termini of the DNA of herpes simplex virus type 1. *J. Virol.* **27**: 374.

Wagner, M.J., J.A. Sharp, and W.C. Summers. 1981. Nucleotide sequence of the thymidine kinase gene of herpes simplex virus type 1. *Proc. Natl. Acad. Sci.* **78**: 1441.

Wagner, M., J. Skare, and W.C. Summers. 1975. Analysis of DNA of defective herpes simplex virus type I by restriction endonuclease cleavage and nucleic acid hybridization. *Cold Spring Harbor Symp. Quant. Biol.* **39**: 683.

Wang, D., D. Leibowitz, and E. Kieff. 1985. An EBV membrane protein expressed in immortalized lymphocytes transforms established rodent cells. *Cell* **43**: 831.

Wathen, M.W., D.R. Thomsen, and M.F. Stinski. 1981. Temporal regulation of human cytomegalovirus transcription at immediate-early and early times after infection. *J. Virol.* **38**: 446.

Watson, R.J. and J.B. Clements. 1978. Characterization of transcription-defective temperature-sensitive mutants of herpes simpelx virus type 1. *Virology* **91**: 364.

———. 1980. A herpes simplex virus type 1 function continuously required for early and late virus RNA synthesis. *Nature* **285**: 329.

Watson, R.J., C.M. Preston, and J.B. Clements. 1979. Separation and characterization of herpes simplex virus type 1 immediate-early mRNAs. J. Virol. **31**: 42.

Watson, R.J., K. Umene, and L.W. Enquist. 1981. Reiterated sequences within the intron of an immediate-early gene of herpes simplex virus type 1. *Nucleic Acids Res.* **16**: 4189.

Watson, R.J., J.H. Weis, J.S. Salstrom, and L.W. Enquist. 1982. Herpes simplex virus type I glycoprotein D gene: Nucleotide sequence and expression in *E. coli. Science* **218**: 381.

Weller, S.K., A. Spadaro, J.E. Schaffer, A.W. Murray, A.M. Maxam, and P.A. Schaffer. 1985. Cloning, sequencing, and functional analysis of ori$_L$, a herpes simplex virus type 1 origin of DNA synthesis. *Mol. Cell. Biol.* **5**: 930.

Whittle, H.C., J. Brown, K. Marsh, B.M. Greenwood, P. Seidelin, H. Tighe, and L. Wedderburn. 1984. T-cell control of Epstein-Barr virus-infected B cells is lost during *P. falciparum* malaria. *Nature* **312**: 449.

Whitton, J.L. and J.B. Clements. 1984a. The junctions between the repetitive and the short unique sequences of the herpes simplex virus genome are determined by the polypeptide-coding regions of two spliced immediate-early mRNAs. *J. Gen. Virol.* **65**: 451.

———. 1984b. Replication origins and a sequence involved in coordinate induction of the immediate-early gene family are conserved in an intergenic region of herpes simplex virus. *Nucleic Acids Res.* **12**: 2061.

Wigdahl, B., C.A. Smith, H.M. Traglia, and F. Rapp. 1984a. Herpes simplex virus latency in isolated human neurons. *Proc. Natl. Acad. Sci.* **81**: 6217.

Wigdahl, B., A.C. Scheck, R.J. Ziegler, E. De Clerq, and F.

Rapp. 1984b. Analysis of the herpes simplex virus genome during in vitro latency in human diploid fibroblasts and rat sensory neurons. *J. Virol.* **49**: 205.

Wigler, M., S. Silverstein, L.-S. Lee, A. Pellicer, Y.-C. Cheng, and R. Axel. 1977. Transfer of purified herpesvirus thymidine kinase gene to cultured mouse cells. *Cell* **11**: 223.

Wilcox, K.W., A. Kohn, E. Sklyanskaya, and B. Roizman. 1980. Herpes simplex virus phosphoproteins. I. Phosphate cycles on and off some viral polypeptides and can alter their affinity for DNA. *J. Virol.* **33**: 167.

Wilkie, N.M. 1976. Physical maps for herpes simplex type I DNA for restriction endonucleases *Hin*dIII, *Hpa*I and *Xba*I. *J. Virol.* **20**: 222

Wilkie, N.M., J.B. Clements, J.C.M. Macnab, and J.H. Subak-Sharpe. 1975. The structure and biological properties of herpes simplex virus DNA. *Cold Spring Harbor Symp. Quant. Biol.* **39**: 657.

Wilkie, N.M., A. Davison, P. Chartrand, N.D. Stow, V.G. Preston, and M.C. Timbury. 1979. Recombination in herpes simplex virus: Mapping of mutations and analysis of intertypic recombinants. *Cold Spring Harbor Symp. Quant. Biol.* **43**: 827.

Wilkinson, G.W.G., A. Ackrigg, and P.J. Greenaway. 1984. Transcription of the immediate early genes of human cytomegalovirus strain AD169. *Virus Res.* **1**: 101.

Wilson, G. and G. Miller. 1979. Recovery of Epstein-Barr virus from nonproducer neonatal human lymphoid cell transformants. *Virology* **95**: 351.

Wolf, H., M. Haus, and E. Wilmes. 1984. Persistence of Epstein-Barr virus in the parotid gland. *J. Virol.* **51**: 795.

Yajima, Y. and M. Nonoyama. 1976. Mechanisms of infection with Epstein-Barr virus. I. Viral DNA replication and formation of non-infectious virus particles in superinfected Raji cells. *J. Virol.* **19**: 187.

Yao, Q.Y., A.B. Rickinson, and M.A. Epstein. 1985a. A re-examination of the Epstein-Barr virus carrier state in healthy seropositive individuals. *Int. J. Cancer* **35**: 35.

Yao, Q.Y., A.B. Rickinson, J.S.H. Gaston, and M.A. Epstein. 1985b. *In vitro* analysis of the Epstein-Barr virus: Host balance in long-term renal allograft recipients. *Int. J. Cancer* **35**: 43.

Yates, J.L., N. Warren, and B. Sugden. 1985. Stable replication of plasmids derived from Epstein-Barr virus in various mammalian cells. *Nature* **313**: 812.

Yates, J., N. Warren, D. Reisman, and B. Sugden. 1984. A *cis*-acting element from the Epstein-Barr viral genome that permits stable replication of recombinant plasmids in latently infected cells. *Proc. Natl. Acad. Sci.* **81**: 3806.

zur Hausen, H., F.J. O'Neill, U.K. Freese, and E. Hecker. 1978. Persisting oncogenic herpesvirus induced by the tumour promoter TPA. *Nature* **272**: 373.

zur Hausen, H., H. Schulte-Holthausen, G. Klein, W. Henle, G. Henle, P. Clifford, and L. Santesson. 1970. EBV DNA in biopsies of Burkitt tumours and anaplastic carcinomas of the nasopharynx. *Nature* **228**: 1056.

# Duplications within Mutated SV40 Enhancers That Restore Enhancer Function

## W. Herr, J. Clarke, B. Ondek, A. Shepard, and H. Fox.
Cold Spring Harbor Laboratory, Cold Spring Harbor, New York 11724

We have analyzed the functional elements within the SV40 enhancer by making directed point mutations that debilitate both SV40 enhancer function and virus viability. We have then selected SV40 growth revertants that overcome the enhancer defects and examined the structure of the revertant enhancer regions responsible for the restored enhancer function. These experiments have identified three functional regions between 15 and 22 bp in length, which we call A, B, and C. When any two of these regions are inactivated by point mutation, enhancer function is restored by tandem duplications that at least span the remaining unmutated element. These results indicate that these three separate elements can functionally compensate for one another. We propose a model in which the three elements A, B, and C each interact with different factors to activate transcription, and these factors differ in DNA-binding specificity but activate transcription by similar mechanisms.

The SV40 early promoter is a prototypical mammalian promoter that contains each of the four known classes of promoter elements: (1) the transcriptional initiation site, (2) the TATA box, (3) upstream promoter elements, and (4) enhancer elements. The TATA box is a common promoter element that is involved in defining where along the DNA molecule transcription begins (e.g., see Gluzman et al. 1980; Benoist and Chambon 1981). Unlike the TATA box element, the upstream promoter and enhancer elements can vary greatly between promoters and have been shown to confer promoter specificity. In the SV40 promoter, the upstream promoter elements lie within two perfect and one imperfect 21-bp tandemly repeated sequence (see the top of Fig. 1 for an illustration of the SV40 early promoter). These sequences interact with the human transcription factor Sp1, which can stimulate transcription from a subset of mammalian promoters, including the SV40 early promoter (see Dynan and Tjian 1985).

Upstream of the Sp1-binding sites lies the enhancer region. Enhancers are unique among eukaryotic promoter elements because they can stimulate transcription of a gene when located several kilobases upstream or downstream of the transcriptional initiation site (see Gluzman and Shenk 1983). These unique properties were first described for the 72-bp repeat region of the SV40 strain 776 early promoter (Banerji et al. 1981; Moreau et al. 1981), a region previously identified as a promoter element by deletion analysis (Benoist and Chambon 1981; Gruss et al. 1981; Fromm and Berg 1982). Since the discovery of the SV40 enhancer, enhancer elements have been shown to be associated with numerous other viral and cellular genes (see Gluzman and Shenk 1983; Gluzman 1985). Comparison of the nucleotide sequences of these different enhancer elements has not identified any sequence common to all enhancers. Nevertheless, a number of short consensus sequences are shared by different sets of enhancers,

and enhancers often contain more than one consensus element. These different consensus sequences include the core consensus sequence $GTGG^A/_T{}^A/_T{}^A/_TG$ (Laimins et al. 1982; Weiher et al. 1983) and stretches of alternating purines and pyrimidines (Pu/Py) (Nordheim and Rich 1983). The SV40 72-bp element contains a core consensus sequence, called the core element (Weiher et al. 1983), and an 8-bp Pu/Py sequence (see Fig. 1). Upstream of the 72-bp element lies a second, different 8-bp Pu/Py sequence that partially overlaps another core consensus sequence.

We have recently described the structure of revertants of two SV40 enhancer mutants: *dpm*12, in which both Pu/Py elements are mutated (Herr and Gluzman 1985), and *dpm*6, in which the core element is mutated (Herr and Clarke 1986). These revertants in each case contain tandem duplications within the enhancer region and these serve to identify three separate elements (called A, B, and C), each of which is apparently capable of compensating for loss of function within the other elements. Here we summarize the structures of these and other revertant enhancer structures and discuss their implications on the mechanism of SV40 enhancer function.

## Experimental Procedures

### Construction of enhancer point mutants

The three sets of double point mutations (*dpm*1, *dpm*2, and *dpm*6; see Figs. 1 and 2) were created by oligonucleotide-directed mutagenesis using an M13 bacteriophage clone of the SV40 control region (mpSV01) as template (see Herr and Gluzman 1985; Herr and Clarke 1986). This SV40 clone contains only a single copy of the 72-bp element (see Fig. 1). We refer to this variant as 1 × 72 and to the wild-type duplicated enhancer as 2 × 72. To construct mutants containing two sets of *dpm* mutations, we used one of the mutant M13 constructs as the template and directed the second set of point mutations with the corresponding oligonucleotide. The mutated enhancers were cloned back into the SV40 vector pK1K1 as described previously (Herr and Gluzman 1985).

*Revertant isolations*
Revertants of the SV40 enhancer mutants were isolated as described previously (Herr and Gluzman 1985; Herr and Clarke 1986). Since the initial isolation of *dpm*12 revertants, we have used the plasmid pK1K1 to propagate the mutant SV40 genomes for transfection into CV-1 cells. This plasmid carries 1.27 copies of the SV40 genome in which nucleotides 346–1782 (*Hpa*II–*Eco*RI) are duplicated. The duplicated 0.27 copies of SV40 sequence allow for excision and replication of the complete SV40 genome after transfection into CV-1 cells. To isolate revertants, large amounts of mutant pK1K1 DNA were transfected into CV-1 cells on multiple (~12 in each case) 25-mm Falcon plates of CV-1 cells. Resulting virus stocks were further passaged once or twice in CV-1 cells, and growth revertants were isolated by plaque purification.

*Structural analysis of revertant enhancer regions*
SV40 DNA from each purified revertant was isolated by Hirt extraction (Hirt 1967) and subjected to structural analysis by digestion with the restriction enzyme *Nco*I. An alteration in the size of the 224-bp SV40 *Nco*I fragment, which spans the 1 × 72 enhancer region, indicates rearrangements within the revertant enhancer region. The generation of a single, new *Nco*I fragment in the absence of the 224-bp parental fragment indicates a pure virus population; impure populations were further purified by plaque isolation until a single viral species was present. Pure revertant viral genomes were further analyzed by double digestion with *Nco*I and *Kpn*I to determine whether the unique SV40 *Kpn*I site had been duplicated. In those cases where the *Kpn*I site was not duplicated, the short *Kpn*I-to-*Bgl*I fragment of the revertant viral genomes was cloned into M13 bacteriophage. Those viral genomes containing duplications of the *Kpn*I site were analyzed by cloning the *Hpa*II-to-*Hin*dIII enhancer fragment into M13 phage. In each case, the exact structure of each revertant enhancer was determined by nucleotide sequence analysis as described previously (Herr and Clarke 1986).

## Results

### Duplications of a mutated SV40 enhancer restore its activity

Figure 1 (top) shows the structure of the SV40 enhancer mutant *dpm*12. This mutant contains a single 72-bp element and two transversions within each of the two (**a** and **b**) 8-bp enhancer Pu/Py elements. (*dpm* is an acronym for *d*ouble *p*oint *m*utant; the double transversion mutations in the **a** Pu/Py sequence are designated *dpm*1 and those in the **b** Pu/Py element are called *dpm*2). The combined *dpm*12 mutations greatly impair SV40 growth in CV-1 cells and weaken the ability of the

enhancer to stimulate transcription of the human β-globin gene as measured in a transient expression assay in HeLa cells. The inability of this virus to grow effectively in CV-1 cells allowed the isolation of a large number of SV40 growth revertants with restored viability; these revertants also showed restored enhancer function (Herr and Gluzman 1985). Figure 1 shows the structure of 18 *dpm*12 revertants; in each case a simple tandem duplication ranging in size from 45 to 135 bp within the mutated enhancer region is responsible for the restored enhancer function. The regions that are duplicated in the revertants are shown in Figure 1 as rectangular boxes (the *r*evertant *d*uplications are referred to by *rd* followed by the size of the duplication in base pairs). No point mutations occurred in the duplicated enhancers; when the duplications span the mutated Pu/Py elements, the mutations are also duplicated.

Examination of the sequences at the junctions of the different revertant *dpm*12 tandem duplications does not identify any consensus sequence that could indicate that these new sequences are responsible for the restored enhancer function; in particular, alternating purine and pyrimidine sequences destroyed by the *dpm*12 mutations are not restored. A consistent feature, however, is the presence in each duplication of a particular 15-bp sequence that we call element C. This sequence lies between the two Pu/Py sequences (its location is shown by the stippled region in each duplication in Fig. 1) and spans the 8-bp SV40 "core" element that Laimins et al. (1982) and Weiher et al. (1983) identified on the basis of sequence homology with other enhancer elements. The sequence of the 15-bp C element, with a bracket identifying the SV40 "core" element, is shown at the bottom of Figure 1. The consistent duplication of the 15-bp C element in the *dpm*12 revertants suggests that duplication of this element can compensate for loss of function within the two Pu/Py elements. These results leave unanswered, however, the relative roles of these various sequences in activating transcription. For example, these results could be explained if the SV40 enhancer contains multiple independent elements, each capable of enhancing transcription or if the 15-bp C element confers a unique and required "core" enhancer function while the sequences mutated in *dpm*12 only serve in an auxiliary stimulatory role(s).

### The SV40 enhancer is composed of multiple functional elements that can compensate for one another

To address the question of what genetic events can compensate for loss of function in the 15-bp C element, we isolated revertants of the core element mutant *dpm*6 (Herr and Clarke 1986). The *dpm*6 mutant contains two point mutations within the SV40 core element and, as in the case of *dpm*12, a single 72-bp element. The structures of 13 *dpm*6 revertants are shown in Figure 2. As in the *dpm*12 revertants, the *dpm*6 revertants contain simple tandem duplications. In the *dpm*6 revertants, however, the duplications consistently duplicate either one or both of the two regions surrounding the Pu/Py elements that are mutated in *dpm*12. We refer to the two regions consistently duplicated in the *dpm*6 revertants as elements A and B. The location and nucleotide sequence of these A and B elements, which are 21 and 22 bp long, respectively, are shown at the bottom of Figure 2. (We differentiate between the 8-bp Pu/Py sequences and the A and B enhancer elements by labeling the former **a** and **b**.) These results suggest that the SV40 enhancer is composed of at least three separate units of enhancer function, elements A, B, and C, and that these units can cooperate with one another or with duplicates of themselves to enhance transcription.

Figure 3 summarizes the commonly duplicated re-

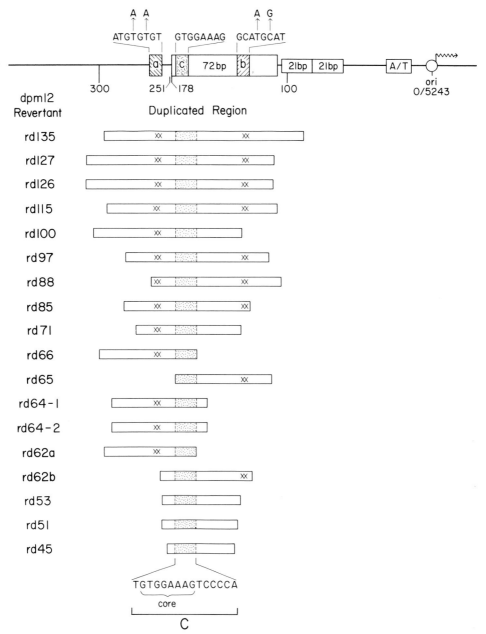

**Figure 1** Regions tandemly duplicated in revertants of the SV40 enhancer mutant, *dpm*12. The diagram at *top* shows the elements within the SV40 control region containing only one copy of the 72-bp element. Shown from right to left are the early transcriptional start site (wavy arrow), the origin of replication (ori), the AT-rich TATA-like element (A/T), the GC-rich 21-bp repeats, and the single 72-bp element. The numbers below the diagram refer to the nucleotide positions in the SV40 sequence. The locations of the three elements discussed in the text, the "Pu/Py" segments (hatched boxes **a** and **b**) and the "core" element (box **c**) are shown with the respective wild-type sequence displayed above each box. The four *dpm*12 transversions are indicated above the two "Pu/Py" segments. The extents of the regions tandemly duplicated within each of the 18 *dpm*12 revertants are indicated by the open rectangular boxes below the diagram of the SV40 control region. The revertants are aligned from top to bottom in order of size. The precise end points of each duplication are described elsewhere (Herr and Gluzman 1985). The series of × × 's identify the positions of the original point mutations where they are included in the duplication. The stippled area within each rectangular box corresponds to the 15-bp C element; the sequence of this region is shown at *bottom* with a bracket outlining the "core" element (Weiher et al. 1983).

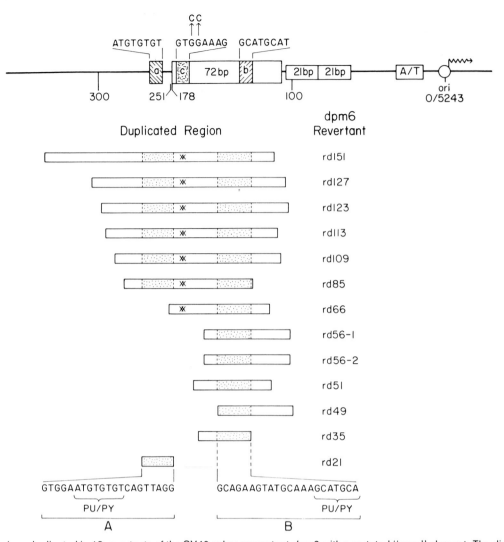

**Figure 2** Regions duplicated in 13 revertants of the SV40 enhancer mutant *dpm*6 with a mutated "core" element. The diagram of the SV40 early promoter shown at *top* is as described in the legend to Fig. 1. The two *dpm*6 transversions are shown above the sequence of the "core" element (box **c**). The extents of the duplicated regions in the *dpm*6 revertants are indicated by the rectangular boxes as described in the legend to Fig. 1; the precise end points are described elsewhere (Herr and Clarke 1986). The stippled regions show the position of the A and B elements. The nucleotide sequences of these two elements are shown at *bottom* with the "Pu/Py" sequences bracketed. (Reprinted, with permission, from Herr and Clarke 1986.)

gions in the *dpm*12 and *dpm*6 revertants, and in two series of revertants of the **a** Pu/Py–core (*dpm*12) and core–**b** Pu/Py (*dpm*26) double mutants (J. Clarke and W. Herr, unpubl.). In addition to the 18 *dpm*12 revertants shown in Figure 1, we isolated 5 *dpm*12 revertants as controls during isolation of the *dpm*6 and other *dpm* revertants (Herr and Clarke 1986). These additional revertants continue to duplicate the 15-bp C element, thus bringing to 23 the total number of *dpm*12 revertants that duplicate this element (see Fig. 3). We have also isolated a second series of 11 *dpm*6 revertants, generating a total of 24 *dpm*6 revertants. Of these revertants, 14 contain duplications spanning both elements A and B, while 8 span the B but not A element and only 2 revertants duplicate the A element but not B element (see Fig. 3).

The paucity of *dpm*6 revertants that duplicate the A

element in the absence of the B domain suggests that this domain is not as capable of compensating for the *dpm*6 mutation in the C element. To show that this region is, however, a bona fide domain, we isolated revertants of the "double" mutant *dpm*26 in which both the B and C domains are mutated. We have isolated 11 revertants of the *dpm*26 mutant; all but one contain a single tandem duplication (one unusual revertant contains two different tandem duplications), and when aligned as in Figure 1, they define a maximal 23-bp region that is contained in each duplication. This region is identified in Figure 3 and contains 20 bp out of the 21-bp A element defined by the *dpm*6 *rd*21 (see Fig. 2). These results strengthen the assignment of the A element as a functional SV40 enhancer element. In a complementary series of 10 revertants of the A- and C-domain "double" mutant *dpm*16, a 27-bp domain is

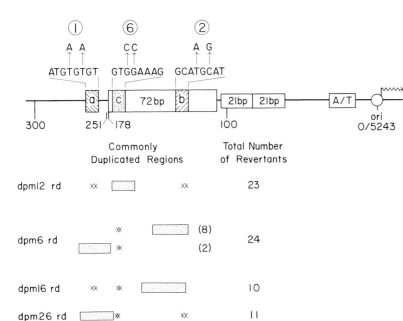

Figure 3 Regions commonly duplicated within revertants of the *dpm*12, *dpm*6, *dpm*16, and *dpm*26 mutants. The stippled boxes identify the regions that are consistently duplicated in revertants of the *dpm* mutants identified at *left*. The × × 's identify the location of the point mutation within each mutant. The numbers to the right of the figure show the total number of revertants analyzed for each mutant. The numbers in parenthesis for the *dpm*6 revertants show the number of revertants that duplicate either the A or B domains exclusively.

consistently duplicated that contains 21 bp of the 22-bp B element shown in Figure 2. These results are all consistent with the hypothesis that the SV40 enhancer is composed of multiple functional elements, each of which is capable of compensating for mutations within the other two elements.

To test whether these elements are individually sufficient to confer enhancer function, we have assayed the ability of synthetic enhancers composed of multiple copies of just one of these three elements to activate transcription. These enhancers have been constructed from multiple, unidirectional tandem copies of a 17-bp SV40 DNA fragment—GGG*TGTGGAAAGTCCCC*—containing 14 bp of the 15-bp C element (italicized; cf. Fig. 1). We have replaced the wild-type SV40 enhancer region of the plasmid πSVHPβΔ128 (see Herr and Gluzman 1985) with four and seven tandem copies of this "C-17mer" and have measured enhancer function in a transient assay of human β-globin gene expression

in both HeLa and CV-1 cells. In both cell lines, the synthetic C-17mer enhancers function effectively; the enhancer with four tandem copies of the C-17mer induces as much β-globin RNA as the wild-type enhancer containing a single copy of the 72-bp element, while the heptameric C-17mer enhancer is nearly as active as the wild-type enhancer with two 72-bp elements (B. Ondek and W. Herr, unpubl.). These experiments show that the C element can function in the complete absence of the A and B elements and other enhancer sequences.

## Discussion

The experiments described indicate that the SV40 enhancer is a composite of different modules of enhancer function that can be used in different combinations to yield a strong enhancer element. Figure 4 shows the relative locations of the three elements A, B, and C within the SV40 enhancer and a comparison of the

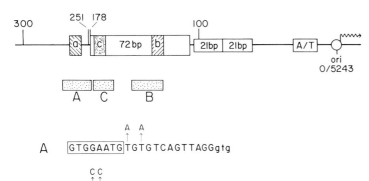

Figure 4 Locations of the three enhancer elements A, B, and C (stippled boxes) within the SV40 early promoter and nucleotide sequence comparison of these three elements. At *top* is shown the SV40 early promoter, which is described in the legend to Fig. 1. The sequence of each element is shown along with the *dpm*1, *dpm*2, and *dpm*6 point mutations. The lower-case letters to the right of the A-element sequence indicate an ambiguity in the sequence of this element caused by the GTG terminal redundancy. The boxes identify the regions of greatest homology between these elements (see text).

nucleotide sequence of each element. The regions of greatest homology between the three elements are boxed. Elements A and C both contain core consensus sequences ($GTGG^A/_T^A/_T^A/_TG$; Weiher et al. 1983), but it is unlikely that the core sequence alone is the entire functional unit within these two elements because the *dpm*1 mutations that affect the A element lie outside the core consensus sequence (see Fig. 4). The B element does not contain a core consensus sequence; the best match is 6 out of 8 bp in the sequence ATGCAAAG. This B-element sequence is instead more homologous with the octamer promoter element ATGCAAAT found in the immunoglobulin enhancer and upstream promoter regions (Falkner and Zachau 1984; Parslow et al. 1984; Mason et al. 1985) and the promoter regions of the chicken histone H2B (Harvey et al. 1982) and *Xenopus* U2 (Mattaj et al. 1985) genes.

Both in vivo and in vitro enhancer competition experiments suggest that *trans*-acting factors are involved in enhancer function (Schöler and Gruss 1984, 1985; Wildeman et al. 1984; Mercola et al. 1985; Sassone-Corsi et al. 1985). The three elements A, B, and C probably represent binding sites for one or more *trans*-acting factors. The simplest way to explain how the three different elements can each independently activate transcription is by proposing that the same *trans*-acting factor (or set of factors) interacts with each element. The basic requirement for strong enhancer function in this model is a sufficient number of binding sites for this factor. Nevertheless, because these elements only share limited sequence homology, we favor a model whereby each element interacts with a unique factor or combination of factors that share common functional domains.

Figure 5 is a diagram of both known and postulated interactions of host-cell transcription factors with the SV40 early promoter. To the right is illustrated a TATA box–binding factor (Davison et al. 1983; Sawadogo and Roeder 1985) and five Sp1 protomers that bind to the 1–3, 5, and 6 CCGCCC boxes within the 21-bp repeats (see Gidoni et al. 1985; Dynan and Tjian 1985). The A, B, and C enhancer elements are shown as stippled boxes of differing shapes because of their unique sequences. In this model, separate factors (labeled in the figure as α, β, and γ) with unique DNA-binding domains interact with each element. These factors may be monomers, as suggested in the illustration; or they may represent protein dimers. In this case, the A and C elements, for example, might interact with the same core-element-specific factor that binds to the core consensus sequence, which these two elements share; a second monomer might be specific for the α and γ factors and bind to the unique sequences within these two elements.

We have demonstrated that the three elements A, B, and C can functionally compensate for one another. Thus, if multiple *trans*-acting enhancer factors with different DNA sequence specificities do exist for these elements, these factors probably activate transcription by similar mechanisms. These factors may, therefore, share common protein domains that interact with other factors in a transcriptional complex. (This is indicated in the figure by the commonly shaped box portion of the putative α, β, and γ enhancer factors.) Such a separation of protein domains for DNA binding and activation of transcription is known to exist in the repressors for *Escherichia coli* lambdoid phages (Hochschild et al. 1983; Wharton et al. 1984).

The potential use of different functionally redundant host-cell factors to activate the SV40 enhancer could, in part, explain how this enhancer can function in a broad range of different cell types and explain why the wild-type enhancer is duplicated. The 72-bp duplication within the wild-type SV40 enhancer creates two copies of each of the B and C domains. The revertants of the *dpm*12 and *dpm*6 mutants (see Fig. 3) show that duplicated B or C elements can effectively enhance transcription when the other element is mutated; thus, the wild-type SV40 enhancer should be capable of enhancing transcription in the presence of either a B or C element-specific *trans*-acting factor alone. This model argues, therefore, that the revertant enhancers, in which one or more elements is mutated, should display a more restricted host-cell preference. Consistent with this conclusion, we have found that, compared with the wild-type enhancer, *dpm*12 revertants exhibit about the same levels of enhancer function in both CV-1 cells and HeLa cells. In the case of the *dpm*6 revertant enhancers we have tested, however, we find that these enhancers are considerably more active in CV-1 cells than in HeLa cells (A. Shepard and W. Herr, unpubl.).

## Acknowledgments

We thank our colleagues in the James and Sambrook laboratories of Cold Spring Harbor Laboratory for many helpful discussions. This work was funded by U.S. Public Health Service grant CA-13106 from the NCI.

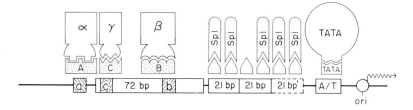

**Figure 5** Illustration of possible interactions between the SV40 early promoter and host-cell factors. The schematic of the early promoter is as described in the legend to Fig. 1. The boxes immediately above the promoter represent sequence elements to which the various transcription factors interact. See the text for a description of the various factors.

# References

Banerji, J., S. Rusconi, and W. Schaffner. 1981. Expression of a β-globin gene is enhanced by remote SV40 DNA sequences. *Cell* **27**: 299.

Benoist, C. and P. Chambon. 1981. *In vivo* sequence requirements of the SV40 early promoter region. *Nature* **290**: 304.

Davison, B.L., J.M. Egly, E.R. Mulvihill, and P. Chambon. 1983. Formation of stable preinitiation complexes between eukaryotic class B transcription factors and promoter sequences. *Nature* **301**: 680.

Dynan, W.S. and R. Tjian. 1985. Control of eukaryotic messenger RNA synthesis by sequence-specific DNA-binding proteins. *Nature* **316**: 774.

Falkner, F.G. and H.G. Zachau. 1984. Correct transcription of an immunoglobulin κ gene requires an upstream fragment containing conserved sequence elements. *Nature* **310**: 71.

Fromm, M. and P. Berg. 1982. Deletion mapping of DNA regions required for SV40 early region promoter function *in vivo. J. Mol. Appl. Genet.* **1**: 457.

Gidoni, D., J.T. Kadonaga, H. Barrera-Saldana, K. Takahashi, P. Chambon, and R. Tjian. 1985. Bidirectional SV40 transcription mediated by tandem Sp1 binding interactions. *Science* **230**: 511.

Gluzman, Y., ed. 1985. *Eukaryotic transcription: The role of cis- and trans-acting elements in initiation.* Cold Spring Harbor Laboratory, Cold Spring Harbor, New York.

Gluzman, Y. and T. Shenk, eds. 1983. *Enhancers and eukaryotic gene expression.* Cold Spring Harbor Laboratory, Cold Spring Harbor, New York.

Gluzman, Y., J.F. Sambrook, and R.J. Frisque. 1980. Expression of early genes of origin-defective mutants of simian virus 40. *Proc. Natl. Acad. Sci.* **77**: 3898.

Gruss, P., R. Dhar, and G. Khoury. 1981. Simian virus 40 tandem repeated sequences as an element of the early promoter. *Proc. Natl. Acad. Sci.* **78**: 943.

Harvey, R.P., A.J. Robins, and J.R.E. Wells. 1982. Independently evolving chicken histone H2B genes: Identification of an ubiquitous H2B-specific 5' element. *Nucleic Acids Res.* **10**: 7851.

Herr, W. and J. Clarke. 1986. The SV40 enhancer is composed of multiple functional elements that can compensate for one another. *Cell* **45**: 461.

Herr, W. and Y. Gluzman. 1985. Duplications of a mutated simian virus 40 enhancer restore its activity. *Nature* **313**: 711.

Hirt, B. 1967. Selective extraction of polyoma DNA from infected mouse cell cultures. *J. Mol. Biol.* **26**: 365.

Hochschild, A., N. Irwin, and M. Ptashne. 1983. Repressor structure and the mechanism of positive control. *Cell* **32**: 319.

Laimins, L.A., G. Khoury, C. Gorman, B. Howard, and P. Gruss. 1982. Host-specific activation of transcription by tandem repeats from simian virus 40 and Moloney murine sarcoma virus. *Proc. Natl. Acad. Sci.* **79**: 6453.

Mason, J.O., G.T. Williams, and M.S. Neuberger. 1985. Transcription cell type specificity is conferred by an immunoglobulin $V_H$ gene promoter that includes a functional consensus sequence. *Cell* **41**: 479.

Mattaj, I.W., S. Lienhard, J. Jiricny, and E.M. De Robertis. 1985. An enhancer-like sequence within the *Xenopus* U2 gene promoter facilitates the formation of stable transcription complexes. *Nature* **316**: 163.

Mercola, M., J. Goverman, C. Mirell, and K. Calame. 1985. Immunoglobulin heavy-chain enhancer requires one or more tissue-specific factors. *Science* **227**: 266.

Moreau, P., R. Hen, B. Wasylyk, R. Everett, M.P. Gaub, and P. Chambon. 1981. The SV40 72 base pair repeat has a striking effect on gene expression both in SV40 and other chimeric recombinants. *Nucleic Acids Res.* **9**: 6047.

Nordheim, A. and A. Rich. 1983. Negatively supercoiled simian virus 40 DNA contains Z-DNA segments within transcriptional enhancer sequences. *Nature* **303**: 674.

Parslow, T.G., D.L. Blair, W.J. Murphy, and D.K. Granner. 1984. Structure of the 5' ends of immunoglobulin genes: A novel conserved sequence. *Proc. Natl. Acad. Sci.* **81**: 2650.

Sassone-Corsi, P., A. Wildeman, and P. Chambon. 1985. A *trans*-acting factor is responsible for the simian virus 40 enhancer activity *in vitro. Nature* **313**: 458.

Sawadogo, M. and R.G. Roeder. 1985. Interaction of a gene-specific transcription factor with the adenovirus major late promoter upstream of the TATA box region. *Cell* **43**: 165.

Schöler, H.R. and P. Gruss. 1984. Specific interaction between enhancer-containing molecules and cellular components. *Cell* **36**: 403.

———. 1985. Cell type-specific transcriptional enhancement *in vitro* requires the presence of *trans*-acting factors. *EMBO J.* **4**: 3005.

Weiher, H., M. Konig, and P. Gruss. 1983. Multiple point mutations affecting the simian virus 40 enhancer. *Science* **219**: 626.

Wharton, R.P., E.L. Brown, and M. Ptashne. 1984. Substituting an α-helix switches the sequence-specific DNA interactions of a repressor. *Cell* **38**: 361.

Wildeman, A.G., P. Sassone-Corsi, T. Grundstrom, M. Zenke, and P. Chambon. 1984. Stimulation of *in vitro* transcription from the SV40 early promoter by the enhancer involves a specific *trans*-acting factor. *EMBO J.* **3**: 3129.

# Identification of Cellular Proteins That Interact with Polyomavirus or Simian Virus 40 Enhancers

## J. Piette, S. Cereghini, M.-H. Kryszke, and M. Yaniv

Department of Molecular Biology, Pasteur Institute, 75724 Paris, France

We have used the genomic-sequencing technique of Church and Gilbert (1984) coupled with nuclease digestion to map salt-labile protein-DNA contacts along the enhancer sequences of polyomavirus and simian virus 40 (SV40). With SV40 we found that roughly 20 bp are protected against micrococcal nuclease around the SphI and EcoRII sites of each of the 72-bp repeats. Another sequence near the PvuII site is also protected. Finally, the 21-bp repeats are occupied by proteins, presumably the Sp1 transcription factor. Similarly, in polyoma, sequences in both enhancer elements A and B are protected against DNase I. These sequences include homologies with the SV40 enhancer, immunoglobulin heavy-chain enhancer, and the adenovirus E1A enhancer. Using extracts of noninfected mouse fibroblasts, we identified proteins that bind specifically to the B enhancer of polyoma, protecting sequences similar to those observed in vivo. Our studies strongly suggest that a multiprotein-DNA complex is formed along the viral enhancers, a complex probably required for its function since it involves all the sequences that were shown to be crucial for its function.

Enhancers are DNA sequences that can stimulate transcription initiation from cellular or viral promoters when placed either 5′ or 3′ in both possible orientations relative to the transcription unit. The noncoding region of papovavirus contains potent enhancers. In the case of simian virus 40 (SV40), these sequences include the 72-bp repeat and some residues on its late side (see Chambon et al; Herr et al; both this volume). In the case of polyomavirus, the enhancer region defined initially by de Villiers and Schaffner (1981) can be divided into two subregions, each one of which behaves like an independent enhancer element (Herbomel et al. 1984; see also Hassell et al., this volume). The major one (A) included in the BclI-PvuII fragment (nucleotides [nt] 5022–5130) shares homology with the enhancer found upstream of the early region 1A (E1A) gene of adenovirus (Hearing and Shenk 1983). The minor enhancer element (B) included in the PvuII-PvuII fragment shares homology with the IgH enhancer (Banerji et al. 1983) and contains the Weiher-Gruss GTGGTTT(T)G consensus sequence (Weiher et al. 1982) as well as a repetition of the CACCA motif found in the bovine papillomavirus (BPV) enhancer (Weiher and Botchan 1984). Despite the large degree of homology in genome organization, in the sequence of the viral proteins, and in the core origin sequences, a certain divergence exists in the noncoding region between SV40 and polyoma. In polyoma the origin sequences are inserted between the enhancer and the early cap site, whereas for SV40 the early cap site(s) coincide with the minimal origin sequences (Tooze 1981). This can explain the fact that polyomavirus replication requires an active enhancer (either A or B, identical with the α and β elements of Muller et al. [1983]) for its replication, whereas SV40 does not.

Studies of SV40 and polyomavirus minichromosomes revealed a particular property of the origin-proximal region. Both in SV40 and polyoma, this region is preferentially cleaved by DNase I or other endonucleases (Scott and Wigmore 1978; Waldeck et al. 1978; Varshavsky et al. 1979; Saragosti et al. 1980; Herbomel et al. 1981). Furthermore, electron microscopy studies have shown that at least in a fraction of the minichromosomes, this region is devoid of nucleosomes (Jakobovits et al. 1980; Saragosti et al. 1980; Jongstra et al. 1984). The exclusion of nucleosomes from the enhancer region is not dependent on the presence of viral proteins. The same DNase I sensitivity is observed when plasmids containing only the origin-enhancer region of SV40 are replicating in COS cells (Innis and Scott 1983; Cereghini and Yaniv 1984). Furthermore, a shift of SV40 tsA mutants to the nonpermissive temperature—a temperature at which the T antigen does not bind to the viral DNA—does not eliminate the hypersensitivity to DNase I (Cereghini et al. 1983). These and other experiments employing SV40 variants with duplications clearly show that the sensitivity to nuclease digestion and the appearance of a nucleosome-free gap is dependent on sequences within or just bordering the sensitive region and not on other distal viral sequences (Fromm and Berg 1983; Jongstra et al. 1984). Such sequences have either an inherent low affinity to bind histones (Wasylyk et al. 1979) or, alternatively, they are interacting with some proteins that interfere with nucleosome formation.

The nuclease digestion studies reported above always showed that the sensitive region is not uniform. Parts of it are highly sensitive, whereas others are highly resistant—a pattern different from that observed with naked DNA. These considerations prompted us to examine the possible association of cellular, nonhistone proteins with the viral enhancer sequences. Two different approaches were used in these studies: (1) genomic sequencing to map diphosphate bonds that are protected against nuclease attack (Church and Gilbert 1984), and (2) search for cellular proteins that interact with the enhancer sequences in vitro.

## Materials and Methods

Nuclei were isolated from SV40-infected CV1 cells 40–41 hr postinfection or from polyoma-infected 3T6 cells 24 hr postinfection as described by Cereghini et al. (1984), except that KCl was replaced by 130 mM NaCl in all solutions. Minichromosomes were extracted from a fraction of the nuclear suspension according to the isotonic procedure of Fernandez-Munoz et al. (1979). They were adjusted to the desired salt concentration, kept for 10 min on ice, and layered onto a sucrose step gradient, consisting of 7 ml of 5% sucrose in 0.025 M Tris · HCl (pH 7.5), 1 mM PMSF and containing the same salt concentration and 3 ml of cushion of 15% sucrose that is in the standard isotonic buffer. After centrifugation for 15 hr at 35,000 rpm, the minichromosome pellet was gently resuspended in the isotonic buffer and digested with nucleases. Digestion of nuclei or chromatin with

microccocal nuclease was done at 18°C, and digestion of free DNA was performed at 0°C. DNase I digestion of nuclei or salt-washed nuclei was done at 35°C. The reactions were arrested and DNA was extracted as described by Cereghini and Yaniv (1984). The extent of digestion was followed by the conversion of the viral DNA into relaxed and linear forms. Samples showing equivalent levels of digestion were used for further studies. The genomic-sequencing procedure of Church and Gilbert (1984) was followed, except that DNA was transferred from the acrylamide gel to the nylon membrane by suction.

Gel retardation assays, BAL-31 deletion, and DNase I footprinting were done as described by Piette et al. (1985).

## Results

### Three different domains along the SV40 enhancer are protected against micrococcal nuclease digestion by salt-washable proteins

As described in Materials and Methods and in the legend to Figures 1 and 2, we prepared nuclei or viral minichromosomes from SV40-infected CVI cells in the beginning of the late phase. The viral minichromosomes

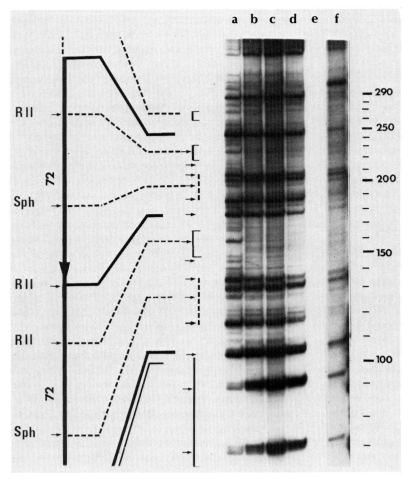

**Figure 1** Protein-DNA contacts along the SV40 enhancer as revealed by micrococcal nuclease footprinting: Autoradiography. Nuclei from SV40-infected cells or minichromosomes extracted from these nuclei and exposed to different salt concentrations were treated with micrococcal nuclease under identical ionic strengths. DNA was extracted, digested with *Hind*III, denatured, and fractionated on a sequencing gel. After the run, the DNA was transferred to a nylon membrane and hybridized with a radioactively labeled, single-stranded M13 probe complementary to the late strand of SV40. In several experiments, a sequence ladder on nonradioactive DNA digested with *Hind*III was run alongside the samples. The location of the protected and hypersensitive sites was obtained by comparing the sequence ladder (not shown) and the nuclease cleavage sites. (Lane *a*) Nuclei; (lane *b*) minichromosomes purified through a sucrose cushion of 0.13 M salt; (lanes *c* and *d*) minichromosomes washed with 0.35 M and 0.55 M salt; (lane *f*) free superhelical SV40 DNA briefly digested with the nuclease. SV40 nucleotide numbers are indicated on the right. The locations of the protected regions (continuous lines), the partially protected segments (discontinuous lines), and the hypersensitive sites along the 21-bp and 72-bp repeats are indicated at left. Also included are the positions of the *Sph*I and *Eco*RII cleavage sites.

were further treated with increasing salt concentrations to remove nonhistone proteins. All the samples were brought back to the same ionic strength and digested for different lengths of time with micrococcal nuclease. DNA was extracted, digested with HindIII, BglI, or NcoI, denatured, fractionated on a sequencing gel, and transferred to a nylon membrane. Uniformly $^{32}$P-labeled, single-stranded probes abutting the restriction site complementary to either the early or late strand of SV40 were used to map the nuclease cleavage sites at the nucleotide level. As can be seen in Figure 1, the enzyme has certain preferred cleavage sites; however, almost every nucleotide is cleaved at a detectable rate. When the cleavage pattern of the salt-washed viral mini-chromosomes or that of free DNA is compared with the products of nuclei digestion, several clear differences appear. A region surrounding the EcoRII site (roughly nt 151–169) on the late strand is fully protected in nuclei with the exception of hypersensitive sites in the middle and in the boundaries of the protected region. Both the

protection and the increased cleavage disappear upon low-salt wash of the minichromosomes and are absent in naked DNA (cf. lane a with lanes b and f, Fig. 1). An identical protected region interrupted by a hypersensitive site was mapped also along the EcoRII site of the second 72-bp element. Another protected region, again restricted to nuclei, is visible along the PvuII site. A partial change in the frequency of cleavage of several bands is seen in both repeats around the SphI sites when lane a is compared with lanes b–f. Cleavage frequency between nt 119 and 139 is increasing, whereas cleavage after residue 140 is decreasing upon salt wash or during purification of minichromosomes (see a scan of the autoradiography in Fig. 2). The cleavage profile of naked DNA is similar to that of salt-washed minichromosomes.

The changes we see upon salt wash are not restricted to the enhancer sequences. Examination of the auto-radiographs clearly showed that the 21-bp repeats are largely protected in nuclei and become exposed upon

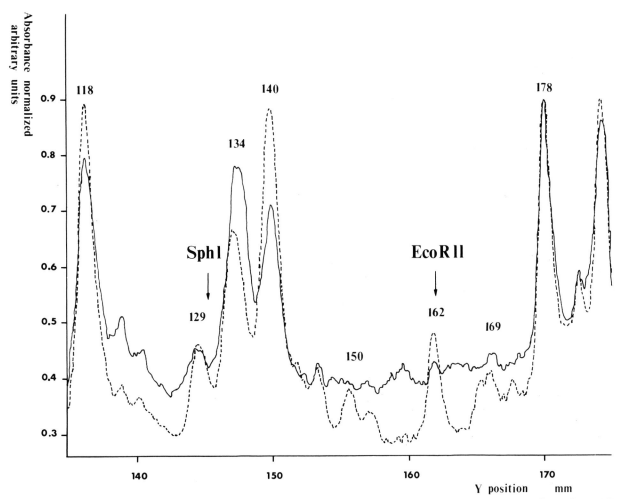

**Figure 2** Protein-DNA contacts along the SV40 enhancer as revealed by micrococcal nuclease footprinting: Scan. Part of the autoradiography reproduced in Fig. 1 corresponding to lanes a (nuclei) (- - - -) and b (purified minichromosomes washed with 0.13 M salt) (———) was scanned and the two profiles were overlapped. Note the relatively equal sensitivity along nt 180–190 between the two lines, the protected and hypersensitive sites between nt 150 and 170, and the change in intensity of the bands around the SphI site (~nt 120–140).

salt wash or in naked DNA (see the bottom part of Fig. 1).

The protection pattern along the early strand is less pronounced than that observed along the late strand. Nevertheless, careful examination of the autoradiographs clearly showed salt-dependent protection of the 21-bp repeats, partial protection of the *Sph*I site, and salt-dependent protection and hypersensitivity of certain bands along the *Eco*RII and *Pvu*II sites. Here again, identical patterns are seen along the first and second 72-bp repeats. Figure 3 summarizes the localization of the different protected or hypersensitized bonds on both strands. For simplicity we assume that each protected domain corresponds to a sequence occupied by a distinct protein. The six GGGCGG motifs found in the

22-bp segment and the 21-bp repeat are probably occupied by the Sp1 protein characterized by Dynan and Tjian (1983). We define the proteins interacting with the *Sph*I, *Eco*RII, and *Pvu*II sites as enhancer-binding proteins (EBP) 1, 2, and 3, respectively. The association of these proteins with the viral chromatin is relatively fragile. They are maintained on the viral minichromosomes after their leakage from the nuclei under isotonic conditions (results not shown). However, the mere sedimentation of the viral chromatin on sucrose gradients under the same salt concentration (0.13 M salt) results in their dissociation. These conditions and even washing the viral minichromosomes at 0.35 M or 0.55 M salt does not dissociate the core histones from the viral DNA. Hence, the protection we see is not the imprint of

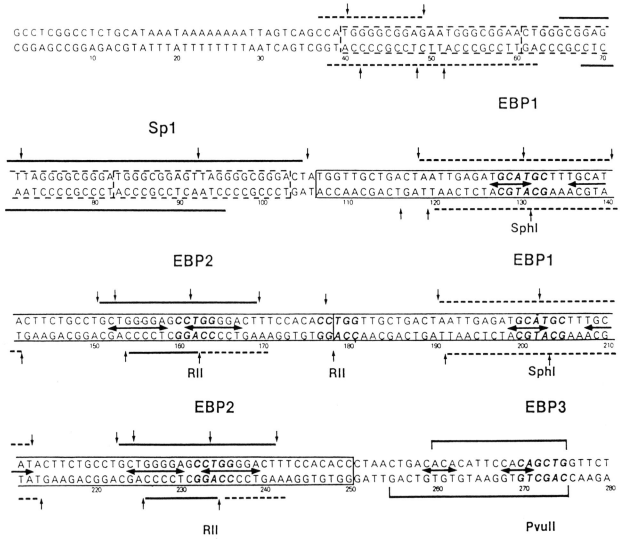

**Figure 3** Summary of the nuclease protection data. The sequence of the SV40 genome from residues 1 to 280 is given. Protected sequences are indicated by a continuous line (full protection) or discontinuous lines (partial protection) above the top strand (late) and the bottom strand (early). Vertical arrows indicate enhanced nuclease cleavage relative to salt-washed chromosomes or to free DNA. The binding sites along the 22-bp region and the two repeats of 21 bp are occupied probably by the Sp1 transcription factor. EBP1, 2, and 3 designate the three different proteins postulated to bind to the two repeats of 72 bp and to the sequences around the *Pvu*II site. Relevant restriction sites are indicated by bold letters. Horizontal, double-headed arrows show the repeated motifs of 5, 7, and 4 nt within EBP1-, EBP2-, and EBP3-protected domains, respectively.

a modified nucleosome present in this position. Furthermore, the fact that we see changes in the sensitivity of this region upon very mild treatment that does not change grossly the topology of the DNA suggests that the changes in the digestion profile we see are not caused merely by conformational transitions in the viral minichromosome.

## Salt-labile contacts along the polyoma enhancers

A rather similar approach to that described in the previous section was undertaken to analyze the structure of the polyoma enhancer in vivo. As described in the legend to Figure 4, nuclei prepared from mouse 3T6 cells infected with polyomavirus were treated with DNase I either before or after washing the nuclei with

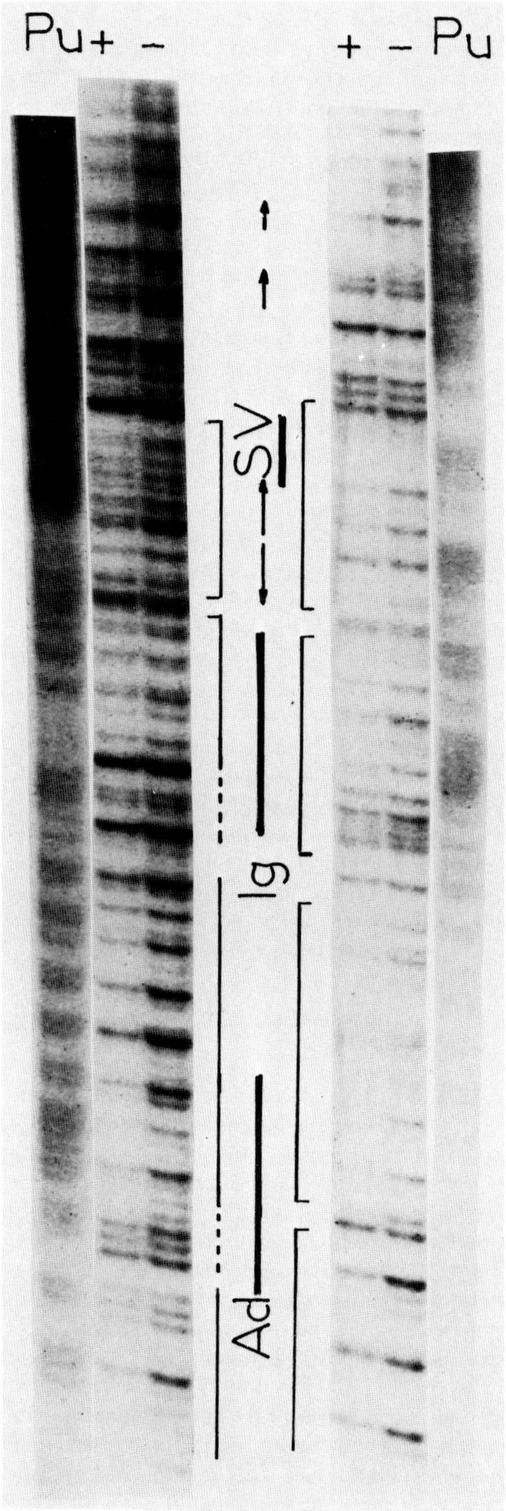

**Figure 4** DNase I footprinting of the enhancer region of polyomavirus in vivo. Nulcei prepared from polyoma-infected 3T6 cells were treated directly or after exposure to salt with DNase I. Total DNA was prepared, cleaved with *Bcl*I, denatured, and electrophoresed on denaturing polyacrylamide gels together with a nonradioactive sequencing ladder starting from the same restriction site. After transfer to nylon membranes, the specific polyoma fragments were revealed with [$\alpha$-$^{32}$P]-labeled M13 probes hybridizing with the early or late strand, respectively. The protection pattern of the late strand is shown on the left and that of the early strand is shown on the right. (Pu) Purine sequence reactions; (+) DNA from nuclei washed with 0.13 M NaCl; (−) DNA from nuclei washed with 0.30 M NaCl. Homologies to other enhancers and characteristic structures of the region are indicated as described further in the legend to Fig. 7. Protected areas are indicated by continuous or discontinuous lines along the left (late strand) or the right (early strand).

0.3 M salt. Total nuclear DNA was purified, cleaved with *Bcl*I, denatured, and fractionated on a sequencing gel. After transfer to nylon membranes, the viral DNA was revealed by hybridization with single-stranded probes specific for the early or late strands. Sequences both in the A and B enhancer elements of polyoma are protected against DNase I by salt-washable proteins. These sequences include the E1A-like core of enhancer A, the IgG, the SV40 homology, and the G + C–rich palindrome of enhancer B. Rather similar results were obtained with both the early and late strands.

### Search for EBPs in cell extracts

It is clear that better understanding of enhancer function will require the identification of cellular proteins that interact with these sequences. As a first step to reach this goal, we searched for a rapid and easy technique to follow specific DNA-protein interactions. To do so we employed the gel retardation procedure introduced by Garner and Revzin (1981) and by Fried and Crothers (1981) for the study of DNA-protein interactions in prokaryotes. More recently, Strauss and Varshavsky (1984) used the same approach to identify a protein in monkey cells that interacts specifically with the α satellite DNA. In essence, a radioactively labeled DNA fragment is mixed with a nuclear extract in the presence of a large excess of nonspecific carrier DNA. The mixture is applied to a nondenaturing polyacrylamide gel. After electrophoresis, the gel is dried and autoradiographed. As shown in Figure 5, when such an experiment was done with a DNA fragment including the polyoma A and B enhancer elements, a fraction of the DNA was retarded as a clear band. Using other fragments containing either the A or the B enhancers of polyoma or pBR322 fragments devoid of enhancer activity clearly showed that the protein-DNA interaction we detect is specific for the polyoma B enhancer. Until now we did not detect in our extracts a binding activity specific for the polyoma A enhancer.

Further competition experiments confirm that the DNA-protein complex we detect is in fact specific for the B enhancer. The radioactive, retarded band disappears when excess cold DNA containing the B enhancer is added to the incubation mixture. On the contrary, equal molarity of nonenhancer DNA fragments, of a fragment containing the A enhancer, and of fragments containing the SV40 origin-promoter-enhancer fragment, the SV40 enhancer, or the IgG enhancer compete only slightly or not at all (see Fig. 5d and Piette et al. 1985). In a certain respect, this is surprising since the polyoma B enhancer shares some homology with both the SV40 and the IgG enhancers.

The degree of retardation of a DNA fragment by bound proteins depends on the size of the protein, on the number of molecules bound, and on the changes in the conformation of the DNA fragment (Wu and Crothers 1984). The decrease in mobility is more pronounced when the bound protein is larger or when several molecules of the same protein are complexed with the same fragment. The frequent observation of several intermediate bands between the free DNA fragment and the final complex (Fig. 5b) strongly suggests that several molecules of the same protein or several different proteins are involved in the formation of the final complex.

The gel retardation assay can also be used to delimit the size of the minimal fragment required for the formation of the specific complex. To do this, we prepared a mixture of unidirectional BAL-31 deletions, starting from either of the two *Pvu*II sites (nt 5130 and 5267). The end-labeled fragments were incubated with the protein fraction and fractionated on a native polyacrylamide gel. After autoradiography, the DNA band corresponding to the bound DNA was eluted and the bulk of the BAL-31–treated DNA was analyzed on a sequencing gel alongside a sequence ladder. Comparison of the free and bound DNA lanes clearly showed that deletions entering from the late proximal *Pvu*II site (nt 5130) abolish complex formation when they reach nt 5150 or thereabout. Deletions from the early proximal *Pvu*II site (nt 5267) seem to decrease the binding when they delete the second CACCA element (nt 5230; the BPV homology). Binding totally disappears when the Weiher-Gruss consensus sequence was deleted (see Fig. 6).

In a similar approach, the position of the interacting protein(s) can be probed with DNase I or with dimethyl sulfate. After formation of a complex with a labeled DNA fragment, we treated the mixture with DNase I, stopped the reaction by the addition of excess EDTA, and applied the sample to a nondenaturing gel. The free and bound DNA bands were eluted from the gel and analyzed as before in a sequencing gel alongside a sequence ladder. Regions of protection and several increased cleavage sites were observed. Along the early strand, the IgG homology and the G + C–rich palindrome were totally protected against DNase I cleavage (results not shown). Along the late strand, we observe protection along the G + C–rich palindrome and the complementary strand of the Weiher-Gruss enhancer consensus sequence. These results, together with the BAL-31 deletion analysis and the in vivo DNase I footprinting, are summarized in Figure 7.

### Discussion and Conclusions

Enhancers are *cis*-acting DNA elements that activate transcription from promoter sequences. In SV40 these sequences are located about 110 bp upstream of the major cap site used in the early phase of the viral lytic cycle. If we suppose that the first 110 bp upstream of this cap site constitute the basic structure of the promoter (Dynan and Tjian 1985), the promoter and enhancer elements are juxtaposed one to the other in SV40. In polyoma the situation is more complex; the enhancer region starts more than 200 bp upstream of the major early cap site (Herbomel et al. 1984). This simple comparison shows that, even in their natural context, viral enhancers can be found in different positions relative to the transcription-initiation site. This example emphasizes the particularity of the enhancer elements. To function,

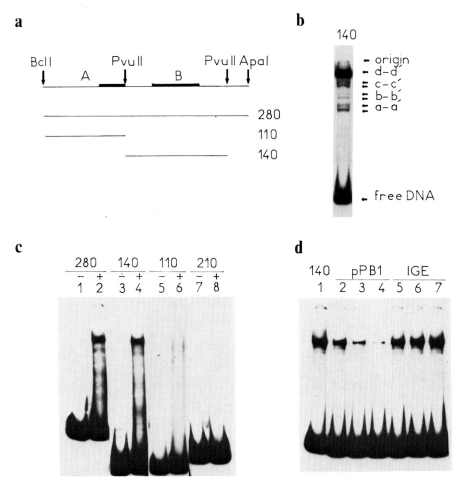

**Figure 5** Band-shifting experiments with the polyoma enhancer. (*a*) Fragments used in the band-shifting experiments. The region extending from the *Bcl*I site at position 5022 to the *Apa*I site at position 5291, containing the A and B enhancers, is indicated; the enhancer core sequences are represented by a black bar. (*b*) Band-shifting pattern obtained with the B enhancer. The 140-bp B-enhancer fragment together with an excess of nonradioactive carrier DNA (sonicated salmon sperm) were incubated with a nuclear extract of 3T6 cells and loaded on a 7.5% polyacrylamide gel. Indicated are the origin of the gel, the DNA fragments complexed by proteins (labeled a-a' to d-d') and the free DNA band. The four bands including d-d' were frequently observed as doublets. (*c*) Band-shifting experiments with different DNA fragments given in *a*. (Lanes *1* and *2*) Fragment 280; (lanes *3* and *4*) fragment 140; (lanes *5* and *6*) fragment 110; (lanes *7* and *8*) 210-bp fragment of prokaryotic origin. The fragments were incubated with (+) or without (−) a nuclear extract of 3T6 cells. (*d*) Competition experiments. A nuclear extract of 3T6 was incubated with the B enhancer in the absence (lane *1*) or presence of linearized pPB1 (a plasmid containing the B enhancer; lanes *2–4*) or linearized IGE (a plasmid containing the mouse heavy-chain gene enhancer; lanes *5–6*). A 7–40 molar excess of competitor was used, as indicated below each lane.

they do not have to be positioned in a constant distance from the transcription-initiation site, where the RNA polymerase is initiating the RNA chain. To explain the fact that enhancers can act at variable distances and in both orientations relative to the promoter, it was suggested that enhancers will serve as entry sites or traps for RNA polymerase. After binding to this site, the enzyme will scan the template until it finds an appropriate promoter (de Villiers et al. 1983; Wasylyk et al. 1983). This hypothesis was founded in part on the previous observations that viral enhancer sequences are included in a nucleosome-free gap at least in a fraction of the SV40 minichromosomes. Furthermore, both SV40, polyoma, and immunoglobulin enhancers were shown to be associated with nuclease-hypersensitive sites. The fine mapping of DNase I–sensitive regions along the SV40

and polyoma enhancers has shown that these regions are not uniformly exposed to the nuclease. A series of hypersensitive and resistant sites were mapped along both enhancers, revealing a pattern different from that observed with free DNA and suggesting that enhancer sequences are interacting with proteins in vivo. In the case of SV40, it was shown that these proteins are of cellular and not of viral origin (Cereghini and Yaniv 1984). These observations prompted us to undertake two complementary approaches for the identification of these proteins and their sites of action—footprinting of protein-DNA contacts in vivo and isolation of specific EBPs in cell extracts.

Using an approach similar to that of Ephrussi et al. (1985), we probed for changes in the sensitivity to DNase I or micrococcal nuclease of the phosphodiester

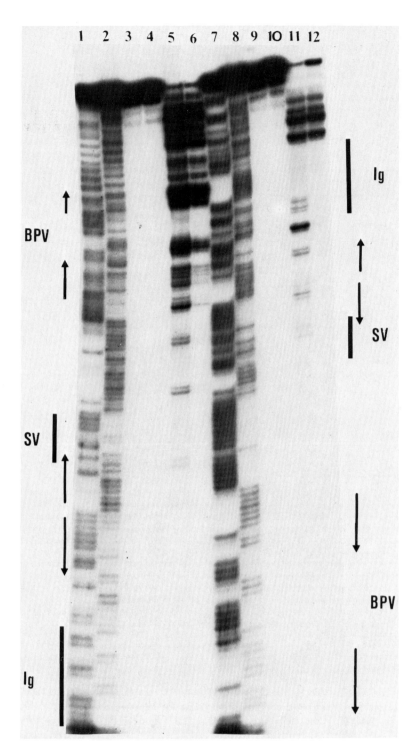

**Figure 6** BAL-31 deletion analysis of the polyoma B enhancer. A mixture of fragments deleted unidirectionally from either the origin-proximal or the late region–proximal PvuII sites were labeled at their invariable 3′ ends and incubated with 0.55 M nuclear extract of 3T6 cells. Complexed fragments were separated from the free DNA, and the DNA was eluted and loaded on a sequencing gel. Deleted but not complexed fragments were taken as controls together with purine or pyrimidine sequence ladders. (Lanes 1–6) DNA fragments labeled at the late side (PvuII site at nt 5130); (lanes 7–12) DNA fragments labeled at the early side (PvuII site at nt 5267); (lanes 1 and 7) purine sequence reactions; (lanes 2 and 8) pyrimidine sequence reactions; (lanes 3 and 6) control fragments; (lanes 4 and 10) retarded fragments of a short BAL-31 digestion; (lanes 5 and 11) control fragments; (lanes 6 and 12) retarded fragments of a long BAL-31 digestion.

bonds along the enhancer sequences upon mild salt wash of nuclei or of viral chromatin. As shown in Figures 1, 2, and 4, clear modifications are observed in the sites of nuclease cleavage upon this treatment. Certain cleavage sites appear or increase in strength upon salt wash, whereas the frequency of cleavage decreases in other positions. Such a profile is characteristic of the protection and of the increased sensitivity of certain phosphodiester bonds upon binding of a protein to specific

DNA sequences (Galas and Schmitz 1978). The genomic sequences that run alongside the cleavage products permitted us to precisely localize the protected regions and the enhanced cleavage sites. The final results are summarized in Figures 3 and 7 for SV40 and polyoma, respectively. In the case of SV40, we observed two protected domains within each of the two 72-bp repeats; one overlapping the SphI site and the other the EcoRII site. In addition, the sequences containing the

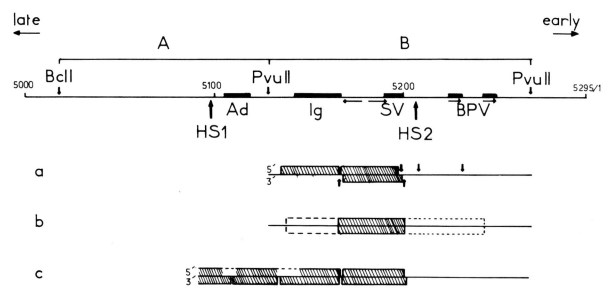

**Figure 7** Summary of the protection and binding data obtained with the polyoma enhancer in vitro and in vivo. The enhancer region of polyomavirus is represented, with homologies to other enhancers indicated by black bars. (Ad) Adenovirus E1A region; (Ig) immunoglobulin heavy chain; (SV) SV40; (BPV) bovine papillomavirus. Double-stranded hypersensitive sites mapped in the chromatin (Herbomel et al. 1981) are indicated by vertical arrows (HS1 and HS2). Also indicated are the G + C–rich palindrome and the short repeats. (a) DNase I footprint obtained in vitro. Protected regions are boxed, and increased digestion is represented by vertical arrows. The results obtained with the early strand are represented above the horizontal line and those with the late strand below. (b) BAL-31 deletion experiments. The limit of the fragments capable of maximal binding are represented by discontinuous lines; the limits of fragments still showing residual weaker binding are indicated by a solid line. (c) DNase I footprint obtained in vivo: (top) early strand; (bottom) late strand.

six GGGCGG motifs in the 21-bp repeats and the region around the *Pvu*II sites are also protected. In the enhancer sequences, almost full protection was observed for the *Eco*RII sites, whereas it is only partial for the *Sph*I sites. It is reasonable to suppose that in most of the intracellular viral chromatin the *Eco*RII domain is occupied by a protein, whereas the *Sph*I domain is occupied only in a fraction of this population. The *Pvu*II domain seems also to be protected in a major fraction of the molecules, although this region was not well resolved in our gels. Finally, the two 21-bp sequences and the 22-bp element are largely protected.

Although we cannot exclude the possibility that a single protein can recognize several different sequences, it is highly improbable. Assuming that each one of the binding domains identified is occupied by a different protein, we designated the species interacting with the *Sph*I, *Eco*RII, and *Pvu*II sites as EBP1, 2, and 3, respectively. The studies of Chambon et al. (this volume) and Sassone-Corsi et al. (1985) show that the SV40 enhancer can be divided into two subdomains, A and B, including the promoter-proximal and -distal parts, respectively (the *Sph*I region and the *Eco*RII and *Pvu*II sequences). The orientation and distance between these subdomains can be variable. Such experiments also tend to exclude the possibility that only one protein species will interact with three different sequences along the enhancer. The three protected domains that we identified correspond rather well to sequences that were shown to be crucial for the function of this element. The *Sph*I and *Pvu*II sites coincide with alternating

purine-pyrimidine blocks that can undergo a B→Z transition in superhelical DNA (Nordheim and Rich 1983). Mutations in these sequences are deleterious for the function of the enhancer (Herr and Gluzman 1985). The protected domain around the *Eco*RII site partially overlaps the Weiher-Gruss GTGGAAAG consensus sequence and includes most of the extended element identified as essential in duplication variants studied by Herr and Gluzman (1985). Each of the protected domains contains a repeated motif of 4–7 bp that is found 10 bp apart. One may wonder if they do not interact with a dimeric protein, recognizing the same face of the double helix 10 bp apart. The hypersensitive site found in the middle of the *Eco*RII domain will be on the opposite site of the double helix. The protein interacting with the *Sph*I domain may be identical with the Z-DNA–binding protein recently identified by Azorin and Rich (1985).

When similar experiments were undertaken with polyomavirus-infected cells, we also found salt-sensitive protection along both the A and B enhancers. Moving from the origin, we observed partial protection from nt 5200 to at least nt 5105. This region includes the Weiher-Gruss consensus sequence, the G + C–rich palindrome, part of the homology with the Ig heavy-chain enhancer, and the E1A enhancer consensus sequence (see Figs. 4 and 7).

Using polyoma DNA, we searched for proteins that interact specifically with the enhancer in vitro. Employing the gel retardation technique, we found proteins in a nuclear extract of noninfected mouse fibroblasts that interact with the B enhancer of polyoma. This interaction is specific, and it is competed only with excess B enhancer fragment and not with excess A enhancer, SV40, or Ig enhancer. By using BAL-31 deletions of the DNA fragment followed by formation and isolation of the complex, we defined the minimal sequences required for the formation of the complex in vitro. Similarly, by treating with DNase I the preformed complex, we mapped the sequences interacting with the cellular proteins. They include residues from nt 5140 to 5200, sequences that contain the SV40 homology, the G + C−rich palindrome, and the Ig enhancer homology (this last sequence was protected mainly along the early strand) (see Fig. 7). As mentioned already, the formation of the complex with the B enhancer is not competed with excess SV40 or IgG enhancer containing DNA fragments, even though the latter two share sequence homology with the protected domains in the B enhancer. One possible explanation is that common sequence motifs—GTGGAAAG in SV40 and GTGGTTTTG in polyoma—are recognized by two homologous but slightly different proteins. Alternatively, if a single factor binds to similar motifs in different enhancers, it is present in large excess, relatively, to limiting amounts of proteins binding to other motifs (e.g., the G + C−rich palindrome).

Our results with both SV40 and polyoma strongly suggest that several cellular proteins interact with the enhancer sequences. The position of the binding sites of these proteins along the DNA is relatively close. We have to suppose that they can interact to form a multi-protein-DNA complex. This complex can be organized linearly along the double helix, perhaps causing bending or even one or two turns of DNA around a central protein core. We have to postulate that such a complex can somehow influence the rate of initiation from a promoter found on the same DNA molecule. When placed in the vicinity, the EBPs can interact directly with proteins bound to the promoter as it may occur in SV40. It is plausible that when placed apart, bending of the DNA double helix can bring the enhancer-bound proteins to interact with the promoter sequences. Alternatively, the protein-DNA complex formed on the enhancer can introduce perturbations in the DNA molecule that will activate promoters found on both sides of the enhancer element.

There is a certain setback in studying these viral structures in the beginning of the late phase of the lytic cycle. Only a small fraction ( < 1%) of the viral mini-chromosomes are actively transcribing in the nuclei at any moment. However, heterogeneity in the viral population is not limited to the late phase. Upon infection or transfection, again only a small fraction of the viral DNA molecules are indeed functional in the nuclei. Most of the experimental work with the SV40 enhancer was done in the absence of DNA replication, under conditions that mimic the early phase. However, recent studies on the function of the SV40 late promoter clearly indicate that the bidirectional enhancer element is also required to enhance late transcription. Deletions or point mutations along the 72-bp element decrease the activity of the viral late promoter even in the presence of DNA replication (Ernoult-Lange et al. 1984, in prep. and pers. comm.). The fact that we observe a visible protection pattern strongly argues that most of the viral mini-chromosomes have an identical structure in vivo. It is possible that all of them are competent for transcription; the cellular RNA polymerase or accessory factors limit the number of active templates. Such a factor(s) can either bind to the promoter-proximal sequences or interact with the protein moieties of the DNA-protein complex along the enhancer. Although we favor this hypothesis, we cannot exclude a more complex scheme. The proteins that we detect in vivo along the enhancer sequences are in fact repressing transcription by binding to crucial enhancer sequences. This repression removes viral minichromosomes from the pool of active templates. Other *trans*-acting, positive factors present in limiting amounts in the cell will bind to the same sequences in a small fraction of the minichromosomes to activate transcription. It is clear that the purification of the EBPs and further development of enhancer-dependent transcription systems in vitro will permit investigators to distinguish between these two models and will help us to understand how such elements work.

## Acknowledgments

We thank F. Arnos and A. Doyen for technical help and J. Ars for the preparation of the manuscript. J. Piette is a recipient of a long-term EMBO fellowship. This work was supported by grants from the CNRS (UA 041149 and ATP biologie moléculaire du gène), the ARC, and the Fondation pour la Recherche Médicale Française.

## References

Azorin, F. and A. Rich. 1985. Isolation of Z-DNA binding proteins from SV40 minichromosomes: Evidence for binding to the viral control region. *Cell* **41**: 365.

Banerji, J., L. Olson, and W. Schaffner. 1983. A lymphocyte-specific cellular enhancer is located downstream of the joining region in immunoglobulin heavy chain genes. *Cell* **33**: 729.

Cereghini, S. and M. Yaniv. 1984. Rapid assembly of transfected DNA into chromatin: Structural changes in the origin-promoter-enhancer region upon replication. *EMBO J.* **3**: 1243.

Cereghini, S., S. Saragosti, M. Yaniv, and D. Hamer. 1984. SV40 α-globin hybrid minichromosomes. Differences in DNase I hypersensitivity of promoter and enhancer sequences. *Eur. J. Biochem.* **144**: 545.

Cereghini, S., P. Herbomel, J. Jouanneau, S. Saragosti, M. Katinka, B. Bourachot, B. de Crombrugghe, and M. Yaniv. 1983. Structure and function of the promoter-enhancer region of polyoma and SV40. *Cold Spring Harbor Symp. Quant. Biol.* **47**: 935.

Church, G.M. and W. Gilbert. 1984. Genomic sequencing. *Proc. Natl. Acad. Sci.* **81**: 1991.

de Villiers, J. and W. Schaffner. 1981. A small segment of polyomavirus DNA enhances the expression of a cloned β-globin gene over a distance at least 1400 base pairs. *Nucleic Acids Res.* **9**: 6251.

de Villiers, J., L. Olson, J. Banerji, and W. Schaffner. 1983. Analysis of the transcriptional enhancer effect. *Cold Spring Harbor Symp. Quant. Biol.* **47**: 911.

Dynan, W. and R. Tjian. 1983. The promoter-specific transcription factor Sp1 binds to upstream sequences in the SV40 early promoter. *Cell* **35**: 79.

———. 1985. Control of eukaryotic messenger RNA synthesis by sequence-specific DNA binding proteins. *Nature* **316**: 774.

Ephrussi, A., G.M. Church, S. Tonegawa, and W. Gilbert. 1985. B lineage specific interactions of an immunoglobulin enhancer with cellular factors in vivo. *Science* **227**: 134.

Ernoult-Lange, M., P. May, P. Moreau, and E. May. 1984. Simian virus 40 late promoter region able to initiate simian virus 40 early gene transcription in the absence of the simian virus 40 origin sequence. *J. Virol.* **50**: 163.

Fernandez-Munoz, R., M. Coca Prados, and M.T. Hsu. 1979. Intracellular forms of simian virus 40 nucleoprotein complexes. *J. Virol.* **29**: 612.

Fried, M. and D.M. Crothers. 1981. Equilibria and kinetics of *lac* repressor interactions by polyacrylamide gel electrophoresis. *Nucleic Acids Res.* **9**: 6505.

Fromm, M. and P. Berg. 1983. SV40 early and late region promoter function are enhanced by the 72 base pair repeat inserted at distant locations and inverted orientations. *Mol. Cell. Biol.* **3**: 991.

Galas, D.J. and A. Schmitz. 1978. DNase footprinting: A simple method for the detection of protein-DNA binding specificity. *Nucleic Acids Res.* **5**: 3157.

Garner, M.M. and A. Revzin. 1981. A gel electrophoresis method for quantifying the binding of proteins in specific DNA regions: Application to components of the *Escherichia coli* lactose operon regulatory system. *Nucleic Acids Res.* **9**: 3047.

Hearing, P. and T. Shenk. 1983. The adenovirus type 5 E1A transcriptional control region contains a duplicated enhancer element. *Cell* **33**: 695.

Herbomel, P., B. Bourachot, and M. Yaniv. 1984. Two distinct enhancers with different specificities coexist in the regulatory region of polyoma. *Cell* **39**: 653.

Herbomel, P., S. Saragosti, D. Blangy, and M. Yaniv. 1981. Sequence rearrangements in polyoma EC mutants affect the DNase I hypersensitive region on the viral chromatin. *Cell* **25**: 651.

Herr, W. and Y. Gluzman. 1985. Duplications of a mutated SV40 enhancer restore its activity. *Nature* **343**: 711.

Innis, J.M. and W.A. Scott. 1983. Chromatin structure of simian virus 40 pBR322 recombinant plasmids in COS-1 cells. *Mol. Cell. Biol.* **3**: 2203.

Jakobovits, E.B., S. Bratosin, and Y. Aloni. 1980. A nucleosome-free region in SV40 minichromosomes. *Nature* **285**: 263.

Jongstra, J., T.L. Reudelhuber, P. Oudet, C. Benoist, C.B. Chae, J.M. Jeltsch, D.J. Mathis, and P. Chambon. 1984. Induction of altered chromatin structures by simian virus 40 enhancer and promoter elements. *Nature* **307**: 708.

Muller, W.J., C.R. Mueller, A.M. Mes, and J.A. Hassell. 1983. Polyomavirus origin for DNA replication comprises multiple genetic elements. *J. Virol.* **47**: 586.

Nordheim, A. and A. Rich. 1983. Negatively supercoiled simian virus 40 DNA contains Z-DNA segments within transcriptional enhancer sequences. *Nature* **303**: 674.

Piette, J., M.-K. Kryszke, and M. Yaniv. 1985. Specific interaction of cellular factors with the B enhancer of polyomavirus. *EMBO J.* **4**: 2675.

Saragosti, S., G. Moyne, and M. Yaniv. 1980. Absence of nucleosomes in a fraction of SV40 chromatin between the origin of replication and the region coding for the late leader RNA. *Cell* **20**: 65.

Sassone-Corsi, P., A. Wildeman, and P. Chambon. 1985. A *trans*-acting factor is responsible for the simian virus 40 enhancer activity in vitro. *Nature* **313**: 458.

Scott, W.A. and D.J. Wigmore. 1978. Sites in SV40 chromatin which are preferentially cleaved by endonucleases. *Cell* **15**: 1511.

Strauss, F. and A. Varshavsky. 1984. A protein binds to a satellite DNA repeat at three specific sites that would be brought into mutual proximity by DNA folding in the nucleosome. *Cell* **37**: 889.

Tooze, J., ed. 1981. *Molecular biology of tumor viruses*, 2nd edition, revised: *DNA tumor viruses*. Cold Spring Harbor Laboratory, Cold Spring Harbor, New York.

Varshavsky, A.J., O.H. Sundin, and M.H. Bohn. 1979. A stretch of ''late'' SV40 viral DNA about 400 bp long which includes the origin of replication is specifically exposed in SV40 minichromosomes. *Cell* **16**: 453.

Waldeck, W., B. Fohring, K. Chowdhury, P. Gruss, and G. Sauer. 1978. Origin of DNA replication in papovavirus chromatin is recognized by endogenous nuclease. *Proc. Natl. Acad. Sci.* **75**: 5964.

Wasylyk, B., P. Oudet, and P. Chambon. 1979. Preferential in vitro assembly of nucleosome cores on some AT-rich regions of SV40 DNA. *Nucleic Acids Res.* **7**: 705.

Wasylyk, B., C. Wasylyk, P. Augereau, and P. Chambon. 1983. The SV40 72 base pair repeat preferentially potentiates transcription starting from proximal natural or substitute promoter elements. *Cell* **32**: 503.

Weiher, H. and M.R. Botchan. 1984. An enhancer sequence from bovine papillomavirus DNA consists of two essential regions. *Nucleic Acids Res.* **12**: 2901.

Weiher, H., M. Konig, and P. Gruss. 1982. Multiple point mutations affecting the simian virus 40 enhancer. *Science* **219**: 626.

Wu, H.M. and D.M. Crothers. 1984. The locus of sequence-directed and protein-induced DNA bending. *Nature* **308**: 509.

# Cis- and Trans-acting Factors Regulating Gene Expression from the Polyomavirus Late Promoter

## F.G. Kern, S. Pellegrini, and C. Basilico
Department of Pathology, New York University School of Medicine, New York, New York 10016

We have studied the regulation of expression of free and integrated polyoma genomes by cis- and trans-acting elements by placing foreign coding sequences whose expression can be easily assayed under the control of the polyoma late promoter. When a cDNA for a mouse dihydrofolate reductase (DHFR) gene or the Tn5 neomycin-resistance-coding sequences are placed under the control of this promoter, cells transfected with these plasmids can be stably transformed to methotrexate or G418 resistance with high efficiency. To determine whether the same elements control both early and late gene expression in this system, we tested the effect of deletions in the polyoma regulatory region. The results indicated that previously identified enhancer sequences are essential for gene expression from both early and late promoters and identified new late transcriptional elements mapping between the origin of replication and the early mRNA cap sites. The effect of polyoma large T antigen on gene expression from the late promoter was tested in a variety of constructs, in both transient and stable transformation assays, under conditions that do not allow plasmid DNA replication. In stable transformation systems, stimulation of late promoter activity is observed only when the target plasmid has a reduced capacity for directing late expression. In contrast, in a transient assay, gene expression is stimulated even when the basal level of activity is not reduced. In both systems, the stimulation of activity does not require the major large T antigen–binding sites A and B nor minor binding sites 1 and 2. These results suggest that the major transcriptional elements in polyomavirus DNA are enhancer sequences and that the effect of trans-activating proteins is exerted, directly or indirectly, on these elements.

The small DNA tumor viruses polyoma and simian virus 40 (SV40) have been invaluable instruments for studying the regulation of gene expression by cis- and trans-acting elements. Not only do these viruses contain in their relatively small genomes a variety of elements that interact with viral and cellular proteins to regulate transcription, but, in addition, they can exist within their host cells either as free, replicating episomes or by being integrated into the host DNA. This allows investigation of whether the same controls operating on free viral genomes modulate the expression of integrated viral genes and the effect of this regulation on the specific expression of the genes that determine the transformed-cell behavior. They therefore provide a plausible model system to study the regulation of transcription of cellular genes.

The polyomavirus genome is organized into two distinct transcriptional units, the early and late regions, that are separated by a noncoding regulatory region including DNA sequences necessary for viral DNA replication and transcription (Tooze 1980). The early region, which encodes the three viral early proteins, is necessary for cell transformation; the late region, whose expression is generally detected only in the late phase of productive infection and is thought to require independent viral DNA synthesis, encodes the viral capsid proteins. Stable transcripts corresponding to the late region are generally not detectable in transformed cells containing only integrated viral genomes (Kamen et al. 1980b; Fenton and Basilico 1981; Kern and Basilico 1985). Transcription of the polyomavirus early genes, both in the free and integrated states, requires the presence of sequences mapping between nucleotides (nt) 5262 and 5021 on the polyoma map. These sequences belong to the category of enhancer sequences and activate in vivo transcription of genes that can be positioned in either polarity with respect to the enhancer (de Villiers and Schaffner 1981; Tyndall et al. 1981; Luthman et al. 1982; Herbomel et al. 1984; Mueller et al. 1984). In addition, one of the viral early proteins, large T antigen, is known to repress early transcription by binding at one or more DNA sites near the early promoter (Fenton and Basilico 1982a; Kamen et al. 1982; Cowie and Kamen 1984; Farmerie and Folk 1984;).

We have sought to determine the requirement for polyoma early and late transcription both in free and integrated genomes but in the absence of viral DNA replication by using a novel system in which foreign genes whose expression can be selected for or otherwise easily measured are put under the control of the late promoter. In this manuscript, we report results indicating that the same enhancer sequences appear to be required for both early and late polyoma gene expression, even in the absence of viral DNA replication. We have also obtained data suggesting that the lack of late gene expression generally observed in transformed cells is due to posttranscriptional regulation. Polyoma large T antigen not only represses early transcription but can be shown to activate, directly or indirectly, transcription from the viral late promoter. Integrated viral genomes, however, appear to be less susceptible to *trans*-activation by large T antigen than "episomal" viral DNA molecules.

## Materials and Methods

### Cells and cell culture

Mouse NIH-3T3 and rat F2408 fibroblasts were cultured in Dulbecco's modified Eagle's medium (DME) containing 10% calf serum and 10% $CO_2$. Details of procedures for transfections with calcium phosphate precipitates and selection in G418-containing medium (Colbere-Garapin et al. 1981; Southern and Berg 1982) have been previously described (Kern and Basilico 1985). For transient chloramphenicol acetyltransferase (CAT) assays, cells were seeded 24 hr prior to transfection into 100-mm dishes at $2 \times 10^5$ cells/dish. Five μg of target plasmid and 5 μg of cotransfecting plasmid were utilized for transfection with rat liver DNA as a carrier to give a final DNA concentration of 20 μg/ml. The calcium phosphate precipitates remained on the cells overnight, and the cells were rinsed and fresh DME containing 10% calf serum was added. Cells were harvested 48 hr after the removal of the calcium phosphate precipitate and analyzed for CAT activity as described by Gorman et al. (1982).

### Plasmid construction

Standard recombinant DNA techniques were utilized. Plasmids pBE27-Neo, pBE21-B27, pBEByNeo, pBE21-By, pBEB8-Neo, and pBE21-B8 were constructed by substitution of *Stu*I-*Eco*RI fragments of previously described plasmids pB27, pBy, and pB8 (Dailey and Basilico 1985) with the same polyoma-containing fragment present in plasmids pPyNeoBE63 or pBE21 (Kern et al. 1985). Plasmids pBEPyCAT, pBE32CAT, pBE21CAT, and p48.19BECAT were constructed by isolation of the 1.5–1.8-kb *Bcl*I-*Eco*RI polyoma-containing fragments of plasmids pBE102, pPyNeoBE63, and pBE21 and of plasmid p48.19, followed by filling in with the Klenow fragment polymerase and adding *Hin*dIII linkers. Fragments containing linkers were then ligated to *Hin*dIII-digested and phosphatase-treated pSV0CAT (Gorman et al. 1982). p48.19 was a generous gift of R. Kamen. It contains an *Xho*I linker inserted between nt 35 and 37, which destroys the ability of the plasmid to replicate (G. Veldman, pers. comm.). Plasmid p48.19BBCAT was constructed by isolating an 830-bp *Bam*HI-*Bst*XI polyoma origin–containing fragment from p48.19, filling in with T4 DNA polymerase, and adding *Hin*dIII linkers. The fragment was then inserted into the pSV0CAT *Hin*dIII-digested vector.

pOriFR-6 was constructed by adding a *Bam*HI linker to a 1.5-kb *Xba*I-*Sal*I fragment of pFR400 (Simonsen and Levinson 1983), which regenerated the *Sal*I site.

This fragment contains a cDNA copy of a mutated mouse dihydrofolate reductase (DHFR) gene and hepatitis B virus surface antigen gene polyadenylation signal. The *Bam*HI-*Sal*I fragment was then inserted into the *Bcl*I-*Sal*I–digested pBgOri3 vector (Kern et al. 1985).

Plasmid pBLTwt12 is an origin-minus plasmid that has an early region coding capability only for the polyoma large T antigen. It was constructed in two steps: A 0.6-kb *Bam*HI-*Hph*I fragment of plasmid pB32 that contains an origin-region deletion of nt 5246–127 (Dailey and Basilico 1985) was ligated to a 1.0-kb *Hph*I-*Eco*RI fragment of pPyLT-1 (Rassoulzadegan et al. 1982). The resulting 1.6-kb *Bam*HI-*Eco*RI fragment was inserted into a *Bam*HI-*Eco*RI double-digested pML vector (Lusky and Botchan 1981) to generate plasmid p1.6BE-3. A 3.4-kb *Eco*RI fragment from p53.A6.6 (Treisman et al. 1981) that contains the carboxyterminal portion of the wild-type A2 strain of the polyoma large T antigen was then inserted into the unique *Eco*RI site. The construction of all other plasmids used in this study has been described in previous publications (Kern and Basilico 1985; Kern et al. 1985).

## Results

Previous work carried out in our laboratory as well as in others had established that polyoma-transformed cells contained integrated viral genomes whose most common configuration is that of a head-to-tail, tandem arrangement of viral DNA molecules (Basilico et al. 1979, 1980; Birg et al. 1979; Lania et al. 1979, 1980; Gattoni et al. 1980). Transformed cells almost invariably expressed only viral early transcripts, even when it could be shown that intact late DNA sequences were present in the cells in question (Kamen et al. 1980b; Fenton and Basilico 1981; Kern and Basilico 1985). The regulation of the expression of early transcripts appeared to be mainly dependent on viral elements. Thus, when a complete early region was present, early mRNAs appeared indistinguishable in their 5′ ends, 3′ ends, and splice junctions from the mRNAs found early in lytic infection (Kamen et al. 1980b, 1982; Fenton and Basilico 1981). No evidence of transcription initiating from host DNA was found, and transcription from integrated genomes appears to require the same viral enhancer sequences necessary for "episomal" gene expression (Tyndall et al. 1981). Furthermore, viral large T antigen, which is known to down-regulate early transcription in the lytic cycle (Cogen 1978; Farmerie and Folk 1984), can exert this effect also on integrated viral DNA molecules (Fenton and Basilico 1982a; Dailey and Basilico 1985). These results indicated that the initiation of early transcription from integrated genomes was regulated in a manner similar to lytic infection. We therefore turned our attention to the expression of late sequences.

### Foreign genes linked to the polyoma late promoter are efficiently expressed in transformed cells

To understand whether the absence of stable late transcripts in polyoma-transformed cells in the absence of extrachromosomal DNA replication was due to transcriptional or posttranscriptional control, we constructed several plasmids in which foreign genes were inserted into the polyoma DNA molecule downstream of the viral *Bcl*I site, about 20 bp upstream of the initiation codon for VP2. Rat cells were transfected with these plasmids, selected for anchorage independence (depending on the expression of the intact early region), and tested for the acquisition of the phenotype conferred by the foreign gene. TK$^-$ cells transfected with chimeric plasmids carrying the herpesvirus thymidine kinase (TK)–coding sequences under the control of the polyoma late promoter were found to have been converted to a TK$^+$ phenotype. When the efficiency of expression of different genes under the control of the late promoter was tested by direct selection, it was found to be similar to or only slightly lower than that of constructs containing the same gene under the control of the early promoter (Kern and Basilico 1985). Plasmids carrying the Tn5 gene for neomycin resistance were capable of transforming cells to a G418-resistant phenotype (Kern and Basilico 1985), and plasmids carrying a mutated DHFR cDNA could transform cells to methotrexate resistance with high efficiency (Table 1). Furthermore, expression of these genes under the control of the late promoter did not require independent viral DNA replication or the presence of the viral early proteins (Table 2 and Kern et al. 1985).

These results indicated that the reason transformed cells generally do not contain viral late mRNAs is unlikely to be due to the lack of transcription initiation but most likely to posttranscriptional mechanisms affecting RNA processing or stability. This hypothesis was supported not only by the fact that the 5′ ends of the chimeric mRNAs map within the polyoma late promoter sequences and are practically indistinguishable from those utilized late in lytic infection, but also by experiments in

which cells were first selected for transformation (growth in agar) and then tested for the acquisition of the TK$^+$ or Neo$^R$ phenotype (Kern and Basilico 1985). This latter experiment rules out the possibility that the simultaneous function of both the early and late promoters cannot take place on the same viral DNA molecule and that it is the initial selection (for early or late functions) that channels the cells into a situation favoring the expression of one or the other promoter. Since some of our transformants contain only a single partial-copy insertion (Fig. 1) of the chimeric plasmid, it can be concluded that both the early and late promoters can be active on the same molecule.

Although the studies bearing more directly on the lack of expression of viral late sequences in transformed cells is discussed here, the observation that foreign genes under the control of the polyoma late promoter are efficiently expressed in the absence of viral DNA replication provided us with a system in which we could easily assess whether deletions or rearrangements in the noncoding regulatory region could affect late transcription and whether the *cis* elements required for early transcription were also required for transcription in the late direction.

**Table 1** Expression of a DHFR cDNA under the Control of the Polyoma Late Promoter

| Plasmid | Promoter | Frequency of colony formation |
|---|---|---|
| pFR400 | SV40 early | $3.4 \times 10^{-4}$ |
| pOriFR-6 | polyoma late | $2.7 \times 10^{-4}$ |
| Carrier | — | $<10^{-6}$ |

$1 \times 10^6$ cells were transfected with 1 μg of plasmid DNA coprecipitated with 19 μg of carrier DNA. 48 hr later, the cells were trypsinized and plated at $2 \times 10^5$ cells per 100-mm dish in DME containing 10% dialyzed fetal calf serum and 100 nM methotrexate. After 2 weeks, colonies were fixed, stained, and counted.

**Table 2** Large T Antigen *Trans*-activation of Gene Expression from the Late Promoter in Stably Transformed Rat F2408 Fibroblasts

| Target plasmid | Regulatory region | Cotransfected plasmid | Input (μg) | Frequency of Neo$^R$ colonies $\times 10^{-5}$ | Stimulation |
|---|---|---|---|---|---|
| pBgNeo3 | wild-type | none | 0.1 | 86 | — |
| | | pBgOri3 | 0.1 : 3.0 | 74 | — |
| | | pBLTwt12 | 0.1 : 3.0 | 259 | 3.0 |
| pBE48.19Neo | *Xho*I linker insertion | none | 0.1 | 221 | — |
| | | pBLTwt12 | 0.1 : 3.0 | 220 | — |
| pBE21 | deletion: 5055−5182 + 5246−127 | none | 1.0 | 0.7 | — |
| | | pBgOri3 | 1.0 : 3.0 | 2.7 | 3.9 |
| | | pBLTwt12 | 1.0 : 3.0 | 16.0 | 23.0 |
| pBE21-B27 | deletion: 5055−5182 + 5−242 | none | 1.0 | 2.5 | — |
| | | pBgOri3 | 1.0 : 3.0 | 3.1 | 1.2 |
| | | pBLTwt12 | 1.0 : 3.0 | 12.4 | 5.0 |

All target plasmids contain neomycin coding sequences linked to the polyoma late promoter at the *Bcl*I site and have a polyoma early polyadenylation signal downstream. pBgNeo3 (Kern et al. 1985) contains the polyoma origin and enhancer region sequences from the *Bcl*I site to the *Hph*I site at nt 153 and thus lacks any early region coding sequences. All other target plasmids contain a truncated early coding region ending at the *Eco*RI site at nt 1560. pBgOri3 (Kern et al. 1985) contains only the polyoma origin and enhancer region sequences from the *Bam*HI site to the *Hph*I site and serves as a control to determine the possible effects of either recombination between the target and cotransfected plasmid during transfection or competition for negative regulatory factors. pBLTwT12 is an origin-minus plasmid that codes only for the polyoma large T antigen.

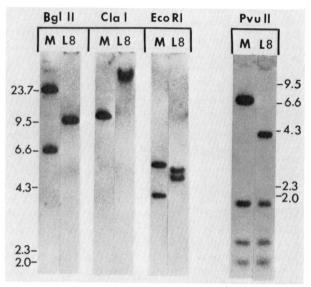

**Figure 1** Southern blot analysis indicating a single, partial insertion of plasmid pPyTK3 in a morphologically transformed cell line capable of growth in HAT medium. Cell line L8 was derived from an agar colony obtained after transfection of rat F2408 cells with plasmid PyTK3, which contains a complete *tsA* polyoma early coding region and the coding region and polyadenylation signal of the HSV TK gene linked to the polyoma late promoter (Kern and Basilico 1985). DNA was extracted from cells grown at 39°C and digested with *Bgl*II (no sites within the pPyTK3 sequences), *Cla*I (single site), *Eco*RI (two sites), or *Pvu*II (multiple sites, with a 1.8-kb fragment being diagnostic of the presence of a complete TK gene). Lanes marked *M* contain 10 pg of plasmid DNA digested with the indicated restriction enzymes. DNA was electrophoresed in 1% agarose gels, transferred to nitrocellulose, and hybridized to a nick-translated, complete plasmid probe as previously described (Kern and Basilico 1985).

## Enhancer sequences are necessary for both early and late gene expression

We constructed a series of chimeric plasmids based on a wild-type parental plasmid, pBE102, that contains the Neo[R] gene under the control of the polyoma late promoter, a truncated polyoma early region terminating at the *Eco*RI site, and deletions within the noncoding regulatory region (Fig. 2). Since the truncated early region is still capable of encoding middle T antigen, which is sufficient for neoplastic transformation in continuous cell lines, we could simultaneously test early and "late" gene expression by transfecting the plasmids into rat F2408 cells and assessing early gene expression by the ability of the cells to form colonies in soft agar and late gene expression by the ability of the cells to grow in G418-containing medium. The fact that the plasmids did not encode a complete large T antigen molecule excluded the possibility that the replicative effect of this protein could have affected the results obtained.

The effect of a number of deletions in the regulatory region on early and late gene activity as assessed by this method is summarized schematically in Figure 3. It can been seen that a large deletion (nt 5247–126; plasmid pPyNeoBE63) that includes the origin core and major T antigen–binding sites A and B had no significant effect on early or late gene expression, and the same was true of a *Pvu*II-4 fragment deletion that includes the so-called B-enhancer box (Herbomel et al. 1984) and a region of homology to the SV40 core enhancer element (Weiher et al. 1982). On the other hand, when this origin deletion was combined with deletions generated by BAL-31 digestion at the *Pvu*II site at nt 5130, different results were obtained. When the second deletion removed sequences containing the adenovirus 5 E1A core enhancer sequence (Hearing and Shenk 1983) or an enhancer box (Herbomel et al. 1984), again there was no effect observed on either early or late gene expression. However when a plasmid (pBE40) containing a second,

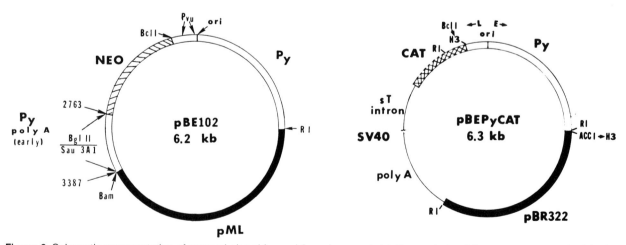

**Figure 2** Schematic representation of parental plasmids used for subsequent deletion analysis of the sequences required for late promoter function and *trans*-activation by the polyoma large T antigen. Plasmid pBE102 contains the coding sequences for neomycin resistance linked to the polyoma late promoter at the *Bcl*I site at nt 5021, with the polyoma early polyadenylation signal inserted downstream. Plasmid pBEPyCAT contains the CAT gene linked to the polyoma late promoter by insertion of a 1.8-kb *Bcl*I-*Eco*RI fragment of polyoma containing *Hin*dIII linkers into the pSV0CAT vector (Gorman et al. 1982).

larger deletion that also removes sequences with homology to the mouse immunoglobulin heavy-chain enhancer (Banerji et al. 1983; Herbomel et al. 1984) was tested, an impairment of both early and late gene activity was observed. We were able to show that this effect is not due to removal of structural elements of the late promoter (Kern et al. 1985). A still larger deletion (pBE21) that now extends in one direction into the G + C–rich inverted repeat (Soeda et al. 1980) and, in the opposite direction, removes part of the reiterated leader sequence (Legon et al. 1979; Flavell et al. 1980; Kamen et al. 1980a; Treisman 1980) had the effect of essentially abolishing both early and late activity when combined with the origin-region deletion.

Interestingly, the origin region, whose deletion per se has no effect on early and late gene expression, appears to contain some elements that increase early and late gene expression. This is shown by the fact that removing this deletion from plasmids pBE21 and pBE40 substantially increases both early and late gene expression. We tried to map these additional transcriptional elements by measuring the effect of other deletions in this region on late gene expression. As expected, these deletions alone do not reduce late gene expression to a significant extent. They are quite effective, however, when combined with the pBE21 enhancer deletion (Fig. 3). An examination of the effects of these deletions on the capacity of the plasmids to confer to cells G418 resistance suggests that these additional late transcriptional elements are dispersed throughout the DNA sequences from nt 5246 to the early cap sites (nt 153–157) and are probably composed of multiple complementing elements.

In conclusion, the results presented above indicate that gene expression from the polyoma late promoter is dependent on regulatory enhancer elements and that these enhancer elements appear to be the same as those necessary for early transcription. Together with other results (Tyndall et al. 1981; Herbomel et al. 1984; Mueller et al. 1984; Veldman et al. 1985), our data indicate that the polyoma regulatory region is a mosaic of complementing transcriptional elements that influence similarly early and late gene expression.

### Polyoma large T antigen activates gene expression from the late promoter

The lack of an absolute requirement for the three viral early proteins for late promoter activity (Table 2 and Kern et al. 1985) does not indicate that expression of the early proteins may not have an effect on late gene expression. We have approached this question by cotransfecting plasmids containing a neomycin-selectable marker and plasmids encoding exclusively polyoma large T antigen (Rassoulzadegan et al. 1982) into rat cells and by measuring the frequency of Neo^R cells obtained in the presence or absence of the large T antigen plasmids. The results shown in Table 2 indicate that when fully competent plasmids are used, there is no detectable effect on the number of Neo^R colonies that are produced by cotransfection with large T antigen–

producing plasmids. A small increase is observed in some experiments using Neo plasmids containing a functional origin of replication. Since this was never observed in origin-defective plasmids, it is likely due to a replication effect promoted by large T antigen. However, the biological activity of plasmids that, because of deletions in the enhancer region, are severely impaired in their transforming ability is effectively increased by large T antigen. Thus, the number of colonies obtained with the pBE21 plasmid is increased about 20-fold, whereas the activity of the pBE21-B27 plasmid is increased about 5-fold.

The ability of the plasmids tested to be *trans*-activated by large T antigen suggests that the major T antigen–binding sites A, B, and C and minor affinity sites (Cowie and Kamen 1984) 1 and 2 are not directly involved in *trans*-activation. It should be noted, however, that although the stimulation of *neo* transformation promoted by large T antigen in these plasmids is quite high (e.g., 23 × for pBE21), the final level of activity obtained is still quite low when compared with that of wild-type plasmids in the presence or absence of large T antigen.

To understand whether the restriction of large T antigen *trans*-activation to defective plasmids was a characteristic of integrated DNA molecules or whether it reflected an intrinsic inability of this protein to activate the promoter in the presence of its essential *cis* transcriptional elements, we measured the effect of T antigen on free plasmid molecules that contained the CAT gene under the control of the polyoma late promoter (Fig. 2). The results shown in Table 3 indicate that in this system, using rat F2408 fibroblasts as recipients, large T antigen can effectively stimulate gene expression from the late promoter. When mouse NIH-3T3 cells are used in the same type of experiment, we observe that large T antigen is a much more potent activator, and cotransfection of origin-defective plasmids with the plasmid encoding large T antigen can result in a stimulation of CAT activity greater than 20-fold (F.G. Kern et al., unpubl.). To facilitate comparison with the stable transformation system, we have shown our results using rat cells. Such a comparison indicates that, in contrast to the stable transformation system, even fully competent plasmids are stimulated by large T antigen, and the results obtained with origin-defective plasmids show that the increased activity does not require replication.

It is interesting to note that two plasmids that are both replication defective, one because of a *Xho*I linker insertion between nt 35 and 37 and the other because of a large deletion including T antigen–binding sites A, B, 1, and 2, are stimulated to approximately the same extent. Therefore, also in these cases most of the T antigen–binding sites do not appear to be required for *trans*-activation. The CAT activity of pBE21CAT plasmids, which is practically zero in the absence of large T antigen, is significantly increased in the presence of this protein, although again the final levels of CAT activity are much lower than those of wild-type plasmids. These results therefore show that polyoma large T antigen significantly activates the viral late promoter and that this

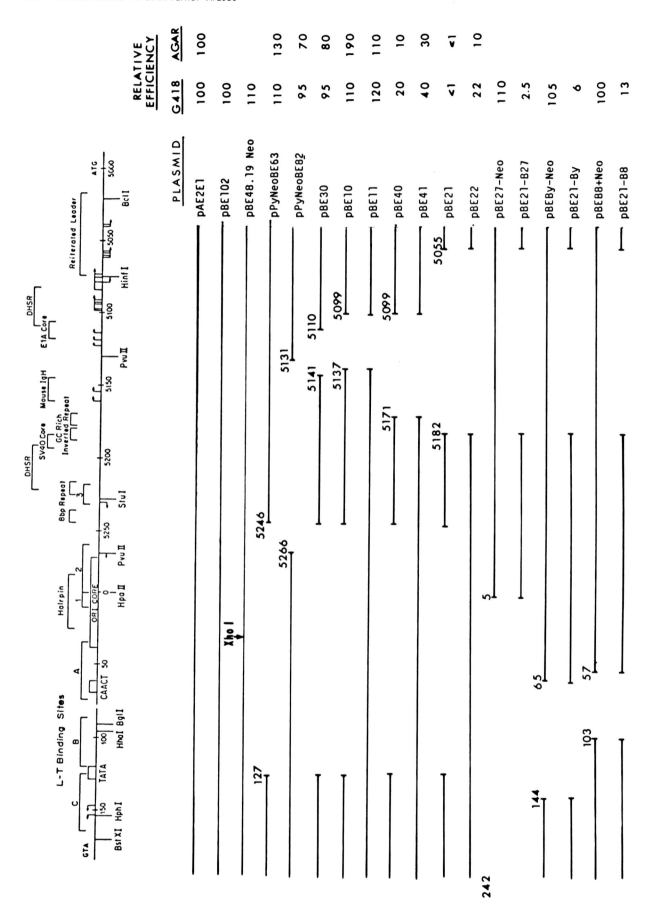

**Figure 3** Effect of enhancer- and origin-region deletions on gene expression from the polyoma early and late promoters. Indicated are the relative efficiencies of colony formation in G418-containing medium or in soft agar for the plasmids containing the deletions indicated by the open spaces based on comparison with the wild-type control plasmids. The nucleotide numbering system is according to Soeda et al. (1980). All plasmids contain the Tn5 gene for neomycin resistance linked to the late promoter at the BclI site and a truncated early coding region ending at the EcoRI site at nt 1560. At the top is a schematic representation of the polyoma A-2 strain control region, showing the location of the major and minor large T antigen–binding sites (Cowie and Kamen 1984), the early region TATA and CAACT boxes (Soeda et al. 1980), sequences comprising the minimal origin of replication (Luthman et al. 1982; Katinka and Yaniv 1983), as well as the location of a reiterated leader sequence generated during lytic infection (Legon et al. 1979; Flavell et al. 1980; Kamen et al. 1980a; Treisman 1980). An area that can form a hairpin loop and the location of DNase-hypersensitive regions (Herbomel et al. 1981), an 8-bp direct repeat, and a G + C–rich inverted repeat are also shown, as are the locations of sequences with homology to the SV40 core enhancer sequence (Weiher et al. 1982), the mouse immunoglobulin heavy-chain enhancer (Banerji et al. 1983; Herbomel et al. 1984), and the adenovirus 5 E1A enhancer consensus sequence (Hearing and Shenk 1983). Arrows indicate the locations of 5′ termini of early (Kamen et al. 1982), late-early (Fenton and Basilico 1982b), late lytic (Treisman 1980; Cowie et al. 1981), or late direction transcripts observed after transfection with chimeric plasmids (Kern and Basilico 1985).

**Table 3** Effect of Large T Antigen on Polyoma Late Promoter Activity in Transient CAT Assays in Rat F2408 Fibroblasts

| Plasmid | Replication | Regulatory region | CAT activity[a] | | | |
|---|---|---|---|---|---|---|
| | | | Promoter | −LT | +LT | Stimulation |
| pBEPyCAT | + | wild-type | polyoma late | 37.2 | 181.3 | 4.9 |
| p48.19BECAT | − | XhoI linker insertion | polyoma late | 30.4 | 171.0 | 5.6 |
| pBE32CAT | − | deletion: 5246−127 | polyoma late | 38.4 | 219.2 | 5.7 |
| pBE21CAT | − | deletion: 5055−5182 + 5246−127 | polyoma late | <0.65 | 1.85 | >2.8 |
| p48.19BBCAT | − | XhoI linker insertion | polyoma early | 85.0 | 27.3 | 0.3 |

F2408 were transfected with the indicated plasmid alone or cotransfected with pBLTwt12, an origin-defective plasmid coding for the polyoma large T antigen, as described in Materials and Methods.
[a] CAT activity is expressed as pmoles converted / min / 100-mm dish.

activation does not require the major T antigen–binding sites deleted in this plasmid. Our results also suggest that the target sequences that interact either directly with large T antigen or indirectly with transcriptional factors affected by large T antigen are located within the enhancer sequences still present in the pBE21CAT plasmid. These sequences include a DNase-hypersensitive region (Herbomel et al. 1981) and regions with homology to the bovine papillomavirus activator element (Lusky et al. 1983) and SV40 core enhancer sequences (Weiher et al. 1982) as well as a low-affinity binding site for large T antigen (Cowie and Kamen 1984). It remains to be determined whether the lack of the other enhancer elements has the effect of diminishing the extent of *trans*-activation. The apparent discrepancy between the susceptibility of integrated and free plasmids to *trans*-activation also remains to be resolved (see Discussion).

## Discussion

The results presented in this paper are part of our current investigation into the regulation of expression of the polyoma genome in the episomal or integrated state. One of the key observations of this work is the finding that foreign DNA sequences can be efficiently expressed from the polyoma late promoter in cells stably transformed by chimeric plasmids. This was somewhat surprising in view of the well-known observation that polyoma- or SV40-transformed cells generally do not contain detectable amounts of stable, polyadenylated late transcripts in the absence of extrachromosomal viral DNA replication. Since excision and free DNA replication lead to the immediate appearance of late RNAs (Kamen et al. 1980b; Fenton and Basilico 1982a; Kern and Basilico 1985), their lack of expression from integrated genomes clearly does not depend on the lack of necessary transcription factors in transformed cells. Rather, it appeared possible that polyoma late gene expression required independent viral DNA replication. Our findings suggest that this is not the case. We have shown that the 5′ ends of the chimeric "late" mRNAs expressed from our integrated plasmids map within the polyoma promoter region at positions similar to or identi-

cal with those utilized by bona fide late lytic transcripts (Kern and Basilico 1985). Furthermore, we have shown that the expression of these sequences depends on the presence of the polyoma transcriptional enhancer elements and can be increased, albeit weakly, by the viral large T antigen. Thus, despite the structural differences that exist between the early and late promoters, it is apparent that transcription from the late promoter is regulated by *cis*- and *trans*-acting factors that also influence transcription from the early promoter. Our results suggest that the polyoma late promoter is functional in an integrated state and that late gene expression in integrated polyoma genomes is not regulated at the level of message initiation but most likely by posttranscriptional events.

These posttranscriptional events could include splicing, mRNA stability, or transport. We presently favor the hypothesis of mRNA stability since it is known that the stable late mRNAs found in the lytic cycle are generated by the splicing of multimeric nuclear transcripts that result from repeated transcription of the circular genome (Acheson 1978). This produces a noncoding leader sequence consisting of a number of repeats of a 57-nt segment (Legon et al. 1979; Flavell et al. 1980; Kamen et al. 1980a; Treisman 1980), and it is possible that this leader may be essential for late mRNA stability. Although in theory these leader sequences could be generated from the primary transcript of multiply repeated head-to-tail insertions, a number of factors would make this unlikely. Such an event would require readthrough of the integrated polyoma late polyadenylation signal, and we have shown previously that, when integrated, this signal can effectively terminate transcription (Kern and Basilico 1985). In addition, this readthrough could be either inhibited by collision with the early region transcriptional machinery or selected against, since such transcription would generate a message that would be antisense with respect to the early messages required for the maintenance of the transformed phenotype (Izant and Weintraub 1984). If the lack of detectable late transcription in transformed cells is indeed related to late mRNA instability, then our results suggest that replacement of the late coding sequences with foreign genes may obviate that need for a repeated leader or, alternatively, that it is the late coding region that confers instability to the mRNA.

Our results on the transient expression of CAT genes linked to the polyoma late promoter, together with other experiments not reported here in which we measured the extent of early and late transcription after transfection of HeLa cells with a variety of chimeric plasmids (A. Velcich et al., unpubl.), also indicate clearly that late gene expression from free genomes does not require viral DNA synthesis and that although the viral late promoter is somewhat weaker than the early, it is still capable of promoting a significant level of gene expression even in the absence of large T antigen *trans*-activation. Thus, a possible explanation for the apparent early-late switch during polyomavirus infection could be the following: When viral genomes are uncoated, transcription initiates from both the early and late promoter. However, as discussed above, it is reasonable to suggest that transcription from the early promoter inhibits the readthrough of the late polyadenylation signal and thus the production of late multimeric transcripts that may be the only ones capable of generating stable late mRNAs. Thus few, if any, stable late transcripts are produced. Early transcription results in the production of large T antigen, which is known to bind at or near the early cap sites and to inhibit early transcription. The inhibition of early transcription and the simultaneous stimulation of the late promoter by large T antigen result in a reversal of promoter strength, the late promoter becoming much stronger than the early (Table 3). This allows efficient multimeric transcription and generates a large number of stable late transcripts. The role of viral DNA synthesis in this model would only be that of increasing copy number and thus allowing the production of a large number of late and early transcripts.

Probably the most intriguing result discussed here is the ability of polyoma large T antigen to stimulate transcription from the viral late promoter. The demonstration that SV40 large T antigen possesses a similar ability (Brady et al. 1984; Keller and Alwine 1984; Alwine 1985; Brady and Khoury 1985) does not make the polyoma finding particularly surprising. Still, this finding increases the number of distinct functions attributable to this viral protein. It would be interesting to determine whether one of the properties of large T antigen—its ability to convert primary cells into continuous cell lines (Rassoulzadegan et al. 1983)—is related to its capacity to activate some promoters.

We have tried to pinpoint the target of T antigen *trans*-activation of the late polyoma promoter and obtained data that suggest that the major T antigen–binding sites A and B and minor sites 1 and 2 are not necessary for *trans*-activation. T antigen–binding site C is also probably not essential, but the data obtained so far do not allow us to rule it out altogether. If the major T antigen–binding sites are not involved in *trans*-activation, it could be concluded that *trans*-activation requires a protein-DNA interaction that is quite different from those involved in repressing early transcription or initiat-

ing DNA synthesis. This could involve the minor binding site 3 or be indirect, resulting from the activation of cellular genes or the removal of inhibitory proteins from the viral DNA.

Finally, we have observed a very different level of *trans*-activation on free or integrated genomes. The observation that the late promoter of integrated plasmids seems to be susceptible to a significant *trans*-activation only when its basal level of activity is reduced by the deletions of *cis* transcriptional elements could be due to trivial reasons, the most likely being that in our neomycin selection system a relatively low level of gene expression is sufficient to ensure cell survival. Increases beyond this level may not produce a significant increase in the number of Neo[R] cells. On the other hand, it is possible that integration leads to changes in the viral chromatin structure within the regulatory region such that a much higher level of basal late promoter activity is maintained. The effect of large T antigen may thus become negligible, perhaps because its main mechanism of action may consist of removing inhibitory proteins that are already present at much lower density on integrated DNA molecules. Work is in progress to distinguish between these possibilities.

## Acknowledgment

This investigation was supported by grants CA-11893 and CA-16239 from the National Cancer Institute.

## References

Acheson, N.H. 1978. Polyoma giant RNAs contain tandem repeats of the nucleotide sequence of the entire viral genome. *Proc. Natl. Acad. Sci.* **75:** 4754.

Alwine, J.C. 1985. Transient gene expression control: Effects of DNA stability and transactivation by viral early proteins. *Mol. Cell. Biol.* **5:** 1034.

Banerji, J., L. Olson, and W. Schaffner. 1983. A lymphocyte specific cellular enhancer is located downstream of the joining region in immunoglobulin heavy chain genes. *Cell* **33:** 729.

Basilico, C., S. Gattoni, D. Zouzias, and G. Della-Valle. 1979. Loss of integrated viral DNA sequences in polyoma transformed cells is associated with an active viral A function. *Cell* **17:** 645.

Basilico, C., S. Gattoni, D. Zouzias, G. Della-Valle, S. Gattoni, V. Colantuoni, R. Fenton, and L. Dailey. 1980. Integration and excision of polyoma virus genomes. *Cold Spring Harbor Symp. Quant. Biol.* **44:** 611.

Birg, F., R. Dulbecco, M. Fried, and R. Kamen. 1979. State and organization of polyomavirus DNA sequences in transformed rat cell lines. *J. Virol.* **29:** 633.

Brady, J. and G. Khoury. 1985. Transactivation of the simian virus 40 late transcription unit by T-antigen. *Mol. Cell. Biol.* **5:** 1391.

Brady, J., J.B. Bolen, M. Radonovich, N. Salzman, and G. Khoury. 1984. Stimulation of SV40 late gene expression by simian virus 40 tumor antigen. *Proc. Natl. Acad. Sci.* **81:** 2040.

Cogen, B. 1978. Virus specific early RNA in 3T6 cells infected by a ts mutant of polyomavirus. *Virology* **85:** 222.

Colbere-Garapin, F., F. Horodniceanu, P. Kourilsky, and A.C. Garapin. 1981. A new dominant hybrid selective marker for higher eukaryotic cells. *J. Mol. Biol.* **150:** 1.

Cowie, A. and R. Kamen. 1984. Multiple binding sites for polyomavirus large T antigen within regulatory sequences of polyomavirus DNA. *J. Virol.* **52:** 750.

Cowie, A., C. Tyndall, and R. Kamen. 1981. Sequences at the capped 5′-ends of polyomavirus late region mRNAs: An example of extreme terminal heterogeneity. *Nucleic Acids Res.* **9:** 6305.

Dailey, L. and C. Basilico. 1985. Sequences in the polyomavirus DNA regulatory region involved in viral DNA replication and early gene expression. *J. Virol.* **54:** 739.

de Villiers, J. and W. Schaffner. 1981. A small segment of polyomavirus DNA enhances the expression of a cloned β-globin gene over a distance of 1,400 base pairs. *Nucleic Acids Res.* **9:** 6351.

Farmerie, W.G. and W.R. Folk. 1984. Regulation of polyomavirus transcription by large tumor antigen. *Proc. Natl. Acad. Sci.* **81:** 6919.

Fenton, R.G. and C. Basilico. 1981. Viral gene expression in polyomavirus–transformed cells and their cured revertants. *J. Virol.* **40:** 150.

———. 1982a. Regulation of polyomavirus early transcription in transformed cells by large T antigen. *Virology* **121:** 1982.

———. 1982b. Changes in the topography of early region transcription during polyomavirus lytic infection. *Proc. Natl. Acad. Sci.* **79:** 7142.

Flavell, A.J., A. Cowie, J.R. Arrand, and R. Kamen. 1980. Localization of three major capped 5′ ends of polyoma late mRNA's within a single tetranucleotide sequence in the viral genome. *J. Virol.* **33:** 902.

Gattoni, S., V. Colantuoni, and C. Basilico. 1980. Relationship between integrated and non-integrated viral DNA in rat cells transformed by polyomavirus. *J. Virol.* **34:** 615.

Gorman, C.M., L.F. Moffat, and B.H. Howard. 1982. Recombinant genomes which express chloramphenicol acetyl transferase in mammalian cells. *Mol. Cell. Biol.* **2:** 1044.

Hearing, P. and T. Shenk. 1983. The adenovirus type 5 E1A transcriptional control region contains a duplicated enhancer element. *Cell* **33:** 695.

Herbomel, P., B. Bourachot, and M. Yaniv. 1984. Two distinct enhancers with different cell specificities coexist in the regulatory region of polyoma. *Cell* **39:** 653.

Herbomel, P., S. Saragosti, D. Blangy, and M. Yaniv. 1981. Fine structure of the origin proximal DNase I–hypersensitive region in wild-type and EC mutant polyoma. *Cell* **25:** 651.

Izant, J.G. and H. Weintraub. 1984. Inhibition of thymidine kinase gene expression by anti-sense RNA: A molecular approach to genetic analysis. *Cell* **36:** 1007.

Kamen, R., J. Favaloro, and J. Parker. 1980a. Topography of the three late mRNAs of polyomavirus which encode the virion proteins. *J. Virol.* **33:** 637.

Kamen, R., P. Jat, R. Treisman, and J. Favaloro. 1982. 5′ termini of polyomavirus early region transcripts synthesized *in vivo* by wild-type virus and viable deletion mutants. *J. Mol. Biol.* **159:** 189.

Kamen, R., J. Favaloro, J. Parker, R. Treisman, L. Lania, M. Fried, and A. Mellor. 1980b. A comparison of polyomavirus transcription in productively infected mouse cells and transformed rodent cell lines. *Cold Spring Harbor Symp. Quant. Biol.* **44:** 189.

Katinka, M. and M. Yaniv. 1983. DNA replication origin of polyomavirus: Early proximal boundary. *J. Virol.* **47:** 224.

Keller, J.M. and J.C. Alwine. 1984. Activation of the SV40 late promoter: Direct effects of T antigen in the absence of viral DNA replication. *Cell* **26:** 381.

Kern, F.G. and C. Basilico. 1985. Transcription from the late promoter in cells stably transformed by chimeric plasmids. *Mol. Cell. Biol.* **5:** 797.

Kern, F.G., L. Dailey, and C. Basilico. 1985. Common regulatory elements control gene expression from polyoma early and late promoters in cells transformed by chimeric plasmids. *Mol. Cell. Biol.* **5:** 2070.

Lania, L., M. Griffiths, B. Cooke, Y. Ito, and M. Fried. 1979. Untransformed rat cells containing free and integrated DNA of a polyoma non-transforming (hrt) mutant. *Cell* **18:** 793.

Lania, L., D. Gandini-Attardi, M. Griffiths, B. Cooke, D. DeCicco, and M. Fried. 1980. The polyomavirus 100K large T antigen is not required for the maintenance of transformation. *Virology* **101:** 217.

Legon, S., A. Flavell, A. Cowie, and R. Kamen. 1979. Amplification of the leader sequences of "late" polyomavirus mRNAs. *Cell* **16:** 373.

Lusky, M. and M. Botchan. 1981. Inhibition of SV40 replication in simian cells by specific pBR322 DNA sequences. *Nature* **293:** 79.

Lusky, M., L. Berg, H. Weiher, and M. Botchan. 1983. Bovine papillomavirus contains an activator of gene expression at the distal end of the early transcription unit. *Mol. Cell. Biol.* **3:** 1108.

Luthman, H., M.G. Nilsson, and G. Magnusson. 1982. Noncontiguous segments of the polyoma genome required in *cis* for DNA replication. *J. Mol. Biol.* **161:** 533.

Mueller, C.R., A.-M. Mes-Masson, M. Bouvier, and J.A. Hassell. 1984. Location of sequences in polyomavirus DNA that are required for early gene expression in vivo and in vitro. *Mol. Cell. Biol.* **4:** 2594.

Rassoulzadegan, M., A. Cowie, A. Carr, N. Glaichenhaus, R. Kamen, and F. Cuzin. 1982. The role of individual polyomavirus early proteins in oncogenic transformation. *Nature* **300:** 713.

Rassoulzadegan, M., Z. Naghashfar, A. Cowie, A. Carr, M. Grisoni, R. Kamen, and F. Cuzin. 1983. Expression of the large T protein of polyomavirus promotes the establishment in culture of "normal" rodent fibroblast cell lines. *Proc. Natl. Acad. Sci.* **80:** 4354.

Simonsen, C.C. and A.D. Levinson. 1983. Isolation and expression of an altered mouse dihydrofolate reductase cDNA. *Proc. Natl. Acad. Sci.* **80:** 2495.

Soeda, E., J.R. Arrand, N. Smolar, J.E. Walsh, and B.E. Griffin. 1980. Coding potential and regulatory signals of the polyomavirus genome. *Nature* **283:** 445.

Southern, P.J. and P. Berg. 1982. Transformation of mammalian cells to antibiotic resistance with a bacterial gene under the control of SV40 early region promoter. *J. Mol. Appl. Genet.* **1:** 327.

Tooze, J., ed. 1980. *Molecular biology of tumor viruses*, 2nd edition: *DNA tumor viruses*. Cold Spring Harbor Laboratory, Cold Spring Harbor, New York.

Treisman, R. 1980. Characterization of polyoma late mRNA leader sequences by molecular cloning and DNA sequence analysis. *Nucleic Acids Res.* **8:** 4867.

Treisman, R., V. Novak, J. Favaloro, and R. Kamen. 1981. Transformation of rat cells by an altered polyomavirus genome expressing only the middle-T protein. *Nature* **292:** 595.

Tyndall, C., G. LaMantia, C.M. Thacker, J. Favaloro, and R. Kamen. 1981. A region of the polyomavirus genome between the replication origin and late protein coding sequences is required in *cis* for both early gene expression and viral DNA replication. *Nucleic Acids Res.* **9:** 6231.

Veldman, G.M., S. Lupton, and R. Kamen. 1985. Polyoma enhancer contains multiple redundant sequence elements that activate both DNA replication and gene expression. *Mol. Cell. Biol.* **5:** 649.

Weiher, H., M. Konig, and P. Gruss. 1982. Multiple point mutations affecting the simian virus 40 enhancer. *Science* **219:** 626.

# Alternate Roles of Simian Virus 40 Sequences in Viral Gene Expression in Different Types of Cells

## C. Prives, L. Covey, T. Michaeli, E.D. Lewis, and J.L. Manley
Department of Biological Sciences, Columbia University, New York, New York 10027

Experiments designed to investigate the role of sequences on the simian virus 40 (SV40) genome that may be involved in gene regulation were carried out. To this end two different studies were undertaken. One of these was designed to determine whether regions of the viral genome can be identified that interact with subnuclear structures that may regulate viral gene expression. The other study was designed to examine the role of sequences at the viral origin of replication in regulating expression from the viral late region.

The association of transcriptionally active and inactive SV40 genes with the nuclear matrix of infected or transfected cells was examined. COS-7 and C6 SV40-transformed moneky cells each contain at least one integrated copy of SV40 DNA in which an intact, transcriptionally active early region is contiguous with a nontranscribed late region. With either cell line, SV40 DNA sequences were found to be asymmetrically distributed within the nuclear soluble and matrix fractions after low-salt extraction followed by extensive restriction enzyme digestion and Southern blotting. Specifically, at least two fragments in the early region were preferentially associated with the nuclear matrix fraction, while sequences in the late coding region, the regulatory *ori* region, and the distal 3' and proximal 5' regions of the early coding sequences were not retained with this fraction. Transfection experiments were carried out to determine whether this asymmetric distribution of viral sequences was related to their transcriptional activity. Surprisingly, we found that transcription of the SV40 early coding region is not a requirement for the preferential association of sequences with the nuclear matrix.

The role of sequences required for DNA replication at the SV40 *ori* region in regulating expression of the viral late region was examined by injecting wild-type and mutant viral DNA into the nucleus of *Xenopus laevis* oocytes. As quantitative and qualitative differences in late RNA 5' ends were observed when SV40 DNA inserted into bacterial plasmids was injected, studies of SV40 gene expression from such recombinant molecules are difficult to interpret in oocytes. Using only the viral portion of recombinant plasmids containing origin-defective mutants, two deletion mutants (6-1 and 8-4) and an additional deletion-insertion mutant (*ori*⁻) were tested. The quantities and patterns of late RNA 5' ends observed upon injection of wild-type and mutant DNAs were virtually identical. Furthermore, synthesis of the capsid proteins were similar in oocytes injected with wild-type or *ori*-defective mutant DNAs. These findings indicate that a functional viral origin of replication is not required for SV40 late gene expression in *X. laevis* oocytes.

Factors governing the regulation of SV40 gene expression have been studied extensively in infected and transfected cells as well as in cell-free systems. These studies have primarily dealt with the analysis of sequences that comprise elements of the early and late promoter regions as well as the viral enhancer sequences. However, it is possible that other regions or sequences of the viral genome are also required for viral expression. Within the host cell, efficient viral gene expression may require an association with intranuclear structures such as the nuclear matrix, considered to be the site of many active processes in eukaryotic gene expression. In addition, since viral late RNA synthesis occurs actively in infected cells primarily after the onset of viral DNA replication, it is possible that sequences necessary for viral DNA replication are also required intrinsically for late gene expression. We have under-

taken two studies designed to investigate the interaction of viral DNA sequences with the nuclear matrix, and the relationship between sequences required for viral DNA replication and late gene expression.

The identification of a residual nuclear structure, termed the nuclear matrix, was first described as the structure remaining after nuclei were extracted and extensively digested with high concentrations of salt, nonionic detergent, and DNase I (Berezney and Coffey 1977). That the nuclear matrix may have a more complex function beyond that of an inert structural framework is suggested by several studies implicating it as a site for DNA replication (Pardoll et al. 1980; Berezney and Buchholtz 1981), RNA transcription (Cook et al. 1982; Robinson et al. 1982; Ciejek et al. 1983), and RNA processing (Herman et al. 1978; Berezney 1979; van Eekelen and van Venrooij 1981; Ben-Ze'ev and Aloni

1983). In addition, several studies in recent years have shown that DNA sequences that are being actively transcribed are preferentially associated with the nuclear matrix (Cook and Brazell 1980). Microscopic studies of chromatin fiber in both metaphase (Adolph et al. 1977; Paulson and Laemmli 1977) and interphase (Benyajati and Worcel 1976) nuclei have shown that it is organized into loops that radiate out from a proteinaceous scaffold. The nuclear lamin proteins are hypothesized to be components of this scaffold structure. Taken together, these data suggest a model of gene transcription that locates actively transcribing sequences to the base of the DNA loops and in association with the matrix scaffold. In contrast, the nontranscribed sequences would be located at some distance from the matrix or scaffold attachment point. Attempts to identify sequences that define potential matrix attachment sites have been ongoing and primarily inconclusive. However, Mirkovitch et al. (1984), using a novel extraction procedure that preserves the looped structure in the absence of histone, recently identified DNA sequences from *Drosophila melanogaster* tissue-culture cells that were retained preferentially with the nuclear scaffold or matrix fraction. Using this protocol, we have analyzed the association of SV40 DNA with the nuclear matrix of both transformed and transfected cells.

Synthesis of SV40 late RNA is temporally regulated during lytic infection, and substantial quantities of late RNA are detected only after the initiation of viral DNA replication following infection of permissive cells. Viral DNA replication is dependent upon a precise DNA sequence at a region approximately 60 bp mapping between nucleotides (nt) 5208–5222 and nt 26–30 that has been defined as the minimal replication origin (DiMaio and Nathans 1978, 1982; Subramanian and Shenk 1978; Bergsma et al. 1982). Deletion or substitution of sequences in this region vastly reduces viral DNA replication. The initiation of SV40 DNA replication also requires the function of the large T antigen (for review, see Tooze 1981). However, an additional, independent effect of large T antigen synthesized in infected monkey cells in stimulating late transcription has been described, in which both sequences in the viral origin of replication and the DNA-binding function of T antigen were implicated (Brady et al. 1984; Brady and Khoury 1985; Keller and Alwine 1985). Because *Xenopus laevis* oocytes do not support the replication of exogenously added DNA (Harland and Laskey 1980) but do accumulate large quantities of SV40 late RNA and proteins, this system can be used for testing the role of both origin sequences and T antigen on late transcription without the complication of ongoing viral DNA replication. We have previously reported that *X. laevis* oocytes do not require the function of T antigen for activation of late RNA synthesis (Michaeli and Prives 1985). In the present study we have asked whether a functional replication origin is required for late transcription in this system. Our results indicate that sequences that are necessary for origin function in permissive cells are not required for the expression of the late promoter in *X. laevis* oocytes.

## Materials and Methods

### Cell culture
C6 (Gluzman et al. 1977) and COS-7 (Gluzman 1981) cells were maintained in Dulbecco's modification of Eagle's medium (DME) supplemented with 10% fetal calf serum. Subconfluent monolayers were obtained from $2 \times 10^7$ cells seeded 24 hr prior to extraction.

### Transfection of HeLa cells
25 µg of supercoiled plasmid DNA was transfected onto $2 \times 10^6$ cells on a 150-mm plate by calcium phosphate coprecipitation (Wigler et al. 1979). Eighteen hr after the addition of the DNA precipitate to the cells, the DNA and medium were removed, and the cells were treated with 10% glycerol for 1.5 min (Frost and Williams 1978). The cells were washed once with TBS (Kimura and Dulbecco 1972), fresh DME with 10% fetal bovine serum was added, and the cells were incubated for a further 30 hr.

### Plasmids
The pφ4SVA and pφ4SVAe plasmids have been described previously (Lewis and Manley 1985). pSVRI is the complete SV40 genome inserted into the *Eco*RI site of pBR322.

### Isolation of nuclear matrix structures
For the high-salt extraction of matrix DNA, a modification of a previously published procedure was used (Covey et al. 1984). Subconfluent monolayers were washed three times with PBS and then swelled in nuclear retention buffer (10 mM NaCl, 20 mM MES [pH 6.0], 1 mM MgCl$_2$) followed by mechanical disruption in a Dounce homogenizer with 15 strokes of an "A" pestle. After centrifugation at 1500 rpm for 10 min, the nuclear pellet was resuspended in isotonic buffer (100 mM NaCl, 50 mM KCl, 20 mM Tris · HCl, 10 mM EDTA [pH 7.4]) (0.3 ml for $5 \times 10^6$ cells) and an equal volume of high-salt buffer (4.0 M NaCl, 20 mM Tris · HCl [pH 7.4]) was added. High-salt-extracted nuclei were allowed to incubate on ice for 10 min, followed by the addition of 30 mM MgCl$_2$ and 30 units/ml DNase I at 37°C for 30 min. The high-salt-extracted nuclei were pelleted through a 15% glycerol cushion in 50% high-salt buffer at 10,000 rpm for 20 min. The supernatant was removed and the pellet was washed extensively with 10 mM Tris · HCl (pH 8.0), 140 mM NaCl, 2 mM EDTA and treated with 0.5 mg/ml proteinase K in the presence of 1% SDS. Released DNA was diluted with TE buffer (10 mM Tris · HCl [pH 8.0], 1 mM EDTA) and initially precipitated with ethanol, followed by deproteinization. Protease inhibitors phenylmethylsulfonyl fluoride (PMSF) (0.5 mM) and L-1-tosylamide-2-phenylmethyl chloromethyl ketone (TPCK) (0.5 mM) were present at each step of the extraction.

For the isolation of low-salt-extracted nuclear matrices, a modified procedure described by Mirkovitch et al. (1984) was used. Briefly, cells were washed three times with isolation buffer (IB) (5.0 mM Tris · HCl [pH 7.4], 0.05 mM spermine, 0.125 mM spermidine, 0.5 mM EDTA/KOH [pH 7.4], 1% [v/v] thiodyglycol, 20 mM

KCl, 10 KIU/ml Trasylol, and 0.1 mM PMSF) at room temperature and pelleted by centrifugation at 900$g$ for 5 min. Nuclei were obtained by resuspending the washed cells in ice-cold IB plus 0.1% digitonin, followed by disruption in a Dounce homogenizer with 15 strokes of an ''A'' pestle. Nuclei were washed two times in the above buffer, followed by centrifugation at 900$g$ for 10 min, and then resuspended on ice in IB plus 0.1% digitonin without EDTA. Nuclei (2.5 $A_{260}$ units) were heated for 20 min at 37°C in 200 $\mu$l of the same buffer. Ten ml of LIS extraction buffer (5 mM HEPES/NaOH [pH 7.4], 0.25 mM spermidine, 2 mM EDTA/KOH [pH 7.4], 2 mM KCl, 0.1% digitonin, 20 mM 3,5-diiodo-salicylic acid, lithium salt) were added to each reaction, followed by incubation at room temperature for 5 min. After centrifugation for 20 min at 2400 rpm and extensive washing with digestion buffer (20 mM Tris · HCl [pH 7.4], 0.05 mM spermine, 0.125 mM spermidine, 20 mM KCl, 50 mM NaCl, 10 mM MgCl$_2$, 0.1% digitonin, 100 KIU/ml Trasylol), LIS-extracted nuclei were digested with the appropriate restriction enzymes until approximately 10–20% of the DNA remained with the matrix fraction. After centrifugation at 2400 rpm for 10 min at 4°C, the supernatant was separated from the pellet and both fractions were digested at 37°C with 0.5 mg/ml proteinase K in the presence of 1% SDS for 16–18 hr.

*DNA analysis*
DNA from both matrix-associated and released fractions was purified by repeated organic extractions and precipitated with ethanol. Samples were resuspended in 100 $\mu$l of 50 mM Tris · Cl (pH 8.0), 10 mM NaCl, 10 mM EDTA and digested with 100 $\mu$g/ml DNase-free RNase A at 37°C for 1 hr. Unless stated, equal amounts of DNA from the matrix-associated and released fractions were separated on a 1.5% agarose gel and transferred to nitrocellulose according to the procedure of Southern (1975). Filters were prehybridized, then hybridized with 100 ng of $^{32}$P-labeled SV40 or pSVRI DNA nick-translated to a specific activity of approximately $5 \times 10^8$ cpm/$\mu$g. DNA bands were visualized by autoradiography using Kodak XAR-5 film.

*Preparation of DNA for microinjection into* X. laevis *oocytes*
pSVRI contains SV40 DNA inserted into the *Eco*RI site of pBR322. Construction of the p6-1 and p8-4 SV40 origin-deletion mutants was previously described by Gluzman et al. (1980a), and pSVRIori$^-$ was described by Chen et al. (1983). Recircularization of viral DNA was as described by Michaeli and Prives (1985), except that vacuum dialysis was used for concentration of DNA samples after ligation.

*Injection into oocyte nuclei and analysis of capsid proteins*
Injection into oocyte nuclei, [$^{35}$S]methionine labeling, protein extraction, and immunoprecipitation of capsid proteins were performed as reported previously (Michaeli and Prives 1985).

*RNA preparation and S1 analysis*
Extraction of RNA and S1 analysis were performed as described (Michaeli and Prives 1985). The probes used were gel-purified *Nae*I-*Bst*XI fragments of p6-1, p8-4, and pSVRI and *Hpa*II-*Taq*I fragments of pSVRI and pSVRIori$^-$. After 5' labeling with polynucleotide kinase, all the probes were denatured and the strand complementary to late RNA was isolated from a 4% polyacrylamide gel.

## Results and Discussion

### The association of SV40 DNA with the nuclear matrix

*High-salt extraction of nuclei does not result in an enrichment of viral DNA sequences with the nuclear matrix*
It was previously reported that viral sequences isolated from SV40-transformed mouse 3T3 cells were enriched with the nuclear matrix after extraction in high-salt buffer followed by limited digestion with DNase I (Nelkin et al. 1980). We attempted to repeat this observation by using two SV40-transformed monkey cell lines, COS-7 and C6, that each contain at least one integrated copy of viral DNA. COS-7 cells express wild-type T antigen but lack a functional viral origin of replication (Gluzman 1981), while C6 cells express a mutant T antigen that lacks specific origin-binding activity (Gluzman et al. 1977; Prives et al. 1983). Using a high-salt extraction procedure (see Materials and Methods), we were unable to detect either any enrichment or altered distribution of viral DNA fragments in the nuclear matrix when compared with the released fraction (Fig. 1). This obser-

**Figure 1** COS-7 SV40 fragments detected in the matrix and released fractions of high-salt-extracted nuclei. Ten $\mu$g of *Hin*dIII-restricted COS-7 matrix (lane *A*) and released (lane *B*) DNA were separated on a 1.5% agarose gel followed by Southern blotting. DNA was hybridized to a $^{32}$P-labeled pSVRI probe. Fragments designated by size correspond to the *Hin*dIII fragment map of SV40 shown in Fig. 2 (top). Fragments 1169 and 526 represent early coding sequences. Late region bands are represented by fragments 1117, which encodes the *ori*, and 447. 1289 is the joint fragment between the early region 3' end and the vector pMK16.

vation was repeated when matrix and released DNA from C6 cells were isolated using the same high-salt extraction procedure (data not shown). Recently, Kirov et al. (1984) reported that the enrichment of an actively transcribing mouse α-globin gene with the nuclear matrix was dependent on whether DNase I digestion of the DNA preceded or followed extraction with high salt and that this enrichment was not found in cell lines that were negative for α-globin gene transcription with either extraction procedure. These results suggest that transcriptionally active complexes may be precipitating under high-salt conditions, resulting in their artifactual association with the nuclear matrix. Since we were unable to detect a preferential attachment of any SV40 DNA with the nuclear matrix, it is possible that the DNA was nicked by an endogenous nuclease prior to high-salt extraction, thus releasing the transcribed sequences from their association with this structure.

*A nonrandom distribution of SV40 sequences is obtained in the matrix and released fractions from LIS-extracted nuclei*

Nuclear matrices were prepared using the procedure described by Mirkovitch et al. (1984) by extraction in a low-salt, spermidine-containing buffer with lithium diiodosalicylate that was reported to remove the majority of the histone and many nonhistone proteins while leaving the DNA in an intact, folded conformation. This method was devised as an alternative to the high-salt method and was used to describe a family of attachment sites located upstream from the *D. melanogaster* heat-shock and histone gene clusters that were preferentially associated with the nuclear scaffold or matrix components. LIS-extracted nuclei from COS-7 cells were digested with restriction enzyme *Hind*III and the matrix-associated DNA was separated from the released DNA as described in Materials and Methods (Fig. 2, lanes A

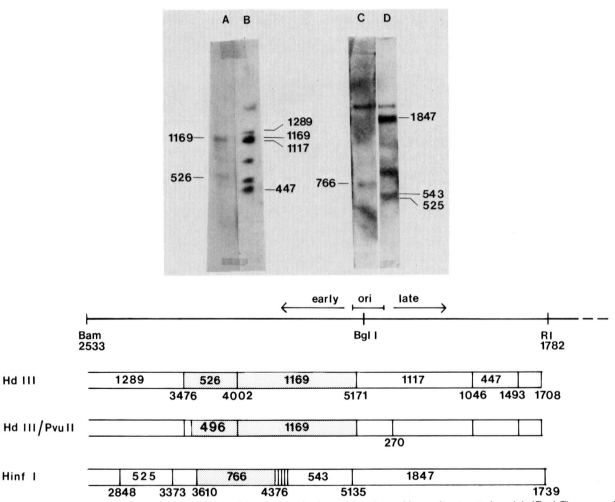

**Figure 2** COS-7 and C6 SV40 fragments detected in matrix and released fractions of low-salt-extracted nuclei. (*Top*) Three μg of COS-7 matrix (lane *A*) and 10 μg of released DNA (lane *B*) that had been separated by *Hind*III digestion and extracted as described in Materials and Methods. After separation on a 1.5% agarose gel and Southern blotting, sequences were hybridized to a [32]P-labeled SV40 probe. Seven μg of *Hinf*I-digested C6 matrix (lane *C*) and released (lane *D*) DNA were processed as in lanes *A* and *B*. (*Bottom*) Diagram showing the relative positions of early, *ori*, and late regions. Numbers within barred regions refer to fragment size. Numbers below barred regions refer to nucleotide numbers on the SV40 map. Stippled fragments indicate preferential association with the nuclear matrix.

and B). Analysis of the DNA by Southern blot hybridization to a [32]P-labeled SV40 DNA probe revealed that two early region fragments of 1169 and 526 bp mapping between nt 4002 and 5171 and between nt 3476 and 4002, respectively, were selectively retained with the nuclear matrix fraction (Fig. 2, lane A). In additional experiments (not shown) the 526-bp fragment could be further cleaved to 496 bp (nt 3506–4002) with *Pvu*II and still retain its association with the nuclear matrix fraction (see Fig. 2, diagram). The same asymmetric distribution of fragments was observed when C6 cells were analyzed in a similar fashion (not shown). The region of attachment was further delineated to a 766-bp fragment containing sequences between nt 3610 and 4376 by cleaving the histone-depleted chromatin of C6 cells with the restriction enzyme *Hin*f-I (Fig. 2, lane C). Sequences directly upstream from this region, lying between nt 4376 and 4592, cannot as yet be identified as being either preferentially associated with the matrix or released fractions since the small size of the fragments (24–109 bp) prevented accurate identification. A 543-bp fragment proximal to this region that encodes the most 5′ part of the gene (nt 4592–5135) and a 525-bp 3′ end early fragment (nt 2848–3373) (Fig. 2, lane D) were preferentially excluded from the nuclear matrix. These experiments indicate that there is a region of the SV40 chromosome, consisting possibly of two distinct sites that can be separated by a *Hin*dIII cleavage site at nt 4002, that has approximate boundaries at nt 3610 and nt 4592 and that attaches to the nuclear matrix. Both putative attachment sites lies within the SV40-transcribed early region, and one and possibly both are located within the large-T-antigen-coding sequences. Because of the uncertainity of the boundary of the most upstream attachment site, this site may include sequences at or near the common 3′ splice site for large and small T antigens. Surprisingly, sequences upstream from the early and late coding region that contain the SV40 origin of replication, the viral enhancer, 21-bp repeats, and both the early and late promoters were

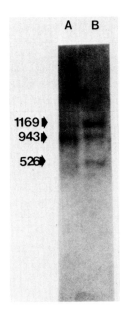

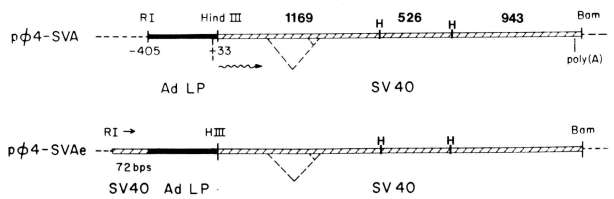

**Figure 3** SV40 fragments detected in matrix and released fractions of low-salt-extracted nuclei from transfected HeLa cells. Subconfluent HeLa cell monolayers were transfected with pφ4SVA DNA 48 hr prior to extraction. Cells were extracted using the LIS method and restricted with *Hin*dIII and *Bam*HI. One μg of pφ4SVA released (lane *A*) and matrix (lane *B*) DNA were separated and hybridized as described in Fig. 2. Numbers on left of figure at *top* refer to SV40-coding sequences indicated in diagram at *bottom*.

found almost exclusively in the released fraction (Fig. 2, lane B, fragment 1117 and lane D, fragment 1847). These results are interesting in light of the fact that this region contains a nucleosome-free region that has been shown to occur in both the integrated (Cremisi 1981; Blanck et al. 1984) and nonintegrated (Waldeck et al. 1978; Varshavsky 1979) states of the SV40 DNA and contain binding sites for T antigen (Tjian 1978; Shalloway et al. 1980) as well as at least one other regulatory factor (Dynan and Tjian 1983). The location of a potential attachment site within a transcribed region is in contrast to published results that located scaffold attachment sites to a region lying upstream from the transcription start sites (Mirkovitch et al. 1984).

*The preferential association of two SV40 early region fragments does not depend on active transcription*
Because the early region is actively transcribed in both COS and C6 cells and because it is within this region that the putative attachment site(s) is located, we were unable to determine whether the enrichment of this sequence with the nuclear matrix in COS and C6 cells was independent of the transcriptional activity of the T antigen gene. Therefore, we chose to look at the distribution of viral sequences in HeLa cells transfected with SV40-containing plasmids that were either positive or negative for T antigen expression in a transient expression assay. HeLa cells transfected with the plasmid pφ4SVA (see Fig. 3, diagram), which contains the adenovirus 2 major late promoter (AMLP) adjacent to SV40 early coding sequences from the *Hin*dIII site at nt 5171 to the *Bam*HI site at nt 2533, have been previously shown not to express T antigen by indirect immunofluorescent antibody staining. This failure of expression has been correlated with a corresponding lack of gene transcription that is dependent on a functional SV40 enhancer (Lewis and Manley 1985). Alternatively, HeLa cells transfected with the pφ4SVAe plasmid, which contains one copy of the SV40 enhancer upstream of the AMLP (see Fig. 3, diagram), is positive for T antigen expression. Forty-eight hr after transfection, nuclei were extracted as described above and the DNA was digested with *Hin*dIII and *Bam*HI. This digest, when hybridized to an SV40 probe, will identify sequences contained only within the SV40 early coding region and will detect neither the enhancer nor the promoter sequences. After extensive digestion of the DNA in LIS-extracted nuclei, which left approximately 15% of the DNA associated with the nuclear matrix, we found an asymmetric distribution of sequences between the matrix and released fractions that corresponded to the pattern seen with the COS and C6 cells. Transfection of HeLa cells with the pφ4SVA plasmid resulted in a distribution pattern in which the *Hin*dIII 1169- and 526-bp fragments were preferentially associated with the nuclear matrix (Fig. 3). The 943-bp fragment, which encodes sequences at the 3′ end of the large-T-antigen-coding sequence (nt 2533–3476), was preferentially distributed with the released fraction (Fig. 3). Similar results were obtained when pφ4SVAe was transfected

into HeLa cells (not shown). These results indicate that the distribution of viral DNA sequences in the nuclear soluble and matrix fractions is similar whether the SV40 genome is integrated into the host genome or is introduced into cells by transfection. Of particular interest is the observation that this distribution is independent of transcriptional activity. Furthermore, cells from two different species exhibited the same pattern of association of the viral DNA with the nuclear matrix. Therefore, the interaction of the SV40 genome with the nuclear matrix appears to be dictated by sequence or structural properties inherent to the DNA molecule and is not a function of cell type or transcriptional activity of the gene.

**Late gene expression of SV40 mutants defective in the viral origin of replication**
*The 5′ ends of the SV40 late RNA transcribed from SV40 DNA I are different from those transcribed from SV40 DNA inserted into bacterial plasmids in X. laevis oocytes*
The alterations in the sequences at the SV40 origin of replication of the mutants employed in this study render the viral DNA unable to replicate in permissive monkey cells (Gluzman et al. 1980b; Chen et al. 1983). Propagation of these mutants is therefore carried out in recombinant plasmids. However, it was observed that injection of viral DNA inserted into bacterial plasmids yields vastly reduced quantities of viral proteins, even though both viral and plasmid-specific RNA can be detected by dot-blot analysis (T. Michaeli and C. Prives, in prep.). To determine whether studies of late RNA synthesis from plasmids containing origin-defective mutants were feasible in oocytes, we compared the pattern of 5′ ends of late RNAs transcribed from viral DNA (SV40 DNA I) to those transcribed from SV40 DNA that was inserted into pBR322 (pSVRI) (Fig. 4). The association of pBR322 DNA sequences with the viral DNA in the recombinant plasmid strongly inhibited the synthesis of late RNA. In addition, a fundamentally altered utilization of initiation sites is evident. Injection of SV40 DNA I yields a very heterogeneous pattern of 5′ ends, mapping upstream of those detected in infected monkey cells and including large transcripts that protect the full length of the probe. However, pSVRI injection yielded two major protected fragments mapping the 5′ ends to the vicinity of nt 225 and nt 300, and a small amount of large transcripts protecting the full length of the probe. Very small quantities of other late viral RNAs were detected upon longer exposure, but virtually none of the major 5′ ends in oocytes that map around nt 290, or the major initiation site in infected monkey cells at nt 325, were detected. Thus, the attachment of pBR322 DNA to SV40 DNA induced quantitative and qualitative differences in the synthesis of late RNA in oocytes. To circumvent this problem, viral DNA was excised from recombinant plasmids. Because only circular DNA serves as a template for polymerase II transcription in oocytes (Miller and Mertz 1982; Harland et al. 1983), viral and plasmid DNAs were subsequently recircularized. The DNA was then treated with *Fnu*DII, a restriction enzyme that does

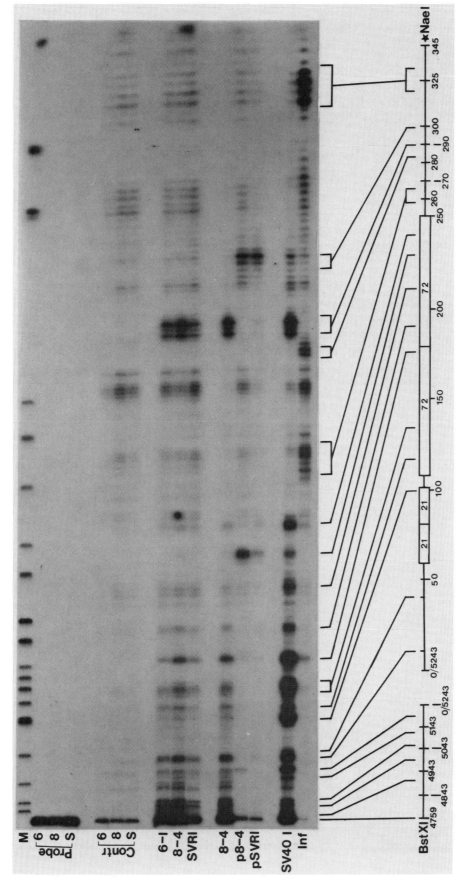

**Figure 4** 5' ends of SV40 late RNA of wild-type and origin-deletion mutants, transcribed from viral DNA and recombinant plasmids, injected into *X. laevis* oocytes. *X. laevis* oocytes were injected with 2.5 ng each of SV40 DNA I (SV40 I), recircularized viral DNAs of recombinant plasmids containing wild-type (SVRI) and origin-deletion mutants 8-4 and 6-1 SV40 DNAs (8-4 and 6-1, respectively), as well as pSVRI and p8-4 DNAs (pSVRI and p8-4, respectively). After 48 hr of incubation, RNA was extracted, treated with DNase I, and hybridized to the 5'-labeled, single-stranded *Nae*I-*Bst*XI fragment of the appropriate injected wild-type, 8-4, or 6-1 DNAs as described in the text. Samples were subsequently digested with nuclease S1, and the protected fragments were resolved on a denaturing 6.5% polyacrylamide/urea gel. RNA from SV40-infected CVI cells (*Inf*) and control oocytes were analyzed, as well (*Contr* lanes with 6 = 6-1, 8 = 8-4, and S = wild-type probes). One-fiftieth the amount of the three probes used is depicted (*Probe* lanes). Molecular-weight markers (*M*) are the fragments generated by *Hpa*II cleavage of pBR322 and are (from the top) 622, 527, 404, 309, 242, 238, 217, 201, 190, 180, 160, 147, 122, 110, 90, 76, 67, 34, 26, and 15 nt in length. A diagram of the late RNA 5' ends detected is shown, including nucleotide numbers and positions of the 21- and 72-bp repeat sequences.

not cut SV40 DNA but cleaves pBR322 and pMK16 into more than 20 fragments. The pattern of late RNA 5' ends transcribed from such templates is very similar to the pattern detected when SV40 DNA I is injected, though their quantities are reduced (Fig. 4). These quantitative differences may be partly the result of minor differences in quantities of DNAs injected or due to the altered superhelicity of the DNA after the ligation process.

*Defects in the viral origin of replication do not affect synthesis of late RNA and proteins in oocytes*
The effects of three different mutations in the origin, generated by in vitro manipulation of DNA, on SV40 late transcription were examined. The mutants 6-1 and 8-4, containing 4- and 6-bp deletions in the origin, respectively, were generated and shown to be replication-defective by Gluzman et al. (1980a). The quantitites of late RNA synthesized upon injection of 6-1 and 8-4 were comparable to those detected after injection of wild-type SV40 DNA, and the patterns of 5' ends were virtually identical (Fig. 4). Viral capsid proteins were synthesized, as well, indicating that functional mRNA was produced (Fig. 5). Variability between groups of oocytes can account for the differences in quantities of capsid proteins detected in these particular analyses. An additional mutant was examined, in which a 4-bp insertion was introduced into the origin that rendered it unable to replicate (*ori*⁻) (Chen et al. 1983). No quantitative or qualitative differences in late RNA 5' ends were detected when RNAs from *ori*⁻ and wild-type SV40–injected oocytes were compared (Fig. 6). Thus, the results obtained with all three mutants analyzed indicate

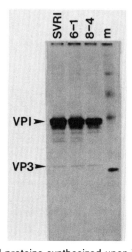

**Figure 5** Capsid proteins synthesized upon injection of SV40 and SV40 origin-deletion mutants into *X. laevis* oocytes. *X. laevis* oocytes were injected with 2.5 ng of recircularized viral DNA of recombinant plasmids containing wild-type (SVRI) and origin-deletion mutants 6-1 (6-1) and 8-4 (8-4) SV40 DNAs. [³⁵S]Methionine-labeled protein extracts were immunoprecipitated with anti–SV40 capsid antiserum and separated by polyacrylamide gel electrophoresis. The locations of the capsid proteins VPI and VP3 are indicated. Protein molecular-weight markers (*m*) are (from the top) 200K, 94K, 69K, 46K, 30K, and 14.3K.

that alterations in the DNA sequence of the SV40 origin of replication do not affect transcription from the late promoter in *X. laevis* oocytes.

The quantitative and qualitative differences induced by the attachment of pBR322 DNA to SV40 DNA were also observed when RNAs from 8-4– and p8-4–injected oocytes were compared (Fig. 4). The plasmid into which 8-4 DNA was inserted is a derivative of pMK16, yet the 5'-ends pattern of p8-4 was virtually identical with that of pSVRI. Thus, different bacterial plasmids have similar effects on utilization of initiation sites of the late promoter. Comparable amounts of late RNA were detected when p8-4– and pSVRI-injected oocytes were compared, offering additional support to the notion that defects in the SV40 origin of replication do not affect late transcription in oocytes.

The observations made in this study contradict a previously published report (Contreras et al. 1982), implicating a direct involvement of the SV40 origin of replication in late RNA synthesis in oocytes. In that study it was reported that early after injection of recombinant plasmids bearing the SV40 regulatory region, the pattern of late RNA 5' ends in oocytes is identical with the one detected in infected monkey cells. This is in contrast to our observation that the 5' ends of late RNA in oocytes are fundamentally different from those in infected cells at all times after microinjection (Michaeli and Prives 1985; Figs. 4 and 5 and unpublished observations). A possible source of this discrepancy is that the constructs used by Contreras et al. (1982) contained small fragments of wild-type SV40 DNA spanning the origin and part of the late leader region in recombinant plasmids that were compared with p8-4, containing the entire SV40 genome with the 8-4 deletion. Deletions within the SV40 late leader sequences have been shown to induce changes in the abundance and pattern of late RNA 5' ends (Ghosh et al. 1982). Our own comparisons of RNA from pSVRI- and p8-4–injected oocytes, both containing the entire SV40 genome, indicated that no inhibitory effects exerted by the defective origin are evident.

The demonstrated independence of late RNA synthesis from sequences in the origin may be unique to oocytes. Keller and Alwine (1985) have demonstrated that 65–75% of the large-T-antigen-mediated activation of the late promoter is independent of both DNA replication and sequences in the origin. However, the remaining 25–35% of the activation mediated by large T antigen is dependent on replication and on an intact origin, as well. In contrast, Brady and Khoury (1985) have demonstrated that the same defects in the origin induced a 20-fold decrease in large-T-antigen-mediated expression of SV40 late genes. Since DNA replication was abolished in their study and it was observed that the pattern of late RNA 5' ends resembled those detected early in lytic infection, it is possible that sequences in the origin have a strong regulatory effect on late RNA synthesis only early in infection. Thus, when viral DNA replication is permitted, late RNA synthesis exhibits a reduced dependence on the sequences in the origin.

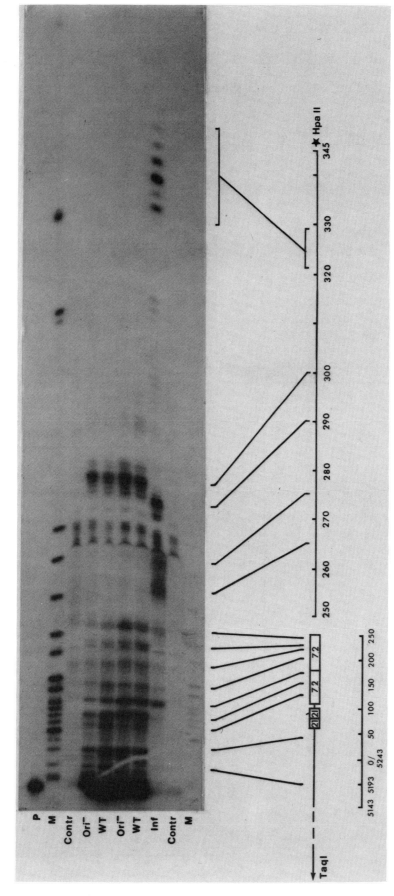

**Figure 6** 5′ ends of SV40 late RNA of wild type and an origin-insertion mutant synthesized in *X. laevis* oocytes. *X. laevis* oocytes were injected with 2.5 ng of recircularized viral DNA of recombinant plasmids containing wild-type and an origin-insertion mutant SV40 DNA (*WT* and *Ori⁻*, respectively). After 48 hr of incubation, RNA was extracted, treated with DNase I, and hybridized to the 5′-labeled, single-stranded *Hpa*II-*Taq*I fragment (*P*) of the viral DNA injected, as described in the text. Samples were subsequently digested with nuclease S1 and the protected fragments were resolved on a denaturing 12% polyacrylamide/urea gel. RNA from SV40-infected CVI cells (*Inf*), and control oocytes (*Contr*) was analyzed, as well. Analyses of two different frogs injected with wild-type and *ori⁻* DNA are shown. Molecular-weight markers are as described in Fig. 4.

In our study, oocytes were presented with a large number of DNA templates, in the absence of DNA replication. Late RNA synthesis in oocytes is neither temporally regulated nor dependent on the function of large T antigen (Michaeli and Prives 1985). The lack of effect of sequences in the origin on late RNA synthesis may be the result of several factors. The high concentration of DNA templates injected into the oocytes may override other regulatory processes, some of which may be dependent on DNA replication and the sequences in the origin. Alternatively, oocytes may contain or lack various activities, so that an altered utilization of the late promoter occurs. The existence of a T-antigen-like activity in oocytes has been postulated previously (Miller et al. 1982), and absence of negative regulation of late gene expression has been proposed, as well (Michaeli and Prives 1985). In addition, the structure of SV40 minichromosomes assembled in oocytes may result in an altered mode of regulation, independent of the origin.

## Acknowledgments

The authors thank John Brady and George Khoury for helpful discussions during the course of these studies and Yakov Gluzman for kindly providing origin-defective mutants. These experiments were supported by grants CA 26905 from the National Institutes of Health and PCM-82-16798 from the National Science Foundation.

## References

Adolph, K.W., S.M. Cheng, and U.K. Laemmli. 1977. Role of nonhistone proteins in metaphase chromosome structure. *Cell* **12**: 805.

Benyajati, C. and A. Worcel. 1976. Isolation, characterization and structure of the folded interphase genome of *Drosophila melanogaster*. *Cell* **9**: 393.

Ben-Ze'ev, A. and Y. Aloni. 1983. Processing of SV40 RNA is associated with the nuclear matrix and is not followed by the accumulation of low-molecular-weight RNA products. *Virology* **125**: 475.

Berezney, R. 1979. In *The cell nucleus* (ed. H. Busch), vol. 7, p. 413. Academic Press, New York and London.

Berezney, R. and L.A. Buchholtz. 1981. Dynamic association of replicating DNA fragments with the nuclear matrix of regenerating liver. *Exp. Cell Res.* **132**: 1.

Berezney, R. and D.S. Coffey. 1977. Isolation and characterization of a framework structure from rat liver nuclei. *J. Cell Biol.* **4**: 559.

Bergsma, D.J., D.M. Olive, S.W. Hartzell, and K.N. Subramanian. 1982. Territorial limits and functional anatomy of the simian virus 40 replication origin. *Proc. Natl. Acad. Sci.* **79**: 381.

Blanck, G., S. Chen, and R. Pollack. 1984. DNase I sensitivity of integrated simian virus 40 DNA. *Mol. Cell. Biol.* **4**: 559.

Brady, J. and G. Khoury. 1985. *Trans*-activation of the simian virus 40 late transcription unit by T-antigen. *Mol. Cell. Biol.* **5**: 1391.

Brady, J., J.B. Bolen, M. Randonovich, N. Salzman, and G. Khoury. 1984. Stimulation of simian virus 40 late gene expression by simian virus 40 tumor antigen. *Proc. Natl. Acad. Sci.* **81**: 2040.

Chen, S., D.S. Grass, G. Blank, N. Hoganson, J.L. Manley, and R.E. Pollack. 1983. A functional simian virus 40 origin of replication is required for the generation of a super T antigen with a molecular weight of 100,000 in transformed mouse cells. *J. Virol.* **48**: 492.

Ciejek, E.M., M. Tsai, and B.W. O'Malley. 1983. Actively transcribed genes are associated with the nuclear matrix. *Nature* **306**: 606.

Contreras, R., D. Gheysen, J. Knowland, A. van de Voorde, and W. Fiers. 1982. Evidence for the direct involvement of the DNA replication origin in synthesis of late SV40 RNA. *Nature* **300**: 500.

Cook, P.R. and I.A. Brazell. 1980. Mapping sequences in loops of nuclear DNA by their progressive detachment from the nuclear cage. *Nucleic Acids Res.* **8**: 2895.

Cook, P.R., J. Lang, A. Hayday, L. Lania, M. Fried, D.J. Chiswell, and J.A. Wyke. 1982. Active viral genes in transformed cells lie close to the nuclear cage. *EMBO J.* **1**: 447.

Covey, L., Y.-S. Choi, and C. Prives. 1984. Association of simian virus 40 T antigen with the nuclear matrix of infected and transformed monkey cells. *Mol. Cell. Biol.* **4**: 1384.

Cremisi, C. 1981. The appearance of DNase I hypersensitive sites at the 5′ end of the late SV40 genes is correlated with the transcriptional switch. *Nucleic Acids Res.* **9**: 5949.

Di Maio, D. and D. Nathans. 1978. Cold sensitive regulatory mutants of simian virus 40. *J. Mol. Biol.* **140**: 129.

———. 1982. Regulatory mutants of simian virus 40: Effect of mutations at a T antigen binding site on DNA replication and expression of viral genes. *J. Mol. Biol.* **156**: 531.

Dynan, W.S. and R. Tjian. 1983. The promoter-specific transcription factor binds to upstream sequences in the SV40 early promoter. *Cell* **35**: 79.

Frost, E. and J. Williams. 1978. Mapping temperature-sensitive and host-range mutations of adenovirus type 5 by marker rescue. *J. Virol.* **91**: 39.

Ghosh, P.K., M. Piatek, J.E. Mertz, S.M. Weissman, and P. Leibowitz. 1982. Altered utilization of splice sites and 5′ termini in late RNAs produced by leader region mutants of simian virus 40 mutants. *J. Virol.* **44**: 610.

Gluzman, Y. 1981. SV40-transformed simian cells support the replication of early SV40 mutants. *Cell* **23**: 175.

Gluzman, Y., R.J. Frisque, and J. Sambrook. 1980a. Origin-defective mutants of SV40. *Cold Spring Harbor Symp. Quant. Biol.* **44**: 293.

Gluzman, Y., J. Sambrook, and R.J. Frisque. 1980b. Expression of early genes of origin-defective mutants of simian virus 40. *Proc. Natl. Acad. Sci.* **77**: 3898.

Gluzman, Y., J. Davison, M. Oren, and E. Winocour. 1977. Properties of permissive monkey cells transformed by UV-irradiated simian virus. *J. Virol.* **22**: 256.

Harland, R.M. and R.A. Laskey. 1980. Regulated replication of DNA microinjected into eggs of *Xenopus laevis*. *Cell* **21**: 761.

Harland, R.M., H. Weintraub, and S.L. McKnight. 1983. Transcription of DNA injected into *Xenopus laevis* oocytes is influenced by template topology. *Nature* **302**: 38.

Herman, R., L. Weymouth, and S. Penman. 1978. Heterogeneous nuclear RNA-protein fibers in chromatin-depleted nuclei. *J. Cell Biol.* **78**: 663.

Keller, J.M. and J.C. Alwine. 1985. Analysis of an activatible promoter: Sequences in the simian virus 40 late promoter required for T-antigen-mediated transactivation. *Mol. Cell. Biol.* **5**: 1854.

Kimura, G. and R. Dulbecco. 1972. Isolation and characterization of temperature-sensitive mutants of simian virus 40. *Virology* **49**: 394.

Kirov, N., L. Djondjurov, and R. Tsanev. 1984. Nuclear matrix and transcriptional activity of the mouse α-globin gene. *J. Mol. Biol.* **180**: 601.

Lewis, E.D. and J.L. Manley. 1985. Control of adenovirus late promoter expression in two human cell lines. *Mol. Cell. Biol.* **5**: 2433.

Michaeli, T. and C. Prives. 1985. Regulation of simian virus 40 gene expression in *Xenopus laevis* oocytes. *Mol. Cell. Biol.* **5**: 2019.

Miller, T.J. and J.E. Mertz. 1982. Template structural requirements for transcription *in vivo* by RNA polymerase II. *Mol. Cell. Biol.* **2**: 1591.

Miller, T.J., D.L. Stephens, and J.E. Mertz. 1982. Kinetics of accumulation and processing of simian virus 40 RNA in

*Xenopus laevis* oocytes injected with simian virus 40 DNA. *Mol. Cell. Biol.* **2:** 1581.

Mirkovitch, J., M.-E. Mirault, and U.K. Laemmli. 1984. Organization of the higher-order chromatin loop: Specific DNA attachment sites on nuclear scaffold. *Cell* **39:** 223.

Nelkin, B.D., D.M. Pardoll, and B. Vogelstein. 1980. Localization of SV40 genes within supercoiled loop domains. *Nucleic Acids Res.* **8:** 5623.

Pardoll, D.M., B. Vogelstein, and D.S. Coffey. 1980. A fixed site of DNA replication in eucaryotic cells. *Cell* **19:** 527.

Paulson, J.R. and U.K. Laemmli. 1977. The structure of histone-depleted metaphase chromosomes. *Cell* **12:** 817.

Prives, C., L. Covey, A. Scheller, and Y. Gluzman. 1983. DNA-binding properties of simian virus 40 T-antigen mutants defective in viral DNA replication. *Mol. Cell. Biol.* **3:** 1958.

Robinson, S.I., B.D. Nelkin, and B. Vogelstein. 1982. The ovalbumin gene is associated with the nuclear matrix of chicken oviduct cells. *Cell* **28:** 99.

Shalloway, D., T. Kleinberger, and D.M. Livingston. 1980. Mapping of SV40 DNA replication origin binding sites for the SV40 T antigen by protection against Exo III digestion. *Cell* **2:** 411.

Southern, E.M. 1975. Detection of specific sequences among DNA fragments separated by gel electrophoresis. *J. Mol. Biol.* **98:** 503.

Subramanian, K. and T. Shenk. 1978. Definition of the origin of DNA replication in simian virus 40. *Nucleic Acids Res.* **5:** 3635.

Tjian, R. 1978. The binding site on SV40 DNA for a T antigen–related protein. *Cell* **13:** 165.

Tooze, J., ed. 1981. *Molecular biology of tumor viruses*, 2nd edition, revised: *DNA tumor viruses*. Cold Spring Harbor Laboratory, Cold Spring Harbor, New York.

van Eekelen, C.A.G. and W.J. van Venrooij. 1981. hnRNA and its attachment to a nuclear protein matrix. *J. Cell Biol.* **88:** 554.

Varshavsky, A.J. 1979. A stretch of late SV40 viral DNA about 400 bp long which includes the origin of replication is specifically exposed in SV40 minichromosomes. *Cell* **16:** 456.

Waldeck, W., B. Fohring, K. Chowdhury, P. Gruss, and G. Sauer. 1978. Origin of DNA replication in papovavirus chromatin is recognized by endogenous endonuclease. *Proc. Natl. Acad. Sci.* **75:** 5964.

Wigler, M., A. Pellicer, S. Silverstein, R. Axel, G. Urlaub, and L. Chasin. 1979. DNA-mediated transfer of the adenine phosphoribosyltransferase locus into mammalian cells. *Proc. Natl. Acad. Sci.* **76:** 1373.

# Enhanced Transcriptional Activity of HeLa Cell Extracts Infected with Adenovirus Type 2

### S.J. Flint and K. Leong*

Department of Molecular Biology, Princeton University, Princeton, New Jersey 08544

The transcriptional activities of whole-cell extracts prepared from normal HeLa cells and HeLa cells infected with adenovirus type 2 (Ad2) harvested after various periods of infection have been compared by using templates that contain both the late (ML) and IVA2 transcriptional control regions. Although the activity displayed by extracts was influenced by such parameters as the KCl concentration or protein-to-DNA ratio, extracts prepared from Ad2-infected cells were more active than their uninfected cell counterparts under all conditions. Maximal stimulation, some 15-fold, of ML and IVA2 transcription in Ad2-infected cell extracts was observed under conditions that were not optimal for transcription in extracts of uninfected cells. Extracts of infected cells harvested during the early phase of infection displayed a greater enhancement of transcriptional activity than those prepared from cells harvested 12–14 hr after infection. Extracts prepared from HeLa cells infected with 150 pfu/cell H5hr1, a virus that cannot express the 289-amino-acid E1A protein (Ricciardi et al. 1981) did not exhibit enhanced transcriptional activity. During chromatography on heparin-agarose, the transcriptional machinery from Ad2-infected cells eluted as a more homogeneous peak at a lower concentration of KCl than did transcriptional activity from uninfected cell extracts. The activities displayed by the fractions recovered from heparin-agarose columns suggest that, in infected cells, RNA polymerase II is more tightly associated with initiation factors or that a greater proportion of the enzyme is associated with such factors.

The late phase of infection of permissive cells by subgroup-C human adenoviruses, such as adenovirus type 2 (Ad2), is characterized by a dramatic inhibition of cellular gene expression (Ginsberg et al. 1967; White et al. 1969; Anderson et al. 1973; Russell and Skehel 1973; Beltz and Flint 1979; for a review, see Flint 1984) mediated posttranscriptionally during mRNA biogenesis (Beltz and Flint 1979; Babich et al. 1983; Flint et al. 1983; Yoder et al. 1983) and translationally (Thimmappaya et al. 1982; Babich et al. 1983; Schneider et al. 1984). Transcription of cellular class-II genes is not inhibited (with the exception of those encoding histones; Flint et al. 1984), although form-II RNA polymerase (Price and Penman 1972; Weinmann et al. 1977) synthesizes an amount of nuclear adenovirus RNA that is approximately equal to the amount of cellular hnRNA present in both uninfected and Ad2-infected cells (Beltz and Flint 1979). The extremely large quantities of viral pre-mRNA made during the late phase of adenovirus infection can be at least partly ascribed to the large number of viral DNA molecules generated during replication, at least 100,000 per cell (Simmons et al. 1974; Flint et al. 1976; Schick et al. 1976), about 5–6% of which have been estimated to serve as templates for transcription of viral genetic information (Wolgemuth and Hsu 1981). Nevertheless, the ability of adenovirus-infected cells to increase production of RNA polymerase II transcripts by close to a factor of two implies either

that HeLa cells possess a large reservoir of untapped transcriptional capacity or that the transcriptional machinery is modified after adenovirus infection.

The latter possibility is of particular interest in view of the well-established role of the Ad2 289-amino-acid (aa) protein encoded by region E1A in stimulating transcription of other viral early regions and of the major late transcriptional unit (Berk et al. 1979; Jones and Shenk 1979; Nevins 1981; Ricciardi et al. 1981; Montell et al. 1982). Moreover, cells in which this protein is made support increased expression of newly introduced cellular genes (e.g., Green et al. 1983; Treisman et al. 1983; Gaynor et al. 1984; Svensson and Akusjärvi 1984) and of the endogenous human hsp70 (Nevins 1982; Kao and Nevins 1983) and β-tubulin (Stein and Ziff 1984) genes.

These properties of adenovirus-infected cells suggest that extracts prepared from them might display increased transcriptional activity in vitro. However, no increases in the transcriptional activity displayed by Ad2-infected cell extracts compared with uninfected HeLa cell extracts have been observed (Fire et al. 1981; Lee and Roeder 1981; Hoeffler and Roeder 1985), using either whole-cell (Manley et al. 1980), S100 (Weil et al. 1979), or nuclear (Dignam et al. 1981) extracts. During the course of an investigation of transcription in whole HeLa cell extracts of the Ad2 IVA2 gene, which does not possess a sequence element related to the "TATA" consensus in the region upstream from its cap sites (Baker and Ziff 1981), we observed that the production of runoff transcripts was stimulated in extracts of Ad2-infected HeLa cells. Here we present the results

*Present address: Department of Microbiology, MBI, University of California, Los Angeles, California 90024.

of a comparison of the abilities of such extracts to transcribe specific adenovirus sequences and of preliminary experiments to elucidate the basis of such enhanced transcriptional activity.

## Materials and Methods

### Cells and virus

HeLa cells were maintained in suspension culture and Ad2 was propagated in them as described previously (Flint et al. 1983).

### Transcription in HeLa cell extracts

The preparation of extracts of HeLa cells, based on the procedure of Manley et al. (1980), has been described (Leong and Flint 1984). Extracts were also prepared from HeLa cells infected with 20–40 pfu/cell Ad2 or with 150 pfu/cell H5hr1 after the periods of infection indicated in the text. Extracts whose activities were to be compared with one another were made at the same time from equal numbers of uninfected or infected cells. Extract protein concentrations were determined by the methods of Lowry et al. (1955) or Bradford (1976). Unless otherwise stated, transcription reactions, 25 or 50 μl, contained 12 mM HEPES, pH 7.9, 0.06 mM EDTA, 1.2 mM DTT, 10% (v/v) glycerol, 2 mM creatine phosphate, 250 μM each of ATP, CTP, and UTP (Schwarz-Mann), 50 μM GTP (Schwarz-Mann) and 5–10 μCi [α-$^{32}$P]GTP (NEN; 760 Ci/mmole), and the concentrations of DNA template, extract protein, MgCl$_2$ and KCl listed in the figure legends. After incubation at 30°C for the periods indicated in the figure legends, the labeled products were purified as described previously (Leong and Flint 1984). Labeled products were denatured (Thomas 1980) and analyzed directly by electrophoresis in 1.4% agarose or 5% polyacrylamide gels as described previously (Leong and Flint 1984). Unlabeled products, synthesized in reactions containing 250 μM of each of the four nucleoside triphosphates, were hybridized to the end-labeled DNA probes listed in the figure legends (Berg and Sharp 1978; Weaver and Weissmann 1979). Hybrids were digested with 100 units of mung bean nuclease (PL Biochemicals), purified, and precipitated with ethanol. The precipitated nucleic acids were collected, dissolved in 20 μl formamide, boiled for 2 min, and electrophoresed in 6% polyacrylamide gels cast in Tris-borate buffer and containing 7 M urea. All gels were dried and exposed to Kodak X/AR or X/RP at −80°C in the presence of intensifying screens.

### Preparation of DNA templates and probes

Transcription from the Ad2 IVA2 and major late (ML) transcriptional control regions was examined using the plasmid pBalE, containing the *Bal*E fragment of Ad2 (14.68–21.24 map units [m.u.]) inserted at the *Bal*I site of pBR322, the kind gift of S.M. Berget. This plasmid was cleaved by appropriate restriction endonucleases (Fig. 1A) under the conditions recommended by the vendors. The DNA was purified by phenol and chloroform extraction followed by ethanol precipitation. Unfractionated DNA fragments were used in transcription reactions. End-labeling of DNA fragments was carried out as described by Maniatis et al. (1982).

## Results

### Enhanced transcriptional activity of extracts prepared from adenovirus-infected cells

During our examination of the transcription of Ad2 IVA2 gene in whole HeLa cell extracts (Leong and Flint 1984), we tested the transcription of templates containing both the IVA2 and ML transcriptional control regions (Fig. 1A) in extracts prepared from uninfected or Ad2-infected HeLa cells. In the experiment shown in Figure 1B, different quantities of extract protein were added to transcription reactions containing the pBalE plasmid cut with *Sma*I and *Sal*I. Initiation of transcription at the ML and IVA2 cap sites would produce runoff transcripts of 0.54 and 1.26 kb, respectively (Fig. 1A). The products of transcription reactions containing either infected or uninfected HeLa cell extracts included species of the predicted size (Fig. 1B, lanes 1–4), whose synthesis was inhibited by low concentrations of α-amanitin (Fig. 1B, lanes 5–8). The likely source of the other labeled products, which included large species comigrating with the DNA fragments of the template whose labeling was not completely eliminated even in the presence of 400 μg/ml α-amanitin, is discussed in the legend to Figure 1. Extracts prepared from infected HeLa cells harvested 14 hr after infection at 40 pfu/cell synthesized larger quantities of both the 1.26-kb and 0.54-kb products but not of the larger labeled species (Fig. 1B, lanes 1–4).

A second example of this phenomenon, with an independently prepared pair of extracts is shown in Figure 1C. In this experiment, increasing amounts of pBalE DNA cleaved with *Sma*I (0.54-kb ML runoff RNA) and *Sst*I (192-nucleotide [nt] IVA2 runoff transcript) were incubated with 4 mg/ml protein of extracts prepared from uninfected or Ad2-infected cells, harvested 14 hr after infection at 20 pfu/cell. In the absence of template DNA, two large, labeled products were observed (Fig. 1C, lanes 1 and 4), presumably the result of labeling of nucleic acids present in the extract. Infected cell extracts synthesized larger quantities of the predicted ML and IVA2 runoff transcripts at both concentrations of DNA tested (Fig. 1C, lanes 2, 3, 5, and 6). Indeed, the synthesis of IVA2 in the uninfected cell extract was so inefficient that it could not be detected unless much longer exposures of the gel shown in Figure 1C were made.

The synthesis of products exhibiting the sizes predicted for IVA2 runoff transcripts from two templates cleaved at different sites downstream from the IVA2 cap sites suggests that initiation of IVA2 transcription in these extracts is accurate. We have previously demonstrated that this is indeed the case (Leong and Flint 1984) and nuclease protection mapping experiments have been performed to confirm the identity of the 0.54-kb ML runoff transcript (data not shown).

The activities of extracts prepared from uninfected

and Ad2-infected HeLa cells have been compared under various conditions. Figure 2A shows typical results obtained when the concentration of extract protein was varied. Synthesis of IVA2 runoff RNA was first stimulated and then inhibited as the concentration of extract protein was increased, a pattern identical with that observed

previously (Leong and Flint 1984). Under all conditions, the infected cell extract produced more IVA2 RNA than the initial cell extract, the greatest stimulation—4-fold to 6-fold in this experiment—being observed at protein concentrations suboptimal for IVA2 transcription in the uninfected cell extract (Fig. 2A). The synthesis of ML

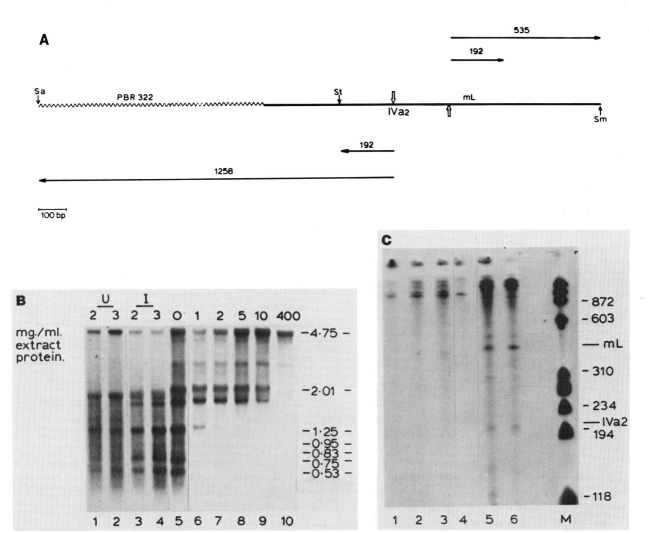

**Figure 1** Synthesis of ML and IVA2 runoff RNA species in extracts of uninfected and Ad2-infected HeLa cells. (*A*) The template DNA used in these, and all subsequent experiments, is represented by the horizontal line in which solid and wavy segments denote Ad2 and pBR322 sequences, respectively. The sites at which this DNA is cleaved by *Sal*I, *Sst*I, *Hind*III, and *Sma*I are indicated by Sa, St, H, and Sm, respectively. The cap sites of the IVA2 and ML transcription units are indicated by the open, vertical arrows, and predicted runoff transcripts are shown by horizontal arrows drawn in the direction of transcription, whose lengths are given in nucleotides. (*B*) The transcription reactions whose products are shown in lanes *1–4* contained 36 μg/ml pBalE DNA cleaved with *Sal*I and *Sma*I, 24 mM KCl, 3 mM MgCl$_2$, other components described in Materials and Methods, and the concentration of uninfected (U) or Ad2-infected (I) HeLa cell extract protein above each lane. The reactions whose products are shown in lanes *5–10* contained 5.0 mg/ml uninfected HeLa cell extract protein, 30 μg/ml template DNA, 5 mM MgCl$_2$, 60 mM KCl, and the 0, 1, 2, 5, 10, or 400 μg/ml α-amanitin, respectively. In all cases, incubation was for 90 min at 30°C and labeled products were purified, denatured, and analyzed in agarose gels as described in Materials and Methods. In addition to the runoff transcripts discussed in the text, other small, labeled products were produced. The full range of these is illustrated in lane *5*. None are synthesized in the presence of low concentrations of α-amanitin. All between 0.7 and 1.0 kb in length are complementary to the l-strand of the IVA2 region and their concentration always reflects that of the 1.25-kb IVA2 runoff RNA. The 0.75-kb species, the most abundant, protects a specific segment of the IVA2 gene, to a site some 30 nt upstream of the IVA2 3′ species site (K. Leong, unpubl.). Thus, these are not nonspecific products but probably are the result of processing of IVA2 transcripts. Their absence from the products of reactions incubated for 30 min (E.H. Postel, unpubl.) is consistent with this conclusion. (*C*) Transcription reactions contained 4 mg/ml protein of an uninfected (lanes *1–3*) or an Ad2-infected (lanes *4–6*) cell extract, 24 mM KCl, 3 mM MgCl$_2$, other components as described in Materials and Methods, and 0 (lanes *1* and *4*), 10 (lanes *2* and *5*), or 20 (lanes *3* and *6*) μg/ml pBalE DNA cut with *Sma*I and *Sst*I. Incubation was for 45 min at 30°C and products were purified, denatured, and analyzed in a 5% polyacrylamide gel. End-labeled, glyoxalated *Hae*III fragments of φX174 DNA were applied to lane *M*.

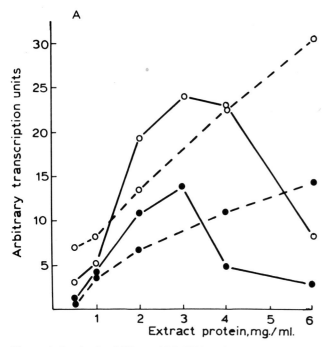

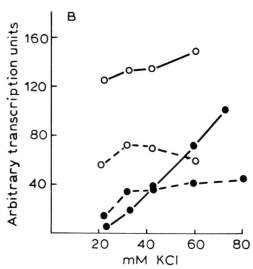

**Figure 2** Synthesis of IVA2 and ML RNA under different in vitro conditions. (*A*) Transcription reactions contained 36 μg/ml pBalE DNA cleaved with *Sma*I and *Sal*I, 3 mM MgCl$_2$, and 24 mM KCl; the concentrations of extract protein are indicated. Reaction products were analyzed as described in Materials and Methods. Autoradiograms obtained by exposure of dried gels to Kodak X/AR (or X.RP) film under conditions established to be in the linear dose-response range were traced using a Joyce-Loebel double-beam recording microdensitometer, and areas under the peaks were excised and weighed. These values are expressed in arbitrary transcription units. (*B*) Transcription reactions contained 25 μg/ml *Sma*I- and *Sal*I-cleaved pBalE DNA, 3 mM MgCl$_2$, 5 mg/ml uninfected or Ad2-infected HeLa cell extract, and the concentrations of KCl indicated. Purification, analysis, and quantitation of products were as described in the legend to Figure 1B. In both panels, synthesis of the IVA2 1.26-kb transcript in uninfected or Ad2-infected cell extracts are shown by ●———● and ○———○, respectively, whereas ●– – – –● and ○– – – –○, respectively, represent ML transcripts (0.54 kb) made in uninfected and Ad2-infected cell extracts. Independent pairs of extracts were used in the experiments shown in *A* and *B*.

RNA, by contrast, increased more or less linearly as a function of extract protein concentration and the infected cell extract was approximately 2.5 times more active at all but the lowest protein concentration examined.

Production of IVA2 and ML runoff transcripts as a function of KCl concentration is illustrated in Figure 2B. Transcription from the IVA2 control region in uninfected HeLa cell extracts was markedly stimulated as the KCl concentration was raised from 24 to 74 mM. Synthesis of the IVA2 runoff transcript in Ad2-infected cell extracts was, by contrast, much less sensitive to KCl concentration, for this RNA species was made at a high level even at the lowest KCl concentrations used (Fig. 2B). Thus, the greatest stimulation of IVA2 transcription—about 15-fold in this experiment—was seen when transcription reactions contained 24 mM KCl. However, the infected cell extract was more active than its uninfected counterpart even at the higher KCl concentrations tested, producing about 2-fold more IVA2 RNA. Synthesis of the 0.54-kb ML transcript was much less dramatically stimulated as the KCl concentration was increased. The greatest difference between the activities of the two extracts was again observed under conditions least optimal for transcription by the uninfected cell extract, in this case 24 mM KCl, when nearly 5-fold more ML RNA was produced by the infected-cell extract.

Enhanced transcriptional activity after Ad2 infection,

assayed as synthesis of IVA2 and ML runoff transcripts in vitro, has been observed with many pairs of uninfected and Ad2-infected HeLa cell extracts. The degree of stimulation is influenced by the conditions under which transcription is assayed (Fig. 2) and also by the time at which infected cells are harvested: Extracts prepared from cells harvested during the early phase of infection were more active than those harvested later in infection (see Fig. 5). The maximal degree of stimulation we have observed was on the order of 15-fold (e.g., Figs. 2 and 5).

A comprehensive analysis of the transcriptional activity of uninfected and Ad2-infected cell extracts upon different DNA templates has not yet been performed. However, extracts have been assayed using a rabbit β-globin DNA template, when 2-fold to 3-fold more 1.1-kb runoff RNA was made by the infected cell extract (data not shown). We do not know whether this value represents the maximal stimulation that can be attained, because transcription of rabbit β-globin RNA has not been examined as a function of parameters other than extract protein concentration.

### Stimulation of transcription in Ad2-infected cell extracts occurs at a step prior to initiation

To investigate which step in the transcription reaction might be accelerated in infected cell extracts, we examined the formation of preinitiation complexes in uninfected and Ad2-infected cell extracts. The formation of stable preinitiation complexes that are resistant to dilution or to competition from a second DNA template was initially reported for RNA polymerase III, its transcription factors, and class-III eukaryotic genes (Bogenhagen et al. 1982; Lassar et al. 1983). Similar complexes that can initiate transcription immediately upon addition of nucleoside triphosphate substrates can be formed when class-II eukaryotic genes are incubated with RNA polymerase II and various transcription factors (Davison et al. 1983; Fire et al. 1984). A preincubation-pulse-chase protocol based on that described by Fire et al. (1984) was employed to assess whether the rate of formation of such complexes was accelerated in Ad2-infected HeLa cell extracts. The pBalE DNA template described in previous sections was preincubated at 30°C with uninfected or Ad2-infected HeLa cell extract, but no nucleoside triphosphates, to permit formation of putative preinitiation complexes. The reactions were then incubated for 5 min at 30°C in the presence of 1 μM [α-$^{32}$P]GTP and 250 μM of each of the other nucleoside triphosphates to permit initiation and limited elongation. Such a pulse was followed by a 15-min chase in the presence of 1 mM GTP to permit elongation and completion of RNA chains initiated during the pulse, in the absence of significant labeling of chains initiated during the chase. At elongation rates measured in an vitro system from a similar source, 400–600 nt/min (Fire et al. 1984), 15 min is clearly sufficient to complete transcription of the 0.54-kb ML RNA transcripts examined in this experiment. The 0.54-kb ML runoff RNA could be detected after as little as 5 min of preincubation with infected cell extracts and was subsequently synthesized at a rate that was more or less a linear function of preincubation time (Fig. 3). By contrast, no ML runoff RNA could be detected until 20 min of preincubation of

the template with the uninfected HeLa cell extract. Such accelerated appearance of ML RNA during preincubation of the Ad2 DNA template with Ad2-infected cell extracts has been observed with four independent pairs of uninfected and Ad2-infected cell extracts. We therefore conclude that a primary consequence of preincubation of the template with Ad2-infected HeLa cell extracts is an accelerated rate of formation of stable preinitiation complexes, capable of initiation of transcription immediately upon addition of ribonucleoside triphosphate substrates; the lag period of about 20 min observed in uninfected cell extracts, presumably required for assembly of preinitiation complexes, was reduced to less than 5 min in Ad2-infected cell extracts (Fig. 3). Moreover, the final differences, about fivefold, in the levels of ML transcripts synthesized in infected compared with uninfected cell extracts by the end of the preincubation period were sufficient to account for the differences typically seen when standard transcription reactions were performed under the same conditions.

## Chromatographic behavior of transcriptional machinery from uninfected and Ad2-infected HeLa cells

As a first step toward characterization of the biochemical basis of the enhanced transcriptional activity shown by extracts of Ad2-infected HeLa cells, we have fractionated pairs of extracts on heparin-agarose columns. Typical results obtained when a pair of columns were eluted with increasing concentrations of KCl are shown in Figure 4. Equal quantities of protein recovered in fractions from uninfected and Ad2-infected cells were added alone and in various combinations to transcription reactions containing pBalE DNA cleaved with *Sma*I and *Sal*I (Fig. 1A). The fractions eluted at 0.2 M and 0.3 M KCl, and to a lesser extent at 0.1 M KCl, from the column to which an uninfected cell extract had been applied

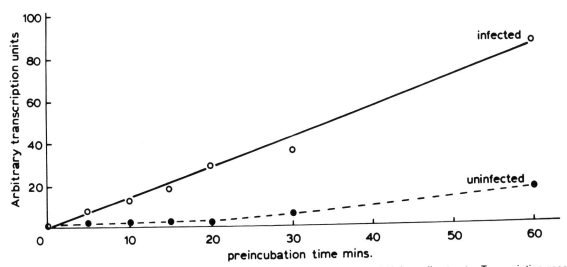

**Figure 3** Preincubation of the Ad2 pBalE template with uninfected or Ad2-infected HeLa cell extracts. Transcription reactions, performed according to the three-step preincubation-pulse-chase protocol discussed in the text, contained 36 μg/ml *Sma*I- and *Sal*I-cleaved pBalE DNA, 24 mM KCl, 3 mM MgCl$_2$, and 4 mg/ml uninfected (●) or Ad2-infected (○) HeLa cell extract protein. Preincubation was at 30°C for the periods indicated in the figure. Transcription products were purified, denatured, and separated in 1.4% agarose gels. The amounts of ML runoff RNA shown were determined as described in the legend to Fig. 2.

contained transcriptional activity (Fig. 4A, lanes 2–5). No activity was recovered at higher salt concentrations (Fig. 4A, lanes 6–8). In some experiments, the 0.04 M KCl fraction stimulated the production of specific transcripts slightly when added to fractions containing RNA

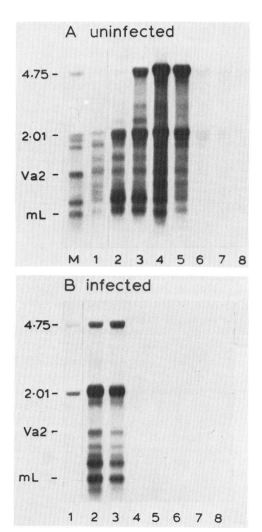

**Figure 4** Heparin-agarose chromatography of uninfected and Ad2-infected HeLa cell extracts. Extracts were prepared as described in Materials and Methods, and 10 mg of protein was fractionated on heparin-agarose (Bio-Rad) columns according to the procedure of Dynan and Tjian (1983). After loading, each column was washed with 0.04 M KCl in HM buffer and eluted serially with HM buffer containing 0.1, 0.2, 0.3, 0.4, and 0.5 M KCl. Peak fractions, based on absorbance at 280 nm, were pooled, dialyzed against 20 mM HEPES (pH 7.9) containing 5 mM MgCl₂, 0.1 mM EDTA, 2 mM DTT, and 20% (w/v) glycerol (HM buffer), concentrated in solid sucrose (Schwarz-Mann, RNase-free) at 4°C and dialyzed against HM buffer. Fractions were assayed in reactions containing 36 μg/ml pBalE DNA cleaved with *Sma*I and *Sal*I, 24 mM KCl, 3 mM MgCl₂, and 10 μg of protein of the 0.04 M KCl fraction or 40 μg of protein of each other fraction. When pairs of fractions were tested, 20 μg of each was added to the reactions. The reactions whose products are shown in lanes *1–8* of each part of the figure contained the following eluates: 0.1 M KCl, 0.2 M KCl, 0.2 M KCl + 0.3 M KCl, 0.3 M KCl, 0.3 M KCl + 0.4 M KCl, 0.4 M KCl + 0.5 M KCl, and 0.5 M KCl, respectively. Reaction products were analyzed as described in the legend to Fig. 1.

polymerase II, although it displayed no transcriptional activity alone (data not shown). Although both the 0.2 M and 0.3 M KCl fractions from an uninfected-cell extract synthesized IVA2 and ML runoff RNA species, the latter fraction supported the synthesis of more nonspecific transcription than did the former (cf. lanes 2 and 4, Fig. 4A). Chromatography, in parallel, of an extract of Ad2-infected cells on heparin-agarose yielded only one fraction, which eluted at 0.2 M KCl, containing transcriptional activity. This fraction contained all components necessary to synthesize IVA2 and ML runoff RNA species and little nonspecific transcription was observed (Fig. 4B). Thus, the components of the infected cell transcriptional machinery monitored with the viral DNA template eluted much less heterogeneously from heparin-agarose than did those of uninfected cells. Similar results have been obtained in several independent experiments. The data shown in Figure 4 suggest that chromatography on heparin-agarose partially separates RNA polymerase II associated with initiation factors (recovered at 0.2 M KCl) from free RNA polymerase II (0.3 M KCl) and thus that the enzyme is more tightly associated with such factors or that greater quantities of factors are present in Ad2-infected cells. Although such inferences remain speculative, it is clear that Ad2 infection induces altered chromatographic behavior of the components required for faithful transcription in vitro.

### Transcriptional activity of extacts of H5hr1-infected cells

The Ad5 mutant H5hr1 carries a single-point mutation in that segment of the E1A region uniquely expressed in the 289-aa protein (Ricciardi et al. 1981) that induces efficient transcription of the viral early and ML transcriptional units (Berk et al. 1979; Jones and Shenk 1979; Nevins 1981; Montell et al. 1982). Extracts prepared from HeLa cells infected with 10 pfu/cell H5hr1 and harvested 14 hr after infection, showed no increase in transcriptional activity (data not shown). This result merely indicates that the manipulations to which cells are subjected during infection do not themselves induce increased transcriptional activity, for all viral mRNA species and proteins are expressed at very low levels under these conditions (Lassam et al. 1978; Berk et al. 1979; Jones and Shenk 1979; Ross et al. 1980). However, cells infected by mutant viruses that cannot express the 289-aa E1A protein eventually produce other viral early products (Nevins 1981; Gaynor and Berk 1983). An extract was therefore prepared from HeLa cells infected with 150 pfu/cell H5hr1 for 16 hr and its activity was compared with those activities displayed by extracts of cells infected with 40 pfu/cell Ad5 for 6 or 12 hr. The results obtained when ML RNA was examined in a nuclease protection assay are shown in Figure 5A. Both extracts made from Ad5-infected cells synthesized greater quantities of specific ML RNA than did an extract of uninfected cells prepared at the same time. The stimulation was, however, far more marked with extracts of cells harvested 6 hr after infection (Fig. 5A, lanes 1–3). By contrast, the H5hr1-infected cell extract pro-

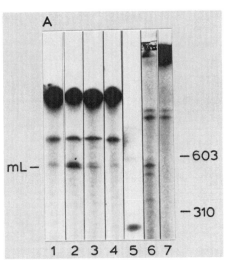

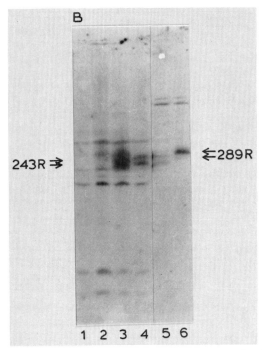

**Figure 5** (*A*) Transcriptional activity of extracts of H5hr1-infected cells. Transcription reactions contained 40 μg/ml *Sma*I-cleaved pBalE DNA, 50 mM (NH₄)₂SO₄, 5 mM MgCl₂ and 4 mg/ml protein of uninfected (lane *1*), Ad2-infected for 6 hr (lane *2*) or 12 hr (lanes *3*, *6* and *7*), and H5hr1-infected (lane *4*) cells. Infections were as described in the text. The reaction whose products are shown in lane *7* also contained 1 μg/ml α-amanitin. Incubation was at 30°C for 30 min. Reaction products were purified and hybridized to Ad2 DNA end-labeled at the *Sma*I site at nt 6575. The position of the fragment protected by ML RNA is indicated by the arrow. The digested products were resolved as described in Materials and Methods. End-labeled *Hpa*II fragments of pBR322 DNA were applied to lane *5*. (*B*) Cells were collected from small portions of the cultures used in the experiment shown in *A*, washed, and lysed in 0.01 M Tris · HCl (pH 7.4), containing 0.1 M NaCl, 1% (w/v) Triton, and 0.5% (w/v) sodium deoxycholate. Samples were sonicated for a total of 2 min in 15-sec bursts, adjusted to 0.1% (w/v) SDS and 1% (v/v) β-mercaptoethanol and boiled for 3 min. 150 μg of each protein sample was electrophoresed in a 10% polyacrylamide-SDS gel. The proteins were then transferred electrophoretically to a nitrocellulose filter (Towbin et al. 1979). The filter was reacted sequentially with anti-E1A antibodies raised in rabbits against a *trpE*-E1A fusion protein (Spindler et al. 1984) produced in *E. coli* and ¹²⁵-labeled protein A (New England Nuclear). Extracts were prepared from (1) uninfected HeLa cells (lane *1*), 20 pfu/cell Ad5-infected HeLa cells at 6 (lane *2*) or 12 (lane *3*) hr after infection, (2) HeLa cells infected with 150 pfu/cell H5hr1 harvested 16 hr after infection (lane *4*) and (3) HeLa cells infected with 20 pfu/cell *dl*347 (lane *5*) or *dl*348 (lane *6*) harvested 6 hr after infection.

duced little ML-specific RNA (Fig. 5A, lane 4) and resembled the uninfected cell extract.

To confirm that the H5hr1-infected cells had indeed entered an infectious cycle, the quantities of viral E1A proteins present in the cells when harvested were compared using a blotting procedure (Towbin et al. 1979), as described in the legend to Figure 5. Extracts of HeLa cells infected by the E1A cDNA viruses *dl*347 and *dl*348 (Winberg and Shenk 1984) were included in this experiment to provide markers for the 243-aa and 289-aa E1A proteins, respectively. The H5hr1-infected cells produced no 289-aa E1A proteins, as expected (cf. lanes 4 and 6, Fig. 5B). They did, however, contain the 243-aa E1A proteins at a concentration significantly greater than that of Ad5-infected cells harvested 6 hr after infection, but lower than that of the proteins in cells harvested 12 hr after infection (Fig. 5B, lanes 2–4). Transcription of the E1A transcription unit is normally autoregulated by its 289-aa products (Berk et al. 1979; Nevins 1981; Osborne et al. 1984; Hearing and Shenk 1985). The production of high levels of the 243-aa E1A proteins in H5hr1-infected cells, but failure of extracts

prepared from them to display enhanced transcriptional activity therefore suggest that the 289-aa E1A protein is required for induction of the enhanced transcriptional activity described in previous sections.

## Discussion

Using DNA templates that contain the sites at which transcription of Ad2 IVA2 and ML RNA initiates, we have routinely observed that whole-cell extracts made from Ad2-infected cells are more active transcriptionally than extracts prepared in parallel from uninfected HeLa cells. The magnitude of the stimulation of transcription depends both upon conditions under which transcription is assayed and the time at which infected cells are harvested, as discussed previously. Such variation may account for previous reports that Ad2-infected cell extracts show little change in transcriptional activity (Fire et al. 1981; Lee and Roeder 1981; Hoeffler and Roeder 1985). Heintz and Roeder (1984) have, however, observed that nuclear extracts of HeLa cells not in S phase synthesize ML runoff transcripts 3-fold to 20-fold more efficiently than do extracts of cells harvested when early in S phase. Since adenovirus infection rapidly induces complete inhibition of cellular DNA synthesis (Ginsberg et al. 1967; Piña and Green 1969), it seems possible that the increased transcriptional activities displayed by extracts of Ad2-infected cells and extracts of non-S-phase HeLa cells might be mediated by similar molecular mechanisms.

The increased transcriptional activity displayed by Ad2-infected cell extracts is accompanied by characteristic changes in the chromatographic behavior of the transcriptional machinery (Fig. 4). The biochemical basis for such altered behavior is not yet known but should be established by further fractionation of infected cell extracts and reconstitution of systems displaying increased transcriptional activity, compared with their counterparts from uninfected cells. It should then be possible to address such questions as whether novel transcription factors are induced, whether larger quantities of specific transcription factors are made, or whether normal transcription factors are modified in Ad2-infected cells.

The results presented in Figure 3 as well as those of other similar experiments using Ad2 and rabbit β-globin DNA templates (data not shown) indicate that stable preinitiation complexes are formed more rapidly in Ad2-infected cell extracts. These complexes have been defined experimentally as those sufficiently stable to survive the preincubation period (up to 60 min) and competent to initiate transcription immediately upon addition of ribonucleoside triphosphates. The accelerated rate of formation of preinitiation complexes in Ad2-infected cell extracts was sufficient to account for the increased activity seen in standard transcription reactions. It therefore seems unlikely that later steps in the transcription reaction, such as elongation, are altered in adenovirus-infected cells. The transcriptional control regions present in the templates used in these experiments share no obvious common features, arguing against a mechanism of stimulation that postulates the association of some factor(s) with a specific DNA sequence. Interactions of the components of the transcriptional machinery with one another, or some process whereby active transcriptional complexes are assembled, not sequence-specific but normally rate-limiting, might be facilitated in adenovirus-infected cells. It is to be hoped that additional studies with transcription systems reconstituted from their component parts will shed some light on the mechanism of stimulation.

Extracts prepared from Ad2-infected cells have recently been reported to transcribe Ad2 VA-RNA and cellular 5S and tRNA genes more efficiently. Such increased RNA polymerase III activity is associated with fractions containing TFIIIC and is dependent on synthesis, in infected cells, of E1A products (Hoeffler and Roeder 1985). The failure of HeLa cells infected with H5hr1 at high multiplicity to display enhanced transcriptional activity when producing close to normal quantities of 243-aa E1A proteins suggests that the stimulation of RNA polymerase II activity reported here is similarly induced by the 289-aa E1A protein. The early increase in transcriptional activity and the apparent lack of sequence specificity are also consistent with this conclusion. It therefore seems likely that this system will facilitate elucidation of the mechanism of the 289-aa E1A protein–induced activation of transcription that characterizes adenovirus-infected cells.

## Acknowledgments

We thank Tom Shenk for the gift of *dl*347 and *dl*348 and Margie Young and Rhonda Lunt for excellent technical assistance. This work was supported by grants from the American Cancer Society (NP 464-F) and from the National Institutes of Health (GM 31419).

## References

Anderson, C.W., P.R. Baum, and R.F. Gesteland. 1973. Processing of adenovirus 2 induced proteins. *J. Virol.* **12**: 241.

Babich, A., L.T. Feldman, J.R. Nevins, J.E. Darnell, and C. Weinberger. 1983. Effects of adenovirus on metabolism of specific host mRNAs: Transport control and specific translational discrimination. *Mol. Cell. Biol.* **3**: 1212.

Baker, C.C. and E. Ziff. 1981. Promoters and heterogeneous 5′ termini of the messenger RNAs of adenovirus 2. *J. Mol. Biol.* **148**: 189.

Beltz, G. and S.J. Flint. 1979. Inhibition of HeLa cell protein synthesis during adenovirus infection: Restriction of cellular messenger RNA sequences to the nucleus. *J. Mol. Biol.* **131**: 353.

Berk, A.J. and P.A. Sharp. 1978. Structure of the adenovirus 2 early mRNAs. *Cell* **14**: 695.

Berk, A.J., F. Lee, T. Harrison, J.F. Williams, and P.A. Sharp. 1979. Pre-early adenovirus 5 gene product regulates synthesis of viral early messenger RNAs. *Cell* **17**: 935.

Bogenhagen, D.F., W.M. Wormington, and D.D. Brown. 1982. Stable transcriptional complexes of *Xenopus* 5S RNA genes: A means to maintain the differentiated state. *Cell* **28**: 413.

Bradford, M.M. 1976. A rapid and sensitive method for the quantitation of microgram quantities of protein utilising the principle of protein-dye binding. *Anal. Biochem.* **72**: 248.

Davison, B.L., J.M. Egly, E.R. Mulvihill, and P. Chambon. 1983. Formation of stable preinitiation complexes between eukaryotic class B transcription factors and promoter sequences. *Nature* **301**: 680.

Dignam, J.D., R.M. Lebowitz, and R.G. Roeder. 1981. Accurate transcription initiation by RNA polymerase II in a soluble extract from isolated mammalian nuclei. *Nucleic Acids Res.* **11**: 1475.

Dynan, W.S. and R. Tjian. 1983. Isolation of transcriptional factors that discriminate between different promoters recognized by RNA polymerase II. *Cell* **32**: 669.

Fire, A., M. Samuels, and P.A. Sharp. 1984. Interactions between RNA polymerase II, factors and template leading to accurate transcription. *J. Biol. Chem.* **259**: 2509.

Fire, A., C.C. Baker, J.L. Manley, E.B. Ziff, and P.A. Sharp. 1981. In vitro transcription of adenovirus. *J. Virol.* **40**: 703.

Flint, S.J. 1984. Adenovirus cytopathology. In *Comprehensive virology* (ed. H. Fraenkel-Conrat and R.R. Wagner), vol. 19, p. 297. Plenum Press, New York.

Flint, S.J., G.A. Beltz, and D.I.H. Linzer. 1983. Synthesis and processing of SV40-specific RNA in adenovirus-infected, SV40-transformed human cells. *J. Mol. Biol.* **167**: 335.

Flint, S.J., S.M. Berget, and P.A. Sharp. 1976. Characterization of the single-stranded viral DNA sequences present during replication of adenovirus types 2 and 5. *Cell* **9**: 559.

Flint, S.J., M. Plumb, U.-C. Yang, J. Stein, and G. Stein. 1984. Effects of adenovirus infection upon histone gene expression. *Mol. Cell. Biol.* **4**: 1363.

Gaynor, R.B. and A.J. Berk. 1983. *Cis*-acting induction of adenovirus transcription. *Cell* **33**: 683.

Gaynor, R.B., D. Hillman, and A.J. Berk. 1984. Adenovirus early region 1A activates expression of nonviral genes introduced into mammalian cells by infection or transfection. *Proc. Natl. Acad. Sci.* **81**: 1193.

Ginsberg, H.S., L.J. Bello, and A.J. Levine. 1967. Control of synthesis of host macromolecules in cells infected with adenovirus. In *The molecular biology of viruses* (ed. J.S. Colter and W. Paranchych), p. 547. Academic Press, New York.

Green, M.R., R. Treisman, and T. Maniatis. 1983. Transcriptional activation of cloned human β-globin genes by viral immediate-early gene products. *Cell* **35**: 137.

Hearing, P. and T. Shenk. 1985. Sequence-independent autoregulation of the adenovirus type 5 E1A transcription unit. *Mol. Cell. Biol.* **5**: 3214.

Heintz, N. and R.G. Roeder. 1984. Transcription of human histone genes in extracts from synchronized HeLa cells. *Proc. Natl. Acad. Sci.* **81**: 2713.

Hoeffler, W.K. and R.G. Roeder. 1985. Enhancement of RNA polymerase III transcription by the E1A gene product of adenovirus. *Cell* **41**: 955.

Jones, N. and T. Shenk. 1979. An adenovirus 5 early gene product function regulates expression of other early viral genes. *Proc. Natl. Acad. Sci.* **76**: 3665.

Kao, H.-T. and J.R. Nevins. 1983. Transcriptional activation and subsequent control of the human heat shock gene during adenovirus infection. *Mol. Cell. Biol.* **3**: 2058.

Lassam, N.J., S.T Bayley, and F.L. Graham. 1978. Synthesis of DNA, late polypeptides and infectious virus by host-range mutants of adenovirus 5 in non-permissive cells. *Virology* **87**: 463.

Lassar, A.B., P.L. Martin, and R.G. Roeder. 1983. Transcription of class III genes: Formation of preinitiation complexes. *Science* **222**: 740.

Lee, D.C. and R.G. Roeder. 1981. Transcription of adenovirus genes in a cell-free system: Apparent heterogeneity of initiation at some promoters. *Mol. Cell. Biol.* **1**: 635.

Leong, K. and S.J. Flint. 1984. Specific transcription of an adenoviral gene that contains no "TATA" sequence homology in extracts of HeLa cells. *J. Biol. Chem.* **259**: 11527.

Lowry, O.H., N.J. Rosenbrough, A.L. Farr, and S.J. Randall. 1955. Protein measurement with the Folin phenol reagent. *J. Biol. Chem.* **193**: 265.

Maniatis, T., E.F. Fritsch, and J. Sambrook. 1982. *Molecular cloning: A laboratory manual.* Cold Spring Harbor Laboratory, Cold Spring Harbor, New York.

Manley, J.L., A. Fire, A. Campo, P.A. Sharp, and M.L. Gefter. 1980. DNA-dependent transcription of adenovirus genes in a soluble, whole-cell extract. *Proc. Natl. Acad. Sci.* **77**: 3855.

Montell, C., E.F. Fisher, M.H. Caruthers, and A.J. Berk. 1982. Resolving the functions of overlapping viral genes by site-specific mutagenesis at a mRNA splice site. *Nature* **295**: 380.

Nevins, J.R. 1981. Mechanism of activation of early viral transcription by the adenovirus E1A gene product. *Cell* **26**: 213.

———. 1982. Induction of the synthesis of a 70,000 dalton mammalian heat-shock protein by the adenovirus E1A gene product. *Cell* **29**: 913.

Osborne, T.F., D.N. Arvidson, E.S. Tyan, M. Dunnsworth-Browne, and A.J. Berk. 1984. Transcriptional control region within the protein-coding portion of adenovirus E1A genes. *Mol. Cell. Biol.* **4**: 1293.

Piña, M. and M. Green. 1969. Biochemical studies on adenovirus multiplication. XIV. Macromolecule and enzyme synthesis in cells replicating oncogenic and non-oncogenic human adenovirus. *Virology* **38**: 573.

Price, R. and S. Penman. 1972. Transcription of the adenovirus genome by an α-amanitin-sensitive RNA polymerase in HeLa cells. *J. Virol.* **9**: 621.

Ricciardi, R.P., R.L. Jones, C.T. Cepko, P.A. Sharp, and B.E. Roberts. 1981. Expression of early adenovirus genes requires a viral-encoded acidic polypeptide. *Proc. Natl. Acad. Sci.* **78**: 6121.

Ross, S.R., A.J. Levine, R.S. Galos, J. Williams, and T. Shenk. 1980. Early viral proteins in HeLa cells infected with adenovirus type 5 host range mutants. *Virology* **103**: 475.

Russell, W.C. and J.J. Skehel. 1973. The polypeptides of adenovirus-infected cells. *J. Gen. Virol.* **15**: 45.

Schick, J., K. Baczko, E. Fanning, J. Groneberg, H. Burger, and W. Doerfler. 1976. Intracellular forms of adenovirus DNA: Integrated form of adenovirus DNA appears early in infection. *Proc. Natl. Acad. Sci.* **73**: 1043.

Schneider, R.J., C. Weinberger, and T. Shenk. 1984. Adenovirus VA1 RNA facilitates the initiation of translation in virus-infected cells. *Cell* **37**: 291.

Simmons, T., P. Heywood, and L.D. Hodge. 1974. Intranuclear site of replication of adenovirus DNA. *J. Mol. Biol.* **89**: 423.

Spindler, K.R., D.S.E. Rosser, and A.J. Berk. 1984. Analysis of adenovirus transformation proteins from early regions 1A and 1B with antisera to inducible fusion antigens produced in *E. coli. J. Virol.* **49**: 132.

Stein, R. and E. Ziff. 1984. HeLa cell β-tubulin gene transcription is stimulated by adenovirus 5 in parallel with viral early genes by an E1A-dependent mechanism. *Mol. Cell. Biol.* **4**: 2792.

Svensson, G. and G. Akusjärvi. 1984. Adenovirus 2 early region 1A stimulates expression of both viral and cellular genes. *EMBO J.* **3**: 789.

Thimmappaya, B., C. Weinberger, R.J. Schneider, and T. Shenk. 1982. Adenovirus VAI RNA is required for efficient translation of viral mRNA at late times after infection. *Cell* **31**: 543.

Thomas, P.S. 1980. Hybridization of denatured RNA and small DNA fragments transferred to nitrocellulose. *Proc. Natl. Acad. Sci.* **77**: 5201.

Towbin, H., T. Staehelin, and J. Gordon. 1979. Electrophoretic transfer of proteins from polyacrylamide to nitrocellulose sheets: Procedures and some applications. *Proc. Natl. Acad. Sci.* **76**: 4350.

Treisman, R., M.R. Green, and T. Maniatis. 1983. *Cis* and *trans*-activation of globin gene expression in transient assays. *Proc. Natl. Acad. Sci.* **80**: 7428.

Weaver, R.F. and C. Weissmann. 1979. Mapping of RNA by a modification of the Berk-Sharp procedure: The 5' termini of 15S β-globin mRNA precursor and mature 10S mRNA have identical map coordinates. *Nucleic Acids Res.* **7**: 1175.

Weil, P.A., D.S. Luse, J. Segall, and R.G. Roeder. 1979. Selective and accurate transcription at the Ad2 major late promoter in a soluble system dependent on purified RNA polymerase II and DNA. *Cell* **18**: 469.

Weinmann, R., H.J. Raskas, and R.G. Roeder. 1977. Role of DNA-dependent RNA polymerase II and III in transcription of the adenovirus genome late in productive infection. *Proc. Natl. Acad. Sci.* **71**: 3426.

White, D.I., M.D. Scharff, and J.V. Maizel. 1969. The polypeptides of adenovirus III synthesis in infected cells. *Virology* **38**: 395.

Winberg, G. and T. Shenk. 1984. Dissection of overlapping functions within the adenovirus type 5 E1A gene. *EMBO J.* **3**: 1907.

Wolgemuth, D.J. and M.-T. Hsu. 1981. Visualization of nascent RNA transcripts and simultaneous transcription and replication in viral nucleoprotein complexes from adenovirus 2-infected HeLa cells. *J. Mol. Biol.* **147**: 247.

Yoder, S.S., B.L. Robberson, E.J. Leys, A.J. Hook, M. Al-Ubaidi, K.-Y. Yeung, R.E. Kellems, and S.M. Berget. 1983. Control of cellular gene expression during adenovirus infection: Induction and shut-off of dihydrofolate reductase gene expression by adenovirus. *Mol. Cell. Biol.* **3**: 819.

# DNA-binding Specificity of USF, a Human Gene-specific Transcription Factor Required for Maximum Expression of the Major Late Promoter of Adenovirus

**M. Sawadogo and R.G. Roeder**
The Rockefeller University, New York, New York 10021

A gene-specific transcription factor, designated USF, is required for maximum expression of the major late promoter of adenovirus. The stimulation by USF requires binding of the transcription factor to the upstream element of the promoter. This binding is stabilized by—and in turn stabilizes—the interaction of the transcription-initiation factor TFIID on the TATA-box region. The DNA sequence requirement for USF binding was determining by using deletion and point mutations in the major late promoter as well as a high-resolution footprint analysis with the small intercalating drug MPE·Fe(II) as the cleaving agent. By these two methods it appears that USF interacts with the small palindromic sequence GGCCACGTGACC located between positions −63 and −52 of the major late promoter, with the central octamer constituting the basic motif of the USF recognition sequence. The possible involvement of USF in other gene systems and in the temporal regulation of adenovirus gene expression is discussed.

The ability to modify isolated genes and analyze their expression in vivo has led to the discovery that transcription of class-II genes in eukaryotes is subject to various levels of control. A basal level of expression is often observed when all but a small region of the promoter, usually centered on the TATA-box sequence, is deleted. Higher levels of expression require the presence of upstream promoter elements and/or (more distal) enhancer elements (for review, see Khoury and Gruss 1983; Dynan and Tjian 1985). These various regulatory elements can be activated in a cell-type or tissue-specific fashion, or they can respond to external stimuli (e.g., hormones, metal, and heat shock).

A major breakthrough for investigation of transcriptional control mechanisms in eukaryotes has been the development of soluble cell-free systems that mediate accurate transcription initiation on purified genes by RNA polymerase II in vitro (Weil et al. 1979; Manley et al. 1980; Dignam et al. 1983). Chromatographic fractionations of these crude extracts have revealed that several components are required for specific transcription initiation by RNA polymerase II (Matsui et al. 1980; Samuels et al. 1982; Parker and Topol 1984). Some of these components are thought to be general transcription factors, i.e., utilized for expression of most or all genes. The transcription factors designated TFIIB, TFIID, and TFIIE belong to this category. These three factors are absolutely required, in addition to the RNA polymerase II, for promoter-dependent transcription initiation in vitro and can even promote transcription from fortuitous TATA-box-like sequences present in bacterial plasmid DNAs (Sawadogo and Roeder 1985a). These same "basic" or general transcription factors are presumably responsible for the basal level of gene ex-

pression observed in vivo when the regulatory promoter elements are deleted or nonfunctional.

The major late (ML) promoter of adenovirus is often used as a model system for in vitro studies of transcription initiation by RNA polymerase II. Recently it has been reported that sequences located upstream of the TATA-box region were necessary for efficient in vivo expression of the adenovirus type-2 (Ad2) ML promoter (Hen et al. 1982). The stimulatory effect of this upstream promoter element could also be observed in vitro in crude extracts (Hen et al. 1982; Miyamoto et al. 1984; Yu and Manley 1984). However, we found that the upstream sequences of the ML promoter were not required for the in vitro reconstituted system composed of transcription factors TFIIB, TFIID, TFIIE, and RNA polymerase II (Sawadogo and Roeder 1985b). We then demonstrated that a different transcription factor can be isolated from uninfected HeLa cells that will stimulate the basal level of transcription obtained in the reconstituted system only when the upstream promoter sequences are present. This new transcription factor was designated USF (for "upstream stimulatory factor") (Sawadogo and Roeder 1985b). We report here an analysis of the DNA-binding specificity of USF and discuss the possibility of its involvement in the temporal regulation of adenovirus gene expression.

## Experimental Procedures

### Transcription reactions

For the analysis of gene specificity, the reaction conditions were as described (Sawadogo and Roeder 1985b) except for the concentrations of the different templates, which were adjusted such that the various promoters were present at the same concentration (2.2 fmoles/ml).

### Footprinting reactions and filter-binding assays

The footprinting reactions and filter-binding assays were described previously (Sawadogo and Roeder 1985b).

### Construction of pMLUS1 and pMLUS2

pMLUS1 and pMLUS2 are pUC13 derivatives that contain ML sequences from −61 to −51 and −61 to −53, respectively. Both plasmids were derived from pML($C_2$AT)19Δ − 61, which contains the ML promoter from position −61 to +10 linked to the ($C_2$AT)19 DNA fragment and cloned in the *Sma*I site of pUC13 (Sawadogo and Roeder 1985b). pMLUS1 was constructed by cleaving the parental plasmid with *Hpa*II endonuclease in a partial reaction leading to one or two *Hpa*II cleavages on average per plasmid molecule. The 3′ recessed ends were filled in using the Klenow fragment of DNA polymerase, followed by intramolecular ligation. For construction of pMLUS2, the parental plasmid was first digested to completion with *Sma*I, then cleaved partially with *Hpa*II (about one cleavage per plasmid molecule). The protruding *Hpa*II ends were removed by S1 nuclease treatment, and the plasmids were recircularized by intramolecular ligation. For both pMLUS1 and pMLUS2 clonings, the ligated DNAs were used to transform JM101 cells that were plated on a

medium containing IPTG and Xgal. Positive clones were selected as giving rise to blue colonies. Restriction digest analysis confirmed that pMLUS1 had an additional 12-bp insert and pMLUS2 had an additional 9-bp insert as compared with pUC13 in the polylinker region of the vector. pMLUS1 also contained (as expected from the construction) a new *Sst*II site between the *Sst*I and *Bam*HI sites of the vector.

## Results

### USF is a gene-specific transcription factor

As illustrated in Figure 1, a number of cellular or viral genes are accurately transcribed by purified RNA polymerase II in an in vitro system reconstituted with partially purified transcription factors TFIIB, TFIID, and TFIIE (Sawadogo and Roeder 1985a). Addition of a chromatographic fraction containing USF (Sawadogo and Roeder 1985b) stimulates drastically the specific transcription driven by the ML promoter (lanes 13 and 14) while having very little effect on the transcription of other adenovirus genes (E3, E4) or of several cellular genes (histone H2B, heat-shock, β-globin, β-actin). USF is therefore a gene-specific transcription factor.

### Site-specific binding of USF to the ML promoter

Using different 5′-deletion mutants in the ML promoter, we have shown that sequences located between −61

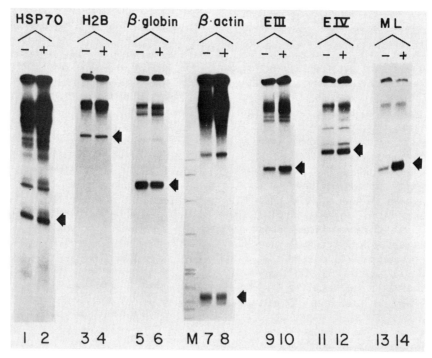

**Figure 1** Gene specificity of transcription stimulation by USF. Transcription reactions were carried out in the reconstituted system (Sawadogo and Roeder 1985a) in the presence (+) or absence (−) of USF. For each template, the arrow indicates the position of the correct-size runoff transcript. Autoradiography of the dried gel was carried out for various periods of time. The 7-min exposure is shown for the ML promoter (lanes *13* and *14*). The 30-min exposure is shown for the human histone H2B gene (lanes *3* and *4*), the mouse β-globin gene (lanes *5* and *6*), and the adenovirus E3 and E4 promoters (lanes *9–10* and *11–12*, respectively). Finally, the 150-min exposure is shown for the *Drosophila* hsp70 promoter (lanes *1* and *2*) and the rat β-actin gene (lanes *7* and *8*). (Reprinted, with permission, from Sawadogo and Roeder 1985b.)

and −50 relative to the cap site are required for stimulation by USF. In addition, we observed that the amount of upstream factor required to reach maximum stimulation was directly proportional to the concentration of the ML promoter present in the transcription reaction (Sawadogo and Roeder 1985b). Taken together, these observations indicated that stimulation by USF was probably dependent upon a stoichiometric interaction of this factor with an upstream DNA element present in the ML promoter but absent in the other nonresponsive genes.

Footprint analyses revealed that USF can actually interact with the ML promoter even in the absence of the other transcription components (Fig. 2). Binding of USF to the ML promoter inhibits DNase I cleavages on both strands over an 18–20-bp region (from position −67 to −48) just upstream of the −45 to +35 region shown previously to interact with the transcription factor TFIID (Sawadogo and Roeder 1985b; N. Nakajima and R.G. Roeder, unpubl.; see also Fig. 4). When methidium-propyl-EDTA·Fe(II) was used instead of DNase I as the cleaving agent, the USF-protected region appeared much smaller, encompassing only 10 bp on each strand (Fig. 2C,D). This smaller size of the MPE footprints most likely reflects the smaller size of the probe itself and therefore offers, in all probability, a more accurate picture of the actual interaction region of the specific DNA-binding protein.

### DNA sequence requirement for USF binding

We have developed a filter-binding assay to analyze the interaction of USF with the ML promoter. In this assay, a small end-labeled DNA fragment containing a USF-binding site (usually 1 or 2 ng of a 500-bp fragment for a 20-μl reaction) is incubated with the USF-containing fraction in the presence of a large excess of nonspecific competitor DNA to titrate out nonspecific binding proteins (usually 1 μg of pUC13 plasmid DNA). After 30-min equilibration at 30°C, the reactions are filtered through nitrocellulose filters and the amount of radioactive fragment specifically retained on the filter by USF can be quantitated. Using this assay, we have determined a dissociation constant ($K_D$) of USF for its specific binding site in the ML promoter of $4 \times 10^{-10}$ M at 30°C in the absence of divalent cations. The $K_D$ is several-fold greater in the presence of magnesium ions ($Mg^{++}$) but does not seem to vary with the topological state of the DNA (linear or supercoiled).

We showed previously that the affinity of USF for the ML promoter is unaltered by progressive 5′ deletions to position −61. A −50 deletion mutant is, however, not recognized (Sawadogo and Roeder 1985b). To determine the 3′ border of the USF recognition sequence, we constructed two new plasmids (pMLUS1 and pMLUS2) that contain only ML sequences from −61 to −51 and −61 to −53, respectively (see Experimental Procedures). USF binds to both of these plasmids. However, although the binding to pMLUS1 is wild type, the $K_D$ is increased twofold to threefold in the case of pMLUS2 (Table 1). We also analyzed a point mutant (pAdCAT-6)

that has the cytosine at position −53 replaced by a thymine residue and that has a wild-type phenotype in vivo (Brunet 1985). Interestingly, whereas the affinity of USF for pAdCAT-6 is decreased about fivefold compared with the wild-type promoter (Table 1), this mutation had almost no effect on the transcriptional ability of the promoter in vitro (result not shown). Therefore, there is no correlation between the independent stability of USF binding and the resulting transcriptional stimulation. This apparent discrepancy might be explained by the stabilization of the USF interaction upon binding of TFIID to the nearby TATA-box region (see below).

### Cooperative interaction between USF and TFIID on the ML promoter

Using the filter-binding assay, we analyzed the dissociation rate of USF from its specific binding site on the ML promoter and found that this dissociation rate was extremely rapid; as shown in Figure 3A, the half-life of the USF/ML complex is only 14 sec at 30°C at the $Mg^{++}$ concentration used in the transcription reactions. (The interaction is noticeably stabilized as the $Mg^{++}$ concentration is decreased; in the absence of divalent cations, the half-life of the complex is 27 min at 30°C.)

It seemed difficult to reconcile this observation of a very rapid dissociation rate of USF under transcription conditions with the large stimulatory effect of the factor on the transcriptional activity of the promoter. We therefore analyzed the stability of the USF interaction in the presence of TFIID by footprint competition analysis (Fig. 3B). USF, TFIID, or a combination of both factors was preincubated with the labeled DNA fragment containing the ML promoter under the same conditions in which footprints had been previously obtained. A 30-fold molar excess of cold ML DNA was then added to the reactions. The dissociation of the transcription factors from their specific binding sites on the labeled DNA fragment was then analyzed by DNase I footprinting at various times after the addition of competitor DNA. As expected from the previous results, the USF footprint had totally disappeared after a 2-min incubation in the presence of competitor DNA. (However, enhanced cleavages at the border of the USF-binding site persisted, reflecting the small proportion of USF molecules bound to the labeled fragment after equilibration of the factor between the labeled and cold competitor DNAs.) In contrast, the TFIID-DNA binding appeared more stable; the footprint was still visible, although attenuated, after a 40-min incubation in the presence of competitor DNA. (A control experiment not shown in this figure indicated that this amount of competitor DNA was quite sufficient to prevent detection of the TFIID footprint when added simultaneously with the labeled fragment.) Interestingly, the USF interaction was greatly stabilized in the presence of TFIID; after a 40-min incubation in the presence of competitor DNA, both the TFIID and USF footprints were unmodified when the two factors were bound together (Fig. 3B, lanes 12–16). In addition, preliminary experiments indicated that the TFIID binding is also stabilized by the presence of USF and that the ternary

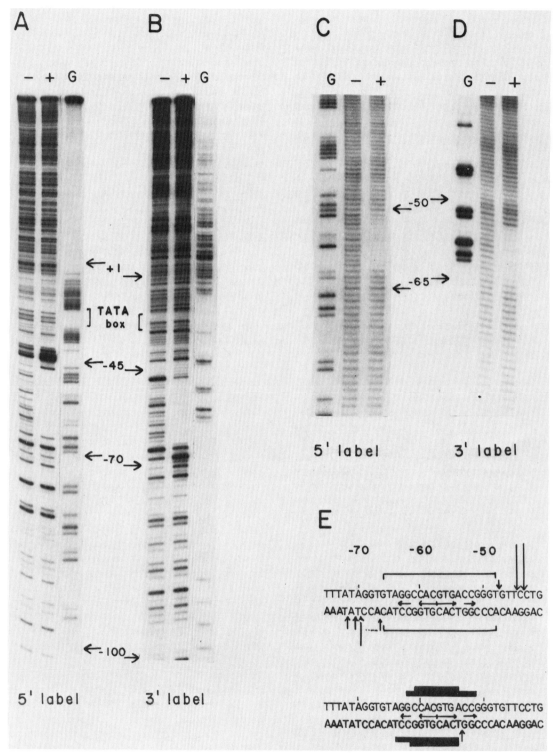

**Figure 2** Footprint analysis of USF on the ML promoter. Footprinting reactions on singly labeled DNA fragments were performed with either DNase I (*A* and *B*) or MPE · Fe(II) (*C* and *D*) as the cleaving agent. The coding strand was 5′-end-labeled and the transcribed strand was 3′-end-labeled. In each case, the control (−) and USF-containing (+) reactions were analyzed along with a G-specific sequencing ladder of the same fragment (lanes *G*). (*E*) A schematic representation of the USF footprints. For the DNase I footprints (*E, top*), the brackets indicate the protected regions and the arrows represent the sites of enhanced cleavages. The arrow length is proportional to the degree of enhancement. The MPE · Fe(II) footprints (*E, bottom*) are represented as histograms. (Reprinted, with permission, from Sawadogo and Roeder 1985b.)

**Table 1** Comparison of USF-binding Sites

| Location | DNA sequence | USF $K_D$ |
|---|---|---|
| Adenovirus 2ML | G G C C A C G T G A C C | $4 \times 10^{-10}$ M |
| pMLUS1 | c c C C A C G T G A C C | $4 \times 10^{-10}$ M |
| pMLUS2 | c c C C A C G T G A C g | $1 \times 10^{-9}$ M |
| pAdCAT-6 | G G C C A C G T G A t C | $2 \times 10^{-9}$ M |

complex is stable for more than 5 hr at 30°C. These observations strongly support the hypothesis that TFIID and USF interact directly when they are bound simultaneously on the ML promoter, although a DNA-transmitted effect cannot be totally excluded.

## Discussion

### The MPE footprints define the minimum sequence requirement for USF binding

It is very interesting to compare the DNA sequences protected by USF against MPE cleavage (Fig. 2) with the minimum sequence requirement for the USF binding as defined by analysis of deletion and point mutations

(Table 1): The MPE footprints indicated an interaction of USF with the small palindromic sequence GGCCACGTGACC, with only the central 8 bp being protected on both strands of the DNA. The deletion analysis indicated that the sequence CCACGTGACC was sufficient to allow wild-type USF binding, with the last cytosine residues being important for stable binding but not for recognition per se. Therefore the octanucleotide CCACGTGA may represent the minimum sequence required for USF binding, with the two bases on either side of this core sequence influencing to some extent the strength of the interaction.

In the case of USF, the MPE footprints give a very precise picture of the DNA residues involved in the interaction. This also seems to be true for the transcription factor TFIIIA interaction with 5S DNA (M. Van Dyke and R.G. Roeder, unpubl.). It will be interesting to determine if this is also true for TFIID, in which case the MPE footprints outlined only the 8–10-bp TATA-box region while the DNase I footprints encompassed more than 80 bp of DNA (Sawadogo and Roeder 1985b; see Fig. 4).

**A**

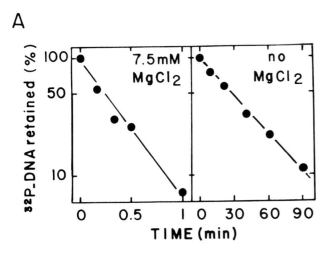

**B**

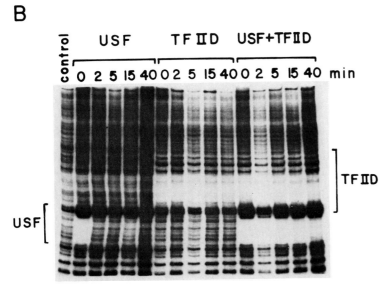

**Figure 3** Dissociation rates of USF and TFIID from their specific binding sites on the ML promoter. (*A*) Dissociation rate of USF measured by filter-binding assay: The dissociation rate of USF from its specific binding site on the ML promoter was followed at different times after addition of a 500-fold molar excess of cold competitor DNA in the presence (*left*) or absence (*right*) of 7.5 mM MgCl$_2$. The results are presented as the logarithm of the percentage of labeled DNA fragment still retained by USF at different times after addition of the competitor DNA. (*B*) Stabilization of USF by TFIID binding: The transcription factors were preincubated for 1 hr at 30°C with a 5′-end-labeled, ML-promoter-containing DNA fragment. A 30-fold molar excess of pML(C$_2$AT)19 was then added in each reaction, and the partial DNase I digestion was carried out at the indicated periods of time after the dilution. The samples were then processed for analysis on a small sequencing gel. (Reprinted, with permission, from Sawadogo and Roeder 1985b).

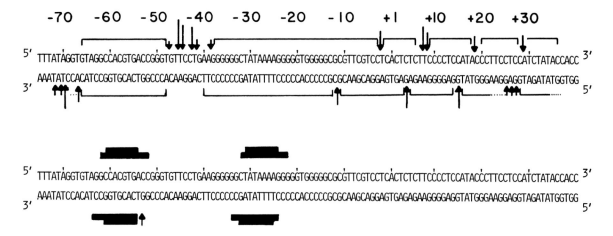

**Figure 4** Schematic representation of the USF and TFIID footprints on the ML promoter. The conventions used to summarize the DNase I (*top*) and MPE · Fe(II) footprints (*bottom*) observed upon binding of USF and TFIID to the ML promoter are described in Fig. 2. (Reprinted, with permission, from Sawadogo and Roeder 1985b.)

## Possible model for the transcriptional stimulation by USF

Because the detailed mechanisms leading to specific transcription initiation by RNA polymerase II are still very poorly understood, it might be somewhat premature to speculate on the mechanism by which a regulatory factor can stimulate the reaction upon binding to an upstream region of the promoter. However, it seems reasonable to postulate that the stimulation would take place by facilitating a step that is otherwise rate-limiting for the overall reaction. Given the proximity of the binding site and the increased stability of USF on the ML promoter when TFIID is present, we have suggested a model in which the two proteins would come in direct contact with each other, as illustrated schematically in Figure 5. If this model is true, the USF interaction could induce a conformational change in TFIID, which might in turn facilitate the subsequent step(s) in preinitiation complex formation or function. The peculiar DNase I cleavage pattern observed in the downstream portion of the TFIID footprint, where protected regions are interrupted with a 10-bp periodicity by regions of enhanced cleavages (Fig. 4), suggests that this portion of the promoter DNA, independently of its nucleotide se-

quence, is in contact with the surface of the TFIID molecule (Sawadogo and Roeder 1985b). A direct interaction between USF and TFIID would result in an even more extensive wrapping of the DNA around the USF-TFIID complex, with the DNA at the junction acting as a hinge (Fig. 5); such a structure could provide a better recognition signal for the interaction of another transcription factor or the binding of the RNA polymerase II itself.

## Involvement of USF in other gene systems

By comparison with the ML sequence, potential USF-binding sites can be found in several other genes (Table 2). The sequence in the LTR region of HTLV-III viruses is almost identical with the ML sequence but is present in the reverse orientation and further away from the cap site (around position −163). This location places it upstream of the enhancer element localized in the HTLV-III promoter between positions −104 and −57 (Rosen et al. 1985). However, it was shown that deletion of sequences from −167 to −120 (which includes the putative USF-binding site) decreases the promoter strength 8-fold to 10-fold as measured by transient assay in various cell types (Rosen et al. 1985), indicating an important role of this promoter region in the early stages of viral infection. It will be particularly interesting to analyze in vitro the effect of USF on the transcription of HTLV-III since the particular positioning of the binding site in this promoter would suggest a different mechanism of action than in ML, perhaps involving an interaction with the protein(s) binding to the downstream enhancer element.

Potential USF-binding sites can also be found in cellular genes, such as the human kininogen (Kitamura et al. 1985). The immunoglobulin enhancer consensus sequence (Church et al. 1985) also presents a striking homology with the ML upstream element (see Table 2). This consensus was derived from the different regions of the immunoglobulin enhancer that have been shown by in vivo footprinting techniques to be associated with a

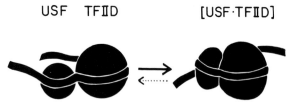

**Figure 5** Possible interpretation of the footprint analyses and dissociation rates for USF and TFIID binding to the ML promoter. The relative sizes of TFIID and USF have been chosen arbitrarily in the absence of data concerning the actual molecular masses of the two proteins. This model, in which the transcription factors interact directly when bound simultaneously on the promoter, implies that the DNA (thick, black string) should be wrapped around the two proteins.

**Table 2** DNA Sequence Homologies with the ML Promoter Upstream Element

| Location | DNA sequence | Reference |
|---|---|---|
| Adenovirus 2 ML promoter (USF-binding site) | GG CCA CGT GA CC | Sawadogo and Roeder (1985b) |
| HTLV-III (LTR) | GG CCA CGT GA t g | Muesing et al. (1985) |
| LAV (LTR) | GG CCA CGT GA t g | Wain-Hobson et al. (1985) |
| Mouse immunoglobulin (heavy-chain enhancer) | t GCCAC at GACC / a c CCA g GT Gg t g | Ephrussi et al. (1985) |
| Immunoglobulin enhancer (consensus sequence) | GCCA g GTG c C | Church et al. (1985) |
| Human kininogen promoter | t GCCACGT GA t t | Kitamura et al. (1985) |
| AAV2 (promoter 1) AAV2 (promoter 3) | a c t CACGT GA CC / a c CCACGT GA t C | Srivastava et al. (1983) |

protein(s) in cells that express the gene (myeloma, B, or pre-B cells) but not in other cell types (Ephrussi et al. 1985). The 1- or 2-nucleotide difference between the immunoglobulin enhancer sequences and the ML promoter upstream element (Table 2) might reflect the involvement of altered forms of the USF protein. Alternatively, these different sequences might bind USF only very weakly, with stable binding occurring only in the presence of another (tissue-specific) protein binding nearby. These various possibilities can be tested once the enhancer-dependent transcription of the immunoglobulin gene will be reproduced in vitro. Also, a major emphasis should be set toward the complete purification of the USF protein, since the availability of antibodies would allow the investigation of the USF concentrations and possible modifications in various tissues or cell types.

## Possible importance of USF in the adenovirus infectious cycle

The strict specificity of USF for the ML promoter versus the other viral genes suggests a possible involvement of this particular host-cell transcription factor in the early-to-late switch of gene expression pattern during the adenovirus infectious cycle. USF might be responsible for the high level of activity of the ML promoter after DNA replication. Because early in infection the ML apparently behaves like all the other early genes (Nevins 1981), it would be of interest to determine whether the ML upstream element is required for the E1A-dependent expression of the promoter. A report by Lewis and Manley (1985) suggests that this is the case when the E1A effect is analyzed by transfection; however, other experiments with a deleted ML promoter reconstructed into the virus indicate that the upstream element is not essential early in infection but is critical for late expression (T. Schenk, pers. comm.). It is therefore possible that the cellular USF protein would be modified (or its concentration increased) by direct or indirect action of an early viral protein to fit better the virus requirements during the late phase of infection. Of possible relevance to this is the observation that potential USF-binding sites can be found in two of the promoters of adenovirus-associated virus 2 (Srivastava et al. 1983; see Table 2).

It is known that this defective virus requires early adenovirus functions for its propagation.

Another interesting question is the role played by USF in IVA2 transcription; the two promoters, IVA2 and ML, direct divergent transcription units with the two cap sites located 210 bp apart. Activation of both promoters requires DNA replication. In vitro analysis of various IVA2 deletion mutants have indicated a requirement for upstream sequences (Natarajan et al. 1985) that appear to coincide with the now well-defined USF-binding site.

## Acknowledgments

We thank L. Brunet for the gift of plasmid pAdCAT-6, R.L. Gilbert for the art work, and C.-G. Balmaceda for technical assistance. This work was supported by research grants to R.G.R. from the National Cancer Institute and from the American Cancer Society, and by a program project grant from the National Cancer Institute to The Rockefeller University.

## References

Brunet, L.J. 1985. "Mutational analysis of the type 5 adenovirus major late promoter." Ph.D thesis, Columbia University, New York.

Church, G.M., A. Ephrussi, W. Gilbert, and S. Tonegawa. 1985. Cell-type-specific contacts to immunoglobulin enhancers in nuclei. *Nature* **313**: 798.

Dignam, J.D., R.M. Lebowitz, and R.G. Roeder. 1983. Accurate transcription initiation by RNA polymerase II in a soluble extract from isolated mammalian nuclei. *Nucleic Acids Res.* **11**: 1475.

Dynan, W.S. and R. Tjian. 1985. Control of eucaryotic RNA synthesis by sequence-specific DNA-binding proteins. *Nature* **316**: 774.

Ephrussi, A., G.M. Church, S. Tonegawa, and W. Gilbert. 1985. B lineage–specific interactions of an immunoglobulin enhancer with cellular factors *in vivo. Science* **227**: 134.

Hen, R., P. Sassone-Corsi, J. Corden, M.P. Gaub, and P. Chambon. 1982. Sequences upstream from the TATA box are required *in vivo* and *in vitro* for efficient transcription from the adenovirus serotype 2 major late promoter. *Proc. Natl. Acad. Sci.* **79**: 7132.

Khoury, G. and P. Gruss. 1983. Enhancer elements. *Cell* **33**: 313.

Kitamura, N., H. Kitagawa, D. Fukushima, Y. Takagaki, T. Miyata, and S. Nakanishi. 1985. Structural organization of the human kininogen gene and a model for its evolution. *J. Biol. Chem.* **260**: 8610.

Lewis, E.D. and J.L. Manley. 1985. Control of adenovirus late promoter expression in two human cell lines. *Mol. Cell. Biol.* **5:** 2433.

Manley, J.L., A. Fire, A. Lang, P.A. Sharp, and M.L. Gefter. 1980. DNA-dependent transcription of adenovirus genes in a soluble whole-cell extract. *Proc. Natl. Acad. Sci.* **77:** 3855.

Matsui, T., J. Segall, P.A. Weil, and R.G. Roeder. 1980. Multiple factors required for accurate initiation of transcription by RNA polymerase II. *J. Biol. Chem.* **255:** 11992.

Miyamoto, N.G., V. Moncollin, M. Wintzerith, R. Hen, J.M. Egly, and P. Chambon. 1984. Stimulation of in vitro transcription by the upstream element of adenovirus-2 major late promoter involves a specific factor. *Nucleic Acids Res.* **12:** 8779.

Muesing, M.A., D.H. Smith, C.D. Cabradilla, C.V. Benton, L.A. Lasky, and D.J. Capon. 1985. Nucleic acid structure and expression of the human AIDS/lymphadenopathy retrovirus. *Nature* **313:** 450.

Natarajan, V., M.J. Madden, and N. Salzman. 1985. Positive and negative control sequences within the distal domain of the adenovirus IVa2 promoter overlap with the major late promoter. *J. Virol.* **55:** 10.

Nevins, J.R. 1981. Mechanism of activation of early viral transcription by the adenovirus E1A gene product. *Cell* **26:** 213.

Parker, C.S. and J. Topol. 1984. A *Drosophila* RNA polymerase II transcription factor contains a promoter-region-specific DNA-binding activity. *Cell* **36:** 357.

Rosen, C.A., J.G. Sodroski, and W.A. Haseltine. 1985. The location of *cis*-acting regulatory sequences in the human T cell lymphotropic virus type III (HTLV-III/LAV) long terminal repeat. *Cell* **41:** 813.

Samuels, M., A. Fire, and P.A. Sharp. 1982. Separation and characterization of factors mediating accurate transcription initiation by RNA polymerase II. *J. Biol. Chem.* **257:** 14419.

Sawadogo, M. and R.G. Roeder. 1985a. Factors involved in specific transcription by human RNA polymerase II: Analysis by a rapid and quantitative *in vitro* assay. *Proc. Natl. Acad. Sci.* **82:** 4394.

———. 1985b. Interaction of a gene-specific transcription factor with the adenovirus major late promoter upstream of the TATA box region. *Cell* **43:** 165.

Srivastava, A., E.W. Lusby, and K.I. Berns. 1983. Nucleotide sequence and organization of the adeno-associated virus 2 genome. *J. Virol.* **45:** 555.

Wain-Hobson, S., P. Sonigo, O. Danos, S. Cole, and M. Alizon. 1985. Nucleotide sequence of the AIDS virus, LAV. *Cell* **40:** 9.

Weil, P.A., D.S. Luse, J. Segall, and R.G. Roeder. 1979. Selective and accurate initiation of transcription at the Ad2 major late promoter in a soluble system dependent on purified RNA polymerase II and DNA. *Cell* **18:** 469.

Yu, Y.T. and J.L. Manley. 1984. Generation and functional analyses for base-substitution mutants of the adenovirus-2 major late promoter. *Nucleic Acids Res.* **12:** 9309.

# The TGGCA Protein Binds In Vitro to DNA Contained in a Nuclease-hypersensitive Region That Is Present Only in Active Chromatin of the Lysozyme Gene

A.E. Sippel,* H.P. Fritton,[†] M. Theisen,* U. Borgmeyer,* U. Strech-Jurk,* and T. Igo-Kemenes[†]

*Zentrum für Molekulare Biologie der Universität Heidelberg (ZMBH), D-6900 Heidelberg 1, Federal Republic of Germany; [†]Institut für Physiologische Chemie, Physikalische Biochemie und Zellbiologie der Universität, D-8000 München 2, Federal Republic of Germany

In vitro DNA-binding properties of the TGGCA protein from chicken cell nuclei and HeLa cell–derived nuclear factor I (NFI), a protein with enhancing function in the in vitro replication of adenovirus DNA, suggest that both factors are members of a ubiquitous, conserved family of proteins.

The position and the fine structure of one out of six 5′-flanking nuclease-hypersensitive sites of the chromatin of the chicken lysozyme gene (HS−6.1) correlates with an in vitro double DNA-binding site for the TGGCA protein. Comparison of the pattern of nuclease-hypersensitive sites in five different states of activity of the lysozyme gene indicates that HS−6.1 harbors a function that is involved in the production and/or maintenance of the active chromatin domain.

In recent years, a number of DNA elements have been defined that are involved in the basic process of transcription initiation by RNA polymerase II (Breathnach and Chambon 1981) and in its regulation (Yaniv 1982; Khoury and Gruss 1983). To understand the molecular mechanism of their action in the cell- and stage-specific activation of eukaryotic genes, it will be necessary to characterize trans-acting factors that interact with these signal sequences and to study their structural organization in chromatin.

Using low-salt protein extracts from isolated nuclei and defined, radioactively labeled DNA fragments in nitrocellulose filter–binding assays (Riggs et al. 1970), we have detected a new sequence-specific DNA-binding activity (Nowock and Sippel 1982). DNA competition experiments show that the same chicken oviduct nuclear protein recognizes DNA sites in the 5′-flanking region of the chicken lysozyme gene (Borgmeyer et al. 1984) and in the adenovirus inverted terminal repeat, the mouse mammary tumor virus LTR, and the BK virus enhancer (Nowock et al. 1985). From these sites, the consensus recognition sequence 5′ YTGGCANNNTGCCAR 3′ could be deduced (Borgmeyer et al. 1984; Nowock et al. 1985). In reference to one half of this palindromic recognition site, the protein was named TGGCA-binding protein (Borgmeyer et al. 1984).

In parallel with our discovery of the TGGCA-binding protein in chicken oviduct nuclei, J. Hurwitz's group has identified a nuclear factor in HeLa cells (Nagata et al. 1982, 1983) that, due to its DNA-binding properties, seems to be a related protein. HeLa cell nuclear factor I (NFI), a protein with enhancer function in the initiation process of adenovirus DNA replication in vitro (Nagata et al. 1982) and in vivo (Hay 1985), recognizes exactly the same DNA site in the inverted terminal repeat of adenovirus DNA (Nagata et al. 1983) as does the chicken TGGCA protein (Nowock et al. 1985), suggesting that both are members of a conserved family of nuclear DNA-binding proteins.

Whereas the viral function of NFI can be conveniently studied in reconstituted adenovirus replication assays, the cellular function of this host factor remains unclear. Since we have found that the TGGCA protein interacts in vitro with high affinity to specific sequences in the 5′-flanking DNA of the chicken lysozyme gene (Borgmeyer et al. 1984), we extended our studies to include the chromatin structure of this gene in order to find clues to its biological function in eukaryotic cells.

The chicken lysozyme gene offers attractive features for the study of cell-specific gene activation. In contrast to most genes, it is expressed in different regulatory modes in two distinct cell types. In the tubular gland cells of chicken oviduct, synthesis of lysozyme mRNA is strictly dependent on steroid hormones (Schütz et al. 1978). Expression of the same gene in mature macrophages is constitutive and totally independent of steroids (Hauser et al. 1981; Sippel et al. 1986).

We have examined the chromatin structure of a 22-kb chromosomal region containing the transcriptional unit of the lysozyme gene. Seven nuclease-hypersensitive sites could be mapped in the flanking regions of the gene in laying-hen oviduct cells, in which the gene is maximally expressed (Fritton et al. 1983). Different sets of DNase I–hypersensitive sites have been found in the promoter-proximal region, depending on whether the gene was constitutively expressed in cultured macrophages or in the steroid hormone–controlled state in the oviduct (Fritton et al. 1984). These data suggest that the pattern of DNase-hypersensitive sites in the 5'-flanking chromatin contributes to the way this gene is transcriptionally regulated. Here we extend the mapping of nuclease-hypersensitive sites to the far distal 5'-flanking region (−8.9 kb to −4.1 kb) and report that the hypersensitivity of the chromatin region at −6.1 kb (HS−6.1) strictly correlates with the active or potentially active state of the lysozyme gene. Recent gene-transfer experiments have demonstrated that DNA sequences containing the HS−6.1 region act as cell-specific enhancer elements in a chicken macrophage cell line (Theisen et al. 1986).

The DNA sequence in the HS − 6.1 region carries two closely positioned binding sites for the TGGCA protein. Because the in vitro double binding site is reflected by a narrow doublet of hypersensitive sites in authentic chromatin, it is possible that the TGGCA protein is a component of the chromatin element at −6.1 kb in the active lysozyme gene domain.

## Experimental Procedures

### Recombinant plasmids

The pBR322-derived plasmids pB2H3 (Borgmeyer et al. 1984), pX1E2, and pE2B1 contain subcloned DNA fragments from λLys-31 (Lindenmaier et al. 1979). There are *Bam*HI, *Hin*dIII, *Xba*I, and *Eco*RI sites in the lysozyme gene region, denoted B, H, X, and E, respectively, in the accompanying figures (Nowock and Sippel 1982). pAd12-*Hin*dIIIG (Schier et al. 1983) was a gift of S. Stabel. For the DNA sequence analysis of the binding-site region of pB2H3, the chemical sequencing method was used on both DNA strands (Maxam and Gilbert 1977).

### Preparation and use of binding-protein functions

TGGCA-binding activity was obtained from laying-hen oviduct and HeLa cell nuclei by extraction with 300 mM NaCl and subsequent precipitation of 40% saturation of ammonium sulfate as described by Borgmeyer et al. (1984). DNase I protection experiments were done according to the procedure described earlier (Borgmeyer et al. 1984) on isolated and 3'-labeled DNA fragment B2H3 from the −6.1 kb region of the lysozyme gene (Borgmeyer et al. 1984) and the adenovirus 12 (Ad12) *Hin*dIIIG fragment (Nowock et al. 1985).

### Preparation of RNA and S1-mapping experiments

Mature chicken macrophages were isolated from blood as described (Beug et al. 1979; Fritton et al. 1984). For preparation of RNA, cells were incubated in Iscove medium supplemented with charcoal-stripped serum. Serum was stripped by stirring with 50 mg/ml charcoal at room temperature for 30 min. After that, serum was sterilized by filtration through a Nalgene filter unit (pore size, 0.2 μm). One day later, cells were supplemented with fresh medium, and diethylstilbestrol (DES) was added to a final concentration of $10^{-8}$ M. Ethanol controls and untreated cultures were incubated in parallel.

Total cellular RNA was prepared from laying-hen oviduct and liver tissue and from $3 \times 10^8$ to $5 \times 10^8$ macrophage cells per tissue-culture flask 24 hr after addition of hormone, according to the method of Kaplan et al. (1979). Cells were lysed in 5 M guanidinium thiocyanate, 50 mM Tris · HCl [pH 7.6], 10 mM EDTA, 5% β-mercaptoethanol. RNA was pelleted through a cushion of 5.7 M CsCl, 0.1 M EDTA in a Beckman SW40 rotor at 29,000 rpm for 22 hr at 20°C. The RNA pellet was resuspended in diethylpyrocarbonate-treated double-distilled water, ethanol precipitated, and stored as an aqueous solution at −20°C. Lysozyme gene-specific transcripts were analyzed by the S1-mapping procedure (Weaver and Weissmann 1979) essentially as described previously (Grez et al. 1981). Total cellular RNA from chicken macrophages, laying-hen oviduct, and laying-hen liver was hybridized in 10 μl of 400 mM NaCl, 40 mM PIPES, pH 6.4, 1 mM EDTA, 80% formamide at 43°C for 12 hr with an excess of radioactively labeled *Bst*NI DNA fragment spanning the promoter region of the chicken lysozyme gene from position +69 to −461. After hybridization, probes were diluted with 250 μl of 300 mM NaCl, 30 mM sodium acetate [pH 4.5], 3 mM $ZnSO_4$, and 100 μg/ml denatured and sonicated salmon sperm DNA and digested with 250 units of S1 nuclease at 30°C for 3.5 hr. After ethanol precipitation, pellets were resuspended in formamide-EDTA dye, and S1-resistant DNA fragments were analyzed on 16% polyacrylamide/urea sequencing gels.

### Preparation of nuclei, DNase I, and endogenous nuclease treatment and hypersensitive-site mapping

Nuclei from oviduct, liver, kidney, and brain of laying hens and nuclei from erythrocytes and total 5-day-old chicken embryos were prepared essentially according to Fritton et al. (1983). Tissues were homogenized in Buffer A (0.15 mM spermine, 0.5 mM spermidine, 60 mM KCl, 15 mM NaCl, 2 mM EDTA, 0.5 mM EGTA, 5 mM Tris · HCl [pH 7.4]), supplemented with 0.5 M sucrose and 1 mM PMSF freshly added from a 0.1 M stock in isopropanol. For preparation of nuclei from macrophages, cells were collected with a rubber policeman and washed with Buffer A supplemented with 0.5 M sucrose and 1 mM PMSF. The pelleted nuclei (2000g for 10 min at 4°C) were washed in the same buffer in the presence of 0.5% Triton X-100 until the nuclei were colorless. To remove the detergent, two additional washing steps were done in Buffer A supplemented with 0.35 M sucrose and 1 mM PMSF. Pelleted nuclei were frozen in liquid nitrogen and stored at −80°C.

Frozen nuclei were thawed and suspended at a DNA concentration of 200–1000 μg/ml in 0.15 mM spermine, 0.5 mM spermidine, 60 mM KCl, 0.15 mM NaCl, 0.2 mM EDTA, 0.2 mM EGTA, 5 mM Tris · HCl (pH 7.4), and freshly added 1 mM PMSF and digested with various amounts of DNase I in the range between 64 and 4096 units/ml at 0°C for 10 min. Digestions were started by the addition of 5 mM MgCl and stopped by the addition of 10 mM EDTA. Digested nuclei were spun down for 1 min at 2000*g* and washed once or twice in buffer A. Endogenous nuclease digestion experiments were performed as described (Fritton et al. 1983). For auto-digestion, nuclei were incubated at 20°C in Buffer A for up to 120 min in the presence of 5 mM MgCl₂. Digestions were terminated as described above. DNA from digested nuclei was isolated by proteinase K treatment, treated with restriction nuclease, and analyzed by gel electrophoresis, blotting, hybridization with nick-translated probes, and autoradiography as described earlier (Fritton et al. 1983).

## Results

### Chicken cells contain a nuclear DNA-binding protein homologous to NFI of HeLa cells

The TGGCA-binding protein was originally detected by us in nuclear extracts of laying-hen oviduct cells (Nowock and Sippel 1982; Borgmeyer et al. 1984). To find the most abundant source for purification of the TGGCA protein, we screened several other chicken tissues and found that the same DNA-binding protein was present in all cell types tested (Borgmeyer et al. 1984). Binding assays with nuclear extracts from mouse liver and human HeLa cells revealed the occurrence of this protein also in other species (Nowock et al. 1985), and we propose that the chicken TGGCA protein is a member of a conserved, ubiquitous family of nuclear DNA-binding proteins. Most suggestive for the biological function of this newly discovered protein-DNA interaction are two specific DNA-binding regions for the TGGCA protein, one on the inverted terminal repeat of adenovirus DNA and the other approximately 6.1 kb 5′ to the chicken lysozyme promoter. Figure 1 outlines the DNA sequence protected by the TGGCA protein against DNase I digestion on the terminal fragments of Ad12 DNA. The protein-covered area is identical with the protected sequence on Ad5 DNA previously seen by footprinting experiments with HeLa cell NFI, discovered by J. Hurwitz's group (Nagata et al. 1982, 1983). Figure 1 also shows the TGGCA-protein-protected DNA in the sequence of the −6.1-kb region of the chicken lysozyme

gene. As can be seen, two binding sites are located close to each other, the centers of the recognition sequences being 92 bp apart.

NFI was shown to enhance replication of adenovirus DNA in a cell-free system with purified proteins (Nagata et al. 1982; Rawlins et al. 1984; Leegwater et al. 1985). In view of the results showing that the chicken TGGCA protein covers exactly the same sequence on the adenovirus inverted terminal repeat as HeLa cell nuclear factor I (Nowock et al. 1985), and that the one can be substituted for the other in reconstituted assay systems of adenovirus replicaton (Leegwater et al. 1986), we consider these nuclear factors to be functionally homologous proteins from different species.

### The structure of the DNase-hypersensitive chromatin site 6.1 kb upstream from the promoter of the chicken lysozyme gene

The biological relevance of a protein that binds in vitro to a specific sequence of cloned genomic DNA is greatly supported by evidence of the same binding in authentic chromatin. DNase-hypersensitive sites in chromatin are believed to be short, nucleosome-free regions that mark sites where *trans*-acting regulatory proteins obtain access to specific DNA sequences (McGhee et al. 1981; Emerson et al. 1985). We have previously seen that one of the 5′ upstream DNase-hypersensitive sites in laying-hen oviduct chromatin could be mapped close to the double binding site of the TGGCA protein around −6.1 kb with respect to the promoter of the lysozyme gene (Fritton et al. 1983). To increase the accuracy of this correlation, we used the indirect end-labeling method (Wu et al. 1979) for mapping of hypersensitive sites with higher resolution in agarose gels. Figure 2 shows the results of these experiments in the respective region of chromatin after treatment of isolated oviduct nuclei with exogenously added DNase I and after self-digestion with endogenous nuclease activity. Whereas DNase I produces a relatively broad band at the respective region, endogenous nuclease creates two clearly separable bands with an approximately 0.1-kb size difference at the same general position (Fig. 2A). It is, in particular, this double structure of the hypersensitive site HS−6.1

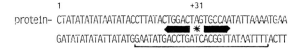

**Figure 1** Chicken oviduct nuclear TGGCA protein—and HeLa cell NFI—binding sites on the Ad12 origin of replication and the chicken lysozyme gene −6.1-kb region. (*Top*) The sequence of the 3′ and 5′ terminal 50 bp of Ad12 DNA (Sugisaki et al. 1980); (*bottom*) the sequence of the chicken lysozyme gene region −6.1 kb (−6110 to −5982 bp) as determined from the B2H3 fragment of pB2H3 (a subclone of λLys-31; Nowock and Sippel 1982). Brackets mark DNA regions protected from DNase I digestion in in vitro DNase protection experiments with TGGCA protein and NFI on the 3′ and 5′ termini of Ad12 DNA (Nowock et al. 1985) and with TGGCA protein on the chicken DNA (Borgmeyer et al. 1984). Black arrows outline the complete or incomplete inverted repeat of the protein recognition sequence, and the asterisks mark the central base pair of protein-binding sites. The covalently coupled adenovirus terminal protein is indicated at the 5′-terminal cytidine.

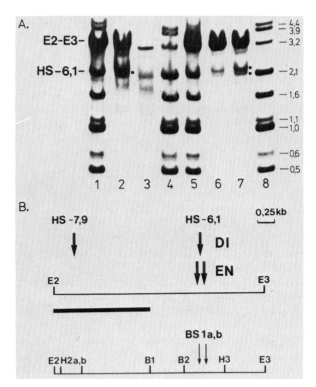

**Figure 2** Mapping of HS−6.1 in chicken oviduct chromatin. (*A*) Laying-hen oviduct nuclei were incubated with two different concentrations of DNase I (lanes *2* and *3*) or were autodigested with endogenous nuclease for two different time periods (lanes *6* and *7*). Isolated DNA was restricted with *Eco*RI, and Southern blots were probed with nick-translated plasmid pE2B1. DNA length markers (lanes *1*, *4*, *5*, and *8*) are given in kilobase pairs. E2−E3 and HS−6.1 mark the positions of the full-sized *Eco*RI DNA fragment and the DNA fragments between E2 and hypersensitive sites, respectively. (*B*) A map of the E2E3 fragment in the 5′ upstream region of chicken lysozyme gene (Nowock and Sippel 1982). (E) *Eco*RI; (H) *Hin*dIII; (B) *Bam*HI; (BS1a,b) TGGCA-protein-binding sites (Borgmeyer et al. 1984). The horizontal bar outlines the position of the radioactive hybridization probe; the arrows mark the positions of hypersensitive sites (HS) with DNase I (DI) and endogenous nuclease (EN).

seen with endogenous nuclease that correlates well with the double binding site for the TGGCA protein (Fig. 2B) in in vitro DNase I protection experiments (Borgmeyer et al. 1984). Even though the accuracy of the applied indirect end-labeling method is not greater than ±50 bp, it is possible that the TGGCA protein is involved in a protein-DNA complex in this far-upstream region. A more accurate chromatin mapping with respect to the nucleotide sequence must await future experiments with methods developed by others for this purpose (Ephrussi et al. 1985; Jackson and Felsenfeld 1985; Nick and Gilbert 1985).

### The constitutive transcription of the lysozyme gene in mature macrophages is initiated at the same promoter active in oviduct cells

Previous Southern blotting experiments have proven that there is only one lysozyme gene per haploid chicken

genome (Lindenmaier et al. 1979; Sippel et al. 1980). Therefore, the same gene that is inducible by steroid hormones in the oviduct tubular gland cells (Schütz et al. 1978) is constitutively expressed in mature macrophages (Hauser et al. 1981). Here, our interest is in confirming the activity of the lysozyme gene in the two modes of regulation by analysis of its transcripts with the S1-mapping procedure (Weaver and Weissmann 1979).

Synthesis of lysozyme mRNA in oviduct is initiated from three distinct start sites (Grez et al. 1981). Since it cannot be excluded that differential control of the lysozyme gene is affected by the use of different start sites, it was essential to know whether the same transcription start sites are used in macrophages. When RNA from laying-hen oviducts and from mature macrophages was analyzed by S1 mapping (Fig. 3), lysozyme transcripts corresponding to the start sites at positions +1, −2, and −24 were mapped. The same bands with equal relative intensities are present in both RNA populations, demonstrating that the same transcription initiation sites are used in similar proportions in both tissues. Although steroid hormones dramatically increase lysozyme gene transcription in oviduct cells (Schütz et al. 1978), incubation of macrophages in medium containing estrogen has no effect on the activity of the gene (cf. lanes 6 and 7 with 10 and 11, Fig. 3). A comparison of the intensities of S1-resistant bands in Figure 3 indicates that total cellular RNA from mature macrophages contains approximately 100 times fewer lysozyme transcripts than total cellular RNA from laying-hen oviduct. No S1-resistant bands were visible when excess liver RNA was used for hybridization (Fig. 3, lane

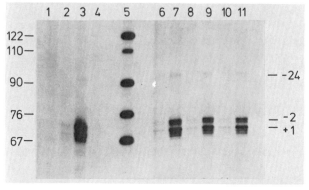

**Figure 3** Analysis of lysozyme gene transcripts in laying-hen oviduct, liver, and macrophage total cellular RNA by S1 mapping. Oviduct RNA (0.02 and 0.5 μg, lanes *2* and *3*), liver RNA (40 μg, lane *4*), macrophage RNA (2 μg, lanes *6*, *8*, and *10*), and macrophage RNA (10 μg, lanes *7*, *9*, and *11*) were hybridized with an excess of *Bst*NI fragment (Grez et al. 1981) labeled at its 5′ ends. Lane *1* contains no RNA. Macrophage RNA was prepared from cells grown with 8% charcoal-stripped fetal calf serum and 2% charcoal-stripped chicken serum that contained no steroid hormone (lanes *6* and *7*), $10^{-8}$ M DES and 0.1% ethanol (lanes *10* and *11*) or only 0.1% ethanol (lanes *8* and *9*). On the autoradiogram, S1-resistant DNA fragments of 69, 71, and 93 bp correspond to the start sites of lysozyme-gene-specific transcripts at positions +1, −2, and −24. Lane *5* contains labeled pBR322 DNA digested with *Hpa*II. The lengths of the marker fragments are indicated on the left.

4), indicating that the lysozyme gene is not transcribed in liver cells. The same is true for most other cell types (data not shown).

## The −6.1-kb chromatin region is nuclease hypersensitive only in active chromatin

The pattern of nuclease-hypersensitive sites in the promoter-proximal region (−4.5 kb to promoter) has previouly been shown to be different in different states of activity of the lysozyme gene (Fritton et al. 1984). Three sites were detected that exhibit tissue specificity (see Fig. 5). HS−1.9 is restricted to the oviduct, whereas HS−0.7 and HS−2.7 are confined to macrophages. The absence of these sites in all inactive tissues tested suggests that their function is limited to either the hormone-dependent or to the constitutive active state. Our cataloging of promoter-proximal hypersensitive sites and their correlation with the various active and inactive states of the gene provided clues to their possible function (Fritton et al. 1984). Here we extended our mapping to the far-upstream chromatin region (−8.9 to −4.1 kb) in order to look for functional indications for the nuclease-hypersensitive site at −6.1 kb.

As can be seen in Figure 4A, HS−6.1 is present in oviduct and macrophage chromatin, but not in chromatin of the inactive tissues such as liver, kidney, brain, and total 5-day embryos, nor in the chromatin of peripheral red blood cells. Figure 5 summarizes the mapping

data of nuclease-hypersensitive sites in the 5′-flanking chromatin of seven different cell types representing five different stages of activity of the lysozyme gene. HS−6.1 is the only one of the six different upstream nuclease-hypersensitive sites that appears coordinately with the hypersensitive site at the promoter region (HS−0.1). Thus, the open structure of the chromatin at −6.1 kb is as characteristic of the active or potentially active state of the gene as is the open chromatin structure at the promoter.

## Discussion

Elucidation of the molecular mechanism of DNA replication in higher eukaryotes is expected to be aided by results produced in the adenovirus replication system. In vitro adenovirus studies have shown that three viral proteins are involved in the protein-priming mechanism in which initiation of DNA synthesis takes place (for review, see Sussenbach and van der Vliet 1983). Reconstituted systems with the purified viral proteins DNA-binding protein (DBP) and DNA polymerase and the precursor of the terminal repeat (pTP) on adenovirus DNA-TP as template sustain a basal level of replication initiation. Fractionation of extracts of uninfected HeLa cells allowed the identification of a nuclear factor (NFI; Nagata et al. 1982, 1983) that enhances initiation and

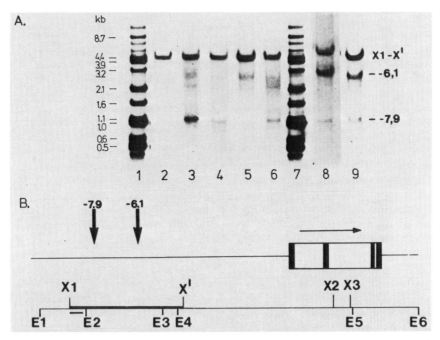

**Figure 4** DNase I–hypersensitive sites in the lysozyme chromatin region −8.9 to −4.1 kb in various chicken tissues. (*A*) Chromatin DNase I–hypersensitive sites were mapped in nuclei from erythrocytes (lane *2*), liver (lane *3*), kidney (lane *4*), brain (lane *5*), total 5-day embryo tissue (lane *6*), mature macrophages (lane *8*), and laying-hen oviduct (lane *9*). X1–X′, −6.1, and −7.9 mark the positions on the autoradiogram of the full-sized DNA fragment X1X′ from −8.9 to −4.1 kb 5′ to the gene, and the respective shorter fragments from X1 to the DNase I cuts as probed by labeled DNA from plasmid pX1E2. DNA size markers (lanes *1* and *7*) are indicated on the left. (*B*) An outline of the mapping procedure. X and E denote *Xba*I and *Eco*RI sites, respectively. The thick horizontal line indicates the region screened for hypersensitive sites with probe X1E2. Arrows mark the position of the hypersensitive sites with respect to the lysozyme gene (bars represent 4 exons). The horizontal arrow indicates the direction of transcription.

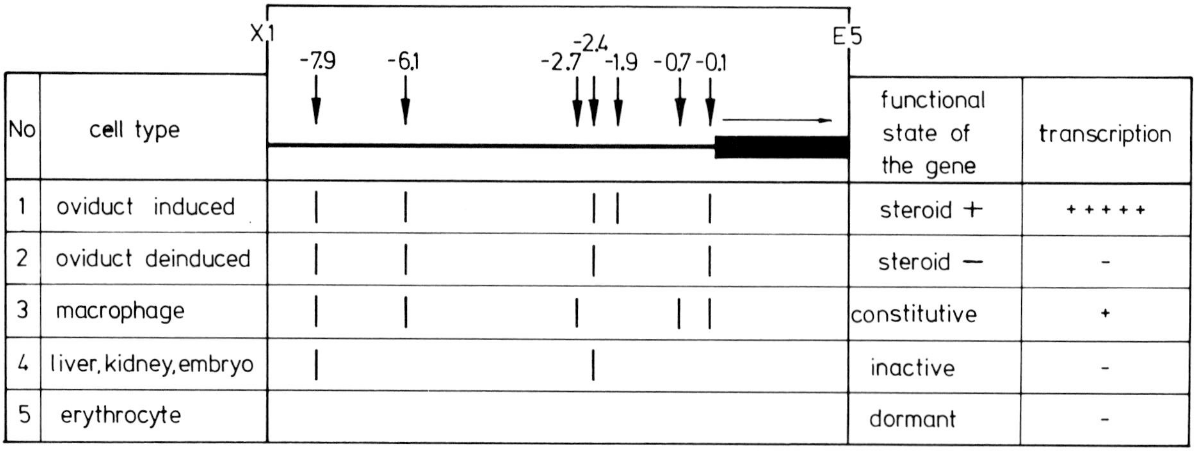

|     |                       | X1 | -7.9 | -6.1 | -2.7 | -2.4 | -1.9 | -0.7 | -0.1 | E5 | functional state of the gene | transcription |
|-----|-----------------------|----|------|------|------|------|------|------|------|----|------------------------------|---------------|
| 1   | oviduct induced       |    | \|   | \|   |      |      | \| \| |     | \|   |    | steroid +                    | + + + + +     |
| 2   | oviduct deinduced     |    | \|   | \|   |      |      | \|    |     | \|   |    | steroid −                    | −             |
| 3   | macrophage            |    | \|   | \|   |      | \|   |      | \|   | \|   |    | constitutive                 | +             |
| 4   | liver, kidney, embryo |    | \|   |      |      | \|   |      |      |      |    | inactive                     | −             |
| 5   | erythrocyte           |    |      |      |      |      |      |      |      |    | dormant                      | −             |

**Figure 5** Alternative sets of 5′-flanking nuclease-hypersensitive sites correlate with the various functional states of the chicken lysozyme gene. Vertical lines in the middle part indicate the pattern of nuclease-hypersensitive sites in the 5′-flanking chromatin in five different functional states of the gene. Mapping data from Fritton et al. (1984) for the promoter-proximal part (HS−2.7 to HS−0.1) are summarized with data obtained from Fig. 4 (HS−7.9, HS−6.1) and transcriptional data from Fig. 3 and Schütz et al. 1978).

subsequent elongation. Evidence that NFI is a site-specific DNA-binding protein that binds to a specific DNA region of the adenovirus origin of replication (Nagata et al. 1983) made it clear to us that, due to its sequence-specific DNA-binding properties, the TGGCA protein that we found in nuclear extracts of chicken cells (Nowock and Sippel 1982; Borgmeyer et al. 1984) must be related to NFI. In in vitro DNA-protection experiments, both proteins covered the same DNA sequence contained in the adenovirus inverted terminal repeat (Nowock et al. 1985). The functional homology was recently extended by results showing that the chicken TGGCA protein can replace HeLa cell NFI in a reconstituted adenovirus DNA replication system (Leegwater et al. 1986). The two proteins are therefore most likely homologous members of the same class of conserved, ubiquitous nuclear proteins.

At present, the cellular function of these DNA-binding proteins is an interesting subject for speculation. Their enhancing function in the initiation process of adenovirus DNA replication is not necessarily equivalent to their cellular function. It was found in other cases that viruses adopt host factors for diverse functions. For example, *Escherichia coli* translation factors Tu and Ts are subunits of Qβ replicase (Blumenthal et al. 1972).

A way of finding the cellular function of TGGCA protein/NFI is expected to come from the identification of genomic binding sites (Nowock and Sippel 1982; Borgmeyer et al. 1984; Gronostajski et al. 1984; Siebenlist et al. 1984; Hennighausen et al. 1985). The colocalization of chromatin nuclease-hypersensitive sites and TGGCA-protein/NFI–binding sites in the chicken lysozyme gene region (Fritton et al. 1983; Borgmeyer et al. 1984), the human c-*myc* gene region (Siebenlist et al. 1984), and the human IgM gene region (Hennighausen et al. 1985) suggests possible sites for factor-DNA interaction in vivo. The finding that the

TGGCA-protein-binding site doublet at approximately 6.1 kb 5′ to the promoter of the lysozyme gene is reflected by a double structure of the nuclease-hypersensitive site at the same position in active chromatin lends further preliminary support to the notion that we have localized an authentic genomic site of action for this protein.

Our experimental results give direct and indirect evidence that the HS−6.1 region functions in the control of gene activity rather than being involved in DNA replication. The open chromatin structure at the far-upstream position is associated with the active state of chromatin in the lysozyme gene domain. Occurrence of the hypersensitive site HS−6.1 is not strictly correlated with the transcriptional activity of the gene but rather with the active conformation of the entire chromatin structure of the gene domain (H.P. Fritton et al., in prep.), as is exemplified in the steroid hormone−deinduced state of the gene (see Fig. 5). From this result we deduce that the −6.1-kb chromatin element harbors a function that is involved in the production and/or maintenance of the active structure of the chromatin domain, a prerequisite for transcriptional activity of the lysozyme promoter. We recently found that, in accordance with our hypothesis derived from structural features of the chromatin, functional assays yielded more-direct arguments. We could show that the DNA element at −6.1 kb acts as a strong cell-specific transcriptional enhancer for lysozyme promoter activity in transient (Theisen et al. 1986) and stable (A. Stief, unpubl.) DNA transfections in the macrophage cell line HBC1 (Beug et al. 1979).

The occurrence of HS−6.1 and its enhancer function cannot be explained by the mere presence of the TGGCA protein. The cellular function of this ubiquitous protein is certainly not sufficient to activate the lysozyme gene domain in a highly cell-specific manner. We are currently looking for other factors interacting with the

DNA element at −6.1 kb and/or the TGGCA protein in order to find general clues to the molecular mechanism of stage- and cell-specific activation of eukaryotic genes.

## Acknowledgments

We thank Rosemary Franklin and Cornelia P. Lohs for the preparation of the manuscript. This work was supported by the Deutsche Forschungsgemeinschaft grant Si 165/4-1 to A.E.S. and grants to T.I.-K. in the Forschungsgruppe "Genomorganisation" and SFB 304.

## References

Beug, H., A. von Kirschbach, G. Döderlein, J.F. Conscience, and T. Graf. 1979. Chicken hematopoietic cells transformed by seven strains of defective avian leukemia viruses display three distinct phenotypes of differentiation. *Cell* **18**: 375.

Blumenthal, T., T.A. Landers, and K. Weber. 1972. Bacteriophage Qβ replicase contains the protein biosynthesis elongation factors Tu and Ts. *Proc. Natl. Acad. Sci.* **69**: 1313.

Borgmeyer, U., J. Nowock, and A.E. Sippel. 1984. The TGGCA-binding protein: A eukaryotic nuclear protein recognizing a symmetrical sequence in double-stranded linear DNA. *Nucleic Acids Res.* **12**: 4295.

Breathnach, T. and P. Chambon. 1981. Organization and expression of eukaryotic split genes coding for proteins. *Annu. Rev. Biochem.* **50**: 349.

Emerson, B.M., C.D. Lewis, and G. Felsenfeld. 1985. Interaction of specific nuclear factors with the nuclease-hypersensitive region of the chicken adult β-globin gene: Nature of the binding domain. *Cell* **41**: 21.

Ephrussi, A., G.M. Church, S. Tonegawa, and W. Gilbert. 1985. B lineage–specific interactions of an immunoglobulin enhancer with cellular factors in vivo. *Science* **227**: 134.

Fritton, H.P., A.E. Sippel, and T. Igo-Kemenes. 1983. Nuclease-hypersensitive sites in the chromatin domain of the chicken lysozyme gene. *Nucleic Acids Res.* **11**: 3467.

Fritton, H.P., T. Igo-Kemenes, J. Nowock, U. Strech-Jurk, M. Theisen, and A.E. Sippel. 1984. Alternative sets of DNase I–hypersensitive sites characterize the various functional states of the chicken lysozyme gene. *Nature* **311**: 163.

Grez, M., H. Land, K. Giesecke, G. Schütz, A. Jung, and A.E. Sippel. 1981. Multiple mRNAs are generated from the chicken lysozyme gene. *Cell* **25**: 743.

Gronostajski, R.M., K. Nagata, and J. Hurwitz. 1984. Isolation of human DNA sequences that bind to nuclear factor I, a host protein involved in adenovirus DNA replication. *Proc. Natl. Acad. Sci.* **81**: 4013.

Hauser, H., T. Graf, H. Beug, I. Greiser-Wilke, W. Lindenmaier, M. Grez, H. Land, K. Giesecke, and G. Schütz. 1981. Structure of the lysozyme gene and expression in the oviduct and macrophages. In *Haematology and blood transfusion* (ed. R. Neth et al.), vol. 26, p. 175. Springer-Verlag, Berlin.

Hay, R.T. 1985. The origin of adenovirus DNA replication: Minimal DNA sequence requirement in vivo. *EMBO J.* **4**: 421.

Hennighausen, L., U. Siebenlist, D. Danner, P. Leder, D. Rawlins, P. Rosenfeld, and T. Kelly, Jr. 1985. High-affinity binding site for a specific nuclear protein in the human IgM gene. *Nature* **314**: 289.

Jackson, P.D. and G. Felsenfeld. 1985. A method for mapping intranuclear protein-DNA interactions and its application to a nuclease hypersensitive site. *Proc. Natl. Acad. Sci.* **82**: 2296.

Kaplan, B.B., S.L. Bernstein, and A.E. Giaio. 1979. An improved method for rapid isolation of brain ribonucleic acid. *Biochem. J.* **183**: 181.

Khoury, G. and P. Gruss. 1983. Enhancer elements. *Cell* **33**: 313.

Leegwater, P.A.J., W. Van Driel, and P.C. van der Vliet. 1985. Recognition site of nuclear factor I, a sequence-specific DNA-binding protein from HeLa cells that stimulates adenovirus DNA replication. *EMBO J.* **4**: 1515.

Leegwater, P.A.J., P.C. van der Vliet, R.A.W. Rupp, J. Nowock, and A.E. Sippel. 1986. Functional homology between the sequence-specific DNA-binding proteins nuclear factor I from HeLa cells and the TGGCA protein from chicken liver. *EMBO J.* **5**: (in press).

Lindenmaier, W., M.C. Nguyen-Huu, R. Lurz, M. Stratmann, N. Blin, T. Wurtz, H.J. Hauser, K. Giesecke, H. Land, S. Jeep, M. Grez, A.E. Sippel, and G. Schütz. 1979. Arrangement of coding and intervening sequences of chicken lysozyme gene. *Proc. Natl. Acad. Sci.* **76**: 6196.

Maxam, A. and W. Gilbert. 1977. A new method for sequencing DNA. *Proc. Natl. Acad. Sci.* **74**: 560.

McGhee, J.D., W.I. Wood, M. Dolan, J.D. Engel, and G. Felsenfeld. 1981. A 200 base pair region at the 5' end of the chicken adult β-globin gene is accessible to nuclease digestion. *Cell* **27**: 45.

Nagata, K., R.A. Guggenheimer, and J. Hurwitz. 1983. Specific binding of a cellular DNA replication protein to the origin of replication of adenovirus DNA. *Proc. Natl. Acad. Sci.* **80**: 6177.

Nagata, K., R.A. Guggenheimer, T. Enomoto, J.H. Lichy, and J. Hurwitz. 1982. Adenovirus DNA replication in vitro: Identification of a host factor that stimulates synthesis of the preterminal protein-dCMP complex. *Proc. Natl. Acad. Sci.* **79**: 6438.

Nick, H. and W. Gilbert. 1985. Detection in vivo of protein-DNA interactions within the *lac* operon of *Escherichia coli*. *Nature* **313**: 795.

Nowock, J. and A.E. Sippel. 1982. Specific protein-DNA interaction at four sites flanking the chicken lysozyme gene. *Cell* **30**: 607.

Nowock, J., U. Borgmeyer, A.W. Püschel, R.A.W. Rupp, and A.E. Sippel. 1985. The TGGCA protein binds to the MMTV-LTR, the adenovirus origin of replication, and the BK virus enhancer. *Nucleic Acids Res.* **13**: 2045.

Rawlins, D.R., P.J. Rosenfeld, R.J. Wides, M.D. Challberg, and T.J. Kelly, Jr. 1984. Structure and function of the adenovirus origin of replication. *Cell* **37**: 309.

Riggs, A.D., S. Burgeois, and M. Cohn. 1970. The *lac* repressor-operator interaction. III. Kinetic studies. *J. Mol. Biol.* **53**: 401.

Schier, P.I., R. Bernhards, R.T.M.J. Vaessen, A. Houeling, and A.J. van der Eb. 1983. Expression of class I major histocompatibility antigens switched off by highly oncogenic adenovirus 12 in transformed rat cells. *Nature* **305**: 771.

Schütz, G., M.C. Nguyen-Huu, K. Giesecke, N.E. Hynes, B. Groner, T. Wurtz, and A.E. Sippel. 1978. Hormonal control of egg white protein messenger RNA synthesis in the chicken oviduct. *Cold Spring Harbor Symp. Quant. Biol.* **42**: 617.

Siebenlist, U., L. Hennighausen, J. Battey, and P. Leder. 1984. Chromatin structure and protein binding in the putative regulatory region of the c-*myc* gene in Burkitt lymphoma. *Cell* **37**: 381.

Sippel, A.E., J. Nowock, M. Theisen, U. Borgmeyer, U. Strech-Jurk, C. Bonifer, T. Igo-Kemenes, and H. Fritton. 1986. Chromatin structure and protein-DNA interactions in the 5'-flanking region of the chicken lysozyme gene. In *Coordinated regulation of gene expression* (ed. R.M. Clayton and D.E.S. Truman). Plenum Press, New York. (In press.)

Sippel, A.E., M.C. Nguyen-Huu, W. Lindenmaier, N. Blin, R. Lurz, H.J. Hauser, K. Giesecke, H. Land, M. Grez, and G. Schütz. 1980. Mechanism of induction of egg white proteins by steroid hormones. In *Steroid induced uterine proteins* (ed. M. Beato), p. 297. Elsevier/North-Holland, Amsterdam.

Sugisaki, H., K. Sugimoto, M. Takanami, K. Shiroki, I. Saito, H. Shimojo, Y. Sawada, Y. Uemizu, S. Uesugi, and K. Fujinaga. 1980. Structure and gene organization in the transforming *Hind*III-G fragment of Ad12. *Cell* **20:** 777.

Sussenbach, J.S. and P. van der Vliet. 1983. The mechanism of adenovirus DNA replication and the characterization of replication proteins. *Curr. Top. Microbiol. Immunol.* **109:** 53.

Theisen, M., A. Stief, and A.E. Sippel. 1986. The lysozyme enhancer: Cell specific activation of the chicken lysozyme gene by a far upstream DNA element. *EMBO J.* **5:** (in press).

Weaver, R.F. and C. Weissmann. 1979. Mapping of RNA by a modification of the Berk-Sharp procedure: The 5′ termini of 155 β-globin mRNA precursor and mature 10S β-globin mRNA have identical map coordinates. *Nucleic Acids Res.* **7:** 1175.

Wu, C., P.M. Bingham, K.J. Livak, R. Holmgren, and S.C.R. Elgin. 1979. The chromatin structure of specific genes. I. Evidence for higher order domains of defined DNA sequence. *Cell* **16:** 797.

Yaniv, M. 1982. Enhancing elements for activation of eukaryotic promoters. *Nature* **297:** 17.

# Regulation of Herpes Simplex Virus 1 Gene Expression: The Effect of Genomic Environments and Its Implications for Model Systems

J.L.C. McKnight,* T.M. Kristie,* S. Silver,*[†] P.E. Pellett,* P. Mavromara-Nazos,*
G. Campadelli-Fiume,[‡] M. Arsenakis,* and B. Roizman*

*The Marjorie B. Kovler Viral Oncology Laboratories, The University of Chicago, Chicago, Illinois;
[‡]The Institute of Microbiology and Virology, The University of Bologna, Bologna, Italy

HSV genes form five major groups, $\alpha$, $\beta_1$, $\beta_2$, $\gamma_1$, and $\gamma_2$, whose expression is coordinately regulated and sequentially ordered in cascade fashion. The promoter-regulatory domains of $\alpha$ genes are transferable to indicator genes (e.g., thymidine kinase gene (TK)). The chimeric gene recombined in the viral genome is regulated as an $\alpha$ gene. When resident in the cellular genome and in transient expression systems, the $\alpha$-TK chimeras are induced by a structural component of the virus designated as the $\alpha$ *trans*-inducing factor ($\alpha$-TIF), whose gene has been mapped and sequenced. $\alpha$-TIF can induce $\alpha$ gene regulation in the absence of other viral proteins if specific *cis*-acting elements contained in $\alpha$ promoter-regulatory domains are linked to the test promoter. However, other viral proteins may enhance $\alpha$-TIF expression.

$\beta$ genes resident in the viral or host genomes require a functional $\alpha4$ protein for maximal expression. (In transient expression systems, $\beta$ genes are induced by $\alpha4$ or $\alpha0$ and only minimally by $\alpha$-TIF.) A DNA fragment encoding only 40% (aminoterminal) of the $\alpha4$ gene is also effective in inducing the $\beta$-TK gene.

The $\gamma_2$-TK chimeric genes are regulated as authentic $\gamma_2$ genes inasmuch as they require viral DNA synthesis for their expression when resident in the viral genome. When resident in host chromosomes, however, these genes are regulated as $\beta$ genes inasmuch as their induction is independent of viral DNA synthesis. Indeed, in transient BHKtk⁻ cell expression systems, the $\gamma_2$ genes cannot be differentiated from $\beta$ genes with respect to the viral *trans*-acting factors capable of stimulating the expression of these genes.

## The pattern of regulation of herpes simplex virus gene expression

Viral gene expression in cells productively infected with herpes simpex virus 1 (HSV-1) appears to be regulated at three different levels:

(1) The major pathway of regulation appears to be at the transcriptional level. HSV-1 genes form at least three transcriptional groups, designated $\alpha$, $\beta$, and $\gamma$, whose expression is coordinately regulated and sequentially ordered in a cascade fashion (Honess and Roizman 1974, 1975). The five $\alpha$ genes, $\alpha0$, $\alpha4$, $\alpha22$, $\alpha27$, and $\alpha47$, encode infected cell proteins (ICP) 0, 4, 22, 27, and 47 (Fig. 1). The $\alpha$ genes are expressed first and are operationally defined as capable of being transcribed in the absence of de novo viral protein synthesis (Honess and Roizman 1974, 1975). The functions of $\alpha$ genes are unknown, but at least three—$\alpha4$, $\alpha22$, and $\alpha27$—have now been shown to play roles in determining the expression of genes expressed later in infection (Kit et al. 1978; Knipe et al. 1978; Dixon and Schaffer 1980; Sacks et al. 1985; Sears et al. 1985). Functional $\alpha$ gene

products, particularly ICP4, are required for the induction of $\beta$ genes (Honess and Roizman 1975; Dixon and Schaffer 1980). The $\beta$ gene products identified to date appear to function in nucleotide metabolism and viral DNA synthesis (for review, see Roizman and Batterson 1985). Unlike the $\alpha$ genes, the $\beta$ genes are not homogeneously expressed. The subgroup designated as $\beta_1$, which includes ICP8 (the major DNA-binding protein [Conley et al. 1981]) and ICP6 (ribonucleotide reductase [Huszar and Bacchetti 1981]), is expressed before the $\beta_2$ subgroup, which includes the gene (TK) specifying thymidine kinase (ICP36) (Honess and Roizman 1974, 1975). One or more $\beta$ gene products, which may include ICP8, shuts off the expression of $\alpha$ genes (Honess and Roizman 1974, 1975; Fenwick and Roizman 1977; Pereira et al. 1977; Dixon and Schaffer 1980; Godowski and Knipe 1985). The $\gamma$ genes are expressed last and, in turn, require prior $\beta$ gene expression. Like the $\beta$ genes, the $\gamma$ genes can be divided into two subgroups. Whereas $\gamma_1$ (or $\beta\gamma$) are expressed, at least in part, in the absence of viral DNA synthesis, the $\gamma_2$ (or true $\gamma$) genes stringently require viral DNA synthesis for their expression (Jones and Roizman 1979; Holland et al. 1980; Conley et al. 1981; Frink et al. 1981).

[†]*Present address*: Department of Microbiology, State University at Stony Brook, Stony Brook, New York 11794.

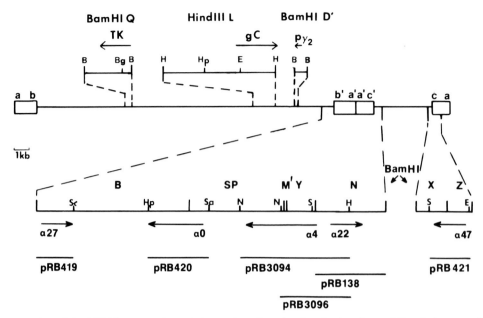

**Figure 1** Representation of the HSV-1 genome in prototype orientation showing the locations of the viral genes discussed in this report. Expansions of the regions of the genome show the transcriptional orientation and restriction maps of specific genes and fragments relevant to the text. Also shown are the plasmid constructs containing either the complete α genes or portions thereof. pRB420, pRB3094, pRB138, pRB419, and pRB421 contain the complete α0, α4, α22, α27, and α47 genes, respectively. pRB3096 contains 40% of the 5′ coding region for ICP4. (Bg) *Bgl*II; (B) *Bam*HI; (H) *Hin*dIII; (Hp) *Hpa*I; (N) *Nar*I; (Sa) *Sac*I; (S) *Sal*I (P. Mavromera-Nazos et al., in prep.).

The gene(s) responsible for the transition from β to γ have not been identified, but as in the case of α-to-β transition, the expression of γ genes signals the shutoff of β gene expression (Honess and Roizman 1974, 1975). HSV-1 genes are transcribed by the host RNA polymerase II (Constanzo et al. 1977).

(2) There is increasing evidence that regulation of gene expression also occurs at the translational level. Thus, the shutoff of host gene expression by a structural component of the virion occurs at a posttranscriptional level (Fenwick and Walker 1978; Silverstein and Engelhardt 1979; Read and Frenkel 1983). The same component appears to affect the expression of α genes (Read and Frenkel 1983). Studies on physically enucleated cells have shown that the shutoff of α gene expression by β gene products is at least in part cytoplasmic, and therefore occurs at a posttranscriptional level (Fenwick and Roizman 1977). Finally, more recent studies have shown that $\gamma_1$ gene mRNA that is capable of being translated in vitro may continue to accumulate in the cell late in infection even though it is no longer translated (Johnson et al. 1984). However, posttranscriptional regulation cannot be ascribed to competition for host or viral translational factors insomuch as experimental conditions have been described in which all three groups of viral genes are efficiently translated at the same time (Honess and Roizman 1975).

(3) The abundance of some viral proteins appears to be regulated by the end product. One example is the report that the abundance of functional ICP4 is autoregulated (Watson and Clements 1978).

These observations suggest the following:

1. The discrimination between α, β, and γ genes by RNA polymerase II must involve both host and viral *trans*-acting factors and *cis*-acting sites within the domains of the viral genes.
2. Posttranscriptional regulation must involve *cis*-acting sites contained within the transcribed domains of the genes and *trans*-acting factors whose origin is not known but may be viral.
3. Autoregulatory genes must contain a domain recognized in *trans* by either the gene product alone or in combination with host or viral factors. The observation that some temperature-sensitive mutants in the α4 gene do not seem to be effective in autoregulating ICP4 abundance at the nonpermissive temperature suggests that autoregulation is conformation-dependent and may involve irreversible binding of the nonfunctional ICP4 to a specific target.

Most of the current work on HSV regulation centers on identification of the myriad *cis*-acting sequences and the *trans*-acting elements involved in the regulation of HSV gene expression. Given the fact that the requirements for expression, temporal pattern, and abundance of expression of each gene are the product of the interaction of several regulatory pathways, the ideal experimental design is to isolate and independently analyze each regulatory event. The key requirement, however, is that the operational definition of the isolated event, whether analyzed in vitro or in the environment of the cell, remain consistent with the real characteristics of the event as it is observed in the productively infected cell.

We summarize here the attempts to define the transcriptional regulation of $\alpha$ and $\gamma_2$ genes both in the infected cell and in cells transfected with subsets of viral genes. There are two major conclusions. First, the systems employed appear to mimic those of productively infected cells in the case of $\alpha$ genes but not in the case of $\gamma_2$ genes. Second, the viral genome and viral gene products may be directed to specific compartments within the cell. Thus, the pattern of regulation of viral genes present in the environment of the cellular genome may be vastly different from that of the same genes in the viral genome.

### How does the infected cell know to transcribe $\alpha$ genes first?

*Experimental design*
When this question first surfaced in 1980, it became obvious that the identification of the *cis*-acting sites responsible for the regulation of $\alpha$ genes would be facilitated by construction of mutants with insertions or deletions in the regulatory domains of $\alpha$ genes. Also, $\alpha$ proteins were not an ideal gene product to measure because of their properties (Wilcox et al. 1980; Ackermann et al. 1984). The decision to construct surrogate $\alpha$ genes using viral thymidine kinase (TK) as an indicator gene product was based on three developments. First, the enzymatic activity of TK is easy to measure and the promoter domains of the gene were well known since it became a model of genes capable of expression in eukaryotic environments (Lin and Munyon 1974; Leiden et al. 1976; McKnight 1980; McKnight et al. 1981). Second, the HSV TK gene is not essential for growth in cell culture, and its coding domains could be interrupted by insertional mutagenesis without inhibiting virus growth (Post et al. 1981). Lastly, the development of antiviral drugs based on specific activation of HSV TK indicated the utility of TK as a powerful selectable marker both for and against recombinant viruses carrying the gene.

In these early studies, putative 5′ promoter-regulatory sequences consisting of the transcription-initiation site and upstream sequences of three $\alpha$ genes ($\alpha 0$, $\alpha 4$, and $\alpha 27$) were fused to the 5′-transcribed noncoding and coding sequences of the TK gene, a $\beta$ gene (Post et al. 1981). The chimeric genes were in the first instance recombined into the viral genomes through homologous flanking sequences by cotransfection with an intact $TK^-$ virus and selection of $TK^+$ recombinants. Subsequently, the chimeric genes were transfected into cells, and these were analyzed directly (transient expression) or selected for stable maintenance of the chimeric gene by selection for $TK^+$ phenotype or for expression of another marker gene cotransfected along with the chimeric gene (Post et al. 1981; Mackem and Roizman 1982a,b,c; Kristie and Roizman 1984).

### $\alpha$-TK chimeric genes in the viral genome are regulated as $\alpha$ genes

Recombinant viruses carrying a TK gene linked to promoter-regulatory domains of the $\alpha$ genes were regulated as $\alpha$ genes (Post et al. 1981). This conclusion was based on the observation that the $\alpha$-TK gene was expressed in the absence of de novo protein synthesis in a fashion analogous to that of authentic $\alpha$ genes and was in contrast to the natural or $\beta$-TK gene, which requires $\alpha$ proteins for its expression.

### Regulation of $\alpha$-TK chimeras resident in the cellular genomes

Because the construction of viral recombinants carrying novel $\alpha$ genes is not a technique readily amenable for mapping the precise domains of the sequence conferring $\alpha$ regulation, the chimeric genes were transfected into $TK^-$ cells and cell lines expressing the $TK^+$ phenotype were selected (Post et al. 1981; Mackem 1982a,b,c). Analyses of the expression of these genes produced two surprises. First, the resident $\alpha$-TK gene, like the resident $\beta$-TK gene in corresponding cells, was inducible by infection of the cells with an HSV-1 $TK^-$ mutant (Post et al. 1981). Because $\alpha$ genes are the first to be transcribed, the expectation was that their transcription would be determined by the affinity of host transcriptional factors for $\alpha$ promoters rather than by factors uniquely available in the infected cell. Second, the resident $\alpha$-TK gene, unlike the $\beta$-TK gene, could be induced in the absence of de novo protein synthesis, suggesting that the induction was caused by a factor introduced into cells during infection (Post et al. 1981). Consistent with this conclusion was the observation that the resident $\alpha$-TK gene could be induced at the nonpermissive temperature by temperature-sensitive mutants in the $\alpha 4$ gene that are incapable of inducing resident $\beta$-TK genes at that temperature (Post et al. 1981). In addition, $\alpha$ promoter-regulatory domains fused to cellular genes (e.g., chick ovalbumin) and stably resident in cells were inducible by HSV infection and regulated as $\alpha$ genes (Post et al. 1982; Herz and Roizman 1983).

### Regulation of $\alpha$-TK chimeras in a transient expression system

Plasmids containing the $\alpha$-TK chimeric genes were tested in BHKtk$^-$ cells infected with an HSV-1 mutant (*ts*502Δ305) carrying a deletion in the TK gene and a temperature-sensitive lesion in the $\alpha 4$ gene. The results were consistent with those observed in the stably transformed cells described in the preceding section (T.M. Kristie, unpubl.).

### Identification of *cis*-acting elements within the domains of $\alpha$ genes

The significant differences in the requirements for enhanced expression of the $\alpha$-TK and $\beta$-TK genes stably residing in cells infected with $TK^-$ virus lent credence to subsequent studies designed to map the regulatory domains of $\alpha$ genes. It is convenient at this point to differentiate between the 5′-nontranscribed $\alpha$ gene sequences that enable the genes to be expressed (promoter domains) from those that confer $\alpha$ gene regulation upon recipient genes. These studies may be summarized as follows:

(1) Experiments similar to those described above involving conversion of Ltk⁻ cells to TK⁺ phenotype with α-TK chimeras followed by *trans* induction with TK⁻ virus demonstrated that the promoter and regulatory α gene sequences are separable and independently movable (Mackem and Roizman 1982b,c). Thus, constructs containing the α promoter alone enable the TK gene to be expressed efficiently in L cells but are not inducible in *trans* by TK⁻ virus. Constructs consisting of the regulatory region fused to a TK gene containing the β promoter element mapped by McKnight are regulated as α genes (Mackem and Roizman 1982b; McKnight 1982; Kristie and Roizman 1984). Results consistent with these observations were obtained by analysis of various 5′-deleted α4-TK chimeras in other systems (Cordingly et al. 1983).

(2) Sequence analysis of the 5′-nontranscribed domains revealed two features common to all α genes (Table 1). The first consisted of numerous repeats containing G + C−rich sequences dispersed throughout the promoter-regulatory domains, whereas the second consisted of one of several homologs of an A + T−rich sequence located primarily in the regulatory domains of α genes (Mackem and Roizman 1982c; Cordingly et al. 1983). The functions of these sequences were tested by fusing short, cloned sequences containing the A + T−rich homologs from the regulatory domains of both the α0 and α27 genes and a G + C−rich series of inverted repeats from the α4 regulatory domain to TK constructs containing either the α4 nonregulated promoter domain fused to the TK gene at +50 or to the β-TK promoter at −80 (Kristie and Roizman 1984). Three series of experiments were done using these constructs. First, Ltk⁻ cells were cotransfected with either control (β-TK) or α-TK chimeras in the presence of a plasmid encoding the neomycin-resistance gene, and the cell lines were selected in the presence of G418 (Southern and Berg 1982). The objective was to measure constitutive (uninduced) TK activity and activity induced by infection with *ts*502Δ305 at the permissive and nonpermissive temperatures. A summary of these experiments is shown in Table 2. Second, Ltk⁻ cells were transfected with control (β-TK) or α-TK chimeras, selected for the TK⁺ phenotype, and assayed as above. Lastly, TK activity of the plasmids was measured in a transient expression system (Kristie and Roizman 1984). These studies revealed the following: (a) The sequences containing the G + C−rich inverted repeats confer an orientation semidependent, high-basal-level expression to both α and β promoters and do not confer the ability to be regulated as an α gene upon the α or β promoter−TK chimeras. (b) The sequences containing the A + T−rich homologs conferred α-specific regulation and consist of two elements, one of which is required to confer α regulation to the α promoter−TK chimera and both of which are required to confer α regulation on chimeric genes linked to β-TK promoter (Kristie and Roizman 1984). Similar elements are present in the common domains of the α22 and α47 genes and, although they have not been individually cloned and tested for function, 5′ deletional analysis of this region in a transient expression system involving induction of an ultraviolet light−inactivated HSV-1(17)*ts*K, which is temperature-sensitive in the α4 gene, has demonstrated loss of similar functions concomitant with increased 5′ deletions (Preston et al. 1984).

**Evidence that the α-*trans*-inducing factor (α-TIF) is the product of a viral gene**

The induction of the α-TK chimeric genes resident in the cells converted to TK⁺ phenotype by infection with TK⁻ virus in the presence of cycloheximide suggested that α gene expression is induced by a virion component introduced into the cell during infection (Post et al. 1981). The properties of this putative virion-associated factor emerged from three series of experiments involving infection of cell lines converted to TK⁺ phenotype with α-TK chimeras. First, TK expression was proportional to the multiplicity of infection within the range of 0.1–10 pfu/cell in TK⁺ cells infected with TK⁻ virus in the presence of cycloheximide. Second, the resident TK gene was induced by ultraviolet light−inactivated virus. Lastly, one of the strongest inducers turned out to be HSV-1(HFEM)*ts*B7 at the nonpermissive temperature (Batterson and Roizman 1983). At that temperature the capsids accumulate at nuclear pores but do not release the viral DNA. These results indicated that the α-TIF is a component of the virion, most probably associated with components located between the capsid and envelope (the tegument) or the envelope.

Cotransfection of plasmids containing α-TK chimeras with plasmids containing DNA fragments representing the entire HSV genome in transient expression assays localized the gene encoding an α-TIF activity to the 8-kb *Bam*HI-F fragment of the HSV-1 genome and, specifically, to the sequences encoding a protein designated as ICP25 (Campbell et al. 1984). The region of the viral DNA encoding the α-TIF activity has been sequenced and its mRNA mapped by S1 analysis. The gene encodes a protein of 54,000 in predicted molecular weight, and its mode of action is not readily discerned from its sequence (Pellett et al. 1985).

**Table 1** Nucleotide Sequence of Specific Domains of the HSV-1(F) α Gene Regulatory Regions

| | −194 | | −135 |
|---|---|---|---|
| G + C−rich | CGGATGGGCGGGGCCGGGGGTTCGACCAACGGCCGCCGGCCAGGGCCCCCCGGCGTGCCG | | |
| α4 gene | GCCTACCCGCCCCGGCCCCCAAGCTGGTTGCCCGGCGCCGGTGCCCGGGGGCCGCACGGC | | |
| | −183 | AT homolog | −134 |
| A + T−rich | CGGAAGCGGAACGGTGTATGTGAT | ATGCTAATTAAATACAT | GCCACGT |
| α27 gene | GCCTTCGCCTTGCCACATACACTA | TACGATTAATTTATGTA | CGGTGCA |

Nucleotide sequence from Mackem and Roizman (1982c) as cloned by Kristie and Roizman (1984).

**Table 2** The Function of Specific Domains of the HSV-1(F) α Gene Regulatory Regions

| α donor sequence and orientation[a] | Recipient promoter and TK gene[a] | $ts502\Delta305$ (cell lines) | | | BamHI-F (transient) | | |
|---|---|---|---|---|---|---|---|
| | | basal level | induction[b] | I:B[c] | basal level | induction[b] | I:B |
| — | −110 α4 | 3.4 | 8.6 | 2.5 | 7.1 | 30 | 4.3 |
| — | −80 β-TK | 0.1 | 0 | — | 8.0 | 8.0 | 1.0 |
| −330 > −110→ | −110 α4 | 90 | 540 | 6.0 | 27 | 1190 | 44 |
| −330 > −110→ | −80 β-TK | 87 | 314 | 3.6 | 30 | 975 | 33 |
| $GC_4$→ | −110 α4 | 186 | 160 | 0.9 | 8.1 | 65 | 8.0 |
| $GC_4$→ | −80 β-TK | 13 | 19 | 1.5 | 7.4 | 30 | 4.1 |
| $AT_{27}$[d]← | −110 α4 | — | — | — | 8.5 | 125 | 15 |
| $AT_{27}$← | −80 β-TK | 3.3 | 21 | 6.4 | 7.2 | 45 | 6.3 |

[a]The first two columns represent the structure of the chimeric genes and have been previously described. The arrows denote the orientation of the donor sequences relative to the recipient promoter-TK genes: (→) direct; (←) inverted.

[b]Levels of basal expression or induced expression in cell lines transformed with the various TK chimeras and selected for G418 resistance was determined by mock infection or infection with $ts502\Delta305$ at the nonpermissive temperature, respectively (Kristie and Roizman 1984). Transient expression assays were done using the intact BamHI-F fragment containing the α-TIF cotransfected with the chimeric TK plasmids as described in Fig. 2. Results are expressed as counts per minute of [$^3$H]thymidine converted to thymidylate per microgram of total cellular protein.

[c]Ratio of induced to basal levels of expression.

[d]This construct was tested only in cells selected for the TK$^+$ phenotype. In this instance, the induction ratio was approximately twofold higher than that obtained with the $AT_{27}$ −80 β-TK construct.

Studies in transient assay systems of the specificity of the α-TIF activity encoded by the BamHI-F fragment indicate that the sequence requirements of the gene product encoded in that fragment for induction of α-TK chimeras is similar to that observed in cell lines transformed to TK$^+$ phenotype with α-TK chimeric genes and infected with TK$^-$ virus (Mackem and Roizman 1982c; Campbell et al. 1984; Kristie and Roizman 1984; Pellett et al. 1985; and J.L.C. McKnight, unpubl.). A summary of these observations is shown in Table 2. In another system, BHKtk$^-$ cells transformed with α-TIF under control of the metallothionein promoter express a higher TK activity when transfected with α-TK chimeras than control cell lines (T.M. Kristie, unpubl.).

## α gene induction may involve additional gene products

Studies on the induction of α-TK chimeras in transient expression systems indicated that the BamHI-F fragment encodes an additional gene (or genes) that enhances the activity of α-TIF. The initial observation was that the intact BamHI-F fragment induced a higher TK activity than the DNA fragment encoding the domain of the gene specifying ICP25 alone. Complementation studies were performed that involved transfections of the plasmids containing the entire α-TIF gene with independently cloned regions of BamHI-F (Fig. 2). These studies localized the enhancement activity to the genes located downstream of the α-TIF gene. Further studies identified two genes 3′ to the α-TIF gene that affect α-TIF activity. One is located immediately 3′ to the α-TIF (Anderson et al. 1981). When this gene is cotransfected with the gene encoding the α-TIF (in different plasmids), repression of α-TIF activity occurs. The second gene is located further 3′ to α-TIF (Anderson et al. 1981). In contrast, when this gene is cotransfected independently

with the α-TIF, the activity of α-TIF is enhanced (Fig. 2). Neither of these genes alone affects the constitutive expression of α-TK chimeras. Studies are currently underway to determine whether gene products are responsible for the observed enhancement and the mechanisms by which they effect the induction of α genes.

## How are HSV-1 γ₂ genes induced?

### Experimental design

The experiment designs employed in the preceding sections indicated that α genes are induced by a structural component of the HSV virion and permitted the identification of a cluster of genes that modulate α gene expression in transient assays systems. Encouraged by the usefulness of these techniques, we applied similar designs to investigate the cis-acting sites and the trans-acting factors involved in γ₂ gene regulation.

The initial studies involved an HSV-1 DNA fragment (BamHI-D′), shown in Figure 1, encoding the promoter-regulatory domain and the transcription-initiation site of a γ₂ gene (Hall et al. 1982). This fragment was fused in the proper transcriptional orientation to the 5′-transcribed noncoding and coding sequences of the TK gene and was either recombined into the viral DNA or used to establish cell lines stably retaining the chimeric gene (Silver and Roizman 1985). Subsequent experiments involved the entire domain of the γ₂ gene specifying the glycoprotein C (gC) gene (Fig. 1). This was cloned into a plasmid carrying the mouse dihydrofolate reductase (DHFR) gene resistant to methotrexate (Simonsen and Levinson 1983) and used to establish L cell lines resistant to the drug (G. Campadelli-Fiume et al., in prep.).

As noted earlier in the text, γ₂ genes are stringently dependent on viral DNA synthesis for their expression

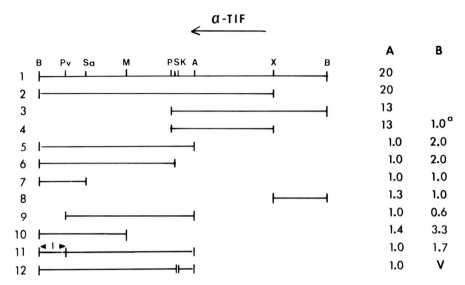

**Figure 2** Complementation studies utilizing transient expression assays to map the α-TIF enhancing activity located within the *Bam*HI-F fragment of HSV-1. In these experiments, 0.05 pmole of an α-TK indicator gene, pRB3354, was cotransfected as described in Fig. 5 with increasing amounts of the complete *Bam*HI-F fragment (line *1*), a subclone containing α-TIF (line *4*), or clones containing α-TIF plus flanking 3′ (line *2*) or 5′ (line *3*) sequences. Column A shows the activities of the α-TK indicator gene in the presence of the individual cloned fragments normalized with respect to that of the α-TK indicator gene alone. Column B shows the activity of the α-TK indicator gene in the presence of the fragment (line *4*) containing α-TIF with the individual fragments shown in lines *5–12* normalized to the activity of the α-TK indicator gene in the presence of the fragment containing α-TIF identified by the superscript a. Line *1*: pRB158 (complete *Bam*HI-F); line *2*: pRB3443 (*Bam*HI to *Xho*I); line *3*: pRB3441 (*Pst*I to *Bam*HI); line *4*: pRB3458 (*Pst*I to *Xho*I); line *5*: pRB3606 (*Bam*HI to ~ 100 bp past the *Asu*II site [*Asu*II + 100]); line *6*: pRB3439 (*Bam*HI to *Sal*I); line *7*: pRB3555 (*Bam*HI to *Sac*I); line *8*: pRB3442 (*Xho*I to *Bam*HI); line *9*: pRB3606 (*Pvu*II to *Asu*II + 100); line *10*: pRB3608 (*Bam*HI to *Mlu*I); line *11*: pRB3609 (*Bam*HI to *Pvu*II fragment inverted relative to the *Pvu*II to *Asu*II + 100 fragment); line *12*: pRB3610 (*Bam*HI to *Asu*II + 100, deleted at the *Kpn*I site). (A) *Asu*II; (B) *Bam*HI; (K) *Kpn*I; (M) *Mlu*I; (P) *Pst*I; (Pv) *Pvu*II; (S) *Sal*I; (Sa) *Sac*I; (X) *Xho*I. (◄ I ►), inverted; (V) variable.

and therefore are readily differentiated from α, β, and γ₁ genes by the sensitivity of their expression to phosphonoacetate (PAA), a selective inhibitor of HSV-1 DNA synthesis (Leinbach et al. 1976).

### γ₂ TK chimeras are regulated as γ₂ genes in the viral genome

In the initial experiment the γ₂ gene constructed by fusion of *Bam*HI-D′ to the TK gene was recombined into the viral genome and tested in TK⁻ cells for the sensitivity of the TK expression to concentrations of PAA known to inhibit viral DNA synthesis (Conley et al. 1981). PAA blocked the expression of the chimeric γ₂-TK gene but not that of the natural β-TK gene, indicating that the chimeric gene was regulated as a bonafide γ₂ gene (Silver and Roizman 1985).

### γ₂-TK chimeras resident in the host genome are regulated as β genes

The initial hypothesis tested in these experiments was that γ₂ genes require template amplification for their expression. Previous studies have shown that HSV-1 DNA fragments consisting of a selectable marker (TK gene) and an origin of HSV DNA synthesis stably associated with the host genome are amplified by infection of the cells with HSV-1 (Mocarski and Roizman 1982). In these experiments the γ₂-TK chimeras both linked and

unlinked to an HSV-1 origin of DNA synthesis were transfected into cells alone or with the gene specifying neomycin resistance. Cell lines carrying the TK constructs were selected either for TK⁺ phenotype or for G418 resistance, infected with TK⁻ virus, and tested for the sensitivity of TK expression to PAA (Silver and Roizman 1985). These studies showed that HSV-1 TK⁻ viruses induced the γ₂-TK gene resident in cells selected for TK⁺ and G418 resistance. S1 analyses indicated that the transcription was initiated from the transcription initiation site in *Bam*HI-D′. Furthermore, as expected, the copy number of the chimeric gene was amplified in constructs containing the origin of DNA replication but not in constructs lacking the origin. However, PAA at inhibitory concentrations did not affect the induction of the γ₂-TK chimeras in any of the constructs.

To test further the regulation of the resident γ₂-TK genes, the cells were infected with *ts*502Δ305 and *ts*HA1Δ305 carrying deletions in the TK gene and temperature-sensitive lesions in the genes specifying ICP4 and ICP8 (a β₁ gene), respectively (Silver and Roizman 1985). At the nonpermissive temperature, *ts*HA1Δ305 induced the resident TK gene, whereas *ts*502Δ305 did not. Since *ts*HA1Δ305 at the nonpermisive temperature does not produce γ₂ proteins (Conley et al. 1981), the results indicated that the induction of γ₂ required functional ICP4 but was not dependent on ICP8 or viral DNA synthesis.

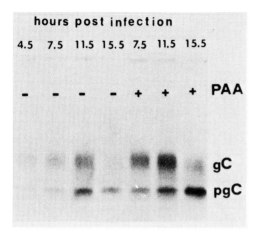

**hours post infection**

| 4.5 | 7.5 | 11.5 | 15.5 | 7.5 | 11.5 | 15.5 | |
|---|---|---|---|---|---|---|---|
| − | − | − | − | + | + | + | PAA |

gC
pgC

**Figure 3** Autoradiograph of electrophoretically separated proteins precipitated with gC-specific monoclonal antibody HC1 from lysates of replicate cultures of L3153$_{28}$ cells infected with HSV-1 and maintained in the presence or absence of PAA. The cells were labeled with [$^{35}$S]methionine for 90 min, beginning at times postexposure of virus to cells shown, and harvested for immunoprecipitation immediately after the labeling interval. The derivation of the cells is described in the legend to Fig. 4. HSV-1(MP) is a spontaneous mutant defective in the production of glycoprotein C (Heine et al. 1974). (gC) Mature form of glycoprotein C; (pgC) precursor form of glycoprotein C.

### An authentic γ$_2$ gene resident in the host genome is also regulated as a β gene

One hypothesis that could explain the data, at least in part, is that the transcribed domain of the TK gene contained in the γ$_2$ construct contains β regulatory

*cis*-acting sites blocked in situ by surrounding genes but effective in the environment of the host genome. To test this hypothesis, a DNA fragment containing the entire domain of the gC gene was cloned into a plasmid carrying the DHFR gene and transfected into Ltk$^-$ cells. The L cells were selected for methotrexate resistance, and clones, selected by end-point dilution, were tested for constitutive and induced gC expression (G. Campadelli-Fiume et al., in prep.). None of the 39 clones tested expressed gC constitutively. However, one of the clones expressed gC after infection with a mutant, HSV-1(MP), that does not express gC (Heine et al. 1974). After infection, this clone produced gC in quantities greater than those made during infection of methotrexate-resistant gC$^-$ cells by wild-type virus. Furthermore, the gC specified by the resident gene was produced earlier in the lytic cycle and its expression was not affected by PAA (Fig. 3). Infection of the cloned line with the temperature-sensitive mutants described above indicated that at the nonpermissive temperature *ts*HA1Δ305 induced the resident gC (Fig. 4), whereas *ts*502Δ305 did not. In all respects tested, the regulation of the authentic γ$_2$-gC gene resident in the L cell clone tested could not be differentiated from that of the chimeric γ$_2$-TK gene described above.

### γ$_2$-TK chimeras are induced by cotransfection with both an intact and truncated ICP4 gene in transient expression systems

The experiments described in the preceding section indicated that the chimeric and authentic γ$_2$ genes resident in the environment of the cellular genome are

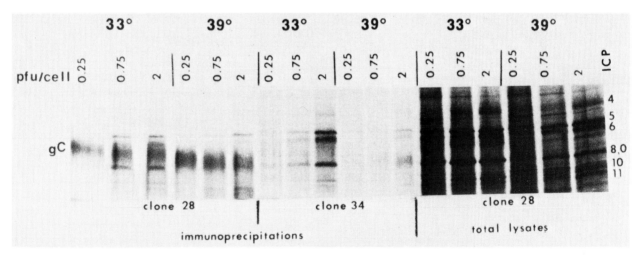

**Figure 4** Autoradiograph of electrophoretically separated total cell lysates and immunoprecipitates obtained with the monoclonal antibody HC1 from lysates of cells infected at permissive and nonpermissive temperatures with HSV-1*ts*HA1. L3153$_{28}$ and L3153$_{34}$ are clonal lines derived from Ltk$^-$ cells transfected with a plasmid containing the methotrexate-resistant mouse dihydrofolate reductase gene and the entire domain of the glycoprotein C (gC) gene contained in the *Hpa*I-*Hind*III subfragment of *Hind*III-L fragment of HSV-1(F) DNA (Fig. 1). Both cell lines are methotrexate resistant, but only the L3153$_{28}$ cells contain a gC gene capable of being induced by infection with HSV. *ts*HA1 contains a lesion in the major DNA-binding protein (ICP8; Conley et al. 1981). At the nonpermissive temperature (39°C), viral DNA replication is blocked and γ$_2$ proteins are not made. In this experiment, replicate cultures of L3153$_{28}$ and L3153$_{34}$ cells were infected with *ts*HA1 at multiplicities of infection shown and incubated at 33°C or 39°C. The cells were labeled with [$^{35}$S]methionine from 1.5 to 10 hr postexposure of virus to cells and harvested at the end of the labeling interval. The autoradiographs of the total cell lysates show that the amounts of viral proteins made in L3154$_{28}$ cells by *ts*HA1 at 10 hr postinfection are small or barely detectable, as would be expected from the very low multiplicities of infection.

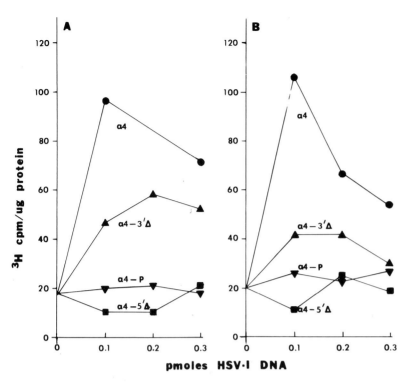

**Figure 5** Transient induction of TK activity in BHKtk⁻ cells. BHKtk⁻ cells were cotranfected with either 0.5 pmole of pRB3488 ($\gamma_2$-TK chimeric gene) (*B*) or 0.1 pmole of pRB3361 DNA ($\beta$-TK gene) (*A*) and increasing amounts of pRB3094 DNA ($\alpha4$; contains the intact $\alpha4$ gene), pRB3096 DNA ($\alpha4$–3'$\Delta$; contains only 40% of the 5' coding sequences of the $\alpha4$ gene), pRB3097 DNA ($\alpha4$–5'$\Delta$; lacks the promoter-regulatory sequence of the $\alpha4$ gene), or pR403 DNA ($\alpha4$–P; contains only the promoter-regulatory sequences). The location of the HSV-1 DNA sequences contained in these plasmids is shown in Fig. 1 and described in P. Mavromara-Nazos et al. (in prep.). The TK activity, assayed 48 hr post-transfection, is expressed as counts per minute of [³H]thymidine converted to thymidilate per microgram of total cellular protein. Time 0 represents the basal-level activity of the TK gene.

regulated as $\beta$ rather than $\gamma_2$ genes. To verify this observation, the expression of the $\gamma_2$-TK chimeric gene was tested in a transient assay system alone or by cotransfection with separate plasmids containing (1) the domain of the intact $\alpha4$, $\alpha0$, $\alpha27$, $\alpha22$, or $\alpha47$ genes, (2) the promoter-regulatory domain of the $\alpha4$ and $\alpha22$ genes, (3) a truncated $\alpha4$ gene lacking the promoter-regulatory domain, and (4) a truncated $\alpha4$ gene lacking 40% of the coding sequence at the 3' end of the gene (Fig. 1). The results of these studies were that only the intact $\alpha4$ and $\alpha0$ genes and the truncated $\alpha4$ gene lacking 40% of its sequences at the 3' end induced the chimeric $\gamma_2$-TK gene (Fig. 5). S1 analyses verified that both the $\gamma_2$-TK and the $\alpha4$ genes capable of inducing it were indeed transcribed from the appropriate transcription-initiation sites. Of particular interest was the observation that $\gamma_2$-TK was induced at relatively low concentrations of the $\alpha4$ genes but not at higher concentrations (P. Mavromara-Nozos et al., in prep.).

## Conclusions: The take-home message
### The obvious
In this report we contrasted the regulation of two sets of HSV genes, $\alpha$ and $\gamma_2$, in similar systems employing chimeric genes. The regulation of these genes resident in the viral genome could not be differentiated from that of authentic $\alpha$ and $\gamma_2$ genes. In the environment of the host genome, the $\alpha$ chimeric genes were unexpectedly found to be inducible by infection of the cells, and a viral gene ($\alpha$-TIF) has been identified whose product effects the induction of the resident gene. The product of this gene, previously identified as ICP25 (Morse et al. 1978; Lemaster and Roizman 1980), has been located in the

envelope-tegument components of the virion. Experiments are underway to determine whether the product of the $\alpha$-TIF gene does in fact play a role in induction of $\alpha$ genes resident in the viral genome. Notwithstanding the remaining uncertainty, the induction of $\alpha$ genes in vitro is consistent with the expression of $\alpha$ genes resident in the viral genome in that neither requires functional ICP4 or de novo protein synthesis for expression.

A very different situation has arisen in the case of $\gamma_2$ genes. In the environment of the cell, both the chimeric $\gamma_2$-TK and the authentic $\gamma_2$-gC genes were regulated as $\beta$ genes. Three conclusions are immediately apparent. First, the results do not support the contention that the 5'-transcribed noncoding and coding sequences of the TK gene carry *cis*-acting sites that confer $\beta$ regulation upon constructs containing these sequences inasmuch as representative genes of all coordinately regulated groups of HSV genes, except $\alpha$ genes, are also regulated as $\beta$ genes. Second, the authentic *cis*-acting sites and *trans*-acting factors involved in the regulation of $\gamma_2$ genes resident in the viral genome cannot be deduced from studies of an isolated $\gamma_2$ gene resident in the environment of the cell by the conventional methods described in these studies. Lastly, the studies described here cast a pale shadow on the conclusions regarding expression and regulation of $\beta$ genes, notably TK, carried out in similar environments and essentially by similar methods. First, if $\beta$ genes are defined operationally as requiring ICP4 but no viral DNA synthesis, as is the case in the environment of the viral genome, then all nonviral genes induced in *trans* by ICP4 must also be classified as $\beta$ genes, a totally nonsensical conclusion, because they are not induced as bonafide β genes after infection. More significant, we cannot differentiate between the hypothesis that β genes are regulated as β and that the induction of all other genes by ICP4 is artifactual and the alternative possibility that a subset of genes resident in the environment of the host genome is induced by ICP4 by a mechanism distinct from that of the authentic *trans*-inducing factor operative in their native genomic environments. Until this is resolved, we cannot be sure whether induction of β genes by ICP4 is by the authentic mechanism operative in the environment of the viral genome or by an entirely different mechanism that reflects conditions that are generated in the environment of the host genome.

*The conjectural*

The results presented in this paper indicate that HSV-1 genes in the environment of cellular genomes fall into two categories only, the α-like and the β-like, and therefore viral gene regulation in that environment does not correspond to that of viral genes in viral genomes. Just because the regulation is different does not mean that the results are devoid of significance. To start with, the results indicate that the infected cell contains two genomic compartments in which genes are regulated differently. Since the transcription of both host and viral genes occurs in the nucleus, the compartmentalization is "environmental." The only datum consistent with environmental segregation of host and viral genomes is the repeated observation that viral DNA in productively infected cells is not organized in nucleosomes (Mouttet et al. 1979; Camerini-Otero and Zasloff 1980; Leinbach and Summers 1980; Sweet et al. 1982). If, in fact, nucleosome organization underlies the difference between the regulation of $\gamma_2$ genes in the two environments, it follows that either (1) the factors necessary to transcribe $\gamma_2$ genes resident in the host genomes are compartmentalized with nucleosomal structures and are not available in the viral genomic compartment until late in infection or (2) $\gamma_2$ gene expression is repressed early in infection, but nucleosomal organization blocks the repressor from reaching the $\gamma_2$ genes in the host genomic compartments.

Both hypotheses rest on the notion that the cell does not require specific viral factors for transcription of $\gamma_2$ genes. In the first hypothesis, the required host factors would be contained in chromatin and generally unavailable for viral DNA transcription until released. Implicit in this hypothesis is the notion that the actual function of viral DNA polymerase (inasmuch as PAA specifically inhibits the polymerase and $\gamma_2$ expression) and not merely the products of β genes that might modify host DNA (e.g., viral DNase and DNA-binding proteins) is required for release of bound factors from the host genomic compartment. This hypothesis differentiates between β and $\gamma_2$ genes in that in the environment of the viral genome the factors necessary for β gene transcription are insufficient for that of $\gamma_2$ genes.

In the second hypothesis, the putative repressor would be a viral gene product that has high affinity for

DNA and that is removed by the initiation of DNA replication and whose synthesis is in turn repressed by $\gamma_2$ proteins. This hypothesis does not discriminate between β-specific and γ-specific transcriptional factors.

Both models postulate a role for ICP4 as a transcriptional activator. The observed range of expression of β genes (e.g., TK) while in the host genomic environment in various systems is consistent with both models (Wigler et al. 1977; McKnight et al. 1981; Kristie and Roizman 1984; El Kareh et al. 1985). It is possible that when cells are selected for TK expression, a modulation of host transcriptional activators results. These activators would then alter the affinity or abundance of transcriptional factors specific for the β-type genes. Implicit in this hypothesis would be that there are also various classes of transcriptional activators, one of which would include ICP4.

The models presented here account for the observations but do not exclude alternatives based on data that is as yet unavailable. The virtue of the models we present is that they are testable.

## Acknowledgments

These studies, carried out at the University of Chicago, were aided by grants from the National Cancer Institute (CA-08494 and CA-19264), U.S. Public Health Service, and the American Cancer Society (MV2T). J.M. is a USPHS postdoctoral trainee (CA-09241). T.K. and P.P. are USPHS predoctoral trainees (CA-192642 and AI-07182, respectively), and S.S. was a USPHS predoctoral trainee (AI-7182). P.N. and M.A. are Damon Runyon postdoctoral trainees (DRG767 and DRG017, respectively). G.C.E. was supported by grants from Progetto Finalizzato Ingeneria Genetica (84.00871.51), Projeto Bilaterale Italia-USA (84.01725.04), and NATO grants (253/83, 265/84).

## References

Ackermann, M., D.K. Braun, L. Pereira, and B. Roizman. 1984. Characterization of herpes simplex virus 1 α proteins 0, 4, and 27 with monoclonal antibodies. *J. Virol.* **52**: 108.

Anderson, K., R. Frink, G. Devi, B. Gaylord, R. Costa, and E. Wagner. 1981. Detailed characterization of the mRNA mapping in the *Hind*III fragment K region of the HSV-1 genome. *J. Virol.* **37**: 1011.

Batterson, W. and B. Roizman. 1983. Characterization of the herpes simplex virion-associated factor responsible for the induction of α genes. *J. Virol.* **46**: 371.

Camerini-Otero, R.D. and M.A. Zasloff. 1980. Nucleosomal packaging of the thymidine kinase gene of herpes simplex virus transferred into mouse cells: An actively expressed single copy gene. *Proc. Natl. Acad. Sci.* **77**: 5079.

Campbell, M.E.M., J.W. Palfreyman, and C.M. Preston. 1984. Identification of herpes simplex virus sequences which encode a transacting polypeptide responsible for stimulation of immediate early transcription. *J. Mol. Biol.* **180**: 1.

Conley, A.J., D.M. Knipe, P.C. Jones, and B. Roizman. 1981. Molecular genetics of herpes simplex virus. VII. Characterization of a temperature-sensitive mutant produced by in vitro mutagenesis and defective in DNA synthesis and accumulation of γ polypeptides. *J. Virol.* **37**: 191.

Constanzo, F., G. Campadelli-Fiume, L. Foa-Tomas, and E. Cassai. 1977. Evidence that herpes simplex virus DNA is transcribed by cellular RNA polymerase II. *J. Virol.* **21:** 996.

Cordingley, M.G., M.E.M. Campbell, and C.M. Preston. 1983. Functional analysis of a herpes simplex virus type 1 promotor: Identification of far upstream regulatory sequences. *Nucleic Acids Res.* **11:** 2347.

Dixon, R.A.F. and P.A. Schaffer. 1980. Fine structural mapping and functional analysis of temperature sensitive mutants in the gene encoding the herpes simplex virus type 1 immediate early protein VP 175. *J. Virol.* **36:** 189.

El Kareh, A., A.J.M. Murphy, T. Fichter, A. Efstratiadis, and S. Silverstein. 1985. "Transactivation" control signals in the promoter of the herpesvirus TK gene. *Proc. Natl. Acad. Sci.* **82:** 1002.

Fenwick, M. and B. Roizman. 1977. Regulation of herpesvirus macromolecular synthesis. VI. Synthesis and modification of viral polypeptides in enucleated cells. *J. Virol.* **22:** 720.

Fenwick, M.L. and M.J. Walker. 1978. Suppression of the synthesis of cellular macromolecules by herpes simplex virus. *J. Gen. Virol.* **41:** 37.

Frink, R.J., K.P. Anderson, and E.K. Wagner. 1981. Herpes simplex virus type 1 *Hind*III L encodes spliced and complementary mRNA species. *J. Virol.* **39:** 559.

Godowski, P.J. and D.M. Knipe. 1985. Identification of a herpes simplex virus function that represses late gene expression from parental viral genomes. *J. Virol.* **55:** 357.

Hall, L.M., K.G. Draper, R.J. Fluck, R.H. Carter, and E.K. Wagner. 1982. Herpes simplex virus mRNA species mapping in *Eco*RI fragment I. *J. Virol.* **43:** 594.

Heine, J.W., R.W. Honess, E. Cassai, and B. Roizman. 1974. Proteins specified by herpes simplex virus. XII. The virion polypeptides of type 1 strains. *J. Virol.* **14:** 640.

Herz, C. and B. Roizman. 1983. The α promotor-ovalbumin chimeric gene resident in human cells is regulated like the authentic α 4 gene after infection with herpes simplex virus 1 mutants in the α 4 gene. *Cell* **33:** 145.

Holland, L.E., K.P. Anderson, C. Shipman, and E.K. Wagner. 1980. Viral DNA synthesis is required for the efficient expression of specific herpes simplex virus type 1 mRNA species. *Virology* **101:** 10.

Honess, R.W. and B. Roizman. 1974. Regulation of herpesvirus macromolecular synthesis. I. Cascade regulation of the synthesis of three groups of viral proteins. *J. Virol.* **14:** 8.

———. 1975. Regulation of herpesvirus macromolecular synthesis: Sequential transition of polypeptide synthesis requires functional viral polypeptides. *Proc. Natl. Acad. Sci.* **72:** 1276.

Huszar, D. and S. Bacchetti. 1981. Partial purification and characterization of the ribonucleotide reductase induced by herpes simplex virus infection of mammalian cells. *J. Virol.* **37:** 580.

Johnson, D.C., M. Wittels, and P.G. Spear. 1984. Binding to cells of virosomes containing herpes simplex virus type I glycoproteins and evidence for fusion. *J. Virol.* **52:** 238.

Jones, P.C. and B. Roizman. 1979. Regulation of herpesvirus macromolecular synthesis. VIII. The transcription program consists of three phases during which both extent of transcription and accumulation of RNA in the cytoplasm are regulated. *J. Virol.* **31:** 299.

Kit, S., D.R. Dubbs, and P.A. Schaffer. 1978. Thymidine kinase activity of biochemically transformed mouse cells after superinfection by thymidine kinase-negative, temperature-sensitive herpes simplex virus mutants. *Virology* **85:** 456.

Knipe, D.M., W.T. Ruyechan, B. Roizman, and I.W. Halliburton. 1978. Molecular genetics of herpes simplex virus: Demonstration of regions of obligatory and nonobligatory identity within diploid regions of the genome by sequence replacement and insertion. *Proc. Natl. Acad. Sci.* **75:** 3896.

Kristie, T.M. and B. Roizman. 1984. Separation of sequences defining basal expression from those conferring α gene recognition within the regulatory domains of herpes simplex virus 1 α genes. *Proc. Natl. Acad. Sci.* **81:** 4065.

Leiden, J.M., R. Buttyan, and P.G. Spear. 1976. Herpes simplex virus gene expression in transformed cells. I. Regulation of the viral thymidine kinase gene in transformed L cells by products of super-infecting virus. *J. Virol.* **20:** 413.

Leinbach, S.S. and W.C. Summers. 1980. The structure of herpes simplex virus type 1 DNA as probed by micrococcal nuclease digestion. *J. Gen. Virol.* **51:** 45.

Leinbach, S.S., J.M. Reno, L.F. Lee, A.F. Isbell, and J.A. Boezi. 1976. Mechanism of phosphonoacetate inhibition of herpesvirus-induced DNA polymerase. *Biochemistry* **15:** 426.

Lemaster, S. and B. Roizman. 1980. Herpes simplex virus phosphoproteins. II. Characterization of the virion protein kinase and of the polypeptides phosphorylated in the virion. *J. Virol.* **35:** 798.

Lin, S.-S. and W. Munyon. 1974. Expression of the viral thymidine kinase gene in herpes simplex virus–transformed L cells. *J. Virol.* **14:** 1199.

Mackem, S. and B. Roizman. 1982a. Regulation of herpes simplex virus: The α27 gene promotor–thymidine kinase chimera is positively regulated in converted L cells. *J. Virol.* **43:** 1015.

———. 1982b. Differentiation between α promotor and regulator regions of herpes simplex virus 1: The functional domains and sequence of a movable α regulator. *Proc. Natl. Acad. Sci.* **79:** 4917.

———. 1982c. Structural features of the herpes simplex virus α gene 4, 0, and 27 promotor-regulatory sequences which confer α regulation on chimeric thymidine kinase genes. *J. Virol.* **44:** 939.

McKnight, S. 1980. The nucleotide sequence and transcript map of the herpes simplex virus thymidine kinase gene. *Nucleic Acids Res.* **8:** 5949.

———. 1982. Functional relationships between transcriptional control signals of the thymidine kinase gene of herpes simplex virus. *Cell* **31:** 335.

McKnight, S.L., E.R. Gavis, R. Kingsbury, and R. Axel. 1981. Analysis of transcriptional regulatory signals of the HSV thymidine kinase gene: Identification of an upstream control region. *Cell* **25:** 385.

Mocarski, E.S. and B. Roizman. 1982. Herpes virus dependent amplification and inversion of a cell-associated viral thymidine kinase gene flanked by viral sequences and linked to an origin of viral DNA replication. *Proc. Natl. Acad. Sci.* **79:** 5626.

Morse, L.S., L. Pereira, B. Roizman, and P.A. Schaffer. 1978. Anatomy of herpes simplex virus (HSV) DNA. X. Mapping of viral genes by analysis of polypeptides and functions specified by HSV-1 × HSV-2 recombinants. *J. Virol.* **26:** 389.

Mouttet, M.E., D. Guetard, and J.M. Bechet. 1979. Random cleavage of intranuclear herpes simplex virus DNA by micrococcal nuclease. *FEBS Lett.* **100:** 107.

Pellett, P.E., J.L.C. McKnight, F.J. Jenkins, and B. Roizman. 1985. Nucleotide sequence and predicted amino acid sequence of a protein encoded in a small herpes simplex virus DNA fragment capable of *trans*-inducing α genes. *Proc. Natl. Acad. Sci.* **82:** 5870.

Pereira, L., M. Wolff, M. Fenwick, and B. Roizman. 1977. Regulation of herpesvirus synthesis. V. Properties of α polypeptides specified by HSV-1 and HSV-2. *Virology* **77:** 733.

Post, L.E., S. Mackem, and B. Roizman. 1981. Regulation of α genes of herpes simplex virus: Expression of chimeric genes produced by fusion of thymidine kinase with α gene promotors. *Cell* **24:** 555.

Post, L.E., B. Norrild, T. Simpson, and B. Roizman. 1982. Chicken ovalbumin gene fused to a herpes simplex virus α promotor and linked to a thymidine kinase gene is regulated like a viral gene. *Mol. Cell. Biol.* **2:** 233.

Preston, C.M., M.G. Cordingly, and N.D. Stow. 1984. Analysis of DNA sequences which regulate the transcription of a herpes simplex virus immediate early gene. *J. Virol.* **50:** 708.

Read, S.G. and N. Frenkel. 1983. Herpes simplex virus mutants defective in the viron-associated shutof of host polypeptide synthesis and exhibiting abnormal synthesis of α (immediate early) viral polypeptides. *J. Virol.* **46:** 498.

Roizman, B. and W. Batterson. 1985. Herpesviruses and their replication. In *Virology* (ed. B.N. Feilds et al.), p. 497. Raven Press, New York.

Sacks, R., C.C. Greene, D.P. Aschman, and P.A. Schaffer. 1985. Herpes simplex virus type 1 ICP27 is an essential regulatory protein. *J. Virol.* **55:** 796.

Sears, A.E., I.W. Halliburton, B. Meignier, S. Silver, and B. Roizman. 1985. Herpes simplex virus 1 mutant deleted in the α22 gene: Growth and gene expression in permissive and restrictive cells and establishment of latency in mice. *J. Virol.* **55:** 338.

Silver, S. and B. Roizman. 1985. $\gamma_2$-thymidine kinase chimeras are identically transcribed but regulated as $\gamma_2$ genes in herpes simplex virus genomes and as β genes in cell genomes. *Mol. Cell. Biol.* **5:** 518.

Silverstein, S. and E.L. Engelhardt. 1979. Alterations in the protein synthetic apparatus of cells infected with herpes simplex virus. *Virology* **95:** 324.

Simonsen, C.C. and A.D. Levinson. 1983. Isolation and expression of an altered mouse dihydrofolate reductase cDNA. *Proc. Natl. Acad. Sci.* **80:** 2495.

Southern, P.J. and P. Berg. 1982. Transformation of mammalian cells to antibiotic resistance with a bacterial gene under control of the SV40 early region promotor. *J. Mol. Appl. Genet.* **1:** 327.

Sweet, R.W., M.V. Chao, and R. Axel. 1982. The structure of the thymidine kinase gene promoter: Nuclease hypersensitivity correlates with expression. *Cell* **31:** 347.

Watson, R.J. and J.B. Clements. 1978. Characterization of transcription-deficient temperature-sensitive mutants of herpes simplex virus type 1. *Virology* **91:** 364.

Wigler, M., S. Silverstein, L.-S. Lee, A. Pellicer, Y.-C. Cheng, and R. Axel. 1977. Transfer of purified herpes virus thymidine kinase gene to cultured mouse cells. *Cell* **11:** 223.

Wilcox, K.W., A. Kohn, E. Sklyanskaya, and B. Roizman. 1980. Herpes simplex virus phosphoproteins. I. Phosphate cycles on and off some viral polypeptides and can alter their affinity for DNA. *J. Virol.* **33:** 167.

# Multiple *Trans*-activating Proteins of Herpes Simplex Virus That Have Different Target Promoter Specificities and Exhibit Both Positive and Negative Regulatory Functions

P. O'Hare,* J.D. Mosca,[†] and G.S. Hayward*

The Virology Laboratories, Departments of *Pharmacology and [†]Oncology, Johns Hopkins University School of Medicine, Baltimore, Maryland 21205

Three *trans*-acting regulatory proteins (VF65, IE175, and IE110) encoded by herpes simplex virus (HSV) have been identified by DNA cotransfection in transient expression assays and shown to have different target promoter specificities. The pre-IE virion factor VF65 (or Vmw65) specifically required the 5'-upstream region of the IE175 (or ICP4) promoter, including several "TAATGARAT" signals for maximal activation. Similarly, stimulation by the IE175 protein showed considerable preference for unknown features of HSV delayed-early (DE) promoters, whereas the promiscuous IE110 (or ICP0) protein activated a variety of heterologous viral and cellular promoters as well as the HSV IE and DE constructs. The IE175 gene product also proved to specifically inhibit expression from its own promoter, with this autoregulation dominating activation by VP65 or IE110. Similarly, in experiments involving integrated forms of the hybrid target genes in neomycin-resistant coselected cell lines, both HSV TK-CAT and IE175-CAT proved to be inducible by the appropriate IE or DE mechanisms after superinfection with HSV. Studies with a hybrid CAT gene driven by the complex IE94 promoter/regulatory region from simian cytomegalovirus (CMV) in transient assay experiments revealed 100-fold stronger basal expression than from IE175-CAT and 3-fold to 5-fold stronger than from SV2-CAT. Furthermore, permanent IE94-CAT cell lines also gave strong basal expression but were not induced by superinfecting HSV or CMV. In contrast, IE94-IFN cell lines gave low basal expression and were strongly induced by either HSV or cycloheximide but not by CMV. This novel HSV *trans*-activation process appeared to represent stabilization of short-half-life IE94-IFN mRNA and required expression of an IE function other than IE175.

The genetic programs for lytic-cycle synthesis of viral proteins of herpes simplex virus (HSV) and cytomegalovirus (CMV) are generally thought to be quite similar, except that the overall process takes much longer in CMV than the 18 to 24 hr for HSV and the permissive host-range for CMV is very narrowly defined in contrast to the extraordinarily broad host-range for HSV. Also, despite its larger genome (240 kb, compared with 150 kb for HSV), CMV is generally thought to rely much more on the utilization of cell-cycle functions for its replication than does HSV. We have been concerned with describing and characterizing the genetic mechanisms involved in the very early regulatory events after infection with these viruses, both as interesting model systems for gene regulation in their own right and with the hope that detailed knowledge of the early lytic-cycle events will also lead to important insights into the nature of the more subtle cell/virus interactions occurring during latency in neurons (HSV) or monocytes and other lymphoid cell types (CMV).

Studies on viral mRNA synthesis in the absence of protein synthesis or with appropriate temperature-sensitive (*ts*) mutant viruses have defined the existence of a group of five immediate-early (IE) genes in HSV that all possess similar *cis*-acting regulatory signals and are transcribed prior to de novo viral protein synthesis (Honess and Roizman 1975; Clements et al. 1979; Mackem and Roizman 1980). Examination of the nature of the IE *cis*-acting signals has revealed four distinct promoters of this type (two of which are arranged in an adjacent, divergent orientation on opposite sides of the *ori* S replication origin). Each possesses multiple, dispersed copies of a characteristic TAATGARAT-like sequence, and some also have multiple G + C−rich Sp1-factor-binding sites (Mackem and Roizman 1982; Lang et al. 1984; Preston et al. 1984; Jones and Tjian 1985). Experiments with DNA-transfected Ltk[+] cell lines containing IE-TK hybrid genes have also revealed that, unlike the 2 × 72-bp SV40 early region enhancers, for example, which mediate a direct response to cellular protein factors, these signals respond preferentially to a factor introduced with the virus particle during infection (Post et al. 1981; Mackem and Roizman 1982; Preston et al. 1984). This process occurs even without entry of virus capsids into the nucleus or release of the input virion DNA and without any requirement for new protein or mRNA synthesis (Batterson and Roizman 1983).

The product of the IE gene encoding a nuclear phosphorylated polypeptide with an apparent molecular weight of 175,000 (also called ICP4) has been known for a number of years to be essential for progression to the second or delayed-early (DE) stage of HSV infection. Viruses that have mutations in this gene and that produce a nonfunctional cytoplasmic form of the protein at nonpermissive temperature (e.g., *tsB2* and *tsK*) fail to synthesize significant amounts of thymidine kinase (TK) or other DE mRNA or proteins (Preston 1979). Furthermore, even after progression to the late stages of infection at 33°C with *tsK* virus, the transcription pattern reverts back to synthesis of IE mRNA after shift up to 39°C (Watson and Clements 1980). Like IE175, all of the IE gene products, except for the smallest (IE12), are also nuclear phosphorylated proteins and have usually been presumed to play some role in gene regulation, although no genetic or other evidence supporting this notion was available until very recently. Similar to the situation for E1A activation of the adenovirus E2 promoter (Imperiale et al. 1985), extensive analysis has shown that in the TK promoter the *cis*-acting signals for response to the DE-activation process virtually overlap at positions −45 to −110 with those that are necessary for basal expression (McKnight et al. 1984; Eisenberg et al. 1985; El Kareh et al. 1985).

In human CMV, a single, abundant IE gene product is known (IE68), and several minor species represent potential additional IE gene products (Stinski et al. 1983). However, the major CMV IE gene that was identified first in our laboratory was that from a strain of African green monkey CMV, and therefore most of our studies have been carried out with the IE94 gene product of CMV (Colburn). In this strain the IE94 protein and mRNA are even more dominant over any potential minor IE species, and the parent virus has a somewhat expanded host-range to the extent that a wide variety of immortalized and transformed primate cell lines are permissive hosts, rather than only diploid fibroblasts as is the case for typical laboratory strains of human CMV (R.L. LaFemina and G.S. Hayward, in prep.). The spliced IE94 gene has been characterized and sequenced, and a large, complex 5′-upstream regulatory protein has been described (Jeang et al. 1984 and in prep.). Both the human IE68 and simian IE94 proteins are phosphorylated and predominantly nuclear in location, and their abundant synthesis immediately after reversal from a cycloheximide block initiated at the time of infection (Jeang et al. 1982) leads to the suspicion that they are also involved at some level in CMV DE gene regulation.

In recent studies described here, we set out to examine the possibilities of using both transient and long-term DNA transfection systems with hybrid chloramphenicol acetyltransferase (CAT) or β-interferon (β-IFN) genes to attempt to achieve the following: (1) reconstruction of the various, different levels of HSV or CMV gene regulation with isolated target promoters; (2) identification and isolation of the individual effector genes involved in *trans*-activation of herpesvirus gene expression; and (3) comparison of the properties, target specificities, and mechanisms of action of the various regulatory gene products.

## Materials and Methods

### Viruses and cells
The growth, preparation, and sources of HSV and CMV virus stocks, mouse Ltk⁻ and Ltk⁺ cells, and VERO cells have been described elsewhere (Jeang et al. 1982; O'Hare and Hayward 1984).

### Plasmid cloning
Construction of the HSV hybrid IFN and CAT plasmids was described by Reyes et al. (1982), O'Hare and Hayward (1984), and Mosca et al. (1985). The hybrid IE94-IFN and IE94-CAT plasmids were constructed by K.-T. Jeang (Jeang et al., in prep.). The HSV effector plasmid clones were all described either by Middleton et al. (1982) or O'Hare and Hayward (1985).

### CAT assays
DNA transfections into VERO cells for transient expression assays were carried out by the calcium precipitation procedure without a glycerol boost and without carrier DNA. Usually 2 μg of effector plasmid DNA or of control pBR322 DNA was cotransfected together with 2 μg of target plasmid DNA into 35-mm dishes. For basal expression and cotransfection assays, cells were harvested at between 40 and 48 hr. Virus superinfection was usually initiated at 24 hr after transfection and the cells were harvested after a further 20 hr. Assays for acetylation of [$^{14}$C]chloramphenicol in the extracts were carried out as described previously (Gorman et al. 1982; O'Hare and Hayward 1984).

### Permanent cell lines
DNA transfection procedures for TK coselection of hybrid IFN genes was described by Reyes et al. (1982) and Mosca et al. (1985). Coselection of hybrid CAT genes into VERO and Ltk⁻ cells with pSV2neo and G418 selection was also described by Mosca et al. (1985). Superinfection was carried out at multiplicities of infection of between 10 and 25 pfu/ml. Cycloheximide was used at 50 μg/ml in VERO cells and 6 μg/ml in L cells. Actinomycin D treatment involved incubation in medium containing 2 μg/ml.

### Oocyte microinjection and IFN assays
These procedures were described previously by Reyes et al. (1982) and Mosca et al. (1985).

### S1 nuclease RNA analysis
Total-cell RNA was prepared by the guanidium thiocyanate/CsCl procedure, and 10-μg quantities from each sample were incubated with S1 nuclease after annealing with an 1800-bp, denatured, *Bgl*II-*Ava*I DNA probe end-labeled with $^{32}$P from pIE94-IFN (pTJ211). The protected hybrids were analyzed by autoradiography after electrophoresis through urea/polyacrylamide gels.

## Results

### Comparison of the basal strengths of the HSV IE175 and CMV IE94 promoter–regulatory regions

In initial studies using hybrid constructs of the CMV IE94 promoter linked to IFN-coding sequences in pBR322-derived plasmids, we observed extraordinarily high yields of released, biologically active, human β-IFN after microinjection of the DNA into *Xenopus laevis* oocytes (K.-T. Jeang et al., in prep.). For example, in experiments in which microinjected TK-IFN DNA and IE175-IFN gave titers of 100 units and 1500 units, respectively, several different IE94-IFN constructs gave at least 25,000 units of β-IFN (Fig. 1B). However, to

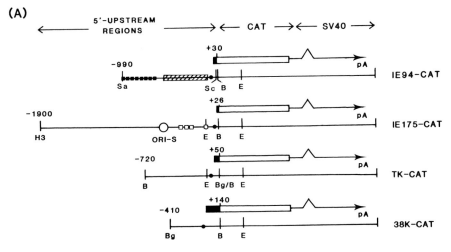

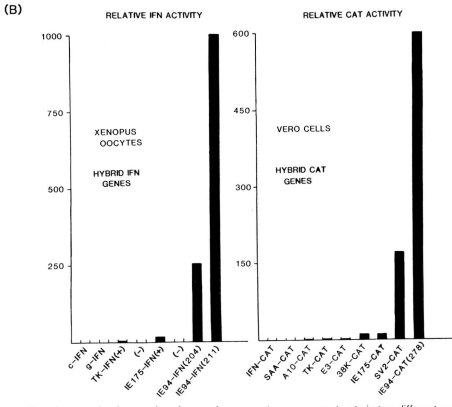

**Figure 1** Comparison of basal expression from various herpesvirus promoter–gene constructs in two different assay systems. (*A*) Structure of the principal HSV and CMV target CAT constructs used throughout these studies. Only the hybrid viral promoter–CAT gene portions are shown. Solid bars indicate the 5′-upstream leader sequence portions of the resulting mRNAs, and open bars represent the CAT-coding sequences. The size of the 5′-upstream sequence present in each construct is given in base pairs above the bars. The positions of TAATGARAT boxes (□), nuclear factor I–binding sites (■), and the CMV enhancer region (▨) are indicated. (B) *Bam*HI; (Bg) *Bgl*II; (E) *Eco*RI; (H3) *Hind*III; (Sa) *Sal*I; (Sc) *Sac*I. (*B*) Histogram of relative amounts of biologically active IFN produced 48 hr after microinjection of plasmid DNA into *Xenopus* oocyte nuclei (*left*) or of CAT enzyme levels produced 44 hr after transfection of plasmid DNA into VERO cell cultures (*right*). Values are relative to TK-IFN and A10-CAT as 1.0.

ensure that the target genes being examined differed only in the nature of their 5'-upstream promoter–regulatory regions, we carried out additional studies with more-uniform constructs containing identical "cassettes" of bacterial CAT-coding sequences plus SV40-derived splicing and poly(A) signals (Gorman et al. 1982). Removal of the 2 × 72-bp enhancer region from the complete SV40 early promoter in SV2-CAT reduces basal CAT enzyme activity produced in transient expression assays by approximately 2 orders of magnitude (Laimins et al. 1982). Therefore, we have used both the complete and minimal SV40 early promoter–CAT constructs as convenient yardsticks for comparison of the basal strengths of the several different herpesvirus promoter constructs (Fig. 1A) and other viral and cellular constructs used in these studies. A summary of the relative levels of CAT expression from equivalent amounts of input DNA over the course of numerous DNA-transfection experiments in VERO cells is given in Figure 1B. On a scale where the yield from the minimal SV40 early promoter construct (A10-CAT) equals 1.0 and that from the complete SV2-CAT plasmid equals 170, we found that two inducible cellular promoter hybrids—SAA-CAT (acute-phase serum amyloid A) and IFN-CAT (human β-IFN)—gave levels of between 0.1 and 0.3. An HSV-1 TK-CAT construct with 720-bp of upstream sequences and an adenovirus E3-CAT early promoter construct each gave values of approximately 1.5. Relative basal CAT levels of about 4 were obtained consistently with both IE175-CAT, which includes all 700 bp of the divergent IE175/$ori_S$/IE68 promoter/regulatory region from HSV-1, and 38K-CAT, which contains the promoter from the gene for a subunit of HSV-2 ribonucleotide reductase. Interestingly, 5' deletions of the IE175-CAT plasmid giving versions containing only 380 bp or 330 bp of upstream IE175 sequences produced a 3-fold increase in basal activity (i.e., average 12-fold stronger that A10-CAT), indicating some apparent negative effects from the presence of the adjacent $ori_S$ or IE68 promoter sequences. In contrast, three CMV IE94-CAT plasmids containing 990 bp, 360 bp, or 260 bp upstream from the cap site all gave greater activity than the complete SV40 promoter, with yields up to 3.5-fold higher than from the SV2-CAT and 600-fold higher than from A10-CAT (Fig. 1B).

### Activation of TK-CAT and IE175-CAT but not IE94-CAT after virus superinfection in the transient CAT assay system

O'Hare and Hayward (1984) reported that expression from both HSV-1 and HSV-2 DE constructs (TK-CAT and 38K-CAT) and from IE175-CAT in transfected VERO cells could be stimulated between 20-fold and 40-fold by superinfection with HSV-1 virus. Furthermore, the HSV-CAT hybrids responded with the expected characteristics of their parental virus genes during viral infection; for example, the activation of IE175-CAT occurred much earlier after infection (within 4 hr) than that of DE constructs and occurred even after infection under nonpermissive conditions with *ts* virus mutants deficient in IE expression. In contrast, little induction of the DE constructs occurred after infection at 39°C with either the HSV-1 *tsB2* or *tsB7* mutants. In control experiments, the levels of CAT product produced by SV2-CAT and A10-CAT plasmids were unaffected by HSV superinfection. These results imply that all three of the HSV promoter–regulatory regions tested possess specific signals for response to viral or cellular factors produced by HSV-1 infection and that the IE and DE *cis*-acting signals respond to different levels of regulatory stimuli, with only the DE-CAT constructs requiring newly synthesized viral gene products. Somewhat surprisingly, CMV (Colburn) infection of DNA-transfected VERO cells produced up to 100-fold activation of HSV DE-CAT and IE175-CAT but it proved to have no significant effect on either SV2-CAT or IE94-CAT expression.

### Evidence for two HSV IE *trans*-activating proteins with different target promoter specificities

Middleton et al. (1982) demonstrated that DNA transfection with cloned plasmids containing several intact IE genes from HSV-2 gave rise to nuclear antigens detectable by immunofluorescence with specific antisera. Therefore, we anticipated that basal expression of the isolated IE gene products in transient assays may be sufficient for study of their transcriptional regulatory action on defined hybrid target genes. Subsequently, O'Hare and Hayward (1985) presented the results of initial cotransfection assays in which the DE targets (TK-CAT or 38K-CAT) were stimulated 20-fold to 40-fold by plasmids containing genes encoding either or both of the HSV-1 IE175 or IE110 proteins (Fig. 2). Linearization of the effector plasmids outside the IE gene-coding regions had no effect on stimulation, whereas cleavage within the coding regions virtually abolished the activity, confirming that the IE gene products themselves were responsible and that possible recombination of the target plasmids with *cis*-acting sequences in the IE plasmids was not involved. The three other isolated intact IE genes—IE68 (or ICP22), IE63 (or ICP27), and IE12 (or ICP47)—gave no activity on DE targets in these assays. Plasmids such as pGR90 (also pGR91 and pGR151), which contained both the HSV-2 IE175 and IE110 genes, gave 30-fold to 60-fold stimulation of the DE constructs but gave no more than 2-fold to 3-fold activation of IE175-CAT, SV2-CAT, or A10-CAT, suggesting considerable target specificity.

Further studies on *trans*-activation by IE175 or IE110 individually revealed markedly different functional properties of the two proteins (Fig. 2B,C). For example, cotransfection with the plasmid containing IE175 alone (p*Xho*I-C) consistently gave 30-fold activation of HSV TK-CAT and 38K-CAT but never gave more than 2-fold to 3-fold activation with A10-CAT, SAA-CAT, E3-CAT, or SV2-CAT and often gave decreased expression from IE175-CAT (see below). In contrast, cotransfection with a plasmid containing IE110 alone (pIGA-15) gave 20-fold to 30-fold activation of nearly all homologous and heterologous targets tested (e.g., TK-CAT, 38K-CAT,

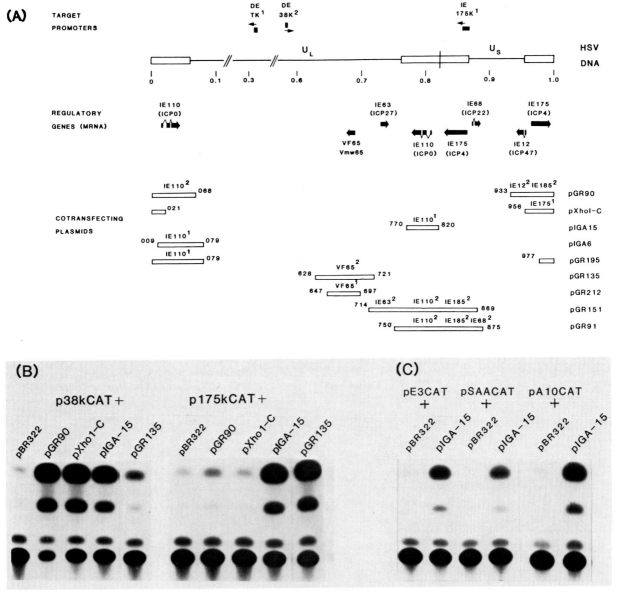

**Figure 2** Map location and target specificity of *trans*-acting regulatory genes from HSV. (*A*) Derivation of the plasmid insert sequences containing intact HSV effector genes that were used for transient cotransfection assays. Open bars and map coordinates define the boundaries of each plasmid relative to the known map location of immediate-early (IE) and virion factor (VF) genes and mRNA (solid arrows). The intact regulatory genes present are named above each plasmid, and their origins from either HSV-1 or HSV-2 are denoted by superscripts. (*B*) Examples of the results from CAT assay experiments showing *trans*-activation of either HSV IE or DE targets by cotranfected plasmid DNA containing single intact genes for the IE175 (pXhoI-C), IE110 (pIGA-15), or VF65 (pGR135) proteins or a mixture of the IE175 and IE110 genes (pGR90). (*C*) Examples of CAT assay experiments demonstrating nonspecific activation of heterologous target constructs by the HSV-1 IE110 gene product (pIGA-15).

IE175-CAT, E3-CAT, SAA-CAT, IFN-CAT, and A10-CAT). The exceptions were SV2-CAT, which was never stimulated greater than 2-fold (but was not inhibited either), and pCATB′, a plasmid containing the CAT-coding region only, without any eukaryotic promoter sequences. Thus IE110 showed the properties of a promiscuous *trans*-activator, whereas IE175 displayed considerable preference for its natural targets, the HSV DE promoters. Addition of an equimolar amount of the IE175 plasmid together with the IE110 effector plasmid

prevented activation of IE175-CAT but increased expression from TK-CAT, thus reproducing the specificity observed originally with the combination of the two IE regulatory genes in pGR90.

## Target specificity for *trans*-activation by the pre-IE virion factor

An HSV-1 DNA fragment containing the gene encoding a function that *trans*-activates a cotransfected HSV IE-TK construct severalfold was identified by Campbell et al.

(1984). This gene produces a 65K late-class phosphorylated protein (Vmw65) that presumably represents the pre-IE virion factor described first by Post et al. (1981) and characterized by Batterson and Roizman (1983). We have examined the target specificity of both the HSV-1 VF65 gene (in plasmid pGR212) and an equivalent function mapping also at 0.67–0.69 in HSV-2 DNA (pGR135). These gene products, despite being driven by a viral late promoter, were synthesized in sufficient abundance in our VERO cell DNA-transfection system to give an average 40-fold stimulation of expression from the cotransfected IE175-CAT target plasmid (Fig. 2B). In contrast, parallel assays with the HSV DE-CAT hybrid genes and heterologous viral and cellular CAT targets failed to give responses greater than 2-fold to 4-fold at best, demonstrating considerable target specificity. Again, the pre-IE stimulatory activity could be mapped by restriction enzyme cleavage to lie within the VF65 gene in the effector plasmids. The use of 5'-deletion plasmids across the *cis*-acting target sequences in the IE175 enhancer–regulatory region (Fig. 3C) confirmed that maximal activation required sequences up to −330, including three of the four TAAT-GARAT-like signals, although the construct deleted down to −160 and containing only one TAATGARAT signal still responded at a 20-fold to 30-fold level (Fig. 3A). Removal of all four of the TAATGARATs by deletion to −108 reduced VF65 activation with the highest

amounts of input effector DNA down to at most 3-fold to 4-fold compared with an equal amount of cotransfected pBR322 DNA (Fig. 3A). Significantly, cotransfection with IE110 could still activate IE175(Δ108)-CAT up to 30-fold (Fig. 3B), an equivalent activation to that obtained with the complete IE175 construct with 1900 bp of upstream sequences, thus emphasizing that IE110 activation apparently does not require the upstream enhancer–regulatory signals that respond to VF65 and further demonstrating the differences in specificity and mechanisms of action of the two HSV effector proteins when acting on the same target gene.

## Negative autoregulatory properties of the IE175 protein

Cotransfection with the plasmid pGR90, which contains both the intact IE175 and IE110 genes, gave no stimulation of IE175-CAT targets despite the fact that cotransfection with the IE110 gene alone could stimulate these constructs 30-fold. To ask whether this effect might be explained by negative regulatory properties of the IE175 protein rather than by complex formation between the two proteins, we examined the effects of cotransfecting increasing amounts of the intact IE175 gene on the basal expression from IE175-CAT. As shown in Figure 4A, this protocol resulted in steadily decreased expression, reaching as much as 30-fold inhibition from the basal level at a 1:1 ratio. This effect could not be

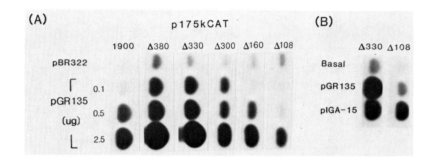

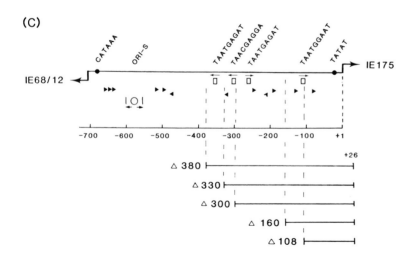

**Figure 3** *Cis*-acting requirements for VF65 and IE110 *trans*-activation within the IE175 promoter–regulatory region. (*A*) Stimulation of complete IE175-CAT and its 5'-deletion variants with different amounts of input VF65 effector DNA (pGR135) in a transient expression assay after cotransfection into VERO cells. Upper samples in each lane show 3-acetyl-chloramphenicol yields after control cotransfections with 2.5 μg of pBR322 DNA. (*B*) Stimulation of IE175 (Δ108)-CAT with cotransfected IE110 DNA (pIGA-15) but not with VF65 DNA. (*C*) Diagram showing locations of the 5'-deletion end-points relative to the TAATGARAT boxes. (□)Sp1-binding sites (◄), and *ori*$_S$ region in the complex, 700-bp, divergent IE68/12-*ori*$_S$-IE175 5'-upstream region.

(A)

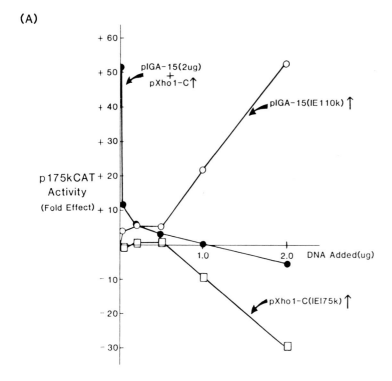

(B)

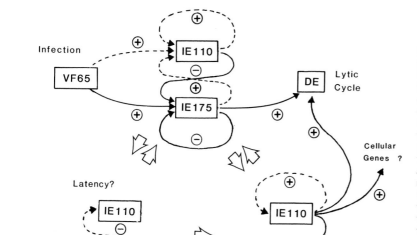

**Figure 4** Positive and negative regulation of the IE175-promoter region by the IE110 and IE175 gene products. (*A*) Effects of the addition of increasing amounts of IE175 (pXhoI-C) or IE110 (pIGA-15) on basal IE175-CAT expression or of IE175 on IE110-stimulated IE175-CAT expression. The diagram shows the fold activation or inhibition of CAT yields produced with different input ratios of cotransfecting DNA (given in micrograms of DNA added per 2 µg of target pIE175-CAT plasmid DNA). (*B*) Model summarizing potential homeostatic transcriptional control interactions between IE175 and IE110 and perturbations of this pattern at different stages of infection or latency: (1) lytic-cycle progression in the presence of input VF65 gene product; (2) predominantly negative interaction in the absence of VF65 during latency in neurons; and (3) possible reactivation of lytic cycle by inactivation of the IE175 negative-regulation mechanism.

accounted for by promoter competition for possibly limiting factors in the cell because the addition of similar amounts of either the IE175 promoter region alone or of a functional IE175-IFN hybrid construct instead of the intact IE175 effector gene gave no more than 2-fold inhibition (not shown). In parallel experiments, the addition of similarly increasing amounts of the IE110 gene plasmid to the IE175-CAT target DNA yielded increased

expression, reaching a maximum of more than 50-fold stimulation at a 1:1 ratio. Therefore, the difference in expression from a standard amount of transfected IE175-CAT target varied over a 1000-fold range, depending on whether the cotransfecting plasmid produced the IE175 or IE110 gene product.

In another experiment, also shown in Figure 4A, we added steadily increasing amounts of the intact IE175 gene (plasmid pXhoI-C) to the 1:1 cotransfection mixture of IE110 and IE175-CAT that gave maximally stimulated expression. Surprisingly, just a trace of the IE175 plasmid (molar ratio, 0.05:1:1) brought IE175-CAT expression back down to near basal levels, and increasing amounts maintained expression at a apparent equilibrium level close to the original basal CAT levels. A similar experiment (not shown), in which low or equimolar amounts of IE175 were added to a maximally stimulated 1:1 cotransfection mix of VF65 and IE175-CAT, gave the same results. Thus, the specific VF65 activation of IE175-CAT was also counteracted by the dominant negative effects of relatively small amounts of IE175. Furthermore, these negative effects of IE175 addition were obtained with VF65-stimulated expression from all of the 5'-deletion mutants down to position −160 and even with IE110-stimulated expression of the IE175(Δ108)-CAT construct. Finally, we found that basal expression from each construct in the IE175-CAT deletion series down to IE175(Δ160)-CAT could be enhanced 10-fold to 15-fold by insertion of the SV40 2× 75-bp repeats in a 5' *cis* position and that these enhancer constructs could also be inhibited up to 20-fold by cotransfection with equimolar amounts of the intact IE175 plasmid. Control experiments examining cotransfection with IE175 showed no negative effects on expression from SV2-CAT.

### Establishment of permanent cell lines containing virus-inducible hybrid CAT target genes

Some of the earliest and most convincing evidence for positive regulation of HSV DE gene expression by IE gene products came from superinfection of Ltk+ cell lines expressing a resident integrated HSV TK gene.

Furthermore, the existence of the pre-IE virion factor was first recognized only after permanent cell lines containing IE-TK hybrids were selected and the resident TK construct was shown to be induced by superinfecting HSV even in the absence of de novo IE protein synthesis (Post et al. 1981). However, the stimulation of the intact TK gene in such cell lines was rarely more than 5-fold to 10-fold, and the studies were complicated by the fact that the integrated target gene had already been selected in HAT medium as one giving relatively high basal TK expression. To systematically compare expression of IE and DE constructs under parallel conditions in permanent cell lines, we carried out several series of studies with stable DNA-transfected cell lines to examine the efficiency and specificity of HSV IE and DE activation mechanisms when the target HSV promoter constructs (containing either IFN- or CAT-coding sequences) were resident in the cell genome in an integrated and presumably chromatin-associated form. In the most extensive studies, we established a relatively large number of cloned G418-resistant (G418^R) cell lines that received pSV2neo plus coselected TK-CAT, IE175-CAT, IE94-CAT, or SV2-CAT constructs and showed either basal or inducible expression of CAT. Essentially identical results were obtained in both VERO cell lines (Fig. 5) and Ltk⁻ cell lines (not shown). The overall findings were that only 5–10% of G418^R cell clones receiving coselected TK-CAT retained inducible CAT enzyme activity and that none yielded significant basal expression. In contrast, between 50% and 80% of all G418^R clones tested from cultures receiving either IE175-CAT, intact and deleted IE94-CAT, or SV2-CAT gave medium to high basal levels of CAT expression. The IE94-CAT cell lines consistently gave the highest

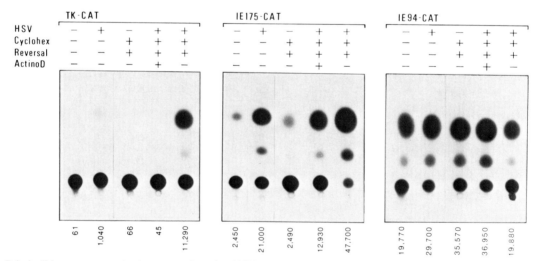

**Figure 5** Inducible versus constitutive expression after HSV superinfection of permanent cell lines containing hybrid CAT genes. Comparison of typical G418^R VERO cell lines containing TK-CAT, IE175-CAT, and IE94-CAT. Parallel cultures were harvested after the following manipulations: (Basal) no treatment; (HSV) 8 hr after infection with HSV-1(MP) virus (w.t.) at 36°C; (Cyclohex plus Reversal) treatment with cycloheximide for 4 hr, followed by incubation in fresh medium for 4 hr; (HSV plus Cyclohex plus Reversal plus ActinoD) infection in the presence of cycloheximide for 4 hr, followed by incubation in fresh medium containing actinomycin D for 4 hr; (HSV plus Cyclohex plus Reversal) infection in the presence of cycloheximide for 4 hr, followed by incubation in fresh medium without actinomycin D for 16 hr.

basal levels initially whereas the IE175-CAT and SV2-CAT lines exhibited a relatively broad range of basal expression. Surprisingly, after subsequent passaging of the cultures, many of the IE94-CAT and SV2-CAT cell lines (but not the IE175-CAT lines) eventually decreased basal expression to either barely detectable or undetectable levels. Evidence that these cells were "turned off" or "repressed" rather than that they had lost the hybrid CAT genes was obtained by treatment with cycloheximide alone, followed by reversal in the absence of actinomycin D, which gave between 5-fold and 10-fold activation in both cell types. In contrast, the IE175-CAT lines were never observed to be inducible by cycloheximide reversal alone.

The yields of CAT enzyme activity were stimulated at least 20-fold to 50-fold in the TK-CAT cell lines at 8 hr after HSV superinfection, and the effect was totally abolished by infection in the presence of cycloheximide for 4 hr, followed by a 4-hr reversal in the presence of actinomycin D. In contrast, the IE175-CAT cell lines were all stimulated at least 10-fold to 20-fold (compared with the steady-state basal levels of CAT accumulated in the cells prior to infection), but the stimulation was unaffected by cycloheximide reversal plus actinomycin D treatment. In both the TK-CAT and IE175-CAT lines, incubation with cycloheximide immediately after infection followed by reversal in the absence of actinomycin D resulted in even greater enhancement in the levels of CAT activity compared with HSV infection in the absence of any inhibitor manipulations. Significantly, HSV superinfection of IE94-CAT or SV2-CAT cell lines, whether of the "high" or "low" basal expression type, failed to yield greater than 2-fold induction in any experiments. The most surprising contrast to the transient assay systems came after superinfection with CMV (Colburn) virus. The same treatment that induced transient TK-CAT and IE175-CAT expression 100-fold failed to induce CAT expression from either construct in permanent VERO cell lines to any detectable degree. Similar results were obtained even with the IE94-CAT lines; neither the "high" nor "low" basal expression type responded to CMV infection.

## Contrast between hybrid target genes producing stable or unstable mRNA

In earlier DNA-transfection studies, we showed that an Ltk[+] cell line that had received a coselected hybrid TK-IFN gene was inducible by HSV superinfection (Reyes et al. 1982). Similar Ltk[+] cell lines receiving IE175-IFN and IE94-IFN genes have also been established. None of these cell cultures gave high-level basal expression of IFN, as assayed by released biological activity or steady-state IFN-specific mRNA levels, although one high-copy-number IE94-IFN line yielded titers of 1000 units per 16-hr collection. After HSV superinfection, a number of TK-IFN lines were stimulated up to 500-fold from 120 units per 16 hr to 60,000 units/16 hr with 16-fold increases in hybrid poly(A)-selected mRNA levels detectable on northern blots. Similarly, a number of IE175-IFN lines could be

activated from 240 units per 16 hr up to 15,000–30,000 units per 16 hr with total IFN-specific RNA increases of 100-fold or greater (Mosca et al. 1985). Poly(IC) plus cycloheximide-reversal treatment did not induce TK-IFN or IE175-IFN expression, although Ltk[+] cell lines that received transfected, intact genomic human β-IFN could be induced by this regimen. Unexpectedly, the IE94-IFN lines could also be induced by HSV superinfection, although not by CMV infection. In particular, a high-copy-number IE94-IFN line could be induced from 1000 units per 16 hr to 60,000 units per 16 hr with increases in northern blot detectable or S1-protected, β-IFN-specific hybrid mRNA of 30-fold to 60-fold (Fig. 6A). Further investigation at the RNA level revealed that β-IFN-specific RNA induction by HSV infection in the presence of cycloheximide was abolished in the TK-IFN cell lines but not in the IE175-IFN lines. These results again indicated correct pre-IE virion factor stimulation of the IE175-IFN gene and the requirement for IE gene expression for stimulation of the DE TK-IFN gene. But how was the IE94-IFN being induced? In the IE94-IFN cells, but not in the IE175-IFN cells, correct, specifically-initiated hybrid IE94-IFN mRNA was also induced by cycloheximide treatment alone (Fig. 6A). Time-course studies suggested that the HSV stimulation of IE94-IFN occurred at DE times in parallel with TK-IFN and considerably later than with IE175-IFN; however, the direct response of IE94-IFN to cycloheximide precluded any definitive statements about the need for viral gene expression. Therefore, as an alternative approach, we turned to a comparison of the pattern of induction in the three cell types, using infection under nonpermissive conditions with either HSV tsB7 (which fails to release virion DNA into the nucleus and fails to synthesize IE mRNA) or HSV tsK and tsB2 (both of which produce all IE mRNAs and proteins but synthesize an inactive, nonnuclear form of the IE175 protein). As expected, TK-IFN was not induced by either mutant virus, whereas IE175-IFN induction did occur under these conditions and even led to overproduction of IFN mRNA, presumably because of a lack of subsequent regulatory "shutoff" events (see Fig. 6B, w.t. panels). In contrast, induction of IE94-IFN, measured at both the total β-IFN-specific RNA and biological activity levels, failed to occur at nonpermissive temperature with tsB7 virus but proceeded normally with the tsK and tsB2 viruses. Therefore, the HSV pre-IE virion factor was not sufficient for IE94-IFN induction, whereas viral IE gene expression was necessary; however, the lack of a requirement for the functional IE175 product implicated a second IE gene product in the process.

A number of features of the comparison between IE94-IFN and IE94-CAT cell lines strongly suggest that mRNA stabilization may be involved in the novel induction of IE94-IFN by HSV infection. First, measurements by S1-hybrid analysis of the amount of remaining mRNA for intact human β-IFN or hybrid IFN-CAT in DNA-transfected mouse Ltk[−] cell lines at different times after poly(IC) plus cycloheximide induction revealed a half-life of approximately 30 min for the β-IFN mRNA

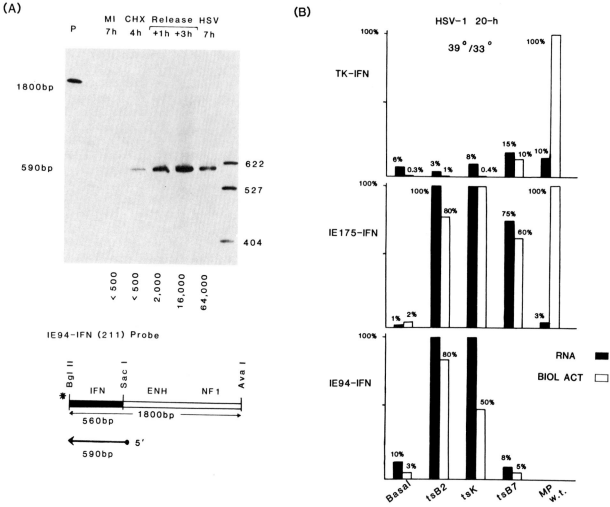

**Figure 6** Induction of coselected CMV IE94-IFN in mouse Ltk+ cells by superinfecting HSV. (*A*) IFN-specific RNA induced by either cycloheximide treatment or HSV-1 superinfection in Ltk+ cells containing the hybrid IE94-IFN (211) gene. The autoradiograph shows the results of urea/polyacrylamide gel electrophoresis of an S1-protected 590-bp hybrid formed with complementary IE94-IFN–specific mRNA present in total cell extracts after treatment with cycloheximide for 4 hr, reversal for 1, 2, or 4 hr, or infection with HSV for 8 hr. Corresponding IFN titers (units/ml) recovered in the supernatants from the same cultures are given below each lane. (*B*) Comparison of the requirements for virion factor and IE gene expression during infection with HSV *ts* mutants in Ltk+ cell lines containing hybrid TK-IFN, IE175-IFN, and IE94-IFN genes. The histogram shows the average percentage ratios of total IFN-specific RNA measured by RNA dot-blot hybridization (solid bars) or biologically active human β-IFN (open bars) produced at 20 hr from cultures at nonpermissive temperature (39.5°C) compared with parallel cultures at permissive temperature (33.5°C) after infection with HSV-1(KOS)*tsB2*, HSV-1(17) *tsK*, HSV(STH2)*tsB7*, or HSV-1(MP) virus (w.t.). Basal percentage levels are expressed relative to maximal induced expression with HSV-1(MP) virus (w.t.) at 36°C. Decreased β-IFN RNA levels in the wild-type samples represent more-rapid shutoff and degradation at 39.5°C compared with 33.5°C. This shutoff did not occur with the *ts* mutants at 39.5°C.

compared with greater than 4–8 hr for CAT mRNA. Second, although both types of constructs gave high-level constitutive expression in microinjected oocytes, coselected IE94-CAT consistently gave high basal expression in both VERO and Ltk⁻ G418ᴿ cells, whereas basal IE94-IFN production in the Ltk+ cells was very weak, suggesting that the IE94-IFN may be transcribed efficiently but gives rise to unstable mRNA. Third, although IE94-IFN could be induced strongly by HSV superinfection, no hint of induction by HSV occurred with any of the IE94-CAT lines. Fourth, preliminary nuclear run-on transcription analysis indicated that no increase in the IE94-IFN transcription rate

occurred after HSV infection although a large increase did occur after cycloheximide treatment. We suggest that an as yet unidentified HSV IE gene function is capable of stabilizing IE94-IFN mRNA in infected cells, whereas direct cycloheximide induction of IE94-IFN (and probably also IE94-CAT and SV2-CAT) involves a genuine transcriptional induction process (perhaps in addition to stabilization).

## Discussion

We have examined various aspects of HSV gene regulation in DNA transfection systems, using hybrid CAT or β-IFN target constructs driven by IE or DE promoter–regulatory regions and either virus superinfection or effector-gene cotransfection protocols. The results have confirmed that these experimental systems permit valid reconstruction assays for efficient and specific *trans*-activation of both IE promoter and DE promoters by HSV-encoded gene products, even when the hybrid targets are resident genes stably integrated into the cellular genome, and that appropriate *cis*-acting transcriptional signals capable of conferring class-specific regulation must lie within the 5′-proximal promoter–regulatory regions or leader sequences included in the target constructs. Our experiments also produced several novel or unexpected observations, including evidence for strong, constitutive expression of the CMV IE94 promoter, the discovery of negative autoregulatory effects of the IE175 protein, the identification of a "promiscuous" *trans*-activating protein, and evidence for an mRNA stabilization factor introduced by HSV infection.

## Specific activation of DE expression by the IE175 gene product

Efficient *trans*-activation of HSV DE hybrid targets occurred in all of the experimental systems that we tested. In contrast to the IE target constructs, this mechanism required either IE gene expression from superinfecting HSV or cotransfection with plasmids containing the intact gene for the IE175 (ICP4) protein. Contrary to the reports and implications from earlier studies with pseudorabies virus and with β-globin that HSV DE activation measured in transient assays may lack specificity (Imperiale et al. 1983; Everett 1984a), both HSV superinfection and IE175 cotransfection in our transient VERO cell assay demonstrated considerable preference for the two HSV DE promoters tested compared with a number of heterologous viral and cellular promoters. Several other investigators have now also directly demonstrated the ability of cotransfected DNA containing the intact gene for the IE175 protein (either alone or in combination with the IE110 gene) to *trans*-activate HSV DE target transcription in a variety of assays (Everett 1984b; Gelman and Silverstein 1985; Quinlan and Knipe 1985). Furthermore, the use of plasmids containing the *tsB2* class of mutations in the IE175 gene abolished activation of DE targets at nonpermissive temperature (DeLuca and Schaffer 1985), and a permanent cell line containing a wild-type IE175 gene has been established that complements the defects of *tsB2* virus for expression of at least some DE genes at 39°C (Persson et al. 1985). Examination of *cis*-acting sequence requirements for HSV superinfection activation of the DE TK promoter have localized the target signals to within the area of the Sp1-factor- and CCAAT-protein-binding sites (Eisenberg et al. 1985; El Kareh et al. 1985; Jones et al. 1985); however, no particular concensus sequences specific to all HSV DE genes have yet been identified, and the multiple Sp1-factor-binding sites within the minimal SV40 early promoter did not provide appropriate response signals for activation by the IE175 gene product.

## Specific activation of IE expression by the HSV pre-IE virion factor

Efficient and specific transcriptional activation of the HSV IE175 hybrid genes by the virion component VF65 (Vmw65) could be accomplished in both transient assay systems and permanent coselected cell lines by superinfection with HSV in the absence of IE protein synthesis or by cotransfection in transient assay systems with the intact VF65 gene of HSV-1 or HSV-2. Preferential *cis*-acting targets occurred within the −330 to −108 region of IE175 corresponding to the location of the TAATGARAT concensus sequences. The levels of IE *trans*-activation obtained in the CAT assay system were 10-fold greater than those observed by Campbell et al. (1984), and this may reflect higher basal expression of the presumably relatively weak late promoter for the VF65 gene in VERO cells.

## Promiscuous *trans*-activation by the IE110K gene product

In contrast to plasmids containing the IE175 or VF65 genes, those containing only the IE110 (ICP0) gene activated (with approximately equal efficiency) all cotransfected target constructs that contained authentic eukaryotic promoters, including HSV IE175-CAT, an IE175-CAT construct lacking all of the TAATGARAT signals and a variety of heterologous viral and cellular hybrid CAT gene constructs. The only exceptions were promoter constructs containing strong enhancer regions such as SV2-CAT and IE94-CAT. Obviously, the presence of IE110 greatly complicates the question of DE specificity both after virus infection and in cotransfection studies when both genes are present. However, in our hands, the isolated IE175 gene was also as active as IE110 on DE targets, and when both genes were present together (e.g., pGR90), the specificity of IE175 dominated over the nonspecificity of IE110. Note that there is no direct evidence as yet that IE110 acts at the transcriptional rather than at a post-transcriptional level, or that it can act on genes in any other state than that presented during transient expression assays. Nonspecific *trans*-activating proteins of this type may turn out to be common within the herpesvirus group. For example, one of the early nuclear antigens of EBV has essentially identical properties in cotransfection assays with those of IE110 (P. Lieberman et al., in prep.).

## Autoregulation at negative *cis*-acting signals in the IE175 promoter region

The strong inhibitory effects that we have described for the IE175 gene product upon expression from IE175-CAT represent the first direct evidence for negative regulation of IE transcription by IE175. There are obviously potential analogies here to SV40-T-antigen expression and autoregulation of early SV40 transcription associated with specific-binding of large T antigen to the origin of DNA replication and adjacent regions. However, neither the minimal SV40 early promoter, with its multiple Sp1-factor- and T-antigen-binding sites, nor the 2 × 72-bp SV40 enhancer contained appropriate signals for response to the negative regulation by IE175. Similarly, although the HSV $ori_S$ signals are an integral part of the divergent IE175 and IE68/12 promoter complex, the negative action of IE175 can be dissociated from any potential role of IE175 in HSV DNA replication because the inhibitory signals lie much closer to the cap site than do either $ori_S$ or the TAATGARAT signals. Except for the "TATAAA-box" signal, the minimal IE175 target region (between −108 and +26) contains a single Sp1-factor-binding site at −70 to −90 and a G + A−rich stretch at −50 to −60, but it has few other notable features. Nevertheless, the IE175 protein does have DNA-binding properties, and a direct sequence-specific interaction within this locus seems to be a reasonable proposition. The interaction between the IE175 protein and its promoter would be expected to have more of an "autoregulatory" role than "shutoff" role to control the levels of IE175 gene expression and as such could potentially have major importance for prevention of the progression to lytic-cycle expression during latency, especially in the absence of any input VF65 protein (Fig. 4B). The dominant, negative autoregulatory action of IE175 during infection could also now provide an explanation for both the observed overproduction of IE transcripts in the presence of cycloheximide and the lack of renewed IE transcription at late times when newly synthesized VF65 presumably accumulates in great abundance.

### Possible interactions between IE175 and IE110

The original discovery of DE *trans*-activation by IE110 in the absence of IE175 was quite unexpected and created something of a dilemma, considering the known inability of temperature-sensitive viruses with mutations mapping in the IE175 gene (e.g., *tsK* and *tsB2*) to permit any DE transcription under nonpermissive conditions. We suggested that in viral infection the IE175 and IE110 proteins may interact directly to form a functional complex, and indeed D. Knipe (pers. comm.) has shown that IE175 and IE110 both remain predominantly in the cytoplasm during infection by *tsB2* at 39°C. At first sight, the inability of plasmids such as pGR90, which contain both genes, to activate IE175-CAT or A10-CAT also supports the idea of an interaction between the two proteins in which the specificity of IE175 for DE targets predominates. However, this result could now also be explained by the IE175 protein negatively regulating expression from the IE110 promoter as well as from the IE175-CAT target. If so, the combined positive action of IE110 and the negative action of IE175 on both promoters could act as a homeostatic mechanism to ensure almost constant and perhaps stoichiometric levels of both proteins when the two genes are present together, as suggested by the data in Figure 4A and the model in Figure 4B. Unfortunately, there is little genetic evidence available as yet that gives any clues about the timing or level of

IE110 action nor about its functional role during virus infection. The IE110 protein could potentially serve to initiate IE175 gene transcription in the absence of VF65 during maintenance of (or reactivation from) the latent state, or it could be a cofactor for the IE175 protein during DE activation that is never normally present in the absence of IE175. Conceivably, IE110 may also function to activate transcription from cellular genes or even to permit transcription of all "open" promoter-like regions at late times after viral DNA replication has occurred.

### Constitutive expression versus *trans*-activation of the CMV IE94 promoter

In comparison with the basal levels of CAT expression from the HSV IE175 constructs, the IE94 promoter–regulatory region from simian CMV (Colburn) was 100-fold more efficient at both driving CAT expression in the transient VERO cell assay system and producing biologically active β-IFN in microinjected oocytes. Boshart et al. (1985) also described part of the equivalent upstream promoter–regulatory region of human CMV IE68 as containing the most powerful enhancer yet discovered. Similarly, in permanent, coselected VERO and Ltk⁻ cells, IE94-CAT was consistently expressed at very high levels initially. However, none of our studies, either in transient or long-term expression assays in permissive VERO cells (and even some experiments carried out in diploid human fibroblasts), have provided any evidence for the existence of positive regulation of IE94 by a virion component of superinfecting CMV (Colburn). In contrast, preliminary studies of IE94-CAT expression in transient assays in nonpermissive mouse Ltk⁻ cells have indicated the existence of negative and possibly autoregulatory effects of CMV superinfection. The absence of either TAATGARAT signals or consensus Sp1-binding sites in the IE94 promoter–regulatory region appears to preclude any cross-reaction with the HSV pre-IE virion factor mechanism. Consequently, in contrast to HSV IE175 and despite a recent report with human CMV IE68 (Stinski and Roehr 1985), the presence of strong virus-independent enhancer signals in the IE94 promoter–regulatory region apparently either compensates for, or precludes the need for, a mechanism for virion factor activation in simian CMV. Nevertheless, although giving strong basal expression in all cell types in transient assays, the IE94-CAT gene tended to become repressed in the integrated form after extensive passaging of the G418-coselected cell lines. Furthermore, the intact CMV IE94 gene is not expressed after virus infection of certain nonpermissive cell types, including Ltk⁻ cells (R.L. LaFemina and G.S. Hayward, in prep.), suggesting the existence of complicated regulatory interactions between this large upstream repeat–enhancer region and both viral and cellular factors.

Evidence for some form of CMV (Colburn) *trans*-activating function was obtained with both HSV IE and DE CAT targets, which were all stimulated even better by CMV superinfection than by HSV in transient assays in VERO cells (O'Hare and Hayward 1984). In the case of IE175-CAT, at least, this CMV induction may be nonspecific because the TAATGARAT region and upstream elements beyond −108 are not required. Interestingly, whatever the mechanism of this activity, it appears to be limited to transient assays because neither the IE175-CAT or TK-CAT nor the IE175-IFN or TK-IFN constructs responded to CMV superinfection when present in an integrated form in permanent cell lines. More recent cotransfection studies with plasmids containing the major IE gene regions of simian and human CMV in transient assay systems have revealed both positive, nonspecific *trans*-activating properties on heterologous promoter-CAT targets and also specific negative autoregulation on IE94-CAT and IE68-CAT constructs (P. O'Hare et al., in prep.).

## Acknowledgments

These studies were funded by DHHS research grants CA22130, CA28473, and CA37314 to G.S.H. from the National Cancer Institute. P.O. was supported by fellowship DRG 714 from the Damon Runyon−Walter Winchell Cancer Fund. J.D.M. was supported through a Regional Oncology Center grant CA06973. G.S.H. is a Faculty Research Awardee of the American Cancer Society (FRA 247).

We thank Dolores Ciufo for excellent technical assistance and Judy DiStefano and Pamela Wright for help with the preparation of the manuscript. We also thank Greg Reyes, Kuan-Teh Jeang, Saul Silverstein, George Khoury, Bruce Howard, John Morrow, Paula Pitha, and Nick Jones for generous gifts of plasmids used in this work.

## References

Batterson, W. and B. Roizman. 1983. Characterization of the herpes simplex virion−associated factor responsible for the induction of α genes. *J. Virol.* **46**: 371.

Boshart, M., F. Weber, G. Jahn, K. Dorsch-Hasler, B. Fleckenstein, and W. Schaffner. 1985. A very strong enhancer is located upstream of an immediate early gene of human cytomegalovirus. *Cell* **41**: 521.

Campbell, M.E.M., J.W. Palfreyman, and C.M. Preston. 1984. Identification of herpes simplex virus DNA sequences which encode a *trans*-acting polypeptide responsible for stimulation of immediate early transcription. *J. Mol. Biol.* **180**: 1.

Clements, J.B., J. McLauchlan, and D.J. McGeoch. 1979. Orientation of herpesvirus type 1 immediate-early mRNAs. *Nucleic Acids Res.* **7**: 77.

DeLuca, N.A. and P.A. Schaffer. 1985. Activation of immediate-early, early, and late promoters by temperature-sensitive and wild-type forms of herpes simplex virus type 1 protein ICP4. *Mol. Cell. Biol.* **5**: 1997.

Eisenberg, S.P., D.M. Coen, and S.L. McKnight. 1985. Promoter domains required for expression of plasmid-borne copies of the herpes simplex virus thymidine kinase gene in virus-infected mouse fibroblasts and microinjected frog oocytes. *Mol. Cell. Biol.* **4**: 1940.

El Kareh, A., A.J.M. Murphy, T. Fichter, A. Efstratiadis, and S. Silverstein. 1985. "Transactivation" control signals in the promoter of the herpesvirus thymidine kinase gene. *Proc. Natl. Acad. Sci.* **82**: 1002.

Everett, R.D. 1984a. A detailed analysis of an HSV-1 early promoter: Sequences involved in *trans*-activation by viral immediate-early gene products are not early-gene specific. *Nucleic Acids Res.* **12**: 3037.

———. 1984b. *Trans*-activation of transcription by herpesvirus products: Requirement for two HSV-1 immediate-early polypeptides for maximum activity. *EMBO J.* **3**: 3135.

Gelman, I.H. and S. Silverstein. 1985. Identification of immediate early genes from herpes simplex virus that transactivate the virus thymidine kinase gene. *Proc. Natl. Acad. Sci.* **82**: 5265.

Gorman, C.M., L.F. Moffat, and B. Howard. 1982. Recombinant genes which express chloramphenicol acetyltransferase in mammalian cells. *Mol. Cell. Biol.* **2**: 1044.

Honess, R.W. and B. Roizman. 1975. Regulation of herpesvirus macromolecular synthesis: Sequential transition of polypeptide synthesis requires functional viral polypeptides. *Proc. Natl. Acad. Sci* **72**: 1276.

Imperiale, M.J., L.T. Feldman, and J.R. Nevins. 1983. Activation of gene expression by adenovirus and herpesvirus regulatory genes acting in *trans* and by a *cis*-acting adenovirus enhancer element. *Cell* **35**: 127.

Imperiale, M.J., R.P. Hart, and J.R. Nevins. 1985. An enhancer-like element in the adenovirus E2 promoter contains sequences essential for uninduced and E1A-induced transcription. *Proc. Natl. Acad. Sci.* **82**: 381.

Jeang, K.-T., G. Chin, and G.S Hayward. 1982. Characterization of cytomegalovirus immediate-early genes. I. Nonpermissive rodent cells overproduce the IE94K protein from CMV (Cloburn). *Virology* **121**: 393.

Jeang, K.-T., M.-S. Cho, and G.S Hayward. 1984. Abundant constitutive expression of the immediate-early 94K protein from cytomegalovirus (Colburn) in a DNA-transfected mouse cell line. *Mol. Cell. Biol.* **4**: 2214.

Jones, K.A. and R. Tjian. 1985. Sp1 binds to promoter sequences and activates herpes simplex virus "immediate-early" gene transcription in vitro. *Nature* **317**: 179.

Jones, K.A., K.R. Yamamoto, and R. Tjian. 1985. Two distinct transcription factors bind to the HSV thymidine kinase promoter in vitro. *Cell* **42**: 559.

Laimins, L.A., G. Khoury, C. Gorman, B. Howard, and P. Gruss. 1982. Host-specific activation of transcription by tandem repeats from simian virus 40 and Moloney murine sarcoma virus. *Proc. Natl. Acad. Sci.* **79**: 6453.

Lang, J.C., D.A. Spandidos, and N.M. Wilkie. 1984. Transcriptional regulation of a herpes simplex virus immediate early gene is mediated through an enhancer-type sequence. *EMBO J.* **3**: 389.

Mackem, S. and B. Roizman. 1980. Regulation of herpesvirus macromolecular synthesis: Transcription-initiation sites and domains of α genes. *Proc. Natl. Acad. Sci.* **77**: 7122.

———. 1982. Differentiation between α promoter and regulatory regions of herpes simplex virus 1: The functional domains and sequence of a movable α regulator. *Proc. Natl. Acad. Sci.* **79**: 4917.

McKnight, S.L., R.C. Kingsbury, A. Spence, and M. Smith. 1984. The distal transcription signals of the herpesvirus tk gene share a common hexanucleotide control sequence. *Cell* **37**: 253.

Middleton, M., G.R. Reyes, D.M. Ciufo, A. Buchan, J.C.M. Macnab, and G.S. Hayward. 1982. Expression of cloned herpesvirus genes. I. Detection of nuclear antigens from herpes simplex virus type 2 inverted repeat regions in transfected mouse cells. *J. Virol.* **43**: 1091.

Mosca, J.D., G.R. Reyes, P.M. Pitha, and G.S. Hayward. 1985. Differential activation of hybrid genes containing herpes simplex virus immediate-early or delayed-early promoters after superinfection of stable DNA transfected cell lines. *J. Virol.* **56**: 867.

O'Hare, P. and G.S. Hayward. 1984. Expression of recombinant genes containing herpes simplex virus delayed-early and immediate-early regulatory regions and *trans*-activation by herpesvirus infection. *J. Virol.* **52**: 522.

———. 1985. Evidence for a direct role for both the 175,000- and 110,000-molecular-weight immediate-early proteins of herpes simplex virus in the transactivation of delayed-early promoters. *J. Virol.* **53:** 751.

Persson, R.H., S. Bacchetti, and J.R. Smiley. 1985. Cells that constitutively express the herpes simplex virus immediate-early protein ICP4 allow efficient activation of viral delayed-early genes in *trans. J. Virol.* **54:** 414.

Post, L.E., S. Mackem, and B. Roizman. 1981. Regulation of α genes of herpes simplex virus: Expression of chimeric genes produced by fusion of thymidine kinase with α gene promoters. *Cell* **24:** 555.

Preston, C.M. 1979. Control of herpes simplex virus type 1 mRNA synthesis in cells infected with wild-type virus or the temperature-sensitive mutant *ts*K. *J. Virol.* **29:** 275.

Preston, C.M., M.G. Cordingley, and N.D. Stow. 1984. Analysis of DNA sequences which regulate the transcription of a herpes simplex virus immediate early gene. *J. Virol.* **50:** 708.

Quinlan, M.P. and D.N. Knipe. 1985. Stimulation of expression of a herpes simplex virus DNA-binding protein by two viral functions. *Mol. Cell. Biol.* **5:** 957.

Reyes, G.R., E.R. Gavis, A. Buchan, N.B. Raj, G.S. Hayward, and P.M. Pitha. 1982. Expression of human β-interferon cDNA under the control of thymidine kinase promoter from herpes simplex virus. *Nature* **297:** 598.

Stinski, M.F. and T.J. Roehr. 1985. Activation of the major immediate early gene of human cytomegalovirus by *cis*-acting elements in the promoter-regulatory sequence and by virus-specific *trans*-acting component. *J. Virol.* **55:** 431.

Stinski, M.F., D.R. Thomsen, R.M. Stenberg, and L.C. Goldstein. 1983. Organization and expression of the immediate early genes of human cytomegalovirus. *J. Virol.* **46:** 1.

Watson, R.J. and J.B. Clements. 1980. A herpes simplex virus type 1 function continuously required for early and late virus RNA synthesis. *Nature* **285:** 329.

# Evidence That Polymerase II Transcription Requires Interaction between Proteins Binding to Control Sequences

**M.R. Loeken, K. Khalili, G. Khoury, and J. Brady**
Laboratory of Molecular Virology, National Cancer Institute, National Institutes of Health, Bethesda, Maryland 20205

The results obtained in independent transcriptional regulatory systems is consistent with the concept that proteins which bind to separate control domains interact to achieve high transcriptional efficiency. Using in vivo competition studies, we examined transcriptional control signals required for T-antigen-dependent *trans*-activation of SV40 late promoter in the absence of DNA replication. Recombinant plasmids containing SV40 late transcriptional regulatory sequences from map position (m.p.) 5171 to m.p. 272 quantitatively inhibited late gene expression in COS-1 cells. Deletion of either the T-antigen-binding sites (m.p. 5171–5243) or the 72-bp tandem repeat (m.p. 128–272) from the competitor plasmid or insertion of increasing lengths of DNA between the T-antigen-binding sites and the enhancer sequences results in markedly less efficient binding of the *trans*-acting factor.

SV40 T antigen and adenovirus E1A stimulate gene expression from the adenovirus E2 early promoter. Basal or *trans*-activated E2 promoter activity requires 79 bp of promoter sequences upstream of the cap site. Using in vivo competition analysis, E2 promoter sequences extending from $-79$ to $+37$ or $-70$ to $+37$ efficiently compete for a limiting transcription factor. In contrast, competition fragments containing promoter sequences $-18$ to $+37$ were significantly less efficient. These experiments demonstrate that the Ad E2 promoter recognition sequences required for binding of a rate-limiting transcription factor(s) for either basal, T-antigen, or E1A *trans*-activation lie within sequences from $-70$ to $-18$.

We have also studied the effect of insertion mutants between the SV40 21-bp repeats and the early-early (EE) TATA sequence. Insertion of 4, 42, or 90 bp of DNA reduced EE transcription, as determined by S1 nuclease analysis, by greater than 10-fold. In contrast, the efficiency of late-early (LE) transcription was increased. These studies suggest that the ability of polymerase II to recognize and utilize the EE TATA transcriptional control sequence requires an interaction with the upstream 21-bp repeats and/or the 72-bp repeats.

The sequences that regulate eukaryotic polymerase II transcription can be divided into essentially three groups. The most highly conserved sequence, the Goldberg-Hogness or TATA box, is located approximately 25 nucleotides upstream of the RNA initiation site and appears to function in vitro and in vivo in positioning the site of initiation of RNA synthesis (Ghosh et al. 1981; Tsai et al. 1981). Additional upstream transcriptional regulatory sequences directly influence the efficiency of RNA synthesis. The pentanucleotide sequence CCAAT plays a positive role in expression of the rabbit β-globin and Maloney murine sarcoma virus long terminal repeat (LTR) (Dierks et al. 1983; Graves et al. 1985). Another upstream element is the guanosine-cytosine (G + C)−rich motif. In simian virus 40 (SV40) this sequence (GGGCGG) is represented twice in each of the three 21-bp repeats and is important for both early and late viral RNA synthesis (see Dynan and Tjian 1985). A similar G + C−rich hexanucleotide is located upstream of the herpesvirus TK RNA initiation site and in several other eukaryotic promoters (McKnight 1982; McKnight et al. 1984; Dynan and Tjian 1985). The third set of

transcriptional control elements, commonly referred to as enhancers (Gruss et al. 1981; Moreau et al. 1981; Banerji et al. 1983), augment the efficiency of transcriptional initiation in a position- and orientation-independent fashion. It has been suggested that enhancers may serve as an entry site for RNA polymerase II or one of its associated subunits.

Recent evidence suggests that the eukaryotic promoter elements described above actually correspond to DNA-binding sites for specific transcription factors. Davison et al. (1983) have demonstrated that HeLa cell extracts contain a transcription factor that binds to the conalbumin and adenovirus major late TATA box, in the absence of RNA polymerase II, to form stable RNA initiation complexes. Dynan and Tjian (1985) have demonstrated that Sp1, a transcription factor isolated from uninfected HeLa cells, binds specifically to the SV40 21-bp repeats, making contact with the hexanucleotide sequence GGGCGG. Evidence for multiple species of Sp1, based on the differential stimulation of early and late SV40 promoters, has been obtained by U. Hansen et al. (pers. comm.). Using in vivo competition analysis,

Schöler and Gruss (1984) first demonstrated that a specific cellular transcription factor binds to the SV40 72-bp enhancer repeat. More recently, these functional binding studies have been extended in vitro (Wildeman et al. 1984).

In the studies presented in this paper, we present evidence suggesting that proteins which recognize separate transcriptional domains interact to achieve high transcriptional efficiency. Efficient binding of a limiting *trans*-acting factor for SV40 late gene expression in COS-1 cells requires the presence, in *cis*, of at least two adjacent SV40 regulatory domains. Using in vivo competition analysis, we have determined that limiting transcription factors required for *trans*-activation of the adenovirus E2 early transcription unit by either T antigen or E1A binds to DNA sequences that are located within E2 promoter sequences −70 to −18. Included in the E2 promoter is an imperfect 14-bp inverted repeat (−75 to −60 and −43 to −30) located on either side of a sequence containing a "CAAT"-like box (−55 to −51). Finally, we demonstrate that insertion of DNA sequences between the SV40 21-bp repeats and TATA box decreases SV40 early-early (EE) TATA-dependent transcription by greater than 10-fold. These results suggest that interaction of the TATA sequence–binding factor and the upstream 21-bp repeats or 72-bp enhancer is important for efficient SV40 EE transcription.

## Materials and Methods

### Cell culture and transfection
CV-1 and COS-1 cells were passed into 10-cm culture dishes with Dulbecco's modified Eagle's medium (DME) supplemented with 10% fetal calf serum (FCS). Twenty-four hr later, when cells were 60–80% confluent, they were transfected by the calcium phosphate precipitation technique. In the in vivo competition experiments, which involved cotransfection of varying concentrations of different plasmid DNAs, the total DNA concentration was kept constant (30–35 µg) with carrier plasmid DNA. Cells were incubated with transfection mixtures for 12 hr, washed, and cultured with fresh media until 40–48 hr posttransfection. In the studies of SV40 late gene expression, cytosine arabinoside (25 µg/ml) was added to prevent viral DNA replication.

### CAT assays
Cells were harvested and assayed for chloramphenicol acetyltransferase (CAT) activity as described by Gorman et al. (1982). One-tenth to one-twentieth of the cell extract was assayed per reaction.

### Western blot analysis of SV40 VP-1
Whole-cell protein extracts were prepared and electrophoretic immunoblot analysis was performed as described previously (Brady et al. 1984).

### S1 nuclease analysis of RNA
RNA was extracted and analyzed by the S1 nuclease technique as described previously (Brady and Khoury 1985).

### Plasmids
Adenovirus E2 promoter and deletion-mutant constructs were the generous gifts of Michael Imperiale and Joseph Nevins (Imperiale and Nevins 1984; Imperiale et al. 1985). Plasmid pEC113 contains 288 bp upstream and 37 bp downstream of the E2 mRNA cap site. 5′-deletion mutants of the E2 promoter were reconstructed for the competition assays. Sequences between *Pvu*II and *Bam*HI were deleted, removing all but 150 bp of CAT-coding sequences. Plasmid p3M*dl*CAT was constructed in an identical manner. Plasmids pRSV-T, in which the SV40 T antigen is expressed under the control of the Rous sarcoma virus LTR, and pE1A were provided by Bruce Howard and Joseph Nevins, respectively. Plasmids containing insertions between the SV40 early TATA box and 21-bp repeats were the kind gift of Jeff Innis and Walter Scott (1984). Using standard recombinant techniques, we transferred a *Bgl*I-*Sph*I fragment (m.p. 0–128) from each mutant to pSV2-CAT. Plasmid DNAs used in the competition experiments for SV40 late gene expression have been described previously (Brady et al. 1985).

## Results

### Binding of *trans*-acting factors is dependent upon spacing between T-antigen-binding sites I and II and the 72-bp enhancer element
In previous studies, we have demonstrated that deletion and base-substitution mutants in the SV40 T-antigen-binding sites or 72-bp repeats decrease T-antigen-mediated SV40 late gene expression (Brady et al. 1984, 1985; Keller and Alwine 1984). The role of these regulatory sequences in binding of positive *trans*-acting factors has been examined by in vivo competition analysis (Brady et al. 1985). This technique examines the ability of molecularly cloned DNA fragments to reduce late gene expression from a cotransfected SV40 template, presumably due to competition for limiting *trans*-acting factors that bind to the competitor DNA. A summary of the competition data is presented in Figure 1. Efficient competition for *trans*-acting factors occurs when the region representing the SV40 T-antigen-binding sites and the 72-bp repeats are linked on the same competitor molecule (Fig. 1, A). Deletion of either the T-antigen-binding sites (m.p. 5171–5243) (Fig. 1, B) or the 72-bp tandem repeats (m.p. 128–272) (Fig. 1, E) from the competitor plasmid results in markedly less efficient binding of the *trans*-acting factor. In addition, neither of the control sequences alone was able to compete for the limiting *trans*-acting factor (Fig. 1, D and F). These findings suggested that efficient binding of the *trans*-acting factors(s) requires the simultaneous interaction between one or more proteins and the two adjacent transcriptional domains.

This hypothesis has been tested directly by analyzing the effect of increasing the distance between the two transcriptional domains on competition efficiency. A series of mutants was recently generated by Innis and Scott (1984) by the insertion of DNA sequences at the SV40 *Nco*I site (m.p. 37). Insertion of 4 bp of DNA at the

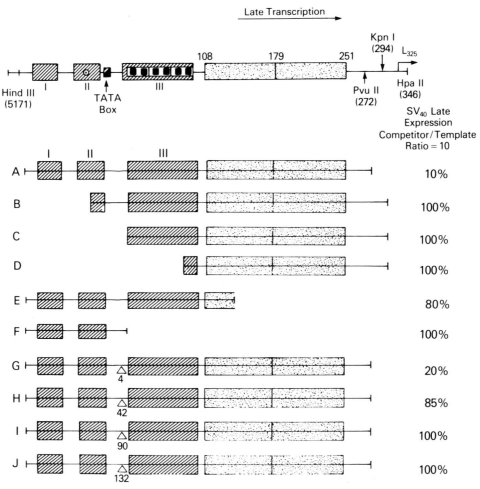

**Figure 1** In vivo competition for *trans*-acting factors in COS-1 cells. The transfection mixtures containing 0.1 μg of supercoiled SV40 template DNA and 1.0 μg of a competitor DNA molecule. Transfected cultures were maintained in DME with 10% FCS and cytosine arabinoside (25 μg/ml). At 40 hr posttransfection, whole-cell protein extracts were prepared and analyzed by immunoblot analysis using anti–SV40 VP-1 antisera (Brady et al. 1984). The SV40 sequences contained in the competition plasmids are as follows: (A) 5171–272, (B) 1–294, (C) 41–294, (D) 95–294, (E) 5171–128, (F) 5171–37, (G) 5171–272 (4-bp insertion at m.p. 37), (H) 5171–272 (42-bp insertion at m.p. 37), (I) 5171–272 (90-bp insertion at m.p. 37), (J) 5171–272 (132-bp insertion at m.p. 37). The level of SV40 late expression was determined by quantitation of bound [$^{125}$I]iodine counts per minute. The level of expression obtained with 0.1 μg of SV40 template DNA, in the absence of competitor DNA, represents 100% (J. Brady et al., pers. comm.).

*Nco*I site had a minimal effect on the competition efficiency, reducing it by approximately 20% (Fig. 1, G). In contrast, pJI-42, pJI-90, and pJI-260 were ineffective competitors, producing little or no decrease in the level of late gene expression (Fig. 1, H, I, and J). These results suggest that efficient binding of the *trans*-acting transcription factors required a precise physical relationship between the DNA sequences and the putative transcriptional factors with which they interact.

*In vivo competition for transcriptional factor(s) that bind to the adenovirus E2 early promoter*
SV40 T antigen, when supplied by cotransfection of plasmid RSV-T, stimulates transcription from the E2 promoter in plasmid pEC113 (Fig. 2) (M. Loeken et al., in prep.). When optimum plasmid concentrations were used, we observed as much as an 80-fold elevation of CAT activity of pEC113 by T antigen. This level of

activation is similar to that observed by cotransfection of a plasmid that codes for the adenovirus E1A gene product. Although factors that mediate T-antigen and E1A *trans*-activation of the E2 promoter may be different, both proteins require transcriptional control sequences 79 bp upstream of the mRNA cap site for full expression of the E2 promoter (Imperiale and Nevins 1984; Imperiale et al. 1985; M. Loeken et al., in prep.).

To determine which promoter regions bind rate-limiting factors responsible for E2 promoter activity, we have used in vivo competition experiments employing various E2 promoter 5′-deletion plasmids as competitors for basal and *trans*-activated pEC113 CAT transcription. In these competition transfections, pEC113 template DNA was transfected alone or with limiting concentrations of pRSV-T or pE1A plus carrier DNA (p3M*dl*CAT$^{(-)}$). Competitor plasmids were added at a twofold or fourfold excess to pEC113. In Table 1, the CAT activity

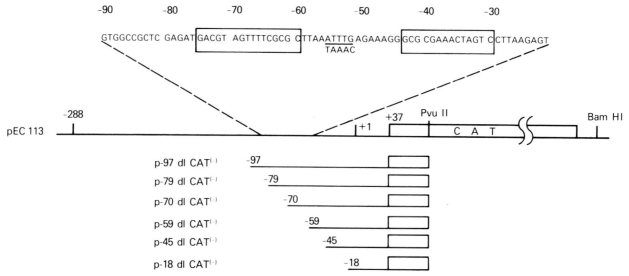

**Figure 2** Adenovirus E2-CAT plasmid constructions. pEC113, containing 288 bp upstream and 37 bp downstream of the E2 mRNA cap site inserted into pCAT3M, is shown with the regulatory sequences expanded. Also shown are 5'-deletion mutants with CAT-coding sequences excised between *Pvu*II and *Bam*HI. These deletion mutants were examined for their ability to inhibit transcription of the CAT gene from pEC113, as described in the text.

obtained in the absence of competitors is expressed as 100%, and the CAT activity in the presence of competitors is expressed relative to the uncompeted levels. There is a 30–60% inhibition of *trans*-activated CAT activity in the presence of twofold excess competitor containing either the entire E2 promoter (−288 to +37) or 5' deletions of the E2 promoter retaining either 97, 79, or 70 bp of the promoter. These results have been confirmed upon examination of steady-state CAT mRNA levels (data not shown). When the competitor levels are increased to a fourfold excess, E1A *trans*-activated CAT activity was further inhibited (Table 1). Similar results were obtained for T-antigen *trans*-acti-

vated E2 expression with a fourfold excess of competitor (data not shown). Basal activity of pEC113 was inhibited up to fivefold in the presence of fourfold excess competitor. When only 18 bp of the E2 promoter were retained on the competing plasmid, competition was significantly less efficient, such that a fourfold excess of the competing promoter fragment was required before a twofold reduction in E1A *trans*-activated CAT activity was observed. We are currently trying to determine the minimal promoter sequences required for competition of basal and *trans*-activated promoter activity with p-59*dl*CAT[(−)] and p-45*dl*CAT[(−)]. Preliminary results suggest that there are probably several protein-binding domains on the E2 promoter.

*Spacing between the SV40 early TATA box and upstream transcriptional control sequences are important for efficient RNA transcription*

We have analyzed transcription of the early region of SV40 to gain insight into some of the mechanisms that regulate EE and late-early (LE) RNA synthesis during the lytic cycle (Ghosh and Lebowitz 1981; Buchman et al. 1984). In these studies, mutants that have insertions between the 21-bp repeats and the early TATA-box promoter were examined using a transient, nonreplicating system. The early SV40 promoter mutants were fused to the bacterial gene coding for CAT. The basic structure of the fusion gene is shown in Figure 3A. As shown in Figure 3B, insertion of 4, 42, and 90 bp between SV40 m.p. 35–37 results in a dramatic decrease in the level of CAT expression driven by the SV40 early promoter. Transfections of pJI-4 resulted in a level of CAT activity 8–10 times less than that observed for control reactions with the wild-type SV40 early promoter (pSV2-CAT).

**Table 1** In Vivo Competition for Adenovirus E2 Transcription Factors

| | CAT activity (% conversion)[a] | | | |
|---|---|---|---|---|
| | T antigen | AdE1A | | basal |
| Competitor | 2× | 2× | 4× | 4× |
| None | 100 | 100 | 100 | 100 |
| pE2*dl*CAT[(−)] | 56 | 66 | 15 | 57 |
| p-97*dl*CAT[(−)] | 37 | 57 | 23 | 36 |
| p-79*dl*CAT[(−)] | 72 | 39 | 26 | 19 |
| p-70*dl*CAT[(−)] | 66 | 48 | 35 | 22 |
| p-18*dl*CAT[(−)] | 121 | 87 | 49 | ND |

[a]CAT activity is expressed relative to the level when pEC113 was transfected with noncompeting carrier DNA (p3M*dl*CAT[(−)]). CAT levels were determined from pEC113 in the absence of viral *trans*-activation (basal) or in the presence of limiting concentrations of pRSV-T (SV40 T antigen) or pE1A. A twofold or fourfold excess of competitor DNA containing 5' deletions of the E2 promoter were used in the competition assays. The actual CAT enzyme levels obtained from pEC113 in the presence of T antigen or E1A was about 16 times greater than from pEC113 alone. (ND) Not determined.

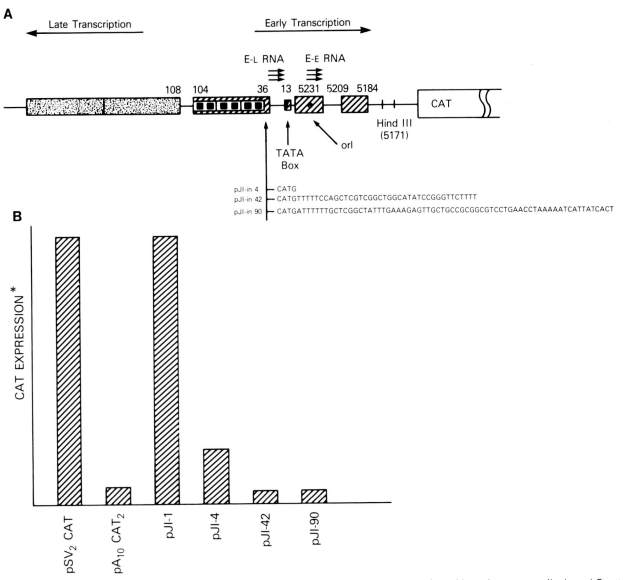

**Figure 3** (*A*) Structure of the SV40-CAT constructs. The *Bgl*I-*Sph*I fragment from a number of insertion mutants (Innis and Scott 1984) was ligated into unique *Sfi*I-*Sph*I restriction sites in pSV2-CAT, yielding pJI-1-CAT, pJI-4-CAT, pJI-42-CAT, and pJI-90-CAT. (*B*) CAT activity in CV-1 cells transfected with pSV2-CAT, pA10-CAT2, and pJI-CAT constructs. The level of CAT activity is expressed relative to the level obtained in extracts from pSV2-CAT–transfected CV-1 cells. The data represent the mean values of at least three separate experiments.

RNA from cells transfected with pJI-1-CAT, pJI-4-CAT, pJI-42-CAT, and pJI-90-CAT was subjected to S1 nuclease analysis to examine the relative steady-state transcript levels and to confirm that transcription initiated at the proper site. As shown in Figure 4A, the major CAT transcript synthesized in pJI-1-CAT and pJI-4-CAT initiated from the SV40 EE promoter. The level of this transcript in the pJI-4-CAT RNA sample was decreased by approximately 10-fold, in good agreement with the CAT expression data. The analysis of pJI-42-CAT and pJI-90-CAT RNA samples revealed a significant level of transcripts whose 5′ ends mapped upstream from the EE RNA initiation site. Primer extension analysis of these transcripts demonstrated that these RNAs were initiated at the SV40 LE start sites

(data not shown). A trace amount of EE RNA was detected in these samples upon longer exposure of the autoradiogram.

## Discussion

The in vivo competition studies contribute important information toward an understanding of the potential interaction of the upstream sequences with putative regulatory molecules. The ability to compete for transcription factors suggests the direct interaction of the factors with the transcriptional control sequences. In the case of SV40 late gene expression, binding to the limiting transcriptional control sequence requires the presence of two transcriptional control sequences in *cis*

A

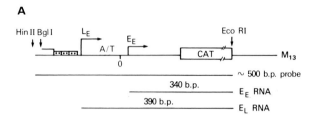

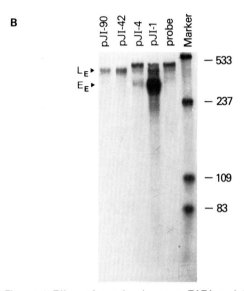

**Figure 4** Effect of spacing between TATA and G + C–rich regions on EE and LE transcription. Transfections were performed with calcium phosphate precipitation, containing 10 μg of each recombinant plasmid and 20 μg of salmon sperm DNA. RNA extraction, hybridization, S1 nuclease digestion, and electrophoresis were performed as described previously (Brady and Khoury 1985). The uniformly labeled DNA probe, synthesized from an M13 recombinant, is shown in *A*.

and at a critical distance from one another (Brady et al. 1985). These results suggest that binding of the limiting transcription factor requires a cooperative interaction between the proteins and the DNA sequences located within these domains.

A minimum of 79 bp of the E2 promoter sequences are essential for optimum basal, T-antigen-mediated, or E1A-mediated transcription. The E2 promoter retaining only 70 bp results in a threefold to fivefold decrease in CAT activity, and deletion to −59 results in a 20-fold decrease in CAT activity, compared with the entire E2 promoter (Imperiale and Nevins 1984; Imperiale et al. 1985; M. Loeken, in prep.). Our in vivo competition studies indicate that a competitor plasmid containing 70 bp of the E2 promoter competes for basal and *trans*-activated activity as effectively as competitor plasmids containing additional upstream sequences. A competitor plasmid containing only 18 bp of the E2 promoter is an inefficient competitor. This suggests that there may be a structural requirement for a minimum of 79 bp of the E2 promoter for transcriptional activity, but binding of rate-limiting transcription factors requires, at a maximum, sequences between −70 and −18. Examination

of these regions of the E2 promoter shows two interesting features (Fig. 2). First, the sequences between −75 and −60 and between −43 and −30 contain an almost perfect inverted repeat. Second, the sequence between −59 and −45 contains a CCAAT-like sequence on its antisense strand. One interpretation of the existing data is that activation of the E2 promoter involves the binding of rate-limiting factors within the inverted repeats. It is possible that another factor may bind between these repeats to sequences between −59 and −45. Binding of all factors may be essential for full activation of the promoter.

It is important to note that both basal and *trans*-activated E2 promoter activity can be inhibited by the same promoter sequences on competing plasmids. This suggests that factors are present in cells in the absence of viral stimulatory factors such as SV40 T antigen or E1A. It is possible that the stimulation by T antigen or E1A is mediated by factors whose affinity for the E2 promoter is increased upon interaction with T antigen or E1A.

The physical structure and spacing of upstream transcriptional elements also plays a major role in the efficiency of SV40 early transcription. Despite the presence of the three major *cis*-acting elements, insertion of even a minimum length of DNA between the 21-bp repeat and TATA region impairs the efficiency of EE transcription and activates the LE promoter. Baty et al. (1984) have recently demonstrated that the G + C–rich repeats I and II, in the divergent 21-bp repeat, play a crucial role in efficient transcription from the EE initiation sites. One interpretation of our results, therefore, is that proteins (Sp1) binding to the divergent 21-bp repeat require interaction with the downstream TATA-box sequences or TATA- box-binding protein for efficient activity. Disruption of this interaction by increasing the length of DNA separating the two promoter elements may cause an inactive complex to form (pJI-4-CAT). Increasing the distance by 42 to 90 bp results in almost total inactivation of the EE promoter and results in an increase in the level of LE promoter activity (pJI-42-CAT and pJI-90-CAT). The switch in promoter activities could be due to either the new configuration of DNA or the chromatin structure that puts the LE promoter in a position more available to transcription factors. Alternatively, those factors that are unable to bind to the EE promoter may associate with the LE promoter.

Additional regulation of the SV40 early mRNA may occur at the translational level. Our analysis of EE and LE RNA by in vitro translation indicates that LE RNA is not an efficient template for translation (K. Khalili et al., in prep.). From these observations, we suggest that both the level and structure of LE RNA contribute in the low level of CAT activity shown in Figure 3.

### References

Banerji, J., L. Olson, and W. Schaffner. 1983. A lymphocyte-specific cellular enhancer is located downstream of the joining region in immunoglobulin heavy chain gene. *Cell* **33**: 729.

Baty, D., H.A. Barrera-Saldana, R.D. Everett, M. Vigneron, and P. Chambon. 1984. Mutational dissection of the 21 bp repeat region of the SV40 early promoter reveals that it contains overlapping elements of the early-early and late-early promoter. *Nucleic Acids Res.* **12:** 915.

Brady, J. and G. Khoury. 1985. *Trans*-activation of the simian virus 40 late transcription unit by T-antigen. *Mol. Cell. Biol.* **5:** 1391.

Brady, J., M.R. Loeken, and G. Khoury. 1985. Interaction between two transcriptional control sequences required for tumor-antigen mediated simian virus 40 late gene expression. *Proc. Natl. Acad. Sci.* **82:** 7299.

Brady, J.N., J. Bolen, M. Radonovich, N. Salzman, and G. Khoury. 1984. Stimulation of simian virus 40 late expression by simian virus 40 tumor antigen. *Proc. Natl. Acad. Sci.* **81:** 2040.

Buchman, A.R., M. Fromm, and P. Berg. 1984. Complex regulation of SV40 early-region transcription from different overlapping promoters. *Mol. Cell. Biol.* **4:** 1900.

Davison, B.L., J.-M. Egly, E. Mulvihill, and P. Chambon. 1983. Formation of stable preinitiation complexes between eukaryotic class B transcription factors and promoter sequences. *Nature* **301:** 680.

Dierks, P., A. van Ooyen, M.D. Cochran, C. Dobkin, J. Reiser, and C. Weissman. 1983. Three regions upstream of the cap site are required for efficient and accurate transcription of the rabbit β-globin gene in mouse 3T6 cells. *Cell* **32:** 695.

Dynan, W.S. and R. Tjian. 1985. Control of eukaryotic messenger RNA synthesis by sequence specific DNA-binding proteins. *Nature* **316:** 774.

Gorman, C.M., L.F. Moffat, and B. Howard. 1982. Recombinant genomes which express chloramphenicol acetyltransferase in mammalian cells. *Mol. Cell. Biol.* **2:** 1044.

Ghosh, P.K. and P. Lebowitz. 1981. Simian virus 40 early mRNAs contain multiple 5′ termini upstream and downstream from a Hogness-Goldberg sequence: A shift in 5′ termini during the lytic cycle is mediated by T-antigen. *J. Virol.* **40:** 224.

Ghosh, P.K., P. Lebowitz, R.J. Frisque, and Y. Gluzman. 1981. Identification of a promoter component involved in positioning the 5′ termini of simian virus 40 early RNAs. *Proc. Natl. Acad. Sci.* **78:** 100.

Graves, B.J., R.N. Eisenman, and S. McKnight. 1985. Delinea-

tion of transcriptional control signals within the Maloney murine sarcoma virus long terminal repeat. *Mol. Cell. Biol.* **5:** 1948.

Gruss, P., R. Dhar, and G. Khoury. 1981. Simian virus 40 tandem repeated sequences as an element of the early promoter. *Proc. Natl. Acad. Sci.* **78:** 943.

Imperiale, M.J. and J. Nevins. 1984. Adenovirus 5 E2 transcription unit: E1A inducible promoter with an essential element that functions independently of position and orientation. *Mol. Cell. Biol.* **4:** 875.

Imperiale, M.J., R.P. Hart, and J. Nevins. 1985. An enhancer-like element in the adenovirus E2 promoter contains sequences essential for uninduced and E1A induced transcription. *Proc. Natl. Acad. Sci.* **82:** 381.

Innis, J. and W. Scott. 1984. DNA replication and chromatin structure of simian virus 40 insertion mutants. *Mol. Cell. Biol.* **4:** 1499.

Keller, J.M. and J. Alwine. 1984. Activation of the SV40 late promoter: Direct effects in the absence of DNA replication. *Cell* **36:** 381.

McKnight, S. 1982. Functional relationships between transcriptional control signals of the thymidine kinase gene of herpes simplex virus. *Cell* **31:** 35.

McKnight, S.L., R. Kingsbury, A. Spence, and M. Smith. 1984. The distal transcription signals of the herpesvirus tk gene share a common hexanucleotide control sequence. *Cell* **37:** 253.

Moreau, P., R. Hen, B. Wasylyk, R. Everett, M.P. Gaub, and P. Chambon. 1981. The SV40 72 base pair repeat has a striking effect on gene expression both in SV40 and other chimeric recombinants. *Nucleic Acids Res.* **9:** 6047.

Schöler, H.R. and P. Gruss. 1984. Specific interaction between enhancer containing molecules and cellular components. *Cell* **36:** 403.

Tsai, S.Y., M.-J. Tsai, and B.W. O'Malley. 1981. Specific 5′ flanking sequences are required for faithful initiation of *in vitro* transcription of the ovalbumin gene. *Proc. Natl. Acad. Sci.* **78:** 879.

Wildeman, A.G., P. Sassone-Corsi, T. Grundstrom, M. Zenke, and P. Chambon. 1984. Stimulation of *in vitro* transcription from the SV40 early promoter by the enhancer involves a specific *trans*-acting factor. *EMBO J.* **3:** 3129.

# Sequences and Factors Involved in E1A-mediated Transcription Activation

## I. Kovesdi, R. Reichel, M. Imperiale,* R. Hart,[†] and J.R. Nevins

The Rockefeller University, New York, New York 10021

The adenovirus E1A gene product activates transcription of five viral transcription units expressed during the early phase of infection. The mechanism of this activation is of general interest because the E1A product also activates several cellular promoters. To elucidate the mechanism of E1A action, we have defined DNA sequences ncecessary for stimulation. In addition, we have begun to identify protein interactions at these sequences, both in vivo as well as in vitro. The data suggest that the viral promoters utilize a cellular transcription factor(s) and that E1A enhances the interaction of the factor with the promoter sequences.

The control of transcription initiation is the basis for many of the phenotypic differences in eukaryotic cells. Therefore, an understanding of the mechanism by which the frequency of transcription of a given gene is controlled is of great importance. One particularly attractive system for the study of transcriptional control is the adenovirus E1A-regulated gene set (for review, see Nevins 1986). Many of the critical features of a system of gene control are available in this system. First, there is a set of viral genes that respond at the level of transcription to E1A activation (Nevins 1981); thus, there is coordinate control. Second, several cellular, chromosomal genes also respond to E1A (Nevins 1982; Kao and Nevins 1983; Stein and Ziff 1984), thus expanding the set of coordinately controlled genes. Finally, the regulatory gene that mediates the activation, the E1A gene, is identified and available (Berk et al. 1979; Jones and Shenk 1979).

Our aim has been to study in detail one particular adenovirus gene that is subject to E1A control in an attempt to elucidate the mechanism of its action. We have made use of the E2 gene of adenovirus, an early gene that encodes the 72K DNA-binding protein. Our studies have been in two directions. First, we have sought to define DNA sequences in the E2 promoter that were required for transcription activation and, second, we have attempted to identify and isolate proteins that interact with these DNA sequences.

## Experimental Procedures

The methods and procedures for the assay of adenovirus gene expression, the generation and analysis of promoter deletions, and the interaction of protein with specific DNA sequences have been described in previous publications (Nevins 1981; Imperiale et al. 1983, 1985; Imperial and Nevins 1984; Wu 1984).

*Present addresses*: *Department of Microbiology and Immunology, University of Michigan School of Medicine, Ann Arbor, Michigan 48109; [†]Department of Zoology and Physiology, Rutgers University, Newark, New Jersey 07102.

## Results

The focus of the work described here is the definition of the critical components involved in E1A-mediated activation of transcription. By this we mean the DNA sequences that are essential and the protein that recognizes these sequences. Over the past several years we have made use of the adenovirus E2A transcription unit for such studies. As shown in Figure 1, we have generated a plasmid version of the E2A transcription unit, termed pE2 (Imperiale and Nevins 1984). This plasmid contains 285 nucleotides (nt) of 5'-flanking sequence and a functional poly(A) addition site. Upon transfection into HeLa cells, there is low activity, and when an E1A plasmid is cotransfected, there is a 20-fold to 40-fold increase in E2 expression (Imperiale et al. 1983). Thus, this plasmid has served as a useful system for investigating activation of transcription by E1A.

## Essential promoter sequences

We previously used the pE2 plasmid to define essential promoter sequences by constructing a series of 5'-deletion mutants. Assays of these mutants in 293 cells (E1A[+]) demonstrated that 79 nt of 5' sequence were sufficient for full activity (Imperiale and Nevins 1984). Deletion to −70 severely impaired activity. To determine the contribution of these sequences in E1A induction, the various promoter deletions were assayed in HeLa cells for comparison with the results in 293 cells, as well as in HeLa cells in the presence or absence of E1A (Fig. 2). As can be seen in Figure 2A, the activity of the plasmids in HeLa cells (E1A[−]) was nearly identical with that in 293 cells (E1A[+]). Furthermore, when assayed in HeLa cells in the presence of E1A (Fig. 2B), the results were nearly the same. Our basic conclusions from these experiments are as follows: First, all of the essential sequences for activity of this promoter are contained within the 79 nt 5' of the gene; second, there appears to be no distinction between sequences required for uninduced promoter activity (HeLa) and those required for E1A-induced activity (293 or HeLa plus E1A); finally, the −70 deletion, which is significantly reduced in function

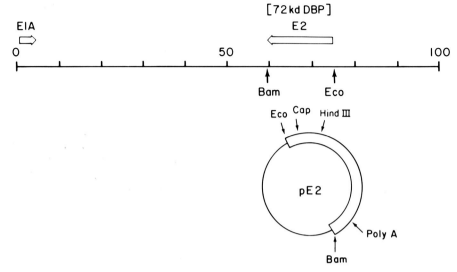

**Figure 1** Map of the adenovirus genome, depicting the location of the E2A transcription unit. Shown below map is the structure of the pE2 plasmid.

but does retain some activity, is inducible by E1A whereas the −59 deletion appears to have lost activity and is not E1A inducible. It would appear that as long as the promoter can be transcribed, even inefficiently as is the case with the −70 deletion, then it is inducible.

These results are very similar to various reports from other laboratories using the same gene (Elkaim et al.

1983). More-extensive analyses, utilizing linker-scanning mutations, have yielded similar conclusions, although with some added detail (Murthy et al. 1985; Zajchowski et al. 1985). In particular, it would appear that there might be two domains to the 5′-flanking sequence. Finally, another study presented similar results but argued that any promoter deletion was induc-

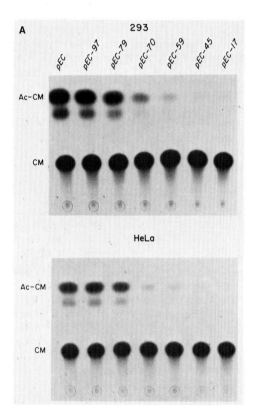

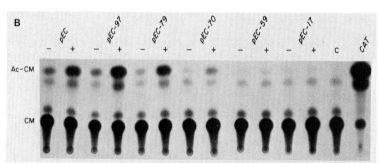

**Figure 2** Functional assays of E2 promoter deletions. (*A*) The various E2 5′ deletions, fused to the CAT gene, were assayed in 293 cells or HeLa cells by transfection. For details, see Imperiale et al. (1985). (*B*) The indicated mutants were assayed in HeLa cells in the presence (+) or absence (−) of the E1A gene. (Reprinted, with permission, from Imperiale et al. 1985.)

ible, even though all 5'-flanking sequences were lacking (Kingston et al. 1984). The basis for the difference is not clear, although in the latter study there was imposed a selection for expression.

### E2 promoter–protein interaction

The above results suggest a sequence element upstream of the E2 promoter that is the site for interaction of a transcription factor. Furthermore, since the same sequences are required before and after E1A, one might argue that the same factor is involved. To eventually define the precise mechanism of E1A transcriptional control, the factor(s) that recognize these sequences must be identified and isolated. We have taken two approaches toward this goal. The first is to map protein binding at this site in vivo during a viral infection, and the second is to identify such a protein in extracts of cells by in vitro binding assays. The in vivo assays have made use of a technique described by Wu (1984) for mapping protein interactions at *Drosophila* heat-shock promoters. Nuclei are prepared from HeLa cells infected with virus at 8 hr postinfection. The nuclei are digested with a restriction enzyme for which there is a cleavage site in the vicinity of the promoter. In the case of the E2 promoter, there is an *Eco*RI site 285 nt 5' to the transcription-initiation site (see Fig. 1). Along with the restriction enzyme is added *Exo*III, a 3'-specific exonuclease that will digest from the free 3' termini created by *Eco*RI. The presence of a protein bound to a specific site on the DNA should then prevent the *Exo*III from further digestion. If the DNA is then purified and cut with a second restriction enzyme, a specific band will have been produced as a result of the *Exo*III stop. This band can be detected by Southern analysis and indirect end-labeling, and the size of the band will predict the position of the *Exo*III stop.

Such an analysis was performed to investigate protein interaction at the E2 promoter, and an example is depicted in Figure 3. Three possible DNA bands could be produced by such an analysis. If the *Eco*RI failed to cut the chromatin near the E2 promoter, then digestion with the second restriction enzyme (*Sst*I) would yield the *Sst*I fragment spanning this region. If the *Eco*RI cut the chromatin but there was no *Exo*III digestion, then an *Eco/Sst* fragment would be produced. Finally, if an *Exo*III stop were generated, then a band will be produced that is shorter than the *Sst/Eco* fragment. The analysis of Figure 3 indeed depicts the *Sst/Sst* and *Sst/Eco* bands as predicted. In the absence of *Exo*III, these are the only bands that are observed. However, in the presence of *Exo*III there is an additional band, shorter in length than the *Sst/Eco* fragment. The size of this band indicates an end point at −85 relative to the E2 transcription-initiation site.

A similar analysis was carried out in the absence of E1A by infecting cells with the E1A mutant *dl*312. When the *Exo*III assay was performed, no *Exo*III stop was found. Thus, under these circumstances it would appear that there is no specific protein interaction at these upstream promoter sequences. Finally, we have pro-

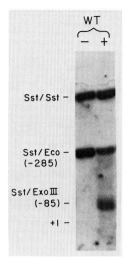

**Figure 3** *Exo*III assay for protein interactions at the E2 promoter. HeLa cells were infected with wild-type (WT) Ad5 and incubated in the presence of cytosine arabinoside for 7 hr. Nuclei were prepared and assayed as described in the text. The (+) and (−) lanes refer to nuclei incubated in the presence or absence of *Exo*III.

duced conditions whereby we observe the same DNA-protein interaction but in the complete absence of E1A. It has previously been shown that there is a slow activation of early viral transcription in the absence of E1A (Gaynor and Berk 1983). By 30–40 hr postinfection, there is a nearly wild-type level of early viral gene expression. An *Exo*III assay performed under these conditions of *dl*312 infection again produced the *Exo*III stop at −85. Thus, even in the complete absence of E1A, we find a protein interaction at the E2 promoter but only under conditions where there is active transcription of the gene. These results would thus imply that the protein is of cellular origin and that the interaction results in a stimulation of transcription.

### Discussion

From the data presented here, we offer the following model for a mechanism of E1A transcription activation. As depicted in Figure 4, we suggest that the upstream sequences of the E2 promoter, which have been deemed critical by mutation studies, are the site for binding of a cellular transcription factor. We further suggest that such a factor is present in uninfected cells and that the role of E1A is to increase the ability of the factor to bind to promoter sequences.

The basis for suggesting such a model is as follows. We have shown that there is very likely an involvement of cellular factors in E1A-mediated induction of transcription. Perhaps the most compelling evidence is that E1A can be effectively replaced by a herpesvirus equivalent, one of the immediate-early genes (Feldman et al. 1982; Imperiale et al. 1983). This appears to argue strongly against a mechanism involving direct sequence recognition by these viral proteins since it would be difficult to envision such specificity between these

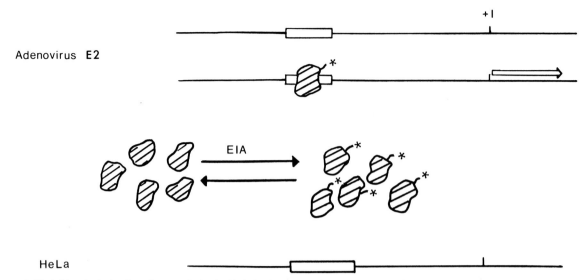

**Figure 4** Model for the induction of transcription by E1A. In this case, it is suggested that E1A catalyzes a modification (*) of a cellular transcription factor, thus increasing its ability to bind to the promoter.

genetically unrelated viruses. Further evidence suggesting cellular factors comes from the observation that viral transcription can take place in the absence of E1A but is dependent on the host cell (Imperiale et al. 1984). There is additional precedent for the use of a cellular factor by a viral promoter since the early SV40 promoter apparently makes use of a factor termed Sp1 (Dynan and Tjian 1983).

The results presented here and elsewhere further argue that the same DNA sequences are required for basal activity as well as induced activity. This suggests that the same factor may be involved in basal transcription and induced transcription. If such a factor were limiting in the cell, then an increase in its concentration could bring about an increase in transcription of the appropriate genes. We suggest that this is the function of E1A; that is, to mediate an increase in the amount or activity of a limiting cellular transcription factor and thus achieve an "induction."

One additional aspect of such a model requires comment. We and others have previously suggested a model wherein the E1A protein counteracted a negative effect imposed by the cell (Katze et al. 1981; Nevins 1981). Although this might imply an involvement of a repressor, there is in fact no evidence for negative action at the level of the gene; i.e., mutation studies all point toward positive control of the promoters. This result does not eliminate negative control, however. If E1A were to regulate the activity of a cellular transcription factor (activation), then a negative component could simply be one that counteracted this effect, i.e., reversed an E1A-induced modification. There are, in fact, examples of such regulated systems, a kinase and phosphatase that act on the same substrate being one such example. The gene-regulatory role of such a system in the cell is of obvious importance, with one possibility being the control of certain aspects of cell-cycle transcription (Kao et al. 1985).

To prove such a model will require the isolation of such a factor and the demonstration that indeed it is a cellular factor. The question then becomes whether the mass of the factor changes as a result of E1A or if the factor is modified in some way. We have recently made progress in this direction by detecting such a protein in extracts of infected cells. Using a gel-band-shift assay (Fried and Crothers 1981; Garner and Revzin 1981), we have detected a protein that interacts with the E2 upstream sequences. The specificity of the interaction has been demonstrated by competition assays with upstream deletion mutants and by footprinting. Furthermore, using a band-shift assay, we have been able to purify the protein through two chromatographic steps. The hope is that with sufficient purification, an antibody to the protein can be produced, thus providing a probe for the factor. Using such a probe, the questions posed above could then be addressed.

## Acknowledgments

M.I. and R.H. were supported by fellowships from the National Institutes of Health; I.K. was supported by a fellowship from the National Science and Engineering Research Council (Canada); R.R. was supported by a fellowship from the Deutsche Forschungsgemeinschaft; J.R.N. was supported by a Research Career Development Award. This work was supported by a grant from the NIH (GM 26765).

## References

Berk, A.J., F. Lee, T. Harrision, J. Williams, and P.A. Sharp. 1979. Pre-early adenovirus 5 gene product regulates synthesis of early viral messenger RNAs. *Cell* **17**: 935.
Dynan, W.S. and R. Tjian. 1983. The promoter-specific transcription factor Sp1 binds to upstream sequences in the SV40 early promoter. *Cell* **35**: 79.

Elkaim, R., C. Goding, and C. Kedinger. 1983. The adenovirus 2 EII2 early promoter: Sequences required for efficient in vitro and in vivo transcription. *Nucleic Acids Res.* **11:** 7105.

Feldman, L.T., M.J. Imperiale, and J.R. Nevins. 1982. Activation of early adenovirus transcription by the herpesvirus immediate early gene: Evidence for a common cellular control factor. *Proc. Natl. Acad. Sci.* **79:** 4952.

Fried, M. and D. Crothers. 1981. Equilibrium and kinetics of *lac* repressor-operator interactions by polyacrylamide gel electrophoresis. *Nucleic Acids Res.* **9:** 6505.

Garner, M. and A. Revzin. 1981. A gel electrophoresis method for quantifying the binding of proteins to specific DNA regions: Application to components of the *E. coli* lactose operon regulatory system. *Nucleic Acids Res.* **9:** 3047.

Gaynor, R.B. and A.J. Berk. 1983. *Cis*-acting induction of adenovirus transcription. *Cell* **33:** 683.

Imperiale, M.J. and J.R. Nevins. 1984. Adenovirus 5 E2 transcription units: An E1A-inducible promoter with an essential element that functions independently of position or orientation. *Mol. Cell. Biol.* **4:** 875.

Imperiale, M.J., L.T. Feldman, and J.R. Nevins. 1983. Activation of gene expression by adenovirus and herpesvirus regulatory genes acting in *trans* and by a *cis*-acting adenovirus enhancer element. *Cell* **35:** 127.

Imperiale, M.J., R.P. Hart, and J.R. Nevins. 1985. An enhancer-like element in the adenovirus E2 promoter contains sequences essential for uninduced and E1A-induced transcription. *Proc. Natl. Acad. Sci.* **82:** 381.

Imperiale, M.J., H.-T. Kao, L.T. Feldman, J.R. Nevins, and S. Strickland. 1984. Common control of the heat shock gene and early adenovirus genes: Evidence for a cellular E1A-like activity. *Mol. Cell. Biol.* **4:** 867.

Jones, N. and T. Shenk. 1979. An adenovirus type 5 early gene function regulates expression of other early viral genes. *Proc. Natl. Acad. Sci.* **76:** 3665.

Kao, H.-T. and J.R. Nevins. 1983. Transcriptional activation and subsequent control of the heat shock gene during adenovirus infection. *Mol. Cell. Biol.* **3:** 2058.

Kao, H.-T., O. Capasso, N. Heintz, and J.R. Nevins. 1985. Cell cycle control of the human HSP70 gene: Implications for the role of a cellular E1A-like activity. *Mol. Cell. Biol.* **5:** 628.

Katze, M.G., H. Persson, and L. Philipson. 1981. Control of adenovirus early gene expression: Post-transcriptional control mediated by both viral and cellular gene products. *Mol. Cell. Biol.* **1:** 807.

Kingston, R.E., R.J. Kaufman, and P.A. Sharp. 1984. Regulation of transcription of the adenovirus EII promoter by E1a gene products: Absence of sequence specificity. *Mol. Cell. Biol.* **4:** 1970.

Murthy, S.C.S., G.P. Bhat, and B. Thimmappaya. 1985. Adenovirus EIIA early promoter: Transcriptional control elements and induction by the viral pre-early E1A gene, which appears to be sequence independent. *Proc. Natl. Acad. Sci.* **82:** 2230.

Nevins, J.R. 1981. Mechanism of activation of early viral transcription by the E1A gene product. *Cell* **26:** 213.

———. 1982. Induction of the synthesis of a 70,000 dalton mammalian heat shock protein by the adenovirus E1A gene product. *Cell* **29:** 913.

———. 1986. Control of cellular and viral transcription during adenovirus infection. *CRC Crit. Rev. Biochem.* **19:** 307.

Stein, R. and E.B. Ziff. 1984. HeLa cell β-tubulin gene transcription is stimulated by adenovirus-5 in parallel with viral early genes by E1a dependent mechanism. *Mol. Cell. Biol.* **4:** 2792.

Wu, C. 1984. Two protein binding sites in chromatin implicated in the activation of heat shock gene. *Nature* **308:** 229.

Zajchowski, D.A., H. Boeuf, and C. Kedinger. 1985. The adenovirus-2 early EIIa transcription unit possesses two overlapping promoters with different sequence requirements for E1A-dependent stimulation. *EMBO J.* **4:** 1293.

# Intricate Sequence Elements Involved in Constitutive and *Trans*-activated Expression from the Adenovirus Early E2A Promoter Region

## H. Boeuf, D. Zajchowski, P. Jalinot, C. Goding, B. Devaux, C. Hauss, and C. Kédinger

Laboratoire de Génétique Moléculaire des Eucaryotes du CNRS, Unité 184 de Biologie Moléculaire et de Génie Génétique de l'INSERM, Université Louis Pasteur, 67085 Strasbourg Cédex, France

To investigate the involvement of viral promoter sequences in the constitutive expression and E1A-mediated stimulation of the adenovirus early E2A transcription unit (EE2A), a series of external (pEII)- and internal-deletion ($\Delta$) and linker-scanning (LS) EE2A promoter mutants were tested in the presence or absence of the E1A gene products in a transient expression system. Transcription efficiency from both the major (+1 or $E_1$E2A) and the minor (−26 or $E_2$E2A) start sites was examined. The results indicate the existence of two overlapping promoters with distinct TATA-like elements and shared upstream components. $E_2$E2A transcriptional stimulation by the E1A products is essentially mediated by the corresponding TATA-like sequence, whereas at least two independent elements are implicated in the case of $E_1$E2A induction, one between −70 and −42 and another upstream of −85.

The entire domain between −111 and −27 behaves as an E1A-inducible enhancer. Within this domain, an essential upstream promoter component (between −90 and −70) is flanked by two sequence elements that exert a negative effect on the enhancer activity in the absence of E1A. The data suggest that the E1A products stimulate the EE2A enhancer by relieving the negative control exerted by these sequences and raise the possibility that at least part of the E1A-mediated induction occurs through this mechanism.

Finally, EE2A transcription was also stimulated by cotransfection with an E4-containing plasmid. However, in contrast to the E1A-responsiveness, the E4-mediated stimulation requires a unique, discrete sequence element in the EE2A promoter, between −48 and −19.

Gene activity in eukaryotes is controlled largely through regulation of transcription. A number of *cis*-acting elements are involved in the regulation of initiation of transcription by RNA polymerase class B (Breathnach and Chambon 1981). Studies of a number of genes have established that faithful and efficient transcription initiation requires both the highly conserved TATA-box sequence, located about 30 nucleotides (nt) upstream from the cap site, and less conserved elements located at variable distances further upstream (Benoist and Chambon 1981; Hen et al. 1982; McKnight and Kingsbury 1982; Dierks et al. 1983; Baty et al. 1984). Another element, known as an enhancer, first discovered in viral promoters but also present in several cellular genes, activates transcription from homologous or heterologous promoters in a manner relatively independent of distance, orientation, and position with respect to these promoters (for references, see Khoury and Gruss 1983; Yaniv 1984; Serfling et al. 1985). An additional type of control element, not necessarily distinct from those required for constitutive transcription, modulates the rate of specific transcription in response to exogenous stimuli such as heat shock (Bienz 1985; Pelham 1985), heavy metal ions (Karin et al. 1984; Searle et al. 1985),

hormones (for references, see Groner et al. 1984), and viral infection (Nevins 1982; Ohno and Taniguchi 1983; Fujita et al. 1985). In some of these cases the responsive sequence components have been reported to exhibit properties of "inducible enhancers" (Goodbourn et al. 1985; Serfling et al. 1985). Accumulating data clearly indicate that most, if not all, classes of promoter elements correspond to DNA-binding sites for specific transcription factors or regulatory components that, together with the RNA polymerase, participate in the formation of active transcription complexes (Davison et al. 1983; Dynan and Tjian 1983; Gidoni et al. 1984; Miyamoto et al. 1984; Parker and Topol 1984; Wildeman et al. 1984; Wu 1984; Sassone-Corsi et al. 1985). In addition to these positive regulatory promoter elements, negatively acting sequences may provide an additional type of gene control. Such elements, which reduce the constitutive level of gene expression, have been described in the upstream promoter region of several yeast genes (for references, see Guarente 1984), the mouse β-major-globin gene (Gilmour et al. 1984), the human β-interferon gene (Zinn et al. 1983), the hormonally regulated chicken ovalbumin gene (M. Gaub et al., pers. comm.), and in the vicinity of the mouse heavy-chain immunoglobulin enhancer (J. Dougherty et al., pers. comm.).

The adenovirus type-2 (Ad2) early E2A (EE2A [or EIIaE]) transcription unit, which encodes a 72,000-dalton DNA-binding protein, provides an interesting example of a regulated promoter. Early in viral infection, its transcription by RNA polymerase B proceeds both from major (+1 or E$_1$E2A [or EIIaE1]) and minor (−26 or E$_2$E2A [or EIIaE2]) start sites, neither of which is preceded by a concensus TATA sequence (Mathis et al. 1981). During lytic infection, efficient transcription from these sites is dependent on the prior expression of the immediate-early E1A (or E1a) transcription unit, which is also required for active expression of the other early regions, E1B (or E1b), E3 (or EIII), and E4 (or EIV) (Jones and Shenk 1979; Flint 1982). As suggested by the sequential activation of these early transcription units (E4 and E3, followed by E1B and EE2A [Nevins et al. 1979]), factors other than those encoded by region E1A are likely to be involved in the temporal control of viral early gene expression. The induction of these early units has been reproduced in transient expression systems with cloned viral genes, confirming the involvement of *trans*-acting E1A products in early gene expression (Elkaim et al. 1983; Imperiale et al. 1983; Jones et al. 1983; Rossini 1983; Weeks and Jones 1983; Leff et al. 1984). In addition, experiments in which EE2A- and E4-containing plasmids were introduced together into HeLa cells revealed that EE2A transcription was stimulated by E4 cotransfection, implying that E4 gene products also act as positive regulators (Goding et al. 1985).

Hence, a study of the Ad2 EE2A transcription unit not only can identify the sequence elements that constitute an atypical class-B promoter, including those that functionally substitute for the TATA box, but also can elucidate the mechanism of transcriptional stimulation by the E1A and E4 products. To define the sequences required for both accurate transcription and E1A- and E4-mediated stimulation, we have constructed an extensive series of EE2A promoter mutants with either 5′-external deletions, internal deletions, or clustered nucleotide substitutions. Using a transient expression assay in HeLa cells, we show that E$_1$E2A and E$_2$E2A transcription is controlled by two overlapping promoters with both common and independent elements. Cotransfection of the EE2A mutants with a plasmid bearing the E1A transcription unit revealed that, although E$_1$E2A stimulation does not involve a unique E1A-inducible sequence, but instead several specific elements, the E$_2$E2A TATA-like element is essential for the E1A-responsiveness of E$_2$E2A transcription (Zajchowski et al. 1985).

Experiments testing for enhancer-like activity revealed that the EE2A upstream promoter domain behaves like an E1A-inducible enhancer composed of a central element flanked by two negative regulatory sequences. Our findings support the hypothesis that the E1A products mediate the E$_1$E2A transcriptional induction by relieving the negative control exerted by these sequences (Jalinot and Kédinger 1986).

Cotransfection experiments in which the EE2A linker-scanning mutants were tested in the presence of an E4-containing plasmid revealed that an EE2A element located between −48 and −19 is essential for maximal E4-mediated stimulation of E$_1$E2A transcription (Goding et al. 1985).

## Experimental Procedures

### Recombinant plasmids

pEII contains the entire E2A transcription unit from −250 to +6800. This plasmid is identical with the pBX recombinant described in Elkaim et al. (1983), except that an *Xba*I linker sequence has been inserted in the *Sma*I site at position −250. The 5′-deletion series constructed by Elkaim et al. (1983) was modified by inserting an *Xba*I linker sequence at the *Bam*HI site present at each deletion end point. The deleted EE2A promoter fragments extending from the *Xba*I site to the *Hin*dIII site (+719) were recloned between the *Xba*I and *Hin*dIII sites of pEII in place of the wild-type sequence, generating a series of unidirectional deletion mutants (see Fig. 1), which were named according to the position of each mutant's deletion end point.

pMTE contains the *Xba*I(−250)−*Hin*dIII(+719) EE2A sequence from pEII, ligated to the rabbit β-globin coding region, which provides splice and polyadenylation sites. The pMTE 5′ deletion series was constructed by replacing the "wild-type" *Xba*I-*Hin*dIII EE2A sequence by the corresponding deleted fragments of the pEII series.

LS WT is the same as pMTE but without the *Xba*I linker sequence at the *Sma*I site. The linker-scanning (LS) derivatives were constructed by the method first developed by McKnight and Kingsbury (1982), and the resulting LS series (Zajchowski et al. 1985) is shown in Figure 2. Internal deletions (Δ series) were generated by in vitro recombination of appropriate LS mutants.

pEIASV is a pBR322 recombinant containing the whole E1A gene (0–4.5 m.u. of the Ad2 genome) ligated to the polyadenylation sites of SV40 (Elkaim et al. 1983).

pEIA 13S is the same as pEIASV but lacks the smallest E1A intron sequence (Leff et al. 1984).

pEIA$^-$ is the same as pEIASV but lacks the E1A coding region downstream from position +129 (Sassone-Corsi et al. 1983).

pEIV contains an Ad2 DNA fragment (89.9−100 m.u.) spanning the entire E4 transcription unit (Goding et al. 1985).

pG is a pBR322-derived plasmid containing the entire rabbit β-globin gene from −425 to +1700 (Jalinot and Kédinger 1986). The pG derivatives contain various EE2A promoter fragments inserted in a polylinker sequence present in front of the globin sequences (see Fig. 3).

### In vivo expression assay

HeLa cells were transfected by calcium phosphate co-precipitation (Banerji et al. 1981) with 2 μg or 5 μg of a given plasmid, and the final DNA concentration per 10-cm petri dish was adjusted to 15 μg with M13mp8 RF DNA. Cytoplasmic RNA was prepared and specific transcripts were analyzed by the S1 nuclease assay previously described (Mathis et al. 1981). Quantification of specific transcription was achieved by determining the intensity of the specific bands by densitometry of autoradiographs similar to that shown in Figure 1. The values were represented as a percentage of the corresponding "wild-type" recombinant in each experiment. Standard deviations varied from 10% to 25% of each value. E1A- and E4-mediated effects on EE2A transription were determined by the ratio of the specific transcription measured in the presence of these plasmids to that measured in their absence.

## Results

### Constitutive EE2A transcription is controlled by two overlapping promoter regions with independent and common sequence requirements

*External-deletion mapping of the EE2A promoter*
Previous results have shown that deletion of EE2A promoter sequences between −250 and −94 did not

significantly affect the level of transcription from both the major (+1 or $E_1E2A$) and minor (−26 or $E_2E2A$) start sites, whereas deletion to position −63 reduced transcription from both sites (Elkaim et al. 1983). A more refined series of 5′ deletions was generated and inserted in two different types of vectors, one containing the E2A coding region (pEII series), the other with the rabbit β-globin coding region in its place (pMTE series). EE2A transcription efficiency from the resulting mutants was determined after transfection into HeLa cells, as described in Experimental Procedures. A typical S1 nuclease analysis of the cytoplasmic RNA is shown in Figure 1 for a set of pEII derivatives. Quantification of $E_1E2A$ transcripts in this and similar experiments is given in Table 1.

Deletion of sequences between −250 and −169 did not significantly impair constitutive $E_1E2A$ promoter function, whereas deletions extending further downstream to positions −146 and −111 reproducibly reduced specific transcription by about 20% and 50%, respectively. Additional deletion to position −88 restored $E_1E2A$ promoter efficiency to approximately wild-type levels. These results indicate that the 5′ limit of the $E_1E2A$ promoter sequences essential for maximal transcription lies around position −169. Furthermore, they

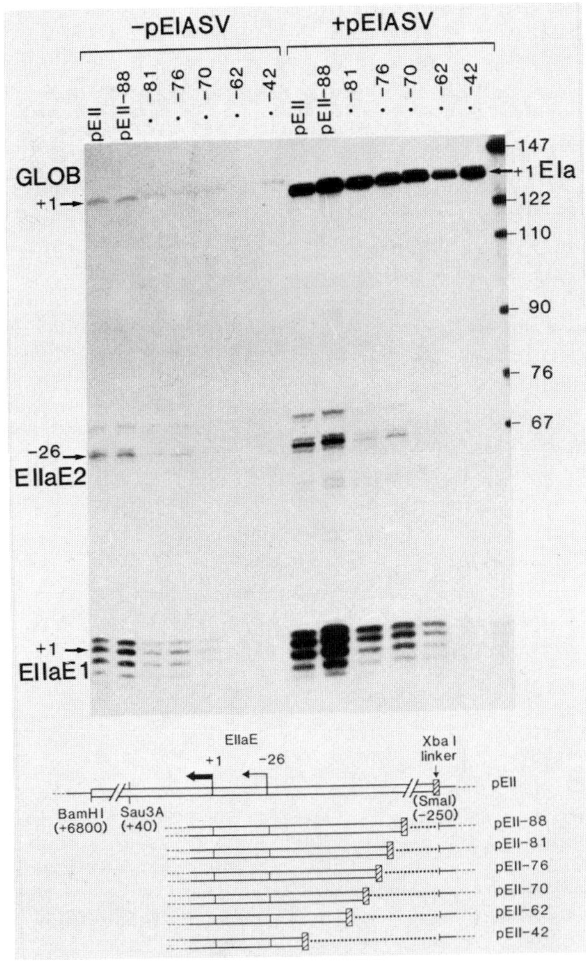

**Figure 1** Quantitative S1 nuclease analysis of cytoplasmic RNA from cells transfected with the pEII series of EE2A 5′-deletion mutants in the presence or absence of a cotransfected E1A-containing plasmid (pEIASV). HeLa cells were transfected with 2 μg of each of the pEII derivatives and either 2 μg of a polyoma β-globin recombinant (pβ[244 + ]β from DeVilliers and Schaffner 1981) used as an internal control of transfection efficiency (lanes −pEIASV) or 1 μg of pEIASV (lanes +pEIASV). Cytoplasmic RNA (10 μg) was analyzed by quantitative S1 nuclease mapping. The probes used to quantitate the specific transcripts were the 5′-end-labeled coding strands of restriction fragments between +40 and −250, +129 and −498, and +137 and −86 of the EE2A (EIIaE), E1A (Ela), and globin (GLOB) genes, respectively. Arrows indicate the probe fragments protected by specific transcripts. Numbers on the right refer to the length in nucleotides of $^{32}$P-labeled *Msp*I fragments of pBR322. The diagram at the bottom depicts the structure of the pEII series used in this experiment, with the parental, undeleted construct on the top. The open box corresponds to the EE2A (EIIaE) sequence, and the hatched box indicates the position of the *Xba*I linker (see Experimental Procedures). The dotted lines represent the deleted portions of the EE2A promoter. The horizontal arrows mark the major (+1) and minor (−26) EE2A cap sites and point in the direction of transcription.

**Table 1** Relative $E_1E2A$ Expression and E1A-mediated Stimulation of the pEII and pMTE Promoter Mutants

| 5′-deletion end point | Constitutive expression (%) pMTE | Constitutive expression (%) pEII | E1A-mediated stimulation (-fold) pMTE | E1A-mediated stimulation (-fold) pEII |
|---|---|---|---|---|
| WT | 100 | 100 | 10 | 10 |
| −169 | 93 | ND | 10 | ND |
| −146 | 81 | ND | 10 | ND |
| −111 | 45 | ND | 10 | ND |
| −88 | 93 | 100 | 8 | 8 |
| −81 | 62 | 90 | 7 | 7 |
| −76 | ND | 80 | ND | 7 |
| −70 | 8 | 45 | 10 | 9 |
| −62 | 2 | 20 | 2–5 | 2–5 |
| −42 | 2 | 15 | 1–2 | 1–2 |
| −39 | ND | 10 | ND | 1 |
| −23 | ND | 2 | ND | 1 |
| −18 | ND | 1 | ND | 1 |

Transcriptional efficiencies of the indicated pEII and pMTE derivatives were measured in the absence and presence of pEIASV in experiments similar to that shown in Fig. 1. WT refers to the parental pEII and pMTE constructs. Quantification of specific transcription and E1A-mediated effects were achieved as described in Experimental Procedures. (ND) Not determined.

suggest that a negative regulatory element exists between positions −111 and −88, the deletion thereof resulting in a slight but significant increase in $E_1E2A$ transcription.

Further deletions removing EE2A sequences between −88 and −62 and between −42 and −18 resulted in a two-step reduction in transcription activity to barely detectable levels. These data suggest that additional promoter elements, required for efficient transcription from the $E_1E2A$ start site, are situated in the −88/−62 and −42/−18 regions.

It is also noteworthy that the effects of EE2A promoter deletions are repeatedly less pronounced in the recombinants bearing the E2A coding region. Although an influence of different, adjacent vector sequences in the two constructs cannot be excluded, these results raise the interesting possibility that the E2A gene product could be involved in its own promoter regulation.

*Efficient $E_1E2A$ constitutive transcription requires a TATA-box substitute at −30 and two separate upstream elements (−48/−39 and −91/−62)*

To define more precisely the sequence components involved in the EE2A promoter function, we have constructed (Zajchowski et al. 1985) a set of mutants with clustered base substitutions in the region between −97 and +1 (see Fig. 2), by employing the LS strategy of McKnight and Kingsbury (1982). Analysis of $E_1E2A$-specific transcription from these recombinants (tabulated values in Fig. 2) clearly revealed three essential regions in the EE2A promoter. Alteration of sequences between −33 and −19 (TATA-like element in Fig. 4) and −48 and −39 (upstream region A in Fig. 4) decreased $E_1E2A$ transcription efficiency by more than 10-fold. Base substitutions in the region between −91 and −62 (upstream region B) less dramatically affected $E_1E2A$ promoter function, reducing it only by 2-fold to 4-fold.

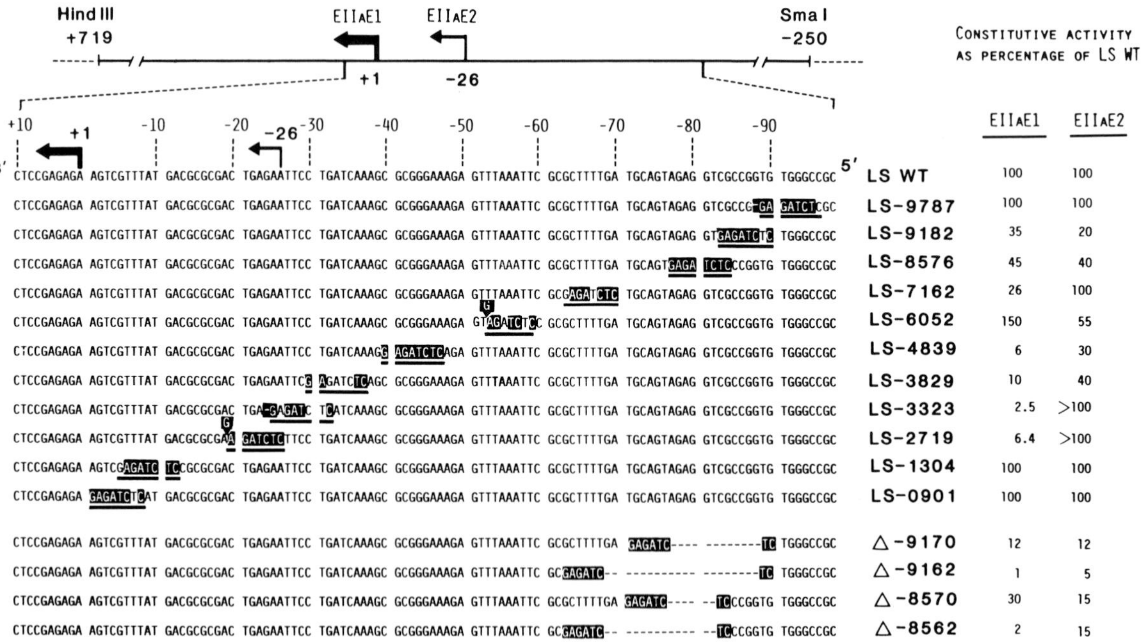


**Figure 2** Nucleotide sequence of the EE2A (EIIaE) LS and Δ mutants and relative $E_1E2A$ (EIIaE1) and $E_2E2A$ (EIIaE2)-specific expression. This scheme depicts the *Sma*I-*Hin*dIII fragment of the EE2A transcription unit and represents the nucleotide sequence between −98 and +10 of the wild-type EE2A promoter region (LS WT), with the linker substitution nucleotides shaded (LS series) and internal deletions dotted (Δ series). Arrows are as described in the legend to Fig. 1. Constitutive transcription from the $E_1E2A$ and $E_2E2A$ sites of the LS and Δ mutants was quantitated, as described in Experimental Procedures, from 5 (or more) different in vivo transfection experiments.

That this region plays an essential role in the $E_1E2A$ promoter was, however, demonstrated by the 100-fold reduction in transcription observed upon its deletion ($\Delta$-9162).

*Maximal $E_2E2A$ constitutive transcription involves a TATA-like element at $-59$ and two separate elements ($-48/-39$ and $-91/-70$)*
Analysis of the effect of the LS and $\Delta$ mutations on the $E_2E2A$ promoter efficiency (tabulated in Fig. 2) also revealed three elements whose alteration markedly reduced $E_2E2A$-specific transcription. One element, between $-60$ and $-52$, corresponds to the TATA-like element located about 30 nt upstream from the start site at $-26$. Another element, between $-48$ and $-39$, that is essential for $E_1E2A$ activity (upstream region A, Fig. 4) is also required for maximal $E_2E2A$ transcription. A more distal element, between $-90$ and $-70$, is part of the upstream region B defined for $E_1E2A$ and is referred to as the upstream region B' (see Fig. 4), essential for optimal $E_2E2A$ promoter activity.

Strikingly, mutations that alter the $E_1E2A$ TATA-like element (LS-3323 and LS-2719) result in an increased $E_2E2A$ promoter efficiency. Similarly, mutation of the $E_2E2A$ TATA-like sequence (LS-6052) stimulated weakly, but repeatedly, transcription from $E_1E2A$. This observation suggests that both TATA-like elements recognize the same or related transcription factors, present in limiting amounts, and that disruption of one element allows more-efficient use of the other one. It is also noteworthy that neither of these TATA-like elements seems to be required for the accuracy of transcription, since the mutated promoters still direct transcription from the same initiating nucleotides as the wild-type ones.

**Multiple and unique specific promoter stretches are required for transcriptional stimulation by the E1A gene products from the $E_1E2A$ and $E_2E2A$ sites, respectively**
Nuclear run-on transcription measurements in similar transient expression assays have demonstrated that the *trans*-activation of EE2A expression by the E1A products corresponds to an increased rate of transcription initiation (Leff et al. 1984). It was, therefore, of interest to localize the promoter sequences involved in this stimulation.

*External deletion analysis identifies an E1A-responsive element in the EE2A promoter between $-70$ and $-42$*
As an initial approach to this goal, we examined the effect of 5' deletions on the E1A-induced level of EE2A transcription. Cells were transfected with the pEII series, together with the E1A-containing plasmid, pEIASV. Cytoplasmic RNA was analyzed as shown in Figure 1, and the results for $E_1E2A$ transcriptional stimulation are given in Table 1. When pEIASV was cotransfected with pEII, the level of transcription from the $E_1E2A$ site was increased and the extent of stimulation varied from 4-fold to 20-fold (Fig. 1), with an

average stimulation of about 10-fold (Table 1). Deletion of EE2A sequences from $-250$ to $-70$ did not markedly affect this stimulation, indicating that the EE2A upstream promoter elements B and B', between $-90$ and $-70$, do not play a major role in the induction mechanism. Further deletion to $-62$ reduced the E1A-mediated effect about 4-fold, whereas deletion to $-42$ generally abolished it. These results suggest the existence of an EE2A element located between $-70$ and $-42$, which is required for maximal E1A responsiveness.

Transcription starting at the $E_2E2A$ site was also stimulated by E1A cotransfection, and the sequence requirement for this induction appeared similar to that of the $E_1E2A$ promoter, as revealed by the transcriptional analysis of the pEII series (Fig. 1).

*None of the LS and $\Delta$ mutations abolish E1A responsiveness of $E_1E2A$ transcription*
Quantification of the effects of the LS and $\Delta$ mutations on the level of transcriptional stimulation from the $E_1E2A$ promoter by the E1A products is provided in Table 2. For any given EE2A recombinant, cotransfection with pEIASV resulted in a 5-fold to 7-fold stimulation from the $E_1E2A$ site (the lower value found for $\Delta$-9162 should be cautiously interpreted since quantification of the very low constitutive transcription level of this mutant was difficult; see Fig. 2). These results, showing that none of these limited substitutions or internal deletions drastically affect the $E_1E2A$ regulation by E1A, suggest that no discrete or unique element is implicated in the E1A-mediated stimulation of $E_1E2A$ transcription. Indeed, the existence of multiple or larger E1A-responsive sequences in the $E_1E2A$ promoter is suggested by the

**Table 2** Fold Increase in Expression of Linker-scanning and Deletional Mutants by pEIASV and E4 Products

| | E1A-mediated stimulation from | | E4-mediated stimulation from |
|---|---|---|---|
| | $E_1E2A$ | $E_2E2A$ | $E_1E2A$ |
| LS WT | 10 | 10 | 9 |
| LS-9787 | 10 | 10 | ND |
| LS-9182 | 10 | 12 | ND |
| LS-8576 | 14 | 14 | ND |
| LS-7162 | 12 | 12 | ND |
| LS-6052 | 10 | 1.5 | 12 |
| LS-4839 | 10 | 16 | 2 |
| LS-3829 | 14 | 20 | 12 |
| LS-3323 | 14 | 10 | 3 |
| LS-2719 | 12 | 10 | 4 |
| LS-1304 | 8 | 8 | 8 |
| LS-0901 | 10 | 10 | ND |
| $\Delta$-9170 | 14 | 7 | ND |
| $\Delta$-9162 | 5 | 10 | 10 |
| $\Delta$-8570 | 10 | 12 | ND |
| $\Delta$-8562 | 14 | 12 | ND |

Transcriptional stimulation by the E1A products of $E_1E2A$ and $E_2E2A$ promoter mutants was determined as described in Fig. 1 and Experimental Procedures. Stimulation by E4 was quantitated by S1 nuclease analysis of RNA from cells transfected with 2 µg per dish of the indicated LS and $\Delta$ mutants or cotransfected with 7.5 µg of pEIV. (ND) Not determined.

observation that an EE2A element (between −70 and −42) becomes essential for the E1A-mediated control, only when the sequences further upstream (from −70) were absent (see above).

### Stimulation of $E_2E2A$ transcription by E1A is dependent upon the TATA-box substitute element around −59 and sequences farther upstream

As shown in Table 2, alteration of the sequence component defined by the LS-6052 mutation severely impaired the inducibility of $E_2E2A$ transcription by the E1A products. In addition, a smaller but reproducible decrease (about twofold) in the extent of EE2A stimulation was observed with the Δ-9170 mutation. Thus it appears that, in contrast to the regulation by E1A of $E_1E2A$ promoter activity, the weaker $E_2E2A$ promoter is composed of discrete sequence elements specific for the E1A response, the most critical control being mediated by the $E_2E2A$ TATA-like element.

### Sequences within the EE2A promoter possess the properties of an enhancer, the activity of which is controlled by flanking E1A-responsive elements

Evidence for the presence of an E1A-inducible enhancer in the EE2A promoter region, between −262 and −21, has previously been reported (Imperiale and Nevins 1984; Imperiale et al. 1985). We have shown, independently, that the minimal EE2A promoter fragment providing maximal, bidirectional, and E1A-inducible enhancer function extends from −111 to −27 (Jalinot and Kédinger 1986). The salient properties of this element are summarized in Figure 3. Insertion in either orientation of

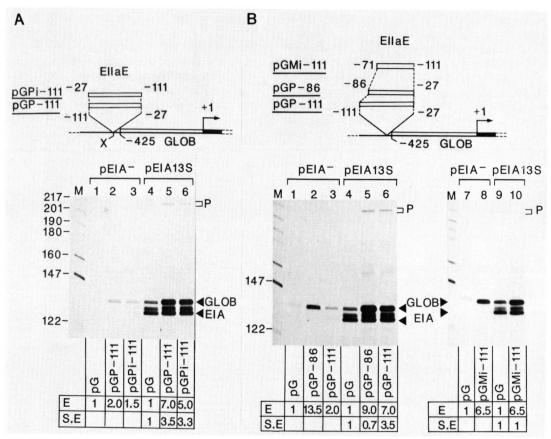

**Figure 3** E1A-inducible and constitutive enhancer activity of EE2A promoter fragments. (*A*) The diagram at the top depicts the pGP-111 and pGPi-111 recombinants, which were obtained by inserting an EE2A (EllaE) fragment extending from −111 to −27, in either orientation, into the *Xba*l site (X) of pG (see Experimental Procedures). The arrow points in the direction of transcription from the globin start sites (+1). Where indicated, 5 μg of pG, pGP-111, or pGPi-111 was cotransfected into HeLa cells with 0.5 μg of either pEIA⁻ (lanes *1–3*) or pEIA13S (lanes *4–6*). The DNA probes used for S1 nuclease mapping of the specific E1A and globin transcripts (EIA and GLOB bands) were the same as in Fig. 1. P refers to the full-length globin probe protected by nonspecific transcripts initiated upstream from the globin-probe end point (−86). Lane *M* contains DNA size markers as in Fig. 1. The E values represent the *enhancement factor* provided by the inserted EE2A fragments and correspond to the globin transcription of the pG derivatives expressed relative to that of the pG parental recombinants, measured under the same conditions. The S.E. values represent the *stimulation factor of the enhancer activity* and correspond to the ratio of E values measured in the presence and absence of the E1A recombinant. (*B*) The pGP-111 recombinant is the same as in *A*. The pGP-86 recombinant contains the EE2A (EllaE) fragment between −86 and −27 inserted into pG in the same orientation as in pGP-111. The pGMi-111 contains the EE2A sequences between −71 and −111 in the same orientation as in pGPi-111 (*A*). Transcription analysis and representation of the results are as described for *A*.

this EE2A promoter segment, at a distance of about 450 nt upstream from the entire rabbit β-globin gene, resulted in a slight (about twofold) increase in globin-specific transcription after transfection into HeLa cells together with an E1A-defective recombinant (pE1A⁻). Cotransfection with an E1A-containing plasmid (pE1A13S) augmented globin transcription fivefold to sevenfold more from the pG derivatives containing the EE2A sequences than from the parent pG vector (cf. E values in Fig. 3A). The E1A-mediated stimulation of the EE2A enhancer activity, represented by the S.E. values, corresponds therefore to a factor of about 3.5 under our transfection conditions.

Deletion of EE2A sequences between −111 and −86 from the pGP-111 plasmid, generating the pGP-86 recombinant, rendered the EE2A enhancer activity independent of the E1A product (cf. E values in lanes 3 and 6 with lanes 2 and 5 in Fig. 3B). Similarly, deletion of the sequences between −27 and −71 from the EE2A element inserted in reverse orientation (yielding pGMi-111), resulted in an E1A-independent enhancer activity (cf. E values in lanes 3 and 6 in Fig. 3A with lanes 8 and 10 in Fig. 3B). A refined set of deletions in pGMi-111 revealed that only a subset of the −27/−71 region, the element between −48 and −71, was responsible for the E1A-dependence of the EE2A enhancer activity in this recombinant (Jalinot and Kédinger 1986). Altogether these results indicate that elements between −111 and −87 and between −71 and −48 are involved in the inhibition of the EE2A enhancer effect in the absence of the E1A products. The fact that deletion of either of these negative regulatory elements activates the enhancer activity only in one direction may be explained by the observation (not shown) that these elements themselves contribute to the enhancer activity, each in an opposite direction: the −71/−48 element is required for maximal enhancement of transcription starting on the −27 side of the enhancer (in the leftward direction; see Fig. 4), whereas the −111/−87 region is involved in the enhancer activity exerted on the −111 side (rightward direction). Bidirectional activation by the intact enhancer element (−111/−27) is obtained in the presence of the E1A products, indicating that these products simultaneously relieve the "repression" exerted by both control regions. The significance of these control elements in the natural context of the EE2A promoter will be discussed below.

### Positive regulation of $E_1E2A$ transcription by the E4 transcription unit is sequence-specific

The contribution of adenovirus early genes, other than the E1A transcription unit, to the transcriptional regulation of EE2A expression was also examined. We have previously shown (Goding et al. 1985) that transfection of HeLa cells with E2A-containing plasmids and cotransfection with distinct plasmids bearing the adenovirus E4 region stimulated specific transcription from the $E_1E2A$ site by 5-fold to 15-fold. The extent of the E4-mediated stimulation was similar to that achieved by E1A under identical conditions.

To determine whether a specific EE2A sequence component was required for this induction, we have analyzed the effect of a selected series of LS and Δ mutations spanning the EE2A promoter region between −91 and −4. The results presented in Table 2 clearly indicate that elements located between −48 and −39 and −33 and −19 are essential for maximal E4-induced stimulation of $E_1E2A$ transcription. Both of these regions correspond to promoter elements required for efficient constitutive transcription from the $E_1E2A$ start site (see Fig. 2), and these results demonstrate their involvement in the E4-mediated control, as well.

## Discussion

### EE2A promoter requirements for constitutive transcription

*$E_1E2A$ and $E_2E2A$ transcription is controlled by overlapping promoters*

It can be deduced from the combined analysis of 5′-deletion, LS, and Δ mutants that the EE2A promoter contains at least five sequence components required for maximal transcription from both the $E_1E2A$ and $E_2E2A$ sites, in the absence of E1A products (see Fig. 4 and Zajchowski et al. 1985). Besides the TATA-like elements around positions −30 and −53 specifically involved in $E_1E2A$ and $E_2E2A$ promoter function, respectively, the other elements (upstream elements A, B, and C; see Fig. 4) are at least partially shared by both promoters. In addition to these positive control elements, the existence of negative regulatory sequences was indicated by the enhancer function studies (Fig. 3) and suggested by the external-deletion analysis (Table 1). One such inhibitory element is located between −111 and −88, and another one is between −70 and −48 (Fig. 4). Deletion of the most upstream of these negative control elements increased EE2A promoter activity (Table 1). That this stimulation corresponds, at least partially, to the activation of an EE2A enhancer function is suggested by the effect of the truncated EE2A sequences, at a distance, on the heterologous globin test gene, in the absence of the E1A products (Fig. 3). The complex structure of the EE2A promoter region appears, therefore, as the juxtaposition and/or superimposition of distinct regulatory sequence components, some of them sharing properties of enhancer elements.

Murthy et al. (1985), who performed a similar LS study of the EE2A promoter, described only two specific transcriptional control sequences: one, between −29 and −21, corresponding to the $E_1E2A$ TATA-like element; the other, between −82 and −66, corresponding to the upstream element B defined here. These authors used the CAT assay to quantitate EE2A promoter efficiency. Under these conditions the CAT values correspond to the cumulated $E_1E2A$ and $E_2E2A$ transcription. This explains why, in their experiments, alteration of the $E_2E2A$ TATA-like element was not detected as a down-mutation, since the effect was probably compensated by the corresponding augmentation of $E_1E2A$ transcription

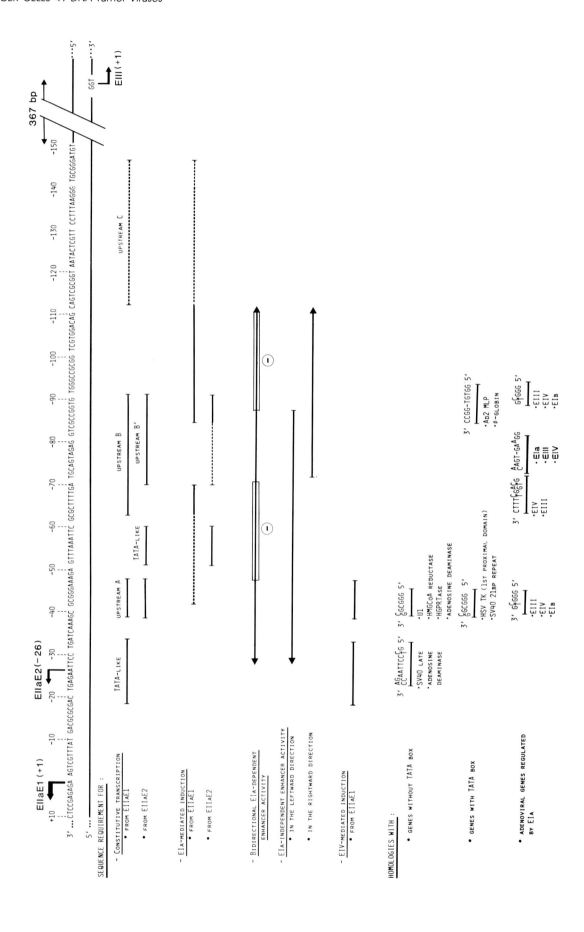

**Figure 4** Summary of the EE2A (EIIaE) promoter analysis. The sequence on the top represents the noncoding strand of the Ad2 EE2A promoter region between positions −150 and +10 relative to the $E_1E2A$ (EIIaE1) (+1) cap site, with symbols as described in the legend to Fig. 2. The initiation site of the E3 (EIII) transcription unit, located 500 nt upstream from the $E_1E2A$ site, is indicated with the arrow showing the direction of E3 transcription. The sequence requirements for constitutive transcription and E1A- or E4 (EIV)-mediated stimulation are indicated by the segments aligned under the relevant nucleotides. Dashes refer to moderately important regions. The position of the minimal E1A-inducible, bidirectional enhancer sequence is shown by the double-arrowed segment, with the open boxes (marked ⊙) corresponding to the negative regulatory elements discussed in the text. The segment below this line refers to the minimal E1A-independent enhancer fragments with the arrowheads pointing in the direction of the enhancer activity (leftward to the EE2A and rightward to the E3 transcriptional directions, respectively). Sequence homologies between EE2A promoter elements and other sequences, located at similar positions in different gene promoters, are also given. The corresponding, tentative consensus sequences are oriented 5′ to 3′, as the noncoding EE2A strand. The origin of the compared gene sequences is as follows: human adenosine deaminase (Valerio et al. 1985), human antithrombine III (Prochownik and Orkin 1984), mouse hypoxanthine guanosine phosphoribosyl transferase (D.W. Melton et al., in prep.), hamster 3-hydroxy 3-methylglutaryl coenzyme A reductase (Reynolds et al. 1984), rabbit β-globin (Dierks et al. 1983), chicken U1 RNA (Earley et al. 1984), SV40 late (Brady et al. 1982) and early promoters (Baty et al. 1984), HSV-1 thymidine kinase (McKnight and Kingsbury 1982), and Ad2 E1A (Hearing and Shenk 1983; Weeks and Jones 1983), E1B (Bos and ten Wolde-Kraamwinkel 1983), E3 (Leff et al. 1985; Weeks and Jones 1985), E4 (Gilardi and Perricaudet 1984), and major late (Hen et al. 1982) promoters.

(see Fig. 2). It is likely also that these authors did not identify the G + C–rich upstream element A (see Fig. 4) as an important promoter element probably because of the high G + C content of the *Bam*HI linker that they used as a substitution sequence.

*Sequence homologies with other eukaryotic promoters*
The TATA-like element around position −30 bears no similarity to the TATA consensus sequence found at the same position in most protein-coding genes. However, this element shares substantial homology with sequences present at equivalent positions in the promoters of a series of genes having no visible, canonical TATA-box sequence (see Fig. 4). Furthermore, a comparison of the nucleotide sequences directly surrounding the consensus TATA box (Breathnach and Chambon 1981) or the substitute TATA sequence reveals, in both cases, a similarly elevated G + C content.

It is interesting to note that the EE2A element involved in the enhancer function (upstream region B) shares sequence homology with an enhancer-like element found in the adenovirus E1A (Hearing and Shenk 1983) and E4 promoter region (Gilardi and Perricaudet 1984). Another sequence homology, which has previously been noted (Hen et al. 1982; Elkaim et al. 1983), concerns an element located around position −89 in the EE2A promoter and also found in the Ad2 major late (position −65) and rabbit β-globin gene (position −79) promoters (Fig. 4). Alteration of the 3′ half of this element (5′ GGCC 3′) reduces the efficiency of all three promoters (Fig. 2 and R. Hen et al., pers. comm.; Dierks et al. 1983; Zajchowski et al. 1985).

Recent evidence suggests, however, that consensus promoter sequences may be only part of the recognition sites for transcription factors and that the presence of a consensus sequence in different promoters does not necessarily imply the involvement of identical factors but rather that of factors belonging to the same family and serving similar functions (Miyamoto et al. 1985; H. Boeuf, unpubl.). Therefore, combinations of footprinting analysis and in vitro transcription studies with purified fractions of cellular extracts will eventually allow the assignment of specific transcription factors to their corresponding promoter element. In the present context, it will be interesting to determine if the $E_1E2A$ and $E_2E2A$ TATA-like functions both implicate the same transcription factor. It will also be of interest to identify the factors contributing to the EE2A enhancer effect and to compare them with those acting on other, more "classical" enhancer elements.

*Different sequence requirements for $E_1E2A$ and $E_2E2A$ transcriptional activation by E1A*
The stimulation of $E_2E2A$ transcription by the E1A products is clearly dependent on $E_2E2A$-specific promoter elements: the TATA-like element at −59 and, to a lesser extent, the upstream region B′ (see Fig. 4). Alteration or deletion of either one impaired the efficiency of induction.

The situation appears different in the case of the $E_1E2A$ transcriptional induction, where several distinct, but apparently equivalent, sequence elements (at least two) appear to be involved in E1A responsiveness. One element could be mapped by external-deletion mutation between −70 and −42, with the most important portion lying between −70 and −62 (Fig. 4). However, alteration of this element alone by the LS-7162 mutation did not reduce the induction level, indicating that other elements, probably located further upstream, are capable by themselves of providing full E1A responsiveness. The fact that a deletion (Δ-8562) that removes upstream sequences, including the −70/−62 element, retains maximal inducibility indicates that such elements are most likely located upstream from −85. Kingston et al. (1984a), using a stable transformation assay, found that none of the tested EE2A external deletions abolished the E1A-mediated stimulation, even those deletions removing the $E_1E2A$ TATA-like element. These results, which suggest, in contradiction with ours, a lack of specific regulatory sequences in the EE2A promoter region, should however be considered with some precaution. The assay itself, which involves a positive selection for the EE2A-driven dihydrofolate reductase expression, could indeed interfere with the mutation analysis.

*Mechanisms of the E1A-mediated activation of transcription*
Many genes have been reported to be stimulated by E1A products in different assay systems (Green et al. 1983; Imperiale et al. 1983; Kao and Nevins 1983; Allan et al. 1984; Courtois and Berk 1984; Gaynor et al. 1984; Kingston et al. 1984b; Stein and Ziff 1984; Chow and Pearson 1985). The level of E1A-mediated stimulation can, however, fluctuate from 3-fold (β-globin and β-tubulin genes) to 20-fold (adenovirus early genes and heat-shock genes).

In the case of the adenovirus genes, Weeks and Jones (1983) have shown in a stable transformation assay with hybrid recombinants that the upstream sequences of early adenovirus genes are implicated in the stimulation process by E1A. Leff and Chambon (1986) have corroborated these results in a transient expression assay. Here we show that for the EE2A transcription unit, two of the sequence components involved in the E1A-mediated induction of $E_1E2A$ transcription roughly overlap with negative regulatory elements identified in the EE2A enhancer region. Since the E1A products most likely activate this enhancer function by relieving the negative effect of these elements, it is tempting to speculate that the induction of $E_1E2A$ transcription by E1A cotransfection directly reflects this mechanism. In the enhancer assay, deletion of the −111/−86 control element resulted in a marked enhancement (~10-fold) of transcriptional activity, which was not stimulated any further in the presence of the E1A products. A 5′ deletion to position −88 in the E2A promoter, removing this same −111/−86 element, increased constitutive EE2A transcription approximately 2-fold compared with the −111 deletion mutant. Cotransfection of this mutant with E1A still activated $E_1E2A$ promoter efficiency, although to a slightly lower extent (Table 1; and Murthy et al. 1985). Hence, these results indicate that the −111/ −86 element exerts less-pronounced effects when tested in its natural position with respect to the transcription start site ($E_1E2A$) and emphasize the role of further downstream control elements (e.g., −70/−42) in E1A responsiveness of the entire transcription unit.

To date, nothing is known about the molecular mechanism of the E1A-mediated induction. In the present example, counteraction of a negative effect on transcription could be interpreted as the inactivation, in the "prokaryotic sense," of DNA-bound repressor molecules by the E1A products (Nevins 1981; Persson et al. 1981; Cross and Darnell 1983) and/or as the catalysis of a more efficient assembly of active transcription complexes (Gaynor and Berk 1983). The latter possibility, which could involve an E1A-mediated activation of transcription factors, is supported by our observation that E1A-responsiveness is provided by promoter elements required for constitutive transcription. It is also supported by the more general, albeit weaker, E1A-caused stimulatory effect on transcription from a number of nonrelated genes in similar short-term transfection experiments (Svensson and Akusjärvi 1984; Berger and Folk 1985; Gaynor et al. 1985).

Finally it is worth recalling that the E1A products exhibit rather pleiotropic functions since, besides their effect on transcription activation, they are also able to repress enhancer-induced stimulation of transcription (Borrelli et al. 1984; Velcich and Ziff 1985), and it is not excluded that these combined effects contribute to the oncogenic transformation by altering the cellular expression program (for references, see Bernards and van der Eb 1984; Branton et al. 1985).

*Mechanism of the E4-mediated activation of transcription*

Activation of $E_1E2A$ transcription by the E4 plasmid requires both the $E_1E2A$ TATA-like region and the upstream element A (Fig. 4). The fact that constitutive promoter elements are involved suggests that E4 cotransfection activates $E_1E2A$ transcription complex formation either by increasing the concentration of transcription factors specifically recognizing these elements or by providing an additional E4-encoded transcription factor(s). Experiments are in progress to identify which of the seven open reading frames in the E4 transcription unit (Rigolet and Galibert 1984; Virtanen et al. 1984) is responsible for the effect.

## Acknowledgments

We wish to thank P. Chambon for stimulating discussions and continuous interest. We are most grateful to M. Acker and B. Heller for tissue culture and C. Aron, C. Kutchis, A. Landman, and C. Werlé for preparing the manuscript. This work was supported by the CNRS (ATP 3582 and 6984) and the Ministère de la Recherche et de la Technologie (82V 1283 and 84V 083).

## References

Allan, M., J. Zhu, P. Montague, and J. Paul. 1984. Differential response of multiple ε-globin cap sites to *cis*- and *trans*-acting control. *Cell* **38**: 399.

Banerji, J., S. Rusconi, and W. Schaffner. 1981. Expression of a β-globin gene is enhanced by remote SV40 DNA sequences. *Cell* **27**: 299.

Baty, D., H.A. Barrera-Saldana, R.D. Everett, M. Vigneron, and P. Chambon. 1984. Mutational dissection of the 21 bp repeat region of the SV40 early promoter reveals that it contains overlapping elements of the early-early and late-early promoters. *Nucleic Acids Res.* **12**: 915.

Benoist, C. and P. Chambon. 1981. *In vitro* sequence requirements of the SV40 early promoter region. *Nature* **290**: 304.

Berger, S.L. and W.R. Folk. 1985. Differential activation of RNA polymerase III–transcribed genes by the polyomavirus enhancer and the adenovirus Ela gene products. *Nucleic Acids Res.* **13**: 1413.

Bernards, R. and A.J. van der Eb. 1984. Adenovirus: Transformation and oncogenicity. *Biochim. Biophys. Acta* **783**: 187.

Bienz, M. 1985. Transient and developmental activation of heat-shock genes. *Trends Biochem. Sci.* **10**: 157.

Borrelli, E., R. Hen, and P. Chambon. 1984. Adenovirus-2 EIA products repress enhancer-induced stimulation of transcription. *Nature* **312**: 608.

Bos, J.L. and H.C. ten Wolde-Kraamwinkel. 1983. The Elb promoter of Ad12 in mouse L tk⁻ cells is activated by adenovirus region Ela. *EMBO J.* **2**: 73.

Brady, J., M. Radonovich, M. Vodkin, V. Natarajan, M. Tgoren, G. Das, J. Janik, and N.P. Salzman. 1982. Site-specific base substitution and deletion mutations that enhance or suppress transcription of the SV40 major late RNA. *Cell* **31**: 625.

Branton, P.E., S.T. Bayley, and F.L. Graham. 1985. Transformation by human adenoviruses. *Biochim. Biophys. Acta* **780**: 67.

Breathnach, R. and P. Chambon. 1981. Organization and expression of eukaryotic split genes coding for proteins. *Annu. Rev. Biochem.* **50**: 349.

Chow, K.C. and G.D. Pearson. 1985. Adenovirus infection elevates levels of cellular topoisomerase I. *Proc. Natl. Acad. Sci.* **82**: 2247.

Courtois, G. and A. Berk. 1984. Adenovirus Ela protein activation of an integrated viral gene. *EMBO J.* **3**: 1145.

Cross, F.R. and J.E. Darnell. 1983. Cycloheximide stimulates early adenovirus transcription if early gene expression is allowed before treatment. *J. Virol.* **45**: 683.

Davison, B.L., J.M. Egly, E.R. Mulvihill, and P. Chambon. 1983. Formation of stable pre-initiation complexes between class B transcription factors and promoter sequences. *Nature* **301**: 680.

DeVilliers, J. and W. Schaffner. 1981. A small segment of polyomavirus DNA enhances the expression of a cloned β-globin gene over a distance of 1400 base pairs. *Nucleic Acids Res.* **9**: 6251.

Dierks, P., A. van Ooyen, M.D. Cochran, C. Dobkin, J. Reiser, and C. Weissman. 1983. Three regions upstream from the cap site are required for efficient and accurate transcription of the rabbit β-globin gene in mouse 3T6 cells. *Cell* **32**: 695.

Dynan, W.S. and R. Tjian. 1983. Isolation of transcription factors that discriminate between different promoters recognized by RNA polymerase II. *Cell* **32**: 669.

Earley, J.M., I. Kenneth, A. Roebuck, and W.E. Stumph. 1984. Three linked chicken U1 RNA genes have limited flanking DNA sequence homologies that reveal potential regulatory signals. *Nucleic Acids Res.* **12**: 7411.

Elkaim, R., C. Goding, and C. Kédinger. 1983. The adenovirus-2 EIIa early gene promoter: Sequences required for efficient *in vitro* and *in vivo* transcription. *Nucleic Acids Res.* **11**: 7105.

Flint, J. 1982. Expression of adenoviral genetic information in productively infected cells. *Biochim. Biophys. Acta* **651**: 175.

Fujita, T., S. Ohno, H. Yasumitsu, and T. Tanigushi. 1985. Delimitation and properties of DNA sequences required for the regulated expression of human interferon-β gene. *Cell* **41**: 489.

Gaynor, R.B. and A.J. Berk. 1983. *Cis*-acting induction of adenovirus transcription. *Cell* **33**: 683.

Gaynor, R.B., L.T. Feldman, and A.J. Berk. 1986. Transcription of class III genes activated by viral immediate early proteins. *Science* **230**: 447.

Gaynor, R.B., D. Hillman, and A.J. Berk. 1984. Adenovirus early region 1A protein activates transcription of a non-viral gene introduced into mammalian cells by infection or transfection. *Proc. Natl. Acad. Sci.* **81**: 1193.

Gidoni, D., W.S. Dynan, and R. Tjian. 1984. Multiple specific contacts between a mammalian transcription factor and its cognate promoters. *Nature* **312**: 409.

Gilardi, P. and M. Perricaudet. 1984. The E4 transcriptional unit of Ad2: Far upstream sequences are required for its transactivation by Ela. *Nucleic Acids Res.* **12**: 7877.

Gilmour, P.S., D.A. Spandidos, J.K. Vass, J.W. Gow, and J. Paul. 1984. A negative regulatory sequence near the mouse β-major globin gene associated with a region of potential Z-DNA. *EMBO J.* **3**: 1263.

Goding, C., P. Jalinot, D. Zajchowski, H. Boeuf, and C. Kédinger. 1985. Sequence specific *trans*-activation of the adenovirus Ella early promoter by the viral EIV transcription unit. *EMBO J.* **4**: 1523

Goodbourn, S., K. Zinn, and T. Maniatis. 1985. Human β-interferon gene expression is regulated by an inducible enhancer element. *Cell* **41**: 509.

Green, M.R., R. Treisman, and T. Maniatis. 1983. Transcriptional activation of cloned human β-globin genes by viral immediate early gene products. *Cell* **35**: 137.

Groner, H., N. Kennedy, P. Skroch, N.E. Hynes, and H. Ponta. 1984. DNA sequences involved in the regulation of gene expression by glucocorticoid hormones. *Biochim. Biophys. Acta* **781**: 1.

Guarente, L. 1984. Yeast promoters: Positive and negative elements. *Cell* **36**: 799.

Hearing, P. and T. Shenk. 1983. The adenovirus type 5 E1A transcriptional control region contains a duplicated enhancer element. *Cell* **33**: 695.

Hen, R., P. Sassone-Corsi, J. Corden, M.P. Gaub, and P. Chambon. 1982. Sequences upstream from the TATA box are required *in vivo* and *in vitro* for efficient transcription from the adenovirus serotype 2 major late promoter. *Proc. Natl. Acad. Sci.* **79**: 7132.

Imperiale, M.J. and J.R. Nevins. 1984. Adenovirus 5 E2 transcription unit: An Ela-inducible promoter with an essential element that functions independently of position or orientation. *Mol. Cell. Biol.* **4**: 875.

Imperiale, M.J., L.T. Feldman, and J.R. Nevins. 1983. Activation of gene expression by adenovirus and herpesvirus regulatory genes acting in *trans* and by *cis*-acting adenovirus enhancer element. *Cell* **35**: 127.

Imperiale, M.J., R.P. Hart, and J.R. Nevins. 1985. An enhancer-like element in the adenovirus E2 promoter contains sequences essential for uninduced and E1A-induced transcription. *Proc. Natl. Acad. Sci.* **82**: 381.

Jalinot, P. and C. Kédinger. 1986. Negative regulatory sequences in the Ela-inducible enhancer of the adenovirus-2 early Ella promoter. *Nucleic Acids Res.* **14**: (in press).

Jones, N.C. and T. Shenk. 1979. Isolation of adenovirus type 5 host range deletion mutants defective for transformation of rat embryo cells. *Cell* **17**: 683.

Jones, N.C., J.D. Richter, D.L. Weeks, and L.D. Smith. 1983. Regulation of adenovirus transcription by an Ela gene in microinjected *Xenopus laevis* oocytes. *Mol. Cell. Biol.* **3**: 2131.

Kao, H.T. and J.R. Nevins. 1983. Transcriptional activation and subsequent control of the human heat-shock gene during adenovirus infection. *Mol. Cell. Biol.* **3**: 2058.

Karin, M., A. Haslinger, H. Holtgreve, R.I. Richards, P. Krauter,

H.M. Westphal, and M. Beato. 1984. Characterization of DNA sequences through which cadmium and glucocorticoid hormones induce human metallothionein-IIA gene. *Nature* **308**: 513.

Khoury, G. and P. Gruss. 1983. Enhancer elements. *Cell* **33**: 313.

Kingston, R.E., R.J. Kaufman, and P.A. Sharp. 1984a. Regulation of transcription of the adenovirus EII promoter by Ela gene products: Absence of sequence specificity. *Mol. Cell. Biol.* **4**: 1970.

Kingston, R.E., P.A. Sharp, and R.J. Kaufman. 1984b. Regulation of gene expression by the adenoviral EIA region and by *c-myc. Cancer Cells* **2**: 539.

Leff, T. and P. Chambon. 1986. Sequence specific activation of transcription by adenovirus Ela products is observed in HeLa cells, but not in 293 cells. *Mol. Cell. Biol.* **6**: 201.

Leff, T., J. Corden, R. Elkaim, and P. Sassone-Corsi. 1985. Sequence elements in the adenovirus 5 EIII promoter required for efficient transcription and for stimulation by Ela gene products. *Nucleic Acids Res.* **13**: 1209.

Leff, T., R. Elkaim, C.R. Goding, P. Jalinot, P. Sassone-Corsi, M. Perricaudet, C. Kédinger, and P. Chambon. 1984. Individual products of the adenovirus 12S and 13S mRNAs stimulate viral Ella and EIII expression at the transcriptional level. *Proc. Natl. Acad. Sci.* **81**: 4381.

Mathis, D.J., R. Elkaim, P. Sassone-Corsi, C. Kédinger, and P. Chambon. 1981. Specific *in vitro* initiation of transcription on the adenovirus type 2 early and late EII transcription units. *Proc. Natl. Acad. Sci.* **78**: 7383.

McKnight, S.L. and R. Kingsbury. 1982. Transcriptional control signals of a eukaryotic protein-coding gene. *Science* **217**: 316.

Miyamoto, N.G., V. Moncollin, J.M. Egly, and P. Chambon. 1985. Specific interaction between a transcription factor and the upstream element of the adenovirus-2 major late promoter. *EMBO J.* **4**: 3563.

Miyamoto, N.G., V. Moncollin, M. Wintzerith, R. Hen, J.M. Egly, and P. Chambon. 1984. Stimulation of *in vitro* transcription by the upstream element of the adenovirus-2 major late promoter involves a special factor. *Nucleic Acids Res.* **12**: 8779.

Murthy, S.C.S., G.P. Bhat, and B. Thimmappaya. 1985. Adenovirus EIIA early promoter: Transcriptional control elements and induction by the viral pre-early EIA gene, which appears to be sequence independent. *Proc. Natl. Acad. Sci.* **82**: 2230.

Nevins, J.R. 1981. Mechanism of activation of early viral transcription by the adenovirus E1A gene product. *Cell* **26**: 213.
———. 1982. Induction of the synthesis of a 70,000 dalton mammalian heat shock protein by the adenovirus Ela gene product. *Cell* **29**: 913.

Nevins, J.R., H.S. Ginsberg, J.M. Blanchard, M.C. Wilson, and J.E. Darnell. 1979. Regulation of the primary expression of the early adenovirus transcriptional units. *J. Virol.* **32**: 727.

Ohno, S. and T. Taniguchi. 1983. The 5′-flanking sequence of human interferon-β$_1$ gene is responsible for viral induction of transcription. *Nucleic Acids Res.* **11**: 5403.

Parker, C.S. and J. Topol. 1984. A *Drosophila* RNA polymerase II transcription factor binds to the regulatory site of an hsp70 gene. *Cell* **37**: 273.

Pelham, H. 1985. Activation of heat-shock genes in eukaryotes. *Trends Genet.* **1**: 31.

Persson, H., M.G. Katze, and L. Philipson. 1981. Control of adenovirus early gene expression: Accumulation of viral mRNA after infection of transformed cells. *J. Virol.* **40**: 358.

Prochownik, E.V. and S.H. Orkin. 1984. *In vivo* transcription of a human antithrombin III ''minigene.'' *J. Biol. Chem.* **259**: 15386.

Reynolds, G.A., S.K. Basu, T.F. Osborne, D.J. Chin, G. Gil, M.S. Brown, J.L. Goldstein, and K.L. Luskey. 1984. HMG CoA reductase: A negatively regulated gene with unusual promoter and 5′ untranslated regions. *Cell* **38**: 275.

Rigolet, M. and F. Galibert 1984. Organization and expression of the E4 region of adenovirus 2. *Nucleic Acids Res.* **12:** 7649.

Rossini, M. 1983. The role of adenovirus early region 1A in the regulation of early regions 2A and 1B expression. *Virology* **131:** 49.

Sassone-Corsi, P., A. Wildeman, and P. Chambon. 1985. A *trans*-acting factor is responsible for the SV40 enhancer activity *in vitro*. *Nature* **313:** 458.

Sassone-Corsi, P., R. Hen, E. Borrelli, T. Leff, and P. Chambon. 1983. Far upstream sequences are required for efficient transcription from the Ad2 EIA transcription unit. *Nucleic Acids Res.* **11:** 8748.

Searle, P.F., G.W. Stuart, and R.D. Palmiter. 1985. Building a metal-responsive promoter with synthetic regulatory elements. *Mol. Cell. Biol.* **5:** 1480.

Serfling, E., M. Jasin, and W. Schaffner. 1985. Enhancers and eukaryotic gene transcription. *Trends Genet.* **1:** 224.

Stein, R. and E.B. Ziff. 1984. HeLa cell β-tubulin gene transcription is stimulated by adenovirus 5 in parallel with viral early genes by an Ela dependent mechanism. *Mol. Cell. Biol.* **4:** 2792.

Svensson, C. and G. Akusjärvi. 1984. Adenovirus 2 early region 1A stimulates expression of both viral and cellular genes. *EMBO J.* **3:** 789.

Valerio, D., M.G.C. Duyvesteyn, B.M.M. Dekker, G. Weeda, T.H.M. Berkvens, L. Van der Voorn, H. van Ormondt, and A.J. Van der Eb. 1985. Adenosine deaminase: Characterisation and expression of a gene with a remarkable promoter. *EMBO J.* **4:** 437.

Velcich, A. and E. Ziff. 1985. Adenovirus E1a proteins repress transcription from the SV40 early promoter. *Cell* **40:** 705.

Virtanen, A., P. Gilardi, A. Naslund, J.M. LeMoullec, U. Pettersson, and M. Perricaudet. 1984. mRNAs from human adenovirus 2 early region 4. *J. Virol.* **51:** 822.

Weeks, D.L. and N.C. Jones. 1983. Ela control of gene expression is mediated by sequences 5′ to the transcriptional starts of the early viral genes. *Mol. Cell. Biol.* **3:** 1222.

———. 1985. Adenovirus E3-early promoter: Sequences required for activation by Ela. *Nucleic Acids Res.* **13:** 5389.

Wildeman, A.G., P. Sassone-Corsi, T. Grundström, M. Zenke, and P. Chambon. 1984. Stimulation of *in vitro* transcription from the SV40 early promoter by the enhancer involves a specific *trans*-acting factor. *EMBO J.* **3:** 3129.

Wu, C. 1984. Two protein-binding sites in chromatin implicated in the activation of heat-shock genes. *Nature* **309:** 229.

Yaniv, M. 1984. Regulation of eukaryotic gene expression by *trans*-activating proteins and *cis*-acting elements. *Biol. Cell.* **50:** 203.

Zajchowski, D.A., H. Boeuf, and C. Kédinger. 1985. The adenovirus-2 early Ella transcription unit possesses two overlapping promoters with different sequence requirements for Ela-dependent stimulation. *EMBO J.* **4:** 1293.

Zinn, K., D. DiMaio, and T. Maniatis. 1983. Identification of two distinct regulatory regions adjacent to the human β-interferon gene. *Cell* **34:** 865.

# Transcriptional Control by the Adenovirus Type-5 E1A Proteins

D.H. Smith, A. Velcich, D. Kegler, and E. Ziff
Department of Biochemistry and Kaplan Cancer Center, New York University Medical Center, New York, New York 10016

The E1A proteins of adenovirus type 5 (Ad5) are best characterized for their role in the activation of transcription of adenovirus early genes and in cell immortalization and transformation. Here we describe novel properties of E1A gene products—the ability to reduce transcription from the SV40 early promoter and potentially to autoregulate their own expression. Our experiments employ vectors that express the 289- and 243-amino-acid (aa) E1A proteins in wild-type or mutant form and contain the 5'-flanking transcription regulatory sequences of the E1A gene as well as the SV40 *ori*, which enables replication in COS-7 cells. Both the 243- and 289-aa E1A proteins repress transcription from the SV40 early promoter, as assayed in HeLa cells, whereas only the 289-aa protein is a transcription activator for Ad5 early genes. We show that under replication conditions in COS-7 cells, a mutant E1A vector that expresses a truncated E1A peptide that lacks the activator and repressor functions substantially overexpresses E1A mRNA, suggesting that wild-type E1A products negatively autoregulate the expression of the E1A gene.

The E1A gene of adenovirus type 5 (Ad5) encodes proteins that are the major regulators of Ad5 transcription. Present evidence suggests the following role for E1A during normal viral infections. *Cis*-acting enhancer elements located upstream from, and within, the E1A gene (Hearing and Shenk 1983; Hen et al. 1983; Imperiale et al. 1983; Weeks and Jones 1983; Osborne et al. 1984) stimulate expression of E1A proteins from the onset of viral infection. Two major E1A proteins, 243 and 289 amino acids (aa) long, are produced that are encoded respectively by 12S and 13S messenger RNAs (mRNAs). The two E1A proteins are amino- and carboxy-coterminal, but since the mRNAs are spliced differently, the larger protein contains a unique 46-aa internal peptide (Perricaudet et al. 1979). Strains of adenovirus that fail to encode E1A peptides, such as the mutant *dl*312, which has a large deletion in the E1A promoter and coding regions (Jones and Shenk 1979a,b), or *hr*1 (Berk et al. 1979), which encodes a truncated 289-aa protein (Ricciardi et al. 1981), fail to activate early promoters, demonstrating a reliance of early transcription upon an E1A-encoded function. The early regions E1B, E2, E3, and E4 all depend upon E1A for their expression, although the kinetics of their activation are not equivalent (Nevins et al. 1979; Nevins 1981). At the close of the early period, the E1A transcription rate, as well as that of the other early transcription units, declines (Shaw and Ziff 1980). E1A transcription increases again with the start of DNA replication and the commencement of the late stage of transcription. However, this increase is less than the magnitude of the increase in transcription of late viral genes or the increase in copy number of adenovirus DNA (Spector et al. 1978; Shaw and Ziff 1980), which raises the possibility that E1A may be negatively regulated during the late stage of infection.

The transcription-activation function of E1A is not confined to viral genes in infected cells. E1A expression plasmids stimulate transcription from viral promoters in cotransfected plasmids (Elkaim et al. 1983; Imperiale et al. 1983; Rossini 1983; Weeks and Jones 1983; Gilardi and Perricaudet 1984). Also, Green et al. (1983) and Svensson and Akusjärvi (1984) showed that cotransfection of an E1A expression plasmid together with a β-globin plasmid stimulated transcription from the cloned β-globin promoter. The Ad5-transformed human embryonic kidney cell line, 293 (Graham et al. 1977), which expresses E1A functions and complements the defects in *dl*312 and *hr*1, also supports transcription from a transfected β-globin promoter (Treisman et al. 1983). Furthermore, E1A directly or indirectly stimulates the expression of the cellular, chromosomal gene for the 70-kD heat-shock protein (Kao and Nevins 1983) and the β-tubulin gene(s) (Stein and Ziff 1984) in HeLa cells. Transcription stimulation by E1A proteins is not confined to genes transcribed by RNA polymerase II inasmuch as Berger and Folk (1985) and Hoeffler and Roeder (1985) have shown that genes transcribed by RNA polymerase III are also stimulated by cotransfection of an E1A expression plasmid.

Experiments employing viruses and plasmids confined to the expression of individual E1A proteins are in agreement that the 289-aa protein can provide the transcription-activation function. In similar experiments transcription activation by the 243-aa protein failed to be observed (Elkaim et al. 1983; Gilardi and Perricaudet 1984; Kingston et al. 1984; Montell et al. 1984; Svensson and Akusjärvi 1984; Winberg and Shenk 1984). However, using an S1 nuclease assay, Leff et al. (1984) found that a 12S cDNA expression plasmid activates transcription from both E2A and E3 vectors, raising the possibility that the 243-aa protein is also a transcription activator. Expression of E1A is a requirement for cell transformation and stimulation of cellular DNA replication by Ad2 and Ad5. Mutant viruses confined to the expression of either the 243- or 289-aa protein will stimulate cellular DNA replication in quiescent human and rodent cells and will transform primary cells or cell lines (Montell et al. 1984; Spindler et al. 1985). Thus it is apparent that both E1A proteins can directly or indirectly influence cellular growth properties. However, both E1A proteins contribute to the full transformed phenotype upon transformation of primary cells (Montell et al. 1984).

We have studied the transcription regulatory properties of the E1A proteins by constructing vectors that encode E1A products in wild-type or mutant form and then determining the effects of the encoded products upon transcription from mutant viral or cloned promoters in transient assays. We have recently reported a novel transcription-repression property of the E1A proteins manifested in their ability to repress the level of transcription from the cloned SV40 promoter (Velcich and Ziff 1985). This repressive function is likely to be related to two previously reported, similar properties of E1A. The first, described by Rossini (1983) and Guilfoyle et al. (1985), negatively regulates the activity of the late promoter of the E2A transcription unit at 72 map units (m.u.). The second, reported by Borrelli et al. (1984), represses transcription from cloned promoters in chimeric plasmids and has been shown to have enhancer elements as its target. In this report, we present evidence that E1A represses transcription from the SV40 early promoter. We also present evidence that, under conditions in which E1A is expressed from a replicative vector, E1A products may repress the E1A promoter and thereby autoregulate the level of their own transcription. We discuss the possibility that this autoregulation operates through a mechanism of enhancer repression. We also discuss the possible significance of E1A autoregulation for the viral life cycle.

## Materials and Methods

### Cells, plasmids, and viruses
COS-7 cells were obtained from Y. Gluzman and grown in Dulbecco's modified minimal essential medium with 10% fetal bovine serum, 100 units/ml penicillin, and 10 μg/ml streptomycin. HeLa cells were grown under identical conditions except that the fetal bovine serum was 5%. pJYM was generously provided by M. Lusky. Construction of wild-type and E1A mutant vectors was described in Smith et al. (1985). The Ad5 mutant, *dl*312, was obtained from T. Shenk (Princeton University) and propagated in 293 cells.

### Transfection and mRNA analysis
COS-7 cells were transfected with 1–20 μg DNA per $10^6$ cells in the presence of 500 μg/ml DEAE-Dextran as described in Smith et al. (1985). HeLa cells were transfected with 5 μg of specific DNA plus 15 μg of pBR322, using the calcium phosphate precipitation technique as described in Velcich and Ziff (1985). Isolation of poly(A)$^+$ mRNA, northern blot analysis, and S1 nuclease analysis were as described previously (Smith et al. 1985).

### Replication assay
100-mm dishes of subconfluent COS cells were transfected with 8 μg of pSVE1a DNA. At the indicated times after transfection, a dish of cells was extracted by the method of Hirt (1967). The DNA obtained was further purified by phenol extraction, precipitated with an equal volume of isopropanol, and an aliquot was cleaved with both *Pst*I and *Mbo*I. The restriction enzyme digests were analyzed on a 1.4% agarose gel, blotted to nitrocellulose, and probed with $^{32}$P-labeled pBR322.

## Results

### Effects of E1A on the expression of SV40 early transcripts
Plasmid vectors that express E1A products were constructed by inserting either the cloned genomic E1A transcription unit or cDNA reconstructions of the E1A gene in wild-type or mutant form into the plasmid pSVOd, a pBR322 variant that contains the SV40 *ori* region and replicates in COS-7 cells. This gave a family of vectors, whose structures are indicated in Figure 1b. The parent vector, pSVE1a, contains genomic adenovirus sequences and encodes both the 289- and 243-aa E1A proteins. Two additional intron-minus vectors related to pSVE1a were pSVN20, a 13S cDNA construction that encodes exclusively the 289-aa E1A protein, and pSVF12, a 12S cDNA construction that encodes the 243-aa E1A protein. Further, a series of mutants in the 289-aa protein were constructed by linearizing pSVN20, introducing small deletions and recircularizing with *Xho*I linkers. Two mutants, described in Figure 5 of Smith et al. (1985), were employed in the current work: (1) pSVXL3, in which Ad5 residues 749–754 were deleted and replaced by an 8-nucleotide-long *Xho*I linker; this shifts the reading frame to give a 70-aa-long truncated peptide. (2) Vector pSVXL105, which has a deletion of 20 residues corresponding to nucleotides 674–693 and an *Xho*I linker replacement; this results in 7 wild-type amino acids being replaced by 3 linker-encoded ones, with no shift of reading frame. The detailed description of the construction of these vectors and mutants is given elsewhere (Smith et al. 1985).

We employed these vectors to examine the effects of E1A proteins on the level of expression of early SV40 transcripts. We transfected the SV40 clone, pJYM, shown in Figure 1a, into HeLa cells in the presence or absence of pSVE1a. The early SV40 promoter in pJYM is constitutively active in HeLa cells by virtue of the 72-bp repeat enhancer elements; and in the absence of pSVE1a, pJYM encoded high levels of SV40 early mRNA, which were detected by northern blotting (Fig. 2A, left lane). To our surprise, when cotransfected with pSVE1a, the level of SV40 early mRNA encoded by

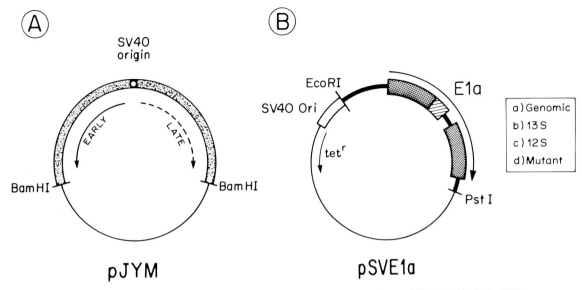

**Figure 1** Structures of pJYM and E1A expression vectors. (A) pJYM, a pBR322 clone of SV40 DNA. (B) pSVE1a, an expression vector that contains genomic sequences from the left end of Ad5. (▓) Sequences common to both 12S and 13S E1A mRNAs; (▨) a sequence exclusive to the 13S mRNA. Variants of pSVE1a, expressing 13S or 12S mRNA cDNA inserts or mutant peptides, are described in the text and in Fig. 5 of Smith et al. (1985).

pJYM was reduced approximately 10-fold (Fig. 2A, right lane). E1A mRNA encoded by the pSVE1a plasmid was, however, readily detectable.

To ascertain the role of SV40 T antigen in the reduction of the levels of early SV40 mRNA, we repeated the experiment using pSV2dhfr, a plasmid in which the

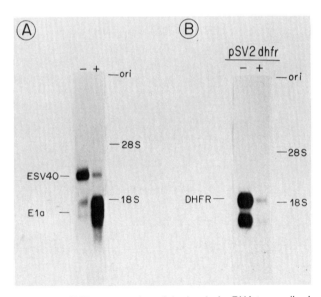

**Figure 2** pSVE1a repression of the level of mRNA transcribed from the SV40 early promoter. Northern blot analyses of the levels of early SV40 mRNA transcribed from pJYM (A) or of DHFR mRNA transcribed from pSV2dhfr (B) in the absence (−) or presence (+) of pSVE1a. Plasmids were introduced into HeLa cells by the calcium phosphate transfection procedure, and mRNA was isolated, fractionated, and probed for E1A and SV40 early region transcripts (A) or DHFR transcripts (B) as described in Materials and Methods and Velcich and Ziff (1985). The positions of 18S and 28S ribosomal RNA markers are given.

SV40 promoter is linked to the dihydrofolate reductase (DHFR) cDNA and which is devoid of T-antigen-coding sequences (Subramani et al. 1981). As shown in the left lane of Figure 2B, DHFR mRNA transcripts were expressed at high levels when pSV2dhfr was transfected on its own. However, when transfected with pSVE1a (Fig. 2B, right lane), the level of DHFR mRNA was depressed approximately 10-fold, as was the case with pJYM-encoded early SV40 mRNA. This experiment rules out any requirement by the repression mechanism for SV40 T antigen. Velcich and Ziff (1985) have examined the ability of the intron-minus and mutant forms of the E1A vectors to repress the SV40 promoter and to complement *dl*312. These results show that whereas the 289- and 243-aa E1A proteins both possess repressor function, only the 289-aa protein is a *dl*312 E2 promoter transcription activator. Vector pSVXL105, the in-frame deletion, linker-substitution mutant, behaves as its wild-type parent, whereas pSVXL3, which encodes the truncated peptide, lacks both activation and repression functions.

We next examined the properties of the E1A expression vectors when transfected into COS-7 cells, an SV40-transformed CV-1 monkey cell line that expresses T antigen from *ori*-defective, integrated SV40 sequences (Gluzman 1981). Vectors that contain the SV40 *ori* replicate in COS-7 cells, and Figure 3 shows this to be the case for pSVE1a. In the assay, vector replication in COS-7 cells yields DNA that lacks bacterial methylations and thus is cleaved by *Mbo*I. Very little of the DNA has replicated at 5 hr posttransfection, since the DNA can be linearized by *Pst*I but is not further cut by *Mbo*I. At 24 hr posttransfection, newly replicated DNA cleaved by *Mbo*I is detectable, and at 48 hr posttransfection all vector DNA in the cells is the product of replication.

We have previously determined the ability of the E1A

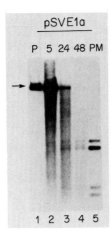

**Figure 3** Southern blot analysis of vector DNA replication after transfection into COS-7 cells. pSVE1a was transfected into COS-7 cells and low-molecular-weight DNA was isolated after 5 hr (lane 2), 24 hr (lane 3), and 48 hr (lane 4), digested with PstI and MboI, fractionated on a 1% agarose gel, and probed with $^{32}$P-labeled HpaI-E fragment (0–4.5 m.u. Ad2). Unreplicated DNA contains bacterial methylations of the MboI recognition sequence that blocks cleavage, but DNA replicated in COS-7 cells is unmethylated and therefore is sensitive to the enzyme. Standards were pSVE1a cleaved with PstI (lane 1) or with both PstI and Sau3A, an isoschizomer of MboI that cleaves despite the bacterial methylations (lane 5).

vectors to complement dl312 in HeLa cells under conditions that do not permit vector replication (Velcich and Ziff 1985). In these assays, transcription activation by E1A products encoded by the vectors results in the production of the E2A-encoded 72-kD DNA-binding protein, whose presence was assayed by indirect immuno-

fluorescence. To determine whether our E1A vectors would complement dl312 in COS-7 cells under conditions of vector replication, we transfected COS-7 cells with the various vectors, and 24 hr later we superinfected with dl312. After an additional period of 24 hr, we harvested cytoplasmic RNA and analyzed the poly(A$^+$) fraction for E2A mRNA by northern blotting. In Figure 4A, lane 2 shows that the wild-type vector pSVE1a stimulated expression of E2A mRNA, whereas lane 1 shows that a control vector lacking the adenovirus insert gave no stimulation. The 13S cDNA reconstruction, pSVN20 (Fig. 4A, lane 3), stimulated E2A mRNA with approximately the same efficiency as the genomic pSVE1a vector. pSVF12, the 12S cDNA reconstruction vector (lane 4), failed to stimulate significantly. pSVXL105, which has an in-frame deletion plus linker substitution near the amino terminus of the 286-aa protein, stimulated E2A mRNA production in lane 5, whereas pSVXL3, the linker-insertion mutant that shifts the reading frame of the 289-aa protein to yield a truncated 70-aa peptide, failed to stimulate production (lane 6). These results closely parallel the dl312 complementation activities of the vectors in HeLa cells (Velcich and Ziff 1985) and are summarized in Table 1.

We also measured the level of expression of E1A mRNA by the vectors by northern blotting (Fig. 4B). In lanes 2–4 of Figure 4B, pSVE1a, pSVN20, and pSVF12 all expressed similar levels of E1A mRNA. Likewise, pSVXL105, which encodes the nearly wild-type 286-aa polypeptide, gave a similar level of E1A mRNA in lane 5. Surprisingly, pSVXL3, which encodes the truncated peptide, greatly overexpressed E1A messenger in lane

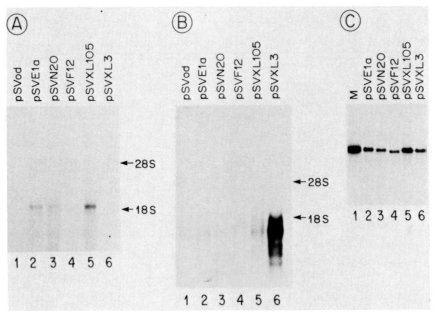

**Figure 4** Levels of mRNAs and plasmids in COS-7 cells transfected with E1A expression vector plasmids. COS-7 cells were transfected with the designated vector superinfected with dl312 24 hr later, and after an additional 24 hr cytoplasmic RNA and low-molecular-weight DNA were harvested. Polyadenylated RNA was isolated from cytoplasmic RNA by poly(U) Sepharose chromatography. A shows a northern blot analysis for E2 mRNA, and B shows a similar analysis for E1A mRNA. In C, low-molecular-weight DNA was linearized by PstI digestion and analyzed for vector sequence by Southern blotting with an E1A DNA probe. Details of procedures are given in Smith et al. (1985).

**Table 1** Properties of Wild-type and Mutant E1A Vectors

| Vector[a] | E1A protein product[a] | E2 activation[b] | SV40 repression[c] | Expression of E1A mRNA[d] |
|---|---|---|---|---|
| pSVE1a | 289 + 243 aa | + | + | + |
| pSVN20 | 289 aa | + | + | + |
| pSVF12 | 243 aa | − | + | + |
| pSVXL105 | 285 aa mutant | + | + | + |
| pSVXL3 | 70 aa mutant | − | − | + + + + + |

[a]Vector structure, construction, and protein-coding capacity are given in Smith et al. (1985).
[b]Activation of the E2 gene in dl312 by vectors in COS-7 cells is from the experiment shown in Fig. 4A.
[c]SV40 early transcript repression by vectors in HeLa cells is from Velcich and Ziff (1985).
[d]Level of expression of E1A mRNA by vectors in COS-7 cells is from Fig. 4B.

6. To determine the level of E1A mRNA per copy of plasmid template, we used the Hirt (1967) procedure to isolate low-molecular-weight DNA from the same cells employed for the northern blot assay and measured plasmid levels by Southern blotting. The result, shown in Figure 4C, show that in this experiment, all vectors were present at equivalent levels, as indicated by comparison of lanes 2–6.

Taken together, these data suggest that under replicative conditions in COS-7 cells, pSVXL3, which encodes the truncated E1A polypeptide, overexpresses E1A mRNA relative to other vectors that encode either wild-type or nearly wild-type E1A polypeptides. This overexpression suggests that the wild-type E1A polypeptide products have a role in establishing the levels of E1A mRNA encoded by the vector templates. One model that explains the overexpression of mRNA by pSVXL3 suggests that the level of E1A mRNA encoded by the vectors is established by a trans-acting repression of the E1A gene by the E1A proteins themselves. If this autoregulatory model for the overexpression by pSVXL3 is correct, when pSVXL3 is transfected into COS-7 cells in the presence of a wild-type vector such as pSVE1a, the pSVXL3 mutant E1A mRNA level will be reduced by wild-type E1A protein products acting upon the mutant gene in trans. To test this prediction, we transfected COS-7 cells with either pSVE1a or pSVXL3 alone, or with a mixture of the two vectors. We then harvested cytoplasmic RNA from the three cultures, isolated the poly(A⁺) fraction, and determined the levels of E1A mRNA encoded by each vector. To determine the levels of mutant and wild-type E1A mRNA species in the presence of each other, we performed an S1 nuclease assay, using a wild-type DNA probe 5′-labeled at a residue within the E1A first exon. When hybridized to wild-type mRNA, the probe is cleaved opposite the cap site, giving a band 319 residues long, whereas the hybrid of the probe to pSVXL3 encoded mRNA is cleaved opposite the linker-encoded mRNA residues, giving a band 68 nucleotides long.

The results of the S1 analysis are given in Figure 5. Lane 1 shows the migration of the undigested probe, while lanes 2 and 4 show the products of digestion of hybrids to pSVXL3-encoded and pSVE1a-encoded mRNAs, respectively. As in the experiment shown in Figure 5, pSVXL3 overexpressed E1A mRNA relative to the pSVE1a wild type. Table 2 shows the quantification of the E1A mRNAs from S1 nuclease hybrid bands from the gel of Figure 5 and the plasmid DNA levels from Southern blots of Hirt extracts (not shown). Overexpression deduced from the levels of the S1 bands was approximately 50-fold when the Hirt analysis of the levels of plasmids in the transfected cells are taken into account. Lane 3 of Figure 5 analyzes the mRNA from the cells cotransfected with the wild-type and mutant vectors. The level of mutant mRNA was reduced 5-fold. This result is consistent with a trans-acting factor encoded by the wild-type pSVE1a vector diminishing the level of mutant E1A mRNA from vector pSVXL3 in the cotransfected cells.

**Table 2** Properties of Wild-type and Mutant E1A Vectors in COS-7 Cells

| Transfected vectors | Vector DNA level | Normalized vector DNA level | Vector mRNA level | Normalized Vector mRNA level | Normalized vector mRNA level corrected for DNA level |
|---|---|---|---|---|---|
| pSVXL3 alone | 9.7 | 5.4 | 514 | >129 | >24 |
| pSVXL3 (+ pSVE1a) | 18 | 1 | 19 | 4.75 | 4.8 |
| pSVE1a (+ pSVXL3) | 44 | 2.4 | ND | ND | ND |
| pSVE1a alone | 34.4 | 1.9 | 4.0 | 1 | 0.5 |

Levels of DNA bands in autoradiograms of Southern blots and mRNA hybrid bands in S1 nuclease analysis were determined by densitometric scanning. The data are from the S1 nuclease experiment shown in Fig. 5. (ND) Not determined.

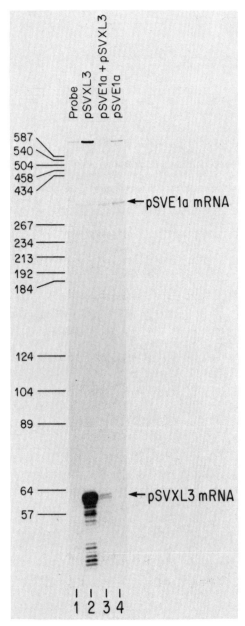

**Figure 5** Effect upon E1A mRNA levels of cotransfection of wild-type and frame-shift mutant E1A expression vectors. COS-7 cells were transfected with plasmids pSVXL3 (frame-shifted vector) and pSVE1a (genome vector), either separately or in combination as indicated, and polyadenylated cytoplasmic RNA was analyzed for E1A mRNA encoded by each vector by means of the S1 mapping procedure. The positions of S1-resistant hybrids to 5'-labeled DNA probe are given for pSVXL3 mRNA and for pSVE1a mRNA. The structure of the probe is given in the text. Migration of *Hae*III-digested pBR322 DNA size markers is indicated.

## Discussion

The Ad5 E1A gene plays a complex role in regulating viral and host gene transcription during infection. The ability of E1A-encoded proteins to activate transcription of both viral and cellular genes has been widely documented. In this paper we present evidence con-

cerning a new activity of E1A, the ability to repress transcription from both homologous and heterologous viral promoters. This activity is likely to be related to that described in reports from two other laboratories. First, Rossini (1983) and Guilfoyle et al. (1985) have shown that E1A can repress transcripts from the E2A late promoter (located at 72 m.u.). Second, Borrelli et al. (1984) have shown that E1A represses the transcription of chimeric plasmid constructs that rely upon enhancer elements for their activity. Using an enhancer competition assay, Borrelli et al. (1984) provided evidence that the enhancer was the target of the E1A repression mechanism.

We have shown that the wild-type SV40 early promoter is susceptible to the repressive activity of the E1A proteins (Velcich and Ziff 1985; and this report). This repression requires neither the SV40 large-T-antigen protein nor replication of the target promoter. It is unlikely that the E1A proteins have evolved to recognize features of the SV40 early promoter that are exclusive to SV40. Because current evidence (Borrelli et al. 1984; Velcich and Ziff 1985) indicates that the target for the repressive action is the SV40 enhancer element, a function of the E1A proteins may be to repress transcription of enhancer-dependent genes during the normal viral life cycle. Hearing and Shenk (1983), Hen et al. (1983), Weeks and Jones (1983), and Osborne et al. (1984) have shown that the E1A gene contains multiple enhancer elements. Thus, one Ad5 gene that relies upon an enhancer is the early region E1A gene itself. Indeed, expression of the E1A gene is likely to be regulated in a complex fashion during the course of viral infection. During the early stages of infection, E1A proteins are reported to stimulate the rate of transcription of the E1A gene itself (Berk et al. 1979; Nevins 1981). However, after the early to late shift, the rate of transcription of E1A sequences increases only modestly compared with the rate of transcription of sequences of the major late transcription unit despite a dramatic increase in the number of potential template molecules (Spector et al. 1978; Shaw and Ziff 1980). Because the vectors employed here replicate in the transfected COS-7 cells, our system provides an analog of the late infected cell in two respects. First, the E1A gene has replicated in the host prior to assay of its transcriptional activity. Replication could influence structural properties of the gene, for example, the physical association of vector DNA with protein. Second, the copy number of the vector is increased, and thus, per cell, both the potential for expression of E1A mRNA and the requirements for transcription factors are elevated.

Table 1 correlates, for each of our E1A vectors, the level of E1A mRNA in COS-7 cells with the transcription activation properties of the vectors. Data for repressor function in HeLa cells from Velcich and Ziff (1985) is also shown. Note that for the several vectors, the level of expression of E1A mRNA in COS-7 cells is directly related to the ability of the vectors to provide the E1A repression function, as assayed in HeLa cells with an SV40 early promoter target. In agreement with the results of Borrelli et al. (1984), we have found that both the 289- and 243-aa E1A proteins are repressors in this assay, although only the larger protein is a transcription activator (Velcich and Ziff 1985). In this report, we show that all vectors encoding at least one of these wild-type products express equivalent levels of E1A mRNA. In particular, vectors encoding exclusively the 289- or 243-aa proteins, pSVN20 and pSVF12, yield equal levels of E1A mRNA. This equivalence suggests that expression of the E1A transcription-activation function is not the determining factor for E1A mRNA levels. However, the vector pSVXL3, which expresses the frame-shifted 70-aa truncated product, a product that lacks both activator and repressor functions (see Table 1), overexpresses E1A mRNA. Therefore we suggest that under the conditions of our experiments, the level of expression of E1A mRNA reflects the ability of the vector to encode E1A protein, which is capable of acting as a repressor product.

A model that accounts for these results is that the E1A proteins, under the conditions of our experiment, are capable of negatively autoregulating the level of their own transcription. The model is based on data obtained with plasmid vectors that express E1A functions under replicative conditions in COS-7 cells. This model as applied to productive infection suggests a hypothetical role for E1A and is shown in Figure 6. The model proposes two roles for E1A proteins in regulation of viral expression, in particular the E1A gene. One is an auto-stimulatory role, as suggested by the results of Berk et al. (1979) and Nevins (1981). The second, which may require high levels of E1A protein and/or vector replication, is a negative autoregulation. The fact that the E1A gene depends on enhancer elements for its transcriptional activity (Hearing and Shenk 1983; Hen et al. 1983; Imperiale et al. 1983; Weeks and Jones 1983;

Osborne et al. 1984) and that E1A proteins display enhancer-repressor function (Borrelli et al. 1984; Velcich and Ziff 1985) further suggests that the mechanism for autoregulation operates by enhancer repression. Borrelli et al. (1984) have suggested a similar, negative autoregulatory role for E1A proteins during infection, which relies on enhancer repression. In a formal sense, the model for E1A autoregulation resembles the negative autoregulation of SV40 early gene expression by SV40 T antigen (Tegtmeyer et al. 1975; Reed et al. 1976; Alwine et al. 1977; Khoury and May 1977; Rio et al. 1980). Like E1A, SV40 T antigen has recently been shown to have a positive-acting as well as a negative-acting role in viral transcription, stimulating synthesis from the late SV40 transcription unit (Khoury and May 1977; Rosenthal and Brown 1977; Brady et al. 1984; Keller and Alwine 1984). Additionally, the herpesvirus immediate-early protein, ICP4, which is a transcription activator (Preston 1979; Watson and Clements 1978), also displays a form of negative autoregulation (Preston 1979; Dixon and Schaffer 1980; DeLuca and Schaffer 1985) by an unknown mechanism. Negative auto-regulation may be a common feature for viral (and perhaps cellular) transcription factors.

Our experiments have utilized a system in which E1A expression vector plasmids replicate to high copy number. We have suggested that vector replication may provide an analog of the late phase of adenovirus infection when templates increase in number through extensive replication. However, we emphasize that neither a requirement nor a precise role for replication has yet been established by our studies. Further, to establish a role for this mechanism during the viral life cycle will require investigation of transcription regulation with viral templates during normal viral infection.

In summary, we have shown that the E1A proteins of

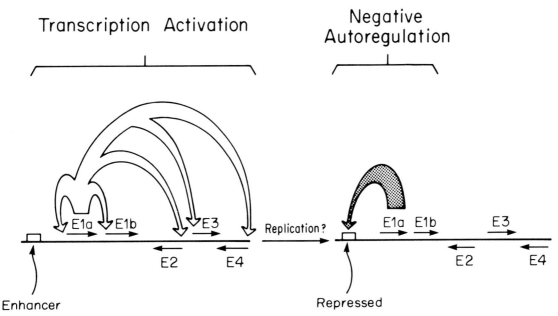

Figure 6 Hypothetical model for positive and negative regulation of the E1A gene by E1A protein products during viral infection.

adenovirus possess a novel transcription-repression activity, and we have proposed a role for this activity in limiting the expression of E1A transcripts through a negative autoregulation mechanism. These results raise the provocative possibility that E1A may serve a complex transcription regulatory role for viral genes during the infectious cycle. Indeed, a similar regulatory activity may extend to the virus's effects on cellular genes and may play a role in the activity of E1A in cell immortalization or transformation.

## Acknowledgments

This work was supported by grants MV-75 from the American Cancer Society, GM 30760 from the U.S. National Institutes of Health, and a Cancer Center Core Support Grant no. P30CA-16087. E.B.Z. is the recipient of a Faculty Research Award FRA 207 from the American Cancer Society. D.H.S. is a predoctoral trainee supported by Public Health Service Institutional Research Service Award 5 T32 GM07238-08 0051 from the National Institutes of Health. We thank R. Stein for supplying *dl*312 and Willy Welch for typing the manuscript. We are grateful to J. Thomas for his assistance in the use of the densitometer.

## References

Alwine, J.C., S.I. Reed, and G.R. Stark. 1977. Characterization of the autoregulation of simian virus gene *A*. *J. Virol.* **24**: 22.

Berger, S.L. and W.R. Folk. 1985. Differential activation of RNA polymerase III–transcribed genes by the polyomavirus enhancer and the adenovirus E1a gene products. *Nucleic Acids Res.* **13**: 1413.

Berk, A.J., F. Lee, T. Harrison, T. Williams, and P.A. Sharp. 1979. Pre-early adenovirus-5 gene product regulates synthesis of early viral messenger RNAs. *Cell* **17**: 935.

Borrelli, E., R. Hen, and P. Chambon. 1984. The adenovirus-2 E1a products repress stimulation of transcription by enhancers. *Nature* **312**: 608.

Brady, J., J.B. Bolen, M. Radonovich, N. Salzman, and G. Khoury. 1984. Stimulation of simian virus 40 late gene expression by simian virus 40 tumor antigen. *Proc. Natl. Acad. Sci.* **81**: 2040.

DeLuca, N.A. and P.A. Schaffer. 1985. Activation of immediate-early, early, and late promoters by temperature-sensitive and wild-type forms of herpes simplex virus type 1 protein ICP4. *Mol. Cell. Biol.* **5**: 1997.

Dixon, R.A.F. and P.A. Schaffer. 1980. Fine structure mapping and functional analysis of temperature-sensitive mutants in the gene encoding the herpes simplex virus type 1 immediate early protein VP175. *J. Virol.* **36**: 189.

Elkaim, R., C. Goding, and C. Kédinger. 1983. The adenovirus-2 EII early gene promoter: Sequences required for efficient *in vitro* and *in vivo* transcription. *Nucleic Acids Res.* **11**: 7105.

Gilardi, P. and M. Perricaudet. 1984. The E4 transcriptional unit of Ad2: Far upstream sequences are required for its transactivation by E1a. *Nucleic Acids Res.* **12**: 7877.

Gluzman, Y. 1981. SV40-transformed simian cells support the replication of early SV40 mutants. *Cell* **23**: 175.

Graham, F.L., J. Smiley, W.C. Russell, and R. Nairu. 1977. Characteristics of a human cell line transformed by DNA from human adenovirus type 5. *J. Gen. Virol.* **36**: 59.

Green, M.R., R. Treisman, and T. Maniatis. 1983. Transcriptional activation of cloned human β-globin genes by viral immediate-early gene products. *Cell* **35**: 137.

Guilfoyle, R., W. Osheroff, and M. Rossini. 1985. Two functions encoded by adenovirus early region 1A are responsible for the activation and repression of the DNA-binding protein gene. *EMBO J.* **4**: 707.

Hearing, P. and T. Shenk. 1983. The adenovirus type 5 E1a transcriptional control region contains a duplicated enhancer element. *Cell* **33**: 695.

Hen, R., E. Borrelli, P. Sassone-Corsi, and P. Chambon. 1983. An enhancer is located 340 bp upstream from the adenovirus-2 E1a capsite. *Nucleic Acids Res.* **11**: 8747.

Hirt, B. 1967. Selective extraction of polyoma DNA from infected mouse cultures. *J. Mol. Biol.* **26**: 365.

Hoeffler, W.K. and R.G. Roeder. 1985. Enhancement of RNA polymerase III transcription by the E1a gene product of adenovirus. *Cell* **41**: 955.

Imperiale, M.J., L.T. Feldman, and J.R. Nevins. 1983. Activation of gene expression by adenovirus and herpesvirus regulatory genes acting in *trans* and by a *cis*-acting adenovirus enhancer element. *Cell* **35**: 127.

Jones, N. and T. Shenk. 1979a. Isolation of adenovirus type 5 host range deletion mutants defective for transformation of rat embryo cells. *Cell* **17**: 683.

———. 1979b. An adenovirus type 5 early gene function regulates expression of other early viral genes. *Proc. Natl. Acad. Sci.* **76**: 3665.

Kao, H.T. and J.R. Nevins. 1983. Transcriptional activation and subsequent control of the human heat shock gene during adenovirus infection. *Mol. Cell. Biol.* **3**: 2058.

Keller, J.M. and J.C. Alwine. 1984. Activation of the SV40 late promoter: Direct effects of T antigen in the absence of viral DNA replication. *Cell* **36**: 381.

Khoury, G. and E. May. 1977. Regulation of early and late simian virus 40 transcription: Overproduction of early viral RNA in the absence of functional T antigen. *J. Virol.* **23**: 167.

Kingston, R., R. Kaufman, and P. Sharp. 1984. Regulation of transcription of the adenovirus EII promoter by E1a gene products: Absence of sequence specificity. *Mol. Cell. Biol.* **4**: 1970.

Leff, T., R. Elkaim, C.R. Goding, P. Jalinot, P. Sassone-Corsi, M. Perricaudet, C. Kédinger, and P. Chambon. 1984. Individual products of the adenovirus 12S and 13S E1a mRNAs stimulate viral E2a and E3 expression at the transcriptional level. *Proc. Natl. Acad. Sci.* **81**: 4381.

Montell, C., G. Courtois, C. Eng, and A. Berk. 1984. Complete transformation by adenovirus-2 requires both E1a proteins. *Cell* **36**: 951.

Nevins, J.R. 1981. Mechanisms of activation of early viral transcription by the adenovirus E1a gene product. *Cell* **26**: 213.

Nevins, J.R., H.W. Ginsberg, J.M. Blanchard, M.C. Wilson, and J.E. Darnell. 1979. Regulation of the primary expression of the early adenovirus transcription units. *J. Virol.* **32**: 727.

Osborne, T.F., D.N. Arvidson, E.S. Tyau, M. Dunsworthe-Brown, and A.J. Berk. 1984. Transcription control region within the protein-coding portion of adenovirus E1a genes. *Mol. Cell. Biol.* **4**: 1293.

Perricaudet, M., G. Akusjärvi, A. Virtanen, and U. Petterson. 1979. Structure of two spliced mRNAs from the transforming region of human subgroup C adenoviruses. *Nature* **281**: 694.

Preston, C.M. 1979. Control of herpes simplex virus type 1 mRNA synthesis in cells infected with wild-type virus or the temperature sensitive mutant, *tsk*. *J. Virol.* **29**: 275.

Reed, S.I., G.R. Stark, and J.C. Alwine. 1976. Autoregulation of simian virus 40 gene *A* by T antigen. *Proc. Natl. Acad. Sci.* **75**: 3083.

Ricciardi, R.P., R.L. Jones, C.P. Cepko, P.A. Sharp, and B.E. Roberts. 1981. Expression of early adenovirus genes requires a viral encoded acidic polypeptide. *Proc. Natl. Acad. Sci.* **78**: 6121.

Rio, D., A. Robbins, R. Myers, and R. Tjian. 1980. Regulation of simian virus 40 early trasncription *in vitro* by a purified tumor antigen. *Proc. Natl. Acad. Sci.* **77**: 5706.

Rosenthal, L.J. and M. Brown. 1977. The control of SV40

transcription during a lytic infection: Late RNA synthesis in the presence of inhibitors of DNA replication. *Nucleic Acids Res.* **4**: 551.

Rossini, M. 1983. The role of adenovirus early region 1A in the regulation of early region 2A and 1B expression. *Virology* **131**: 49.

Shaw, A.R. and E.B. Ziff. 1980. Transcripts from the adenovirus-2 major late promoter yield a single family of 3' co-terminal mRNAs during early infection and five families at late times. *Cell* **22**: 905.

Smith, D.H., D.M. Kegler, and E.B. Ziff. 1985. Vector expression of the adenovirus-5 E1a proteins: Evidence for E1a autoregulation. *Mol. Cell. Biol.* **5**: 2684.

Spector, D.J., M. McGrogan, and H.J. Raskas. 1978. Regulation of the appearance of cytoplasmic RNAs from region 1 of the adenovirus 2 genome. *J. Mol. Biol.* **126**: 395.

Spindler, K.R., C.Y. Eng, and A.J. Berk. 1985. An adenovirus early region 1a protein is required for maximal viral DNA replication in growth-arrested human cells. *J. Virol.* **53**: 742.

Stein, R. and E.B. Ziff. 1984. HeLa cell β-tubulin gene transcription is stimulated by adenovirus-5 in parallel with viral early genes by an E1a-dependent mechanism. *Mol. Cell. Biol.* **4**: 2792.

Subramani, S., R. Mulligan, and P. Berg. 1981. Expression of the mouse dihydrofolate reductase complementary deoxyribonucleic acid in simian virus 40 vectors. *Mol. Cell. Biol.* **2**: 854.

Svensson, C. and G. Akusjärvi. 1984. Adenovirus 2 early region 1A stimulates expression of both viral and cellular genes. *EMBO J.* **3**: 789.

Tegtmeyer, P., M. Schwartz, J.K. Collins, and K. Rundell. 1975. Regulation of tumor antigen synthesis by simian virus 40 gene A. *J. Virol.* **16**: 168.

Treisman, R., M.R. Green, and T. Maniatis. 1983. *Cis* and *trans* activation of globin gene transcription in transient assays. *Proc. Natl. Acad. Sci.* **80**: 7428.

Velcich, A. and E. Ziff. 1985. Adenovirus E1a proteins repress transcription from the SV40 early promoter. *Cell* **40**: 705.

Watson, R.J. and J.B. Clements. 1978. Characterization of transcription deficient temperature-sensitive mutants of herpes simplex type 1. *Virology* **9**: 364.

Weeks, D.L. and N.C. Jones. 1983. E1a control of gene expression is mediated by sequences 5' to the transcriptional start of the early viral genes. *Mol. Cell. Biol.* **3**: 1222.

Winberg, G. and T. Shenk. 1984. Dissection of overlapping functions within the adenovirus type 5 E1a gene. *EMBO J.* **3**: 1907.

# Identification of the DNA Sequences in the Adenovirus E2A Late Promoter That Are Responsive to E1A Repression

**R. Guilfoyle,\* M.R. Fontana,[†] M. Mora,[†] W. Osheroff,\*[††] and M. Rossini[†]**

\*Cold Spring Harbor Laboratory, Cold Spring Harbor, New York 11724; [†]Sclavo Research Center, Molecular Biology Laboratory, 53100 Siena, Italy

Using different transient expression assays, we have studied the interaction between adenovirus 2 early regions 1A (E1A) and 2A (E2A). We have been able to assign two different functions to the products encoded by the E1A gene. The 289-amino-acid (aa) protein encoded by the E1A 13S mRNA is responsible for the activation of the E2A early promoter, while the 243-aa protein encoded by the 12S mRNA represses transcription from the E2A late promoter.

To identify the sequences at the E2A late promoter region that are responsive to the repression, we have constructed a set of deletion mutants by BAL-31 digestion and analyzed them by a microinjection assay. The results of these experiments suggest that a DNA region from nucleotide position −87 to −51 from the E2A late cap site contains the E1A regulatory interaction site.

In human cells infected with adenovirus, the expression of early genes depends upon products encoded by the viral early region 1A (E1A) transcription unit, which acts in *trans* to activate transcription from the other viral early promoters (Berk et al. 1979; Jones and Shenk 1979a; Nevins 1981; Montell et al. 1982; Rossini 1983; Guilfoyle et al. 1985), as well as certain cellular genes (Kao and Nevins 1983; Stein and Ziff 1984). E1A is also required for transformation of rodent cells by adenovirus (Graham et al. 1978; Jones and Shenk 1979b; Houwelling et al. 1980). In addition, E1A is responsible for establishment functions and enables the oncogene T24 Ha-*ras* to transform primary BRK cells (Ruley 1983).

Two differentially spliced mRNAs (12S and 13S) are transcribed from E1A during the early phase of adenovirus lytic infection (Berk and Sharp 1978; Chow et al. 1979; Kitchingman and Westphal 1980). Their translation products, which are 243 and 289 amino acids (aa) long, respectively (Perricaudet et al. 1979; Ricciardi et al. 1981; Gaynor et al. 1982), have the amino- and carboxyterminal portions in common. The principal role of the larger peptide is promoter activation. Both products have a role in transformation (Solnick and Anderson 1982; Montell et al. 1984) and the ability to reduce transcription from the early SV40 promoter by interaction with the *cis*-acting SV40 enhancer elements (Borrelli et al. 1984; Velcich and Ziff 1985).

As described in this report, we have analyzed the regulatory interaction of E1A with early region 2A, which codes for the 72K DNA-binding protein (DBP). The E2A gene is transcribed from different promoters during the infectious cycle. At early times, the majority of transcripts have 5' ends at map coordinate 75. At late times,

the DBP mRNA is transcribed primarily from its late promoter, located at coordinate 72 (Westphal and Lai 1977; Chow et al. 1979).

We have previously reported that these two promoters are differentially regulated by the E1A product(s). The E1A gene product(s) activates the expression of DBP transcribed from the early promoter and inhibits the expression of DBP from its late promoter (Rossini 1983; Guilfoyle et al. 1985).

Using E1A plasmids derived from different mutant viruses (Ad5*hr*1 and Ad2/5 pm975), we have been able to assign a function for each of the E1A products in the E2A regulatory interaction. The 289-aa polypeptide is responsible for the activation of the E2A early promoter, and the product encoded by the 12S mRNA (243 aa long) represses transcription from the late promoter. By testing a series of deletion mutants in the E2A late promoter region, we suggest that the DNA sequences between coordinates −87 and −51 from the E2A late cap site are responsive to E1A repression and that a region between coordinates −51 and −33 is essential for the synthesis of DBP from its late promoter.

## Experimental Procedures

### Cell culture
Thymidine kinase–deficient hamster fibroblasts, TK⁻ ts13 (Jonak and Baserga 1980), and HeLa cells were maintained in Dulbecco's modified Eagle's medium supplemented with 10% calf serum or 10% fetal calf serum, respectively.

### Plasmids
Plasmid L-251.1, containing the late promoter, was constructed by inserting the Ad2 *Bam*HI-*Hind*III fragment from map coordinates 59.5 to 72.8 into the corresponding sites in pBR322.

[††]*Present address*: Department of Biochemistry and Nutrition, University of North Carolina at Chapel Hill, Chapel Hill, North Carolina 27514.

Plasmid E-252.1, which contains only the early promoter, was constructed by insertion of the Ad2 *Hind*III-*Eco*RI fragment (coordinates 72.8–75.9) upstream of the *Bam*HI-*Eco*RI fragment (coordinates 59.5–70.7) cloned in pBR322 at the *Bam*HI and *Hind*III sites. The vector was modified by introducing a *Hind*III linker at the *Eco*RI site (coordinate 70.7) to facilitate the construction.

The E2A early promoter–CAT plasmid (p2CAT) was kindly given to us by Dr. N. Jones (Weeks and Jones 1983). For the construction of the 251CAT plasmid, the L-251.1 plasmid was digested with *Kpn*I (coordinate 71.9) and treated briefly with BAL-31 nuclease to remove the presumptive donor splice junction of the E2A late 5′ leader sequence. *Hind*III linkers were ligated to the blunt ends. Subsequent cleavage with *Hind*III released a segment that also contained 420 bp of the E2A late promoter region up to the *Hind*III site located at coordinate 72.8. The deleted fragments were purified and inserted into the *Hind*III site of pSV0CAT (Gorman et al. 1982) in both orientations. The isolate used in this report contains the promoter in the correct orientation to drive expression of the CAT gene. As expected, plasmid 251CAT contains a small deletion (5–10 nt) at the 3′ end of the E2A late leader, the extent of which has not been determined precisely. In pSV0CAT reside the CAT gene, SV40 small T intron, and the SV40 early polyadenylation site.

Plasmid DNAs were prepared according to the method of Birnboim and Doly (1979) and purified by cesium chloride–ethidium bromide equilibrium density gradient centrifugation (Radloff et al. 1967).

*Site-directed mutagenesis*
The E2A late promoter/leader fragment (*Eco*RI-*Hind*III, coordinates 70.7–72.8) was cloned in pBR322. The plasmid was then linearized by digestion with *Hind*III (i.e., at the 5′ side of the promoter). The linearized plasmid was digested for various times with BAL-31 nuclease as described by Guilfoyle and Weinmann (1981). *Hind*III synthetic linkers were joined to the 5′ end of deleted fragments using T4 DNA ligase.

The deletion fragments were released from the vector by digestion with *Eco*RI and *Hind*III, purified by agarose gel electrophoresis, and eluted using the NaI glass method (Vogelstein and Gillespie 1979).

The L-251.1 plasmid described above (see Fig. 1) was digested with *Eco*RI and *Hind*III to remove the wild-type promoter fragment. The large fragment containing the DBP coding sequences and the first internal leader was purified by agarose gel electrophoresis and eluted, then incubated with the purified 5′-end deletion fragments in the presence of T4 DNA ligase and used to transfect *Escherichia coli* strain HB101. The exact end point of the deletions was determined by sequence analysis using the Maxam and Gilbert method (1980).

*Microinjection and immunofluorescence*
Nuclear microinjection was performed into TK⁻ts13 cells according to the method of Graessmann et al. (1980). TK⁻ts13 cells were cultured in the presence of glass slides (22 × 30 mm) numbered and marked with circles. A different DNA sample was injected into each circle. About 200 copies/cell of plasmid DNA suspended in 10 mM Tris·HCL (pH 7.2) was injected.

Twenty hr after microinjection, TK⁻ts13 cells were fixed for 15 min with methanol at −20°C. The Ad2 72K DBP was visualized in microinjected cells by indirect immunofluorescence, using hamster anti-72K serum as the first antibody and fluorescein-isothiocyanate-labeled goat anti-hamster globulin as the second antibody.

*DNA transfection*
HeLa cells were plated at a density of $10^4$ cells/cm² in 100-mm plates on the day before transfection in Dulbecco's modified Eagle's medium containing 10% fetal calf serum. Four hr before transfection, the cells were refed with fresh medium. Calcium phosphate–DNA precipitates were prepared according to the procedure of Wigler et al. (1978). Precipitations were carried out for 10–15 min and contained the amounts of DNA indicated in the legend to Figure 4. One ml of the precipitate was added to each plate of cells, which were then covered

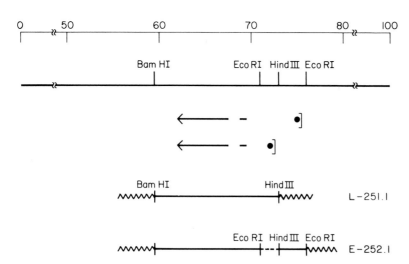

**Figure 1** Plasmids containing the DBP gene associated with its early or late promoter. A simplified depiction of transcription from E2A is shown. At early times during infection, the majority of mRNAs have 5′ ends at map coordinate 75 and, at late times, at coordinate 72. The plasmid L-251.1 contains sequences encoding the DBP late transcription unit. The plasmid E-252.1 contains sequences comprising the DBP early transcription unit that have been deleted in the late promoter region between coordinates 70.7 and 72.8. The construction of these plasmids is described in Experimental Procedures. (Reprinted, with permission, from Guilfoyle et al. 1985.)

with 2 ml of medium and incubated for 15 min at room temperature. The volume of medium was brought to 10 ml and the cells were incubated at 37°C for 36–48 hr. Fresh medium was added at 24 hr posttransfection.

### Assay for CAT activity in HeLa cells

The assay for chloramphenicol acetyltransferase (CAT) activity was performed essentially as described by Gorman et al. (1982), using some minor modifications. The washed cell pellets were suspended in 100 μl of 0.25 M Tris (pH 8.7) and sonicated briefly. The sonicated cells were centrifuged for 15 min at 4°C in an Eppendorf microfuge. The entire supernatant was assayed for CAT activity by the addition of 1 μCi of [$^{14}$C]chloramphenicol (40 mCi/mmole; New England Nuclear), 0.5 μl of 40 mM acetyl Coenzyme A, and 100 μl of water. The reactions were carried out for 60 min at 37°C and then extracted with 1 ml of ethyl acetate. The organic layer was dried, taken up in 20 μl of ethyl acetate, and spotted onto silica gel thin-layer chromatography plates (Eastman Kodak Co., no. 13179). The plates were developed in chloroform/methanol (95:5).

## Results

### Expression of the DBP gene

To determine the effect of the E1A products on the expression of the DBP gene, the following plasmids were used for nuclear microinjection into hamster TK$^-$ts13 cells:

1. *E2A plasmids*: E-252.1 contains the DBP gene associated with its early promoter; L-251.1 contains the DBP gene associated with its late promoter (Fig. 1).

2. *E1A plasmids*: HE4 (coordinate 0–4.5) is an E1A wild-type plasmid described by Stow (1981). p*hr*1, derived from AD5*hr*1 virus, has a single-base-pair mutation, which results in a reading frame shift and the termination of the protein coded by the 13S mRNA, whereas the protein coded by the 12S mRNA is not altered (Esche et al. 1980; Ricciardi et al. 1981). pEKpm975 plasmid has a single-base transversion at nucleotide (nt) 975, which eliminates the 12S mRNA splicing but leaves the 13S mRNA and its product unchanged (Montell et al. 1982).

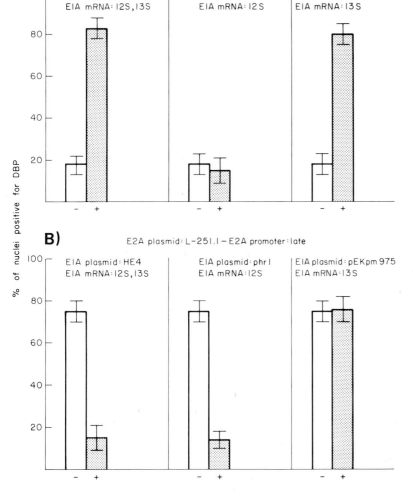

**Figure 2** Effect of E1A mutations on the expression of E2A plasmids containing the DBP gene associated with the early or the late promoter. The E2A plasmids E-252.1 (*top*) or L-251.1 (*bottom*) described in the text were injected into TK$^-$ts13 cell nuclei in the absence (open bar) or in the presence (stippled bar) of the E1A wild-type or mutant plasmids. The wild-type mRNAs and proteins produced by each E1A plasmid are also indicated. About 200 copies/cell of each plasmid were injected. The ordinate indicates the percentage of nuclei that stain positively for the E2A DBP. The average percentages of positive nuclei and standard deviations were calculated from 5 to 10 separate experiments. (Reprinted, with permission, from Guilfoyle et al. 1985.)

Cell nuclei were injected with the E2A plasmids in the absence or in the presence of E1A wild-type or mutant plasmids. The presence of DBP in injected cells was analyzed by indirect immunofluorescence, using specific antibodies.

The results of these experiments are shown in Figure 2. As already described, the expression of the DBP gene and its response to the E1A-mediated regulation is dependent on the presence of the E2A early or late promoter (Rossini 1983). The expression of DBP after injection of the E2A early promoter clone E-252.1 (Fig. 2, A) was very poor (18% of injected cells express DBP), and the presence of the E1A wild-type plasmid increases the proportion of positive cells (80%). The same increase is seen when the E2A early promoter plasmid was coinjected with the mutant pEKpm975, which produces only the 13S mRNA. In contrast, the mutant p*hr*1, in which the mutation affects the 13S mRNA product but not the product of the 12S mRNA, has lost the ability to stimulate the synthesis of DBP. The percentage of positive nuclei (15%) is comparable to the basal level observed in the absence of E1A.

Microinjection of the E2A late promoter plasmid L-251.1 (Fig. 2, B) in the absence of E1A resulted in a high (85%) level of expression of DBP. Consistent with our previous finding, a decrease in the number of cells producing DBP was observed after coinjection of E1A DNA.

The E1A mutant p*hr*1, by which the effect of 12S mRNA product was tested, maintained the wild-type property of repressing DBP production from the late promoter, whereas coinjection of pEKpm975 with the E2A late promoter plasmid had no effect on the production of DBP.

The results suggest that the product of the 13S mRNA, but not the protein encoded by the 12S mRNA, is responsible for the activation of DBP from its early promoter, whereas the 12S mRNA product is sufficient for the repression of DBP synthesis from its late promoter.

## Expression of the CAT gene from the E2A promoters

To determine more precisely whether the E1A-mediated regulation was acting at the E2A promoter regions, we used plasmids p2CAT and 251CAT, which contained the E2A early and late promoters associated with the bacterial CAT gene, respectively, to transfect HeLa cells (251CAT is described in Fig. 3). HeLa cells were transfected with these plasmids in the absence or presence of plasmids containing the wild-type or mutant E1A

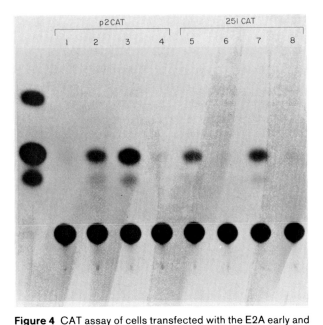

**Figure 4** CAT assay of cells transfected with the E2A early and late promoter CAT plasmids. HeLa cells were transfected with p2ACAT or 251CAT, in the absence or presence of the wild-type or mutant E1A plasmids. The calcium phosphate–DNA precipitate contained 20 μg/plate of the CAT plasmids and an equimolar amount of the E1A plasmid DNAs (40 μg/plate). For transfections of CAT plasmid alone, pBR322 DNA was substituted for E1A. After 48 hr, cell extracts were assayed for CAT activity and analyzed by thin-layer chromatography as described in Experimental Procedures. Lanes *1–4* represent transfections with p2CAT; lanes *5–8* represent transfections with 251CAT. (Lanes *1* and *5*) CAT plasmids alone; (lanes *2* and *6*) cotransfections with HE4; (lanes *3* and *7*) cotransfections with pEKpm975; (lanes *4* and *8*) cotransfections with p*hr*1. The $^{14}$C spots in the autoradiograph corresponding to the acetylated forms of chloramphenicol were cut out of the silica gel and counted: (lane *1*) 1,709 cpm; (lane *2*) 19,288 cpm; (lane *3*) 39,780 cpm; (lane *4*) 4,324 cpm; (lane *5*) 18,033 cpm; (lane *6*) 2,208 cpm; (lane *7*) 23,031 cpm; (lane *8*) 3,768 cpm. (Reprinted, with permission, from Guilfoyle et al. 1985.)

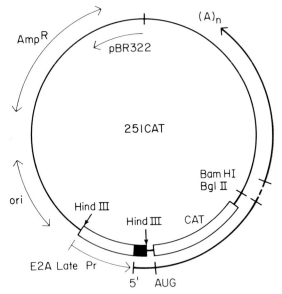

**Figure 3** A plasmid containing the CAT gene linked to the E2A late promoter. In the plasmid shown, 251CAT, a 420-bp fragment comprising a major portion of the E2A late leader and 5′-flanking promoter region was inserted into the *Hind*III site of pSV0CAT. The CAT gene in pSV0CAT is associated with the 3′-flanking sequences that contain the SV40 small T intron and polyadenylation signals (Gorman et al. 1982). The construction of plasmid 251CAT is described in Experimental Procedures. (Reprinted, with permission, from Guilfoyle et al. 1985.)

gene. Forty-eight hr after transfection, cell extracts were prepared and the expression of the chimeric gene was measured by determining the proportion of chloramphenicol converted to the acetylated form (CAT activity). The results of these experiments are shown in Figure 4.

Cells transfected with the E2A early promoter–CAT chimeric plasmid (p2CAT) show a very low level of CAT activity (Fig. 4, lane 1). This activity increased 10-fold to 20-fold in cells cotransfected with the same CAT plasmid in combination with either the wild-type (HE4) or the mutant (pEKpm975) E1A plasmids (lanes 2 and 3), both of which produced the 13S mRNA product. In contrast, the presence of plasmid p*hr*1, which codes for the 12S mRNA product, had no effect on the expression of the early promoter–CAT gene (lane 4). HeLa cells transfected with the E2A late–CAT gene (251CAT) show a high level of CAT activity (lane 5). Consistent with the results obtained in microinjection experiments, cotransfection of the same plasmid with the E1A wild-type (HE4) or p*hr*1 mutant plasmids results in a 10-fold to 15-fold reduction of the CAT activity (lanes 6 and 7). The high level of CAT activity was not altered in extracts of cells transfected with the 251CAT and the E1A pEKpm975 mutant plasmids (lane 7).

From these results we conclude that the sequences responsive to the positive or negative control exerted by the 13S and 12S mRNA products are located at the 5' end of the DBP early and late transcription units.

## Construction and analysis of deletion mutants in the E2A late promoter region

To define the DNA sequences in the E2A late promoter region responsive to the E1A-mediated repression, we introduced deletions into this region. A plasmid containing the E2A late promoter region was cut at the unique *Hin*dIII site and treated with BAL-31 exonuclease. A *Hin*dIII linker was then inserted at the deletion end point, using T4 DNA ligase. The deletion fragments released from the vectors were inserted into a plasmid containing the DBP coding sequences and first internal leader. The precise location of the deletion end points, ranging from −126 to −3 (Fig. 5) from the E2A late cap site (+1) were then determined by the sequencing method of Maxam and Gilbert (1980).

To determine the effect of deletion on the expression and regulation of DBP, these plasmids were microinjected into TK⁻ts13 cell nuclei in the absence or in the presence of E1A DNA. The results of such experiments are given as the percentage of injected cells positive for the 72K DBP, as determined by indirect immunofluorescence (see Fig. 5).

In the absence of E1A DNA, deletion to coordinate −51 does not alter the expression of DBP, but deletion

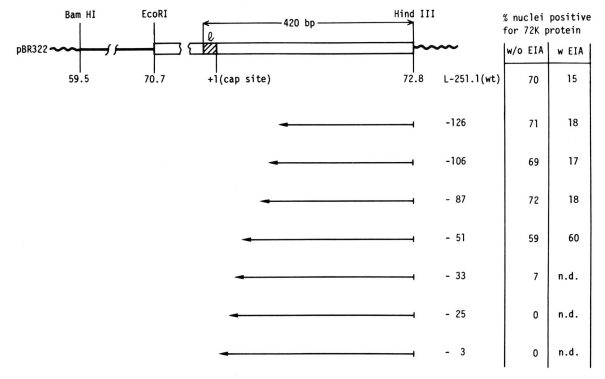

**Figure 5** Effect of deletion on the E2A late promoter region on the expression and regulation of DBP. The structures of the DBP gene linked to its late promoter (L-251.1) and deletion-mutant derivatives that affect the promoter region are shown. The sequences upstream of the cap site that are deleted and substituted by pBR322 (arrows) are illustrated. The nucleotide corresponding to the end point of deletion in each mutant plasmid is indicated to the right side of each diagram (the 5' capped nucleotide of the mRNA being at nucleotide position +1). On the far right are the results obtained after microinjection of these plasmids in the absence or presence of E1A DNA, reported as percentages of nuclei that are positive for the DBP.

of sequences between −51 and −31 dramatically reduces the synthesis of DBP.

The ability of deletion mutants to respond to E1A repression was tested by coinjection of the promoter deletion mutants with the E1A plasmids. Deletion to coordinate −87 responded to E1A repression in a way similar to the response of the wild type. In contrast, repression was not observed when the mutant with deletion to coordinate −51 was coinjected with E1A DNA. The percentages of positive cells were indeed comparable in cells injected with or without E1A DNA (∼70%).

These results strongly suggest that DNA sequences between coordinates −87 and −51 from the E2A late cap site contain the regulatory interaction site for E1A-mediated repression. Furthermore, deletion of sequences between coordinates −51 and −33 almost completely abolishes the synthesis of DBP.

## Discussion

The mechanism of action of E1A has been intensely studied in the past few years, and in particular, the role played by the 289-aa protein in the regulation of transcription has been analyzed. It has become clear that this protein has an essential role in the activation of early adenovirus transcription units and therefore is necessary for the viral reproductive cycle (Berk et al. 1979; Carlock and Jones 1981; Ricciardi et al. 1981; Montell et al. 1982; Ferguson et al. 1984; Leff et al. 1984; Guilfoyle et al. 1985). The 13S mRNA product also affects the expression of certain cellular genes (Kao and Nevins 1983; Stein and Ziff 1984).

The structural difference between the 289- and 243-aa products resides in the internal 46 aa unique to the larger product, but the 12S product fails to induce transcription from viral or cellular genes (Berk et al. 1979; Ricciardi et al. 1981; Gaynor et al. 1984; Svensson and Akusjärvi 1984).

In this report we have shown that E1A repression of the DBP late promoter is a function encoded by the 12S mRNA. This negative control is promoter-dependent and exerted only on the late and not the early DBP promoter, suggesting a control at the level of transcription. In addition, we suggest that this interaction is mediated by specific DNA sequences located between −87 and −51 upstream from the DBP late cap site. Computer analysis of the DNA sequences within this region has revealed an interesting feature that consists of a duplicated GCGG sequence flanking a potential stem-loop, which can also be regarded as a partial inverted repeat sequence (Fig. 6). Perhaps binding of the 12S product to this region blocks entry of RNA polymerase II or translocation to the preferred initiation site. Alternatively, binding may induce a change in the DNA or chromatin conformation that renders this region transcriptionally inactive.

Recently it has been shown E1A can repress transcription of a rabbit β-globin gene linked to a polyoma enhancer element (Borrelli et al. 1984) as well as SV40 early transcription by acting on the enhancer sequences

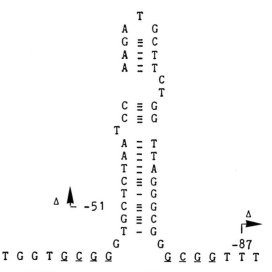

**Figure 6** DNA sequences in the DBP late promoter region. The DNA sequence between nucleotides −88 and −40 located upstream from the E2A late transcription initiation site (Galibert et al. 1979). The dyad axis of near symmetry found in this region is shown diagrammatically as a hairpin structure. The nucleotides substituted with pBR322 sequences in deletion mutants LΔ−87 and LΔ−51 are indicated. The deletion 5′ to nucleotide −51, but not the deletion 5′ to nucleotide −87, abolishes the E1-mediated repression, as indicated in Fig. 5.

(Velcich and Ziff 1985). This repression is a function of both 13S and 12S mRNA products, suggesting that the repressor function resides in a domain common to both proteins. The specific repression of the DBP late promoter by the 243-aa protein found in our system seems to be somewhat in contrast with the other reports. The interaction with unrelated regulatory elements such as the SV40 or polyoma enhancers may lead to disruption of transcription by a different mechanism.

Clearly, in the adenovirus system these two products act differently. Recently it has been shown that infection of primary BRK cells with a virus expressing the 13S and not the 12S mRNA product results in cell death typical of productive infection, whereas cells infected with a virus expressing the 12S mRNA and not the 13S mRNA product undergo cell proliferation (M. Quinlan and T. Grodzicker, pers. comm.). This feature may indicate that the 243-aa protein has its own function, not shared with the larger 289-aa protein. The 12S mRNA product may play an important role in cell transformation by acting as a specific repressor factor on the synthesis of DBP required for viral DNA replication and progression to the late phase. Furthermore, by interacting with specific cellular genes that regulate the cell cycle, the 12S repressor factor could be responsible for cell immortalization and transformation.

## Acknowledgments

We are grateful to Dr. A. Lewis, Jr. for supplying antibodies and Drs. A. Berk, N. Jones, and N. Stow for providing the pEKpm975, p2CAT, and HE4 plasmids, respectively. We also thank Ms. Vanna Pieri for typing this manuscript.

This work was partially performed while three of the authors (R.G., W.O., and M.R.) were working at Cold Spring Harbor Laboratory. Financial support was obtained from American Cancer Society Research Grants to M.R. (MV-147A) and to L.T. Chow (MV-144) and an NIH/NCI Program Project Grant to Cold Spring Harbor Laboratory; R.G. was supported by an NIH Training Grant (CA 90311-03).

## References

Berk, A.J. and P.A. Sharp. 1978. Structure of the adenovirus 2 early mRNAs. *Cell* **14**: 695.

Berk, A.J., F. Lee, T. Harrison, J. Williams, and P.A. Sharp. 1979. Pre-early adenovirus 5 gene product regulates synthesis of early viral messenger RNAs. *Cell* **17**: 935.

Birnboim, H.C. and J. Doly. 1979. A rapid alkaline extraction procedure for screening recombinant plasmid DNA. *Nucleic Acids Res.* **7**: 1513.

Borrelli, E., R. Hen, and P. Chambon. 1984. Adenovirus-2 E1A products repress enhancer-induced stimulation of transcription. *Nature* **312**: 608.

Carlock, L.R. and N. Jones. 1981. Transformation-defective mutant of adenovirus type 5 containing a single altered E1A mRNA species. *J. Virol.* **40**: 657.

Chow, L.T., T.R. Broker, and J.B. Lewis. 1979. Complex splicing patterns of RNAs from the early regions of adenovirus 2. *J. Mol. Biol.* **134**: 265.

Esche, H., M.B. Mathews, and J.B. Lewis. 1980. Proteins and messenger RNAs of the transforming region of wild-type and mutant adenoviruses. *J. Mol. Biol.* **142**: 399.

Ferguson, B., N. Jones, J. Richter, and M. Rosenberg. 1984. Adenovirus E1A gene product expressed at high levels in *Escherichia coli* is functional. *Science* **224**: 1343.

Galibert, F., J. Herisse, and G. Courtois. 1979. Nucleotide sequence of the *Eco*RI-F fragment of adenovirus 2 genome. *Gene* **6**: 1.

Gaynor, R.B., D. Hillman, and A.J. Berk. 1984. Adenovirus early region 1A protein activates transcription of a nonviral gene introduced into mammalian cells by infection or transfection. *Proc. Natl. Acad. Sci.* **81**: 1193.

Gaynor, R.B., A. Tsukamoto, C. Montell, and A.J. Berk. 1982. Enhanced expression of adenovirus transforming proteins. *J. Virol.* **44**: 276.

Gorman, C.M., L.F. Moffat, and B.H. Howard. 1982. Recombinant genomes which express chloramphenicol acetyltransferase in mammalian cells. *Mol. Cell. Biol.* **2**: 1044.

Graessmann, A., M. Graessmann, and C. Mueller. 1980. Microinjection of early SV40 DNA fragments and T antigen. *Methods Enzymol.* **65**: 816.

Graham, F.L., T. Harrison, and J. Williams. 1978. Defective transforming capacity of adenovirus type 5 host-range mutants. *Virology* **86**: 10.

Guilfoyle, R. and R. Weinmann. 1981. Control region for adenovirus VA RNA transcription. *Proc. Natl. Acad. Sci.* **78**: 3378.

Guilfoyle, R., W. Osheroff, and M. Rossini. 1985. Two functions encoded by adenovirus early region 1A are responsible for the activation and repression of the DNA-binding protein gene. *EMBO J.* **4**: 707.

Houwelling, A., P. van den Elsen, and A.J. van der Eb. 1980. Partial transformation of primary rat cells by the leftmost 4.5% fragment of adenovirus 5 DNA. *Virology* **105**: 537.

Jonak, G.J. and R. Baserga. 1980. The cytoplasmic appearance of three functions expressed during the $G_0 \rightarrow G_1 \rightarrow S$ transition is nucleus-dependent. *J. Cell. Physiol.* **105**: 347.

Jones, N. and T. Shenk. 1979a. An adenovirus type 5 early

gene function regulates expression of other early viral genes. *Proc. Natl. Acad. Sci.* **76**: 3665.

———. 1979b. Isolation of adenovirus type 5 host range mutants defective for transformation of rat embryo cells. *Cell* **17**: 683.

Kao, H.-T. and J. Nevins. 1983. Transcriptional activation and subsequent control of the human heat shock gene during adenovirus infection. *Mol. Cell. Biol.* **3**: 2058.

Kitchingman, G.R. and H. Westphal. 1980. The structure of adenovirus 2 early nuclear and cytoplasmic RNAs. *J. Mol. Biol.* **13**: 23.

Leff, T., R. Elkaim, C.R. Godin, P. Jalinot, P. Sassone-Corsi, C. Perricaudet, C. Kedinger, and P. Chambon. 1984. Individual products of the adenovirus 12S and 13S E1A mRNAs stimulate viral E2A and E3 expression at the transcriptional level. *Proc. Natl. Acad. Sci.* **81**: 4381.

Maxam, A.M. and W. Gilbert. 1980. Sequencing end-labeled DNA with base-specific chemical cleavages. *Methods Enzymol.* **67**: 499.

Montell, C., G. Courtois, C. Eng, and A.J. Berk. 1984. Complete transformation by adenovirus 2 requires both E1A proteins. *Cell* **36**: 951.

Montell, C., E.F. Fisher, M.H. Caruthers, and A.J. Berk. 1982. Resolving the functions of overlapping viral genes by site-specific mutagenesis at a mRNA splice site. *Nature* **295**: 380.

Nevins, J.R. 1981. Mechanism of activation of early viral transcription by the adenovirus E1A gene product. *Cell* **26**: 213.

Perricaudet, M., G. Akusjärvi, A. Virtanen, and U. Pettersson. 1979. Structure of two spliced mRNAs from the transforming region of human subgroup C adenoviruses. *Nature* **281**: 694.

Radloff, R., W. Bauer, and J. Vinograd. 1967. A dye buoyant-density method for detection and isolation of closed circular duplex DNA: The closed circular DNA in HeLa cells. *Proc. Natl. Acad. Sci.* **57**: 1514.

Ricciardi, R.P., R.L. Jones, C.L. Cepko, P.A. Sharp, and B.E. Roberts. 1981. Expression of early adenovirus genes requires a viral encoded acidic polypeptide. *Proc. Natl. Acad. Sci.* **78**: 6121.

Rossini, M. 1983. The role of adenovirus early region 1A in the regulation of early regions 2A and 1B expression. *Virology* **131**: 49.

Ruley, E.H. 1983. Adenovirus early region 1A enables viral and cellular transforming genes to transform primary cells in culture. *Nature* **304**: 602.

Solnick, D. and M.A. Anderson. 1982. Transformation deficient adenovirus mutant defective in expression of region 1A but not region 1B. *J. Virol.* **42**: 106.

Stein, R. and E.B. Ziff. 1984. HeLa cell β-tubulin gene transcription is stimulated by adenovirus 5 in parallel with viral early genes by an E1a-dependent mechanism. *Mol. Cell. Biol.* **4**: 2792.

Stow, N.D. 1981. Cloning of a DNA fragment from the left-hand terminus of the adenovirus type 2 genome and its use in site-directed mutagenesis. *J. Virol.* **37**: 171.

Svensson, C. and G. Akusjärvi. 1984. Adenovirus 2 early region 1A stimulates expression of both viral and cellular gene. *EMBO J.* **3**: 789.

Velcich, A. and E. Ziff. 1985. Adenovirus E1A proteins repress transcription from the SV40 early promoter. *Cell* **40**: 705.

Vogelstein, B. and D. Gillespie. 1979. Preparative and analytical purification of DNA from agarose. *Proc. Natl. Acad. Sci.* **76**: 615.

Weeks, D.L. and N.C. Jones. 1983. E1A control of gene expression is mediated by sequences 5′ to the transcriptional starts of the early viral genes. *Mol. Cell. Biol.* **3**: 1222.

Westphal, H. and S.-P. Lai. 1977. Quantitative electron microscopy of early adenovirus RNA. *J. Mol. Biol.* **116**: 525.

Wigler, M., A. Pellicer, S. Silverstein, and R. Axel. 1978. Biochemical transfer of single copy eukaryotic genes using total cellular DNA as donor. *Cell* **14**: 729.

# A Genetic Analysis of Bovine Papillomavirus Type-1 Transformation and Plasmid-maintenance Functions

**M.S. Rabson, Y.C. Yang, and P.M. Howley**
Laboratory of Tumor Virus Biology, National Cancer Institute, National Institutes of Health, Bethesda, Maryland 20892

A genetic analysis of the bovine papillomavirus type-1 (BPV-1) genome has been performed to evaluate the role of individual open reading frames (ORFs) in plasmid maintenance and transformation. Deletion and translational termination linker–insertion mutants have been evaluated for transforming ability and for their ability to remain extrachromosomal. Mutations in the E2 ORF significantly reduce efficiency of focus formation and invariably result in the integration of the viral DNA into the cellular DNA, indicating a requirement for the putative product of the E2 ORF for efficient transformation and for plasmid maintenance. The plasmid maintenance function can be provided in *trans* to complement the E2 mutant DNAs in a cell line expressing the BPV-1 3'-ORF gene functions, including E2. The effect of inserting the translational termination linker into E1, E5, and E7 ORFs has also been evaluated. The E1 ORF is essential for plasmid maintenance, whereas the E7 ORF is not. The insertion of the premature translational termination linker in the E5 ORF after the first methionine codon drastically reduced the efficiency of transformation, implicating the putative product of the E5 ORF in cellular transformation.

The bovine papillomavirus type 1 (BPV-1) causes benign fibropapillomas in cattle and can induce fibroblastic tumors in a variety of heterologous hosts (Lancaster and Olson 1982). Intact virus, cloned BPV-1 DNA, and a specific subgenomic BPV-1 fragment comprising 69% of the viral genome ($BPV_{69T}$) can each morphologically transform susceptible mouse cells in vitro (Lowy et al. 1980). In these transformed mouse cells, the viral DNA is maintained as an extrachromosomal, multicopy plasmid (Law et al. 1981), in a way analogous to the viral extrachromosomal DNA maintenance observed in naturally occurring, benign tumors.

The genome of BPV-1 is a 7945-bp double-stranded, circular molecule, which has been sequenced (Chen et al. 1982). The genomic organization has been established from analysis of the sequence and from transcriptional studies of BPV-1–transformed cells (Heilman et al. 1982) and of productively infected bovine fibropapillomas (Amtmann and Sauer 1982; Engel et al. 1983). Eight open reading frames (ORFs), designated E1 through E8 (Fig. 1), have been identified on the same DNA strand within the 69% subgenomic transforming segment, which contains all of the viral gene required for transformation and for plasmid replication and maintenance. All of the viral RNAs detected in BPV-1–transformed cells map to $BPV_{69T}$ (Heilman et al. 1982). Analysis of the viral RNAs in transformed cells by cDNA mapping (Yang et al. 1985a) and electron microscopy (Stenlund et al. 1985) have demonstrated multiple species of spliced and unspliced RNAs. These viral RNA species are all polyadenylated at BPV-1 nucleotide (nt) 4203 but differ in the location of their 5' ends and patterns of splicing. Two transcriptional regulatory elements with

the properties of enhancers have been identified in the $BPV_{69T}$. One, which maps to the 1000-bp BPV-1 noncoding region (NCR) upstream of the eight early ORFs, is an inducible enhancer that can be activated in *trans* by the E2 ORF gene product (Spalholz et al. 1985a). Another enhancer element mapping to the distal end of the $BPV_{69T}$ region has also been described (Campo et al. 1983; Lusky et al. 1983). The role of these enhancer elements in the life cycle of the virus has not yet been determined (Howley et al. 1985; Spalholz et al. 1985a).

A number of studies have been conducted to characterize and map the regions of the viral genome involved in cellular transformation. Mutagenesis studies (Sarver et al. 1984; Lusky and Botchan 1985) and studies with subgenomic viral DNA fragments expressed from surrogate promoters (Nakabayashi et al. 1983; Schiller et al. 1984) have shown that at least one BPV-1 transforming function maps to the 3' early ORFs (E2, E3, E4, and E5). The E5 ORF downstream from the first 5' ATG methionine codon (nt 3879) appears to be a critical domain for transformation since frameshift mutations or the introduction of an in-frame termination codon into this region has a marked effect on the transformation efficiency of the viral DNA (DiMaio et al. 1985; Yang et al. 1985b; Schiller et al. 1986; D. Groff and W. Lancaster, in prep.). The integrity of the E2 ORF is also critical for efficient transformation when the entire BPV-1 genome is analyzed for transformation (Sarver et al. 1984; DiMaio et al. 1985; Lusky and Botchan 1985). An additional transforming function that is required for the fully transformed phenotype maps to the 5' E6 and E7 regions of the genome (Sarver et al. 1984). Expression of the E6 ORF itself behind a surrogate promoter or

**235**

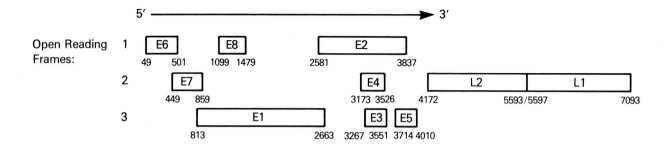

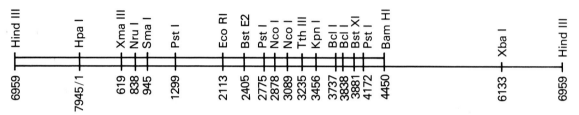

**Figure 1** The genomic organization of BPV-1 DNA. The full-length molecule (7945 bp) of BPV-1 opened at the unique *Hind*III site (nt 6959) is marked off with restriction endonuclease sites and base pairs at the bottom of the figure. The BPV-1 subgenomic fragment that contains all the sequences necessary for autonomous plasmid replication (Law et al. 1981) and transformation (Lowy et al. 1980) is indicated by the open bar at the bottom. The direction of transcription and the region transcribed in transformed cells is indicated by the arrow at top (Heilman et al. 1982). The open bars represent the ORFs and, therefore, the potential coding sequences in each of the three translation frames predicted from the sequence (Chen et al. 1982). ORFs within the transforming region have been designated E1 to E8. The numbers beneath the ORFs designate the first and last nucleotides of an ORF.

expression of the E6 cDNA is sufficient for morphologic transformation of mouse C127 cells (Schiller et al. 1984; Yang et al. 1985a).

Genetic analyses of BPV-1 have also defined regions of the genome involved in plasmid replication and maintenance. Two independent, *cis*-essential elements designated as plasmid-maintenance sequences have been defined as being independently capable of supporting extrachromosomal plasmid maintenance in the presence of diffusible viral replication and plasmid-maintenance factors provided in *trans* (Lusky and Botchan 1984). The putative E1 gene product is essential for plasmid replication and maintenance. Mutations in the E1 region invariably result in the integration of viral DNA into the host chromosome (Sarver et al. 1984; Lusky and Botchan 1985), and this function can be complemented in *trans* (Lusky and Botchan 1985). From mutagenesis of the E7 ORF and complementation studies with mutant DNAs, the E7 ORF product has been ascribed a role in regulating plasmid copy number (Lusky and Botchan 1985).

The role of the E2 ORF product in plasmid maintenance has been unclear. In our studies (Sarver et al. 1984) as well as in those of D. DiMaio (DiMaio et al. 1985), mutagenesis of the E2 ORF has resulted in the loss of plasmid maintenance. Lusky and Botchan, however, have described mutants deleted of part or all of the E2 ORF that are able to replicate as stable extrachromosomal plasmids (Lusky and Botchan 1984, 1985).

A limitation of the genetic analyses that we performed previously (Sarver et al. 1984) derived from our use of mutants with large deletions that, because of the overlapping ORFs in BPV-1, affected the integrity of more than one ORF. The current study represents a further genetic analysis of BPV-1 using DNAs with targeted mutations that affect specific ORFs or groups of ORFs. We describe experiments examining additional deletion mutants as well as a set of mutant DNAs generated by the insertion of a translational termination linker (TTL) at specific sites in the BPV-1 genome.

The results of our analysis demonstrate that in a wild-type genome background, the integrity of the E2 ORF is essential for stable plasmid maintenance and is important for efficient transformation. Mutants affected only in the E2 ORF transformed C127 cells at efficiencies of less than 5% of that observed for the full genome, confirming the role of E2 in inducing efficient transformation. Furthermore, these E2 mutants were found to integrate into chromosomal DNA when they were selected after transfection for either focus formation or by an independent drug-resistance marker, indicating a role for the putative E2 gene product in plasmid maintenance. The E2 ORF mutants were found to replicate as stable plasmids in cells expressing the E2 gene product from an integrated cDNA, suggesting that the E2 gene product can act in *trans* to allow extrachromosomal plasmid maintenance. The effect of inserting the translational termination into each of the E1, E5, and E7 ORFs has also been evaluated.

## Materials and Methods

### Cell lines

Mouse C127 cells (Dvoretzky et al. 1980) and C127 cells containing BPV-1 DNAs were maintained in Dulbecco's modified Eagle's medium supplemented with 10% fetal bovine serum, penicillin (100 μg/ml), and streptomycin (100 μg/ml).

### Transfections

DNA transfections of C127 cells were performed as previously described, using a modification of the calcium phosphate coprecipitation method (Sarver et al. 1982). To each 60-mm plate were added 10 μg of precipitate containing 1 μg of BPV-1 DNA and 9 μg of carrier salmon sperm DNA. When the cells were to be selected for drug resistance to G418, 1 μg of a plasmid in which the Tn5 neomycin-resistance gene is expressed from the mouse metallothionein (pMMTneo[302-3]) (Law et al. 1983) was added and the amount of carrier DNA was reduced accordingly. Four hr after transfection, the cells were treated with 15% glycerol for 2 min and then washed. The cells were incubated in complete media with 5 mM sodium butyrate for 24 hr (Gorman and

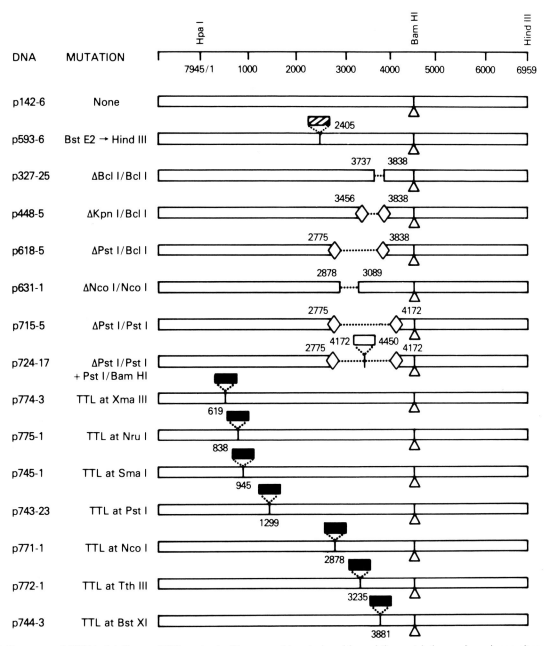

**Figure 2** Structure of BPV-1 deletion and TTL mutants. The recombinant plasmids and the restriction endonuclease sites used in generating the various mutants are listed on the left. In each case, the plasmid consisted of BPV-1 DNA sequences linked to the pML2d vector sequences at the *Bam*HI site (△). The BPV-1 sequences (▭) are covalently linked at the sites indicated. (◇) *Xho*I linker elements; ( · · · · ) deleted sequences; (■) the TTL element. In p593-6, the *Hin*dIII linker (▨) has been inserted into the *Bst*E2 site. In p724-17, the BPV-1 fragment nt 4172–4450 (▭) has been inserted.

Howard 1983) and each 60-mm plate was split into two 100-mm dishes. For drug-resistance selection, G418 (400 μg/ml) was added to the dishes 72 hr after transfection, and the cells were maintained continuously under selection (Colbere-Garapin et al. 1981).

*Cellular DNA*
Total cellular DNA was extracted from C127 cells according to a method described previously (Law et al. 1981). Restriction endonuclease digestion, gel electrophoresis, and Southern blot analysis of the DNAs were performed by standard methods (Maniatis et al. 1982).

*Construction of BPV-1 mutant DNAs*
In most cases, except where otherwise noted, the starting plasmid for all BPV-1 mutant constructions was pdBPV-1 (p142-6), which contains the entire BPV-1 genome cloned at the *Bam*HI site of pML2d (Sarver et al. 1982). Most of the BPV-1 DNAs used in this study are diagramed in Figure 2.

The BPV-1 mutant DNAs p593-6, p620-7, p327-25, p618-5, and p448-5 have been described in detail previously (Sarver et al. 1984). The plasmids p593-6 and p620-7 are mutated only in the E1 ORF; p593-6 was generated by the insertion of a *Hind*III linker at the unique *Bst*E2 site (nt 2405) in BPV-1 DNA, and p620-7 is deleted of sequences between the *Eco*RI site (nt 2113) and the *Bst*E2 site (nt 2405). The mutants p327-5, p448-5, and p618-5 each contain deletions extending upstream from the *Bcl*I site at nt 3838: p327-5 has a 101-bp deletion between the *Bcl*I sites at nt 3737 and 3838, which affects the 3′ end of the E2 ORF and a portion of the E5 ORF 5′ to the first methionine codon; p448-5 is deleted of the 383 bases between the *Kpn*I site (nt 3455) and the *Bcl*I site (nt 3838); and p618-5 is deleted of 1063 bp between the *Pst*I site at nt 2775 and the *Bcl*I site at nt 3838. Each of the mutants contains an *Xho*I linker at the site of their deletions.

For these experiments, three additional deletion mutants of BPV-1 DNA were constructed. The mutant p631-1, which is affected only in the E2 ORF, was made by cleaving p142-6 DNA with *Nco*I and religating, thereby removing a 211-bp *Nco*I fragment. Two additional gross-deletion mutants that removed all of the E3, E4, and E5 ORFs and most of E2 were made. p715-5 lacked the sequences between the *Pst*I site at nt 2775 and the *Pst*I site at nt 4172 and contained an *Xho*I linker at the site of the deletion. p724-17 is similar to p715-5, having the same deletion but containing a tandem duplication of the distal BPV-1 enhancer. This construction was made by converting the *Bam*HI site at the 3′ end of the genome in p715-5 to an *Xho*I site by adding a linker to create the intermediate p722-1. The 278-bp *Xho*I fragment (BPV-1 nt 4172–4450) from p722-1 was then cloned into the *Xho*I site of p715-5 to create p724-17, which is deleted of sequences from nt 2775 to 4172 and has tandem head-to-tail duplication of the sequence nt 4172–4450.

To target mutations that would lead to premature translational termination of the products of specific ORFs, a different approach was developed. An oligonucleotide linker element (TTL) containing a *Hpa*I cleavage site was synthesized and inserted into the full-length clone of BPV-1 DNA (p142-6) at a variety of different restriction enzyme sites. This 12-bp element was designed to contain a translation stop codon in each of the three reading frames when inserted, regardless of the orientation (Fig. 3). The termination linker was inserted at two sites in the E1 ORF at the *Sma*I site at nt 945 (p745-1) and at the *Pst*I site nt 1299 (p743-23). The latter mutation also affects the E8 ORF. The plasmids p774-3 and p775-1 contained the termination element within the E7 ORF at the *Xma*III site (nt 619) and the *Nru*I site (nt 838), respectively. The E5 ORF was mutated by insertion of the termination linker into the *Bst*XI site of p142-6 at nt 3881 (p744-3). Two BPV-1 DNAs that were mutated in the E2 ORF were generated by moving restriction fragments containing the termination linker from the background of the C59 cDNA, which contains the E2 ORF intact, into the p142-6 background. The cDNAs, C59-2878 and C59-3235, containing the termination linkers have been described previously (Yang et al. 1985b); each was cleaved with *Bst*E2 and *Kpn*I and the resultant 1500-bp BPV-1 DNA fragments were exchanged into the p142-6 plasmid that had been deleted of the analogous element. Ligation of these DNA fragments generated the plasmids p771-1 and p772-1, which contain the termination linker at nt 2878 (*Nco*I) and nt 3235 (*Tth*III), respectively, in the full-length BPV-1 genome.

**Results**

These experiments were initiated as part of an analysis of the genetic elements involved in BPV-1 transformation, in plasmid replication, and in plasmid-maintenance functions. Our previous analysis of deletion mutants that

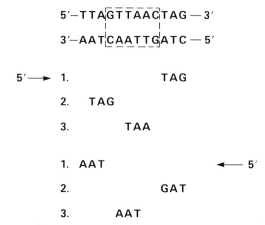

**Figure 3** DNA sequence of the TTL used for mutagenesis. The location of the stop codon in each reading frame is shown. The nucleotide sequence enclosed within the dotted lines is the recognition sequence for the restriction endonuclease *Hpa*I.

affected the E2 ORF suggested that the product of this ORF had a role in both of these functions. We observed that deletions that affected the E2 ORF, particularly in the 5' region, diminished transforming efficiency and that cells selected for transformation by these mutants contained the viral DNA integrated into the host chromosome (Sarver et al. 1984). These results differ from those of Lusky and Botchan (1984, 1985), who found that the E2 ORF was not absolutely required for plasmid replication and plasmid maintenance in mouse C127 cells. The first experiments presented were performed to examine in more detail the role of the putative E2 gene product in plasmid replication and maintenance.

### Is expression of the 5' ORFs sufficient for plasmid maintenance?

Previous genetic analyses of BPV-1 functions involved in plasmid maintenance have identified critical sequences in the BPV-1 5' ORFs (E1, E6, E7, and E8). Two independent, *cis*-acting plasmid-maintenance sequences termed PMS-1 and PMS-2 have been identified in the BPV-1 NCR and in the E1 ORF, respectively (Lusky and Botchan 1984). A *trans*-acting factor necessary for plasmid maintenance has also been mapped to the E1 ORF (Sarver et al. 1984; Lusky and Botchan 1985). Another *trans*-acting factor, involved in regulating plasmid copy number, has been mapped to the E7 ORF (Lusky and Botchan 1985).

We were interested in determining whether the BPV-1 5' ORFs contained all the genetic information necessary for plasmid maintenance or whether sequences mapping to the 3' ORFs (E2, E3, E4, and E5) were also involved. Two deletion mutants affected in the 3' ORFs were assayed for their ability to replicate as plasmids. These two mutants, p715-5 and p724-17, were deleted of sequences between the *Pst*I sites at nt 2775 and 4172. Each retained the BPV-1 distal enhancer and early region polyadenylation signals but were deleted of the E3, E4, and E5 ORFs and all but 194 bp of the E2 ORF. The mutant p715-5 is closely analogous to the Bal-15 mutant described by Lusky and Botchan (1984), which has been reported to be capable of plasmid replication and maintenance. The mutant p724-17 is identical with p715-5 except that it contains a tandem duplication of a 278-bp fragment (nt 4172–4450) containing the distal enhancer and 3' transcriptional polyadenylation signals. The mutant p724-17 was constructed to assess whether duplication of the distal enhancer element might have a positive effect on plasmid replication in a manner analogous to what has been found for polyomavirus, where multiple copies of an enhancer segment can activate DNA replication (Veldman et al. 1985).

When transfected into C127 cells, neither p715-5 nor 724-17 induced transformation. Plasmid replication was assessed by examining the DNA from cell lines containing the DNAs introduced by cotransfection with pMMTneo(302-3) and selected for resistance to the drug G418. The viral DNA of both these mutants was found to be integrated (data not shown), suggesting that some sequences found in the 3' ORFs are necessary for plasmid maintenance.

### Analysis of a deletion mutant affecting only E2

Due to the overlapping ORFs at the 3' end of the BPV-1 transforming region, each of the deletion mutants that we previously had studied were mutated in more than one ORF. Since our previous results had implicated the E2 ORF as playing a role in plasmid maintenance, we decided to generate mutants that were affected only in the E2 ORF. A deletion mutant, p631-1, which removed a 211-bp *Nco*I fragment, was constructed that affected only the E2 ORF. This mutant DNA was capable of transforming C127 cells with an efficiency of approximately 1–5% that of p142-6 DNA when linked to pML2d (Table 1). Cell lines were established from transformed foci induced by p631-1 and from individual neomycin-resistant colonies established by cotransfection of C127 cells with this mutant DNA and pMMTneo(302-3). Total cellular DNA from these cells was examined by Southern blot analysis and in both types of lines the viral DNA was found to be integrated into the cellular DNA (data not shown). These results are in agreement with those of DiMaio et al. (1985) who found that an identical deletion mutant integrates into the host chromosome regardless of whether cells containing the DNA are selected by transformation or by expression of an unlinked second marker. These results differ, however, from those reported by Lusky and Botchan (1985), who used an analogous mutant deleted of this *Nco*I fragment. They observed integrated viral DNA in lines expanded from transformed foci but found extrachromosomal viral DNA in lines established by selection for drug resistance to G418 provided by a second unlinked marker.

### Analysis of TTL insertion mutants in the E2 ORF

Although the mutation in p631-1 is localized only to the E2 ORF, we were concerned about other effects the relatively large 211-bp deletion in this mutant might have on the expression of other viral genes, possibly by affecting the stability of various mRNAs. We, therefore, developed a strategy that would affect the translational product of targeted ORFs that would not grossly alter the viral genome, or presumably, the mRNAs. A linker element was designed that contained a premature translational termination codon in each of the possible reading frames regardless of the orientation in which the linker was inserted (Fig. 3). The introduction of this linker, which could be confirmed by the presence of the *Hpa*I cleavage site contained within it, led to the premature truncation of any polypeptide encoded by an ORF into which the linker element has been inserted.

Two mutant DNAs, p771-1 and p772-1, were generated that contained the termination linker in the E2 ORF at nt 2878 and 3235, respectively. The presence of the linker at each of these sites was confirmed by DNA sequencing (data not shown). The introduction of the TTL at nt 3235 in p772-1 also affects the E4. The E2 mutant p771-1 was capable of transformation at a rate

**Table 1** Transformation of Mouse C127 Cells by Mutated BPV-1 DNAs

| Plasmid | ORFs affected | Transformation efficiency[a] | State of DNA[b] |
|---|---|---|---|
| p142-6 | w.t. | 4+ | E |
| p593-6 | E1 | 4+ | I |
| p745-1 | E1 | 3+ | I |
| p743-23 | E1, E8 | 3+ | I |
| p631-1 | E2 | 1+ | I |
| p771-1 | E2 | 1+ | I |
| p772-1 | E2, E4 | 0 | I |
| p448-5 | E2, E3, E4, E5 | 0 | I |
| p327-25 | E2, E5 | 4+ | I |
| p744-3 | E5 | 1+ | ND |
| p775-1 | E7 | 3+ | E |
| p774-3 | E7 | 3+ | E |

[a]Transformation efficiency was determined from a focus-forming assay, using the intact mutant BPV-1 DNA in pML2d on C127 cells. The wild-type BPV-1 DNA (p142-6) induced approximately 100 foci per 0.5 μg of DNA in a 60-mm dish. The following scale was used for expressing the transformation data: (4+) 50–100% of p142-6; (3+) 25–49% of p142-6; (2+) 10–24% of p142-6; (1+) 1–9% of p142-6; (0) <1% of p142-6.

[b]State of viral DNA in C127 cells transfected with BPV-1 DNA linked to pML2d vector. (E) Extrachromosomal; (I) integrated; (ND) not determined.

approximately 5% that of p142-6, linked or unlinked from the prokaryotic pML2d sequences (Table 1). Interestingly, the mutant p772-1, affecting E2 as well as E4, was completely negative in our transformation assay, linked or unlinked to pML2d DNA (Table 1). When the total DNA from cell lines established from foci (p771-1) and G418-resistant cotransformants (p771-1 and p772-1) was examined, the viral DNA was found to be integrated into the host DNA (data not shown). These data confirm a role for the putative gene product of the E2 ORF in plasmid replication and maintenance in C127 cells.

## The E2 function for plasmid replication can be complemented in *trans*

Having established that E2 mutants are not able to replicate in mouse C127 cells in the transformation assays utilized, we next assessed whether the E2 function necessary for plasmid replication and maintenance could be provided in *trans*. For these studies, we cotransfected BPV-1 mutant DNAs with pMMTneo(302-3) into a transformed C127 cell line containing an integrated copy of the BPV-1 cDNA, C59. This cDNA contains the BPV-1 E2, E3, E4, and E5 ORFs intact, under control of the SV40 early promoter (Yang et al. 1985a). The YC-C59 cell line was established from a single focus induced by C59 DNA and has been shown to encode a function that can activate the BPV-1 NCR enhancer in *trans* (Spalholz et al. 1985a).

After pMMTneo(302-3) cotransfection of the YC-C59 cell line with various mutant BPV-1 DNAs and G418 selection, mutiple drug-resistant colonies were pooled, total cellular DNA was prepared, and the state of the viral DNA was examined. The results of this experiment are presented in Figure 4. Lane c contains uncleaved DNA from nontransfected YC-C59 cells, which demon-

strates the integrated BPV-1 cDNA sequences. Lane h represents the positive control in which wild-type p142-6 DNA was transfected; abundant extrachromosomal DNA forms are apparent. Lane d is the negative control

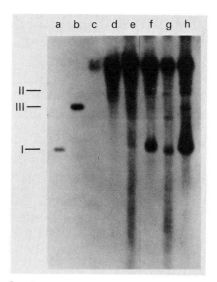

**Figure 4** Southern blot analysis of cellular DNAs from lines YC-C59 cells transfected with BPV-1 DNAs. The YC-C59 cell line, a C127 cell containing the C59 cDNA (Yang et al. 1985a), was examined for its ability to support plasmid replication for the mutant E1 and E2 DNAs. YC-C59 cells were transfected with BPV-1 DNAs and pMMTneo(302-3), which carries the Tn5 gene for neomycin resistance. After selection in G418, drug-resistant colonies were pooled. Lanes *a* and *b* are marker lanes and contain 10 pg of p142-6 DNA uncleaved (*a*) and cleaved (*b*) with XbaI, respectively. Lanes *c*–*h* each contain 10 μg of cellular DNA cleaved with ApaI, which does not recognize any sites in BPV-1 DNA. Lane *c* contains YC-C59 cells that have not been transfected. Lanes *d*–*h* contain DNA from pooled foci from YC-C59 cells cotransfected with p593-6 (*d*), p327-25 (*e*), p771-1 (*f*), p448-5 (*g*), and p142-6 (*h*).

in which the E1 ORF mutant p593-6 introduced by cotransfection with pMMTneo(302-3) is found integrated. Three E2 ORF mutants were tested in this experiment. The E2 deletion mutants p327-25 (lane e) and p448-5 (lane g) were each found to remain as extrachromosomal plasmids in the YC-C59 cells, as does the termination linker mutant p771-1 (lane f). Thus, the YC-C59 cell line provides a factor that permits the E2 but not the E1 mutant to remain extrachromosomal. Since this cell line contains a cDNA corresponding to the full E2 ORF and has independently been shown to express the NCR *trans*-activation function mapped to the E2 ORF (Spalholz et al. 1985b). we postulate that this complementation is due to a diffusible factor encoded by the E2 ORF. Proof that this complementation is due to the E2 gene product, however, requires analysis in cell lines transformed by this cDNA mutated within the E2 ORF. These experiments are in progress.

## TTL insertion mutants in E1

Two E1 mutant DNAs, p745-1 and p743-23, which contained the TTL at nt 945 (*Sma*I) and nt 1299 (*Pst*I), respectively, were studied. Each of these DNAs transformed at a level only slightly reduced from that of p142-6 DNA when linked to the pML2d vector sequences (Table 1). Each of these mutant DNAs was found to be integrated into the cellular DNA in selected transformants. These results are in agreement with previous studies establishing the essential role of the putative E1 gene product for plasmid maintenance and plasmid replication (Sarver et al. 1984; Lusky and Botchan 1985).

## TTL insertion mutants in E7

The termination linker was also inserted at two sites in the E7 ORF. The plasmids p774-3 and p775-1 contained the linker at the unique BPV-1 *Xma*III (nt 619) and the *Nru*I (nt 838) sites, respectively. Each of these mutations is located distal to the splice-acceptor site at nt 527 and would consequently affect the spliced E6/E7 viral gene product that was predicted (Yang et al. 1985a). The biological properties of both of these E7 mutants were similar; each transformed at a frequency approximately 85% of that of wild-type p142-6 DNA (Table 1), and each remained extrachromosomal when linked to pML2d or cleaved away. The state of the viral DNA was examined in three independent cell lines established from individual foci transformed by each DNA. In each of the cell lines expanded from a single focus, the copy number was found to be approximately 100 per diploid genome (Fig. 5). These results agree with those of Lusky and Botchan (1985) in that both sets of E7 mutants remained extrachromosomal, indicating that the product of the E7 ORF is not essential for plasmid maintenance or transformation. Our results differ, however, in that each of our E7 mutants transform C127 cells at a much higher efficiency than the two E7 mutants, *dl*54 and *dl*576, examined by Lusky and Botchan (1985). Furthermore, the plasmid copy number in cell

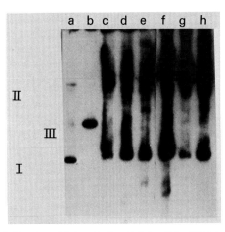

**Figure 5** Southern blot analysis of total cellular DNA from cells transfected with E7 TTL mutants p774-1 and p775-1. Lane *a* contains 10 pg of uncleaved p142-6 marker DNA, and lane *b* has 10 pg of *Xba*I-cut p142-6 marker DNA. Lanes *c–h* each contain 10 μg of DNA cleaved with *Xho*I, which does not cut the plasmids p774-3 or p775-1; (lane *c*) MR 775-1A; (lane *d*) MR 775-1 C; (lane *e*) MR775-1 D; (lane *f*) MR 774-1A; (lane *g*) MR 774-1 B; (lane *h*) MR 774-1 D. All six of these lines were expanded from individual BPV-1–transformed foci. This filter was hybridized with a nick-translated probe composed of the full BPV-1 genome. The Roman numerals indicate the migration of forms I, II, and III of marker p142-6 DNA. Note that all six cell lines contain extrachromosomal BPV-1 DNA in high copy number.

lines established with the E7 mutants *dl*576 and *dl*54 were reported to contain only 1–5 plasmid copies per cell in contrast to the 100 copies per cell detected in cell lines we established after p774-3 and p775-1 transfections.

## TTL insertion mutant in E5

The TTL was also inserted into the E5 ORF at the *Bst*XI site at nt 3881 (p744-3). This mutant was dramatically reduced in transforming ability, forming loci with an efficiency 5% or less of that of p142-6 DNA (Table 1). This result is in agreement with those obtained by several laboratories (DiMaio et al. 1985; Yang et al. 1985b; Schiller et al. 1986; D. Groff and W. Lancaster, unpubl.), which indicate the importance of the sequences downstream from the first AUG in the E5 ORF in cellular transformation.

## Discussion

The experiments described here extend our previous genetic analyses directed at identifying regions of the BPV-1 genome involved in transformation and in plasmid replication and maintenance. Data based on the study of deletion and TTL mutants of BPV-1 DNA indicate that the putative E2 gene product plays an important role in the extrachromosomal maintenance of BPV-1 DNA as well as in transformation. Several questions concerning the role of the E2 gene product in plasmid maintenance remain to be resolved. One is whether or not there is an absolute requirement for the E2 product in plasmid maintenance and replication. Although we have found the E2 product to be essential for this function in our assays outlined in this paper, Lusky and Botchan (1984) have described a replication-competent deletion mutant (Bal-15) that is deleted of viral sequences between nt 2694 and 4172, which includes almost the complete E2 ORF. None of our E2 mutants, even p715-5, which has a deletion quite analogous to that in Bal-15, remain extrachromosomal in C127 cells. Another discrepancy between our results and those of Lusky and Botchan (1985) pertains to NcoI deletion mutant p631-1. In our assay, this mutant integrated into the host chromosome regardless of the mode of selection; however, Lusky and Botchan found that although the identical mutant integrated when foci were selected for focus formation, the DNA did persist as a stable plasmid in cells selected for G418 resistance when tested by using an unlinked cotransfection assay. These differences suggest that the role of the E2 gene product in plasmid replication is indirect and that its function can possibly be circumvented under some experimental conditions. It is possible that under certain growth conditions a cellular factor functionally analogous to E2 could be induced and permit plasmid maintenance of the E2 mutants. At this time, therefore, we do not know whether the E2 ORF product is acting directly or indirectly in plasmid maintenance. Spalholz et al. (1985a) have recently demonstrated that the E2 gene product can act in *trans* to induce the BPV-1 NCR enhancer. It is possible that the role of the E2 product in plasmid maintenance is entirely indirect through the activation of the NCR enhancer, leading to expression of critical viral replication functions such as E1. However, we are not yet prepared to rule out the possibility of a direct role for the putative E2 gene product in plasmid maintenance; further studies will be necessary to clarify this.

It is also not clear whether the role of the E2 gene product in transformation is direct or indirect. Although it has been previously suggested that the E2 gene might encode a transforming protein (Sarver et al. 1984), it now seems likely that the role of E2 in transformation is indirect, in view of the evidence that the E2 product can act as a transcriptional regulatory protein (Spalholz et al. 1985a). Most of the E2 mutants analyzed in this study have a similar phenotype in that they are significantly impaired in their ability to transform, are replication-incompetent, and unable to *trans*-activate the NCR enhancer (Sarver et al. 1984; Spalholz et al. 1985a,b; Yang et al. 1985b). The *Bcl*I deletion mutant (p327-25), however, has a different phenotype. Although this mutant is replication incompetent in C127 cells and is unable to *trans*-activate the NCR enhancer (Yang et al. 1985b), it transforms at an efficiency comparable to that of the wild-type BPV-1 plasmid, p142-6 (Table 1) (Sarver et al. 1984). Therefore, the inefficient transformation character of many of the E2 mutants cannot be fully explained by implicating an indirect role for the E2 gene product in viral transcriptional control. It should be stressed, however, that a direct transforming function for the E2 gene has not yet been identified. When expressed from either the SV40 early promoter or a

Harvey sarcoma virus long terminal repeat (LTR), the primary transforming function of the 3' ORFs (E2, E3, E4, and E5) maps to E5 downstream from the first AUG at nt 3879, not to the E2 ORF (Yang et al. 1985b; Schiller et al. 1986). These transformation experiments were carried out by assaying for focus formation on the immortalized cell lines C127 and NIH-3T3. It remains possible that a direct role for E2 in transformation could be demonstrated in alternative assays using nonimmortalized cells or epithelial cells.

Our results further suggest a role for the E4 ORF in transformation. The mutant p772-1 affects both the E2 and E4 ORFs by the insertion of the TTL at nt 3235. We have been unable to induce any transformed foci with this mutant whether it is linked or unlinked from the bacterial vector DNA (Table 1). In contrast, p771-1, which is mutated only in E2 by the insertion of the termination linker at nt 2878, retains at least some low level of transforming ability (Table 1). Evidence that E4 is a coding ORF comes from the structural analysis of cDNAs of viral mRNAs in BPV-1–transformed cells (Yang et al. 1985a). One class of RNA has a splice from nt 304 to nt 3224, bringing in frame the E6 and E4 ORFs as the primary ORF in the cDNA. Direct genetic analysis of the E4 ORF, however, is difficult since it overlaps both the E2 and E3 ORFs. Studies currently in progress are directed toward studying the role of the E4 ORF by generating E4 mutants that are unaffected in E2 and E3 ORFs.

Two termination linker mutants altered in the E7 ORF were also examined in this study. We found, in agreement with Lusky and Botchan (1985), that E7 mutants remain extrachromosomal, suggesting that the E7 ORF product is not essential for plasmid replication and maintenance. However, unlike the E7 mutants of Lusky and Botchan (1985), which transformed with low efficiency and were present in low copy numbers (1–5 copies/cell), each of our E7 mutants transformed with high efficiency (Table 1) and were present in high copy numbers per cell (Fig. 5). The site of the deletion and consequent frame-shift mutations in *dl*576 and *dl*54, described by Lusky and Botchan (1985), map more 5' in E7 ORF than the sites of either of the TTL insertion mutants described here. It is possible that the shortened E6/E7 and/or E7 proteins encoded by p774-3 and p775-1 function normally in plasmid copy control. These differences suggest that the carboxyl terminus of the E7 ORF product may be dispensable in regulating plasmid copy number of BPV-1–transformed cells.

Results obtained from examination of the mutant p774-3, which contains the termination linker at the *Bst*XI site (nt 3881) in the E5 ORF, suggests strongly that the E5 gene product is involved in transformation. This result confirms studies of similar mutants in the full-length BPV-1 background (DiMaio et al. 1985; D. Groft and W. Lancaster, in prep.). Studies of cDNAs mutated in either the E2 or E5 ORFs (Yang et al. 1985b) and mutagenesis studies of subgenomic fragments of BPV-1 DNA cloned behind the Harvey sarcoma virus LTR have also indicated the important role of E5 in transformation (Schiller et al. 1986). Indeed, when a fragment of BPV-1 DNA (nt 3838–4450) containing only the E5 ORF was cloned behind the Moloney murine sarcoma virus LTR, this DNA fragment was capable of transforming C127 cells, although at a low frequency (M. Rabson and P. Howley, unpubl.). Taken together, these studies suggest that the 44-amino-acid hydrophobic protein, potentially encoded by the E5 ORF product, plays a major role in transformation by BPV-1.

Examination of E1 TTL mutants in this study produced results consistent with previous observations (Sarver et al. 1984; Lusky and Botchan 1985) of BPV-1 deletion mutants, indicating that a protein critical for plasmid maintenance or replication is encoded by the E1 ORF, since the complementation studies of Lusky and Botchan (1985) have shown that this function can act in *trans*. The failure to find a cDNA with the full E1 ORF in a cDNA library from transformed cells (Yang et al. 1985a) or by electron microscopic analysis of viral mRNAs in transformed cells (Stenlund et al. 1985) suggests that the message that encodes this protein is rare. The mechanism of action of the E1 protein is unknown but similarities identified in amino acid structure between certain domains of the putative E1 ORF and those domains of the polyoma and SV40 large T antigens (Clertant and Seif 1984) suggest that the role of E1 in plasmid replication may be direct.

## Acknowledgments

We are grateful to Joe Bolen for a critical reading of this manuscript and to Nan Freas for its preparation. This work was supported in part by grant 1597 from the Council for Tobacco Research, U.S.A., Inc. to MSR.

## References

Amtmann, E. and G. Sauer. 1982. Bovine papillomavirus transcription: Polyadenylated RNA species and assessment of the direction of transcription. *J. Virol.* **43:** 59.

Campo, M.S., D.A. Spandidos, J. Lang, and N.M. Wilkie. 1983. Transcriptional control signals in the genome of bovine papillomavirus type 1. *Nature* **303:** 77.

Chen, E.Y., P.M. Howley, A. Levinson, and P. Seeberg. 1982. The primary structure and genetic organization of the bovine papillomavirus type 1 genome. *Nature* **299:** 529.

Clertant, P. and I. Seif. 1984. A common function for polyomavirus large-T and papillomavirus E1 proteins. *Nature* **311:** 276.

Colbere-Garapin, F., F. Horodniceanu, P. Kourilsky, and A.C. Garapin. 1981. A new dominant hybrid selective marker for higher eukaryotic cells. *J. Mol. Biol.* **150:** 1.

DiMaio, D., J. Metherall, K. Neary, and D. Guralsk. 1985. Genetic analysis of cell transformation by bovine papillomavirus. In *Papilloma viruses: Molecular and clinical aspects* (ed. P.M. Howley and T.R. Broker), p. 437. A.R. Liss, New York.

Dvoretzky, I., R. Shober, S.K. Chattopadhy, and D.R. Lowy. 1980. A quantitative *in vitro* focus assay for bovine papillomavirus. *Virology* **103:** 369.

Engel, L., C.A. Heilman, and P.M. Howley. 1983. Transcriptional organization of the bovine papillomavirus type 1. *J. Virol.* **47:** 516.

Gorman, C.M. and B. Howard. 1983. Expression of recombinant plasmids in mammalian cells is enhanced by sodium butyrate. *Nucleic Acids Res.* **11:** 7631.

Heilman, C.A., L. Engel, D.R. Lowy, and P.M. Howley. 1982. Virus-specific transcription in bovine papillomavirus–transformed mouse cells. *Virology* **119:** 22.

Howley, P.M., E.T. Schenborn, E. Lund, J.C. Byrne, and J.E. Dahlberg. 1985. The bovine papillomavirus distal "enhancer" is not *cis*-essential for transformation or for plasmid maintenance. *Mol. Cell. Biol.* **5:** 3310.

Lancaster, W.D. and C. Olson. 1982. Animal papillomaviruses. *Microbiol. Rev.* **46:** 191.

Law, M.-F., J.C. Byrne, and P.M. Howley. 1983. A stable bovine papillomavirus hybrid plasmid that expresses a dominant selective trait. *Mol. Cell. Biol.* **3:** 2110.

Law, M.-F., D.R. Lowy, I. Dvoretzky, and P.M. Howley. 1981. Mouse cells transformed by bovine papillomavirus contain only extrachromosomal viral DNA sequences. *Proc. Natl. Acad. Sci.* **78:** 2727.

Lowy, D.R., I. Dvoretzky, R. Shober, M.-F. Law, L. Engel, and P. Howley. 1980. In vitro tumorigenic transformation by a defined subgenomic fragment of bovine papillomavirus DNA. *Nature* **287:** 72.

Lusky, M. and M. Botchan. 1984. Characterization of the bovine papillomavirus plasmid maintenance sequences. *Cell* **36:** 391.

———. 1985. Genetic analysis of bovine papillomavirus type 1 *trans*-acting replication factors. *J. Virol.* **53:** 955.

Lusky, M., L. Berg, H. Weiher, and M. Botchan. 1983. Bovine papillomavirus contains an activator of gene expression at the distal end of the early transcription unit. *Mol. Cell. Biol.* **3:** 1108.

Maniatis, T., E.F. Fritsch, and J. Sambrook. 1982. *Molecular cloning: A laboratory manual.* Cold Spring Harbor Laboratory, Cold Spring Harbor, New York.

Nakabayashi, Y., S. Chattopadhyay, and D.R. Lowy. 1983. The transformation function of bovine papillomavirus DNA. *Proc. Natl. Acad. Sci.* **80:** 5832.

Sarver, N., J.C. Byrne, and P.M. Howley. 1982. Transformation and replication in mouse cells of bovine papillomavirus/pML2 plasmid vector that can be rescued in bacteria. *Proc. Natl. Acad. Sci.* **79:** 7147.

Sarver, N., M.S. Rabson, Y.-C. Yang, J.C. Byrne, and P.M. Howley. 1984. Localization and analysis of bovine papillomavirus type 1 transforming functions. *J. Virol.* **52:** 377.

Schiller, J., W.C. Vass, and D.R. Lowy. 1984. Identification of a second transforming region of bovine papillomavirus. *Proc. Natl. Acad. Sci.* **81:** 7880.

Schiller, J., K. Vousden, W.C. Vass, and D.R. Lowy. 1986. E5 open reading frame of bovine papillomavirus type 1 encodes a transformed gene. *J. Virol.* **57:** 1.

Spalholz, B.S., Y.-C. Yang, and P.M. Howley. 1985a. Transactivation of a bovine papillomavirus transcriptional regulatory element by the E2 gene product. *Cell* **42:** 183.

———. 1985b. Identification of new enhancer within the noncoding region of BPV-1 which is *trans*-activated by BPV-1 early gene products. In *Papilloma virus: Molecular and clinical aspects* (ed. P.M. Howley and T.R. Broker), p. 343. A.R. Liss, New York.

Stenlund, A., J. Zabielski, H. Ahola, J. Moreno-Lopez, and U. Pettersson. 1985. The messenger RNAs from the transforming region of bovine papillomavirus type 1. *J. Mol. Biol.* **182:** 541.

Veldman, G.M., S. Lupton, and R. Kamen. 1985. Polyomavirus enhancer contains multiple redundant sequence elements that activate both DNA replication and gene expression. *Mol. Cell. Biol.* **5:** 649.

Yang, Y.-C., H. Okayama, and P.M. Howley. 1985a. Bovine papillomavirus contains multiple transforming genes. *Proc. Natl. Acad. Sci.* **82:** 1030.

Yang, Y.-C., B.A. Spalholz, M.S. Rabson, and P.M. Howley. 1985b. Dissociation of transforming and transactivation functions for bovine papillomavirus type 1. *Nature* **318:** 575.

# Analysis of Splicing of mRNA Precursors In Vitro

## P.A. Sharp

Center for Cancer Research and Department of Biology, Massachusetts Institute of Technology, Cambridge, Massachusetts 02139

Splicing of mRNA precursors occurs through the formation of a lariat RNA containing a branch site. Sequences at the branch site do not appear to play a critical role in specifying the splice sites utilized during splicing. An intermediate in splicing is found exclusively in a large, multicomponent complex. This multicomponent complex ("spliceosome") probably contains U1 RNP, U2 RNP, and hnRNP proteins. It is hypothesized that formation of such a complex by two substrate RNAs results in an observed *trans*-splicing reaction. *Trans*-splicing is the joining of exons from two substrate RNAs.

The study of DNA tumor viruses has contributed significantly to our understanding of gene expression in mammalian cells. Perhaps the most startling contribution was the discovery of the role of RNA splicing in the synthesis of mRNAs (Berget et al. 1977). It is now clear that the heterogeneous nuclear RNA (hnRNA) transcribed from most cellular genes contains introns that are excised by splicing. For viral as well as cellular genes, the question of the nature of the signals specifying the precise sequences to be excised remains an enigma. Introns of thousands of nucleotides in length are specifically processed and yet the critical sequences directing splicing seem to be confined to short tracts at the 5' splice site, 3' splice site, and perhaps the branch site. Constraints on the nature of these sequences are reflected in their consensus relationship to one another and the effects of nucleotide changes on the activity of splice sites. The consensus sequence for the 5' splice site extends only 6 nucleotides into the intron (AG:GUAAGU), whereas the 3' splice site seems to possess comparable specificity in the form of a pyrimidine tract ending in a conserved CAG trinucleotide sequence [(U/C$_{16}$NCAG:G] (Sharp 1981). The consensus sequences of splice sites in viral genes are indistinguishable from those of cellular genes, suggesting a common splicing mechanism (Sharp 1984). In addition, 5' splice sites from viral genes can be accurately joined to 3' splice sites from cellular genes when constructed into transcription units with chimeric introns (Berkner and Sharp 1984).

The nature of sequences important in splicing became more interesting with the discovery of the role of branch formation in this process (Grabowski et al. 1984; Padgett et al. 1984; Ruskin et al. 1984). This insight followed from the development of in vitro soluble reactions that accurately spliced RNA in a reproducible fashion (Hernandez and Keller 1983; Padgett et al. 1983; Krainer et al. 1984). As shown in Figure 1, using as an example a substrate containing the first (L1) and second (L2) leaders of the major late transcription unit of adenovirus, an intermediate is generated during splicing (Grabowski et al. 1984). The intermediate contains two RNAs, the L1 exon with a 3'-hydroxyl group and a lariat RNA where the guanosine at the 5' splice site is covalently linked to an adenosine group upstream of the 3' splice site. A 2'−5' phosphodiester bond links the branch site to the 5' splice site (Konarska et al. 1985b). This intermediate accumulates during splicing in vivo (Domdey et al. 1984; Zeitlin and Efstratiadis 1984) and in vitro and is subsequently processed to mature spliced product L1-L2 and the excised intervening sequence. The latter RNA is produced as a lariat RNA and is rapidly degraded in vivo. As indicated in Figure 1, sequences at the branch site of this substrate are strikingly complementary to sequences at the 5' splice site. A similar but less extensive complementarity is found in other introns. Note that since the consensus sequence at the 5' splice site is complementary to both the branch site and sequences at the 5' end of U1 RNA, the two latter sequences are homologous. A consensus sequence can be formed with sequences encompassing the branch sites of introns (Keller and Noon 1984; Ruskin et al. 1984). This consensus is quite limited, suggesting that the branch site may not be a highly constrained sequence element in the splicing of most introns (Fig. 2).

## Results and Discussion

Perhaps the first indication that sequences at branch sites were not critical for splicing was the observation of Wieringa et al. (1984) that deletions to within 24 nucleotides of the 3' splice site of the large intron of the rabbit β-globin yielded mutants that were active in vivo for synthesis of mature mRNA. We have analyzed the splicing in vitro of RNAs transcribed from both mutant DNAs as well as the wild-type β-globin gene (Padgett et al. 1985). As previously reported, the wild-type intron is processed with the formation of a branch site at 32 nucleotides from the 3' splice site (Zeitlin and Efstratiadis 1984). RNAs transcribed from mutants deleted in this branch site sequence are spliced at about 1/5 the efficiency as RNA from the wild-type gene. Branch formation occurs during splicing of these mutant

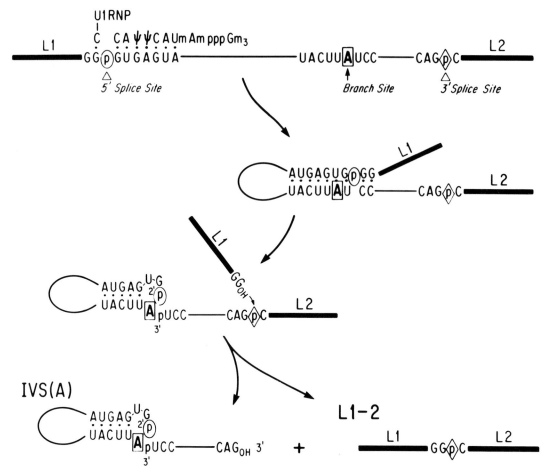

**Figure 1** The splicing process. The base pair complementarity between the sequences at the 5′ splice site and branch site is indicated by ·. Phosphate moieties from the 5′ splice site (Ⓟ) and 3′ splice site (◇Ⓟ) are retained in the product RNA during processing. The intermediate in splicing is schematically drawn on the third line. As discussed in the text, the L1 exon and lariat intervening sequence–L2 exon are found in a multicomponent complex.

RNAs at a site within the *Xho*I linker sequence utilized in construction of the deletions. As shown in Figure 2, the sequence encompassing the branch site in the mutants has limited homology to either previously proposed consensus sequences or to a sequence complementary to the 5′ splice site. Others have also analyzed the effect of deleting the normal branch site from introns (Reed and Maniatis 1985; Ruskin et al. 1985). In general, the same results were obtained. A surrogate adenosine was used for branch formation, and kinetics of splicing was slower for RNAs from the mutant genes as compared with the wild-type gene. These observations suggest that sequences at the branch site do not have an essential role in specifying splicing of introns in mammalian cells. As a corollary, it is unlikely that an essential component in splicing directly recognizes sequences at the branch site.

As mentioned previously, the intermediate in splicing contains two RNAs, the 5′ exon and a lariat RNA composed of the intervening sequence and 3′ exon. It was anticipated that these two RNAs would be stably juxta-

posed in a splicing complex for further reaction. Such a complex can be easily resolved by velocity sedimentation in glycerol gradients where the RNAs characteristic of the intermediate cosediment in a 60S peak (Brody and Abelson 1985; Frendewey and Keller 1985; Grabowski et al. 1985). This complex has been referred to as a spliceosome. It is interesting to compare the size of the spliceosome with another ribonucleoprotein structure, the 60S ribosome subunit. The latter contains 5300 nucleotides of RNA and 45 proteins, which sums to a mass of $3 \times 10^6$ daltons. The precursor RNA used to form the spliceosome was 454 nucleotides in length (Grabowski et al. 1985). Thus, several cellular components must also be present in the complex. Specific antisera to either U1 RNP or U2 RNP will immunoprecipitate the splicing complex, and thus these small nuclear ribonucleoprotein components are likely also to be present (Grabowski et al. 1985). This is consistent with the recent finding that both U1 RNP and U2 RNP are essential for splicing (Black et al. 1985; Krainer and Maniatis 1985). The total mass of U1 RNP, U2 RNP, and

```
RβG wild type IVS-2                        UGCUAAC

RβG LIVS 3' SS-24                          UCUAGAG

RβG mini LIVS 38/129                       GCUAGAG

RβG mini LIVS 38/102                       GAGAGAG

Consensus from Keller & Noon (1984)        CUGAC

Consensus from Ruskin et al. (1984)        U U A U
                                           C X C U G A A C
```

**Figure 2** Sequences at the branch site of mutants of the second intron (IVS-2) of the rabbit β-globin (RβG) gene were determined by both primer extension analysis and direct RNA analysis (Padgett et al. 1985). The sequence structure of the mutants has previously been described (Wieringa et al. 1984) as well as their activity in vivo. The adenosine at the site of branch formation is underlined. Two consensus sequences for branch sites are given. The branch sites for the mutant genes poorly conform to either consensus sequence.

precursor RNA does not account for a 60S structure unless multiple copies of these species are present. Thus, it is highly likely that other components are present in this complex.

In the nucleus of cells, newly synthesized precursor RNA is rapidly assembled into heterogeneous nuclear ribonucleoprotein particles, hnRNPs (LeStourgeon et al. 1981). Light nuclease digestion of hnRNP releases most of the newly synthesized RNA in the form of 30S–40S particles. These particles contain a set of basic proteins, $A_1$ and $A_2$, $B_1$ and $B_2$, $C_1$ and $C_2$, which are tightly associated with the RNA and other more loosely associated proteins (Dreyfuss et al. 1984). Since these basic proteins are associated with nascent RNA transcripts, it was reasonable that they played a structural role in splicing. To test this possibility, two specific monoclonal antibodies that react with the $C_1$

and $C_2$ doublet were tested for inhibition of splicing in the in vitro extract (Choi et al. 1986). Surprisingly, one of the monoclonal antibodies totally inhibited splicing. The step blocked by inhibition was before the formation of the intermediate RNAs. The second monoclonal antibody, which was added at an equivalent titer, did not affect splicing. The lack of inhibition by a monoclonal antibody is difficult to interpret. For example, the binding of the antibody to a particular epitope might not inhibit the function of the protein. These results suggest that proteins related to the 30S RNP particle are also important in splicing and may be components of the spliceosome.

A schematic of the current resolution of the splicing process as now understood is shown in Figure 3. The first step in the reaction is the formation of a 60S spliceosome structure (Grabowski et al. 1985). In this complex, the intermediate RNAs are generated through cleavage at the 5' splice site and formation of an RNA branch. This bipartite intermediate upon further reaction yields the spliced RNA product and the excised intervening sequence. In vivo, the latter RNA is probably released from the complex and rapidly degraded. The spliced RNA product is transported to the cytoplasm.

The above scheme for splicing does not provide an obvious answer to the question of the sequence specificity directing the process. The consensus sequences at the 5' and 3' splice sites seem inadequate to specify the excision of introns thousands of bases in length. As mentioned previously, neither specific sequences at the branch site nor at other sites within the intron are apparently critical in the reaction. A possible solution to the apparent, limited set of sequences specifying splicing is the formation on the RNA of a highly ordered multicomponent structure. In this structure, a limited set of sequences would be available for

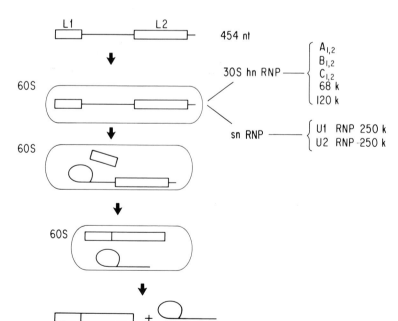

**Figure 3** The spliceosome. Schematic of the splicing process showing the involvement of splicing complex formation. Multiple nuclear factors likely to be present in the complex are listed.

reacting with other splice sites. Such a hypothetical structure would have to be formed on the precursor RNA in a unique fashion to approximately align the correct splice sites. The nature of the signals for formation of such a structure and whether it exists at all will require further study.

If cellular components recognized different sequences in the precursor RNA and then coalesce into a complete spliceosome, the process could potentially utilize two RNA substrates and trans-splice exons from two distinct precursors. To test whether trans-splicing occurred in extracts in vitro, we added two RNA substrates, one containing the L1 exon with the 5' splice site and part of the intervening sequence, and the other containing the remainder of the intervening sequence, the 3' splice site and 3' exon (Konarska et al. 1985a). Splicing of the L1 and L2 exons was detected at a low level. Trans-splicing was dependent upon the addition of ATP. Experiments designed to inhibit end-to-end ligation of RNA molecules suggested that the trans-splicing observed was not due to ligation of the two substrate RNAs prior to splicing. The efficiency of trans-splicing of the two RNAs could be markedly improved if the two RNAs could associate through complementary sequences (Konarska et al. 1985a; Solnick 1985). Assuming that the process of trans-splicing involves formation of a splicing complex, then the above results suggest that two distinct RNAs can participate in this reaction and that two recognition processes occur. One step would recognize the 3' splice site and another would recognize the 5' splice site. Each step would result in the formation of a partial complex at the splice site. Defining the components responsible for these partial complexes should permit reconstruction of the spliceosome complex.

## Acknowledgments

This work was supported by grants from the National Institutes of Health (RO1-GM32467 and RO1-GM 34277), partially from a National Cancer Institute Cancer Center Support (core) (P30-CA14051), and from the National Science Foundation (DCB-8502718) to P.A.S.

## References

Berget, S.M., C. Moore, and P.A. Sharp. 1977. Spliced RNA segments at the 5'-terminus of late adenovirus 2 mRNA. Proc. Natl. Acad. Sci. 74: 3171.

Berkner, K.L. and P.A. Sharp. 1984. Expression of dihydrofolate reductase, and of the adjacent Elb region, in an Ad5-dihydrofolate reductase recombinant virus. Nucleic Acids Res. 12: 1925.

Black, D.L., B. Chabot, and J.A. Steitz. 1985. U2 as well as U1 small nuclear ribonucleoproteins are involved in premessenger RNA splicing. Cell 42: 737.

Brody, E. and J. Abelson. 1985. The "splicesome": Yeast premessenger RNA associates with a 40S complex in a splicing-dependent reaction. Science 228: 963.

Choi, Y.D., P.J. Grabowski, P.A. Sharp, and G. Dreyfuss. 1986. Inhibition of RNA splicing by an antibody to proteins of the pre-mRNA-ribonucleoprotein (hnRNP) particle. Science (in press).

Domdey, H., B. Apostol, R.-J. Lin, A. Newman, E. Brody, and J. Abelson. 1984. Lariat structures are in vivo intermediates in yeast pre-mRNA splicing. Cell 39: 611.

Dreyfuss, G., Y.D. Choi, and S.A. Adam. 1874. Characterization of heterogeneous nuclear RNA–protein complexes in vivo with monoclonal antibodies. Mol. Cell. Biol. 4: 1104.

Frendewey, D. and W. Keller. 1985. Stepwise assembly of a pre-mRNA splicing complex requires U-snRNPs and specific intron sequences. Cell 42: 355.

Grabowski, P.J., R.A. Padgett, and P.A. Sharp. 1984. Messenger RNA splicing in vitro: An excised intervening sequence and a potential intermediate. Cell 37: 415.

Grabowski, P.J., S.R. Seiler, and P.A. Sharp. 1985. A multicomponent complex is involved in the splicing of messenger RNA precursors. Cell 42: 345.

Hernandez, N. and W. Keller. 1983. Splicing of in vitro synthesized messenger RNA precursors in HeLa cell extracts. Cell 35: 89.

Keller, E.B. and W.A Noon. 1984. Intron splicing: A conserved internal signal in introns of animal pre-mRNAs. Proc. Natl. Acad. Sci. 81: 7417.

Konarska, M.M., R.A. Padgett, and P.A. Sharp. 1985a. Trans-splicing of mRNA precursors in vitro. Cell 42: 165.

Konarska, M.M., P.J. Grabowski, R.A. Padgett, and P.A. Sharp. 1985b. Characterization of the branch site in lariat RNAs produced by splicing of mRNA precursors. Nature 313: 552.

Krainer, A.R. and T. Maniatis. 1985. Multiple factors including the small nuclear ribonucleoproteins U1 and U2 are necessary for pre-mRNA spliced in vitro. Cell 42: 725.

Krainer, A.R., T. Maniatis, B. Ruskin, and M.R. Green. 1984. Normal and mutant human β-globin pre-mRNAs are faithfully and efficiently spliced in vitro. Cell 36: 993.

LeStourgeon, W.M., L. Lothstein, B.W. Walker, and A.L. Beyer. 1981. The composition and general topology of RNA end protein in monomer 40S ribonucleoprotein particles. In The cell nucleus (ed. H. Busch), vol. 9, p. 49. Academic Press, New York.

Padgett, R.A., S.F. Hardy, and P.A. Sharp. 1983. Splicing of adenovirus RNA in a cell free transcription system. Proc. Natl. Acad. Sci. 80: 5230.

Padgett, R.A., M.M. Konarska, P.J. Grabowski, S.F. Hardy, and P.A. Sharp. 1984. Lariat RNAs as intermediates and products in the splicing of mRNA precursors. Science 225: 898.

Padgett, R.A., M.M. Konarska, M. Aebi, H. Hornig, C. Weissmann, and P.A. Sharp. 1985. Non-consensus branch site sequences in the in vitro splicing of transcripts of mutant rabbit beta-globin genes. Proc. Natl. Acad. Sci. 82: 8349.

Reed, R. and T. Maniatis. 1985. Intron sequences involved in lariat formation during pre-mRNA splicing. Cell 41: 95.

Ruskin, B., J.M. Greene, and M.R. Green. 1985. Cryptic branch point activation allows accurate in vitro splicing of human β-globin intron mutants. Cell 41: 883.

Ruskin, B., A.R. Krainer, T. Maniatis, and M.R. Green. 1984. Excision of an intact intron as a novel lariat structure during pre-mRNA splicing in vitro. Cell 38: 317.

Sharp, P.A. 1981. Speculations on RNA splicing. Cell 23: 643.
———. 1984. Adenovirus transcription. In The adenoviruses (ed. H.S Ginsberg), p. 173. Plenum Press, New York.

Solnick, D. 1985. Trans splicing of mRNA precursors. Cell 42: 157.

Wieringa, B., E. Hofer, and C. Weissmann. 1984. A minimal intron length but no specific internal sequence is required for splicing the large rabbit β-globin intron. Cell 37: 915.

Zeitlin, S. and A. Efstratiadis. 1984. In vivo splicing products of the rabbit β-globin pre-mRNA. Cell 39: 589.

# Analysis of mRNA Splicing Using Fractions Isolated from HeLa Cells and Mutant Adenovirus Pre-mRNAs

### K.K. Perkins, H.M. Furneaux, G.A. Freyer, J. Arenas, L. Pick, and J. Hurwitz

Graduate Program in Molecular Biology, Sloan-Kettering Institute for Cancer Research, New York, New York 10021

Three fractions (Ia, Ib, and II) have been isolated from a nuclear extract of HeLa cells that are required for splicing of adenovirus precursor mRNA in vitro. Incubation of fractions Ia, Ib, and II with precursor mRNA in the presence of ATP yielded four RNA species that have been identified as spliced RNA, intron lariat, intron-exon lariat, and the 5′ exon. When fractions Ib and II were combined, only cleavage at the 5′ splice site of precursor mRNA and formation of the intron-exon lariat occurred. Incubation of precursor mRNA with either fraction alone did not yield any ATP-dependent RNA species. The activity in fraction II was abolished by preincubation with micrococcal nuclease, indicating that this fraction contained an essential nucleic acid component. When fraction II was incubated with precursor RNA in the presence of ATP and the resulting products were sedimented through sucrose gradients, a 30S complex was detected that contained precursor RNA. The combination of fractions Ib and II resulted in the production of a 55S complex that contained the 5′ exon as a prominent RNA species. Combination of fractions I (containing Ia and Ib) and II yielded the 55S complex and spliced RNA sedimenting between 40S and 20S. Two mutant pre-mRNAs containing base changes at the site of lariat attachment and at the 3′ splice site did not yield spliced RNA but were cleaved at the 5′ splice site and formed intron-exon lariat stuctures.

The elimination of intervening sequences from primary transcripts by RNA splicing is essential for the correct expression of many eukaryotic genes (Abelson 1979). Studies of the kinetics of RNA splicing in vitro and characterization of the likely reaction intermediates suggest that RNA splicing proceeds by at least two discrete steps (Grabowski et al. 1984; Krainer et al. 1984; Ruskin et al. 1984). Early in the reaction, precursor mRNA (pre-mRNA) is cleaved at the 5′ splice site, and the 5′-terminal phosphate of the intron becomes esterified to the 2′-hydroxyl groups of an adenosine residue within the intron, resulting in the formation of an intron-exon lariat structure (Padgett et al. 1984; Ruskin et al. 1984; Konarska et al. 1985). At a later stage, the 5′ exon species and the intron-exon lariat are thought to interact to form spliced RNA, concomitant with the release of the intron in a lariat configuration. The discovery of lariat structures in vivo (Domdey et al. 1984; Rodriquez et al. 1984; Zeitlin and Efstratiadis 1984) lends considerable support to the validity of the pathway described above.

To gain further insight into the mechanism of this reaction, we have concentrated on the isolation of enzymes required for RNA splicing and have previously reported the isolation of two fractions (I and II) from a HeLa cell nuclear extract that were both required for RNA splicing (Furneaux et al. 1985). Here we report the separation of fraction I into two components (Ia and Ib) and their ability to form ATP-dependent splicing complexes with pre-mRNA.

Although a complete understanding of the mechanism of RNA splicing will undoubtedly result from the purification of the various components involved, one can use the partially purified system to examine the effects of specific sequence changes in the pre-mRNA. Comparison of a large number of pre-mRNA sequences has revealed the existence of highly conserved regions surrounding the intron-exon junctions (Mount 1982). Almost all pre-mRNAs have an invariant GU dinucleotide at the 5′ end of the intron and an invariant AG dinucleotide at the 3′ end. Other regions of interest include four highly conserved residues adjacent to the 5′ GU dinucleotide and a polypyrimidine tract near the 3′ end of the intron. The critical importance of both the GU and AG dinucleotides has been well documented by studies done in vivo (Montell et al. 1982; Treisman et al. 1982) and in vitro (Krainer et al. 1984; Reed and Maniatis 1985).

Although other sequences in the intron were not previously thought to be essential for accurate splicing (Wieringa et al. 1984), this issue has been reexamined in light of the discovery of the lariat structure. Recent studies (Keller and Noon 1984; Ruskin et al. 1984; Zeitlin and Efstratiadis 1984; Konarska et al. 1985) have in fact indicated a possible consensus sequence for the lariat branch point (PyXPyUPuAPy) close to the 3′ end of the intron. Deletion of this consensus sequence or mutation of the adenosine residue showed that splicing of the mutated pre-mRNA was relatively unaffected but that a cryptic lariat attachment site was utilized (Ruskin et al. 1985).

In our studies we have employed an adenovirus pre-mRNA that contains 16 nucleotides (nt) of SP6-promoter-proximal sequence followed by 41 nt of the major late first-leader exon, an abbreviated 86-nt intron, and 38 nt of the second-leader exon (Furneaux et al. 1985). This intron includes the 6-nt conserved region at the 5' end, the lariat branch point mapped by Grabowski et al. (1984), the polypyrimidine tract, and the AG dinucleotide at the 3' end. This short pre-mRNA is therefore an excellent substrate for site-directed mutagenesis studies, and in this report we have examined the effect of changing residues at the 3' end of the intron.

## Experimental Procedures

### Materials

Bacteriophage SP6 RNA polymerase fraction V (Blue Dextran–Sepharose pool) was purified by the method of Butler and Chamberlin (1982). Restriction enzymes were obtained from New England Biolabs; Bio-Rex 70 was from Bio-Rad Laboratories; micrococcal nuclease was from P-L Biochemicals; vanadyl ribonucleoside was from Bethesda Research Laboratories; and $[\alpha\text{-}^{32}P]GTP$ was from New England Nuclear. $^{32}P$-labeled 40S ribosomal subunit prepared from *Artemia salina* was the kind gift of Dr. D. Tabarini of this Institute.

### Plasmid DNA

Plasmid pKT1 was constructed by inserting a modified DNA sequence encoding the Ad2 major late RNA into the SP6 RNA polymerase promoter containing plasmid pSP65 (Furneaux et al. 1985). A subclone of Ad2 DNA (pBalE) containing the *Bal*IE fragment (nt positions 5360–7751) cloned into pBR322 with *Bam*HI linkers was cleaved with *Mbo*II (nt pos. 6035) and treated with the large fragment of DNA polymerase I (Klenow) and dNTPs to produce a blunt-ended molecule. This was followed by digestion with *Hin*dIII. The *Mbo*II-*Hin*dIII fragment was gel-purified. In a separate reaction, pBR322 DNA was digested with *Eco*RI, incubated with DNA polymerase I large fragment plus dNTPs, and digested with *Hin*dIII. The digested pBR322 DNA and the *Mbo*II-*Hin*dIII fragment were ligated with T4 DNA ligase and transformed into *E. coli* mm294. The resulting plasmid, pVM7, was cleaved with *Sca*I (nt pos. 6086) and *Eco*RI (which was regenerated during the ligation), and the DNA fragment was purified by agarose gel electrophoresis. This fragment contains the first-leader exon (41 nt long) and the first 6 nt of the intron. pBalE DNA was then digested with *Bst*EII (nt pos. 7020), incubated with DNA polymerase I large fragment and dNTPs, and then digested with *Bam*HI. This fragment contains the 80 nt at the 3' end of the first intron, the second exon (72 nt), and 600 nt at the 5' end of the second intron. This *Bst*EII-*Bam*HI fragment was gel purified and mixed with the *Eco*RI-*Sca*I DNA fragment and joined at the *Sca*I-*Bst*EII sites in a reaction containing T4 DNA ligase and 15% polyethylene glycol (Zimmerman and Pheiffer 1983). After extraction with phenol-chloroform (1:1), the product was cleaved with

*Eco*RI and *Bam*HI and the resulting fragment was ligated into *Eco*RI-*Bam*HI–digested pSP65.

### In vitro RNA splicing reaction

Adenovirus major late pre-mRNA transcripts were synthesized using SP6 RNA polymerase as previously described (Furneaux et al. 1985). RNA splicing reaction mixtures (0.05 ml) containing 20 mM HEPES-KOH buffer (pH 7.6), 3 mM $MgCl_2$, 2 mM dithiothreitol, 0.4 mM ATP, 20 mM creatine phosphate, 2% PEG-6000, $[\alpha\text{-}^{32}P]GTP$-labeled RNA transcripts (expressed as fmoles of molecules, as indicated in figure legends), and various enzyme fractions (as indicated in legends) were incubated at 30°C for 2 hr. The reaction was terminated by the addition of 0.2 ml of SDS buffer (200 mM Tris-HCl [pH 7.5], 25 mM EDTA, 300 mM NaCl, 2% SDS) and water to 0.4 ml. After extraction with phenol/chloroform (1:1) and precipitation with ethanol, the RNA products were dissolved in formamide and then analyzed by polyacrylamide/urea gel electrophoresis and visualized by autoradiography.

### Sucrose gradient centrifugation analysis of mRNA splicing complexes

Splicing reactions (0.10 ml) were stopped by quick-freezing in a dry ice/ethanol bath. The samples were then thawed and applied to 5-ml 10–30% sucrose gradients containing 100 mM KCl, 3 mM $MgCl_2$, 10 mM HEPES-KOH buffer (pH 7.6). Centrifugation was at 48,000 rpm in a Sorvall AH-650 rotor at 4°C for 195 min.

### Preparation of fractions

A HeLa cell nuclear extract prepared by the method of Dignam et al. (1983) was separated into two fractions (I and II) by DEAE-cellulose chromatography, as previously described (Furneaux et al. 1985). Fraction I (DEAE-cellulose flowthrough, 43 ml, 11.3 mg/ml protein) was dialyzed against 50 mM KCl in 20 mM HEPES-KOH buffer (pH 7.6), 1 mM dithiothreitol, 0.1 mM EDTA/10% (v/v) glycerol (Buffer A) and applied to a Bio-Rex 70 column (5.5 × 2.5 cm) previously equilibrated with 50 mM KCl/Buffer A. The material that passed through the column was collected and pooled to yield fraction Ia (30 ml, 2.7 mg/ml protein). The column was then eluted with Buffer A containing 1 M KCl. The eluate was pooled to yield fraction Ib (18 ml, 7.5 mg/ml protein) and dialyzed against Buffer A containing 0.1 M KCl.

## Results

We demonstrated previously that two fractions (I and II) isolated from HeLa cell nuclear extracts were capable of accurately splicing adenovirus pre-mRNA in the presence of ATP (Furneaux et al. 1985). Analysis of reaction products by 6% polyacrylamide/urea gel electrophoresis revealed two species with apparent mobilities corresponding to nucleotide lengths of 94 and 56 that were identified as spliced RNA and the 5' exon. No attempt was made to analyze putative lariat structures. No spliced RNA was detected when either fraction was incubated alone (Furneaux et al. 1985). Fraction I has now been further separated into two distinct fractions (Ia and Ib) by Bio-Rex chromatography (as described in Experimental Procedures). The combination of fractions Ia, Ib, and II catalyzed the ATP-dependent formation of four RNA species from adenovirus pre-mRNA that migrated with apparent mobilities corresponding to nucleotide lengths of 430, 255, 94, and 56 on 15% polyacrylamide/urea gels (Fig. 1A, lanes 6–10). The 94-nt species was identified as spliced RNA by hybridization with the M13 single-stranded cDNA derived from the adenovirus major late mRNA (Padgett et al. 1983) and protection from RNase T1 digestion as described previously (Furneaux et al. 1985). Quantitation (by gel excision followed by liquid scintillation spectroscopy; Fig. 1B) revealed that in the presence of the three fractions approximately 15% of the input pre-mRNA had been converted to spliced RNA after 4 hr at 30°C.

No spliced RNA was detected when fractions Ib and II were combined (Fig. 1A, lanes 1–5, and Fig. 1B). However, the RNA species of 430, 255, and 56 nt were observed. The 430- and 255-nt species have been iden-

tified as the intron-exon lariat and the intron lariat, respectively, on the basis of their anomalous electrophoretic mobilities, by their RNase T1 digestion patterns, and by the observation that they yielded trinucleotides upon digestion with nuclease P1 (data not shown). In addition, incubation of these species with HeLa cytoplasmic extracts containing the lariat-debranching enzyme described by Ruskin and Green (1985), which specifically cleaves 2′,5′-phosphodiester bonds at lariat attachment sites, resulted in their conversion to linear RNA species with mobilities corresponding to nucleotide lengths of 125 and 86 when analyzed by 15% polyacrylamide/urea gel electrophoresis (data not shown). The 56-nt species has been identified as the 5′ exon on the basis of its size and RNase T1 digestion pattern (data not shown). Therefore the combination of fractions II and Ib was sufficient to yield putative intermediates but not spliced RNA.

The amount of spliced RNA formed with the purified fractions was compared with the amount formed by the HeLa cell nuclear extract. The combination of fractions Ia (27 μg of protein), Ib (37 μg of protein), and II (46 μg of protein) yielded 0.73 fmole of spliced RNA in 2 hr at 30°C, representing 10% of the input adenovirus pre-

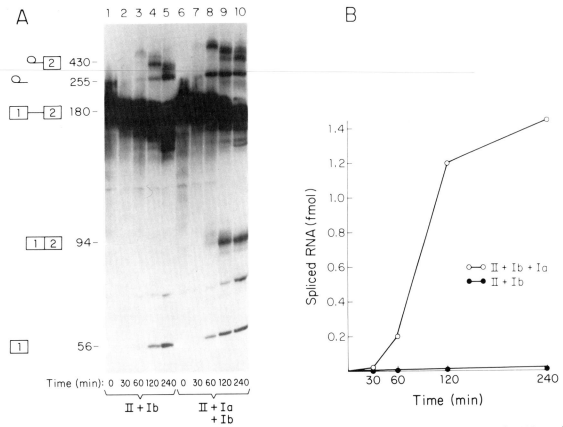

**Figure 1** Kinetics of the splicing reaction. Reaction mixtures contained SP6 adenovirus transcript (14 fmoles ends, 120 cpm/fmole), ATP (0.4 mM) and creatine phosphate (20 mM), fraction II (46 μg protein), fraction Ib (37 μg protein), and fraction Ia (27 μg protein as indicated above. (A) Lanes 1–5: Fractions Ib and II were combined and incubated for the indicated time periods. Lanes 6–10: Fractions Ia, Ib, and II were combined and incubated for the indicated time periods. (B) Spliced RNA from A was quantitated by liquid scintillation spectroscopy of gel fragments. Results are expressed graphically as the number of molecules of spliced RNA formed during incubation for the indicated time periods at 30°C.

mRNA. Under the same conditions, the combination of fractions I (226 μg of protein) and II (46 μg of protein) yielded 1.6 fmoles of spliced RNA (20% of the input pre-mRNA), and the nuclear extract (300 μg of protein) yielded 1.0 fmole of spliced RNA (12% of the input pre-mRNA). Addition of increasing amounts of fraction Ib resulted in increased cleavage at the 5' splice site; however, due to an exoribonuclease present in this fraction, the intron-exon lariat was extensively degraded and less spliced RNA was produced. Therefore, the amount of fraction Ib used in these reactions was limiting and is likely to account for the lower yield of spliced RNA observed.

The amount of cleavage at the 5' splice site that had occurred in the absence (Fig. 1A, lanes 1–5) and in the presence (Fig. 1A, lanes 6–10) of fraction Ia was determined by measuring the amount of free 5' exon observed and the 5' exon content of spliced RNA. Incubation of fractions Ia, Ib, and II with pre-mRNA for 1.0, 2.0, and 4.0 hr resulted in 5%, 11%, and 15% cleavage at the 5' site. In contrast, incubation of fractions Ib and II with pre-mRNA for 1.0, 2.0, and 4.0 hr resulted in 1%, 4%, and 9% cleavage at the 5' splice site.

The amount of intron-exon lariat formed could not be quantitated due to digestion of this species by a 3'-exonuclease activity present in fraction Ib (Fig. 1A, lanes 1–5). We believe that the species with electrophoretic mobility similar to the intron lariat produced by incubation with fractions II and Ib (Fig. 1A, lanes 1–5) was formed by digestion of the intron-exon lariat (to a specific site) by this 3' exonuclease. This product is probably not formed by endonucleolytic cleavage at the 3' splice site since the 3' exon species (38 nt in length)

has not been observed. In addition, the intron lariat was not detected when β-globin pre-mRNA was incubated with fractions II and Ib (Perkins et al. 1986).

## Properties of the three fractions

We have previously demonstrated that fraction II is sensitive to micrococcal nuclease when human β-globin pre-mRNA was used as a substrate (Furneaux et al. 1985). Pretreatment of fraction II with micrococcal nuclease in the presence of $CaCl_2$ (1 mM) abolished its ability to yield spliced RNA or any of the ATP-dependent RNA species when combined with fraction I and adenovirus pre-mRNA (Fig. 2, lanes 3 and 4). In contrast, preincubation of fraction I with micrococcal nuclease in the presence of $CaCl_2$ had no effect on its ability to catalyze ATP-dependent formation of spliced RNA when combined with fraction II (Fig. 2, lanes 5 and 6). As a control, fraction II was preincubated with micrococcal nuclease in the presence of both $CaCl_2$ (1 mM) and EGTA (3 mM), and this had no effect on the activity of fraction II (data not shown). The RNA splicing reaction was not affected by the addition of pancreatic DNase (data not shown). These results indicate that fraction II contains an RNA component essential for an early step in the splicing reaction.

The activities in fractions Ia and Ib were distinguished on the basis of their sensitivities to heat and N-ethylmaleimide (5 mM). The activity in fraction Ia (measured by quantitation of spliced RNA formed after 2 hr at 30°C) was reduced 50% by N-ethylmaleimide (5 mM) and was reduced 50% and 100% by heating at 55°C and 60°C, respectively, for 10 min. The activity in fraction Ib (measured by quantitation of the 5' exon formed

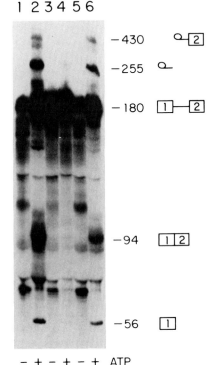

**Figure 2** Effect of micrococcal nuclease on the splicing activity of fractions I and II. Splicing reactions containing SP6 adenovirus pKT1 transcript (23 fmoles ends, 210 cpm/fmole), fraction I (226 μg protein), and fraction II (46 μg protein) were incubated at 30°C for 2 hr in the absence or presence of ATP (0.4 mM) and creatine phosphate (20 mM) as indicated. Fraction I (565 μg protein) or fraction II (230 μg protein) was preincubated with micrococcal nuclease (40 units/ml) and $CaCl_2$ (1 mM) for 30 min at 30°C, followed by the addition of EGTA (3 mM). (Lanes *1* and *2*) No preincubation with micrococcal nuclease; (lanes *3* and *4*) preincubation of fraction II with micrococcal nuclease; (lanes *5* and *6*) preincubation of fraction I with micrococcal nuclease. The various RNA structures are indicated schematically on the right: Boxes represent exons (1 and 2); lines stand for introns.

after 2 hr at 30°C) was inhibited 100% by *N*-ethyl-maleimide (5 mM). The activity of fraction Ib was inhibited by 60% and 98% by heating at 45°C and 55°C, respectively, for 10 min.

### Analysis of macromolecular complexes formed by fractions IA, IB, and II

Earlier studies on mRNA splicing in vitro suggested that the first step in the reaction might be the assembly of a ribonucleoprotein complex (Konarska et al. 1985). The existence of such complexes has recently been observed in both yeast and mammalian systems (Brody and Abelson 1985; Frendewey and Keller 1985; Grabowski et al. 1985). It was therefore of interest to establish which, if any, of the protein fractions described above was capable of forming a complex with precursor RNA. Incubation (2 hr at 30°C) of the adenovirus pre-mRNA with fractions I and II followed by sucrose gradient sedimentation analysis revealed the ATP-dependent formation of material much larger than the 6S precursor RNA (Fig. 3A). The sedimentation coefficients of the two main peaks of radioactivity were determined to be 55S and 20S, using a 40S ribosomal subunit and 6S pre-mRNA as markers. The specificity of this interaction was investigated using a precursor RNA that could not undergo cleavage at the 5′ splice site, the first observable

step in mRNA splicing. A truncated adenovirus pre-mRNA was synthesized from plasmid pKT1 digested with restriction enzyme *Hha*I. This pre-mRNA contained the first exon and only 25 nt of the intron. The essential role of 3′ intron sequences in the cleavage of the 5′ exon has been documented (Reed and Maniatis 1985). Incubation of this pre-mRNA with fractions I and II did not result in cleavage at the 5′ splice site (data not shown) and therefore provided a useful control transcript to ascertain RNA-protein interactions that are involved in mRNA splicing. Incubation of the truncated pre-mRNA with fractions I and II did not result in the formation of the fast-sedimenting complexes described above (Fig. 3A). This result, taken in conjunction with the stringent ATP requirement, suggests that the fast-sedimenting material is a product of the RNA splicing system and not merely the nonspecific binding of the RNA to large components in the reaction mixture.

The structure of the RNA found in these ATP-dependent complexes was analyzed by 15% polyacrylamide/urea gel electrophoresis. Precursor RNA was found in almost all fractions but was visibly concentrated in the 55S peak (Fig. 3B). Spliced RNA was also found in the 55S peak but was predominant in an area between 40S and 20S. The intron lariat was present in the 55S peak, but the majority of this species was found between 55S

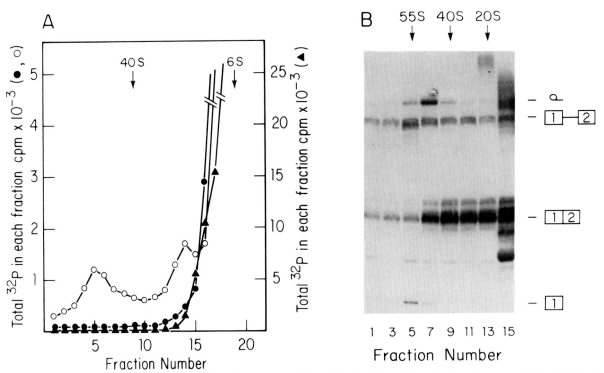

**Figure 3** Sedimentation analysis of macromolecular complexes formed during the splicing of adenovirus pre-mRNA. (*A*) Fraction I (452 μg protein) and fraction II (92 μg protein) were incubated with adenovirus pre-mRNA (84 fmoles ends, 120 cpm/fmole) in the presence (○) or absence (●) of ATP (0.4 mM) and creatine phosphate (20 mM). The truncated pre-mRNA (▲) was also incubated with ATP (0.4 mM) and creatine phosphate (20 mM) under the same conditions. The resulting macromolecular complexes were analyzed by sucrose gradient centrifugation as described in Experimental Procedures. The indicated markers of 40S and 6S refer to the sedimentation positions of 40S ribosomal subunit and adenovirus pre-mRNA, respectively (*B*). The radioactive material in the indicated gradient fractions was extracted with phenol/chloroform, precipitated by addition of ethanol, and analyzed by 15% polyacrylamide/urea gel electrophoresis. The various RNA structures are indicated schematically on the right.

and 40S. In contrast, the 5' exon species was located almost exclusively in the 55S peak. Although not clearly shown here, in other experiments the intron-exon lariat species was also found to be exclusively associated with the 55S complex, albeit present in a low amount.

A time course of complex formation (data not shown) showed that the 55S complex was formed after 30 min of incubation. There was no apparent increase in the amount of radioactivity associated with the 55S complex upon further incubation for 60 min and 120 min. Gel analysis of individual fractions revealed that although the amount of 5' exon associated with the 55S complex remained constant throughout this time period, the amount of spliced RNA in the 20S–40S area both markedly increased and progressively sedimented more slowly.

We next investigated which if any of the protein fractions described above was capable of forming a complex with pre-mRNA. Neither fraction I nor its derivative fractions Ia and Ib (data not shown) formed any discernible (> 10S) complex with the adenovirus pre-mRNA (Fig. 4A). On the other hand, fraction II, the microccoccal-nuclease-sensitive fraction, did form a 30S complex, which on analysis was found to contain precursor RNA (Fig. 4A). This interaction is also likely to be involved in mRNA splicing since it required ATP (Fig. 4A) and was not manifested with the truncated precursor (data not shown).

We then determined which combination of fractions could yield the 55S complex. Figure 4B shows that the addition of fraction Ia to fraction II yielded only the 30S complex. In contrast, incubation of fractions Ib and II yielded a 55S complex that by gel analysis was found to contain the 5' exon and to a lesser extent the intron-exon lariat species. Since the combination of fractions Ib and II yielded the first observable event in mRNA splicing (see above), this strengthens the conclusion that the 55S complex is involved in the splicing of RNA.

## Analysis of mutations in the adenovirus pre-mRNA

To gain a better understanding of the structural requirements for mRNA splicing, we made base substitutions within the intron of the adenovirus pre-mRNA (pKT1) (as described in Experimental Procedures). Two mutants are presented here. One mutant (pUSA) was constructed that contained two single base changes: a G residue was substituted for the A residue at the lariat attachment site (24 nt upstream from the 3' splice site) as well as for the A residue 6 nt upstream of the lariat attachment site (Fig. 5). pUSA pre-mRNA and pKT1 pre-mRNA were incubated with the HeLa cell nuclear extract for various time periods, and the reaction products were analyzed by 15% polyacrylamide/urea gel electrophoresis (Fig. 5). The most striking observation was that spliced RNA was not formed (<1.0%) when pUSA pre-mRNA was used as a substrate (Fig. 5, lanes 1–6). In contrast, approximately 30% of pKT1 pre-mRNA was spliced in 2 hr at 30°C (Fig. 5, lane 12).

The amount of cleavage at the 5' splice site of pKT1 pre-mRNA (Fig. 5, lanes 7–12) was compared with the amount of cleavage at the 5' splice site of pUSA pre-mRNA (Fig. 5, lanes 1–6) by measuring the level of free 5' exon observed and the 5' exon content of spliced RNA. Incubation of the nuclear extract with pKT1 pre-mRNA for 0.5, 1.0, and 2.0 hr resulted in 6.0%, 15.0%, and 21.0% cleavage at the 5' splice site. In contrast, incubation of pUSA pre-mRNA with the nuclear extract for 0.5, 1.0, and 2.0 hr resulted in 0.3%, 2.0%, and 5.0% cleavage at the 5' splice site. Therefore, both the rate and yield of cleavage at the 5' splice site of pUSA pre-mRNA are lower than that observed with wild-type pre-mRNA.

Analysis of the pUSA intron-exon lariat revealed that the 5'-terminal phosphate of the intron was esterified to the 2'-hydroxyl group of the new G residue that had been substituted for the A residue at the lariat attachment site. RNase T1 digestion of purified pUSA intron-

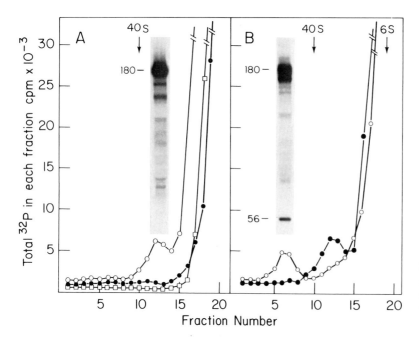

**Figure 4** The formation of a 30S complex by fraction II and a 55S complex by fraction II and fraction Ib. (*A*) Fraction I (□) or fraction II (○) were incubated (2 hr, 30°C) with pre-mRNA (84 fmoles ends, 120 cpm/fmole) in the presence of ATP (0.4 mM) and creatine phosphate (20 mM), and resulting complexes were analyzed by sucrose gradient centrifugation. Fraction II was also incubated in the absence of ATP and creatine phosphate (●). (*B*) Fraction II and fraction Ia (●) or fraction II and fraction Ib (○) were incubated (2 hr, 30°C) with adenovirus pre-mRNA (84 fmoles ends, 120 cpm/fmole), and resulting complexes were analyzed by sucrose gradient centrifugation. The inserts show the RNA compositions of fractions 12 and 6, respectively. Amounts of fractions used were: fraction I, 452 μg; fraction II, 92 μg; fraction Ia, 54 μg; fraction Ib, 75 μg.

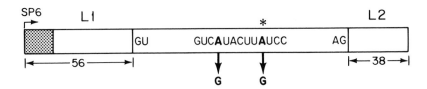

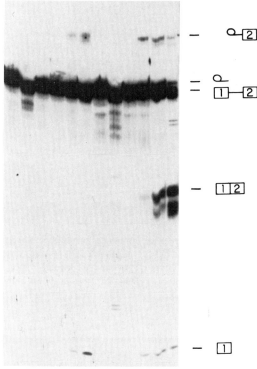

**Figure 5** Effect of base changes at the lariat attachment site. Splicing reactions contained either adenovirus pUSA pre-mRNA (22 fmoles ends, 90 cpm/fmole, lanes *1–6*) or adenovirus pKT1 pre-mRNA (22 fmoles ends, 90 cpm/fmole, lanes *7–12*), HeLa cell nuclear extract (280 μg of protein), and where indicated, ATP (0.4 mM) and creatine phosphate (20 mM); splicing reactions were incubated at 30°C for the indicated time periods. Reaction products were analyzed by 15% polyacrylamide/urea gel electrophoresis and visualized by autoradiography.

exon lariat produced a unique oligonucleotide of 12 nt in length that was not present in the pre-mRNA. This oligonucleotide arises as a consequence of the 2′,5′-phosphodiester bond at the G residue, rendering it resistant to RNase T1 digestion. Both this oligonucleotide as well as the corresponding oligomer from wild-type lariat generated by RNase T1 were purified and digested with nuclease P1. Two-dimensional chromatographic analysis was consistent with the mutant lariat RNA nuclease P1 digestion product being

pG
  pG
pU

. Incubation of either the pKT1 intron-exon lariat or the pUSA intron-exon lariat with a HeLa cytoplasmic extract containing the lariat-debranching enzyme described by Ruskin and Green (1985) resulted in their conversion to species with the electrophoretic mobilities of 125 nucleotides, the size expected for the linearized intron-exon lariat. These results indicate that the A-to-G change at the lariat attachment site produces a substrate that can be cleaved at the 5′ exon and can form an

intron-exon lariat RNA but is unable to produce either intron lariat or spliced RNA.

A second mutant (pIEC) was constructed in which the dinucleotide AG at the 3′ splice site was substituted with the dinucleotide AC. When pIEC pre-mRNA was incubated with the HeLa cell nuclear extract, no spliced RNA (<1.0%) was formed (Fig. 6, lanes 1–5). A comparison with pKT1 pre-mRNA is shown (Fig. 6, lanes 6–10) in which 17% of the input pre-mRNA was spliced after 4 hr at 30°C.

Cleavage at the 5′ splice site and intron-exon lariat formation did occur when pIEC was used as a substrate. Although the intron-exon lariat formed with pIEC is not clearly shown here, we have verified by primer extension analysis that this species was produced (data not shown). The amount of cleavage at the 5′ splice site of pIEC pre-mRNA was compared with the amount of cleavage at the 5′ splice site of pKT1 pre-mRNA. Incubation of pIEC pre-mRNA with the nuclear extract for 0.5, 1.0, 2.0, and 4.0 hr resulted in 0.8%, 1.5%, 2.0%, and 1.2% cleavage at the 5′ splice site, respectively. In

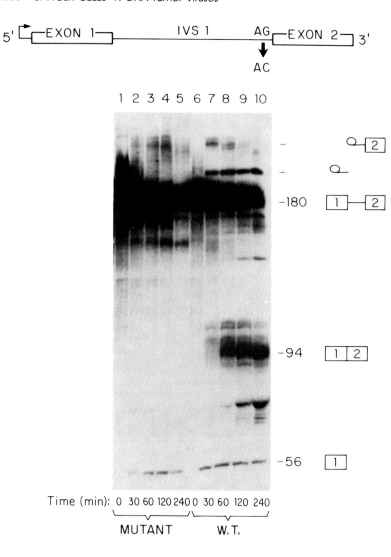

**Figure 6** Effect of base substitution at the 3′ splice site. Splicing reactions containing either adenovirus pIEC pre-mRNA (13.9 fmoles ends, 36 cpm/fmole, lanes *1–5*) or adenovirus pKT1 pre-mRNA (9.6 fmoles ends, 36 cpm/fmole, lanes *6–10*), HeLa cell nuclear extract (298 μg protein), ATP (0.4 mM), and creatine phosphate (20 mM) were incubated at 30° for the indicated time periods. Reaction products were analyzed by 15% polyacrylamide/urea gel electrophoresis and visualized by autoradiography.

contrast, incubation of pKT1 pre-mRNA with the nuclear extract for 0.5, 1.0, 2.0, and 4.0 hr resulted in 5.0%, 13.0%, 14.0%, and 13.0% cleavage at the 5′ splice site, respectively. These results demonstrate that both the rate and yield of cleavage at the 5′ splice site of pIEC pre-mRNA are lower than the rate and yield of cleavage at the 5′ splice site of pKT1 pre-mRNA.

## Discussion

We previously reported the isolation and characterization of two fractions (I and II) from nuclear extracts of HeLa cells that were required for ATP-dependent splicing of both adenovirus and β-globin pre-mRNAs. In this report we have described the separation of fraction I into two further components (Ia and Ib), which have been distinguished on the basis of their sensitivities to *N*-ethylmaleimide and heat. The combination of fractions II, Ia, and Ib was required to yield spliced RNA and the intron lariat. Two of these fractions (II and Ib), however, were sufficient for cleavage at the 5′ splice site and intron-exon lariat formation, apparently uncoupling the proposed two-stage splicing mechanism. Quantitation

of the reaction products revealed that when fractions Ib and II were combined, the amount of cleavage at the 5′ splice site was less than the amount of cleavage occurring when fractions Ia, Ib, and II were combined. These results suggest that the addition of fraction Ia (which results in the formation of spliced RNA) enhances further cleavage at the 5′ splice site of another pre-mRNA. We have reported that fraction Ib was added in limiting amounts to the splicing reactions due to the presence of an exoribonuclease activity in this fraction that degraded the intron-exon lariat. It is unlikely that the amount of 5′ exon observed in the absence of Ia was decreased by this exonuclease since a time-dependent degradation of this species was never observed.

In this report we have also demonstrated the effect of mutations at the 3′ end of the intron on RNA splicing. Earlier reports of mutations in the lariat attachment site of β-globin pre-mRNA demonstrated the activation of cryptic lariat attachment site, which permitted the formation of spliced RNA, and although the yield of spliced RNA was significantly reduced, it was not abolished (Ruskin et al. 1985). A mutation (A to C) at the lariat attachment site within the conserved yeast sequence

TACTAAC prevented cleavage at the 5' splice site and intron-exon lariat formation as well as the production of spliced RNA in vitro, suggesting that the conserved yeast sequence was essential for substrate recognition (Newman et al. 1985). Here we have shown that a mutant containing the A-to-G substitution at the lariat attachment site was not spliced, although cleavage did occur at the 5' splice site and intron-exon lariat formation occurred at the new guanosine residue. This is the first demonstration that a nucleotide substituted at the lariat attachment site was utilized for intron-exon lariat formation and suggests that recognition of the nucleotides in the branched structure is critical for the formation of spliced RNA.

The 3' splice site has also been shown to play a critical role in the RNA splicing process. A substitution (AG to GG) at the 3' splice site of the human β-globin gene resulted in cleavage at the 5' splice site and intron-exon lariat formation, but spliced RNA was not formed (Reed and Maniatis 1985). This result is similar to our finding that the adenovirus pIEC pre-mRNA, which contains a substitution (AG to AC) at the 3' splice site, is capable of 5' splice site cleavage and intron-exon lariat formation but does not undergo formation of spliced RNA. Taken together, these results indicate that the consensus sequence elements at the 3' end of the intron are critical for ligation of exon 1 to exon 2, producing spliced RNA.

We also demonstrated that RNA splicing products formed with the fractionated system were associated with fast-sedimenting complexes. Our observations are similar to those of Frendewey and Keller (1985) and Grabowski et al. (1985) in that putative splicing intermediates were found to be associated with a smaller complex and that pre-mRNA was found to be associated with a smaller complex. These groups have provided evidence that the smaller complex was formed as a precursor to a larger complex (Frendewey and Keller 1985) and that complex formation requires the U-snRNPs (Frendewey and Keller 1985; Grabowski et al. 1985).

In this report we have established which fractions are required for complex formation. Incubation of fraction II with pre-mRNA resulted in the ATP-dependent formation of a 30S complex that contained pre-mRNA. Since we know that fraction II is sensitive to micrococcal nuclease (Furneaux et al. 1985), we believe that RNA components may be involved in the formation of this complex. Attractive candidates for these RNA components are the U-snRNPs as suggested by other investigators (Frendewey and Keller 1985; Grabowski et al. 1985). Incubation of fractions II and Ib with pre-mRNA in the presence of ATP resulted in the formation of a 55S complex that contained the 5' exon and the intron-exon lariat as well as pre-mRNA.

We speculate that 30S complex formation is the first step in the splicing reaction and may be a precursor to the 55S complex. Our argument is strengthened by the observation that a mutant pre-mRNA in which the 3' splice junction has been changed from CAG to CAC

formed very little 55S complex (11% of wild type) and formed predominantly 30S material when incubated with fractions Ib and II (data not shown). At present we cannot ascertain whether the component(s) in fraction Ib bind directly to the pre-mRNA or to the 30S complex, nor can we etablish whether ATP is required for the proposed transition from 30S to 55S. The isolation of an active 30S complex that could be reincubated with fraction Ib to yield the putative intermediates would be likely to resolve these questions.

We have also demonstrated that when fraction I (which contains Ia and Ib) was incubated with fraction II, the 55S complex was formed, which contained the putative intermediates as well as spliced RNA. However, the majority of the spliced RNA produced under these conditions sedimented between 40S and 20S. Thus it appeared that spliced RNA was released from the 55S complex. The amount of the 5' exon contained within the 55S complex formed with fractions I and II did not increase with time, although increasing amounts of spliced RNA were released.

Our observations therefore suggest a model for RNA splicing in which the first step is the ATP-dependent association of pre-mRNA with U-snRNPs and as yet unidentified proteins to form a 30S complex. In the presence of the appropriate splicing factors (fraction Ib), a 55S complex is formed that produces the putative intermediates suitably juxtaposed for ligation of exon 1 to exon 2. In the presence of fraction Ia, spliced RNA is formed, resulting in the dissociation of the 55S complex. The components of the 55S complex are now available to reassemble on a new pre-mRNA molecule, resulting in cleavage at the 5' splice site.

## Acknowledgments

We are indebted to Ms. Barbara Philips and Ms. Nilda Belgado for their supberb technical assistance. This work was supported by grant DMB8415369 from the National Science Foundation, and grant NP89 from the American Cancer Society. H.F. was supported by an Exxon Fellowship Training Grant.

## References

Abelson, J. 1979. RNA processing and the intervening sequence problem. *Annu. Rev. Biochem.* **48:** 1035.

Brody, E. and J. Abelson. 1985. The "Spliceosome": Yeast pre-messenger RNA associates with a 40S complex in a splicing-dependent reaction. *Science* **228:** 963.

Butler, E.T. and M.J. Chamberlin. 1982. Bacteriophage SP6-specific RNA polymerase I. Isolation and characterization of the enzyme. *J. Biol. Chem.* **257:** 5772.

Dignam, J.D., R.B. Lebovitz, and R.G. Roeder. 1983. Accurate transcription initiation by RNA polymerase II in a soluble extract from isolated mammalian nuclei. *Nucleic Acids Res.* **11:** 1475.

Domdey, H., B. Apostol, R.-J. Lin, A. Newman, E. Brody, and J. Abelson. 1984. Lariat structures are *in vivo* intermediates in yeast pre-mRNA splicing. *Cell* **39:** 611.

Frendewey, D. and W. Keller. 1985. Stepwise assembly of a pre-mRNA splicing complex requires U-snRNPs and specific intron sequences. *Cell* **42:** 355.

Furneaux, H.M., K.K. Perkins, G.A. Freyer, J. Arenas, and J. Hurwitz. 1985. Isolation and characterization of two fractions from HeLa cells required for *in vitro* RNA splicing. *Proc. Natl. Acad. Sci.* **82:** 4355.

Grabowski, P.J., R.A. Padgett, and P.A. Sharp. 1984. Messenger RNA splicing *in vitro*: An excised intervening sequence and a potential intermediate. *Cell* **37:** 415.

Grabowski, P.J., S.R. Seiler, and P.A. Sharp. 1985. A multicomponent complex is involved in the splicing of messenger RNA precursors. *Cell* **42:** 345.

Keller, E. and W. Noon. 1984. Intron splicing: A conserved internal signal in introns of animal pre-mRNA. *Proc. Natl. Acad. Sci.* **81:** 7417.

Konarska, M.M., P.J. Grabowski, R.A. Padgett, and P.A. Sharp. 1985. Characterization of the branch site in lariat RNAs produced by splicing of mRNA precursors. *Nature* **313:** 552.

Krainer, A.R., T. Maniatis, B. Ruskin, and M.R. Green. 1984. Normal and mutant human β-globin pre-mRNAs are faithfully and efficiently spliced *in vitro*. *Cell* **36:** 993.

Montell, C., E.F. Fisher, M.H. Caruthers, and A.J. Berk. 1982. Resolving the functions of overlapping viral genes by site-specific mutagenesis at a mRNA splice site. *Nature* **295:** 380.

Mount, S.M. 1982. A catalogue of splice junction sequences. *Nucleic Acids Res.* **10:** 459.

Newman, A.J., R.-J. Lin, S.-C. Cheng, and J. Abelson. 1985. Molecular consequences of specific intron mutations on yeast mRNA splicing in vivo and in vitro. *Cell* **42:** 335.

Padgett, R.A., S.F. Hardy, and P.A. Sharp. 1983. Splicing of adenovirus RNA in a cell-free transcription system. *Proc. Natl. Acad. Sci.* **80:** 5230.

Padgett, R.A., M.M. Konarska, P.J. Grabowski, S.F. Hardy, and P.A. Sharp. 1984. Lariat RNAs as intermediates and products in the splicing of messenger precursors. *Science* **225:** 898.

Perkins, K.K., H.M. Furneaux, and J. Hurwitz. 1986. RNA splicing products formed with isolated fractions from HeLa cells are associated with fast-sedimenting complexes. *Proc. Natl. Acad. Sci.* (in press).

Reed, R. and T. Maniatis. 1985. Intron sequences involved in lariat formation during pre-mRNA splicing. *Cell* **41:** 95.

Rodriquez, J.R., C.W. Pikielny, and M. Rosbach. 1984. *In vivo* characterization of yeast mRNA processing intermediates. *Cell* **39:** 603

Ruskin, B. and M.R. Green. 1985. An RNA processing activity that debranches RNA lariats. *Science* **229:** 135.

Ruskin, B., J.M. Greene, and M.R. Green. 1985. Cryptic branch point activation allows accurate *in vitro* splicing of human β-globin intron mutants. *Cell* **41:** 833.

Ruskin, B., A.R. Krainer, T. Maniatis, and M.R. Green. 1984. Excision of an intact intron as a novel lariat structure during pre-mRNA splicing *in vitro*. *Cell* **38:** 317.

Treisman, R., N.J. Proudfoot, M. Shander, and T. Maniatis. 1982. A single base change at a splice site in a β⁰-thalassemic gene causes abnormal splicing. *Cell* **29:** 903.

Wieringa, B., E. Hofer, and C. Weissmann. 1984. A minimal intron length but no specific internal sequence is required for splicing the large rabbit β-globin intron. *Cell* **37:** 915.

Zeitlin, S. and A. Efstratiatis. 1984. *In vivo* splicing products of the rabbit β-globin pre-mRNA. *Cell* **39:** 589.

Zimmerman, S.B. and B.H. Pheiffer. 1983. Macromolecular crowding allows blunt-end ligation by DNA ligases from rat liver or *Escherichia coli*. *Proc. Natl. Acad. Sci.* **80:** 5852.

# The Pathway of SV40 Early mRNA Splicing

J.L. Manley, J.C.S. Noble, M. Chaudhuri, X.-Y. Fu, T. Michaeli, L. Ryner, and C. Prives
Department of Biological Sciences, Columbia University, New York, New York 10027

Addition of a precursor RNA containing SV40 early splicing signals to HeLa nuclear extracts resulted in the production of an RNA corresponding to correctly spliced large T mRNA. However, an RNA corresponding to small T mRNA could not be detected. This appears to be due to a lack of utilization of the small T 5' splice site, because the predicted products of cleavage of the precursor RNA at this site have not been observed. The precursor RNA used in these experiments can be used to produce small T mRNA, however, when microinjected into Xenopus laevis oocytes. Additionally, HeLa cells are capable of utilizing the small T splice site in vivo, as shown by DNA transfection experiments. The small-T-related RNA that would be produced by in vitro splicing appears not to be preferentially degraded, because an RNA synthesized from a small T cDNA template shows the same stability as other RNAs when incubated in the nuclear extract. This RNA was found not to be a substrate for RNA processing in vitro, suggesting that small T mRNA cannot serve as a precursor to large T mRNA.

The development of in vitro systems capable of splicing mRNA precursors has resulted in a considerable advance in our understanding of the mechanism of pre-mRNA splicing (Kole and Weissman 1982; Hernandez and Keller 1983; Padgett et al. 1983; Krainer et al. 1984). From studies with such systems, we have learned that splicing of adenovirus late leader and human β-globin mRNAs proceeds by a pathway that involves cleavage of the pre-mRNA at the 5' splice site and formation of an intermediate in which the 5' end of the intron is covalently joined to an adenosine residue just upstream of the 3' splice site by an unusual 2'-5' phosphodiester bond. The pre-mRNA is then cleaved at the 3' splice site and the two exons are joined together, releasing the intron in the form of a lariat (Grabowski et al. 1984; Padgett et al. 1984; Ruskin et al. 1984). Evidence that this pathway is utilized in splicing of a number of pre-mRNAs (Reed and Maniatis 1985), including the SV40 early pre-mRNA that encodes the large tumor (T) antigen (Noble et al. 1986), has been obtained.

The nature of the factors required for pre-mRNA splicing remains essentially obscure. A role for the U1 RNA containing a small nuclear ribonucleoprotein (snRNP) particle is strongly supported by experiments demonstrating a requirement for this particle for splicing in vitro (Padgett et al. 1983; Kramer et al. 1984) and in vivo (Fradin et al. 1984). U1 snRNP appears to function at the 5' splice site in a mechanism presumably involving base pairing (Lerner et al. 1980; Rogers and Wall 1980; Mount et al. 1983). Recent experiments have also suggested a role for the U2 snRNP particle in splicing (Black et al. 1985), via interaction with the pre-mRNA in the region of the lariat branch point-3' splice site. Various scenarios involving base pairing between U2 RNA and pre-mRNAs have been suggested (Ohshima et al. 1981; Fradin et al. 1984; Keller and Noon 1984).

We have been interested in SV40 early pre-mRNA splicing because it offers an excellent system to begin to learn about the mechanism and control of splicing of pre-mRNAs displaying alternative splicing pathways. The splicing pathways of all other RNAs that have been analyzed in vitro are simple; i.e., one 5' splice site is utilized in conjunction with one 3' splice site. Thus, it is important to determine, as a first step, whether the same pathway is utilized in splicing RNAs with alternative splice sites. SV40 early pre-mRNA contains two 5' splice sites that are utilized in vivo in conjunction with a common 3' splice site to produce either large T or small T mRNA. Here we describe the patterns of splicing of this pre-mRNA in vivo and in vitro in the context of possible mechanisms for alternative splice site utilization.

## Materials and Methods

### Template construction

A 45-bp EcoRI-HindIII fragment containing a synthetic promoter with the consensus sequence for E. coli RNA polymerase (Rossi et al. 1983) cloned into plasmid vector pEA300 (obtained from J. Brosius) was fused to the 1169-bp HindIII B fragment of SV40 early region (nt 5171–4002) to create pYSVHdB. pYLSVt is identical, except that the HindIII fragment was from a plasmid containing small T cDNA (a gift of M. Botchan).

### Preparation of precursor RNA

Precursor RNA was prepared by run-off transcription and contained 2 nt and 15 nt at the 5' and 3' ends, respectively, of bacterial sequences. Transcription reactions contained, in 50 µl, 0.25 pmole of template and 2 pmoles of E. coli RNA polymerase (the gift of S. Beychok) in a buffer comprising 20 mM Tris·HCl (pH 7.9), 0.4 M KCl, 10 mM $MgCl_2$, 0.1 mM EDTA, 2.5 mM dithiothreitol, 1 mM ATP, CTP, and UTP, 500 µM GTP, and 25 μCi [α-$^{32}$P]GTP. To synthesize capped RNAs, the concentration of ATP was reduced to 100 μM, and $^7$mGpppA was added to a final concentration of 1 mM (Contreras et al. 1982). After 1 hr at 30°C, transcription was terminated by adding rifampicin at 500 μg/ml and DNase I at 20 μg/ml and incubating for 10 min at 37°C. RNA was then extracted with phenol/chloroform (1:1) and precipitated with ethanol. Transcriptions performed using this protocol yielded 2–4 pmoles of precursor RNA.

### In vitro splicing reaction

Nuclear extracts were prepared from HeLa cells by using a modification of the method of Dignam et al. (1983). Standard splicing reactions contained, in 25 μl, 0.15 pmole of precursor RNA and 5 μl of nuclear extract in a buffer comprising 4 mM Tris·HCl (pH 7.9), 4% (v/v) glycerol, 20 mM KCl, 2 mM MgCl$_2$, 40 μM EDTA, 0.2 mM dithiothreitol, 4 mM creatine phosphate (discoidium salt), and 500 μM ATP. After 3 hr at 30°C, processing was terminated and RNA was extracted as described previously (Manley et al. 1980).

### S1 nuclease analysis

S1 nuclease mapping of processed RNAs was carried out by the method of Berk and Sharp (1978), using as a probe the 1169-bp *Hind*III B fragment of SV40 early region (SV40 nt 5171–4002) labeled at either 5′ or 3′ ends. Typically, 10–25 ng of probe was used for one-fifth of the products of a splicing reaction. Hybridization was carried out for 4–5 hr at 45°C, and S1 nuclease digestion was performed using $4 \times 10^3$ units/ml S1 for 1 hr at 40°C. After digestion, products were analyzed under denaturing conditions on 5% polyacrylamide–8.3 M urea sequencing-type gels (Maxam and Gilbert 1980).

### X. laevis *microinjection*

Excised ovaries of *X. laevis* mature females were defolliculated by treatment with 0.15% Collagenase (Worthington) at 20°C and maintained in modified Barth's solution (Fradin et al. 1984). Ooctyes were spun in an IEC clinical centrifuge at 800$g$ for 12 min immediately prior to injection. Micropipettes were calibrated to deliver 20 sequential microinjections at 25 nl each. Five μg of capped precursor RNA was injected per oocyte, and groups of 10 to 20 oocytes were harvested at the indicated times. RNA was extracted as described (Fradin et al. 1984), and approximately 1% of the RNA was used for S1 nuclease analysis.

### DNA transfection

Twenty-five μg of plasmid DNA was used to transfect $1.8 \times 10^6$ HeLa cells on a 150-mm plate, as previously described (Lewis and Manley 1985). Forty-eight hr after addition of DNA, the cells were harvested and total cytoplasmic RNA was prepared.

## Results

### The small T mRNA 5′ splice site is not used in a HeLa nuclear extract

Figure 1A shows the relevant features of the plasmid used to produce substrates for RNA processing reactions. Capped RNAs were synthesized by including $^7$mGpppA in the transcription reaction, and uncapped RNAs were prepared by omitting this dinucleotide (see Materials and Methods). Also indicated in Figure 1A are the predicted sizes of some of the intermediates or products expected to arise from splicing of this RNA. Figure 1B is a schematic representation of the products of in vitro splicing reactions that we have previously characterized (Noble et al. 1986). The point most relevant to the experiments described here is that none of the intermediates or products expected to result from utilization of the small T mRNA 5′ splice site have been observed among the products of in vitro splicing reactions.

An example of the evidence that led to the conclusion that the small T 5′ splice site is not utilized in HeLa nuclear extracts is shown in Figure 2A, which displays the results of an S1 nuclease analysis of the RNAs present after increasing times of incubation. The DNA probe employed (3′-end labeled) differentiates between utilization of the large and small T 5′ splice sites. The results show that by 1 hr, RNAs cleaved at the 5′ splice site were readily detectable. After 3 or 6 hr of incubation, RNAs cleaved at this splice site were the most abundant products of the in vitro processing reaction. In contrast, no evidence for utilization of the small T 5′ splice site is apparent. This can be seen in the analysis shown here only after longer times of incubation, because of degradation products present at early times. After 3 or 6 hr of incubation, however, when large T 5′-splice-site cleavage is most prominent, there was no indication of cleavage at the small T 5′-splice-site. (Note in Fig. 2 the small amount of an S1-resistant species approximately the size of the one corresponding to small T 5′-splice-site cleavage. Close inspection of the autoradiogram shown here, as well as of numerous others not shown, revealed that this species is significantly larger than that produced by authentic small T mRNA.) Further evidence that the small T 5′ splice site is not utilized has been obtained by direct analysis of the products of RNA processing reactions on denaturing polyacrylamide gels, by additional S1 nuclease mapping, and by analysis of cDNA products synthesized by primer extension of processed RNAs (Noble et al. 1986).

### Small T mRNA is not selectively degraded in vitro

A substantial amount of the RNA added to processing reactions was degraded during incubations (e.g., see Fig. 2). We have found the amount of degradation to be variable (between 30% and 80% of the input RNA was degraded), for unknown reasons. However, capped RNAs were found to be considerably more stable than uncapped precursors.

The fact that a significant fraction of the RNA added to

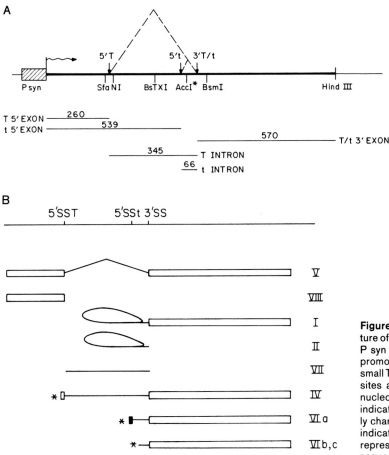

**Figure 1** Products of SV40 early RNA splicing. (*A*) Structure of DNA template used to synthesize RNA precursors. P syn represents the synthetic *E. coli* RNA polymerase promoter (Rossi et al. 1983). Splice sites for large and small T antigens (T and t) are indicated. Several restriction sites are shown below the solid line, and the sizes (in nucleotides) of the expected products of splicing are also indicated. (*B*) Products of in vitro splicing. RNAs previously characterized as products of processing reactions are indicated schematically (Noble et al. 1986). Rectangles represent exon sequences, and lines represent intron sequences. Stars denote species produced by a 5′-to-3′ exonuclease activity (see text).

in vitro processing reactions was degraded was consistent with the formal possibility that RNAs that had undergone cleavage at the small T mRNA 5′ splice site were selectively degraded during processing. Although no evidence exists suggesting that specific degradation of certain RNAs can occur in vitro, we decided to test this notion in the following manner. An SV40 *Hind*III fragment obtained from a small T mRNA cDNA clone was inserted into the *E. coli* expression plasmid shown in Figure 1A. RNA transcribed from this template was identical with that obtained from a wild type–containing template, except that RNA sequences corresponding to the small T intron were deleted; i.e., the RNA has the same structure that the wild-type RNA would have if the small T splice were made. This RNA was synthesized (in both capped and uncapped forms), and its fate after incubation in the nuclear extract was determined and compared with that of a wild-type precursor. Figure 3 shows the results of an S1 nuclease analysis (with a 5′-end-labeled probe) of uncapped processed RNAs. Although a substantial fraction of both precursors was degraded (~60–70% in the experiment shown here), both RNAs were degraded to the same extent. Essentially identical results were obtained with capped precursors (data not shown). These results argue strongly against the idea that an RNA corresponding to spliced

small T mRNA was not detected among the products of an in vitro processing reaction because it was selectively lost or degraded.

An interesting sidelight of the experiment shown in Figure 3 emerges when one notes that the pYLSVt-generated transcript failed to serve as a substrate for RNA processing, as judged by the absence of an S1 nuclease–resistant species other than the intact precursor. This result is significant because it suggests that small T mRNA is not a substrate for further splicing; i.e., large T mRNA cannot be formed from small T mRNA. In fact, an RNA from which the small T intron had been removed not only could not be spliced, but also was not cleaved at the large T 5′ splice site (Fig. 3, although this is seen more clearly with a 3′-end-labeled DNA probe; results not shown). This finding suggests that sequences at the 3′ end of the large T intron are required for cleavage at the large T 5′ splice site.

**Small T mRNA splicing in vivo**

As alternate explanations for our failure to detect small T mRNA splicing in vitro, we considered two additional possibilities. First, the structure of the precursor RNA might be such that it would have been impossible to use the small T 5′ splice site, even in vivo. Second, HeLa

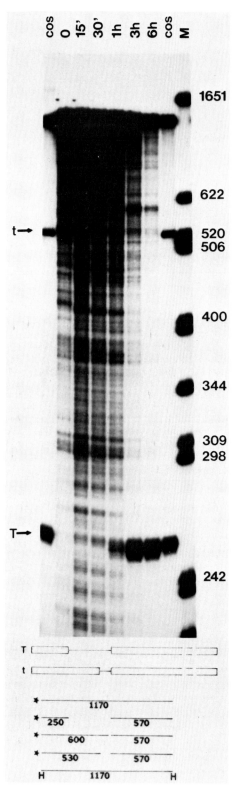

Figure 2 S1 nuclease analysis of processed RNAs. Uncapped precursor RNAs were incubated in vitro for the times indicated. S1 mapping was performed using a single-stranded, 3'-end-labeled *Hin* dIII B fragment probe (SV40 nt 5175–4003), and the products were analyzed as described in Materials and Methods. (cos) Total cytoplasmic RNA from an SV40-transformed monkey cell line; (M) size markers. A diagram of expected S1 nuclease–resistant probe fragments is shown at bottom.

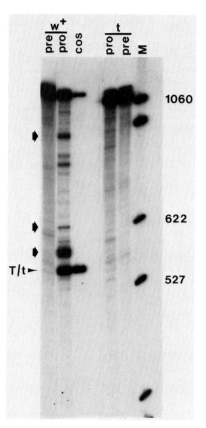

Figure 3 Incubation of a small-T-related RNA in vitro. Uncapped RNAs transcribed from pYSVHdB (denoted w$^+$) or pYLSVt (denoted t) were incubated in vitro, and the RNA products were analyzed by S1 nuclease mapping, using either a wild-type or a small-T-related 5'-end-labeled DNA probe, as appropriate. (pre) Precursor RNA; (pro) processed RNA; (cos) RNA from COS monkey cells and the wild-type probe in the S1 analysis; (M) DNA size markers; (T/t) an S1-resistant fragment indicative of utilization of the common 3' splice site; (arrowheads) fragments that reflect protection.

cells might lack a factor required for carrying out the small T splice.

To test the first possibility, capped precursor RNA was synthesized and microinjected into *X. laevis* oocytes. At various times afterwards, oocytes were harvested, and RNA was extracted and subjected to S1 nuclease analysis. The results indicate that the small T 5' splice site in this precursor can be utilized (Fig. 4). In fact, the small T splice site appears to be used more efficiently than the large T splice site in oocytes. This ratio contrasts with that observed not only in permissive monkey cells (see Fig. 5), but also in most, if not all, other cell types that have been examined. However, a similar ratio was observed previously for RNA transcribed from SV40 DNA microinjected into *X. laevis* oocytes (Fradin et al. 1984).

The possibility that HeLa cells cannot make the small T splice is not a completely trivial one, because, to our knowledge, it has not been demonstrated that SV40-infected HeLa cells synthesize small T antigen. To test this, we introduced a plasmid containing an intact SV40 early region (pSTER; Lewis and Manley 1985) into HeLa

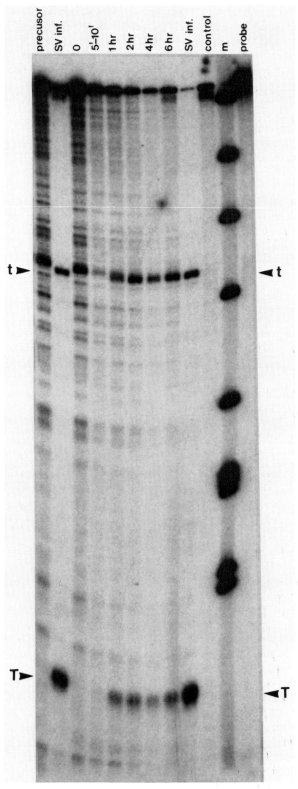

**Figure 4** Splicing of SV40 precursor RNA in *X. laevis* oocytes. Capped RNA transcribed from pYSVHdB was injected into oocyte nuclei as described in Materials and Methods. At the indicated times, oocytes were harvested, and RNA was extracted and analyzed by S1 nuclease analysis, using a single-stranded 3′-end-labeled DNA probe. (T) A DNA fragment produced by hybridization of the probe to an RNA splice using the large T 5′ splice site; (t) an RNA splice using the small T 5′ splice site of precursor RNA.

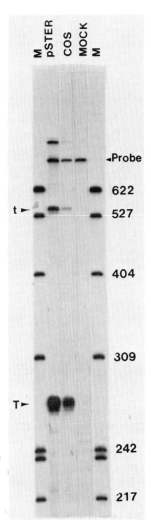

**Figure 5** SV40 early RNA splicing in HeLa cells. S1 nuclease analysis of cytoplasmic RNA (25 μg) extracted from HeLa cells that had been transfected with the plasmid pSTER or no DNA (mock). A 3′-end-labeled *Sty*I (SV40 nt 4409–5187) was utilized as probe. The COS lane displays the products of S1 analysis of 25 μg of RNA from COS monkey cells.

cells by calcium phosphate coprecipitation and analyzed the SV40-specific mRNAs present 48 hr posttransfection by S1 nuclease analysis. The results (Fig. 5) indicate that the small T 5′ splice site can be used in HeLa cells, with the same relative efficiency as it is in COS cells, an SV40-transformed monkey cell line (Gluzman 1981).

## Discussion

Incubation of an SV40 early pre-mRNA in a HeLa nuclear extract results in removal of the large T mRNA intron and formation of a spliced large-T-like RNA (Noble et al. 1986). However, no evidence that the small T 5′ splice site could be utilized was obtained. The pathway followed in the formation of large T mRNA RNA appears to be similar to, if not identical with, that deduced for adenovirus late (Grabowski et al. 1984) and globin (Ruskin et al. 1984) precursor RNAs. Thus, the first detectable step is cleavage of the RNA at the large T 5′ splice site, in a reaction that may be coupled with formation of a lariat, or branched structure, involving the 5′ end of the intron and one of two adenosine residues located 18 or 19 nt upstream of the 3′ splice site. The second step, which may also take place in a coupled reaction, consists of cleavage of the RNA at the 3′ splice site and ligation of the two exons, resulting in release of the lariat-form intron. No evidence of either 5′ splice site cleavage uncoupled from lariat formation or 3′ splice site cleavage uncoupled from ligation has been obtained.

In contrast, injection of an aliquot of the capped precursor RNA used in these experiments into frog ooctyes revealed that the small T 5′ splice site was indeed functional in this precursor. Apart from ruling out trivial explanations for our failure to detect small T splicing in vitro (mutations introduced during cloning or miscopying during in vitro transcription), this experiment indicates that this precursor RNA, which is only a fragment of the authentic SV40 early pre-mRNA that would exist in vivo, contained all the nucleotide sequences required for small T splicing, and it was able to form whatever higher-order structure might be required for this reaction. Additionally, we have used precursor RNAs in in vitro splicing reactions that very closely approximate the precursor that would exist in vivo and failed to detect any evidence for small T splicing (results not shown). Finally, the presence or absence of a 5′ cap structure and/or a 3′ polyadenylation signal (which can be efficiently utilized under the conditions employed for in vitro splicing) failed to influence the utilization of splice sites (results not shown).

It was originally surprising to us that splicing of small T mRNA could not be detected in vitro. The fact that this splice could occur in microinjected *X. laevis* oocytes even when other SV40 splices were inhibited (by microinjection of anti–U1 RNP antibodies) and that the size of the intron is only 66 nt (perhaps allowing easier juxtapositioning of 5′ and 3′ splice sites) had raised the possibility that this splice might occur efficiently in vitro. With respect to the second point, however, a number of recent experiments suggest that the opposite may indeed be true; i.e., the small size of the intron may negatively affect its utilization. When the sizes of a number of different introns were reduced to 75–80 nt, in each case the efficiency of splicing was dramatically reduced, both in vivo (Wieringa et al. 1984; Ulfendahl et al. 1985) and in vitro (Ruskin et al. 1985). However, the small T splice obviously is made in vivo, and with a reasonable efficiency. To explain this discrepancy, we propose that a specific factor exists that is required for a class of introns (perhaps all small ones) exemplified by small T and that this factor is lost or inactivated in preparation of HeLa nuclear extracts. This idea is consistent with our recent observations showing that, although the ratio of small T to large T 5′-splice-site utilization in vivo was dramatically increased in mutants in which the size of the small T intron had been lengthened, small T splicing remained virtually undetect-

able in vitro when precursor RNAs synthesized from such mutants were utilized as substrates (X.Y. Fu et al., in prep.). We are currently testing a variety of cell types and fractionation procedures to determine whether a factor capable of activating the small T 5′ splice site can be uncovered. One interesting source of cells to test are *X. laevis* oocytes, because of their demonstrated ability to utilize the small T splice site more efficiently than do other cell types (e.g., see Fig. 4).

We have previously characterized, among the products of in vitro splicing reactions, several RNAs that result from blockage of an endogenous 5′-to-3′ exonuclease by binding of splicing components to the SV40 early precursor RNA (Noble et al. 1986). These species are designated IV, VIa, and VIb,c in Figure 1B and are apparent in the S1 nuclease analysis shown in Figure 3, where they are designated by arrows. Relevant to the current discussion are products IV and VIa, which result from blockage of the nuclease at points 8–10 nt upstream of the large and small T 5′ splice sites, respectively. Recent experiments have shown that the appearance of these products can be inhibited if the U1 RNA in the nuclear extract is destroyed by oligonucleotide-targeted degradation (J.L. Manley et al., unpubl.). These findings thus suggest that U1 snRNP particles can recognize and bind to the small T mRNA 5′ splice site and raise several intriguing possibilities. First, binding of a U1 snRNP may be a necessary first step in the 5′-splice-site cleavage reaction, but it is not by itself sufficient to bring about splicing (see Mount et al. 1983). Second, the binding of a U1 snRNP to the small T splice site may be nonproductive, perhaps because a required factor has been dissociated from the particle, or because a different type of U1 RNA–containing particle, which has been depleted from the extract, is required. Finally, and more speculatively, perhaps binding of U1 snRNPs to the small T 5′ splice site actually prevents utilization of that site. By this model, a factor might be envisioned that in vivo prevents binding of U1 snRNPs to the small T splice site. We note that the latter two ideas are consistent with our previous observations that small T splicing is resistant to inactivation of U1 snRNPs by lupus anti–U1 RNP antibodies in microinjected *X. laevis* oocytes (Fradin et al. 1984).

In summary, we have shown that, although large T mRNA splicing can occur at moderate efficiency in HeLa nuclear extracts, small T mRNA splicing cannot be detected in vitro. The defect appears to be at the level of 5′-splice-site cleavage, since we have failed to observe any of the predicted intermediates. From a number of facts, we propose that the inability of such extracts to utilize the small T splice site is due to loss or inactivation of a specific factor required for small T splicing. Determining the nature and identity of this factor will be the goal of future work.

## Acknowledgments

We thank R. Mathews and W. Ehrman for technical assistance. This work was supported by National Institutes of Health grant CA 33620 and National Science Foundation grant PCM-82-16798.

## References

Berk, A.J. and P.A. Sharp. 1978. Spliced early pre mRNAs of simian virus 40. *Proc. Natl. Acad. Sci.* **75:** 1274

Black, D.L., B. Chabot, and J.A. Steitz. 1985. U2 as well as U1 small nuclear ribonucleoproteins are involved in pre-messenger RNA splicing. *Cell* **42:** 737.

Contreras, R., H. Cheroutre, O.W. Degrave, and W. Fiers. 1982. Simple, efficient *in vitro* synthesis of capped RNA useful for direct expression of cloned eukaryotic genes. *Nucleic Acids Res.* **10:** 6353.

Dignam, J.D., R.M. Lebovitz, and R.G. Roeder. 1983. Accurate transcription initiation by RNA polymerase II in a soluble extract from isolated mammalian nuclei. *Nucleic Acids Res.* **11:** 1475.

Fradin, A., R. Jove, C. Hemenway, H.D. Keiser, J.L. Manley, and C. Prives. 1984. Splicing pathways of SV40 mRNAs in *X. laevis* ooctyes differ in their requirements for snRNPs. *Cell* **37:** 927.

Gluzman, Y. 1981. SV40-transformed simian cells support the replication of early SV40 mutants. *Cell* **23:** 175.

Grabowski, P.J., R.A. Padgett, and P.A. Sharp. 1984. Messenger RNA splicing *in vitro*. An excised intervening sequence and a potential intermediate. *Cell* **37:** 415.

Hernandez, N. and W. Keller. 1983. Splicing of *in vitro* synthesized messenger RNA precursors in HeLa cell extracts. *Cell* **35:** 89.

Keller, E.B. and W.A. Noon. 1984. Intron splicing: A conserved internal signal in introns of animal pre mRNAs. *Proc. Natl. Acad. Sci.* **81:** 7417.

Kole, R. and S.M. Weissman. 1982. Accurate *in vitro* splicing of human β-globin RNA. *Nucleic Acids Res.* **10:** 5429.

Krainer, A.R., T. Maniatis, B. Ruskin, and M.R. Green. 1984. Normal and mutant human β-globin pre-mRNAs are faithfully and efficiently spliced *in vitro*. *Cell* **36:** 993.

Kramer, A., W. Keller, B. Appel, and R. Luhrmann. 1984. The 5′ terminus of the RNA moiety of $U_1$ small nuclear ribonucleoprotein particles is required for the splicing of messenger RNA precursors. *Cell* **38:** 299.

Lerner, M.R., J.A. Boyle, S.M. Mount, S.L. Wolin, and J.A. Steitz. 1980. Are snRNPs involved in splicing? *Nature* **283:** 220.

Lewis, E.D. and J.L. Manley. 1985. Repression of simian virus 40 early transcription by viral DNA replication in human 293 cells. *Nature* **317:** 172.

Manley, J.L., A. Fire, A. Cano, P.A. Sharp, and M.L. Gefter. 1980. DNA-dependent transcription of adenovirus genes in a soluble whole-cell extract. *Proc. Natl. Acad. Sci.* **77:** 3855.

Maxam, A.M. and W. Gilbert. 1980. Sequencing end-labeled DNA with base-specific chemical cleavages. *Methods Enzymol.* **65:** 499.

Mount, S.M., I. Petterson, M. Hinterberger, A. Karmas, and J.A. Steitz. 1983. The $U_1$ small nuclear RNA protein complex selectively binds a 5′ splice site *in vitro*. *Cell* **33:** 509.

Noble, J.C.S., C. Prives, and J.L. Manley. 1986. *In vitro* splicing of simian virus 40 early pre mRNA. *Nucleic Acids Res.* (in press).

Ohshima, Y., M. Itoh, N. Okada, and T. Miyata. 1981. Novel models for RNA splicing that involve a small nuclear RNA. *Proc. Natl. Acad. Sci.* **78:** 4471.

Padgett, R.A., S.M. Mount, J.A. Steitz, and P.A. Sharp. 1983. Splicing of messenger RNA precursors is inhibited by antisera to small nuclear ribonucleoprotein. *Cell* **35:** 101.

Padgett, R.A., M.M. Konarska, P.J. Grabowski, S.F. Hardy, and P.A. Sharp. 1984. Lariat RNAs as intermediates and products in the splicing of messenger RNA precursors. *Science* **225:** 898.

Reed, R. and T. Maniatis. 1985. Intron sequences involved in lariat formation during pre-mRNA splicing. *Cell* **41:** 95.

Rogers, J. and R. Wall. 1980. A mechanism for RNA splicing. *Proc. Natl. Acad. Sci.* **77:** 1877.

Rossi, J.J., X. Soberon, Y. Marumoto, J. McMahon, and K. Itakura. 1983. Biological expression of an *Escherichia coli* consensus sequence promoter and some mutant derivatives. *Proc. Natl. Acad. Sci.* **80:** 3203.

Ruskin, B., J.M. Greene, and M.R. Green. 1985. Cryptic branch point activation allows accurate *in vitro* splicing of human β-globin intron mutants. *Cell* **41:** 833.

Ruskin, B., A.R. Krainer, T. Maniatis, and M.R. Green. 1984. Excision of an intact intron as a novel lariat structure during pre-mRNA splicing *in vitro*. *Cell* **38:** 317.

Ulfendahl, P.J., U. Pettersson, and G. Akusjarvi. 1985. Splicing of the adenovirus-2 E1A 13S mRNA requires a minimal intron length and specific intron signals. *Nucleic Acids Res.* **13:** 6299.

Wieringa, B., F. Meyer, J. Reiser, and C. Weissmann. 1983. Unusual splice sites revealed by mutagenic inactivation of an authentic splice site of the rabbit β-globin gene. *Nature* **301:** 38.

# Adenovirus as a Model for Transcription-termination Studies

**E. Falck-Pedersen,\* J. Logan,† T. Shenk,† G. Galli,‡ and J.E. Darnell, Jr.‡**

\*Department of Microbiology, Cornell Medical College, New York, New York 10021; †Department of Molecular Biology, Princeton University, Princeton, New Jersey 08544; ‡The Rockefeller University, New York, New York 10021

We have placed the sequences of the mouse β-globin gene responsible for RNA polymerase II termination into the adenovirus E1A transcription unit. When transcription is scored by a combination of in vivo and in vitro techniques, termination occurs in the new viral chromosome as it does in the mouse cell. We propose that the termination event requires two sequence-specific elements, the first a functional poly(A) site and the second a sequence involved directly in the termination (gF). In addition to a termination event, we have observed a secondary effect attributed to termination of the E1A transcription unit in the newly constructed viral genome: The E1B transcription unit is no longer functional. This observation indicates a dependence of E1B transcription on neighboring activities within the E1A transcription unit.

Transcription termination in prokaryotic and eukaryotic cells is a poorly understood component of the transcriptional machinery. The importance of the termination event in bacterial gene regulation is well established (Rosenberg and Court 1979), whereas in eukaryotic transcription units the termination event has only recently been characterized, and its role in the scheme of mRNA production has yet to be defined.

The termination event has been demonstrated for all three classes of eukaryotic transcription units by direct and indirect methods (Bogenhagen and Brown 1981; Hofer and Darnell 1981; Bakken et al. 1982). The first characterization of a class-II termination event was carried out by $^3$H labeling of the mouse β-major-globin transcription unit, demonstrating that approximately 1500 nucleotides (nt) downstream from the poly(A) site the transcription rate for the β-globin transcription unit is reduced 10-fold (Hofer and Darnell 1981). Recently a number of additional transcription units have been analyzed in a similar manner, all demonstrating transcription well past the poly(A) site, followed by a loss of transcription as measured by either in vivo or in vitro analysis of nascent RNA chains (Frayne et al. 1984; Lemur et al. 1984; Mather et al. 1984). When the RNA products have been analyzed for site-specific termination as occurs in the Rho independent bacterial system, a collection of pause/termination sites were revealed, with no outstanding site or sequences presenting themselves as site-specific termination candidates (Citron et al. 1984).

To define the role of various sequences in transcription termination, we wished to transfer the putative terminator elements to another genomic site and still score the termination event at the level of transcription. We have used the adenovirus as a transcription vector that under appropriate conditions would allow us to score nascent RNA production before and after the termination region in question. We have inserted sequences from the 3'-end transcription unit of the mouse β-major-globin into the E1A region, where it is possible to score transcription of E1A sequences 5' and 3' to the site of insertion. This allows a direct examination of the effect of the inserted sequences in termination. The findings indicate that termination for this polymerase II transcription unit results from a collaboration of two independent sequence elements. In addition, the termination element exerts a negative *cis* effect on transcription of the adjacent E1B gene.

## Materials and Methods

### Cells and viral stocks

Human 293 monolayer cells (Graham et al. 1977) maintained in Eagle's medium containing 10% fetal calf serum were used for cotransfections, plaque purification, and initial virus stock lysates. HeLa cells and spinner 293 cells were maintained in Joklik's modified minimal essential medium containing 5% fetal calf serum, and virus stocks were grown and purified as previously described (Nevins 1980).

### Plasmid and virus constructions

The mouse genomic insert from λgtWES · MβG2 (Tilghman et al. 1977) was subcloned into the plasmid pUC18-MβG2 for further subcloning. Using DNA polymerase I (Klenow), blunt ends were created on restriction fragments gDEF from the BalI site (1233 nt from cap site + 1) to the BglII site (2790 nt from the cap site) or gF from the XbaI site (nt 1985) to the AvaI site (nt 2488). The fragments were then ligated into the unique XbaI site (nt 1336) of plasmid pMLP6 (Logan and Shenk 1984), which contains 0–15 map units of adenovirus type-5 DNA. After transfection into *Escherichia coli* strain HB101, colonies were selected and inserts were analyzed for size and orientation by the alkaline minilysate technique. Adenovirus subclones of the E1A and E1B regions were made by standard cloning methods; actin, globin, and additional adenovirus subclones were previously described.

Viruses *dl*309, *dl*118, and *sub*360-LO have been described (Jones and Shenk 1978; Babiss and Ginsberg 1984; Logan and Shenk 1984). The reconstructed, globin-containing viruses were generated by the method of overlap recombination in 293 cells (Chinnadurai 1979; Stow 1981). After cotransfection of the *Eco*RI-linearized plasmid and the large *Xba*I fragment from *dl*309, individual plaques were picked, and viral DNA was prepared from $5 \times 10^7$ cells and verified by restriction digestion and Southern analysis. Having verified the virus constructs, large virus stocks were prepared in spinner 293 cells as described above.

*Preparation and hybridization of labeled nuclear RNA*
HeLa cells were harvested either 4 or 18 hr after infection with virus at $10^3$ particles/cell (25 pfu), and nuclei were isolated by Dounce homogenization for production of [α-$^{32}$P]UTP-labeled nuclear RNA. Nuclei from $5 \times 10^7$ cells were incubated for 10 min at 30°C in the presence of 0.5 mM ATP, CTP, and GTP with 2.5 μM UTP at a specific activity of 40 μCi/nmole. Labeled nuclear RNA was extracted by hot phenol/SDS extraction, and after ethanol precipitation RNA was purified by 2 M LiCl precipitation ($5 \times 10^7$ cells/ml) at 4°C (Nevins 1980) and unincorporated triphosphates were removed by TCA precipitations. After mild alkali cleavage (0.1 M NaOH at 0°C for 10 min) and neutralization by 2.0 M HEPES, $10^7$ cpm in 1.5 ml of 2 × TESS (RNA from ~ $2.5 \times 10^7$ cells) were used for each hybridization with early RNA; for late infections, $10^6$ cpm/ml in 1.5 ml (RNA from $2 \times 10^6$ nuclei) was used for hybridization. The RNA hybridizations to a DNA dot matrix were performed for 36 hr at 65°C. Filters were washed and digested with RNase prior to exposure to Kodak XAR-7 film.

Nitrocellulose DNA dots were prepared as previously described (Kafatos and Efstratiadis 1979). Each DNA dot represents 5–7 μg of DNA denatured in 0.2 M NaOH at 100°C for 5 min, neutralized with 2.0 M HEPES, and applied to a nitrocellulose grid in 5 × SSC.

In vivo labeling of adenovirus-infected HeLa cells and hybridization of nuclear RNA was essentially as previously described (Nevins 1980).

*Northern analysis of cytoplasmic poly(A)$^\pm$ RNA*
Cytoplasmic supernatants from a nuclear pellet were treated as previously described (Nevins 1980), and poly(A)$^+$ RNA was selected by three passages over oligo(dT) cellulose (Calbiochem). Usually 2 μg of early-infected poly(A)$^+$ RNA (or 0.5 μg of late-infected RNA) was loaded onto a 1.4% agarose/3% formaldehyde gel and electrophoresed in a 50 mM borate buffer (pH 8.3). RNA was transferred to nitrocellulose and hybridized to nick-translated DNA probes.

## Results

Because of the extreme instability of the RNA that is transcribed downstream from a poly(A) site, the charac-

terization of the termination event requires a quantitative assay of nascent RNA transcription of the sequences in question. Measurement of steady-state nuclear RNA is not an accurate representation of transcribed RNA and thus does not allow valid conclusions about termination. To simplify the analysis of the termination event identified in the mouse β-major-globin transcription unit, we have elected to implant the terminator sequences into the E1A gene of adenovirus. The technique for these constructions requires all sequence manipulations to be carried out in a plasmid (mIP6), and after plasmid characterization they are cotransfected with the large *Xba*I fragment (3.8–100 m.u.) of the adenovirus variant *dl*309 into 293-HEK cells, where homologous recombination generates mutant virus that will propagate in these cells. The use of this system allows us to manipulate the inserted DNA sequence with relative ease, and by reconstruction back into adenovirus, we can measure the effect of the insertion in a new (viral) transcriptional context under both early and late (before and after replication) stages of viral infection.

The sequences we chose to insert are taken from the 3′ region of the globin transcription unit (Fig. 1A), and the region where transcription begins its decline in vivo is within the fragment designated F (gF). The series of termination constructs we are analyzing in these experiments (Fig. 1B) includes the insertion of 500 nt of the 5′ gF sequence (*sub*360-gF$^+$ and *sub*360-gF$^-$) identified with termination in vivo of erythroleukemia mouse cells and a pair of constructs that include the globin poly(A) site present in the gD fragment (*sub*360-gDEF and *sub*360-gFED). To assay for transcription, we have subcloned portions of the E1A and E1B transcription units (1A1, 1A2, 1A3, 1B5′, and 1B3′) as well as globin fragments D, E, and F. Because the site of insertion at nt 1336 separates 1A2 and 1A3 genomic subclones, the termination effect is accurately identified by comparing the relative transcription rates of these two fragments.

## Transcription termination occurs in the new virus constructs

HeLa cells were infected with one of six viruses, and after a 4-hr incubation, cells were harvested and nuclei isolated for nuclear sample assays. The purified RNA from each sample was hybridized to a nitrocellulose DNA dot-blot matrix (Fig. 2). The control transcriptions are represented by grids labeled *dl*309 and *sub*360-LO (the only difference between these viruses is the substitution of the major late transcriptional control regions, MLTCR, in *sub*360 5′ to the E1A cap site). Comparing hybridization signals 1A2 and 1A3 for each of these viruses demonstrates equimolar transcription, as expected. For viruses *sub*360-gF$^+$ and gF$^-$, in which the terminator sequence was inserted in the plus and minus orientations, we see transcription of the gF sequence for both constructs as well as transcription of both 1A2 and 1A3 genomic sequences. For these constructs, insertion of the gF element alone does not induce the termination event. The transcription analysis of the *sub*360-gFED and *sub*360-gDEF viruses demonstrates

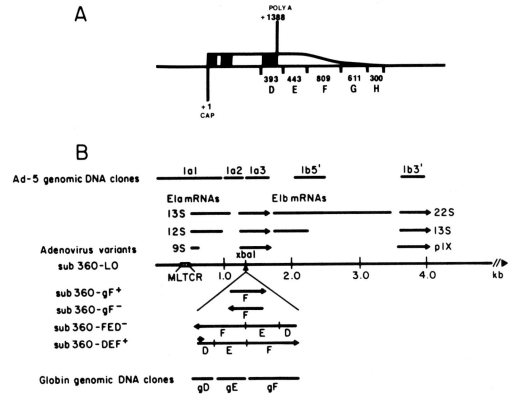

Figure 1 (*A*) Schematic representation of the mouse β-major-globin transcription unit (Konkel et al. 1978). Black areas represent the three globin exons; open regions represent either intervening sequences or transcribed sequences downstream of the poly(A) site. Transcription proceeds equally from the cap site through the poly(A) site (1390 nt). From analysis of nuclear RNA (Hofer and Darnell 1981; Citron et al. 1984), there is a decrease in transcription in the F fragment and complete loss of transcription between the G and H genomic fragments (indicated by convergence of the top line with the bottom line). (*B*) Plasmid pMLP-6, used for inserting globin termination sequences into the E1A second exon. Three transcription units are contained in this plasmid: E1A initiates at nt 498; its polyadenylation site is at nt 1630. E1B initiates at nt 1699; its polyadenylation is at nt 4061; polypeptide IX mRNA is initiated at nt 3576, and its polyadenylation site is at nt 4061. Globin fragments termed DEF (nt 1233–2790) and F (nt 1985–2488) were inserted into the unique *Xba*I site at position 1336 in the second exon of E1A. ( + and − ) The orientations relative to the original mouse genomic orientation. Genomic subclones used for transcriptional analysis are 1A1, 1A2, 1A3 gD, gE, gF, 1B5′, and 1B3′. Viruses were made by cotransfecting linearized, reconstructed plasmid with a large *Xba*I fragment from *dl*309 in 293 cells (see Methods).

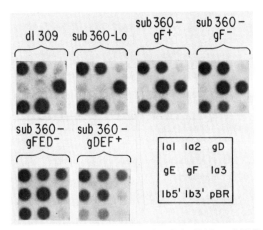

Figure 2 Transcription-rate analysis of the E1A and E1B transcription units early in infection with globin-virus constructs. HeLa cells were infected with 1000 particles of virus (*dl*309, *sub*360-LO, *sub*360-gF⁺, *sub*360-gF⁻, *sub*360-gFEF⁻, or *sub*360-gDEF⁺), and nuclei were isolated after 4 hr. Nascent $^{32}$P-labeled RNA (~ 10⁷ cpm) elongated in isolated nuclei was hybridized to a nitrocellulose filter containing plasmid DNAs (5 μg/dot) as indicated in the key.

for the gDEF construct (which contains the poly[A] signal sequence in addition to the downstream terminator element), a 10-fold to 15-fold reduction of transcriptional activity comparing the intensity of $^{32}$P-labeled RNA hybridized to the 1A2 and 1A3 DNA dots. The virus that contains these same sequences in the opposite orientation (*sub*360-gFED) transcribes through the 1A2 and 1A3 DNA regions in an equimolar fashion. This demonstrates an orientation (and therefore sequence) specificity for the termination reaction and provides a viral control for sequence insertions of this size into the E1A second exon. We also note in the *sub*360-gDEF virus a strong suppression of transcription in the neighboring E1B transcription unit.

To analyze for the termination event in vivo, the adenovirus is again an exceptional model for in vivo [³H]uridine pulse-labeling of nascent RNA. Because of the efficiency of viral templates for transcription, we are able to score nascent RNA transcription after a very brief (5 min) [³H]uridine pulse. By relying on the short pulse, we reduce the effect of differential half-lives of exon and nonexon sequences present in a given transcription unit. After a 4-hr infection with *sub*360-LO, *sub*360-gFED, and the *sub*360-gDEF viruses, cells were labeled with [³H]uridine and the isolated RNA was hybridized to genomic subclones on nitrocellulose filters (Fig. 3). In the control *sub*360-gFED virus, we see equal levels of transcription for all sequences present in the E1 transcription unit (both E1A and E1B) and strong transcription from two additional early adenovirus transcription units, E2 and E4, and from the cellular actin gene. The terminator virus, *sub*360-gDEF, shows a strong reduction of nascent labeling from (and including gF) continuing through the entire 1A3 and E1B sequences. The in vivo pulse not only shows termination of the 1A3 downstream sequence but again indicates a strong suppression of E1B transcription, presumably at the level of RNA chain initiation. The E2, E4, and actin

controls are transcribed as they were in the *sub*360-gFED virus.

## Termination in E1A causes a negative *cis* effect on E1B transcription

The architecture of the majority of eukaryotic transcription units (class II) is such that considerable DNA sequence lies between the 3' end of one gene and the 5' cap site of its nearest neighbor. Adenovirus, presumably due to size restrictions, packages its transcription units very efficiently, thus reducing the amount of intergenic sequence. For the E1 transcription unit, we have the extreme example of 69 nt separating the 3' end of E1A from the 5' end of E1B. Regulation of E1B is distinct from E1A and is dependent on the production of E1A in the early stages of viral infection (Berk et al. 1979; Jones and Shenk 1979; Nevins 1981). The DNA sequence elements involved in controlling E1B transcription have not been characterized, and it has been difficult to separate transcription from E1A and E1B because of potential readthrough from the E1A promoter into the E1B transcription unit (Wilson et al. 1979). The suppression of E1B transcription rates (Figs. 2 and 3) and steady-state mRNA levels (data not shown) indicate

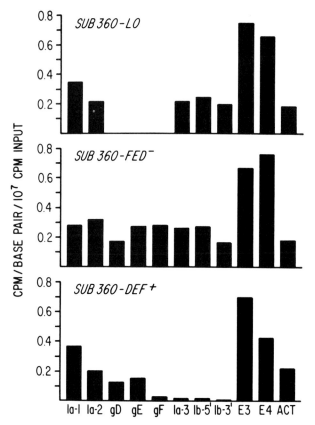

**Figure 3** Transcription rate analysis of in vivo [³H]uridine-labeled RNA (5-min pulse) from HeLa cells infected with *sub*360-LO, *sub*360-gFED⁺, or *sub*360-gDEF. For each infection, 10⁸ HeLa cells were infected at 10³ particles/cell for 4 hr, concentrated to 10⁷ cells/ml, and incubated for 5 min with 300 μCi/ml of [³H]uridine. [³H]RNA was extracted as described and hybridized to 25 μg of DNA on a nitrocellulose filter (∼ 0.8 cm²). (*Top*) 1.6 × 10⁷ cpm of [³H]RNA from *sub*360-LO-infected HeLa cells. (*Middle*) 1.6 × 10⁷ cpm of [³H]RNA from *sub*360-gFED⁻-infected HeLa cells. (*Bottom*) 2.1 × 10⁷ cpm of [³H]RNA from *sub*360-gDEF⁺-infected HeLa cells. Hybridized counts per minute have been normalized for input counts per minute, and the length of each subclone has been normalized to the value cpm/bp/10⁷ input cpm. pBR background in the hybridizations was below 55 cpm and was subtracted from each sample before normalization.

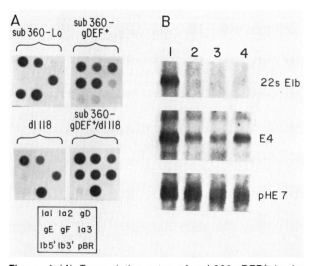

**Figure 4** (*A*) Transcription rates of *sub*360-gDEF⁺ in the presence of wild-type E1A. Nuclei were isolated and labeled nascent RNA from parallel infections of HeLa cells with 1.5 × 10³ particles/cell of *sub*360-LO, *sub*360-gDEF⁺, *dl*118, or both *sub*360-gDEF⁺ and *dl*118 was prepared and hybridized to nitrocellulose filters containing 5 μg of plasmid DNA dots as indicated in the key. 5 × 10⁶ cpm was hybridized for the *sub*360-LO, *sub*360-gDEF⁺, and *dl*118 RNAs, and 10⁷ cpm of RNA from the *sub*360-gDEF⁺/*dl*118 coinfection was used. (*B*) Steady-state RNA from E1B is not detectable in the *sub*360-gDEF⁺/*dl*118 coinfection. Cytoplasmic poly(A)⁺ RNA was prepared from the infections described in *A*, and 5 μg of each was resolved on a 1.4% gel (see Materials and Methods). (Lane *1*) *sub*360-LO; (lane *2*) *sub*360-gDEF⁺; (lane *3*) *dl*118; (lane *4*) *dl*118/*sub*360-gDEF⁻. The filters were hybridized to each of the following nick-translated DNA plasmids: E1B5' plasmid; the adenovirus *Eco*RI-C genomic fragment corresponding to the E4 transcription unit; and the pHe7 plasmid, a cDNA that serves as a cellular gene.

that the insertion of the termination element, but not simply the insertion of additional DNA sequences, inhibits transcription of the E1B region. This might be caused by loss of *trans* activation provided by the E1A protein or by a *cis* effect. Such a *cis* effect might suggest that the E1B promoter is only accessible to transcription-complex formation if a polymerase reads across it.

To separate these *cis* and *trans* possibilities, we carried out a coinfection of the *sub*360-gDEF virus with a mutant E1B virus *dl*118 (Babiss and Ginsberg 1984). This virus contains all the wild-type E1A functions and carries a deletion in the 5′ region of E1B. If the wild-type E1A protein products from *dl*118 were able to complement transcription in *sub*360-gDEF, an increase in the transcription signal of the E1B 5′ DNA dot should occur. This sequence is present in *sub*360-gDEF but not present in *dl*118. In addition, there should be an increase in the level of processed E1B mRNA, arising from transcription of the intact E1B transcription unit of *sub*360-gDEF. The results from this experiment (Fig. 4A,B) indicate that *dl*118 does not complement E1B at the level of transcription or steady-state mRNA production.

We are therefore left to conclude that by insertion of the functioning terminator element into the E1A transcription unit, we are causing a *cis* inactivation of E1B transcription. In late infections, termination is not as efficient (data not shown) and E1B transcription is recovered.

## Discussion

The present studies on transcription termination effected by sequences from the mouse β-major-globin gene indicate a two-step process for transcription termination in poly(A)-containing transcription units. The termination event occurs in only one of several mutant viruses, *sub*360-gDEF, which contained both the globin poly(A) site and the downstream termination element (gF). When the fragment gF was inserted by itself in the second exon of E1A, transcription continued through the downstream 1A3 fragment in equimolar fashion. This indicates that a sequence upstream of gF, presumably one with an active poly(A) site, is also required to reproduce the termination event.

In a recent series of virus constructs that we are using

DNA SEQUENCE FOR GLOBIN D AND F TERMINATOR ELEMENTS

```
        3360       3370       3380       3390       3400
CCACTGCCTT GGCTCACAAG TACCACTAAA CCCCCTTTCC TGCTCTTGCC
GGTGACGGAA CCGAGTGTTC ATGGTGATTT GGGGGAAAGG ACGAGAACGG
        3410       3420       3430       3440       3450
TGTGAACAAT GGTTAATTGT TCCCAAGAGA GCATCTGTCA GTTGTTGGCA
ACACTTGTTA CCAATTAACA AGGGTTCTCT CGTAGACAGT CAACAACCGT
        3460       3470       3480       3490       3500
AAATGATAGA CATTTGAAAA TCTGTCTTCT GACAAATAAA AAGCATTTAT
TTTACTATCT GTAAACTTTT AGACAGAAGA CTGTTTATTT TTCGTAAATA
                                    _____ globin polyA signal
        3510       3520       3530       3540       3550
GTTCACTGCA ATGATGGTTTT AAATTATTTG TCTGTGTCAT AGAAGGGTTT
CAAGTGACGT TACTACAAAA TTTAATAAAC AGACACAGTA TCTTCCCAAA
  • globin polyA site
        3560       3570       3580       3590       3600
ATGCTAAGTT TTCAAGATAC AAAGAAGTGA GGGTTCAGGT CTGACCTTGG
TACGATTCAA AAGTTCTATG TTTCTTCACT CCCAAGTCCA GACTGGAACC
        3610       3620       3630       3640       3650
GGAAATAAAT GAATTACACT TCAAATTGTG TTGTCAGCTA AGCAGCAGTA
CCTTTATTTA CTTAATGTGA AGTTTAACAC AACAGTCGAT TCGTCGTCAT
    _____ second polyA signal in gD
        3660
GCCACAGATC ------gE sequence---------- begin gF -->
CGGTGTCTAG
        4110       4120       4130       4140       4150
CTAGAGAGAA TAAGAATATC TAGTTTTTAA GGCTCATTAC TGGGGTCTTA
GATCTCTCTT ATTCTTATAG ATCAAAAATT CCGAGTAATG ACCCCAGAAT
        4160       4170       4180       4190       4200
TGAAATTTCC ATAATACCCT GTAAATGGAA GCATTTATTT TTTCAATAAA
ACTTTAAAGG TATTATGGGA CATTTACCTT CGTAAATAAA AAAGTTATTT
                                               _____
        4210       4220       4230       4240       4250
TCTATCTTGA ATATCCAGTG TGGGTTAGGA TTAAATCTCT CCTTCATACA
AGATAGAACT TATAGGTCAC ACCCAATCCT AATTTAGAGA GGAAGTATGT
        4260       4270       4280       4290       4300
GTTGGACTGC TTTTATTTAT ATGGAGTTAC TAGAGTTAAC ACAATAAGTA
CAACCTGACG AAAATAAATA TACCTCAATG ATCTCAATTG TGTTATTCAT
        4310       4320       4330       4340       4350
ATATACCCTT GATTTGTTTT TCTTTCCATA ACCACCAGGT TATGCGCAAT
TATATGGGAA CTAAACAAAA AGAAAGGTAT TGGTGGTCCA ATACGCGTTA
        4360       4370       4380       4390       4400
TCCGGAAATA AAATGTGTGT TCCAAGAGTT CTTTACGCTA CTCTCTGGTA
AGGCCTTTAT TTTACACACA AGGTTCTCAA GAAATGCGAT GAGAGACCAT
    _____
        4410       4420       4430       4440       4450
CAGTTTTAGT GAGATTTTGA AATGACTACA TATAATAAGT GGCCTTTAAT
GTCAAAATCA CTCTAAAACT TTACTGATGT ATATTATTCA CCGGAAATTA
        4460       4470       4480       4490       4500
TACAGAATGG TTTGTGTAGG TACAGAATAA AATACACCAA ATATTATGAG
ATGTCTTACC AAACACATCC ATGTCTTATT TTATGTGGTT TATAATACTC
```

**Figure 5** Sequences of the terminator fragments. DNA sequences of the two elements (gD and gF) that cause transcription termination in the E1A transcription unit. The sequence numbering originates from the 5′ *Eco*RI end of the parent genomic clone (Tilghman et al. 1977). The underlined sequences show the positions of the poly(A) signal hexamer AATAAA in both the gD and gF fragments.

to pinpoint the sequence elements involved in transcription termination, the following preliminary results support this two-step model. Using the same site of insertion and sequences involved in the first constructions, we have demonstrated that the globin EF sequences, which start 146 nt 3′ to the globin poly(A) site (Fig. 1) when inserted into the E1A second exon, do not cause termination before the 1A3 region. In addition, the globin poly(A) site, when inserted without the additional 3′ sequences, also does not cause termination. Preliminary results from a pair of virus constructs, *sub*360-gDF$^+$ and *sub*360-gDF$^-$, which have the globin poly(A) site followed by the gF fragment in either the plus or minus orientation (relative to the mouse transcription unit), reproduce the termination event when the gF fragment is in the plus orientation. This demonstrates the sequence-specific nature of the downstream gF termination element, which functions only in the presence of the upstream gD fragment containing the functional globin poly(A) site. We cannot determine from these experiments the type of interaction between these two elements and how they are affecting the transcription complex. We can rule out a specific spacing requirement for the two elements since in the gDF construct we have eliminated 450 nt of the gE sequence.

The termination region has been sequenced (Citron et al. 1984), and the two elements that are acting together (gD and gF) show little sequence homology or secondary structure reminiscent of bacterial terminators (Fig. 5). The gD fragment contains two hexamers, AAUAAA, which are part of the sequence requirements for polyadenylation. The gF sequence does contain some sequence components that may prove interesting for termination. In the first 400 nt of the gF fragment, there are three perfect poly(A) signal sequences (AATAAA), several TTTT stretches (associated with yeast and 5S termination), and a small region of dyad symmetry at position 4250 of gF. We are now trying to determine which of these components interact with the upstream sequence in gD to bring about termination of the class-II transcription unit.

The second conclusion from the work presented here regards transcriptional control of E1B. The negative *cis* effect of E1A termination on the transcription of the downstream E1B unit indicates a possible dependence of E1B transcription on the activities in the E1A transcription unit. By coinfecting the terminator *sub*360-gDEF with the *dl*118 (wild type in E1A), we have eliminated a faulty *sub*360-gDEF E1A protein as the lesion in E1B transcription. We also know that *sub*360-gFED, which has the same sequence in the opposite orientation and which does not terminate, does not inhibit E1B transcription. This indicates that the inactivation of E1B is not due to separation of enhancer-promoter elements (possibly in E1A) from the E1B promoter. One of two reasons can be proposed for the effect of E1B transcription. A requirement for functional transcription complexes reading into the E1A poly(A) site, inducing an active E1B promoter or the terminator element present in gF, induces a DNA conformation that is capable of extend-ing an effect downstream that interferes with formation of initiation complexes for the E1B transcription unit. From the preliminary results of the *sub*360-gEF virus, which does not terminate but has the identical downstream sequence of the *sub*360-gDEF virus, we see E1B transcriptional activity. This implies that the suppression of E1B transcription in *sub*360-gDEF is in fact due to termination of E1A complexes before their reaching the 1A3 downstream sequence adjacent to E1B and that opening the E1B promoter requires transcriptional transit through the region.

## Acknowledgments

This work was supported by grants from the National Institutes of Health (CA16006-12, CA18213-10) and the American Cancer Society (CD123N).

## References

Babiss, L.E. and H.S. Ginsberg. 1984. Adenovirus type 5 early region 1b gene product is required for efficient shutoff of host protein synthesis. *J. Virol.* **50:** 202.

Bakken, A., G. Morgan, B. Sollner-Webb, J. Roan, S. Busby, and R.H. Reeder. 1982. Mapping of transcription initiation and termination signals on *Xenopus laevis* ribosomal DNA. *Proc. Natl. Acad. Sci.* **79:** 56.

Berk, A.J., F. Lee, T. Harrison, J. Williams, and P.A. Sharp. 1979. Pre early adenovirus 5 genome product regulates synthesis of early viral messenger RNAs. *Cell* **17:** 935.

Bogenhagen, D.F. and D.D. Brown. 1981. Nucleotide sequences in *Xenopus* 5S DNA required for transcription termination. *Cell* **24:** 261.

Chinnadurai, G., S. Chinnadurai, and J. Brusca. 1979. Physical mapping of a large-plaque mutation of adenovirus type 2. *J. Virol.* **32:** 623.

Citron, B., E. Falck-Pedersen, M. Salditt-Georgieff, and J.E. Darnell, Jr. 1984. Transcription termination occurs within a 1000 base pair region downstream from the poly(A) site of the mouse β-globin (major) gene. *Nucleic Acids Res.* **12:** 8723.

Frayne, E.G., E.J. Leys, G.F. Crouse, A.G. Hook, and R.F. Kellems. 1984. Transcription of the mouse dihydrofolate reductase gene proceeds unabated through seven polyadenylation sites and terminates near a region of repeated DNA. *J. Mol. Cell. Biol.* **4:** 2921.

Graham, F.L., J. Smiley, W.C. Russel, and R. Nairn. 1977. Characteristics of human cell line transformed by DNA from human adenovirus type 5. *J. Gen. Virol.* **36:** 59.

Hofer, E. and J.E. Darnell, Jr. 1981. The primary transcription unit of the mouse β-major globin gene. *Cell* **23:** 585.

Jones, N. and T. Shenk. 1978. Isolation of deletion and substitution mutants of adenovirus type 5. *Cell* **13:** 181.

———. 1979. An adenovirus type 5 early gene function regulates expression of other early viral genes. *Proc. Natl. Acad. Sci.* **76:** 3665.

Kafatos, F.C. and A. Efstratiadis. 1979. Determination of nucleic acid sequence homologies by a dot hybridization procedure. *Nucleic Acids Res.* **7:** 1541.

Konkel, D.A., S.M. Tilghman, and P. Leder. 1978. The sequence of the chromosomal mouse β-globin major gene: Homologies in capping splicing and poly(A) sites. *Cell* **15:** 1125.

Lemur, M.A., B. Galliot, and P. Gerlinger. 1984. Termination of the ovalbumin gene transcription. *EMBO J.* **3:** 2779.

Logan, J. and T. Shenk. 1984. Adenovirus tripartite leader sequence enhances translation of mRNAs late after infection. *Proc. Natl. Acad. Sci.* **81:** 3655.

Mather, E.L., K.J. Nelson, J. Haimovich, and R.P. Perry. 1984. Mode of regulation of immunoglobulin μ and δ chain expression varies during B lymphocyte maturation. *Cell* **81:** 328.

Nevins, J. 1980. Definition and mapping of adenovirus-2 nuclear transcription. *Methods Enzymol.* **65:** 765.

Rosenberg, M. and D. Court. 1979. Regulatory sequences involved in the promotion and termination of RNA transcription. *Annu. Rev. Genet.* **13:** 319.

Stow, N.D. 1981. Cloning of a DNA fragment from the left-hand terminus of the adenovirus type 2 genome and its use in site directed mutatgenesis. *J. Virol.* **37:** 171.

Tilghman, S.M., F. Polsky, M.H. Edgall, J.G. Seidman, A. Leder, L.W. Enquist, B. Norman, and P. Leder. 1977. Cloning specific segments of the mammalian genome: Bacteriophage containing mouse globin and surrounding gene sequences. *Proc. Natl. Acad. Sci.* **74:** 4406.

Wilson, M.C., N.W. Fraser, and J.E. Darnell, Jr. 1979. Mapping of RNA initiation sites by high doses of UV irradition: Evidence for three independent promoters within the left 11% of the Ad-2 genome. *Virology* **94:** 175.

# The Adenovirus E1B 495R Protein Plays a Role in Regulating the Transport and Stability of the Viral Late Messages

**J. Williams,\* B.D. Karger,\* Y.S. Ho,[†] C.L. Castiglia,[‡] T. Mann,[‡] and S.J. Flint[‡]**
\*Department of Biological Sciences, Carnegie-Mellon University, Pittsburgh, Pennsylvania 15213;
[‡]Department of Molecular Biology, Princeton University, Princeton, New Jersey 08544

The type-5 host-range (*hr*) mutants *hr*6 and *hr*[cs]13 possess mutations that specifically alter the E1B 495R protein but leave the E1B 175R and 155R proteins unchanged. These mutants grow efficiently on 293 cells at all temperatures tested, but are cold-sensitive for growth on HeLa cells. At restrictive temperatures on HeLa cells, these mutants display normal viral early gene expression and DNA replication, but at late times in infection, the viral late mRNAs and proteins do not accumulate in the cytoplasm to wild-type levels. Further, these mutant-infected cells show an incomplete inhibition of host-cell protein synthesis, by comparison with the total shutdown seen in the wild-type infection. The phenotype results from atypical transport of RNA from nucleus to cytoplasm in mutant-infected cells. Cellular mRNAs continue to exit, whereas the transport of viral late mRNAs is defective, and those messages that do move out to the cytoplasm show reduced stability.

During the early phase in the productive infection of permissive human cells by group-C adenoviruses, rightward transcription from the E1B promoter at 4.6 map units (m.u.) generates two major mRNA species, which possess identical 5′ and 3′ termini but are of unequal size as a result of differential splicing. The larger mRNA (22S) encodes two proteins from overlapping reading frames (Bos et al. 1981); one of these frames begins at the 5′-proximal AUG and encodes a 175R protein, while the other begins at the second AUG and encodes a 495R protein. The smaller (13S) message encodes only the 175R protein. Two minor, intermediately sized mRNA species have been identified, one of which encodes a 155R protein that is coterminal with the 495R species (Anderson et al. 1984; Virtanen and Pettersson 1985). A physical map illustrating these mRNAs and their coding regions is shown in Figure 1.

There is a large body of evidence to suggest that E1B products play an important role in the infectious cycle and are required for complete transformation of rodent cells (for review, see Graham 1984; Shenk and Williams 1984; Williams 1985). Currently, however, very little is known about the functional roles of these E1B gene products in either infection or transformation. In this paper we describe experiments, using two mutants with specific alterations in the E1B 495R protein, that indicate that this product regulates the transport and stability of viral late mRNA species in infected cells.

## Materials and Methods

### Cells and viruses
The HeLa cells were those previously used for characterization of *ts* and *hr* mutants (Williams et al. 1971;

[†]*Present address*: Department of Medicine, Duke University Medical Center, Durham, North Carolina 27710.

Harrison et al. 1977) and were maintained in Dulbecco's modified Eagle's medium (DME), supplemented with 7% calf serum. Wild-type adenovirus 5 (Ad5) stocks were propagated, and infectivity was measured by plaque assay on these cells (Williams 1970). The 293 cells were those originally obtained from Dr. Frank Graham and used for isolation and characterization of *hr* mutants (Harrison et al. 1977; Ho et al. 1982). The mutants, *hr*6 (Harrison et al. 1977) and *hr*[cs]13 (Ho et al. 1982) were grown and titrated by plaque assay on these cells.

### DNA replication
Cells infected at an input multiplicity of 30 pfu/cell were harvested at various times (up to 30 hr at 38.5°C and 60 hr at 32.5°C), and lysed in buffer containing 0.5% SDS (Wilkie et al. 1973). The total DNA extracted from these lysates was treated with RNase A (Sigma), subsequently digested with *Hind*III, and several dilutions were electrophoresed along with *Hind*III-digested pure Ad5 DNA as a standard. The electrophoresed DNA was transferred to nitrocellulose (Southern 1975), and the blots were hybridized to nick-translated Ad5 DNA (Rigby et al. 1977) and autoradiographed.

### Analysis of protein synthesis
Cells infected at an input multiplicity of 30 pfu/cell were labeled at various times with [35S]methionine (5 μCi/ml). After a 2-hr pulse at 38.5°C or a 3-hr pulse at 32.5°C, the cells were lysed in SDS buffer, and aliquots containing equal numbers of TCA-precipitable counts were analyzed by electrophoresis on SDS/10% acrylamide gels (Laemmli 1970). Subsequently, the gels were fixed, fluorographed (Bonner and Laskey 1974), and exposed to X-ray film.

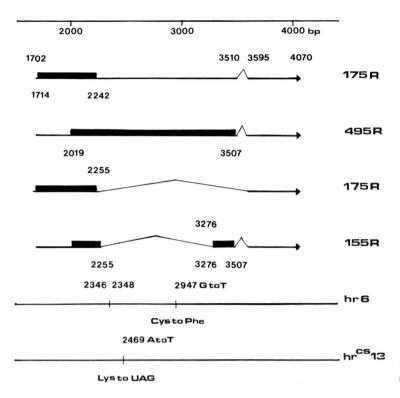

**Figure 1** Physical map of E1B mRNAs, coding regions, and mutations.

*Determination of the steady-state levels of RNA*

Total cytoplasmic RNA was prepared by the isotonic NP-40 lysis method of Kumar and Lindberg (1972), treated with proteinase K (BRL), and precipitated with ethanol. Nuclei were resuspended in 0.01 M Tris · HCl (pH 7.4) containing 0.01 M MgCl$_2$ with NaCl added to a final concentration of 0.5 M, digested for 30 min at 37°C with 20 mg/ml DNase I (RNase-free, Worthington), then similarly deproteinized and precipitated with ethanol. Poly(A)$^+$ mRNA or nuclear RNA was prepared from total RNA by oligo(dT)-cellulose chromatography (Aviv and Leder 1972), denatured by either glyoxal (McMaster and Carmichael 1977) or formaldehyde (Boedtker 1971; Rave et al. 1979), and electrophoresed in 0.8% agarose gels. The RNAs were transferred to nitrocellulose (Thomas 1980), and the resulting blots were hybridized with nick-translated $^{32}$P-labeled, region-specific Ad5 or Ad2 DNA probes obtained from appropriate plasmids. The following probes were used: E1A + E1B—the Ad2 *Bal*I fragment (0.7–5.7 m.u.); E1B + IVa$_2$—the Ad2 *Hin*dIII-C fragment (7.7–17.1 m.u.); L1—the Ad2 *Hin*dIII-I fragment (31.5–37.3 m.u.); L2—the Ad5 *Hin*dIII-D fragment (37.3–50.1 m.u.) or the Ad2 *Hin*dIII fragment (41.8–50.9 m.u.); L3, L4, and E2A—the Ad5 *Hin*dIII-A fragment (50.1–72.8 m.u.); L5 and E4—the Ad2 *Hin*dIII-F fragment (89.5—97.1 m.u.).

*Analysis of nascent RNA*

For in vivo labeling, infected cells were preincubated in DME containing 2% calf serum and 20 mM glucosamine for 30 min prior to addition of [$^3$H]uridine. Pulses of 10 min were made with 0.75 mCi/ml of [$^3$H]uridine (39 Ci/mmole, ICN) at each temperature. Total nuclear RNA was purified as described above and hybridized to increasing amounts of Ad5 DNA immobilized on nitrocellulose filters.

Nuclei were also isolated from wild type–infected and mutant-infected cells in the late phase of infection and permitted to elongate the chains initiated in vivo in the presence of [α-$^{32}$P]GTP (600 Ci/mmole, NEN). The RNA labeled in isolated nuclei was purified as described above and hybridized to either Ad5 DNA or to nitrocellulose filters to which restriction endonuclease fragments of Ad2 had been transferred (Southern 1975). The probes used were *Bal*I-E fragment (14.7–21.5 m.u.), *Hin*dIII-I fragments (31.5–37.3 m.u.), *Hin*dIII-D fragment (41.5–50.1 m.u.), *Hin*dIII-A fragment (50.1–72.8 m.u.) digested with *Eco*RI, *Hin*dIII-H fragment (72.8–79.9 m.u.) digested with *Sma*I, and *Hin*dIII-F fragment (89.5–97.1 m.u.). Prehybridization, hybridization, and posthybridization conditions were as described previously (Flint et al. 1983, 1984). Filters were exposed to Kodak X/AR or X/RP film, and autoradiograms were scanned with a Joyce-Loebel recording microdensitometer for calculation of the 38.5°/32.5° ratios of hybridized RNA.

*Measurement of the stability of viral mRNAs*

Cells were labeled with 200–300 μCi/ml [$^3$H]uridine for 1 hr at each temperature, washed twice with DME containing 2% calf serum and 200 mM unlabeled uridine, and preequilibrated to the appropriate temperature. The cultures were incubated in this medium for various periods, with medium changes every 2 hr. Cytoplasmic

RNA was isolated as described above and hybridized to nitrocellulose filters loaded with either denatured salmon sperm DNA, M13 DNA bearing an insert of the r-strand segment from 0 to 4.5 m.u. (E1A), or M13 carrying the r-strand sequence from 52.6 to 56.9 m.u. (hexon). Both were the generous gift of Faye Eggerding. Hybridization was in 2.5 ml of 0.01 M Tris · HCl (pH 7.4) containing 1 mM EDTA, 1 M NaCl, and 0.1% (w/v) SDS at 68°C for 30–40 hr; after this procedure, the filters were washed with two 500-ml changes of 2 × SSPE containing 0.1% SDS at 65°C for 1 hr, rinsed with 2 × SSPE, and digested with 20 μg/ml RNase in 2 × SSPE for 1 hr at room temperature. Subsequently, the filters were washed extensively in 2 × SSPE, dried, and counted in Econofluor (NEN). The amounts of DNA added to filters, 20 μg in the case of the M13-hexon DNA and 10 μg in all other cases, were previously shown by experiment to be in excess of the RNA to be hybridized.

*Measurement of the transport of newly synthesized RNA*
Infected cells were pulse-labeled with 0.75 mCi/ml [³H]uridine for 10 min and the label was then chased in the presence of 200 mM unlabeled uridine as described above. Cytoplasmic RNA was isolated and hybridized, as described in the preceding section, to filters loaded with salmon sperm DNA (20 μg), M13-hexon DNA (20 μg) or human β-actin cDNA cloned in pBR322 (10 μg), clone LK221 (the gift of P. Ponte, R. Gunning, and L. Kedes).

## Results

### Nature of the E1B mutations

The mutants used in the present work both exhibit a cold-sensitive, host-range phenotype in HeLa cells (Ho et al. 1982). Their mutations were previously mapped by complementation, recombination, deletion, and marker-rescue analyses (Galos et al. 1980; Ho et al. 1982) to the region between 5.6 and 8.5 m.u., strongly suggesting that they lay within the sequence encoding the 495R E1B protein. However, these limits include about 200 bp encoding the carboxyterminal end of the 175R protein, and the sequence encoding the aminoterminal part of the 155R protein (Fig. 1). To establish that the mutations alter only a single E1B protein, we have sequenced the entire E1B region (from the HpaI site at nt 1571 to the HpaII site at nt 3679) of both hr6 and hr^cs 13, using the dideoxy chain termination method (Sanger et al. 1977). The details of the cloning and sequence analysis have been given elsewhere (B.D. Karger et al., in prep.), and are simply summarized in Figure 1. In each case the mutations specifically change the 495R protein but do not alter the 175R and 155R moieties. Presumably, the hr6 leftward mutation (the deletion) determines the phenotype of this mutant, and the rightward one has no effect. By segregating the two hr6 mutations, however, we have shown that the latter, when present singly in the 495R gene, still generates a host-range phenotype that is less cold-sensitive than that of hr6 or the hr6 leftward segregant.

### Early events—presence of cytoplasmic early mRNA species in mutant-infected cells

Northern blot analysis of the cytoplasmic, poly(A)-containing RNA from wild type— and hr mutant—infected HeLa cells grown at 32.5°C and 38.5°C, using various restriction endonuclease fragments as probes (see Materials and Methods), revealed no marked shifts in the levels of the major viral early messages as a result of the mutations. The results are given in more detail elsewhere (B.D. Karger et al. in prep.), and are summarized here in Table 1.

The hr mutants make more or less normal (wild-type) levels of the E1B message (22S, 2.38 kb) encoding the 495R protein. Previous tests using tumor antisera and a 495R monolconal antibody indicated that these mutants did not produce immunoprecipitable protein under restrictive or permissive conditions (P. Sarnow, A. Levine, Y. Ho, B. Karger, and J. Williams, unpubl.). In view of the mutational lesions reported here, this is not surprising, since hr6 is likely to produce only a truncated aminoterminal protein of 188 residues, whereas hr^cs 13 should make a similar product of 151 residues. Preliminary immunoprecipitation tests using an antiserum specific for the amino terminus of the 495R moiety (kindly provided by Dr. P. Branton) have failed to reveal the truncated product in extracts of infected HeLa cells (B. Karger, unpubl.). Thus, these products are either not made at all, are made but degraded, or are made but exist in an immunologically nonreactive form. In contrast, tests on the same extracts with a 72K (E2A) monoclonal (kindly provided by B. Miller) show that this protein is made normally at early times by the E1B mutants at both temperatures (B. Karger, unpubl.).

### Viral DNA synthesis

Clearly, many of the early events that precede viral DNA replication are not severely perturbed as a result of the E1B 495R mutations, but it was of interest to determine whether the alterations to that protein resulted in aberrant viral DNA synthesis. We find that both hr6 and hr^cs 13 make essentially wild-type levels of DNA at both 32.5°C and 38.5°C. A comparison of wild-type and hr6 levels is shown in Figure 2. Analysis of [³H]thymidine-labeled, infected-cell DNA by CsCl gradient analysis led us to the same conclusion (data not shown). These results, in conjunction with those discussed above, can

**Table 1** Relative Concentrations of Early mRNA Species at 32.5°C and 38.5°C in Ad5 Mutant and Wild Type–infected HeLa Cells

| Region (est. size in kb) | RNA concentration[a] | | |
|---|---|---|---|
| | hr6 | hr^cs 13 | wild type |
| E1A (0.91 + 0.79) | 2.91 | ND | 3.53 |
| E1B (2.38) | ND | 0.73 | 0.78 |
| E1B (1.32) | 2.94 | 3.89 | 1.24 |
| E2A (2.03) | 0.46 | 0.51 | 0.56 |
| E3 (1.78) | ND | 1.14 | 1.02 |
| E4 (1.96) | 1.82 | 1.85 | 1.70 |

[a]Ratio of RNA concentration at 32.5°C: concentration at 38.5°C. (ND) Not done.

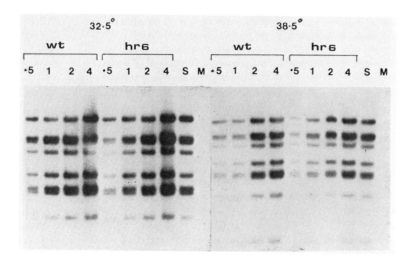

**Figure 2** Viral DNA synthesis by *hr*6 and wild-type viruses in HeLa cells. Lanes were loaded with from 0.5 μg to 4 μg of DNA. (Lane *S*) Pure viral DNA standard; (lane *M*) mock-infected extract.

be interpreted to mean that either the 495R protein is not required for the process of viral DNA replication, or, less likely, that the aminoterminal region of that protein provides the necessary function. On the other hand, at least one E1B gene function is probably required for viral DNA replication, since the E1B mutant *hr*7 is defective in this respect (Harrison et al. 1977; Lassam et al. 1978) and shows delayed synthesis (Stillman 1983).

### Viral late protein synthesis
Viral early events, including DNA synthesis, are clearly not severely disrupted in HeLa cells infected with the mutants at either 32.5°C or 38.5°C, yet the viral yields attained at low temperature are greatly depressed in comparison with those of wild-type virus (Ho et al.

1982). This suggested that a late event (or events) is altered as a result (direct or indirect) of the mutations in the 495R gene, and to test this view, we first examined late protein synthesis in HeLa cells infected at 32.5°C and 38.5°C with *hr*6 and *hr*$^{cs}$13. Preliminary tests, in which HeLa cells infected with wild-type virus were pulse-labeled with [$^{35}$S]methionine, established that viral late protein synthesis was optimal between 20 and 25 hr at 38.5°C and between 45 and 55 hr at 32.5°C (data not shown). In striking contrast to the wild type, however, the mutants make greatly reduced levels of all late proteins (undetectable in most cases) in these cells at 32.5°C, although they make close to wild-type levels of most at 38.5°C. One example of this effect is shown in Figure 3, in which cells were pulsed at 20 and 28 hr

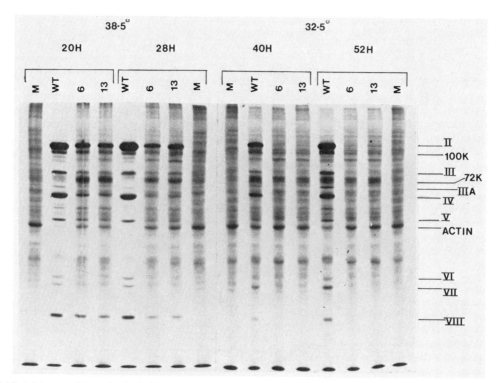

**Figure 3** Viral late protein synthesis in HeLa cells infected at 32.5°C and 38.5°C with *hr*6, *hr*$^{cs}$13, and wild-type viruses.

(38.5°C), and at 40 and 52 hr (32.5°C). In other experiments, we have found that incubation for up to 120 hr at 32.5°C induced little or no leak through of synthesis of these proteins at later times, although in some cases small amounts of hexon and 100K were visible. It is also apparent that although viral late proteins are made at reduced levels in mutant-infected cells at 32.5°C, the E2A 72K protein is made in both mutant- and wild type–infected cells at that temperature. At 38.5°C, this protein is not made by wild-type virus at late times but continues to be made at quite high levels in mutant-infected cells.

The inhibition of cellular protein synthesis, which is typically observed with wild-type virus, is incomplete at 38.5°C in HeLa cells infected with the mutants. Likewise, at 32.5°C the shutoff is incomplete, even with wild-type virus, although at somewhat later times (60 hr) the wild-type virus is more effective at shutting down cellular protein synthesis. Thus, the 495R protein may play a role, either directly or indirectly, in controlling the inhibition of host-cell protein synthesis normally observed in adenovirus infections of human cells.

### Steady-state levels of viral late mRNA species

To determine the basis for the reduced levels in viral late protein synthesis observed with the *hr* mutants, we first measured the relative levels of the L1–L5 mRNA species in the cytoplasm of HeLa cells infected at 32.5°C and 38.5°C. The results of a complete northern blot analysis, carried out as described in Materials and Methods, are shown in Figure 4. Clearly, there is no temperature-dependent effect on the levels of any of the

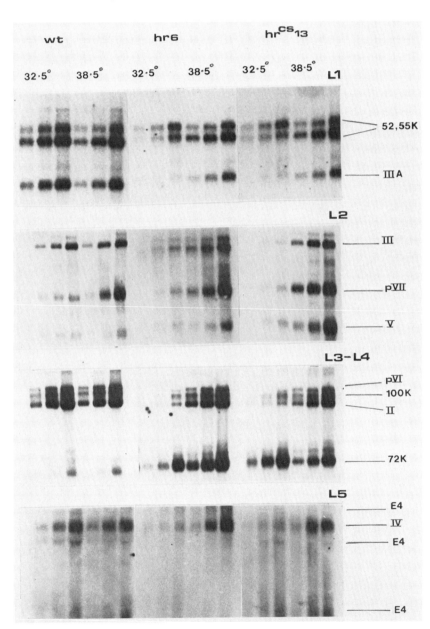

**Figure 4** Steady-state levels of viral late mRNA species in the cytoplasm of HeLa cells infected at 32.5°C and 38.5°C. From *left* to *right*, at each temperature, lanes were loaded with 0.25, 0.50 and 1.00 μg of cytoplasmic poly(A)-containing RNA, respectively.

late mRNA species in wild-type infections. In marked contrast, the levels of all late messages (encoded in L1–L5) are greatly reduced in cells infected at 32.5°C by both *hr*6 and *hr*$^{cs}$13. The magnitude of the effect ranges from about 4-fold to 5-fold for the fiber mRNA, through about an 8-fold effect upon the hexon mRNA, to more than a 10-fold effect on the L1 and L2 messages (densitometry data not shown). In contrast, the early L1 messages and the E2A (72K) message are not dramatically reduced at low temperatures in mutant-infected cells. Thus, the decreased levels of the viral late proteins at 32.5°C can be explained at least in part by a depression of viral late message levels.

### Transcription of viral late mRNA sequences in mutant-infected cells

As described in Materials and Methods, transcription of viral late RNA sequences was examined by two procedures. In the first, infected cells were pulse-labeled with uridine for 10 min after incubation for 18 hr at 38.5°C or 36 hr at 32.5°C, and the total nuclear RNA was purified and hybridized to Ad5 DNA attached to nitrocellulose filters, giving the results summarized in Table 2. It is clear that there are no temperature-dependent differences in the fractions of virus-specific RNA labeled in this short pulse period. Furthermore, *hr*6-infected cells synthesize the large fraction of the total nuclear RNA that is characteristic of the late phase of adenovirus infections (see Flint 1982).

In the second procedure, nuclei were isolated from infected cells during the late phase of infection and incubated at 37°C in the presence of [α-$^{32}$P]GTP, as described in Materials and Methods. When this RNA, labeled in vitro, was hybridized to Ad5 DNA, results similar to those described for the first procedure were obtained. For example, the ratios of viral RNA transcribed in nuclei from cells maintained at 32.5°C were 1.26 and 1.34 for wild type–infected and *hr*$^{cs}$13-infected cells, respectively. Some of this $^{32}$P-labeled RNA (from equal numbers of nuclei) was also hybridized to filters carrying specific fragments of the viral genome (either segments of the major late transcriptional unit or early units) as described in Materials and Methods, generating the results shown in Table 3. In all cases the increase in the ratios for the mutant was less than 2-fold, too small to account for the decreased levels of cytoplasmic, viral mRNAs described in the previous section. Thus, we conclude that the decreased accumulation of viral late mRNAs in the cytoplasm at 32.5°C does not

**Table 3** Transcription of Viral RNA in Mutant-infected HeLa Cells: Hybridization of RNA $^{32}$P-labeled in Isolated Nuclei to Viral DNA Fragments

| Region of Ad5 genome | Hybridized RNA[a] | |
|---|---|---|
| | wild-type | *hr*6 |
| L1 | ND | 1.0 |
| L2 | 1.3 | 1.8 |
| L3 | 1.0 | 1.8 |
| L4 | 1.5 | ND |
| E2A + L3 | 0.9 | 1.6 |
| E4 | 1.2 | 1.1 |

[a]Ratio of hybridized RNA at 38.5°C: 32.5°C.

result primarily from reduced transcription of the major late transcriptional unit.

### Processing of nuclear RNA

In view of the fact that defects in transcription of late RNA could not account for the failure of mutant-infected cells to accumulate viral late cytoplasmic mRNA species at 32.5°C, we next examined processing of nuclear RNA. Nuclear poly(A)-containing RNA was isolated from infected cells harvested during the late phase of infection at 38.5°C or 32.5°C and analyzed by northern blotting, using the *Hind*III fragment D of Ad2 DNA (41.8–50.9 m.u.) as a probe for L2. As shown in Figure 5, three major mRNA species, which comigrate with the cytoplasmic L2 mRNA species, were observed at 32.5°C and 38.5°C. Correcting the amounts of RNA added to the gel, the 38.5°C/32.5°C ratio for *hr*6-infected cell nuclear species is estimated to be 1.5. By comparison, the ratios for the cytoplasmic L2 mRNA species from the same culture examined in parallel was about 10-fold (data not shown, but see Fig. 4). Thus, the decrease in the amount of processed viral RNA seen in the nuclei of *hr*6-infected cells is similar in magnitude to the decrease in transcription and significantly smaller than the reduction in steady-state levels of viral late mRNA species found in the cytoplasm of mutant-infected cells at 32.5°C.

### The stability of viral mRNA species

Since the amounts of viral late cytoplasmic mRNAs in mutant-infected cells are apparently not determined at the levels of transcription or nuclear processing, we next examined the stability of the viral late mRNA species at permissive and restrictive temperatures. Rather than

**Table 2** Transcription of Viral RNA in Mutant-infected HeLa Cells: Hybridization of RNA Pulse-labeled In Vivo to Ad5DNA

| Infection | Hybridized cpm | Input cpm | % Virus-specific RNA | 38.5°C: 32.5°C |
|---|---|---|---|---|
| Wild-type, 32.5°C | 191.4 | 1579.0 | 12.1 | |
| Wild-type, 38.5°C | 1373.7 | 11071.5 | 12.4 | 1.0 |
| *hr*6, 32.5°C | 265.9 | 1562.5 | 17.0 | |
| *hr*6, 38.5°C | 758.1 | 4019.4 | 18.8 | 1.1 |

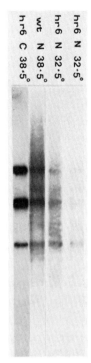

**Figure 5** Processing of nuclear RNA in *hr*6-infected HeLa cells. (N) Nuclear RNA; (C) cytoplasmic RNA. From *left* to *right*, lanes were loaded with 0.45 μg, 0.90 μg, and 0.70 μg of nuclear and 0.78 μg of cytoplasmic poly(A)-containing RNA, respectively.

use a protocol of continuous labeling and measurement of accumulation to steady state of cytoplasmic RNA species to measure stability (Wilson and Darnell 1981), we chose to measure the turnover rate of viral late mRNAs directly, using the pulse-chase protocol described in Materials and Methods. It should be noted that no [³H]uridine was incorporated into RNA during the chase period, when fresh, cold uridine was supplied at regular intervals (data not shown). Cytoplasmic RNA was purified from cells harvested after pulse-labeling and, after various periods of chase, added to filters containing an excess of M13-hexon or M13-E1A DNA; the amounts of labeled RNA hybridized were then determined. The half-lives of the mRNAs determined from these measurements are listed in Table 4. In Ad5-infected cells, hexon mRNA exhibited a turnover time ($t_{1/2}$) of about 2 hr at either temperature. This value is in reasonable agreement with those 75–100 min made

**Table 4** Turnover Times of Hexon and E1A mRNAs in Mutant-infected HeLa Cells

| | Turnover time (hr) | | | |
|---|---|---|---|---|
| | hexon | | E1A | |
| Virus | 32.5°C | 38.5°C | 32.5°C | 38.5°C |
| Wild-type | 2.0, 1.6 | 2.0, 2.0 | 1.0, ND | 1.3, 1.2 |
| *hr*$^{cs}$13 | 0.5, 0.6, 0.5 | 1.3, 1.3, 1.4 | 1.2, ND | 1.4, 0.8 |

previously for the mRNA species produced from the adenovirus major late transcriptional unit (Nevins and Darnell 1978). Hexon mRNA species were found to be more unstable in *hr*$^{cs}$13-infected cells, with $t_{1/2}$s of 0.53 hr and 1.33 hr at 32.5°C and 38.5°C, respectively. Thus, at 32.5°C, the hexon mRNA made in mutant-infected cells is about 4-fold less stable than that made in wild type–infected cells. This effect seems to be specific for late species, for no similar alteration was found for the $t_{1/2}$ of the E1A mRNA in mutant-infected cells at either temperature. The E1A $t_{1/2}$s we have found are in good agreement with those previously reported by others (Wilson and Darnell 1981). The stability of late mRNA sequences is clearly decreased at nonpermissive (and to a lesser extent at permissive) temperatures in mutant-infected cells, although again this is of insufficient magnitude to account completely for the reduction in the steady-state level of the hexon and other late mRNA species.

## Transport of newly synthesized viral and cellular RNA to the cytoplasm

We (Beltz and Flint 1979; Castiglia and Flint 1983; Flint et al. 1983) and others (Babich et al. 1983; Yoder et al. 1983) have previously established that transport from the nucleus to the cytoplasm of newly synthesized cellular mRNA and 28S ribosomal RNA is inhibited during the late phase of adenovirus infection, when viral late messages are efficiently moving out of the nucleus. We thus wished to find out if the reduced concentrations of viral late cytoplasmic mRNA species observed in mutant-infected cells at 32.5°C might result in part from a failure to induce selective transport. Labeling and chase times of 10 min each were chosen (see Materials and Methods), on the basis of previous estimates of transcription and exit times of Ad2 late mRNAs (Nevins and Darnell 1978), and to minimize the effects of turnover in mutant-infected cells at 32.5°C, where the $t_{1/2}$ for hexon mRNA is just over 30 min. The results are summarized in Figure 6. It is clear that in *hr*$^{cs}$13-infected cells, smaller quantities of newly synthesized late RNA sequences enter the cytoplasm, compared with the levels entering in wild-type infection. The amounts of RNA entering the cytoplasm are given as percentages of the total RNA, so these changes do not simply reflect the lower rates of transcription observed at lower temperatures. We conclude that newly synthesized late mRNA sequences are not transported to the cytoplasm as efficiently in *hr*$^{cs}$13-infected cells as in wild type–infected cells and that the mutant-infected cells show a temperature-dependent defect in transport. Accompanying the transport defect of viral late mRNA, there is continued transport of cellular mRNA species, exemplified in this experiment by a β-actin message. As expected, in Ad5-infected cells, no newly synthesized actin mRNA entered the cytoplasm, despite transcription of the β-actin species at these late times (data not shown). In contrast, *hr*$^{cs}$13-infected cells continued to transport newly synthesized actin mRNA to the cytoplasm, and the extent of this continued transport was

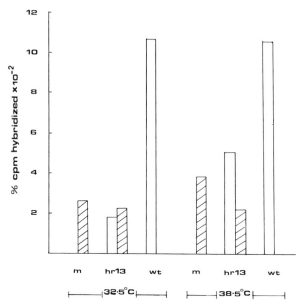

**Figure 6** Transport of viral late and cellular mRNAs from nucleus to cytoplasm. Amounts of RNAs entering cytoplasm are given as percentages of total RNA. (□) RNA hybridizing to filter containing M13-hexon DNA; (▨) RNA hybridizing to filter hybridizing β-actin DNA; (m) mock-infected.

inversely proportional to the degree to which transport of new viral late mRNA sequences was inhibited. The mutant-infected cells show a similar continued transport of newly synthesized 28S rRNA (data not shown).

## Discussion

In HeLa cells infected by $hr^{cs}$ mutants, which do not encode a complete E1B 495R protein, the viral late mRNAs and proteins do not accumulate in the cytoplasm to wild-type levels at restrictive temperatures. The most likely explanations for this phenomenon are that transport of viral late messages from the nucleus is defective in mutant-infected cells and that the stability of the messages that do exit to the cytoplasm is reduced.

Normally, at the onset of the late phase of infection in adenovirus-infected cells, there is induction of selective transport of viral mRNA sequences and a parallel inhibition of cellular mRNA and 28S rRNA transport (Beltz and Flint 1979; Babich et al. 1983; Castiglia and Flint 1983; Flint et al. 1983; Yoder et al. 1983). In contrast, E1B mutant-infected cells continue to transport newly synthesized actin mRNA sequences (Fig. 6) and 28S rRNA at levels similar to those found in noninfected cells. Concomitant with this continued transport of cellular RNA sequences, there is a 5-fold reduction in the transport of newly synthesized viral (L3) mRNA sequences to the cytoplasm (Fig. 6). As discussed above, this aberrant transport pattern does not reflect any difference in the transcription of actin or viral L3 sequences, and it is known that the processing of actin mRNA occurs normally in wild type–infected cells (Babich et al. 1983), as does processing of viral late

transcripts in mutant-infected cells (Fig. 5). Furthermore, the labeling protocol used to measure RNA transport was chosen to minimize effects upon cytoplasmic RNA concentration of the more rapid turnover of viral late mRNA sequences in mutant-infected cells compared with wild type–infected cells (Table 4). We therefore conclude that the breakdown in selective transport of viral mRNA sequences in mutant-infected cells is largely responsible for the much lower steady-state concentrations seen at the restrictive temperature (Fig. 4). The 4-fold reduction in transport of viral L3 mRNA sequences in $hr^{cs}$13-infected cells at the restrictive temperature compared with the permissive temperature (Fig. 6) and the 2.5-fold reduction in L2 mRNA stability (Table 4) accounts very well for the 7-fold to 8-fold reductions in the concentration of the cytoplasmic L3 mRNA species observed (Fig. 4).

The nuclear location of the E1B 495R moiety in virally infected cells is consistent with a role for this protein in the selective transport of viral mRNA species from the nucleus to the cytoplasm. However, the mechanism by which the 495R protein induces selective transport of viral mRNA sequences is not yet understood, in part because the normal transport mechanism(s) is poorly characterized. The phenotypes of the *hr* mutants described here and other mutants bearing deletions in the 495R-coding sequence (Babiss and Ginsberg 1984; Logan et al. 1984; T. Shenk, pers. comm.) do seem to rule out a simple competition mechanism. Clearly, the production of mature, viral late mRNA species in the nucleus is insufficient to induce selective transport. At the same time, the mere synthesis and presence of the E1B 495R protein does not alter transport, since rodent and human transformants (for review, see Flint 1982; Graham 1984; Williams 1985) that produce this protein have been isolated. Presumably, such cell lines would not survive were the E1B protein to inhibit transport of cellular mRNA and 28S rRNA from the nucleus. Thus, it is quite possible that the 495R entity mediates selective transport indirectly, via another viral early or late protein that is not produced efficiently in the absence of the E1B protein. The E4 34K protein, which physically complexes with the E1B component (Sarnow et al. 1984), is a good candidate for this second early protein. In this respect, it is of interest that mutants defective for production of the E4 protein have a lytic-cycle phenotype similar in some aspects to that described for the E1B mutants (Challberg and Ketner 1981; T. Shenk, pers. comm.), although it has yet to be shown that the E4 mutant phenotypes result from aberrant transport of viral or cellular mRNAs. The observed correlation between the efficiency of synthesis of late proteins and the degree of selective transport observed in our experiments (see Figs. 3 and 6) suggests that one or more viral late protein may be involved in the inhibition of cellular RNA transport that develops late in infection. However, it has been shown that cells infected by mutants that cannot express their VAI-RNA genes and that produce reduced levels of viral late proteins are capable of transporting viral late mRNA species efficiently (Thimmappaya et al. 1982; Schneider et al. 1984).

The 495R protein may act indirectly to influence transport via a cellular component. It is known, for example, that the 495R protein forms physical complexes with the p53 cellular tumor antigen (Sarnow et al. 1982), and it is possible that the transport defect is mediated by this protein. Alternatively, the E1B protein may act directly with a cellular component of the transport machinery but can only do so with an appropriate substrate. Consistent with (but not proving) the view that there is either direct or indirect interaction with a cellular factor, is the fact that these E1B mutants demonstrate a cold-sensitive, host-range phenotype in HeLa cells but grow productively at both high and low temperatures in normal human embryonic kidney cells and certain human tumor cell lines (Harrison et al. 1977; J. Williams, unpubl.).

In addition to displaying atypical transport of viral late and cellular messages, the stability of those late messages that do exit to the cytoplasm is reduced in cells infected at the restrictive temperature. Although these defects may be independent, there is a good chance that the second is linked to the first. An aberrant transport mechanism could result in temporal and/or compartmental perturbations of late messages in the cytoplasm, so that a late message that is not in the correct place at the correct time cannot be translated and is more prone to nuclease digestion. Such misplacement of message would explain why, despite the apparent presence of detectable amounts of viral late mRNA species in mutant-infected HeLa cells at 32.5°C (see Fig. 4), little if any of the corresponding proteins are made under these conditions (Fig. 3). Alternatively, the E1B protein may play a more direct role in the translation of viral mRNA species. In this respect, we do know that the translational effect does not result from aberrant production by the mutants of the viral VAI-RNA species (C. Castiglia and S.J. Flint, unpubl.), which is known to facilitate translation of viral late messages (Thimmappaya et al. 1982).

Lastly, it should be mentioned that these E1B *hr* mutants are defective for transformation of rat cells at 32.5°C, 35.5°C, 37.0°C, and 38.5°C (Ho et al. 1982). At present, however, we do not know if any of the defects observed in the lytic cycle is responsible for the transformation defect or if the latter is independent of those.

## Acknowledgments

We gratefully acknowledge the excellent technical assistance of R. Lunt, M. Williams, and M. Yonge. This work was funded by grants from the National Institutes of Health to J.W. (CA21375) and S.J.F. (AI17265).

## References

Anderson, C.W., R.C. Schmitt, J.E. Smart, and J.B. Lewis. 1984. Early region 1B of adenovirus 2 encodes two coterminal proteins of 495 and 155 amino acid residues. *J. Virol.* **50**: 387.

Aviv, H. and P. Leder. 1972. Purification of biologically active globin messenger RNA by chromatography on oligothymidylic acid-cellulose. *Proc. Natl. Acad. Sci.* **69**: 1408.

Babich, A., L.T. Feldman, J.R. Nevins, J.E. Darnell, and C. Weinberger. 1983. Effects of adenovirus on metabolism of specific host mRNAs: Tranport control and specific translational discrimination. *Mol. Cell. Biol.* **3**: 1212.

Babiss, L.E. and H.S. Ginsberg. 1984. Adenovirus type 5 early region 1B gene product is required for efficient shutoff of host protein synthesis. *J. Virol.* **50**: 202.

Beltz, G. and S.J. Flint. 1979. Inhibition of HeLa cell protein synthesis during adenovirus infection: Restriction of cellular messenger RNA sequences to the nucleus. *J. Mol. Biol.* **131**: 353.

Boedtker, H. 1971. Conformation-independent molecular weight determinations of RNA by gel electrophoresis. *Biochim. Biophys. Acta* **240**: 448.

Bonner, W.M. and R.A. Laskey. 1974. A film detection method for tritium-labelled proteins and nucleic acids in polyacrylamide gels. *Eur. J. Biochem.* **46**: 83.

Bos, J.L., L.J. Polder, R. Bernards, P.I. Schrier, P.J. van den Elsen, A.J. van der Eb, and H. van Ormondt. 1981. The 2.2 kb E1B mRNA of human Ad12 and Ad5 codes for two tumor antigens starting at different AUG triplets. *Cell* **27**: 121.

Castiglia, C.L. and S.J. Flint. 1983. Effects of adenovirus infection on rRNA synthesis and maturation in HeLa cells. *Mol. Cell. Biol.* **3**: 662.

Challberg, S.S. and G. Ketner. 1981. Deletion mutants of adenovirus 2: Isolation and initial characterization of virus carrying mutations near the right end of the viral genome. *Virology* **114**: 196.

Flint, S.J. 1982. Expression of adenoviral genetic information in productively-infected cells. *Biochim. Biophys. Acta* **651**: 175.

Flint, S.J., G.A. Beltz, and D.I.H. Linzer. 1983. Synthesis and processing of simian virus 40-specific RNA in adenovirus-infected, simian virus 40-transformed human cells. *J. Mol. Biol.* **167**: 335.

Flint, S.J., M.A. Plumb, U.-C. Yang, G.S. Stein, and J.L. Stein. 1984. Effect of adenovirus infection on expression of human histone genes. *Mol. Cell. Biol.* **4**: 1363.

Galos, R.S., J. Williams, T. Shenk, and N. Jones. 1980. Physical location of host range mutations of adenovirus type 5: Deletion and marker-rescue mapping. *Virology* **104**: 510.

Graham, F.L. 1984. Transformation by and oncogenicity of human adenoviruses. In *The viruses: Adenoviruses* (ed. H.S. Ginsberg), p. 339. Plenum Press, New York.

Harrison, T.J., F.L. Graham, and J. Williams. 1977. Host range mutants of adenovirus 5 defective for growth in HeLa cells. *Virology* **77**: 310.

Ho, Y.S., R. Galos, and J. Williams. 1982. Isolation of type 5 adenovirus mutants with a cold-sensitive host range phenotype: Genetic evidence of an adenovirus transformation maintenance function. *Virology* **122**: 109.

Kumar, A. and U. Lindberg. 1972. Characterization of messenger ribonucleoprotein and messenger RNA from KB cells. *Proc. Natl. Acad. Sci.* **69**: 681.

Laemmli, U.K. 1970. Cleavage of structural proteins during the assembly of the head of bacteriophage T4. *Nature* **227**: 680.

Lassam, N.J., S.T. Bayley, and F.L. Graham. 1978. Synthesis of DNA, late polypeptides and infectious virus by host-range mutants of adenovirus 5 in nonpermissive cells. *Virology* **87**: 463.

Logan, J., S. Pilder, and T. Shenk. 1984. Functional analysis of adenovirus type 5 early region 1B. *Cancer Cells* **2**: 527.

McMaster, G.K. and G.G. Carmichael. 1977. Analysis of single- and double-stranded nucleic acids on polyacrylamide and agarose gels by using glyoxal and acridine orange. *Proc. Natl. Acad. Sci.* **74**: 4835.

Nevins, J.R. and J.E. Darnell. 1978. Groups of adenovirus type 2 mRNAs derived from a large primary transcript: Probable nuclear origin and possible common 3′ ends. *J. Virol.* **25**: 811.

Rave, N., R. Crkvenjakov, and H. Boedtker. 1979. Identification of procollagen mRNAs transferred to diazobenzyloxymethyl paper from formaldehyde agarose gels. *Nucleic Acids Res.* **6**: 3559.

Rigby, P.W.J., M. Dieckmann, C. Rhodes, and P. Berg. 1977. Labelling deoxyribonucleic acid to high specific activity *in vitro* by nick translation with DNA polymerase I. *J. Mol. Biol.* **113:** 237.

Sanger, F., S. Nicklen, and A.R. Coulson. 1977. DNA sequencing with chain-terminating inhibitors. *Proc. Natl. Acad. Sci.* **74:** 5463.

Sarnow, P., Y.S. Ho, J. Williams, and A.J. Levine. 1982. Adenovirus E1B-58kd tumor antigen and SV40 large tumor antigen are physically associated with the same 54kd cellular protein in transformed cells. *Cell* **28:** 387.

Sarnow, P., P. Hearing, C.W. Anderson, D.N. Halbert, T. Shenk, and A.J. Levine. 1984. Adenovirus early region 1B 58,000-dalton tumor antigen is physically associated with an early region 25,000-dalton protein in productively infected cells. *J. Virol.* **49:** 692.

Schneider, R.J., C. Weinberger, and T. Shenk. 1984. Adenovirus VAI RNA facilitates the initiation of translation in virus-infected cells. *Cell* **37:** 291.

Shenk, T. and J. Williams. 1984. Genetic analysis of adenoviruses. *Curr. Top. Microbiol. Immunol.* **111:** 1.

Southern, E.M. 1975. Detection of specific sequences among DNA fragments separated by gel electrophoresis. *J. Mol. Biol.* **98:** 503.

Stillman, B.W. 1983. The replication of adenovirus DNA. *UCLA Symp. Mol. Cell. Biol.* **10:** 381.

Thimmappaya, B., C. Weinberger, R.J. Schneider, and T. Shenk. 1982. Adenovirus VAI-RNA is required for efficient translation of viral mRNAs at late times after infection. *Cell* **31:** 543.

Thomas, P.S. 1980. Hybridization of denatured RNA and small DNA fragments transferred to nitrocellulose. *Proc. Natl. Acad. Sci.* **77:** 5201.

Virtanen, A. and U. Pettersson. 1985. Organization of early region 1B of human adenovirus type 2: Identification of four differentially spliced mRNAs. *J. Virol.* **54:** 383.

Wilkie, N.M., S. Ustacelebi, and J.F. Williams. 1973. Characterization of temperature-sensitive mutants of adenovirus type 5: Nucleic acid synthesis. *Virology* **51:** 499.

Williams, J.F. 1970. Enhancement of adenovirus plaque formation on HeLa cells by magnesium chloride. *J. Gen. Virol.* **9:** 251.

———. 1985. Adenovirus genetics. In *Developments in molecular virology* (ed. W. Doerfler), vol. 5, p. 247. Martinus Nijhoff, The Hague.

Williams, J.F., M. Gharpure, S. Ustacelebi, and S. McDonald. 1971. Isolation of temperature-sensitive mutants of adenovirus type 5. *J. Gen. Virol.* **11:** 95.

Wilson, M.C. and J.E. Darnell. 1981. Control of messenger RNA concentration by differential cytoplasmic half-life: Adenovirus messenger RNAs from transcription units 1A and 1B. *J. Mol. Biol.* **148:** 231.

Yoder, S.S., B.L. Robberson, E.J. Leys, A.G. Hook, M. Al-Ubaidi, K.-Y. Yeung, R.E. Kellems, and S.M. Berget. 1983. Control of cellular gene expression during adenovirus infection: Induction and shutoff of dihydrofolate reductase gene expression by adenovirus type 2. *Mol. Cell. Biol.* **3:** 819.

# Functional Analysis of the Adenovirus E1B 55K Polypeptide

S. Pilder, K. Leppard, J. Logan, and T. Shenk

Department of Molecular Biology, Princeton University, Princeton, New Jersey 08544

The adenovirus type-5 mutant H5*dl*338 lacks 524 bp within early region 1B (E1B). The mutation removes a portion of the region encoding the related E1B 55K and 17K polypeptides but does not disturb the E1B 21K coding region. The virus is defective for growth in HeLa cells, where its final yield is reduced about 100-fold compared with wild-type virus. The site of the *dl*338 defect was studied in HeLa cells. Early gene expression and DNA replication appeared normal. Late after infection, mRNAs coded by the major late transcription unit accumulated to reduced levels. At a time when transcription rates and steady-state nuclear RNA species were normal, the rate at which late mRNA accumulated in the cytoplasm was markedly reduced. Further, in contrast to the wild-type case, transport and accumulation of cellular mRNAs continued late after infection with *dl*338. The E1B product appears to facilitate transport and accumulation of viral mRNAs late after infection while blocking the same processes for cellular mRNAs.

The adenovirus type-5 (Ad5) E1B transcription unit is located between about 4.5 and 11.5 map units on the viral chromosome (Fig. 1). It encodes three mRNAs early after infection. The largest mRNA (22S) can code both 21K and 55K polypeptides, whereas the smallest (13S) codes only the 21K species (Bos et al. 1981). The 55K coding region overlaps the carboxyterminal 40% of the 21K region in a second reading frame. Anderson et al. (1984) have identified an additional E1B-coded polypeptide (17K) synthesized from an intermediate-sized mRNA. The 17K moiety is encoded in the 55K reading frame and comprises the amino- and carboxyterminal segments of the larger protein. The synthesis of a fourth mRNA (9S) is directed by a control element within the E1B region that becomes active late after infection. This small, unspliced mRNA encodes a 14K structural polypeptide (IX). All of the mRNAs discussed so far are coded by the same DNA strand. There are also several small open reading frames on the opposite DNA strand that may be utilized (Gingeras et al. 1982; Katze et al. 1983).

The E1B region provides functions required both for oncogenic transformation and productive growth in cultured cells (for review, see Graham 1984; Shenk and Williams 1984). To elucidate the function of the E1B 55K polypeptide, we have constructed a viral mutant, H5*dl*338, which carries a deletion within the E1B region. This mutant was shown to be defective for transformation of rat cells and for growth in human cells (Logan et al. 1984). Here we report the physiological consequences of the *dl*338 mutation. Our results indicate that the E1B 55K polypeptide functions to simultaneously facilitate transport and accumulation of late viral mRNAs while blocking the same processes for cellular mRNAs. Similar conclusions have been reached by two other research groups using different mutants (Babiss et al. 1985; Williams et al., this volume).

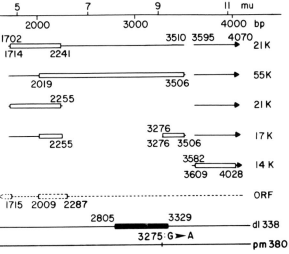

**Figure 1** Physical map of E1B mRNAs, coding regions, and mutations. Scale (*top* line): Map units (mu) and nucleotide sequence position (bp) relative to the left end of the viral chromosome. Key: mRNAs, lines; introns, spaces; coding region, open rectangles; open reading frame (ORF) encoded by the opposite DNA strand, open rectangle outlined with broken line; segment deleted in *dl*338, solid rectangle. The nucleotide sequence numbers of the base pairs bracketing the deletion are indicated.

## Experimental Procedures

### Viruses and cells

H5*dl*309 serves as the wild-type Ad5 parent in these studies. This virus was generated from H5*wt*300, which is a plaque-purified derivative of a virus stock obtained from H. Ginsberg. H5*dl*309 was selected as a variant that contains only one *Xba*I cleavage site, located 1339 bp from the left end of the chromosome, and it displays a wild-type phenotype (Jones and Shenk 1979). H5*dl*338 carries a 524-bp deletion located between nucleotide sequence positions 2805 and 3329, and H5*pm* 380 contains a G-to-A substitution at sequence position 3275. Mutations were originally constructed in a recombinant plasmid, pA5XhoI-C (Hearing and Shenk 1983), and alterations were verified by nucleotide sequence analysis. The mutated Ad5 segments were rebuilt into an intact viral chromosome (*dl* 309) by overlap recombination (Chinnadurai et al. 1979).

The 293 cell line (a human embryonic kidney cell line transformed with a DNA fragment carrying the left 11% of the Ad5 genome) was obtained from H. Young and has been described by Graham et al. (1977). Cells were maintained in medium containing 10% calf serum. Monolayer culture HeLa cells were obtained from the American Type Culture Collection and grown in medium supplemented with 7.5% calf serum. All infections were at a multiplicity of 20 pfu/cell.

*DNA replication*
DNA replication rates were assayed by labeling for 60 min with [$^3$H]thymidine (100 μCi/ml, 50 Ci/mmole) at various times after infection of HeLa cells. Total DNA was extracted and analyzed by electrophoresis after digestion with the *Hin*dIII endonuclease.

*RNA preparation and analyses*
Protocols for cytoplasmic poly(A)$^+$ RNA isolation from HeLa cells and RNA blot analyses have been described (Hearing and Shenk 1983). The Ad5 region L5–specific probe DNA was a recombinant M13 DNA carrying the segment of the viral chromosome from 89.1 to 92 map units. The actin probe was a recombinant plasmid carrying chicken sequences (pA1; Cleveland et al. 1980).

Transcription rates were measured in isolated nuclei essentially as described by Hofer and Darnell (1981) and Groudine et al. (1981). Nuclei prepared from infected cells were incubated for 15 min at 30°C in the presence of [$^{32}$P]UTP (750 μCi/ml, 410 Ci/mmole), and nuclear RNA was isolated, degraded by treatment with 0.2 N NaOH for 10 min at 4°C, and hybridized to region-specific, single-stranded DNA probes bound to nitrocellulose filters (100-μg genome equivalents/filter) by the method of McKnight and Palmiter (1979). The probe DNAs were as described above. After the first round of hybridization, a second DNA-containing filter was added to each mix to ensure quantitative results.

Cytoplasmic accumulation of mRNAs was measured using previously described procedures (Greenberg 1972; Babich and Nevins 1981; Babich et al. 1983). In brief, infected cells were labeled with [$^3$H]uridine (200 μCi/ml, 50 Ci/mmole) in the presence of added unlabeled uridine (14 μM). One infected monolayer culture was harvested at each 30-min interval, and cytoplasmic RNA was prepared using the guanidinium isothiocyanate procedure of Chirgwin et al. (1979), degraded by treatment with 0.2 N NaOH at 4°C for 10 min, and then hybridized to the region-specific, single-stranded DNA probes as described for nuclear RNAs.

*Analysis of polypeptides*
One hour before labeling, infected HeLa cells were placed in medium lacking methionine and supplemented with 2% calf serum. Cultures were labeled with [$^{35}$S]methionine (50 μCi/ml, 1100 Ci/mmole) for 30 min unless otherwise noted. Preparation of cellular extracts, immunoprecipitations, and SDS-polyacrylamide gel electrophoresis were carried out as described by Sarnow et al. (1982b). The E1A- and E2A-specific monoclonal antibodies were gifts of E. Harlow and A. Levine, respectively.

## Results

### The H5*dl* 338 phenotype is due to alteration of the E1B 55K but not 17K polypeptide
The alteration in *dl* 338 prevents synthesis of the E1B 55K and 17K polypeptides but not the E1B 21K species (data not shown). To determine whether the missing 17K polypeptide contributes to the *dl* 338 phenotype, a second variant was constructed. *dl* 380 carries a single base-pair change (G to A at sequence position 3275), which destroys the 3' splice site required for synthesis of the E1B 17K-coding mRNA (Fig. 1) without affecting the amino acid sequence of the 55K polypeptide. This mutant grows indistinguishably from wild-type virus (data not shown). Apparently, the 17K polypeptide encodes no function required for growth in cultured cells that is not carried out by the related 55K species. It is possible that the 17K product could at least partially compensate for the lack of the 55K polypeptide, but our present mutants do not speak to that issue. Consistent with our results, Montell et al. (1984) have mutated the 5' splice site required for synthesis of the mRNA encoding the 17K polypeptide and found the variant to be phenotypically wild type.

### H5*dl* 338 early gene expression and DNA replication are normal
To assay early gene expression, RNAs were prepared from HeLa cells at 6 hr after infection with either *dl* 338 or *dl* 309. RNA blot analysis revealed no differences in the variety or levels of E1A-, E2A-, or E4-specific mRNAs in mutant as compared with wild-type virus–infected cells (data not shown). Further, the E2A 72K polypeptide, as judged by immunoprecipitation, was produced in normal quantities early after infection of HeLa cells with *dl* 338 (data not shown). Thus, it appears that early gene expression proceeds normally in the absence of the E1B 55K and 17K polypeptides.

Viral DNA replication was measured by labeling cells with [$^3$H]thymidine for short periods at various times after infection, extracting and cleaving total DNA with a restriction endonuclease, and subjecting the resulting fragments to electrophoresis. Radioactivity in viral DNA fragments was then detected by fluorography (Fig. 2). By this assay, mutant and wild-type DNA replication are indistinguishable.

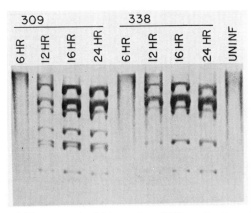

**Figure 2** Mutant and wild-type virus DNA replication. HeLa cells were labeled for 60 min with [³H]thymidine at the times indicated after infection. Total cellular DNA was prepared, cleaved with HindIII, and subjected to electrophoresis in an agarose gel. DNA fragments were visualized by fluorography.

## Accumulation of H5dl338-encoded mRNAs is aberrant

Steady-state levels of cytoplasmic mRNAs were monitored by blot analysis at 12, 16, and 24 hr after infection. At all times tested, dl338-infected cells contained reduced levels of mRNAs coded by the viral major late transcription unit. Some late mRNA families were more severely affected than others. For example, the L4 unit was reduced only twofold or threefold (data not shown) whereas the L5 unit was reduced by a factor of about 15 (Fig. 3A).

To determine the basis for abnormal mRNA levels observed in dl338-infected cells, transcription rates were measured in isolated nuclei. Nascent RNA synthesized in isolated nuclei was labeled with [³²P]UTP, and total nuclear RNA was prepared and hybridized to an L5-specific probe DNA (Fig. 4A). Transcription rates for the L5 mRNA family were identical at 12 and 16 hr, but at 20 and 24 hr the late RNAs were transcribed at about fourfold higher levels in wild-type than in mutant-infected cells. Reduced transcription rates can at least partially account for the lower mRNA levels in dl338-infected

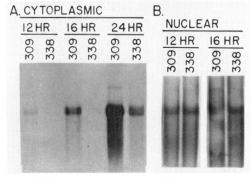

**Figure 3** RNA blot analysis of viral mRNA species present in HeLa cells after infection with mutant or wild-type viruses. Polyadenylated RNA was prepared at various times after infection from either cytoplasm (A) or nucleus (B) and analyzed using an L5-specific probe DNA.

cells at 24 hr after infection. However, transcription rates are normal at 12 and 16 hr when cytoplasmic steady-state mRNA levels are already reduced. Therefore, the earliest perturbation of late mRNA levels in dl338-infected cells must be due to a posttranscriptional event.

To elucidate the nature of the posttranscriptional event, the rate at which newly synthesized mRNAs accumulated in the cytoplasm of virus-infected cells was monitored. Cultures were labeled continuously with [³H]uridine from 12 to 14.5 hr after infection, and the rate of appearance of labeled RNA in the infected-cell cytoplasm was monitored by hybridization to a specific probe DNA (Fig. 4B). Transcription rates are identical for mutant and wild-type virus at this time (Fig. 4A), so the slopes of accumulation plots are a direct measure of relative accumulation rates. L5 mRNAs were transported at a threefold-reduced rate in dl338 as compared with wild-type virus–infected cells.

The reduced rate of late mRNA transport in dl338-infected cells could result from a primary defect in nuclear RNA processing. If polyadenylation were compromised, less RNA should be present in the poly(A)⁺ nuclear fraction; if splicing were inhibited, less mature-sized RNA should be present and large precursors might be evident. Blot analysis of polyadenylated nuclear RNA using an L5-specific probe DNA indicates that the same quantities of similarly sized RNA species are present in mutant as compared with wild-type virus–infected cells (Fig. 3B). The presumptive nuclear species cannot be cytoplasmic contaminants since there is little L5-specific cytoplasmic RNA available to contaminate the nuclear fraction in dl338-infected cells. Thus, the reduced rate of cytoplasmic accumulation is not due to a processing defect.

Finally, the apparent reduction in accumulation rate of L5 mRNAs within dl338-infected cells could result from a reduced mRNA half-life. The accumulation plot in Figure 4B does not reach a plateau during the labeling period, suggesting that the mRNAs under study had a half-life in excess of 150 min. However, it is difficult to reach definitive conclusions about viral mRNA half-lives on the basis of the accumulation data since the rate of viral transcription was increasing during the labeling period. Therefore, a pulse-chase experiment was performed to compare the half-lives of mRNAs produced in mutant and wild-type virus–infected cells (Fig. 4C). Cells were radiolabeled for 30 min with [³H]uridine, and the chase was carried out in medium containing high levels of unlabeled uridine and glucosamine. The stability of L5 mRNAs was monitored at 45 min intervals by hybridization of cytoplasmic, polyadenylated RNA to an L5-specific probe DNA. L5 mRNAs displayed the same half-life (~3.2 hr) in dl338- and dl309-infected cells. Therefore, it is unlikely that the dl338 phenotype is due to a primary effect on mRNA half-lives.

Polyadenylation, splicing, and cytoplasmic L5 mRNA half-lives are normal in dl338-infected cells at 12–16 hr after infection. We conclude that the abnormal viral mRNA accumulation rates observed at this time in mu-

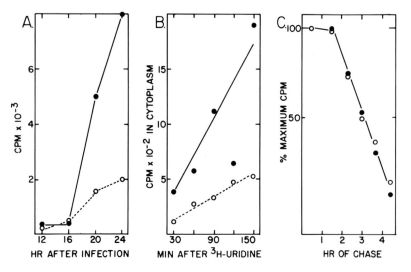

**Figure 4** L5-specific RNA metabolism in mutant or wild-type virus–infected HeLa cells. (*A*) Analysis of transcription rates. Nuclei were prepared from infected cells at the indicated times after infection and incubated for 15 min at 30°C in the presence of [$^{32}$P]UTP, and nuclear RNA was prepared. (*B*) Kinetics of mRNA accumulation in the cytoplasm. Cultures were labeled with [$^3$H]uridine beginning at 12 hr after infection, portions were harvested every 30 min, and cytoplasmic RNA was prepared. (*C*) Half-life determination of mRNAs. Cells were labeled with [$^3$H]uridine for 30 min at 12 hr after infection, chased in the presence of excess uridine and glucosamine, and cytoplasmic, polyadenylated RNA was isolated. All experiments utilized an L5-specific, single-stranded probe DNA. (●) *dl*309; (○)*dl*338.

tant-infected cells result from a defect in either transport or stabilization of mRNA as it first reaches the cytoplasm.

## Accumulation of cellular mRNA is not shut off in H5*dl*338-infected cells

Normally, host-cell mRNAs continue to be transcribed but fail to appear in the cytoplasm of adenovirus-infected cells (Beltz and Flint 1979). As expected, transcription of the cellular actin gene continued in both wild-type- and *dl*338-infected cells (Fig. 5A). However, actin mRNA continued to accumulate in the cytoplasm of *dl*338-infected cells after its appearance had ceased in the case of the wild type (Fig. 5B). Thus, the E1B 55K polypeptide plays a role in host-cell shutoff. This con-

clusion is consistent with the work of Babiss and Ginsberg (1984), who have described E1B mutants similar to *dl*338 and demonstrated their failure to shut off host-cell protein synthesis.

## Polypeptide synthesis late after *dl*338 infection

In general, the pattern of polypeptide synthesis mimicked late mRNA levels. Late viral polypeptides were reduced and cellular products continued to be synthesized in virus-infected cells (data not shown). However, even though early mRNA levels were normal late after infection, at least some (E1A and E2A were tested) and possibly all early polypeptides were synthesized at elevated levels (~eightfold) late after infection with the mutant virus (Fig. 6).

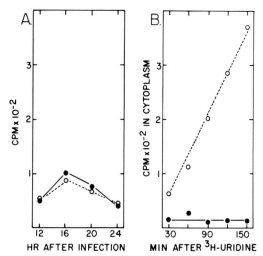

**Figure 5** Actin-specific RNA metabolism in mutant or wild-type virus–infected HeLa cells. (*A*) Analysis of transcription rates. (*B*) Kinetics of mRNA accumulation in the cytoplasm. Accumulation was measured 16–18.5 hr after infection. Experiments utilized an actin-specific (pA1) probe DNA. (●) *dl*309; (○)*dl*338.

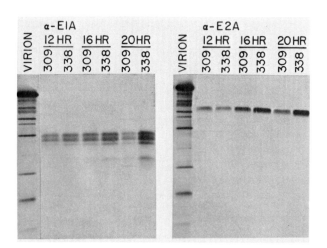

**Figure 6** Electrophoretic analysis of E1A- and E2A-specific polypeptides immunoprecipitated from infected HeLa cells. Extracts were prepared at indicated times and subjected to immunoprecipitation using a monoclonal antibody specific to either the E1A or E2A polypeptide. Electrophoresis was in a 12.5% polyacrylamide gel containing SDS.

## Discussion

The *dl*338 phenotype is complex. Early gene expression and DNA replication appear normal. Transcription rates across the major late unit are normal at 12 and 16 hr after infection with *dl*338 but fall off sharply relative to the wild type at 20 and 24 hr (Fig. 4A). Although nuclear mRNA processing (Fig. 3B) and mRNA half-lives (Fig. 4C) appear normal, the rate at which late mRNAs are transported and accumulate in the cytoplasm is reduced as early as 12 hr after infection (Fig. 4B). Not surprisingly, then, steady-state levels of mRNAs encoded by the major late transcription unit are reduced in *dl*338 as compared with wild-type virus–infected cells at all times tested (Fig. 3A). The picture is quite the opposite for the cellular actin mRNA (Fig. 5). In wild-type virus–infected cells, transport of this mRNA is terminated at 16 hr after infection. In contrast, transport continues within mutant-infected cells. Finally, translation appears to faithfully track mRNA transport and accumulation for both late viral mRNAs (synthesis of late polypeptides is reduced) and cellular mRNAs (host-cell translation is not shut off in *dl*338-infected cells). At least some early polypeptides are overproduced by 20 hr after infection with *dl*338 (Fig. 6), even though their mRNA levels are normal. Presumably, the effects of these aberrations on the dynamics of virus multiplication are additive since no one alteration appears to be of sufficient magnitude to account for the 100-fold reduction in *dl*338 yield subsequent to infection of HeLa cells.

What is the primary defect responsible for the *dl*338 phenotype? Transport and accumulation of both viral and host-cell mRNAs were clearly abnormal in *dl*338-infected cells at 12 hr after infection, when transcription rates for the RNAs and all other parameters of the virus infection tested were still normal. Thus, the temporal sequence in which abnormalities of *dl*338 metabolism appear indicates that the E1B 55K polypeptide functions both to facilitate accumulation of late viral mRNAs and to prevent accumulation of host-cell mRNAs. The nuclear plus perinuclear location of the E1B 55K polypeptide (Sarnow et al. 1982a; Yee et al. 1983) is consistent with this role. Several other research groups (Babiss et al. 1985; Williams et al., this volume) have studied different adenovirus mutants and similarly concluded that E1B products are required for efficient cytoplasmic accumulation of late viral mRNAs.

There are two possible explanations for the reduced rate at which late viral mRNAs accumulate in the cytoplasm of *dl*338-infected cells. Possibly, the rate of transport itself is reduced. Since late viral RNAs do not accumulate to abnormally high levels in the nucleus, they would necessarily be degraded within the nucleus. Alternatively, mRNAs may be transported at a normal rate but not stabilized and immediately degraded within the cytoplasm. If degradation occurs within the cytoplasm, it cannot simply result from the generalized activation of a cytoplasmic nuclease since other mRNAs (early viral and host cell) accumulate to normal levels. Degradation could result from incorrect transport or positioning of

the late mRNAs within the cytoplasm or from activation of a nuclease that specifically degrades late viral mRNAs. In fact, the E1B 55K polypeptide might redirect this nuclease to cellular mRNAs.

The interrelationship of late viral and host-cell mRNA transport and accumulation is intriguing. The E1B 55K polypeptide may play a direct role in both processes, actively facilitating viral and inhibiting host-cell mRNA accumulation. It is also possible that the viral product directly mediates one process whereas the other is modulated indirectly. For example, facilitated transport of viral mRNAs may lead to a dramatic reduction in host-cell mRNA accumulation due to competition for limited transport machinery. This notion is consistent with the observation that adenovirus must enter the late phase of its growth cycle for shutoff to occur.

The accumulation of some late viral mRNAs in *dl*338-infected cells is more severely affected than others (e.g., the accumulation gradient is L5 mRNA < L3 mRNA < L4 mRNA). Curiously, the mRNAs most reduced encode polypeptides whose production normally begins latest during the growth cycle of the wild-type virus. Translation of the L4 100K polypeptide turns on early in the late phase, whereas the L5 IV species turns on later (Bablanian and Russell 1974). Clearly, some mRNAs are more dependent than others on E1B 55K polypeptide function for transport and accumulation.

On the basis of the time of appearance, the reduced rates of late mRNA transcription observed at 20 and 24 hr after infection with *dl*338 as well as overexpression of early polypeptides are likely indirect effects of the *dl*338 mutation. The basis for altered transcription rates is unclear. Transcription and transport might somehow be coupled and inefficient transport could feed back on the transcription process. It is also possible that an overproduced early gene product or underproduced late product generates the transcriptional abnormality. The overproduction of early polypeptides must occur at the level of translation since early mRNA levels are normal. Increased synthesis likely results from the lack of competition by late viral mRNAs. E2A 72K polypeptide overproduction has been noted previously in E1B mutant–infected cells (Lassam et al. 1978; Ross et al. 1980).

We have constructed Ad5 mutants that fail to produce the E4 34K polypeptide (Halbert et al. 1985) and display lytic growth phenotypes similar in many respects to that described here for *dl*338. The similarity makes sense because the E1B 55K and E4 34K polypeptides exist in a physical complex (Sarnow et al. 1984). The similar phenotypes indicate that these two polypeptides probably do function as a complex. It is interesting to note that wild-type adenovirus displays a growth defect in semipermissive mouse cells that appears similar to that of *dl*338 in human cells (Cheng and Praszkier 1982). Conceivably, the poor growth in mouse cells is due to the presence of the cellular tumor antigen, p53, which complexes with the E1B 55K polypeptide. This could preclude formation of sufficient E1B-55K/E4-34K complex. We have noted one difference between *dl*338 and E4 mutant phenotypes: E4 mutants exhibit reduced levels of viral DNA replication. Although *dl*338 appears normal, an E1B mutant designated *hr*7 (Harrison et al. 1977) exhibits a delayed DNA replication phenotype (Stillman 1983). These observations raise the possibility that the E1B 55K polypeptide plays a role in DNA replication. If this is true, the function may reside in the aminoterminal E1B 55K fragment, which *dl*338 can, in theory, produce.

Mutant *dl*338 is markedly defective for transformation of rat cells (Logan et al. 1984). It is not clear whether its transforming defect is mechanistically related to its lytic defect. It is certainly possible that the E1B polypeptide exerts its oncogenic activity by perturbing transport and accumulation of host-cell mRNAs. Alternatively, its function may be modified by association with the p53 tumor antigen of rodent cells (Sarnow et al. 1982b).

## Acknowledgments

We thank E. Harlow and A. Levine for gifts of E1A- and E2A-specific monoclonal antibodies, respectively. This work was supported by grants from the American Cancer Society (MV-45) and the National Cancer Institute (CA 38965). K.L. is the recipient of an EMBO Postdoctoral Fellowship, and J.L. was supported by a Postdoctoral Fellowship from the Ohtsuka Pharmaceutical Company.

## References

Anderson, C.W., R.C. Schmitt, J.E. Smart, and J.B. Lewis. 1984. Early region 1B of adenovirus 2 encodes two coterminal proteins of 495 and 155 amino acid residues. *J. Virol.* **50:** 387.

Babich, A. and J.R. Nevins. 1981. The stability of early adenovirus mRNA is controlled by the viral 72 kd DNA-binding protein. *Cell* **26:** 371.

Babich, A., L.T. Feldman, J.R. Nevins, J.E. Darnell, Jr., and C. Weinberger. 1983. Effect of adenovirus on metabolism of specific host mRNAs: Transport control and specific translational discrimination. *Mol. Cell. Biol.* **3:** 1212.

Babiss, L.E. and H.S. Ginsberg. 1984. Adenovirus type 5 early region 1b gene product is required for efficient shut off of host protein synthesis. *J. Virol.* **50:** 202.

Babiss, L.E., H.S. Ginsberg, and J.E. Darnell, Jr. 1985. Adenovirus E1B proteins are required for accumulation of late viral mRNA and for effects on cellular mRNA translation and transport. *Mol. Cell. Biol.* **5:** 2552.

Bablanian, R. and W.C. Russell. 1974. Adenovirus polypeptide synthesis in the presence of non-replicating poliovirus. *J. Gen. Virol.* **24:** 261.

Beltz, G.A. and S.J. Flint. 1979. Inhibition of HeLa cell protein synthesis during adenovirus infection: Restriction of cellular mRNA sequences to the nucleus. *J. Mol. Biol.* **131:** 353.

Bos, J.L., L.J. Polder, R. Bernards, P.I. Schrier, P.J. van den Elsen, A.J. van der Eb, and H. van Ormondt. 1981. The 2.2 kb E1B mRNA of human Ad12 and Ad5 codes for two tumor antigens starting at different AUG triplets. *Cell* **27:** 121.

Cheng, C. and J. Praszkier. 1982. Regulation of type 5 adenovirus replication in murine teratocarcinoma cell lines. *Virology* **123:** 45.

Chinnadurai, G., S. Chinnadurai, and J. Brusca. 1979. Physical mapping of a large-plaque mutation of adenovirus type 2. *J. Virol.* **32:** 623.

Chirgwin, J.M., A.E. Przybyla, R.J. MacDonald, and W.J. Rutter. 1979. Isolation of biologically active RNA from sources enriched in ribonuclease. *Biochemistry* **18:** 5294.

Cleveland, D.W., M.A. Lopata, R.J. MacDonald, N.J. Cowan, W.J. Rutter, and M.W. Kirschner. 1980. Number and evolutionary conservation of alpha- and beta-tubulin and cytoplasmic beta- and gamma-actin genes using specific cloned cDNA probes. *Cell* **20:** 95.

Darnell, J.E. 1982. Variety in the level of gene control in eukaryotic cells. *Nature* **297:** 365.

Gingeras, T.R., D. Sciaky, R.E. Gelina, J. Bing-Dong, C.E. Yen, M.M. Kelly, P.A. Bullock, B.L. Parsons, K.E. O'Neill, and R.J. Roberts. 1982. Nucleotide sequences from the adenovirus 2 genome. *J. Biol. Chem.* **257:** 13457

Graham, F.L. 1984. Transformation by and oncogenicity of human adenoviruses. In *The adenoviruses* (ed. H.S. Ginsberg), p. 339. Plenum Press, New York.

Graham, F.L., J. Smiley, W.C. Russell, and R. Nairu. 1977. Characteristics of a human cell line transformed by DNA from human adenovirus type 5. *J. Gen. Viol.* **36:** 59.

Greenberg, J.R. 1972. High stability of mRNA in growing cultured cells. *Nature* **240:** 102.

Groudine, M., M. Peretz, and H. Weintraub. 1981. Transcriptional regulation of hemoglobin switching in chicken embryos. *Mol. Cell. Biol.* **1:** 281.

Harrison, T., F. Graham, and J. Williams. 1977. Host range mutants of adenovirus type 5 defective for growth in HeLa cells. *Virology* **77:** 319.

Hearing, P. and T. Shenk. 1983. The adenovirus type 5 E1A transcriptional control region contains a duplicated enhancer element. *Cell* **33:** 695.

Hofer, E. and J.E. Darnell. 1981. The primary transcription unit of the mouse β-major globin gene. *Cell* **23:** 585.

Jones, N. and T. Shenk. 1979. Isolation of Ad5 host range deletion mutants defective for transformation of rat embryo cells. *Cell* **17:** 683.

Katze, M.G., H. Persson, and L. Philipson. 1983. A novel mRNA and a low molecular weight polypeptide encoded in the transforming region of adenovirus DNA. *EMBO J.* **1:** 783.

Lassam, N.J., S.T. Bayley, and F.L. Graham. 1978. Synthesis of DNA late polypeptides, and infectious virus by host-range mutants of adenovirus 5 in nonpermissive cells. *Virology* **87:** 463.

Logan, J., S. Pilder, and T. Shenk. 1984. Functional analysis of adenovirus type 5 early region 1B. *Cancer Cells* **2:** 527.

McKnight, G.S. and R.D. Palmiter. 1979. Transcriptional regulation of the ovalbumin and conalbumin genes by steroid hormones in chick oviduct. *J. Biol. Chem.* **254:** 9050.

Montell, C., E.F. Fisher, M.H. Caruthers, and A.J. Berk. 1984. Control of adenovirus E1B mRNA synthesis by a shift in the activities of RNA splice sites. *Mol. Cell. Biol.* **4:** 966.

Ross, S.R., A.J. Levine, R.S. Galos, J. Williams, and T. Shenk. 1980. Early viral proteins in HeLa cells infected with adenovirus type 5 host range mutants. *Virology* **103:** 475.

Sarnow, P., C.A. Sullivan, and A.J. Levine 1982a. A monoclonal antibody detecting the Ad5 E1B-58K tumor antigen. Characterization of the E1B-58K tumor antigen in adenovirus-infected and transformed cells. *Virology* **120:** 510.

Sarnow, P., Y.S. Ho, J. Williams, and A.J. Levine. 1982b. Adenovirus E1B-58kd tumor antigen and SV40 large tumor antigen are physically associated with the same 54kd cellular protein in transformed cells. *Cell* **28:** 387.

Sarnow, P., P. Hearing, C.W. Anderson, D.N. Halbert, T. Shenk, and A.J. Levine. 1984. Adenovirus early region 1B 58,000-dalton tumor antigen is physically associated with an early region 4 25,000-dalton protein in productively infected cells. *J. Virol.* **49:** 692.

Shenk, T. and J. Williams. 1984. Genetic analysis of adenoviruses. *Curr. Top. Microbiol. Immunol.* **111:** 1.

Stillman, B. 1983. The replication of adenovirus DNA. *UCLA Symp. Mol. Cell. Biol.* **10:** 381.

Yee, S.-P., D.T. Rowe, M.L. Tremblay, M. McDermott, and P.E. Branton. 1983. Identification of human adenovirus early region 1 products by using antisera against synthetic peptides corresponding to the predicted carboxy termini. *J. Virol.* **46:** 1003.

# The Control of Protein Synthesis by Adenovirus VA RNA

R.P. O'Malley,* T.M. Mariano,[†] J. Siekierka,[†‡] W.C. Merrick,[§] P.A. Reichel,* and M.B. Mathews*

*Cold Spring Harbor Laboratory, Cold Spring Harbor, New York 11724; [†]Department of Biochemistry, The Roche Institute of Molecular Biology, Nutley, New Jersey 07110; [§]Department of Biochemistry, Case Western Reserve University, Cleveland, Ohio 44106

VA RNA$_I$ is required for efficient protein synthesis in adenovirus-infected cells. The Ad5 deletion mutant dl331 fails to synthesize VA RNA$_I$ and shows a 10-fold reduction in growth as a result of an inhibition of protein synthesis at late times of infection. The defect is due to the activation of a protein kinase that phosphorylates the α subunit of initiation factor eIF-2. The phosphorylated form of eIF-2 traps a second factor that is needed for eIF-2 recycling and blocks the earliest stage of polypeptide chain initiation. We have identified the kinase responsible for the translational defect as DAI, the double-stranded RNA (dsRNA)–activated inhibitor of protein synthesis. This enzyme is present in HeLa cells at a basal level in an inactive form and becomes activated in the presence of low concentrations of dsRNA. DAI synthesis can be induced by interferon, but interferon does not appear to be involved in the protein synthesis block in dl331-infected HeLa cells. During the late phase of dl331 infection, kinase activity increases dramatically, and RNA isolated from infected cells is capable of activating the enzyme. The source of the presumed dsRNA activator appears to be the symmetrical transcription of the adenovirus genome. In extracts of interferon-treated HeLa cells, VA RNA$_I$ blocks activation of DAI by dsRNA. We propose that although VA RNA$_I$ cannot activate DAI, its duplex regions are of sufficient length to allow it to interact with the enzyme and compete with dsRNA, thereby preventing activation of the enzyme. A model for DAI activation is described.

The adenovirus genome contains two genes that encode the virus-associated (VA) RNAs, VA RNA$_I$ and VA RNA$_{II}$ (Reich et al. 1966; Mathews 1975; Pettersson and Philipson 1975; Mathews and Pettersson 1978; Akusjärvi et al. 1980). The genes are arranged in tandem and are located between map units 29 and 31 on the viral chromosome (Mathews 1975; Mathews and Pettersson 1978; Akusjärvi et al. 1980). Unlike the other viral transcription units, which are transcribed by RNA polymerase II, VA RNA$_I$ and VA RNA$_{II}$ are synthesized by RNA polymerase III (Price and Penman 1972; Weinmann et al. 1974; Söderlund et al. 1976). Both are present from early times of infection, but the synthesis of VA RNA$_I$ increases dramatically with the onset of the late phase and this species becomes the predominant cytoplasmic RNA of infected cells (Söderlund et al. 1976). Although synthesis of VA RNA$_{II}$ continues, it is less pronounced than VA$_I$ and the final levels are only 2–5% of the level of VA RNA$_I$. The VA RNAs are approximately 160 nucleotides in length and are capable of extensive intramolecular base-pairing, which gives rise to stable secondary structures (Akusjärvi et al. 1980; Monstein and Philipson 1981). The VA RNAs can also form intramolecularly paired structures with late viral mRNA

(Mathews 1980). A small percentage ($<5\%$) of the VA RNAs is found in ribonucleoprotein particles with the La antigen (Lerner et al. 1981; Francoeur and Mathews 1982), a cellular protein that binds to short runs of 3'-terminal U residues characteristic of RNA polymerase III transcripts (Mathews and Francoeur 1984; Stefano 1984). The significance of these interactions is not yet understood.

VA RNA$_I$ is required for the efficient translation of both viral and cellular mRNAs at late times of infection. The adenovirus type-5 (Ad5) mutant dl331, which fails to make VA RNA$_{II}$, shows a 10-fold depression in protein synthesis at 24 hr postinfection (hpi). The mutant virus proceeds through the early phase of infection but fails to make late viral proteins, even though undiminished levels of late viral mRNAs are present. These mRNAs are normal in structure and are capable of efficient translation in a heterologous cell-free system, suggesting that the defect lies in the translational apparatus (Thimmappaya et al. 1982). Subsequent work by Schneider et al. (1984) excluded defects in polypeptide chain elongation or termination and led to the conclusion that the defect occurs at the level of chain initiation. The same conclusion was reached by Reichel et al. (1985) from studies conducted in vitro. In this report we describe the biochemical basis for the translational defect and recent findings that elucidate the role of VA RNA$_I$ in regulating viral mRNA translation.

Present address: Merck, Sharp and Dohme Research Laboratories, Department of Immunology, Box 2000, Rahway, New Jersey 07065.

## Materials and Methods

### Cells and viruses

HeLa cells were grown in suspension culture, infected with *dl*331 or wild-type (Ad2) virus, and labeled with [$^{35}$S]methionine as previously described (Reichel et al. 1985). Where indicated, cells were treated for 24 hr with α-interferon at 200 units/ml to induce DAI.

### Preparation of extracts

Extracts for protein synthesis were prepared from HeLa cells grown in suspension culture as described previously (Reichel et al. 1985). Endogenous RNA was removed where indicated by treating the extracts with 75 units/ml of micrococcal nuclease in the presence of 1 mM CaCl$_2$ for 15 min at 20°C. The nuclease was then inactivated by the addition of EGTA to a final concentration of 3 mM. Initiation of protein synthesis was blocked as indicated by the addition of edeine to a final concentration of 25 μM. HeLa extracts intended for kinase assays only were prepared by lysis with Nonidet P-40 instead of homogenization. Cell-free translation reactions using the micrococcal nuclease–treated reticulocyte lysate system (Pelham and Jackson 1976; Dunn et al. 1978) were primed with mRNA present in HeLa cell extracts or with mRNA isolated from HeLa cell extracts by deproteinization. In the former case, in the small amounts used, any activated kinase present in the extract did not interfere with protein synthesis (Reichel et al. 1985; Siekierka et al. 1985). Protein concentrations were determined by the method of Bradford (1976).

### Phosphopeptide and phosphoamino acid analysis

The α subunit of purified eIF-2 (eIF-2α) was phosphorylated with [γ-$^{32}$P]ATP in vitro by the kinase present in *dl*331 extracts or by crude heme-controlled inhibitor present in heated reticulocyte lysates, using the standard kinase assay conditions (see below). Phosphorylated eIF-2α was isolated from a 10% SDS-polyacrylamide gel by homogenization, elution, and precipitation with trichloroacetic acid. The pellet was washed with acetone, resuspended in 100 μl of 10 mM Tris · HCl (pH 7.4), 1 mM EDTA and aliquots were partially digested with *Staphylococcus aureus* V8 protease (Cleveland et al. 1977). For phosphoamino acid analysis, phosphorylated eIF-2α was hydrolyzed with 5.7 M HCl and subjected to two-dimensional thin-layer electrophoresis using the procedure of Cooper et al. (1983).

### Assays for eIF-2 kinase and oligoadenylate synthetase

Two types of protein kinase were used, the standard assay (Siekierka et al. 1985) and a modified version of a cellulose–poly(I):poly(C) binding assay of Hovanessian and Kerr (1979), which has been described in detail elsewhere (O'Malley et al. 1986). Purified eIF-2 (Benne et al. 1976) was added as indicated. 2′,5′-Oligoadenylate synthetase assays were performed following the protocol of Revel et al. (1981) with the following exceptions: After diluting each reaction mixture to 400 μl with 1 M glycine-HCl (pH 2), 200 μl was applied to a 0.3-ml alumina column equilibrated with this buffer. Each column was washed with 3 ml of 1 M glycine (pH 2) to elute the $^{32}$P-labeled 2′,5′-oligoadenylate; 200 μl of the eluate was added to 20 ml of Aquasol (New England Nuclear) and counted in a Beckman LS7000 scintillation counter.

### Isolation of cytoplasmic RNA and VA RNA$_I$

Cytoplasmic RNA was isolated either from cells or from cell extracts by phenol extraction (Anderson et al. 1974). Northern blot analysis and isolation of VA RNA$_I$ and tRNA were as described by O'Malley et al. (1986).

## Results

### The mechanism of translational inhibition

In the absence of VA RNA$_I$, protein synthesis is severely depressed at late times in the infection of HeLa cells by adenovirus, as illustrated in Figure 1A. Protein synthesis appeared normal during the early phase of infection with *dl*331, but with the onset of late transcription at approximately 12 hpi, protein synthesis in *dl*331-infected cells began to decline when compared with Ad2-infected cells and was only 10% of the level observed in wild-type-infected cells at 24 hpi. In contrast, the mRNA present in cytoplasmic extracts prepared from *dl*331-infected HeLa cells was as efficient in programming translation in the reticulocyte lysate system as the mRNA present in extracts of wild-type-infected cells (Fig. 1B). Thus, the viral mRNA present in *dl*331-infected cells at late times of infection is normal in structure and abundance (Thimmappaya et al. 1982) and is not irreversibly sequestered (Reichel et al. 1985). In cells infected with wild-type virus, viral mRNAs are translated, whereas the translation of cellular messages is largely inhibited. During *dl*331 infection, viral protein synthesis is blocked but the inhibition of host mRNA translation is not relieved. The inhibition of protein synthesis affects both viral and cellular protein synthesis (Reichel et al. 1985).

Our first step in defining the translational defect was the development of a cell-free protein-synthesizing system from adenovirus-infected HeLa cells (Reichel et al. 1985). Extracts of *dl*331-infected cells were only 1–5% as efficient in protein synthesis from endogenous mRNA as extracts prepared from Ad2-infected or uninfected cells (Fig. 2, lanes 13, 7, and 1). The pattern and relative extent of protein synthesis was similar to that seen when the corresponding cultures were pulse-labeled in vivo (Fig. 2, lanes 19, 20, and 21). After removal of endogenous mRNA with micrococcal nuclease, RNA from Ad2-infected cells programmed translation in mock and Ad2 extracts much more efficiently than in *dl*331 extracts (Fig. 2, lanes 5, 11, and 17). It should be noted that none of these HeLa cell systems, especially the nuclease-treated ones, approached the reticulocyte lysate in translational efficiency. The residual protein synthesis observed in extracts of *dl*331-infected cells was largely dependent on reinitiation as shown by sensitivity to edeine, an inhibitor of polypeptide chain initiation. Less than 50% of [$^{35}$S]methionine incorporation was edeine-sensitive in mock- and Ad2-infected cells (Fig. 2, lanes 3

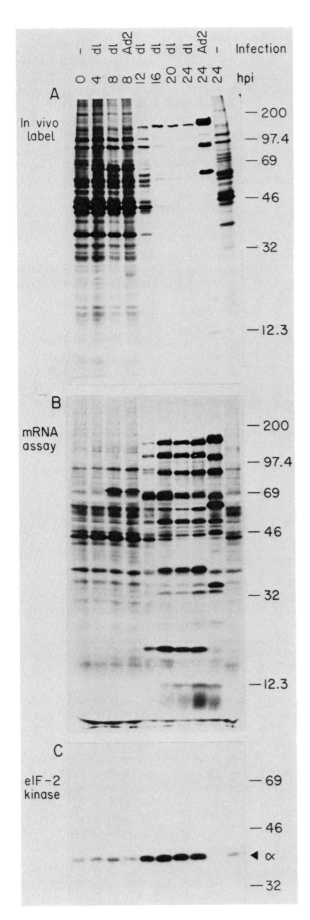

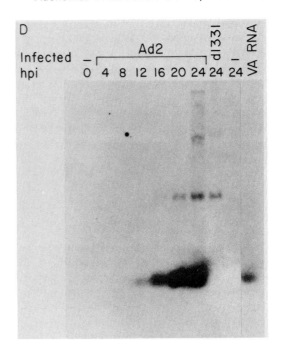

**Figure 1** Kinetics of protein synthesis, mRNA accumulation, kinase activation, and VA RNA$_I$ appearance in infected HeLa cells. HeLa cells in suspension culture were mock-infected or infected with Ad2 or *dl*331 at a multiplicity of infection of 5. At the times shown, portions of the cultures were removed for labeling with [$^{35}$S]methionine or preparation of cell extracts (Reichel et al. 1985). (*A*) Proteins labeled in vivo, analyzed by SDS-gel electrophoresis. (*B*) Polypeptides synthesized in vitro in 12.5 μl of mRNA-dependent reticulocyte lysate translation reactions primed with 0.5 μl of HeLa cell extract. (*C*) eIF-2 kinase assayed in the standard reaction using 0.5 μl of extract (6.6–9.4 μg protein) per reaction. (*D*) VA RNA$_I$ analysis by northern blot, probed with pBalM (see O'Malley et al. 1986). The lane marked VA RNA contained 20 ng of purified VA RNA$_I$. (Reprinted, with permission, from O'Malley et al. 1986.)

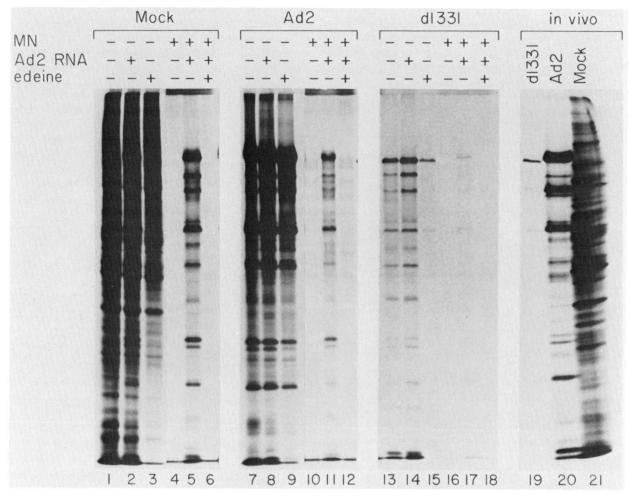

**Figure 2** Properties of cell-free systems. Proteins synthesized in vitro by extracts of mock-infected (lanes *1–6*), Ad2-infected (lanes *7–12*), and *dl*331-infected (lanes *13–18*) HeLa cells are compared with protein labeled in vivo in mock-infected, Ad2-infected, and *dl*331-infected HeLa cells (lanes *19–21*). Extracts were pretreated with micrococcal nuclease, and Ad2 mRNA and edeine were added as indicated. Lanes *1–3* and *7–8* represent a lighter photographic print than the other lanes.

and 9), whereas more than 75% of the incorporation in *dl*331 extracts was edeine-sensitive (Fig. 2, lane 15), suggesting that the translational defect in *dl*331-infected cells is at the level of polypeptide chain initiation. This conclusion is supported by the observation that, in contrast to Ad2-infected cells, only a small percentage of the mRNA in *dl*331-infected cells occurs on polysomes (Schneider et al. 1984; our unpublished results).

The defect was not rescued by addition of purified VA RNA or of infected-cell extract. Furthermore, when the *dl*331 extract was mixed with translationally active Ad2 or mock extracts, translation was depressed below the levels expected for either of these extracts. Likewise, the *dl*331 extract depressed protein synthesis in rabbit reticulocyte lysates to a much greater extent than uninfected or Ad2-infected HeLa cell extract. These findings argue for the appearance, in the absence of VA RNA$_{\mathrm{I}}$, of an inhibitor capable of blocking the function of some essential initiation factor (Reichel et al. 1985; Siekierka et al. 1985). Despite these inhibitory effects, fractions prepared from translationally active HeLa or rabbit reticulocyte lysate translation systems were able to rescue

translation in *dl*331 extracts (Reichel et al. 1985). The ability of HeLa cell extracts to rescue translation was low in comparison with reticulocyte lysates, and most of the stimulatory activity sedimented with the ribosomes. In contrast, whereas a significant percentage (20%) of the stimulatory activity of the reticulocyte lysate sedimented with the ribosomes, the majority of the activity was found in the postribosomal supernatant (S100). The stimulatory activity associated with the ribosomes was separated from the ribosomal subunits by salt treatment. Since this behavior is characteristic of protein synthesis initiation factors, we tested individual initiation factors for their ability to rescue translation in *dl*331 extracts. Of the seven initiation factors examined in purified form and the larger number in less-pure mixtures, only eIF-2 was able to rescue translation in *dl*331 extracts (Reichel et al. 1985). A small fraction of the eIF-2 in reticulocyte lysate is present in the S100 fraction in a complex with a second initiation factor, the guanine nucleotide exchange factor (GEF; also known as eIF-2B). This complex was about 15-fold more effective than eIF-2 in rescuing translation in extracts of *dl*331-infected HeLa

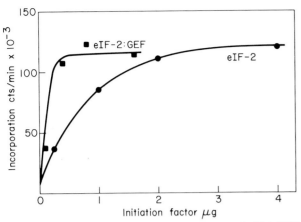

**Figure 3** Rescue of protein synthesis by eIF-2 and eIF-2:GEF complex. Increasing amounts of eIF-2 (●) or eIF-2:GEF complex (■) were titered into 25-μl reactions containing 6.25 μl of *dl*331 extracts. Translation was measured by [$^{35}$S]methionine incorporation. (Reprinted, with permission, from Reichel et al. 1985.)

cells (Fig. 3). When this complex was dissociated and its components separated on a sucrose gradient, most of the stimulatory activity migrated with GEF. Apparently, the function of one or both of these initiation factors is altered during adenovirus infection if VA RNA$_I$ is absent. Similar experiments by Schneider et al. (1984, 1985) lead to the same conclusion.

eIF-2 is required for the earliest step in polypeptide chain initiation (for review, see Jagus et al. 1982; Ochoa 1983). Normally, eIF-2 interacts with GTP and methionyl-tRNA$_F$ to form a ternary complex that binds to a 40S ribosomal subunit, forming a 40S initiation complex (Fig. 4). Subsequent reactions involving additional initiation factors result in mRNA binding and the joining of the 60S ribosomal subunit to the 40S initiation complex. During subunit joining, the ribosome-bound GTP is hydrolyzed to GDP and released with eIF-2 as an eIF-2:GDP complex. In this form eIF-2 is inactive for further rounds of initiation until the GDP is replaced by GTP, since methionyl-tRNA$_F$ binds only to eIF-2:GTP. The exchange reaction is catalyzed by GEF (Clemens et al. 1982; Siekierka et al. 1982; Konieczny and Safer 1983; Ochoa 1983; Panniers and Henshaw 1983). GEF activity is regulated by the phosphorylation state of the α subunit of eIF-2 because phosphorylated eIF-2α traps GEF in an inactive complex, preventing the GEF-catalyzed exchange of GTP for GDP (Fig. 4).

Siekierka et al. (1985) showed that the inhibition of translation in *dl*331-infected cell extracts is due to phosphorylation of eIF-2α and the consequent sequesteration of GEF. GEF activity was measured directly by assaying its ability to stimulate ternary complex formation from eIF-2:GDP, methionyl-tRNA$_F$, and GTP. Extracts of mock-infected and Ad-2 infected cells exhibited comparable levels of GEF, whereas extracts of *dl*331-infected cells had little detectable GEF activity. Further analysis demonstrated the presence in *dl*331 extracts of an eIF-2α kinase capable of inhibiting ternary complex formation. These results indicate that in the absence of

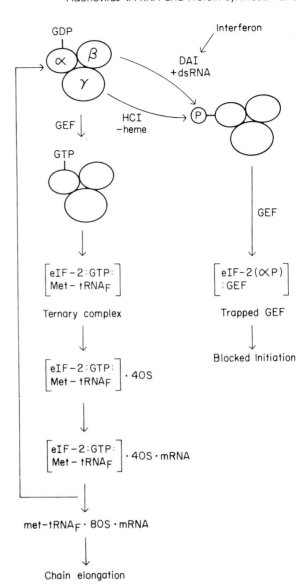

**Figure 4** The roles of eIF-2 and GEF in the initiation of translation. The α subunit of eIF-2 can be phosphorylated by the heme-controlled inhibitor (HCI) or the dsRNA-activated inhibitor (DAI), leading to sequestration of GEF. Subsequently, eIF-2 molecules can participate no more than once in ternary complex formation. (Reprinted, with permission, from Reichel et al. 1985.)

VA RNA$_I$, a kinase becomes activated and, by phosphorylating eIF-2α, sequesters GEF and thereby blocks polypeptide chain initiation.

### Identification of the *dl*331 kinase as DAI

Two protein kinases are known to phosphorylate the α subunit of eIF-2. Both of them phosphorylate eIF-2α on the same serine residue but under different conditions (Farrell et al. 1977, 1978; Ernst et al. 1980). The first is activated by many stimuli, including heme deprivation, and is called the heme-controlled inhibitor (HCI). The second kinase, DAI, can be activated by a low concentration of dsRNA (see Fig. 9A) or by heparin (Galabru and Hovanessian 1985). To compare the substrate specificity of the *dl*331 kinase with these kinases, we incubated eIF-2 with [γ-$^{32}$P]ATP in the presence of HCl or *dl*331 extract. The $^{32}$P-labeled α subunit was isolated, eluted, and digested partially with *S. aureus* V8 protease. Figure 5A shows that essentially identical peptide patterns were derived from the two preparations. Similarly, phosphoserine was the only labeled amino acid detected after acid hydrolysis (Fig. 5B). These data suggest that the *dl*331 kinase and HCl phosphorylate eIF-2α at the same site. By extension, this is also the site phosphorylated by DAI.

In earlier reports it was proposed that the *dl*331

kinase is DAI (Reichel et al. 1985; Schneider et al. 1985; Siekierka et al. 1985). We confirmed this hypothesis by comparing the known properties of HCl and DAI with those of the *dl*331 kinase (O'Malley et al. 1986). Incubation of interferon-treated cell extracts in the presence of dsRNA and [γ-$^{32}$P]ATP leads to the phosphorylation of two polypeptides (Johnston and Torrence 1984). The smaller phosphoprotein migrates with an apparent molecular weight of 38,000 (38K) and has been identified as eIF-2α. The larger phosphoprotein migrates between 68K and 72K, depending on its phosphorylation state and species of origin (Petryshyn et al. 1983; Johnston and Torrence 1984), and has been identified as DAI or one of its subunits (Petryshyn et al. 1983; Galabru and Hovanessian 1985). DAI activity is roughly proportional to the phosphorylation state of the ~70K protein (Petryshyn et al. 1983), so the appearance of the phosphorylated form of this protein serves as a marker for DAI activity. Unlike HCl, DAI and the *dl*331 kinase bind agarose–poly(I):poly(C) and phosphorylate eIF-2α while bound to this matrix (Fig. 6, lanes

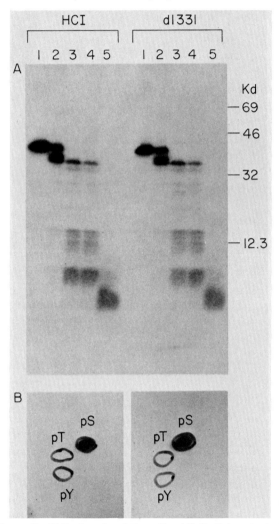

**Figure 5** eIF-2 phosphorylation site. (*A*) Partial proteolysis products of eIF-2α labeled with $^{32}$P by HCl (left lanes) or the *dl*331 kinase (right lanes). The α subunit, isolated by gel electrophoresis, was incubated for 15 min at 37°C with an equal volume of *S. aureus* V8 protease at 0, 1, 10, 100, 1000 μg/ml (lanes *1–5*, respectively). Digestion products were separated in a 20% SDS-polyacrylamide gel and detected by autoradiography. The positions of molecular mass markers are shown. (*B*) Phosphoamino acids released by hydrolysis of the $^{32}$P-labeled α-subunit preparations were resolved by two-dimensional electrophoresis and detected autoradiographically. The first dimension (pH 1.9) is shown left to right and the second (pH 3.5) is shown bottom to top. Marker amino acids (pS, phosphoserine; pT, phosphothreonine; pY, phosphotyrosine) were added to the digests prior to electrophoresis and their positions, revealed by ninhydrin staining, are circled.

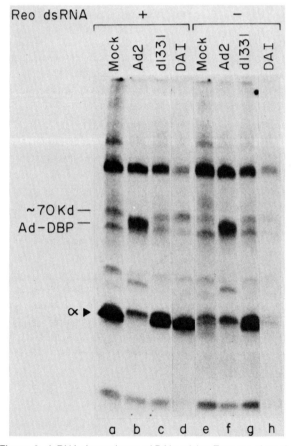

**Figure 6** dsRNA-dependence of DAI activity. Extracts of mock-infected, wild-type, Ad2-infected, or *dl*331-infected HeLa cells (125 μg of protein) or of uninfected, interferon-treated HeLa cells (DAI; 50 μg of protein) were preincubated with ATP, with or without 10 ng/ml reovirus dsRNA as indicated (lanes *a–d* and *e–h*, respectively). eIF-2 kinase activity was assayed by the agarose–poly(I):poly(C) method. The positions of the α subunit of eIF-2, ~70K DAI-associated polypeptide, and the Ad2 DNA-binding protein (Ad-DBP) are shown.

d and c). When the extract of *dl*331-infected cells was incubated with [γ-$^{32}$P]ATP, a polypeptide of ∼70K was labeled that bound to agarose–poly(I):poly(C) (Fig. 6, lanes c and g). This polypeptide comigrated with the ∼70K protein of DAI-treated cell extract (Fig. 6, lane d), and both were immunoprecipitated with antiserum raised against DAI (our unpublished results). In addition, the *dl*331 kinase, like DAI, was inhibited by concentrations of the sulfhydryl reagent *N*-ethyl maleimide, which stimulates HCl (Gross and Rabinovitz 1972; Grosfeld and Ochoa 1980). This evidence identifies the *dl*331 kinase as DAI.

## Activation of DAI

DAI is one of several enzymes involved in establishing an antiviral state in response to interferon (for reviews, see Revel 1979; Baglioni 1979; Lengyel 1982; Johnston and Torrence 1984). The enzyme is present at a basal level in a latent form in many cell lines; its synthesis is induced by interferon, and interferon production is induced by virus infection. The activation of the enzyme by dsRNA follows a characteristic bell-shaped concentration curve (Fig. 9A; Hunter et al. 1975; Farrell et al. 1977, 1978). Activation of DAI in vivo is often attributed to the presence of dsRNA replication intermediates in the case of RNA viruses and to symmetrical transcription of the viral genome in the case of DNA viruses (Johnston and Torrence 1984).

Unlike extracts of untreated or interferon-treated cells, which require preincubation with dsRNA for DAI activity (Fig. 6, lanes a and d), *dl*331 extracts contained DAI in an activated state (Fig. 6, lane c). Preincubation with dsRNA did not significantly increase the activity of DAI in *dl*331 extracts (Fig. 6, lane g), implying that DAI in this extract is largely active. The level of kinase activity in *dl*331 extracts was equivalent to the activity observed in uninfected extracts preincubated with dsRNA, but lower than the activity seen in dsRNA-activated extracts of interferon-treated cells. These data are consistent with the idea that the kinase activity in *dl*331-infected cells represents the activation of the basal levels of DAI present in HeLa cells. However, as yet, we have not measured the actual levels of DAI and so the possibility of enzyme induction cannot be excluded.

Preincubation with dsRNA and [γ-$^{32}$P]ATP did not activate significant amounts of eIF-2α kinase in extracts of cells infected with wild-type virus, but a slightly faster-moving polypeptide (∼65K) was heavily labeled (Fig. 6, lanes b and f). This species was not precipitated with antibody to DAI but reacted with antibody to the adenovirus DNA-binding protein (Ad-DBP), a known phosphoprotein (data not shown). Further experiments confirmed the identity of the ∼65K species as Ad-DBP: Purified Ad-DBP comigrated with the ∼65K species, was phosphorylated in the uninfected HeLa extract, and also bound to agarose–poly(I):poly(C). Also, as expected for Ad-DBP, the protein was selectively removed by pretreatment of the adenovirus-infected cell extract with single-stranded DNA–cellulose (not shown). The Ad-DBP was present in *dl*331-infected cells, although at a

much lower level than in cells infected with wild-type virus (Thimmappaya et al. 1982; and our unpublished results), and was barely detectable in the mutant extract by the agarose–poly(I):poly(C) binding assay.

Since DAI was activated in *dl*331-infected cells, it appeared likely that dsRNA was produced during adenovirus infection. When preparations of cytoplasmic RNA from mock-, Ad2- and *dl*331-infected HeLa cells were tested for their ability to activate DAI, only *dl*331 RNA activated DAI and produced the characteristic bell-shaped activation curve (O'Malley et al. 1986). RNA from mock extracts failed to activate DAI at any concentration tested, but RNA isolated from extracts of Ad2-infected cells gave a small response. As discussed below, in the case of Ad2 RNA, DAI activation appears to be blocked by the presence of high concentrations of VA RNA$_I$ (O'Malley et al. 1986). Presumably, the dsRNA component is generated by the symmetrical transcription of the viral genome during the late phase of infection (Pettersson and Philipson 1974; Sharp et al. 1975).

If dsRNA is generated by symmetrical transcription of the viral genome, DAI activation and the shut-off of protein synthesis should be coupled to the onset of late transcription in *dl*331-infected cells. Analysis of the kinetics of Ad2 and *dl*331 infections support the model that dsRNA generated by symmetrical transcription of the viral genome activates DAI. In the experiment illustrated in Figure 1, late transcription began between 8 and 12 hpi; as determined by the appearance of late viral mRNA (Fig. 1B). Even though functional viral transcripts continued to accumulate throughout the late phase (Fig. 1B), protein synthesis progressively decreased (Fig. 1A). Kinase activation followed the same kinetics as late transcription and protein synthesis shut-off in *dl*331 infection (Fig. 1C). Further experiments in which the protein synthesis inhibitor cycloheximide or the DNA replication inhibitor hydroxyurea were used to block progression of viral infection into late phase showed that DAI activity, as well as late viral transcription, depend on the transition from the early to the late phase (O'Malley et al. 1986).

Conceivably, interferon could be produced in response to adenovirus infection and might limit infection of HeLa cells in a fashion that is prevented by VA RNA$_I$. To test this hypothesis, we monitored the activity of 2',5'-oligoadenylate synthetase, another interferon-induced enzyme. During infection of HeLa cells with *dl*331, enzyme activity rose slightly until 12–16 hpi and then fell at later times, while treatment of uninfected cells with interferon gave a 20-fold increase (Fig. 7). Thus, the activity of the synthetase did not parallel that of the kinase during infection with *dl*331. Furthermore, in preliminary experiments, the presence of anti-α-interferon with antibody before and during infection of HeLa cells with *dl*331 did not prevent the protein synthesis block. In addition, pretreatment of HeLa cells with α-interferon did not hamper infection with Ad2 (data not shown). Although these experiments are not conclusive, they do not implicate interferon in the pathway leading to protein synthesis inhibition in *dl*331 infection.

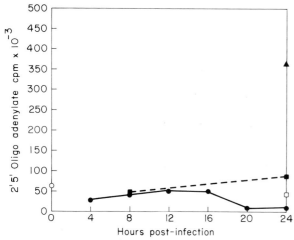

**Figure 7** 2',5'-Oligoadenylate synthetase activity in infected and interferon-treated cells. HeLa cells in spinner culture were infected with *dl*331 (●) or Ad2 (■), mock-infected (□), treated with α-interferon (▲), or untreated (○). At the times indicated, cells were harvested and extracts were assayed as described in Materials and Methods.

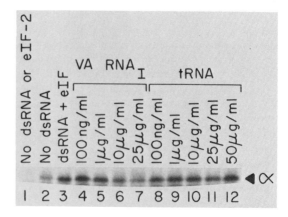

**Figure 8** VA RNA₁ blocks activation of DAI by dsRNA. Interferon-treated HeLa cell extracts (11.7 μg) were assayed for DAI activity by the standard assay without dsRNA or eIF-2 (lane 1), without dsRNA (lane 2), and with dsRNA (25 ng/ml) and increasing concentrations of VA RNA (lanes 4–7) or tRNA (lanes 8–12).

## VA RNA₁ blocks activation of DAI

Although the genetic experiments of Thimmappaya et al. (1982) clearly established that the translational defect is due to the absence of VA RNA₁, our early work demonstrated that the effect of VA RNA₁ on translation in adenovirus-infected cells is indirect. Over a wide range of concentrations, addition of purified VA RNA₁ to extracts of *dl*331-infected HeLa cells did not restore translation. Likewise, addition of VA RNA₁ failed to stimulate translation of *dl*331 mRNA in micrococcal nuclease–treated HeLa extracts or reticulocyte lysates (Reichel et al. 1985). However, the discovery that the translation defect in *dl*331-infected cells is due to the activation of DAI suggested a possible role for VA RNA₁. We proposed that in the Ad2-infected cell, VA RNA₁ blocks the activation of DAI by competing with dsRNA (Reichel et al. 1985). This hypothesis was supported by the kinetics of VA RNA₁ synthesis in Ad2-infected cells. VA RNA₁ accumulation to high levels began at 12 hpi and continued until 24 hpi (Fig. 1D), corresponding to the onset of late transcription, the production of the presumptive dsRNA by symmetrical transcription of the adenovirus genome, and the activation of DAI in the absence of VA RNA₁ (O'Malley et al. 1986). To test the hypothesis directly, we incubated VA RNA₁ with extracts of interferon-treated HeLa cells prior to the addition of activating concentrations of reovirus dsRNA and eIF-2 in the standard kinase assay. VA RNA₁ at high concentrations (~10–25 μg/ml) blocked activation of DAI, whereas tRNA over the same concentration range failed to block DAI activation (Fig. 8). Similar results have been obtained using the agarose–poly(I):(C) assay. VA RNA₁ can block activation of DAI when added before or at the same time as dsRNA but does not if added after dsRNA (O'Malley et al. 1986). We conclude that VA RNA₁ blocks activation of DAI but not the activity of the activated enzyme.

## Discussion

HeLa cells infected with the Ad5 deletion mutant *dl*331, which does not produce VA RNA₁, fail to translate mRNA efficiently at late times of infection (Thimmappaya et al. 1982; Fig. 1). This defect can be reproduced in cell-free translation systems prepared from adenovirus-infected HeLa cells (Schneider et al. 1984; Reichel et al. 1985; Fig. 2). Translation in *dl*331 extracts is restored by the addition of initiation factor eIF-2 or of much smaller quantities of GEF (Reichel et al. 1985; Schneider et al. 1985; Fig. 3). Furthermore, the ability of GEF to recycle eIF-2 is severely restricted in extracts of *dl*331-infected cells in comparison with extracts of mock- or Ad2-infected cells (Siekierka et al. 1985). GEF activity can be regulated by the phosphorylation state of eIF-2: When the α subunit of eIF-2 is phosphorylated, it sequesters GEF in a tight complex and prevents it from catalyzing GTP/GDP exchange on eIF-2 (Fig. 4; Clemens et al. 1982; Siekierka et al. 1982; Konieczny and Safer 1983; Ochoa 1983; Panniers and Henshaw 1983). Activation of an eIF-2 kinase occurs during the late phase of *dl*331 infection (Schneider et al. 1985; Siekierka et al. 1985). Recently, the *dl*331 kinase was identified as DAI. In wild-type Ad2-infected cells, activation of DAI is prevented by the presence of large quantities of VA RNA₁ (O'Malley et al. 1986).

We identified the *dl*331 kinase as DAI by comparing the biochemical properties of the two enzymes. The *dl*331 kinase resembles DAI in its sensitivity to NEM and in its ability to bind agarose–poly(I):poly(C) and phosphorylate eIF-2α while bound to this matrix (O'Malley et al. 1986). Activation of the *dl*331 kinase is accompanied by the phosphorylation of an ~70K polypeptide that corresponds to an ~70K phosphoprotein present in dsRNA-activated, interferon-treated cell extracts (Fig. 6, lanes c, g, and d). This protein copurifies with DAI activity and seems to be the enzyme or one of its subunits (Revel 1979; Lengyel 1982; Johnston and Torrence 1984; Galabru and Hovanessian 1985; Laurent et al. 1985). In addition, the 70K polypeptide of *dl*331-infected cells is precipitated by antibody directed against DAI (our unpublished results). Furthermore, cell lines GM2767A and RD114, which have little or no DAI activity, are permissive for *dl*331 growth and protein synthesis (Schneider et al. 1985; our unpublished results).

Activation of DAI in extracts of mock-infected cells and interferon-treated cells requires preincubation with ATP and dsRNA (Fig. 6, lanes a and d). However, in extracts of *dl*331-infected cells, the kinase is already activated (Fig. 6, lane c). Several lines of evidence support the notion that DAI activation in *dl*331-infected cells is due to the generation of dsRNA by symmetrical transcription of the adenovirus genome at late times of infection. Analysis of the kinetics of *dl*331 infection reveals that activation of DAI coincides with the onset of late transcription (Fig. 1) and is dependent on transition into the late phase of infection (O'Malley et al. 1986). Direct proof for the production of dsRNA by adenovirus infection comes from experiments that examine the ability of RNA isolated from extracts of infected and uninfected cells to activate DAI. RNA isolated from *dl*331 extracts is capable of activating DAI and exhibits the characteristic bell-shaped activation curve of authentic dsRNA. RNA from mock-infected cell extracts did not stimulate DAI at any concentration tested, but Ad2 RNA stimulated DAI slightly, probably because of the presence of VA RNA$_I$, which blocks activation of DAI both in vivo and in vitro (O'Malley et al. 1986).

We propose that the inhibitory action of VA RNA$_I$ is due to its partially double-stranded nature. VA RNA$_I$ contains several short duplex regions and accumulates to high levels ($\sim$ 100 µg/ml) at late times of infection. At high concentrations VA RNA$_I$ is capable of blocking activation of DAI by reovirus dsRNA (Fig. 8, lanes 4–7), but VA RNA cannot itself activate the enzyme (O'Malley et al. 1986). Thus it seems likely that VA RNA$_I$ prevents the interaction of authentic dsRNA because its duplex regions are equivalent to the short or imperfectly base-paired dsRNA molecules that at high concentrations can block activation of DAI (Minks et al. 1979; Baglioni et al. 1980; Torrence et al. 1981).

The interaction of DAI with dsRNA exhibits a number of intriguing features. First, the activation of DAI is dependent on dsRNA concentration: Low concentrations of dsRNA activate DAI whereas high concentrations of dsRNA prevent activation. Second, after incubation with an appropriate low concentration of dsRNA, activation is not readily reversed by addition of excess dsRNA. Third, only dsRNA containing 50 bp or more of perfect duplex can activate the enzyme; molecules with shorter or imperfectly base-paired duplex regions fail to activate DAI and at high concentrations prevent activation of DAI by authentic dsRNA (Fig. 9; Minks et al. 1979; Baglioni et al. 1980; Torrence et al. 1981). These characteristics can be explained if it is assumed that activation requires that two polypeptide subunits of the enzyme interact with one another to form a stable, active

complex while bound to the same molecule of dsRNA. As illustrated in Figure 9, at very low dsRNA concentrations the enzyme will be in excess and activity will be low. Likewise, at high dsRNA concentrations enzyme will be limiting and dimerization on the RNA duplex will not tend to occur frequently. Only at intermediate concentrations will cooperative interactions between the subunits be favored.

This model can also explain the inhibitory behavior of short and imperfect RNA duplexes and of VA RNA, which should bind the enzyme without permitting dimerization. Recent studies of DAI structure indicate that the

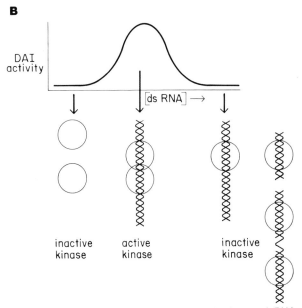

**Figure 9** Model for dependence of DAI activation on dsRNA. (*A*) To demonstrate the dependence of DAI activation on dsRNA concentration, 12.3 µg of extract from interferon-treated cells was preincubated without reovirus dsRNA (lane *1*) or with increasing concentrations of reovirus dsRNA (lanes *3–8*) and assayed for eIF-2 kinase activity by the agarose–poly(I):poly(C) binding assay. (*B*) Activation of DAI is supposed to require the interaction of two protein subunits (circles) bound to a dsRNA molecule of > 50 bp. Shorter and imperfect duplexes, such as present in VA RNA$_I$, block DAI activation. Active and inactive conformations are illustrated under the schematic dsRNA concentration curve.

enzyme has a molecular weight of 100–120K and that it comprises two polypeptide chains, the ~ 70K polypeptide discussed above and an additional 48K polypeptide (Levin et al. 1980; Galabru and Hovanessian 1985). Both of these moieties are needed for enzyme activity, and Galabru and Hovanessian (1985) have suggested that the smaller protein is a kinase that phosphorylates the larger one. It is attractive to speculate that the juxtaposition of these moieties on dsRNA and the subsequent phosphorylation of the ~ 70K polypeptide are the critical features for activation of DAI.

## Acknowledgments

The work described has been supported by grants CA 13106 to M.B.M. and GM 26796 to W.C.M. from the National Institutes of Health. We are indebted to Drs. Tim Hunt, Tony Hunter, Richard Jackson, and Hugh Robertson for discussions that led to the cartoon in Figure 9.

## References

Akusjärvi, G., M.B. Mathews, P. Andersson, B. Vennström, and U. Pettersson. 1980. Structure of genes for virus-associated RNA$_I$ and RNA$_{II}$ of adenovirus type 2. *Proc. Natl. Acad. Sci.* **77:** 2424.

Anderson, C.W., J.B. Lewis, J.F. Atkins, and R.F. Gesteland. 1974. Cell-free synthesis of adenovirus 2 proteins programmed by fractionated messenger RNA: A comparison of polypeptide products and messenger RNA lengths. *Proc. Natl. Acad. Sci.* **71:** 2756.

Baglioni, C. 1979. Interferon induced enzymatic activities and their role in the antiviral state. *Cell* **17:** 255.

Baglioni, C., S. Benvin, P.A. Maroney, M.A. Minks, T.W. Nilsen, and D.K. West. 1980. Interferon-induced enzymes: Activation and role in the antiviral state. *Ann. N.Y. Acad. Sci.* **350:** 497.

Benne, R., C. Wong, M. Luedi, and J.W.B. Hershey. 1976. Purification and characterization of initiation factor IF-E2 from rabbit reticulocytes. *J. Biol. Chem.* **23:** 7675.

Bradford, M.M. 1976. A rapid and sensitive method for the quantitation of microgram quantities of protein utilizing the principle of protein-dye binding. *Anal. Biochem.* **72:** 248.

Clemens, M.J., V.M. Pain, S.T. Wong, and E.C. Henshaw. 1982. Phosphorylation inhibits guanine nucleotide exchange on eukaryotic initiation factor 2. *Nature* **296:** 93.

Cleveland, D.W., S.G. Fischer, M.W. Kirschner, and U.K. Laemmli. 1977. Peptide mapping by limited proteolysis in sodium dodecyl sulfate and analysis by gel electrophoresis. *J. Biol. Chem.* **252:** 1102.

Cooper, J.A., B.M. Sefton, and T. Hunter. 1983. Detection and quantification of phosphotyrosine in proteins. *Methods Enzymol.* **99:** 387.

Dunn, A.R., M.B. Mathews, L.T. Chow, J. Sambrook, and W. Keller. 1978. A supplementary adenoviral leader sequence and its role in messenger translation. *Cell* **15:** 511.

Ernst, V., D. Levin, A. Leroux, and J.M. London. 1980. Site-specific phosphorylation of the α subunit of eukaryotic initiation factor eIF-2 by the heme-regulated and double-stranded RNA-activated eIF-2α kinases from rabbit reticulocyte lysates. *Proc. Natl. Acad. Sci.* **77:** 1286.

Farrell, P.J., T. Hunt, and R.J. Jackson. 1978. Analysis of phosphorylation of protein synthesis initiation factor eIF-2 by two-dimensional gel electrophoresis. *Eur. J. Biochem.* **89:** 517.

Farrel, P.J., K. Balkow, T. Hunt, R.J. Jackson, and H. Trachsel. 1977. Phosphorylation of initiation factor eIF-2 and the control of reticulocyte protein synthesis. *Cell* **11:** 187.

Francoeur, A.M. and M.B. Mathews. 1982. Interaction between VA RNA and the lupus antigen La: Formation of a ribonucleoprotein particle *in vitro*. *Proc. Natl. Acad. Sci.* **79:** 6772.

Galabru, J. and A.G. Hovanessian. 1985. Two interferon-induced proteins are involved in the protein kinase complex dependent on double-stranded RNA. *Cell* **43:** 685.

Grosfeld, H. and S. Ochoa. 1980. Purification and properties of the double-stranded RNA-activated eukaryotic initiation factor 2 kinase from rabbit reticulocytes. *Proc. Natl. Acad. Sci.* **77:** 6526.

Gross, M. and M. Rabinovitz. 1972. Control of globin synthesis by hemin. Factors influencing formation of an inhibitor of globin chain initiation in reticulocyte lysates. *Biochim. Biophys. Acta* **287:** 340.

Hovanessian, A.G. and I.M. Kerr. 1979. The (2'-5') oligoadenylate (pppA2'-5'A2'-5'A) synthetase and protein kinase(s) from interferon-treated cells. *Eur. J. Biochem.* **93:** 515.

Hunter, T., T. Hunt, R.J. Jackson, and H.D. Robertson. 1975. The characteristics of inhibition of protein synthesis by double-stranded ribonucleic acid in reticulocyte lysates. *J. Biol. Chem.* **250:** 409.

Jagus, R., D. Crouch, A. Konieczny, and B. Safer. 1982. The role of phosphorylation in the regulation of eukaryotic initiation factor 2 activity. *Curr. Top. Cell. Regul.* **21:** 35.

Johnston, M.I. and P.F. Torrence. 1984. The role of interferon-induced proteins, double-stranded RNA and 2',5' oligoadenylate in the interferon-mediated inhibition of viral translation. In *Interferon 3 mechanism of production and action* (ed. R.M. Friedman), p. 189. Elsevier/North-Holland, Amsterdam.

Konieczny, A. and B. Safer. 1983. Purification of the eukaryotic initiation factor 2—eukaryotic initiation factor 2B complex and characterization of its guanine nucleotide exchange activity during protein synthesis initiation. *J. Biol. Chem.* **258:** 3402.

Laurent, A.G., B. Krust, J. Galabru, J. Svab, and A.G. Hovanessian. 1985. Monoclonal antibodies to an interferon-induced $M_r$ 68,000 protein and their use for the detection of double-stranded RNA-dependent protein kinase in human cells. *Proc. Natl. Acad. Sci.* **82:** 4341.

Lengyel, P. 1982. Biochemistry of interferons and their actions. *Annu. Rev. Biochem.* **51:** 251.

Lerner, M.R., J.A. Boyle, J.A. Hardin, and J.A. Steitz. 1981. Two novel classes of small ribonucleoproteins detected by antibodies associated with lupus erythematosus. *Science* **211:** 400.

Levin, D.H., R. Petryshyn, and I.M. London. 1980. Characterization of double-stranded-RNA-activated kinase that phosphorylates α subunit of eukaryotic initiation factor 2 (eIF-2α) in reticulocyte lysates. *Proc. Natl. Acad. Sci.* **77:** 832.

Mathews, M.B. 1975. Genes for VA-RNA in adenovirus 2. *Cell* **6:** 223.

———. 1980. Binding of adenovirus VA RNA to mRNA: A possible role in splicing. *Nature* **285:** 575.

Mathews, M.B. and M. Francoeur. 1984. La antigen recognizes and binds to the 3'-oligouridylate tail of a small RNA. *Mol. Cell. Biol.* **4:** 1134.

Mathews, M.B. and U. Pettersson. 1978. The low molecular weight RNA of adenovirus 2-infected cells. *J. Mol. Biol.* **119:** 293.

Minks, M.A., D.K. West, S. Benvin, and C. Baglioni. 1979. Structural requirements of double-stranded RNA for the activation of 2',5'-oligo(A) polymerase and protein kinase of interferon-treated HeLA cells. *J. Biol. Chem.* **254:** 10180.

Monstein, B. and L. Philipson. 1981. The conformation of adenovirus VAI-RNA in solution. *Nucleic Acids Res.* **9:** 4239.

Ochoa, S. 1983. Regulation of protein synthesis initiation in eukaryotes. *Arch. Biochem. Biophys.* **223:** 325.

O'Malley, R.P., T.M. Mariano, J. Siekierka, and M.B. Mathews. 1986. A mechanism for the control of protein synthesis by adenovirus VA RNA$_I$. *Cell* **44:** 391.

Panniers, R. and E.C. Henshaw. 1983. A GDP/GTP exchange factor essential for eukaryotic initiation factor 2 cycling in

Ehrlich ascites tumor cells and its regulation by eukaryotic initiation factor 2 phosphorylation. *J. Biol. Chem.* **258**: 7928.

Pelham, H.R.B. and R.J. Jackson. 1976. An efficient mRNA-dependent translation system from reticulocyte lysates. *Eur. J. Biochem.* **67**: 247.

Petryshyn, R., D.H. Levin, and I.M. London. 1983. Double-stranded RNA-dependent eIF-2α protein kinase. *Methods Enzymol.* **99**: 346.

Pettersson, U. and L. Philipson. 1974. Synthesis of complementary RNA sequences during productive adenovirus infection. *Proc. Natl. Acad. Sci.* **71**: 4887.

———. 1975. Location of sequences on the adenovirus genome coding for the 5.5S RNA. *Cell* **6**: 1.

Price, R. and S. Penman. 1972. A distinct RNA polymerase activity, synthesizing 5.5S, 5S and 4S RNA in nuclei from adenovirus 2-infected HeLa cells. *J. Mol. Biol.* **70**: 435.

Reich, P.R., B.G. Forget, S.M. Weissman, and J.A. Rose. 1966. RNA of low molecular weight in KB cells infected with adenovirus type 2. *J. Mol. Biol.* **17**: 428.

Reichel, P.A., W.C. Merrick, J. Siekierka, and M.B. Mathews. 1985. Regulation of a protein synthesis initiation factor by adenovirus virus-associated RNA₁. *Nature* **313**: 196.

Revel, M. 1979. Molecular mechanisms involved in the antiviral effects of interferon. In *Interferon 1* (ed. I. Gresser), p. 102. Academic Press, New York.

Revel, M., D. Wallach, G. Merlin, A. Schattner, A. Schmidt, D. Wolf, L. Shulman, and A. Kimchi. 1981. Interferon-induced enzymes: Micro-assays and their applications; purification and assay of (2′-5′)-oligoadenylate synthetase and assay of 2′ phosphodiesterase. *Methods Enzymol.* **79**: 149.

Schneider, R.J., C. Weinberger, and T. Shenk. 1984. Adenovirus VAI RNA facilitates the initiation of translation in virus infected cells. *Cell* **37**: 291.

Schneider, R.J., B. Safer, S.M. Munemitsu, C.E. Samuel, and T. Shenk. 1985. Adenovirus VAI RNA prevents phosphorylation of the eukaryotic initiation factor 2 α subunit subsequent to infection. *Proc. Natl. Acad. Sci.* **82**: 4321.

Sharp, P.A., P.H. Gallimore, and S.J. Flint. 1975. Mapping of adenovirus 2 RNA sequences in lytically infected cells and transformed cell lines. *Cold Spring Harbor Symp. Quant. Biol.* **39**: 457.

Siekierka, J., L. Mauser, and S. Ochoa. 1982. Mechanism of polypeptide chain initiation in eukaryotes and its control by phosphorylation of the α subunit of initiation factor 2. *Proc. Natl. Acad. Sci.* **79**: 2537.

Siekierka, J., T.M. Mariano, P.A. Reichel, and M.B. Mathews. 1985. Translational control by adenovirus: Lack of virus-associated RNA₁ during adenovirus infection results in phosphorylation of initiation factor eIF-2 and inhibition of protein synthesis. *Proc. Natl. Acad. Sci.* **82**: 1959.

Söderlund, H., U. Pettersson, B. Vennström, L. Philipson, and M.B. Mathews. 1976. A new species of virus coded low molecular weight RNA from cells infected with adenovirus type 2. *Cell* **7**: 585.

Stefano, J.E. 1984. Purified lupus antigen La recognizes an oligouridylate stretch common to the 3′ termini of RNA polymerase III transcripts. *Cell* **36**: 145.

Thimmappaya, B., C. Weinberger, R.J. Schneider, and T. Shenk. Adenovirus VAI RNA is required for efficient translation of viral mRNA at late times after infection. *Cell* **31**: 543.

Torrence, P.F., M.I. Johnston, D.A. Epstein, H. Jacobsen, and R.M. Freedman. 1981. Activation of human and mouse 2–5A synthetases and mouse protein p₁ kinase by nucleic acids. *FEBS Lett.* **130**: 291.

Weinmann, R., H.J. Raskas, and R.G. Roeder. 1974. Role of DNA dependent RNA polymerase II and III in transcription of the adenovirus genome late in productive infection. *Proc. Natl. Acad. Sci.* **71**: 3426.

# Changes in the Expression of Cellular Genes in Cells Immortalized or Transformed by Polyomavirus

N. Glaichenhaus,* P. Léopold,* P. Masiakowski,[†] P. Vaigot,[‡] M. Zerlin,[§]
M. Julius,[§] K.B. Marcu,[§] and F. Cuzin*

*INSERM U273, Centre de Biochimie, Université de Nice, 06034 Nice, France; [†]Laboratoire de Génétique
Moléculaire des Eucaryotes du CNRS (INSERM U184), 67000 Strasbourg, France; [‡]Centre International de
Recherches Dermatologiques, Sophia Antipolis, 06565 Valbonne, France; [§] Department of Biochemistry,
State University of New York, Stony Brook, New York 11794

Differential screening of cDNA libraries yielded nine clones (pIL plasmids) characteristic of rat
cellular mRNAs whose cytoplasmic levels increase upon expression of the large T protein of
polyomavirus. One of them was identified as a mitochondrial transcript (COII), and, in fact, levels
of all mitochondrial transcripts were found to be increased in cells immortalized by the polyoma
large T protein as compared with rat embryo fibroblasts. Analysis of the expression of four other
pIL RNAs throughout the cell cycle demonstrated that they accumulated during the $G_2$ period of
the cycle in normal FR3T3 cells but were expressed mostly in $G_1$ in polyoma-transformed and
immortalized cell lines.

A two-step transformation process was evidenced upon transfection of rodent embryo fibroblast (REF) cells with genes of DNA tumor viruses and cloned cellular oncogenes. As an initial step, induced by the *plt* gene of polyomavirus (large T protein) (Rassoulzadegan et al. 1982, 1983), by the E1A genes of adenoviruses (van den Elsen et al. 1982; Ruley 1983), and by the viral and rearranged forms of the *myc* gene (Land et al. 1983; Mougneau et al. 1984; Ruley et al. 1984)—"group-I" oncogenes—REF cells acquire the ability of long-term growth and of growth at low serum concentration in culture and become reactive to tumor promoters (for review, see Connan et al. 1985; Glaichenhaus et al. 1985). These cells could be further transformed by transfer of other oncogenes (e.g., polyoma *pmt* and mutated *ras* genes), which by themselves could not transform REF cells. A possible clue for elucidating the molecular mechanisms involved was given by the observation that at least two of the proteins encoded by group-I oncogenes act in the regulation of gene expression at the transcriptional level. The polyoma large T protein, as its SV40 equivalent, was shown to repress the early viral promoter and to activate late gene expression (for review, see Hand 1981; Brady et al. 1984; Keller et al. 1984; G. Khoury and M. Yaniv, pers. comm.). E1A is required for transcription of the other early viral promoters, activates the expression of various cellular genes, but represses enhancer-dependent transcription (Borrelli et al. 1984 and references therein). One may therefore consider the hypothesis that oncogenes of group 1 act by modulating, either positively or negatively, the expression of a series of cellular genes critical for the regulation of division and growth. We searched cDNA libraries for cellular transcripts accumulated in larger amounts in cells that express the polyoma large T protein than in the corresponding normal cells.

## Only a limited number of individual RNA species are increased in amount in polyoma-transformed rat fibroblast cells

Libraries of cDNA-pBR322 recombinant clones were prepared according to standard procedures (Maniatis et al. 1982; Masiakowski et al. 1982) from FR3T3-LT1 and MTT4 cells grown at low serum concentration (0.5%). Both cells lines were derived (Rassoulzadegan et al. 1982) after transfer into early-passage FR3T3 rat fibroblasts of the modified polyoma genes constructed by Kamen and colleagues (Treisman et al. 1981; Zhu et al. 1984) and were shown to express only one early polyoma protein, large T (FR3T3-LT1) or middle T (MTT4). They both grew in low-serum medium with an apparently normal phenotype—FR3T3-LT1 because these cells lack the genetic information for the middle T protein, and MTT4 cells because they require an exogenous supply of serum factors. In this way we hoped to minimize the secondary effects on gene expression that one would expect in phenotypically transformed cells. This procedure, however, implied that differential screening would be performed only between FR3T3-LT1 and MTT4 cells, with no possibility of a direct comparison with normal REF or FR3T3 cells, which cannot be maintained in low-serum medium. About 2500 clones from each library were transferred into microtitration plates and used to prepare colony replicas on Whatman 540 filter paper (Gergen et al. 1979). The filters were hybridized with [32]P-labeled cDNA probes from FR3T3 and MTT4, again grown in low-serum medium, and from chicken oviduct RNA as a control (data not shown). About 100 clones were selected for a second screening, which eventually led to the identification of nine cDNA clones, designated pIL, which produced a stronger hybridization signal with FR3T3-LT1 probes. The fraction of the total library thus selected (0.2%) is in good agreement

**303**

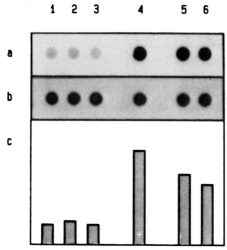

Figure 1 Steady-state levels of pIL3 RNA in cell lines with isolated polyomavirus genes. Cytoplasmic RNAs (20 μg) extracted from different polyoma transformants were spotted onto nitrocellulose paper and hybridized with [32]P-labeled pIL3 (*a*) or pHGAPDH-4 (Dani et al. 1984) plasmid DNA (*b*). Cell lines MTT1 (*1*), MTT2 (*2*), and MTT4 (*3*) expresses only middle T; FR3T3-LT1 (*4*) expresses large T only; MTT4-LT1 (*5*) expresses large T and middle T; and PYT21 (*6*) expresses the three early proteins. Ratios of densitometer measurements performed on the corresponding spots in *a* and *b* are shown in *c*.

with previous estimates by the same approach of the fraction of the total cellular transcripts, which show differential levels of accumulation during transformation (Scott et al. 1983).

The level of hybridization, measured by computer reading of the intensity of the spots, increased by 1.5-fold to 3-fold in FR3T3-LT1 as compared with MTT4 cells (Fig. 1). Although of a limited amplitude, this increase was reproducible, among polyoma transformants, in cells that expressed the large T protein and only in these cells. For instance, probes prepared from the two other "middle-T-only" lines, MTT1 and MTT2, produced the same result as MTT4. On the other hand, cell line 4-MT-LT1, derived from MTT4 cells by selection for growth in suspension in a low-serum medium after transfer of the *plt* gene (Rassoulzadegan et al. 1982), behaved like FR3T3-LT1, as did a cell line transformed with the complete polyoma early region (Fig. 1). The levels of the nine pIL RNAs increased when MTT4 cells were propagated in medium containing 10% calf serum, where they grew with a fully transformed phenotype (Rassoulzadegan et al. 1982).

To check whether the changes we observed were a peculiarity of FR3T3 cells and their derivatives, the same analysis was performed on REF cells and on two cell lines derived from REF, one by colony formation at low cell input (immortalization assay) after transfer of the *plt* gene (RAT-LT1) and the other by transformation with the complete polyoma early region (RAT-PYWT1) (Rassoulzadegan et al. 1983). Levels of the cytoplasmic RNAs detected by the pIL1 and pIL7 probes were augmented (Fig. 2).

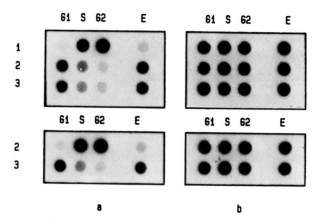

Figure 2 Cell-cycle-dependent regulation of pIL2 RNA level. Actively growing cells of lines FR3T3 (*1*), MTT4 (*2*), and MTT4-LT1 (*3*) were sorted as a function of their DNA content according to the method of Imbert et al. (1984). Cytoplasmic RNA was extracted from fractions corresponding to the $G_1$, S and $G_2$ DNA content, and 5 μg of RNA was spotted onto nitrocellulose and hybridized with [32]P-labeled pIL2 (*a*) and pB2-m2 DNA (β$_2$-microglobulin; Daniel et al. 1983) (*b*). Cells were grown in medium containing either 10% (upper panels) or 0.5% (lower panels) newborn calf serum (SEROMED).

## Cell-cycle dependence of the level of pIL RNAs

Analysis of expression at various phases of the cell cycle demonstrated that the moderate increase in the levels of several pIL RNAs observed during exponential growth reflected a change of a larger amplitude but limited to a particular period of the cell cycle. Rather than synchronizing cells by procedures that might result in artifactual alteration of gene expression, we used automated cell sorting to fractionate actively growing populations on the basis of their DNA content. As previously described by Imbert et al. (1984), RNA of the purified $G_1$, S, and $G_2$ populations could be analyzed for the presence of specific transcripts. As many as $5 \times 10^6$ cells in a given phase could be separated without apparent deleterious effect, and the homogeneity of the fractions was checked by rerunning aliquots of the suspensions of sorted cells through the cytofluorometer. The cytoplasmic and nuclear RNA fractions were analyzed by reverse transcription and hybridization of the [32P]cDNA probes with reference DNA clones. Alternatively, RNA deposited onto a nitrocellulose paper could be hybridized with nick-translated DNA probes. We used the expression of either β$_2$-microglobulin or glyceraldehyde-3P dehydrogenase as internal standards for constitutive expression throughout the cycle.

An unexpected result was observed for four pIL RNAs (pIL2, 6, 8, and 10). As exemplified for pIL2 in Figure 3, these RNAs were found to accumulate during the $G_2$ phase of the cycle ($G_2/G_1$ ratio of 8:15) in untransformed FR3T3 cells and in MTT4 cells cultivated in low-serum medium, where they exhibit a normal phenotype (Rassoulzadegan et al. 1982). By contrast, in fully transformed cells, the same RNAs were present in amounts fivefold to eightfold greater in $G_1$ than in $G_2$. This was the case of cell lines expressing both early proteins (4-MTLT-1) and of MTT4 cells cultivated in

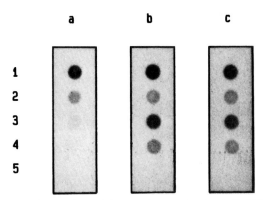

a    b    c

1
2
3
4
5

**Figure 3** Comparison of the steady-state levels of pIL RNAs in REF cells and in cell lines derived after transfer of polyomavirus genes. Cytoplasmic RNA extracted from REF cells (*a*), RAT-LT1 (large T only) cells (*b*), and RAT-PY1 cells (*c*) (all three polyoma early proteins) (Rassoulzadegan et al. 1983) were used as a template for oligo(dT)-primed synthesis by reverse transcriptase of $^{32}$P-labeled probes (Masiakowski et al. 1982). Probes were hybridized with an excess (1 µg) of various plasmid DNAs immobilized on nitrocellulose paper; (*1*) pB2-m2; (*2*) pHGAPDH-4; (*3*) pIL2; (*4*) pIL1; (*5*) pBR322.

high-serum medium (Fig. 3). Expression of these pIL genes was also shown to occur during the $G_2$ phase of the synchronous cycle induced by stimulation of $G_0$-arrested FR3T3 cells with a high concentration of serum (not shown).

## pIL1 and pIL7 RNAs are also overexpressed in cells containing *myc* or E1A oncogenes; one is a mitochondrial mRNA

Two of the pIL mRNAs, pIL1 and pIL7, were present in increased amounts, not only in polyoma transformants, but also in rat fibroblast cells immortalized by adenovirus E1A genes (van den Elsen et al. 1982) or by rearranged *myc* genes (Mougneau et al. 1984). In agreement with the latter observation, increased levels of these two RNA species were observed in plasmacytoma cell lines with translocated *myc* genes (not shown).

Each of these inserts was subcloned in vector M13mp10 (Messing and Vieira 1982), and nucleotide sequences were established by the method of Sanger et al. (1977). Sequences were compared with those in the rat, mouse, and human files of the GENBANK and EMBL libraries, using the Wilbur-Lipman algorithm ("IFIND" program, BIONET™). Positive identification was obtained for pIL7 sequences. A computer search revealed extensive homologies between the cDNA insert and the coding region of the rat mitochondrial (mt) COII gene (subunit II of cytochrome oxydase; see Attardi et al. 1982; Clayton 1984) (Fig. 4); at least part of the differences between these two sequences, obtained from Sprague-Dawley and Fischer rats, respectively, was expected from the known high rate of divergence of mtDNA (Brown and Simpson 1982). The size of the corresponding mRNA, estimated by northern blot analysis, was that expected from the literature, 750 nt (see Clayton 1984). COII mRNA, as all the mitochondrial RNAs, is produced by processing of a full-length primary transcript of the H strand of mtDNA (for reviews, see Attardi et al. 1982; Clayton 1984). A nearly full length rat mtDNA clone was isolated after screening with the pIL7 probe a genomic library prepared in the λEMBL4 vector. When restriction fragments from this DNA were used as probes in northern blot hybridization, they revealed the expected series of mtRNAs, including the two prominent ribosomal RNA bands. Amounts of all these transcripts were elevated in a comparable manner in large-T-immortalized cells as compared with REF cells (Fig. 5). This coordinated increase of the levels of various transcripts, in spite of their different half-lives and steady-state amounts (Attardi et al. 1982), clearly suggests a common transcriptional control, characteristic of an early stage of tumor progression in REF cells. However, pIL7 expression was not increased by transfection of the established Rat-2 cell line with pSVc-*myc* plasmid (Land et al. 1983), even though the resulting "*myc*-transformed" Rat-2 clones showed increased tumorigenicity. These results would collectively

```
PIL7COR--RATMTCYOS

       X      10        20        30        40        50
       AGATTAATAACCCAGTTCTAACAGTAAAAACTATAGGACTACCATTGATA
       ::::::::::::::::::::::::::::::::::::::::::  ::::  :::::
ATAATAGACGAGATTAATAACCCAGTTCTAACAGTAAAAACTATAGGAC-ACCAATGATA
       2380      2390      2400      2410      2420      2430

           60        70        80        90       100
       CTGAAGCTATGA-TATACTGACTATGAAGACCTATGCTATGACTCCTACATAATCCCAAC
       ::::::::::::: ::::::::::::::::::::::::::: :::::::::::::::::::::
       CTGAAGCTATGAATATACTGACTATGAAGACCTATGCTTTGACTCCTACATAATCCCAAC
           2440      2450      2460      2470      2480      2490

           120       130         X
       CAATGACCTAA-ACCAGGTGA-CTTCGTCTAT
       :::::::::::: :: ::::::  :::::   :::
       CAATGACCTAAAAACTAGGTGAACTTCGCTTATTAGAAGTTGA
           2500      2510      2520      2530
```

**Figure 4** Part of the pIL7 nucleotide sequence (top line, pIL7COR) and comparison with the sequence of the gene encoding subunit II of cytochrome oxydase (bottom line, RATMTCYOS).

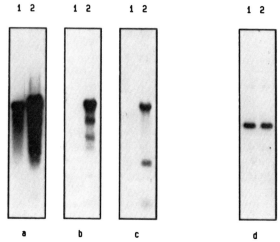

**Figure 5** Steady-state levels of mitochondrial transcripts in REF cells and in cells immortalized by large T antigen. Cytoplasmic RNA and total genomic DNA were extracted from REF (*1*) and from RAT-LT1 cells (*2*) (Rassoulzadegan et al. 1983). Twenty μg of RNA was fractionated on an agarose gel containing formaldehyde, tranferred to nitrocellulose, and hybridized with three different ³²P-labeled subgenomic fragments of the rat mitochondrial genome obtained by *Eco*RI digestion of a mtDNA clone in λEMBL4 vector (*a, b,* and *c*). In addition, 20 μg of total genomic DNA was digested with *Eco*RI, fractionated on an agarose gel, transferred to nitrocellulose, and hybridized with ³²P-labeled pIL7 plasmid DNA.

suggest that activation of the pIL1 gene and mtDNA (pIL7) is an early event in cellular transformation that already occurred in some established cells lines, like Rat-2, that, unlike FR3T3, have been grown in culture for large numbers of generations.

The pIL1 transcript was identified as a unique mRNA species of about 3000 nt on northern blots, and a simple pattern of fragments was seen on Southern blots, which probably corresponds to a unique nuclear gene. Screening of a library of full-length cDNAs led to the isolation of several clones in the expected size range. Their structural and biological properties are now under study, as well as are those of several clones corresponding to the cell-cycle-dependent pIL2 RNA.

## Acknowledgments

We thank J.J. Lawrence for introducing us to the cell-sorter technology, P. Chambon for helpful discussion and laboratory facilities during the initial part of this work, and M. Bordes for help with computer-assisted reading of autoradiograms. We are indebted to L. Carbone, M.J. Gonzalez, and F. Tillier for skilled technical help. This work was made possible by grants from the Association pour la Recherche sur le Cancer and the Fondation pour la Recherche Médicale (France), from the North Atlantic Treaty Organization, and from the Public Health Service (U.S.A.). Computer resources used to carry out our studies were provided by the BIONET™ National Computer Resource for Molecular Biology, whose funding is provided by the Biomedical Research Technology Program, Division of Research Resources, National Institutes of Health, grant 1U41RR-01685-02.

## References

Attardi, G., P. Cantatore, A. Chomyn, S. Crews, R. Gelfand, C. Merkel, J. Montoya, and D. Ojala. 1982. A comprehensive view of mitochondrial gene expression in human cells. In *Mitochondrial genes* (ed. P. Slonimski et al.), p. 51. Cold Spring Harbor Laboratory, Cold Spring Harbor, New York.

Borrelli, E., R. Hen, and P. Chambon. 1984. Adenovirus-2 E1A products repress enhancer-induced stimulation of transcription. *Nature* **312:** 608.

Brady, J., J.B. Bolen, M. Radonovich, N. Salzman, and G. Khoury. 1984. Stimulation of simian virus 40 late gene expression by simian virus 40 tumor antigen. *Proc. Natl. Acad. Sci.* **81:** 2040.

Brown, G.G. and M.V. Simpson. 1982. Novel features of animal mtDNA evolution as shown by sequences of two rat cytochrome oxidase subunit II genes. *Proc. Natl. Acad. Sci.* **79:** 3246.

Clayton, D.A. 1984. Transcription of the mammalian mitochondrial genome. *Annu. Rev. Biochem.* **53:** 573.

Connan, G., M. Rassoulzadegan, and F. Cuzin. 1985. Focus formation in rat fibroblasts exposed to a tumour promoter after transfer of polyoma *plt* and *myc* oncogenes. *Nature* **314:** 277.

Dani, C., J.M. Blanchard, M. Piechaczyk, L. El Sabouty, L. Marty, and P. Jeanteur. 1984. Extreme instability of *myc* mRNA in normal and transformed human cells. *Proc. Natl. Acad. Sci.* **81:** 7046.

Daniel, F., D. Morello, O. Le Bail, P. Chambon, Y. Cayre, and P. Kourilsky. 1983. Structure and expression of the mouse β₂-microglobulin gene isolated from somatic and non-expressing teratocarcinoma cells. *EMBO J.* **2:** 1061.

Gergen, J.P., R.H. Stern, and P.C. Wensink. 1979. Filter replicas and permanent collections of recombinant DNA plasmids. *Nucleic Acids Res.* **7:** 2115.

Glaichenhaus, N., E. Mougneau, G. Connan, M. Rassoulzadegan, and F. Cuzin. 1985. Cooperation between multiple oncogenes in fodent embryo fibroblasts: An experimental model of tumor progression. *Adv. Cancer Res.* **45:** 291.

Hand, R. 1981. Functions of T antigens of SV40 and polyomavirus. *Biochim. Biophys. Acta* **651:** 1.

Imbert, J., J.J. Lawrence, F. Coulier, E. Jeunet, V. Billotey, and F. Birg. 1984. Cell cycle-dependent expression of early viral genes in one group of simian virus 40-transformed rat cells. *EMBO J.* **3:** 2587.

Keller, J.M. and J.C. Alwine. 1984. Activation of the SV40 late promoter: Direct effects of T antigen in the absence of DNA replication. *Cell* **36:** 381.

Land, H., L.F. Parada, and R.A. Weinberg. 1983. Tumorigenic conversion of primary embryo fibroblasts requires at least two cooperating oncogenes. *Nature* **304:** 596.

Maniatis, T., E.F. Fritsch, and J. Sambrook. 1982. *Molecular cloning: A laboratory manual.* Cold Spring Harbor Laboratory, Cold Spring Harbor, New York.

Masiakowski, P., R. Breathnach, J. Bloch, F. Gannon, A. Krust, and P. Chambon. 1982. Cloning of cDNA sequences of hormone-regulated genes form the MCF-7 human breast cancer cell line. *Nucleic Acids Res.* **10:** 7895.

Messing, J. and J. Vieira. 1982. A new pair of M13 vectors for selecting either DNA strand of double digest restriction fragments. *Gene* **19:** 269.

Mougneau, E., L. Lemieux, M. Rassoulzadegan, and F. Cuzin. 1984. Biological activities of the v-*myc* and of rearranged c-*myc* oncogenes in rat fibroblast cells in culture. *Proc. Natl. Acad. Sci.* **81:** 5758.

Rassoulzadegan, M., A. Cowie, A. Carr, N. Glaichenhaus, R. Kamen, and F. Cuzin. 1982. The roles of individual polyomavirus early proteins in oncogenic transformation. *Nature* **300:** 713.

Rassoulzadegan, M., Z. Naghashfar, A. Cowie, A. Carr, M. Grisoni, R. Kamen, and F. Cuzin. 1983. Expression of the large T protein of polyomavirus promotes the establishment in culture of "normal" rodent fibroblast cells. *Proc. Natl. Acad. Sci.* **80:** 4354.

Ruley, H.E. 1983. Adenovirus early region 1A enables viral and cellular transforming genes to transform primary cells in culture. *Nature* **304:** 602.

Ruley, H.E., J.F. Moomaw, and K. Maruyama. 1984. Avian myelocytomatosis virus *myc* and adenovirus early region E1A promote the in vitro establishment of cultured primary cells. *Cancer Cells* **2:** 481.

Sanger, F., S. Nicklen, and A.R. Coulson. 1977. DNA sequencing with chain-terminating inhibitors. *Proc. Natl. Acad. Sci.* **74:** 5463.

Scott, M.R.D., K.H. Westphal, and P.W.J. Rigby. 1983. Activation of mouse genes in transformed cells. *Cell* **34:** 557.

Treisman, R., U. Novak, J. Favaloro, and R. Kamen. 1981. Transformation of rat cells by an altered polyomavirus genome expressing only the middle-T protein. *Nature* **292:** 595.

van den Elsen, P.J., S. de Pater, A. Houweling, J. van der Veer, and A. van der Eb. 1982. The relationship between region E1a and E1b of human adenoviruses in cell transformation. *Gene* **18:** 175.

Zhu, A., G.M. Veldman, A. Cowie, A. Carr, B. Schaffhausen, and R. Kamen. 1984. Construction and functional characterization of polyoma virus genomes that separately encode the three early proteins. *J. Virol.* **51:** 170.

# The Organization and Expression of Integrated Polyomavirus DNA in Transformed Cells

## M. Fried, J.B. Wilson, T. Williams, and C. Norbury

Department of Tumour Virus Genetics, Imperial Cancer Research Fund, Lincoln's Inn Fields, London WC2A 3PX, United Kingdom

Polyomavirus DNA is found integrated into the cellular chromosome in transformed cells. A number of properties of polyoma integration are described, including the illegitimate recombination events and gross and local rearrangements of cellular DNA sequences that occur at polyoma integration sites. The transformed phenotype is mediated by the polyoma oncogene—middle T antigen (TAg). Three unstable, flat-cell revertants of a polyoma-transformed cell line have been found to contain a deletion of one of nine consecutive cytidine residues between polyoma nucleotides 1239 and 1247 in a region of the polyoma early region where there are two translated reading frames, one coding for the carboxyl terminus of middle TAg and the other for the central region of large TAg. This frameshift mutation results in the shuffling of the different domains of the polyoma TAg's. The 43K protein generated by this shuffling contains the aminoterminal 75% of middle TAg, has an associated protein kinase activity, and forms a complex with c-src but is unable to induce a transformed phenotype. Transient frameshifting during normal polyoma gene expression as a mechanism to give the virus greater genetic flexibility is discussed.

Polyomavirus can transform cells to a neoplastic state in vitro and induce tumors in susceptible animals. The transforming ability is mediated by the viral early region, which encodes the three (large, middle, and small) tumor antigens (TAg's). The three TAg's are translated from three different mRNAs, which, although transcribed from the same early region DNA sequence, vary as a result of differential splicing (Tooze 1981) (Fig. 1). The large TAg, which has a molecular weight of 100,000 (100K) and is predominantly located in the nucleus, is a phosphoprotein with DNA-binding activity. The large TAg is required for the initiation of viral DNA replication (Fried 1970; Francke and Eckhart 1973) and can negatively regulate transcription of the polyoma early region mRNAs (Cogen 1978). It is also capable of "immortalizing" primary cells (Rassoulzadegan et al. 1983), and its activity is required for the initiation, but not the maintenance, of the transformed phenotype after virus infection (Fried 1965). The 56K middle TAg, which is the polyoma oncogene (Treisman et al. 1981), has a hydrophobic carboxyl terminus and is found associated with membranes (Ito et al. 1977; Silver et al. 1978). Middle TAg has an associated protein kinase activity (Eckhart et al. 1979; Schaffhausen and Benjamin 1979; Smith et al. 1979), which may be the result of the kinase activity of the c-src protein with which it can form a complex (Courtneidge and Smith 1983, 1984). Little is known of the biological activity of the 22K small TAg, which appears to be located in the cytoplasm. In polyoma-transformed cells the viral DNA is invariably found integrated into the host-cell chromosome. In this report we present some of the properties of the organization and expression of the integrated polyoma sequences in transformed cells.

## Results and Discussion

### Properties of polyomavirus integration

A great deal of information concerning polyomavirus integration has been generated in our laboratory from studies of the integrated polyoma and adjacent cellular DNA sequences of the five polyoma-transformed cell lines shown in Figure 2. These five cell lines (82-Rat, 53-Rat, 7axT, 7axB, and TSA-3T3) all contain a single insert of polyoma DNA, and in some cases the viral insert and adjacent cellular DNA sequences have been molecularly cloned and analyzed (Hayday et al. 1982; Ruley et al. 1982, 1986; Ruley and Fried 1983). A number of the properties of polyoma integration are presented in Table 1. From an inspection of virus-host joins (Figs. 2 and 3), it can be seen that integration can occur at many different locations in the host-cell DNA or viral DNA.

In contrast to retroviruses such as avian leukemia viruses and mouse mammary tumor virus, the transformed phenotype induced by polyomavirus does not appear to be caused by integration in specific regions of the cellular genome and consequent activation of an adjacent cellular gene (for review, see Varmus 1984). Neither does polyoma transformation appear to occur as the result of the inactivation of a cellular gene, because an insertion in only one cellular chromosomal allele (haploid insertion) is sufficient for expression of the transformed phenotype (Hayday et al. 1982). Polyoma

## POLYOMA VIRUS

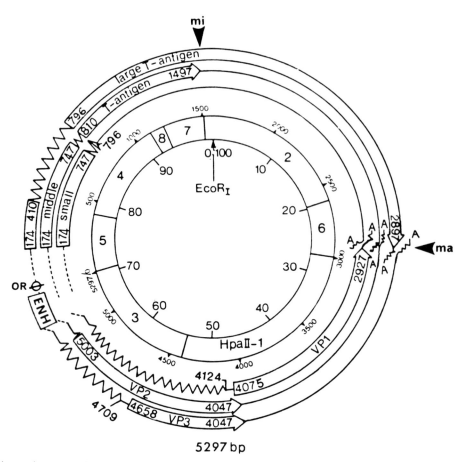

**5297 bp**

**Figure 1** The polyomavirus map of the A-2 strain (5297 bp). The early and late region mRNAs (thin lines) and their coding regions (boxed regions) are shown relative to the *Hpa*II physical map. Map units are shown inside and nucleotide numbers outside the physical map. The nucleotide numbers at the starts and ends of the coding regions and at the splice donors and acceptors are shown. Introns are indicated by jagged lines. The positions of the origin of DNA replication (OR), the enhancer region (ENH), and the major (ma) and minor (mi) early region polyadenylation sites are shown. Modified from Tooze (1981).

transformation seems, rather, to be the result of the addition of virally specified genetic information to the cellular genome. In transformed cells the polyoma early region is always expressed whereas the polyoma late region is poorly expressed, if at all. Although after virus infection a functional polyoma large TAg is required for the initiation of transformation (Fried 1965), upon continued growth of the transformed cells both in vitro (Basilico et al. 1979, 1980) and in vivo (Lania et al. 1981) there appears to be a selection for a nonfunctional large TAg by either deletion or mutation of the early region DNA sequences that are unique to the coding region of large TAg (polyoma nucleotides [nt] 1498–2897; see Fig. 1). These sequences are truncated in the polyoma-transformed cell lines 82-Rat, 53-Rat, 7axT, 7axB, and TSA-3T3 as a result of recombination between cellular and viral sequences (Fig. 2). The intact early regions of 7axB contain a single-base-change mutation in the sequences unique to large TAg that inactivates large TAg function (Hayday et al. 1983), whereas one of the early regions of 82-Rat contains a deletion-

insertion in the sequences unique to large TAg (Ruley and Fried 1983; Ruley et al. 1986). Cells containing integrated polyoma host-range transformation (*hr-t*) mutant DNA, capable of synthesizing a functional large TAg but not a functional middle TAg, have a normal phenotype (Lania et al. 1979), indicating that integration per se or the expression of a functional large TAg is not sufficient for the induction of a transformed phenotype. In contrast, the transformed phenotype requires an intact, functional middle TAg coding region. An intact middle TAg coding region is always present and is always expressed in polyoma-transformed cells (Lania et al. 1980) (Fig. 2), and a middle TAg cDNA will efficiently transform established cells (Treisman et al. 1981). In a number of cell lines, the recombinant join has resulted in the loss of the 3′ end of the polyoma early region and its associated polyadenylation signal. This can result in transcripts initiated at the start of the viral early region continuing into the adjacent host DNA where their polyadenylation is dependent on a cellular processing signal (Ruley et al. 1982).

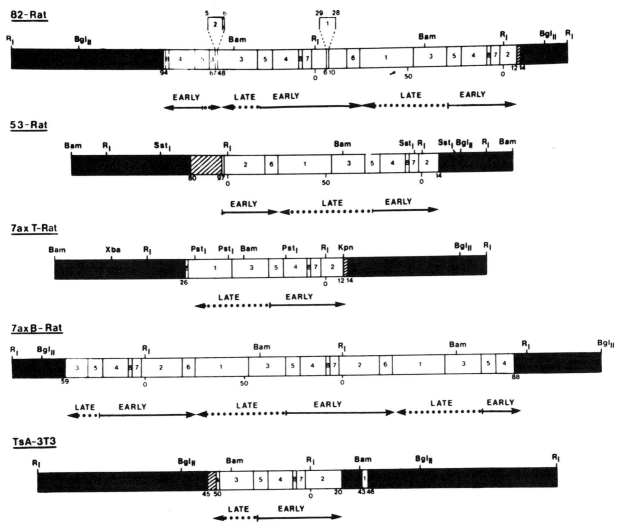

**Figure 2** The physical maps of the viral and adjacent cellular DNA of five polyoma-transformed cell lines. The *Hpa*II maps of the integrated viral sequences are shown in white (see Fig. 1), and the flanking cellular sequences are in black for the cell lines 82-Rat (Ruley and Fried 1983; Ruley et al. 1986), 53-Rat (Ruley et al. 1986), 7axT (Hayday et al. 1982), 7axB (Hayday et al. 1982), and TSA-3T3 (Ruley et al. 1982). Regions of uncertainty at the virus-host joins are indicated by hatching. The viral early and late regions are shown.

**Table 1** Polyomavirus Integration

1. Integration appears to take place in many different regions of the host cell and viral DNA. There is no evidence of integration in specific regions of cellular DNA.

2. A haploid insertion of viral DNA is sufficient for the transformed phenotype.

3. Cellular DNA adjacent to the viral insert does not contain transforming activity; viral transformation does not appear to activate cellular oncogenes.

4. The transformed phenotype requires an intact, functional middle TAg coding region. An intact large TAg coding region is not essential for the maintenance of the transformed phenotype after viral integration, although it is required for initiation of transformation by the infecting virus.

5. There appears to be selection for loss of functional large TAg both in vitro and in vivo either by mutation or deletion.

6. There are no common sequence or structural features at the sites of integration.

7. Integration appears to be the result of an illegitimate recombination event. There are no large regions of homology between viral and cellular DNA at the integration sites, although very small regions of homology (2–5 bp) may be found at virus-host joins.

8. Integration causes a rearrangement (probably a deletion) of host DNA at the integration site.

9. Initial transformants usually contain tandem arrays of viral genomes.

10. In some cases transcription initiated in viral sequences extends into host sequences, resulting in hybrid mRNAs and proteins. Processing of these mRNAs is under host control.

```
           VIRUS              CELL

TSA1    TGTTGGATTTCACCT | GGCTTTCCTTCCCAG
                  2643
TSA2    CTCAAGCTGAGGCAT | TTAATCCCTCTGCTC
                  3833
TSA3    CCAGCACCTCCATAC | TCTGGCTCATGCCTT
                  3974
53-RAT  ACAGGCCTAGCCGCG | GGGACAGCAGTGCCA
                  2312
82-RAT  GTACCCCAGCTCATC | AATAACTCATAACTT
                  1239
7B-L    ATGGAATGATTTCTT | GCTGTTAATTGCTAA
                   916
7B-R    TACCTACACAACGAA | TGAGTGCTATGAAAT
                  4682
7T-L    ATATCTGTATTTCCT | TTAAAAAATAATAAT
                  2932
7T-R    GAAAATGTGCCAAAA | GAACTTAAAGGGTTG
                  2252
```

**Figure 3** The nucleotide sequence at the virus-cell joins in polyoma-transformed cell lines. The nucleotide number beneath each sequence indicates the viral nucleotide at the virus-cell join (see Fig. 2) (Hayday et al. 1982; Ruley et al. 1982, 1986).

No sequence specificity or common structural features have been detected at the virus-host joins in polyoma-transformed cells (Hayday et al. 1982; Ruley et al. 1982, 1986; Ruley and Fried 1983) (Fig. 3). The integration of polyoma sequences appears to involve nonspecific, illegitimate recombination events. No large regions of homology between viral and cellular sequences are found at the integration sites, although between 2 and 5 bp of homology may exist at the recombinant join (Stringer 1982; Ruley and Fried 1983; Williams and Fried 1986). The cellular DNA sequences on either side of the viral insert in the transformed cells are not found adjacent to each other in untransformed cells (Hayday et al. 1982; Stringer 1982), demonstrating that integration has caused a gross rearrangement (deletion or chromosomal translocation) of cellular DNA sequences at the integration site.

## An inverted duplication-transposition of cellular DNA at a virus-host join

In addition to the gross rearrangement of cellular sequences at the polyoma integration site, a perturbation has been found at the left virus-host join in 7axB cells. Thirty-seven extraneous nt (filler DNA), derived from neither the viral nor directly adjacent host sequences, were found at the join when the viral sequences were compared with the cellular DNA of the unoccupied site (cellular sequences prior to the polyoma integration) (Hayday et al. 1982). By using a synthetic oligomer of these 37 nt as a probe, it was demonstrated that the filler DNA is an exact inverted duplication of a single-copy cellular sequence found 650 bp upstream from the virus-host join (Williams and Fried 1986) (Fig. 4). Thus, in addition to the large deletion or chromosomal translocation of cellular DNA at the integration site in 7axB cells, a more local inverted duplication-transposition of upstream cellular DNA has occurred, generating 37 bp

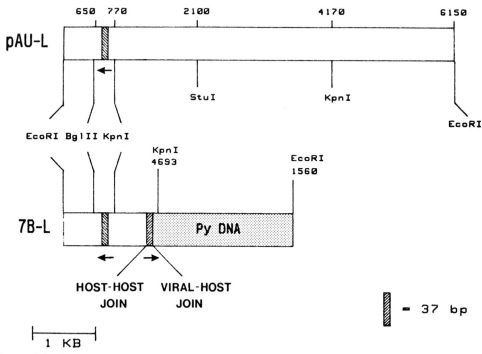

**Figure 4** An inverted duplication-transposition of 37 bp at the left virus-host join in 7axB cells. Shown is a diagram of the normal upstream location of the 37 bp (hatched region) in both the unoccupied-site fragment (pAU-L) from untransformed RAT-1 cells and in the left virus-host junction fragment (7B-L) from 7axB cells. A transposed copy of the 37 bp is also found in an inverted position at the virus-host join in 7B-L. (Reprinted, with permission, from Williams and Fried 1986.)

of filler DNA at the left virus-host junction. In contrast, the right 7axB join showed an abrupt transition from viral to host sequences with 2 bp of homology at the join (Williams and Fried 1986).

## Cellular sequences govern the 3′ processing of early region mRNAs initiated in the integrated polyoma DNA

Both a strong (major) and a weak (minor) polyadenylation signal are found in the polyoma early region (Fig. 1) (Kamen et al. 1980). In many polyoma-transformed cell lines, the viral sequences specifying the major polyadenylation site are lost (Fig. 2), and the vast majority of early region transcripts initiated in the viral sequences do not utilize the polyoma minor polyadenylation signal but continue past the virus-host join and are polyadenylated in the cellular sequences. In TSA-3T3 cells such viral-cellular hybrid transcripts were found to utilize a cellular polyadenylation signal (Ruley et al. 1982). The polyoma major polyadenylation signal is also absent from the transcribed early region in 53-Rat cells (Fig. 2) (Kamen et al. 1980; Ruley et al. 1986). The processed polyoma transcripts initiated in the polyoma early region of 53-Rat differ from most transcripts in other polyoma-transformed cells in that they utilize the inefficient minor polyoma polyadenylation signal instead of a cellular polyadenylation signal. Therefore the cellular DNA downstream from the polyoma insert in 53-Rat contains sequences that govern the use of the polyoma minor

polyadenylation signal. These cellular sequences could contain either a transcriptional termination or processing site so that a cellular polyadenylation signal is never reached. Alternatively, transcription may proceed very slowly through these cellular sequences, allowing adequate time for the use of the inefficient minor polyoma polyadenylation signal.

## Definition of the domains of the polyoma TAg's by a frameshift mutation in a polyoma-transformed cell line

Three independent flat-cell revertants isolated from the polyoma-transformed cell line 7axT (Fig. 2) have been analyzed (Wilson et al. 1986). All three revertants were observed to spontaneously retransform at a high frequency. All the revertants produced two novel TAg species of 43K and 37K instead of the 56K polyoma middle TAg and the 75K truncated large TAg synthesized by the parental 7axT cells and the spontaneous retransformants (Fig. 5). All the revertant and transformed cells also produced the polyoma 22K small TAg. A number of properties of the 43K and 37K TAg's are compared with those of polyoma large and middle TAg's in Table 2. The 43K protein shares a number of properties with the polyoma oncogene middle TAg, including its associated protein kinase activity and its ability to form a complex with c-src. The 43K protein differs from middle TAg in being unreactive with an anti-peptide serum, termed MTC, directed against the carboxyl ter-

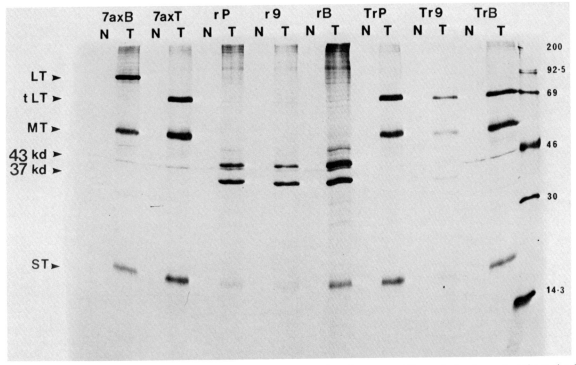

**Figure 5** TAg's in transformed and revertant cell lines. Immunoprecipitation of extracts of the control polyoma-transformed cell line 7axB, the polyoma-transformed parental cell line 7axT, the three flat-cell revertants (rP, r9, and rB), and their three spontaneous retransformants (TrP, Tr9, TrB) with polyoma anti-T (T) or normal (N) serum. The positions of the following TAg's are indicated: large TAg (LT), middle TAg (MT), small TAg (ST), the 75K truncated large TAg of 7axT and the retransformants (tLT), and the novel 37K and 43K TAg species of the flat-cell revertants. (Reprinted, with permission, from Wilson et al. 1986.)

**Table 2** Comparison of Properties of the Normal TAg's and the Two Truncated Proteins

| Property | LT | 37K | MT | 43K |
|---|---|---|---|---|
| In vitro phosphate acceptor | − | − | +++ | + |
| Association with c-*src* | − | − | + | + |
| In vivo phosphorylation | +++ | ++ | + | − |
| Antibody reactivity: | | | | |
|   monoclonal LT1 | + | + | − | − |
|   monoclonal MT16 | − | − | + | + |
|   anti-peptide Glu4 | − | − | + | + |
|   monoclonal LT4 | + | − | − | − |
|   anti-peptide MTC | − | + | + | − |
| Location[a] | N | M | M | C |

Data taken from Wilson et al. (1986).
[a](N) Nucleus; (M) membrane; (C) cytoplasm.

minus of middle TAg, in being predominantly located in the cytoplasm instead of being membrane-associated, and in being insufficient to induce a transformed phenotype. The 37K protein resembles the polyoma large TAg in a number of respects but differs from large TAg in being unreactive with the anti-large TAg monoclonal LT-4, reacting with the middle TAg anti-peptide serum MTC and in being predominantly membrane-associated instead of being localized in the nucleus. The generation of the two novel proteins is the result of a frameshift mutation created by a deletion of one of nine consecutive cytidine residues between polyoma nt 1239 and 1247 (Fig. 6). In this region there are two open reading frames; one coding for the carboxyl terminus of middle TAg and the other for the central region of large TAg. The frameshift mutation results in the shuffling of different domains of the polyoma TAg's, generating the

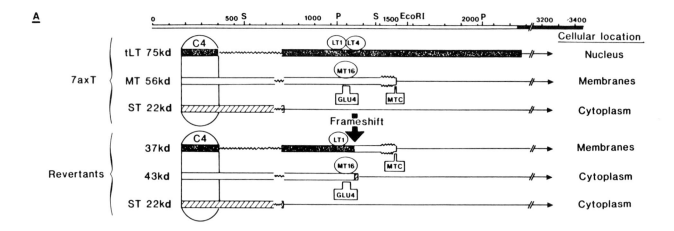

**Figure 6** Generation of the 37K and 43K novel TAg's by a frameshift at a mutational hot spot in the polyoma early region sequences encoding the carboxyl terminus of middle TAg and the central region of large TAg. (*A*) At *top* is shown the integrated polyoma early region (open band) and adjacent cellular DNA sequences (filled band) in 7axT. Below this are shown the proteins produced by 7axT, indicating the epitopes recognized by the monoclonal antibodies and the antipeptide sera specific for the polyoma TAg's. Beneath this is shown the generation of the revertant cell 37K and 43K novel TAg's as a result of frameshift mutation caused by the deletion of any one of nine cytidines between nt 1239 and 1247 (see *B*). This results in the large TAg reading frame (filled bands in *A*, cross-hatched bands in *B*) switching into the middle TAg reading frame (open bands) to generate the 37K protein and the middle TAg reading frame switching into the small TAg reading frame (hatched band) to generate the 43K protein (see *B*). (*B*) The deletion of a cytidine residue at a mutational hot spot in the polyoma early region causes a frameshift, generating the flat-cell revertant phenotype. The *top* half of the figure shows the large TAg and middle TAg reading frames above the DNA sequence of 7axT (which is the same as the sequence of the spontaneous retransformants). The *bottom* half of the figure shows the DNA sequence of the flat-cell revertants, which is generated by the deletion of any one of the nine consecutive cytidines between polyoma nt 1239 and 1247 (underlined). This results in the frameshift that generates the revertant 37K and 43K novel TAg's shown beneath the sequence (see *A*). The 37K TAg contains the amino terminus of large TAg and the carboxyl terminus of middle TAg, and the 43K TAg contains the amino terminus of middle TAg and seven amino acids at its carboxyl terminus (diagonal hatching) from the small TAg reading frame (after the termination of the small TAg protein). (Reprinted, with permission, from Wilson et al. 1986.)

37K hybrid protein with the amino terminus of large TAg and the carboxyl terminus of middle TAg and the hybrid 43K protein with the amino terminus of middle TAg and a carboxyl terminus encoded by a normally untranslated region of the small TAg reading frame (Fig. 6). These results indicate that the hydrophobic carboxyl terminus of middle TAg is responsible for membrane association even when attached to the amino terminus of large TAg, which is thought to contain a nuclear localization signal (Richardson et al. 1986). In addition, these results show that the aminoterminal 75% of middle TAg is sufficient for its associated protein kinase activity and the ability to form a complex with c-src but is not sufficient to induce a transformed phenotype. Thus, some other function of intact middle TAg is required for transformation.

### The possible use of frameshifting for the generation of other polyoma early region proteins

The spontaneous retransformants all contain an identical back mutation, which is a precise correction of the revertant mutation by the addition of a cytidine residue, regenerating nine cytidine residues between polyoma nt 1239 and 1247 (Fig. 6). As well as the previously described spontaneous retransformants (Wilson et al. 1986), nine other spontaneous retransformants (three derived from each of the three revertants) show the same phenotype and polyoma proteins as 7axT and presumably also contain nine cytidines in this region. The high frequency of generation of spontaneous retransformants ($3 \times 10^{-5}$) must be due, at least in part, to the ability to precisely correct the revertant mutation by the addition of a cytidine residue at any one of nine different positions in the sequence (Fig. 6). The ease of isolation of deletions (revertants) and additions (spontaneous retransformants) at the cytidine run indicates that it is a hot spot for frameshift mutations. Consecutive runs of a single nucleotide have been found to be hot spots for frameshift mutations in both prokaryotes (Streisinger et al. 1967; Okada et al. 1972; Pribnow et al. 1981) and recently in eukaryotes (Baumann et al. 1985). Additions or deletions presumably arise by slippage or stuttering of the DNA polymerase at the homonucleotide run or by misalignment of the replicating and parental DNA strands, as proposed by Streisinger and Owen (1985).

In a number of organisms, a low level of normal expression from frameshift mutants has been observed, resulting in "leaky" phenotypes (Atkins et al. 1972, 1983; Fox and Weiss-Brummer 1980). There is also accumulating evidence that certain organisms, particularly viruses, utilize translational frameshifting during normal expression of their intact genomes and that this may even have a regulatory function (Kastelein et al. 1982; Dunn and Studier 1983; Craigen et al. 1985; Mellor et al. 1985; Varmus 1985). The frameshifting in viruses may be achieved with the normal translational machinery, unlike the mechanism that exists in yeast and bacteria, which relies on suppressor tRNAs containing mutant anticodons (Roth 1981). Another possible mechanism to produce transient frameshifting would be at the

transcriptional level. Certain DNA sequences may cause the RNA polymerase to generate addition or deletion errors in the mRNAs, in the same manner as frameshift mutations arise during DNA replication.

It is interesting to speculate that the region of polyoma containing nine consecutive cytidines, which is a frameshift mutational hot spot for DNA replication, might also cause similar errors during transcription and/or translation of the mRNA. The ability of portions of the polyoma early region to encode hybrid proteins may endow the virus with greater genetic flexibility than would be achieved by encoding just two proteins (large and middle TAg's) in the same region of the DNA sequence. The exact polyoma gene products involved in the induction of the polyoma tumor-specific transplantation antigen (Tooze 1981) or the immortalization of primary cells (Rassoulzadegan et al. 1983) remain unclear and could be mediated by proteins made in small quantities as a result of frameshifting at either the translational or transcriptional level. Minor TAg species (37–43K) have been noted by others (Hutchinson et al. 1978; Schaffhausen et al. 1978; Silver et al. 1978; Simmons et al. 1979; Ito and Spurr 1980) during the lytic cycle and in transformed cells. In one case a 39K TAg species appeared to contain the aminoterminal amino acid sequence from large TAg and the carboxyterminal amino acid sequence of middle TAg (Ito and Spurr 1980), which is similar to the 37K TAg species produced by the revertants (Wilson et al. 1986). These minor TAg species could be specific gene products, breakdown products, or hybrid molecules of the polyoma middle and large TAg's caused by frameshifting.

### Acknowledgments

The authors wish to thank Mrs. G. Briody for her help in the typing and preparation of parts of this manuscript.

### References

Atkins, J.F., D. Elseviers, and L. Gorini. 1972. Low activity of β-galactosidase in frameshift mutants of *Escherichia coli. Proc. Natl. Acad. Sci.* **69**: 1192.

Atkins, J.F., B.P. Nichols, and S. Thompson. 1983. The nucleotide sequence of the first externally suppressible-1 frameshift mutant, and of some nearby leaky frameshift mutants. *EMBO J.* **2**: 1345.

Basilico, C., S. Gattoni, D. Zouzias, and G. Della Valle. 1979. Loss of integrated viral DNA sequences in polyoma transformed cells is associated with an active viral A function. *Cell* **17**: 645.

Basilico, C., D. Zouzias, G. Della Valle, S. Gattoni, V. Colantuoni, R. Fenton, and L. Dailey. 1980. Integration and excision of polyomavirus genomes. *Cold Spring Harbor Symp. Quant. Biol.* **44**: 611.

Baumann, B., M.J. Potash, and G. Kohler. 1985. Consequences of frameshift mutations at the immunoglobulin heavy chain locus of the mouse. *Embo J.* **4**: 351.

Cogen, B. 1978. Virus-specific early RNA in 3T6 cells infected by a *ts* mutant of polyomavirus. *Virology* **85**: 222.

Courtneidge, S.A. and A.E. Smith. 1983. Polyomavirus transforming protein associated with the product of the c-src cellular gene. *Nature* **303**: 435.

———. 1984. The complex of polyomavirus middle-T antigen and pp60$^{c-src}$. *EMBO J.* **3**: 585.

Craigen, W.J., R.G. Cook, W.P. Tate, and C.T. Caskey. 1985. Bacterial peptide chain release factors: Conserved primary structure and possible frameshift regulation of release factor 2. *Proc. Natl. Acad. Sci.* **82:** 3616.

Dunn, J.J. and F.W. Studier. 1983. Complete nucleotide sequences of bacteriophage T7 DNA and the locations of T7 genetic elements. *J. Mol. Biol.* **166:** 477.

Eckhart, W., M.A. Hutchinson, and T. Hunter. 1979. An activity phosphorylating tyrosine in polyoma T-antigen immunoprecipitates. *Cell* **18:** 925.

Fox, T.D. and B. Weiss-Brummer. 1980. Leaky +1 and −1 frameshift mutations at the same site in yeast mitochondrial gene. *Nature* **288:** 60.

Francke, B. and W. Eckhart. 1973. Polyoma gene function required for viral DNA synthesis. *Virology* **55:** 127.

Fried, M. 1965. Cell-transforming ability of a temperature-sensitive mutant of polyomavirus. *Proc. Natl. Acad. Sci.* **53:** 486.

———. 1970. Characterization of a temperature-sensitive mutant of polyomavirus. *Virology* **40:** 605.

Hayday, A.C., F. Chaudry, and M. Fried. 1983. Loss of polyomavirus infectivity as a result of a single amino acid change in a region of polyomavirus large T-antigen which has extensive amino acid homology with simian virus 40 large T-antigen. *J. Virol.* **45:** 693.

Hayday, A., H.E. Ruley, and M. Fried. 1982. Structural and biological analysis of integrated polyomavirus DNA and its adjacent host sequences cloned from transformed rat cells. *J. Virol.* **44:** 67.

Hutchinson, M.A., T. Hunter, and W. Eckhart. 1978. Characterization of T antigens in polyoma-infected and transformed cells. *Cell* **15:** 65.

Ito, Y. and N. Spurr. 1980. Polyomavirus T antigens expressed in transformed cells: Significance of middle T antigen in transformation. *Cold Spring Harbor Symp. Quant. Biol.* **44:** 149.

Ito, Y., J. Brocklehurst, and R. Dulbecco. 1977. Virus-specific proteins in the plasma membrane of cells lytically infected or transformed by polyomavirus. *Proc. Natl. Acad. Sci.* **74:** 4666.

Kamen, R., J. Favaloro, J. Parker, R. Treisman, L. Lania, M. Fried, and A. Mellor. 1980. Comparison of polyomavirus transcription in productively infected mouse cells and transformed rodent cells. *Cold Spring Harbor Symp. Quant. Biol.* **44:** 63.

Kastelein, R.A., E. Remaut, W. Fiers, and J. Van Duin. 1982. Lysis gene expression of RNA phase MS2 depends on a frameshift during translation of the overlapping coat protein gene. *Nature* **295:** 35.

Lania, L., A. Hayday, and M. Fried. 1981. Loss of functional large T-antigen and free viral genomes from cells transformed *in vitro* by polyomavirus after passage *in vivo* as tumor cells. *J. Virol.* **39:** 422.

Lania, L., M. Griffiths, B. Cooke, Y. Ito, and M. Fried. 1979. Untransformed rat cells containing free and integrated DNA of polyoma non-transforming (hr-t) mutant. *Cell* **18:** 793.

Lania, L., D. Gandini-Attardi, M. Griffiths, B. Cooke, D. De Cicco, and M. Fried. 1980. The polyomavirus 100K large T-antigen is not required for the maintenance of transformation. *Virology* **101:** 217.

Mellor, J., S.M. Fulton, M.J. Dobson, W. Wilson, S.M. Kingsman, and A.J. Kingsman. 1985. A retrovirus-like strategy for expression of a fusion protein encoded by yeast transposon Ty1. *Nature* **313:** 243.

Okada, Y., G. Streisinger, J.E. Owen, and J. Newton. 1972. Molecular basis of a mutational hot spot in the lysozyme gene of bacteriophage T4. *Nature* **236:** 236.

Pribnow, D., D.C. Sigurdson, G. Gold, B.S. Singer, and C. Naroli. 1981. rII cistrons of bacteriophage T4 DNA sequence around the intercistronic divide and positions of genetic landmarks. *J. Mol. Biol.* **149:** 337.

Rassoulzadegan, M., Z. Naghashfar, A. Cowie, A. Carr, M. Grisoni, R. Kamen, and F. Cuzin. 1983. Expression of the large T protein of polyomavirus promotes the establishment in culture of "normal" rodent fibroblast cell lines. *Proc. Natl. Acad. Sci.* **80:** 4354.

Richardson, W.D., B.L. Roberts, and A.E. Smith. 1986. Nuclear location signals in polyomavirus large-T. *Cell* **44:** 77.

Roth, J. 1981. Frameshift suppression. *Cell* **24:** 601.

Ruley, H.E. and M. Fried. 1983. Clustered illegitimate events in mammalian cells which involve very short sequence homologies. *Nature* **304:** 181.

Ruley, H.E., L. Lania, F. Chaudry, and M. Fried. 1982. Use of a cellular polyadenylation signal by viral transcripts in polyomavirus transformed cells. *Nucleic Acids Res.* **10:** 4515.

Ruley, H.E., F. Chaudry, L. Lania, M. Griffiths, and M. Fried. 1986. Correlation of the organization and expression of the integrated polyomavirus DNA in transformed 82-Rat and 53-Rat cells. *J. Virol.* (in press).

Schaffhausen, B.S. and T.L. Benjamin. 1979. Phosphorylation of polyoma T-antigens. *Cell* **18:** 935.

Schaffhausen, B.S., J.E. Silver and T.L. Benjamin. 1978. Tumor antigen(s) in cells productively infected by wild-type polyomavirus and mutant NG-18. *Proc. Natl. Acad. Sci.* **75:** 79.

Silver, J., B. Schaffhausen, and T. Benjamin. 1978. Tumor antigens induced by nontransforming mutants of polyomavirus. *Cell* **15:** 485.

Simmons, D.T., C. Chang, and M.A. Martin. 1979. Multiple forms of polyomavirus tumor antigens from infected and transformed cells. *J. Virol.* **29:** 881.

Smith, A.E., R. Smith, B.E. Griffin, and M. Fried. 1979. Protein kinase activity associated with polyomavirus middle T-antigen *in vitro*. *Cell* **18:** 915.

Streisinger, G., Y. Okada, J. Emrich, J. Newton, T. Tsugita, E. Terzaghi, and M. Inouye. 1967. Frameshift mutations and the genetic code. *Cold Spring Harbor Symp. Quant. Biol.* **31:** 77.

Streisinger, G. and J.E. Owen. 1985. Mechanisms of spontaneous and induced frameshift mutation in bacteriophage T4. *Genetics* **109:** 633.

Stringer, J.R. 1982. DNA sequence homology and chromosomal deletion at a site of SV40 DNA integration. *Nature* **296:** 363.

Tooze, J., ed. 1981. *Molecular biology of tumor viruses*, 2nd edition, revised: *DNA tumor viruses*. Cold Spring Harbor Laboratory, Cold Spring Harbor, New York.

Treisman, R.H., V. Novak, J. Favaloro, and R. Kamen. 1981. Transformation of rat cells by an altered polyomavirus genome expressing only the middle T protein. *Nature* **292:** 595.

Varmus, H.E. 1984. The molecular genetics of cellular oncogenes. *Annu. Rev. Genet.* **18:** 533.

———. 1985. Reverse transcriptase rides again. *Nature* **314:** 583.

Williams, T. and M. Fried. 1986. An inverted duplication-transposition event in mammalian cells at an illegitimate recombination join. *Mol. Cell. Biol.* **6:** 2179.

Wilson, J.B., A. Hayday, S. Courtneidge, and M. Fried. 1986. Frameshift in polyoma early region generates two new proteins that define T-antigen functional domains. *Cell* **44:** 477.

# The Role of Adenovirus E1A in Transformation and Oncogenicity

R.T.M.J. Vaessen,* A.G. Jochemsen,* J.L. Bos,* R. Bernards,* A. Israël,[†]
P. Kourilsky,[†] and A.J. van der Eb*

*Department of Medical Biochemistry, Sylvius Laboratories, 2300 RA Leiden, The Netherlands; [†]Unité de Biologie Moléculaire du Gène, Institut Pasteur, 75724 Paris, France

In this summary we report results on the mechanism of suppression of class-I major histocompatibility complex genes by adenovirus (Ad) E1A and the role of region E1A in transformation of primary rat cells. The results can be summarized as follows:

(1) Suppression of class-I gene activity by Ad12 is caused by the switching off of gene expression.

(2) All class-I loci are affected to the same extent.

(3) Introduction of region E1A of Ad5 in Ad12-transformed cells causes reappearance of class-I expression. A transfected class-I H-2 gene is also differentially regulated by Ad5 or Ad12 E1A.

(4) A transfected class-I H-2 gene is also differentially regulated by Ad5 or Ad12 E1A.

(5) The immortalizing activity of E1A is localized in the aminoterminal part of the E1A protein.

(6) Region E1B does not seem to play a major role in transformation. Hence, region E1A is probably mainly responsible for this property.

Oncogenic transformation by adenovirus is mediated by the concerted activity of at least three viral gene products, encoded by the early regions E1A and E1B, which together constitute region E1 of the viral genome. Each of the two subregions specifies a set of coterminal mRNAs, which are translated into a number of partially overlapping proteins. In transformed cells, region E1A is expressed into two mRNAs of 12S and 13S, which specify related polypeptides of about 26 kD and 32 kD, respectively. Region E1B yields a single mRNA of 22S in transformed cells, which is translated into two unrelated proteins of about 19 kD and 55 kD (references in Pettersson and Akusjärvi 1983; Van Ormondt and Galibert 1984). Results from a number of studies have shown that region E1 contains all the information required for oncogenic transformation (although other early regions of the viral genome may also contribute to the oncogenic phenotype). The smallest DNA segment still possessing transforming activity is region E1A. Primary rodent cells transformed by this region only show a partially transformed phenotype and are nononcogenic, but they have become established into immortal cell lines (Houweling et al. 1980). Expression of E1A products in E1A-transformed cells is at least 20-fold lower than in cells transformed by E1A plus E1B, suggesting that region E1B is somehow required for high expression of region E1A (Van den Elsen 1983a,c). Conversely, region E1A is required for efficient expression of E1B, a phenomenon attributed to an E1A-mediated activation of E1B transcription (Berk et al. 1979; Jones and Shenk 1979; Bos and ten Wolde-Kraamwinkel 1983). Region E1B has no detectable transforming activity, even when it is fully expressed under the influence of a heterologous promoter (Van den Elsen et al. 1983b). Since region E1A is expressed to very low levels in E1A-immortalized cells and E1B alone does not seem to

have transforming activity, it cannot be excluded that region E1A actually contains most, if not all, of the information needed for morphological transformation and that E1B is only required for full expression of E1A (Van den Elsen et al. 1983a). Although this interpretation is suggested by the data mentioned above, evidence has also been found that region E1B must have a role in determining the degree of oncogenicity of transformed cells (Bernards et al. 1983). Thus, the precise role of E1A and E1B in oncogenic transformation is still unresolved.

Recently, a new property has been identified for the E1A region of oncogenic Ad12. Cells transformed by the E1 region of this virus differ from cells transformed by the corresponding region of the nononcogenic Ad5 in that they lack expression of the class-I major histocompatibility complex (MHC) antigens on their cell membrane (Schrier et al. 1983). Class-I MHC antigens consist of a heavy chain of approximately 45 kD, which is noncovalently bound to $\beta_2$-microblogulin, the 12-kD light chain. Since killing of cells expressing foreign antigens by cytotoxic T lymphocytes (CTL) will only occur if the cells also express their class-I MHC antigens, the absence or strong reduction of these antigens on Ad12-transformed cells will render them more or less resistant to this type of cellular immune defense (cf. Bernards et al. 1983). This may explain, at least in part, why Ad12-transformed cells are as oncogenic in T-cell-deficient nude mice as in immunocompetent syngeneic rats. Since it was found that suppression of class-I MHC genes is a property of the E1A region of Ad12, it appeared of interest to study this interaction with cellular gene expression in more detail.

Subsequent work has shown that suppression of class-I genes in Ad12-transformed cells occurs at the level of mRNA accumulation, that is, the cells contain strongly reduced amounts not only of class-I proteins but also of cytoplasmic mRNA. $\beta_2$-Microglobulin synthesis does not appear to be strongly affected. The Ad12 product responsible for the effect is encoded by the largest E1A mRNA (13S) and more specifically by its first exon, implying that the polypeptide specified by the nucleotide stretch that is unique for the 13S mRNA plays an important role in the suppression (Jochemsen et al. 1984). Interestingly, the Ad5 E1 region appears to counteract the effect of the Ad12 E1 region: BRK cells cotransformed by the Ad5 E1A and the Ad12 E1 region show no reduction of class-I expression, although both transforming regions are expressed to normal levels. This indicates that Ad5 E1A can somehow prevent Ad12 E1A from switching off expression of class-I genes. In the first part of this paper we report the results of further studies on the interaction of adenovirus genes with expression of the class-I MHC genes. In the second part results are presented of a study on the role of region E1A in oncogenic transformation in combination with the EJ *ras* oncogene.

## Experimental Procedures

### Transfections
All DNA transfections were performed as described by van der Eb and Graham (1980). Primary cultures of baby rat kidney (BRK) and baby mouse kidney (BMK) cells were prepared from 1-week-old WAG/RIJ rats and 1-week-old BALB/c or C57BL mice, respectively, obtained from the Radiobiological Institute, Rijswijk, The Netherlands.

### RNA analysis
Standard procedures were used for S1 nuclease analysis (Bernards et al. 1982), isolation of total cytoplasmic RNA, and northern blotting analysis (Schrier et al. 1983).

### Immunoprecipitation
Cell labeling with [$^{35}$S]methionine and subsequent immunoprecipitation was carried out as described previously (Schrier et al. 1979).

### Fluorescence-activated cell sorting (FACS)
FACS analysis was performed by an Ortho Diagnostic System 50-HH and computer MP-2150.

### Oncogenicity
Tumorigenicity of transformed cells was investigated by injection of $10^7$ cells into adult nude mice or 4-day-old syngeneic WAG/RIJ rats.

## Results

### Interaction of region E1A with expression of class-I MHC genes

*Ad5 E1 can restore expression of class-I genes in Ad12-transformed cells*
Our previous results have shown that primary BRK cells

transformed by the E1 region of highly oncogenic Ad12 have a strongly reduced expression of the class-I MHC antigens on their plasma membrane, as opposed to cells transformed by nononcogenic Ad5. When the E1A regions of both Ad5 and Ad12 are present in the same transformed cells, expression of the class-I genes is normal, indicating that Ad5 E1A can prevent the effect associated with Ad12 E1A (Bernards et al. 1983; Schrier et al. 1983). To further investigate the apparently opposite effects of Ad5 and Ad12 E1A, we have transfected class-I-negative, Ad12-transformed BRK cells with the Ad5 E1 region, in the presence of a dominant selectable marker (R.T.M.J. Vaessen, in prep.). Figure 1 shows that the level of class-I gene expression in the Ad5 E1-supertransfected cells was restored to the level found in Ad5-transformed cells. This indicated that Ad5 E1 can reactivate expression of the previously suppressed class-I genes and that the Ad12-transformed cells are intrinsically capable of forming normal amounts of class-I gene products. Experiments with hybrid Ad5/Ad12 E1A regions and mutant Ad5 E1 plasmids expressing exclusively either the 13S or the 12S E1A mRNA have shown that the dominant effect of Ad5 is a function of the product encoded by the first exon of the 13S E1A mRNA (Bernards et al. 1983; Jochemsen et al. 1984). This indicates that the class-I MHC-suppressing effect of Ad12 E1A and the opposing, dominant effect of Ad5 E1A on this suppression are functions of the same domain of the respective 13S mRNA products. It also suggests that the opposite activities may be due to a similar interaction of the E1A products with a common cellular receptor or other target and that the Ad5 product has a higher affinity for this receptor and competes out the Ad12 product.

*Is the Ad12 E1A effect a general phenomenon?*
To investigate whether the reduction of class-I MHC expression observed in Ad12-transformed BRK cells also occurs in cells from other organs or other species, we have transformed baby mouse kidney (BMK) cells, hamster embryo (HE) cells, rat embryo brain (REB) cells, and human embryonic retinoblast (HER) cells with Ad5 or Ad12 E1 plasmids. In all Ad12-transformed cells, a significant reduction of class-I MHC expression was observed as measured by northern blotting analysis and immunoprecipitation (not shown). The suppression was also found in cells transformed by highly oncogenic Ad31 but not in cells transformed by weakly oncogenic Ad7. Surprisingly, however, suppression was not observed in Ad12-transformed cells that were derived from established cell lines, that is, in cells that had already been established into immortal cells lines long before the Ad12 E1 region was introduced (e.g., rat 3Y1 cells; not shown). The results indicate that reduced expression of class-I MHC genes (the reduction factor varies from 4× to more than 10×) is a common feature of cells derived from a variety of animal species when they are transformed as primary or secondary cultures by highly oncogenic adenoviruses.

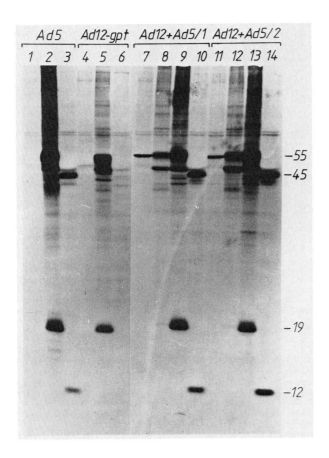

Ad5  Ad12-gpt  Ad12+Ad5/1  Ad12+Ad5/2

1 2 3  4 5 6  7 8 9 10  11 12 13 14

−55
−45

−19

−12

**Figure 1** SDS-polyacrylamide gel electrophoresis of proteins immunoprecipitated from extracts of [$^{35}$S]methionine-labeled cells with normal rat serum (lanes *1, 4, 7,* and *11*), Ad5 anti-T serum (lanes *2, 9,* and *13*), Ad12 anti-T serum (lanes *5, 8,* and *12*), and a mouse monoclonal antibody against rat class-I MHC heavy chains (lanes *3, 6, 10,* and *14*). The following cell lines were used: Ad5 (BRK cells transformed by Ad5 E1), Ad12-gpt (Ad12 E1–transformed BRK cells transfected with pSV2gpt [Mulligan and Berg 1981]), Ad12 + Ad5 (Ad12 E1–transformed BRK cells transfected with Ad5 E1 and pSV2gpt; 1 and 2 represent different cell lines). The molecular sizes (in kD) of the adenovirus small and large T antigens (19 and 55), the class-I MHC heavy chain (45), and $\beta_2$-microglobulin are indicated.

*Evidence that the Ad12 E1A-induced suppression is due to active switching-off*

As mentioned above, inhibition of class-I expression by Ad12 region E1 and the activation of expression after introduction of Ad5 E1 are both functions of the first exon product of the 13S E1A mRNAs. This suggested that the suppression of class-I gene expression by Ad12 E1A may be due to an active switching-off phenomenon. However, the primary cell cultures used for the transformation assays presumably consist of heterogeneous cell populations, and it cannot be excluded that Ad5 preferentially transforms cells expressing high levels of class-I antigens, whereas Ad12 selects other cells that have low expression of class-I genes. This interpretation is supported by the observation showing that Ad12 transforms primary cell cultures with a much lower frequency than Ad5 (Bernards et al. 1982), and that Ad12 may preferentially transform cells of neural origin (Mukai 1976; Gallimore and Paraskeva 1980), which are known to express low levels of class-I antigens (Vitetta and Capra 1978; Williams et al. 1980).

To distinguish between active switching-off of MHC genes and selective transformation of cells expressing low levels of MHC antigens, we have made use of a BRK cell line transformed by a mutant Ad12 E1 region, R11. This mutant plasmid carries a 109-bp deletion in region E1A so that it can only code for a 15-kD amino-terminal truncated protein (Bos et al. 1983). The R11 mutant plasmid is defective in transformation unless the SV40

enhancer sequences are inserted upstream of the R11 region (SVR11). BRK cells transformed by the chimeric SVR11 plasmid have become immortal and express the E1B region to normal levels but are nononcogenic in nude mice. Interestingly, the SVR11-transformed cells do not show a reduction of class-I gene expression, in contrast to cells transformed by wild-type Ad12 E1. Therefore, we were interested to investigate the effect of introduction of a wild-type Ad12 E1A region on class-I expression in these cells. Transfection of an Ad12 E1 plasmid in the presence of a dominant selectable marker (pSV2neo) into SVR11-transformed cells resulted in the isolation of clones expressing both the R11 E1A and the wild-type Ad12 E1A region. One of the lines, SVR11/Ad12-C8, expressed levels of Ad12 E1A mRNAs and proteins comparable to that in Ad12 E1-transformed control cells, but the second line, SVR11/Ad12-C6, contained only small amounts of wild-type E1A mRNA and no detectable E1A proteins (Fig. 2 and Fig. 3, lanes 12 and 16). Both lines contained normal amounts of R11 E1A mRNA (Fig. 2). Interestingly, immunoprecipitation of cell extracts from the supertransfected cells with an anti–rat class-I alloantiserum showed that expression of class-I antigens was drastically reduced in the C8 line, which showed high expression of wild-type Ad12 E1A (Fig. 3, lane 16), but was not reduced in the C6 line, in which the Ad12 E1A expression was low (Fig. 3, lane 12). Similar results were obtained for class-I-specific mRNA, using northern blotting analysis. This experiment

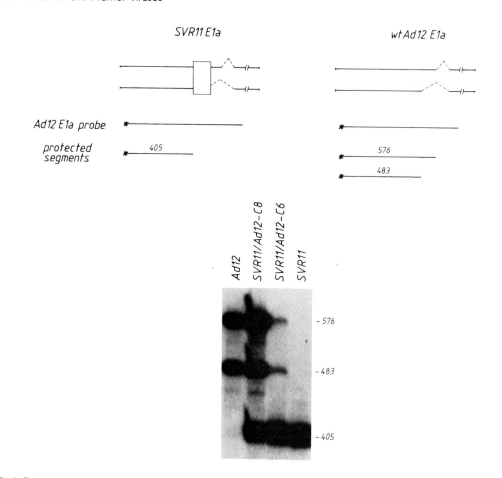

**Figure 2** (*Top*) Schematic representation of the S1 nuclease analysis performed to verify the expression of wild-type Ad12 E1A mRNAs after transfection of Ad12 E1 into SVR11-transformed BRK cells. The E1A mRNAs from SVR11 and wild-type Ad12 are shown. The box in SVR11 E1A represents a 109-bp deletion. A *Nar*I-*Dde*I fragment isolated from pAd12Acc (Jochemsen et al. 1984) and 3′-end-labeled at the *Nar*I position was used as a probe. The lengths of the protected segments are indicated. The autoradiograph shows the S1-resistant fragments run on a 5% acrylamide–7 M urea gel after hybridization of the DNA probe to 20 μg of total cytoplasmic RNA at 55°C.

indicates that expression of class-I MHC products is only reduced when wild-type Ad12 E1A is expressed at a sufficiently high level, which strongly suggests that Ad12 E1A products can indeed actively suppress class-I MHC genes.

*Expression of all class-I loci is suppressed in Ad12-transformed cells*
Although the level of class-I MHC antigens was strongly reduced in most Ad12-transformed primary cell cultures tested so far, the suppression was never complete since residual amounts of class-I mRNA and protein usually could be detected. To investigate whether the residual class-I mRNA and protein is due to the fact that certain class-I genes are partially suppressed and others completely, we have also investigated the expression of the individual class-I loci in Ad5- and Ad12-transformed cells from BALB/c mice. S1 nuclease analysis of cytoplasmic mRNA, using locus-specific DNA probes, showed that the level of RNA corresponding to all three loci, *K*, *L*, and *D*, is somewhat elevated in Ad5-transformed cells as compared with untransformed cells, but

that the RNA concentration from all three loci is considerably reduced in the Ad12-transformed cells (not shown). Similar results were obtained for the H-2 antigens specified by the class-I genes in the same transformed cells as well as in transformed BMK cells from C57BL mice, as detected by FACS with specific alloantisera. The results obtained with the C57BL BMK cells, which lack genes of the *H-2L* locus, are presented in Figure 4. The results show that the suppression of class-I gene products in Ad12 E1-transformed cells affects all class-I loci approximately to the same extent.

*Evidence that transfected class-I MHC genes are also differentially regulated by Ad5 and Ad12 region E1*
In an attempt to identify which DNA sequences in or around class-I genes are involved in the inhibition of gene expression, we have first investigated whether cloned class-I genes, when introduced into Ad5- or Ad12-transformed cells, are subject to the same regulation as the endogenous genes. Ad5- and Ad12-transformed BALB/c BMK cells (*d* haplotype) were transfected with a cloned H-2 gene of the *b* haplotype (kindly

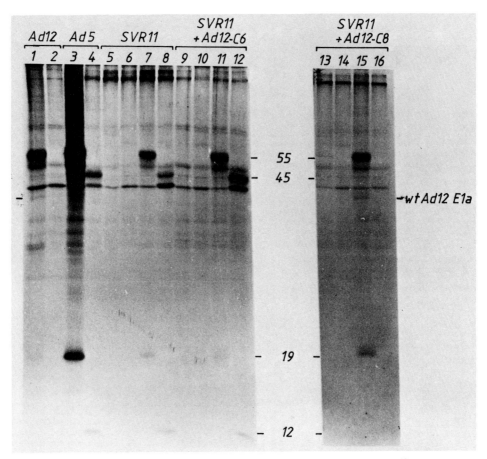

**Figure 3** SDS-polyacrylamide gel electrophoresis of proteins immunoprecipitated from extracts of [$^{35}$S]methionine-labeled cells with normal rat serum (lanes *5, 9,* and *13*), Ad5 anti-T serum (lanes *3, 6, 10,* and *14*), Ad12 anti-T serum (lanes *1, 7, 11,* and *15*), and a Lew anti-Wag rat alloantiserum (Schrier et al. 1983) (lanes *2, 4, 8, 12,* and *16*). The following cell lines were used: Ad5 (BRK cells transformed by Ad5 E1), Ad12 (BRK cells transformed by Ad12 E1), SVR11 (BRK cells transformed by plasmid SVR11), SVR11 + Ad12 (SVR11 cells transfected with Ad12 E1 and pSV2neo; C6 and C8 refer to individual G418-resistant cell lines. wtAd12 E1A refers to wild-type Ad12 E1A protein precipitated in lanes *1* and *15*. The molecular sizes (in kD) of the adenovirus small and large T antigens (19 and 55), the class-I MHC heavy chain (45), and $\beta_2$-microglobulin are indicated.

provided by Dr. R.A. Flavell, Biogen) in the presence of a dominant selection marker. This *H-2* clone contains about 4000 bp upstream from the cap site. The expression of the transfected *H-2* gene was determined by S1 analysis using a DNA probe that specifically recognizes the transfected gene. Figure 5 shows that the transfected *H-2* gene is efficiently expressed in Ad5-transformed cells, but to a much lower level in Ad12-transformed cells. This suggests that expression of transfected genes may indeed be regulated by Ad5 or Ad12 E1A in the same way as the endogenous genes. Experiments are in progress to determine whether specific DNA sequences in the 5′ upstream region of the gene are involved in the inhibition by Ad12 E1A.

### Interaction of region E1A with the T24 *ras* oncogene

To further define the role of region E1A in transformation and oncogenicity, we have studied the interaction of this region with the T24 *ras* oncogene. As was first demon-

strated by Ruley (1983) and Land et al. (1983), primary rat cell cultures can be oncogenically transformed by the combination of Ad2 region E1A and the EJ *ras* oncogene. As such, the *ras* oncogene can functionally replace E1B in transformation. To investigate the effect on oncogenicity of different E1A regions in transformation with the *ras* oncogene, we have transformed BRK cultures with EJ *ras* plus either Ad5 E1A, Ad12 E1A, or the mutant Ad12 E1 region from pSVR11. As described in a previous section, R11 is an Ad12 E1 region with a 109-bp deletion in region E1A. The R11 plasmid was capable of causing transformation only when the SV40 enhancer was inserted upstream of the R11 E1 region (Bos et al. 1983).

As can be seen in Table 1, all three E1A regions could transform BRK cells in the presence of EJ *ras*. The efficiency of transformation with wild-type E1A plus *ras* was considerably lower than that of E1A plus E1B, and in the case of Ad5 E1A plus *ras* the transformation efficiency approached that of E1A alone. Analysis of the

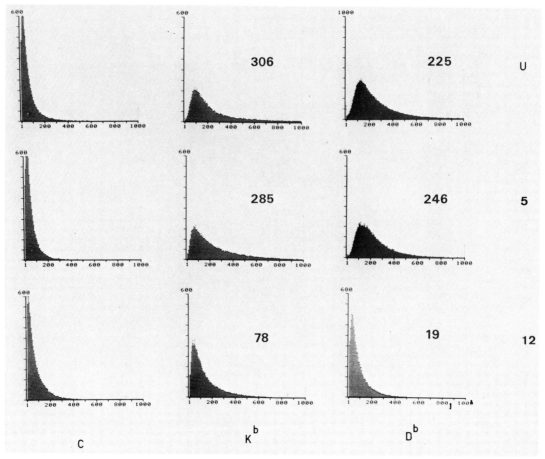

**Figure 4** FACS analysis of untransformed, Ad5 E1− and Ad12 E1−transformed C57/B1 (*H-2*$^b$) BMK cells. Cells were incubated with specific anti-*K*$^b$ or -D$^b$ antiserum and fluorescent conjugate, respectively, or conjugate only (*C*). Numbers indicate the average fluorescence value for each sample, which was calculated as follows: $m_{antibody}/m_{conjugate}^{-1}$.

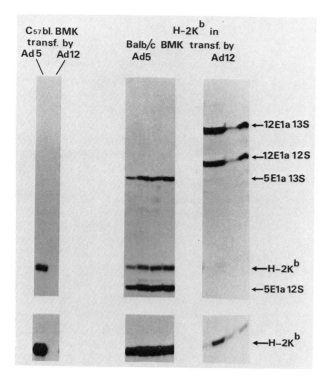

**Figure 5** S1 nuclease analysis of the *H-2K*$^b$ expression in supertransfected Ad5 E1− and Ad12 E1−transformed BALB/c BMK cells. As a reference, the E1A expression level in these cells was also determined. Probes: Ad5 E1A (3′-end-labeled *Hin*fI-fragment [nt 761–1372]); Ad12 E1A (3′-end-labeled *Nar*I-*Dde*I fragment (nt 1193–1255]); *H-2K*$^b$ (3′-end-labeled *Sin*I-fragment [600 bp], spanning the third exon–third intron junction); C57/bl (Ad5 E1− or Ad12 E1−transformed *b* haplotype BMK cells). (*Bottom*) Longer exposure of the same gel, with only *H-2K*$^b$ expression showing.

**Table 1** Properties of BRK Cells Transformed by Various E1A Regions in the Absence or Presence of Either Ad12 E1B or the EJ *ras* Oncogene

| DNA used for transformation | Avg. no. foci/μg genome equivalent E1A | Oncogenicity of transformed cells[a] | | E1A expression | EJ *ras* expression[b] |
| --- | --- | --- | --- | --- | --- |
| | | nude mice | syngeneic rats | | |
| Ad5 E1A | 0.36 | 0/12 | n.d. | low | − |
| Ad5 E1A + Ad12 E1B | 7.6 | 18/18 (35 days) | 0/26 | high | − |
| Ad5 E1A + EJ *ras* | 0.34 | 15/15 (15 days) | 9/9 (10 days) | low | + + |
| Ad12 E1A | (0)[c] | — | — | — | − |
| Ad12 E1A + Ad12 E1B | 0.55 | 23/23 (45 days) | 18/18 (110 days) | high | − |
| Ad12 E1A + EJ *ras* | 0.02 | 5/15 (45 days) | 2/9 (90 days) | low | + + |
| SVR11 E1A | 0.7 | 0/9 | ND | high | − |
| SVR11 E1A + Ad12 E1B | 1.0 | 0/12 | ND | high | − |
| SVR11 E1A + EJ *ras* | 0.8 | 9/15 (30 days) | ND | high | + + |

Data are taken from Bernards et al. 1982, 1983; Van den Elsen et al. 1982; Bos et al. 1983; R. Bernards et al.; A.G. Jochemsen et al.; both in prep.

[a]Oncogenicity is indicated as the number of animals with tumors per the number of animals injected. The average latency period of a tumor formation is indicated in parentheses. (ND) Not done.

[b]The c-Ha-*ras* expression in the E1A + *ras*–transformed cells is comparable to the expression level in the T24 bladder carcinoma cell line.

[c]The transformation frequency of Ad12 E1A probably is below our detection level, or the immortalized cells are not distinguishable between untransformed cells. Others have reported immortalizing activity of Ad12 E1A (Gallimore et al. 1984).

expression level of E1A RNA in the transformed cells showed that the RNA concentration of wild-type E1A was very low in all cases except when region E1B was present in addition to E1A (Table 1). Apparently, the presence of region E1B is required for efficient expression of region E1A (Van den Elsen et al. 1982). Thus, the efficiency of transformation appears to be proportional to the level of E1A mRNA. Furthermore, the results summarized in Table 1 show that the EJ *ras* gene can substitute for region E1B in conferring a fully transformed phenotype to primary rat cells as well as the ability to form tumors, but that it cannot enhance the level of E1A mRNA, as region E1B can. Together, the results suggest that the major role of region E1A in the transformation with *ras* is to immortalize the primary rat cells (as it also does in the absence of other transforming genes). Morphological transformation and oncogenicity could then be functions of EJ *ras*. The question can then be asked which gene product(s) are responsible for the fully transformed phenotype when cells are transformed by E1A plus E1B and region E1A is expressed at high levels. Since we have not been able to detect any transforming activity for region E1B alone (Van den Elsen et al. 1983b), there is no clear indication that the acquisition of morphological transformation is a direct function of region E1B product(s). Therefore, there may be more reasons to believe that this property is largely a function of region E1A, provided it is expressed at a sufficiently high level (cf. Van den Elsen et al. 1983a). Major functions of E1B would then include the stimulation of expression of region E1A and a role in oncogenicity (Bernards et al. 1982, 1983). That region E1B may have no important role in transformation is also suggested by the observation that BRK cells transformed by SVR11 E1A alone or SVR11 E1A plus Ad12 E1B are very similar in phenotype and hardly appeared transformed (but nevertheless are immortalized), showing that additional expression of E1B has little or no effect. In combination with the EJ *ras* oncogene, the SVR11 E1A region can induce a fully transformed phenotype. The results with the SVR11 mutant also show that the aminoterminal part (15 kD) of the E1A protein(s) encoded by the 13S mRNA has immortalizing potential. The additional presence of the carboxyterminal part would then be required for transformation. Attempts to prove the assumption that morphological transformation is largely a function of region E1A have failed because it has not been possible to isolate E1A-transformed primary cells with a high expression of E1A. Possibly, high levels of E1A proteins are toxic in the absence of E1B protein, but other interpretations cannot be excluded.

An interesting observation was that cells transformed by Ad5 E1A plus *ras* strongly differ from cells transformed by Ad12 E1A plus *ras*. The former have a rounded-cell morphology (possibly due to the absence of fibronectin; H. van Kranen and J.L. Bos, pers. comm.), are anchorage independent, and highly oncogenic, whereas the latter have a flat appearance, adhere strongly to substrate (there is only a partial reduction of fibronectin mRNA), and are only weakly oncogenic. Thus, the cells containing the E1A region of the nononcogenic virus are much more strongly oncogenic than those containing the E1A region of the oncogenic virus. Since the expression levels of the E1A regions and of *ras*, respectively, were equal, a possible explanation for the more highly oncogenic and transformed phenotype of the cells transformed by Ad5 E1A plus *ras* may be the stronger gene expression–modulating activity of the Ad5 E1A region (cf. Bos and ten Wolde-Kraamwinkel 1983). Hence, E1A fulfills more functions in cells transformed by E1A plus *ras* than immortalization alone. If this interpretation is correct, it would imply that oncogenic transformation by adenoviruses is due to E1A-induced alterations of cellular gene expression.

## Discussion

In this paper we report the results on the functional activities of the E1A regions of human adenovirus types 5 and 12 in oncogenic transformation of primary rat cell cultures. Functions that have been associated with region E1A include:

1. immortalization (Houweling et al. 1980);
2. transcriptional activation of region E1B and other early regions of the viral genome (Berk et al. 1979; Jones and Shenk 1979) and of certain cellular genes (such as the hsp70 heat-shock gene; Nevins 1979);
3. determination of the efficiency of transformation of primary rat cells (Ad5 E1A causes at least a 10× higher transformation frequency than Ad12 E1A; Bernards et al. 1982);
4. determination of the morphology of transformed colonies (Van den Elsen et al. 1983a);
5. differential modulation of class-I MHC expression (Schrier et al. 1983). The latter phenomenon provides at least a partial explanation for the fact that Ad12-transformed cells are oncogenic in immunocompetent animals.

In this paper we report results of additional experiments on the effect of region E1A on class-I MHC gene expression and on the interaction of region E1A with the EJ ras gene in oncogenic transformation of primary rat cell cultures. The observation that the suppressing effect of Ad12 E1A on class-I expression and the "dominant," activating effect of Ad5 E1A on class-I expression are functions of the product encoded by the first exon of the respective 13S mRNAs suggests that the two effects may be based on a similar interaction of the E1A products with a common cellular receptor or other target. The Ad5 E1A product would bind more strongly to this receptor than the Ad12 E1A product and could even compete-out the Ad12 product. The nature of this putative receptor is unknown, but possible candidates would be, for example, transcription complexes of certain genes. Evidence has indeed been presented that adenovirus E1A products directly or indirectly participate in viral gene transcription and that Ad5 E1A is a more efficient transcriptional activator than Ad12 E1A (Bos and ten Wolde-Kraamwinkel 1983). If one assumes that E1A products can also participate in the regulation of transcription of certain cellular genes, as has in fact been shown for the hsp70 gene (Kao and Nevins 1983), it is conceivable that Ad5 E1A products indeed stimulate transcription but that Ad12 products inhibit or reduce transcription due to the formation of inactive or weakly active transcription complexes. It is clear, however, that this mechanism can only be valid if suppression of class-I activity indeed occurs at the level of gene transcription. That this is still uncertain was recently suggested by nuclear run-off experiments that indicated that the rate of transcription of class-I genes is the same in Ad5- and Ad12-transformed human cells (our unpublished observations). Thus, it is still uncertain whether reduced expression of class-I genes by Ad12 E1A is caused by inhibition of class-I transcription or by a decrease in the rate of posttranscriptional maturation of mRNA (e.g., as a result of decreased stability of the primary transcripts). Our finding that a transfected genomic clone of a H-2 gene is differentially expressed in Ad5- and Ad12-transformed cells may be of considerable help in distinguishing between these two possibilities. That the Ad12 effect is indeed due to inhibition of class-I gene expression and not to selective transformation of target cells in the primary cultures that express low levels of class-I products was indicated by the supertransfection experiment of the (class-I-positive) SVR11-transformed cells.

In the second part of this study we show that different E1A regions can cooperate with the EJ ras oncogene in transformation of primary rat cells. The efficiency of transformation with the combinations of these two was very low and approximately equal to that of the E1A region alone. High transformation efficiencies were only obtained when region E1B was present in addition to E1A. The frequency of transformation appeared to be correlated with the level of expression of the E1A region, that is, when E1A was expressed at high levels, the frequency of transformation was also high, and when E1A expression was low, transformation was also low. Thus, the ras oncogene can replace the E1B region in oncogenic transformation with E1A, but it cannot stimulate the frequency of transformation or enhance the expression of E1A. Since the effect of region E1A alone in primary cells is to cause immortalization and, at the most, partial transformation, it is reasonable to conclude that the major contribution of region E1A in the transformation with ras is to confer immortality to the primary cells.

Then the question arises which role should be attributed to region E1A when it is expressed at the high levels found in cells transformed by E1A and E1B. The experimental evidence obtained so far suggests that region E1A may have a major role in morphological transformation when it is expressed at high levels, because region E1B alone has no detectable transforming activity (Van den Elsen et al. 1983a) and because cells transformed by SVR11 E1A alone or by SVR11 E1A plus E1B do not differ in phenotype, and, in fact, both appear only semitransformed. Attempts to prove the assumption that high levels of E1A products alone cause full transformation have failed so far since we have not been able to isolate E1A-transformed cells with a high expression level of E1A. If this assumption will prove to be correct, the concept that E1A belongs to the myc-like oncogenes and E1B to the ras-like oncogenes may require reconsideration.

The surprising observation that cells transformed by Ad5 E1A plus ras appear more highly transformed and are more strongly oncogenic than cells transformed by Ad12 E1A plus ras implies that the role of E1A is not only to confer immortality but also to induce transformed properties, even when expressed at a low level. Since the E1A regions of the two adenoviruses have been shown to differ in their ability to stimulate transcription of the E1B gene (Bos and ten Wolde-Kraamwinkel 1983), it is tempting to speculate that the differences in transformed phenotype just mentioned are due to differences in activity to modulate cellular gene expression.

## Acknowledgments

We thank Rita Veeren for typing the manuscript, our colleagues Ada Houweling, Ineke de Wit, and Henk van Kranen for their expert technical assistance, and P. Metezeau and "Service du Trieur de cellules de l'Institut Pasteur" for help with the cell sorter.

This work was supported in part by the Netherlands Organization for the Advancement of Pure Research (ZWO) through the Foundation for Fundamental Medical Research (FUNGO).

## References

Berk, A.J., F. Lee, T. Harrison, J. Williams, and P. Sharp. 1979. Pre-early adenovirus 5 gene products regulate synthesis of early viral messenger RNAs. *Cell* **17**: 935.

Bernards, R., A. Houweling, P.I. Schrier, J.L. Bos, and A.J. van der Eb. 1982. Characterization of cells transformed by Ad5/Ad12 hybrid early region 1 plasmids. *Virology* **120**: 422.

Bernards, R., P.I. Schrier, A. Houweling, J.L. Bos, A.J. van der Eb, M. Zijlstra, and C.J.M. Melief. 1983. Tumorigenicity of cells transformed by adenovirus type 12 by evasion of T-cell immunity. *Nature* **305**: 776.

Bos, J.L. and H.C. ten Wolde-Kraamwinkel. 1983. The E1b promoter of Ad12 in mouse Ltk⁻ cells is activated by adenovirus region E1a. *EMBO J.* **2**: 73.

Bos, J.L., A.G. Jochemsen, R. Bernards, P.I. Schrier, H. van Ormondt, and A.J. van der Eb. 1983. Deletion mutants of region E1a of Ad12 E1 plasmids: Effect on oncogenic transformation. *Virology* **129**: 393.

Gallimore, P.H. and C. Paraskeva. 1980. A study to determine the reasons for the difference in the tumorigenicity of rat cell lines transformed by adenovirus 2 and adenovirus 12. *Cold Spring Harbor Symp. Quant. Biol.* **44**: 703.

Gallimore, P., P. Byrd, R. Grand, J. Whittaker, D. Breiding, and J. Williams. 1984. An examination of the transforming and tumor-inducing capacity of a number of adenovirus type 12 early-region 1, host range mutants and cells transformed by subgenomic fragments of Ad12 E1 region. *Cancer Cells* **2**: 519.

Houweling, A., P.J. van den Elsen, and A.J. van der Eb. 1980. Partial transformation of primary rat cells by the leftmost 4.5% fragment of adenovirus 5 DNA. *Virology* **105**: 537.

Jochemsen, A.G., J.L. Bos, and A.J. van der Eb. 1984. The first exon of region E1a genes of adenoviruses 5 and 12 encodes a separate functional protein domain. *EMBO J.* **3**: 2923.

Jones, N. and T. Shenk. 1979. Isolation of adenovirus type 5 host range deletion mutants defective for transformation of rat embryo cells. *Cell* **17**: 683.

Kao, H.T. and J.R. Nevins. 1983. Transcriptional activation and subsequent control of the human heat shock gene during adenovirus infection. *Mol. Cell. Biol.* **3**: 2058.

Land, H., L.F. Parada, and R.A. Weinberg. 1983. Tumorigenic conversion of primary embryo fibroblasts requires at least two cooperating oncogenes. *Nature* **304**: 596.

Mukai, M. 1976. Human adenovirus induced embryonic neuronal tumor phenotypes in rodents. *Prog. Neuropathol.* **3**: 89.

Mulligan, R.C. and P. Berg. 1981. Selection for animal cells that express the *Escherichia coli* gene coding for xanthine guanine phosphoribosyltransferase. *Proc. Natl. Acad. Sci.* **78**: 2072.

Nevins, J.R. 1979. Induction of the synthesis of a 70,000 dalton mammalian heat shock protein by the adenovirus E1A gene product. *Cell* **29**: 913.

Pettersson, U. and G. Akusjärvi. 1981. Molecular biology of adenovirus transformation. *Adv. Viral Oncol.* **3**: 83.

Ruley, H.E. 1983. Adenovirus region E1A enables viral and cellular transforming genes to transform primary cells in cultures. *Nature* **304**: 602.

Schrier, P.I., P.J. Van den Elsen, J.J.L. Hertoghs, and A.J. van der Eb. 1979. Characterization of tumor antigens in cells transformed by fragments of adenovirus type 5 DNA. *Virology* **99**: 372.

Schrier, P.I., R. Bernards, R.T.M.J. Vaessen, A. Houweling, and A.J. van der Eb. 1983. Expression of class I major histocompatibility antigens switched off by highly oncogenic adenovirus 12 in transformed rat cells. *Nature* **305**: 771.

Van den Elsen, P.J., A. Houweling, and A.J. van der Eb. 1983a. Morphological transformation of human adenoviruses is determined to a large extent by gene products of region E1a. *Virology* **131**: 242.

———. 1983b. Expression of region E1b of human adenoviruses in the absence of region E1a is not sufficient for complete transformation. *Virology* **128**: 377.

Van den Elsen, P.J., S. de Pater, A. Houweling, J. van der Veer, and A.J. van der Eb. 1982. The relationship between region E1a and E1b of human adenoviruses in cell transformation. *Gene* **18**: 175.

Van den Elsen, P.J., B. Klein, B. Dekker, H. van Ormondt, and A.J. van der Eb. 1983c. Analysis of virus-specific mRNAs present in cells transformed with restriction fragments of adenovirus type 5 DNA. *J. Gen. Virol.* **64**: 1079.

van der Eb, A.J. and F.L. Graham. 1980. Assay of transforming activity of tumor virus DNA. *Methods Enzymol.* **65**: 826.

Van Ormondt, H. and F. Galibert. 1984. Nucleotide sequences of adenovirus DNAs. *Curr. Top. Microbiol. Immunol.* **110**: 73.

Vitetta, E.S. and J.D. Capra. 1978. The protein products of the murine 17th chromosome: Genetics and structure. *Adv. Immunol.* **26**: 147.

Williams, K., D. Hart, J. Fabre, and P. Morris. 1980. Distribution and quantitation of HLA-A,B,C and DR (Ia) antigens on human kidney and other tissue. *Transplantation* **29**: 274.

# Production of a Cell Proliferation Factor by Baby Rat Kidney Cells Infected with Adenovirus Type-5 12S Virus

## M.P. Quinlan and T. Grodzicker

Cold Spring Harbor Laboratory, Cold Spring Harbor, New York 11724

Primary baby rat kidney (BRK) cultures consist of fibroblasts and epithelial cells. Infection of these cultures with Ad5 12S virus results in a rapid increase in ($^3$H)thymidine incorporation and proliferation of the epithelial cell population. Infection with Ad5 dl312 or mock infection results in the eventual loss of the epithelial cells from the culture. The conditioned medium from 12S-infected cells contains stimulatory factor(s) that are able to produce the same proliferative response when placed on new, uninfected BRK cells. The continued presence of the 12S-conditioned medium is required to maintain epithelial cell proliferation. Preliminary analyses have characterized some of the physical properties of this factor(s).

The E1A region of adenovirus encodes products important in both the lytic cycle of the virus and in oncogenic transformation, although their biochemical functions are not fully understood. E1A is required for efficient transcription of viral early genes (Berk et al. 1979; Jones and Shenk 1979; Nevins 1981; Hearing and Shenk 1985). Their expression also increases the transcription of several cellular genes either resident within the genome or carried on plasmids or viral vectors (Nevins 1982; Green et al. 1983; Kao and Nevins 1983; Treisman et al. 1983; Gaynor et al. 1984; Leff et al. 1984; Stein and Ziff 1984). Although no direct interaction between DNA and E1A protein has been demonstrated, there are sequences upstream of genes that are required for trans-activation by E1A. However, such upstream sequences vary for several different promoters (Green et al. 1983; Weeks and Jones 1983; Imperiale and Nevins 1984). The E1A region is required for an efficient, productive adenovirus infection of quiescent permissive cells and the stimulation of cellular DNA synthesis (Montell et al. 1984; Bellett et al. 1985; Spindler et al. 1985; Stabel et al. 1985). The induction of $G_1$-arrested cells to progress into S phase after infection by adenovirus is also a function of the E1A region (Shimojo and Yamashita 1968; Strohl 1969; Takahashi et al. 1969; Stich and Yohn 1970; Younghusband et al. 1979; Pochran et al. 1980; Braithwaite et al. 1981, 1983; Cheetham and Bellett 1982). The cell-cycle progression of growing cells is also altered by the expression of the E1A region (Bellett et al. 1982; Murray et al. 1982a,b).

Together with the E1B region, E1A is capable of complete transformation of primary rat cells (Gallimore et al. 1974; Graham et al. 1975). In collaboration with other viral and cellular oncogenes, E1A is capable of transforming primary cells (Ruley 1983). E1A expresses activities that extend the growth potential of primary cells, resulting in cell lines that are capable of growing in

vitro as established cell lines that show morphological alterations (Houweling et al. 1980; Ruben et al. 1982; Zerler et al. 1986).

Two E1A mRNAs (12S and 13S) and their corresponding proteins are produced at early times after productive infection and in transformed cells. The 12S and 13S mRNAs are generated by differential splicing and encode proteins that are, respectively, 243 and 289 amino acids (aa) in length. Since the splicing process maintains the same reading frame in both RNAs, the 289-aa protein differs only by the presence of an additional 46-aa internal stretch.

The 12S and 13S proteins act with equal efficiency in conjunction with polyoma middle T or T24 ras gene to produce transformed baby rat kidney (BRK) cells that are tumorigenic in nude mice (Zerler et al. 1986). Several independent studies have shown that the 12S protein has only a subset of the activities responsible for affecting transcription of viral and cellular genes (Harrison et al. 1977; Esche et al. 1980; Carlock and Jones 1981; Ricciardi et al. 1981; Montell et al. 1982, 1984; Borrelli et al. 1984; Haley et al. 1984; Svensson and Akusjärvi 1984; Hurwitz and Chinnadurai 1985; Krippl et al. 1985; Velcich and Ziff 1985; Moran et al. 1986; Zerler et al. 1986). The 13S protein is capable of trans-activating viral early genes. Whether the 12S protein possesses this function remains controversial and may depend upon the experimental system employed. The 12S and 13S proteins both seem to stimulate DNA synthesis when microinjected into growth-arrested 3T3 cells (Stabel et al. 1985). The 12S protein stimulates virus replication in quiescent cells to a greater extent than the 13S protein (Montell et al. 1984; Spindler et al. 1985), suggesting that the 12S product may play a role in stimulating cell proliferation responses.

Because the 12S cDNA virus (Moran et al. 1986) can be used to efficiently infect all of the cells in a culture and because it is not cytotoxic, we have used it to study the early effects of the 12S gene product on primary BRK cells. We have found that the 12S product stimulates proliferation of the epithelial cell population in the absence of serum. Furthermore, the conditioned medium from 12S-infected cells induces quiescent, uninfected primary epithelial cells to enter a cell-cycle phase and proliferate.

## Experimental Procedures

### Materials
Collagenase, dispase (Boehringer Mannheim), hydrocortisone, insulin, prostaglandin $E_1$, selenium, transferrin, triiodothyronine (Sigma), and 6-[$^3$H]thymidine (Amersham, 20 Ci/mmole) were purchased from the suppliers indicated.

### Cells and viruses
Primary BRK cells were prepared as described previously (Ruley 1983) from 2- or 6-day-old Fisher rats (Taconic Farms). The medium, Dulbecco's modified Eagle's medium (DME) plus 5% fetal calf serum (FCS) or K1-1, was replaced twice a week with fresh medium. Medium K1-1 (Taub 1984) was prepared by combining an equal volume of medium F12 and DME (Gibco) and supplementing with hydrocortisone (40 ng/ml), insulin (5 μg/ml), prostaglandin $E_1$ (250 ng/ml), selenium ($10^{-8}$ M), transferrin (5 μg/ml), and triiodothyronine (4 ng/ml). Viral stocks were propagated on 293 cells. Viral infections and titrations were performed as described (Moran et al. 1986).

Conditioned medium was harvested after 2, 3, or 4 days as indicated, filtered through a 0.2-μm filter (Nalgene), and stored at 4°C. Conditioned K1-1 and DME without serum were supplemented with an equal volume of the appropriate fresh medium before adding to BRK cells. Conditioned medium was replaced twice a week with fresh conditioned medium. Treatments of conditioned medium were as described (De Larco and Todaro 1978). Cultures were fixed with methanol and stained with 2% Giemsa (Fisher).

### [$^3$H]Thymidine labeling
Cultures were incubated with a 5 μCi of [$^3$H]thymidine per milliliter of DME plus 5% FCS or K1-1 for the time periods indicated. The medium was removed, and the cells were washed three times with PBS at 4°C and then lysed with 0.3 N NaOH and scraped into a tube. The lysates were neutralized by adding Tris · HCl (pH 7.6) to 0.15 M and an equal volume of 0.3 N HCl. Equal aliquots were assayed for total or incorporated counts after trichloroacetic acid precipitation and liquid scintillation counting.

## Results

### Proliferation of epithelial cells
Primary BRK cultures isolated from 2- or 6-day-old rats and plated in DME plus 5% FCS consist of a mixture of fibroblasts and epithelial cells. Because FCS optimizes fibroblast and not epithelial cell growth, the epithelial cell population decreases rapidly until only a sparse, nonproliferating monolayer of fibroblasts remains. However, if these cultures are infected 2–3 days postplating with adenovirus type-5 (Ad5) 12S virus at a multiplicity of infection (moi) of 10, there is rapid cellular proliferation (Fig. 1). Microscopic examination of these infected cultures reveals that it is the epithelial subpopulation that is exhibiting the proliferative response. In mock-infected cells or cells infected with Ad5 *dl*312 (which contains a large E1A deletion; Jones and Shenk 1979), the epithelial cells decrease in number and disappear by about 5–6 days postplating (Fig. 1). Epithelial cell proliferation is first detectable in the 12S-infected cells between 24 and 36 hr postinfection. Initially, the epithelial cells form tightly delineated islands, with tightly packed cells. By about 3 days postinfection, the islands enlarge to form a monolayer. These rapidly growing areas may obscure any effect on the fibroblasts in the population. By about 8–10 days postinfection, large numbers of cells detach from the dish, although epithelial cell growth continues. Those cells that detach from the dish early after infection will continue to proliferate in suspension for 3–4 days (data not shown). By 3–4 weeks after infection, foci of cells are found, presumably representing those that have stably integrated adenoviral DNA into their genomes. The 12S-infected cells retain their epithelial phenotype; they are difficult to trypsinize and exhibit a very poor plating efficiency.

The degree and rapidity of the proliferative response is dependent on the moi (data not shown). When cells are infected with an moi of less than 1 plaque-forming unit (pfu) per cell, the proliferative response is not seen, although at a low frequency some foci may be observed after several weeks. In general, the best response is seen at an moi of 5, 10, or 20. At an moi of 50–100, there is extensive cellular proliferation, but some cytotoxic effects are observed. Infection of the cells with the 13S or wild-type virus results in extensive cell death starting at 24–36 hr postinfection, although small numbers of transformed foci are obtained with wild-type virus at a low moi (<0.01 pfu/cell) (data not shown).

### Cellular DNA synthesis
To examine levels of cellular DNA synthesis, cultures of primary BRK cells infected with Ad5 *dl*312 or Ad5 12S were incubated with [$^3$H]thymidine for increasing time periods for up to 24 hr. Figure 2 illustrates the results of an experiment performed at 4 days postinfection. The intracellular levels and the amount of [$^3$H]thymidine incorporated into DNA is always greater in 12S-infected cultures than in *dl*312-infected cells (Fig. 2). Maximal levels of [$^3$H]thymidine are observed after 12 hr of labeling. Although some [$^3$H]thymidine is taken up by cells infected with *dl*312, it is not incorporated into TCA-precipitable material (Fig. 2). Thus, the 12S-infected cultures not only exhibited increased levels of [$^3$H]thymidine uptake, but also enhanced efficiency of incorporation of [$^3$H]thymidine into cellular DNA. Cultures were also labeled for 24-hr intervals every day for a

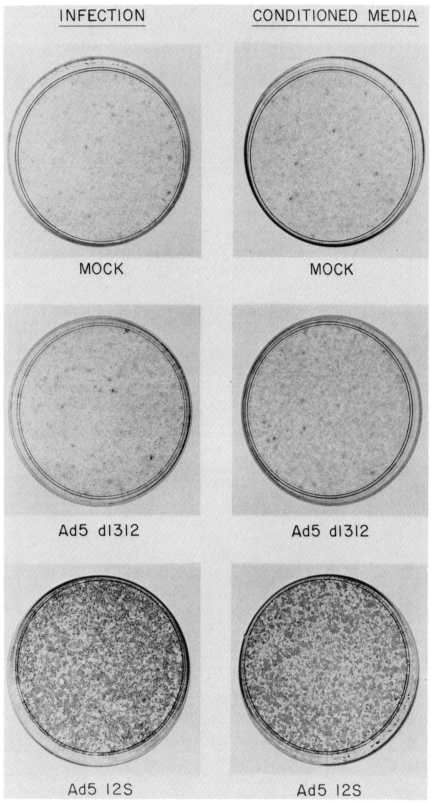

**Figure 1** Proliferation of primary BRK cells infected with an E1A 12S cDNA virus. Primary BRK cells from 2-day-old rats were mock-infected or infected with Ad5 *dl*312 or the Ad5 12S virus (*left* column) or treated with conditioned medium generated by BRK cells infected with the indicated viruses (*right* column) at 2 days postplating. The cells were fixed 3 days posttreatment and stained with Giemsa.

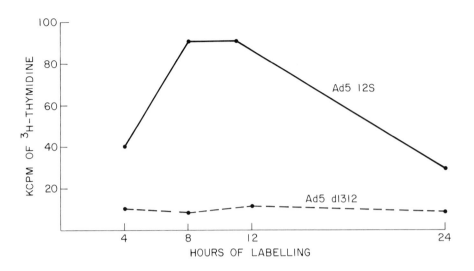

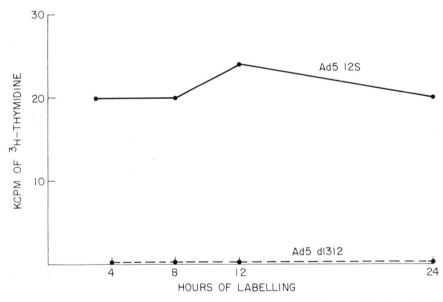

**Figure 2** Increased cellular DNA synthesis in cells infected with the 12S virus. At 3 days postinfection with 12S virus (———) or *dl*312 (– – –) virus, BRK cells were incubated with [³H]thymidine for the times indicated. At the end of the labeling period, the cells were lysed and the levels of [³H]thymidine uptake and incorporation were determined.

week. A 5-fold to 10-fold increase in [³H]thymidine incorporation in 12S-infected cultures compared with *dl*312-infected cultures is observed at all times examined (data not shown). No viral DNA replication was detected (see below). Thus, the increase in DNA synthesis is due to cellular DNA replication.

**Proliferation in the absence of serum**
The ability of cells infected with the 12S virus to grow in a serum-free, hormonally defined media, K1-1, was examined. K1-1 has been developed to optimize epithelial

cell growth (Taub 1984) and is inhibitory to fibroblast growth. This medium contains insulin as a growth factor. Cells were plated in DME plus FCS, infected with 12S or *dl*312 or mock-infected, and thereafter maintained in K1-1. The epithelial cells in 12S-infected cultures proliferated rapidly. The growth and life spans of the epithelial cells in uninfected or *dl*312-infected cultures were still only 4–5 days. BRK cells infected with 12S virus are also able to synthesize cellular DNA at a similar rate in K1-1 media (data not shown) as in DME plus FCS. If BRK cells are initially plated in K1-1, no fibroblasts grow. Thus, the ability of the 12S-infected cells in K1-1 to proliferate in the absence of fibroblasts eliminated the possible effects due to cross-feeding.

To determine whether infected BRK cells could survive in DME in the absence of serum, BRK cells were plated in DME plus FCS. Subsequent to infection, they were maintained for two days in the presence of serum and then switched into DME without serum. We observed that the epithelial cells continued to proliferate in 12S-infected cultures. The number of fibroblasts decreased rapidly in the absence of serum (data not shown). Thus, the presence of the 12S gene product enabled the epithelial cells to proliferate in the absence of serum.

### 12S-conditioned media induced proliferation

The rapid cellular proliferation induced after 12S virus infection raised the possibility that the infected BRK cells produce a growth factor that induces cellular DNA synthesis and subsequent cell division. To investigate this, primary BRK cells were first infected at an moi of 10 with different adenoviruses or were mock-infected. At 3 days after infection, the conditioned media from the respective cultures were removed and filtered to remove cells and debris. The conditioned medium from each infection was added to new, uninfected BRK cells 2–3 days after plating. The cellular response to the presence of the conditioned media was similar to that induced by infection (Fig. 1). The proliferation-inducing potential of 12S-conditioned media generated in DME in the absence of serum (Fig. 3) or in K1-1 (data not shown) was also tested. Both conditioned media are equally competent at stimulating epithelial cell proliferation (Fig. 3 and data not shown). The minimal nature of these media results in the depletion of required components. Hence, conditioned media was supplemented with an equal volume of fresh media before adding to BRK cells. The rapid decrease in the fibroblast population, upon addition of conditioned medium from Ad5 12S-infected BRK cells, together with the rapid proliferation of the epithelial cells, suggests the presence of an epithelial cell–specific proliferation factor(s). To determine how long the conditioned medium was capable of mediating epithelial cell proliferation, uninfected primary BRK cells were placed in 12S- or *dl*312-conditioned medium. The conditioned medium was replaced with fresh conditioned medium twice a week. Epithelial cell proliferation was observed for at least 6 weeks in the presence of fresh conditioned medium. These cells retained their epithelial cell characteristics.

The kinetics of production of the proliferation factor(s) by infected BRK cells were determined. BRK cells were infected with 12S virus at an moi of 10. At different times postinfection, conditioned medium was collected and replaced with fresh DME. The conditioned medium was tested for the presence of stimulatory factor(s). Proliferation of epithelial cells was detected in response to media taken from 12S-infected cells as early as 6 hr after infection and as late as 5 days after infection. Media harvested after 5 days was negative. This is about

the time that viral DNA is rapidly lost from infected cultures and cells begin to be lost from cultures. It is therefore possible that the concentration of stimulatory factor(s) drops below the level that can be detected in our bioassay. Alternatively, the expression of the factor(s) may be altered. It was difficult to test whether the 13S gene product expressed from a 13S cDNA virus or from wild-type virus induced similar stimulatory factor(s) because of its cytotoxic effect.

### Cellular DNA synthesis induced by 12S-conditioned medium

The levels of DNA synthesis in cultures receiving conditioned media from mock-, 12S-, or *dl*312-infected cells were examined. The intracellular levels and the amount of [$^3$H]thymidine incorporated into DNA are substantially greater in cultures receiving conditioned media from 12S-infected cultures than from mock- or *dl*312-infected cultures (Fig. 4). The 12S-conditioned medium harvested from infected cells 4 days after infection could be diluted 1:20 and yield a similar response, both with respect to observable cell proliferation and [$^3$H]thymidine incorporation (data not shown).

### Requirement for the continued presence of conditioned media

To determine whether there was a need for the continuous presence of the proliferation factor(s), we performed reversal experiments. All media were replaced with the appropriate fresh media every 3 days. Primary BRK cells were placed in 12S-conditioned media. After 6 days, one set of cells was changed into and maintained in fresh DME plus FCS for another 6 days (Fig. 5, bottom panel). The other set was maintained in fresh 12S-conditioned medium for another 6 days (Fig. 5, middle panel). Cellular proliferation ceased in the cultures changed into DME, and the epithelial cells gradually disappeared from the population (Fig. 5, bottom panel) so that it resembled cultures that had been maintained in DME throughout the experiment (Fig. 5, top panel). Growth-curve analyses showed that cells maintained in 12S-conditioned medium for shorter time periods (i.e., 2–3 days) ceased cell division within 24 hr after removing the 12S-conditioned medium and resembled mock-treated cultures within 2 days (data not shown). We have no direct evidence whether the factor(s) responsible for epithelial cell proliferation is secreted into the medium or is released due to cell lysis.

The ability of the 12S-conditioned medium to complement T24 Ha-*ras* in the production of transformed loci of primary cells was also tested. Primary BRK cells were transfected with the Ha-*ras* gene with and without the E1A 12S gene. After transfection, the cells were plated in DME plus FCS or 12S-conditioned media. Many foci appeared in those dishes in which the E1A 12S and Ha-*ras* genes were cotransfected, consistent with previous observations (Ruley 1983). No permanently transformed foci appeared in the cultures transfected with the Ha-*ras* gene plus 12S-conditioned medium. Thus, the presence of a proliferation factor(s) does not seem

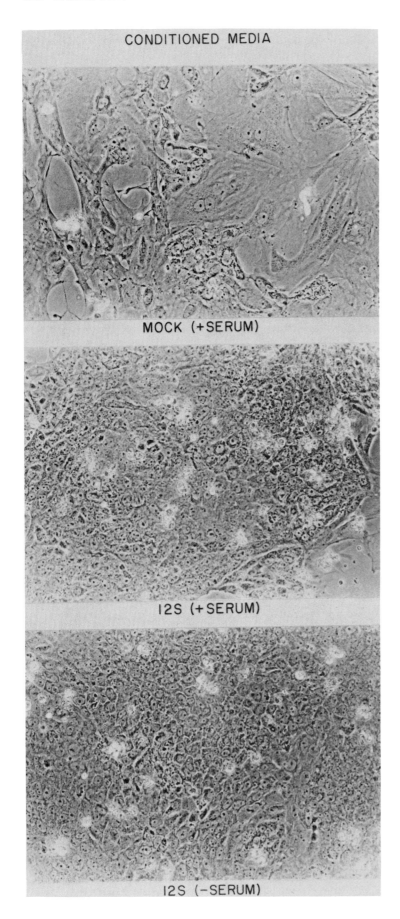

**Figure 3** Proliferation of primary epithelial cells in the presence of 12S-conditioned medium. At 2 days postplating, primary BRK cells received medium conditioned by mock- or 12S-infected BRK cells that had been maintained for 2 days in DME with and without serum. Photographs were taken 4 days after the addition of the conditioned medium.

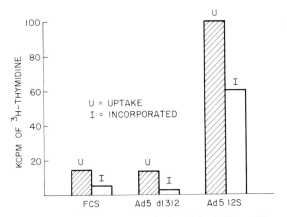

Figure 4 Cellular DNA synthesis induced by 12S-conditioned medium. Primary BRK cells were incubated with the indicated conditioned medium and [³H]thymidine for a 24-hr period. The cells were lysed and the levels of [³H]thymidine were determined.

to effect transformation by *ras* and suggests that the induction of a growth factor(s) is not the sole function of the 12S product. It cannot be ruled out that the concentration of the factor(s) in the conditioned medium of infected cells is sufficient to induce cellular DNA synthesis and proliferation but is insufficient to produce additional responses.

Conditioned media from cell lines containing integrated E1A regions and expressing E1A gene products were negative for the induction of epithelial cell proliferation (data not shown). Thus, the constitutive expression of the 12S gene need not, per se, result in the production of high levels of stimulatory factor(s) (see Discussion).

### Absence of viral infection

The above results suggest that the induction of cellular DNA synthesis and cellular proliferation by 12S-conditioned medium was not due to reinfection by residual viral particles in the conditioned medium. To substantiate this, several control experiments were performed. The conditioned medium from 12S- and dl312-infected BRK cells were titered on human 293 cells, which contain integrated E1A sequences and complement the growth of E1A mutant viruses. The residual virus titers in the conditioned medium would enable an moi of 0.01–0.001. A proliferative response was not detected in cultures infected with an moi of less than 1.0 (see above). No viral DNA was detected in Hirt extracts of cells treated with conditioned medium, although reconstruction experiments showed that viral DNA would have been detected in cells infected with an moi of 0.05–0.1. Cells receiving conditioned medium were also examined for adenovirus gene expression. No viral proteins (E1A, 72K, or hexon) could be detected in immunoprecipitates from these cells (data not shown).

### Characteristics of the factor

To further characterize the stimulatory factor(s) in 12S-conditioned medium, several physical properties were determined. We tested the stability of 12S-conditioned media to heat treatment at different temperatures and for increasing time periods. Treated medium was then diluted with an equal volume of fresh DME plus FCS to counteract any negative effect of the heat treatment on other components of the medium and tested for proliferative activity on uninfected BRK cells. The activity was found to be stable up to 43°C, some loss of activity was detected at 46°C, and the activity was totally inactivated after treatment at 50°C. Trypsin sensitivity was determined by incubating conditioned medium with trypsin for 2 hr at 37°C. After dilution with fresh medium, it was tested for stimulatory activity. The trypsin-treated, conditioned medium was negative for proliferative activity. If the trypsin was pretreated with soybean trypsin inhibitor, the proliferative activity was retained. The soybean trypsin inhibitor, by itself, had no effect on the proliferation-stimulating activity of the conditioned medium.

The stimulatory factor(s) was retained after dialysis with membranes with a molecular-weight cutoff (MWCO) of 12,000–14,000 (12–14K). Conditioned medium was subjected to centrifugation at 100g for 60 min at 4°C, and the pellet was reconstituted with DME plus FCS. The reconstituted pellet and the supernatant were then added to primary BRK cells. The stimulatory activity was found to be associated with the pellet fraction. The pellet fraction was then treated with increasing amounts of NaCl, again subjected to centrifugation, and dialyzed against DME. With increasing salt concentration, the activity was gradually released into the supernatant, until at 4 M NaCl, all the stimulatory activity was found in the supernatant fraction. However, only 10% of the residual virus was present in the active fraction, whereas 90% was retained in the inactive pellet fraction. These results suggest that the protein factor(s) may be found as a complex with a large macromolecule(s). Consistent with this observation was that the factor in the 12S-conditioned medium was retained by Amicon filters with a MWCO of 100K. The activity is stable for 4–6 weeks at 4°C. The factor(s) is sensitive to acetone precipitation and acetic acid extraction but is stable to lyophilization.

### Discussion

Adenovirus normally infects terminally differentiated, noncycling epithelial cells (Tooze 1981) that are arrested in $G_0$. To undergo replication, the virus makes use of the DNA synthetic machinery of the host. The ability of the 12S gene product to induce cellular DNA synthesis and proliferation may serve both the lytic and the immortalization/transformation pathways of adenovirus. It has been shown that the 12S product is necessary for maximal viral replication in quiescent, permissive cells (Montell et al. 1984; Spindler et al. 1985) but not in growing permissive cells (Montell et al. 1982). Primary

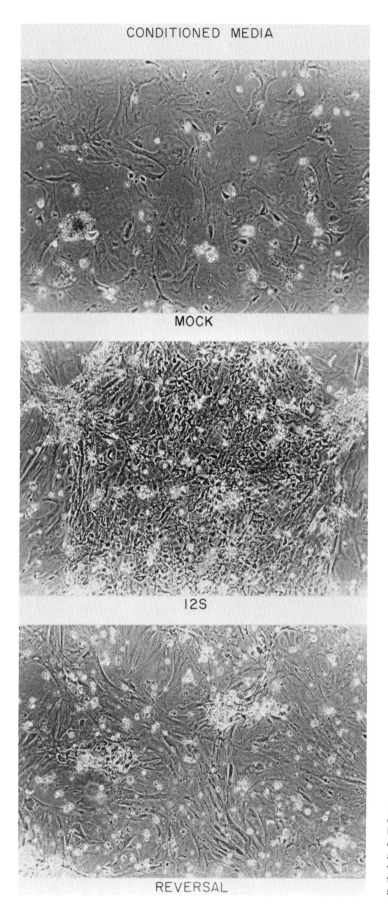

**Figure 5** Reversible nature of the effect of 12S-conditioned medium on epithelial cell proliferation. Primary BRK cells were incubated with mock-infected (*top* panel) or 12S-conditioned medium (*middle* and *bottom* panels). After 6 days, one set of cultures was switched into DME (*bottom* panel). See Results for details. Photographs were taken 6 days after the switch.

BRK epithelial cells in culture cease growth and eventually die, even in the presence of serum. Infection with the 12S virus induces these cells to enter a proliferative phase. This suggests that the presence of the 12S gene overcomes quiescence and shifts cells into a proliferative state, either directly or indirectly.

Polypeptide mitogenic growth factors act to induce morphological and mitogenic changes in their target cells (for reviews, see Heldin and Westermark 1984; James and Bradshaw 1984). The close relationship between certain oncogenes, proteins, growth factors, and their receptors emphasizes the role of growth factors in neoplastic and normal cell growth (for review, see Heldin and Westermark 1984; Hunter and Cooper 1985; Sporn and Roberts 1985).

Proliferation in the absence of serum is a common phenotypic property of malignant cells derived from tumors and cells that have been transformed with viruses and/or chemicals in vitro. The reduced requirements can be due to (1) alterations in growth-factor receptors, (2) the production of growth factors by the cells, or (3) constitutive alterations in growth-factor-controlled processes. The conditioned medium from 12S-infected and not *dl*312-infected BRK cells was able to induce a proliferative response, suggesting that these cells are producing a mitogenic factor(s). The induction of cellular DNA synthesis by E1A could then be due to the presence of the growth factor(s). In this context, it is interesting to note that the stimulatory activity was detected in the medium as early as 6 hr postinfection, prior to the onset of cellular DNA synthesis. The factor(s) seems to be specific for epithelial cells, although we cannot, at this time, eliminate the possibility of an effect on the fibroblasts.

There is no direct evidence to suggest that the activity found in the medium is the 12S protein itself. It has recently been shown that vaccinia virus codes for a protein that has a segment homologous to epidermal growth factor (EGF) and type-$\alpha$ transforming growth factor (TGF-$\alpha$) and that vaccinia virus–infected cells release a growth factor with mitogenic activity (Brown et al. 1985; Stroobant et al. 1985). We have assayed conditioned medium for the 12S protein by immunoprecipitation analysis, with negative results. However, it is possible that the levels are too low to be detected, because even the intracellular levels of E1A proteins during an infection are very low. EGF and TGF-$\alpha$ seemed to be possible candidates for the 12S-induced factor due to the epithelial cell specificity of the activity. However, with respect to heat stability and acetic acid resistance, at least a component of the 12S-induced factor(s) seems quite distinct from TGF-$\alpha$ and EGF (De Larco and Todaro 1978). We are presently trying to determine whether this is a previously identified growth factor or a novel one whose expression is induced by the 12S virus.

No stimulatory activity was detected in conditioned medium from transformed cells that constitutively express the 12S protein. Several explanations for this are, of course, possible. Transformed cells may have undergone many changes and lost many original epithelial cell characteristics such that they no longer produce the same factor(s) that they did initially. Most transformed epithelial cells replicate very well in the presence of serum. The production of the growth factor(s) may be an early event in the transformation process and may not be required for the maintainence of the transformed state. The 12S products may provide other functions necessary for transformation. Alternatively, transformed cells may produce the stimulatory factor(s) at a level too low to be detected in the bioassay used. In this regard, it should be noted that the levels of E1A proteins are higher in infected cells than in transformed cells. It is not known whether continued high levels of E1A expression are toxic to cells and transformed cells are selected that have lower E1A levels and thus lower levels of growth factors. Alternatively, the stimulatory factor(s) induced by 12S may not be directly related to its transforming ability but may simply induce infected cells to enter a proliferative phase to support viral DNA replication. However, an initial round of host-cell DNA synthesis after infection appears to be essential for the establishment of adenovirus transformation (Casto 1973; Tooze 1981).

Whether the production of the growth factor(s) is the primary function of 12S protein and the other effects of the 12S product observed are mediated by the growth factor(s) remain to be determined. The inability of 12S-conditioned medium to complement Ha-*ras* to fully transform BRK cells suggests otherwise. However, it is possible the level of factor(s) was insufficient for this purpose. Further studies are needed to determine the role of the growth factor(s) in transformation.

## Acknowledgments

We are grateful to Dolly Chao for excellent technical assistance. We would like to thank Dr. Sri Sharma for helpful discussions and critical reading of the manuscript, and we appreciate the helpful suggestions of Dr. Neil Sullivan. We are grateful to Carrie O'Loughlin for the careful animal work. We thank Jean Michaelis, David Greene, and Mike Ockler for excellent photographic and art work. This work was supported by the National Institutes of Health (Cancer Research grant CA 13106). M.Q. is a recipient of a National Institutes of Health postdoctoral fellowship (F32 CA 97676) and funds from an institutional grant from the American Cancer Society (ACS IN153).

## References

Bellett, A.J.D., L.K. Waldron-Stevens, A.W. Braithwaite, and B.F. Cheetham. 1982. Spermine and amino-guanidine protect cells from chromosome alterations induced by adenovirus during the $G_2$ phase of the cell cycle. *Chromosoma* **84**: 571.

Bellett, A.J.D., P. Li, E.T. David, E.J. Mackey, A.W. Braithwaite, and J.R. Cutt. 1985. Control functions of adenovirus transformation region E1A products in rat and human cells. *Mol. Cell. Biol.* **5**: 1933.

Berk, A.J., F. Lee, T. Harrison, J. Williams, and P.A. Sharp. 1979. Pre-early adenovirus 5 gene product regulates the synthesis of early viral messenger RNAs. *Cell* **17**: 935.

Borrelli, E., R. Hen, and P. Chambon. 1984. Adenovirus-2 E1A products repress enhancer-induced stimulation of transcription. *Nature* **312**: 608.

Braithwaite, A.W., J.D. Murray, and A.J.D. Bellett. 1981. Alterations to controls of cellular DNA synthesis by adenovirus infection. *J. Virol.* **39**: 331.

Braithwaite, A.W., B.F. Cheetham, P. Li, C.R. Parish, L.K. Waldron-Stevens, and A.J.D. Bellett. 1983. Adenovirus induced alterations of the cell growth cycle: A requirement for the expression of E1A but not of E1B. *J. Virol.* **45**: 192.

Brown, J.P., D.R. Twardzik, H. Marquardt, and G.J. Todaro. 1985. Vaccinia virus encodes a polypeptide homologous to epidermal growth factor and transforming growth factor. *Nature* **313**: 491.

Carlock, L.R. and N.C. Jones. 1981. Transformation-defective mutant of adenovirus type 5 containing a single altered E1A mRNA species. *J. Virol.* **40**: 657.

Casto, B.C. 1973. Biologic parameters of adenovirus transformation. *Prog. Exp. Tumor Res.* **18**: 166.

Cheetham, B.F. and A.J.D. Bellett. 1982. A biochemical investigation of the adenovirus-induced $G_1$ to S phase transition: Thymidine kinase, ornithine decarboxylase and inhibitors of polyamine biosynthesis. *J. Cell. Physiol.* **110**: 114.

De Larco, J.E. and G.J. Todaro. 1978. Growth factors from murine sarcoma virus-transformed cells. *Proc. Natl. Acad. Sci.* **75**: 4001.

Esche, H., M.B. Mathews, and J.B. Lewis. 1980. Proteins and messenger RNAs of the transforming region of wild type and mutant adenoviruses. *J. Mol. Biol.* **142**: 399.

Gallimore, P.H., P.A. Sharp, and J. Sambrook. 1974. Viral DNA in transformed cells. III. A study of the sequences of adenovirus 2 DNA in nine lines of transformed rat cells using specific fragments of the viral genome. *J. Mol. Biol.* **8**: 49.

Gaynor, R.B., D. Hillman, and A.J. Berk. 1984. Adenovirus E1A protein activates transcription of a non-viral gene introduced into mammalian cells by infection or transfection. *Proc. Natl. Acad. Sci.* **81**: 1193.

Graham, F.L., P.J. Abrahams, C. Mulder, H.L. Heijneker, S.O. Warnaar, F.A.J. deVries, W. Fiers, and A.J. van der Eb. 1975. Studies on in vitro transformation by DNA and DNA fragments of human adenovirus and simian virus 40. *Cold Spring Harbor Symp. Quant. Biol.* **39**: 637.

Green, M.R., R. Treisman, and T. Maniatis. 1983. Transcriptional activation of cloned β-globin genes by viral immediate early gene products. *Cell* **35**: 137.

Haley, K.P., G. Overhauser, L.E. Babiss, H.S. Ginsberg, and N.C. Jones. 1984. Transformation properties of type 5 adenovirus mutants that differentially express the E1A gene products. *Proc. Natl. Acad. Sci.* **81**: 5734.

Harrison, T., F. Graham, and J. Williams. 1977. Host range mutants of adenovirus type 5 defective for growth in HeLa cells. *Virology* **77**: 319.

Hearing, P. and T. Shenk. 1985. Sequence-independent autoregulation of the adenovirus type 5 E1A transcription unit. *Mol. Cell. Biol.* **5**: 3214.

Heldin, C.-H. and B. Westermark. 1984. Growth factors: Mechanism of action and relation to oncogenes. *Cell* **37**: 9.

Houweling, A., P.J. van der Elsen, and A.J. van der Eb. 1980. Partial transformation of primary rat cells by the leftmost 4.5% fragment of adenovirus 5 DNA. *Virology* **105**: 537.

Hunter, T. and J.A. Cooper. 1985. Protein-tyrosine kinases. *Annu. Rev. Biochem.* **54**: 897.

Hurwitz, D.R. and G. Chinnadurai. 1985. Evidence that a second tumor antigen coded by adenovirus early region E1a is required for efficient cell transformation. *Proc. Natl. Acad. Sci.* **82**: 163.

Imperiale, M.J. and J.R. Nevins. 1984. Adenovirus 5 E2 transcription unit: An E1A inducible promoter with an essential element that functions independently of position or orientation. *Mol. Cell. Biol.* **4**: 875.

James, R. and R.A. Bradshaw. 1984. Polypeptide growth factors. *Annu. Rev. Biochem.* **53**: 259.

Jones, N. and T. Shenk. 1979. An adenovirus type 5 early gene function regulates expression of other early viral genes. *Proc. Natl. Acad. Sci.* **76**: 3665.

Kao, H.-T. and J.R. Nevins. 1983. Transcriptional activation and subsequent control of the human heat shock gene during adenovirus infection. *Mol. Cell. Biol.* **3**: 2058.

Krippl, B., B. Ferguson, N. Jones, M. Rosenberg, and H. Westphal. 1985. Mapping of functional domains in adenovirus E1 proteins. *Proc. Natl. Acad. Sci.* **83**: 7480.

Leff, T., R. Elkaim, C.R. Goding, P. Jalinot, P. Sassone-Corsi, M. Perricaudet, C. Kédinger, and P. Chambon. 1984. Individual products of the adenovirus 12S and 13S E1A mRNAs stimulate viral EIIA and EIII expression at the transcriptional level. *Proc. Natl. Acad. Sci.* **81**: 4381.

Montell, C., G. Courtois, C. Eng, and A. Berk. 1984. Complete transformation by adenovirus 2 requires both E1A proteins. *Cell* **36**: 951.

Montell, C., E.F. Fisher, M.H. Caruthers, and A.J. Berk. 1982. Resolving the functions of overlapping viral genes by site specific mutagenesis at a mRNA splice site. *Nature* **295**: 380.

Moran, E., T. Grodzicker, R.J. Roberts, M.B. Mathews, and B. Zerler. 1986. Lytic and transforming functions of individual products of the adenovirus E1A gene. *J. Virol.* **57**: 765.

Murray, J.D., A.W. Braithwaite, I.W. Taylor, and A.J.D. Bellett. 1982a. Adenovirus-induced alterations of the cell growth cycle: Effects of mutations in early regions E2A and E2B. *J. Virol.* **44**: 1072.

Murray, J.D., A.J.D. Bellett, A.W. Braithwaite, L.K. Waldron, and I.W. Taylor. 1982b. Altered cell cycle progression and aberrant mitosis in adenovirus-infected rodent cells. *J. Cell. Physiol.* **111**: 89.

Nevins, J.R. 1981. Mechanism of activation of early viral transcription by the adenovirus E1A gene product. *Cell* **26**: 213.

———. 1982. Induction of the synthesis of a 70,000 dalton mammalian heat shock protein by the adenovirus E1A gene product. *Cell* **29**: 913.

Pochron, A., M. Rossini, Z. Darzynkiewcz, F. Traganos, and R. Baserga. 1980. Failure of accumulation of cellular RNA in hamster cells stimulated to synthesize DNA by infection with adenovirus 2. *J. Biol. Chem.* **255**: 4411.

Ricciardi, R.P., R.L. Jones, C.L. Cepko, P.A. Sharp, and B.E. Roberts. 1981. Expression of early adenovirus genes requires a viral encoded acidic polypeptide. *Proc. Natl. Acad. Sci.* **78**: 6121.

Ruben, M.S., S. Bachetti, and F.L. Graham. 1982. Integration of viral DNA in cells transformed by host range mutants of adenovirus type 5. *J. Virol.* **41**: 674.

Ruley, H.E. 1983. Adenovirus early region 1A enables viral and cellular transforming genes to transform primary cells in culture. *Nature* **304**: 602.

Shimojo, H. and T. Yamashita. 1968. Induction of DNA synthesis by adenovirus in contact-inhibited hamster cells. *Virology* **36**: 422.

Spindler, K.R., C.Y. Eng, and A.J. Berk. 1985. An adenovirus early region 1A protein is required for maximal viral DNA replication in growth arrested human cells. *J. Virol.* **53**: 742.

Sporn, M.B. and A.B. Roberts. 1985. Autocrine growth factors and cancer. *Nature* **313**: 745.

Stabel, S., P. Argos, and L. Philipson. 1985. The release of growth arrest by microinjection of adenovirus E1A DNA. *EMBO J.* **4**: 2329.

Stein, R. and E.B. Ziff. 1984. HeLa cell β-tubulin gene transcription is stimulated by adenovirus 5 in parallel with viral early genes by an E1a-dependent mechanism. *Mol. Cell. Biol.* **4**: 2792.

Stich, H.F. and D.S. Yohn. 1970. Viruses and chromosomes. *Prog. Med. Virol.* **12**: 78.

Strohl, W.A. 1969. The response of BHK21 to infection with type 12 adenovirus. II. Relationship of virus-stimulated DNA synthesis to other viral functions. *Virology* **39**: 653.

Stroobant, P., A.P. Rice, W.J. Gullick, D.J. Cheng, and I.M. Kerr. 1985. Purification and characterization of vaccinia virus growth factor. *Cell* **43:** 383.

Svensson, C. and G. Akusjärvi. 1984. Adenovirus 2 early 1A stimulates expression of both viral and cellular genes. *EMBO J.* **3:** 789.

Takahashi, M., T. Ogino, K. Baba, and M. Onaka. 1969. Synthesis of deoxyribonucleic acid in human and hamster kidney cells infected with human adenovirus types 5 and 12. *Virology* **37:** 513.

Taub, M. 1984. Growth of primary and established kidney cell cultures in serum-free media. In *Methods of serum-free culture of epithelial and fibroblastic cells* (ed. D.W. Barnes et al.), p. 3. A.R. Liss, New York.

Tooze, J., ed. 1981. *Molecular biology of tumor viruses*, 2nd edition, revised: *DNA tumor viruses*. Cold Spring Harbor Laboratory, Cold Spring Harbor, New York.

Treisman, R., M.R. Green, and T. Maniatis. 1983. *Cis*- and *trans*-activation of globin transcription in transient assays. *Proc. Natl. Acad. Sci.* **80:** 7428.

Velcich, A. and E. Ziff. 1985. Adenovirus E1A proteins repress transcription from the SV40 promoter. *Cell* **40:** 705.

Weeks, W. and N.C. Jones. 1983. E1A control of gene expression is mediated by sequence 5′ to the transcriptional starts of the early viral genes. *Mol. Cell. Biol.* **3:** 1222.

Younghusband, H.B., C. Tyndall, and A.J.D. Bellett. 1979. Replication and interaction of virus DNA and cellular DNA in mouse cells infected by a human adenovirus. *J. Gen. Virol.* **45:** 455.

Zerler, B., B. Moran, K. Maruyama, J. Moomaw, T. Grodzicker, and H.E. Ruley. 1986. Adenovirus E1A coding sequences that enable *ras* and *pmt* oncogenes to transform cultured primary cells. *Mol. Cell. Biol.* **6:** 887.

# Studies on Adenovirus Type-12 E1 Region: Gene Expression, Transformation of Human and Rodent Cells, and Malignancy

P.H. Gallimore,* J. Williams,[†] D. Breiding,[†] R.J.A. Grand,* M. Rowe,* and P. Byrd*

*Cancer Research Campaign Laboratories, Department of Cancer Studies, The Medical School, University of Birmingham, Birmingham B15 2TJ, England; [†]Department of Biological Sciences, Carnegie-Mellon University, Pittsburgh, Pennsylvania 15213

We have isolated a series of adenovirus type-12 (Ad12) early region 1 (E1) mutants that have been sequenced and characterized for kinetics of virus protein synthesis, effect on cellular gene expression, transformation, and tumorigenicity. In addition, we have been examining complementation of adenovirus E1A region by *ras* genes in the transformation of primary human cells. Results from both of these studies have provided a clearer insight to the specific properties of E1 proteins.

Adenovirus genes located in early region 1 (E1) play a dominant role in the transformation of rodent cells and in adenovirus oncogenesis. However, the precise biochemical activity of individual E1 proteins has yet to be determined. Up to now, two approaches, namely DNA transfection and the use of adenovirus mutants, have provided some insight into the roles the E1A and E1B proteins play in infection and transformation (for reviews, see Bernards and van der Eb 1984; Graham 1984; Gallimore et al. 1985a).

Over the past four years we have isolated a number of mutants of the oncogenic human adenovirus type 12 (Ad12) that map to the E1A 266- and 235-amino-acid (aa) proteins and large E1B 482-aa (54K) protein. In these Ad12 mutants, the DNA sequence has been determined and the genetic lesions have been identified. In the studies described here, we report the effect these mutations have had on transformation, tumorigenicity, and the expression of adenovirus and cell proteins. In addition, we describe a human tissue culture system in which adenovirus E1 genes transform almost as efficiently as on baby rat kidney (BRK) cells. Using primary human embryo retinoblasts (HERs), we show that adenovirus E1A and human activated *ras* genes can complement for transformation, as originally demonstrated by Ruley (1983) for rat cells, and describe some of the features of transformed human retinoblasts.

## Experimental Procedures

### Plasmids

The following plasmids were used in this study: pAsc2 (Ad12 E1A + E1B; Byrd et al. 1982b); pAsc4.7 (Ad12 E1A; Byrd et al. 1982b); pLB205 (Ad5 E1A + E1B; obtained from Lee Babiss); pR29 (Ad2 E1A; obtained from Nigel Stow); pN-*ras* (activated human N-*ras*; obtained from Alan Hall); and pHO6T1 (containing the activated human H-*ras* and the G418-resistance gene; Spandidos and Wilkie 1984).

### Virus stocks

Ad12 Huie strain was plaque-purified on primary human embryo kidney (HEK) cells. Low-passage seed stocks were grown up on HEK cells, purified using CsCl gradients, and plaque-titrated on HEK cells. Ad12 mutants were constructed either by recombinant DNA technology (Ad12 *in*600, *in*601, *in*602, and *del*620) or by nitrous acid mutagenesis (Ad12 *hr*700 and *hr*703) and isolated on Ad12 HER-3 cells (Byrd 1983; Gallimore et al. 1984). Stocks of mutant virus were grown up on Ad12 HER-3 cells, purified using CsCl gradients, and titrated on Ad12 HER-3 cells.

### Transformation assays

1. Baby rat kidney cell cultures (BRK) and baby mouse kidney cells (BMK) were prepared as previously described (Gallimore et al. 1985b). Exponentially growing cultures were infected over a range of virus doses for wild-type virus of 0.1–10 pfu/cell and for mutant viruses of 10–200 pfu/cell. Transformation experiments were carried out in quadruplicate and at three temperatures— 33°C, 37°C, and 38.5°C. Dishes were fixed and stained at either 4 or 6 weeks postinfection.

2. HER cultures were prepared as described by Byrd et al. (1982a). DNA transfection was carried out using the glycerol boost modification (Frost and Williams 1978) of the original method described by Graham and van der Eb (1973). HER cultures were transfected with a total of 10 µg of DNA per dish or 10 + 10 µg in cotransfection experiments, and transformed foci were counted microscopically at 4–9 weeks after transfection. Individual transformed foci were picked using finely drawn pasteur pipettes at 6–10 weeks posttransfection. The transforming activities of the different plasmids were related to that of Ad2 E1A, correction being made for the differences in the sizes of the plasmids. The relative transforming activities of E1A plus *ras* mixtures were calculated using the concentrations and sizes of the E1A plasmids only.

*Tumorigenicity studies*
One-day-old Hooded Lister rats were inoculated intraperitoneally with 0.2 ml of virus over the range of $1 \times 10^5$ to $1 \times 10^7$ pfu/rat for wild-type virus and $1 \times 10^8$ to $1 \times 10^9$ pfu/rat for the Ad12 mutants (at least 10 animals per virus dose).

*Western blotting and immunoprecipitation*
Western blotting was carried out essentially as described by Gallimore et al. (1984). Samples were prepared for polyacrylamide gel electrophoresis as follows: Infected or transformed cells were washed in ice-cold saline and harvested using a rubber policeman. Cell pellets were suspended in 9 M urea, 50 mM Tris · HCl (pH 7.5), 0.15 M β-mercaptoethanol and sonicated for 30 sec. Aliquots (25 μg of protein) were mixed with SDS and subjected to polyacrylamide gel electrophoresis. After western blotting and autoradiography, the gels were scanned using an Isco gel scanner and quantitation was relative to Ad12 protein IX for structural protein studies and to the level of major histocompatibility complex (MHC) class I in primary HER cells for MHC class-I studies. The procedure for immunoprecipitation was that of Paraskeva et al. (1982).

*Thymidine kinase assay*
Confluent 9-cm dishes of primary HEK cells were infected at multiplicities of infection of 50 pfu/cell (wild-type virus) or 200 pfu/cell (mutant viruses). After appropriate times, the infected cells were washed twice in ice-cold saline and harvested using a rubber policeman. The cell pellet was resuspended in 20 mM Tris · HCl, pH 7.4 (300 μl), and sonicated for 30 sec, and after centrifugation the supernatant was used for the determination of thymidine kinase (TK) activity. The assay mixture (250 μl) contained 50 mM Tris · HCl (pH 7.5), 2 mM DTT, 10 mM MgCl$_2$, 5 mM ATP, 1 mg BSA, 10 mM NaF, and 50 μl of cell supernatant. The reaction was started by the addition of 0.25 mM thymidine containing 1 μCi of [$^{14}$C]thymidine. After a 30-min incubation at 37°C, aliquots (50 μl) were removed and spotted onto Whatman DE-81 paper discs. The discs were washed five times (5 min each wash with agitation) with 90% ethanol, dried, and counted in scintillation fluid.

*Fluorescence-activated cell sorting*
Single cell suspensions were prepared either directly before assaying by using versene (transformed HER lines) or by trypsinization followed by overnight culture in methylcellulose/Dulbecco's modified Eagle's medium plus 10% fetal calf serum semisolid medium (primary HERs). Cells were reacted with either normal mouse serum or the monoclonal antibody W6/32 (Barnstable et al. 1978; purchased from Sera Lab) at 4°C for 1 hr. The second antibody step used anti-mouse IgG labeled with fluorescein isothiocyanate at 4°C for 1 hr. After extensive washing, cells were fixed in 1% formol saline and processed on a Becton Dickinson FACS 440. Cells were analyzed at 1000 cells/sec at a stimulatory wavelength of 488 nm. Antibodies were diluted in saline containing 20% goat serum to quench nonspecific binding.

## Results

### The effect of Ad12 E1 mutations on the expression of Ad12 genes and specific cellular genes
Ad12 E1 and structural protein synthesis was studied by the western blotting technique and/or immunoprecipitation, using tumor-bearer serum with antibodies to Ad12 E1 proteins (Gallimore et al. 1985b) or SDS-disrupted, purified Ad12 virion antiserum (Paraskeva et al. 1982). HEK cells were infected with Ad12 wild-type or mutant virus at 200 pfu/cell, and the cells were collected over a 48-hr infection period. Table 1 provides a description of the lesion sustained by each mutant virus and a summary of protein data from a number of experiments. Ad12 hr700, an E1A mutant, was the least disabled mutant in that this virus replicated on HEK cells, producing titers only 10-fold lower than wild-type virus. Infection of HEK cells by this mutant led to the expression of E1A 235-aa and 266-aa proteins, E1B 163-aa and 482-aa proteins, and structural proteins to wild-type levels, but the infection cycle was up to 6 hr behind wild-type virus. No E1A products were identified in Ad12 in601–infected cells nor was the expected, truncated 13S mRNA product observed in all experiments after Ad12 in600 infection. For the latter mutant, synthesis of the 235-aa protein (translated from the 12S E1A mRNA) was delayed and never reached the wild-type level throughout the 48-hr time course. For both Ad12 in600 and in601, synthesis of E1B proteins was markedly delayed; however, the presence of the E1A 235-aa protein in Ad12 in600–infected cells permitted the accumulation of wild-type levels of the E1B 163-aa protein. Identical phenotypes were displayed by the three E1B 54K mutants Ad12 in602, del620, and hr703; no E1B 482-aa protein was detected over the 48-hr time course, and expression of both of the E1A proteins and the E1B 163-aa protein was delayed and observed only at reduced levels. Mutants Ad12 in600, in601, in602, hr703, and del620 are very tight host-range mutants.

In light of the observation by Ledinko (1967) that Ad12 significantly increased the expression of the cellular TK gene, experiments were carried out with the Ad12 mutants to examine their ability to stimulate expression of TK after infection of HEK cells. Table 2 shows relative levels of TK activity in infected cells as compared with mock-infected HEK controls. Thirty hr after infection, Ad12 wild type stimulated expression of TK 8-fold, while mutants in602 and hr700 increased TK activity 5-fold. No stimulation was observed after infection by in600 or in601.

In western blotting studies using an antibody directed against the heat-shock 70K protein (HS70K) isolated from 293 cells (α-hsp70; obtained from H. Pelham) and a polyclonal rabbit serum with antibodies to p53 (designated Rba/3 gal C19 p53; obtained from D. Lane), no change in the overall levels of p53 or HS70K were observed in HEK cells after Ad12 wild-type or mutant

**Table 1** The Lesions in the Ad12 E1 Mutants and Their Predicted and Observed Effects on E1A, E1B, and Structural Protein Synthesis

| Mutant | Lesion | Predicted effect | Observed effects in E1A, E1B, and structural protein expression in infected HEK cells | | | | |
| --- | --- | --- | --- | --- | --- | --- | --- |
| | | | E1A 266 aa | E1A 235 aa | E1B 163 aa | E1B 482 aa | structural proteins |
| Ad12 in601 | Xbal linker inserted at Rsal site at nt 751 | premature termination of E1A 266- and 235-aa proteins after residue 84 with 67 additional aa | ND | ND | D/R | D/R | D/R |
| Ad12 hr700 | G→A transition at nt 949 | Asp→Asn at residue 150 common to both E1A proteins | 6-hr delay compared with wild-type virus | | | | |
| Ad12 in600 | two Bcll linkers inserted at Rsal site at nt 1004 | premature termination of the E1A 266-aa protein after residue 168 with 3 additional aa | T/R | D/R | D | D/R | D/R |
| Ad12 hr703 | C→T transition at nt 2234 | premature termination of E1B 482-aa protein at residue 134 | D/R | D/R | D/R | ND | D/R |
| Ad12 in602 | Xbal linker inserted at Rsal site at nt 2247 | premature termination of E1B 482-aa protein after residue 134 with 10 additional aa | D/R | D/R | D/R | ND | D/R |
| Ad12 del620 | in-frame deletion 2129→2825 | deletion of 232 internal aa of 482-aa protein | D/R | D/R | D/R | ND | D/R |

(T) Truncated protein; (D) delayed synthesis; (R) reduced levels compared with wild type over 48-hr infection period; (ND) not detected; (aa) amino acid.

**Table 2** Effect of E1 Mutations on the Stimulation of Cellular Gene Expression, Transformation, and Tumorigenicity by Ad12

| Mutant | Location | TK induction[a] | Transformation | | Virus dose (pfu) to give TPD$_{50}$[b] |
| | | | BMK cells | BRK cells | |
|---|---|---|---|---|---|
| Ad12 *in*601 | E1A | 1.2 | − | − | $>1 \times 10^9$ |
| Ad12 *hr*700 | E1A | 5.6 | R[c] | R[c] | $5 \times 10^8$ |
| Ad12 *in*600 | E1A | 1.2 | − | − | $>1 \times 10^9$ |
| Ad12 *hr*703 | E1B | ND | − | − | $>1 \times 10^9$ |
| Ad12 *in*602 | E1B | 5.1 | − | − | $>1 \times 10^9$ |
| Ad12 *del*620 | E1B | ND | − | − | $>1 \times 10^9$ |
| Ad12 wild type | − | 8.0 | + | + | $7 \times 10^5$ |
| Mock-infected | NA | 1.0 | − | − | NA |

[a]HEK cells infected for 30 hr; mutant viruses at 200 pfu/cell, and wild type at 50 pfu/cell.

[b]TPD$_{50}$ (tumor-producing dose) assayed in 1-day-old Hooded Lister rats.

[c]Reduced 10-fold to 20-fold compared with wild-type virus.

(NA) Not applicable; (ND) not determined.

virus infections (R.J. Grand and P.H. Gallimore, unpubl.).

## The effects of Ad12 E1 mutations on transformation and tumorigenicity

In transformation experiments carried out on both BMK and BRK cells at 37°C, Ad12 wild-type virus at 1 pfu/cell induced an average of 40 foci and 1 foci, respectively, per 5-cm dish. The mutants Ad12 *in*600, *in*601, *in*703, *in*602, and *del*620 all failed to transform either cell type at the three incubation temperatures of 33°C, 37°C, and 38.5°C and at a maximum input virus dose of 200 pfu/cell. Ad12 *hr*700 transformed both cell types at a reduced frequency; i.e., at 80 pfu/cell and 37°C an average of 4 foci/dish developed on BRK cells and 60 foci/dish on BMK cells. Overall, Ad12 *hr*700 transformed primary rodent cells 10-fold to 20-fold less efficiently than did wild-type virus (Table 2).

In tumorigenesis studies, carried out using 1-day-old Hooded Lister rats (see Table 2), Ad12 wild-type virus produced a 50% tumor incidence at a virus dose equivalent to $7 \times 10^5$ pfu. To obtain the same tumor incidence, $5 \times 10^8$ pfu of Ad12 *hr*700 had to be inoculated. Solitary adenovirus tumors were produced by both Ad12 *in*600 and *del*620 at a virus inoculation dose of $1 \times 10^9$ pfu/rat. Both of these tumors were shown by Southern and western blotting analyses to have been induced by these mutants (P.J. Bryd et al., unpubl.).

## Transformation of human embryo retinoblasts

In 1982 we described the development of a primary human tissue culture system in which a plasmid containing Ad12 E1A plus E1B induced malignant transformation (Byrd et al. 1982a). We have now carried out a large number of transformation experiments using HER cells with plasmids containing the E1 or E1A regions of oncogenic and nononcogenic adenoviruses. Cotransfection studies with adenovirus E1A plasmids in combination with activated human *ras* genes (N-*ras* with an amino acid substitution at position 61 and H-*ras* with a substitution at position 12) have also been performed.

Table 3 shows comparative transformation data (see Experimental Procedures) for all the plasmids or combination of plasmids used so far. pLB205 DNA (Ad5 E1A + E1B) transformed HER cells efficiently and microscopic foci could be identified as early as 14 days posttranfection. The vast majority of these foci had a distinctive border (Fig. 1) and an overall morphology distinguishable from both Ad12 E1 transformants (see Byrd et al. 1982b) and those produced by cotransfection with E1A plasmids and *ras* genes (Fig. 1). The Ad12 E1A + E1B plasmid transformed 20-fold less efficiently than Ad5 E1 DNA; the frequency of 0.1 transformations per microgram of Ad12 E1 was 4-fold to 5-fold lower than that on BRK cells for the same plasmid (Gallimore et al. 1985a). The Ad12-induced HER foci were always observed 2 to 3 weeks after Ad5 foci. A transformation frequency comparable with Ad12 E1 DNA was seen when either Ad2 E1A or Ad12 E1A DNAs were cotransfected with activated N-*ras* (Table 3). These foci were morphologically different (Fig. 1), and the time of their appearance in the HER cultures was determined by which adenovirus E1A DNA was used. The Ad2 E1A + N-*ras* foci appeared 14 to 20 days posttransfection, whereas the Ad12 E1A + N-*ras* foci were first observed 2 weeks later. In two experiments where H-*ras* was cotransfected with the adenovirus E1A plasmids, only the Ad2 E1A + H-*ras* combination produced transformed foci. The time of appearance and morphology of these foci were identical with that of Ad2 E1A + N-*ras* transformants. The adenovirus E1A plasmids produced only rare transformed foci, and no foci were recorded in HER cultures exposed to either of the activated *ras* genes. Parallel cultures transfected with H-*ras* constructs were also subjected to geneticin (G418) selection. Many G418-resistant colonies were observed 12 days postselection, but very few colonies survived beyond 21 days. No cell lines could be developed from either individual foci or by passaging all the G418-resistant colonies in individual dishes (P.J. Byrd and P.H. Gallimore, unpubl.). Primary HER cells split 1:3 can be passaged 3 to 6 times, depending on the fetal specimen.

**Table 3** Summary Table of HER Experiments

| Transfecting DNA | Relative transformation frequency | No. lines established/ no. foci picked | No. lines developed | % Entering crisis | Tumourigenicity in nude mice[a] | Mean tumor latent period | Total tumor positives/ total mice injected |
|---|---|---|---|---|---|---|---|
| Ad12 E1A | 1.7 | 0/6 | 1[b] | 0 | 0/1 | >150 Days | 0/15 |
| Ad2 E1A | 1.0 | 0[c] | 0 | — | — | — | — |
| Ad12 E1A + E1B | 33.4 | 22/22 | 21 | 9.1 | 21/21 | 42 Days | 95/95 |
| Ad5 E1A + E1B | 724.3 | 10/10 | 10 | 0 | 4/6 | 185 Days | 8/23 |
| Ad12 E1A + N-ras | 44.7 | 8/9 | 4 | 11 | 3/3 | 30 Days | 19/20 |
| Ad12 E1A + H-ras | 0.0 | — | — | — | — | — | — |
| Ad2 E1A + N-ras | 37.1 | 0/10 | 5[b] | 100 | 3/5 | 66 Days | 7/20 |
| Ad2 E1A + H-ras | 29.8 | 0/7 | ND | — | ND | ND | ND |
| N-ras | 0.0 | — | — | — | — | — | — |
| H-ras | 0.0 | — | — | — | — | — | — |

[a]No. of lines positive per no. of lines injected; 1–2-week-old athymic nude mice injected intracerebrally with $1 \times 10^5$ cells/mouse.
[b]Lines developed by allowing transformed cells to overgrow the dish.
[c]No foci picked.
(ND) Not done.

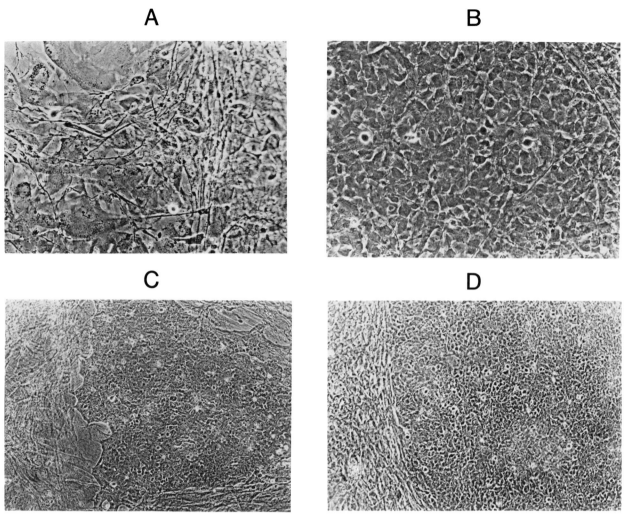

**Figure 1** Phase-contrast photomicrographs of transformed HER foci. (*A*) Normal HER cells 9 weeks in culture; (*B*) Ad12 E1A + N-*ras*–transformed focus; (*C*) Ad5 E1–transformed focus; and (*D*) Ad2 E1A + N-*ras*–transformed focus. *A* and *B* are 2 × the magnification of *C* and *D*.

No spontaneous foci have been observed in HER cultures (produced from more than 40 fetuses) even when held in culture for up to a year.

Individual foci picked from HER cultures tranfected with either Ad12 E1 or Ad5 E1 plasmid DNA developed, albeit slowly, into cell lines. None of the Ad5 lines entered crisis, whereas only one of the Ad12 lines failed to establish; two lines entered crisis around passage 15, one of which recovered after a further 3 months in culture. Cell lines were also easily isolated when individual Ad12 E1A + N-*ras* foci were picked. Of the eight out of nine foci producing cell lines, four were frozen in liquid nitrogen and four others, selected at random, have been passaged 14 times with no crisis. Individual foci picked from experiments in which Ad2 E1A was cotransfected with either H-*ras* or N-*ras* did not establish (Table 3). However, cell lines were developed by allowing the transformants to overgrow the majority of the dish and then, after mild versene treatment, passaging the whole culture. Therefore, the cell lines were theoretically the products of multiple foci.

Only the transformed lines produced from the combination of Ad2 E1A + N-*ras* have been extensively studied, and all five successfully traversed multiple crisis periods before developing into rapidly growing cell lines. Only one adenovirus E1A transformation event produced a continuously growing line. Ad12 E1A/HER-1 was established by allowing the primary transformation to expand into a massive colony on the original dish because three early attempts to isolate a cell line by picking the focus were unsuccessful. Ad12 E1A/HER-1 has a doubling time of 7 days, has been passed 23 times in 2 years, and has not entered crisis.

**Tumorigenicity of transformed human retinoblast cell lines**

Table 3 provides tumorigenicity data for studies on the transformed HER cell lines after intracerebral inoculation of athymic nude mice with $1 \times 10^5$ cells/mouse. All of the Ad12 E1 cell lines tested (21/21) produced rapidly growing tumors. Not all the Ad5 E1 lines were tumorigenic, and only one of these cell lines produced tumors in all the inoculated animals. The latent period was also very long, some tumors arising 250 days after inoculation, with a mean of 185 days. Similarly, the Ad2 E1A + N-*ras* cell lines were not all tumorigenic and only one produced a 100% tumor incidence. In contrast, the three Ad12 E1A + N-*ras* cell lines were highly tumorigenic, having a short latent period and a high tumor incidence. The one adenovirus E1A–transformed line, Ad12 E1A/HER-1, was nontumorigenic (Table 3).

### Expression of viral and cellular genes in adenovirus E1–transformed human cell lines

We have also examined the adenovirus E1–transformed human cell lines for the expression of adenovirus E1A– and E1B–encoded proteins and for the level of expression of the class-I MHC antigens (MHC-I), HS70K, and p53. With two exceptions, the adenovirus E1–transformed lines produced the expected E1A and E1B proteins at high ( > 40) and low ( < 10) passages. The two exceptions were Ad12 HER-1 and Ad12 HER-9A (Byrd 1983), which expressed all the E1 proteins at early passage levels but at some as yet undefined stage switched off the expression of the E1B 54K protein. No obvious change in cell morphology or behavior accompanied the down-regulation of this E1B product.

Figure 2 illustrates the profiles obtained by fluorescence-activated cell sorting of normal primary HERs, two Ad5 E1 cell lines, and one Ad12 E1 cell line that were labeled with a monoclonal antibody that recognizes all of the MHC-I molecules. This illustration is representative of a number of experiments carried out on two batches of primary HER cells, five Ad5 E1 and four Ad12 E1 transformed lines. It can be seen from Figure 2 that the Ad12 HER-2 line only contains cells with low levels of MHC-I molecules on the cell surface. The two Ad5 lines shown in Figure 2 had similar profiles to the normal HER cells. Overall, a clear pattern emerged, namely that Ad12 HER cell lines have significantly reduced levels of

MHC-I antigen on their cell surface, but that the Ad5 transformants were a very heterogeneous group in that they overlapped both the high end of the Ad12 spectrum and the normal HER range. In experiments examining total MHC-I levels by western blotting, using a rabbit antibody against papain-digested human MHC-I (designated K455; obtained from S. Paabo), both Ad12 E1– and Ad5 E1–transformed cell lines expressed significantly reduced levels of MHC-I compared with the level seen in primary HER cells. Data obtained from gel scans indicated that both Ad5 lines (five examined) and Ad12 lines (six examined) contained less than 30% of the MHC-I expressed by normal cells.

Evidence has been presented that the large E1B protein of Ad5 binds the mouse p53 protein in Ad5-transformed rodent cell lines (Sarnow et al. 1982). We therefore examined the expression of p53 in the adenovirus-transformed human cell lines. Figure 3 shows an autoradiograph of western blot in which a number of cell extracts from adenovirus cell lines have been reacted with a polyclonal rabbit serum with antibodies to p53. In those cell lines expressing the Ad12 E1B 54K protein, increased levels of p53 were observed. The Ad5 E1 transformants and Ad12 HER-1 and -9A (at high passage) were indistinguishable from primary HEK and HER cells.

Nevins (1982) reported that Ad5 infections of HeLa cells induced high levels of the HS70K protein early in infection. Also, the Ad5-transformed cell line 293 was shown by Nevins and his colleagues to overproduce HS70K (Nevins et al. 1984). Using the HS70K antibody α-hsp70, we noted a wide variation in the level of expression of this cellular protein (Fig. 4). However, all of the adenovirus-transformed cell lines showed higher levels of HS70K than primary HEK (Fig. 3) or HER cells (data not shown). Overproduction of this cell protein was most pronounced in 293 cells, Ad5 HER-B and -C, and Ad12 HER-9A.

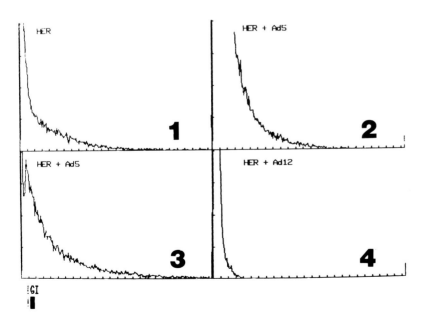

**Figure 2** MHC-I cell-surface labeling. (*1*) Normal HER cells; (*2*) Ad5 HER-B; (*3*) Ad5 HER-A; (*4*) Ad12 HER-2. (*x* axis) Fluorescence intensity; (*y* axis) cell number.

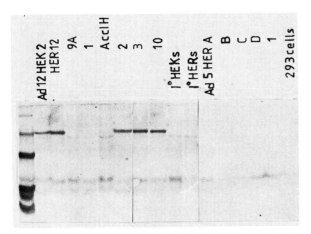

**Figure 3** Western blot showing the level of p53 expression in Ad5- and Ad12-transformed human cell lines. Cells were processed as described in Experimental Procedures and subjected to SDS-polyacrylamide gel electrophoresis and western blotting, using an antibody against p53. The following cell lines were examined: one Ad12-transformed HEK line (Ad12 HEK-2, passage 31); six Ad12 HER lines (HER-12, passage 14; HER-9A, passage 93; HER-1, passage 82; Ad12 E1A/HER-1, passage 17; HER-2, passage 71; HER-3, passage 32; HER-10, passage 20); primary HEKs and HERs; five Ad5 HER lines (HER-A, passage 37; HER-B, passage 22; HER-C, passage 47; HER-D, passage 54; HER-1, passage 5); and an Ad5-transformed HEK line (293, passage 90). The leftmost lane contains molecular-weight markers.

## Discussion

A number of biological characteristics distinguish the oncogenic adenovirus serotype 12 from the nononcogenic serotypes 2 and 5. Very few human cell types are absolutely permissive for Ad12 infection, whereas rodent cells are nonpermissive (Doerfler 1969; P.H. Gallimore, unpubl.). In contrast, rat (Gallimore 1974) and mouse cells are semipermissive for Ad2 and Ad5, whereas hamster cells support the lytic cycles of these viruses. Additionally, at similar input multiplicities, the replication cycle of Ad2 and Ad5 in HEK cells precedes Ad12 by 4–6 hr, a difference that is a reflection of the appearance of the respective E1A proteins (i.e., Ad2 E1A 2 hr and Ad12 E1A 6 hr postinfection). From studies of mutant viruses, it is clear that Ad2 and Ad5 E1A regions have profound regulatory functions on both viral (Berk et al. 1979; Jones and Shenk 1979) and cellular gene expression, the latter including the β-tubulin (Stein and Ziff 1984) and HS70K (Nevins 1982) genes. So far we have been unable to detect an analogous stimulation of HS70K by Ad12 in a permissive infection. However, the levels of this protein were found to be elevated in both Ad5- and Ad12-transformed HER cell lines (Fig. 4).

The Ad12 E1A mutant *in*600, which expresses a wild-type 12S mRNA, has an identical phenotype with the Ad5 *hr* mutants *hr*3 and *hr*5 (i.e., a tight host-range and nontransforming phenotype). Ad12 *hr*700 is a unique E1A mutant in that it grows well on HEK cells but has a significantly reduced (temperature-independent) transforming capacity on primary rodent cells

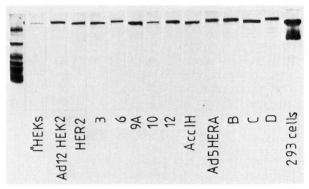

**Figure 4** Western blot showing the level of heat-shock protein expression in Ad5- and Ad12-transformed human cell lines. Cells were processed as described in Experimental Procedures and subjected to SDS-polyacrylamide gel electrophoresis and western blotting, using an antibody to HS70K. The following cells were used: primary HEKs; Ad12-transformed HEK line (Ad12 HEK-2); seven Ad12-transformed HER lines (HER-2, HER-3, HER-6, HER-9A, HER-10, HER-12, and Ad12 E1A/HER-1); four Ad5-transformed HER lines (HER-A, HER-B, HER-C, and HER-D); and one Ad5-transformed HEK line (293 cells). The leftmost lane contains molecular-weight markers.

(Table 2.) The single-amino-acid substitution in *hr*700 (Asp → Asn) affects both E1A proteins (residue 150) and causes a dramatic reduction in the oncogenic potential of this virus (Table 2). In common with the analogous Ad5 mutant *hr*6 (Graham et al. 1978; Ho et al. 1982), the Ad12 E1B 54K mutants Ad12 *hr*703, *in*602, and *del*620 are nontransforming and essentially nontumorigenic (Table 2). However, unlike Ad5 *hr*6 (see Williams et al., this volume), these Ad12 mutants did not grow in HEK cells, failing to make any E1B 54K protein and synthesizing reduced levels of the E1A, E1B 19K, and viral structural proteins. Since wild-type levels of E1B 19K protein are made by these Ad12 mutants in cells that express E1A products constitutively (i.e., Ad12 E1A/HER-1) (P.H. Gallimore and R.J.A. Grand, unpubl.), the reduced levels of E1B 19K protein synthesis in HEK cells infected by these mutants is due to a paucity of the regulatory E1A proteins.

The efficiency with which Ad5 E1 and Ad12 E1 DNAs transformed primary HERs (Table 3) is not too dissimilar from data previously published for BRK cells (van den Elsen et al. 1982), showing a differential of approximately 20:1 in favor of Ad5 E1. In addition, the activated human N-*ras* gene was able to complement both Ad2 E1A and Ad12 E1A for transformation in the HER cell system, in an analogous manner to that previously reported for rat cells by Ruley (1983). In cotransfection experiments, H-*ras* and N-*ras* complemented Ad2 E1A to the same degree, whereas only N-*ras* was shown to complement Ad12 E1A (Table 3). The reason for this is unclear at present. Similarly, we are unable to explain why transformants produced by the Ad12 E1A plus N-*ras* combination developed easily into cell lines when single foci were picked, whereas foci transformed by Ad2 E1A plus N-*ras* or H-*ras* did not. In previously reported DNA transfection studies on normal rat cells, E1A of both Ad5 (Houweling et al. 1980) and Ad12 (Gallimore et al. 1985a) were capable of producing transformed foci from which cell lines were developed. Although both Ad12 and Ad2 E1A were capable of inducing foci of transformed HER cells, immortalization was an exceedingly rare event.

Tumorigenicity studies carried out in nude mice with the adenovirus-transformed HER cell lines provided data similar to those obtained in previous studies carried out with rodent cells (Gallimore and Paraskeva 1980; Bernards et al. 1982). Cell lines expressing Ad12 E1A in the presence of E1B or N-*ras* were highly malignant, whereas Ad12 E1A/HER-1, which expresses E1A only, was nontumorigenic (Table 3). The Ad5 E1A + E1B and Ad2 E1A + N-*ras* lines were not all tumorigenic and generally did not produce a 100% tumor incidence.

Compared with normal HERs, the total MHC-I expression was significantly reduced in both Ad12 and Ad5 HER cell lines. This finding appears to contradict the observations reported for rat cells (Schrier et al. 1983) and is most probably a reflection of the different cell system. From cell-surface-labeling studies, it can be concluded that both Ad5 and Ad12 HER lines form heterogeneous groups, as do rat cell lines transformed by these viruses (Mellow et al. 1984). There was a clear overlap for the number of cells positive for MHC-I between Ad12 and Ad5 lines, although in general Ad12 lines had lower levels of MHC-I on the cell surface. The examination of these cell lines for MHC-I expression and the parallel tumorigenicity study were not aimed at resolving the underlying mechanism(s) by which adenovirus-transformed cells escape immune detection (Bernards et al. 1983; Mellow et al. 1984; Lewis and Cook 1985). However, they have provided additional support for the concept that adenovirus E1A in as yet some undefined way interacts with MHC-I expression at two levels, namely, an initial up-regulation in Ad12- and Ad5-infected mouse cells (Rosenthal et al. 1985) and subsequent down-regulation (Schrier et al. 1983) in Ad12-transformed rat cells.

Heterogeneity was also observed for the level of expression of the HS70K protein, but again there was a clear tendency for the Ad12 HER lines to be lower than the Ad5 lines (Fig. 4). A rather surprising result was obtained from studies on p53 expression, namely, that an elevated level was only observed in Ad12 lines expressing E1B proteins (Fig. 3). There are no reports that Ad12 E1B 54K protein binds p53 as is the case for Ad5 (Sarnow et al. 1982), and we do not know whether the increased abundance of p53 is a reflection of increased expression or increased stability afforded by the Ad12 E1B protein.

The Ad12 mutants reported here allow us to come to a number of conclusions:

1. The E1A 12S mRNA product in the absence of the 13S mRNA products does not permit replication in the permissive cells or neoplastic transformation.
2. A lesion defined by Ad12 *hr*700, which is common to the first exon of both the E1A 12S and 13S mRNAs

(G → A at nt 949), did not make this virus host range in HEK cells (whereas it does in HeLa cells) but significantly reduced transformation and tumorigenicity (Table 2). The importance of the first exon in determining the frequency of transformation has also been identified in a study of the transforming activities of adenovirus E1 plasmids that contain chimeric E1A regions (Jochemsen et al. 1984).
3. Three Ad12 E1B 54K mutants provided evidence that this Ad12 E1B protein influences the expression of E1A in permissive cells. These mutants were also nontransforming and nontumorigenic. It seems probable that the E1B 54K mutations would have a more dramatic effect on E1A expression in nonpermissive cells, and this may account for the nontransforming phenotype. However, we have found that expression of E1A is at a constant level in transformed cell lines irrespective of the presence or absence of E1B proteins (Gallimore et al. 1985a).

A number of differences in the phenotypes of the A-group and C-group adenoviruses are defined by their E1 regions in spite of conservation of genetic information. The different host-range phenotypes noted for some Ad12 and Ad5 E1B mutants, together with the differential effects on cellular gene expression, provide an insight into the complexity of function of these adenovirus transforming proteins.

## Acknowledgments

The authors acknowledge the following for supplying either plasmids or sera: Drs. D. Spandidos, N. Stow, L. Babiss, A. Hall, S. Paabo, D. Lane, and H. Pelham. We thank Paul Reeve, Liz Fletcher, Val Nash, Peter Grabham, Paul Biggs, and Ann Maguire for excellent technical assistance and Debbie Williams for superb secretarial skills.

This study was supported by the Cancer Research Campaign, England, and by the National Cancer Institute (grant CA-21375).

## References

Barnstable, C.J., W.F. Bodmer, G. Brown, G. Galfré, C. Milstein, A.F. Williams, and A. Ziegler. 1978. Production of monoclonal antibodies to group A erythrocytes, HLA and other human cell surface antigens—New tools for genetic analysis. *Cell* **14:** 9.

Berk, A.J., F. Lee, T. Harrison, J. Williams, and P.A. Sharp. 1979. Pre-early adenovirus 5 gene product regulates synthesis of early viral messenger RNAs. *Cell* **17:** 935.

Bernards, R. and A.J. van der Eb. 1984. Adenovirus: Transformation and oncogenicity. *Biochim. Biophys. Acta* **783:** 187.

Bernards, R., A. Houweling, P.I. Schrier, J.L. Bos, and A.J. van der Eb. 1982. Characterization of cells transformed by Ad5/Ad12 hybrid early region I plasmids. *Virology* **120:** 422.

Bernards, R., P.I. Schrier, A. Houweling, J.L. Bos, A.J. van der Eb, M. Zijlstra, and C.J.M. Melief. 1983. Tumourigenicity of cells transformed by adenovirus type 12 by evasion of T cell immunity. *Nature* **305:** 776.

Byrd, P.J. 1983. "Transformation of mammalian cells by human adenovirus DNA." Ph.D. thesis, University of Birmingham, England.

Byrd, P.J., K.W. Brown, and P.H. Gallimore. 1982a. Malignant transformation of human embryo retinoblasts by cloned adenovirus 12 DNA. *Nature* **298:** 69.

Byrd, P.J., W. Chia, P.W.J. Rigby, and P.H. Gallimore. 1982b. Cloning of DNA fragments from the left end of the adenovirus type 12: Transformation by cloned early region 1. *J. Gen. Virol.* **60:** 279.

Doerfler, W. 1969. Nonproductive infection of baby hamster kidney cells (BHK21) with adenovirus 12. *Virology* **38:** 587.

Frost, E. and J. Williams. 1978. Mapping temperature sensitive and host-range mutations of adenovirus type 5 by marker rescue. *Virology* **91:** 39.

Gallimore, P.H. 1974. Interactions of adenovirus type 2 with rat embryo cells. Permissiveness, transformation and *in vitro* characteristics of adenovirus transformed rat embyro cells. *J. Gen. Virol.* **25:** 263.

Gallimore, P.H. and C. Paraskeva. 1980. A study to determine the reasons for differences in the tumorigenicity of rat cell lines transformed by adenovirus 2 and adenovirus 12. *Cold Spring Harbor Symp. Quant. Biol.* **44:** 703.

Gallimore, P.H., P.J. Byrd, and R.J.A. Grand. 1985a. Adenovirus genes involved in transformation. What determines the oncogenic phenotype? *Symp. Soc. Gen. Microbiol.* **37:** 125.

Gallimore, P.H., P.J. Byrd, J.L. Whittaker, and R.J.A. Grand. 1985b. The properties of rat cells transformed by DNA plasmids containing adenovirus type 12 DNA or specific fragments of the E1 region: A comparison of transforming frequencies. *Cancer Res.* **45:** 2670.

Gallimore, P., P. Byrd, R. Grand, J. Whittaker, D. Breiding, and J. Williams. 1984. An examination of the transforming and tumor-inducing capacity of a number of adenovirus type 12 early-region 1, host-range mutants and cells transformed by subgenomic fragments of Ad12 E1 region. *Cancer Cells* **2:** 519.

Graham, F.L. 1984. Transformation by and oncogenicity of human adenoviruses. In *The adenoviruses* (ed. H.S. Ginsberg), p. 339. Plenum Press, New York.

Graham, F.L. and A.J. van der Eb. 1973. A new technique for the assay of infectivity of human adenovirus 5 DNA. *Virology* **52:** 456.

Graham, F.L., T. Harrison, and J. Williams. 1978. Defective transforming capacity of adenovirus type 5 host range mutants. *Virology* **86:** 10.

Ho, Y.S., R. Galos, and J. Williams. 1982. Isolation of type 5 adenovirus mutants with a cold-sensitive host range phenotype: Genetic evidence of an adenovirus transformation maintenance function. *Virology* **122:** 109.

Houweling, A., P.J. van den Elsen, and A.J. van der Eb. 1980. Partial transformation of primary rat cells by the left-most 4.5% fragment of adenovirus 5 DNA. *Virology* **105:** 537.

Jochemsen, A.G., J.L. Bos, and A.J. van der Eb. 1984. The first exon of region E1a genes of adenoviruses 5 and 12 encodes a separate functional protein domain. *EMBO J.* **3:** 2923.

Jones, N. and T. Shenk. 1979. An adenovirus type 5 early gene function regulates expression of the other early viral genes. *Proc. Natl. Acad. Sci.* **76:** 3665.

Ledinko, N. 1967. Stimulation of DNA synthesis and thymidine kinase activity in human embryonic kidney cells infected by adenovirus 2 or 12. *Cancer Res.* **27:** 1459.

Lewis, A.M. and J.L. Cook. 1985. A new role for DNA virus early proteins in viral carcinogenesis. *Science* **227:** 15.

Mellow, G.H., B. Fohring, J. Dougherty, P.H. Gallimore, and K. Raska. 1984. Tumorigenicity of adenovirus-transformed rat cells and expression of class I major histocompatibility antigen. *Virology* **134:** 460.

Nevins, J.R. 1982. Induction of the synthesis of a 70,000 dalton mammalian heat shock protein by the adenovirus E1A gene product. *Cell* **29:** 913.

Nevins, J.R., M.J. Imperiale, L.T. Feldman, and H.-T. Kao. 1984. Role of adenovirus transforming gene (E1A) in the general control of gene expression. *Transplant. Proc.* **16:** 438.

Paraskeva, C., K.W. Brown, A.R. Dunn, and P.H. Gallimore. 1982. Adenovirus type 12—transformed rat embryo brain and rat liver epithelial cell lines: Adenovirus type 12 genome content and viral protein expression. *J. Virol.* **44:** 759.

Rosenthal, A., S. Wright, K. Quade, P.H. Gallimore, H. Cedar, and F. Grosveld. 1985. MHC H-2K gene transcription in cultured mouse embryo cells is increased following adenovirus infection. *Nature* **315:** 579.

Ruley, H.E. 1983. Adenovirus early region 1a enables viral and cellular transforming genes to transform primary cells in culture. *Nature* **304:** 602.

Sarnow, P., Y.S. Ho, J. Williams, and A.J. Levine. 1982. Adenovirus E1b-58kd tumor antigen and SV40 large tumor antigen are physically associated with the same 54 kd cellular protein in transformed cells. *Cell* **28:** 387.

Schrier, P.I., R. Bernards, R.T. Vaessen, A. Houweling, and A.J. van der Eb. 1983. Expression of class I major histocompatibility antigens switched off by highly oncogenic adenovirus in transformed rat cells. *Nature* **305:** 771.

Spandidos, D.A. and N.M. Wilkie. 1984. Malignant transformation of early passage rodent cells by a single mutated human oncogene. *Nature* **310:** 469.

Stein, R. and E.B. Ziff. 1984. HeLa cell β-tubulin gene transcription is stimulated by adenovirus-5 in parallel with viral early genes by an E1a dependent mechanism. *Mol. Cell. Biol.* **4:** 2792.

van den Elsen, P.J., S. de Pater, A. Houweling, J. van der Veer, and A.J. van der Eb. 1982. The relationship between region E1a and E1b of human adenoviruses in cell transformation. *Gene* **18:** 175.

# The E1 Region of Human Adenovirus Type 12 Determines the Sites of Virally Induced Chromosomal Damage

**D.M. Durnam, P.P. Smith, J.C. Menninger, and J.K. McDougall**

Fred Hutchinson Cancer Research Center, Program in Experimental Pathology, Seattle, Washington 98104

Infection of mouse/human hybrid cells with an Ad5/Ad12 recombinant virus and transfection with Ad12 E1 DNA sequences has shown that the nonrandom effect of Ad12 on band q21-q22 of human chromosome 17 is determined by a gene(s) in the E1 region.

Infection of human cells with the highly oncogenic adenovirus type 12 (Ad12) at a multiplicity of infection (moi) of 1–10 results in chromosomal damage primarily in the q21-q22 region of chromosome 17 and the p32, q21, and q42 regions of chromosome 1 (zur Hausen 1967; McDougall 1970, 1979). In typical experiments, 30–50% of the metaphase spreads from Ad12-infected human cells show damage in the 17q21-q22 region; 10–20% of the spreads show damage in chromosome 1 (McDougall 1971a). No preferential integration of viral sequences has been detected at any of the sites of virally induced chromosome damage (zur Hausen 1967; McDougal et al. 1972). Unlike Ad12, the nononcogenic (class C) human adenoviruses, such as Ad2 and Ad5, do not cause specific chromosome damage on chromosomes 1 and 17 but rather cause the appearance of random damage throughout the human genome (McDougall 1971b).

The chromosomal sites affected by Ad12 are also common sites of chromosomal rearrangements in human cancers. For example, the 17q21-q22 region is thought to be the critical site of rearrangements in myeloproliferative leukemias (Yamada et al. 1983; Knuutila et al. 1984). The most common rearrangement at this site occurs in acute promyelocytic leukemia and involves the transfer of chromosomal material between chromosomes 15 and 17: t(15;17)(q21;q22) (Rowley et al. 1977). Rearrangements in chromosomes 1 and 17 are also frequently associated with neuroblastomas (Gilbert et al. 1984). Although there is no direct linkage between adenoviral infection and these cancers, it is possible that an understanding of the mechanism by which the virus induces chromosome damage may yield clues as to how cancer-related aberrations occur in these regions, as well.

Little is currently known about the mechanism by which Ad12 induces specific chromosome damage. From studies of thymidine kinase activity, it was proposed that an alteration in the transcription of genes in the 17q21-q22 region accounts for the susceptibility of the region to gap formation and breakage (McDougall 1971b). Since that time, a significant number of additional genes have been mapped in the 17q21-q22 region, including the galactose kinase gene (Elsevier et al. 1974), the pro-$\alpha$1(I) collagen gene (Church et al. 1980), the erbA1 oncogene (Spurr et al. 1984) and, more recently, the neu oncogene (Schechter et al. 1985), the homeo box gene (Rabin et al. 1985), and the U2 small nuclear RNA gene family (Lindgren et al. 1985). It is not clear at the present time where any of these genes map relative to the site of Ad12-induced chromosome damage or whether Ad12 affects their transcriptional activity.

Faced with this accumulation of information, which shows that this chromosomal site (17q21-q22) is a hot spot for rearrangement and gene mapping, we have returned to an analysis of the nonrandom virus interaction. Several lines of evidence suggest that the Ad12 gene (or genes) responsible for the induction of specific chromosome damage is an early viral gene. First of all, the time course of the appearance of chromosome damage correlates with the time course of early gene expression (Raska and Strohl 1972). Secondly, Ad12 infection of mouse/human hybrid lines that have retained human 17q sequences results in specific 17q damage (McDougall et al. 1973) in spite of the fact that neither late gene transcription nor viral replication (MacKinnon et al. 1966) occurs in rodent cells. In this report we describe the chromosome damage induced by an Ad5/Ad12 recombinant virus (RC15GT), which has the Ad12 E1 region inserted in place of Ad5 E1 (Bernards et al. 1984). In separate experiments an Ad12 E1 region clone has been transfected into mouse/human hybrid cells and surviving clones have been examined for chromosomal changes. The results show that both the RC15GT recombinant virus and the isolated E1 region of Ad12 induce chromosome damage primarily at the 17q21-q22 region.

## Experimental Procedures

### Cell culture and chromosome preparation

All cells were grown in Dulbecco's modified Eagle's medium (DME) supplemented with 10% fetal bovine serum. Colcemid (0.1 $\mu$g/ml) was added to the culture medium 1 hr prior to chromosome preparation. Chromosomes were prepared by incubating trypsinized cells in hypotonic KCl (0.075 M) for 5 min, followed by fixation in methanol/acetic acid (3:1) for at least 24 hr. Chromosome preparations were stored in fixative until analyzed. For light microscope analysis, chromosomes were stained in 4% Giemsa (Biomedical Specialties, Santa Monica) in Gurr's buffer (Biomedical Specialties) for 2.5 min at room temperature.

### Preparation of virus stocks

Ad12 and Ad5 were gifts from M. Green and J. Lewis, respectively. RC15GT was kindly provided by R. Bernards; its construction has been described (Bernards et al. 1984). Virus stocks were amplified by infecting human embryo kidney cells with Ad12, Ad5, or RC15GT at an moi of 1–10. After 2 days, the cells were freeze-thawed three times, the cell debris was removed by centrifugation, and the virus containing supernatant was quick-frozen and stored at −70°C. Virus stocks prepared in this manner were titered both in infectivity and by the ability to cause chromosome damage. For chromosome analysis, cells were plated 24 hr prior to infection and washed three times prior to the addition of virus at an moi of 1–500. Virus was adsorbed for 90 min before readdition of culture media. Chromosomes were prepared 16–24 hr postinfection.

### Transfection assay

The Ad12 E1 region clone, pC-1, was prepared by cloning the EcoRI C fragment of Ad12 (0–16.5 map units) into the plasmid vector pUC12, using the protocol described by Byrd et al. (1982). The identity of the plasmid was confirmed both by restriction mapping and by its ability to transform primary rat embryo cells (D.M. Durnam, unpubl.). In all experiments, a clone containing the hygromycin B phosphotransferase gene, pUC18SV2hygA, was cotransfected as a selectable marker (pUC18SV2hygA was a gift from P. Berg). The methods for transfection have been described (Wigler et al. 1978). Briefly, we mixed 1 µg of pC-1, 1 µg of pUC18SV2hygA and 10 µg of carrier calf thymus DNA in 200 µl of TE (1 mM Tris, pH 7.9, 0.1 mM EDTA) and added 250 µl of 2× HEPES-buffered saline (280 mM NaCl, 50 mM HEPES, 1.5 mM $Na_2HPO_4$, pH 7.1) before slowly adding 30 µl of 2 M $CaCl_2$. The precipitate was allowed to form for 30 min at room temperature before being added to a 60 mM culture of WL 24a-2-A cells in 5 ml of medium. Cells were plated 1 day before transfection and were fed fresh medium 4 hr prior to the addition of the DNA precipitate. The cells were washed and the medium was changed 4 hr after the addition of the DNA precipitate. One day after transfection, the cells were split 1:3. Hygromycin (300–500 µg/ml) was added 24 hr later. Selective medium was changed approximately every 3 days. Individual clones were selected after 2–3 weeks and expanded in selective medium.

### DNA and RNA analysis

Total nucleic acids were isolated by lysing cells in 1× SET (1% SDS, 5 mM EDTA, 10 mM Tris, pH 7.5) in the presence of 50 µg/ml proteinase K. After incubation at 55°C for 1–2 hr, samples were extracted with phenol/ chloroform, precipitated with ethanol, and redissolved in TE. DNA concentrations were determined by the fluorescence assay described by Labarca and Paigen (1980) with a Perkin-Elmer Model 650 10S fluorescence spectrophotometer. Restriction enzyme digestions were carried out according to the manufacturers specifications. Gel electrophoresis, Southern blotting techniques, nick translation, and hybridization conditions were carried out essentially as described by Palmiter et al. (1982). For RNA analysis, total nucleic acid samples were treated with RNase-free DNase and spotted onto nitrocellulose as described by Norstedt and Palmiter (1984). Conditions for hybridization were identical with those used for Southern analysis.

### Scanning electron microscope analysis

Chromosome spreads were examined and photographed by light microscopy before further processing for scanning electron microscope (SEM) analysis. For SEM, specimens were treated with 0.25% trypsin in 0.85% NaCl for 5–8 sec before being fixed in 3% glutaraldehyde in 0.1 M Sorenson's phosphate buffer (pH 7.4) for 30 min. After rinsing in Sorenson's buffer, specimens were: (1) fixed in 1% osmium tetroxide in Sorenson's buffer for 10 min; (2) rinsed three times in distilled $H_2O$; (3) dehydrated through a graded ethanol series (30–100%); and (4) dried by the critical point method from liquid carbon dioxide with 100% ethanol as transitional fluid. Specimens were then mounted onto SEM stubs, sputter-coated with approximately 50 nM gold/palladium, and examined in a JEOL JSM-35C electron microscope at an accelerating voltage of 17–21 kV and a tilt of 30°.

## Results

### Chromosome damage induced by recombinant RC15GT

The role of the Ad12 E1 region was initially tested by comparing the specificity of the chromosome damage induced by Ad12, Ad5, and the recombinant virus RC15GT. RC15GT has the Ad12 E1 region substituted for the Ad5 E1 region; the rest of the viral genome is Ad5. Initial experiments were conducted using human embryo lung (HEL) cells since they routinely show 17q21-q22 damage in a high percentage (30–40%) of cells after Ad12 infection. An example of the type of chromosome damage induced by Ad12 in HEL cells is shown in Figure 1A. In this example, the 17q21-q22 region is the only site of chromosome damage. Unlike Ad12, Ad5 and RC15GT induced chromosome damage in very few of the HEL cells in spite of the fact that parallel infections led to the production of high viral titers. Similar results with Ad5 have been reported by zur Hausen (1967) and are thought to result from the fact that class-C adenoviruses often turn off host-cell DNA synthesis too rapidly to allow the examination of infected-cell chromosomes. In the metaphases that did show RC15GT-induced chromosome damage, 17q21-q22 damage was frequently apparent; 17q damage was

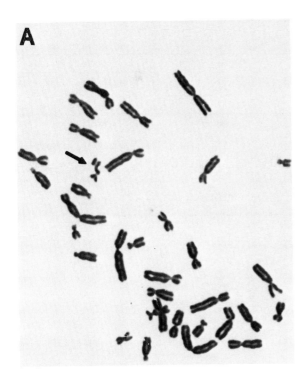

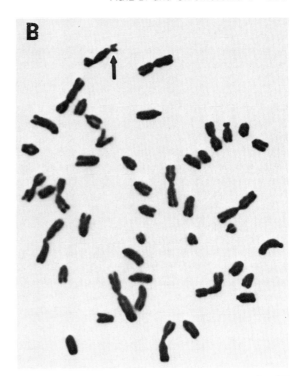

**Figure 1** Chromosome 17q21-q22 aberrations induced by Ad12. (*A*) Ad12-induced damage in human embryo lung cells; (*B*) Ad12-induced damage in WL24A-2-A cells. In both panels, 17q21-q22 damage is indicated by an arrow.

never observed after Ad5 infection, although in all experiments with Ad5 and RC15GT the number of metaphases exhibiting damage was too low to be statistically analyzed.

To circumvent the problems associated with human cells, all subsequent experiments were conducted using the mouse/human hybrid WL24a-2-A, a cell line that is semipermissive for class-C adenovirus replication (McDougall et al. 1975). These cells have retained the long arm of chromosome 17 as their sole human component (Boone et al. 1972) and readily allow the appearance of 17q21-q22–specific damage after Ad12 infection (McDougall et al. 1973) (Fig. 1B). In these cells, the human 17q sequences have been translocated to a mouse chromosome yet remain identifiable in routine Giemsa-stained preparations. The data in Table 1 show that there is a distinct difference in the specificity of the damage induced by Ad12 and Ad5 in WL 24a-2-A cells. Ad12 causes 17q21-q22 gaps and breaks in 89% of the metaphases showing virally induced damage. In the majority of the metaphases with damage, the 17q aberration is either the sole site or one of few sites of

chromosome damage. An example of a 17q21-q22 chromosome gap induced by Ad12 in these cells is shown in Figure 1B. In contrast to Ad12, Ad5 induces random chromosomal damage, showing no obvious preferential sites on either the mouse or human chromosomes (Table 1). No 17q gaps or breaks were detected after Ad5 infection in spite of the fact that the average number of Ad5-induced breaks per cell was approximately the same as the number induced by Ad12.

Examination of RC15GT damage shows that of the 100 metaphases examined, 40% have virally induced damage and, of these, 62% have 17q damage (Table 1). The appearance of gaps and breaks indistinguishable from those induced by Ad12 is observed after RC15GT infection. We infer from these experiments that a gene or genes in the Ad12 E1 region must be responsible for dictating the specificity of virally induced chromosome damage.

### Transfection of Ad12 E1 into WL24a-2-A cells

To further establish the role of Ad12 E1 expression in the induction of the specific chromosomal aberration,

**Table 1** Comparison of Chromosome Aberrations Induced in WL24A-2-A Cells by Ad12, Ad5, and RC15GT

| Virus | No. of cells analyzed | No. of cells with aberrations | Average no. of aberrations/cell | No. of cells with 17q gaps or breaks |
|---|---|---|---|---|
| Ad12 | 100 | 44 | 2.5 | 39 |
| Ad5 | 90 | 15 | 2.1 | 0 |
| RC15GT | 110 | 42 | 3.2 | 26 |

we cotransfected an Ad12 E1 region containing plasmid pC-1 together with a plasmid, pUC18SV2hygA, which confers resistance to the drug hygromycin, into WL24a-2-A cells. Cells were selected by growth in hygromycin medium, and surviving colonies were picked and replated for further growth. Preparations of DNA and RNA were made from confluent cultures, and chromosome preparations were made from parallel, subconfluent dishes. The cloned cell lines were examined by Southern blot hybridization for integration of Ad12 E1 DNA and by spot hybridization for virus-specific RNA (Fig. 2). Approximately 80% of the hygromycin-resistant colonies contained Ad12 E1 DNA sequences. Four colonies positive for E1 region DNA and four colonies lacking E1 sequences were selected for in-depth analysis. Figure 2 shows that all four of the colonies positive for E1 DNA sequences also show evidence of E1 region RNA expression. In contrast, no E1 region RNA is detected in the colonies scoring negative for E1 region DNA. When these cloned cell lines were examined cytogenetically, in all of the cell lines positive for E1 DNA and RNA, between 28% and 43% of the cells contained a visible aberration in the human chromosome 17q21-q22 portion of the translocated mouse/human chromosome, but none of the E1-DNA-negative clones were found to have this altered chromosome region above background level (2–4%) (Table 2). The observed effect on the 17q21-q22 region varies from minor to extensive "uncoiling" (Fig. 3A) and often results in the formation of chromatid and isochromatid gaps and breaks.

To further investigate the structure of the virus-

**Table 2** Transfection or WL24A-2-A Cells with Ad12 E1 Region DNA

| Transfectant clone | Positive for E1 | | Chromosome aberrations[a] | |
|---|---|---|---|---|
| | DNA | RNA | cells with 17q lesion (%) | cells with random damage (%) |
| 1B5-4 | + | + | 39 | 8 |
| 1B5-6 | + | + | 37 | 7 |
| 2B5-7 | + | + | 43 | 13 |
| 2B3-1 | + | ± | 28 | 7 |
| 2A2-1 | − | | 3 | 5 |
| 1A4-1 | − | | 4 | 6 |
| 2B2-1 | − | | 2 | 6 |
| 2B4-3 | − | | 3 | 7 |

[a]100 metaphases analyzed per clone.

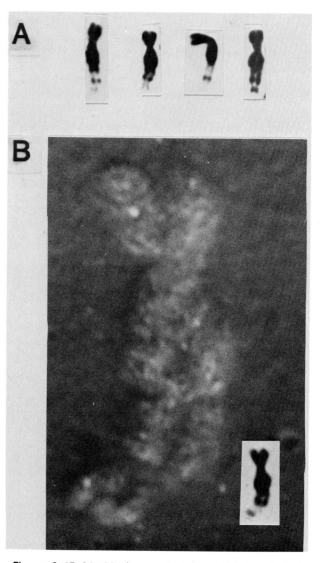

**Figure 3** 17q21-q22 damage in WL24A-2-A transfectants positive for E1 region DNA. (*A*) Light microscope analysis of 17q21-q22–specific damage. (*B*) Specific 17q21-q22 damage as visualized by SEM (magnification, × 40,000). (Insert) Corresponding light microscope observation showing uncoiling in the 17q21-q22 region.

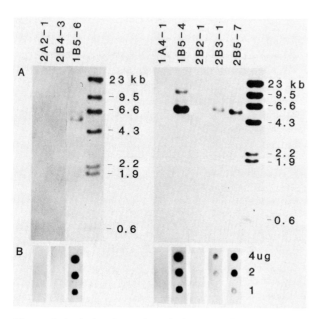

**Figure 2** Analysis of transfected clones for Ad12 E1 region DNA and RNA. (*A*) Southern blot analysis of eight transfectants. DNA was digested with *Eco*RI, and blots were probed with E1 region DNA released from pC-1. (*B*) Analysis of E1 region RNA. Spot hybridizations are aligned with corresponding lanes of the Southern analysis in *A*. RNA spots were hybridized with E1 region DNA released from pC-1.

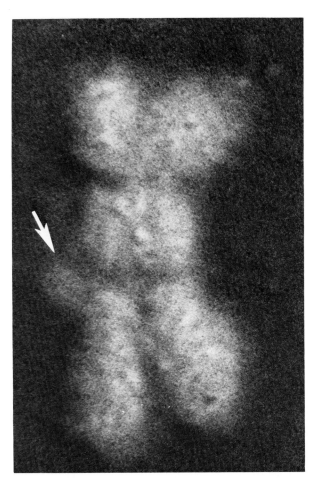

**Figure 4** SEM analysis of 17q21-q22 damage induced in human cells by Ad12. Uncoiled DNA at the site of 17q21-q22 damage is indicated by an arrow (magnification, × 40,000).

induced chromosome lesion, we commenced SEM studies. These studies, although still in the preliminary stages of development, confirmed our belief that the major effect of E1 region gene expression is the uncoiling of DNA sequences in the 17q21-q22 region. A SEM photograph of a chromosome from an Ad12 E1 RNA–expressing WL24a-2-A cell line is shown in Figure 3B. In this example, the SEM analysis showed "uncoiled" DNA in the "gapped" region identifiable by light microscopy. Evidence of uncoiling can also be seen at Ad12-induced gaps on chromosome 17q21-q22 in human cells as well (Fig. 4).

## Discussion

The results presented here assign the previously described, nonrandom effect of Ad12 on human chromosomes to genes mapping in the E1 region. Comparison of the functions of Ad5 and Ad12 E1 region genes could provide clues about why the two viruses differ in the specificity of the chromosome damage they induce. The E1 region encodes two transcription units termed E1A and E1B (Flint 1980). Both Ad5 and Ad12 E1A regions encode several related gene products that regulate the

transcription of adenoviral (Jones and Shenk 1979) and cellular (Schrier et al. 1983) genes and may also have limited transforming capabilities (Shiroki et al. 1979). The functions of the Ad5 and Ad12 E1A gene products are not identical, however, since these products also account for the difference in the oncogenic potential of the two viruses (Schrier et al. 1983; Tanaka et al. 1985). The E1B regions of the viruses are also similar but not identical and encode at least two early gene products that are primarily involved in determining the transforming capabilities of the viruses (Bernards et al. 1984). It has been clearly demonstrated that Ad12 and Ad5 E1A region genes can have different effects on the transcription of the major histocompatibility class-I gene (Schrier et al. 1983), but it remains to be seen whether any of the host genes mapping in or close to the site of Ad12-induced chromosome damage are differently affected by Ad5 and Ad12. We have preliminary evidence that suggests that at least the erbA1 gene maps in the region uncoiled by Ad12 since it has an increased susceptibility to nuclease digestion after Ad12 infection (D.M. Durnam, unpubl.).

Further experiments are required to identify the specific Ad12 E1 region gene that determines the site of virally induced chromosome damage. Analysis of Ad12 viruses carrying mutations in specific E1 region genes is perhaps the simplest aproach to this question. We have initiated this type of analysis by studying the chromosome damage induced by the Ad12 mutant, dl205, a virus carrying a mutation in the E1B gene for the 19K product (Fukui et al. 1984). Our preliminary data indicate that dl205 is capable of inducing a level of 17q damage equivalent to that induced by Ad12 in HEL cells, which suggests that the 19K product neither selects the site of nor affects the type of chromosome damage we have observed. Analysis of the effects of additional Ad12 mutants and cloned fragments of the E1 region introduced into cells by transfection will resolve this issue and allow definition of the gene in Ad12 that induces specific aberrations.

The results of the transfection experiments indicate not only that the E1 region is responsible for the selectivity of the induced damage, but also that no other viral genes are required for either site selection or uncoiling process. Along these same lines, it is interesting to note that random chromosome damage on mouse chromosomes is much lower in the transfected cell lines than in WL24a-2-A cells infected by Ad12. This may suggest that nonrandom and random chromosome effects may be occurring by different mechanisms. The transfection data also suggest that expression of E1 RNA is quantitatively linked to frequency of chromosome aberration, because clone 2B3-1 has both the lowest level of E1 RNA and the lowest percentage of 17q21-q22 abnormalities. The fact that none of the cloned cell lines expressing E1 show 100% of the specific aberration was unexpected. However, although the effect may be universal, careful analysis suggests that the extent of visible "uncoiling" varies and that the result could in fact reflect our depth of resolution.

It has been tempting to view this virus-induced aberration as being similar to the chromosome "puff" found in, for example, *Drosophila* and *Chironomus*. The chromosome shown in Figure 4 is representative of a number found in our preliminary electron microscope studies in which there is electron-dense material visible in the chromosome gap. Further studies using electron microscopy and in situ hybridization should confirm this observation.

## Acknowledgments

We thank R. Bernards for giving us the recombinant virus RC15GT; M. Green and J. Lewis for giving us Ad12 and Ad5 virus stocks, respectively; P. Berg for the plasmid pUC18SV2hygA; and Toni Higgs for preparation of the manuscript. This work was supported in part by National Institute of Health grant CA07188 (to D.M.D.), American Cancer Society grant IN-26Z (to D.M.D.), and March of Dimes grant I769 (to J.M.).

## References

Bernards, R., M.G.W. de Leew, M.J. Vaessen, A. Houweling, and A.J. van der Eb. 1984. Oncogenicity by adenovirus is not determined by the transforming region only. *J. Virol.* **50:** 847.

Boone, C., T. Chen, and F.H. Ruddle. 1972. Assignment of three human genes to chromosomes (LDH-A to 11, TK to 17 and IDH to 20) and evidence for translocation between human and mouse chromosomes in somatic cell hybrids. *Proc. Natl. Acad. Sci.* **69:** 510.

Byrd, P.J., W. Chia, P.W.J. Rigby, and P.H. Gallimore. 1982. Cloning of DNA fragments from the left end of the adenovirus type 12 genome; transformation by the cloned early region 1. *J. Gen. Virol.* **60:** 279.

Church, R.L., N. SunderRaj, and J.K. McDougall. 1980. Regional chromosome mapping of the human skin type 1 procollagen gene using adenovirus 12 fragmentation of human/mouse somatic cell hybrids. *Cytogenet. Cell Genet.* **27:** 24.

Elsevier, S.M., R.C. Kucherlapati, E.A. Nicholls, K. Willecke, R.P. Creagan, R.E. Giles, F.H. Ruddle, and J.K. McDougall. 1974. Assignment of the gene for human galactokinase to chromosome E17 band q21-22. *Nature* **252:** 633.

Flint, S.J. 1980. Structure and genomic organization of adenoviruses. In *The molecular biology of tumor viruses*, 2nd edition: *DNA tumor viruses* (ed. J. Tooze), p. 383. Cold Spring Harbor Laboratory, Cold Spring Harbor, New York.

Fukui, Y., I. Saito, K. Shiroki, and H. Shimojo. 1984. Isolation of transformation-defective, replication-non-defective early region 1B mutants of adenovirus 12. *J. Virol.* **49:** 154.

Gilbert, F., M. Feder, G. Balaban, D. Brangman, D.K. Lurie, R. Podolsky, V. Rinaldt, N. Vinikoor, and J. Weisband. 1984. Human neuroblastomas and abnormalities of chromosomes 1 and 17. *Cancer Res.* **44:** 5444.

Jones, N. and T. Shenk. 1979. An adenovirus type 5 early gene function regulates expression of other early viral genes. *Proc. Natl. Acad. Sci.* **76:** 3665.

Knuutila, S., T. Ruutu, R. Kovanen, M. Klblom, and A. de la Chapelle. 1984. Critical chromosome rearrangement in acute promyelocytic leukemia. *Cancer Genet. Cytogenet.* **11:** 473.

Labarca, C. and K. Paigen. 1980. A simple, rapid and sensitive DNA assay procedure. *Anal. Biochem.* **102:** 344.

Lindgren, V., M. Ares, Jr., A.M. Weiner, and U. Francke. 1985. Human genes for U2 small nuclear RNA map to a major adenovirus 12 modification site on chromosome 17. *Nature* **314:** 115.

MacKinnon, E., I.V. Kalnins, H.F. Stich, and D.S. Yohn. 1966. Viruses and mammalian chromosomes. V. Comparative karyologic and immunofluorescent studies on Syrian hamster and human amnion cells infected with human adenovirus type 12. *Cancer Res.* **26:** 612.

McDougall, J.K. 1970. Effects of adenoviruses on the chromosomes of normal human cells and cells trisomic for an E group chromosome. *Nature* **225:** 456.

———. 1971a. Adenovirus-induced chromosome aberrations in human cells. *J. Gen. Virol.* **12:** 43.

———. 1971b. Spontaneous and adenovirus type 12 induced chromosome aberrations in Fanconi's anemia fibroblasts. *Int. J. Cancer* **7:** 526.

———. 1979. The interactions of adenovirus with host cell gene loci. *Cytogenet. Cell Genet.* **25:** 183.

McDougall, J.K., A.R. Dunn, and P.H. Gallimore. 1975. Recent studies on the characteristics of adenovirus-infected and transformed cells. *Cold Spring Harbor Symp. Quant. Biol.* **39:** 591.

McDougall, J.K., A.R. Dunn, and K.W. Jones. 1972. In situ hybridization of adenovirus RNA and DNA. *Nature* **236:** 345.

McDougall, J.K., K.S. Kucherlapati, and F.H. Ruddle. 1973. Localization and induction of the human thymidine kinase gene by adenovirus type 12. *Nat. New Biol.* **245:** 172.

Norstedt, G. and R. Palmiter. 1984. Secretory rhythm of growth hormone regulates differentiation of mouse liver. *Cell* **36:** 805.

Palmiter, R.D., H.Y. Chen, and R.L. Brinster. 1982. Differential regulation of metallothionein-thymidine kinase fusion genes in transgenic mice and their offspring. *Cell* **29:** 701.

Rabin, M., C.P. Hart, A. Ferguson-Smith, W. McGinnis, M. Levine, and F.H. Ruddle. 1985. Two homeo box loci mapped in evolutionarily related mouse and human chromosomes. *Nature* **314:** 175.

Raska, K. and W.A. Strohl. 1972. The response of BHK21 cells to infection with type 12 adenovirus. *Virology* **47:** 734.

Rowley, J.D., H.M. Golomb, J. Vardiman, S. Fukuharo, C. Dougherty, and D. Potter. 1977. Further evidence for a non-random chromosome abnormality in acute promyelocytic leukemia. *Int. J. Cancer* **20:** 869.

Schechter, A.L., M.C. Hung, L. Vaidyanathan, R.A. Weinberg, T.L. Yang-Feng, U. Francke, A. Ulrich, and L. Coussens. 1985. The *neu* gene: An *erb* B homologous gene distinct from and unlinked to the gene encoding the EGF receptor. *Science* **229:** 976.

Schrier, P.I., R. Bernards, R.T.M.J. Vaessen, A. Houweling, and A.J. van der Eb. 1983. Expression of class I major histocompatibility antigens switched off by highly oncogenic adenovirus 12 in transformed rat cells. *Nature* **305:** 771.

Shiroki, K., H. Shimojo, Y. Sawada, Y. Uemizu, and K. Fujinaga. 1979. Incomplete transformation of rat cells by a small fragment of adenovirus 12 DNA. *Virology* **95:** 127.

Spurr, N.K., E. Solomon, M. Jansson, D. Sheer, P.N. Goodfellow, W.F. Bodmer, and B. Vennstrom. 1984. Chromosomal localization of the human homologues to the oncogenes *erb* A and B. *EMBO J.* **3:** 159.

Tanaka, K., K.J. Isselbacher, G. Khoury, and G. Jay. 1985. Reversal of oncogenesis by the expression of a major histocompatibility complex class 1 gene. *Science* **228:** 26.

Wigler, M., A. Pellicer, S. Silverstein, and R. Axel. 1978. Biochemical transfer of single copy eukaryotic genes using total cellular DNA as donor. *Cell* **14:** 725.

Yamada, K., E. Sugimoto, M. Amano, Y. Imamura, T. Kubota, and M. Matsumoto. 1983. Two cases of acute promyelocytic leukemia with variant translocations: The importance of chromosome no. 17 abnormality. *Cancer Genet. Cytogenet.* **9:** 93.

zur Hausen, H. 1967. Induction of specific chromosomal aberrations by adenovirus type 12 in human embryonic kidney cells. *J. Virol.* **1:** 1174.

# Herpes Simplex Virus and Cytomegalovirus: Unconventional DNA Tumor Viruses

D.A. Galloway, F.M. Buonaguro, C.R. Brandt, and J.K. McDougall

Fred Hutchinson Cancer Research Center, Seattle, Washington 98104

Transfection of rodent cells with defined fragments of DNA has identified regions of the genomes of herpes simplex virus type 2 (HSV-2) and cytomegalovirus (CMV) that are capable of malignant transformation; however, these fragments do not specify viral proteins. Contained within these fragments are sequences that resemble insertion-like elements. One hypothesis is that the transforming fragments activate the expression of cellular genes. When the CMV fragment was linked upstream of the HSV-1 thymidine kinase gene or upstream of a chloramphenicol acetyltransferase gene containing a simian virus 40 promoter, transcription was significantly enhanced. Analysis of transcription in a CMV-transformed cell line, using as probe the cellular sequences flanking the integrated CMV, showed that transcription was elevated threefold to fivefold in the transformant over the normal NIH-3T3 cells. In contrast, we could not demonstrate transcriptional enhancing activity for the HSV-2 transforming region. However, in a mutagenesis assay designed to test an alternative hypothesis, the HSV-2 fragment was capable of inducing mutations, a high proportion of which were gene rearrangements.

Since the initial reports that herpes simplex virus types 1 and 2 (HSV-1 and HSV-2) could morphologically transform rodent cells (Duff and Rapp 1971, 1973), there have been a number of reports confirming and extending those observations, using a variety of types of inactivated virus, viral DNA, and subgenomic fragments (for review, see Minson 1984). Attempts to define the transforming sequences gave surprising results. First, experiments with defined fragments of HSV-1 place the transforming sequences (mtrI) between map units (m.u.) 0.30 and 0.42 (Camacho and Spear 1978; Reyes et al. 1980), whereas with HSV-2 the transforming region (mtrII) was located between positions 0.58 and 0.63 (Reyes et al. 1980; Galloway and McDougall 1981), a rather surprising finding given the colinearity of the genomes, including the two transforming regions (Hayward and Reyes 1983; our unpublished results). Second, another region of the HSV-2 genome (m.u. 0.43–0.58) was able to convert cells to a tumorigenic phenotype in a continuous passage assay (Jariwalla et al. 1980). Third, despite the ability of at least six laboratories to transform cells with mtrII, as defined by the BglII N fragment, neither the retention of viral sequences nor the expression of viral proteins was consistently, or even frequently, observed (Reyes et al. 1980; Galloway and McDougall 1981; Bejcek and Conley 1985; Cameron et al. 1985; M.K. Howett and A. Kessous, pers. comm.).

The overall organization of the HSV-2 BglII N fragment has been determined, including mapping of the viral mRNA transcripts (McLaughlan and Clements 1983a; Jenkins and Howett 1984), identification of the protein products by in vitro translation (Galloway et al. 1982), and partial DNA sequence analysis (McLaughlan and Clements 1983b; Galloway and Swain 1984; Swain

et al. 1985). Several lines of evidence, not reviewed here, show that genes at the left end of the fragment encode two subunits of ribonucleotide reductase; the gene spanning the right end of the fragment encodes glycoprotein C; and the functions of the other proteins are not known. By constructing deleted fragments of mtrII in vitro and transfecting them into rodent cells, we localized the transforming region to a 794-bp fragment, designated pBC24. The transforming fragment lies within the coding region for a 61K protein but does not appear to specify a viral polypeptide (Galloway et al. 1984). Contained within the transforming fragment are sequences that can be drawn as a stem-loop structure flanked by direct repeats, similar to an insertion sequence–like element. These results are summarized in Figure 1.

Human cytomegalovirus (CMV) was also shown to transform rodent cells (Albrecht and Rapp 1973) and attempts to define the mtr have also given unconventional results for a DNA tumor virus. Using defined fragments of CMV, the mtr was localized to a 2.9-kb region of the genome that is expressed as immediate early RNA (Nelson et al. 1982). The structure of this region has been determined, including mapping of the RNAs (Stinski et al. 1983; Jahn et al. 1984) and fine structure analysis of the gene encoding the major immediate early gene product (Stenberg et al. 1984) and its regulatory region (Thomsen et al. 1984; Boshart et al. 1985). Using a protocol identical with that described for HSV-2, a 558-bp fragment of CMV designated pCM4127 was shown to have transforming activity (Nelson et al. 1984). The fragment falls within a 5.0-kb unspliced RNA that by DNA sequence analysis reveals no significant open reading frame (Nelson et al. 1984; B. Fleckenstein, pers. comm.). An insertion-like structure was also found

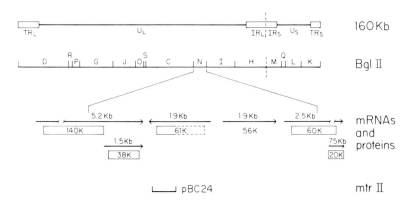

**Figure 1** The organization of the HSV-2 *mtr*. The overall structure of the HSV-2 strain 333 genome is depicted, as are the fragments generated by cleavage with *Bgl*II. Shown are the mRNAs that can be selected by hybridization with *Bgl*II N and the products they translate in vitro. Solid boxes for the proteins indicate confirmation by DNA sequence analysis. The sequences encoded by BC24 were shown to have morphological transforming activity.

in the CMV *mtr*I. These results are summarized in Figure 2.

The precedent for tumor viruses that do not carry their own transforming genes has been established by the avian leukosis viruses that transform cells by activating cellular oncogenes (Neel et al. 1981; Payne et al. 1981). To determine whether this is a possible mechanism for herpesvirus transformation, the *mtr* genes were linked to test genes to assay their ability to promote or enhance transcription. Additionally, the ability of the sequences to alter transcription of cellular genes in transformed cells was examined by using a phage that contained the integrated CMV plasmid and flanking cellular sequences. To investigate an alternative hypothesis, we examined the ability for the transforming fragments to induce mutations at the HGPRT locus. UV and neutral-red-inactivated HSV-1 have been shown to induce mutations (Schlehofer and zur Hausen 1982) and to selectively amplify DNA in a way analogous to chemical carcinogens (Schlehofer et al. 1983). Our experiments have led us to conclude that HSV-2 and CMV may transform cells by different mechanisms, with HSV-2 having the potential to induce mutations by gene rearrangement and with CMV having the potential to activate cellular genes.

## Materials and Methods

### Construction of plasmids

A series of plasmids was constructed to link the *mtr*s of HSV-2 and CMV to the HSV-1 thymidine kinase (TK)

gene. The plasmids pBC24 (Galloway et al. 1984) and pCM4127 (Nelson et al. 1984) were digested with *Hind*III and *Bam*HI and the 794-bp and 555-bp fragments, respectively, were purified from low-melting-point agarose gels. The TK deletion plasmids were obtained from S. McKnight (Carnegie Institute, Baltimore) and have been described previously (McKnight and Gavis 1980). Plasmids 5′ − 6, 5′ + 36, and 5′ + 109 were cleaved with *Bam*HI and the *mtr* fragments were ligated to them. The plasmid was then cleaved with *Hind*III to release the *mtr*-TK construct and that was ligated into *Hind*III-cleaved, phosphatase-treated pBR322. The plasmids 5′ − 4, 5′ + 35, and 5′ + 480 were first cleaved with *Hind*III and treated as above except that the second cleavage was done with *Bam*HI.

Another series of plasmids was constructed to link the *mtr*s to the chloramphenicol acetyltransferase (CAT) gene. The plasmid pSV2CAT (Gorman et al. 1982) was digested with *Sph*I and *Acc*I to remove the SV40 72-bp repeats and filled in with Klenow polymerase, and *Sal*I linkers were ligated on to create the vector pSVSalCAT. The *mtr* fragments were purifed from low-melting-point agarose after digestion by *Hind*III and *Bam*HI. The fragments were filled in, *Sal*I linkers were added, and the fragments were ligated into pSVSalCAT. The orientation of the insert in the vector was determined by restriction analysis. In the plasmids pBC24CAT1 and p4127CAT2 the *Hind*III site of the fragment is closest to the CAT gene, and in pBC24CAT2 and p4127CAT1 the *Bam*HI site is closest to the CAT gene. Constructs linking the pCM4127 insert to the 3′ end of the CAT

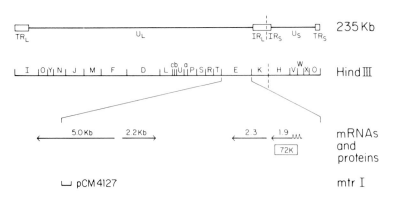

**Figure 2** The organization of the CMV *mtr*. The overall structure of the CMV strain AD169 genome is shown, as are the fragments generated by cleavage with *Hind*III. The mRNAs that can be identified by hybridization with the *Hind*III E fragment are shown, as is the 72K major immediate-early protein. The sequences encoded by pCM4127 were shown, using deleted fragments, to have morphological transforming activity.

gene were made by purifying the insert, filling in with Klenow polymerase, adding BamHI linkers to the fragment, and ligating the fragment to BamHI-linearized, phosphatase-treated pSVSalCAT.

The restriction enzymes used were purchased from Bethesda Research Laboratories (Bethesda), T4 ligase was obtained from New England Nuclear (Cambridge [USA]), and Klenow polymerase from Boehringer Mannheim (Indianapolis). The cloning procedures were performed using protocols of Maniatis et al. (1982).

*Transfections, TK transformants, and the CAT assay*
The TK plasmids were tested for their ability to transform Ltk$^-$ cells to a TK$^+$ phenotype. Ltk$^-$ cells were seeded into 60-mm dishes at a concentration of $5 \times 10^5$ cells 24 hr before transfection. Calcium phosphate precipitates containing 1 μg of plasmid DNA and 20 μg of salmon sperm DNA were formed, transfected onto cells, and the medium replaced 4 hr later. After 24 hr the cells were passaged 1:3 and changed to DME, 10% serum, and HAT, and the medium was changed every 3 days. Colonies were counted after 21 days. This method has been described in detail previously (McDougall et al. 1980).

The CAT plasmids were tested for their ability to enhance the transcription of the CAT gene in transient expression assays. The method followed was modified from Lopata et al. (1984). Twenty-four hr prior to transfection, a total of $1.5 \times 10^6$ cells were seeded per 100-mm dish and medium was changed 20 hr later. Sample DNAs, 8–10 μg of plasmid, were complexed to 4.5 ml of DEAE-Dextran (200 μg/ml; Pharmacia, m.w. 500,000) in DME with 50 mM Tris·HCl, pH 7.3. After rinsing twice with serum-free medium, the complexed DEAE-DNA samples were added to the subconfluent dishes. Plates were incubated at 37°C for 4 hr, followed by a shock with 10% DMSO in 1 × HBS for 2 min. Cells were rinsed twice with DME and maintained at 37°C in medium containing 10% fetal calf serum. Cells were harvested approximately 72 hr after transfection and assayed. Extracts were normalized for total protein as determined by the Bradford method. After incubation for up to 6 hr with 17.3 μM [$^{14}$C]chloramphenicol (NEN; 57.8 mCi/mM, 0.1 mCi/ml), samples were spotted onto a thin-layer silica gel plate. Products were separated from substrate by development in 95:5 chloroform/methanol. Chromatograms were visualized by exposure to Kodak XAR-5 film. Comparison of acetylated versus nonacetylated derivatives was made by liquid scintillation in a Beckman LS-7000 counter.

*Analysis of RNA from transformed cells*
The characterization of cells transformed by pCM4127 and their tumor-derived cell lines will be described elsewhere (F. Buonaguro et al., in prep.). Total cytoplasmic RNA was prepared from ten 100-mm dishes of NIH-3T3 cells and from a transformant TN4 2/2 according to Scott et al. (1983). The RNA was quantitated by spectrophotometric reading and by staining on an agarose gel containing ethidium bromide. The RNA was

transferred onto nitrocellulose using a "slot-blot" apparatus by following the instructions of Schleicher and Schuell. The most concentrated sample contained 25 μg, and 1:3 dilutions were used. The nitrocellulose filters were hybridized with a $^{32}$P-labeled actin gene and a histone probe (gifts of M. Groudine, Fred Hutchinson Cancer Research Center, Seattle) to further determine that equivalent amounts of RNA were placed on each slot. The filters were then hybridized with a $^{32}$P-labeled probe designated pλE2.3, which contained a 2.3-kb EcoRI fragment from a λ clone containing the NIH-3T3 sequences flanking the integrated pCM4127. Hybridization and preparations of nick-translated probes followed the protocols of Maniatis et al. (1982). Comparison of the RNA hybridization was done with a LKB scanning densitometer.

*Mutagenesis of the HGPRT gene*
NIH-3T3 cells were passed in HAT medium for 2 weeks to reduce spontaneous HGPRT$^-$ cells. Five 100-mm plates per sample were then seeded with $3.0 \times 10^6$ cells/plate and transfected 24 hr later using the calcium phosphate coprecipitation method. Twenty-four hr after transfection, the plates were trypsinized and the cells were seeded into roller bottles in DME with 10% fetal calf serum. The cells were passed twice during the next 8 days, and a minimum of $1.5 \times 10^7$ cells were reseeded at each pass. After the 8-day expression time, the cells were plated in selective medium (DME with 10% FCS and 0.1 mM 6-thioguanine). For each sample $1.5 \times 10^7$ cells were seeded at a density of $5 \times 10^5$ cells/plate (30 plates/sample). Resistant mutants were counted 2 to 4 weeks after plating in selective medium.

## Results and Discussion

### The enhancing activity of the CMV *mtr*
To test the hypothesis that transformation with herpesvirus may result from the ability of the *mtr* regions to alter the expression of cellular genes, these fragments were linked to test genes. In the first instance the fragments were linked to the HSV-1 TK gene by using a series of defined TK deletions created by Steve McKnight (McKnight and Gavis 1980). The plasmids 5′ − 6, 5′ + 36, and 5′ + 109 contain the sequences as indicated, relative to the start of the transcription, and are bounded by a BamHI site; the former two lack a functional promoter. The plasmids 5′ − 4, 5′ + 35, and 5′ + 480 are bounded by a HindIII site and contain the sequences upstream of TK as indicated. These TK fragments allowed the *mtr* regions to be placed in either orientation (by virtue of the *mtr* HindIII and BamHI ends) upstream of the TK genes in two positions where they could serve as a promoter and in both orientations further upstream of a gene containing its own promoter, where they could serve as an enhancer. The *mtr*-TK plasmids were transfected into Ltk$^-$ cells, and TK$^+$ colonies were selected as summarized in Table 1. The promoter-negative TK deletions were not able to synthesize TK, nor were any of the constructions containing the HSV-2 or CMV *mtr*

**Table 1** The Ability of *mtr*-TK Constructs to Transform Ltk⁻ Cells

| DNA[a] | No. of colonies[b] | DNA[a] | No. of colonies[b] | DNA[a] | No. of colonies[b] |
|---|---|---|---|---|---|
| 5' − 6 | 0 | BC24B-1 | 0 | 4127B1 | 0 |
| 5' + 36 | 0 | BC24B-2 | 0 | 4127B2 | 0 |
| 5' + 109 | 78 | BC24B-3 | 86 | 4127B3 | 350 |
| 5' − 4 | 0 | BC24H-4 | 0 | 4127H4 | 0 |
| 5' + 35 | 0 | BC24H-5 | 0 | 4127H5 | 0 |
| 5' + 480 | 112 | BC24H-6 | 90 | 4127H6 | 194 |

[a]Transfection contained 1 μg plasmid DNA and 20 μg salmon sperm per 60-mm dish Ltk⁻ cells.
[b]Total number from two 60-mm dishes that were transfected.

regions (here shown as BC24 or 4127, respectively). The TK deletions containing a promoter were able to give rise to TK⁺ transformants. No increase in the number of colonies was observed when the HSV-2 *mtr* was located upstream of the promoter; however, when the CMV *mtr* was located upstream of the promoter, the number of colonies was increased approximately four-fold in one orientation and twofold in the other orientation. From these results we conclude that neither the HSV-2 *mtr* nor the CMV *mtr* are likely to function as a promoter, and that the CMV *mtr* shows characteristics that are consistent with its functioning as a transcriptional enhancer (Banerji et al. 1981; Gruss 1984).

To further examine the possibility that the CMV *mtr* might act as an enhancer, the fragment was linked to the CAT gene. A plasmid pSVSalCAT was constructed that contained the SV40 TATA homology, the 21-bp repeats, a *Sal*I site in place of the *Sph* site, and no other upstream SV40 sequences. Both the HSV-2 *mtr*, pBC24, and the CMV *mtr*, pCM4127, were inserted in either orientation into the *Sal*I site. The plasmids were transfected into a variety of species cell types (e.g., mouse 3T3 cells, hamster BHK cells, monkey CVI cells, and human foreskin fibroblasts) and in transient expression assays were measured for their ability to synthesize CAT. The results in two cell lines are shown in Figure 3. In either orientation or cell type, BC24 did not enhance the expression of CAT above that of pSV1CAT alone. On the other hand, the CMV *mtr* could enhance the level of CAT activity to that of the SV40 72-bp enhancer sequences, as shown in comparison with pSV2CAT. Figure 3 shows a dramatic example of the directionality preference exhibited by this fragment. In most experiments weak enhancing activity, 2-fold to 3-fold, could be observed with pCM4127 in one orientation; however, when the *Hin*dIII border of the fragment was placed nearest the CAT gene, enhancement of 5-fold to 10-fold was observed. Data to be presented more fully elsewhere (F. Buonaguro et al., in prep.) show that the CMV *mtr* can enhance transcription when placed at the 3' end of the CAT gene and that it does not function in *trans*. Unpublished data of B. Fleckenstein, W. Schaffner, and colleagues have shown that pCM4127 can substitute for the SV40 72-bp repeats in an enhancer trap protocol. From these data we conclude that sequences within the CMV *mtr* pCM4127 can serve as a transcriptional activator. The preference for a given orientation is unusual for a classic enhancer, and experiments are in progress to determine whether there are sequences at one end of pCM4127 that are inhibitory to enhancing activity and that, when deleted, will allow the element to function in a bidirectional fashion.

**Transcriptional activation in CMV-transformed cells**
The demonstration that the CMV *mtr* can enhance the transcription of linked test genes is suggestive of the fact, but certainly not proof, that CMV induces morphological transformation by increasing the expression of cellular genes. In data to be presented elsewhere (F. Buonaguro et al.), CMV-transformed cell lines were obtained by cotransformation with pSV2neo and selection in both G418 and low serum. These cell lines were analyzed for retention of CMV sequences by Southern blot analysis, and a tumor derivative retaining only a single copy of pCM4127 was selected for further study. A bacteriophage λ library containing the transformed-

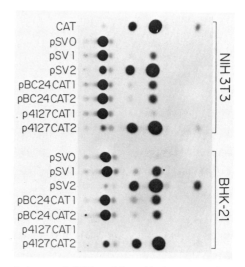

**Figure 3** Assay of CAT activity with the *mtr* regions placed upstream of the SV40 promoter. Plasmid DNAs are transfected into either NIH-3T3 cells or BHK-21 cells. After 48–72 hr extracts are prepared and assayed for their ability to acetylate [¹⁴C]chloramphenicol, and the products are separated on thin-layer chromatography. The plasmids pSV0, pSV1, and pSV2 contain no transcriptional elements, the SV40 promoter, and the SV40 promoter and enhancer linked to the CAT genes, respectively. The other plasmids contain the HSV-2 (pBC24) and CMV (pCM4127) sequences inserted in either orientation of a *Sal*I site located upstream of the SV40 promoter. CAT is the commercially available enzyme.

cell DNA was constructed and a phage was isolated that contained the integrated pCM4127 plasmid and flanking NIH-3T3 cellular sequences. From this phage, a 2.3-kb *Eco*RI fragment containing mouse sequences was isolated and used to probe slot-blots containing RNA from normal NIH-3T3 cells and the transformants, as shown in Figure 4. After spectrophotometric analysis to determine that equivalent amounts of RNA were loaded, the blot was hybridized with an actin probe (as shown in Fig. 4) and a histone probe (not shown) to confirm that the concentration of RNA in the two samples was equivalent. When the flanking cellular sequences of the transformant were used as probe, Figure 4 demonstrates that these sequences are transcribed in normal 3T3 cells and that the level of this RNA is increased threefold to fivefold in the transformant. From these results, we conclude that the CMV *mtr* can augment the expression of cellular genes.

Several questions about CMV transformation remain unanswered. The first is, what is the up-regulated gene in this transformant? When the *Eco*RI fragment was hybridized to a battery of oncogenes, no hybridization was detected (data not shown). Whether the transcription of this gene is altered in other CMV transformants or in transformed cells in general is not known.

Second, if CMV transformation occurs by a fragment enhancing the transcription of cellular genes, one would expect that a general feature of transformed cells would be the retention of viral sequences in proximity to important cellular genes. In fact, many CMV transformants do not apparently retain viral DNA, as judged by hybridization with pCM4127 (J. Nelson; F. Buonaguro; B. Fleckenstein; all unpubl.). Either only a very small segment of the *mtr* is consistently retained, not detectable by hybridization, or the fragment need only be integrated transiently. Another possibility is that the transformant observed in this study is unusual, and to test the generality of the model that CMV enhances transcription of cellular genes, additional transformants will have to be examined.

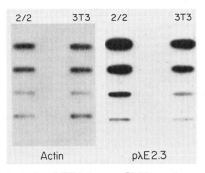

**Figure 4** Analysis of RNA from a CMV-transformed cell line. The lanes 3T3 and 2/2 contain total cytoplasmic RNA from normal NIH-3T3 cells and from a tumor-derived CMV transformant TN4 2/2. RNA was quantitated and transferred to a filter by using a slot-blot apparatus in threefold dilutions, starting with 25 μg. The panel on the left was hybridized with an actin probe, and the panel on the right was hybridized to a plasmid containing a 2.3-kb *Eco*RI fragment from a λ phage containing the mouse sequences flanking the inserted pCM4127 DNA.

Third, if transcriptional enhancers can cause morphological transformation, why is it impossible to induce transformation with other CMV enhancers, such as the strong element located upstream of the major immediate early gene (Boshart et al. 1985) or other known viral enhancers? Although there has been one report that retroviral long terminal repeats (LTRs) alone can transform cells (Muller and Muller 1984), this phenomenon has not been widely observed, though perhaps it has not been widely examined. One explanation may be that in addition to enhancing activity, the CMV *mtr* may have other features that facilitate its integration into, and perhaps excision from, cellular genes.

## Mutagenic potential of the HSV-2 *mtr*

Since the HSV-2 *mtr* did not exhibit any ability to act as transcriptional regulatory elements, other mechanisms to explain its transforming potential were examined. The hypothesis that HSV may act as a mutagen was first suggested by its ability to damage chromosomes (Hampar and Ellison 1961; O'Neill and Rapp 1971) and by experiments that showed that UV- or neutral-red-inactivated HSV-1 could induce mutations in the HGPRT gene (Schlehofer and zur Hausen 1982). This hypothesis has been reviewed recently (Galloway and McDougall 1983; zur Hausen et al. 1984). To test this idea, the HSV-2 *mtr*, pBC24, was transfected into NIH-3T3 cells, and the cells were selected for resistance to 6-thioguanine as described in Materials and Methods and in more detail elsewhere (C. Brandt et al., in prep.). As controls, cells were treated with the mutagen *N*-methyl-*N*-nitro-*N*-nitrosoguanidine and mock-transfected or transfected with pBR322 or the CMV *mtr* pCM4127. The results from one experiment are shown in Table 2. The HSV-2 fragment shows a mutation frequency enhanced 11-fold above mock-transfected cells, a value in agreement with that of Schlehofer and zur Hausen when comparing UV-inactivated HSV-1 with mock infection. However, when another plasmid such as pBR322 was used, an increased mutation frequency was observed, a variable not controlled for in the experiments of Schlehofer and zur Hausen (1982). Therefore we would conclude that in this assay, the HSV-2 *mtr* acts as a weak mutagen.

Interesting results were obtained when the nature of the mutations was analyzed by probing Southern blots of mutant-cell DNAs with a cDNA of the HGPRT gene (a gift of T. Caskey) (see Fig. 5). Of 21 mutants obtained from control transfections, 9 of which were transfected with pBR322, only one showed any alteration in the HGPRT gene. Of 6 mutants obtained by transfection with pBC24, 3 showed alterations in the HGPRT restriction pattern. More transformants are being analyzed to confirm the statistical significance of this observation; however, the first impression of these results is that the HSV-2 *mtr* is qualitatively different from other mutagens in that it may readily induce gene rearrangement in an as yet undefined manner. This is an intriguing hypothesis in light of the fact that activation of many oncogenes occurs by gene rearrangements.

In summary, the data presented in this study suggest

**Table 2** Frequencies of 6-Thioguanine-resistant Mutants

| Sample | DNA transfected per plate | Mutants | Surviving fraction | Mutation frequency | Enhancement over control |
|---|---|---|---|---|---|
| Control | — | 1 | 0.446 | $1.4 \times 10^{-7}$ | — |
| pBR322/8 | 8 | 3 | 0.297 | $6.7 \times 10^{-7}$ | 4.78 |
| pBC24/25 | 25 | 3 | 0.200 | $1 \times 10^{-6}$ | 7.14 |
| pBC24/50 | 50 | 5 | 0.197 | $1.6 \times 10^{-6}$ | 11.4 |
| pBC24/200 | 200 | 3 | 0.210 | $9.5 \times 10^{-7}$ | 6.78 |
| p4127/6 | 6 | 3 | 0.442 | $4.5 \times 10^{-7}$ | 3.21 |
| p4127/30 | 30 | 5 | 0.377 | $8.8 \times 10^{-7}$ | 6.29 |

that although there are similarities between the *mtr* regions of HSV-2 and CMV, in that neither appears to encode viral proteins and both contain IS-like elements, they may induce transformation by different mechanisms. In the case of CMV, the *mtr* can augment transcription when linked to the test genes TK or CAT and can in at least one instance enhance the expression of a cellular gene in a morphologically transformed cell. In contrast, no transcriptional activation by the HSV-2 *mtr* could be demonstrated. Preliminary evidence suggests that pBC24 acts as a weak mutagen when transfected into cells and that its mutagenic potential may come from its ability to induce gene rearrangements. Furthermore, the data presented here and elsewhere demonstrate that HSV and CMB are unconventional DNA tumor viruses because they do not encode their own oncogenes. Their ability to transform cells in culture stems from their tendency to alter cellular genes, in a broad sense as mutagenic agents. If these viruses contribute to the etiology of human cancers, it is by virtue of their subtle mutagenic potential rather than by the action of a dominant oncogene, a concept that is unusual for virology (and virologists) but fits well with our current understanding of the multistep process leading to human cancers.

## Acknowledgments

We are grateful for helpful discussions with J. Nelson and B. Fleckenstein; plasmid constructions from J. Nelson and T. Hill; gifts of plasmids from S. McKnight, C. Gorman, M. Groudine, and T. Caskey; dedicated technical assistance of A. Taeschner; artwork by P. Su; and preparation of the manuscript by T. Higgs. This work was supported by grants from the National Cancer Institute to D.A.G. (CA 26001) and J.K.M. (CA 29350). C.R.B. is supported by a Public Health Service fellowship (GM 09125), and F.M.B. was initially supported by a fellowship from the World Health Organization.

## References

Albrecht, T. and F. Rapp. 1973. Malignant transformation of hamster embryo fibroblasts following exposure to ultraviolet irradiated human cytomegalovirus. *Virology* **55:** 53.

Banerji, J., S. Rusconi, and W. Schaffner. 1981. Expression of a β-globin gene is enhanced by remote SV40 DNA sequences. *Cell* **27:** 299.

Bejcek, B. and A.J. Conley. 1985. A rat oncogene isolated as an extra-chromosomal element from herpes simplex virus 2 transformed cells. *Virology* (in press).

Boshart, M., F. Weber, G. Jahn, K. Dorsch-Hasler, B. Fleckenstein, and W. Schaffner. 1985. A very strong enhancer is located upstream of an immediate early gene of human cytomegalovirus. *Cell* **41:** 521.

Camacho, A. and P.G. Spear. 1978. Transformation of hamster embryo fibroblasts by a specific fragment of the herpes simplex virus genome. *Cell* **15:** 993.

Cameron, I.R., M. Park, B.M. Dutia, A. Orr, and J.C.M. Macnab. 1985. Herpes simplex virus sequences involved in the initiation of oncogenic morphological transformation of rat cells are not required for maintenance of the transformed state. *J. Gen. Virol.* **66:** 517.

Duff, R. and F. Rapp. 1971. Oncogenic transformation of hamster cells after exposure to herpes simplex virus type 2. *Nat. New Biol.* **233:** 45.

———. 1973. Oncogenic transformation of hamster embryo cells after exposure to herpes simplex virus type 1. *J. Virol.* **12:** 209.

Galloway, D.A. and J.K. McDougall. 1981. Transformation of rodent cells by a cloned DNA fragment of herpes simplex virus type 2. *J. Virol.* **38:** 749.

———. 1983. The oncogenic potential of herpes simplex viruses: Evidence for a "hit-and-run" mechanism. *Nature* **302:** 21.

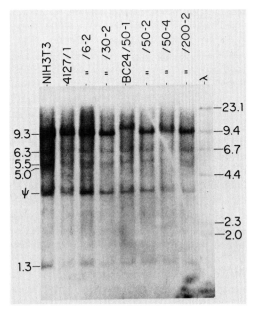

**Figure 5** Analysis of the HGPRT gene in cells mutagenized by transfection of *mtr* sequences. DNA from NIH-3T3 cells and from HGPRT⁻ mutant cell lines was digested with *Eco*RI and analyzed by Southern blot analysis with a probe containing a cDNA of the HGPRT gene.

Galloway, D.A. and M.A. Swain. 1984. Organization of the left-hand end of the herpes simplex virus type 2 *Bgl*II N fragment. *J. Virol.* **49:** 724.

Galloway, D.A., L.C. Goldstein, and J.B. Lewis. 1982. Identification of proteins encoded by a fragment of herpes simplex virus type 2 DNA that has transforming activity. *J. Virol.* **45:** 530.

Galloway, D.A., J.A. Nelson, and J.K. McDougall. 1984. Small fragments of herpesvirus DNA with transforming activity contain insertion sequence-like structures. *Proc. Natl. Acad. Sci.* **81:** 4736.

Gorman, C.M., L.F. Moffat, and B.H. Howard. 1982. Recombinant genomes which express chloramphenicol acetyl transferase in mammalian cells. *Mol. Cell. Biol.* **2:** 1044.

Gruss, P. 1984. Eukaryotic enhancers. *DNA* **3:** 1.

Hampar, B. and S.A. Ellison. 1961. Chromosomal aberrations induced by an animal virus. *Nature* **192:** 145.

Hayward, G.S. and G.R. Reyes. 1983. Biochemical aspects of transformation by herpes simplex viruses. *Adv. Viral Oncol.* **3:** 271.

Jahn, G., E. Knust, H. Schmolla, T. Sarre, J.A. Nelson, J.K. McDougall, and B. Fleckenstein. 1984. Predominant immediate early transcripts of human cytomegalovirus AD 169. *J. Virol.* **49:** 363.

Jariwalla, R.J., L. Aurelian, and P.O.P. Ts'o. 1980. Tumorigenic transformation induced by a specific fragment of DNA from herpes simplex virus type 2. *Proc. Natl. Acad. Sci.* **77:** 2279.

Jenkins, F.J. and M.K. Howett. 1984. Characterization of mRNAs that map in the *Bgl*II N fragment of the herpes simplex virus type 2 genome. *J. Virol.* **52:** 99.

Lopata, M.A., D.W. Cleveland, and B. Sollner-Webb. 1984. High level transient expression of a chloramphenicol acetyl transferase gene by DEAE-dextran mediated DNA transfection coupled with a dimethyl sulfoxide or glycerol shock treatment. *Nucleic Acids Res.* **12:** 5707.

Maniatis, T., E.F. Fritsch, and J. Sambrook. 1982. *Molecular cloning: A laboratory manual.* Cold Spring Harbor Laboratory, Cold Spring Harbor, New York.

McDougall, J.K., T. Massey, and D. Galloway. 1980. Location and cloning of the herpes simplex virus type 2 thymidine kinase gene. *J. Virol.* **33:** 1221.

McKnight, S.L. and E.R. Gavis. 1980. Expression of the herpes thymidine kinase gene in *Xenopus laevis* oocytes: An assay for the study of deletion mutants constructed *in vitro. Nucleic Acids Res.* **8:** 5931.

McLaughlan, J. and J.B. Clements. 1983a. Organization of the herpes simplex virus type 1 transcription unit encoding two early proteins with molecular weights of 140,000 and 40,000. *J. Gen. Virol.* **64:** 997.

———. 1983b. DNA sequence homology between two co-linear loci on the HSV genome which have different transforming abilities. *EMBO J.* **2:** 1953.

Minson, A.C. 1984. Cell transformation and oncogenesis by herpes simplex virus and human cytomegalovirus. *Cancer Surv.* **3:** 91.

Muller, R. and D. Muller. 1984. Co-transfection of normal NIH3T3 DNA and retroviral LTR sequences: A novel strategy for the detection of potential c-*onc* genes. *EMBO J.* **3:** 1121.

Neel, B.G., W.S. Hayward, H.L. Robinson, J. Fang, and S. Astrin. 1981. Avian leukosis virus-induced tumors have common proviral integration sites and synthesize discrete RNAs: Oncogenesis by promoter insertion. *Cell* **23:** 323.

Nelson, J.A., B. Fleckenstein, D.A. Galloway, and J.K. McDougall. 1982. Transformation of NIH3T3 cells with cloned fragments of human cytomegalovirus strain AD169. *J. Virol.* **43:** 83.

Nelson, J.A., B. Fleckenstein, G. Jahn, D.A. Galloway, and J.K. McDougall. 1984. Structure of the transforming region of human cytomegalovirus AD169. *J. Virol.* **49:** 109.

O'Neill, F.J. and F. Rapp. 1971. Early events required for the induction of chromosomal abnormalities in human cells by herpes simplex virus. *Virology* **44:** 544.

Payne, G.S., S.A. Courtneidge, L.B. Crittenden, A.M. Fadley, J.M. Bishop, and H.E. Varmus. 1981. Analysis of avian leukosis virus DNA and RNA in bursal tumors: Viral gene expression is not required for maintenance of the tumor state. *Cell* **23:** 311.

Reyes, G.R., R. LaFemina, S.D. Hayward, and G.S. Hayward. 1980. Morphological transformation by DNA fragments of human herpesviruses: Evidence for two distinct transforming regions in herpes simplex virus types 1 and 2 and lack of correlation with biochemical transfer of the thymidine kinase gene. *Cold Spring Harbor Symp. Quant. Biol.* **44:** 629.

Schlehofer, J.R. and H. zur Hausen. 1982. Induction of mutations within the host cell genome by partially inactivated herpes simplex virus type 1. *Virology* **122:** 471.

Schlehofer, J.R., L. Gissman, B. Matz, and H. zur Hausen. 1983. Herpes simplex virus-induced amplification of SV40 sequences in transformed hamster embryo cells. *Int. J. Cancer* **32:** 99.

Scott, M.R.D., K.-H. Westphal, and P.W.J. Rigby. 1983. Activation of mouse genes in transformed cells. *Cell* **34:** 557.

Stenberg, R.M., D.R. Thomsen, and M.F. Stinski. 1984. Structural analysis of the major immediate early gene of human cytomegalovirus. *J. Virol.* **49:** 190.

Stinski, M.F., D.R. Thomsen, R.M. Stenberg, and L.C. Goldstein. 1983. Organization and expression of the immediate early genes of human cytomegalovirus. *J. Virol.* **46:** 1.

Swain, M.A., R.W. Peet, and D.A. Galloway. 1985. Characterization of the gene encoding herpes simplex virus 2 glycoprotein C and comparison with the type 1 counterpart. *J. Virol.* **53:** 561.

Thomsen, D.R., R.M. Stenberg, W.F. Goins, and M.F. Stinski. 1984. Promoter-regulatory region of the major immediate early gene of human cytomegalovirus. *Proc. Natl. Acad. Sci.* **78:** 1441.

zur Hausen, J., L. Gissman, and J.R. Schlehofer. 1984. Viruses in the etiology of human genital cancer. *Prog. Med. Virol.* **30:** 170.

# Expression of Simian Virus 40 Oncogenes in F9 Embryonal Carcinoma Cells, in Transgenic Mice, and in Transgenic Embryos

F. Kelly,[*] O. Kellermann,[†] F. Mechali,[*] J. Gaillard,[‡] and C. Babinet[§]

[*]Institut de Recherches Scientifiques sur le Cancer, Centre National de la Recherche Scientifique, 94800 Villejuif, France; [†]Département de Biologie Moléculaire, [‡]Département de Physiologie Expérimentale, et [§]Département d'Immunologie, Institut Pasteur, 75015 Paris, France

The expression of simian virus 40 (SV40) oncogenes was examined in F9 embryonic carcinoma cells and in transgenic mice carrying an integrated recombinant plasmid, pK4, in which the SV40 early region is linked to the enhancer-promoter sequences of the E1A transcription unit of adenovirus type 5. F9-K4B2, a pK4-transformed F9 cell line, expresses the SV40 T antigens when induced to differentiate in vitro, and early embryonic derivatives, in particular cells with characteristics of parietal endoderm, have been derived with high efficiency from the differentiated cell population. In pK4 transgenic mice, expression of the viral genes is observed in several tissues, and some mice develop glioblastomas at 3–4 months of age. Permanent cell lines and, in particular, glial-like cells have also been established from 12-day embryos. The production of a strain of pK4 transgenic mice should allow the study of the expression of T antigens in the early mouse embryo and their role in the development of brain tumors.

One approach to understanding mechanisms involved in cellular determination in the mouse embryo has been to deal with in vitro systems. Many cell lines have been isolated from various tissues and retain their initial characteristics. On the one hand, multipotential embryonic carcinoma (EC) cells, which resemble early mouse embryo cells, have been isolated from spontaneous teratocarcinomas and have been widely used as a paradigm to study early mouse embryogenesis (Silver et al. 1983). EC cells can now be derived very efficiently from early embryos grown in vitro (Evans and Kaufman 1981; Martin 1981). On the other hand, many cell lines with restricted possibilities are useful tools for the study of terminal differentiation. In contast, only a few cell lines corresponding to early intermediate stages of differentiation have so far been isolated (Nicolas et al. 1981).

Simian virus 40 (SV40) has served extensively in the establishment of cells lines that are derived from a variety of highly differentiated tissues and that retain their differentiated properties (Tooze 1980; for more-recent references, Tanigawa et al. 1983). Recently, transgenic mice have been obtained that express the SV40 oncogenes either in the brain (Brinster et al. 1984) or in the pancreas (Hanahan 1985), depending on the enhancer-promoter sequences used to drive the viral genes. These results indicate that this system may be an extremely powerful tool to derive specific cell lines through the use of adequate enhancer sequences.

The resistance of early embryonic cells to SV40 (for review, see Kelly and Condamine 1982) precludes the use of the wild-type virus to immortalize early embryonic derivatives. Although the mechanism of resistance is not fully understood, several experiments suggest that the enhancer-promoter region of the virus is not efficiently recognized in these cells (Sleigh 1985).

We decided to examine whether the early region of SV40, which carries the transforming and immortalizing genes of the virus, might be expressed in early embryonic cells when placed under the control of a heterologous promoter active in EC cells. Adenovirus types 2 and 5 (Ad2 and Ad5) represent potential sources of such promoters since, in contrast to SV40, they are able to infect EC cells (Kelly and Boccara 1976; Cheng and Praszkiev 1982).

A recombinant plasmid pK4 (Fig. 6A) was therefore constructed (O. Kellermann and F. Kelly, in prep.) in which the early region of SV40 was linked to the enhancer-promoter region of the early transcription unit 1A (E1A) of Ad5 (Tooze 1980).

This plasmid was introduced into F9 EC cells by cotransfer with the neo gene (Southern and Berg 1982) and into mice by microinjection of fertilized eggs.

The results presented here show that immortalized derivatives from early embryonic cells, in particular cells with characteristics of parietal endoderm, were derived with high efficiency from a pK4-transformed F9 cell line and expressed the SV40 oncogenes. In pK4 transgenic mice, expression of the viral genes has been observed in several tissues, and some mice develop brain glioblastomas at 3–4 months of age. Permanent cell lines, and in particular glial-like cells, were also derived from 12-day embryos.

## Materials and Methods

### Cell culture

All cultures were performed in Dulbecco's modified Eagle's medium (DME) containing 10% fetal calf serum. F9 cells were grown as described (Nicolas et al. 1976) on gelatin-coated tissue-culture dishes.

Tissues from adult mice were minced finely and placed in tissue-culture dishes with medium. Cells were fixed and stained for the presence of SV40 T antigen after 3–5 days, when sufficient outgrowth had developed. Twelve-day embryos were first minced and cells were further dissociated by pipetting. Cells were replated at lower density when the dishes became confluent and divided 1:5 for successive passages. The cloning efficiency was determined on dishes seeded with $10^3$ and $10^4$ cells, fixed, and stained with Giemsa after 3 weeks. Colonies were isolated from duplicate dishes using a pipetman tip. They were plated, then expanded, in 5-cm dishes.

*Transformation of F9 cells by pK4*
pK4 and the plasmid pSVtk-neoβ were cotransfected into F9 cells by the calcium phosphate coprecipitation technique (Graham and van der Eb 1973). pSVtk-neoβ carries a bacterial resistance gene expressed efficiently in EC cells under the control of a composite SV40 early–herpes simplex virus I thymidine kinase promoter (SVtk) (Nicolas and Berg 1983). After cotransfection, cells were grown in the presence of Geneticin (G418) at a concentration of 400 μg/ml. G418-resistant colonies were isolated 10–15 days later and subcloned. The frequency of transformation was approximately $10^{-5}$.

*Differentiation of F9 cells*
F9 cells and pK4-transformed F9 cells were seeded at low density ($10^4$ cells per 3-cm dish) and treated with $10^{-7}$ M retinoic acid (Sigma) and $10^{-3}$ M dibutyryl-cAMP (Sigma) (Strickland and Madhavi 1978; Strickland et al. 1980). Medium was changed every 48 hr.

*Microinjection into mouse eggs*
Fertilized eggs were from (C57Bl6 × SJL/J) F2 or female F1 (BALB/c × SJL/J) × male SJL/J.

They were obtained from superovulated mice. Approximately 500 copies of the *Eco*RI-linearized pK4 plasmid were injected into one of the pronuclei (Gordon et al. 1980). The eggs were transferred into the oviducts of pseudopregnant females for further development.

*Southern blot hybridization of genomic DNA*
Genomic DNAs were extracted from cell monolayers, tail biopsies, and 12-day embryos. Aliquots (10 μg) were digested with restriction endonucleases, electrophoresed in 0.7% agarose gels, and blot-transferred using zetabind membranes (AMF cuno) that were hybridized with a pK4 nick-translated probe ($3.10^8$ cpm/μm) using an Amersham kit and the conditions suggested by the supplier.

*Detection of SV40 early proteins*
Viral T antigens were detected by indirect immunofluorescence on cells grown on culture dishes, fixed with methanol/acetone (3:7), and exposed to mouse monoclonal antibody 412 specific for SV40 tumor an-

tigens (Gurney et al. 1980), followed by fluorescein-conjugated rabbit anti-mouse immunoglobulin (Miles).

*Histological procedures*
Brains were fixed in situ by addition of Bouin's fixative into the cranial cavity and further incubation of the whole head for 6 days, cut sagittally at 5 μm, and stained with hematoxylin-eosin.

## Results

### Expression of SV40 early genes in pK4-transformed F9 cells

*pK4 genomes integrated into F9 cells are expressed only upon in vitro differentiation of the cells*
Several G418-resistant sublines were obtained after cotransfection of F9 cells with pK4 and pSVtk-neoβ and subsequent growth in the presence of G418. They were identical with the F9 parent in their morphology and tumorigenicity. Analysis of cellular DNA by blot hybridization showed the presence of intact and multiple copies (2–50) of plasmid pK4 (results not shown). None of the clones expressed the viral early proteins, as shown by indirect immunofluorescence. However, in the case of one clone, F9-K4B2, T-antigen-positive cells could be obtained reproducibly when differentiation was induced by treatment with retinoic acid and dibutyryl-cAMP. A heterogeneous population containing EC and endoderm-like cells was obtained. The endoderm-like cells were shown to express the SV40 T antigens, whereas the EC cells remained T-antigen-negative (results not shown).

*Permanent cell lines can be isolated from an F9-K4B2–differentiated cell population*
The mixed population derived from F9-K4B2 cells as described above had a cloning efficiency of approximately 50%. Two types of colonies were obtained with a comparable efficiency and consisted of either EC or differentiated cells. Several clones of differentiated cells were isolated and expanded. Most of them have an epithelial morphology characteristic of that observed for endoderm cells (Fig. 1A) and display stable traits characteristic of parietal endoderm. In particular, they have been shown by the use of specific antibodies and indirect immunofluorescence staining to express high levels of laminin and type-IV collagen, two basement membrane proteins synthesized in large amounts in parietal endoderm (Adamson et al. 1979; Hogan 1980; Howe and Solter 1980). The cells in these clones also present a network of intermediate filaments recognized by the monoclonal antibodies TROMA-1, which reacts with both visceral and parietal endoderm, and TROMA-3, which reacts with parietal endoderm (Boller and Kemler 1983). They do not express an EC-cell-specific surface antigen recognized by monoclonal antibody ECMA-7 (Kemler et al. 1981).

Another type of clone was also observed that consisted of small, rounded cells with bipolar extensions rem-

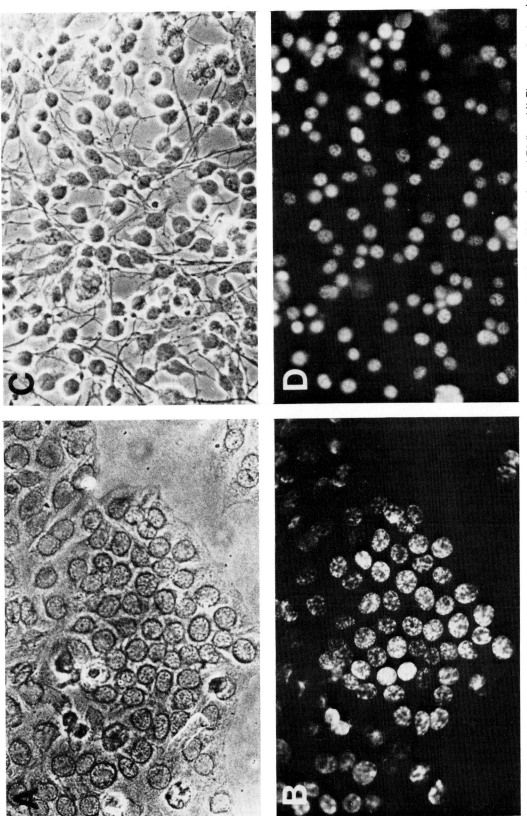

**Figure 1** Morphology and immunofluorescent staining for SV40 T antigens of differentiated cell lines derived from F9-K4B2, a pK4-transformed F9 line. (A) Phase contrast and (B) fluorescence of K4B2-IVF6, a parietal endoderm line; (C) phase contrast and (D) fluorescence of K4B2-IC11, a nonendodermal line.

iniscent of neuronal cells (Fig. 1C). Cells from these clones failed to express any of the markers specific for EC or parietal endoderm cells and might represent an intermediate stage in a differentiation pathway of F9 cells distinct from that leading to endodermal derivatives. Cells from all differentiated clones express the SV40 T antigens (Fig. 1, B and D), are tumorigenic in irradiated syngeneic mice, and have maintained their properties upon successive passages both in vivo and in vitro. These cell lines will be described in more detail elsewhere (O. Kellermann and F. Kelly, in prep.).

## Expression of the SV40 early region in transgenic mice

### Introduction of pK4 into mice and analysis of the genomic DNA

The *Eco*RI-linearized plasmid pK4 was microinjected into fertilized eggs, and the offspring carrying plasmid sequences were identified by Southern blot hybridization of the genomic DNA isolated from tail biopsies. The DNA samples were first digested with *Eco*RI, which cuts only once in pK4. Out of a total of 50 offspring, 11 carried integrated sequences. A band migrating like the linearized plasmid was observed in most cases. By comparison with the SVT2 line of SV40-transformed fibroblasts, which contains 5–6 copies of the early region of SV40 (Kelly and Sambrook 1975), the number of copies of pK4 in the transgenic mice was estimated to vary from approximately 1 to 10–20 copies per genome. When the DNA samples were cut with *Eco*RI and *Pvu*II, three fragments of 3.6, 1.8, and 0.46 kb were observed in most cases, indicating the presence of an intact pK4 plasmid (Fig. 2A). Additional bands were seen in several instances.

In another series of experiments, embryos were removed from the uterus at day 12 of embryonic development and cut transversally, and DNA was extracted from the posterior half. Four of the 10 embryos carried apparently intact pK4 sequences (Fig. 2B).

### Phenotype of the transgenic mice

An unexpected phenotype has emerged in the transgenic mice. Four of the 11 mice (P11, P16, P17, and P23) were noticeably smaller than their littermates at 2–3 weeks of age. Two of them developed neurological disorders: One (P11) showed an altered equilibrium; the other (P23) was prostrated and appeared to have a partial paralysis of the hind legs. Both animals were killed shortly after the appearance of the neurological signs at 4 months and 3 months of age, respectively, and autopsied. In both animals, a brain tumor was observed in the posterior part of the cerebral cortex in front of the cerebellum (Fig. 3A). Histological examination identified these tumors as multiform glioblastomas (Fig. 3B) (Zimmermann and Innes 1979). No other macroscopic abnormalities were observed. Several tissues were cultured in vitro and cells were examined for the presence of nuclear T antigen 2–4 days after the onset of the culture. Skin fibroblasts from mouse P11 and epithelial kidney cells from mouse P23 expressed nuclear T antigen in 100% of the cells. A third animal (P16) died at 4 months of age and was not autopsied. The fourth "small" transgenic mouse (P17) and the 7 of "normal" size are still alive and apparently normal at 6 months of age.

Mouse P23 has transmitted the pK4 sequences to 2 offspring out of 15. P11 appeared to be sterile, and P16 died before breeding.

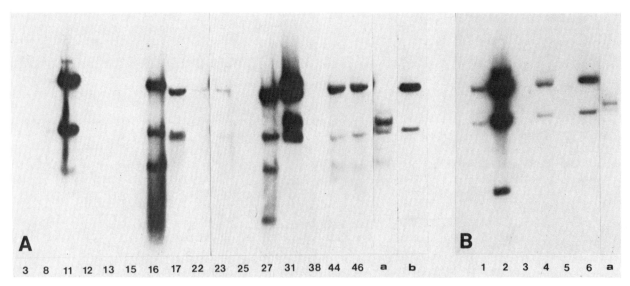

**Figure 2** Southern hybridization analysis of pK4 transgenic mice and embryos. DNA was extracted from tail biopsies (*A*) or from the posterior half of 12-day embryos (*B*). DNA samples were digested with *Eco*RI and *Pvu*II, electrophoresed through 0.7% agarose gels, transferred to zetabind membranes, and hybridized with a pK4 probe. (*A*) Lane numbers at bottom correspond to the number of each individual mouse (i.e., 3 = P3); (lane *a*) SVT2; (lane *b*) pK4. (*B*) Lanes *1–5* correspond to DNA from embryos 001, 004, 010, 008, and 005; (lane *a*) SVT2.

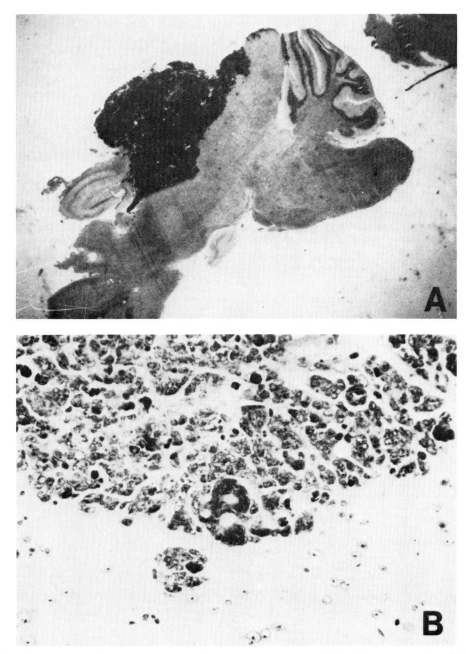

**Figure 3** Histological appearance of a spontaneous brain tumor occurring in mouse P23. (*A*) Low-power view of a sagittal section of the brain, showing the localization of the tumor in the cerebral cortex in front of the cerebellum; (*B*) high-power view of the tumor, showing the histological aspect and the invasion of the normal brain by tumor islands.

*In vitro growth properties of cells from 12-day embryos*
In view of the previous results that showed that the SV40 early region can be expressed in some pK4 transgenic mice (~1/4), it was of interest to search for an earlier expression.

A sensitive method to detect the presence of rare cells expressing the SV40 T antigens in a given population is to examine its growth properties. We have therefore examined several growth parameters of cells from transgenic embryos compared with cells from normal embryos. Embryos derived from pK4 microinjected eggs

were removed from the uterus at day 12 of embryonic development. Four embryos out of 10 were found to be transgenic, as shown above. Cells from the anterior half of the embryos were dissociated and cultured in vitro. We first examined the potential of the cell population for continuous growth. Primary cultures were divided 1:5 every 5 days. Cultures from normal embryos and from three transgenic embryos stopped multiplying after several passages (~5–6). In contrast, cells derived from one of the transgenic embryos (embryo 001) retained their ability to divide, and a population of permanently growing cells was obtained. Expression of T antigens was examined on cell cultures from the successive passages. T-antigen-positive cells were first detected in cultures at the fourth passage and represented 1 or less in $10^4$ of the cell population. The number of positive cells increased rapidly afterwards and reached 80–100% after the ninth passage. No T-antigen-positive cells were detected in cultures from the other 3 transgenic embryos.

The ability of embryonic cells to form colonies was then examined in cells from early passages. Embryo 001 was shown to have a higher cloning efficiency than the other 9 embryos, as demonstrated in Figure 4. Whereas cells from embryos 002 and 006 (which do not carry pK4 sequences) and transgenic embryo 004 gave rise only to rare colonies when seeded at $10^4$ cells/dish, cells from embryo 001 had a cloning efficiency of approximately 1%.

*Permanent cell lines can be established from transgenic embryo 001*
Cells derived from early passages of 001 embryo cultures were seeded at $10^3$ cells/dish. Colonies were isolated and expanded. Most colonies consisted of cells that had a fibroblastic morphology and expressed T antigen. Some colonies ($\sim 1/10$) had a different morphology. They were smaller than the fibroblast-like colonies and contained two cell types, flat cells with large nuclei and small, round cells with little cytoplasm and bi- or tripolar extensions forming an interconnected network (Fig. 5A). This aspect is reminiscent of glial cells

cultivated in vitro. Several such colonies were isolated and have given rise to permanent cell lines. T antigen could not be detected in these cells at early passages (Fig. 5B) but was eventually expressed efficiently after several passages (Fig. 5D). The cells have maintained their morphology for more than 3 months of continuous culture (Fig. 5C). The large, flat cells and the small, round cells could not be separated by subcloning, which suggests that one type may be a precursor to the other. Experiments are in progress to characterize these cells.

### Expression of the SV40 early region is accompanied by demethylation of the pK4 sequences

DNA from embryo 001 was analyzed by Southern blot hybridization after digestion with *Hpa*II, which is sensitive to DNA methylation, and *Pst*I, which is not. *Pst*I cuts twice in the pK4 plasmid, resulting in two bands of 3.3 and 2.5 kb (Fig. 6A). *Hpa*II cuts pK4 once in the E1A enhancer-promoter region, whereas the pML2 sequences contain approximately 25 sites. After *Hpa*II digestion, the 2.5-kb *Pst*I fragment will be degraded into a large number of small fragments, whereas the 3.3-kb fragment will give rise to a 2.3-kb fragment containing the E1A-SV40 junction as well as smaller fragments. Digestion of genomic DNA with *Pst*I and *Hpa*II therefore allows the determination of the extent of methylation of the plasmid sequences.

In Figure 6B (lanes 1 and 2) the DNA from embryo 001 is compared with DNA extracted from a T-antigen-positive cell line derived from cultured cells. Although

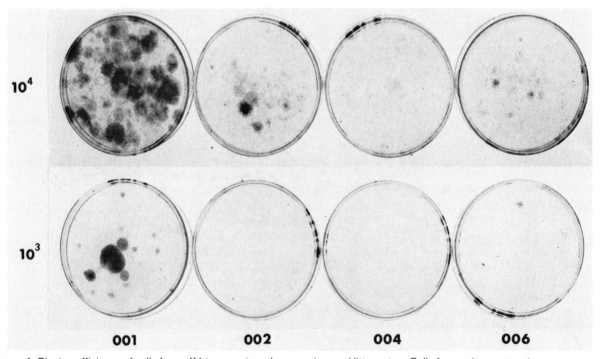

**Figure 4** Cloning efficiency of cells from pK4 transgenic embryos and normal littermates. Cells from cultures at early passages were seeded as indicated at $10^4$ and $10^3$ cells per 5-cm dish. Colonies were fixed and stained with Giemsa 3 weeks later. Numbers beneath the dishes correspond to individual embryos: 001 and 004 carry pK4 sequences; 002 and 006 do not.

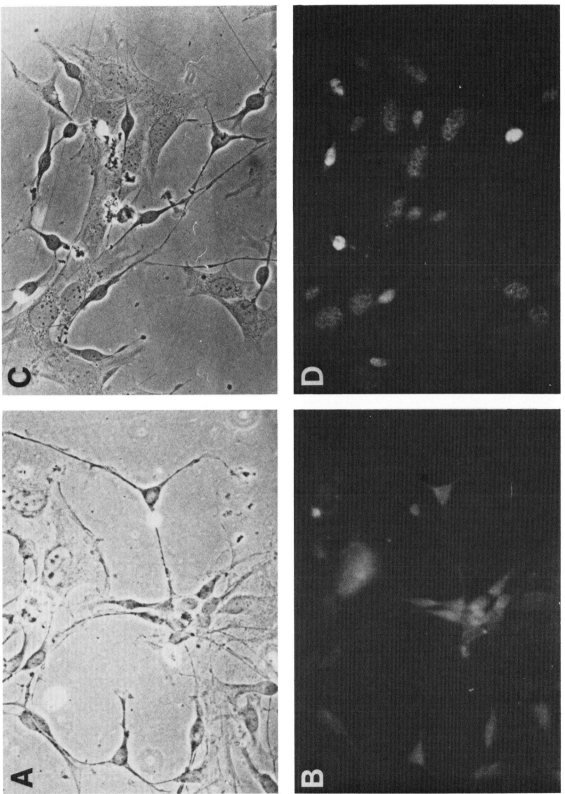

**Figure 5** Glial-like cells derived from embryo 001; morphology and immunofluorescent staining for SV40 T antigen. Cells from early passages are shown in phase contrast (A) and fluorescence (B). Little if any specific staining can be seen at this stage. Cells from late passages are shown in phase contrast (C) and fluorescence (D). The initial morphology is retained and cells now express the SV40 T antigen.

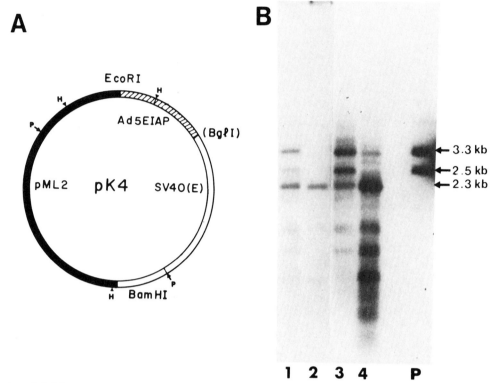

**Figure 6** Southern hybridization analysis showing the level of methylation of pK4 DNA in embryo 001 and cell line F9-K4B2. (A) plasmid pK4. The 2.6-kb BglI-BamHI fragment spanning the early region of SV40 (open area) is linked to the proximal 495 bp of Ad5 containing the enhancer-promoter region of the E1A transcription unit (hatched area) in a pML2 vector (black area). (P) PstI site; (H) HpaII site. One HpaII site only is present in the adeno-SV40 hybrid gene, whereas there are approximately 25 sites in the pML2 vector, the two extremes of which only are indicated here. (B) DNA samples were digested with PstI and HpaII, electrophoresed through 0.7% agarose gels, transferred to zetabind membranes, and hybridized with a pK4 probe. (Lane 1) Embryo 001; (lane 2) T-antigen-positive fibroblastic cell line derived from embryo 001; (lane 3) T-antigen-negative F9-K4B2 cell line; (lane 4) K4B2-IVF6, a T-antigen-positive parietal endoderm cell line; (lane P) pK4 digested with PstI alone.

the DNA from the cell line is fully unmethylated, approximately 50% of the HpaII sites and, in particular, the site present in the E1A enhancer-promoter region are methylated in the DNA from embryo 001. A similar pattern is observed when the T-antigen-negative F9-K4B2 cell line is compared with one of the differentiated T-antigen-positive cell lines derived from it (Fig. 6B, lanes 3 and 4).

## Discussion

The aim of the experiments reported here has been to obtain early mouse embryonic cells carrying heritable SV40 oncogenes and to define conditions in which the viral genes might be expressed efficiently so that a variety of permanent cell lines corresponding to early stages of differentiation could be established.

We have used a recombinant plasmid pK4 in which the SV40 early region is under the control of the enhancer-promoter region of the E1A transcription unit of Ad5. The reason to prefer the E1A enhancer-promoter region to that of SV40 is that adenovirus is expressed in a large variety of cells like SV40 but, in addition, can infect early mouse embryonic cells (Kelly and Boccara 1976; Cheng and Praszkiev 1982). F9-K4B2, a pK4-

transformed clone of F9 cells, was obtained as well as several pK4 transgenic mice and embryos. Results reported in this paper indicate that these materials may indeed be very useful for the immortalization of various cell types. Parietal endoderm cell lines as well as another type of early embryonic derivative could be obtained with high efficiency from F9-K4B2 induced to differentiate in vitro into parietal endoderm, as described by Strickland et al. (Strickland and Madhavi 1978; Strickland et al. 1980). These cell lines will be described in more detail elsewhere (O. Kellermann and F. Kelly, in prep.). Due to the versatility of F9 cells, which can be engaged by varying the culture conditions into different pathways of differentiation (e.g., visceral endoderm [Hogan et al. 1981] or neuronal cells [Liesi et al. 1983]), one may hope to obtain other immortalized derivatives from the F9-K4B2 cell line. Cell lines with a different tissue type, either fibroblast-like or glial-like, were derived from one 12-day pK4 transgenic embryo. A better evaluation of the possibilities offered by this material requires a systematic analysis of the various tissues where the SV40 oncogenes are expressed at different times in the life of the animal. For this purpose, we are now trying to produce strains of pK4 transgenic mice.

Contrary to what we expected, the plasmid pK4 is not expressed in the undifferentiated F9-K4B2 cell line nor in early cultures derived from 12-day transgenic embryos. Thus the E1A promoter is not in itself sufficient to allow proper expression of the SV40 genes. Various mechanisms might be responsible for the restriction of the SV40 genes in early embryonic cells and have been discussed elsewhere (O. Kellermann and F. Kelly, in prep.). Our results on the extent of methylation of the pK4 sequences in embryo 001, F9-K4B2 cells, and their derivatives suggest that demethylation accompanies the activation of the viral genes, and we are now investigating the effect on these cells of conditions capable of inducing demethylation, such as azacytidine treatment (Jones 1985).

It is interesting that some of the transgenic mice (2/11) develop brain tumors, which were shown histologically to be glioblastomas. Although we have not examined here the expression of the SV40 genes in the tumor tissues, the occurrence of the glioblastomas is likely to be related to the presence of the pK4 sequences, since in more than 50 transgenic mice carrying various plasmids in the same genetic background no tumor has been found (C. Babinet, unpubl.). R.L. Brinster and colleagues have recently reported the occurrence of brain tumors characterized as choroid plexus papillomas or carcinomas in transgenic mice carrying a plasmid containing the SV40 early region under the control of its own enhancer-promoter sequence and another gene (the thymidine kinase gene of herpesvirus or the human growth hormone gene) under the control of the metallothionein promoter (Brinster et al. 1984). In this case the tissue-specific expression of the SV40 oncogenes was shown to be linked to the presence of the SV40 enhancer sequence (Palmiter et al. 1985). The occurrence in the experiments reported here of brain tumors different from those described by Brinster et al. (1984) might be related to the use of the E1A enhancer-promoter sequences. This interpretation is supported by the fact that Ad12, which has a sequence very similar to Ad5, can induce tumors of neuronal origin. In particular, tumors derived from neuronal precursor cells (Mukai and Kobayashi 1972) and retinoblastomas (Kobayashi and Mukai 1974) have been described. Obtaining a strain of pK4 transgenic mice able to develop glioblastomas would be very useful for examining the mechanism of activation and the role of the pK4 sequences in the development of tumors. The glial-like permanent cell lines established from a 12-day embryo may also be a useful in vitro system to approach these questions. We already known that these cells can be cloned and expanded at a time when no T antigen can be detected by our immunofluorescent technique. It is only after a certain time in culture that the cells start expressing efficiently the viral genes. We are now examining the expression of SV40 as well as other cellular genes in these cultures at different passages.

From the results reported, it appears that the introduction of SV40 oncogenes into EC cells and in early mouse embryos constitutes a potent tool for the establishment of cell lines corresponding to early stages of differentiation. In addition, the production of a strain of pK4 transgenic mice carrying the potential to develop brain glioblastomas should be useful in the study of events involved in tumor formation.

## Acknowledgments

This work was supported by grants from the Centre National de la Recherche Scientifique (ER 272), and the Association pour la Recherche sur le Cancer (no. 6312).

## References

Adamson, E.D., S.J. Gaunt, and C.F. Graham. 1979. The differentiation of teratocarcinoma stem cells is marked by the types of collagen which are synthesized. *Cell* **17:** 469.

Boller, K. and R. Kemler. 1983. In vitro differentiation of embryonal carcinoma cells characterized by monoclonal antibodies against embryonic cell markers. *Cold Spring Harbor Conf. Cell Proliferation* **10:** 39.

Brinster, R.L., H.Y. Chen, A. Messing, T. Van Dyke, A.J. Levine, and R.D. Palmiter. 1984. Transgenic mice harboring SV40 T-antigen genes develop characteristic brain tumors. *Cell* **37:** 367.

Cheng, C. and J. Praszkiev. 1982. Regulation of type 5 adenovirus replication in murine teratocarcinoma cell lines. *Virology* **123:** 45.

Evans, M.J. and M.H. Kaufman. 1981. Establishment in culture of pluripotential cells from mouse embryos. *Nature* **292:** 154.

Gordon, J.W., G.A. Scangos, D.J. Plotkin, J.A. Barbosa, and F.H. Ruddle. 1980. Genetic transformation of mouse embryos by microinjection of purified DNA. *Proc. Natl. Acad. Sci.* **77:** 7380.

Graham, F.L. and A.J. van der Eb. 1973. A new technique for the assay of infectivity of human adenovirus 5. *Virology* **52:** 456.

Gurney, E.G., R.O. Harrison, and J. Fenno. 1980. Monoclonal antibodies against simian virus 40 T antigens: Evidence for distinct subclasses of large T antigen and for similarities among non-viral T antigens. *J. Virol.* **34:** 752.

Hanahan, D. 1985. Heritable formation of pancreatic β-cell tumors in transgenic mice expressing recombinant insulin/simian virus 40 oncogenes. *Nature* **315:** 115.

Hogan, B.L.M. 1980. High molecular weight extracellular proteins synthesized by endoderm cells derived from mouse teratocarcinoma cells and normal extraembryonic membrane. *Dev. Biol.* **76:** 275.

Hogan, B.L.M., A. Taylor, and E.D. Adamson. 1981. Cell interactions modulate embryonal carcinoma cell differentiation into parietal and visceral endoderm. *Nature* **291:** 235.

Howe, C.C. and D. Solter. 1980. Identification of non-collagenous basement membrane glycopolypeptides synthesized by mouse parietal endoderm and an endodermal cell line. *Dev. Biol.* **77:** 480.

Jones, P.A. 1985. Altering gene expression with 5-azacytidine. *Cell* **40:** 485.

Kelly, F. and M. Boccara. 1976. Susceptibility of teratocarcinoma cells to adenovirus type 2. *Nature* **262:** 409.

Kelly, F. and H. Condamine. 1982. Tumor viruses and early mouse embryos. *Biochim. Biophys. Acta* **651:** 105.

Kelly, F. and J. Sambrook. 1975. Variants of simian virus 40–transformed mouse cells resistant to cytochalasin B. *Cold Spring Harbor Symp. Quant. Biol.* **39:** 345.

Kemler, R., P. Brulet, and F. Jacob. 1981. Monoclonal antibodies as a tool for the study of embryonic development. *Immune Syst.* **1:** 102.

Kobayashi, S. and N. Mukai. 1974. Retinoblastoma-like tumors induced by human adenovirus type 12 in rats. *Cancer Res.* **34:** 1646.

Liesi, P., L. Reichardt, and J. Wartiovaara. 1983. Nerve growth factor induces adrenergic neuronal differentiation in F9 teratocarcinoma cells. *Nature* **306:** 265.

Martin, G.R. 1981. Isolation of a pluripotent cell line from early mouse embryos cultured in medium conditioned by teratocarcinoma cells. *Proc. Natl. Acad. Sci.* **78:** 7634.

Mukai, N. and S. Kobayashi. 1972. Undifferentiated intraperitoneal tumors induced by human adenovirus type 12 in hamsters. *Am. J. Pathol.* **69:** 331.

Nicolas, J.F. and P. Berg. 1983. Regulation of expression of genes transduced into embryonal carcinoma cells. *Cold Spring Harbor Conf. Cell Proliferation* **10:** 469.

Nicolas, J.F., H. Jacob, and F. Jacob. 1981. Teratocarcinoma derived cell lines and their use in the study of differentiation. In *Functionally differentiated cell lines* (ed. G. Sato), p. 185. A.R. Liss, New York.

Nicolas, J.F., P. Avner, J. Gaillard, J.L. Guenet, H. Jakob, and F. Jacob. 1976. Cell lines derived from teratocarcinomas. *Cancer Res.* **36:** 4224.

Palmiter, R.D., H.Y Chen, A. Messing, and R.L. Brinster. 1985. SV40 enhancer and large T-antigen are instrumental in development of choroid plexus tumors in trangenic mice. *Nature* **316:** 457.

Silver, L.M., G.R. Martin, and S. Strickland, eds. 1983. *Cold Spring Harbor conferences on cell proliferation; Teratocarcinoma stem cells*, vol. 10. Cold Spring Harbor Laboratory, Cold Spring Harbor, New York.

Sleigh, M.J. 1985. Virus expression as a probe of regulatory events in early mouse embryogenesis. *Trends Genet.* **1:** 17.

Southern, P.J. and P. Berg. 1982. Transformation of mammalian cells to antibiotic resistance with a bacterial gene under the control of the SV40 early region promoter. *J. Mol. Appl. Genet.* **1:** 327.

Strickland, S. and V. Madhavi. 1978. The induction of differentiation in teratocarcinoma stem cells by retinoic acid. *Cell* **15:** 393.

Strickland, S., K.K. Smith, and R.R. Marotti. 1980. Hormonal induction of differentiation in teratocarcinoma stem cells: Generation of parietal endoderm by retinoic acid and dibutyryl cAMP. *Cell* **21:** 347.

Tanigawa, T., H. Takayama, A. Takagi, and G. Kimura. 1983. Cell growth and differentiation *in vitro* in mouse macrophages transformed by a tsA mutant of simian virus 40. *J. Cell. Physiol.* **116:** 303.

Tooze, J., ed. 1980. *Molecular biology of tumor viruses*, 2nd edition: *DNA tumor viruses.* Cold Spring Harbor Laboratory, Cold Spring Harbor, New York.

Zimmermann, H.M. and J.R.M. Innes. 1979. Tumours of the central and peripheral nervous system. *IARC Sci. Publ.* **23:** 644.

# Transformation-associated Domains of Polyomavirus Middle T Antigen

## W. Markland, S.H. Cheng, B.L. Roberts, R. Harvey, and A.E. Smith

Protein Engineering Group, Integrated Genetics, Inc., Framingham, Massachusetts 01701

Putative transformation-associated domains of polyomavirus middle T antigen were investigated using site-directed and site-specific mutagenesis. Three sites were chosen: (1) tyrosines 315 and 250, two major in vitro phosphorylation sites; (2) a carboxyterminal hybrophobic sequence, the putative membrane-binding domain; and (3) amino acids 177–180, the site of the lesion in the NG59 mutant. Results indicate that tyrosine residues 315 and 250 are not an absolute requirement for middle-T-induced transformation. By contrast, the integrity of the hydrophobic domain of middle T antigen is essential for both transformation and for association with cellular membranes. Furthermore, $pp60^{c-src}$ binding and transformation are sensitive to mutations within the region around amino acids 177–180 of middle T antigen. Interestingly, the presence of a predicted β-turn at this location appears to correlate with the ability of middle T antigen to transform.

Polyomavirus middle T antigen transforms the growth of established rodent cells in vitro (Treisman et al. 1981) by a mechanism that is yet to be established. Middle T antigen is present in the membrane fraction of cells and has an associated tyrosine-specific kinase activity that can be detected in vitro (Eckhart et al. 1979; Schaffhausen and Benjamin 1979; Smith et al. 1979). The presence of the in vitro kinase activity correlates well with the transforming ability of different species of middle T antigen (Smith and Ely 1983; Courtneidge and Smith 1984). All attempts to demonstrate that the kinase activity is intrinsic to middle T antigen have proved unsuccessful (Schaffhausen et al. 1985), and it is probably that the protein itself does not possess intrinsic kinase activity (Courtneidge and Smith 1983). The discovery that middle T antigen forms a complex with $pp60^{c-src}$ provided an alternative explanation that suggests that the activity measured in vitro is a property of the associated cellular tyrosine kinase.

The presence of the middle-T:$pp60^{c-src}$ complex in cell extracts also suggests a possible model for the transforming activity of middle T antigen. This proposes that complex formation induces a change in activity or specificity of $pp60^{c-src}$. Consistent with this, the interaction between transformation-competent species of middle T antigen and $pp60^{c-src}$ in virally infected or transformed cells, leads to an increase in the $pp60^{c-src}$ tyrosyl kinase–specific activity (Bolen et al. 1984) and a change in the phosphorylated state of $pp60^{c-src}$ (Courtneidge 1985; Yonemoto et al. 1985).

Although the normal cellular function of $pp60^{c-src}$ is unknown, because it is strongly conserved in vertebrates and because the overexpression of $pp60^{v-src}$ in virally infected cells generally causes transformation, it is assumed that $pp60^{c-src}$ is involved in some way in the control of cellular proliferation.

To investigate middle T antigen both as a transforming protein and as a component in the middle-T:$pp60^{c-src}$ complex, several sites within the protein were selected for specific mutagenesis. The sites chosen were regions of the gene encoding (1) two major tyrosine phosphorylation sites, (2) the putative membrane-binding domain of middle T antigen, and (3) amino acids 177–180 (the site of the NG59 lesion).

The findings in this report confirm the importance of the middle-T:$pp60^{c-src}$ complex in the transformation process and reiterate the observation that although complex formation is necessary for middle-T-mediated transformation, it is not always sufficient.

## Experimental Procedures

### Cells and assays

Purified plasmid DNAs were introduced into Rat-1 cells or NIH-3T3 cells by the calcium phosphate transfection method (Graham and van der Eb 1979). Nontransforming mutant plasmids of polyomavirus middle T antigen were cotransfected onto NIH-3T3 cells with the Tn5-encoded phosphotransferase (neo) gene (pSV2Neo) in the ratio of 10:1 by the calcium phosphate method. Cellular clones resistant to the antibiotic G418 were analyzed for the presence and subcellular location of middle T antigen and for middle-T-antigen-associated in vitro kinase activity.

Conditions for labeling with [$^{35}$S]methionine, preparation of cell lysates, immunoprecipitation, and the in vitro kinase assays have all been previously described (Smith et al. 1979; Courtneidge and Smith 1983). Cellular fractionation was carried out as described in Smith et al. (1979) and Templeton and Eckhart (1984).

### Mutagenic procedures

Several strategies were utilized in the mutagenesis of the three regions described. These involved linker insertion/deletion mutagenesis, site-directed mutagenesis using sodium bisulfite, and site-specific mutagenesis using single and mixed oligonucleotides. The details of these procedures have been published elsewhere (Kalderon et al. 1982; Oostra et al. 1983; Cheng et al. 1986; W. Markland et al., in prep.).

Mutations were verified by DNA sequencing and, where appropriate, by restriction enzyme analysis. Some mutants that generated a transformation-defective phenotype were reverted to a DNA sequence encoding the wild-type amino acid sequence using oligonucleotide mutagenesis to ensure that the mutant phenotype was a consequence of the characterized mutation.

## Results

### Phosphorylation sites of middle T antigen

Polyomavirus middle T antigen contains several tyrosine residues that accept phosphate in vitro. These residues lie in the carboxyterminal half of the molecule. Tyrosine 315 has been demonstrated to be a major phosphate-acceptor site (Schaffhausen and Benjamin 1981; Smith and Ely 1983), whereas tyrosines 250, 297, and 322 are minor sites (Harvey et al. 1984; Hunter et al. 1984). Deletion mutants that remove tyrosines 297 and 322 do not affect transformation, whereas several that delete tyrosine 315 are severely reduced in transforming ability. However, Mes-Masson et al. (1984) reported a deletion mutant lacking tyrosines 315 and 322 that retained 20% of wild-type transforming ability on F111 cells. Furthermore, Oostra et al. (1983) have shown by using point mutants of middle T antigen that tyrosine 315 and the surrounding acidic tract are not essential for middle-T-antigen-mediated transformation. Middle T antigen species that lack tyrosine 315 are still able to accept phosphate in vitro, in which case tyrosine 250 becomes a major acceptor site (Harvey et al. 1984). To investi-

gate further the role of middle T antigen phosphorylation in transformation, DNA sequences encoding tyrosine 250 and the surrounding regions were mutated and a combined mutant encoding a middle T antigen lacking both tyrosines 315 and 250 was created.

Mutation of middle T antigen residue 250 from tyrosine to phenylalanine, produced a molecule (T250) that was handicapped when compared with wild-type middle T antigen (Table 1), without being totally defective in any of the parameters investigated. In a focus-formation assay, T250 produced fewer foci with an increased lag period when compared with wild-type middle T antigen. A similar handicap was noted in both colony formation in agar and in tumor formation in rats. T250 middle T antigen had reduced phosphate acceptor activity in in vitro kinase assays, and the tryptic fingerprint of the phosphate-labeled protein gave a pattern consistent with that predicted for a middle T antigen molecule that lacked tyrosine 250 (W. Markland et al., in prep.). Extensive mutagenesis of the DNA encoding the amino acids surrounding tyrosine 250 had little effect on the transforming ability of the mutant middle T antigen species generated. Reversion of the DNA sequences encoding phenylalanine at residue 250 in T250 and at 315 in pAS131 to tyrosine (T250 w.t. and T315 w.t., respectively) resulted in wild-type phenotypes for all the parameters studied.

Somewhat surprisingly, combining the mutations in which tyrosines 315 and 250 had been converted to phenylalanine into one middle T antigen (TH) produced a protein that was totally defective in transformation but still retained a reduced phosphate acceptor activity in an in vitro kinase assay (Table 1). This assay was carried out using cell lysates of neomycin-resistant cell lines expressing the doubly mutated middle T antigen.

Taken together, these results suggest that although

**Table 1** Properties of the Tyrosine Mutants of Polyomavirus Middle T Antigen

| Plasmid | Focus formation | Tumor induction | Middle T phosphate acceptor activity |
|---|---|---|---|
| pAS101 (wild type) | + + + | + + + | + + + |
| pAS131 (Tyr-315 to Phe) | + + | + + + | + + + |
| pAS131 w.t. (Phe-315 to Tyr) | + + + | ND | + + + |
| pT250 (Tyr-250 to Phe) | + | + | + + |
| pT250 w.t. (Phe-250 to Tyr) | + + + | ND | + + + |
| pTH (Tyr-315 to Phe and Tyr-250 to Phe) | − | ND | + |

Transformation was measured by focus formation using the calcium phosphate transfection method of plasmid DNA onto a monolayer of Rat-1 cells. Tumor induction was determined by the injection of transformed Rat-1 cells into Fisher rats, as described in Oostra et al. (1983). Phosphate acceptor activity was estimated by in vitro kinase assays utilizing cell lysates (normalized for total protein concentration) made from transformed Rat-1 cells or, in the case of TH, from neomycin-resistant NIH-3T3 cell lines expressing this transformation-defective middle T antigen. + + + represents wild-type levels of the properties assayed, while + + and + represent reduced levels and − defective levels for the middle-T species when compared with wild-type middle T. (ND) Not determined.

tyrosines 315 and 250 can be individually altered in middle T antigen without abolishing the ability to transform, the simultaneous removal of both phosphate acceptor sites results in the creation of a mutant middle T antigen that is still able to associate with pp60$^{c\text{-}src}$ but is unable to transform.

## Putative membrane-binding domain

Middle T antigen has a distinct, uncharged, and predominantly hydrophobic 22-amino-acid domain near the carboxyl terminus (Soeda et al. 1980) that is likely to be the membrane-binding domain of the molecule. Early reports suggested that middle T antigen was associated with the plasma membrane fraction; however, more recent investigations (Zhu et al. 1984) reported it to be also associated with the internal membrane fractions of the cell.

Deletions within the region of the genome coding for the hydrophobic domain (Novak and Griffin 1981) or mutations causing the premature termination of middle T antigen translation upstream of the hydrophobic domain lead to transformation-defective species of middle T antigen. The truncated molecules also lack an associated in vitro kinase activity (Carmichael et al. 1982; Templeton and Eckhart 1982). The putative membrane-binding domain of middle T antigen has been exchanged for the transmembrane anchorage domain of vesicular stomatitis virus glycoprotein G (VSV-G) (Templeton et al. 1984). Although this chimeric protein (MT/G1) was shown both to be located to the membrane fractions of cells and to possess in vitro kinase activity, it was unable to transform cultured rodent cells. Further point and deletion mutants have been generated within the DNA coding for the putative membrane-binding domain of middle T antigen to undertake a more detailed investigation into the effect of lesions within this region.

Mutagenesis of the putative membrane-binding domain yielded an extensive series of mutants, called RX mutants (W. Markland et al., in prep.). These could be categorized into three groups depending upon their biological and biochemical properties, as shown in Table 2. One group (RX38, RX51, and RX77) with deletions outside but proximal to the hydrophobic domain displayed an essentially wild-type middle T antigen phenotype. They had the ability to transform established rodent cells, and they associated with cellular membrane fractions and possessed associated kinase activity. A second set of RX mutants either had extensive deletions within the DNA encoding the hydrophobic domain (RX26, RX69) or contained premature termination signals within or proximal to it (RX13 and RXT). These mutants failed to transform, did not associate with cellular membrane fractions, and failed to display associated kinase activity. Finally, the third group of RX mutants had small mutations within the hydrophobic domain (RX2, RX67, and RX68). They retained the biochemical properties associated with wild-type middle T antigen but were unable to transform Rat-1 cells or NIH-3T3 cells.

To determine the presence of RX mutant–encoded middle T antigen species in transfected cells, cells were metabolically labeled, and immunoprecipitates were prepared from either transformed Rat-1 cells or, in the case of nontransforming RX mutants, from neomycin-resistant NIH-3T3 cells lines and examined by SDS-polyacrylamide gel electrophoresis. The size of the middle T antigen mutants was consistent with their expected lesions.

These results illustrate that the integrity of the hydrophobic domain is crucial to the transforming ability of middle T antigen and that this domain is required for membrane association.

**Table 2** Mutations within the Hydrophobic Domain of Polyomavirus Middle T Antigen

| Plasmid | Mutation deletion (amino acid numbers) | Mutation insertion (amino acids) | Transformation | Kinase | Cellular fraction |
|---------|------------------------|---------------------------|----------------|--------|-------------------|
| pAS101 | | Wild type | + | + | M |
| RX38 | 388–391 | Pro · Ser · Arg | + | + | M |
| RX51 | 382–391 | Pro · Ser · Arg | + | + | M |
| RX77 | 359–391 | Pro · Ser · Arg | + | + | M |
| RX13 | 398T | Leu · Glu · Glu | − | − | C |
| RX26 | 403–417 | Leu · Asp · Gly | − | − | C |
| RX69 | 405–417 | Glu · Gly | − | − | C |
| RXT | 388T | Pro · Ser · Thr · Ser | − | − | C |
| RX2 | 404–405 | Glu · Val | − | + | M |
| RX67 | 404 | Ser · Ser · Arg | − | + | M |
| RX68 | 404 | Pro · Arg · Gly · Ala · Ala Leu · Leu. | − | + | M |

Transformation was measured by focus formation on monolayers of Rat-1 cells. In vitro kinase assays were carried out using cell lysates from transformed Rat-1 cells or neomycin-resistant NIH-3T3 cell lines expressing the nontranforming RX mutant middle T antigens. Cellular fractionation was carried out using metabolically labeled cells expressing the mutant middle T antigens, which were subsequently lysed and fractionated. Mutant middle-T species were found to associate with the membrane pellet (M) or the supernatant cytosol (C). T indicates a termination mutant. The presence and absence of the properties assayed is indicated by + and −, respectively.

### NG59 site of middle T antigen

The host-range transformation-defective (hrt) mutant NG59 contains one of the most subtle mutations within the middle T antigen, which results in a transformation defect (Carmichael and Benjamin 1980). The lesion in NG59 results in an insertion of an isoleucine residue between amino acids 178 and 179 together with a transition of aspartic acid at 179 to asparagine. It generates a molecule defective in transforming ability that lacks associated kinase activity.

Association between middle T antigen and $pp60^{c-src}$ implies that each molecule contains complementary binding sites. Results to date indicate that any variant of middle T antigen with a mutation that results in a failure to associate with $pp60^{c-src}$ is also transformation-defective. Due to the subtle nature of the amino acid changes in NG59 middle T antigen and the finding that this mutated middle T antigen fails to associate actively with $pp60^{c-src}$, this region of the middle T antigen molecule was considered to be a potential binding site for $pp60^{c-src}$. This site was mutated further to investigate this possibility and to test the hypothesis that middle T antigen species unable to bind $pp60^{c-src}$ are unable to transform (Cheng et al. 1986).

Deletion mutants (AB1–AB5), which collectively deleted amino acids 157–185 (Cheng et al. 1986), were found to be transformation-defective as judged by the focus-formation assay. A more detailed mutagenic analysis of the region was subsequently undertaken. Initially to ascertain the relative contribution of the insertion and transition components of the NG59 lesion, two individual mutations were generated to create separately the two differences in amino acid sequence in NG59 middle T antigen (Table 3). In NS1 the transition from aspartic acid (179) to asparagine was produced, and in NS2 the isoleucine insertion between amino acids 178 and 179 was introduced. NS1 middle T antigen was wild type in transforming ability and exhibited associated kinase activity, whereas NS2 was defective in both, being indistinguishable in phenotype from the parent NG59 mutant. Thus the transformation defect of the NG59 lesion can be attributed entirely to the insertion component of the mutation rather than to the transition.

To examine further the importance of this region of middle T antigen, point mutants (called the MG series) were generated in the region of middle T antigen between amino acids 177 and 180. Although the MG series of mutants were clustered to the region of middle T antigen corresponding to the NG59 lesion, they differed in their ability to transform established rodent cells (Table 3). MG2, MG3, MG6, and MG8 displayed a wild-type transforming ability, whereas MG1 and MG14 were transformation-defective. MG12 and MG13 were intermediate in transforming ability, being unable to transform Rat-1 cells but promoting anchorage-independent growth when expressed in NIH-3T3 cells. Of the MG series of mutants, only those middle T antigen species that were able to transform exhibited associated in vitro kinase activity. This is consistent with previous data that indicate that there is a direct correlation between the ability of a mutant middle T antigen to transform and the possession of associated kinase activity.

The ability of the mutant middle T antigen species to stimulate the kinase activity of $pp60^{c-src}$ (as measured by the phosphorylation of an exogenous substrate) also reflected the transforming abilities of the mutants. Transformation-competent mutants of middle T antigen expressed a high (14-fold to 20-fold) activation of $pp60^{c-src}$ phosphorylation, whereas transformation-defective mutants failed to activate $pp60^{c-src}$. The mutants intermediate in their transformation capabilities displayed intermediate (5-fold) levels of activation. Thus, the MG series of mutants demonstrated an exact correlation between transforming ability, the presence of associated in vitro kinase activity, and the activation of $pp60^{c-src}$ upon complex formation.

**Table 3** Mutations Surrounding the NG59 Region of Polyomavirus Middle T Antigen

| 176–Met Pro Ile Asp Trp Leu Asp Leu–183 |  | Transformation | Kinase | Increased $pp60^{c-scr}$ activity | Predicted $\beta$ turn |
|---|---|---|---|---|---|
| \|------------------------\| $\beta$ turn | | | | | |
| --------------------------------------------------- | pAS101 | + | + | + | + |
| -----------------Ile Asn-------------------- (*) | NG59[b] | − | − | − | − |
| -------------------Asn------------------------ | NS1 | + | + | + | + |
| -----------------Ile-------------------------- | NS2 | − | − | − | − |
| ---------------------------Arg---------------- | MG1 | − | − | − | − |
| ---------------------Arg--------Tyr------------ | MG14 | − | − | − | − |
| ----------Thr---------------------------------- | MG12 | (+) | + | (+) | − |
| ----------------Arg-------Cys----------------- | MG13 | (+) | + | (+) | + |
| ----------------Arg---------------------------- | MG2 | + | + | + | + |
| --------------------- Asn--Cys---------------- | MG3 | + | + | ND | + |
| ----------------Lys---------------------------- | MG6 | + | + | ND | + |
| --------------------------- Cys---------------- | MG8 | + | + | ND | + |

Transforming activity was measured by the ability of neomycin-resistant NIH-3T3 cell lines expressing the middle-T variants to promote anchorage-independent growth in agar; + represents an intermediate ability to form colonies in this assay. In vitro kinase and quantitative kinase assays using enolase were performed using cell lysates from neomycin-resistant NIH-3T3 cell lines expressing the middle-T mutants; + indicates an intermediate ability to activate the specific kinase activity of $pp60^{c-src}$. NG59b was created for this study on the basis of the lesion originally reported by Carmichael and Benjamin (1980).
* Indicates an insertion of Ile between Ile-178 and Asp-179.

## Discussion

A current working model for transformation by polyomavirus middle T antigen proposes that it forms a complex with pp60$^{c\text{-}src}$ and thereby produces a pp60$^{c\text{-}src}$ species with enhanced tyrosine kinase activity. This, it is supposed, triggers the processes that eventually result in the uncontrolled growth of cells (Bolen et al. 1984; Courtneidge and Smith 1984; Courtneidge 1985; Yonemoto et al. 1985).

The role of phosphorylation of middle T antigen on tyrosine residues in the transformation process remains unclear. There is no compelling evidence of such phosphorylation in vivo. Individual mutants in which either tyrosine 315 (Oostra et al. 1983) or tyrosine 250 (W. Markland et al., in prep.) have been converted to phenylalanine or in which tyrosines 315 and 322 have been deleted (Mes-Masson et al. 1984) show that there is no absolute requirement for these residues in the transformation process, at least when assayed by transfection of plasmid DNAs onto established rodent cells. However, removal of these tyrosine residues reduces the transforming efficiency of middle T antigen, as judged by the number and time of appearance of foci, when compared with the wild-type protein. A middle T antigen mutant in which both tyrosines 315 and 250 are changed to phenylalanine in the same molecule is transformation-defective while still retaining the ability to associate with pp60$^{c\text{-}src}$. It is possible that this species of middle T antigen may affect the substrate specificity of pp60$^{c\text{-}src}$ or fail to activate the tyrosine kinase activity of pp60$^{c\text{-}src}$ above the threshold level required to induce transformation. This possibility is under investigation. It is interesting to note that the sequence immediately adjacent to and including tyrosine 315 in the murine polyomavirus middle T antigen is conserved in the recently described hamster papovavirus middle T antigen (Delmas et al. 1985), whereas the surrounding regions share little homology between the two molecules (Fig. 1).

The hydrophobic tail mutants also illustrate that transformation requires more than a simple association between middle T antigen and pp60$^{c\text{-}src}$. The first group of RX mutants, with deletions outside but proximal to the hydrophobic domain, are wild type for transforming ability and display an associated kinase activity. RX mutants with deletions within the hydrophobic sequence or that have premature terminations within or prior to this domain are unable to transform, presumably because they fail to associate with cellular membrane fractions and with pp60$^{c\text{-}src}$. They are equivalent to the previously described mutants of Carmichael et al. (1982) and Templeton and Eckhart (1982). The more interesting set of RX mutants are those with small changes within the hydrophobic sequence. These mutant middle T antigen species possess associated kinase activity and associate with pp60$^{c\text{-}src}$ but are unable to transform. This third set of RX mutants are equivalent to the MT/G1 mutant of Templeton et al. (1984). Several mutants that are able to associate with pp60$^{c\text{-}src}$ yet are unable to transform have now been described and include *dl*23, *dl*1015, and *dl*2208. This phenotype demonstrates that although middle T antigen–associated kinase activity may be necessary for transformation, it is not sufficient. The RX mutant–encoded middle-T-antigen species with small changes in the hydrophobic domain are being examined for their effect on pp60$^{c\text{-}src}$ tyrosyl kinase activity.

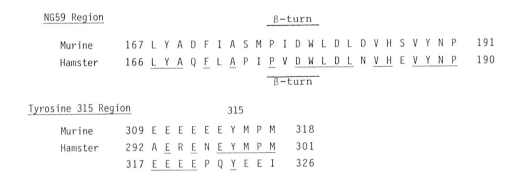

**Figure 1** Homologous amino acid sequences in murine polyomavirus middle T antigen and hamster papovavirus middle T antigen are shown here using the single-letter amino acid code. The amino acid sequence numbers are indicated at the extremities of the regions considered, and conserved amino acid residues are underlined. The amino acid sequence of the membrane anchorage domain of vesicular stomatitis virus glycoprotein G (VSV-G) is also shown to demonstrate the lack of homology between this region and either middle T antigen putative membrane-binding domains.

The hydrophobic domains of the hamster and murine middle T antigens are strongly conserved in amino acid sequence (Fig. 1). This is in contrast to adjacent sequences and a comparison with the anchorage domain of VSV-G that demonstrates little homology, and it may imply a need for a particular amino acid sequence and not simply a tract of hydrophobic amino acids. This hypothesis is supported by the phenotype of the middle-T/VSV-G chimeric protein. It is possible that this region of middle T antigen is a specific binding site for some other component important in transformation.

Mutants with lesions in and around the site of the NG59 mutation show that this region of middle T antigen is important in the binding and activation of $pp60^{c-src}$. Middle T antigen species containing mutations within this region are either fully transforming, in which case the protein associates with and activates $pp60^{c-src}$, or nontransforming, in which case the middle T antigen species fail to associate with and hence do not activate $pp60^{c-src}$. The mutants intermediate in transforming ability associate with $pp60^{c-src}$ but only partially activate it. This series of mutants demonstrates a close correlation between transformation and the active association of middle T antigen with $pp60^{c-src}$.

Whether this site within middle T antigen is part of the $pp60^{c-src}$ binding site remains to be determined. The fact that certain point mutations abolish $pp60^{c-src}$ association while others do not does not rule out this possibility. Analysis of the predicted secondary structure of this region of middle T antigen (using the Chou and Fasman [1978] program) suggests that the four amino acids, 177–180, form a β turn between two α helixes. There is an almost perfect correlation between the predicted presence of this β turn and the ability of each MG mutant middle T antigen to transform. A possible explanation for the observed phenotypes is that the removal of a β turn in middle T antigen at this site causes a gross conformational change that interferes, at a distant site, with $pp60^{c-src}$ association. Alternatively, the NG59 region in wild-type middle T antigen could be part of a $pp60^{c-src}$ binding site. The equivalent region in hamster middle T antigen shares sequence homology and is also predicted to comprise a β turn (Fig. 1).

There is still no known mutant of middle T antigen that is able to transform without associating with $pp60^{c-src}$. Thus the ability of middle T antigen to associate with $pp60^{c-src}$ and to active its tyrosyl kinase activity remains the best correlate of polyomavirus–induced transformation. However, there are an increasing number of middle T antigen mutants that while associating with $pp60^{c-src}$ fail to transform. An analysis of the middle-T:$pp60^{c-src}$ complex (Courtneidge and Smith 1983) estimates that the complex has a molecular weight of about 200,000. This implies that it may contain more than one copy of each of the two proteins or possibly another component. One interpretation of the multiple phenotypes of the mutants reported here is that, in addition to the interaction between $pp60^{c-src}$ and middle T antigen and with membranes, the complex does contain other proteins and that these are important in transformation. So far there are no suggestions as to what these proteins might be.

## References

Bolen, J.B., C.J. Thiele, M.A. Israel, W. Yonemoto, L.A. Lipsich, and J.B. Brugge. 1984. Enhancement of cellular *src* gene product associated tyrosyl kinase activity following polyomavirus infection and transformation. *Cell* **38:** 767.

Carmichael, G.G. and T.L. Benjamin. 1980. Identification of DNA sequence changes leading to loss of transforming activity in polyomavirus. *J. Biol. Chem.* **255:** 230.

Carmichael, G.G., B.S. Schaffhausen, D.I. Dorsky, D.B. Oliver, and T.L. Benjamin. 1982. Carboxy terminus of polyoma middle-sized tumor antigen is required for attachment to membranes, associated protein kinase activities and cell transformation. *Proc. Natl. Acad. Sci.* **79:** 3579.

Cheng, S.H., W. Markland, A.F. Markham, and A.E. Smith. 1986. Mutations mapping to the region around the NG59 lesion of polyomavirus middle-T antigen indicate this region is important in $pp60^{c-src}$ association and cell transformation. *EMBO J.* (in press).

Chou, P.Y. and G.D. Fasman. 1978. Empirical predictions of protein conformation. *Annu. Rev. Biochem.* **47:** 251.

Courtneidge, S.A. 1985. Activation of the $pp60^{c-src}$ kinase by middle-T antigen binding or by dephosphorylation. *EMBO J.* **4:** 1471.

Courtneidge, S.A. and A.E. Smith. 1983. Polyomavirus transforming protein associated with the product of the c-*src* cellular gene. *Nature* **303:** 435.

————. 1984. The complex of polyomavirus middle-T antigen and $pp60^{c-src}$. *EMBO J.* **3:** 585.

Delmas, V., C. Bastien, S. Scherneck, and J. Feunteun. 1985. A new member of the polyomavirus family: The hamster papovavirus. Complete nucleotide sequence and transformation properties. *EMBO J.* **4:** 1279.

Eckhart, W., M.A. Hutchinson, and T. Hunter. 1979. An activity phosphorylating tyrosine in polyoma T antigen immunoprecipitates. *Cell* **18:** 925.

Graham, F.L. and A.J. van der Eb. 1979. A new technique for the assay of infectivity of human adenovirus 5 DNA. *Virology* **52:** 456.

Harvey, R., B.A. Oostra, G.J. Belsham, P. Gillett, and A.E. Smith. 1984. An antibody to a synthetic peptide recognizes polyomavirus middle-T antigen and reveals multiple *in vitro* tyrosine phosphorylation sites. *Mol. Cell. Biol.* **7:** 1334.

Hunter, T., M.A. Hutchinson, and W. Eckhart. 1984. Polyoma middle-T antigen can be phosphorylated on tyrosine at multiple sites *in vitro*. *EMBO J.* **3:** 73.

Kalderon, D., B.A. Oostra, B.K. Ely, and A.E. Smith. 1982. Deletion loop mutagenesis: A novel method for the construction of point mutants using deletion mutants. *Nucleic Acids Res.* **10:** 5161.

Mes-Masson, A.M., B. Schaffhausen, and J.A. Hassell. 1984. The major site of tyrosine phosphorylation in polyomavirus middle-T antigen is not required for transformation. *J. Virol.* **52:** 457.

Novak, U. and B.E. Griffin. 1981. Requirement for the C-terminal region of middle-T antigen in cellular transformation by polyomavirus. *Nucleic Acids Res.* **9:** 2055.

Oostra, B.A., R. Harvey, B.K. Ely, A.F. Markham, and A.E. Smith. 1983. Transforming activity of polyomavirus middle-T antigen probed by site-directed mutagenesis. *Nature* **304:** 456.

Schaffhausen, B.S. and T.L. Benjamin. 1979. Phosphorylation of polyoma T antigen. *Cell* **18:** 935.

————. 1981. Comparison of phosphorylation of two polyomavirus middle-T antigens *in vivo* and *in vitro*. *J. Virol.* **40:** 184.

Schaffhausen, B., T.L. Benjamin, J. Lodge, D. Kaplan, and T.M. Roberts. 1985. Expression of polyoma early gene products in *E. coli. Nucleic Acids Res.* **13:** 501.

Smith, A.E. and B.K. Ely. 1983. The biochemical basis of transformation by polyomavirus. *Adv. Viral Oncol.* **3:** 3.

Smith, A.E., R. Smith, B. Griffin, and M. Fried. 1979. Protein kinase activity associated with polyoma middle-T antigen *in vitro. Cell* **18:** 915.

Soeda, E., J.R. Arrand, N. Smolar, J.E. Walsh, and B.E. Griffin. 1980. Coding potential and regulatory signals of the polyomavirus genome. *Nature* **283:** 445.

Templeton, D. and W. Eckhart. 1982. Mutation causing premature termination of the polyomavirus medium T antigen blocks cell transformation. *J. Virol.* **41:** 1014.

———. 1984. N-terminal amino acids sequences of the polyoma middle-sized antigen are important for protein kinase activity and cell transformation. *Mol. Cell. Biol.* **4:** 817.

Templeton, D., A. Voronova, and W. Eckhart. 1984. Construction and expression of a recombinant DNA gene encoding a polyomavirus middle-sized tumor antigen with the carboxyl terminus of vesicular stomatitis virus glycoprotein G. *Mol. Cell. Biol.* **4:** 282.

Treisman, R., U. Novak, J. Favaloro, and R. Kamen. 1981. Transformation of rat cells by an altered polyomavirus genome expressing only the middle-T protein. *Nature* **292:** 595.

Yonemoto, W., M. Yarvis-Morar, J.S. Brugge, J.B. Bolen, and M. Israel. 1985. Tryosine phosphorylation within the amino-terminal domain of pp60$^{c-src}$ molecules associated with polyomavirus middle-sized tumor antigen. *Proc. Natl. Acad. Sci.* **82:** 4568.

Zhu, J., G.M. Veldman, A. Cowie, A. Carr, B. Schaffhausen, and R. Kamen. 1984. Construction and functional characterization of polyomavirus genomes that separately encode the three early proteins. *J. Virol.* **51:** 170.

# DNA Rearrangement and the Role of Viral Origin in SV40-transformed Mouse Cells

## S. Chen and R. Pollack

Department of Biological Sciences, Columbia University, New York, New York 10027

In simian virus 40 (SV40)–transformed mouse cells, a 100-kD super T antigen is always found in addition to the wild-type 94-kD and 17-kD T antigens. We have determined the coding sequence for this super T antigen, which includes two separate partial repeats of the SV40 genome. The downstream repeat contained the complete coding sequence for the SV40 large T antigen, whereas the upstream repeat was a truncated copy of the same gene, which varied in length in different clones.

We also have shown that in SV40-transformed mouse cells, a functional origin of replication is not needed for either partial repeat formation or subclonal rearrangement of integrated viral DNA.

Simian virus (SV40) encodes two early proteins, large T and small T antigens, with molecular masses of 94 kD and 17 kD, respectively (Tooze 1981). Both proteins are derived from a single transcript whose 5′ end lies near the viral origin of replication.

The large T antigen has been shown to be involved in transformation, in some cases including maintenance of certain transformed phenotypes in nonpermissive cells (Tooze 1981). T antigen also participates directly in many biochemical events, including initiation of viral DNA replication in lytically infected permissive cells (Tegtmeyer 1975), down-regulation of early RNA synthesis (Tegtmeyer et al. 1975; Reed et al. 1976), activation of host ribosomal genes (Soprano et al. 1980), stimulation of cellular DNA synthesis (Henry et al. 1966), provision of adenovirus helper function (Cole et al. 1979), high-affinity binding to specific DNA sequences at the SV40 origin of replication (Tjian 1978), in vitro ATPase activity (Tjian and Robbins 1979), and formation of a stable complex with a cellular 54-kD phosphoprotein (Lane and Crawford 1979).

Infection of rodent cells by SV40 frequently results in the integration of tandem arrays of viral DNA into host chromosomes (Bender and Brockman 1981; Clayton and Rigby 1981; Blanck et al. 1982). A comparison of the amplified arrays of viral sequences in several SV40 mouse transformants reveals that they share some consistent, structured features. Usually they contain one or more copies of a full-length viral genome, in addition to several partial copies (Bender and Brockman 1981; Clayton and Rigby 1981; Sager et al. 1981; Blanck et al. 1982). Generation of variant-size T antigens probably occurs through viral DNA rearrangements since tandem arrays of SV40 DNA encode and express variant-size T antigens (Chang et al. 1979; Kress et al. 1979; Smith et al. 1979; McCormick et al. 1980; Chen et al. 1981). Those that are larger than 94 kD T are termed "super T antigens" (Chang et al. 1979; Kress et al. 1979; Smith et al. 1979; McCormick et al. 1980; Chen et al. 1981).

One such super T antigen, 100 kD, is ubiquitous in cell lines transformed by either SV40 virus infection or SV40 DNA transfection. This 100-kD T antigen does not appear in mouse transformants generated by a variant SV40 DNA virus that lacks a functional viral replication origin (Chen et al. 1983). This is true for two different mutants; one has a 6-bp in-phase deletion (Gluzman et al. 1980), and the other has a 4-bp insertion at the origin of replication (Chen et al. 1983).

## SV40 DNA Rearrangement

Multiple copies of foreign DNA that are introduced into a cell by infection, transfection, or microinjection often integrate in a tandem array, and significant perturbations occur in the host sequences adjacent to sites of integration (Botchan et al. 1980; Stringer 1982; Kopchick and Stacey 1984). The input foreign DNA can be of either supercoiled or linear form and still be efficiently ligated together in head-to-tail polymers (Wilson et al. 1982). Although these DNA concatemers are relatively stable, subsequent rearrangements also take place.

Rearrangements of SV40 DNA can occur both before integration and after integration. Chia and Rigby (1981) have demonstrated that prior to integration SV40 replicates in the form of large polymers and that double-crossover events can account for some multiple insertions of viral DNA. However, further amplification, excisions, and recombinations also occur after the initial integration event.

Rearrangements of SV40 DNA have been reported in many mouse transformants (Bender and Brockman 1981; Clayton and Rigby 1981; Blanck et al. 1982). We have shown that selection for phenotypic reversion from the transformed state to a more normal serum or anchorage-dependent state is accompanied by rearrangement of integrated SV40 DNA. Anchorage-dependent revertants preferentially lose defective viral DNA while retaining an intact SV40 early region and the ability to express lytic-size large and small T antigens. A considerable amount of viral DNA rearrangement also occurs in some anchorage-dependent revertant cell lines, such as the line LS₁ isolated from the fully transformed mouse line SV101. In semisolid medium, anchorage-independent "relapsed" subclones can be isolated from LS₁. These have undergone amplification of their integrated viral DNA inserts (Blanck et al. 1982). Similar results were reported in rat transformants: Rat transformants selected for growth in semisolid medium contain amplified viral sequences, while those isolated from nonselective growth only contain a single insert (Mougneau et al. 1980).

In SV40-transformed mouse cells, DNA rearrangements also occur with serial cell passage in culture. Recently, we and others (Sager et al. 1981; Chen et al. 1983) have shown that the SV40-transformed BALB/c-3T3 cell line SVT2 contains only one variant early region of SV40, about 4.4 kb in length. The only detectable virus-specific protein made by this region is a 100-kD super T antigen. Upon subcloning or passaging in culture, a normal-size early region of 2.7 kb and lytic-size 94-kD large T and 17-kD small T antigens arise in SVT2 cells in addition to the super T antigen (Chen et al. 1983).

### SV40 origin of replication is not required for DNA rearrangements

It has been assumed that SV40 DNA rearrangement depends on a functional viral origin of replication, because this has been shown to be the case for polyoma rat transformants (Basilico et al. 1979; Pellegrini et al.

1984). However, subclones derived from mouse transformants generated either with wild-type or origin-defective SV40 show DNA rearrangements (Fig. 1). This is in direct contrast to results obtained from polyoma rat transformants. In the case of SV40, mouse transformants can have partial repeats of SV40 DNA and DNA rearrangements can occur, despite the absence of a functional viral origin of replication (G. Blanck et al., in prep.). The difference in the two results may be due to the differences in host cells and/or virus types.

### The coding region of super T antigens

Some super T antigens are encoded by variant early region sequences that contain internal in-phase duplications. For instance, the sequence that encodes a 145-kD super T antigen in mouse transformants contains a perfect, in-phase duplication of 1212 bp within the large T second exon from nucleotides 4103 to 2892 (Lovett et al. 1982). Similarly, May et al. (1981) have determined that the template sequence for a 115-kD super T antigen in rat transformants includes a duplication of 572 bp in the same region (4116–3544). The duplicated sequences are not adjacent, however; they are separated by a 93-bp segment that is a nearly perfect inversion of a sequence located within the large T intron.

Two other super antigens from mouse transformants contain internal duplications that span the acceptor splice junction between the large T intron and the second exon. The 130-kD super T antigen in mouse transformants contains a duplication of 788 bp, which extends from nucleotides 4588 to 3800 (Lovett et al. 1982). The 100-kD super T antigen in SVT2 has a very

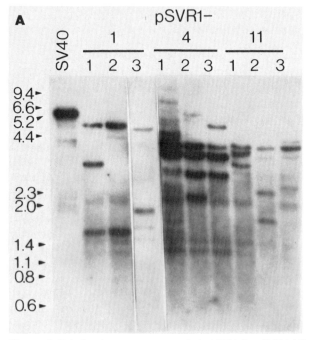

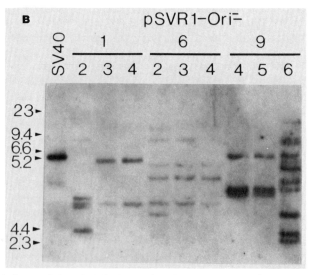

**Figure 1** Subclonal rearrangement of viral DNA in pSVR1 (A) subclones: 1-1, 1-2, 1-3, 4-1, 4-2, 4-3, 11-1, 11-2, and 11-3 and pSVR1-*ori⁻* (B) subclones: 1-2, 1-3, 1-4, 6-2, 6-3, 6-4, 9-4, 9-5, and 9-6 transformants. High-molecular-weight DNA was extracted from the subclones, digested with *Taq*, and probed with ³²P-labeled SV40 DNA. Comparison of the subclones indicates the rearrangement of integrated SV40 DNA.

large duplication of 1750 bp that also spans the acceptor splice junction (Sager et al. 1981). It is still not known whether the duplicated part of the intron is used as a coding sequence or a new splice pattern is involved.

Both the 115-kD and 145-kD super antigens have been cloned and transfected into rodent and monkey cells, and both of these super T antigens are able to transform rodent cells, but neither one is able to support lytic replication (May et al. 1981; Clayton et al. 1982; Lovett et al. 1982).

### The 100-kD super T antigen

We have been studying the 100-kD super T antigen that is found in fully transformed, anchorage-independent mouse cell lines and lost specifically in subclones that have regained serum and/or anchorage requirements (Chen et al. 1981).

We have identified the coding sequences for the 100-kD super T antigen in two different SV40 mouse transformants (SV101 and SV3T3 CIM). Unlike the other super T antigens, the 100-kD T antigen is not coded by internal duplication of SV40 DNA. In both clones, one complete SV40 early region is preceded by a truncated copy of the early region (Levitt et al. 1985). The truncated early region can be of different lengths but includes the first exon, the intron of large T antigen, and part of the second exon. The full-length early region can be separated from the truncated early region by as much as 600 bp as in pSV3T3-M-A, or it can be immediately adjacent as in p100D. pSV3T3-M-A was one of the clones from a library of genomic clones of SV3T3 CIM made by Clayton and Rigby (1981). Although p25B was made from SV101 after virus rescue by fusion with monkey cells, p100D was constructed in vitro from wild-type SV40 by placing a truncated copy of the early region upstream from a full-length early region.

All three plasmids when transfected into monkey cells express the 100-kD super T antigen (Fig. 2). Both p25B and pSV3T3-M-A have been mapped extensively with restriction enzymes. The upstream, truncated early region is of two different sizes; however, both contain a complete SV40 early region downstream (Fig. 3). The truncated early region in both cases starts before the origin/control region and continues past the acceptor splice site for the large T antigen. Only about 100 bp of mouse DNA separate the two inserts in p25B, whereas in pSV3T3-M-A about 600 bp of host DNA are present between the upstream partial repeat and the downstream early region.

For the 100-kD super T antigen to be transcribed, the two insertions must be colinear; that is, if the partial repeat upstream is separated from the complete early region, then only wild-type 94-kD T antigen is produced. Therefore the partial and full-length early regions must be *cis* for 100-kD super T antigen expression (Levitt et al. 1985).

Mouse transformants generated from origin-defective SV40 fail to express the 100-kD super T antigen (Chen et al. 1983). Partial repeats of integrated SV40 DNA occur in these transformants as examined by three

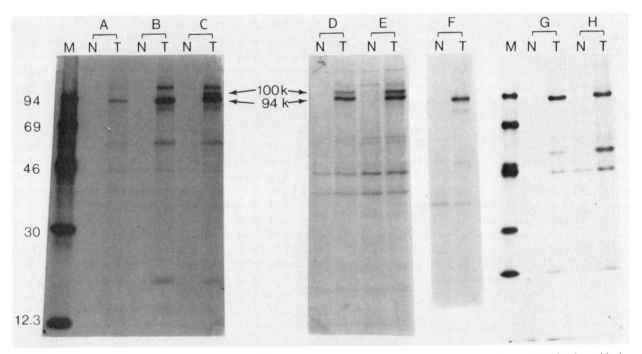

**Figure 2** Immunoprecipitation of viral proteins in [$^{35}$S]methionine-labeled monkey cell extracts at 48 hr posttransfection with the following plasmid DNA and analyzed on a 10–20% SDS-polyacrylamide gel: wild-type SV40 pSVR1 (*A*), origin-defective SV40 pSV ori⁻ (*B*), p25B (*C*), pSV3T3-M-A (*D*), p100D (*E*), p25B-ori⁻ (*F*). Lanes *G* and *H* are [$^{35}$S]methionine-labeled cell extracts from mouse transformants generated by cDNA clones of large T pSVT-2 (lane *G*) and pSVT-5 (lane *H*).

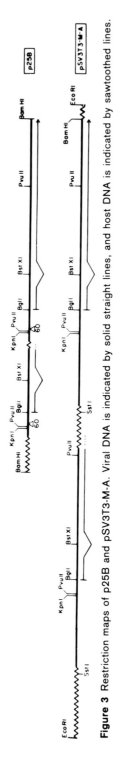

**Figure 3** Restriction maps of p25B and pSV3T3-M-A. Viral DNA is indicated by solid straight lines, and host DNA is indicated by sawtoothed lines.

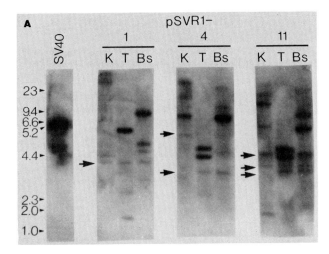

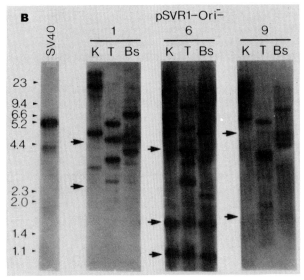

**Figure 4** Partially repeated viral DNA in pSVR1 and pSVR1-*ori*⁻ mouse transformants. Tandem repeats of the viral DNA can be identified by digesting the high-molecular-weight transformant DNA with several enzymes that cut the SV40 DNA in one place. Each digest is run in a separate, but parallel, gel lane. Fragments that contain SV40 DNA and comigrate most likely represent viral DNA repeats, since a tandem duplication of a set of viral enzyme sites creates the same size fragments when the DNA is cut off at any one of those sites. If the repeat has suffered a deletion between the duplicated enzyme sites, then comigrating fragments will be smaller than the wild-type SV40 DNA length of 5.2 kb. Although it is possible that fragments detected in this way represent some other structure, more-detailed work has shown that ~80% or more of the fragments so identified do in fact represent partial duplications (Blanck et al. 1982). Here, the enzymes *Kpn* (K), *Taq* (T), and *Bst*XI (Bs) were used. The arrowheads indicate comigration of bands in the separate digests: (*A*) pSVR1 transformants; (*B*) pSVR1-*ori*⁻ transformants.

one-cut restriction enzymes (Fig. 4). Thus, formation of tandem arrays of integrated SV40 can occur in the absence of a functional origin of replication.

If the downstream, full-length SV40 early region is replaced by the origin-defective SV40, then no 100-kD super T expression is seen (Fig. 2, lane F). The sequences at the origin of replication in the downstream full-length region are crucial for the 100-kD super T antigen expression. The origin may serve as a splice signal rather than coding region for the 100-kD protein.

Either one or both of the acceptor and donor splice sites of the large T antigen are needed for the generation of 100-kD T antigen. Mouse transformants generated only by the cDNA clone of large T antigen express the wild-type 94-kD antigen even after many passages in culture (Fig. 2, lanes G and H). Small T antigen deletion mutant 884 generated mouse transformants that express the 100-kD super T antigen (Chen et al. 1983).

The transcripts for the wild-type 94-kD and 17-kD antigen probably initiate at the downstream, full-length early region, whereas transcription for the 100-kD super T antigen probably initiates at the partial repeat upstream and continues through the downstream, full-length early region. The primary transcript is then removed by splicing to generate the 100-kD super T antigen.

Since the 100-kD super T antigen is ubiquitous in SV40-transformed mouse cells, we would like to directly test the functions of this protein. This will only become possible when we obtain a clone that produces the 100-kD super T antigen exclusively.

## Acknowledgments

We wish to thank Dr. George Blanck and Dr. Alexandra Levitt for helpful discussions. This work was supported by Public Health Service grant CA 336207 and New Investigator Research Award CA 36319 (to S.C.) from the National Institutes of Health.

## References

Basilico, C., S. Gattoni, D. Zouzias, and G. Della Valle. 1979. Loss of integrated viral DNA sequences in polyoma-transformed cells is associated with an active viral A function. *Cell* **17:** 645.

Bender, M. and W. Brockman. 1981. Rearrangements of integrated viral DNA sequences in mouse cells transformed by SV40. *J. Virol.* **38:** 872.

Blanck, G., S. Chen, and R. Pollack. 1982. Integration, loss and reacquisition of defective viral DNA in SV40-transformed mouse cell lines. *Virology* **126:** 413.

Botchan, M., J. Stringer, T. Mitchison, and J. Sambrook. 1980. Integration and excision of SV40 DNA from the chromosome of a transformed cell. *Cell* **20:** 143.

Chang, C., D.T. Simmons, M.A. Martin, and P.T. Mora. 1979. Identification and partial characterization of new antigens from simian virus 40-transformed mouse cells. *J. Virol.* **31:** 463.

Chen, S., G. Blanck, and R. Pollack. 1983. Reacquisition of a functional early region by a mouse transformant containing only defective viral DNA. *Mol. Cell. Biol.* **4:** 666.

Chen, S., M. Verderame, A. Lo, and R. Pollack. 1981. Nonlytic simian virus 40-specific 100K phosphoprotein is associated with anchorage-independent growth in simian virus 40-transformed and revertant mouse cell lines. *Mol. Cell. Biol.* **1:** 994.

Chen, S., D.S. Grass, G. Blanck, N. Hoganson, J.L. Manley, and R.E. Pollack. 1983. A functional simian virus 40 origin of replication is required for the generation of a super T antigen with a molecular weight of 100,000 in transformed mouse cells. *J. Virol.* **48:** 492.

Chia, W. and P. Rigby. 1981. Fate of viral DNA in nonpermissive cells infected with SV40. *Proc. Natl. Acad. Sci.* **78:** 6638.

Clayton, C. and P. Rigby. 1981. Cloning and characterization of the integrated viral DNA from three lines of SV40-transformed mouse cells. *Cell* **25:** 547.

Clayton, C.E., M. Lovett, and P.W.J. Rigby. 1982. Functional analysis of a simian virus 40 super T antigen. *J. Virol.* **44:** 974.

Cole, C., L. Crawford, and P. Berg. 1979. Simian virus 40 mutants with deletions at the 3′ end of the early region are defective in adenovirus helper function. *J. Virol.* **30:** 683.

Gluzman, Y., R. Frisque, and J. Sambrook. 1980. Origin-defective mutant of SV40. *Cold Spring Harbor Symp. Quant. Biol.* **44:** 293.

Henry, P., P.H. Black, M.N. Oxman, and S.M. Weissman. 1966. Stimulation of DNA synthesis in mouse cell line 3T3 by simian virus 40. *Proc. Natl. Acad. Sci.* **56:** 1170.

Kopchick, J.J. and D.W. Stacey. 1984. Differences in intracellular DNA ligation after microinjection and transfection. *Mol. Cell. Biol.* **4:** 240.

Kress, M., E. May, R. Cassingena, and P. May. 1979. Simian virus 40-transformed cells express new species of proteins precipitable by anti-simian virus 40 tumor serum. *J. Virol.* **31:** 472.

Lane, D.P. and L. Crawford. 1979. T antigen is bound to a host protein in SV40 transformed cells. *Nature* **278:** 261.

Levitt, A., S. Chen, G. Blanck, D. George, and R. Pollack. 1985. Two integrated partial repeats of simian virus 40 together code for a super T antigen. *Mol. Cell. Biol.* **5:** 742.

Lovett, M., C.E. Clayton, D. Murphy, P. Rigby, A.E. Smith, and F. Chaudry. 1982. Structure and synthesis of a simian virus 40 super T antigen. *J. Virol.* **44:** 963.

May, E., J.M. Jeltsch, and F. Gannon. 1981. Characterization of a gene encoding a 115 K super T antigen expressed by an SV40-transformed rat cell line. *Nucleic Acids Res.* **9:** 4111.

McCormick, F., F. Chaudry, R. Harvey, R. Smith, P.W.J. Rigby, E. Paucha, and A.E. Smith. 1980. T antigens of SV40-transformed cells. *Cold Spring Harbor Symp. Quant. Biol.* **44:** 171.

Mougneau, E., F. Birg, M. Rassoulzadegan, and F. Cuzin. 1980. Integration sites and sequence arrangement of SV40 DNA in a homogeneous series of transformed rat fibroblast lines. *Cell* **22:** 917.

Pellegrini, S., L. Dailey, and C. Basilico. 1984. Amplification and excision of integrated polyoma DNA sequences require a functional origin of replication. *Cell* **36:** 943.

Reed, S.J., G.R. Stark, and J.C. Alwine. 1976. Autoregulation of simian virus 40 gene by T antigen. *Proc. Natl. Acad. Sci.* **73:** 3083.

Sager, R., A. Anisowicz, and N. Howell. 1981. Genomic rearrangement in a mouse cell line containing integrated SV40 DNA. *Cell* **23:** 41.

Smith, A.E., R. Smith, and E. Paucha. 1979. Characterization of different tumor antigens present in cells transformed by simian virus 40. *Cell* **18:** 335.

Soprano, K.J., V. Dev, C.M. Croce, and R. Baserga. 1980. Reactivation of silent rRNA genes by simian virus 40 in human-mouse hybrid cells. *Proc. Natl. Acad. Sci.* **76:** 3885.

Stringer, J.R. 1982. DNA sequence homology and chromosome deletion at a site of SV40 integration. *Nature* **296:** 363.

Tegtmeyer, P. 1975. Function of simian virus 40 gene A in transforming infection. *J. Virol.* **15:** 613.

Tegtmeyer, P., M. Schwartz, J.K. Collins, and K. Rundell. 1975. Regulation of tumor antigen synthesis by simian virus 40 gene A. *J. Virol.* **16:** 168.

Tjian, R. 1978. The binding site on SV40 DNA for a T-antigen-related protein. *Cell* **13:** 165.

Tjian, R. and A. Robbins. 1979. Enzymatic activities associated with a purified SV40 T antigen related protein. *Proc. Natl. Acad. Sci.* **76:** 610.

Tooze, J., ed. 1981. *Molecular biology of tumor viruses*, 2nd edition, revised: *DNA tumor viruses*. Cold Spring Harbor Laboratory, Cold Spring Harbor, New York.

Wilson, J.H., P.B. Berget, and J.M. Pipas. 1982. Somatic cells efficiently join unrelated DNA segments end-to-end. *Mol. Cell. Biol.* **2:** 1258.

# Monoclonal Antibody Analysis of the SV40 Large T Antigen–p53 Complex

## D.P. Lane* and J. Gannon*

Cancer Research Campaign, Eukaryotic Molecular Genetics Research Group, Department of Biochemistry, Imperial College of Science and Technology, London SW7 2AZ, United Kingdom

A wide range of monoclonal antibodies to SV40 large T antigen (large TAg) and the host-coded p53 protein have been used to analyze the antigenic structure of the large TAg–p53 complex. Several anti–large TAg antibodies directed to discrete epitopes failed to efficiently immuno-precipitate the large TAg–p53 complex from SV40-transformed or abortively infected mouse fibroblasts. These results were confirmed and extended using solid-phase radioimmunoassays and an in vitro system of complex formation. The binding of one group of antibodies to large TAg was inhibited as p53 binding occurred. Another set of antibodies reacted weakly with the in vitro complex but were uniquely able to inhibit its in vitro assembly. Immunocytochemical experiments using these reagents suggest that all the large TAg in abortively infected cells is complexed to p53 while transformed cells contain free large TAg molecules as well as complexed ones.

Large T antigen (large TAg), the transforming protein of SV40, is complexed to a host protein, p53, in SV40-transformed cells (Lane and Crawford 1979). The importance of this interaction for the biology of SV40 transformation is not yet clear. Primary cultures of rodent cells can be immortalized by overexpression of the p53 gene (Jenkins et al. 1984) and can be transformed by transfection of the p53 gene and an activated ras gene (Eliyahu et al. 1984; Jenkins et al. 1984; Parada et al. 1984). The level of p53 protein found in a wide range of transformed cells of both rodent and human origin is much higher than in the equivalent normal cells (for review, see Crawford 1983), and the p53 gene undergoes frequent rearrangement (Mowat et al. 1985) and possibly chromosomal translocation (LeBeau et al. 1985) in some neoplasia. The p53 protein has a very short half-life in nontransformed cells, and the protein is therefore present in minute concentrations. The protein's half-life is greatly extended in a range of transformed cells, and this may be achieved by complexing to large TAg in SV40-transformed cells (Oren et al. 1981). Mutations in the p53 gene that increase its product's half-life render the gene more effective as an immortalizing factor in transfection experiments (Jenkins et al. 1985). These findings raise a critical question about the nature of SV40 transformation: Is it achieved solely, or partly, through large TAg stabilizing p53? We have addressed this question here by using a wide range of antibodies to large TAg and p53 to analyze the formation of the large TAg–p53 complex in vivo and in vitro.

## Materials and Methods

### Eukaryotic cells and virus stocks

293 cells were obtained from Dr. P. Gallimore, and SVA31E7 cells were originally from Dr. Y. Ito. T3T3 cells, a spontaneously transformed mouse cell line derived from a culture of 3T3 cells, were isolated at Imperial College by Ms. C. Paul.

Ad5-SVRIII virus was from Dr. Y. Gluzman and SV40 virus (strain 830) was from Dr. P. Rigby. All cells were grown at 37°C in a humidified incubator gassed at 10% v/v $CO_2$ in air. Cells were grown in Dulbecco's modification of Eagle's medium supplemented with 10% fetal calf serum, 500 units/ml penicillin, and 100 µg/ml streptomycin.

### Monoclonal antibodies

The PAb4 series of antibodies were from Dr. E. Harlow (Harlow et al. 1981). The PAb2 series antibodies PAb203, PAb204, and PAb205 have been described (Clark et al. 1981). PAb250 and PAb251 are two new antibodies raised against the HindIII D fragment of SV40 large TAg (amino acids 271–447) fused to the carboxyl terminus of the lacZ gene (Mole and Lane 1985). 1D9 is an antibody raised against SV40 small TAg (Montano and Lane 1984) that reacts with the amino terminus of large and small TAgs. CAT-2 is a monoclonal antibody to chloramphenicol acetyltransfer-ase and was used as a control antibody. The anti-p53 antibody 200.47 was from Dr. Dippold (Dippold et al. 1981) and another anti-p53 antibody, RA3 2C2, was from Dr. Rotter (Rotter et al. 1981). The monoclonal cell lines were grown as described above, and the antibodies they secreted were purified by protein A–Sepharose chromatography (Ey et al. 1978). Purified antibodies were iodinated using the iodogen-coated tube method to specific activities of 1–2 µCi/µg.

*Present address: Imperial Cancer Research Fund, Clare Hall Laboratories, South Mimms, Potters Bar, Herts EN6 3LD, United Kingdom.

### Cell extracts

Cell monolayers (SVA31E7, T3T3, or 293 cells, the latter having been infected 24 hr earlier with Ad5-SVRIII virus at 5 pfu/cell) were rinsed in TD (25 mM Tris · HCl, pH 7.4, 136 mM NaCl, 5.7 mM KCl, 0.7 mM $Na_2HPO_4$) and harvested by incubating in TD containing 10 mM EDTA until cells detached from the dishes. The cells were collected, counted, and centrifuged (2000$g$ for 5 min), and the pellets were snap-frozen in a bath of ethanol and solid $CO_2$. The pellets were stored at −70°C. Extracts were prepared by thawing the cell pellets rapidly and resuspending them in extraction buffer (150 mM NaCl, 5 mM EDTA, 50 mM Tris · HCl, pH 8.0, 2 mM PMSF, 1% NP-40) at a ratio of 1 ml of buffer per $4 \times 10^7$ cells. The extracts were kept on ice for 30 min and then clarified by centrifugation at 10,000$g$ for 30 min. The supernatants were used immediately.

### Immunoprecipitation

[$^{35}$S]Methionine-labeled extracts of infected 3T3 cells were immunoprecipitated using the conditions described by Mole and Lane (1985).

### Solid-phase radioimmunoassay

This is a modification of that employed by Benchimol et al. (1982).

Purified antibody (50 μl) at a concentration of 30 μg/ml in 10 mM phosphate buffer (pH 7.5) was pipetted into each well of a flexible plastic microtiter dish (Falcon) and allowed to absorb overnight in a humidified chamber. The plate was rinsed in phosphate-buffered saline (PBS) and blocked in 3% bovine serum albumin BSA) in PBS for 3 hr. Plates were finally rinsed in PBS before either immediate use or storage at −20°C.

A volume of 50 μl of clarified cell lysate was applied to each well of the antibody-coated plate, incubated overnight, and then rinsed in PBS. $^{125}$I-labeled antibody (50 μl) was pipetted into the wells, incubated for 3 hr, and rinsed in PBS. Each well was counted in an LKB Wallac 80,000 gamma counter.

### In vitro assay of p53–SV40 large TAg association

This assay is based on the solid-phase radioimmunoassay (RIA). Clarified cell lysate (50 μl) from T3T3 and 50 μl of a given dilution of an Ad5-SVRIII–infected 293 cell lysate were incubated together for 3 hr. The mixture was transferred to either an RA32C2 antibody–coated well in a plastic microtiter plate or a PAb205 antibody–coated well, allowed to incubate overnight, and then rinsed in PBS. Fifty μl of $^{125}$I-labeled PAb419, PAb414, or PAb250 were added to the wells, incubated for 3 hr, and rinsed in PBS, and the individual wells were cut out and counted as described. For titration purposes, the T3T3 or infected 293 cell extracts were diluted in extraction buffer.

### Antibody inhibition of large TAg–p53 complex assembly in vitro

Microtiter plates were coated with the anti–large TAg antibody PAb419 as described. Large TAg was then added as a 50-μl extract of Ad5-SVRIII–infected 293 cells ($4 \times 10^7$ cells/ml) and incubated overnight. The plates were washed and the inhibitor antibodies were added at a range of concentrations in 50 μl of NET buffer. One hr later, the p53 was added as a 50-μl extract of T3T3 cells ($3.5 \times 10^7$ cells/ml) in NET (extraction buffer minus PMSF and NP-40) containing calcium and magnesium ions (5 mM each). The plates were again incubated overnight and washed, and 50 μl of iodinated anti-p53 (200.47) was added (50,000 cpm/well). The plates were incubated overnight, washed, and counted as above.

## Results

Figure 1 shows the result of immunoprecipitations of a [$^{35}$S]methionine-labeled extract of 3T3 cells abortively infected with SV40 virus. The majority of the anti–large TAg monoclonal antibodies used and the anti-p53 antibody PAb421 (lane 24) clearly immunoprecipitate a band of large T and a band of p53 absent from the

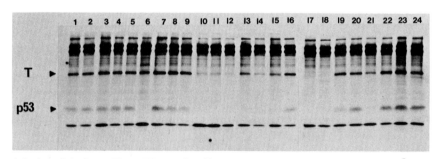

**Figure 1** Immunoprecipitation of the large TAg–p53 complex. Ten 9-cm plates of BALB/c-3T3 cells ($10^7$/plate) were infected with SV40 at an m.o.i. of 50. Forty-eight hr after infection, the plates were labeled with [$^{35}$S]methionine (50 μCi/plate) for 3 hr and a soluble extract was prepared. The extract was immunoprecipitated with a range of monoclonal antibodies and the precipitated protein was analyzed on a 15% SDS gel. The antibodies used are: PAb419 (lane *1*); 1D9 (lane *2*); PAb416 (lane *3*); PAb438 (lane *4*); PAb430 (lane *5*); PAb204 (lane *6*); PAb423 (lane *7*); PAb205 (lane *8*); PAb203 (lane *9*); PAb250 (lane *10*); PAb251 (lane *11*); PAb402 (lane *12*); PAb414 (lanes *13–15*); PAb404 (lane *16*); control, CAT-2 (lane *17*); PAb420 (lane *18*); PAb433 (lane *19*); PAb405 (lane *20*); PAb413 (lane *21*); PAb435 (lane *22*); PAb419 (lane *23*); PAb421 (lane *24*). The autoradiograph was exposed for 3 weeks.

control in lane 17. Some of the anti–large TAg antibodies immunoprecipitate either very little or no p53
and often reduced amounts of large TAg. This is notice-
able in lane 6 (PAb204), lane 10 (PAb250), lane 11
(PAb251), lane 12 (PAb402), lanes 13–15 (PAb414),
lane 18 (PAb420), and lane 21 (PAb413). All of the
anti–large TAg antibodies were used in excess and had
been recently checked for activity in a range of assays.
Broadly similar results were seen when the same antibodies were used to immunoprecipitate an extract of
SV40-transformed cells (SVA31E7), though here the
analysis was complicated by the presence of a lower-
molecular-weight TAg species recognized by all the
antibodies except PAb250 and PAb251. This lower-
molecular-weight TAg band did not seem to be associated with p53 since it was efficiently immunoprecipitated by PAb414, for instance, while no p53 was present
in this lane (data not shown). The lack of immunoprecipitation of the p53–large T complex by PAb414 was
further investigated over a wide range of antibody concentrations and using a variety of preparations with the
same result.

Solid-phase RIAs using a more restricted group of
antibodies confirmed and extended the immunoprecipitation data. In Figure 2, large TAg and the large TAg–
p53 complex present in an extract of SVA31E7 cells is
titrated in four RIAs. When the anti-large TAg antibody
PAb205 is used as the solid phase (coating the microtiter wells), large TAg is detected using either PAb414
(Fig. 2A) or PAb419 (Fig. 2B) as the radiolabeled probe.
The titrations with the two probes are broadly similar

over a five $\log_{10}$ range of antigen concentration. However, when the large TAg–p53 complex present in the
extracts is measured using the anti-p53 antibody
RA32C2 as a solid phase, the titration obtained with
PAb414 (Fig. 2C) is very different to that seen with
PAb419 (Fig. 2D). Simply, PAb414 fails to detect any
large TAg immobilized through p53, whereas the titration with PAb419 looks very similar to that obtained
using an anti–large TAg solid phase (Fig. 2B). Thus
PAb414 appears able only to recognize large TAg
molecules that are not complexed to p53 in the extracts.

The ability to assemble the large TAg–p53 complex in
vitro (Lane et al. 1985; Simanis and Lane 1985; Yewdell
et al. 1986) was then used to look for antigenic alterations in large TAg resulting from complex formation.
Again, solid-phase RIAs were used, and some typical
results are shown in Figure 3. In these assays, complex
formation was established using a fixed concentration of
p53 and a 10,000-fold range of large TAg concentrations. The same range of large TAg concentrations were
also measured after incubation in a control buffer to
quantitate the epitopes on free large TAg. The basis of
the assay and the formation of the complex is illustrated
by the positive control graphs in Figure 3, C and F. In
Figure 3C, large TAg is measured in the presence and
absence of complex formation using as solid phase the
anti–large TAg PAb205 and as label the anti–large TAg
PAb419. In Figure 3F, the same extracts and mixtures
are titrated, but instead the anti-p53 solid phase is used
with the same anti–large TAg, PAb419, as label. Formation of the complex is seen in Figure 3F where large TAg

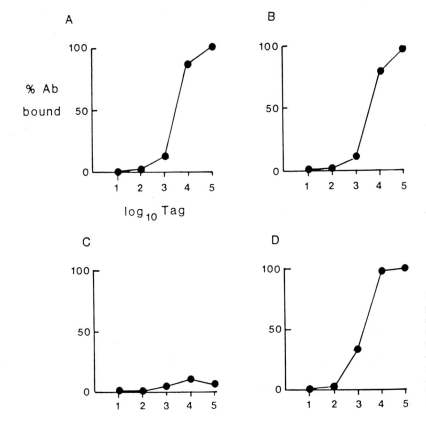

**Figure 2** Solid-phase RIA of the large TAg–
p53 complex from SV40-transformed cells. A
soluble extract of SVA31 E7 cells was titrated
in four different RIAs. In *A* and *B* the solid-
phase antibody is the anti–large TAg monoclonal antibody PAb205. In *C* and *D* the solid
phase is the anti-p53 antibody RA32C2. In *A*
and *C* the $^{125}$labeled anti–large TAg antibody
probe is PAb414, while in *B* and *D* it is
PAb419.

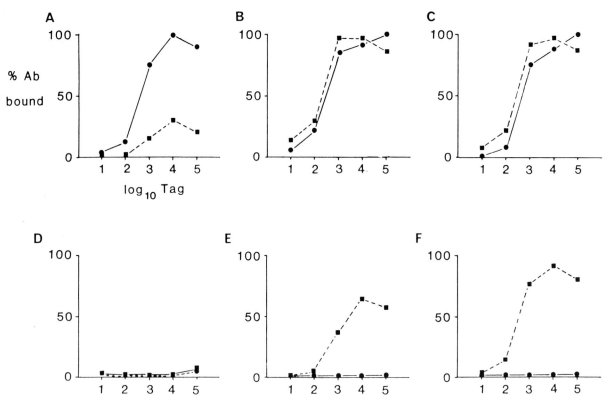

**Figure 3** Antigenic analysis of the in vitro–assembled large TAg–p53 complex. The effects of complex formation with p53 on the antigenic structure of large TAg were analyzed in a series of RIAs. A five $\log_{10}$ range of large TAg–containing extract was incubated either with a soluble p53-containing extract (■) or with a control buffer (●). After incubation at room temperature for 3 hr, aliquots of the mixtures were transferred to antibody-coated plates for RIA. The solid-phase antibody in *A*, *B*, and *C* is the anti–large TAg antibody PAb205. In *D*, *E*, and *F* the solid phase is the anti-p53 antibody RA32C2. The labeled anti–large TAg antibodies are PAb250 (*A* and *D*), PAb 414 (*B* and *E*), and PAb419 (*C* and *F*). Antibody binding is expressed as a percentage of the maximum binding seen with that labeled antibody within the experiment to aid comparison between graphs (see text).

is immobilized through the p53 it has bound in the in vitro incubation. In the absence of p53, no binding of large TAg to the solid phase is seen. In Figure 3C, broadly similar titrations are seen in the presence or absence of p53 complex formation. Strikingly different results are seen using the anti–large TAg PAb250 as label. In Figure 3D, it is clear that PAb250 fails to bind to large TAg that is complexed to p53. (It should be emphasized that all the titrations used aliquots of the same large TAg plus p53 extract or control mixture.) In Figure 3A, substantially less large TAg is able to bind PAb250 in the presence of p53 than in the absence of p53 and complex formation. Thus, it appears that the PAb250 epitope on large TAg is occluded by p53 binding either by direct stearic blocking or perhaps by a p53-induced conformational shift. The reduction in PAb250 binding is not simply due to breakdown of large TAg since the PAb419 control (Fig. 3C) shows no such effect. PAb251 gave identical results with those obtained with PAb250 in this assay.

The behavior of the PAb414 epitope is also distinct. As with PAb419 (Fig. 3C), PAb414 binding to large TAg is not markedly altered by the presence of p53 and complex formation. However, PAb414 is slightly less efficient than PAb419 at binding to p53-complexed large

TAg as seen in Figure 3E. These results obtained with the in vitro–assembled complex differ from those obtained using the in vivo complex from SVA31E7 cells in that some binding of PAb414 to the complex is detected in the former but not in the latter (cf. Fig. 2C and Fig. 3E).

In the next stage of the analysis, the ability of a range of anti–large TAg antibodies to block the in vitro assembly of the large TAg–p53 complex was measured. For these experiments, the in vitro complex assay was modified so that rather than assemble the complex in free solution as in the earlier studies (Lane et al. 1985; Simanis and Lane 1985; Yewdell et al. 1986), a solid-phase approach was used. Anti–large TAg–coated wells were first incubated with large TAg–containing extract, washed, and then incubated with a p53-containing extract. The assembly of the complex proceeded efficiently and could be measured using a radiolabeled anti-p53 antibody. Control experiments established that no p53 bound the plates in the absence of large TAg. If the solid-phase large TAg was incubated with anti–large TAg antibody before the addition of p53, some, but not all, antibodies inhibited p53 binding to the immobilized large TAg. The data for PAb251, PAb414, and a control antibody, CAT-2, are shown in Figure 4; clearly PAb414

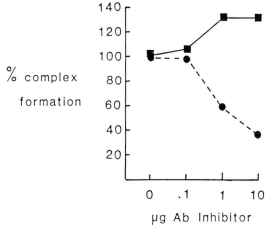

**Figure 4** Antibody inhibition of large TAg–p53 complex formation in vitro. The ability of a range of anti–large TAg antibodies to inhibit the formation of the large TAg–p53 complex was titrated in a two-stage solid-phase assay. Plates were coated with the anti–large TAg PAb419 and then incubated with a soluble large TAg–containing extract. After washing, a range of concentrations of inhibiting antibody was added, followed by a soluble p53 extract after 1 hr. After incubation, p53 attached to the solid phase was measured using an iodinated anti-p53 antibody (200.47). The results are expressed as a percentage of complex formation (200.47 binding) relative to that found in the absence of inhibitor (= 100%). (●) PAb414 anti–large TAg; (■) PAb251 anti–large TAg and CAT-2 anti-CAT control antibody.

but not PAb251 or CAT-2, inhibits complex formation. The effect is not particularly potent, since quite high antibody concentrations are required, but it is clearly specific.

To try to determine whether the effects of p53 binding seen in vitro on large TAg's epitopes were also occurring in vivo, an immunocytochemical analysis of SV40 abortive infection and transformation has been initiated. Subconfluent 3T3 cells were infected at a high multiplicity of SV40 virus and then fixed and stained with a range of different anti–large TAg and anti-p53 antibodies at timed intervals postinfection. The most striking result seen was the absence of staining of the infected cells with PAb414, whereas PAb250, PAb419, PAb423, and a range of other antibodies efficiently stained large TAg in about 50% of the cells present in the culture. The PAb414 was not inactive since it stained SV40-transformed 3T3 cells very efficiently even when used at a 100-fold lower concentration.

## Discussion

The locations of the epitopes recognized by the anti–large TAg antibodies used in these studies are listed in Table 1. The antibodies that immunoprecipitate very little large TAg–p53 complex from the abortively infected cells recognize epitopes located within the central region of large T antigen, whereas antibodies directed against amino and carboxyl termini of the molecule efficiently immunoprecipitate the complex (Table 2).

The precise locations of the epitopes of the antibodies to the middle region are not known in many cases, but it is clear that not all antibodies directed against this area fail to recognize the complex. Thus, PAb433 and PAb435, which can immunoprecipitate the 56,000-m.w. (56K) but not the 42K Ad2ND2 protein, both efficiently immunoprecipitate the complex. The antibodies PAb204, PAb414, PAb413, PAb420, and PAb402 all immunoprecipitate the Ad2ND2 42K protein but not the AD2ND1 28K protein, and all fail to recognize the complex. However, other antibodies with this phenotype do recognize the complex, in particular PAb203 and PAb205. The new antibodies PAb250 and PAb251 also fail to recognize the complex in immunoprecipitation experiments; their binding sites have been mapped between nucleotides (nt) 4002 and 3716 (S.E. Mole and D.P. Lane, unpubl.), but their ability to recognize the Ad2ND2 42K protein has not yet been measured. Thus, the majority of antibodies that fail to recognize the complex have binding sites that lie between nt 3476 and 3245. This includes the antibody PAb204, which has been mapped to bind between nt 3476 and 3290. Antibodies in this group do not recognize the same epitope because PAb250 and PAb204 have their binding sites mapped to discrete linear sequences, whereas the other antibodies have been shown by solid-phase RIAs to recognize at least three noncompeting epitopes. Antibodies PAb250 and PAb251 compete, as do PAb414 and PAb413; these two sets do not cross-compete with each other or with PAb204. The competition patterns of PAb402 and PAb420 have not yet been fully assessed, but neither of them cross-competes with the antibodies in the other three sets. The complex-recognizing antibodies PAb203 and PAb205 compete with each other but do not cross-compete with the other four sets. The picture that emerges is that of an antigenic "hot spot" on large TAg that displays multiple epitopes and is potentially involved in p53 binding.

The RIA data have confirmed and extended these conclusions for two sets of antibodies—(1) PAb250 and

**Table 1** Monoclonal Antibodies to T Antigen That Are Defective in Immunoprecipitating the T-p53 Complex from Abortively Infected Cells

| Antibody | Minimum epitope coordinates | Maximum epitope coordinates |
|---|---|---|
| PAb204 | 3476–3290 | 3476–3290 |
| PAb250, PAb251 | 4002–3716 | 4002–3716 |
| PAb413, PAb414, PAb402, PAb420 | 3820–3476 | 3820–2727 |

**Table 2** Monoclonal Antibodies to T Antigen That Efficiently Immuno-precipitate the T-p53 Complex from Abortively Infected Cells

| Antibody | Minimum epitope coordinates | Maximum epitope coordinates |
|---|---|---|
| PAb419, PAb430, PAb438, 1D9 | 5163–4918 | 5163–4918 |
| PAb416 | 5163–4475 | 5163–4002 |
| PAb435 | 4571–4002 | 4571–4002 |
| PAb433 | 4081–3820 | 4002–2727 |
| PAb203, PAb205 | 3820–3476 | 3820–2727 |
| PAb423, PAb405, PAb404 | 3245–2727 | 3245–2727 |

Epitope locations have been determined by using fusion proteins that express parts of large T antigen or by using recombinant adenoviruses that produce T antigen fragments. For maximum epitope coordinates, only positive reactions with fragments are used. For minimum coordinates, the antibodies failure to react with a given fragment is also taken into account. Coordinates are given as nucleotide numbers on the SV40 sequence. Details are found in Clarke et al. 1981; Harlow et al. 1981; Mole and Lane 1985.

PAb251 and (2) PAb413 and PAb414. In the test that most directly relates to the immunoprecipitation data, both antibodies recognized free large TAg but did not bind to the p53-complexed large TAg present in extracts of SV40-transformed cells. Thus, the in vivo complex in both RIA and immunoprecipitation analyses lacks these large TAg epitopes. The remaining studies all use an in vitro complex assay where the free large TAg and the free p53 are obtained from different crude cell extracts. The results obtained are more complicated and suggest that complex formation may involve two discrete steps or processes. The in vitro complex completely lacks the PAb250/251 epitope, which maps between amino acids 271 and 366, but somewhat surprisingly still displays some of the PAb414 epitope. Conversely, PAb414 but not PAb250/251 can inhibit in vitro complex formation. Complex formation may, therefore, involve a two-stage process consisting of an initial binding event involving areas around the PAb414 site followed by some conformational shift that obscures the PAb250/251 site and also exposes the PAb414 site in vitro but not in vivo. Clearly, the in vitro complex does not yet precisely mirror the in vivo complex. This may be due to an absence of modification to either large TAg or p53 that occurs only when they assemble together in vivo, or more prosaically, it could be due to the relatively low concentrations of the proteins attained in vitro compared with in vivo conditions (large TAg has a concentration of ~ 2 mg/ml in the nucleus of the transformed cell; D. Lane, unpubl.). Further study should distinguish between these alternatives.

The immunocytochemical study of abortive infection and transformation that we have just begun has provided very provocative results in that PAb414 fails to stain the nucleus of abortively infected 3T3 cells but does stain the nucleus of SV40-transformed cells. This leads to a tentative hypothesis that all the large TAg in the abortively infected cell can be complexed by p53, whereas in the transformed cell free large TAg is present.

We are now trying to determine whether there is a selection for free large TAg–containing cells in the process of SV40 transformation. It would fit well with recent ideas of complementary groups of oncogenes if SV40 large TAg exerted two discrete transforming functions, one in its free state and the other in a complex with p53 (for a brief discussion, see Lane 1984).

## Acknowledgments

We thank Drs. E. Harlow, V. Rotter, and W. Dippold for their gifts of monoclonal antibody producing cell lines and Dr. Y. Gluzman for his gift of virus. The research was supported by a grant from the Cancer Research Campaign.

## References

Benchimol, S., D. Pim, and L. Crawford. 1982. Radioimmunoassay of the cellular protein p53 in mouse and human cell lines. *EMBO J.* **1**: 1055.

Clark, R., D. Lane, and R. Tjian. 1981. Use of monoclonal antibodies as probes of simian virus 40 T antigen ATPase activity. *J. Biol. Chem.* **256**: 11854.

Crawford, L. 1983. The 53,000 dalton cellular protein and its role in transformation. *Int. Rev. Exp. Pathol.* **25**: 1.

Dippold, W.G., G. Jay, A.B. DeLeo, G. Khoury, and L.J. Old. 1981. p53 transformation-related protein: Detection by monoclonal antibody in mouse and human cells. *Proc. Natl. Acad. Sci.* **78**: 1695.

Eliyahu, D., A. Raz, P. Gruss, D. Givol, and M. Oren. 1984. Participation of p53 cellular antigen in transformation of normal embryonic cells. *Nature* **312**: 646.

Ey, P.L., S.J. Prowse, and C.R. Jenkin. 1978. Isolation of pure IgG1, IgG2a, and IgG2b. Immunoglobulins from mouse serum using protein A sepharose. *Immunochemistry* **15**: 429.

Harlow, E., L.V. Crawford, D.C. Pim, and N.M. Williamson. 1981. Monoclonal antibodies specific for simian virus 40 tumor antigens. *J. Virol.* **39**: 861.

Jenkins, J.R., K. Rudge, and G.A. Currie. 1984. Cellular immortalization by a cDNA clone encoding the transformation-associated phosphoprotein p53. *Nature* **312**: 651.

Jenkins, J.R., K. Rudge, P. Chumakov, and G.A. Currie. 1985. The cellular oncogene p53 can be activated by mutagenesis. *Nature* **317**: 816.

Lane, D.P. 1984. Cell immortalization and transformation by the p53 gene. *Nature* **312:** 596.

Lane, D.P. and L.V. Crawford. 1979. T antigen is bound to a host protein in SV40-transformed cells. *Nature* **278:** 261.

Lane, D.P., V. Simanis, R. Bartsch, J. Yewdell, J. Gannon, and S. Mole. 1985. Cellular targets for SV40 large T-antigen. *Proc. R. Soc. Lond. B* **226:** 25.

LeBeau, M.M., C.A. Westbrook, M.D. Diaz, J.D. Rowley, and M. Oren. 1985. Translocation of the p53 gene in t(15;17) in acute promyelocytic leukaemia. *Nature* **316:** 826.

Mole, S.E. and D.P. Lane. 1985. Use of simian virus 40 large T-β galactosidase fusion proteins in an immunochemical analysis of simian virus 40 large T antigen. *J. Virol.* **54:** 703.

Montano, X. and D.P. Lane. 1984. Monoclonal antibody to simian virus 40 small T. *J. Virol.* **51:** 760.

Mowat, M., A. Cheng, N. Kimura, A. Bernstein, and S. Benchimol. 1985. Rearrangements of the cellular p53 gene in erythroleukaemic cells transformed by Friend virus. *Nature* **314:** 633.

Oren, M., W. Maltzman, and A.J. Levine. 1981. Post-translational regulation of the 54K cellular tumor antigen in normal and transformed cells. *Mol. Cell. Biol.* **1:** 101.

Parada, L.F., H. Land, R.A. Weinberg, D. Wolf, and V. Rotter. 1984. Cooperation between gene encoding p53 tumor antigen and *ras* in cellular transformation. *Nature* **312:** 649.

Rotter, V., M.A. Boss, and D. Baltimore. 1981. Increased concentration of an apparently identical cellular protein in cells transformed by either Abelson murine leukemia virus or other transforming agents. *J. Virol.* **38:** 336.

Simanis, V. and D.P. Lane. 1985. An immunoaffinity purification procedure for SV40 large T antigen. *Virology* **144:** 88.

Yewdell, J.W., J.V. Gannon, and D.P. Lane. 1986. Monoclonal antibody analysis of p53 expression in normal and transformed cells. *J. Virol.* (in press).

# Identification and Characterization of the Sites Phosphorylated in the Cellular Tumor Antigen p53 from SV40-transformed 3T3 Cells and in the DNA-binding Protein from Adenovirus 2

C.W. Anderson,* A. Samad,[†] and R.B. Carroll[†]

*Biology Department, Brookhaven National Laboratory, Upton, New York 11973; [†]Department of Pathology and Rita and Stanley H. Kaplan Cancer Center, New York University Medical School, New York, New York 10016

The cellular tumor antigen p53, isolated from SV40-transformed 3T3 cells, is phosphorylated at at least four distinct sites on the 390-amino-acid protein. Two sites have been identified by radiochemical protein sequence analyses as Ser-312 and Ser-389. At least two additional phosphorylation sites, one involving serine, the other threonine, have not been identified, but both probably reside in the hydrophobic aminoterminal segment of p53. The peptide containing Ser-312 is resistant to dephosphorylation by alkaline hydrolysis and by alkaline phosphatase, suggesting that the Ser-312 phosphate is in a phosphodiester linkage. Alkaline hydrolysis of the peptide containing Ser-389 liberated the four ribonucleoside monophosphates, suggesting that Ser-389 may be linked to RNA. Phosphorylation of Ser-389 is markedly reduced in p53 isolated from *tsA58*-transformed cells labeled at the nonpermissive temperature; thus, the phosphorylation of this site is dependent upon native T antigen and may play a role in cell transformation. The aminoterminal phosphorylation sites appear to be simple phosphate monoesters. A similar analysis of the adenovirus DNA-binding protein (DBP) is in progress. Seven phosphorylation sites in the aminoterminal domain of DBP have tentatively been identified.

Protein phosphorylation/dephosphorylation is the major posttranslational modification used to regulate biochemical function (Krebs 1983). Although not all protein phosphorylations have obvious regulatory roles, many do. Among the recognized phosphoproteins are viral tumor antigens, other nonstructural viral proteins, and the cellular tumor antigen p53. The role played by phosphorylation in regulating the function of these viral proteins and p53 is not known; in most instances, the actual sites for phosphorylation have not yet been mapped. To elucidate the role phosphorylation may have with respect to their function, we have begun to map and characterize the phosphorylated sites of several adenovirus early proteins and of the tumor antigen p53.

p53 is the product of a cellular proto-oncogene. Its expression is elevated in cells transformed by a wide variety of agents, including the DNA tumor viruses simian virus 40 (SV40), adenovirus 2 (Ad2), Epstein-Barr virus (EBV), RNA tumor viruses such as Abelson murine leukemia virus (Ab-MLV), chemical carcinogens, and ionizing radiation (for review, see Klein 1982). A variety of spontaneous tumors also have elevated levels of p53. The cloned mouse p53 gene has been shown to immortalize transfected primary cells (Jenkins et al. 1984) and to render these cells tumorigenic when cotransfected with the *ras* gene of Harvey sarcoma virus (Eliyahu et al. 1984; Parada et al. 1984).

In SV40-transformed cells, p53 is complexed with the SV40 large T antigen (TAg) (Lane and Crawford 1979). p53 also forms complexes with virus-specific tumor antigen E1B-58K found in adenovirus-transformed cells (Sarnow et al. 1982). Complex formation with SV40 large TAg depends upon the integrity of large TAg since complexes do not form at the nonpermissive temperature in cells transformed with SV40 mutants defective at the *A* (large TAg) locus (Greenspan and Carroll 1981). The half-life of p53 is significantly increased in SV40-transformed 3T3 cells (Oren et al. 1981; Reich et al. 1983), and the p53 concentration in SV3T3 cells is 25 to 100 times higher than in nontransformed 3T3. The increased stability of p53 is thought to result from the formation of p53–large TAg complexes (Reich et al. 1983); however, the factors that regulate p53 stability are not yet well understood.

The DNA-binding protein (DBP) of adenovirus 2 (Ad2) is a multifunctional, nonstructural phosphoprotein with an apparent molecular weight of 72,000. Ad2 DBP is synthesized in large amounts beginning about 8 hr after infection (Anderson et al. 1973) and at late times is present in both the nucleus and cytoplasm of infected cells. Ad2 DBP binds single-stranded DNA (van der Vliet and Levine 1973) and the termini of double-stranded DNA fragments (Fowlkes et al. 1979; Schechter et al. 1980). DBP is required for adenovirus DNA replication (van der Vliet and Sussenbach 1975; Nagata et al. 1982; Friefeld et al. 1983), may play a role in adenovirus virion assembly (Nicolas et al. 1983), has been implicated in the regulation of early adenovirus transcription (Carter and Blanton 1978; Handa et al. 1983), and plays an important but poorly understood role in adenovirus host-range (Klessig and Grodzicker 1979; Anderson et al. 1983). Ad2 DBP is not required for adenovirus-induced transformation; in fact, viruses defective in DBP function transform at a somewhat higher frequency and integrate larger segments of the viral genome (Mayer and Ginsberg 1977) than wild-type virus.

The aminoterminal segment of the DBP, which represents a physically distinct domain from the carboxyterminal DNA-binding segment (Schechter et al. 1980), is believed to be phosphorylated at about a dozen sites involving both serine and threonine residues (Klein et al. 1979; Linné and Philipson 1980). As yet no function has been ascribed to DBP phosphorylation (Linné and Philipson 1980); however, mutants that allow productive adenovirus growth in monkey cells occur in the aminoterminal segment of DBP.

We report here the identification and preliminary characterization of two phosphorylation sites in the cellular tumor antigen p53. A preliminary analysis of several phosphorylation sites in the Ad2 DBP is also presented.

## Experimental Procedures

### Preparation of radiolabeled p53 and Ad2 DBP

Mouse cells transformed by wild-type SV40 and by the *tsA* mutant *tsA58* were obtained from G. Khoury and P. Tegtmeyer, respectively. The origin of our HeLa cells and Ad2 stocks have been described previously (Anderson et al. 1983). Cell lines were grown in Dulbecco's modified Eagle's medium (DME) containing 10% calf serum (GIBCO) and 1% antibiotic-antimycotic (ABAM, GIBCO). The mouse hybridoma line PAb416, which produces antibodies specific for SV40 TAg, was obtained from E. Harlow (Harlow et al. 1985); the DBP-specific mouse hybridoma line H2-19 is described by Rowe et al. (1984). Ascites fluid from mice injected with the appropriate hybridoma cell line was used as the source of specific antisera.

Cell monolayers in 100-mm petri dishes were labeled with inorganic phosphate or a tritiated amino acid as indicated in the text. Extracts containing 600 μg/ml RNAsin (Promega Biotec) were prepared, clarified, processed for immunoprecipitation, and fractionated by SDS-polyacrylamide gel electrophoresis as previously described (Greenspan and Carroll 1981). Polypeptides were recovered from gels by electroelution. All tubes were siliconized and sterilized.

### Protein sequence analysis and phosphoamino acid analysis

Our procedures for tryptic peptide fractionation, radiochemical sequence analysis, and phosphoamino acid analysis are described in greater detail elsewhere (Anderson 1982; Samad et al. 1986). Briefly, the desired proteins were precipitated from extracts of radiolabeled cells with a monoclonal antiserum, and the immunoprecipitate was further fractionated by SDS-polyacrylamide gel electrophoresis. Eluted radiolabeled proteins were precipitated with ice-cold trichloroacetic acid and extracted with ethanol:ether to remove SDS. DBP preparations were reduced and alkylated with iodoacetamide after immunoprecipitation and prior to gel electrophoresis. p53 preparations were oxidized with performic acid prior to trypsin digestion. Trypsin digestion was in 50 mM $NH_4HCO_3$ (pH 8). After lyophilization, the peptides were dissolved in 0.1% trifluoroacetic acid (TFA) and applied to a $0.39 \times 30$-cm μBondapak C18 reverse-phase column (Waters, Inc.) equilibrated with 0.1% TFA (pH 2). Peptides were eluted with a nonlinear gradient of 0.1% TFA, 80% acetonitrile, 20% water at a flow rate of 1 ml/min, and 1-ml fractions were collected. Radiolabeled peptides were detected by Cerenkov or scintillation counting. For protein sequence analysis, appropriate fractions were pooled, dried, dissolved in 20% formic acid, and applied to a Beckman 890C protein sequencer together with 3 mg of polybrene and 2 mg of apomyoglobin (Anderson 1982).

Phosphorylated amino acids were identified by two-dimensional thin-layer electrophoresis after acid hydrolysis of individual peptides (Greenspan et al. 1982). The nature of the phosphate bond was examined by electrophoresis of the products produced by base hydrolysis or alkaline phosphatase treatment (Samad et al. 1986). Digested samples were spotted directly on Whatman 3MM paper and electrophoresed in 5% acetate, 0.5% pyridine, pH 3.5. The air-dried paper was exposed to Kodak X-omat film and an intensifying screen (Dupont) at −70°C.

## Results

### Phosphorylation of a p53 peptide requires SV40 TAg

The complex between p53 and SV40 TAg does not form at the nonpermissive temperature (39.5°C) in cells transformed by *tsA58*, a mutant defective in SV40 TAg function; however, complex formation is observed at 32°C. To determine the effect of complex formation on the phosphorylation of p53, we prepared p53 from wild-type SV40-transformed 3T3 cells and from 3T3 cells transformed by the TAg mutant *tsA58*. Both transformants were labeled with [$^{32}$P]phosphate at 32°C and 24 hr after a shift to 39.5°C. Phosphorylated p53 was isolated from cell extracts by immunoprecipitation with a monoclonal antiserum specific for SV40 TAg followed by SDS-polyacrylamide gel electrophoresis. Each p53 preparation was then digested with trypsin, and the resulting phosphorylated tryptic peptides were fractionated by reverse-phase high-performance liquid chromatography (HPLC).

After 5 hr of phosphate labeling, four major radioactive components were reproducibly observed upon fractionation of p53 tryptic digests isolated from wild-type SV40-transformed cells labeled at either 32°C or 39.5°C (data not shown). These peptides have been designated P1 (peak 1) through P4, in order of increasing hydrophobicity (Samad et al. 1986). Component P1 was not retained by the column and thus is likely to consist of multiple components. Peak 2 eluted as a relatively broad peak and may only be retarded by the column. Peak 3 eluted as a sharp peak at an acetonitrile concentration of about 9% and was followed by a small shoulder (designated P3b) that is probably due to a related peptide, as discussed below. The last component eluted at approximately 40–50% acetonitrile and was frequently observed as a group of three peaks designated P4a, P4b, and P4c. These rather hydrophobic peptides may be related and are considered collectively in discussions below.

When *tsA58*-transformed 3T3 cells were labeled at 32°C (Fig. 1A), the profile of p53 phosphopeptides was essentially identical with that found for wild-type SV40-transformed 3T3 cells labeled at 37°C. In contrast, p53 isolated from SV40 *tsA58*-transformed 3T3 cells labeled at 39.5°C showed a marked reduction in phosphopeptide P2 (Fig. 1B). The relative amounts of components P3 and P4 were not significantly affected by the temperature shift. The apparent shift of label among the individual components of P4 may have resulted from small differences in digestion conditions since such differences were observed in different preparations labeled under similar conditions.

## Analysis of p53 phosphate bonds

Although phosphoproteins frequently contain simple phosphomonoesters, a variety of other phosphate-containing posttranslational modifications are known. Furthermore, at least five different amino acid residues may be phosphorylated. As a step toward identification of the actual posttranslation modifications of p53, we subjected each of the components shown in Figure 2 to several chemical and enzymatic digestions.

Trypsin digestion of p53 should produce about 40 peptides. Therefore, we first examined the four phosphorylated HPLC fractions (P1–P4) obtained from trypsin-digested p53 to see if they were homogeneous. Each of the four components was subjected to high-voltage paper electrophoresis at pH 3.5 (Fig. 2). All yielded two or more phosphorylated components except for component P4, which did not migrate from the origin during electrophoresis. P1 yielded three major phosphate-containing species designated P1-1, P1-2, and P1-3 in order of increasing mobility toward the anode (+). P2 yielded one major component (P2-1) that remained near the origin and a smear of material migrating

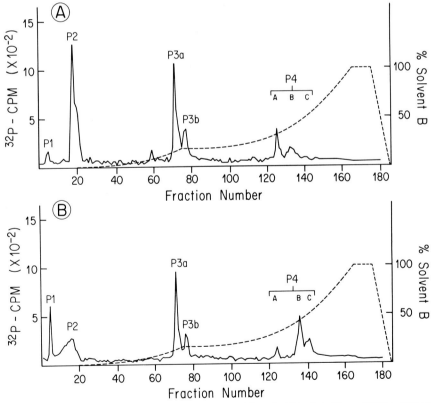

**Figure 1** Reverse-phase HPLC fractionation of the tryptic phosphopeptides of p53 isolated from SV40 *tsA58*-transformed 3T3 cells. SV40 *tsA58*-transformed 3T3 cells were labeled with [$^{32}$P]PO$_4$ for 5 hr at 32°C (*A*) or for 5 hr beginning 24 hr after a shift to 39.5°C (*B*). Labeled p53 was isolated by immune precipitation and SDS-polyacrylamide gel electrophoresis as described in Experimental Procedures. After performic acid oxidation and digestion with trypsin, the peptides were fractionated by reverse-phase C18 chromatography. Radiolabeled peptides were detected after addition of a scintillant to each fraction. Approximately 15,000 cpm of each sample was applied to the reverse-phase column. Peptide peaks (P1–P4) are numbered in order of increasing hydrophobicity.

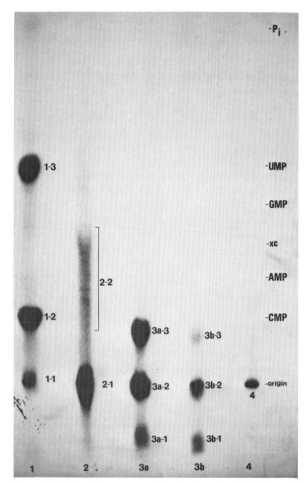

**Figure 2** High-voltage paper electrophoresis of p53 tryptic phosphopeptides. Fractions from a reverse-phase separation of p53 tryptic phosphopeptides similar to those shown in Fig. 1 (except these were from wild-type SV40-transformed 3T3 mouse cells) were pooled, dried, redissolved in 1% acetic acid, and applied to Whatman 3MM paper for high-voltage electrophoresis at pH 3.5 Phosphopeptides were detected by exposure of the dried paper to Kodak X-omat film. Individual components have been numbered according to the HPLC peak from which they were derived and in order of increasing mobility toward the anode (+, top) during paper electrophoresis. The positions of 3'-ribonucleotide standards and of xylene cyanol dye are indicated.

toward the anode (P2-2). P3 yielded three species (P3-1, P3-2, and P3-3), and the shoulder, P3b, gave three species of precisely the same mobility.

Phosphoamino acid content was determined by two-dimensional electrophoresis after acid hydrolysis (data not shown). With the exception of P4, each component yielded phosphoserine and inorganic phosphate ($P_i$). No phosphotyrosine was observed. Phosphohistidine and phospholysine, if present, would have been lost during performic acid oxidation of the protein. P4 (a–c pooled) yielded phosphoserine and a lesser amount of phosphothreonine. Previously, intact p53 was shown to contain approximately 92% phosphoserine and 8% phosphothreonine (Greenspan et al. 1982).

Next each of the components depicted in Figure 2

was treated with KOH, with hydroxylamine, or with alkaline phosphatase to determine the nature of the phosphate bond. P4 was found to be sensitive to base hydrolysis but resistant to hydroxylamine as expected for simple phosphate monoesters of serine or threonine (data not shown). In contrast, components P3-1 and P3-2 were substantially resistant to base and alkaline phosphatase hydrolysis, whereas P3-3 was sensitive to these agents. Components P3-1 and P3-2 thus have the properties expected for phosphate diester bonds. We believe that P3-1, P3-2, and P3-3 probably represent differently modified forms of the same peptide (see below); however, the precise nature of these modifications remains to be determined.

Surprisingly, base hydrolysis of components P2-1 and P2-2 yielded material with the mobility of the four ribonucleoside monophosphates (Fig. 3). Nucleoside monophosphates were also liberated by RNase A and T2 digestion of P2-1 and P2-2, but P2-1 and P2-2 were largely resistant to digestion by alkaline phosphatase (Fig. 4). The properties of P2 suggest that it may contain RNA covalently coupled to a peptide. We are reluctant to draw this conclusion, however, until we have actually identified an oligoribonucleotide chain clearly bound to peptidyl serine.

P1 was also found to be largely resistant to hydrolysis by alkaline phosphatase, and it was also moderately alkali-resistant. Base hydrolysis of P1-1 liberated $P_i$ and material with the mobility of P1-2 and P1-3, whereas hydrolysis of P1-2 yield primarily material with the mobility of P1-3. P1-3 has a mobility similar to UMP whereas the mobility of P1-2 is rather similar to that of CMP. As noted above, P1-1 and P1-2 were shown to contain phosphoserine (P1-3 was not analyzed). We show below that P1 and P2 are related in sequence; however, the precise chemical relationship between these two components has yet to be determined.

### Sequence identification of p53 tryptic phosphopeptides

To identify the different amino acid residues phosphorylated in p53, and, in particular, the residue whose phosphorylation is affected by p53-TAg complex formation, we analyzed each of the four phosphorylated HPLC-fractionated components by radiochemical protein sequence analysis. Samples for sequence analysis were prepared by labeling SV3T3 cells with [$^{32}$P]PO$_4$ and one tritiated amino acid at a time. The labeled p53 was then isolated, digested, and the tryptic peptides were fractionated as described in Figure 1. Identical preparations were made with several different labeled amino acids, including [$^3$H]glycine, [$^3$H]proline, [$^3$H]leucine, [$^3$H]lysine, and [$^{35}$S]cysteine. Each preparation was sequenced separately. The profiles of tritium (or $^{35}$S) and [$^{32}$P]phosphate released in each Edman cycle were then compared with the predicted sequence of p53 tryptic peptides (Pennica et al. 1984). Phosphate release was expected to coincide with a prediction of serine (for P1–P3) in the appropriate residue after a trypsin cleavage site (trypsin cleaves after arginine or

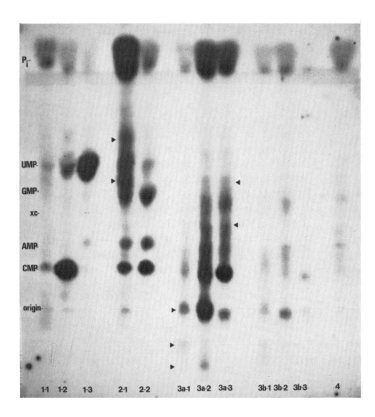

**Figure 3** Aklaline hydrolysis of p53 phosphopeptides. Each of the phosphorylated species shown in Fig. 2 was eluted with 50 mM ammonium bicarbonate, subjected to hydrolysis in 0.3 M KOH at 37°C for 20 hr, and electrophoresed again as described in Fig. 2. The positions of unlabeled standards are indicated as in Fig. 2.

lysine residues). Unambiguous identification of the phosphorylation site was then accomplished by a similar comparison of predicted sequence with the position of other amino acids determined from separate analyses.

P1-1 has been identified as the carboxyterminal tryptic peptide Lys-Val-Gly-Pro-Asp-Ser-Asp that begins at Lys-384. This identification is based primarily upon the identification of glycine at residue 3 (Fig. 5C), proline at residue 4 (Fig. 5B), and phosphate at residue 6 (Fig. 5A). The release of phosphate at residue 6 is consistent with the phosphorylation of Ser-389. No other site in the predicted sequence of p53 can be accommodated with these results. Not unexpectedly, the column flowthrough (P1) also contains several other peptides (detected with other amino acid labels) but none of these appear to be phosphopeptides.

P2-1 has also been identified as the carboxyterminal tryptic peptide. This identification is based upon the release of tritium in residue 1 from [$^3$H]lysine-labeled P2-1 (Fig. 5F), the release of tritium in residue 3 from [$^3$H]glycine-labeled P2-1 (data not shown), tritium in residue 4 from [$^3$H]proline-labeled P2-1 (Fig. 5E), and $^{32}$P in residue 6 from phosphate-labeled P2-1 (Fig. 5D). Careful analysis of the sequence data suggests that P2 probably contains a smaller quantity of a second peptide, Val-271 to Arg-277. This second peptide does not contain serine or threonine and cannot be phosphorylated.

P3 was found to have leucine at residue 2 (data not shown), proline at residues 3, 10, and 11 (Fig. 5H), cysteine at residue 5 (Fig. 5I), and phosphate at residue 9 (Fig. 5G). The presence of proline at residues 3, 10,

and 11 indicates P3 contains the tryptic peptide Ala-Leu-Pro-Thr-Cys-Thr-Ser-Ala-Ser-Pro-Pro-Gln-Lys beginning at Ala-304. Inspection of the sequence reveals that p53 residues 310 and 312 are both serine; however, the pattern of phosphate release indicates that only residue 312 is phosphorylated under the conditions used for labeling. The peptide also contains two unphosphorylated threonine residues. Since sequence analysis of P3 provided evidence for only one phosphorylated peptide, we believe that the three charged species observed by electrophoretic analysis (Fig. 2) result from different modifications of this peptide. At the present time we cannot be sure whether these modifications occurred in vivo or in vitro during the procedures required for isolation of the peptide. The carboxyterminal Lys-316 of this peptide is immediately followed by the sequence Lys-317—Lys-318—Pro-391—Leu-320, etc. We suggest that P3 ends with Lys-316 because in our experience trypsin cleaves most efficiently after the first of several consecutive basic amino acids (Lewis and Anderson 1983). Furthermore, Lys-Pro bonds are relatively resistant to trypsin cleavage. We further suggest that P3b is likely to result from cleavage after Lys-317 or Lys-318 instead of Lys 316.

Component P4 is rather hydrophobic and is probably not homogeneous. We have not yet identified P4 by sequence analysis, but predictions of retention times for p53 tryptic peptides suggest that P4 may correspond to one (or both) of two large peptides, Leu-28—Arg-62 (t2) and Val-63—Lys-98 (t3) derived from the aminoterminal segment of p53. The first of these peptides (t2) from mouse p53 is not predicted to contain threonine (but the

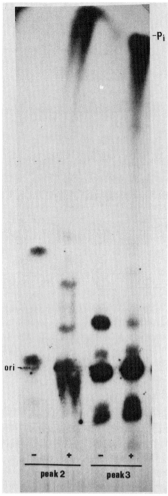

**Figure 4** Alkaline phosphatase digestion of p53 components P2 and P3. p53 tryptic peptides were isolated by reverse-phase HPLC as described in the legend to Fig. 1. Each component was divided into two portions, which were dried and redissolved in 20 mM Tris · HCl, 1 mM EDTA, pH 8. To one portion of each sample was added 1 unit of calf intestinal alkaline phosphatase (CIP). Both portions were then incubated for 1 hr at 37°C, and the products were spotted on 3MM paper for electrophoresis as described in the legend to Fig. 2. Plus and minus signs denote the CIP-treated and mock-treated samples, respectively.

equivalent peptide from human p53 does). Since phosphothreonine was found in P4, almost certainly Thr-74, Thr-76, or Thr-86 is phosphorylated. Peptide t2 contains two serines and peptide t3 contains four serines. Because of the large sizes of t2 and t3, unambiguous identification of the P4 phosphorylation sites will require redigestion with another proteinase and isolation of the resulting subdigested peptides.

## A preliminary analysis of Ad2 DBP phosphorylation sites

Recently we have begun to analyze the phosphorylation sites of several Ad2 proteins, including those of the Ad2 DBP. The phosphorylation of Ad2 DBP is intriguing because previous studies have suggested that it may be

phosphorylated at as many as 12 different sites that are predominantly and perhaps exclusively located in the aminoterminal domain of the protein. This domain is not required for binding to single-stranded DNA but may be required for other functions.

Ad2-infected HeLa cells were labeled with [³H]proline from 24 to 42 hr, and with $^{32}PO_4$ from 43 to 48 hr after infection. DBP was then isolated by immune precipitation and gel electrophoresis as described for p53 except that DBP was reduced and alkylated prior to gel electrophoresis. After trypsin digestion, the resulting peptides were separated by reverse-phase HPLC (Fig. 6). Seven phosphate-labeled peaks and 14 proline-containing peptides were observed. Each of the phosphate-containing peaks has been subjected to phosphoamino acid analysis and protein sequence analysis. Our preliminary analyses indicate that several of these peaks represent peptide mixtures. Several peptides also appear to be phosphorylated at multiple sites. The existence of peptide mixtures coupled with the large number of contiguous basic amino acids makes it difficult to definitively identify most DBP phosphorylation sites at this time. However, we can draw a few tentative conclusions.

Fraction 8 contains a relatively hydrophilic peptide mixture that has phosphothreonine as its third residue and proline as residue 4. Proline was also found at residues 1 and 2, but in lesser amounts. Inspection of the DBP sequence suggests that fraction 8 probably contains the peptide Glu-Thr-Thr-Pro-Glu-Arg beginning with Glu-10. Thus Thr-12 is probably phosphorylated whereas Thr-11 is probably not. However, to be certain of this conclusion, a second-dimension separation will be required to show that the proline at residue 4 and the phosphothreonine at residue 3 are in the same peptide.

Fraction 79 contains phosphothreonine and phosphoserine; protein sequence analysis yielded phosphate in residues 9 and 10, and a lesser amount in residue 3. The phosphate peak at fraction 79 eluted between two major proline peaks that elute at fractions 78 and 82; however, proline was observed from residues 2, 4, and 6. These data suggest that fraction 79 contains the peptide Thr-Pro-Ser-Pro-Arg-Pro-Ser-Thr-Ser-Thr-Ala-Asp-Leu-Ala-Ile-Ala-Ser-Lys beginning with Thr-68. Our data suggest that the serines at positions 3 and 9 and the threonine at position 10 in this peptide are phosphorylated; the threonines at 1 and 8 and the serine at 7 are probably not phosphorylated. Further experiments will be required to confirm these conclusions since fraction 79 probably contains at least one other peptide of unknown sequence.

Fraction 135 yielded phosphate in residues 4, 9, 11, 12, 18, and 19 and was shown to contain phosphoserine and phosphothreonine. Proline was found at residues 3, 5, 7, 10, 11, and 13; thus, this fraction contains the peptide Lys-Arg-Pro-Ser-Pro-Lys-Pro-Glu-Arg-Pro-Pro-Ser-Pro-Glu-Val-Ile-Val-Asp-Ser-Glu-Glu-Glu-Arg beginning with Lys-89. All three serines in this peptide are probably phosphorylated, but fraction 135 must also contain at least one additional phosphorylated peptide

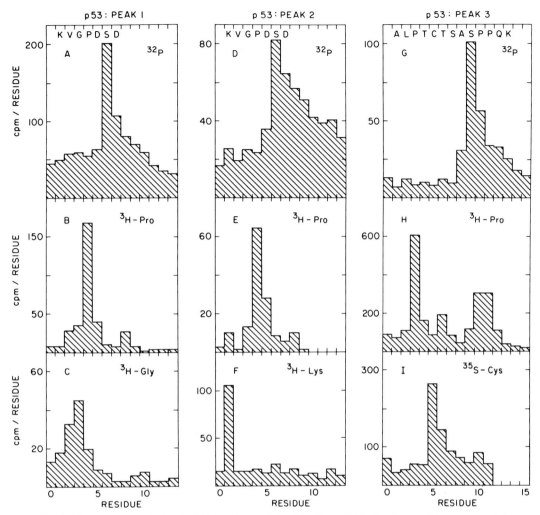

**Figure 5** Radiochemical sequence analysis of p53 phosphopeptides P1, P2, and P3. Tryptic peptides, prepared from p53 labeled with [$^{32}$P]PO$_4$ and a single tritiated amino acid, were fractionated as described in the legends to Figs. 1 and 2. Components P1-1 and P2-1 were obtained after high-voltage electrophoresis; component P3 was the HPLC component without fractionation by paper electrophoresis. The indicated peptide components were applied to a Beckman 890C protein sequencer, and the total radioactivity released after each Edman cycle has been plotted; the cycle labeled 0 was performed without addition of the coupling reagent, phenylisothiocynate. (*A*) $^{32}$P-labeled P1-1, 2200 cpm applied; (*B*) [$^3$H]proline-labeled P1-1, 1030 cpm applied; (*C*) [$^3$H]glycine-labeled P1-1, 770 cpm applied; (*D*) $^{32}$P-labeled P2-1, 1000 cpm applied; (*E*) [$^3$H]proline-labeled P2-1, 530 cpm applied; (*F*) [$^3$H]lysine-labeled P2-1, 560 cpm applied; (*G*) $^{32}$P-labeled P3, 570 cpm applied; (*H*) [$^3$H]proline-labeled P3, 3700 cpm applied; (*I*) [$^{35}$S]cysteine-labeled P3, 3000 cpm applied. The predicted sequence of each peptide is given in the top panels using the single-letter amino acid code.

to account for the release of phosphate in residues 9 and 11.

Analyses of the phosphate in peptides from HPLC fractions 76, 82, and 132 suggest these peaks probably represent yet other phosphorylated peptides rather than modifications of the three phosphopeptides tentatively identified. Thus, substantially more work will be required before all of the DBP phosphorylation sites can be identified.

## Discussion

Protein phosphorylation has been shown to activate or to inhibit the function of a variety of enzymes, but in only a few cases are the effects of protein phosphorylation well established. Not infrequently, proteins become phosphorylated at multiple sites. This is clearly the case for p53, SV40 TAg (Scheidtmann et al. 1982), and the Ad2 DBP; the Ad2 E1A proteins are probably also multiply phosphorylated (our unpublished results). Although some phosphorylations may be fortuitous, we would be greatly surprised if phosphorylation at one or more sites of p53, SV40 TAg, or the Ad2 DBP and E1A proteins did not profoundly affect the activity of these proteins.

At least four sites in the cellular tumor antigen p53 were shown to be phosphorylated. Peptide sequence analysis has unambiguously identified Ser-312 and Ser-389 as phosphorylation sites. Both of these sites are conserved in the sequence of human p53 (Harlow et al.

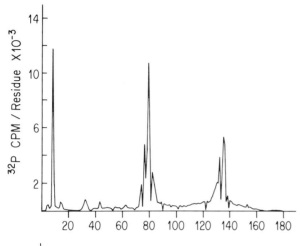

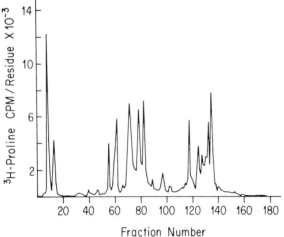

**Figure 6** Tryptic phosphopeptides and proline-containing peptides of Ad2 DBP fractionated by reverse-phase liquid chromatography. DBP was isolated from Ad2-infected HeLa cells labeled with [³H]proline and [³²P]PO₄ as described in the text and harvested 48 hr after infection. The protocol for DBP isolation and trypsin digestion was similar to that used for p53 (see Fig. 1); reverse-phase chromatography was as described in the legend to Fig. 1. Phosphate and proline-containing peptides were detected using a liquid scintillation counter after mixing an aliquot (100 μl) of each fraction with Aquasol II (New England Nuclear). Phosphate-labeled components were found in fractions 8, 76, 79, 82, 132, and 135 (*top*). The principal proline-labeled components were found in fractions 8, 13, 55, 61, 71, 78, 82, 97, 117, 124, 126, 132, and 134 (*bottom*).

1985; Zakut-Houri et al. 1985), but we have not yet examined the phosphorylation pattern of human p53. In contrast, the sequences of the aminoterminal peptides t2 and t3, which probably contain two other phosphorylation sites, are poorly conserved. None of the threonines contained in mouse t3 are conserved in the predicted sequence of t3 from human p53.

Both Ser-312 and Ser-389 appear to involve phosphodiester bonds rather than simple phosphate monoesters. We have not yet identified the moiety linked to Ser-312; however, Ser-389 appears to be coupled to ribonucleotides. All four ribonucleotides were liberated from the carboxyterminal tryptic peptide containing Ser-

389 that eluted as peak 2 when this peptide was digested with alkali or with RNase A and T2. These results suggest that Ser-389 may be covalently attached to RNA. P2 eluted as a skewed, relatively broad peak; a portion of the carboxyterminal tryptic peptide was also found in the column flowthrough. The carboxyterminal peptide thus behaved as a heterogeneously modified peptide, as might be expected if it were attached to RNA. Only very few instances of RNA covalently attached to proteins are known (Flanegan et al. 1977; Lee et al. 1977). Recently the suggestion was made that SV40 TAg may be attached to RNA (Khandjian et al. 1982). Before we can conclude that p53 is covalently attached to an RNA molecule, the released ribonucleotides must be shown to be derived from an oligoribonucleotide that was covalently attached to Ser-389.

When 3T3 cells transformed with a temperature-defective SV40 TAg gene were labeled with [³²P]PO₄ at 39.5°C, less radioactivity was incorporated at Ser-389 of p53 relative to other p53 phosphorylation sites than when these cells were labeled at 32°C. Since p53 phosphorylated at the high temperature was isolated by virtue of a continued association with TAg, and since new p53-TAg complexes do not form at this temperature (Greenspan and Carroll 1981), the labeling pattern observed may reflect phosphate turnover rates in p53-TAg complexes formed prior to the temperature shift. The lower labeling rate of Ser-389 at 39.5°C suggests that functional TAg may be required for the metabolic turnover of ribonucleotides associated with this serine. Alternatively, only newly synthesized p53 may function as a substrate for ribonucleotide addition. We have not yet examined the rate of phosphate turnover for either free p53 or p53 associated with TAg. The phosphorylation of p53 complexed with the adenovirus E1B-58K protein will also be of interest.

Our results suggest that p53 phosphorylation at Ser-389 may be important for transformation. It remains to be seen if the increased stability of p53 in transformed cells results from complex formation with viral tumor antigens, from specific phosphorylations, or from other changes in cell metabolism. Nevertheless, elucidation of the precise nature of the Ser-389 modification may provide an important clue as to the biochemical role of p53 in cell growth control.

The adenovirus DBP is a more extreme example of a multiply phosphorylated protein. Previous studies have suggested that about a dozen sites in the aminoterminal domain of DBP become phosphorylated. Our preliminary data are consistent with this estimate, and we have tentatively identified two threonine and five serine residues in the DBP aminoterminal domain that become phosphorylated. Unlike p53, most of the DBP residues that we have examined appear to be simple phosphate monoesters. In particular, the DBP phosphopeptide found in fraction 8 had the properties expected of a phosphate monoester. We note that the p53 ribonucleotide-associated carboxyterminal peptide eluted at a similar position.

The aminoterminal sequence of the DBP is unusual (Kruijer et al. 1982). It is very rich in proline and both basic (Lys, Arg) and acidic (Asp, Glu) amino acids. Two results suggest that the aminoterminal DBP domain is not involved in binding to DNA. It has been shown that the carboxyterminal 44,000-dalton chymotrypsin fragment binds efficiently to single-stranded DNA (Klein et al. 1979), and dephosphorylated DBP is reported to bind to single-stranded DNA as well as does phosphorylated DBP (Linné and Philipson 1980). The only functions thus far ascribed to the aminoterminal domain of DBP is that of participation in the extended host range of adenovirus in monkey cells (Klessig and Grodzicker 1979; Anderson et al. 1983). Ad2-infected monkey cells fail to express several late mRNAs and proteins efficiently unless they are coinfected with SV40 (i.e., the carboxyterminal domain of SV40 TAg is expressed) or residue 130 of the DBP is altered from histidine to tyrosine. The block to efficient Ad2 growth in monkey cells appears to involve both transcriptional and translational defects. If DBP influences transcription and translation in monkey cells, it is likely to do so in human cells as well. One possible explanation for the failure of late gene expression in monkey cells could be an altered phosphorylation of DBP in monkey cells. DBP phosphorylation is probably mediated by host-cell enzymes, and altering histidine to tryosine at position 130 might functionally substitute for an altered phosphorylation. Because DBP is phosphorylated at many sites, an altered DBP phosphorylation pattern might be difficult to recognize. Obviously, to evaluate these concepts, it will be necessary to both map and alter individual DBP phosphorylation sites.

## Acknowledgments

We thank A. Blum, R. Feldman, D. Hardy, and S. Lamb for excellent technical assistance, and Dr. V. Nussenzweig (NYU) for use of his HPLC facilities. This work was supported by Research Grants CA16239 and CA20802 (to R.B.C.), Training Program Grant CA09161 (to A.S.), and by the U.S. Department of Energy (to C.W.A.). R.B.C. is a Scholar of the Leukemia Society of America.

## References

Anderson, C.W. 1982. Partial sequence determination of metabolically labeled radioactive proteins and peptides. In *Genetic engineering: Principles and methods* (ed. J.K. Setlow and A. Hollaender), vol. 4, p. 147. Plenum Press, New York.

Anderson, C.W., P.R. Baum, and R.F. Gesteland. 1973. Processing of adenovirus 2-induced proteins. *J. Virol.* **12**: 241.

Anderson, C.W., M.M. Hardy, J.J. Dunn, and D.F. Klessig. 1983. Independent, spontaneous mutants of adenovirus type 2–simian virus 40 hybrid Ad2⁺ND3 that grow efficiently in monkey cells possess identical mutations in the adenovirus type 2 DNA-binding protein. *J. Virol.* **48**: 31.

Carter, T.H. and R.A. Blanton. 1978. Possible role of the 72,000 dalton DNA binding protein in regulation of adenovirus type 5 early gene expression. *J. Virol.* **25**: 664.

Eliyahu, D., A. Raz, P. Gruss, D. Givol, and M. Oren. 1984. Participation of p53 cellular tumour antigen in transformation of normal embryonic cells. *Nature* **312**: 646.

Flanegan, J., R. Petterson, V. Ampros, M. Hewlett, and D. Baltimore. 1977. Covalent linkage of a protein to a defined nucleotide sequence at the 5′-terminus of virion and replicative intermediate RNAs of poliovirus. *Proc. Natl. Acad. Sci.* **74**: 961.

Friefeld, B.R., M.D. Krevolin, and M.S. Horwitz. 1983. Effects of the adenovirus H5ts125 and H5ts107 DNA binding proteins on DNA replication *in vitro*. *Virology* **124**: 380.

Fowlkes, D.M., S. Lord, T. Linné, U. Pettersson, and L. Philipson. 1979. Interaction between the adenovirus DNA-binding protein and double-stranded DNA. *J. Mol. Biol.* **132**: 163.

Greenspan, D.S. and R.B. Carroll. 1981. Complex of simian virus 40 large tumor antigen and 48,000-dalton host tumor antigen. *Proc. Natl. Acad. Sci.* **78**: 105.

Greenspan, D.S., A.S. Blum, and R.B. Carroll. 1982. Host nuclear phosphoproteins that complex simian virus 40 large T antigen. *Adv. Viral Oncol.* **2**: 103.

Handa, H., R.E. Kingston, and P.A. Sharp. 1983. Inhibition of adenovirus early region IV transcription *in vitro* by a purified DNA binding protein. *Nature* **302**: 545.

Harlow, E., N.M. Williamson, R. Ralston, D.M. Helfman, and T.E. Adams. 1985. Molecular cloning and in vitro expression of a cDNA clone for human cellular tumor antigen p53. *Mol. Cell. Biol.* **5**: 1601.

Jenkins, J.R., K. Rudge, and G.A. Currie. 1984. Cellular immortalization by a cDNA clone encoding the transformation-associated phosphoprotein p53. *Nature* **312**: 651.

Khandjian, E., M. Loche, J.L. Darlix, R. Carmer, H. Turler, and R. Weil. 1982. Simian virus 40 large tumor antigen: A "RNA binding protein?" *Proc. Natl. Acad. Sci.* **70**: 1139.

Klein, G., ed. 1982. Transformation associated cellular p53 protein. *Adv. Viral Oncol.* **2**: 180.

Klein, H., W. Maltzman, and A.J. Levine. 1979. Structure-function relationships of the adenovirus DNA-binding protein. *J. Biol. Chem.* **254**: 11051.

Klessig, D.F. and T. Grodzicker. 1979. Mutations that allow human Ad2 and Ad5 to express late genes in monkey cells map in the viral gene encoding the 72K DNA binding protein. *Cell* **17**: 957.

Krebs, E.G. 1983. Historical prespectives on protein phosphorylation and a classification system for protein kinases. *Philos. Trans. R. Soc. Lond. B* **302**: 3.

Kruijer, W., F.M. A. van Schaik, and J.S. Sussenbach. 1982. Nucleotide sequence of the gene encoding adenovirus type 2 DNA binding protein. *Nucleic Acids Res.* **10**: 4493.

Lane, D. and L.V. Crawford. 1979. T antigen is bound to a host protein in SV40-transformed cells. *Nature* **278**: 261.

Lee, Y.K., A. Nomoto, B. Detjen, and E. Wimmer. 1977. The genome-linked protein of picornaviruses. I. A protein covalently linked to poliovirus genome. *Proc. Natl. Acad. Sci.* **74**: 59.

Lewis, J.B. and C.W. Anderson. 1983. Proteins encoded near the adenovirus late messenger RNA leader segments. *Virology* **127**: 112.

Linné, T. and L. Philipson. 1980. Further characterization of the phosphate moiety of the adenovirus type 2 DNA-binding protein. *Eur. J. Biochem.* **103**: 259.

Mayer, A.J. and H.S. Ginsberg. 1977. Persistence of type 5 adenovirus DNA in cells transformed by a temperature sensitive mutant, H5ts125. *Proc. Natl. Acad. Sci.* **74**: 785.

Nagata, K., R.A. Guggenheimer, T. Enomoto, J.H. Lichy, and J. Hurwitz. 1982. Adenovirus DNA replication *in vitro*: Identification of a host factor that stimulates synthesis of the pre-terminal protein-dCMP complex. *Proc. Natl. Acad. Sci.* **79**: 6438.

Nicolas, J.C., P. Sarnow, M. Girard, and A.J. Levine. 1983. Host range temperature conditional mutants in the adenovirus DNA binding protein are defective in the assembly of infectious virus. *Virology* **126**: 228.

Oren, M., W. Maltzman, and A.J. Levine. 1981. Post-transcriptional regulation of the 54K cellular tumor antigen in normal and transformed cells. *Mol. Cell. Biol.* **1:** 101.

Parada, L.F., H. Land, R.A. Weinberg, D. Wolf, and V. Rotter. 1984. Cooperation between gene encoding p53 tumour antigen and *ras* in cellular transformation. *Nature* **312:** 649.

Pennica, D., D.V. Goddel, J.S. Hayflick, N.C. Reich, C.W. Anderson, and A.J. Levine. 1984. The amino acid sequence of murine p53 determined from a c-DNA clone. *Virology* **134:** 477.

Reich, N.C., M. Oren, and A.J. Levine. 1983. Two distinct mechanisms regulate the levels of a cellular tumor antigen, p53. *Mol. Cell. Biol.* **3:** 2143.

Rowe, D.T., P.E. Branton, and F.L. Graham. 1984. The kinetics of synthesis of early viral proteins in KB cells infected with wild-type and transformation-defective host-range mutants of human adenovirus type 5. *J. Gen. Virol.* **65:** 585.

Samad, A., C.W. Anderson, and R.B. Carroll. 1986. Mapping of phosphomonoesters and apparent phosphodiester bonds of the oncogene product p53 from SV40-transformed 3T3 cells. *Proc. Natl. Acad. Sci.* **83:** 897.

Sarnow, P., Y. Ho, J. Williams, and A.J. Levine. 1982. The adenovirus E1b-58K tumor antigen and SV40 large tumor antigen are physically associated with the same 54K cellular protein. *Cell* **28:** 387.

Schechter, N.M., W. Davies, and C.W. Anderson. 1980. Adenovirus coded deoxyribonucleic acid binding protein. Isolation, physical properties, and effects of proteolytic digestion. *Biochemistry* **19:** 2802.

Scheidtmann, K.-H., B. Echle, and G. Walter. 1982. Simian virus 40 large T antigen is phosphorylated at multiple sites clustered in two separate regions. *J. Virol.* **44:** 116.

van der Vliet, P.C. and A.J. Levine. 1973. DNA binding proteins specific for cells infected by adenovirus. *Nat. New Biol.* **246:** 170.

van der Vliet, P.C. and J. Sussenbach. 1975. An adenovirus type 5 gene function required for initiation of viral DNA replication. *Virology* **67:** 415.

Zakut-Houri, R., B. Bienz-Tadmor, D. Givol, and M. Oren. 1985. Human p53 cellular tumor antigen: cDNA sequence and expression in COS cells. *EMBO J.* **4:** 1251.

# The Role of Small T Antigen in the Simian Virus 40 Transforming Mechanism

I. Bikel,* H. Mamon,* E.L. Brown,[†] J. Boltax,* M. Agha,* and D.M. Livingston*

*The Dana-Farber Cancer Institute and the Departments of Medicine, Brigham and Women's Hospital and The Harvard Medical School, Boston, Massachusetts 02115; [†]Genetics Institute, Cambridge, Massachusetts 02140

SV40 small T antigen is a protein of 174 amino acids that has long been known to assist SV40 large T antigen in the transformation of rodent cells. Mutations Cys-111 → Trp and Cys-111 → STOP, introduced into the small T–unique segment of the protein by site-directed mutagenesis, have led to the production of small T species defective in their ability to cooperate with large T in the acute transformation of mouse cells. These mutant proteins are also defective in promoting actin cable dissolution in rat cells. Therefore, native structure of the small T–unique region is essential to its transforming activity. Moreover, there is a positive correlation between the actin cable perturbation and the transforming function(s) of this oncogenic product of papovavirus.

The simian virus 40 (SV40) early region is a 2500-nucleotide (nt) segment governed by two overlapping early promoters. Its primary transcript is processed to two spliced messengers. The products of these RNAs are large T antigen and small T antigen. The aminoterminal 82 residues of these proteins are identical, whereas the carboxyterminal 92 residues of small T and 626 residues of large T are unique. In tissue-culture cells, only large T is required for effective viral replication, although there is evidence that, in the absence of small T antigen, viral bursts are lower than normal (Shenk et al. 1976). Since virtually all of the papovaviruses of the nonpapilloma type encode a small T antigen that is significantly homologous with that of SV40, it is likely that this protein has functions that were present in the common progenitor of these various small T species. Moreover, in view of the fact that it has been preserved over millions of years of viral evolution, it has presumably played an essential, albeit presently unknown, role in natural viral propagation.

There is abundant evidence suggesting that SV40 can transform cells in culture and generate tumors in certain newborn animals in the absence of functional small T antigen (Lewis and Martin 1979; Martin et al. 1979). By contrast, there are also data which indicate that, in some cells cultivated under particular conditions, small T is either absolutely required for the expression of the SV40-transformed phenotype or can facilitate its appearance (Sleigh et al. 1978; Fluck and Benjamin 1979; Rubin et al. 1982; Sompayrac and Danna 1983). The fact that large T can serve a transforming role in the absence of small T has led to the belief that some cells synthesize one or more proteins that can do what small T does.

One example of a cell line that requires both small and large T antigens for the immediate expression of an SV40-transformed phenotype after virus infection is BALB/c-3T3 Cl A31. Soon after SV40 infection of these cells at high multiplicity, one observes their relatively efficient growth in semisolid medium for several replication rounds (Rubin et al. 1982). By contrast, comparable infection by an SV40 mutant that encodes large T but not small T, or vice versa, led to little or no acute transforming response. These findings have been shown to constitute the basis for an acceptable assay of the transforming function of small T, which, alone, is unable to transform such cells.

Until now, there has been little or no understanding of which residues or segments of the small T primary sequence are required for its transforming function(s). In those cases where deletion mutants of small T have been studied, the rate of synthesis and/or the stability of the truncated products have been sufficiently low that it has been difficult to ascribe functional significance to the absent protein sequences. Given this gap in knowledge, we have mutated selected regions of the small-T-protein coding unit by site-directed mutagenesis and then analyzed the effects of these alterations. The long-term goal of this work is to learn which elements of protein structure are critical to each of the antigen's functions and, thereby, to increase one's appreciation of how certain in vitro functions of the protein contribute to its activity in vivo.

## Materials and Methods

### Cells and viruses

Cells were grown on plastic surfaces in Dulbecco's modified Eagle's medium at 37°C in a 10% $CO_2$-containing atmosphere. SV402, a large T$^-$/small T$^+$ (T$^-$/t$^+$) derivative of SV40, was routinely grown in COS-1 cells (Rubin et al. 1982). SVt-$Trp_{111}$ and SVt-$STOP_{111}$ are derivatives of SV402 containing a TGG and TGA mutation, respectively, in the triplet encoding Cys-111. These viruses were also propagated in COS-1 cells, and all viruses were titered by VP$_1$ induction, as measured in a specific immunofluorescence assay (Rubin et al. 1982). Plaque-forming assays were performed in CV-1P cells. BALB/c-3T3 Cl A31 and Rat-1 cells are immortal rat and mouse fibroblast lines that are otherwise untransformed. REF cells are secondary Fisher rat embryo fibroblasts.

### Plasmids

Established techniques for isolating plasmid DNA, cleavage with restriction enzymes, and molecular cloning were employed in these studies (Maniatis et al. 1982).

pMEAt is a partially deleted derivative of pHR402, a plasmid that encodes intact small T but not large T. It also encodes an ~25-kD truncated large T species (Rubin et al. 1982). Unlike pHR402, pMEAt lacks all sequences from the large T unique coding region and cannot encode the 25-kD large T species. Nevertheless, it contains an intact small T cistron, early promoter, and polyadenylation sequence. In its construction, all residual large T coding sequences of pHR402 between nt 4568 and 2666 were deleted. In CV-1P cells, it was responsible for the synthesis of full-size SV40 small T but no detectable intact or truncated large T (I. Bikel and M.E. Agha, unpubl.).

### Immunofluorescent and anti−large T/small T (T/t) immunoprecipitation analyses

These were carried out by established methods (Rubin et al. 1982). Rabbit anti-SDS gel band−purified SV40 small T antibody was used in these experiments. This antibody reacts with both small and large T antigens in specific immunological assays and does not appear to contain any antibody that reacts uniquely with small T (Bikel et al. 1983).

### Abortive transformation tests

These tests were performed by a published method (Rubin et al. 1982).

### Microinjection of various cloned DNAs

Each DNA to be microinjected was dissolved in 0.01 M $NaPO_4$ (pH 7.4), 0.14 M NaCl at 100 µg/ml. Aliquots of 20–40 pl were injected into each target cell as described previously (Rubin et al. 1982).

### Actin cable analyses

Actin cable analyses were performed with Rhodamine-conjugated phalloidin as probe by a previously described method (Pollack et al. 1975). The phalloidin derivative was the generous gift of S. Chen and R. Pollack.

### Introduction of specific mutations into the SV40 small T coding unit

pHR402-STOP$_{111}$ (pt-STOP$_{111}$) and pHR402-Trp$_{111}$ (pt-Trp$_{111}$) are identical with pHR402 except for the presence of a TGC → TGA and a TGC → TGG change in small T codon 11, respectively. They were constructed by the method of Oostra et al. (1983). Specifically,

an ~1:1 mixture of a 19-member oligodeoxyribonucleotide, synthesized in an Applied Biosystems 380A instrument and representing the antisense strand of the SV40 sequence

$$5' \; d(GCAAGCATAT^T_C CAGTTAGC) \; 3'$$

was annealed to its complementary sequence in the 412-nt single-stranded gap of a circular heteroduplex structure, as described in Figure 1B. The above-noted gap extended from the HindIII site at nt 5171 to the BstX site at nt 4759. A 314-bp fragment extending from nt 5176 (see legend to Fig. 1) to the HaeIII site at nt 4862 was also annealed to this gap. The target nucleotide for mutagenesis (G$_{4827}$ in the antisense strand) was the complement of the 3′ base (i.e., C) of the SV40 E-strand triplet, TGC, encoding small T Cys-111. It is also a component of an NdeI recognition site (CATATG), as shown in Figure 1A. After annealing the 5′ phosphorylated synthetic oligodeoxyribonucleotide mixture and the adjoining restriction fragment, as noted above, the single-stranded regions abutting these fragments were filled by incubation of the heteroduplex with the Klenow fragment of DNA polymerase I and four deoxynucleoside triphosphates, as described (Oostra et al. 1983). After ligation of the residual strand interruption with T4 DNA ligase, Escherichia coli HB101 cells were transformed with the DNA product. Colonies were screened for the presence of the predicted TGC → TGG and TGC → TGA mutations by parallel hybridization of replicate, colony-imprinted nitrocellulose filters with two 5′-end-labeled ($^{32}$P), 13-member oligodeoxyribonucleotide probes—5′ d(CATAT*T*CAGTTAG) 3′, $T_m = 34°C$ (for the STOP mutant; Fig. 2, left filter), and 5′ d(CATAT*CC*AGTTAG) 3′, $T_m = 36°C$ (for the Trp mutant; Fig. 2, right filter) (spec. act. of each ≈ 10$^7$ cpm/µg)—each bearing one of the above-noted third-base substitutions. Hybridization of groups of four filters was performed at 29°C in 8 ml of 0.1% Ficoll, 0.1% polyvinylpyrollidone, 0.1% BSA, 10% Dextran sulfate, 0.4 M NaCl, 0.09 M Tris·HCl (pH 7.5), 6 mM EDTA, 0.5% SDS. The filters were rinsed three times in 25 ml of 4× SSC (SSC = 0.01 M sodium citrate, 0.14 M NaCl) at room temperature. These rinses were followed by progressive 4-hr rinses at increasing temperatures in order to take advantage of the relative thermal stability of perfect oligonucleotide:single-strand DNA duplexes compared with the relative instability of duplexes containing single-base-pair mismatches (Wallace et al. 1979). All rinses were performed in a circulating water bath containing 3.5 liters of 4 × SSC. After drying, each filter was exposed to X-ray film in the presence of an intensifying screen. Colonies hybridizing to one but not the other probe were picked (cf. Fig. 2) and replated for repeat analysis. Plasmid DNA from a suspected clonal member of each mutant population was isolated, and the presence of the suspected mutation was tested first by assessing the sensitivity of the relevant region to digestion with NdeI, followed by direct DNA sequencing of the region from the BstX site at nt 4759 to nt 4840 by the method of Maxam and Gilbert (1980).

## (A) Nucleotide Sequence of the Site

5′-GCAAGCATAT $_{C}^{T}$ CAGTTAGC-3′ 19-mer oligo mix (approx.1:1 T:C)
3′-CGTTCGTATAC GTCAATCG-5′ SV40 Early strand

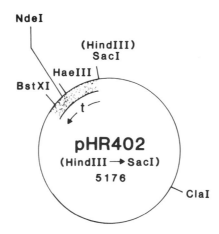

## (B) DNA Fragments Used for Obtaining Gapped Heteroduplex Molecules for Mutant Construction

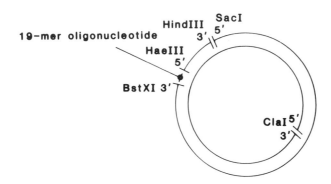

**Figure 1** Site-directed mutagenesis of the C · G base pair at the *Nde*I site (nt 4827) of the pHR402 small-T-antigen gene. Two μg of the plasmid pHR402 (Rubin et al. 1982) were partially digested with *Hin*dIII and the products were separated by electrophoresis in a 1% agarose gel. Full-length linear DNA was isolated from this gel, and its overhanging 5′ ends were filled by incubation with 1 unit of the Klenow fragment of DNA polymerase I and 2.5 mM of each of the four standard deoxyribonucleoside triphosphates in a 20-μl reaction mixture. A 5′ phosphorylated *Sac*I linker was then introduced at this site by incubation with 20 units of T4 DNA ligase. The DNA product was used to transform *E. coli* HB101, and colonies bearing plasmids containing a *Sac*I rather than a *Hin*dIII site at nt 5171 (denoted 5176 in this figure) were identified. The *Sac*I linker–containing derivative was denoted pHR402S. Five μg of pHR402S were: (1) digested with *Cla*I and the resulting ends were dephosphorylated with calf intestine alkaline phosphatase, phenol-extracted, and ethanol-precipitated (see Materials and Methods); (2) digested with *Sac*I (at nt 5176) and *Bst*XI (at nt 4759) and the resulting 4826-bp fragment was isolated; and (3) digested with *Sac*I (at nt 5176) and *Hae*III (at nt 4862) and the resulting 314-bp fragment was isolated. The fragments isolated from these digestions were mixed, denatured, and allowed to reanneal, as described in Materials and Methods. This protocol was also followed to anneal the 1:1 mixture of synthetic oligodeoxyribonucleotides shown in the figure to the resulting gapped heteroduplex and to close the remaining gaps with the DNA polymerase I Klenow fragment and T4 DNA ligase. Mutagenized plasmids were used to transform *E. coli* HB101, and possible mutants were identified by parallel screening with end-labeled oligonucleotides of known sequence, as described in Materials and Methods (see also Fig. 2). (Reprinted with permission, from Bikel et al. 1986.)

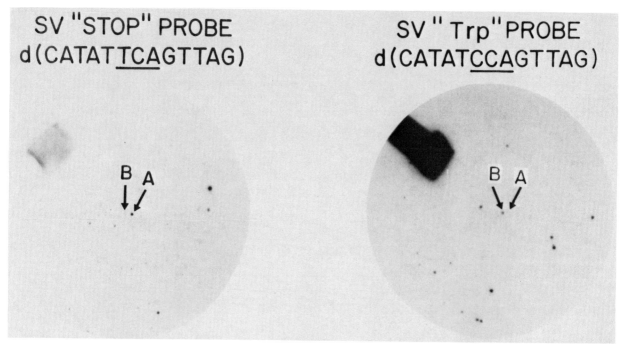

**Figure 2** Mutant colony hybridization analysis. Shown are autoradiograms of replicate nitrocellulose filters containing the same, initial *E. coli* HB101 ampicillin-resistant colonies resulting from transformation with in vitro–mutagenized pHR402. The filters were hybridized with the indicated probes, as described in Materials and Methods. When analyzed further, the clones labeled A and B proved to contain the predicted mutations—pt*STOP*$_{111}$ and pt-*Trp*$_{111}$. (Reprinted, with permission, from Bikel et al. 1986.)

## Results

### Isolation of mutant small T genomes with substitutions at codon 111

Two defined-sequence oligodeoxyribonucleotides, one leading to the introduction of a G and the other of an A in the third position of the TGC triplet encoding small-T-antigen Cys-111, were annealed to the relevant single-stranded region of a wild-type copy of the small T gene and ligated into a suitable plasmid, and the appropriate mutant derivatives were isolated by an established technique (Wallace et al. 1979; Oostra et al. 1983; Fig. 1). The genome containing the TGC-111 → TGG-111 substitution was denoted pt-*Trp*$_{111}$ and that of the TGC-111 → TGA-111 was denoted pt-*STOP*$_{111}$. The mutation in each case was identified by direct DNA sequencing of the relevant region in the small T cistron.

Construction of these mutants was performed on a pHR402 plasmid template (Rubin et al. 1982). The SV40 early region contained in this plasmid bears an intact small T cistron but has sustained an ~ 1.5-kb deletion of large T–unique coding sequences. Indeed, it retains only 380 nt of the second exon of the large T gene. Hence, it cannot encode intact large T. However, under some conditions, cells bearing this plasmid can synthesize small amounts of an extremely unstable, ~ 25-kD truncated species of large T (X. Montano et al., unpubl.). pHR402 also contains a complete viral late region.

The viral sequences contained in both mutant plasmids and in pHR402 were excised from the backbone

plasmid and individually transfected onto COS-1 cells. After serial passage of lysates of the original transfected cultures, infectious virus stocks appeared as reflected by the fact that lysates of infected COS-1 cells contained an agent that was capable of causing a cytopathic effect on subsequent cultures of COS-1 cells, led to SV40 VP$_1$ synthesis in these cells, and led to SV40 small T synthesis in infected CV-1P and BALB/c-3T3 Cl A31 cells. Titration of these viruses was performed by immunofluorescent assay for VP$_1$ synthesis after infection of COS-1 cells at serial dilution, as described previously (Rubin et al. 1982).

### Detection of mutant SV40 small T gene products in infected and transfected cells

Equal amounts of SV402, SVt-*Trp*$_{111}$ and SVt-*STOP*$_{111}$ were applied to identically sized cultures of CV-1P cells. One day later, each culture was labeled with [$^{35}$S]methionine. After extraction, each cell lysate was immunoprecipitated with monospecific rabbit anti–small T serum (Bikel et al. 1983). Aliquots of cultures of the same cells microinjected with each of the two mutant plasmids (pt-*Trp*$_{111}$ and pt-*STOP*$_{111}$) and suitable positive and negative controls were also analyzed for small T cell location by specific immunofluorescence, using the above noted serum. The results of the former experiments are shown in Figure 3. The data indicate that wild-type and both mutant-infected cultures produce comparable amounts of readily identified, anti–small T reactive polypeptide. Specifically, SVt-*Trp*$_{111}$ and

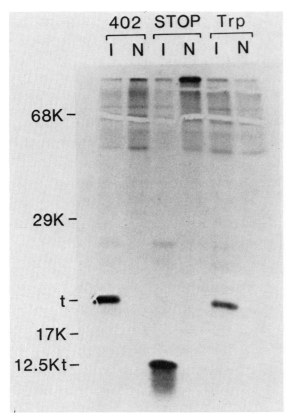

**Figure 3** Immunoprecipitation of [$^{35}$S]methionine-labeled CV-1P cell extracts 24 hr after infection with SV402, SVt-STOP$_{111}$, and SVt-Trp$_{111}$. Parallel cultures of subconfluent CV-1P cells (~75% confluent, 100-mm dishes) were infected by each of the above-noted viruses at a multiplicity of ~10. Twenty-four hr later, each culture was labeled with [$^{35}$S]methionine (100 μCi/ml) for 4 hr. After lysis of each, anti–small T immunoprecipitation was performed, as described in Materials and Methods. SDS gel electrophoresis was performed in a 15% polyacrylamide gel, and a fluorographic exposure (3 days) of the dried and stained gel is shown. Half of each extract (700 μl) was reacted with rabbit antiserum raised against gel-band-purified small T antigen (lanes *I*), and the other half was reacted with preimmune rabbit serum (lanes *N*). (Reprinted, with permission, from Bikel et al. 1986.)

SV402 infection resulted in the synthesis of comigrating 20-kD small T bands, as expected. By contrast, SVt-STOP$_{111}$ infection resulted in the appearance of a ~12-kD immunoreactive band, which corresponds to the predicted size of the truncated small T molecule (cf. Fig. 1). The relative yields of wild-type and mutant small T proteins in a series of steady-state labeling experiments, such as those described in Figure 3, were generally indistinguishable with, at most, twofold to threefold variations in the abundance of wild-type and the two mutant proteins in any given experiment. Furthermore, microinjection of plasmids encoding wild-type and each of the two mutant small T antigens resulted in typical nuclear and cytoplasmic fluorescence, as demonstrated in Figure 4. By contrast, large T fluorescence was found to be exclusively nuclear in the same cell line. No signal was observed in suitable negative controls. Therefore, both mutant genomes encode stable, mutant small T proteins that are, grossly, distributed in the same manner in an SV40-permissive monkey cell line as the wild-type molecule. Importantly, the results shown in Figure 4 include an immunofluorescent staining pattern obtained in CV-1P cells microinjected with a new, small-T-only plasmid, pMEAt. The pattern of staining observed in these microinjected cells was identical with that observed after SV402 infection of the same cells. Importantly, no ~20-kD truncated protein was synthesized in such cells. Given the fact that the immunofluorescent staining patterns were identical in these two cultures and in cells containing pt-Trp$_{111}$ and pt-STOP$_{111}$, it appears that the intracellular distribution of the two mutant small T species is, by this relatively superficial method of analysis, indistinguishable from that of wild-type small T.

## Comparison of the biological properties of mutant and wild-type small T species

The effect of the wild-type and each mutant small T antigen on the stability of the actin cable cytoskeleton of a continuous, but otherwise untransformed, rat fibroblast line, Rat-1, and of primary REF cells was tested

**Table 1** Effect on Rat-1 and REF Actin Cable Structures of Wild-type and Mutant Small T Antigens

| DNA injected | Large T/ small T (T/t) | Percent cells with | | | |
| --- | --- | --- | --- | --- | --- |
| | | intact cables | | dissolved cables | |
| | | Rat-1 | REF | Rat-1 | REF |
| pHR401 | T, t | 39 | ND | 61 | ND |
| pHR402 | t | 30 | 19 | 70 | 81 |
| pt-STOP$_{111}$ | t-STOP$_{111}$ | 83 | 81 | 17 | 19 |
| pt-Trp$_{111}$ | t-Trp$_{111}$ | 86 | 89 | 14 | 11 |
| pBR322 | — | 88 | 77 | 12 | 23 |

Growing Rat-1 cells or REFs were injected with the various cloned DNAs, and the cells were incubated at 37°C for 24 hr and then stained for actin cables, as described in Materials and Methods. (ND) Not determined. (Reprinted, with permission, from Bikel et al. 1986.)

after microinjection of the relevant, parental plasmids. As noted previously (Graessmann et al. 1980; Rubin et al. 1982), significant perturbation of the actin cable architecture followed microinjection of the wild-type plasmid pHR402 (Table 1). However, under the conditions employed, neither mutant plasmid was capable of disturbing the actin cytoskeleton in either cell line. Furthermore, when the abortive transforming activity of the wild-type and two mutant viruses was tested

(Table 2), replicate cultures of BALB/c-3T3 Cl A31 cells formed agar microcolonies with much higher efficiency after mixed infection by $dl$883 ($T^+/t^-$) and SV402 ($T^-t/t^+$) than by mixtures of $dl$883 and either mutant virus. The need for mixed infection by $dl$883 and SV402 for A31 microcolony formation has been noted previously (Rubin et al. 1982). These results indicate that both small T mutants are defective in their ability to promote actin cable dissolution of two species of cul-

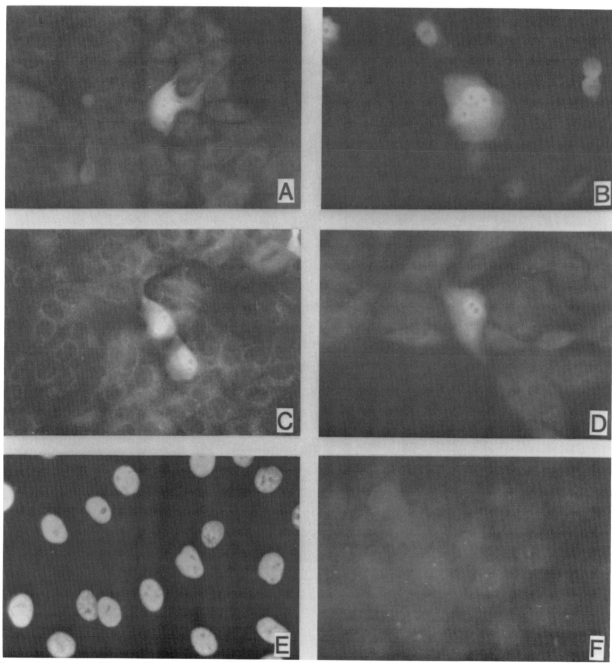

**Figure 4** Anti–small T immunofluorescence of CV-1P cells 48 hr after microinjection or infection with the following plasmids or viruses: pMEAt injection (*A*); SV402 infection (*B*); pt-*STOP*$_{111}$ injection (*C*); pt-*Trp*$_{111}$ injection (*D*); $dl$883 infection (*E*). In *F*, the cells were mock-infected with an extract of COS-1 cells. (Reprinted, with permission, from Bikel et al. 1986.)

**Table 2** Abortive Transformation of Large T–containing BALB/c-3T3/Cl A31 Cells by Wild-type and Mutant Small T Antigens

| Infect with | Percent microcolonies |
|---|---|
| — | <0.5 |
| SV40 | 12 |
| dl883 | 1.8 |
| SV402 | <0.5 |
| SVt-$Trp_{111}$ | <0.5 |
| SVt-$STOP_{111}$ | <0.5 |
| dl883+SV402 | 10.8 |
| dl883+t-$STOP_{111}$ | 2.6 |
| dl883+t-$Trp_{111}$ | 2.5 |

Growing cells ($\sim1.3\times10^5$ per 35-mm dish) were infected with the various viruses. Eight, serial 45-min absorption periods at 37°C with 1.5-ml aliquots of the indicated virus stock(s) were employed. After a 24-hr incubation at 37°C, the cells were trypsinized, seeded in agar, and incubated at 37°C for up to 2 weeks. Similar infection multiplicities (200 per absorption) were used in all single virus infections. In coinfections (with dl883), the same amount of each virus tested in the singly infected cultures was employed. (Reprinted, with permission, from Bikel et al. 1986.)

tured rat fibroblasts as well as the abortive transformation of a large T–containing, immortal mouse cell line.

## Discussion

The cysteine codon (Cys-111) that serves as the amino-terminal boundary of a six-residue canonical sequence (Cys-X-Cys-XX-Cys), which is repeated twice in the small T gene of multiple papovaviruses, has been mutagenized in two ways, and the effects of mutagenesis on the behavior of the small T gene product of each mutant cistron have been studied. The above-noted canonical sequence is widely represented in the genomes encoding another growth-promoting protein family—the glycopeptide tropic hormones (Friedmann et al. 1978). Cys-111 also lies within the unique segment of the small T gene, a region already suspected of contributing to the transformation maintenance function of this protein (Sleigh et al. 1978). Although both mutant genomes encode stable polypeptides that could be readily detected in immunofluorescence and immunoprecipitation experiments, they also direct the synthesis of small T antigens that are defective in their ability to promote two small-T-specific functions—the dissolution of the actin cable cytoskeleton in rodent cells and the complementation of SV40 large T function in the abortive transformation of BALB/c-3T3 and A31 cells. Thus, more than the aminoterminal 111 residues are required for both functions. Moreover, the dual defectiveness of both mutants opens the possibility that an intact small T–unique region contributes significantly to these two biological functions, perhaps even operating as a semiindependent domain, given the segmental molecular architecture of the viral early region. The data further

suggest that the cytoskeletal perturbation effect of small T antigen is linked to its function in transformation maintenance. The absence of actin cable–perturbing activity and abortive transforming activity of both mutant genomes is consistent with this hypothesis. Nevertheless, the results do not show that the actin cable effect is directly linked to the biochemical change that underlies expression of a neoplastic phenotype. Rather, one could argue equally well that the perturbation of the actin cytoskeleton is an effect as opposed to a cause of the development of transforming behavior and is, therefore, a post facto marker of this change.

Given the results described here, it will be interesting to assess the individual roles of all six of the cysteines of the twice-repeated motif, noted above. One obvious question is whether these residues are disulfide-bonded in parallel, thereby creating a stable, 33-amino-acid segment that functions as a minidomain. One prediction of such a model is that the 21-residue intervening segment might have special functional properties that would be perturbed equally well by destroying its disulfide-bonded base as by altering its intrinsic sequence. Experiments designed to test this hypothesis have been initiated.

## References

Bikel, I., H. Mamon, E.L. Brown, J. Boltax, M. Agha, and D.L. Livingston. 1986. The t-unique coding domain is important to the transformation maintenance function of the simian virus 40 small t antigen. *Mol. Cell. Biol.* (in press).

Bikel, I., T.M. Roberts, M.T. Bladon, R. Green, E. Amann, and D.M. Livingston. 1983. Purification of biologically active simian virus 40 small tumor antigen. *Proc. Natl. Acad. Sci.* **80:** 906.

Fluck, M.M. and T.L. Benjamin. 1979. Comparisons of two early gene functions essential for transformation in polyomavirus and SV40. *Virology* **96:** 205.

Friedmann, T., R.F. Doolittle, and G. Walter. 1978. Amino acid sequence homology between polyoma and SV40 tumor antigens deduced from nucleotide sequences. *Nature* **274:** 291.

Graessmann, A., M. Graessmann, R. Tjian, and W.C. Topp. 1980. Simian virus 40 small t protein is required for loss of actin cable networks in rat cells. *J. Virol.* **33:** 1182.

Lewis, A.M., Jr. and R.G. Martin. 1979. Oncogenicity of simian virus 40 deletion mutants that induce altered 17-kilodalton t-proteins. *Proc. Natl. Acad. Sci.* **76:** 4299.

Maniatis, T., E.F. Fritsch, and J. Sambrook. 1982. *Molecular cloning: A laboratory manual.* Cold Spring Harbor Laboratory, Cold Spring Harbor, New York.

Martin, R.G., V.P. Setlow, C.A.F. Edwards, and D. Vembu. 1979. The roles of the simian virus 40 tumor antigens in transformation of Chinese hamster lung cells. *Cell* **17:** 635.

Maxam, A. and W. Gilbert. 1980. Sequencing end-labeled DNA with base-specific chemical cleavage. *Methods Enzymol.* **65:** 499.

Oostra, B.A., R. Harvey, B.K. Ely, A.F. Markham, and A.E. Smith. 1983. Transforming activity of polyomavirus middle T antigen probed by site-directed mutagenesis. *Nature* **304:** 454.

Pollack, R., M. Osborn, and K. Weber. 1975. Patterns of organization of actin and myosin in normal and transformed cultured cells. *Proc. Natl. Acad. Sci.* **72:** 994.

Rubin, H., J. Figge, M.T. Bladon, L.B. Chen, M. Ellman, I. Bikel, M.P. Farrell, and D.M. Livingston. 1982. Role of small t antigen in the acute transforming activity of SV40. *Cell* **30:** 469.

Shenk, T.E., J. Carbon, and P. Berg. 1976. Construction and analysis of viable deletion mutants of simian virus. *J. Virol.* **18:** 664.

Sleigh, M.J., W.C. Topp, R. Hanich, and J.F. Sambrook. 1978. Mutants of SV40 with an altered small t protein are reduced in their ability to transform cells. *Cell* **14:** 79.

Sompayrac, L. and K.J. Danna. 1983. A simian virus 40 dl884/tsA58 double mutant is temperature sensitive for abortive transformation. *J. Virol.* **46:** 620.

Wallace, R.B., J. Shaffer, R.F. Murphy, J. Bonner, T. Hirose, and K. Itakura. 1979. Hybridization of synthetic oligodeoxyribonucleotides to φX174 DNA: The effect of single base pair mismatch. *Nucleic Acids Res.* **6:** 3543.

# Growth-stimulating Activity of Polyomavirus Small T Antigen

Y. Ito, T. Noda, M. Satake, T. Robins, and M. Gonzatti-Haces

National Cancer Institute–Frederick Cancer Research Facility, Litton Bionetics, Inc.–Basic Research Program, Frederick, Maryland 21701

Using a murine sarcoma virus shuttle vector, NIH-3T3 cells that express only the small T antigen of polyomavirus were isolated and characterized. Immunofluorescent staining of these cells with monoclonal antibodies directed against the aminoterminal region of the T antigens showed that the small T antigen is localized in the cell nucleus. These cells, expressing only the small T antigen gene, displayed a normal microfilament bundle pattern. Comparison of the growth characteristics of parental NIH-3T3 cells and cells that express the small T antigen gene revealed a mitogenic activity that is directly or indirectly associated with the expression of polyoma small T antigen. In addition, small T antigen was found to cooperate with middle T antigen in cell transformation by promoting cell growth, via an unidentified middle T antigen–independent mechanism.

The process of carcinogenesis seems to require multistep genetic alterations, and there is increasing evidence that oncogenes are responsible for at least some of these alterations. Consistent with this idea, it has been observed that at least two distinctly different oncogenes are required for the expression of the phenotype of cancer cells (Land et al. 1983; Ruley 1983). By introducing a pair of oncogenes into primary rat embryo fibroblasts (REFs), one can observe whether they cooperate in cell transformation. Polyomavirus is unique in that it has three distinct transforming genes and that all three genes appear to be required for full expression of the transformed phenotype in REFs (Cuzin 1984). Two of the genes, for large and middle T antigen, have been extensively characterized. These two genes have been shown to cooperate with either heterologous viral and/or cellular oncogenes in the REF transformation assay (Land et al. 1983; Ruley 1983).

Large T antigen is a DNA-binding protein and is primarily present in nuclei. In the lytic cycle of the virus, it is required for viral DNA replication and for the regulation of RNA transcription, but in transformation, only the aminoterminal 40% of large T antigen is required. Large T antigen has been shown to promote the establishment of permanent cell lines from primary cell cultures (Rassoulzadegan et al. 1983). Middle T antigen is associated with membranes (Ito 1979) and is primarily responsible for inducing the transformed phenotype (Ito et al. 1980). A portion of the middle T antigen present in transformed cells is associated with an activity that phosphorylates middle T antigen on tyrosine residues. It has been suggested that this tyrosine kinase activity is due to $pp60^{c\text{-}src}$, a cellular tyrosine kinase (Courtneidge and Smith 1983). There is a strong correlation between the level of this $pp60^{c\text{-}src}$–middle T antigen function and the extent of expression of the transformed phenotype (Smith et al. 1979). In addition, this tyrosine kinase activity is enhanced when extracellular levels of epidermal growth factor are increased (Segawa and Ito 1983).

Less is known about the third polyomavirus transforming gene, the small T antigen gene. The entire small T antigen polypeptide chain, except for the four carboxy-terminal amino acids, is included in the aminoterminal portion of middle T antigen (Fig. 1), raising the question of whether there is a specific function for small T antigen that is not shared by middle T antigen. We have, therefore, isolated NIH-3T3 cells expressing only small T antigen. This has made it possible to study the subcellular localization of small T antigen, its effect on microfilaments, the effect of small T antigen on growth characteristics of cells, and the cooperation of the small T antigen gene with other transforming genes. In this communication we will describe some of the properties of small T antigen–expressing cells so far examined and will discuss the possibility that small T antigen represents the third complementation group in cell transformation.

## Materials and Methods

### Construction of a retrovirus vector for expression of the small T antigen gene

The BamHI/EcoRI fragment of small T antigen gene of pST-1 (Treisman et al. 1981) was inserted into the murine sarcoma virus shuttle vector pGV16, constructed by T. Robins. Details of the construction of this small T antigen expression vector, pGVST, will be described elsewhere.

### Helper-free defective murine retrovirus

Mouse cells containing murine retrovirus lacking a packaging sequence and expressing all the viral proteins necessary for virus assembly, $\psi2$ (Mann et al. 1983), were transfected with pGVST. Populations of G418-resistant cells were obtained that express small T antigen. Culture supernatant fluids of these $\psi2$-pGVST cells containing defective virus were collected and used

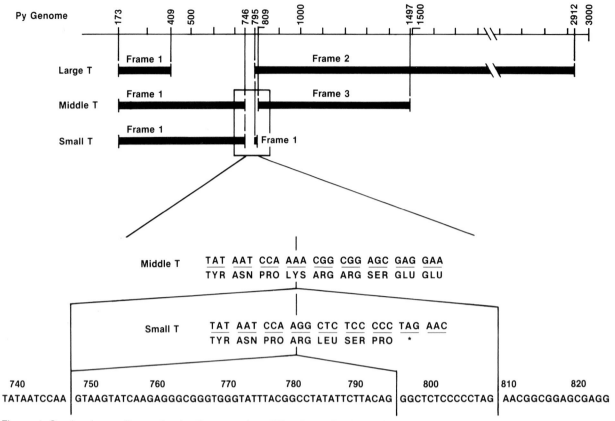

**Figure 1** Overlapping coding region for three species of T antigen of polyomavirus.

to infect NIH-3T3 cells. This virus stock was found to contain about $10^4$ infectious units/ml and was able to transfer G418 resistance to the recipient cells.

*Immunoprecipitation of [$^{35}$S]methionine-labeled small T antigen*

Labeling of cells with [$^{35}$S]methionine, extraction of small T antigen, and immunoprecipitation of T antigen using rat anti-tumor serum were all carried out as described previously (Ito 1979).

## Results

### Isolation of cells expressing small T antigen of polyomavirus

The small T antigen gene was introduced into NIH-3T3 cells using helper-free defective retrovirus containing small T antigen and neomycin genes as described in Materials and Methods. About half of the G418-resistant cell clones isolated express small T antigen. Figure 2 shows immunoprecipitation of small T antigen from one of those cell clones.

### Subcellular localization of small T antigen of polyomavirus

In past cell-fractionation experiments, we recovered small T antigen from the soluble cytoplasmic fraction

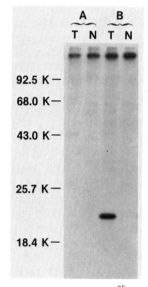

**Figure 2** Immunoprecipitation of [$^{35}$S]methionine-labeled small T antigen of polyomavirus using rat anti-tumor serum (T) or control serum (N). (Lanes *A*) NIH-3T3 cells; (lanes *B*) NIH-3T3 cells infected with helper-free defective retrovirus containing the small T antigen gene.

(Segawa and Ito 1982). In contrast, immunofluorescent staining of the G418-selected, small T antigen–positive clones with two independent monoclonal antibodies that recognize the aminoterminal region of T antigens (Dilworth and Griffin 1982) demonstrated that small T antigen was localized in the nuclei. The staining was observed mainly in nucleoplasm, not in nucleolei, and there was no detectable staining in the cytoplasm or cell surface (Fig. 3). Thus, this staining pattern is indistinguishable from that of large T antigen. A possible explanation for the recovery of small T antigen in the soluble cytoplasmic fraction in cell fractionation experiments is that small T antigen may be soluble in the nucleus and, therefore, leaks out during cell fractionation.

## Effect of each of the three T antigens of polyomavirus on microfilament bundles

Cells transformed by polyomavirus have reduced microfilament bundles. For the first time, we are in a position to ask which of the three T antigens is responsible for this phenomenon, by using a full set of cells uniquely expressing each of the three T antigens. As shown in Figure 3B, cells expressing middle T antigen (Fig. 3B, b) are the only cells that have an altered

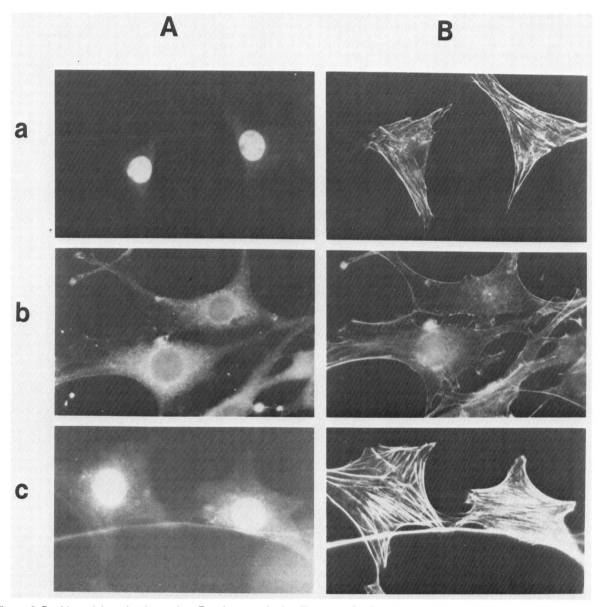

**Figure 3** Double staining of polyomavirus T antigens and microfilaments of cells expressing large, middle, or small T antigen. (*A*) Immunofluorescent staining of T antigens using monoclonal antibodies that recognize the aminoterminal common region of the T antigens and are conjugated with fluorescein isothiocyanate; (*B*) staining of microfilaments with rhodamine-conjugated phalloidin. (*a*) Cells expressing large T antigen; (*b*) cells expressing middle T antigen; (*c*) cells expressing small T antigen. With middle T antigen–expressing cells, the monoclonal antibodies used here do not show any specific staining.

microfilament staining pattern. In these cells, the staining is mainly seen in ruffles at the edge of cells and very little bundling is seen. On the contrary, large and small T antigens seem to have little effect, if any, on microfilament bundle formation (Fig. 3B, a and c). We conclude that middle T antigen alone is necessary and sufficient for altering the microfilament bundle pattern observed in polyomavirus–transformed cells.

## Mitogenic activity of small T antigen of polyomavirus

To examine the effect of small T antigen on cells, we compared the growth characteristics of NIH-3T3 cells expressing small T antigen and those of parental NIH-3T3 cells. The comparison was made by analyzing the growth curves at two different serum concentrations, the saturation density, the incorporation of [³H]thymidine into cells at confluency, and colony formation in soft agar.

Small T antigen–producing cells were morphologically indistinguishable from control cells. At confluency, cell-to-cell contact appeared unchanged compared with control cells, and they formed a flat cell-sheet. However, cell division appeared to continue beyond the confluency stage, and they formed a more dense cell-sheet. The density of cells under such conditions varied from clone to clone, and there was some correlation between the amount of small T antigen expressed in these cells and cell density (data not shown). Moreover, small T antigen–expressing cells formed flat, dense foci when these cells were mixed with a large excess of parental cells and allowed to grow until confluency.

When cells expressing small T antigen were suspended in soft agar, they formed small colonies under conditions in which the control cells did not divide more than a few times. The size of the colonies was generally smaller than that induced by cells expressing only middle T antigen. The proportion of cells that formed such colonies per given number of cells suspended was rather small and, depending upon the clone, varied from about 0.1% to 10%.

## Cooperation of middle and small T antigens of polyomavirus in the transformation of NIH-3T3 cells

We have compared the phenotype of parental NIH-3T3 cells and of cells expressing small T antigen after transfection of middle T antigen genes into the two types of cells. Dense foci appeared in about 2 weeks on monolayers of NIH-3T3 cells. In the case of small T antigen–expressing cells, it was not easy to obtain dense foci on monolayers because, although the cells formed a flat monolayer, they continued to grow beyond the saturation density of NIH-3T3 cells and came off the plates due to overgrowth, not too long after the dense foci could be recognized. A few of these foci were picked and immunoprecipitation was carried out to confirm that, after transfection, cells isolated from the NIH-3T3 cell culture expressed only middle T antigen, whereas those isolated

from small T antigen–expressing cells expressed both small and middle T antigens (Fig. 4).

No morphological differences were observed on the monolayers of these two sets of cells. However, when they were suspended in soft agar, a dramatic difference was observed; cells expressing only middle T antigen grew moderately well in soft agar, whereas those expressing both T antigens grew much faster. As an additional control for these results, we performed these experiments in reverse order: pGVST or pGV16 was introduced into a cell clone expressing middle T antigen and the appropriate cells were isolated after selection for resistance to G418. The advantage of this procedure was that middle T antigen, which is a known transforming protein, was present at exactly the same level in both cultures and, therefore, any effect seen after the introduction of the small T antigen gene could be attributed to the function of that gene. Immunoprecipitation experiments revealed that about half of the cell clones isolated from the culture that received pGVST contained both small and middle T antigens. When these cell clones were suspended in soft agar, the cells expressing both small T antigen and middle T antigen grew to large colonies, whereas the clones that received only pGV16 or pGVST but did not express small T antigen grew to medium-sized colonies (Fig. 5). The amount of middle T antigen expressed in these clones is not appreciably different. This suggests that the enhancing effect of the small T antigen on anchorage-independent growth is not due to the increased expression of middle T antigen.

It has been suggested that there is a strong correlation between the tyrosine kinase activity associated with middle T antigen of polyomavirus and the expression of the transformed cell phenotype (Smith et al. 1979). Preliminary results suggest that there is no appreciable difference in the tyrosine kinase activity associated with

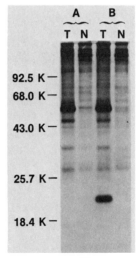

**Figure 4** Immunoprecipitation of [³⁵S]methionine-labeled T antigens of polyomavirus. (Lanes *A*) NIH-3T3 cells containing only the middle T gene; (lanes *B*) NIH-3T3 cells containing both small and middle T genes.

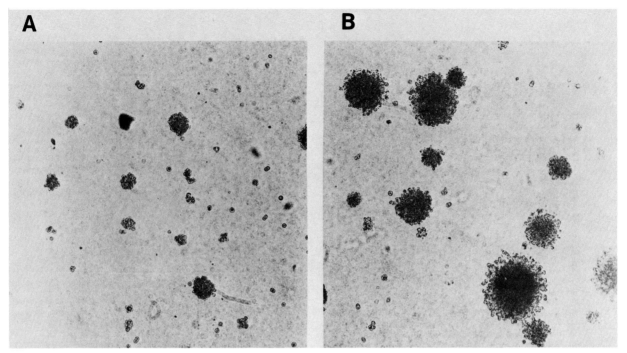

**Figure 5** Cooperation of middle and small T antigens for anchorage-independent growth. Cells expressing only middle T antigen (*A*) or those expressing both middle and small T antigens (*B*) were suspended in 0.33% agar and incubated for 2 weeks.

middle T antigen in cells expressing only middle T antigen and those expressing both middle and small T antigens. The results strongly suggest that small T antigen possesses its own unique function and that it contributes to expression of the phenotype of transformed cells independently of the functions of middle T antigen and perhaps also of large T antigen.

## Discussion

We have shown in this communication that small T antigen of polyomavirus has growth-stimulating activity and cooperates with middle T antigen in the transformation of NIH-3T3 cells. Small T antigen appears to cooperate with middle T antigen not by enhancing the function of middle T antigen, but rather by adding independent effects on cells. These results suggest that small T antigen has its own unique function and that small T antigen represents a third independent complementation group. Our results also suggest that even though the entire small T antigen polypeptide sequence is contained at the amino terminus of middle T antigen, this moiety does not express small T antigen function. A possible explanation for this would be that the carboxy-terminal region of middle T antigen, which is responsible for its association with membranes, might affect the function of the aminoterminal region and prevent its mitogenic activity. We can test this interpretation by examining whether several transformation-negative middle T antigen mutants lacking the membrane anchorage domain but containing an intact aminoterminal region are localized in nuclei and whether they are mitogenic.

In the present studies, we did not examine whether large T antigen has any additional effect on the cooperative effect of middle T and small T antigens. Large T antigen has been shown not to have any effect on cell morphology or growth in established cell lines (Schlegel and Benjamin 1978; Lania et al. 1979). In addition, biochemical and biological properties of large T antigen such as DNA binding, regulation of viral RNA transcription, *trans*-activation of some cellular genes, and establishment of permanent cell lines from primary culture fibroblasts suggest that large and small T antigens, although they are both present in nuclei, are distinctively different. Therefore, it is likely that small and large T antigens will have their own unique contribution to cell transformation. It is necessary to test directly whether the small T antigen gene represents the third complementation group in cell transformation. From this point of view, it is also important to see whether the small T antigen gene would cooperate with heterologous viral or cellular genes in cell transformation.

## Acknowledgments

The research described was sponsored by the National Cancer Institute under contract NO1-CO-23909 with Litton Bionetics, Inc. The contents of this publication do not necessarily reflect the views or policies of the Department of Health and Human Services, nor does mention of trade names, commercial products, or organizations imply endorsement by the U.S. Government.

## References

Courtneidge, S. and A.E. Smith. 1983. Polyomavirus transforming protein associates with the product of the c-*src* cellular gene. *Nature* **303:** 435.

Cuzin, F. 1984. The polyomavirus oncogenes. Coordinated functions of three distinct proteins in the transformation of rodent cells in culture. *Biochim. Biophys. Acta* **781:** 193.

Dilworth, S.M. and B.E. Griffin. 1982. Monoclonal antibodies against polyomavirus tumor antigens. *Proc. Natl. Acad. Sci.* **79:** 1059.

Ito, Y. 1979. Polyomavirus—specific 55K protein isolated from plasma membrane of productively infected cells is virus-coded and important for cell transformation. *Virology* **98:** 261.

Ito, Y., N. Spurr, and B.E. Griffin. 1980. Middle T antigen as primary inducer of full expression of the phenotype of transformation by polyomavirus. *J. Virol.* **35:** 219.

Land, H., L.F. Parada, and R. Weinberg. 1983. Tumorigenic conversion of primary embryo fibroblasts requires at least two cooperating oncogenes. *Nature* **304:** 596.

Lania, L., M. Griffiths, B. Cooke, Y. Ito, and M. Fried. 1979. Untransformed rat cells containing free and integrated DNA of a polyoma non-transforming (Hr-t) mutant. *Cell* **18:** 793.

Mann, R., R.C. Mulligan, and D. Baltimore. 1983. Construction of a retrovirus packaging mutant and its use to produce helper-free defective retrovirus. *Cell* **33:** 153.

Rassoulzadegan, M., Z. Naghashfar, A. Cowie, A. Carr, M. Grisoni, R. Kamen, and F. Cuzin. 1983. Expression of the large T protein of polyomavirus promotes the establishment in culture of "normal" rodent fibroblast cell lines. *Proc. Natl. Acad. Sci.* **80:** 4354.

Ruley, H.E. 1983. Adenovirus early region 1A enables viral and cellular transforming genes to transform primary cells in culture. *Nature* **304:** 602.

Schlegel, R. and T.L. Benjamin. 1978. Cellular alterations dependent upon the polyomavirus Hr-t function: Separation of mitogenic from transforming capacities. *Cell* **14:** 587.

Segawa, K. and Y. Ito. 1982. Differential subcellular localization of in vivo—phosphorylated and nonphosphorylated middle-sized tumor antigen of polyomavirus and its relationship to middle-sized tumor antigen phosphorylating activity in vitro. *Proc. Natl. Acad. Sci.* **79:** 6812.

———. 1983. Enhancement of polyomavirus middle T antigen tyrosine phosphorylation by epidermal growth factor. *Nature* **304:** 742.

Smith, A.E., R. Smith, B.E. Griffin, and M. Fried. 1979. Protein kinase activity associated with polyomavirus middle T antigen in vitro. *Cell* **18:** 915.

Treisman, R., U. Novak, J. Favaloro, and R. Kamen. 1981. Transformation of rat cells by an altered polyomavirus genome expressing only the middle T protein. *Nature* **292:** 595.

# Nuclear Localization and Gene Activation in *Trans* Are Mediated by Separate Domains of the Adenovirus E1A Proteins

B. Krippl,* B. Ferguson,[†] N. Jones,[‡] M. Rosenberg,[†] and H. Westphal*

*Laboratory of Molecular Genetics, National Institute of Child Health and Human Development, Bethesda, Maryland 20892; [†]Department of Molecular Genetics, Smith Kline and French Laboratories, Philadelphia, Pennsylvania 19101; [‡]Department of Biological Sciences, Purdue University, West Lafayette, Indiana 47907

The adenovirus E1A gene specifies nuclear functions that modulate initiation of transcription from certain viral and nonviral promoters. Insertion of E1A coding sequences in *Escherichia coli* expression vectors allowed us to obtain purified E1A proteins in quantities sufficient to initiate a detailed functional analysis of this important class of regulators of gene activity. We microinjected the proteins into mammalian cells and measured their ability to localize to the nucleus and to complement the adenovirus E1A deletion mutant H5*dl*312. The *E. coli*–expressed E1A 12S and 13S mRNA products both localize quantitatively to the nucleus within 30 min after cytoplasmic injection, and both induce H5*dl*312 gene expression. By introducing various deletions in the E1A coding sequence, we obtained mutant proteins that allowed us to map the two parameters tested to nonoverlapping domains present in both the 12S and the 13S mRNA products. The domain required for efficient nuclear localization was mapped to the carboxyl terminus of the proteins, whereas information for H5*dl*312 complementation was located in an internal region, comprising sequences encoded within both exons of the E1A gene.

It has been suggested that the products of the adenovirus E1A gene and of the *myc* gene bear certain structural and functional relationships. We obtained a c-*myc* product that had been shown to be functional to act as a competence factor in the cell cycle and promote the progression of $G_0$-arrested cells to S phase. This c-*myc* product was unable to complement H5*dl*312 even when injected at high concentrations.

As an alternative to microinjection, E1A proteins can also be transferred by protoplast fusion between the induced bacteria and the mammalian cell. We show that the E1A 13S mRNA products of both Ad2/5 and Ad12, transferred by this method, retain their ability to complement H5*dl*312.

Some transforming proteins, like the tumor antigens of papovaviruses, the herpesvirus immediate-early gene product, and the E1A proteins of adenoviruses have attracted considerable attention because of their ability to modulate transcription (for review, see Velcich and Ziff 1984; Kingston et al. 1985). The human adenovirus E1A gene is transcribed into two mRNAs, 12S and 13S, which share common 5′ and 3′ termini but differ internally by the size of the intron removed. The 12S RNA encodes a protein of 243 amino acids (aa), which differs from the 289-aa 13S mRNA product by an internal deletion of 46 aa (Perricaudet et al. 1979). The E1A proteins *trans*-activate gene expression by stimulating the rate of initiation of transcription from a number of viral and cellular promoters. During the early phase of the lytic cycle, E1A induces transcription from five viral promoters (Berk et al. 1979; Jones and Shenk 1979a; Nevins 1981). Nonviral genes, such as the rabbit β-globulin gene, the human ε-globin gene, and the pre-proinsulin gene are also activated by E1A when newly introduced into cells along with the E1A gene (Green et al. 1983; Treisman et al. 1983; Allan et al. 1984; Gaynor et al. 1984; Svensson and Akusjärvi 1984). Certain endogenous cellular genes such as the human gene for the 70-kD heat-shock protein (Kao and Nevins 1983) and the β-tubulin gene (Stein and Ziff 1984) also appear to be stimulated by E1A. In certain cases, E1A has been noted to suppress the ability of viral enhancer elements to stimulate transcription (Borelli et al. 1984; Velcich and Ziff 1985). This suppression may be due to a competition between E1A and viral enhancers for a limited supply of transcriptional factors (Kingston et al. 1985). In addition to modulating RNA polymerase II–directed transcription, E1A can stimulate RNA polymerase III–transcribed genes (Berger and Folk 1985; Hoeffler and Roeder 1985). The E1A proteins also play an important role in cell immortalization and oncogenic transformation (Bernards et al. 1983b). The E1A gene alone is sufficient to immortalize primary cells (Houweling et al. 1980; van den Elsen et al. 1982). In combination with adenovirus E1B or certain other oncogenes, E1A produces a fully transformed phenotype (Ruley 1983). It is not known whether the role of E1A in cell transformation is associated with its transcription-modulating activity.

Detailed analysis of the E1A gene products has been

difficult since the proteins are difficult to isolate from adenovirus-infected or transformed cells. To obtain sufficient quantities of the proteins to study their role in the regulation of gene expression and in the transformation process, we have constructed plasmid expression vectors that permit the regulated and high-level expression in *Escherichia coli* of the products of human adenovirus serotype C E1A gene (a serotype 2/5 hybrid gene) (Ferguson et al. 1984). The E1A 13S mRNA product (an Ad2/5 hybrid protein) and the 12S mRNA product were purified and characterized by microinjection into intact cells (Krippl et al. 1984; Ferguson et al. 1985a,b). The proteins rapidly and quantitatively localize to the nucleus. Both the 12S and 13S proteins activate expression of the E2A gene and stimulate expression of the adenovirus type-5 (Ad5) E1A deletion mutant H5*dl*312.

Here, we report on a study of E1A mutant proteins that were generated by introducing various deletions in the E1A gene. We find that rapid nuclear localization of both E1A products requires the presence of an intact carboxyl terminus. In contrast, an internal region is required for the activation of the H5*dl*312 gene expression. We also report on functional properties of an Ad12 E1A gene product, and on the direct transfer, by protoplast fusion, of functional E1A proteins from bacteria to mammalian cells.

## Experimental Procedures

### Plasmid constructions

The plasmid expression vector pAS1 (Rosenberg et al. 1983) (Fig. 1A) was used for the expression in *E. coli* of E1A-derived proteins. The unique *Bam*HI restriction site in pAS1 allows the inserted E1A coding sequence to be fused precisely to the translation initiation codon on the vector (Fig. 1B). The DNA segments of the E1A coding sequence that have been inserted into pAS1 are shown schematically in Figure 1C. The construction of pAS1-E1A410, which encodes the E1A 13S protein, has been described (Ferguson et al. 1984). The details of the construction of the pAS1 derivatives that encode E1A 12S protein, the E1A deletion-mutant proteins, and the Ad12 E1A protein are given elsewhere (Ferguson et al. 1985a; Krippl et al. 1985, 1986).

### Expression in E. coli and isolation of E1A proteins

The expression in *E. coli* of E1A proteins encoded by derivatives of pAS1 was tightly controlled by temperature induction, using the phage $\lambda$ $p_L$ promoter and the $\lambda$ lysogenic host, N5151(*cl ts*857) (Rosenberg et al.

1983; Ferguson et al. 1984). Expression was also controlled by using nalidixic acid (Mott et al. 1985). E1A-derived proteins were purified from *E. coli* as described previously (Krippl et al. 1984).

### Cells and viruses

VERO, an African green monkey kidney cell line permissive for adenovirus growth, was propagated as previously described (Richardson et al. 1980). The E1A deletion mutant H5*dl*312 was grown and titered in 293 cells (Jones and Shenk 1979b).

### Microinjection and immunofluorescent staining

Proteins were microinjected via glass capillaries into the nucleus or cytoplasm of VERO cells grown in monolayers (Krippl et al. 1984). Immunofluorescent staining of cells, using rabbit antiserum specific to *E. coli*–expressed E1A 13S protein and monoclonal antibody specific to the E2A DNA-binding protein (DBP) was carried out as described previously (Krippl et al. 1984).

### Assay for expression of H5dl312 major late transcription unit

VERO cells were infected with H5*dl*312 at a multiplicity of infection (moi) of approximately 200 pfu/cell. Every cell in a defined area was microinjected with E1A protein or control protein, either before or after H5*dl*312 infection. Cells were labeled with [$^{35}$S]methionine for 2 hr at 22 hr postinfection, and the major viral coat proteins (hexon, penton base, and fiber) were immunoprecipitated and analyzed by SDS-polyacrylamide gel electrophoresis (PAGE) (Krippl et al. 1984).

### Transfer of protein into tissue-culture cells by protoplast fusion

*E. coli* strain N5151 cells carrying pAS1-E1A410 (Ferguson et al. 1984) or pAS-A12N (Krippl et al. 1986) were grown at 32°C to an OD$_{650}$ of 0.8. Expression from the vector systems was induced by temperature shift of the cultures to 42°C. After incubation at 42°C for 90 min, bacteria were collected and converted to protoplasts as described (Waldman and Milman 1984). Induction by nalidixic acid of the E1A proteins in the defective lysogenic host is described elsewhere (Mott et al. 1985; Krippl et al. 1986). Protoplasts were fused to H5*dl*312-infected VERO cells (200 pfu/cell) in the presence of polyethylene glycol at a ratio of $3 \times 10^4$:1 (Sandri-Goldin et al. 1983). Fused cells ($3 \times 10^3$) were analyzed for expression of H5*dl*312 coat proteins as described (Krippl et al. 1984).

**Figure 1** (*see facing page*) Plasmid vectors for the expression of E1A proteins in *E. coli*. (*A*) The plasmid expression vector pAS1 (Rosenberg et al. 1983); transcription from the vector is controlled by the $\lambda$ $p_L$ promoter. (Amp$^R$) β-lactamase gene. (*B*) The E1A coding sequence was inserted into pAS1, using the *Bam*HI restriction endonuclease site, in-frame with the translation initiation site on the vector. (cII R.B. site) $\lambda$ cII gene ribosome-binding site; (fMet) $\lambda$ cII translation initiation codon. (*C*) Schematic of insertion- and deletion-mutant gene products expressed in *E. coli*. Plasmid expression vectors encoding insertion- and deletion-mutant E1A proteins were constructed as described (Krippl et al. 1985). The number of the first and last amino acids encoded by segments of E1A coding sequence are indicated by open thick lines. Amino acids unique to the 13S mRNA product (stippled area) as well as missense (hatched area) and deleted E1A-specific (dashed line) amino acids are indicated. Also shown are the regions of E1A protein encoded by the first and second exons, the region unique to the 13S mRNA product, restriction endonuclease sites used in the construction of the plasmid expression vectors, and the amino acid lengths of the E1A-derived polypeptides expressed in *E. coli*.

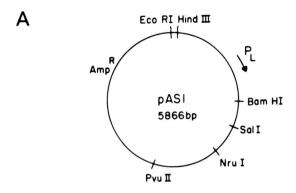

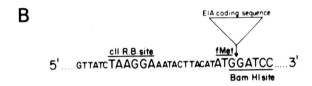

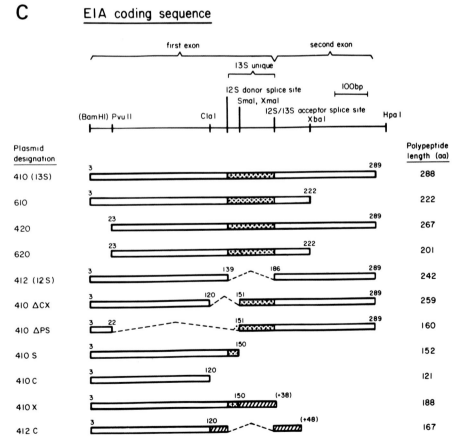

**Figure 1** (*See facing page for legend.*)

## Results and Discussion

### Expression of E1A proteins in *E. coli*

We have used a derivative of the plasmid vector pAS1 (Rosenberg et al. 1983) (Fig. 1A) to express in *E. coli* the products of the Ad2/5 E1A 13S mRNA and 12S mRNA (Ferguson et al. 1985a) and the 13S mRNA product of Ad12. In addition, derivatives of pAS1 were constructed that encode and express several deletion-mutant proteins (Fig. 1C). After temperature induction, each of the E1A-derived proteins was expressed at a high level (5–10% of total *E. coli* protein) and purified as described (Krippl et al. 1984). The purification procedure we use has proved useful for the isolation, in biologically functional form, of other eukaryotic proteins similarly expressed in *E. coli* (Ferguson et al. 1985c; Watt et al. 1985). Figure 2 shows an immunoblot analysis of the partially purified E1A-derived polypeptides, using polyclonal antibody specific to the E1A 13S protein. Each of the truncated E1A polypeptides clearly contain determinants that are recognized by this polyclonal E1A-specific antibody. As shown in Figure 2, each of the partially purified E1A-derived proteins is obtained predominantly as a single major immunoreactive polypeptide species.

### Nuclear accumulation

Purified E1A proteins were injected into the cytoplasm of VERO cells, and the cells were subsequently analyzed by E1A-specific immunofluorescent staining. By examining the location of the E1A proteins at various times after injection into the cell cytoplasm, we found that the 12S and 13S proteins were rapidly and quantitatively localized to the cell nucleus. Within 30 min, the injected cells showed predominately nuclear E1A-specific immunofluorescence (Fig. 3A,B,C). Both E1A proteins could be detected up to 12 hr after injection, attesting to their stability within the cell.

Next, we analyzed the effects of structural alterations of the E1A gene products on their ability to localize to the nucleus. Each of the mutant proteins was injected into the cytoplasm of VERO cells, and the cells were analyzed at various times after injection. The results are shown in Table 1 and Figure 3. The products of pAS1-E1A420, 410ΔPS, and 410ΔCX, like the full-length E1A 12S and 13S proteins, localized rapidly (within 30 min) to the nucleus after cytoplasmic microinjection. These variant E1A proteins contained aminoterminal or internal deletions but had unaltered carboxyl termini. In contrast, each of the E1A variants with carboxyterminal deletions, including the products of pAS1-E1A610, 620, 410X, 412C, 410S, and 410C, localized to the nucleus very slowly. For example, the products of pAS1-E1A610 and 620, which have 67 carboxyterminal amino acid residues deleted, took 7–8 hr to fully localize to the nucleus after cytoplasmic microinjection. Our results indicate that sequences of the carboxyterminal portion of the E1A protein are essential for rapid nuclear accumulation. Removal of these sequences, however, did not prevent the truncated proteins from eventually localizing to the nucleus. The fact that all of the E1A-derived proteins that lack the carboxyl terminus are detectable in the cell nucleus, albeit many hours after cytoplasmic injection, suggests that they retain affinity for nuclear components.

These findings are consistent with an active mechanism for translocation of E1A into the nucleus that depends on recognition of a signal located in the carboxyl

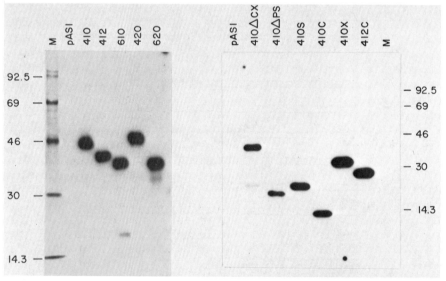

**Figure 2** Isolation and immunoblot analysis of *E. coli*–expressed insertion and deletion-mutant E1A proteins. Insertion- and deletion-mutant E1A proteins (see Fig. 1) were expressed in *E. coli* and partially purified as described previously (Krippl et al. 1985). Samples containing 1–3 ng of E1A-derived protein (except 410ΔPS, where 30 ng was analyzed) were analyzed by SDS-PAGE and immunoblot analysis using rabbit antiserum specific to the *E. coli*–expressed E1A 13S mRNA product (Krippl et al. 1984). The position and size (in kilodaltons) of marker proteins (Amersham) are indicated. Each insertion- and deletion-mutant E1A protein is designated according to the plasmid expression vector by which it is encoded (see Fig. 1).

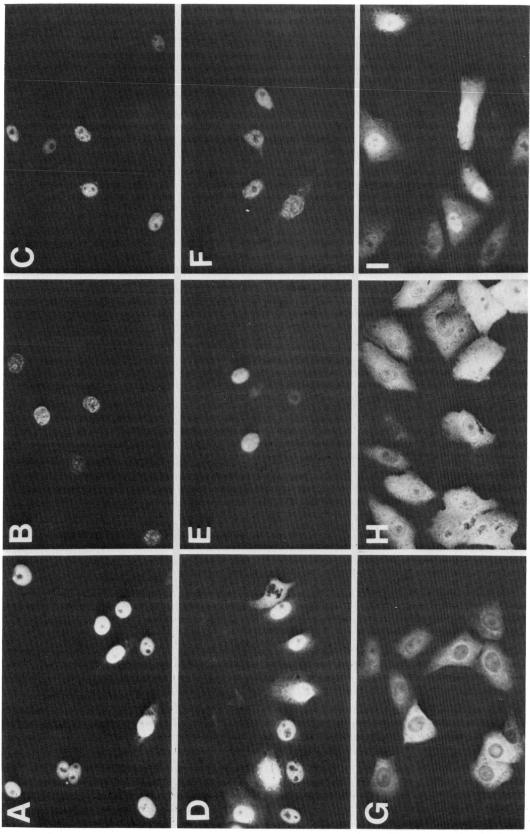

**Figure 3** Subcellular location of microinjected E1A-derived proteins. About $2 \times 10^{-11}$ ml of a solution containing 1 mg/ml of the protein products of pAS1-E1A410 (*A–C*), 410ΔPS (*D–F*), or 410C (*G–I*) was microinjected into the cytoplasm of VERO cells. The microinjected cells were analyzed by E1A-specific immunofluorescent staining (Krippl et al. 1984) at 30 min (*A, D, G*), 4 hr (*B, E, H*), and 6 hr (*C, F, I*) after injection.

**Table 1** Intracellular Location of Microinjected E1A Proteins as a Function of Time after Cytoplasmic Microinjection

| Protein | Hours after injection | | | | | | | | | |
|---|---|---|---|---|---|---|---|---|---|---|
| | 1/4 | 1/2 | 1 | 2 | 3 | 4 | 5 | 6 | 7 | 8 |
| 410 | NC | N | N | N | N | N | N | N | N | N |
| 412 | NC | N | N | N | N | N | N | N | N | N |
| 420 | NC | | N | N | N | N | N | N | N | N |
| 410ΔCX | NC | NC | N | | | N | N | | N | N |
| 410ΔPS | NC | NC | N | N | N | N | N | N | | |
| 610 | | NC | | NC | NC | | NC | NC | | N |
| 620 | | C | | C | | NC | NC | NC | N | N |
| 410X | | NC | NC | | NC | | NC | | NC | |
| 412C | | NC | | | | | NC | | NC | |
| 410S | C | C | C | | C | C | NC | | NC | NC |
| 410C | C | C | C | C | C | C | NC | NC | NC | NC |

E1A-derived proteins (see Fig. 1C) were injected into the cytoplasm of VERO cells. At given times thereafter, cells were analyzed by E1A-specific immunofluorescent staining. Positive staining is symbolized as N (nuclear), C (cytoplasmic), or NC (both nuclear and cytoplasmic).

terminus of the protein. Analogous information has been detected in other nuclear proteins (Kalderon et al. 1984; Ellis 1985). In SV40 T antigen, for example, there is a short, internal region that functions as a signal for transport across the nuclear membrane (Kalderon et al. 1984). A similar internal region that includes the amino acid sequence Lys-Arg-Pro-Arg has recently been identified in polyoma large T antigen (Richardson et al. 1986). This same sequence is also found at the carboxyl terminus of both Ad5 and Ad12 (Sukisaki et al. 1980). Since all of our mutant proteins that migrate rapidly to the nucleus contain an intact carboxyl terminus, we predict that this terminus indeed contains the signal for nuclear transport.

## Gene activation functions

The following experiments were designed to show that E1A products made in *E. coli* substitute for genuine E1A proteins in the adenovirus life cycle. One of the known E1A functions in this regard is their ability to turn on other early adenovirus genes (Berk et al. 1979; Jones and Shenk 1979a; Nevins 1981). Accordingly, we microinjected into VERO cells the 12S and 13S E1A proteins together with the early Ad2 E2A gene and measured DBP expression. Whereas less than 1% of the cells injected with the E2A gene alone displayed immunofluorescent staining using DBP antibody, this figure rose to 30% when either 12S or 13S E1A protein were coinjected (Ferguson et al. 1985c). Both proteins, therefore, were able to stimulate DBP expression to a comparable degree.

A more sensitive and also more stringent and comprehensive assay of E1A functional activity in the virus life cycle is the H5*dl*312 induction assay (Krippl et al. 1984). The defective adenovirus H5*dl*312 is unable to synthesize an E1A product since the E1A promoter region and most of the coding region have been deleted. The defect results in very low levels of early viral transcripts, of DNA replication, and of late gene expression in a variety of primary and established cell lines, includ-

ing VERO. In the H5*dl*312 assay, therefore, we measure the ability of the E1A proteins to stimulate sufficient early gene transcription to allow viral replication and subsequent expression of the major late transcription unit. The results in Figures 4 and 5 show that the level of H5*dl*312 hexon protein expression was proportional to the concentration of either E1A 13S (protein 410) or 12S (protein 412) microinjected over a range of 0.1–2.0 mg/ml. Both the 12S and the 13S E1A products activate H5*dl*312 with comparable efficiencies in this assay system. This does not exclude the possibility that both may function differently when expressed in the context of a genuine viral infection (Montell et al. 1984). Our findings do show, however, that both E1A proteins have the potential of activating other viral genes.

To map functional domains within the E1A genes that are responsible for gene activation, we tested each of the mutant E1A proteins (Figs. 1 and 2) in our H5*dl*312 assay. Late gene expression of the deletion mutant was determined as a function of the amount of protein applied to the cell (Figs. 4 and 5). The E1A mutant product 610, which lacks 67 carboxyterminal residues, was able to induce H5*dl*312 expression nearly as efficiently as the full-length E1A 13S protein (Fig. 5), even though this protein locates to the nucleus much more slowly than the E1A 13S or 12S proteins. Deletion of 21 aminoterminal residues from E1A (mutant product 420) only slightly reduced its ability to induce H5*dl*312 expression (Fig. 5). Moreover, this defect was only observed when the protein was injected at low concentration (0.2 mg/ml). Full induction of H5*dl*312 expression was achieved when the protein was injected at concentrations of 0.5–1 mg/ml. We interpret this to mean that deletion of the first 21 aminoterminal residues only partially impairs the function of E1A and that this defect is overcome by increasing the concentration of the defective protein. Essentially identical results were obtained with the E1A mutant product 620, which is missing both its aminoterminal 21 and carboxyterminal 67 residues. Again, although nuclear localization is significantly retarded, this

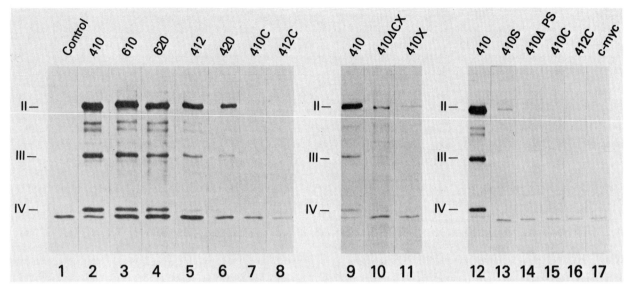

**Figure 4** Adenovirus coat protein expression in VERO cells infected with 200 pfu/cell (titered on 293 cells) of H5*dl*312 and injected with E1A-derived proteins. Protein was injected at 2–4 hr postinfection, cells were labeled for 2 hr with [$^{35}$S]methionine at 22 hr postinfection, and the major viral coat proteins were immunoprecipitated and analyzed by SDS-PAGE (Krippl et al. 1984). The indicated E1A-derived proteins (see Fig. 1C) were injected at a concentration of 1 mg/ml (lanes *2–9*) or 2 mg/ml (lanes *10–16*). The control lane represents cells injected with an extract of *E. coli* carrying the plasmid pAS1 (Ferguson et al. 1984). Cells in lane *17* were injected with a solution containing 3 mg/ml of purified *E. coli*–expressed human c-*myc* protein (Watt et al. 1985). Each lane represents an immunoprecipitate from approximately 50 cells. The major viral coat proteins hexon (II), penton base (III), and fiber (IV) are indicated.

variant also shows only a partial defect in gene activation that is compensated for by injecting higher amounts.

In an effort to further define domains of the E1A protein that are essential for H5*dl*312 complementation, we analyzed two mutant E1A proteins that lack 139 carboxyterminal residues (410X, 410S). These proteins are unable to generate a hexon signal when injected at a concentration of 1.0 mg/ml or less. However, low levels of H5*dl*312 hexon expression were observed upon in-

jection of these proteins at 2.0 mg/ml. Mutant E1A proteins containing a larger deletion of 169 carboxyterminal residues (410C, 412C) completely lost any detectable H5*dl*312-inducing activity, even when injected into cells at concentrations up to 3 mg/ml.

In addition to the E1A mutants with aminoterminal and carboxyterminal deletions, we have examined the H5*dl*312 hexon-inducing activity of two E1A mutants with internal deletions in the aminoterminal half of the

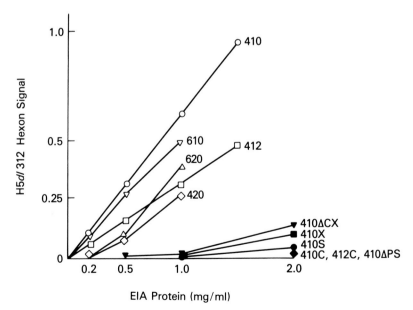

**Figure 5** H5*dl*312 hexon protein expression as a function of the concentration of E1A-derived protein injected. VERO cells were infected with H5*dl*312 and microinjected with E1A-derived protein, and viral coat protein expression was analyzed as described in the legend to Fig. 4. Hexon protein (band II) expression was quantitated by densitometric scanning of autoradiograms. The level of hexon, given in arbitrary units, is the mean value of three experiments. The E1A-derived proteins microinjected are indicated (see Fig. 1C).

molecule (410ΔCX, 410ΔPS). Both of these mutant E1A proteins were unable to induce H5*dl*312 expression when injected at concentrations of 1 mg/ml or less. However, the protein with the smaller deletion (410ΔCX) induced low levels of H5*dl*312 gene expression when injected at 2 mg/ml (Figs. 4 and 5).

In summarizing our observations of the E1A deletion proteins, we note that each of the mutant E1A proteins (610, 620, and 420) that have an unaltered sequence between aa 23 and 222 exhibit activities similar to the full-length E1A products. We conclude therefore that a central region of the E1A gene is required for the effective regulatory function of E1A in the lytic cycle. These findings are in agreement with studies on adenovirus mutants that have shown that E1A retains its regulatory function after deletion of 14 aa at the amino terminus or 87 aa at the carboxyl terminus (Jones and Shenk 1979b; Osborne et al. 1982; Downey et al. 1984).

Mutant E1A proteins that lack part of this central region are inactive in the H5*dl*312 complementation assay when tested at a concentration of 1 mg/ml. However, some of these (410S, 410X, 410ΔCX) clearly exhibit low activity when tested at higher concentrations. Protein 410ΔCX carries a small deletion comprising aa 121–151. This protein does not efficiently induce H5*dl*312 gene expression, implying that sequences within this region are crucial for gene activation. Since aa 140–151 are deleted in the E1A 12S protein and since the activity of the 12S protein is similar to that of the 13S protein, the critical segment might be even smaller, that is, from aa 121–139. A portion of the second exon of the E1A protein probably also contributes to the ability of the protein to complement H5*dl*312 virus. This is demonstrated by the fact that E1A mutant proteins 410S and 410X, which have unaltered first-exon-encoded regions, induce hexon protein synthesis poorly. As shown above, deletion of 67 aa from the carboxyl terminus of the E1A protein does not impair its ability to induce hexon protein expression. Therefore, the information that the second exon of E1A supplies to H5*dl*312 complementation is most likely contained within a region that maps immediately downstream of the 12S/13S acceptor splice site. The first exon of E1A encodes a domain that is clearly sufficient to induce low levels of H5*dl*312 gene expression (see above). The first exon of E1A also encodes a domain sufficient for the *trans*-activation of the Ad5 E3 promoter in *Xenopus* oocytes (Richter et al. 1985). Recent studies indicate that the first exon also contains information sufficient for the immortalization of primary rat cells, as measured by the E1A/*ras* cotransformation assay (Ruley 1983; N. Haley and N. Jones, unpubl.). These results are therefore consistent with the model that the transcription-modulating functions of E1A proteins are required for their transforming activities. Previous reports have also indicated that the first exon may encode a distinct functional domain (Solnick and Anderson 1982; Bos et al. 1983; Jochemsen et al. 1984).

### E. coli–expressed E1A 13S protein is unable to induce hsp70 and β-tubulin gene expression in VERO cells

It has been suggested that the transcription of two endogenous genes, namely, the genes encoding a 70-kD heat-shock protein (hsp70) (Kao and Nevins 1983) and β-tubulin (Stein and Ziff 1984), may be stimulated by E1A. We therefore tested whether the *E. coli*–expressed E1A protein could induce hsp70 and β-tubulin gene expression in microinjected VERO cells. VERO cells were injected with E1A 13S protein. At 1, 2, 4, 6, and 12 hr postinjection, the cells were labeled with [$^{35}$S]methionine and cell lysates were immunoprecipitated with either a rabbit antibody specific to the human 70-kD heat-shock protein (kindly provided by J. Nevins) or a β-tubulin–specific monoclonal antibody (Amersham). The microinjected E1A protein had no effect on the level of hsp70 and β-tubulin protein expression in VERO cells (data not shown). In a control experiment, expression of hsp70 protein was induced greater than 10-fold by temperature shift of VERO cells from 37°C to 41.5°C. It appears that under our assay conditions, the E1A protein is unable to stimulate transcription of the endogenous hsp70 and β-tubulin genes in the absence of other viral functions. Our results indicate that the stimulation of the hsp-70 and β-tubulin gene transcription observed in response to adenovirus infection (Kao and Nevins 1983; Stein and Ziff 1984) is not a direct effect of the E1A protein alone.

### E1A and c-*myc*

There are a number of apparent similarities between the adenovirus E1A and the human c-*myc* gene products. The proteins share structural homologies and both can complement the product of the T24 H-*ras* gene in transformation of primary cells in culture (for review, see Kingston et al. 1985). The product of the human c-*myc* gene has been expressed in *E. coli* (Watt et al. 1985) and is active as a cofactor in stimulating cellular DNA synthesis (Kaczmarek et al. 1985). We tested the *E. coli*–made c-*myc* protein in the H5*dl*312 assay. The protein did not complement H5*dl*312, even at high concentrations (Fig. 4), nor could it compete with the 13S E1A product or help the mutant protein E1A-410ΔPS in this assay (Krippl et al. 1985). Therefore, if there are indeed functional relationships between E1A and c-*myc*, they have not become apparent in our analysis.

### E1A protein transfer by protoplast fusion

As has been demonstrated in the experiments described so far, the microinjection technique is well suited for directing known quantities of functional proteins to the cell compartment of choice, the cytoplasm or the nucleus. However, in experiments to be described elsewhere in detail (Richter et al. 1985), the protocol called for reextraction of E1A protein from the cells some time after transfer in order to study secondary modifications. In this case, we opted instead for the protoplast fusion technique, which allows transfer of DNA or protein to a large number of mammalian cells (Sandri-Goldin et al. 1983). The efficiency of E1A protein transfer into VERO cells is demonstrated in the following experiment. *E. coli* cells containing Ad2/5 or Ad12 E1A 13S cDNA expression vectors (Ferguson et al. 1984; Krippl et al. 1986) were induced by temperature shift or by the addition of nalidixic acid to produce large amounts of E1A proteins (Ferguson et al. 1984). The bacteria were subsequently converted to protoplasts by the addition of lysozyme and fused to H5*dl*312-infected VERO cells in the presence of polyethylene glycol. At 22 hr after fusion, cells were labeled with [³⁵S]methionine and examined for the presence of hexon antigen. As shown in Figure 6, the E1A 13S mRNA products of Ad2/5 and of Ad12 complemented H5*dl*312 equally well. Our controls indicate that the E1A proteins (as opposed to the vector encoding them) were transferred from the bacteria to the VERO cells in this experiment. The hexon protein that was synthesized appeared only in cells that were fused with induced bacteria (Fig. 6, lanes 2, 3, 5, and 6). Fusion of uninduced bacteria (lanes 1 and 4) failed to result in H5*dl*312 complementation.

Apart from demonstrating that E1A proteins can indeed be transferred in a functionally active form directly from bacteria to mammalian cells, our experiment confirms the previously established fact (Bernards et al. 1983a) that Ad12 E1A can substitute for Ad5 E1A in complementing the Ad5 deletion mutant. Although the protoplast fusion technique bypasses the need for protein purification prior to transfer, it has its own limitations. For instance, there is the potential for deleterious effects from bacterial components cotransferred with the gene product of choice. The two quite different techniques will enable us to adapt the method of protein transfer to the particular needs of our future experiments.

## Acknowledgment

We thank Rosemary Watts for the purified c-*myc* product.

## References

Allan, M., J. Zhu, P. Montague, and J. Paul. 1984. Differential response of multiple ε-globin cap sites to *cis*- and *trans*-acting controls. *Cell* **38**: 399.

Berger, S.L. and W.R. Folk. 1985. Differential activation of RNA polymerase III–transcribed genes by the polyomavirus enhancer and the adenovirus E1A gene products. *Nucleic Acids Res.* **13**: 1413.

Berk, A.F., F. Lee, T. Harrison, J. Williams, and P.A. Sharp. 1979. Pre-early adenovirus 5 gene product regulates synthesis of early viral messenger RNAs. *Cell* **17**: 935.

Bernards, R., J. Vaessen, A. van der Eb, and J.S. Sussenbach. 1983a. Construction and characterization of an adenovirus type 5/adenovirus type 12 recombinant virus. *Virology* **131**: 30.

Bernards, R., P.I. Schrier, A. Houweling, J.L. Bos, and A.J. van der Eb. 1983b. Tumorigenicity of cells transformed by adenovirus type 12 by evasion of T-cell immunity. *Nature* **305**: 776.

Borelli, E., R. Hen, and P. Chambon. 1984. Adenovirus-2 E1A products repress enhancer-induced stimulation of transcription. *Nature* **312**: 608.

Bos, J.L., A.G. Jochemsen, R. Bernards, P.I. Schrier, H. van Ormondt, and A.J. van der Eb. 1983. Deletion mutants of region E1a of Ad12 E1 plasmids: Effect on oncogenic transformation. *Virology* **129**: 393.

Downey, J.F., C.M. Evelegh, P.E. Branton, and S.T. Bayley. 1984. Peptide maps and N-terminal sequences of polypeptides from early region 1A of human adenovirus 5. *J. Virol.* **50**: 30.

Ellis, J. 1985. Genetic engineering. Eukaryotic proteins retargeted among cell compartments. *Nature* **313**: 353.

Ferguson, B., N. Jones, J. Richter, and M. Rosenberg. 1984. Adenovirus E1A gene product expressed at high levels in *Escherichia coli* is functional. *Science* **224**: 1343.

Ferguson, B., B. Krippl, O. Andrisani, N. Jones, H. Westphal, and M. Rosenberg. 1985a. E1A 13S and 12S mRNA products made in *E. coli* both function as nuclear-localized transcriptional activators but do not directly bind DNA. *Mol. Cell. Biol.* **5**: 2653.

Ferguson, B., B. Krippl, N. Jones, J. Richter, H. Westphal, and M. Rosenberg. 1985b. Functional characterization of purified adenovirus E1A protein expressed in *Escherichia coli*. *Cancer Cells* **3**: 265.

Ferguson, B., L.M. Pritchard, J. Feild, D. Rieman, R.G. Greig, G. Poste, and M. Rosenberg. 1985c. Isolation and analysis of an Abelson murine leukemia virus–encoded tyrosine-specific kinase produced in *Escherichia coli*. *J. Biol. Chem.* **260**: 3652.

Gaynor, R.B., D. Hillman, and A.J. Berk. 1984. Adenovirus early region 1A protein activates transcription of a nonviral gene introduced into mammalian cells by infection or transfection. *Proc. Natl. Acad. Sci.* **81**: 1193.

Green, M.R., R. Treisman, and T. Maniatis. 1983. Transcrip-

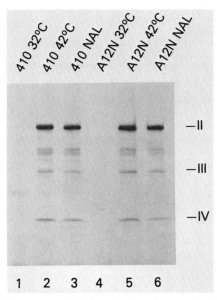

**Figure 6** Immunoprecipitation of adenovirus coat proteins from H5*dl*312-infected VERO cells after protoplast fusion with *E. coli* cells carrying the plasmid pAS1-E1A410 (lanes *1–3*) and pAS-A12N (lanes *4–6*). Induction was achieved by temperature shift (lanes *2* and *5*) and by nalidixic acid (lanes *3* and *6*). VERO cells in lane *1* and *4* were fused with *E. coli* prior to temperature induction. Each lane represents an immunoprecipate from approximately 3 × 10³ cells.

tional activation of cloned human β-globin genes by viral immediate-early gene products. *Cell* **35**: 137.

Hoeffler, W.K. and P.G. Roeder. 1985. Enhancement of RNA polymerase III transcription by the E1A gene product of adenovirus. *Cell* **41**: 955.

Houweling, A., P.J. van den Elsen, and A.J. van der Eb. 1980. Partial transformation of primary rat cells by the left-most 4.5% fragment of adenovirus 5 DNA. *Virology* **105**: 537.

Jochemsen, A.G., J.L. Bos, and A.J. van der Eb. 1984. The first exon of region E1A genes of adenovirus 5 and 12 encode a separate functional protein domain. *EMBO J.* **3**: 2923.

Jones, N. and T. Shenk. 1979a. An adenovirus type 5 early gene function regulates expression of other early viral genes. *Proc. Natl. Acad. Sci.* **76**: 3665.

———. 1979b. Isolation of adenovirus type 5 host range deletion mutant defective for transformation of rat embryo cells. *Cell* **17**: 683.

Kaczmarek, L., R. Watt, M. Rosenberg, and R. Baserga. 1985. Microinjected c-*myc* as a competence factor. *Science* **288**: 1313.

Kalderon, D., B.L. Roberts, W.D. Richardson, and A.E. Smith. 1984. A short amino acid sequence able to specify nuclear location. *Cell* **39**: 499.

Kao, H.T. and J.R. Nevins. 1983. Transcriptional activation and subsequent control of the human heat shock gene during adenovirus infection. *Mol. Cell. Biol.* **3**: 2058.

Kingston, R.E., A.S. Baldwin, and P.A. Sharp. 1985. Transcriptional control by oncogenes. *Cell* **41**: 3.

Krippl, B., B. Ferguson, M. Rosenberg, and H. Westphal. 1984. Functions of purified E1A protein microinjected into mammalian cells. *Proc. Natl. Acad. Sci.* **81**: 6988.

Krippl, B., B. Ferguson, N. Jones, M. Rosenberg, and H. Westphal. 1985. Mapping of functional domains in adenovirus E1A proteins. *Proc. Natl. Acad. Sci.* **82**: 7480.

———. 1986. Nuclear localization and gene activation in *trans* are mediated by separate domains of the adenovirus E1A proteins. *J. Virol.* (in press).

Montell, C., G. Courtois, C. Eng, and A.J. Berk. 1984. Complete transformation by adenovirus 2 requires both E1A proteins. *Cell* **36**: 951.

Mott, J.E., R.A. Grant, Y.-S. Ho, and T. Platt. 1985. Maximizing gene expression from plasmid vectors containing the λ P_L promoter: Strategies for overproduction transcription termination factor ρ. *Proc. Natl. Acad. Sci.* **82**: 88.

Nevins, J.R. 1981. Mechanism of activation of early viral transcription by the adenovirus E1A gene product. *Cell* **26**: 213.

Osborne, T.F., R.B. Gaynor, and A.J. Berk. 1982. The TATA homology and the mRNA 5′ untranslated sequence are not required for expression of essential adenovirus E1A functions. *Cell* **29**: 139.

Perricaudet, M., G. Akusjärvi, A. Virtanen, and U. Pettersson. 1979. Structure of two spliced mRNAs from the transforming region of human subgroup C adenoviruses. *Nature* **281**: 694.

Richardson, W.D., B.J. Carter, and H. Westphal. 1980. VERO cells injected with adenovirus type 2 mRNA produce authentic viral polypeptide patterns: Early mRNA promotes growth of adenovirus-associated virus. *Proc. Natl. Acad. Sci.* **77**: 931.

Richardson, W.D., B.L. Roberts, and A.E. Smith. 1986. Nuclear location signals in polyoma virus large-T. *Cell* **44**: 77.

Richter, J.D., P. Young, N. Jones, B. Krippl, M. Rosenberg, and B. Ferguson. 1985. A first exon-encoded domain of E1A sufficient for posttranslational modification, nuclear-localization and induction of adenovirus E3 promoter expression in *Xenopus* oocytes. *Proc. Natl. Acad. Sci.* **82**: 1.

Rosenberg, M., Y. Ho, and A. Shatzman. 1983. The use of pKC30 and its derivatives for controlled expression of genes. *Methods Enzymol.* **101**: 123.

Ruley, H.E. 1983. Adenovirus early region 1A enables viral and cellular transforming genes to transform primary cells in culture. *Nature* **304**: 602.

Sandri-Goldin, R.M., A.L. Goldin, M. Levine, and J. Gloriso. 1983. High-efficiency transfer of DNA into eukaryotic cells by protoplast fusion. *Methods Enzymol.* **101**: 27.

Solnick, D. and M.A. Anderson. 1982. Transformation-deficient adenovirus defective in expression of region 1A and not region 1B. *J. Virol.* **42**: 106.

Stein, R. and E.B. Ziff. 1984. HeLa cell β-tubulin gene transcription is stimulated by adenovirus 5 in parallel with viral early genes by an E1a-dependent mechanism. *Mol. Cell. Biol.* **4**: 2792.

Sukisaki, H., K. Sugimoto., M. Takanami, K. Shiroki, J. Saito, H. Shimojo, Y. Sawada, Y. Uemizu, S. Uesugi, and K. Fujinaga. 1980. Structure and gene organization in the transforming *Hind*III-G fragment of Ad12. *Cell* **20**: 777.

Svensson, C. and G. Akusjärvi. 1984. Adenovirus 2 early region 1A stimulates expression of both viral and cellular genes. *EMBO J.* **3**: 789.

Treisman, R., M.R. Green, and T. Maniatis. 1983. *Cis*- and *trans*-activation of globin gene transcription in transient assays. *Proc. Natl. Acad. Sci.* **80**: 7428.

van den Elsen, P.J., S. de Peter, A. Houweling, J. van der Veer, and A.J. van der Eb. 1982. The relationship between region E1a and E1b of human adenoviruses in cell transformation. *Gene* **18**: 175.

Velcich, A. and E. Ziff. 1984. Gene regulation: Repression of activators. *Nature* **312**: 594.

———. 1985. Adenovirus E1a proteins repress transcription from SV40 early promoter. *Cell* **40**: 705.

Waldman, A.S. and G. Milman. 1984. Transfer of herpes simplex virus thymidine kinase synthesized in bacteria by a high-expression plasmid to tissue culture cells by protoplast fusion. *Mol. Cell. Biol.* **4**: 1644.

Watt, R., A. Shatzman, and M. Rosenberg. 1985. Expression and characterization of the human c-*myc* DNA-binding protein. *Mol. Cell. Biol.* **5**: 448.

# Heterogeneity of Adenovirus E1A Proteins Is Due to Posttranslational Modification of the Primary Translation Products of the 12S and 13S mRNAs

C. Stephens, B.R. Franza, Jr., C. Schley, and E. Harlow

Cold Spring Harbor Laboratory, Cold Spring Harbor, New York 11724

The early region 1A (E1A) of adenovirus 5 codes for proteins of 289 and 246 amino acids. These proteins are translated from the 13S and 12S mRNAs, respectively. Using monoclonal antibodies, the E1A proteins were immunoprecipitated from extracts of infected HeLa cells and separated on two-dimensional, isoelectric-focusing, SDS-polyacrylamide gels. In these analyses the E1A proteins were resolved into about 40 polypeptide species, with approximately 20 polypeptides arising from translation of the 12S mRNA and 20 from the 13S mRNA. Pulse-chase experiments were performed to identify the primary translation products of both mRNAs. After synthesis, the primary translation products are extensively modified, and this posttranslational processing accounts for the heterogeneity of the E1A proteins.

The early region 1A (E1A) of adenovirus codes for proteins that are involved in the regulation of viral gene expression and transformation (for reviews, see Flint 1984; Branton et al. 1985; Akusjärvi et al. 1986; Jones 1986). The E1A proteins are the first viral proteins expressed after cells are infected with adenovirus, and one of the functions of these proteins is to activate transcription of other viral early genes (Berk et al. 1979; Jones and Shenk 1979a). The E1A proteins also stimulate transcription from some cellular genes, including the 70K heat-shock and β-tubulin genes (Nevins 1982; Kao and Nevins 1983; Stein and Ziff 1984). Recently, it has been shown that the E1A proteins repress transcription of genes found adjacent to SV40 or polyoma enhancers (Borelli et al. 1984; Velcich and Ziff 1985). In addition, the infection of primary rodent cultures with adenovirus often leads to transformation (Gallimore et al. 1974; Graham et al. 1974, 1978; Flint et al. 1976; Jones and Shenk 1979b). The E1A proteins allow these cells to overcome senescence and, together with the E1B proteins, provide all the viral information necessary for full transformation (Houweling et al. 1980; Ruley 1983).

Three mRNA species are synthesized from the E1A region (Berk et al. 1979; Chow et al. 1979; Kitchingman and Westphal 1980). These three mRNAs, referred to by their sedimentation values of 13S, 12S, and 9S, arise from differential splicing of a common RNA precursor (Perricaudet et al. 1979; Svensson et al. 1983; Virtanen and Pettersson 1983; Roberts et al. 1985). They possess the same 5′ and 3′ termini and differ only by the size of their excised introns. The 12S and 13S mRNAs are seen throughout infection, whereas the 9S mRNA is found only late after infection (Chow et al. 1979; Spector et al. 1980b). The 13S and 12S mRNAs code for polypeptides of 289 and 246 amino acids, respectively

(Perricaudet et al. 1979). Because these two proteins are translated from the same reading frame, they are identical except for an additional 46 amino acids within the 289-amino-acid protein. The sequence of the 9S mRNA predicts that its product will be a protein of 55 amino acids. The first exon of the 9S mRNA is translated in the same reading frame as the 13S and 12S mRNA, while the second exon uses a different reading frame.

The E1A proteins have been studied by several different methods, including in vitro translation of hybrid-selected mRNAs (Halbert et al. 1979; Ricciardi et al. 1979; Esche et al. 1980; Spector et al. 1980a; Smart et al. 1981; Lupker et al. 1981) and immunoprecipitation with specific antibodies (Feldman and Nevins 1983; Rowe et al. 1983; Yee et al. 1983; Scott et al. 1984; Spindler et al. 1984; Harlow et al. 1985). When the E1A proteins are analyzed on one-dimensional SDS polyacrylamide gels, the proteins migrate slower than would be expected for their molecular sizes. Several workers have suggested that this property may be due to the high proline and acidic amino acid content. The products of the 12S and 13S mRNAs each run as two polypeptide bands on one-dimensional gels. However, when the 12S and 13S proteins are separated on two-dimensional gels, these bands can be resolved into approximately 20 different protein products from each mRNA (Harlow et al. 1985). We show here that this heterogeneity is due to posttranslational modification of the E1A primary translational products.

## Materials and Methods

### Cells and viruses

HeLa cells were grown in Dulbecco's modified Eagle's medium (DME) supplemented with 5% fetal bovine serum. When the cultures were near confluency, the cells were infected with adenovirus type 5 at a multiplicity of infection of 20 plaque-forming units per cell.

*Radiolabeling and immunoprecipitation*
Beginning at 14 hr postinfection, cultures were labeled for 4 hr in methionine-free DME with 0.5 mCi of [$^{35}$S]methionine (New England Nuclear). Cells were lysed in buffer containing 250 mM NaCl, 0.1% Nonidet P-40, and 50 mM HEPES, pH 7.0 (lysis buffer), and the lysates were precleared with *Staphylococcus aureus* Cowan I as described by Harlow et al. (1985). The E1A proteins were immunoprecipitated with monoclonal antibodies M58 or M73 (Harlow et al. 1985). For the denaturation experiment, extracts were split, and one-half was made with 2% SDS and boiled for 10 min. Prior to immunoprecipitation, both the boiled and untreated samples were diluted 20-fold in lysis buffer.

For pulse-chase experiments, cells were starved for methionine for 5 min in DME without methionine. Each plate of cells was labeled for 1 min in methionine-free DME containing 1 mCi of [$^{35}$S]methionine. At the end of the 1-min pulse, the cells were rinsed with DME containing a 5-fold excess of cold methionine and then chased for different periods in DME with a 5-fold excess of methionine. At the end of each chase period, the proteins were extracted with lysis buffer, and the lysates were processed as described above.

*Synthesis of the 13S and 12S transcripts and their in vitro translation*
SP6 transcription vectors were generated by cloning cDNA copies of the 12S or 13S mRNA into pSP64

(Melton et al. 1984). SP6 transcription was carried out as described by Melton et al. (1984), and RNA was translated in vitro by using rabbit reticulocyte lysates (Pelham and Jackson 1976). The translated proteins were then immunopreciptated and separated on two-dimensional gels.

*Two-dimensional gel electrophoresis*
Two-dimensional gel electrophoresis was performed as described by Garrels (1983). Immunoprecipitated proteins were stripped from the protein A–Sepharose by incubation with sample buffer (0.3% SDS, 9.95 M urea, 4% Nonidet P-40, 2% ampholytes [pH 3.5–10], 100 mM dithiothreitol). Samples were then loaded onto pH range 3.5–10 isofocusing gels. Second-dimension gels were 10% polyacrylamide.

## Results

When E1A proteins from adenovirus-infected HeLa cells were immunoprecipitated with the M58 monoclonal antibody and separated on two-dimensional gels, the 12S and 13S proteins were resolved into approximately 40 polypeptides (Fig. 1). Similar experiments were performed using recombinant adenoviruses that contain a cDNA copy of either the 12S or 13S mRNA in place of the wild-type E1A coding region. These viruses can only synthesize 12S- or 13S-related products, and approximately 20 polypeptides can be assigned to each mRNA (Harlow et al. 1985). There are a number of explanations for the extensive heterogeneity of these polypeptides, including: (1) considerable posttranslational modifica-

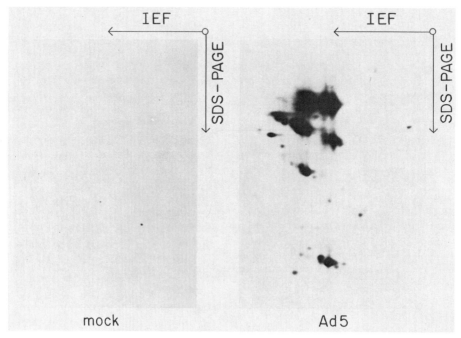

**Figure 1** Two-dimensional gel analysis of E1A proteins precipitated from mock-infected and Ad5-infected HeLa cells. At 14 hr postinfection, cells were labeled for 4 hr with [$^{35}$S]methionine. Lysates were immunoprecipitated with anti-E1A monoclonal antibodies (M58) and precipitated proteins were run on a two-dimensional gel. The region of the gel in which the E1A proteins migrate is shown. The E1A proteins run in the molecular-weight range of 42,000–58,000 and focus at a pH of between 4 and 6.

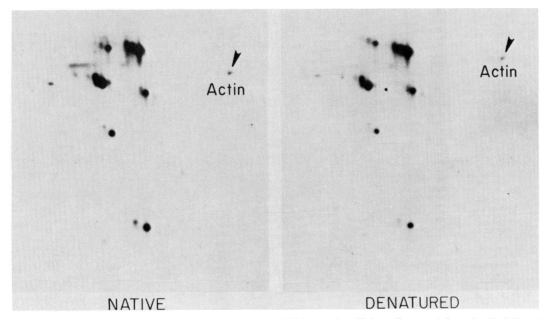

**Figure 2** Two-dimensional gel analysis of native and of denatured E1A proteins. HeLa cells were infected with Ad5 and at 14 hr postinfection were labeled with [$^{35}$S]methionine for 4 hr. Cells were lysed and half the sample was adjusted to 2% SDS and boiled for 10 min. Both samples were diluted 20-fold and precipitated with the anti-E1A monoclonal antibody M73. The region of the gel in which the E1A proteins migrate is shown. The E1A proteins run in the molecular-weight range of 42,000–58,000 and focus at a pH of between 4 and 6.

tion of a single primary translation product, (2) multiple initiation or termination events that lead to multiple translation products, (3) proteolytic breakdown, or (4) immunoprecipitation of cellular proteins that share epitopes with E1A or that are complexed to E1A. To distinguish between these possibilitites or combinations thereof, we have analyzed the individual proteins in greater detail.

## The immunoprecipitated proteins are not encoded by host DNA

Two experimental approaches were used to show that the polypeptides from Figure 1 were not cellular proteins precipitated because they shared epitopes with E1A or were associated with E1A. First, none of these polypeptides was seen when extracts of uninfected HeLa cells were immunoprecipitated with the monoclon-

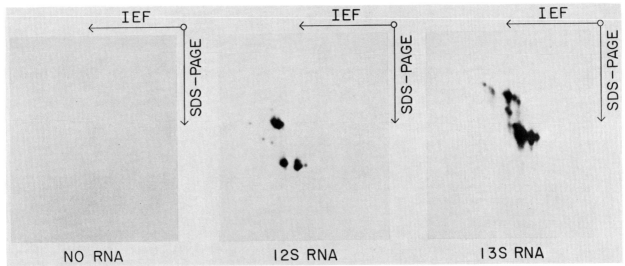

**Figure 3** Two-dimensional gel analysis of E1A proteins synthesized in vitro. SP6-synthesized 12S and 13S E1A RNAs were translated in vitro using rabbit reticulocyte lysates. Proteins were immunoprecipitated with the anti-E1A monoclonal antibody M73 and were run on two-dimensional gels. The region of the gel in which the E1A proteins migrate is shown.

al antibodies (Fig. 1). Second, when extracts from infected HeLa cells were denatured by boiling in SDS prior to immunoprecipitation, the multiple species were still detected (Fig. 2). These conditions should disrupt most noncovalent associations, and only proteins that are recognized directly by the monoclonal antibodies will be precipitated. These conditions have been shown to dissociate the known E1A–cellular protein complexes (Harlow et al. 1986). The cellular proteins that complex with E1A were not seen in our experiments because of their low levels and because host-protein synthesis is inhibited during the late phase of adenovirus infections.

## The immunoprecipitated proteins are products of 12S and 13S mRNA translation

To analyze the protein products of tne E1A mRNAs, we subcloned cDNA copies of the 12S and 13S mRNAs (kindly provided by E. Moran and B. Zerler; Moran et al. 1986) behind the SP6 promoter and synthesized RNA complementary to the 12S and 13S DNAs. When these RNAs were used to program rabbit reticulocyte lysates, authentic E1A proteins were synthesized. The products of these in vitro translations were analyzed by immunoprecipitation and two-dimensional gel electrophoresis (Fig. 3). Results from these experiments agree with the

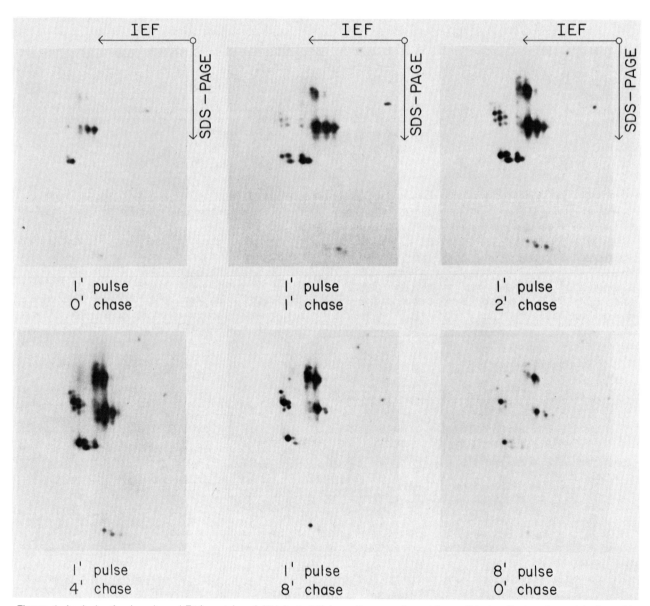

**Figure 4** Analysis of pulse-chased E1A proteins. Ad5-infected HeLa cells were starved for methionine 5 min prior to labeling with [$^{35}$S]methionine. Cells were pulse-labeled for 1 min and then chased with DME containing a 5-fold excess of cold methionine. At the end of each chase period, cells were lysed and proteins were immunoprecipitated with an anti-E1A antibody. The region of the gel in which the E1A proteins migrate is shown. The E1A proteins run in the molecular-weight range of 42,000–58,000 and focus at a pH of between 4 and 6.

in vivo virus infections and indicate that most of the precipitated proteins can be placed into two groups, those arising from translation of the 12S mRNA and those from translation of the 13S mRNA. In addition, these experiments have shown that the majority of the polypeptide species can be detected after in vitro translation, suggesting that their generation is not specific for infected cells.

### Multiple forms of E1A are due to posttranslational modification

Three possibilities still remain as a means to explain the heterogeneity of the E1A polypeptides: The proteins may be proteolytic breakdown products; they may result from the synthesis of multiple primary translation products; or they may arise from posttranslational modification. To determine which of these possibilities is correct, we performed the pulse-chase experiments shown in Figure 4. Translation of the 12S or 13S mRNAs yields a single product, and, with time, this polypeptide becomes more acidic and moves more slowly in the sizing dimension. This indicates that most of the complicated patterns seen with E1A immunoprecipitations can be explained by posttranslational modification of a single translation product.

In preparation for the experiment shown in Figure 4, a number of pulse-chase experiments with variations in pulse times were carried out (data not shown). A pulse time of longer than 1 min results in multiple protein spots on two-dimensional gels, making it difficult to determine the primary translation product. This also implies that the E1A modifications are taking place at a very rapid rate. The 1-min-pulse/1-min-chase time point in Figure 4 contains more counts than the 1-min pulse because [$^{35}$S]methionine probably has not been cleared from the intracellular amino acid pools until the end of the 1-min chase.

### Discussion

Using anti-E1A antibodies, we have precipitated the E1A 12S and 13S proteins from adenovirus-infected HeLa cells. On two-dimensional gels these proteins resolve into a set of about 40 polypeptide species. This heterogeneity is due to posttranslational modification of the primary translation products of the 12S and 13S E1A mRNAs. The addition of modifications results in the E1A proteins migrating as more-acidic and apparently larger moieties in two-dimensional gels.

The complexity of the E1A 12S and 13S proteins might lead one to believe that there are 20 different modifications, but two other factors may contribute to the complicated patterns. First, it is possible that subsets of the E1A proteins have different modifications. Second, if the modifications are not added in a strict temporal order, heterogeneity can be introduced by changes in the sequence of addition. Understanding which of these alternatives is correct must await the detailed analysis of the modification events themselves.

We and others (Yee et al. 1983; Spindler et al. 1984) have demonstrated that the E1A proteins are phosphorylated. In many cases posttranslational modification has been shown to play an important role in determining proper protein function. For example, proteins are known to be regulated by phosphorylation (for review, see Cohen 1982), and in some cases the correct subcellular localization of a protein has been shown to be due to the addition of certain modifications. An example is the myristylation of the *src* protein (Garber et al. 1983), which is necessary for the proper localization of *src* in the cellular membrane. As of yet, little is known about the modification of the E1A proteins. Future experiments will be directed at determining what residues of the E1A molecules are modified and the role such modification plays in the function of the E1A proteins.

### Acknowledgments

We thank Tim Adams, Karen Buchkovich, and Peter Whyte for helpful discussions. As well, we thank H. Sacco for technical assistance in running the two-dimensional gels. This work was supported by U.S. Public Health Service grant CA13106 from the National Cancer Institute.

### References

Akusjärvi, G., U. Pettersson, and R.J. Roberts. 1986. Structure and function of the adenovirus-2 genome. In *Adenovirus DNA* (ed. W. Doerfler), p. 53. Academic Press, New York.

Berk, A.J., F. Lee, T. Harrison, J. Williams, and P.A. Sharp. 1979. Pre-early adenovirus 5 gene product regulates synthesis of early viral messenger RNAs. *Cell* **17**: 935.

Borrelli, E., R. Hen, and P. Chambon. 1984. Adenovirus-2 E1A products repress enhancer-induced stimulation of transcription. *Nature* **312**: 608.

Branton, P.E., S.T. Bayley, and F.L. Graham. 1985. Transformation by human adenoviruses. *Biochim. Biophys. Acta* **780**: 67.

Chow, L.T., T.R. Broker, and J.B. Lewis. 1979. Complex splicing patterns of RNAs from the early regions of adenovirus-2. *J. Mol. Biol.* **134**: 265.

Cohen, P. 1982. The role of protein phosphorylation in neural and hormonal control of cellular activity. *Nature* **296**: 613.

Esche, H., M.B. Mathews, and J.B. Lewis. 1980. Proteins and messenger RNAs of the transforming gene of wild-type and mutant adenoviruses. *J. Mol. Biol.* **142**: 399.

Feldman, L.T. and J.R. Nevins. 1983. Localization of the adenovirus E1A protein, a positive-acting transcriptional factor, in infected cells. *Mol. Cell. Biol.* **3**: 829.

Flint, S.J. 1984. Cellular transformation by adenovirus. *Pharmacol. Ther.* **26**: 59.

Flint, S.J., J. Sambrook, J.F. Williams, and P.A. Sharp. 1976. Viral nucleic acid sequences in transformed cells. IV. A study of the sequences of adenovirus 5 DNA and RNA in four lines of adenovirus 5-transformed rodent cells using specific fragments of the viral genome. *Virology* **72**: 456.

Gallimore, P.H., P.A. Sharp, and J. Sambrook. 1974. Viral DNA in transformed cells. II. A study of the sequences of adenovirus 2 DNA in nine lines of transformed rat cells using specific fragments of the viral genome. *J. Mol. Biol.* **89**: 49.

Garber, E., J.G. Kruger, H. Hanafusa, and A.R. Goldberg. 1983. Only membrane-associated RSV *src* proteins have amino-terminally bound lipid. *Nature* **302**: 161.

Garrels, J.I. 1983. Quantitative two-dimensional gel electrophoresis of proteins. *Methods Enzymol.* **100**: 411.

Graham, F.L., T. Harrison and J. Williams. 1978. Defective transforming capacity of adenovirus type 5 host range mutants. *Virology* **86:** 10.

Graham, F.L., A.J. van der Eb, and H.L. Heijneker. 1974. Size and location of the transforming region in human adenovirus type 5 DNA. *Nature* **251:** 687.

Halbert, D.H., D.J. Spector, and H.J. Raskas. 1979. In vitro translation products specified by the transforming region of adenovirus type 2. *J. Virol.* **31:** 621.

Harlow, E., B.R. Franza, and C. Schley. 1985. Monoclonal antibodies specific for adenovirus early region 1A proteins: Extensive heterogeneity in early region 1A proteins. *J. Virol.* **55:** 553.

Harlow, E., P. Whyte, B.R. Franza, and C. Schley. 1986. Association of the adenovirus early region 1A proteins with cellular polypeptides. *Mol. Cell. Biol.* **6:** 1579.

Houweling, A., P.J. Van den Elsen, and A.J. van der Eb. 1980. Partial transformation of primary rat cells by the leftmost 4.5% fragment of adenovirus 5 DNA. *Virology* **105:** 537.

Jones, N. 1986. The viral genome and its expression. In *Adenovirus DNA* (ed. W. Doerfler), p. 161. Academic Press, New York.

Jones, N. and T. Shenk. 1979a. An adenovirus type 5 early gene function regulates expression of other early viral genes. *Proc. Natl. Acad. Sci.* **76:** 3665.

———. 1979b. Isolation of adenovirus type 5 host range deletion mutants defective for transformation of rat embryo cells. *Cell* **17:** 683.

Kao, H. and J.R. Nevins. 1983. Transcriptional activation and subsequent control of human heat shock gene during adenovirus infection. *Mol. Cell. Biol.* **3:** 2058.

Kitchingman, G.R. and H. Westphal. 1980. The structure of adenovirus 2 early nuclear and cytoplasmic RNAs. *J. Mol. Biol.* **137:** 23.

Lupker, J.H., A. Davis, H. Jochemsen, and A.J. van der Eb. 1981. In vitro synthesis of adenovirus type 5 T antigens. I. Translation of early region 1–specific RNA from lytically infected cells. *J. Virol.* **37:** 524.

Melton, D.A., P.A. Krieg, M.R. Rebagliati, T. Maniatis, K. Zinn, and M.R. Green. 1984. Efficient in vitro synthesis of biologically active RNA and RNA hybridization probes from plasmids containing a bacteriophage SP6 promoter. *Nucleic Acids Res.* **12:** 7035.

Moran, E., T. Grodzicker, R. Roberts, M. Mathews, and B. Zerler. 1986. Lytic and transforming functions of individual products of the adenovirus E1A gene. *J. Virol.* **57:** 765.

Nevins, J.R. 1982. Induction of the synthesis of a 70,000 dalton mammalian heat shock protein by the adenovirus E1A gene product. *Cell* **29:** 913.

Pelham, R.B. and J. Jackson. 1976. Efficient mRNA-dependent translation system in reticulocyte lysates. *J. Biochem.* **67:** 247.

Perricaudet, M., G. Akusjärvi, A. Virtanen, and U. Pettersson. 1979. Structure of two spliced mRNAs from the transforming region of human subgroup C adenoviruses. *Nature* **281:** 694.

Ricciardi, R.P., J.S. Miller, and B.E. Roberts. 1979. Purification and mapping of specific mRNAs by hybridization-selection and cell-free translation. *Proc. Natl. Acad. Sci.* **76:** 4927.

Roberts, B., J. Miller, D. Kimelman, C. Cepko, I. Lemischka, and R. Mulligan. 1985. Individual adenovirus type 5 early region 1A gene products elicit distinct alterations of cellular morphology and gene expression. *J. Virol.* **56:** 404.

Rowe, D.T., S. Yee, J. Otis, F.L. Graham, and P.E. Branton. 1983. Characterization of human adenovirus type 5 early region 1A polypeptides using antitumor sera and an antiserum specific for the carboxy terminus. *Virology* **127:** 253.

Ruley, H.E. 1983. Adenovirus early region 1A enables viral and cellular transforming genes to transform primary cells in culture. *Nature* **304:** 602.

Scott, M.O., D. Kimelman, D. Norris, and R.P. Ricciardi. 1984. Production of monospecific antiserum against the early region 1A proteins of adenovirus 12 and the adenovirus 5 by an adenovirus 12 early region 1A–β-galactosidase fusion protein antigen expressed in bacteria. *J. Virol.* **50:** 895.

Smart, J.E., J.B. Lewis, M.B. Mathews, M.L. Harter, and C.W. Anderson. 1981. Adenovirus type 2 early proteins: Assignment of the early 1A proteins synthesized in vivo and in vitro to specific mRNAs. *Virology* **112:** 703.

Spector, D.J., M. McGrogan, and H.J. Raskas. 1980a. Regulation of appearance of cytoplasmic RNAs from region 1 of the adenovirus 2 genome. *J. Mol. Biol.* **126:** 395.

Spector, D.J., L.D. Crossland, D.N. Halbert, and H.J. Raskas. 1980b. A 28K polypeptide is the translation product of 9S RNA encoded by region 1A of adenovirus 2. *Virology* **102:** 218.

Spindler, K.R., D.S.E. Rosser, and A.J. Berk. 1984. Analysis of adenovirus transforming proteins from early region 1A and 1B with antisera to inducible fusion antigens produced in *Escherichia coli*. *J. Virol.* **49:** 132.

Stein, R. and E.B. Ziff. 1984. HeLa cell tubulin gene transcription is stimulated by adenovirus 5 in parallel with viral early genes by an E1a-dependent mechanism. *Mol. Cell. Biol.* **4:** 2792.

Svensson, C., U. Pettersson, and G. Akusjärvi. 1983. Splicing of adenovirus 2 early region 1A mRNAs is non-sequential. *J. Mol. Biol.* **165:** 475.

Velcich, A. and E. Ziff. 1985. Adenovirus E1A proteins repress transcription from the SV40 early promoter. *Cell* **40:** 705.

Virtanen, A. and U. Pettersson. 1983. The molecular structure of the 9S mRNA from early region 1A of adenovirus serotype 2. *J. Mol. Biol.* **165:** 496.

Yee, S., D.T. Tremblay, M. McDermott, and P.E. Branton. 1983. Identification of human adenovirus early region 1 products by using antisera against synthetic peptides corresponding to the predicted carboxy termini. *J. Virol.* **46:** 1003.

# Experimental Induction of Autocrine Synthesis of GM-CSF Results in Leukemogenicity

T.J. Gonda,* D. Metcalf,[†] R.A. Lang,* N.M. Gough,* and A.R. Dunn*

*Melbourne Tumour Biology Branch, Ludwig Institute for Cancer Research, and [†]The Walter and Eliza Hall Institute of Medical Research, P.O. Royal Melbourne Hospital, Victoria 3050, Australia

We have examined the role that autocrine production of a hemopoietic growth factor, the granulocyte-macrophage colony–stimulating factor (GM-CSF), might play in the development of leukemia. A GM-CSF retrovirus has been constructed and used to infect a factor-dependent nonleukemogenic cell line FDC-Pl. A number of cell lines have been isolated that grow autonomously, express virally encoded GM-CSF, and are leukemogenic in syngeneic mice. Thus, in this system, we have experimentally induced autocrine growth regulation and have shown that this represents a key event in the development of leukemia.

The growth and multiplication of mammalian cells depends on the availability of a complex and diverse range of molecules. In addition to general nutrients, cells from particular tissues often have a requirement for specific polypeptide growth factor(s) that bind to corresponding receptors located at the cell surface. It is thought that this interaction activates other proteins that in turn deliver intracellular signals mediating DNA synthesis and cell division. Although our knowledge of these processes is scant, it is well-established that tumor cells often have quite different requirements for growth factors compared with their normal cellular progenitors.

Much of our understanding of the genetic basis of malignant transformation has been derived from studies with transforming viruses. In particular, acute transforming retroviruses are known to carry oncogenes whose expression both initiates and maintains the transformed phenotype of target cells. It is now firmly established that retroviral oncogenes represent transduced copies of cellular genes (proto-oncogenes); furthermore, there is accumulating evidence that at least some of the proteins encoded by viral and cellular oncogenes correspond to specific growth factors (Doolittle et al. 1983; Waterfield et al. 1983) and growth factor receptors (Downward et al. 1984). Thus, it has been suggested that genetic alterations involving these genes may result in the unscheduled synthesis of the intracellular signals that initiate DNA synthesis. This would effectively uncouple proliferation from growth factor regulation and result in uncontrolled growth. Indeed, there are several examples where cells have become independent of external growth factors and where such autonomous growth is considered to be an important stage in the progression to a tumorigenic phenotype (De Larco and Todaro 1978; Kaplan et al. 1982).

In our laboratory we are interested in the regulation of growth factors controlling blood cell development. Hemopoietic cells are dependent on a group of specific growth factors, the colony-stimulating factors (CSFs), for survival, proliferation, differentiation, and in some cases functional end-cell activation (Metcalf 1985).

Interestingly, primary leukemias of the granulocyte-macrophage (GM) lineage retain dependence on exogenous CSF for proliferation in vitro, but upon prolonged maintenance in culture or upon repeated transplantation, such leukemic cells become autonomous and in many cases can be shown to constitutively synthesize one of the CSFs.

Although most hemopoietic cells grown in the presence of CSF in vitro proliferate, differentiate, and ultimately die, a number of laboratories have isolated cell lines that retain their dependence on CSF for proliferation and are not leukemogenic (Dexter et al. 1980; Greenberger et al. 1983). There are two reports of CSF-independent variants of a factor-dependent parent cell line whose autonomous growth coincided with leukemogenicity (Hapel et al. 1981; Schrader and Crapper 1983), and it has been proposed that autocrine synthesis of CSF was responsible for conversion to a malignant phenotype.

The availability of molecular clones corresponding to various murine CSFs (Fung et al. 1984; Gough et al. 1984; Dunn et al. 1985) has allowed us to directly test whether autonomous growth of factor-dependent cell lines results in tumorigenicity. To this end we have constructed a retrovirus containing a murine GM-CSF cDNA and infected a cell line, FDC-P1 (FD) (Dexter et al. 1980; Hapel et al. 1984), whose proliferation depends on GM-CSF (or Multi-CSF). A number of independent cell lines have been isolated from the infected cells that synthesize CSF, grow autonomously, and induce leukemias in mice.

## Materials and Methods

### DNA transfections

$\psi$2 cells (Mann et al. 1983) were transfected with 1 μg of vector DNA and 8 μg of carrier DNA per 50-mm petri dish by using the calcium phosphate coprecipitation technique (Graham and van der Eb 1973). Forty-eight hr after transfection, cells were trypsinized and replated at a 1:20 dilution in medium containing G418 (400 μg/ml).

Drug-resistant colonies were picked and expanded after 2 weeks.

### Maintenance of cell lines

Cells were routinely grown in Dulbecco's modified Eagle's medium supplemented with 10% fetal calf serum. Where applicable for growth and clone selection, medium was supplemented with G418 at 400 μg/ml for ψ2 cells and BALB/c-3T3 fibroblasts and at 1 mg/ml for FD cells. FD cells (Dexter et al. 1980; Hapel et al. 1984) were maintained with Multi-CSF using WEHI-3B(D −)-conditioned medium at 10% culture volume. Individual G418-resistant and CSF-independent colonies were maintained in 20-ml flat-bottomed culture bottles and the medium was changed twice weekly.

### Viral infections

Infectious virus was collected from ψ2 cells as a 24-hr culture supernatant. The virus titer was determined on the basis of G418-resistant cfu/ml of supernatant on BALB/c-3T3 cells.

Nonadherent FD cells were infected either by exposure to ψ2 clone culture supernatant or by cocultivation with adherent ψ2 cells.

### Plasmid construction

pGMV was constructed by first ligating BamHI linkers to the GM-CSF cDNA clone pGM5′Δ7 (Gough et al. 1985) and subsequent ligation into the BamHI site of pZIPNeoSV(x)1 (Cepko et al. 1984). Enzymic manipulations of DNA and preparation of plasmid DNA were performed according to standard procedures (Maniatis et al. 1982).

### Determination of tumorigenicity

GMV-infected FD cells were washed and $1 \times 10^6$ cells were injected subcutaneously to groups of eight 3-month-old DBA/2 mice. Mice were monitored twice weekly, and, when moribund, killed and sectioned.

Cells from tumor tissue, spleen, and bone marrow were cultured in 1-ml agar medium cultures lacking added GM-CSF to determine the frequency of clonogenic autonomous FD cells.

## Results

### Isolation and expression of GM-CSF cDNA

We have previously reported the isolation of a partial cDNA copy of the murine GM-CSF (Gough et al. 1984). More recently we have isolated an essentially complete GM-CSF cDNA clone (pGM3.2) from a library established using mRNA from lectin-activated T cells (Gough et al. 1985). When this cDNA was installed in a eukaryotic expression vector, pJL (Fig. 1A), and transfected into COS cells, an activity was secreted with the same biological properties as GM-CSF purified from cell lines or mouse tissues (Dunn et al. 1985; Gough et al. 1985).

Since the nucleotide sequence of pGM3.2 revealed the presence of two potential translational initiation codons, we analyzed the sequences required for the synthesis of secreted GM-CSF by constructing a set of mutants with deletions extending from the 5′ terminus of the cDNA clone using BAL-31 nuclease (Gough et al. 1985). During the course of these experiments, it became clear that the media from COS cells transfected with various pJL/cDNA constructions varied considerably in the levels of steady-state GM-CSF assayed at 2 days (Fig. 1B). One such mutant, pGM3.2Δ7, reproducibly directed the expression of relatively high levels of

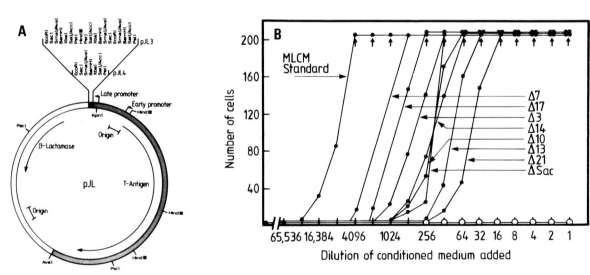

**Figure 1** (A) Map of eukaryotic expression vectors pJL3 and pJL4 (for details of structure, see Gough et al. 1985). (B) Stimulation of factor-dependent FD cells by COS-cell-conditioned media from transfections using DNA from various GM-CSF cDNA mutants (Gough et al. 1985). The titration curves (●) for each mutant (Δ) are indicated. Note that transfection with vector alone (○) did not result in GM-CSF synthesis.

GM-CSF and accordingly was selected for insertion into a retroviral vector.

## Construction of a GM-CSF retrovirus

The pGMV plasmid was constructed by inserting pGM3.2Δ7 into the Moloney murine leukemia virus–based vector pZIPNeoSV(x)1 (Cepko et al. 1984) as shown schematically in Figure 2. pGMV was subsequently transfected into the packaging cell line, ψ2, where it was expected to be transcribed to give two overlapping mRNAs originating from the retroviral promoter located in the long terminal repeat. The longer, genome-length mRNA should be translated to give GM-CSF, whereas the shorter, spliced subgenomic mRNA should encode the selectable marker for neomycin resistance (Neo$^R$; Fig. 2). From these transfections a number of drug-resistant colonies developed after exposure to G418 that were subsequently developed into cell lines (GMV-FD cells). The titers of virus produced by individual clones ranged between $10^4$ and $2 \times 10^5$ G418-resistant cfu/ml when assayed in BALB/c-3T3 cells. These G418-resistant, GMV-infected 3T3 cells secreted GM-CSF into the medium, whereas 3T3 cells derived from infection with ZIPNeo virus lacked GM-CSF activity (data not shown).

## Infection of FD cells with GMV

Having established that GMV could induce both GM-CSF production and G418 resistance in infected fibroblasts, we proceeded to infect CSF-dependent FD cells with this virus. This was achieved either by exposing the FD cells to virus-containing culture supernatants from the GMV-producing ψ2 clones or by cocultivating the FD cells with the GMV-producing cells. The FD cells were then plated in soft agar cultures containing GM-CSF and G418 to select for drug resistance or in cultures lacking CSF to select for autonomously growing clonogenic cells. Whereas normal FD cells failed to proliferate in the presence of G418 or in the absence of GM-CSF, GMV-infected FD cells gave rise to colonies under both selective conditions. Figure 3 indicates the number of colonies obtained under each selective regime after infection with GMV or ZIPNeo viruses for

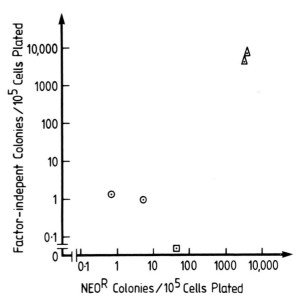

**Figure 3** Coincidence of factor-independent and drug-resistant colonies. Number of factor-independent and G418-resistant colonies obtained after GMV infection of FD cells. FD cells were infected either by exposure to GMV-containing medium (○), by cocultivation with GMV-producing ψ2 cells (△), or by cocultivation with ZIPNeo virus–producing ψ2 cells (□). Each point represents a single experiment, and its position indicates the number of colonies obtained on selection for G418 resistance or factor independence.

each of several separate experiments. It can be seen that GMV infection resulted in similar numbers of G418-resistant and factor-independent colonies in each case, whereas, as expected, ZIPNeo infection gave rise to G418-resistant colonies but not to factor-independent colonies. The large difference in colony numbers between the two sets of GMV-infection experiments (circles and triangles, respectively, in Fig. 3) reflects the greater efficiency of the cocultivation regime for infecting FD cells; in the latter experiments, up to 10% of clonogenic FD cells became factor-independent (data not shown). Single colonies of GMV-infected, factor-independent FD cells were removed from the agar plates and established as continuous cell lines in

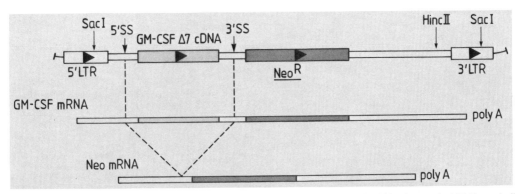

**Figure 2** Structure and transcripts of GMV. Cleavage sites of *Sac*I, *Bam*HI, and *Hinc*II within the proviral DNA are indicated as are 5′- and 3′-splice-site (SS) positions. Large arrowheads indicate the direction of transcription. The predicted structures of two viral mRNAs are indicated.

**Table 1** Colony-stimulating Activity of Media Conditioned by GMV-FD Cells

| Cell line | GM-CSF production (units/ml) | Percent colonies[a] | | | |
|---|---|---|---|---|---|
| | | G | GM | M | Eo |
| A1.1 | 2000 | 16 | 18 | 64 | 2 |
| A1.3 | 4000 | 18 | 46 | 36 | 0 |
| B5.1 | 4000 | 14 | 50 | 33 | 3 |
| B5.3 | 2000 | 10 | 38 | 50 | 2 |
| Purified recombinant GM-CSF 4000 units | | 28 | 26 | 41 | 5 |
| Purified native GM-CSF 4000 units | | 40 | 23 | 34 | 3 |

[a]Cultures of 75,000 C57Bl bone marrow cells were stimulated by 0.1. ml of test material diluted to contain 400 units of CSF. At 7 days, cultures were fixed and stained with Luxol Fast Blue–hematoxylin, and the frequency of various colony types was determined. Control cultures were stimulated using purified GM-CSF (400 units/culture) from bacterially expressed GM-CSF cDNA or from medium conditioned by concanavalin A–primed LB3 cells (a T-lymphocyte blast cell line). (G) Granulocyte; (GM) granulocyte-macrophage; (M) macrophage; (Eo) eosinophil.

medium lacking exogenous CSF. Thus we could conclude that GMV-infection of FD cells resulted in their ability to proliferate in the absence of added CSF.

### Analysis of CSF activity in GMV-FD cell lines

The media conditioned by the growth of GMV-FD cell lines was assayed for colony-stimulating activity in semisolid cultures of mouse bone marrow cells. The levels of CSF produced by $2 \times 10^6$ cells from various cloned cell lines ranged from 1000 to more than 4000 units/ml during a 5-day incubation period. As shown in Table 1, our analysis showed that GMV-FD–conditioned medium stimulated the formation of granulocyte and/or macrophage colonies and also small numbers of eosinophilic colonies. The relative frequencies of these colonies were similar to that observed after stimulation by comparable concentrations of purified GM-CSF either of natural or recombinant origin. Since the GMV-

FD–conditioned media stimulated the proliferation of FD cells (responsive to GM-CSF and Multi-CSF) but not of 32D cells (responsive to Multi-CSF; data not shown), we conclude that the activity corresponds to GM-CSF.

### Expression of GMV in FD cells

To establish that the GMV was transcriptionally active in individual GMV-FD cell lines, we examined poly(A)-containing mRNA from these lines by northern blotting. Hybridization to a GM-CSF cDNA probe (Fig. 4B) revealed that the GMV-infected FD cell lines each contained a single hybridizing species of RNA of 5.6 kb corresponding to the full-length genomic transcript (lanes 2–5). No mRNAs containing GM-CSF sequences were detected in the two FD cells lines established using ZIPNeo virus (lanes 6 and 7). When mRNA from the same GMV-infected FD cells was examined using a Neo[R] probe (Fig. 4A), two different size-classes

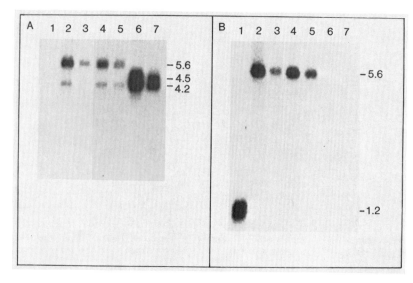

**Figure 4** Detection of virus-specific mRNAs encoding GM-CSF and the Neo[R] gene product. Polyadenylated RNA from the following cell lines were fractionated on a denaturing gel and analyzed by northern blotting with a Neo probe (*A*) (Southern and Berg 1982) and a GM-CSF cDNA probe (*B*) (Gough et al. 1984): LB3 cells stimulated with concanavalin A (lane *1*) (Kelso et al. 1984); GMV-infected cell lines B5.3, B5.1, A1.3, and A1.1, respectively (lanes *2–5*); ZIPNeo virus–infected cell lines E2.4 and E2.3, respectively (lanes 6 and 7).

of mRNA were observed (lanes 2 and 5), the 5.6-kb, full-length genomic mRNA and the 4.2-kb spliced mRNA. The ZIPNeo-infected FD cells contained two Neo$^R$ hybridizing mRNAs, the 4.2-kb spliced mRNA and the 4.5-kb, full-length genomic transcript (lanes 6 and 7). Thus, GM-CSF produced by GMV-FD cells is due to translation of viral mRNA.

## GMV-FD cells are leukemogenic

To establish whether the GMV-FD cell lines were leukemogenic, 10$^6$ cells from each of four GMV-FD cell lines were injected subcutaneously into syngeneic DBA/2 mice. As revealed in Table 2, all 32 mice that received these cells developed tumors at the site of injection (see Fig. 5A) and progressively growing leukemias, which killed the animals within 5–8 weeks after transplantation. By contrast, neither a ZIPNeo virus–infected cell line nor the parental FD cells were leukemogenic under an identical experimental regime during an observation period of 26 weeks.

To establish that the leukemic cells retained their ability for autonomous proliferation after growth in vivo, we plated 10$^4$ cells from tumor, spleen, and bone marrow tissues in agar cultures and determined the fraction of cells able to form colonies in the absence of GM-CSF. As shown in Table 2, we were able to recover clonogenic cells from the tumor masses themselves and in most cases from the tissues that showed pathological evidence of invading tumor cells (e.g., see Fig. 5B). Where tested, all of the clonogenic cells recovered in this way were cytogenetically indistinguishable from the parental FD cell lines containing eight characteristic metacentric chromosomes (data not shown). Thus we conclude that the disseminating leukemia arises through the unrestrained proliferation and spread of autonomous GMV-FD cells.

## Discussion

The conversion of normal cells into their tumorigenic counterparts is thought to be a multistage process involving several compounding genetic events. In recent years a number of cellular and viral genes have been identified and classified according to their ability to immortalize primary cells or to superimpose malignant characteristics on populations of immortalized cells (Land et al. 1983; Ruley 1983). Thus, it is assumed that one of the stages in neoplasia might be the emergence of premalignant immortalized cells; such cells might subsequently undergo genetic alterations affecting the genes encoding a growth factor on which the cells, proliferation depends (Todaro et al. 1977; Sporn and Todaro 1980; Sporn and Roberts 1985). Indeed, some transformed cells have been shown to secrete and respond to tumor growth factors (De Larco and Todaro 1978; Kaplan et al. 1982). Similarly, simian sarcoma virus (SSV) encodes a growth factor that is identical with platelet-derived growth factor (Doolittle et al. 1983; Waterfield et al. 1983), and transformation of cells infected with SSV is attributed, in part, to the autocrine production of PDGF, which stimulates their continuous proliferation.

Among populations of normal hemopoietic cells, the processes of growth and differentiation are tightly coupled and therefore the autonomous production of CSF by otherwise normal progenitor cells could not by itself lead to the emergence of a leukemic clone. This would only occur if there existed a nonleukemic progenitor cell that had a full capacity for self-generative proliferation. Autonomous production of CSF would then be a secondarily acquired characteristic of the leukemic clone. Indeed, GMV-infected bone marrow cells do not form permanent cell lines, and the infected cells are not leukemogenic upon injection into the syngeneic mouse (our unpublished results with Dr. G. Johnston).

Little is known about the development of immortalized hemopoietic cells in vivo, and clonogenic cells with extensive self-renewal potential have not been identified in primary bone marrow cultures. Consequently it is unclear what time interval elapses between the emergence of such clonogenic cells and the subsequent appearance of a leukemic clone. Nonetheless, it has proved relatively simple to establish continuous cell lines from ostensibly normal long-term marrow cultures in the presence of Multi-CSF. These cell lines are often blast cell–like and are characterized by an absolute dependence on CSF and by being nonleukemogenic. In recent years there have been a number of reports of factor-

**Table 2** Development and Clonogenicity of GMV-FD-induced Tumors

| Cell line | No. of tumors/ no. injected[a] | Time to death (weeks) | Metacentric chromosomes[b] | Recovery of clonogenic tumor cells (mean frequency/10$^4$ cells plated)[c] | | |
|---|---|---|---|---|---|---|
| | | | | tumor | spleen | bone marrow |
| A1.1 | 8/8 | 5–8 | + | 580 | 12 | 5 |
| A1.3 | 8/8 | 6 | + | 790 | 0 | 0 |
| B5.1 | 8/8 | 5–7 | + | 980 | 29 | 13 |
| B5.3 | 8/8 | 6–8 | + | 1980 | 3 | 50 |
| E2.1 | 0/8 | – | – | – | – | – |
| Normal FD | 0/22 | – | – | – | – | – |

[a]10$^6$ cells infected subcutaneously into DBA/2 mice.
[b]Scored on metaphase chromosome spreads prepared from tumor-derived GMV-FD colonies.
[c]Cell suspensions from the tumor mass, spleen, and bone marrow were plated in soft agar cultures in the absence of exogenous CSF, and colonies were counted after 7 days of incubation.

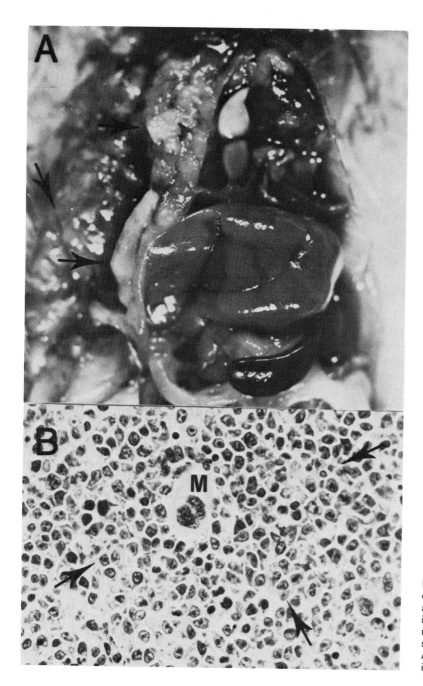

**Figure 5** Tumors caused by GMV-infected FD cells. (*A*) Mouse with transplanted leukemia after subcutaneous injection of 1 × 10⁶ GMV-infected FD cells. Note infiltrating subcutaneous tumor (arrowed). (*B*) Section of spleen showing massive infiltration of GMV-FD cells. A single surviving normal megakaryocyte (M) is indicated in the center.

independent subclones of hemopoietic cell lines that have arisen by spontaneous mutation (Schrader and Crapper 1983) or by transformation with Abelson murine leukemia virus (Ab-MLV) (Cook et al. 1985; Oliff et al. 1985; Pierce et al. 1985). In these cases factor independence arose concomitantly with the potential to grow as transplantable leukemic cells. Interestingly, the spontaneously arising mutants and the Ab-MLV–transformed cells must have achieved factor independence by different mechanisms since the former cells were shown to synthesize CSF, whereas no factor could be demonstrated in the media conditioned by the growth of the Ab-MLV transformants, and furthermore no CSF transcripts could be detected by northern blotting. Thus,

expression of the v-*abl* oncogene must short-circuit the requirement for interaction between CSF and its receptor and directly activate molecules involved in the post-receptor cascade.

The factor-independent cell lines generated in the present study have clearly become autonomous by the constitutive expression of viral GM-CSF sequences. Importantly, release from dependence on exogenous growth factors, as in the case of the spontaneous and Ab-MLV transformants, accompanies the conversion to a highly leukemogenic phenotype and therefore represents a final event in the progression to a fully leukemogenic state.

Could autonomous growth represent an important

stage in the development of human myeloid leukemia? Although many murine and human leukemic cells that have been passaged in vitro are factor-independent, primary cultures of human myeloid leukemic cells exhibit absolute dependency on exogenous CSF for proliferation in vitro. As such, myeloid leukemias do not represent examples of autocrine stimulation of growth. However, it is not clear that the leukemic clones growing in conventional cultures in response to CSF stimulation are derived from the true leukemic stem cells since the leukemic cells within such colonies do not always possess extensive proliferative capacity. Alternatively, if the true leukemic stem cells are CSF-dependent, then they may proliferate in microenvironments that contain sufficient levels of CSF. By contrast, the parental FD factor-dependent cell lines used in the present study might not gain access to these privileged sites after injection into the animal and as such fail to develop into tumors. In this scheme the factor-dependent FD cells may represent potentially malignant cells that fail to grow as tumors because of the unavailability of sufficient levels of circulating CSF to sustain their proliferation. We are presently trying to distinguish between these possibilities.

## Acknowledgments

We thank Richard Mulligan for providing the parental vector and ψ2 cell line, Tony Burgess for his continued support and encouragement, and Joanne Wright for her invaluable help in the preparation of the manuscript.

## References

Cepko, C.L., B.E. Roberts, and R.C. Mulligan. 1984. Construction and applications of a highly transmissible retrovirus shuttle vector. *Cell* **37**: 1053.

Cook, W.D., D. Metcalf, N.A. Nicola, A.W. Burgess, and F. Walker. 1985. Malignant transformation of a growth factor–dependent myeloid cell line by Abelson virus without evidence of an autocrine mechanism. *Cell* **41**: 667.

De Larco, J.E. and G.J. Todaro. 1978. Growth factors from murine sarcoma virus–transformed cells. *Proc. Natl. Acad. Sci.* **75**: 4001.

Dexter, T.M., J. Garland, D. Scott, E. Scolnick, and D. Metcalf. 1980. Growth of factor-dependent hemopoietic precursor cell lines. *J. Exp. Med.* **152**: 1036.

Doolittle, R.F., M.W. Hunkapiller, S.G. Devare, K.C. Robbins, S.A. Aaronson, and H.N. Antoniades. 1983. Simian sarcoma virus *onc* gene v-sis is derived from the gene (or genes) encoding a platelet-derived growth factor. *Science* **221**: 275.

Downward, J., Y. Yarden, E. Mayes, G. Scrace, N. Totty, P. Stockwell, A. Ullrich, J. Schlessinger, and M.D. Waterfield. 1984. Close similarity of EGF receptor and v-*erbB* oncogene protein sequences. *Nature* **307**: 521.

Dunn, A.R., D. Metcalf, E. Stanley, D. Grail, J. King, E.C. Nice, A.W. Burgess, and N.M. Gough. 1985. Biological characterization of regulators encoded by cloned hematopoietic growth factor gene sequences. *Cancer Cells* **3**: 227.

Fung, M.C., A.J. Hapel, S. Ymer, D.R. Cohen, R.M. Johnson, H.D. Campbell, and I.G. Young. 1984. Molecular cloning of cDNA for murine interleukin-3. *Nature* **307**: 233.

Gough, N.M., D. Metcalf, J. Gough, D. Grail, and A.R. Dunn. 1985. Structure and expression of the mRNA for murine granulocyte-macrophage colony-stimulating factor. *EMBO J.* **4**: 645.

Gough, N.M., J. Gough, D. Metcalf, A. Kelso, D. Grail, N.A. Nicola, A.W. Burgess, and A.R. Dunn. 1984. Molecular cloning of cDNA encoding a murine haematopoietic growth regulator, granulocyte-macrophage colony-stimulating factor. *Nature* **309**: 763.

Graham, F.L. and A.J. van der Eb. 1973. A new technique for the assay of infectivity of human adenovirus 5 DNA. *Virology* **52**: 456.

Greenberger, J.S., M.A. Sakakeeny, R.K. Humphries, C.J. Eaves, and R.J. Eckner. 1983. Demonstration of permanent factor-dependent multipotential (erythroid/neutrophil/basophil) hematopoietic progenitor cell lines. *Proc. Natl. Acad. Sci.* **80**: 2931.

Hapel, A.J., H.S. Warren, and D.A. Hume. 1984. Different colony-stimulating factors are detected by the "interleukin-3" dependent cell lines FDC-P1 and 32D cl-23. *Blood* **46**: 786.

Hapel, A.J., J.C. Lee, W.C. Farrar, and J.N. Ihle. 1981. Establishment of continuous cultures of Thy 1.2+ Lyt 1+2- T cells with purified interleuken 3. *Cell* **25**: 179.

Kaplan, P.L., M. Anderson, and B. Ozanne. 1982. Transforming growth factor(s) production enables cells to grow in the absence of serum: An autocrine system. *Proc. Natl. Acad. Sci.* **79**: 485.

Kelso, A., H.R. MacDonald, K.A. Smith, J.-C. Cerottini, and K.T. Brunner. 1984. Interleukin 2 enhancement of lymphokine secretion by T lymphocytes: Analysis of established clones and primary limiting dilution microcultures. *J. Immunol.* **132**: 2932.

Land, H., L.F. Parada, and R.A. Weinburg. 1983. Tumorigenic conversion of primary embryo fibroblasts requires at least two cooperating oncogenes. *Nature* **304**: 596.

Maniatis, T., E.F. Fritsch, and J. Sambrook. 1982. *Molecular cloning: A laboratory manual.* Cold Spring Harbor Laboratory, Cold Spring Harbor, New York.

Mann, R., R.C. Mulligan, and D. Baltimore. 1983. Construction of a retrovirus packaging mutant and its use to produce helper-free defective retrovirus. *Cell* **33**: 153.

Metcalf, D. 1985. Molecular control of granulocyte and macrophage production. In *Hemoglobin switching* (ed. G. Stamatoyannopoulos and A. Nienhuis). Academic Press, New York. (In press.)

Oliff, A., O. Agranovsky, M.D. McKinney, V.V.S. Murty, and R. Banchwitz. 1985. Friend murine leukemia virus–immortalized myeloid cells are converted into tumorigenic cell lines by Abelson leukemia virus. *Proc. Natl. Acad. Sci.* **82**: 3306.

Pierce, J.H., P.P. Di Fione, S.A. Aaronson, M. Potter, J. Pumphrye, A. Scott, and J.N. Ihle. 1985. Neoplastic transformation of mast cells by Abelson MuLV: Abrogation of IL-3 dependence by nonautocrine mechanism. *Cell* **41**: 685.

Ruley, H.E. 1983. Adenovirus early region 1A enables viral and cellular transforming genes to transform primary cells in culture. *Nature* **304**: 602.

Schrader, J.W. and R.M. Crapper. 1983. Autogenous production of a hemopoietic growth factor "P cell stimulating factor" as a mechanism for transformation of bone marrow-derived cells. *Proc. Natl. Acad. Sci.* **80**: 6892.

Southern, P.J. and P. Berg. 1982. Transformation of mammalian cells to antibiotic resistance with a bacterial gene under control of the SV40 early region promoter. *J. Mol. Appl. Genet.* **1**: 327.

Sporn, M.B. and A.B. Roberts. 1985. Autocrine growth factors and cancer. *Nature* **313**: 745.

Sporn, M.B. and G.J. Todaro. 1980. Autocrine secretion and malignant transformation of cells. *N. Engl. J. Med.* **303**: 878.

Todaro, G.J., J.E. De Larco, S.P. Nissley, and M.M. Rechler. 1977. MSA and EGF receptors on sarcoma virus–transformed cells and human fibrosarcoma cells in culture. *Nature* **267**: 526.

Waterfield, M.D., G.T. Scrace, N. Whittle, P. Stroobant, A. Johnson, A. Wasteson, B. Westermark, C.-H. Heldin, J.S. Huang, and T.F. Deuel. 1983. Platelet-derived growth factor is structurally related to the putative transforming protein p28$^{sis}$ of simian sarcoma virus. *Nature* **304**: 35.

# DNA Affinity Purification of Nuclear Factor I and Characterization of Its Recognition Site

**P.J. Rosenfeld,\* R.J. Wides,\* M.D. Challberg,[†] and T.J. Kelly\***

\*Department of Molecular Biology and Genetics, The Johns Hopkins University School of Medicine, Baltimore, Maryland 21205; [†]Laboratory of Viral Diseases, National Institute of Allergy and Infectious Diseases, National Institutes of Health, Bethesda, Maryland 20892

Nuclear factor I (NFI) is a cellular, sequence-specific DNA-binding protein that enhances the initiation of adenovirus DNA replication in vitro. We have developed a purification method for NFI based upon the high-affinity interaction between NFI and its recognition site. The essential feature of the method is a two-step column chromatographic procedure involving fractionation on a nonspecific DNA affinity matrix followed by a specific DNA recognition-site affinity matrix. The specific affinity matrix is prepared from a plasmid that contains 88 concatenated NFI-binding sites arranged exclusively in a direct head-to-tail configuration. Using this purification scheme, we have obtained a 2400-fold purification of NFI from crude HeLa nuclear extract with a 57% recovery of specific DNA-binding activity.

The essential base pairs that constitute the specific NFI recognition site have been identified by studying the affinity of NFI for DNAs that contain base substitutions. In conjunction with studies that compared numerous viral and cellular NFI recognition sequences, we have identified the idealized NFI-binding site as the symmetrical sequence TTGGCNNNNNGCCAA.

Initiation of adenovirus DNA replication takes place at the ends of the viral genome and involves the formation of a covalent complex between a dCMP residue and the virus-encoded preterminal protein (pTP) (Challberg et al. 1980, 1982; Lichy et al. 1981; Pincus et al. 1981; Tamanoi and Stillman 1982). Analysis of the initiation reaction in vitro and in vivo has established that formation of the pTP-dCMP complex requires the presence of specific nucleotide (nt) sequence elements at the terminus of the template (Tamanoi and Stillman 1982, 1983; van Bergen et al. 1983; Challberg and Rawlins 1984; Guggenheimer et al. 1984; Lally et al. 1984; Rawlins et al. 1984; Hay et al. 1985; Pearson and Wang 1985). These sequences, which collectively constitute the viral origin of DNA replication, are contained within the first 67 bp of the adenovirus genome. Moreover, the origin can be functionally divided into at least two functionally distinct domains (Guggenheimer et al. 1984; Ralwins et al. 1984). The first 18 bp constitute the minimal origin of DNA replication, and nt 19–67 are required for optimal levels of replication.

Both virus-encoded and cell-encoded proteins are required for initiation of adenovirus DNA replication. The viral initiation proteins include the 80-kD pTP and the 140-kD adenovirus DNA polymerase (Challberg et al. 1980; Enomoto et al. 1981; Lichy et al. 1982; Stillman et al. 1982; Ostrove et al. 1983). These two proteins are sufficient for initiation of replication, but in the absence of additional factors the reaction proceeds at a suboptimal rate. The addition of nuclear extract from uninfected HeLa cells greatly enhances the efficiency of initiation (Enomoto et al. 1981; Nagata et al. 1982, 1983; Stillman et al. 1982; Ostrove et al. 1983). Nagata et al. (1982)

partially purified a stimulatory factor from such extracts that they designated nuclear factor I (NFI).

The enhancement of initiation by NFI appears to be mediated via its interaction with a specific sequence within the adenovirus origin of DNA replication (Nagata et al. 1983; Guggenheimer et al. 1984; Rawlins et al. 1984; de Vries et al. 1985; Leegwater et al. 1985). DNase I protection experiments of bound NFI revealed a single footprint within the region of the origin required for optimal replication efficiency (between nt 19 and 67). The ability of NFI to recognize various mutant replication origins containing base substitutions or deletions has been shown to correlate directly with its ability to enhance the initiation reaction in vitro.

We have developed a purification method based upon the high-affinity interaction between NFI and its recognition site. The approach may be generally applicable to the purification of other site-specific DNA-binding proteins. An affinity matrix was prepared using plasmid DNA that contains 88 copies of the NFI-binding site. This plasmid was constructed by means of a novel cloning strategy that generated concatenated NFI-binding sites arranged exclusively in the direct head-to-tail orientation. Purification of NFI was effected by a two-step column chromatographic procedure in which proteins were first fractionated on a matrix consisting of nonspecific (*Escherichia coli*) DNA and then on the recognition-site affinity matrix. During the first step, NFI activity coeluted with proteins that have a similar general affinity for DNA. During the second step, NFI eluted at a much higher ionic strength than the contaminating, nonspecific DNA-binding proteins. These two steps, combined with one stage of ion-exchange chromatography, resulted in a 2400-fold purification of NFI from crude HeLa nuclear extract with a final yield of 57%. Such highly purified NFI had not been previously available for detailed biochemical analysis. We have analyzed the physical and biochemical properties of the purified protein and continued to study its role in adenovirus DNA replication. In addition, binding studies with DNA molecules containing a variety of base-substitution mutants have allowed the identification of the specific nucleotide pairs that constitute the NFI recognition site.

## Methods

### Purification of NFI

#### Buffers

The buffers used in the purification of NFI contained the protease inhibitors pepstatin A, chymostatin, and antipain at a concentration of 1 $\mu$g/ml. The composition of Buffer W is 137 mM NaCl, 10 mM $Na_2HPO_4$, 2 mM $KH_2PO_4$, 2.7 mM KCl, 0.5 mM $MgCl_2$. Buffer H is 25 mM HEPES (pH 7.5), 5 mM KCl, 1 mM $MgCl_2$, 1 mM DTT, 1 mM PMSF. Buffer E is 25 mM HEPES (pH 7.5), 10% sucrose, 0.01% NP-40, 1 mM DTT, 1 mM PMSF. Buffer D is 25 mM HEPES (pH 7.5), 40% glycerol, 0.01% NP-40, 1 mM DTT, 2 mM EDTA, 0.1 mM PMSF. Buffer S is 25 mM HEPES (pH 7.5), 20% glycerol, 0.01% NP-40, 1 mM DTT, 1 mM EDTA, 0.1 mM PMSF.

#### Nitrocellulose filter-binding assay

The DNAs used in the parallel nitrocellulose filter-binding assays were pKP45 and pKR67. The plasmid pKR67 contains the specific recognition sequence for NFI, and the plasmid pKP45 contains only vector sequences that serve as nonspecific DNA. Plasmids were linearized with *Eco*RI and labeled at their 3' termini ($4.0 \times 10^4$ cpm/fmole) by incubating with [$\alpha$-$^{32}$P]dATP and [$\alpha$-$^{32}$P]dTTP (3000 Ci/mmole) in the presence of *Micrococcus luteus* polymerase. The standard filter-binding assay used in the purification of NFI was performed in parallel with labeled pKP45 and pKR67. Each assay (50 $\mu$l) contained 25 mM HEPES (pH 7.5), 150 mM NaCl, 5 mM $MgCl_2$, 1 mM DTT, 2% glycerol, 5 $\mu$g BSA, 5 $\mu$g sheared *E. coli* DNA, and 0.1 nM $^{32}$P-labeled DNA. Protein fractions were diluted (1:50–1:200) in Buffer S containing 25 mM NaCl, and 5 $\mu$l of the dilution was added to reaction mixtures containing $^{32}$P-labeled pKR67 or $^{32}$P-labeled pKP45. The binding assays were incubated at 4°C for 30 min and then filtered at a rate of 25 ml/hr through a nitrocellulose membrane mounted on a Schleicher and Schuell Minifold apparatus. Each sample was washed once with 0.5 ml of 25 mM HEPES (pH 7.5), 150 mM NaCl, 2% glycerol, 5 mM $MgCl_2$, 1 mM DTT. Filters were dried and radioactive DNA bound to the filter was quantitated by liquid scintillation counting. Each assay was performed in duplicate and the average values were plotted. Duplicate samples varied by less than 5%.

The competition nitrocellulose filter–binding assay was performed using the purified preparation of NFI as

previously described (Rawlins et al. 1984) with the following modifications. Radioactive DNA was prepared as described previously for the standard nitrocellulose filter–binding assays. The reaction mixtures contained 50 pM $^{32}$P-labeled pKR67 and form-I plasmid DNA was used as competitor.

#### Preparation of DNA-celluloses

DNA was adsorbed to cellulose according to the method of Albert and Herrick (1971) with the following modifications. The nonspecific DNA-cellulose was prepared from high-molecular-weight *E. coli* DNA (Miura 1967). The specific affinity matrix was prepared from pKB67-88 DNA linearized with *Pst*I. DNA (1 mg/ml) was mixed with cellulose (0.5 g/ml DNA) and air-dried for 48 hr, followed by lyophilization for 12 hr. The matrices were hydrated in Buffer S containing 2 M NaCl and stored as a frozen slurry (−20°C). The amount of DNA bound to cellulose (1 mg DNA/ml packed matrix) was estimated by $A_{260}$ measurements of the column effluent during the packing and equilibration steps.

#### Preparation of HeLa nuclear extract

S-3 HeLa cells were propagated at 37°C in suspension culture in Eagle's minimal essential medium supplemented with 5% horse serum. Each liter of cells was grown to a density of $4 \times 10^5$ to $5 \times 10^5$ cells/ml and collected by centrifugation at 3000$g$ for 5 min and washed twice with 15 ml of Buffer W. The washed cell pellets were frozen at −70°C. A total of $6 \times 10^{10}$ cells (120 g) were used in the purification. Each gram of HeLa cells was thawed at 4°C and resuspended in 5 ml of Buffer H containing 0.2% NP-40. Cells were completely disrupted by 10 strokes with a tight-fitting dounce homogenizer, and nuclei were collected by centrifugation at 1000$g$ for 5 min. The nuclear pellet was washed once with 5 ml of Buffer H containing 0.01% NP-40 and once with 5 ml of Buffer E. Nuclei were resuspended in 2.0 ml of Buffer E containing 0.35 M NaCl and incubated on ice for 60 min. The residual nuclei were removed by centrifugation at 10,000$g$ for 30 min and an equal volume of Buffer D was added to the supernatant.

#### Bio-Rex 70 chromatography

The diluted nuclear extract (900 ml, 4590 mg) was loaded onto a Bio-Rex 70 column (200 ml, 5 cm × 10 cm) that had been preequilibrated with 200 mM NaCl in Buffer S. The matrix was washed with 200 mM NaCl in Buffer S (600 ml), and the DNA-binding activity was eluted with a linear gradient from 200 to 600 mM NaCl in Buffer S (800 ml). The peak of specific DNA-binding activity eluted at 350 mM NaCl. Fractions were pooled between 300 and 400 mM NaCl (215 ml) and diluted with an equal volume of Buffer S.

#### E. coli DNA–cellulose chromatography

The diluted pool of activity from the Bio-Rex 70 column (430 ml, 550 mg) was loaded onto an *E. coli* DNA–cellulose column (60 ml, 2.5 cm × 12 cm) that had been preequilibrated with 200 mM NaCl in Buffer S. The matrix was washed with 200 mM NaCl in Buffer S (200 ml) and the DNA-binding activity was eluted with a linear gradient from 200 to 500 mM NaCl in Buffer S (300 ml). The gradient elution was followed by a 2 M NaCl step elution. The peak of specific DNA-binding activity eluted between 280 and 350 mM NaCl. Fractions were pooled (75 ml) and diluted with an equal volume of Buffer S.

### pKB67-88 DNA–cellulose chromatography

The diluted eluate from the *E. coli* DNA–cellulose column (62.5 mg, 75 ml) was loaded onto the specific DNA-cellulose column (5 ml, 1 cm × 6.3 cm) that had been preequilibrated with 200 mM NaCl in Buffer S. The column was washed with 200 mM NaCl in Buffer S (20 ml), and proteins were eluted with a linear gradient of 200–500 mM NaCl (30 ml). The specific DNA-binding activity was recovered in the 2 M NaCl step elution (5.28 ml, 2.1 mg). This eluate was diluted with 9 volumes of Buffer S and reapplied to the same pKB67-88 DNA–cellulose column that had been preequilibrated with 200 mM NaCl in Buffer S. The column was washed with 0.2 M NaCl (10 ml) followed by step elutions at 450 mM NaCl (20 ml) and 2 M NaCl (20 ml). The specific DNA-binding activity eluted with the 2 M NaCl step (4.28 ml, 1.1 mg). The eluate was dialyzed against 100 mM NaCl in Buffer S and stored at −70°C.

## Results

### Construction of a plasmid containing multiple copies of the adenovirus origin of replication

An attractive approach to the purification of a site-specific DNA-binding protein is chromatography on a DNA affinity matrix that contains the specific recognition sequence. This approach has not been used in practical purification procedures, but previous studies have provided evidence of its feasibility. Herrick (1980) demonstrated that *lac* repressor bound more tightly to a column matrix prepared from a plasmid containing a single *lac* operator site than to a similar matrix prepared with nonspecific DNA. Oren et al. (1980) obtained similar results with SV40 T antigen, a site-specific DNA-binding protein that recognizes a sequence within the SV40 origin of DNA replication. Buffers with higher ionic strength were required to elute T antigen from SV40 DNA–cellulose than from calf thymus DNA–cellulose. Moreover, the proportion of specifically bound T antigen that eluted at the highest ionic strength was increased significantly by preparing the affinity matrix from the DNA of an SV40 variant that contained five origins of replication. This observation suggested that a useful affinity matrix should contain a high ratio of specific to nonspecific binding sites. In the present study we maximized this ratio by engineering a plasmid that contains multiple copies (88) of the NFI-binding site. By amplifying the number of binding sites in a plasmid, we also reduced the molar amount of plasmid that would be required to prepare an affinity matrix of a given capacity.

The strategy used to generate a plasmid with multiple copies of the NFI-binding site is outlined in Figures 1

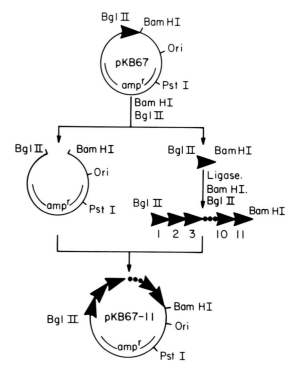

**Figure 1** Construction of a plasmid containing 11 copies of the NFI-binding site. The multiple-binding-site plasmid was constructed from the plasmid pKB67, which contains nt 1–67 of the Ad2 genome inserted between cleavage sites for the restriction enzymes *Bgl*II and *Bam*HI. After cleavage of pKB67 with *Bgl*II and *Bam*HI, the 73-bp fragment containing the NFI-binding site was purified and incubated with T4 DNA ligase in the presence of *Bgl*II and *Bam*HI. Concatemers containing 73-bp fragments oriented in a direct head-to-tail configuration were reinserted between the *Bgl*II and *Bam*HI sites of the parental plasmid. The largest insert detected during subsequent screening contained 11 direct repeats and was designated pKB67-11. (Reprinted, with permission, from Rosenfeld and Kelly 1986.)

and 2. The starting point for the construction was the plasmid pKB67, which contains the terminal 67 bp of the adenovirus type-5 (Ad5) genome inserted between cleavage sites of the restriction enzymes *Bgl*II and *Bam*HI. Digestion of pKB67 with *Bgl*II and *Bam*HI released a 73-bp fragment, which served as the basic repeat unit for the multiple-binding-site plasmid. Since cleavage with *Bgl*II and *Bam*HI yields fragments with complementary 5′ termini, incubation of the 73-bp fragment with T4 ligase yielded three types of joints: *Bgl*II/*Bgl*II, *Bam*HI/*Bam*HI, and *Bgl*II/*Bam*HI. Inclusion of the restriction enzymes *Bgl*II and *Bam*HI in the ligation mixture resulted in the cleavage of *Bgl*II/*Bgl*II and *Bam*HI/*Bam*HI joints but not the *Bgl*II/*Bam*HI hybrid joints. Thus, the ligation products contained the 73-bp fragments oriented exclusively in the direct head-to-tail configuration. This configuration is known to be more stable in *E. coli* than the inverted configuration. Using this protocol, we initially constructed the plasmid pKB67-11, which contains 11 direct repeats of the NFI-binding site. The insert in pKB67-11 was subsequently amplified in twofold steps as described in Figure

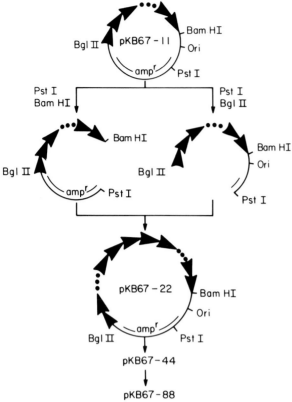

**Figure 2** Construction of a plasmid containing 88 copies of the NFI-binding site. The multiple direct repeats in the plasmid pKB67-11 were amplified twofold by cleaving pKB67-11 with *Pst*I followed by *Bam*HI or *Bgl*II. The *Pst*I/*Bam*HI and the *Pst*I/*Bgl*II fragments containing the concatemers were isolated and ligated together. The appropriate recombinant containing 22 tandem repeats (pKB67-22) was identified by restriction analysis after transformation of HB101 and ampicillin selection. Two additional rounds of amplification resulted in the construction of pKB67-88, which contained 88 copies of the NFI-binding site. (Reprinted, with permission, from Rosenfeld and Kelly 1986.)

1 to obtain a plasmid with 88 repeats (pKB67-88). The insert in pKB67-88 was quite stable when the plasmid was propagated in HB101, a *recA* host. However, the addition of chloramphenicol prior to the isolation of plasmid DNA results in accumulation of deletion mutations within the repeated sequences. For this reason, plasmid DNA was purified from bacterial cultures grown to saturation without chloramphenicol amplification.

## Purification of NFI

To monitor DNA-binding activity during purification, we developed a rapid and quantitative nitrocellulose filter-binding assay (Rosenfeld and Kelly 1986). Each fraction was assayed in two parallel assay mixtures, one containing radioactive pKR67, a plasmid that harbors the terminal 67 bp of the adenovirus genome, and the other containing radioactive pKP45, a plasmid that harbors no adenovirus sequences. Both assay mixtures contained a 1000-fold excess of nonradioactive, sheared *E. coli* DNA that served as competitor for nonspecific DNA-binding proteins in crude nuclear-extract and partially purified fractions. Specific NFI-binding activity was calculated by subtracting the binding activity observed with pKP45 from that observed with pKR67. The usefulness of this approach is demonstrated by the elution profile from the Bio-Rex 70 column, the first fractionation step in the purification procedure (Fig. 3). Despite high levels of nonspecific binding, the fractions containing the specific NFI-binding activity were easily detected. Bio-Rex 70 chromatography resulted in an 8.7-fold purification with an apparent recovery of specific DNA-binding activity of 100%.

The NFI activity recovered from the Bio-Rex 70 column was loaded onto an *E. coli* DNA-cellulose column and eluted with a linear gradient of NaCl (200–500 mM). In this step the NFI activity cochromatographed with DNA-binding proteins that had a similar affinity for nonspecific DNA (Fig. 4). DNA-binding proteins with very high affinity for nonspecific DNA were effectively removed, thus improving the efficiency of the subsequent recognition-site affinity step. *E. coli* DNA-cellulose chromatography resulted in a 6.5-fold purification of NFI with an 83% yield. The fractions containing NFI activity were loaded onto a column of pKB67-88 DNA-cellulose, the recognition site affinity matrix. The column was developed under the same conditions as the *E. coli* DNA-cellulose column, and the bulk of the protein eluted at the same position in the gradient (Fig. 5). However, the specific DNA-binding activity was shifted to the 2 M NaCl step elution, resulting in a large (25-fold) purification of NFI. Chromatography of the active fractions on the recognition-site affinity matrix a second time provided an additional 1.7-fold purification. The overall purification of NFI achieved by the method described here was about 2400-fold. The overall yield was 57% (Table 1).

**Table 1** Affinity Purification of Nuclear Factor I

| Fraction | Total protein (mg) | Volume (ml) | Sp. act.[a] ($\times 10^{-3}$) | Purification (-fold) | Yield (%) |
|---|---|---|---|---|---|
| I. HeLa nuclear extract[b] | 4590 | 900.0 | 3.1 | 1.0 | 100 |
| II. Bio-Rex 70 | 550 | 215.0 | 27.1 | 8.7 | 104 |
| III. *E. coli* DNA cellulose | 65.2 | 75.0 | 181.0 | 58.4 | 83 |
| IV. pKB67-88 DNA cellulose I | 2.1 | 5.25 | 4510.0 | 1455 | 67 |
| V. pKB67-88 DNA cellulose II | 1.1 | 4.28 | 7517.0 | 2425 | 57 |

[a]Specific $^{32}$P-labeled DNA bound (fm)/mg protein.
[b]$6 \times 10^{10}$ (120 g) HeLa cells.

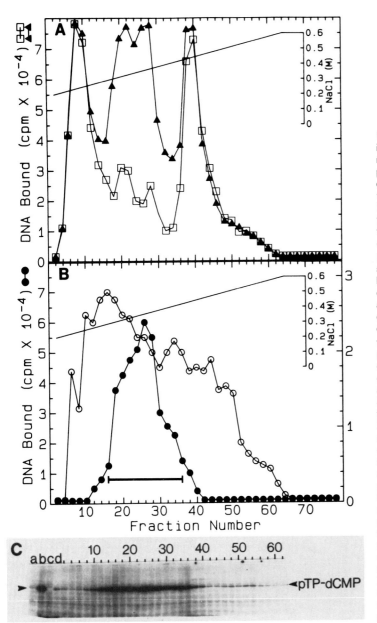

**Figure 3** Fractionation of HeLa nuclear extract by ion-exchange chromatography on Bio-Rex 70. HeLa nuclear extract was prepared as described in Experimental Procedures and loaded onto a Bio-Rex 70 column at 200 mM NaCl. The column was developed with a linear gradient between 200 mM and 600 mM NaCl. The elution position of NFI DNA-binding activity was determined by parallel nitrocellulose filter–binding assays. The difference in the amount of radioactivity bound to nitrocellulose filters in the presence of $^{32}$P-labeled specific (pKR67) and nonspecific (pKP45) DNAs represented specific NFI binding activity. Additionally, the elution of NFI-dependent adenovirus DNA replication activity was assayed by the formation of pTP-dCMP initiation complexes. (*A*) The DNA-binding activity was detected by assaying aliquots (0.1 μl) of the column eluate with $^{32}$P-labeled pKP45 (□) and pKR67 (▲). (*B*) The profile of specific NFI DNA-binding activity (●) was calculated by subtracting the DNA-binding activity observed with nonspecific DNA from the binding activity associated with specific DNA. Fractions 16–36 were pooled for subsequent purification. The protein concentrations (○) were determined by the Bradford dye assay. (*C*) The stimulation of pTP-dCMP complex formation was assayed by incubating cytoplasmic extract from Ad5-infected HeLa cells, adenovirus DNA–protein complex (except in lane c), and [α-$^{32}$P]dCTP as the only deoxynucleoside triphosphate with uninfected nuclear extract or fractions eluted from the Bio-Rex 70 column. The pTP-dCMP complexes were analyzed by SDS-polyacrylamide gel electrophoresis followed by autoradiography. (Lane a) No addition of protein from uninfected cells; (lane b) addition of uninfected crude nuclear extract (1 μl); (lane c) addition of uninfected crude nuclear extract with deproteinized adenovirus DNA as template; (lane d) flowthrough fraction; (lanes 2–62) elution fractions from the Bio-Rex 70 column. (Reprinted, with permission, from Rosenfeld and Kelly 1986.)

## Characterization of purified NFI

Throughout the purification procedure, the specific DNA-binding activity copurified with an activity that enhanced the efficiency of initiation of adenovirus DNA replication in vitro (see Figs. 3C, 4B, and 5B). The initiation assay monitors the formation of a covalent linkage between pTP and [α-$^{32}$P]dCTP (Challberg et al. 1980, 1982; Lichy et al. 1981; Pincus et al. 1981; Tamanoi and Stillman 1982). Thus, the purified protein has the biological activity expected of NFI.

SDS-polyacrylamide gel electrophoresis of the purified protein revealed a population of polypeptides, most of which fell in a narrow relative-molecular-mass range of 52–66 kD (Rosenfeld and Kelly 1986). The major polypeptides were shown to be related in primary sequence by partial proteolysis experiments, and they all

cosedimented with the specific DNA-binding activity in glycerol density gradients (data not shown). The native molecular mass of NFI was deduced from gel filtration and sedimentation analyses performed in buffers containing 0.5 M NaCl. Under these conditions the sedimentation coefficient was 4.0S and the Stokes radius was 33 Å. Assuming a partial sequence volume of 0.725 cm³/gm, the calculated molecular mass was 55 kD, in good agreement with the results obtained by SDS-polyacrylamide gel electrophoresis.

The DNA-binding properties of affinity-purified NFI were compared with those described previously for NFI purified by conventional methods (Nagata et al. 1982, 1983; Rawlins et al. 1984). DNase I protection analysis revealed a footprint within the adenovirus origin of replication (data not shown). The protein protected the seg-

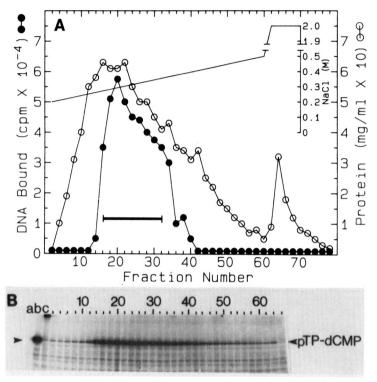

**Figure 4** Chromatography of NFI on *E. coli* DNA–cellulose. The Bio-Rex 70 elution fractions that contained the specific DNA-binding activity were pooled, diluted to 200 mM NaCl, and loaded onto an *E. coli* DNA–cellulose column as described in Experimental Procedures. The elution profiles of the specific DNA-binding activity and the pTP-dCMP stimulatory activity were determined. (*A*) (●) The profile of specific DNA-binding activity, as determined by parallel nitrocellulose filter–binding assays. Fractions *16–32* were pooled for further purification. (○) The protein concentrations, as determined by the Bradford dye assay. (*B*) Fractions were assayed for their ability to stimulate formation of the pTP-dCMP complex. (Lane *a*) No addition of protein from uninfected cells; (lane *b*) addition of uninfected, crude nuclear extract; (lane *c*) flow-through fraction; (lanes *2–66*) elution fractions from the *E. coli* DNA–cellulose column. (Reprinted, with permission, from Rosenfeld and Kelly 1986.)

ment from nt 21 to nt 44 on the viral strand whose 5′ end lies at the terminus of the genome. The protected region on the other strand extended from nt 19 to nt 42 (see Fig. 7). This protected region closely approximates the previously identified NFI-binding site. The affinity of NFI for its recognition site was determined by Scatchard analysis. The measured dissociation constant for specific binding was $2.1 \times 10^{-11}$ M.

### Analysis of the NFI recognition site

As an approach to defining the base pairs required for specific binding of NFI, we measured the effect of

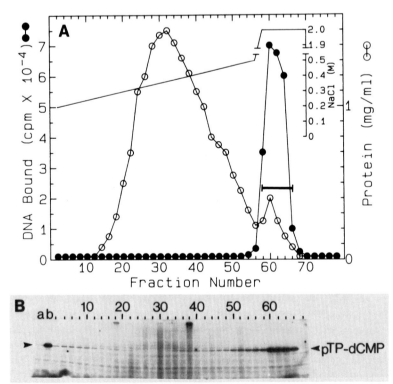

**Figure 5** Purification of NFI by DNA recognition-site affinity chromatography. The fractions containing specific DNA-binding activity from the *E. coli* DNA–cellulose column were pooled, diluted to 200 mM NaCl, and loaded onto a pKB67-88 DNA–cellulose column as described in Experimental Procedures. (*A*) The bulk of the protein (○) eluted from the recognition-site DNA affinity column at a position similar to its elution position from the *E. coli* DNA–cellulose column. The specific DNA-binding activity (●) eluted at 2 M NaCl. (*B*) Fractions were assayed for their ability to stimulate the formation of the pTP-dCMP initiation complex. (Lane *a*) No addition; (lane *b*) addition of uninfected, crude nuclear extract; (lanes *2–68*) fractions from the pKB67-88 DNA–cellulose column. (Reprinted, with permission, from Rosenfeld and Kelly 1986.)

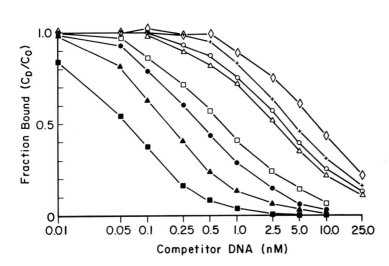

**Figure 6** Determination of the relative affinity of NFI for various mutant DNA molecules of competition binding assays. Fixed concentrations of purified NFI and $^{32}$P-labeled pKB67 DNA (wild-type origin of replication) were incubated with various concentrations of unlabeled competitor DNA, and the extent of binding was determined by a quantitative nitrocellulose filter assay (Rawlins et al. 1984; Rosenfeld and Kelly 1986). The concentration of radioactive DNA bound to NFI at each competitor concentration ($C_D$) was normalized to the concentration of radioactive DNA bound to protein in the absence of competitor ($C_o$). The structures of the mutants studied in this experiment are summarized in the legend to Fig. 7. For simplicity of presentation the mutants are divided into several classes. The wild-type DNA competition profile (▲) was identical with the profile observed with the point mutants $pm17$, $pm18$, $pm19$, $pm20$, $pm21$, $pm30$, $pm31$, $pm32$, $pm33$, $pm40$, and $pm42$. The nonspecific DNA competition profile (pUC9, ◇) was identical with the profile observed with the point mutants $pm26$, $pm35$, and $pm36$. Identical competition curves were obtained with DNAs containing point mutations at positions 24 (+) and 25. The competition profiles of $pm27$ (○), $pm34C$, and $pm37$ were superimposable. The remaining competition profiles are as follows: $pm28$ (■); $pm29$ (●); $pm34T$ (□); $pm38$ (△).

base-substitution mutations on the affinity of the protein for its recognition site. The mutations were generated by the synthetic oligonucleotide procedure (Norris et al. 1983; Zoller and Smith 1983, 1984) and were targeted to the region of the adenovirus origin that is protected by DNase I cleavage by NFI (see Fig. 7). The effects of the mutations were quantitated by a competition filter-binding assay (Rawlins et al. 1984; Rosenfeld and Kelly 1986). In this assay fixed quantities of NFI and a radioactive plasmid DNA (pKR67) containing the wild-type adenovirus origin were incubated with various concentrations of unlabeled competitor DNA. At each competitor concentration the fraction of the radioactive DNA bound to NFI was determined, and the resulting data were used to generate a competition curve (Fig. 6). By comparing the amounts of two competitor DNAs required to decrease the binding of the radioactive pKR67 DNA to a given value, one can obtain a good measure of the relative affinities of NFI for the two DNAs. As shown in Figure 6, the various mutant DNAs yielded several different types of competition curves. A number of the mutant DNAs (e.g., $pm17$, $pm18$, $pm19$, $pm20$, $pm21$, $pm30$, $pm31$, $pm32$, $pm33$, $pm40$, and $pm42$) competed for binding to NFI with an efficiency identical with the wild-type DNA ($dl67$). Thus, the base substitutions present in these mutants do not alter the affinity of NFI for its recognition site. Another class of mutants displayed competition curves identical with those observed with nonspecific DNA (e.g., pUC9). The base substitutions present in these mutants (located at nt 26, 35, and 36 of the adenovirus genome) abolish detectable, specific NFI binding, and thus they must affect important contacts for the protein. Most of the remaining mutants (e.g., $pm24$, $pm25$, $pm27$, $pm29$, $pm34$, $pm35$, $pm36$, $pm37$, and $pm38$) competed for NFI binding with ef-

ficiencies intermediate between those of wild-type DNA and nonspecific DNA. The base substitutions in these mutants also define base pairs involved in specific binding by NFI, but the particular mutations tested did not result in complete loss of specificity. Finally, one mutant ($pm28$) bound NFI with an affinity greater than the wild-type adenovirus origin of DNA replication. It is of interest that the base substitution in $pm28$ increases the symmetry of the NFI recognition site.

The results of the binding studies that have been performed to date are summarized in Figure 7. Mutations that significantly affect the affinity of NFI for the viral origin of replication lie in two clusters of 5 bp (half-sites) separated by a spacer segment of 5 bp. Mutations in the spacer segment had minimal ($\leq$ 2-fold) effect on NFI binding. The two half-sites are about one turn of the helix apart and display approximate dyad symmetry. As described above, the affinity of NFI binding can be increased by increasing the symmetry of the site. These results suggest that NFI binds to its recognition site as a dimer with one protein monomer interacting with each half-site.

## Discussion

In this report we have described a rapid, high-yield purification scheme for NFI (DNA recognition-site affinity chromatography) that may be generally applicable to other sequence-specific DNA-binding proteins. The method is useful for the isolation of nuclear regulatory proteins that are present in low abundance and requires only that the recognition sequence be present in a DNA fragment suitable for amplification. The success of the method depends upon the relative affinity of the protein for its recognition site versus unrelated DNA sequences.

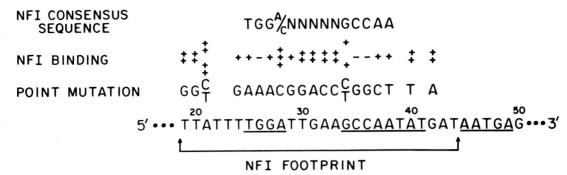

**Figure 7** Summary of effects of base-substitution mutations on NFI binding. At *bottom* is the 5′ sequence of the terminus of the adenovirus genome from nt 19 to nt 50. The composite footprint from both strands is shown by the bracket. Sequence elements that are conserved among adenovirus serotypes are underlined. Each base-substitution mutation is shown above the terminal sequence together with an indication of its effect on the binding affinity of NFI. (+ +) The mutation has no effect on binding affinity; (−) the mutation abolishes specific binding; (+) NFI binds to the mutant DNA with intermediate affinity; (+ + +) NFI binds the mutant DNA with greater affinity than wild-type DNA. The consensus sequence shown at *top* is derived from comparison of a number of viral and cellular NFI-binding sites (Borgmeyer et al. 1984; Siebenlist et al. 1984; Gronostajski et al. 1985; Hennighausen et al. 1985; Nowock et al. 1985).

To take advantage of this affinity difference, it is important that the specific affinity matrix contain a high ratio of specific to nonspecific sequences. In the present case this requirement was satisfied by constructing a plasmid that contained 88 tandem copies of the adenovirus origin of replication. Alternatively, one could achieve similar results by ligating together chemically synthesized fragments containing the appropriate recognition sequence.

Using the purification method described here, we have purified NFI about 2400-fold with a 57% recovery from crude nuclear extract. The protein present in the most highly purified fractions bound specifically to the NFI recognition site in the adenovirus origin of replication. Moreover, the purified protein was capable of enhancing the initiation of adenovirus DNA replication in vitro. SDS-polyacrylamide gel electrophoresis followed by silver staining revealed several related polypeptides with molecular masses between 52 kD and 66 kD. We do not yet know the reason for the heterogeneity of the peptides in our most-purified NFI preparation. One obvious possibility is that NFI undergoes partial proteolysis during purification. However, there are a number of other possible explanations as well, including the intriguing possibility that these polypeptides are products of a gene family that encodes several site-specific DNA-binding proteins that recognize a similar sequence.

The estimate of molecular mass that we have obtained for NFI is not in agreement with a previous study of Nagata et al. (1982). The latter workers reported the presence of a major polypeptide of 47 kD in partially purified fractions containing NFI activity. A polypeptide of this molecular size was not observed in preparations of NFI that we have purified by recognition-site affinity chromatography. Since the purity of the latter preparations is more than 10-fold greater than that obtained by Nagata et al., it seems likely that the 47-kD protein represented a contaminant or possibly a proteolytic product of NFI.

NFI binds to its specific recognition site with an apparent dissociation constant of $2 \times 10^{-11}$ M. In contrast, the affinity of the protein for nonspecific sequences of a similar length is several orders of magnitude lower. To define the determinants of this binding specificity, we have examined the effects of base-substitution mutations on the affinity of NFI for its recognition site. Our studies have so far identified 10 bp within the viral origin that are important components of the NFI recognition site. These base pairs are organized as two 5-bp half-sites separated by a 5-bp spacer sequence. The two half-sites are related by twofold rotational symmetry, although in the case of the adenovirus origin the symmetry is not perfect. The observation that a mutation that increases the symmetry of the NFI recognition site also increases binding affinity strongly suggests that NFI binds DNA as a dimer. We have demonstrated that NFI is largely monomeric in solution under conditions of relatively high ionic strength. It is possible that the protein exists as a dimer in solution under other conditions or forms a multimeric complex upon binding to its recognition site.

Our data are in good agreement with the consensus recognition sequence (TGGA/CNNNNNGCCAA) derived by comparing NFI-binding sites from various viral and cellular sources (Borgmeyer et al. 1984; Siebenlist et al. 1984; Gronostajski et al. 1985; Hennighausen et al. 1985; Nowock et al. 1985). As suggested by this consensus sequence, most naturally occurring NFI-binding sites are not perfectly symmetrical and therefore do not bind NFI with the highest possible affinity. It is possible that the biological function of NFI is incompatible with extremely tight binding (e.g., the protein must be able to dissociate from DNA at some minimal rate). Further studies are underway to define the biological effects of the mutations described in this paper.

## Acknowledgments

We thank our colleagues Dan Rawlins, Gary Ketner, Ed O'Neill, Mark Bolanowski, Harvey Ozer, Joachim Li,

James Sherley, and Steve Desiderio for useful discussions. This study was supported by research grant CA 16519 from the National Cancer Institute. P.J.R. is a trainee of the Medical Scientist Training Program (GM 07309).

## References

Albert, B. and G. Herrick. 1971. DNA-cellulose chromatography. *Methods Enzymol.* **21**: 198.

Borgmeyer, U., J. Nowock, and A.E. Sippel. 1984. The TGGCA-binding protein: Recognizing a symmetrical sequence on double-stranded linear DNA. *Nucleic Acids Res.* **12**: 4295.

Challberg, M.D. and D.R. Rawlins. 1984. Template requirements for the initiation of adenovirus DNA replication. *Proc. Natl. Acad. Sci.* **81**: 100.

Challberg, M.D., S.V. Desiderio, and T.J. Kelly, Jr. 1980. Adenovirus DNA replication in vitro: Characterization of a protein covalently linked to nascent DNA strands. *Proc. Natl. Acad. Sci.* **77**: 5105.

Challberg, M.D., J.M. Ostrove, and T.J. Kelly, Jr. 1982. Initiation of adenovirus DNA replication: Detection of covalent complexes between nucleotide and the 80kd terminal protein. *J. Virol.* **41**: 265.

de Vries, E., W. van Driel, M. Tromp, J. van Boom, and P.C. Van der Vliet. 1985. Adenovirus DNA replication *in vitro*: Site-directed mutagenesis of the nuclear factor I binding site of the Ad2 origin. *Nucleic Acids Res.* **13**: 4935.

Enomoto, T., J.H. Lichy, J.E. Ikeda, and J. Hurwitz. 1981. Adenovirus DNA replication *in vitro*: Purification of the terminal protein in a functional form. *Proc. Natl. Acad. Sci.* **78**: 6779.

Gronostajski, R.N., S. Adhya, K. Nagata, R.A. Guggenheimer, and J. Hurwitz. 1985. Site-specific DNA binding of nuclear factor I: Analyses of cellular binding sites. *Mol. Cell. Biol.* **5**: 964.

Guggenheimer, R.A., B.W. Stillman, K. Nagata, F. Tamanoi, and J. Hurwitz. 1984. DNA sequences required for the in vitro replication of adenovirus DNA. *Proc. Natl. Acad. Sci.* **81**: 3069.

Hay, R.T. 1985. The origin of adenovirus DNA replication: Minimal DNA sequence requirement in vivo. *EMBO J.* **4**: 421.

Hennighausen, L., U. Siebenlist, D. Danner, P. Leder, D. Rawlins, P. Rosenfeld, and T.J. Kelly, Jr. 1985. High-affinity binding site for a specific nuclear protein in the human IgM gene. *Nature* **314**: 289.

Herrick, G. 1980. Site-specific DNA affinity chromatography of the *lac* repressor. *Nucleic Acids Res.* **8**: 3721.

Lally, C., T. Dorper, W. Groger, G. Antoine, and E.-L. Winnacker. 1984. A size analysis of the adenovirus replicons. *EMBO J.* **3**: 333.

Lichy, J.H., M.S. Horwitz, and J. Hurwitz. 1981. Formation of a covalent complex between the 80,000 dalton adenovirus terminal protein and 5'-dCMP in vitro. *Proc. Natl. Acad. Sci.* **78**: 2678.

Lichy, J.H., J. Field, M.S. Horwitz, and J. Hurwitz. 1982. Separation of the adenovirus terminal protein precursor from its associated DNA polymerase: Role of both proteins in the initiation of adenovirus DNA replication. *Proc. Natl. Acad. Sci.* **79**: 5225.

Leegwater, P.A.J., W. van Driel, and P.C. Van der Vliet. 1985. Recognition site of nuclear factor I, a sequence-specific DNA-binding protein from HeLa cells that stimulates adenovirus DNA replication. *EMBO J.* **4**: 1515.

Miura, K. 1967. Preparation of bacterial DNA by the phenol-pH-pN9-RNAses methods. *Methods Enzymol.* **12A**: 543.

Nagata, K., R.A. Guggenheimer, and J. Hurwitz. 1983. Specific binding of a cellular DNA replication protein to the origin of replication of adenovirus DNA. *Proc. Natl. Acad. Sci.* **80**: 6177.

Nagata, K., R.A. Guggenheimer, T. Enomoto, J.H. Lichy, and J. Hurwitz. 1982. Adenovirus DNA replication in vitro: Identification of a host factor that stimulates synthesis of the preterminal protein-dCMP complex. *Proc. Natl. Acad. Sci.* **79**: 6438.

Norris, K., F. Norris, L. Christiansen, and N. Fill. 1983. Efficient site-directed mutagenesis by simultaneous use of two primers. *Nucleic Acids Res.* **11**: 5103.

Nowock, J., U. Borgmeyer, A.W. Puschel, R.A.W. Rupp, and A.E. Sippel. 1985. The TGGCA protein binds to the MMTV-LTR, the adenovirus origin of replication, and the BK virus enhancer. *Nucleic Acids Res.* **13**: 2045.

Oren, M., E. Winocour, and C. Prives. 1980. Differential affinities of simian virus 40 large tumor antigen for DNA. *Proc. Natl. Acad. Sci.* **77**: 220.

Ostrove, J.M., P.J. Rosenfeld, J. Williams, and T.J. Kelly, Jr. 1983. In vitro complementation as an assay for purification of adenovirus DNA replication proteins. *Proc. Natl. Acad. Sci.* **80**: 935.

Pearson, G.D. and K. Wang. 1985. Adenovirus sequences required for replication in vivo. *Nucleic Acids Res.* **13**: 5173.

Pincus, S., W. Robertson, and D.M.K. Rekosh. 1981. Characterization of the effect of aphidicolin on adenovirus DNA replication: Evidence in support of a protein primer model of initiation. *Nucleic Acids Res.* **9**: 4919.

Rawlins, D.R., P.J. Rosenfeld, R.J. Wides, M.D. Challberg, and T.J. Kelly, Jr. 1984. Structure and function of the adenovirus origin of replication. *Cell* **37**: 309.

Rosenfeld, P.J. and T.J. Kelly, Jr. 1986. Purification of nuclear factor I by DNA recognition site affinity chromatography. *J. Biol. Chem.* **261**: 1398.

Siebenlist, U., L. Hennighausen, J. Battey, and P. Leder. 1984. Chromatin structure and protein binding in the putative regulatory region of the c-*myc* gene in Burkitt lymphoma. *Cell* **37**: 381.

Stillman, B.W., F. Tamanoi, and M.D. Mathews. 1982. Purification of the adenovirus-coded DNA polymerase that is required for initiation of DNA replication. *Cell* **31**: 613.

Tamanoi, F. and B.W. Stillman. 1982. Function of the adenovirus terminal protein in the initiation of DNA replication. *Proc. Natl. Acad. Sci.* **79**: 2221.

———. 1983. Initiation of adenovirus DNA replication in vitro requires a specific DNA sequence. *Proc. Natl. Acad. Sci.* **80**: 6446.

van Bergen, B.G.M., P.A. Van der Ley, W. van Driel, A.D.M. van Mansfeld, and P.C. Van der Vliet. 1983. Replication of origin containing adenovirus DNA fragments that do not carry the terminal protein. *Nucleic Acids Res.* **11**: 1975.

Zoller, M.J. and M. Smith. 1983. Oligonucleotide-directed mutagenesis of DNA fragments cloned into M13-derived vectors. *Methods Enzymol.* **100**: 468.

———. 1984. Oligonucleotide-directed mutagenesis: A simple method using two oligonucleotide primers and a single-stranded DNA template. *DNA* **3**: 479.

# DNA-Protein Interactions at the Replication Origins of Adenovirus and SV40

**B. Stillman, J.F.X. Diffley, G. Prelich, and R.A Guggenheimer**
Cold Spring Harbor Laboratory, Cold Spring Harbor, New York 11724

Biochemical studies on the proteins that interact with the origin of replication in both adenovirus and simian virus 40 (SV40) DNA are described. One of two cellular proteins that are essential for the initiation of adenovirus DNA replication in vitro was previously identified as a 47,000-dalton, site-specific DNA-binding protein called nuclear factor I (NFI). We have developed a rapid filter-binding method that is suitable for the detection of NFI in crude cellular extracts and this assay was used for its subsequent purification. The purified protein, which has an apparent molecular mass of 160,000 daltons, binds specifically to the origin of adenovirus DNA replication and stimulates by approximately 20-fold the formation of the protein-dCMP initiation complex. In contrast to adenovirus, SV40 DNA replication requires only one virus-encoded protein, the large tumor (T) antigen, for DNA replication and thus relies upon many host-cell proteins for complete DNA synthesis. Studies with SV40 T-antigen mutants have defined at least three separate functions of the T antigen that are required for the initiation of DNA replication in vitro. Furthermore, the minimal origin required for replication in vitro has been defined by deletion analysis. Finally, we demonstrate that the kinetics of SV40 DNA synthesis commences with a lag of 10–15 min at 37°C and that it can be overcome by prior incubation of the template DNA with T antigen and cellular proteins.

The value of DNA tumor viruses as agents for investigating basic biological phenomena in mammalian cells is obvious from even a brief survey of the work presented in this volume. As has been the case for studies on gene regulation and expression, DNA viruses have allowed significant advances in our understanding of the diverse mechanisms and control of DNA replication and cell proliferation. Two major handicaps that essentially prevent a direct attack on many aspects of the replication of mammalian cell DNA are the lack of genetics to probe cellular functions and the large number of origins contained within the cell's chromosomes. The use of the smaller and less complex DNA genomes of adenovirus and simian virus 40 (SV40) circumvent some of these problems. Furthermore, the development of cell-free replication systems for these DNAs has provided a starting point for probing the function and regulation of cellular proteins that may be required for cell chromosome replication.

### Adenovirus DNA replication

The adenovirus replicon consists of a linear, double-stranded DNA of 36,000 bp that has origins of replication at each end of the genome contained within an inverted, terminally repetitive sequence (Challberg and Kelly 1982; Stillman 1983). DNA replication proceeds from either origin by displacement of the nontemplate strand as a single-stranded DNA, which is subsequently replicated from the other origin (for reviews, see Stillman 1983; Friefeld et al. 1984; Kelly 1984). Five proteins have been identified and characterized that are required for the complete replication of adenovirus DNA

in vitro. Three of these are encoded by the virus genome and include the single-stranded DNA-binding protein (DBP) (Kaplan et al. 1979; Ostrove et al. 1983; Stillman et al. 1984; Prelich and Stillman 1986), the adenovirus DNA polymerase (Adpol) (Lichy et al. 1982, 1983; Stillman et al. 1982), and the terminal protein precursor (pTP) (Challberg et al. 1980, 1982; Lichy et al. 1981; Stillman et al. 1981). All of these proteins are synthesized from a single transcription unit in the adenovirus genome (Stillman et al. 1982), the expression of which is regulated by other adenovirus gene products (Sharp 1984). In addition to these virus-encoded proteins, fractionation of the replication extracts has revealed the existence of two cellular proteins that are essential for the complete replication of adenovirus DNA in vitro. Nuclear factor I (NFI) is required for initiation of adenovirus DNA replication and binds to a specific sequence located within a domain of the origin of DNA replication (Nagata et al. 1982, 1983a; Guggenheimer et al. 1984; Rawlins et al. 1984; Leegwater et al. 1985). The other cellular protein, nuclear factor II (NFII) is not required for initiation of DNA synthesis, nor for partial elongation, but is required for completion of the nascent DNA strand to yield full-length adenovirus DNA (Nagata et al. 1983b). This protein has been identified as a type-I DNA topoisomerase (Nagata et al. 1983b).

The mechanism of priming adenovirus DNA replication at the origin proved to be a novel and unexpected method for synthesis of nascent DNA chains. The protein-priming model, first proposed by Bellett and co-workers (Rekosh et al. 1977), has been substantiated by subsequent biochemical studies and has been suggested as the method of choice in a number of other replication systems. The pTP and Adpol exist as a functional complex and, together with NFI, recognize sequences at the origin of DNA replication and form a priming complex. This reaction is stimulated by the presence of DBP (Nagata et al. 1982; de Vries et al. 1985) and requires specific DNA sequences within the origin (Tamanoi and Stillman 1982, 1983; van Bergen et al. 1983; Guggenheimer et al. 1984; Challberg and Rawlins 1984; Lally et al. 1984; Rawlins et al. 1984). The first nucleotide in the nascent chain, dCMP, is covalently bound to the pTP in a reaction that requires Adpol and NFI (Lichy et al. 1981; Pincus et al. 1981; Challberg et al. 1982; Tamanoi and Stillman 1982). The 3'-hydroxyl of the covalently bound dCMP is then utilized as the primer for subsequent DNA chain elongation, a reaction that is absolutely dependent upon the DBP (Kaplan et al. 1979; Enomoto et al. 1981; Ikeda et al. 1981; Challberg et al. 1982; Friefeld et al. 1983; Ostrove et al. 1983; Stillman et al. 1984; Prelich and Stillman 1986). The pTP that remains covalently bound to the replicated DNA plays a role in the initiation of the next round of DNA synthesis, but it is not essential (Challberg and Kelly 1979; Tamanoi and Stillman 1982). A proteolytically cleaved product of pTP is found attached to the 5' end of mature virion DNA (Robinson et al. 1973; Rekosh et al. 1977; Challberg and Kelly 1981; Stillman et al. 1981). These studies demonstrating the protein-priming mechanism for initiation of adenovirus DNA replication not only illuminated a new method for replicating DNA, but also provided a mechanism for replication of the ends of a linear DNA molecule.

## SV40 DNA replication

In contrast to the mode of replication of adenovirus DNA, the mechanism of SV40 DNA replication more closely parallels that of the host cell's chromosomes. The SV40 genome exists in a chromatin structure similar to that of cellular chromatin and contains a unique sequence that constitutes the origin of bidirectional replication, much like an origin within a replicon of a linear chromosome (for review, see DePamphilis and Wassarman 1982). DNA synthesis terminates when the replication forks meet at a point approximately 180° from the origin on the circular DNA, and the two daughter molecules segregate into two minichromosomes (Sundin and Varshavsky 1980, 1981; Weaver et al. 1985).

Only one virus-encoded protein, the SV40 tumor (T) antigen, is required for SV40 DNA replication (Tegtmeyer 1972). The T antigen binds specifically to three sites within the origin region and is required for initiation of DNA replication (DePamphilis and Wassarman 1982). In addition to this origin binding, an associated ATPase activity appears to be required for DNA replication (Clark et al. 1981, 1983; Stillman et al. 1985). Recently, Stillman et al. (1985) provided evidence that another function of T antigen, perhaps a specific protein-protein interaction, is required for the initiation of DNA replication. This suggestion was based upon two observations. First, a defective T antigen (Manos and

Gluzman 1984, 1985) that retains both ATPase activity and origin-binding activities, fails to support SV40 DNA replication in vitro and in vivo, and second, defective T antigens that do not bind to the origin retain the ability to compete with wild-type T antigen in an in vitro replication system.

Because SV40 encodes only one replication protein, the preponderance of factors required to replicate SV40 DNA in cells must be of cellular origin and in all probability are factors that replicate cellular chromosomes. This makes SV40 replication an attractive model system for the analysis of eukaryotic replication and its control, and the recent development of a cell-free system for the replication of SV40 DNA in monkey cell extracts (Li and Kelly 1984) will enable biochemical studies to advance rapidly. We have demonstrated that SV40 DNA replicates efficiently in extracts prepared from human 293 cell extracts but less efficiently in similarly prepared extracts from human HeLa cells (Stillman and Gluzman 1985). SV40 DNA replication in both monkey and human cell extracts is dependent upon purified SV40 T antigen and SV40 origin sequences, and replication proceeds bidirectionally from the origin in a manner analogous to that observed in vivo (Li and Kelly 1984, 1985; Stillman and Gluzman 1985; Stillman et al. 1985; Wobbe et al. 1985).

In this paper, we describe our recent studies on cellular factors required for the replication of both adenovirus and SV40 DNA. The cellular protein, NFI, appears to be a 160,000-dalton (160 kD) protein that binds specifically to the adenovirus origin of DNA replication and stimulates the initiation reaction approximately 20-fold. Although the identities of cellular factors that are required for SV40 DNA replication are not known, we demonstrate that cellular factors, in addition to SV40 T antigen, are required in a presynthesis reaction on SV40 DNA prior to elongation of DNA chains.

## Materials and Methods

### Cells and viruses

Human HeLa and 293 (Graham et al. 1977) cells were grown in suspension cultures as described by Stillman and Gluzman (1985). The recombinant adenoviruses that overproduce wild-type and mutant T antigens were obtained from Y. Gluzman (pers. comm.) and have been briefly described (Stillman and Gluzman 1985; Stillman et al. 1985).

### Purified enzymes and proteins

The adenovirus-encoded pTP and Adpol (native DNA-cellulose fraction) were purified as described previously (Stillman et al. 1982). The adenovirus single-stranded DBP was purified to apparent homogeneity by the method of Schechter et al. (1980). The cellular protein NFI was purified by the method described by Nagata et al. (1983a) with modifications described by Diffley and Stillman (1986). The protein was detected during purification by a rapid filter-binding assay (Diffley and Stillman 1986). Wild-type and mutant SV40-encoded T antigens were synthesized in recombinant adenovirus-infected HeLa or 293 cells (Stillman and Gluzman 1985) and were purified to apparent homogeneity using the method of Simanis and Lane (1985). The *Escherichia coli* DNA polymerase I and restriction enzymes were purchased from New England Biolabs.

## Adenovirus DNA replication

Initiation of adenovirus DNA replication in vitro was measured by the covalent linkage of dCMP to the pTP, essentially as described previously (Nagata et al. 1982). Reactions (20 μl) contained 25 mM HEPES/KOH (pH 7.5), 5 mM MgCl$_2$, 2 mM DTT, 3 mM ATP, 0.8 μM [α-$^{32}$P]dCTP, 60 ng of adenovirus DNA-protein complex (Challberg and Kelly 1979), 70 ng of the pTP-Adpol fraction, 1.25 μg of DBP, and 0.4 μg of NFI were indicated. Reactions were incubated for 1 hr at 30°C, and the formation of the pTP-dCMP complex was measured by electrophoresis in SDS-polyacrylamide gels.

## SV40 DNA replication

Extracts from uninfected human 293 cells were prepared by a modification of the Li and Kelly (1984) procedure described by Stillman and Gluzman (1985). The 293 cell cytosol extract was adjusted to 0.1 M NaCl and centrifuged for 1 hr at 100,000*g* at 4°C. The supernatant (293 cell cytosol S100) contained all the DNA replication activities. DNA synthesis reactions were essentially as described by Stillman and Gluzman (1985) except that the 50-μl reactions contained approximately 200 μg of 293 cell cytosol S100 fraction and 1 μg of SV40 T antigen, which typically incorporated approximately 150–200 pmoles of dAMP in 1 hr at 37°C using 300 ng of pSV40 form-I DNA as template. This was about twice the level of synthesis observed with the crude cytosol. The amount of DNA synthesis is expressed as pmoles of dAMP incorporated into acid-precipitable form using 300 ng of template DNA and optimal amounts of T antigen and extract.

## Fractionation of 293 cell cytosol

Solid ammonium sulfate was added to the 293 cell S100 fraction to a final concentration of 45% and the precipitate was collected by centrifugation. The pellet (C45 fraction) was dissolved in Buffer A (20 mM Tris [pH 7.5], 1 mM DTT, 1 mM EDTA, 10% glycerol, and 10$^{-4}$ M PMSF) containing 50 mM NaCl. Solid ammonium sulfate was added to the supernatant fraction to a final concentration of 60%, and this precipitate was collected by centrifugation (C60 fraction). The C60 fraction was also resuspended in Buffer A containing 50 mM NaCl. Both fractions were dialyzed against Buffer A containing 50 mM NaCl.

## DNA-binding (footprint) assay

The specific binding of NFI to the origin of adenovirus DNA replication was determined by the deoxyribonuclease (DNase) footprint technique (Galas and Schmitz 1978). The 72-bp fragment from pLAS108 (Tamanoi and Stillman 1983) containing the origin of adenovirus

DNA replication was labeled at one end with $^{32}$P by treatment with polynucleotide kinase and the labeled fragment was digested with DNase I in the absence or presence of purified NFI protein. The reaction products were separated by polyacrylamide gel electrophoresis (sequencing gel) together with marker DNA fragments produced by partial DNA cleavage (Maxam and Gilbert 1980).

## DNA polymerase and DNA primase assays

Enzyme activities were measured in crude extracts as described previously (Field et al. 1984).

# Results

## Characterization of a cellular protein required for initiation of adenovirus DNA replication

The cellular protein NFI was first identified as a 47-kD cellular protein that binds to the origin of adenovirus DNA replication and promoted the formation of the pTP-dCMP primer of nascent DNA chains (Nagata et al. 1982, 1983a; Guggenheimer et al. 1984; Rawlins et al. 1984). NFI was purified using a reconstitution of adenovirus DNA replication assay in reactions that contained partially purified pTP and Adpol, as well as the DBP (Nagata et al. 1982). Since this assay might have selectively detected a subset of the TGGCA-binding proteins from cells that function in DNA synthesis (see Discussion), we have developed a rapid filter-binding assay that can detect NFI or other site-specific DNA-binding proteins in crude cellular extracts and have used this assay to follow the purification of NFI (Diffley and Stillman 1986). The DNA-binding activity of NFI cofractionated with the stimulation of DNA replication activity during all steps in the purification. An example is shown in Figure 1A, which shows the coelution of the NFI DNA binding and replication activities from a phosphocellulose column. The filter-binding assay used to detect NFI in the phosphocellulose column fractions is shown in Figure 1B. Two DNA fragments were used for each assay. Fragment *f* 114, containing a deleted NFI-binding site (nucleotides [nt] 1–35 of adenovirus DNA) and *f* 107, containing the entire NFI-binding site (nt 1–48 of adenovirus DNA), were labeled with $^{32}$P at their 5' termini and mixed with proteins present in the phosphocellulose column fractions that had been previously bound to competitor DNA. The DNA-protein complexes formed were detected by filtration through a nitrocellulose filter, which was then subjected to autoradiography (Fig. 1B) and scintillation counting. The specific DNA-binding (Fig. 1A) is the difference between the cpm bound with *f* 107 and the cpm bound with *f* 114.

When NFI was purified using the procedures described by Nagata et al. (1982) and protease inhibitors were included in all buffers, a major polypeptide of apparent molecular mass of 160 kD was observed in SDS-polyacrylamide gels stained with silver nitrate, with no bands visible in the region of 47 kD (Fig. 2A). This highly purified protein stimulated the formation of the pTP-dCMP complex 20-fold in the presence of DBP and

A.

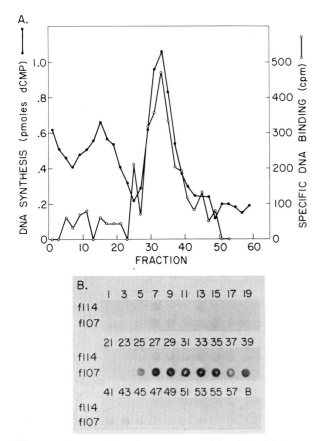

B.

**Figure 1** Fractionation of extracts containing NFI. (*A*) Phosphocellulose chromatography of a pooled fraction from a DEAE-cellulose column and elution of NFI with a NaCl gradient. The open circles show specific DNA binding and the closed circles show the DNA synthesis assay, done in the presence of purified pTP, Adpol, and DBP. The peak of DNA-binding activity eluted at a concentration of 300 mM NaCl. (*B*) An autoradiogram of the nitrocellulose filter after detecting specific DNA-binding proteins with ³²P-labeled *f*114 (no NFI-binding site) and *f*107 (containing an NFI-binding site) probes (Diffley and Stillman 1986). The specific binding described in *A* was determined from the cpm present using the specific DNA probe versus the cpm present using the nonspecific DNA probe. Odd-numbered fractions were assayed, together with an assay with no added protein (*B*).

pTP-Adpol (Fig. 2B) and also bound specifically to the NFI-binding site identified previously (Fig. 2C) (Nagata et al. 1983a; Guggenheimer et al. 1984; Rawlins et al. 1984; Leegwater et al. 1985). These results suggest that NFI is a 160-kD polypeptide that can bind specifically to the origin of adenovirus DNA replication and promote the formation of the primer for DNA synthesis.

## Initiation of DNA synthesis at the origin of SV40 DNA replication

### Definition of the origin sequences

It was previously demonstrated that SV40 DNA replication in vitro starts at a unique origin and proceeds in both directions around the DNA (Li and Kelly 1985; Stillman and Gluzman 1985; also see Fig. 5 below). The se-

quences that constitute the origin of SV40 replication had been defined by studies in vivo (for review, see DePamphilis and Wassarman 1982), and we have demonstrated that the origin sequences required for efficient DNA synthesis in vitro are essentially the same as these (Stillman et al. 1985). A summary of the results is presented in Figure 3, with the addition of the plasmid pRG53 (R. Gerard, pers. comm.). This plasmid has combined the deletion end points present in pSV0*dl*3 and pS1*dl*5 to create a minimal origin of DNA replication of 65 bp that replicates in vitro as efficiently as both parental deletion plasmids (Fig. 3).

Four subregions have been defined within the origin of SV40 DNA replication by these studies. The first consists of the region of T-antigen-binding site I, which affects the level of replication. Second, a region between the end points for pSV0*dl*3 and pSV0*dl*6 is a critical cutoff point for the ability to support DNA replication in vitro and overlaps with a part of the origin that Hay and DePamphilis (1982) have recognized as the transition point between leading strand (continuous) and lagging strand (discontinuous) synthesis in vivo. This region may play a critical role in initiation of bidirectional replication. A third element within the origin that is necessary for any DNA synthesis consists of T-antigen-binding site II, which is centered within a 27-bp palindrome. Finally, a stretch of 17 A·T base pairs that lie next to the palindrome is required for replication. This stretch can be shortened to 16 A·T residues, but not to 12 A·T residues (cf. pS1*dl*5 and pS1*dl*6 in Fig. 3).

### Requirement for the SV40 T antigen

Since T antigen is required for DNA replication in vitro (Li and Kelly 1984; Stillman and Gluzman 1985), we have recently examined the ability of purified mutant T antigens (Gluzman and Ahrens 1982; Manos and Gluzman 1984, 1985) to (1) support DNA synthesis in vitro, (2) bind to origin sequences, and (3) contain the T-antigen-associated ATPase activity. The results, summarized in Table 1, suggest that both ATPase activity and origin binding are important functions for the replication activity of T antigen. Moreover, the defective T antigen C8A, which retains ATPase and origin-binding activities, does not support replication in vitro (Table 1;

**Table 1** Biochemical Functions of Wild-type and Defective SV40 T Antigens

| T antigen | Origin binding | ATPase activity | DNA synthesis in vitro |
|---|---|---|---|
| R112 (wild-type) | + | + | + |
| C2 | − | + | − |
| C6.2 | − | + | − |
| C8A | + | + | − |
| C11 | + | − | − |
| T22 | − | + | − |

Results are summarized from data obtained from Manos and Gluzman (1984, 1985) and Stillman et al. (1985).

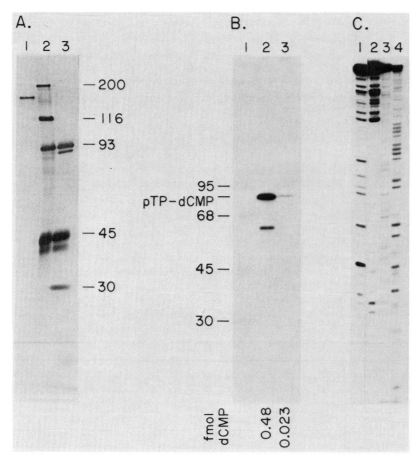

**Figure 2** Characterization of purified NFI. (*A*) (Lane *1*) SDS-polyacrylamide gel electrophoresis of the most highly purified preparation of NFI (specific DNA-cellulose fraction; Diffley and Stillman 1986); (lanes *2* and *3*) molecular mass markers, with the apparent molecular mass indicated in daltons ( × 10³). (*B*) Stimulation of pTP-dCMP complex formation by purified NFI. (Lane *1*) Molecular-mass markers; (lanes *2* and *3*) reactions containing purified pTP, Adpol, DBP, and adenovirus DNA-protein complex and [α-³²P]dCTP in the presence (lane *2*) or absence (lane *3*) of NFI; the reaction products were separated by SDS-polyacrylamide gel electrophoresis. (*C*) Footprint analysis with purified NFI. The 72-bp *Eco*RI-*Sal*I fragment present in pLAS108 (Tamanoi and Stillman 1983) was labeled with [γ-³²P]ATP and polynucleotide kinase at the *Sal*I site and subjected to DNase I digestion in the absence (lane *1*) or presence (lane *2*) of purified NFI. The region protected from DNase (lane *2*) corresponds to the NFI-binding site reported by Nagata et al. (1983a). Lanes *3* and *4* are DNA sequencing reactions as markers.

Stillman et al. 1985) and in vivo (Manos and Gluzman 1985). We have suggested that an additional function of the SV40 T antigen, perhaps a specific protein-protein interaction, is required for DNA replication in vitro (Stillman et al. 1985). This proposal is supported by the observation that all the defective SV40 T antigens described in Table 1 compete with wild-type T antigen and reduce DNA synthesis in reactions where both wild-type and mutant proteins are added together (Stillman et al. 1985).

*Characterization of a presynthesis reaction*
A time course of DNA synthesis using the crude 293 cell cytosol fraction indicated that a lag of 10–15 min occurred prior to the incorporation of dAMP into nascent DNA chains (Fig. 4A and Stillman and Gluzman 1985). DNA synthesis then proceeded linearly for another 30 min before stopping about 2 hr later (Fig. 4A). The lag in the replication reaction can be eliminated by a 15-min preincubation of the reaction mixture in the absence of precursor deoxynucleoside triphosphates (dNTP) (Fig. 4B, complete reaction). Under these conditions, which include preincubation of the reaction mixture containing template DNA (pSV40), T antigen, 293 cytosol extract, and ATP at 37°C, the rate of synthesis immediately increases linearly for at least 30 min after addition of the

dNTP precursors (at 0 min). The presynthesis reaction requires T antigen, template DNA, and factors in the crude extract since omission of any of these components results in the reappearance of the lag (Fig. 4B). The presynthesis reaction is also critically dependent upon temperature, since only a 15-min incubation at 37°C, but not 30°C, 25°C, or 4°C, will eliminate the lag in synthesis (Fig. 4C). These results suggest that a complex reaction involving interactions between the origin DNA, T antigen, and cellular factors is required prior to the synthesis of nascent DNA chains.

We have demonstrated that there is not extensive DNA synthesis during the presynthesis reaction since the origin-proximal DNA is the first to be labeled after addition of the dNTPs (Fig. 5). Using pSV01 (Myers and Tjian 1980) as template DNA in this experiment, instead of pSV40, reaction mixtures without dNTPs were incubated for 15 min at 37°C, and then labeled dNTPs were added. At 0.5, 1, 2, 5, and 10 min after addition of the dNTPs, the reaction was terminated and the DNA isolated. The DNA was then digested with a combination of the *Dde*I, *Pvu*II, and *Sal*I restriction enzymes and the fragments were separated by electrophoresis through a 4% polyacrylamide gel (Maniatis et al. 1975), shown in Figure 5B. The bands were excised from the gel, counted, and the counts per minute in each fragment relative

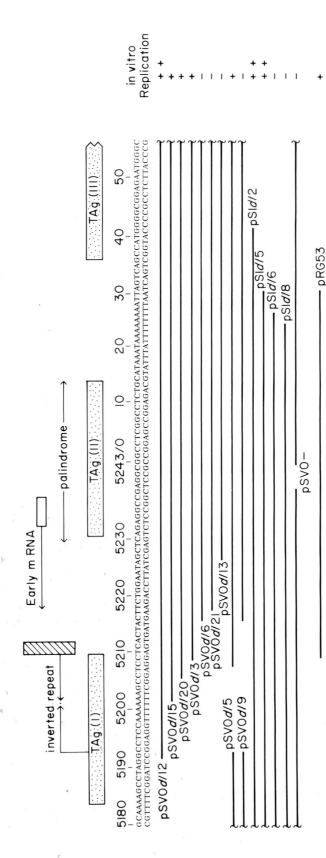

**Figure 3** Definition of the SV40 DNA replication origin. The nucleotide sequence surrounding the SV40 origin region is shown. Above the sequence, the three T-antigen-binding sites (TAgI, II, and III) are indicated, together with the start site for early gene transcription (□) and the transition point between leading- and lagging-strand synthesis defined by Hay and DePamphilis (1982) (▨). Below the line are the deletion end points for plasmids in the pSV0 and pS1 series, together with plasmids pSV0− and pRG53. The construction and analysis of these plasmids was described previously (Stillman et al. 1985). The solid line represents sequences present in the plasmid DNAs. On the right-hand side, the ability of these plasmids to support SV40 DNA replication in vitro is indicated. (+ +) 100% of wild-type activity; (+) 40–50% of wild-type activity; (−) no detectable activity.

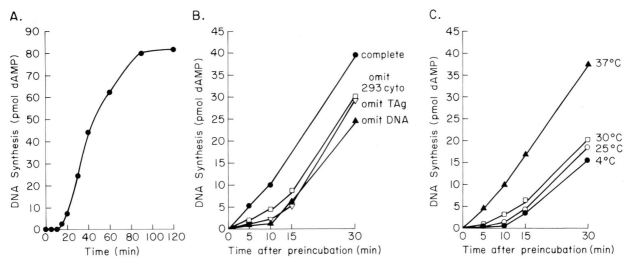

**Figure 4** The effect of preincubation on the time course of SV40 DNA synthesis. (*A*) Time course of DNA synthesis with pSV40 form I as template DNA and no preincubation. (*B*) Time course of DNA synthesis with a 15-min preincubation at 37°C in the presence (●) of T antigen (0.5 μg), 293 cell cytosol extract, and pSV40 or in reactions in which the indicated components were omitted from the preincubation. In the latter cases, the omitted component was added at 0 min with dNTPs. (*C*) Time course of DNA synthesis following a 15-min preincubation at each of the indicated temperatures, followed by DNA synthesis at 37°C. In *B* and *C*, dNTPs were omitted during the preincubation and were added at 0 min.

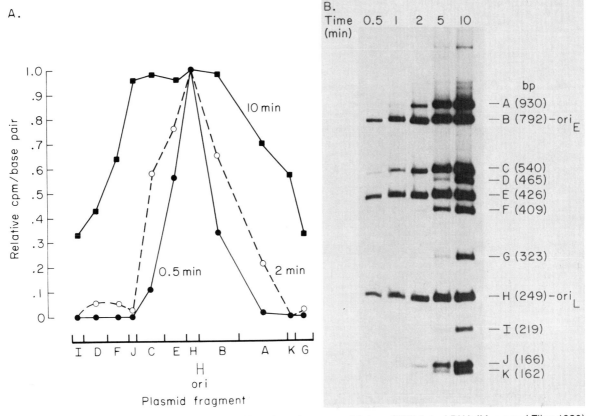

**Figure 5** Extent of DNA replication after preincubation. Reaction mixtures containing pSV01 form-I DNA (Myers and Tjian 1980), T antigen, and 293 cell cytosol extract were preincubated for 15 min at 37°C and DNA synthesis was started at 0 min by addition of dNTPs. At the indicated times, reactions were terminated and DNA was isolated as described previously (Stillman and Gluzman 1985). The DNA was then digested with a combination of *Dde*I, *Pvu*II, and *Sal*I restriction enzymes and the DNA fragments were separated by polyacrylamide gel electrophoresis (*B*). The bands were excised from the gel, and the cpm in each band is expressed relative to its length in nucleotides. The H fragment was arbitrarily given a value of 1.0. The restriction fragment map and position of the origin are indicated.

to its length in base pairs has been graphed in Figure 5A. Under presynthesis conditions, the first fragments to become labeled are in the B, E, and H fragments. Fragments B and H make up the origin sequences since a *Dde*I site is present at nt 5228 within the functional origin (see Fig. 1), whereas the E fragment is located adjacent to H on the late transcription side of the origin. As the time of replication after presynthesis is increased, the distribution of fragments that are labeled shifts in both directions from the origin at approximately equal rates. This analysis demonstrates that extensive replication does not occur during the presynthesis reaction and that subsequent DNA synthesis after addition of dNTPs is bidirectional from the origin in a population of DNA molecules.

*Fractionation of cellular factors required for the presynthesis reaction*

In addition to T antigen and the template DNA, factors present in the 293 cell extract are required for the presynthesis reaction. We have begun biochemical fractionation of these extracts and have tested each fraction to determine if it will support the presynthesis reaction. An example of this kind of analysis is shown in Figure 6. The 293 cell cytosol S100 extract was separated into two fractions by ammonium sulfate precipitation. The resultant fractions, C45 (0–45% $(NH_4)_2SO_4$ pre-

cipitate) and C60 (45–60% $(NH_4)_2SO_4$ precipitate) showed little (C45) or no (C60) DNA synthesis when used alone, but when combined efficient synthesis was restored (Fig. 6A). The C45 fraction had some residual DNA replication activity, but the level of this activity varied from preparation to preparation. Addition of the C60 fraction stimulated DNA synthesis at least fivefold.

The C60 fraction was able to partially eliminate the lag period for DNA synthesis when preincubated with T antigen and pSV40 DNA for 15 min prior to addition of dNTPs and the C45 fraction (Fig. 6B). Thus, even though the C60 fraction alone was inactive for replication, it contained factors that increase the rate of dAMP incorporation after the preincubation. A similar experiment was not possible with the C45 fraction since it supported significant DNA replication by itself, probably due to contamination of the fraction with factors present in the C60 fraction.

The partial elimination of the lag period for DNA replication by the C60 fraction is interesting since there was no detectable DNA primase activity in this fraction (Table 2), although the C60 fraction does retain some aphidicolin-sensitive DNA polymerase activity. This polymerase activity may represent polymerase α core protein that has become dissociated from the primase under the conditions used. The absence of detectable DNA primase activity in the C60 fraction suggests that

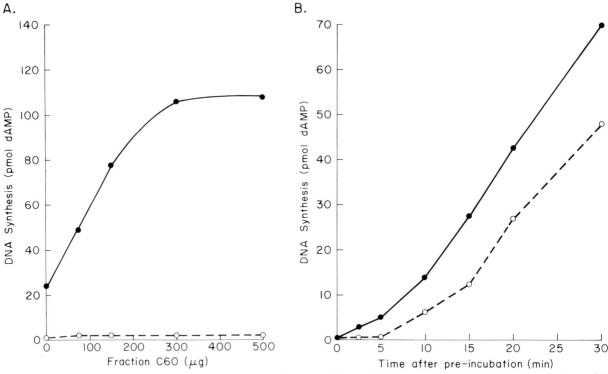

**Figure 6** Fractionation of the 293 cell cytosol S100 extract. The 293 cell cytosol S100 extract was divided into two fractions, C45 and C60, as described in Materials and Methods. (*A*) Reactions were performed with pSV40 plasmid DNA and T antigen in the absence (○) or presence (●) of the C45 fraction (optimal amount, 300 μg) and increasing amounts of the C60 fraction as indicated. Reactions were terminated after 60 min. (*B*) Reactions containing pSV40 and T antigen in the absence (○) or presence (●) of the C60 fraction were preincubated for 15 min at 37°C and then the C45 fraction was added with dNTPs at 0 min. Reactions were terminated at the indicated times after preincubation.

**Table 2** DNA Polymerase and DNA Primase Activities in Fractions C45 and C60

| Fraction | DNA polymerase[a] (units/mg) | | DNA primase (units/mg) |
|---|---|---|---|
| | − aphidicolin | + aphidicolin | |
| C45 | 5.4 | 0.4 | 180 |
| C60 | 0.54 | 0.08 | <1 |

[a] 1 unit of DNA polymerase or DNA primase is defined as the incorporation of 1 nmole of dAMP into acid-insoluble cpm in 15 min at 37°C under standard conditions. DNA polymerase reactions were performed in the absence (−) or presence (+) of aphidicolin (40 μg/ml), a concentration that inhibits DNA polymerase α.

this enzyme, which is believed to function in the formation of the primer for DNA synthesis, is not required during the presynthesis reaction.

## Discussion

A number of origins of DNA replication have an obligatory requirement for a sequence-specific, double-stranded DNA–binding protein to recognize the origin prior to DNA synthesis. The *dnaA* gene product in *E. coli*, which binds specifically to the *oriC* replication origin, is required for initiation of DNA replication in vitro (Fuller et al. 1984; Kaguni and Kornberg 1984). Similarly, the bacteriophage λ O protein binds to the origin of phage λ replication and is required for the initiation reaction (Tsurimoto and Matsubara 1981; Anderl and Klein 1982; Wold et al. 1982). Three proteins that bind to eukaryotic replication origin sequences have been identified, two of which have been described above, i.e., T. antigen and NFI. A third protein, the Epstein-Barr virus–encoded nuclear antigen 1 (EBNA-1) has recently been shown to bind to a putative origin of replication from Epstein-Barr virus DNA (Rawlins et al., this volume). Both the SV40 T antigen and the NFI protein are required to bind to DNA prior to initiation of DNA replication at their respective origins, but it is premature to conclude anything about the involvement of the EBNA-1 protein in initiation.

NFI was first identified and purified based on its ability to stimulate the formation of the pTP-dCMP priming complex and thus DNA synthesis (Nagata et al. 1982). The purified protein recognized sequences that form a part of the adenovirus DNA replication origin and that lie adjacent to an essential A·T-rich domain (Tamanoi and Stillman 1982, 1983; Nagata et al. 1983a; Challberg and Rawlins 1984; Guggenheimer et al. 1984; Rawlins et al. 1984; de Vries et al. 1985; Hay 1985; Leegwater et al. 1985; Wang and Pearson 1985). Concomitant with these studies on NFI, two groups have identified proteins (called TGGCA-binding proteins) that specifically recognized DNA sequences similar to the NFI-binding site and are present in crude cellular extracts from a wide variety of species and tissues (Nowock and Sippel 1982; Borgmeyer et al. 1984; Siebenlist et al. 1984; Nowock et al. 1985; Sippel et al., this volume). Although

these proteins have not been purified, it is most likely that they are related to NFI.

We have used a DNA-binding assay (Diffley and Stillman 1986) to purify HeLa cell proteins that bind to the adenovirus replication origin NFI-binding site and have found a single polypeptide of 160 kD. This protein is probably related to the 47-kD NFI reported by Nagata et al. (1982), since it binds to the same region of adenovirus DNA and stimulates DNA replication in vitro, satisfying the only known activities of NFI. It is possible that the 47-kD protein is a proteolytic breakdown product of the 160-kD polypeptide that retains biological activity. Since the only protein found in the most highly purified preparation of NFI is the 160-kD protein and the protein was detected during purification using the DNA-binding assay, it is likely that NFI and the TGGCA-binding proteins are the same molecules.

In contrast to adenovirus, SV40 encodes only one protein that is required for initiation of DNA replication. The binding site II for T antigen, which lies within the minimal origin of DNA replication, is a necessary but not sufficient element for DNA replication. The T-antigen-binding site II lies adjacent to a stretch of A·T-rich DNA sequence that is also necessary for origin function. Thus, despite the divergent nature of the adenovirus and SV40 replication mechanisms, some features of the overall structure of the replication origins are very similar. The A·T region in the adenovirus origin has been implicated as an essential domain and may be the site of pTP and Adpol entry onto the DNA for primer formation (Rijinders et al. 1983; Tamanoi and Stillman 1983; Challberg and Rawlins 1984). Although the nature and mechanism of formation of the primer for SV40 DNA replication has not been described, it is possible that the A·T-rich region in the SV40 origin plays a similar function.

In this report, we have identified a reaction that occurs prior to synthesis of SV40 DNA in vitro and have demonstrated that the template DNA, SV40 T antigen, and cellular proteins are required. It appears that the reaction, which is temperature-dependent, does not require the cellular DNA primase enzyme that is associated with DNA polymerase α, since the lag in DNA replication could be partially eliminated by preincubation with the C60 fraction, which lacks primase activity. Since both SV40 DNA and T antigen are required for the presynthesis reaction, we conclude that a complex set of interactions occur on the template DNA prior to formation of the initial primers for leading-strand synthesis.

A similar presynthesis reaction has recently been observed in the enzymatic replication from the *E. coli oriC* region in vitro (van der Ende et al. 1985). Reactions that were dependent upon DNA primase for initiation, rather than RNA polymerase, show a lag of 10–15 min, which could be eliminated by prior incubation of the reaction without dNTPs. Interestingly, the presynthesis reaction did not require primase, but was temperature-dependent and required the origin-binding *dnaA* protein, which may have a similar function to T antigen (van der

Ende et al. 1985). Since the prepriming reaction at *oriC* required DNA gyrase and other initiation complex enzymes (i.e., *dnaB*, *dnaC*, SSB), it is possible that a prepriming complex is formed on the template DNA and that this requires topological strain in the circular DNA. We have previously demonstrated (Stillman and Gluzman 1985) that circular, but not linear, DNAs containing the SV40 origin are required for the initiation of SV40 DNA replication in vitro, which suggests that topological constraint in the template DNA may be a requirement for the prepriming event at the SV40 *ori* sequence. It is clear that further analysis of this reaction will indicate which cellular functions are required for initiation of SV40 DNA replication in vitro.

## Acknowledgments

We thank B. Faha for excellent technical assistance and Y. Gluzman and R. Gerard for valuable discussions. This research was supported by a Cancer Center Grant to Cold Spring Harbor Laboratory (CA-13106) and by grants to B.S. from the National Institutes of Health (AI-20460) and the Rita Allen Foundation.

## References

Anderl, A. and A. Klein. 1982. Replication of λ dv DNA in vitro. *Nucleic Acids Res.* **10**: 1733.

Borgmeyer, Y., J. Nowock, and A.E. Sippel. 1984. The TGGCA-binding protein: A eukaryotic nuclear protein recognizing a symmetrical sequence on double-stranded virus DNA. *Nucleic Acids Res.* **12**: 4295.

Challberg, M.D. and T.J. Kelly, Jr. 1979. Adenovirus DNA replication *in vitro*. *Proc. Natl. Acad. Sci.* **76**: 655.

———. 1981. Processing of the adenovirus terminal protein. *J. Virol.* **38**: 272.

———. 1982. Eukaryotic DNA replication: Viral and plasmid model systems. *Annu. Rev. Biochem.* **51**: 901.

Challberg, M.D. and D.R. Rawlins. 1984. Template requirement for the initiation of adenovirus DNA replication. *Proc. Natl. Acad. Sci.* **81**: 100.

Challberg, M.D., S.V. Desiderio, and T.J. Kelly, Jr. 1980. Adenovirus DNA replication *in vitro*: Characterization of a protein covalently linked to nascent DNA strands. *Proc. Natl. Acad. Sci.* **77**: 5105.

Challberg, M.D., J.M. Ostrove, and T.J. Kelly, Jr. 1982. Initiation of adenovirus DNA replication: Detection of covalent complexes between nucleotide and the 80Kd terminal protein. *J. Virol.* **41**: 265.

Clark, R., D.P. Lane, and R. Tjian. 1981. Use of monoclonal antibodies as probes of simian virus 40 T antigen ATPase activity. *J. Biol. Chem.* **256**: 11854.

Clark, R., K. Peden, J. Pipas, D. Nathans, and R. Tjian. 1983. Biochemical activities of T-antigen proteins encoded by simian virus 40 A gene deletion mutants. *Mol. Cell. Biol.* **3**: 220.

DePamphilis, M.L. and P.M. Wassarman. 1982. Organization and replication of papovavirus DNA. In *Organization and replication of viral DNA* (ed. A.S. Kaplan), p. 37. CRC Press, Boca Raton, Florida.

de Vries, E., W. van Driel, M. Tromp, J. van Boom, and P.C. van der Vliet. 1985. Adenovirus DNA replication *in vitro*: Site-directed mutagenesis of the nuclear factor I binding site of the Ad2 origin. *Nucleic Acids Res.* **13**: 4935.

Diffley, J.F.X. and B. Stillman. 1986. Purification of a cellular, double stranded DNA binding protein required for initiation of adenovirus DNA replication using a rapid filter binding assay. *Mol. Cell. Biol.* (in press).

Enomoto, T., J.H. Lichy, J.E. Ikeda, and J. Hurwitz. 1981.

Adenovirus DNA replication *in vitro*: Purification of the terminal protein in a functional form. *Proc. Natl. Acad. Sci.* **78**: 6779.

Field, J., R.M. Gronostajski, and J. Hurwitz. 1984. Properties of the adenovirus DNA polymerase. *J. Biol. Chem.* **259**: 9487.

Friefeld, B.R., J.H. Lichy, J. Hurwitz, and M.S. Horwitz. 1983. Evidence for an altered adenovirus DNA polymerase in cells infected with the mutant H5*ts*149. *Proc. Natl. Acad. Sci.* **80**: 1589.

Friefeld, B.R., J.H. Lichy, J. Field, R.M. Gronostajski, R.A. Guggenheimer, M.D. Krevolin, K. Nagata, J. Hurwitz, and M.S. Horwitz. 1984. The *in vitro* replication of adenovirus DNA. *Curr. Top. Microbiol. Immunol.* **110**: 221.

Fuller, R.S., B. Funnell, and A. Kornberg. 1984. The DNA A protein complex with the *E. coli* chromosomal replication (ori C) and other DNA sites. *Cell* **38**: 889.

Galas, D.J. and A. Schmitz. 1978. DNAase footprinting: A simple method for the detection of protein-DNA binding specificity. *Nucleic Acids Res.* **5**: 3157.

Gluzman, Y. and B. Ahrens. 1982. SV40 early mutants that are defective for viral DNA synthesis but competent for transformation of cultured rat and simian cells. *Virology* **123**: 78.

Graham, F.L., J. Smiley, W.C. Russell, and R. Nairu. 1977. Characteristics of a human cell line transformed by DNA from human adenovirus type 5. *J. Gen. Virol.* **36**: 59.

Guggenheimer, R.A., B.W. Stillman, K. Nagata, F. Tamanoi, and J. Hurwitz. 1984. DNA sequences required for the *in vitro* replication of adenovirus DNA. *Proc. Natl. Acad. Sci.* **81**: 3069.

Hay, R.T. 1985. The origin of adenovirus DNA replication: Minimal DNA sequence requirement *in vivo*. *EMBO J.* **4**: 421.

Hay, R.T. and M.L. DePamphilis. 1982. Initiation of SV40 DNA replication *in vivo*: Location and structure of 5′ ends of DNA synthesized in the *ori* region. *Cell* **28**: 767.

Ideka, J.-E., T. Enomoto, and J. Hurwitz. 1981. Replication of the adenovirus DNA-protein complex with purified proteins. *Proc. Natl. Acad. Sci.* **78**: 884.

Kaguni, J.M. and A. Kornberg. 1984. Replication initiated at the origin (ori C) of the *E. coli* chromosome reconstituted with purified enzymes. *Cell* **38**: 183.

Kaplan, L.M., H. Ariga, J. Hurwitz, and M.S. Horwitz. 1979. Complementation of the temperature sensitive defect in H5*ts*125 adenovirus DNA replication *in vitro*. *Proc. Natl. Acad. Sci.* **76**: 5534.

Kelly, T.J., Jr. 1984. Adenovirus DNA replication. In *The adenoviruses* (ed. H.S. Ginsberg), p. 271. Plenum Press, New York.

Lally, C., T. Dorper, W. Groger, G. Antoine, and E.-L. Winnacker. 1984. A size analysis of the adenovirus replicon. *EMBO J.* **3**: 333.

Leegwater, P.A.J., K. van Driel, and P.C. van der Vliet. 1985. Recognition site of nuclear factor I, a sequence specific DNA-binding protein from HeLa cells that stimulates adenovirus DNA replication. *EMBO J.* **4**: 1515.

Li, J.J. and T.J. Kelly, Jr. 1984. Simian virus 40 DNA replication *in vitro*. *Proc. Natl. Acad. Sci.* **81**: 6973.

———. 1985. SV40 DNA replication *in vitro*: Specificity of initiation and evidence for bidirectional replication. *Mol. Cell. Biol.* **5**: 1238.

Lichy, J.H., M.S. Horwitz, and J. Hurwitz. 1981. Formation of a covalent complex between the 80,000 dalton adenovirus terminal protein and 5′-dCMP *in vitro*. *Proc. Natl. Acad. Sci.* **78**: 2678.

Lichy, J.H., J. Field, M.S. Horwitz, and J. Hurwitz. 1982. Separation of the adenovirus terminal protein precursor from its associated DNA polymerase: Role of both proteins in the initiation of adenovirus DNA replication. *Proc. Natl. Acad. Sci.* **79**: 5225.

Lichy, J., T. Enomoto, J. Field, R.A. Guggenheimer, J.-E. Ikeda, K. Nagata, M. Horwitz, and J. Hurwitz. 1983. Isolation of proteins involved in the replication of adenovirus DNA. *Cold Spring Harbor Symp. Quant. Biol.* **47**: 731.

Maniatis, T., A. Jeffrey, and H. van de Sande. 1975. Chain

length determination of small double- and single-stranded DNA molecules by polyacrylamide gel electrophoresis. *Biochemistry* **14**: 3787.

Manos, M.M. and Y. Gluzman. 1984. Simian virus 40 large T-antigen point mutants that are defective in viral DNA replication but competent in oncogenic transformation. *Mol. Cell. Biol.* **4**: 1125.

————. 1985. Genetic and biochemical analysis of transformation-competent, replication-defective simian virus 40 large T antigen mutants. *J. Virol.* **53**: 120.

Maxam, A. and W. Gilbert. 1980. Sequencing end-labeled DNA with base-specific chemical cleavages. *Methods Enzymol.* **65**: 499.

Myers, R.M. and R. Tjian. 1980. Construction and analysis of simian virus 40 origins defective in tumor antigen binding and DNA replication. *Proc. Natl. Acad. Sci.* **77**: 6491.

Nagata, K., R.A. Guggenheimer, and J. Hurwitz. 1983a. Specific binding of a cellular DNA replication protein to the origin of replication of adenovirus DNA. *Proc. Natl. Acad. Sci.* **80**: 6177.

————. 1983b. Adenovirus DNA replication *in vitro*: Synthesis of full-length DNA with purified proteins. *Proc. Natl. Acad. Sci.* **80**: 4266.

Nagata, K., R.A. Guggenheimer, T. Enomoto, J.H. Lichy, and J. Hurwitz. 1982. Adenovirus DNA replication *in vitro*: Identification of a host factor that stimulates synthesis of the preterminal protein-dCMP complex. *Proc. Natl. Acad. Sci.* **79**: 6438.

Nowock, J. and A.E. Sippel. 1982. Specific protein-DNA interaction at four sites flanking the chicken lysozyme gene. *Cell* **30**: 607.

Nowock, J., V. Borgmeyer, A.W. Püschel, R.A.W. Rupp, and A.E. Sippel. 1985. The TGGCA protein binds to the MMTV-LTR, the adenovirus origin of replication, and the BK virus enhancer. *Nucleic Acids Res.* **13**: 2045.

Ostrove, J.M., P. Rosenfeld, J. Williams, and T.J. Kelly, Jr. 1983. *In vitro* complementation as an assay for purification of adenovirus DNA replication proteins. *Proc. Natl. Acad. Sci.* **80**: 935.

Pincus, S., W. Robertson, and D.M.K. Rekosh. 1981. Characterization of the effect of aphidicolin on adenovirus DNA replication: Evidence in support of a protein primer model of initiation. *Nucleic Acids Res.* **9**: 4919.

Prelich, G. and B.W. Stillman. 1986. Functional characterization of thermolabile DNA-binding proteins that affect adenovirus DNA-replication. *J. Virol.* **57**: 883.

Rawlins, D.R., P.J. Rosenfeld, R.J. Wides, M.D. Challberg, and T.J. Kelly, Jr. 1984. Structure and function of the adenovirus origin of replication. *Cell* **37**: 309.

Rekosh, D.M.K., W.C. Russell, A.J.D. Bellett, and A.J. Robinson. 1977. Identification of a protein linked to the ends of adenovirus DNA. *Cell* **11**: 283.

Rijinders, A.W.M., B.G.M. van Bergen, P.C. van der Vliet, and J.S. Sussenbach. 1983. Specific binding of the adenovirus terminal protein precursor-DNA polymerase complex to the origin of DNA replication. *Nucleic Acids Res.* **11**: 8777.

Robinson, A.J., H.B. Younghusband, and A.J.D. Bellett. 1973. A circular DNA-protein complex from adenoviruses. *Virology* **56**: 54.

Schechter, N.M., W. Davies, and C.W. Anderson. 1980. Adenovirus coded deoxyribonucleic acid binding protein: Isolation, physical properties, and effects of proteolytic digestion. *Biochemistry* **19**: 2802.

Sharp, P.A. 1984. Adenovirus transcription. In *The adenoviruses* (ed. H.S. Ginsberg), p. 173. Plenum Press, New York.

Siebenlist, U., L. Henninghausen, J. Battey, and P. Leder. 1984. Chromatin structure and protein binding in the putative regulatory region of c-*myc* in Burkitt lymphoma. *Cell* **37**: 381.

Simanis, V. and D.P. Lane. 1985. An immunoaffinity purification procedure for SV40 large T antigen. *Virology* **144**: 88.

Stillman, B.W. 1983. The replication of adenovirus DNA with purified proteins. *Cell* **35**: 7.

Stillman, B.W. and Y. Gluzman. 1985. Replication and supercoiling of SV40 DNA in cell free extracts from human cells. *Mol. Cell. Biol.* **5**: 2051.

Stillman, B.W., F. Tamanoi, and M.B. Mathews. 1982. Purification of an adenovirus-coded DNA polymerase that is required for initiation of DNA replication. *Cell* **31**: 613.

Stillman, B.W., E. White, and T. Grodzicker. 1984. Independent mutations in Ad2*ts*111 cause degradation of cellular DNA and defective viral DNA replication. *J. Virol.* **50**: 598.

Stillman, B., R.D. Gerard, R.A. Guggenheimer, and Y. Gluzman. 1985. T antigen and template requirements for SV40 DNA replication *in vitro*. *EMBO J.* **4**: 2933.

Stillman, B.W., J.B. Lewis, L.T. Chow, M.B. Mathews, and J.E. Smart. 1981. Identification of the gene and mRNA for the adenovirus terminal protein precursor. *Cell* **23**: 497.

Sundin, O. and A. Varshavsky. 1980. Terminal stages of SV40 DNA replication proceed via multiply intertwined catenated dimers. *Cell* **21**: 103.

————. 1981. Arrest of segregation leads to accumulation of highly intertwined catenated dimers: Dissection of the final stages of SV40 DNA replication. *Cell* **25**: 659.

Tamanoi, F. and B.W. Stillman. 1982. Function of the adenovirus terminal protein in the initiation of DNA replication. *Proc. Natl. Acad. Sci.* **79**: 2221.

————. 1983. Initiation of adenovirus DNA replication *in vitro* requires a specific DNA sequence. *Proc. Natl. Acad. Sci.* **80**: 6446.

Tegtmeyer, P. 1972. Simian virus 40 deoxyribonucleic acid synthesis: The viral replicon. *J. Virol.* **10**: 591.

Tsurimoto, T. and K. Matsubara. 1981. Purified bacteriophage lambda γ protein binds to four repeating sequences at the lambda replication origin. *Nucleic Acids Res.* **9**: 1789.

van Bergen, B.G.M., P.A. van der Ley, W. van Driel, A.D.M. van Mansfeld, and P.C. van der Vliet. 1983. Replication of origin containing adenovirus DNA fragments that do not carry the terminal protein. *Nucleic Acids Res.* **11**: 1975.

van der Ende, A., T.A. Baker, T. Ogawa, and A. Kornberg. 1985. Initiation of enzymatic replication of the origin of the *Escherichia coli* chromosome: Primase as the sole priming enzyme. *Proc. Natl. Acad. Sci.* **82**: 3954.

Wang, K. and G.D. Pearson. 1985. Adenovirus sequences required *in vivo*. *Nucleic Acids Res.* **13**: 5173.

Weaver, D.T., S.C. Fields-Berry, and M.L. DePamphilis. 1985. The termination region for SV40 DNA replication directs the mode of separation for the two sibling molecules. *Cell* **41**: 565.

Wobbe, C.R., F. Dean, L. Weissbach, and J. Hurwitz. 1985. In vitro replication of duplex circular DNA containing the simian virus 40 DNA origin site. *Proc. Natl. Acad. Sci.* **82**: 5710.

Wold, M.S., J.B. Mallory, J.D. Roberts, J.H. Lebowitz, and R. McMacken. 1982. Initiation of bacteriophage lambda DNA replication *in vitro* with purified lambda replication proteins. *Proc. Natl. Acad. Sci.* **79**: 6176.

# In Vitro Replication of DNA Containing the Simian Virus 40 Origin

## Y. Murakami, C.R. Wobbe, L. Weissbach, F.B. Dean, and J. Hurwitz

Graduate Program in Molecular Biology and Virology, Sloan-Kettering Institute for Cancer Research, New York, New York 10021

The in vitro replication of DNA containing the simian virus 40 (SV40) origin has been carried out using cell-free extracts of HeLa cells. The in vitro system required T antigen, DNA containing the SV40 origin, ATP, and an ATP-regenerating system. The role of DNA polymerase α and DNA primase in the in vitro replication system was examined. Aphidicolin, a known, specific inhibitor of DNA polymerase α in vitro, inhibited T-antigen ori$^+$ DNA–dependent replication. Removal of DNA polymerase α and DNA primase activities from HeLa cell crude extract using an anti-DNA polymerase α immunoaffinity column resulted in the loss of replication activity. The addition of purified DNA polymerase α–primase complex isolated from HeLa cells restored the replication activity of depleted extracts. Both activities were required to restore DNA synthesis; the addition of DNA polymerase α or DNA primase alone supported replication poorly. Under conditions where DNA polymerase α–primase complex isolated from HeLa cells and from monkey cells restored replication by the depleted extract, the DNA polymerase α–primase complex from mouse cells and from calf thymus did not. *Escherichia coli* DNA polymerases I and III did not restore replication activity. In addition, crude extracts have been resolved into multiple fractions that are inactive by themselves but are active in combination.

The replication of SV40 DNA is an important model system for studying eukaryotic DNA replication because it requires only one virus-coded protein, the SV40 large T antigen; all of the other components involved in this process are supplied by host cells. Our interest is in the isolation of factors that are involved in mammalian DNA replication. For this purpose, we have developed an in vitro replication system that is analogous to the systems described by other laboratories (Ariga and Sugano 1983; Li and Kelly 1984; Stillman and Gluzman 1985), using a salt extract of exponentially growing HeLa cells and plasmid DNA containing the SV40 origin sequence (Wobbe et al. 1985). In this communication we will describe the requirements of this replication system, the role of DNA polymerase α and DNA primase in the replication system and the fractionation of the replication activity of HeLa cell crude extract.

DNA polymerase α is known to play a key role in mammalian DNA replication (Sugino and Nakayama 1980; Murakami et al. 1985). We have found that DNA polymerase α and DNA primase activity are essential for the in vitro replication of DNA containing the SV40 origin. We have also isolated multiple protein fractions that, in addition to the DNA polymerase α–primase complex and T antigen, support this DNA replication system.

## Experimental Procedures

### In vitro replication system of SV40 ori$^+$ DNA

Reaction conditions as well as methods used for the preparation of HeLa cell extracts (Wobbe et al. 1985) and SV40 T antigen (Simanis and Lane 1985) were described previously. Reaction mixtures contained 30 mM Tris · HCl buffer (pH 8.5), 7 mM MgCl$_2$, 0.5 mM DTT, 4 mM ATP, 200 μM each of CTP, UTP, and GTP, 100 μM each of dATP, dCTP, and dGTP, 25 μM [methyl-$^3$H]dTTP, 40 mM creatine phosphate, 20 μg/ml creatine phosphokinase, 0.3 μg RFI plasmid (pSV01ΔEP3, ori$^+$), 200–400 μg of HeLa extract, and 0.5–1 μg of T antigen. The ori$^+$ DNA consisted of the 311-bp origin-containing *Eco*RII fragment of SV40 DNA cloned into a 2481-bp pBR322 derivative described earlier (Wobbe et al. 1985). Reactions were incubated at 30°C as indicated, and the amount of acid-insoluble material formed was measured. Products obtained in the replication reaction were analyzed by 1.5% agarose gel electrophoresis.

### Depletion of DNA polymerase α–primase complex

HeLa cell extracts were applied to anti-DNA polymerase α immunoaffinity columns prepared with SJK287-37 hybridoma (source ATCC) supernatant and anti-mouse IgG–Sepharose by the method previously described (Chang et al. 1984); flowthrough fractions were used as the source of DNA polymerase α–primase-depleted extract.

### Purification of DNA polymerase α–primase complex and separation of these activities

Purification of DNA polymerase α–primase complex was carried out as described earlier (Gronostajski et al. 1984). The fraction obtained after the second phosphocellulose column chromatographic step was used as the purified enzyme. Mouse mammary carcinoma FM3A cells were used as the source for the isolation of the mouse DNA polymerase α–primase complex. The calf thymus enzyme was purchased from Pharmacia. DNA polymerase α activity without detectable DNA primase activity was obtained by DEAE-cellulose column chromatography in the presence of 50% ethylene glycol (Suzuki et al. 1985). DNA primase activity was obtained from the anti–DNA polymerase α immunoaffinity column by elution with 50% ethylene glycol. Such DNA primase fractions contained 2.5% of the DNA polymerase α activity normally found in DNA polymerase α–primase complex preparations. The assay systems for DNA polymerase α and DNA primase activity were described earlier (Gronostajski et al. 1984).

*Fractionation procedure*
HeLa cell crude extract prepared as described earlier (Wobbe et al. 1985) was fractionated by ammonium sulfate precipitation into two fractions (0–40% and 40–65%). The 40% ammonium sulfate fraction was further fractionated by Bio-Rex chromatography into three fractions: fraction I (80 mM NaCl flowthrough), fraction II (250 mM NaCl eluate), and fraction III (1 M NaCl eluate). The details of this procedure will be published elsewhere.

## Results

### Requirements for replication of SV40 *ori* $^+$ DNA
We investigated the requirements of the replication system using crude HeLa extracts. As shown in Figure 1,

**Table 1** Requirements for Replication of SV40 *ori* $^+$ DNA

| Component omitted | DNA synthesis (%) |
|---|---|
| None | 100 |
| T antigen | <2 |
| DNA | <2 |
| DNA + *ori* $^-$ DNA | <2 |
| DNA + topoisomerase I–treated *ori* $^+$ DNA | 83 |
| DNA + *Pst*I RFIII of *ori* $^+$ DNA | 3 |
| ATP | 35 |
| Creatine phosphate and creatine phosphokinase | 9 |
| dATP, dGTP, dTTP and CTP, UTP, GTP | 3 |
| dATP, dGTP, dTTP | 3 |
| CTP, UTP, GTP | 80 |
| None + aphidicolin | |
|   100 μM | <2 |
|   400 μM | <2 |
| None + camptothecin | |
|   100 μM | 58 |
|   500 μM | 42 |

Reaction mixtures (50 μl) were as described in Experimental Procedures, with 0.3 μg of pSV01ΔEP (*ori* $^+$ DNA) or 0.3 μg of pBR322ΔEP (*ori* $^-$ DNA), 380 μg of HeLa extract protein, and 25 μM [α-$^{32}$P]dCTP (2000 cpm/pmole). Reactions were carried out for 2 hr at 37°C. The 100% value was equivalent to the incorporation of 128 pmoles of dCMP into an acid-insoluble form.

replication occurred only in the presence of DNA containing the SV40 origin and the SV40 T antigen. In the absence of T antigen or with DNA lacking the SV40 origin, replication did not occur. Table 1 shows the further analysis of the system. Topoisomerase I–relaxed DNA was as active as RFI DNA, whereas linearized duplex DNA was inactive. ATP and an ATP-regenerating system were required. Replication was completely dependent on the addition of dNTPs but was marginally stimulated by CTP, UTP, and GTP. Aphidicolin, a known inhibitor of DNA polymerase α in vitro (Huberman 1981), inhibited replication. Camptothecin, a drug that inhibits eukaryotic DNA replication in vivo (Horowitz et al. 1971), also reduced replication 60% maximally. We have shown that this drug inhibits eukaryotic topoisomerases I and II in vitro (Dean et al. 1985). Inhibition of SV40 DNA replication by this drug suggests the involvement of topoisomerases in this system.

### The role of DNA polymerase α and DNA primase
To study the role of DNA polymerase α and DNA primase, we prepared DNA polymerase α–DNA primase–depleted extracts by passing the crude extract through an anti-DNA polymerase α immunoaffinity column. This procedure removed about 70% of DNA polymerase α and 90% of DNA primase activity present in the extract. Further recycling of the depleted extract through immunoaffinity columns did not remove the residual DNA polymerase α activity. When the depleted extract was examined in the replication system, it was inactive (Table 2). This suggests that the DNA polymerase α–primase complex is essential for the replication system. This was

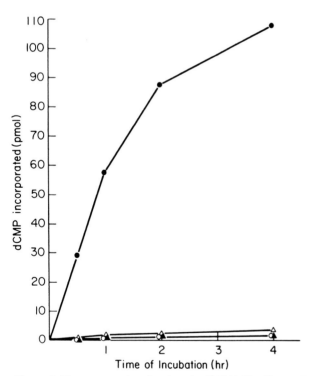

**Figure 1** Time course of replication of *ori* $^+$ DNA (●, ○) or *ori* $^-$ DNA (▲, △) in the presence (●, ▲) or absence (○, △) of 0.6 μg of SV40 T antigen. Reaction conditions were as described in Experimental Procedures. At indicated times, aliquots were assayed for acid-insoluble radioactivity.

**Table 2** Depletion of DNA Polymerase α–Primase from HeLa Cells

| Fraction analyzed | DNA polymerase α (units/mg protein) | Activity measured | |
| --- | --- | --- | --- |
| | | DNA primase (units/mg protein) | replication (pmoles dTMP incorp./60 min) |
| HeLa crude extract | 6.90 | 22.3 | 37.6 |
| Depleted extract | 1.96 | 2.7 | 0.5 |

The removal of DNA polymerase α–primase activities from HeLa crude extract was as described in Experimental Procedures. The definition of DNA polymerase and DNA primase are: 1 unit of DNA polymerase α activity catalyzes the incorporation of 1 nmole of [$^3$H]dTMP in 30 min at 30°C in the DNA polymerase α assay and 1 unit of DNA primase resulted in the incorporation of 1 nmole of [$^3$H]dAMP in 30 min at 30°C in the DNA primase assay. Replication reactions were carried out using 0.3 μg pSV01ΔEP (*ori*$^+$ DNA) and 0.6 μg T antigen.

verified by the reconstitution of the replication activity with highly purified DNA polymerase α–primase complex (Fig. 2). The extent of replication depended upon the amount of DNA polymerase α–primase activity added; all incorporation with the reconstituted system required T antigen and DNA containing the SV40 origin. Crude extract containing 1 unit of DNA polymerase α activity supported the incorporation of 60 pmoles of dTMP in 60 min. When 0.64 unit of DNA polymerase α (60% of the DNA polymerase α activity present in the untreated crude extract) was added to depleted extacts, about 60% of the original replication activity was observed. We also analyzed the replication products obtained with reconstituted extracts (Fig. 3). The original extracts yielded RFI, topoisomers, RFII, and high-molecular-weight product only in the presence of T antigen. Identical replication products were formed with the depleted extract in the presence of exogenous DNA polymerase α–primase complex and T antigen.

The experiments described above were carried out with the DNA polymerase α–primase complex isolated from HeLa cells. We investigated whether DNA polymerase α–primase complex from other species could restore replication activity of the depleted extract of HeLa cells. Under conditions where the addition of HeLa DNA polymerase α–primase supported DNA synthesis (Table 3), neither the mouse nor the calf thymus enzyme preparations did. On the other hand, monkey DNA polymerase α was as efficient as the HeLa enzyme in supporting replication. In addition, *E. coli* DNA polymerase I and DNA polymerase III holoenzyme (plus *dnaN*) were inactive in this system.

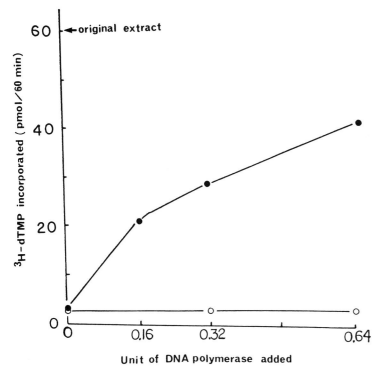

**Figure 2** Reconstitution of depleted extract with HeLa DNA polymerase α–primase complex. Indicated amounts of purified DNA polymerase α–primase complex were added to DNA polymerase α–depleted extract. After 60 min of incubation, [$^3$H]dTMP incorporation into acid-insoluble material was determined. Reactions were carried out in the absence (○) or presence (●) of 0.6 μg T antigen.

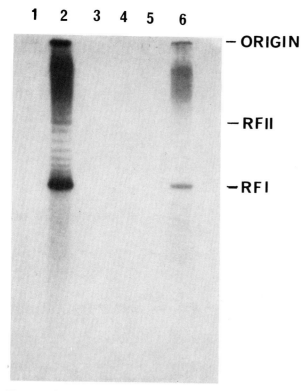

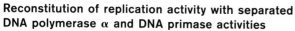

1  2  3  4  5  6

— ORIGIN

— RFII

— RFI

**Figure 3** Analysis of the products formed in the reconstitution experiment. Reaction mixtures contained HeLa cell extract (lanes *1* and *2*) or depleted extract (lanes *3–6*) with (lanes *2, 4* and *6*) or without (lanes *1, 3,* and *5*) SV40 T antigen. Lanes *5* and *6* contained depleted extracts supplemented with HeLa DNA polymerase α–primase. All reactions were incubated and processed for gel electrophoresis as described in Experimental Procedures.

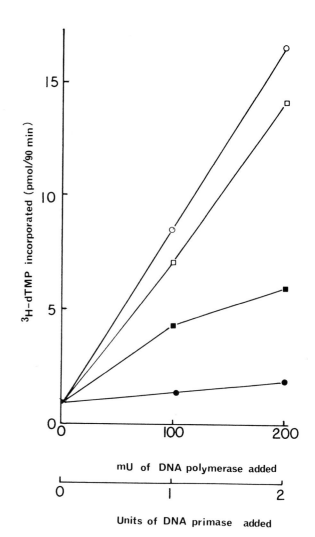

mU of DNA polymerase added

Units of DNA primase added

**Figure 4** Response to DNA polymerase α, DNA primase, and DNA polymerase α + DNA primase. Reaction mixtures containing DNA polymerase α–primase complex (○), a mixture of DNA polymerase α and DNA primase (□), DNA primase alone (■), or DNA polymerase α alone (●) added to the depleted extract were incubated for 90 min and then assayed for acid-insoluble radioactivity.

## Reconstitution of replication activity with separated DNA polymerase α and DNA primase activities

As described in the previous section, we could restore the replication reaction with depleted extract by addition of HeLa cell DNA polymerase α–primase complex. We examined whether both activities were essential for rep-

**Table 3** The Response of Depleted Extract to Various DNA Polymerases

| Fraction tested | Replication activity (pmoles dTMP incorp./60 min) | |
|---|---|---|
| | TAg (−) | TAg (+) |
| 1. Original HeLa extract | 1.47 | 76.8 |
| 2. pol α-depleted extract | 1.46 | 2.46 |
| 3. No. 2 + 0.4 unit of HeLa pol α–primase complex | 1.61 | 26.1 |
| 4. No. 2 + 0.4 unit of monkey pol α–primase complex | 1.41 | 23.4 |
| 5. No. 2 + 0.4 unit of mouse pol α–primase complex | 1.49 | 3.75 |
| 6. No. 2 + 0.4 unit of calf thymus pol α–primase complex | 1.47 | 1.87 |
| 7. No. 2 + 1.0 unit of *E. coli* pol I | 1.04 | 2.07 |
| 8. No. 2 + 1.0 unit of *E. coli* pol III | 1.38 | 1.87 |

Replication reactions were carried out in the presence or absence of 0.6 μg of T antigen for 60 min and the amount of acid-insoluble radioactivity was determined.

lication. For this purpose, we have separated HeLa DNA polymerase α activity from DNA primase activity. These isolated activities were added to depleted extract and assayed for replication activity (Fig. 4). Under conditions where the DNA polymerase α–primase complex supported replication, the combination of DNA polymerase α and DNA primase also worked well. However, DNA polymerase α alone was inactive and DNA primase alone possessed low activity. The low activity observed upon addition of DNA primase alone could be due to minor contamination of the DNA primase preparation with DNA polymerase as well as the presence of DNA polymerase α activity remaining in the depleted extract. These results suggest that both DNA polymerase α activity and DNA primase activity are essential for replication.

### Fractionation of HeLa crude extract

To isolate the protein factors essential for the replication system, we have resolved the HeLa cell crude extract into multiple fractions. Using ammonium sulfate precipitation, two fractions (0–40% and 40–65%) were obtained from crude extract. The 0–40% fraction alone contained detectable replication activity that could be increased fivefold by the addition of the 40–65% fraction. The latter fraction was inactive by itself. We further fractionated the 0–40% fraction into three components using Bio-Rex column chromatography (Table 4).

The complete system, containing the Bio-Rex I, II, and III and 40–65% ammonium sulfate fractions, supported extensive DNA replication. Omission of any one of the fractions resulted in the loss of replication activity. The Bio-Rex II fraction stimulated the reaction, but it was not absolutely essential for replication. In addition, a fifth fraction, isolated from nuclear extracts, stimulated incorporation. This fraction was inactive by itself.

The requirements for DNA synthesis with partially purified fractions were also examined (Table 5). The system was dependent upon origin-containing DNA, T antigen, ATP, and an ATP-regenerating system. Omission of UTP, CTP, and GTP did not affect incorporation. The replication system was sensitive to DNase I and

**Table 4** Requirements of Multiple Fractions for SV40 Replication

| Fractions added | dTMP incorporated (pmoles/70 min) |
|---|---|
| Bio-RexI + II + III + AS$_{40-65}$ | 37.4 |
| omit Bio-Rex I | 3.4 |
| omit Bio-Rex II | 11.4 |
| omit Bio-Rex III | 4.7 |
| omit As$_{40-65}$ | 2.0 |
| Bio-RexI + II + III + AS$_{40-65}$ + nuclear fraction | 60.6 |
| omit nuclear fraction | 27.4 |
| nuclear fraction alone | <1.0 |

Replication reactions were carried out as described in Experimental Procedures for 70 min at 37°C and scored for acid-insoluble ratioactivity.

**Table 5** Requirements for Replication with Separated Protein Fractions

| Additions | dTMP incorporated (pmoles/60 min) |
|---|---|
| Bio-Rex I + II + III + As$_{40-65}$ | 35.9 |
| Omit ATP + CrP + CrK | 1.4 |
| Omit DNA | <1.0 |
| Omit T Ag | <1.0 |
| Use ori$^-$ DNA in place of ori$^+$ DNA | <1.0 |
| Omit UTP, CTP, and GTP | 34.0 |
| Complete + DNase I μg | 1.3 |
| Complete + RNase 4 μg | <1.0 |
| Complete + aphidicolin (6.7 × 10$^{-5}$ M) | <1.0 |
| Complete + pol α–primase complex | 39.9 |
| Complete + nuclear extract (AS$_{40-50}$) | 53.4 |

Replication reactions were as described in Experimental Procedures, using the multiple fractions described above. (CrP) Creatine phosphate; (CrK) creatine phosphokinase.

RNase A. It is also highly sensitive to aphidicolin, indicating involvement of DNA polymerase α in this system. The addition of the DNA polymerase α–primase complex had little effect on the reaction nor did it replace any of the fractions. The Bio-Rex fractions contained substantial amounts of DNA polymerase α and primase activities.

### Discussion

We have used an anti–DNA polymerase α monoclonal antibody to analyze the role of DNA polymerase α in the replication of DNA containing the SV40 origin. Since DNA primase forms a tight complex with DNA polymerase α in higher eukaryotes (Conaway and Lehman 1982; Gronostajski et al. 1984), this procedure also removed DNA primase. The depleted extracts were inactive in the replication reaction. The addition of purified DNA polymerase α–primase complex restored this activity. In contrast, DNA polymerase α devoid of DNA primase activity did not support replication; DNA primase, containing trace amounts of DNA polymerase α, also did not efficiently restore replication activity. The combination of DNA polymerase α and DNA primase activities worked as well as the DNA polymerase α–primase complex, indicating that both activities are essential for the replication system.

The depleted system enabled us to test different DNA polymerase α–primase complex preparations for their ability to support replication in the in vitro system. DNA polymerase α–primase complex purified from exponentially growing mouse cells and from calf thymus, neither of which is a natural host for the propagation of SV40 virus, were inactive, suggesting that enzyme from cells that can support replication of SV40 DNA is required. On the other hand, the DNA polymerase α–primase complex purified from monkey COS cells, a natural host for SV40, could reactivate the depleted extract. Human cells are not a natural host for SV40 virus, but they are permissive for replication of SV40 DNA (Stillman and Gluzman 1985). Therefore, it is possible that the permissiveness of a given cell centers on the ability of T antigen to interact with the cellular DNA polymerase $\alpha$–primase complex. Further work comparing the different systems is now under way.

In view of the complexity of prokaryotic bidirectional replication, it is likely that multiple factors are also required for the bidirectional replication of SV40 DNA in vitro. To isolate host factors that are necessary for the replication system, we have fractionated HeLa cell crude extract and have now obtained five fractions other than DNA polymerase $\alpha$–primase and T antigen, which were required for replication. Elucidation of the mechanism by which replication of SV40 origin–containing DNA occurs will require the identification of the specific reactions carried out by these fractions. It is clear that the SV40 system is amenable to purification and will provide important insights into chromosomal replication.

## Acknowledgments

We are indebted to Ms. Claudette Turck and Ms. Nilda Belgado for their technical assistance. This work was supported by grant GM-34559 from the National Institutes of Health. C.R.W. is a postdoctoral fellow of the Damon Runyon Foundation, and L.W. is a postdoctoral fellow of NIH.

## References

Ariga, H. and S. Sugano. 1983. Initiation of simian virus 40 DNA replication *in vitro*. *J. Virol.* **48:** 481.

Chang, L.M.S., E. Rafter, C. Angle, and F.J. Bollum. 1984. Purification of a DNA polymerase–DNA primase complex from calf thymus glands. *J. Biol. Chem.* **259:** 14679.

Conaway, R.C. and I.R. Lehman. 1982. A DNA primase activity associated with DNA polymerase $\alpha$ from *Drosophila melanogaster* embryos. *Proc. Natl. Acad. Sci.* **79:** 2523.

Dean, F.B., C.R. Wobbe, L. Weissbach, Y. Murakami, and J. Hurwitz. 1985. *In vitro* replication of DNA containing the SV40 origin sequence. *UCLA Symp. Mol. Cell. Biol. New Ser.* **32:** (in press).

Gronostajski, R., J. Field, and J. Hurwitz. 1984. Purification of a primase activity associated with DNA polymerase $\alpha$ from HeLa cells. *J. Biol. Chem.* **259:** 9479.

Horowitz, S.B., C.K. Chang, and A.P. Grollman. 1971. Studies on camptothecin. I. Effects on nucleic acid and protein synthesis. *Mol. Pharmacol.* **7:** 632.

Huberman, J.A. 1981. New views of biochemistry of eukaryotic DNA replication revealed by aphidicolin, an unusual inhibitor of DNA polymerase $\alpha$. *Cell* **23:** 647.

Li, J.J. and T.J. Kelly. 1984. Simian virus 40 DNA replication *in vitro*. *Proc. Natl. Acad. Sci.* **81:** 6973.

Murakami, Y., H. Yasuda, H. Miyasawa, F. Hanaoka, and M. Yamada. 1985. Characterization of a temperature-sensitive mutant of mouse FM3A cells defective in DNA replication. *Proc. Natl. Acad. Sci.* **82:** 1761.

Simanis, V. and D.P. Lane. 1985. An immunoaffinity purification procedure for SV40 large T antigen. *Virology* **144:** 88.

Stillman, B.W. and Y. Gluzman. 1985. Replication and supercoiling of simian virus 40 DNA in cell extracts from human cells. *Mol. Cell. Biol.* **5:** 2051.

Sugino, A. and K. Nakayama. 1980. DNA polymerase $\alpha$ mutants from a *Drosophila melanogaster* cell line. *Proc. Natl. Acad. Sci.* **77:** 7049.

Suzuki, M., T. Enomoto, F. Hanaoka, and M. Yamada. 1985. Dissociation and reconstruction of a DNA polymerase $\alpha$–primase complex. *J. Biochem.* **98:** 581.

Wobbe, C.R., F.B. Dean, L. Weissbach, and J. Hurwitz. 1985. *In vitro* replication of duplex circular DNA containing the SV40 DNA origin site. *Proc. Natl. Acad. Sci.* **82:** 5710.

# A New Biochemical Property of SV40 Large T Antigen

## E. Fanning and B. Vogt

Institute for Biochemistry, 8000 Munich 2, Federal Republic of Germany

Specific binding of simian virus 40 (SV40) large tumor antigen (TAg) to sequences in the viral origin of DNA replication is essential for initiation of viral DNA replication and control of viral transcription. In the present communication, we present evidence that sequence-specific DNA binding of immunopurified large TAg is markedly inhibited by low concentrations of purine nucleoside triphosphates. The inhibition is reversible after removal of the nucleotide, suggesting that simple nucleotide binding rather than a covalent modification of large TAg in the presence of ATP is responsible. ATP fails to inhibit DNA binding of a large TAg-related protein encoded by the hybrid virus Ad2$^+$D2. The results suggest that large TAg may assume either of two conformations: one able to bind to nucleotides but not SV40 DNA and another able to bind DNA but not nucleotides. The possible significance of this new biochemical property of large TAg in viral DNA replication and transcription is discussed.

Simian virus 40 (SV40) DNA replication and transcription are controlled by a multifunctional viral protein, the large tumor antigen (TAg) (for review, see Rigby and Lane 1983). Genetic and biochemical studies have shown that sequence-specific binding of large TAg to two major sites in the viral origin of replication is required for this regulation. In addition, large TAg exhibits several other biochemical activities, including ATPase, nonspecific binding to single- and double-stranded DNA, and ATP binding. Although ATPase and possibly ATP binding appear to be essential for viral DNA replication (Clark et al. 1983; Clertant et al. 1984), the biochemical roles of these activities in replication are not understood.

The biochemical properties of a number of prokaryotic proteins involved in DNA replication and recombination have been more extensively investigated. DNA-binding activity of several of these proteins has been shown to be modulated by nucleotide binding. For example, binding of *Escherichia coli* dnaB protein to single-stranded DNA is allosterically activated by binding to ATP and other ribonucleoside triphosphates (Arai and Kornberg 1981). More recently, it was demonstrated that the presence of ATP is required for sequence-specific binding of the Tn*3* transposase to the inverted repeats of the transposon (Wishart et al. 1985).

These studies raised the question whether the DNA-binding activity of large TAg might also be influenced by its nucleotide-binding activity. In the present communication, we present evidence that binding of large TAg to purine nucleoside triphosphates allosterically inhibits its origin-DNA-binding activity.

## Experimental Procedures

### Immunoprecipitation of large TAg

Extracts of COS-1 cells, which constitutively express wild-type large TAg (Gluzman 1981), were incubated with 5 μg of purified monoclonal antibody directed against large TAg and fixed *Staphylococcus aureus* as described previously (Huber et al. 1985). The epitopes recognized by each antibody have been localized as shown in Figure 1A (Gurney et al. 1980; Harlow et al. 1981; Ball et al. 1984; E. Gurney et al., in prep.). Immune complexes were washed with NET (50 mM Tris [pH 7.5], 5 mM EDTA, 150 mM NaCl, 0.05% NP-40) with or without 500 mM LiCl before use in DNA-binding assays. D2 protein was immunopurified in the same fashion from extracts of Ad2$^+$D2-infected HeLa cells (Hassell et al. 1978).

### SV40 DNA binding

Cloned SV40 DNA (pSV-wt) was cleaved with HindIII and 5'-end-labeled as described (Fanning et al. 1982). Immunopurified large TAg (~ 100 ng) was taken up in 0.15 ml of binding buffer (10 mM HEPES [pH 7.8], 80 mM KCl, 0.5 mM MgCl$_2$, 1 mM DTT, 1 mM PMSF, 0.2 mg/ml glycogen, 1 mg/ml bovine serum albumin) with or without 20 mM ATP (pH 7.8). After incubation for 30 min at 0°C, 0.25 μg of end-labeled DNA fragments were added. After equilibrium was reached (1–2 hr at 0°C), unbound DNA was washed away with NET, and bound DNA was analyzed by agarose gel electrophoresis and autoradiography (Fanning et al. 1982).

## Results

### Large TAg binding to SV40 origin DNA is reversibly inhibited by ATP

COS-1 large TAg was immunopurified on each of the monoclonal antibodies whose binding sites are illustrated in Figure 1A. D2 large TAg, a 107-kD protein whose amino terminus is encoded by adenovirus sequences and whose carboxy terminus consists of residues 115–708 of SV40 large TAg (Baumann et al. 1985) (Fig. 1A), was immunopurified on Pab405 and

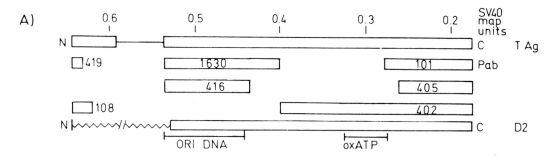

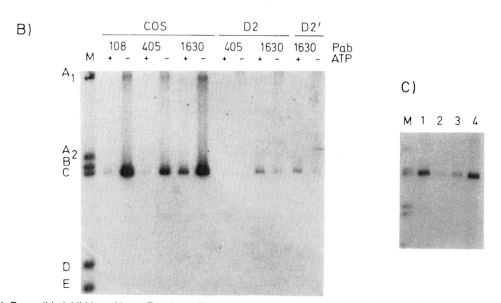

**Figure 1** Reversible inhibition of large T antigen–DNA binding in the presence of ATP. (*A*) The SV40 sequences for wild-type large TAg and the hybrid D2-large TAg (T Ag and D2) (Baumann et al. 1985), the monoclonal antibody–binding sites for each antibody used for immunopurification (Pab) (Gurney et al. 1980; Harlow et al. 1981; Ball et al. 1984; E. Gurney et al., in prep.), and the functional domains for binding to origin DNA (ORI DNA) (Rigby and Lane 1983) and periodate-oxidized ATP (oxATP) (Clertant et al. 1984) are shown. (*B*) D2 and COS-1 large TAg's immunopurified on Pab405, 1630, or 108 were incubated with (+) or without (−) ATP and then assayed for origin-DNA binding. Marker DNA (lane *M*) was 12.5 ng of the input DNA. In the lanes marked D2', D2 immune complexes were treated with ATP, then washed and resuspended with or without ATP. (*C*) COS-1 large TAg immunopurified on Pab108 was preincubated with ATP (lane *1*) or without ATP (lanes *2–4*). Immune complexes were washed and resuspended with ATP (lane *2*) or without ATP (lanes *1*, *3*, and *4*). After 30 min, pSV-wt DNA fragments were added to all samples and incubated for 1 hr. After 30 min, ATP was added to one sample (lane *3*).

1630. Binding of these large TAg's to SV40-origin DNA was assayed in the presence and absence of ATP (Fig. 1B). DNA binding of COS-1 large TAg was markedly reduced in the presence of ATP, regardless of the antibody used for purification (Fig. 1B; B. Vogt and E. Fanning, unpubl.). Binding to both site I and site II in the origin was decreased in the presence of ATP (B. Vogt and E. Fanning, unpubl.). Similar inhibition was observed with large TAg's prepared from productively infected cells and nine other SV40-transformed cell lines (Huber et al. 1985; B. Vogt and E. Fanning, unpubl.). D2 large TAg–DNA binding, on the other hand, was not inhibited by ATP (Fig. 1B). In fact, slightly more origin DNA was bound in the presence of ATP. Thus, COS-1 and several other large TAg's exhibit a new biochemical property not shared by the hybrid D2 protein.

To ascertain whether an ATP-dependent modification of large TAg, such as phosphorylation, might be re-

sponsible for the effects of ATP on DNA binding, we preincubated D2 (Fig. 1B, lanes D2') and large TAg (Fig. 1C, lane 1) with ATP. The ATP was then removed by washing the immune complexes and DNA binding was measured in the presence and absence of ATP. The amounts of origin DNA bound by the ATP-treated proteins did not differ significantly from that bound by the controls that had not been exposed to ATP (Fig. 1B, lanes D2, and Fig. 1C, lane 4). Thus, ATP inhibition of DNA binding is readily reversible. Furthermore, the increase in DNA binding of D2 in the presence of ATP does not appear to be due to a stable modification of the protein.

## Nucleotide inhibition of large TAg–DNA binding: Specificity and concentration dependence

Various nucleotides were assayed for their ability to inhibit specific binding of large TAg to SV40 DNA (Fig. 2A). Purine ribo- and deoxyribonucleoside triphos-

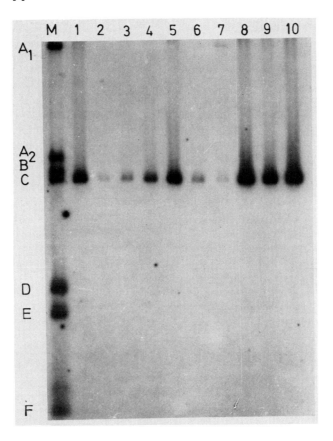

**A**

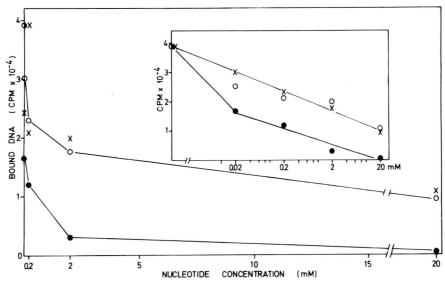

**B**

**Figure 2** Specificity and concentration dependence of nucleotide inhibition of large TAg–DNA binding. (A) Binding of immunopurified COS-1 large TAg to pSV-wt DNA was tested in the presence of various nucleotides, each at 20 mM. (Lane 1) Without nucleotides; (lane 2) ATP; (lane 3) dATP; (lane 4) CTP; (lane 5) dCTP; (lane 6) GTP; (lane 7) dGTP; (lane 8) dTTP; (lane 9) UTP; (lane 10) NADH. (B) DNA binding of COS-1 large TAg was assayed in the presence of ATP (●), ADP (○), or AMP-PNP (×) at the concentrations indicated. The bands of bound HindIII-C fragment were isolated from the gels and counted. (Inset) Semilog plot of cpm bound as a function of nucleotide concentration.

phates, ATP-γ-S, and AMP-PCP were about equally effective as inhibitors, whereas nucleoside diphosphates and AMP-PNP were less effective. Little or no inhibition of DNA binding was observed in the presence of pyrimidine nucleotides, nucleoside monophosphates, cAMP, or NADH (Fig. 2A; B. Vogt and E. Fanning, unpubl.).

The ability of several nucleotides to inhibit DNA binding was measured as a function of nucleotide concentration (Fig. 2B). DNA-binding activity of large TAg dropped sharply as nucleotide concentration increased. An ATP concentration of about 10 μM and an ADP or AMP-PNP concentration of about 1 mM were sufficient to inhibit DNA binding by 50% (Fig. 2B, inset).

## Discussion

Specific binding of large TAg to the SV40 origin of replication is markedly reduced in the presence of purine nucleoside triphosphates (Fig. 1). The inhibition is completely reversible upon removal of nucleotides. These results suggest that the effects of ATP on DNA binding are probably not due to covalent modification of large TAg in the presence of ATP, but rather to simple nucleotide binding. The fact that ATP inhibits COS-1 large TAg–DNA binding in the presence of excess EDTA or EGTA provides further support for the notion that an enzymatic reaction is not involved (B. Vogt and E. Fanning, unpubl.).

That the monoclonal antibody bound to large TAg is somehow responsible for the effects of ATP on DNA binding seems improbable since the inhibition was observed with antibodies directed against many epitopes distributed throughout the molecule (Fig. 1; B. Vogt and E. Fanning, unpubl.). Moreover, ATP and AMP-PNP inhibit origin-DNA binding of large TAg in crude cell extracts, as tested by immunoprecipitation of large TAg–DNA complexes formed in the absence of antibody (E. Fanning, unpubl.).

It is conceivable that a large TAg-associated protein may act to block origin-DNA-binding activity in the presence of ATP. Although we cannot rule out this possibility, we feel it is unlikely for two reasons. First, a mutant large TAg, D2, showed no inhibition of DNA binding in the presence of ATP, though it retains ATPase and periodate-oxidized ATP (oxATP) binding activities (Rigby and Lane 1983; B. Vogt and E. Fanning, unpubl.). Second, stringent washing of immunopurified large TAg with 0.5 M LiCl to remove weakly associated proteins failed to abolish the inhibitory effect of ATP (B. Vogt and E. Fanning, unpubl.).

### A model for nucleotide inhibition of large TAg–DNA binding

Assuming that simple binding of nucleotides to large TAg causes inhibition of DNA binding, ATP could compete with origin DNA for the same binding site or it could induce a conformational change in the protein. Since the affinity of large TAg for ATP must be much lower than its affinity for origin DNA, as judged by the relative

$$T_N \cdot ATP \underset{-ATP}{\overset{+ATP}{\rightleftharpoons}} T_N \rightleftharpoons T_D \overset{ORI\text{-}DNA}{\rightarrow} T_D \cdot ORI$$

**Figure 3** Model for allosteric control of large TAg–origin DNA binding by purine nucleoside triphosphates.

stabilities of the two ligands during washing (Fig. 1C), it seems unlikely that ATP could compete effectively with origin DNA for the same binding site. Furthermore, D2 protein retains the origin-DNA-binding domain and yet its DNA-binding activity is not inhibited by ATP (Fig. 1B). We therefore favor the idea that ATP binds to a second site on large TAg and allosterically inhibits its origin-DNA-binding activity.

Thus, we propose that large TAg may exist in either of two stable conformations, one (N) that binds nucleotides but not origin DNA, and a second (D) that binds origin DNA but not nucleotides (Fig. 3). Binding of either ligand to large TAg would stabilize it in the corresponding conformation. We speculate that the N form of large TAg may predominate, at least under the conditions of our studies, because simultaneous addition of ATP and origin DNA to large TAg results in efficient inhibition of DNA binding (B. Vogt and E. Fanning, unpubl.). Work is in progress to localize the allosteric nucleotide-binding site on large TAg and test this model for the mechanism of inhibition.

The role of allosteric regulation of large TAg–DNA binding in vivo is an open question. It may be that large TAg synthesized early in infection binds ATP rather than the SV40 template, which would presumably lead to premature autoregulation of early transcription. Allosteric regulation might thus allow accumulation of a certain threshold concentration of large TAg before the onset of autoregulation, initiation of DNA replication, and the late phase of infection.

Allosteric regulation could also be more directly involved in initiation of viral DNA replication. To our knowledge, initiation of SV40 DNA by the D2 protein has not been demonstrated so far, raising the question whether loss of allosteric regulation might be correlated with an inability to replicate SV40 DNA. Curiously, the observed nucleotide specificity of allosteric inhibition (Fig. 2A) overlaps with that of the substrates utilized by mammalian primases for initiation of primer synthesis (Eliasson and Reichard 1978; Kaufmann and Falk 1982). Perhaps large TAg, complexed with cellular replication proteins, plays a role in priming either at the origin, as suggested by Bradley et al. (1984), or possibly during elongation (Stahl et al. 1985).

## Acknowledgments

We thank S. Dehde, A. Schmid, and U. Markau for excellent technical assistance, E. Baumann for Ad2$^+$D2, and E. Harlow, R. Ball, E. Gurney, and W. Deppert for hybridomas. The financial support of the Deutsche Forschungsgemeinschaft and Fonds der Chemischen Industrie is gratefully acknowledged.

# References

Arai, K. and A. Kornberg. 1981. Mechanism of dnaB protein action. III. Allosteric role of ATP in the alteration of DNA structure by dnaB protein in priming replication. *J. Biol. Chem.* **256:** 5260.

Ball, R., B. Siegl, S. Quellhorst, G. Brandner, and D.G. Braun. 1984. Monoclonal antibodies against simian virus 40 large T tumor antigen. Epitope mapping, papova virus cross-reaction and cell surface staining. *EMBO J.* **3:** 1485.

Baumann, E.A., C.-P. Baur, M. Baack, and S. Beck. 1985. DNA sequence of the leftward junction in the adenovirus—simian virus 40 hybrid Ad2$^+$D2 and determination of the structure of the D2-T antigen. *J. Virol.* **54:** 882.

Bradley, M., J. Hudson, M. Villanueva, and D.M. Livingston. 1984. Specific in vitro adenylation of the simian virus 40 large tumor antigen. *Proc. Natl. Acad. Sci.* **81:** 6574.

Clark, R., K. Peden, J. Pipas, D. Nathans, and R. Tjian. 1983. Biochemical activities of T-antigen proteins encoded by simian virus 40 gene A deletion mutants. *Mol. Cell. Biol.* **3:** 220.

Clertant, P., P. Gaudray, E. May, and F. Cuzin. 1984. The nucleotide binding site detected by affinity labeling in the large T proteins of polyoma and SV40 viruses is distinct from their ATPase catalytic site. *J. Biol. Chem.* **259:** 15196.

Eliasson, R. and P. Reichard. 1978. Replication of polyoma DNA in isolated nuclei. Synthesis and distribution of initiator RNA. *J. Biol. Chem.* **253:** 7469.

Fanning, E., K.-H. Westphal, D. Brauer, and D. Cörlin. 1982. Subclasses of simian virus 40 large T antigen: Differential binding of two subclasses of T antigen from productively infected cells to viral and cellular DNA. *EMBO J.* **1:** 1023.

Gluzman, Y. 1981. SV40-transformed simian cells support the replication of early SV40 mutants. *Cell* **23:** 175.

Gurney, E.G., R.O. Harrison, and J. Fenno. 1980. Monoclonal antibodies against simian virus 40 T antigens: Evidence for distinct subclasses of large T antigen and for similarities among nonviral T antigens. *J. Virol.* **34:** 752.

Harlow, E., L.V. Crawford, D.C. Pim, and N.M. Williamson. 1981. Monoclonal antibodies specific for simian virus 40 tumor antigens. *J. Virol.* **39:** 861.

Hassell, J., E. Lukanidin, G. Fey, and J. Sambrook. 1978. The structure and expression of two defective adenovirus 2/simian virus 40 hybrids. *J. Mol. Biol.* **120:** 209.

Huber, B., E. Vakalopoulou, C. Burger, and E. Fanning. 1985. Identification and biochemical analysis of DNA replication-defective large T antigens from SV40-transformed cells. *Virology* **146:** 188.

Kaufmann, G. and H. Falk. 1982. An oligoribonucleotide polymerase from SV40-infected cells with properties of a primase. *Nucleic Acids Res.* **10:** 2309.

Rigby, P. and D. Lane. 1983. The structure and function of SV40 large T antigen. *Adv. Viral Oncol.* **3:** 31.

Stahl, H., P. Dröge, H. Zentgraf, and R. Knippers. 1985. A large-tumor-antigen-specific monoclonal antibody inhibits DNA replication of simian virus 40 minichromosomes in an in vitro elongation system. *J. Virol.* **54:** 473.

Wishart, W., J. Broach, and E. Ohtsubo. 1985. ATP-dependent specific binding of Tn*3* transposase to Tn*3* inverted repeats. *Nature* **314:** 556.

# A Sensitive and Rapid Assay for DNA-binding Proteins Exemplified by a Functional Analysis of the Inverted Terminal Repetition of Adenovirus DNA

R. Schneider,* T. Dörper,* I. Gander,* R. Mertz,[†] and E.L. Winnacker*

*Institut für Biochemie der Universität München, D-8000 München 2, Federal Republic of Germany;
[†]Genzentrum der Universität München, D-8033 Martinsried, Federal Republic of Germany

This paper describes a rapid and sensitive assay for DNA-binding proteins that interact with specific and defined binding sites. It exploits the observation that complexes of proteins and small synthetic DNA fragments (40 bp) containing the protein-binding site can enter native polyacrylamide gels and remain stably associated during electrophoresis under nondenaturing conditions. The assay was applied to nuclear factor I (NFI), to its identification and purification from porcine liver, to an analysis of its binding site on adenovirus type-5 DNA, and to an exploration of other potential binding sites for DNA-binding proteins within the inverted terminal repetition (ITR) of adenovirus DNA. The extreme sensitivity of the assay, which surpasses that of conventional footprint assays by two orders of magnitude, permitted the identification of NFI-like activities in *Saccharyomyces cerevisiae* and in higher plants.

The linear, double-stranded genome of the adenoviruses is characterized by two structural features, a 55-kD terminal protein (TP) attached to both 5' ends of the DNA molecule and the inverted terminal repetition (ITR). Varying in length between 63 and 164 bp (in different serotypes), the ITR (Fig. 1) has been shown to represent the origin of DNA replication (for reviews, see Challberg and Kelly 1982; Fütterer and Winnacker 1984). In vitro assays (Stillman 1983) for the initiation reaction of DNA replication have shown that this reaction requires the presence of a highly conserved region between nucleotides (nt) 9 and 18 (the "9–18 box") of the ITR as well as of a binding site for nuclear factor I (NFI). Footprint analyses have placed the latter binding sites between positions 17 and 49 or 19 and 42 within

the ITR (Nagata et al. 1983; Rawlins et al. 1984). In this paper we present studies of the role of the conserved region between nt 9 and 19 and show that not only its presence but also its orientation with respect to the termini is essential for the initiation of DNA replication. In addition, we have developed a sensitive and rapid assay for NFI (and DNA-binding proteins, in general) that permitted not only a mutational analysis of its binding site but also the identification of a third functional domain within the ITR of adenovirus.

## Materials and Methods

### Plasmid construction and isolation

Plasmid DNA was isolated according to the SDS-NaOH lysis procedure of Birnboim and Doly (1979). Plasmids pTD19, pTD38, and pAD5-I have been described previously (Lally et al. 1984). Oligonucleotides were synthesized on an Applied Biosystems Model 308A DNA synthesizer using the chemistry and purification methods described by Dörper and Winnacker (1983). The identity of the desired sequences was confirmed by plasmid DNA sequencing (Chen and Seeburg 1985) after cloning of appropriate oligonucleotide pairs into linearized pBR327 DNA.

### NFI-binding assay

Oligonucleotides were 5'-labeled with [$^{32}$P]ATP by T4 polynucleotide kinase (Boehringer Mannheim) to a specific activity of 50–100 Bq/fmole DNA (1–2 nCi/fmole DNA). Six pmoles of single-stranded oligonucleotide were added to 10 μl of doubly distilled water together with 18 pmoles of [$^{32}$P]ATP (50 μCi or 2.5 × 10$^6$ Bq), 2 μl (10 units) of polynucleotide kinase, and 1

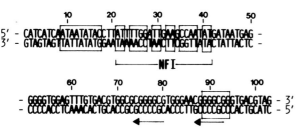

**Figure 1** Structure of the ITR of Ad5 DNA. A highly conserved region is located between positions 9 and 18 (9–18 box). The NFI-binding site consists of a region with dyad symmetry centered around base pair 31 and extending from position 21 to 41. A GGGCGG box between positions 89 and 94 may represent a binding site for the transcription factor Sp1 (Gidoni et al. 1984). A 6-bp-long repetition is indicated by two arrows (from Fütterer and Winnacker 1984). DNA fragments A1/B1 and K1/K2 contain base pairs 17–51 or 60–94, respectively. They were synthesized with protruding *Eco*RI and *Hin*dIII termini (see Fig. 4).

μl of kinase buffer (Maniatis et al. 1982). After 10 min at 37°C, the reaction was stopped by addition of 5 μl of 0.5 M EDTA, and the reaction mixture was purified by chromatography on DE-52. Oligonucleotides were always labeled separately and only mixed later in appropriate pairs by heating for 30 sec to 100°C followed by a cooling step for 15 min at room temperature.

Between 10 and 20 fmoles of a DNA fragment were incubated for 15 min at room temperature together with 2–5 μg of protein (e.g., from a crude nuclear extract) in 25 mM HEPES (pH 7.5), 1 mM EDTA, 5 mM DTT, 10% glycerol, 150 mM NaCl, and 0.5 mM PMSF (Serva, Heidelberg). The total volume per assay was 20 μl. The nuclear extract was added last. The reaction mixture was subsequently subjected to gel electrophoresis on 11% polyacrylamide gels (acrylamide/bisacrylamide = 44:0.8) in 0.375 M Tris-glycine (pH 8.8). Gels were electrophoresed at 15–25 mA with 40 mM Tris-glycine (pH 8.5) in the buffer chambers. Samples were applied in 5% glycerol, 10 mM DTT, and 0.05% bromophenol blue. After electrophoresis, gels were soaked in 5% glycerol, dried, and autoradiographed.

*Preparation of nuclear extracts from porcine liver*
All steps were performed at 4°C. A fresh liver from the slaughter house (1.3 kg) was broken up in a Waring Blendor. The cell homogenate was taken up in 3 ml/g liver of a solution containing 50 mM HEPES (pH 7.5), 340 mM sucrose, and 4 mM CaCl. After homogenization in a "magic stick" homogenizer (ESGE AG, Mettlen, Switzerland) and a centrifugation step for 10 min at 600*g*, the pellet was resuspended in 50 mM HEPES (pH 7.5), 250 mM sucrose, and 3 mM HEPES (pH 7.5), 340 mM sucrose, and 4 mM CaCl$_2$. Subsequent centrifugation in a Sorvall SS-34 rotor for 1 hr at 20,000 rounds/min yielded a pellet that was washed three times by resuspension in 10 volumes of 50 mM Tris · HCl (pH 7.5), 240 mM sucrose, and 3 mM CaCl$_2$ and centrifugation for 10 min at 1500*g*. A total of $5 \times 10^{10}$ nuclei were obtained per liver. Nuclear extracts were prepared by resuspension of the nuclei pellet in 10 volumes of 50 mM HEPES (pH 7.5), 240 mM sucrose, 3 mM CaCl$_2$, 2 mM DTT, 200 mM NaCl, and 0.5 mM PMSF. After incubation for 30 min at 4°C, the suspension was centrifuged for 20 min at 12,000*g*. The NaCl concentration in the supernatant was raised to 300 mM. The extract, which could be kept frozen at this stage, was either used directly for binding studies or for the purification of NFI.

*Purification of NFI*
To purify NFI, the nuclear extract (20.2 mg protein/ml) was subjected to DEAE-cellulose chromatography (Whatman DE-52) in 50 mM HEPES (pH 7.5), 1 mM EDTA, 2 mM DTT, 10% glycerol, 0.5 mM PMSF, 300 mM NaCl. All NFI activity, which was assayed fraction by fraction with the electrophoretic assay described above, was found in the flowthrough. In a second purification step, this material was loaded onto a phosphocellulose column in the same buffer described above and eluted with 1 column volume of 600 mM NaCl in this buffer. The

NFI activity eluted at 450 mM NaCl. This material was estimated to be enriched by a factor of approximately 100 from the crude nuclear extract but is still only 0.01% pure. It was used as such in most of the binding studies described in this paper. The purification protocol is based on the procedure of Nagata et al. (1982) with some modifications required due to the different source of starting material.

*Assay for in vitro DNA replication*
Reaction mixtures (25 μl) contained 25 mM HEPES/KOH (pH 7.5), 3 mM ATP, 2 mM DTT, 5 mM MgCl$_2$, 50 μM ddATP, 0.6 μM dCTP, 2 μCi (α-$^{32}$P)dCTP (Amersham; 3000 Ci/mmole), 0.5–2 μg of plasmid DNA, and 8–12 μl of nuclear extract (8–15 mg protein/ml) from adenovirus type-5 (Ad5)–infected HeLa cells. For the formation of the elongated product, ddATP was omitted and replaced by dATP, dTTP, and ddGTP, each at a final concentration of 40 μM. After incubation for 60 min at 30°C, the reaction was stopped by addition of sample buffer (50 mM Tris · HCl [pH 6.8], 10% glycerol, 3% β-mercaptoethanol, 3% SDS). Samples were heated for 3 min at 100°C and separated by electrophoresis in 10% polyacrylamide-SDS gels. $^{14}$C-labeled adenovirus was used as a molecular-weight marker.

## Results

### The orientation of the conserved region between positions 9 and 18 within the ITR of adenoviruses
To study the effect of the orientation of a conserved region (the 9–18 box) close to the adenovirus DNA termini on the initiation reaction, two 30-nt-long complementary oligonucleotides were synthesized (Fig. 2).

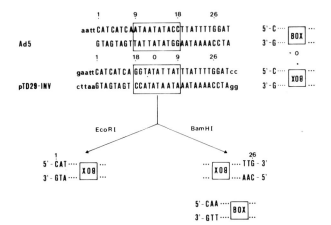

**Figure 2** Construction and in vitro DNA replication activity of Ad5-specific termini with the 9–18 box in an inverted orientation. The uppermost line shows the terminal 26 bp of Ad5 DNA. The line labeled pTD29-INV shows the sequence of an insert within plasmid pBR327 containing the 9–18 box in an inverted orientation. The insert was synthesized as two oligonucleotides with protruding 5′ *Eco*RI and *Bam*HI termini and cloned into *Bam*HI/*Eco*RI-linearized pBR327. Capital letters indicate the Ad5-specific sequences; lower-case letters indicate the pBR327 derived sequences. Cleavage of pTD29-INV with *Eco*RI or *Bam*HI yields termini with the 9–18 box in an inverted or a correct orientation, respectively.

The DNA fragment with protruding *Eco*RI and *Bam*HI termini formed after mixing of the two oligonucleotides contains the first 29 bp of Ad5 DNA with the 9–18 box in the reverse orientation. The plasmid pTD29-INV, formed after cloning the synthetic DNA fragment into an *Eco*RI/*Bam*HI-linearized pBR327, could subsequently be cleaved by either *Eco*RI or *Bam*HI digestion. *Eco*RI cleavage results in a terminus containing the terminal 8 bp of Ad5 DNA as well as the 9–18 box, the latter in a reverse orientation. Cleavage with *Bam*HI, in turn, leads to an end that contains the conserved region in the proper orientation with respect to the terminus but with nucleotides 20–29 of the ITR separating it from the molecular end (Fig. 2). When pTD29-INV DNA molecules linearized by either *Eco*RI or *Bam*HI digestion were subjected to an in vitro initiation or partial elongation assay, only the *Bam*HI-cleaved molecule was active (Fig. 3, lanes 5 and 6). Its activity was comparable to a control terminus derived from plasmid pTD19 (Fig. 3, lanes 3 and 4), whereas the *Eco*RI-linearized pTD29-INV probe was totally inactive (Fig. 3, lanes 7 and 8) although it carries the genuine terminal 8 bp from Ad5 DNA. It is thus both the nature of the conserved region and its orientation that determine the ability of a DNA terminus to function in adenovirus DNA replication. The insignificance of the terminal 8 bp apart from the

terminal dC residue had already been proven previously through deletion analyses (Tamanoi and Stillman 1983).

## A sensitive and rapid assay for NFI

Protein-DNA interactions have in the past been analyzed by sucrose gradient centrifugation and/or nitrocellulose filtration. Fried and Crothers (1981, 1984) have shown that DNA-protein complexes as exemplified by *lac* operator and *lac* repressor can jointly enter native polyacrylamide gels and remain stably associated during electrophoresis. The protein, the *lac* repressor in this case, can be identified on the gels through autoradiography if the DNA fragment is appropriately labeled. The unexpected stability of such complexes is not entirely understood but is attributed, at least in part, to a caging effect mediated by the gel matrix. We have exploited this observation in order to develop a rapid assay for NFI. The DNA fragment A1/B1, composed of two complementary synthetic oligonucleotides, carries the NFI-binding site between positions 17 and 51 of the adenovirus ITR; a control fragment K1/K2 extends from positions 60 to 94 of the ITR (see Fig. 4). Both fragments carry protruding 5' *Eco*RI and *Bam*HI termini, respectively. The 5' $^{32}$P-labeled DNA fragments are mixed with protein extracts (either crude or enriched) containing NFI, incubated for 15 min at room temperature, and subjected to polyacrylamide gel electrophoresis. The small size of the oligonucleotide-derived DNA fragments permits electrophoresis even in high-percentage polyacrylamide gels (up to 15%). Under our conditions (see Materials and Methods) the unbound DNA fragments move with the front while the protein-bound fragments are retained and produce a characteristic band (Fig. 5, lane 1, arrow 2) upon autoradiography. This band is missing in lane 2 with the control DNA fragment, which lacks the appropriate binding site. The origin of a band specific for the control fragment K1/K2 (Fig. 5, lane 2, arrow 1) will be dis-

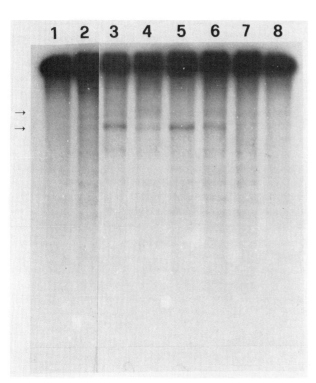

**Figure 3** Analysis of in vitro DNA replication products. Plasmids as described in Fig. 2 were subjected to in vitro initiation and elongation assays for adenovirus DNA replication. (Lanes *1* and *2*) *Eco*RI-linearized pBR327, initiation and elongation; (lanes *3* and *4*) *Eco*RI-linearized pTD19, initiation and elongation; (lanes *5* and *6*) *Bam*HI-linearized pTD29-INV, initiation and elongation; (lanes *7* and *8*) *Eco*RI-linearized pTD29-INV, initiation and elongation.

```
A1/B1   5'--aattCCTTATTTTGGATTGAAGCCAATATGATAATGAGG    --3'
        3'--    GGAATAAAACCTAACTTCGGTTATACTATTACTCCctag --5'


K1/K2   5'--aattCTTGTGACGTGGCGCGGGGCGTGGGAACGGGGCGG    --3'
        3'--    GAACACTGCACCGCGCCCCGCACCCTTGCCCCGCCctag --5'


myc1    5'--tcgaCCTCCCCACCTTCCCCACCCTCCCCACCCTCCCCAT    --3'
myc2    3'--    GGAGGGGTGGAAGGGGTGGGAGGGGTGGGAGGGGTAgatc--5'


U1/U2   5'--agctTAACTGCTTCTTAATTTGCATACCCTCACTGCATCG    --3'
        3'--    ATTGACGAAGAATTAAACGTATGGGAGTGACGTAGCttaa--5'
```

**Figure 4** Structure of oligonucleotides used in the gel assay. The pair A1/B1 extends from base pairs 17 to 51 of the ITR of adenovirus DNA; fragment K1/K2 from base pairs 60 to 94 of the adenovirus ITR (Fütterer and Winnacker 1984). Bold letters represent NFI- and transcription factor Sp1–binding sites, respectively. Fragments myc1/myc2 represent base pairs 2175–2214 of the human c-*myc* gene promoter (Gazin et al. 1984); fragments U1/U2 represent a sequence from the recombined mouse κ light-chain gene from positions 13 to 50 (Seidman et al. 1979). Lower-case letters are protruding termini characteristic of various restriction enzyme recognition sites.

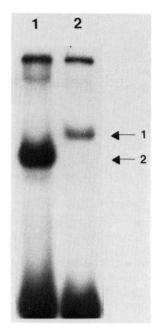

**Figure 5** Assay for NFI. A crude nuclear extract was incubated for 15 min at room temperature with $^{32}$P-labeled DNA fragments A1/B1 (lane 1) or K1/K2 (lane 2) and subjected to gel electrophoresis on a native, 11% polyacrylamide gel. Radioactive bands were identified by autoradiography. Fragment-specific bands are indicated by arrows and labeled 1 or 2, respectively, for the control fragment K1/K2 or the NFI-binding site containing fragment A1/B1. The assay was performed in a volume of 20 μl with 10 fmoles of the respective fragments (4000 cpm/fmole) and with 10 μl of a crude nuclear extract from porcine liver.

cussed below. The bands in Figure 5 disappear when the assay mixtures are supplemented with increasing concentrations of the identical but unlabeled DNA fragment (see below). Addition of control DNA fragments to the assay mixtures does not interfere with the banding pattern, indicating that, indeed, these bands arise through specific interactions of distinct proteins (see below; Figs. 6 and 7). The assay has been used for the purification of NFI from various sources, including rat and porcine liver (Schneider et al. 1986).

## Sensitivity and specificity of the gel assay

The sensitivity of the gel assay for NFI as well as the amount of NFI in protein extracts were determined from competition experiments (Fig. 6). A constant amount of the $^{32}$P-labeled fragment A1/B1 was mixed with increasing amounts of the unlabeled fragment (10-, 100-, 500-, and 1000-fold the amount of the labeled fragment) and incubated with NFI containing crude nuclear extract from porcine liver under conditions of protein excess. After electrophoresis, the NFI-containing bands could be identified through autoradiography. From the radioactivity present in each band and the known specific radioactivity of the $^{32}$P-labeled fragment, it was possible to calculate the amount of bound oligonucleotide in each sample. A reciprocal representation of bound fragment versus the total fragment concentration permits an ex-

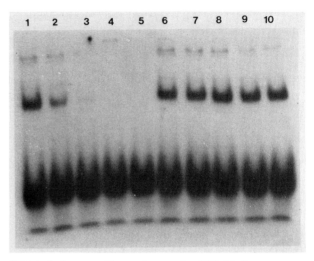

**Figure 6** A competition gel assay for NFI. Each assay was performed with 10 μl of a crude nuclear extract from porcine liver in the presence of 6 fmoles of the $^{32}$P-labeled DNA fragment A1/B1 with a specific activity of 4000 cpm/fmole. Lanes 1 and 6 did not contain any additional DNA, whereas lanes 2–5 contained additions of 60, 600, 3000, and 6000 fmoles, respectively, of the unlabeled fragment A1/B1, and lanes 7–10 contained equivalent amounts of the unlabeled control fragment K1/K2.

trapolation to saturation conditions. Under the assumption that 2 moles of protein are bound to every mole of DNA fragment at saturation (see Discussion), we could then calculate that 10 μl of a crude nuclear extract from porcine liver contains approximately 50 fmoles of NFI. This would amount to 2–3 nmoles of protein (100–150 μg) per liver.

The lower limits of sensitivity of the gel assay can be estimated from these assays as well. The amount of 50 fmoles of NFI, representing approximately 2.5 ng of protein, were detected with 6 fmoles of a $^{32}$P-labeled oligonucleotide. It has been possible to reduce the amount of oligonucleotide to 1 fmole (at a specific activity of 4000 cpm/fmole) and still to detect approximately 1 fmole (50 pg) of protein. Figure 8, lane 8, for example, displays an assay with 5 fmoles of NFI. The assay is thus at least 2 orders of magnitude more sensitive than a typical DNase I footprint assay (cf. Gidoni et al. 1984; Gronostajski et al. 1985), not considering the comparable ease with which it is performed.

To determine the specificity of the assay, experiments were performed in the presence of constant amounts of the $^{32}$P-labeled fragment A1/B1 and increasing concentrations of DNA fragments with different sequences but of identical size.

The fragments used in the experiment illustrated in Figure 7 are derived from a more internal region of the adenovirus ITR (fragment K1/K2 of Fig. 3), from the human c-myc gene promoter, and from the mouse immunoglobulin κ light-chain gene (cf. Fig. 3). These fragments were employed in excesses of up to 800-fold. No decrease in the binding of the specific, homologous fragment A1/B1 could be detected under these conditions (see also Fig. 6, lanes 6–10), indicating that the

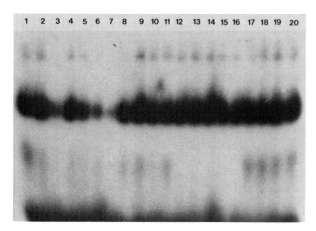

**Figure 7** An extended competition assay for NFI with various control fragments. The assays were performed under conditions similar to those described in the legend to Fig. 6. Lanes *1* and *11* represent controls containing only 6 fmoles of the $^{32}$P-labeled fragment A1/B1. Lanes *2-7* show increasing concentrations (100 fmoles, 4 pmoles, 500 fmoles, 1 pmole, and 2 and 9 pmoles) of the homologous, unlabeled fragment A1/B1, displaying the expected competition. Lanes *8, 9,* and *10* represent additions of 1, 4, and 8 pmoles of fragment U1/U2, respectively (cf. Fig. 4); lanes *12-16* contained additions of 500 fmoles, and 1, 2, 4, and 8 pmoles of fragment myc1/myc2, respectively (cf. Fig. 4); and lanes *17-20* represent additions of 1, 2, 4, and 8 pmoles of fragment K1/K2, respectively. No competition could be observed for these fragments, either by autoradiography or scintillation counting (the latter is not shown).

assay is indeed specific for a DNA fragment carrying the NFI-binding site.

## A mutational analysis of the NFI-binding site

We have exploited the above-mentioned assay to map the NFI-binding site in Ad5 DNA by the introduction of specific mutations. A total of seven pairs of oligonucleotides were synthesized comprising the region from positions 17 to 51 of the Ad5 genome and containing the transitions and transversions indicated in Figure 8A. Appropriate pairs were 5'-end-labeled, incubated with a purified NFI preparation, and subjected to the electrophoresis assay. As shown in Figure 8B, six of these oligonucleotide pairs showed no band, even upon prolonged autoradiography. From the amount of radioactivity, however, observed by scintillation counting at the expected positions of the bands, we estimate a 250- to 400-fold reduction in the binding activity as compared with fragment A1/B1.

The transversions and transitions in these mutated fragments invariably reside in areas that define a region of dyad symmetry centered around base pair 31, as indicated by the brackets in Figures 1 and 8. In contrast to these six fragments, mutant fragment $TT_1$ displays considerable activity. As demonstrated convincingly in lanes 6 (mutant) and 8 (wild type) of Figure 8, this activity amounts to approximately 50% of the wild type. Mutation $TT_1$, which simultaneously introduces four transitions immediately around the central G·C base pair at position 31, does not represent by itself an

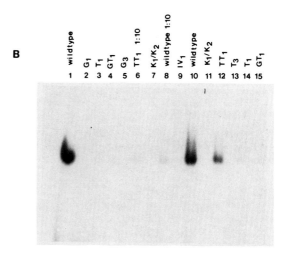

**Figure 8** Electrophoretic analysis of mutations in the NFI-binding site. Mutant fragments were synthesized as oligonucleotide pairs, 5' labeled with $^{32}$P, and incubated with crude nuclear extracts from porcine liver. The assay conditions are described in Materials and Methods. "Wildtype" refers to a DNA fragment containing the original Ad5-derived sequence, which is outlined in *A*, together with the various mutant fragments. Wild-type and mutant fragments were subjected to electrophoresis on an 11% polyacrylamide gel under native conditions. K1/K2 refers to incubations with the control fragment K1/K2 (cf. Figs. 1 or 4); "1:10" in lanes 6 and 8 indicates 1:10 dilutions of the nuclear extract. Mutations were always introduced into both strands.

entirely symmetrical arrangement. The binding site, however, appears to tolerate these changes around the center of symmetry, whereas it is extremely sensitive toward any changes in the more distant regions of the binding site. We conclude not only that the NFI-binding site, as has been shown previously, resides around positions 17-41 of the viral genome, but also that the symmetrical arrangement as presented in Figure 8A appears to be of particular significance.

## An additional protein-binding site in the ITR of Ad5 DNA

The electrophoretic assay described above requires an oligonucleotide pair containing a binding site, as exemplified by the pair A1/B1, as well as a control fragment.

In our case this was chosen to represent the region from positions 60 to 94 of the adenovirus genome. As shown in Figure 5, lane 2, this DNA fragment creates a different band with a slower mobility as compared with the fragment containing the NFI-binding site when exposed to a purified NFI preparation. This band is considerably weaker than the NFI-specific band, indicating that this protein is either present in comparatively lower concentrations or that it displays only a weak binding interaction with this DNA fragment. The nature of this protein, which elutes at a lower NaCl concentration from phosphocellulose as compared with NFI (350 vs. 450 mM), has not been elucidated unequivocally as yet. However, a sequence comparison of the interior, G + C-rich portion of the ITR as represented in the control fragment indicates a characteristic sequence motif—GGGCGG—between positions 89 to 94 (Fig. 1; Fütterer and Winnacker 1984). This sequence occurs repeatedly within the early promoter region of SV40 DNA and has been shown to represent a binding site for transcription factor Sp1 (Gidoni et al. 1984). Although it is thus tempting to conclude that the protein binding to the DNA fragment K1/K2 represents transcription factor Sp1, a final answer as to the identity of this protein, however, will only be obtained from footprint analyses and from the purification of this factor.

## Discussion

This paper describes two approaches to a functional analysis of the ITR of Ad5 DNA, a region representing the origin of DNA replication. The first strategy involves the use of synthetic oligonucleotides in a study of the role of the conserved region from positions 9 to 17 of the adenovirus genome. Its presence has previously been shown to be absolutely required for in vitro DNA replication. We are now able to add that it is not only its presence but also its orientation with respect to the termini of the linear Ad5 DNA molecule that is of significance. A DNA molecule carrying this sequence motif in the inverse orientation but otherwise indistinguishable from a wild-type terminus is totally inactive in an in vitro initiation assay. This may not be surprising since this region does not display the usual dyad symmetry of a typical protein-binding site. In fact, this binding site may have to interact with a monomeric protein that must approach this terminus in a particular orientation to properly initiate DNA replication. It remains to be seen whether it is the 80-kD precursor to the terminal protein or another protein that recognizes this highly conserved region in the adenovirus DNA ITR.

A second strategy for a functional analysis of the adenovirus ITR involves a novel assay for NFI. It exploits previous observations (Fried and Crothers 1981, 1984; Garner and Revzin 1981; Strauss and Varshavsky 1984) according to which a DNA-protein complex can stably enter a polyacrylamide gel matrix upon gel electrophoresis under nondenaturing conditions. The approach by Fried and Crothers requires low-percentage gels (3–4%) since they use comparatively long DNA fragments (200 bp). Our conditions with 40-bp-long oligonucleotides permit a wide range of polyacrylamide concentrations to be employed. In addition, we recommend buffers with comparatively high ionic strengths in order to reduce unspecific binding and to permit analyses of crude extracts in the absence of unspecific, competing DNA (e.g., *E. coli* DNA). Our assay thus may appear somewhat similar in its concept but is in fact very different in its actual design and performance compared with previously known procedures. Salt concentrations, pH dependence, and polyacrylamide cross-linking have been optimized as described in Materials and Methods and should be strictly adhered to.

The sensitivity of the assay is, of course, dependent on the binding constants of complex formation between the protein in question and the oligonucleotide carrying the binding site. In the case of NFI, a lower limit of sensitivity of 0.1–1.0 fmole could easily be achieved, surpassing the footprinting methods by at least 2 orders of magnitude.

The absolute amount of NFI is very different in extracts from different sources. The best source in our hands, nuclear extracts from porcine liver, contains about 250 ng/ml of NFI at a protein concentration of 20 mg/ml, indicating that it would require an almost 80,000-fold enrichment to purify NFI to homogeneity. In HeLa cells, the concentration of NFI is only a fifth of that in porcine liver.

One pertinent question refers to the number of molecules of NFI that bind to each binding site. This question can easily be addressed by this assay since addition of a second molecule of NFI would be expected to lead to the formation of a new complex with a higher molecular weight and with greatly reduced mobility. All experiments described in this paper were performed at protein-DNA concentrations that lead to formation of a 1:1 complex. However, upon reduction of the DNA concentrations, it is possible to demonstrate that indeed a 2:1 complex can be formed (not shown). The exact concentrations and conditions required for this transition are currently under investigation.

The present assay permitted a mutational analysis of the NFI-binding site. According to our results, the binding site displays dyad symmetry centered around base pair 31 in the Ad5 genome. It extends approximately 10 bp to either side of this central G·C pair and does not tolerate any transitions or transversions in its flanking regions. Mutations in the most outward symmetry region covered by base pairs 40 and 41 have been tested by us, not in the electrophoretic assay, but in the in vitro DNA replication assay. The relevant sequence, exemplified by plasmid pTD38, contains the correct base pair at position 41 but carries transitions at positions 39 and 40 that affect the proposed pattern of symmetry (Fig. 8). As compared with a wild-type sequence, this construct displays only one-fourth to one-fifth of the in vitro initiation activity. Using filter-binding assays, Leegwater et al. (1985) conclude that DNA fragments containing the first 40 bp are bound normally, whereas the first 38 bp are insufficient to sustain binding. These results confirm our observations and permit us to conclude that the terminal 40 bp of Ad5 DNA carry the minimal recognition site for NFI. The significance of a hairpin structure formed by nucleotides 17–51 as a prerequisite for NFI binding thus appears questionable (Guggenheimer et al. 1984).

In the course of the purification of NFI, as assayed by the electrophoretic assay, we observed the binding of another protein of a different and reduced mobility to a control fragment covering Ad5 positions 60–94. A sequence comparision with the SV40 early promoter region leads us to suspect that this region contains a binding site for transcription factor Sp1 and that the observed protein may, in fact, be factor Sp1. This is under investigation. In the present context, the identification of a protein binding to the G + C-rich portion of the Ad5 DNA ITR served only to demonstrate that there may, in fact, be three functional domains in the ITR and that the assay described may be of general significance for the identification of DNA-binding proteins. Preliminary results already established its usefulness in the characterization of proteins binding to the SV40 DNA enhancer, the mouse IgG enhancer, and of proteins from *S. cerevisiae* binding to NFI-binding sites.

## Acknowledgments

The authors wish to thank Renate Föckler and Sabine Stelzig for excellent technical assistance. This work was supported by the Deutsche Forschungsgemeinschaft (SFB 304).

## References

Birnboim, H.C. and J. Doly. 1979. A rapid alkaline extraction procedure for screening recombinant plasmid DNA. *Nucleic Acids Res.* **7**: 1513.

Challberg, M.D. and T.J. Kelly, Jr. 1982. Eukaryotic DNA replication: Viral and plasmid model systems. *Annu. Rev. Biochem.* **51**: 901.

Chen, E.Y. and P.H. Seeburg. 1985. Supercoil sequencing: A fast and simple method for sequencing plasmid DNA. *DNA* **4**: 165.

Dörper, T. and E.L. Winnacker. 1983. Improvements in the phosphoramidite procedure for the synthesis of oligodeoxyribonucleotides. *Nucleic Acids Res.* **11**: 2575.

Fried, M. and D.M. Crothers. 1981. Equilibria and kinetics of *lac* repressor-operator interactions by polyacrylamide gel electrophoresis. *Nucleic Acids Res.* **9**: 6505.

———. 1984. Equilibrium studies of the cyclic AMP receptor protein-DNA interaction. *J. Mol. Biol.* **172**: 241.

Fütterer, J. and E.L. Winnacker. 1984. Adenovirus DNA replication. *Curr. Top. Microbiol. Immunol.* **111**: 41.

Garner, M.M. and A. Revzin. 1981. A gel electrophoresis method quantifying the binding of proteins to specific DNA regions. *Nucleic Acids Res.* **9**: 3047.

Gazin, C., S. Dupont de Dinechin, A. Hampe, J.M. Masson, P. Martin, D. Stehelin, and F. Galibert. 1984. Nucleotide sequence of the human c-*myc* locus: Provocative open reading frame within the first exon. *EMBO J.* **3**: 383.

Gidoni, D., W.S. Dynan, and R. Tjian. 1984. Multiple specific contacts between a mammalian transcription factor and its cognate promoters. *Nature* **312**: 409.

Gronostajski, R.M., S. Adhya, K. Nagata, R. Guggenheimer, and J. Hurwitz. 1985. Site-specific DNA binding of nuclear factor I: Analyses of cellular binding sites. *Mol. Cell. Biol.* **5**: 964.

Guggenheimer, R.A., B.W. Stillman, K. Nagata, F. Tamanoi, and J. Hurwitz. 1984. DNA sequences required for the *in vitro* replication of adenovirus DNA. *Proc. Natl. Acad. Sci.* **81**: 3069.

Lally, C., T. Dörper, W. Gröger, G. Antoine, and E.L. Winnacker. 1984. A size analysis of the adenovirus replicon. *EMBO J.* **3**: 333.

Leegwater, P.A.J., W.V. Driel, and P.C. van der Vliet. 1985. Recognition site of nuclear factor I, a sequence-specific DNA binding protein from HeLa cells that stimulates adenovirus DNA replication. *EMBO J.* **4**: 1515.

Maniatis, T., E.F. Fritsch, and J. Sambrook. 1982. *Molecular cloning: A laboratory manual.* Cold Spring Harbor Laboratory, Cold Spring Harbor, New York.

Nagata, K., R.A. Guggenheimer, and J. Hurwitz. 1983. Specific binding of a cellular DNA replication protein to the origin of replication of adenovirus DNA. *Proc. Natl. Acad. Sci.* **80**: 6177.

Nagata, K., R.A. Guggenheimer, T. Enomoto, J.H. Lichy, and J. Hurwitz. 1982. Adenovirus DNA replication *in vitro*: Identification of a host factor that stimulates synthesis of the preterminal protein dCMP complex. *Proc. Natl. Acad. Sci.* **79**: 6438.

Rawlins, J.R., P.J. Rosenfeld, R.J. Wides, M.D. Challberg, and T.J. Kelly, Jr. 1984. Structure and function of the adenovirus origin of replication. *Cell* **37**: 309.

Schneider, R., I. Gander, U. Muller, R. Mertz, and E.L. Winnacker. 1986. A sensitive and rapid gel retention assay for nuclear factor I and other DNA-binding proteins in crude nuclear extracts. *Nucleic Acids Res.* **14**: 1303.

Seidman, J.G., E.E. Max, and P. Leder. 1979. A κ immunoglobulin gene is formed by site-specific recombination without further somatic mutation. *Nature* **280**: 370.

Stillman, B.W. 1983. The replication of adenovirus DNA with purified proteins. *Cell* **35**: 7.

Strauss, F. and A. Varshavsky. 1984. A protein binds to a satellite DNA repeat at three specific sites that would be brought into mutual proximity by DNA folding in the nucleosome. *Cell* **37**: 889.

Tamanoi, F. and B.W. Stillman. 1983. Initiation of adenovirus DNA replication *in vitro* requires a specific DNA sequence. *Proc. Natl. Acad. Sci.* **80**: 6446.

# Studies on the Adenovirus DNA-binding Protein

D.F. Klessig,*[†] S.A. Rice,* V. Cleghon,*[†] D.E. Brough,*[†] J.F. Williams,[‡] and K. Voelkerding*[†]

*Department of Cellular, Viral and Molecular Biology, University of Utah, Medical School, Salt Lake City, Utah 84132; [‡]Department of Biological Sciences, Carnegie-Mellon University, Pittsburgh, Pennsylvania 15213

Genetic and biochemical approaches were employed to help decipher the mechanism(s) by which the adenovirus-encoded DNA-binding protein (DBP) carries out its myriad functions. Using site-directed mutagenesis of plasmids carrying the DBP gene, followed by in vivo recombination to introduce the altered genes into the viral genome, several deletion mutants of DBP have been constructed. The defective DBP mutants were propagated by growth in cell lines that express an integrated copy of the DBP gene under the control of a dexamethasone-inducible promoter. Characterization of two of these mutants indicated that DBP (1) is absolutely essential for viral DNA replication, (2) does not play a major role in regulating early gene expression, and (3) normally has little effect on the efficiency by which this DNA tumor virus transforms cells in culture.

The multifunctional nature of DBP as well as the host-range phenotypes of a number of DBP mutants suggest that this protein interacts with different cellular macromolecules and/or structures. Using a combination of immunofluorescence and in situ fractionation, two distinct nuclear subclasses of DBP have been defined. The first subclass was represented by a diffuse nuclear staining pattern and could be released by 1% NP-40, 150 mM NaCl. The second subclass was sequestered in globular structures and required a much higher salt concentration (1–2 M) for extraction. The transition from the diffuse state to the globular state was concurrent with the onset of viral DNA replication and entry into the late phase. Viral DNA appears to be associated with these globular structures since (1) inhibitors of DNA replication blocked the transition and (2) viral DNA was localized to similar structures by in situ hybridization.

Previous characterizations of DBP mutants suggested that DBP might interact with RNA. This interaction has been directly demonstrated in a variety of ways, including cross-linking of DBP to RNA in vivo by UV irradiation.

The DNA-binding protein (DBP) encoded by early region 2A (E2A) of the human adenovirus genome is a 72-kilodalton (kD) phosphoprotein (Linne et al. 1977; Klein et al. 1979) that is synthesized both early and late during the infectious cycle (Levinson and Levine 1977; Axelrod 1978). A large number of functions have been ascribed to this protein based primarily on analyses of DBP mutants; these include roles in viral DNA replication, viral early and late gene expression, and virion assembly.

DBP's role in viral DNA replication is well-established. The prototype DBP temperature-sensitive mutant, Ad5ts125, provided the first genetic evidence for this function. At the nonpermissive temperature Ad5ts125 is defective for viral DNA synthesis (Ensinger and Ginsberg 1972) and encodes a thermolabile DBP (van der Vliet et al. 1975). With the advent of the in vitro adenovirus DNA replication systems, this function was directly demonstrated (Horwitz 1978). DBP is required for the strand-elongation reaction of viral DNA synthesis (Friefeld et al. 1983; van Bergen and van der Vliet 1983)

and may play a role in the initiation reaction, as well (van der Vliet and Sussenbach 1975; Nagata et al. 1982).

This polypeptide has also been implicated in the regulation of adenovirus early gene expression. Studies involving the microinjection of purified adenovirus genes or mRNAs into cells have suggested that DBP may stimulate gene expression of both early region 1B (E1B) (Rossini 1983) and early region 4 (E4) (Richardson and Westphal 1981). It has been proposed that DBP also has a negative influence on early gene expression at intermediate and late times after infection. Evidence for this comes primarily from studies with the temperature-sensitive E2A mutant, Ad5ts125. This mutant, unlike wild-type adenovirus, fails to turn down early gene expression as viral infection proceeds and thus accumulates higher amounts of early mRNAs than wild-type virus (Carter and Blanton 1978a,b). The negative regulation mediated by the DBP is thought to affect both the cytoplasmic stabilities of early region 1A (E1A) and E1B mRNAs (Babich and Nevins 1981) as well as the rate of transcription of E4 (Nevins and Jensen-Winkler 1980). Repression of E4 transcription by DBP has also been demonstrated in vitro (Handa et al. 1983).

Further support for DBP's multifunctional nature

---

[†]Present address: Waksman Institute, Rutgers University, P.O. Box 759, Piscataway, New Jersey 08854.

comes from studies of abortive infections of monkey cells by human adenovirus. Wild-type adenovirus fails to multiply in these cells due to a complex block to viral late gene expression. This block includes a reduction in the rate of transcription of late genes (Johnston et al. 1985), alterations in the pattern of mRNA splicing for the fiber polypeptide (Klessig and Chow 1980; Anderson and Klessig 1984), and poor utilization of this mRNA in vivo (Anderson and Klessig 1983). Host-range mutants of adenovirus, which overcome these blocks and thus are capable of productive growth in monkey cells, contain alterations in the DBP gene (Klessig and Grodzicker 1979; Anderson et al. 1983; Brough et al. 1985).

Finally, evidence for DBP's role in virion assembly is based on work with a revertant of the DNA-replication-negative mutant, Ad5*ts*107. The revertant, R(*ts*107)202, produces apparently normal amounts of late viral structural proteins but is deficient at assembling virions in human 293 cells at the nonpermissive temperature (Nicolas et al. 1983).

Physical mapping of the two general classes of DBP mutations suggest that this polypeptide contains at least two functionally distinct domains (Kruijer et al. 1981; Klessig and Quinlan 1982; Anderson et al. 1983; Brough et al. 1985). Whereas temperature-sensitive mutants reside in the 3′ two-thirds of the gene, the host-range mutants are located in the 5′ portion of the gene.

Moreover, digestion of purified DBP with a variety of proteases generates a carboxyterminal fragment of approximately 44 kD and an aminoterminal fragment of approximately 26 kD (Klein et al. 1979; Schechter et al. 1980). The 44-kD carboxyterminal fragment is capable of binding single-stranded DNA (Klein et al. 1979) and can substitute for intact DBP in an in vitro DNA replication system (Ariga et al. 1980). The 26-kD aminoterminal

fragment, which contains most of the protein's sites of phosphorylation (Klein et al. 1979), retains its activity(s) required for normal late gene expression even when DBP's DNA replication function is perturbed by temperature-sensitive mutations in the carboxyl terminus (Rice and Klessig 1984). Together these results suggest that DBP contains at least two functionally and physically separable domains (Fig. 1).

The small number of temperature-sensitive and host-range mutants isolated and characterized to date have been invaluable in ascribing functions to the DBP. It is apparent, however, that a larger collection of mutants will be needed to understand this multifunctional protein in more detail. Unfortunately, since several of the activities of DBP are essential for virus growth, many potential E2A mutants are expected to have a nonconditionally lethal phenotype and thus cannot be isolated under normal circumstances. To circumvent this problem, we have constructed several human cell lines (designated gmDBP cells) that contain and express integrated copies of the E2A gene (Klessig et al. 1984a). Because we suspected that expression of DBP was toxic to cells (Klessig et al. 1984b), the E2A gene in these cell lines was placed under the control of a glucocorticoid hormone–inducible promoter. In the presence of dexamethasone, a synthetic glucocorticoid hormone, DBP synthesis is induced 50-fold to 200-fold and reaches levels that are approximately 10% of that observed during peak synthesis in adenovirus-infected HeLa cells. Utilizing these cell lines, several DBP deletion mutants have been isolated (Rice and Klessig 1985). Salient features of several of these mutants are described.

The pleiotropic nature of DBP implies that there may be several subclasses of protein that carry out different functions. Defining such potential subclasses of DBP

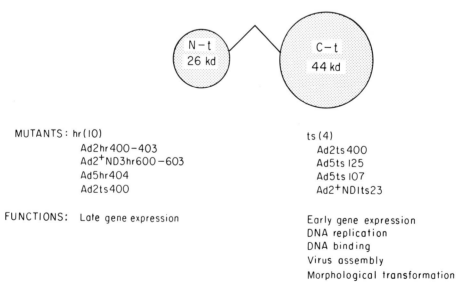

**Figure 1** Separate physical and functional domains of DBP. At top are represented the two physical domains of DBP connected by a protease-sensitive hinge region. Cleavage with a variety of proteases, including chymotrypsin, generates an aminoterminal (N-t) 26-kD fragment and a carboxyterminal (C-t) 44-kD fragment. Listed below each domain is a set of mutants mapped to and functions ascribed to each domain.

might be achieved by demonstrating that they associate with different macromolecules and structures in the cell. Delineating and isolating these subpopulations would, moreover, provide a useful framework upon which to compare and contrast the distribution and interactions of mutant DBP proteins. To this end, we have examined the subcellular distribution of DBP during adenovirus type-5 (Ad5) infection of HeLa cells as well as the association of the protein with RNA in vitro and in vivo.

## Experimental Procedures

### Construction and general characterization of DBP deletion mutants
The construction and general characterization of DBP deletion mutants are detailed in Rice and Klessig (1985).

### Analysis of early viral RNA
For comparison of steady-state levels of early adenovirus mRNAs, HeLa monolayers were infected with virus at a multiplicity of 50 pfu/cell. The overlay media included 10 mM hydroxyurea (Calbiochem) to block DNA replication. Cytoplasmic RNA was prepared from the cells by phenol and chloroform extraction of the 0.5% Nonidet P-40 (NP-40) supernatant fractions followed by ethanol precipitation (Klessig and Anderson 1975). For dot-blot analysis, 5 μg of each cytoplasmic RNA preparation was denatured in formaldehyde (Maniatis et al. 1982), mixed with an equal volume of cold 20× SSC, (1× SSC is 0.15 M NaCl, 0.015 M sodium citrate) and aspirated onto a nitrocellulose filter through a 96-well dot-blot manifold (Bethesda Research Laboratories). DNA probes specific for adenovirus early region were labeled by nick-translation with [$^{32}$P]dCTP to a specific activity of $1 \times 10^8$ to $5 \times 10^8$ cpm/μg. pBR322-derived plasmids containing the BalI-L (coordinates 6.0–7.7) and EcoRI-B (coordinates 58.5–70.7) fragments were used as probes for E1A and E2A, respectively. The filter was hybridized with nick-translated $^{32}$P-labeled Ad5 probes at 42°C in 50% formamide, 0.4 M NaCl, 0.1 M HEPES (pH 8.0), 5 mM EDTA, 0.2% SDS, 5× Denhardt solution (1× Denhardt solution is 0.2% [w/v] bovine serum albumin, Ficoll [m.w. 400,000], and polyvinyl pyrrolidone), 100 μg/ml denatured salmon sperm DNA. The filters were washed twice in 2× SSC, 0.1% SDS at room temperature, and three times in 0.1× SCC, 0.1% SDS at 55°C. For quantitation of the dot-blot data, individual dots were cut out and counted in 3 ml of Aquasol-2 (New England Nuclear). The radioactivity bound to duplicate dots differed by less than 20% and usually by less than 10%. The number of counts per minute (cpm) binding to dots containing 5 μg of HeLa cell RNA was considered background and subtracted before quantitation.

### Transformation of rat embryo cells
Primary rat embryo cultures were prepared from 16–18-day embryos taken from pregnant inbred Fisher rats. Embryos were eviscerated, washed to remove blood,

chopped into small pieces, and trypsinized to give a single cell suspension. Cells seeded in 60-mm dishes ($2 \times 10^6$ cells/dish) in Dulbecco's modified Eagle's medium (DME) supplemented with 7% calf serum formed confluent monolayers around 4 days after seeding and at that time were infected with either wild-type or mutant virus at an input multiplicity of 10 pfu/cell unless otherwise stated. After infection, cultures were washed with 0.02% EDTA in PBS-A solution and were trypsinized to release cells. Replicate dishes were seeded with $10^6$ in DME + 7% CS and incubated at the appropriate temperature. At 4 days after seeding, the medium was removed and replaced with calcium-free DME + 7% calf serum, and for the duration of the experiment this medium was replaced at 3–4-day intervals.

### Immunofluorescence microscopy
For immunofluorescence microscopy studies, mock- or virus-infected cells grown on coverslips were washed briefly in PBS (13.38 mM Na$_2$HPO$_4$, 1.47 mM KH$_2$PO$_4$, 2.68 mM KCl, 154 mM NaCl), fixed for 20 min in −20°C methanol, then rehydrated in PBS. Primary antibody incubations were at 37°C for 30 min using either a mouse anti-DBP monoclonal antibody or a polyclonal rabbit antisera raised against purified DBP. After three 5-min PBS washes, secondary antibody reactions using either FITC-conjugated sheep anti-mouse IgG (Sigma) or FITC-conjugated goat anti-rabbit (Meloy Laboratories) were performed at 37°C for 30 min. Coverslips were then washed three times in PBS, once in distilled H$_2$O, and mounted on glass slides with 90% glycerol + 10% PBS. Slides were viewed on a Zeiss IM35 microscope equipped with epifluorescence and recorded on Kodak Plus-X pan film.

In situ fractionation of cell monolayers or coverslip cultures was as described by Staufenbiel and Deppert (1984) with the following modifications. Cells were washed briefly, first with cold PBS, then with cold 10 mM MES (pH 6.2), 1.5 mM MgCl$_2$. Cells were first extracted with two sequential incubations for 5 min and 10 min on ice in 1% NP-40, 150 mM NaCl, 1 mM EGTA, 5 mM DTT, 1.5 mM MgCl$_2$, 10 mM MES, pH 6.2. Cells were then subjected to two successive 5-min extractions on ice in 2 M NaCl, 1 mM EGTA, 5 mM DTT, 1.5 mM MgCl$_2$, 10 mM MES (pH 6.2). Solubilization of residual cell structures was achieved by incubation for 30 min on ice with 1% Empigen BB (Albright and Wilson, Inc.), 25 mM KCl, 1 mM EGTA, 5 mM DTT, 40 mM Tris (pH 9.0). For IF microscopy analysis, residual cell structures, remaining after the various extraction steps, were briefly rinsed in cold 10 mM MES (pH 6.2), 1.5 mM MgCl$_2$, fixed for 20 min in −20°C methanol, rehydrated in PBS, and incubated with primary and secondary antibodies as described above.

### In situ hybridization
In situ hybridization was performed using a biotinylated Ad2 probe supplied in an Adenovirus 2 Patho-Gene™

Identification Kit (Enzo Biochem., Inc.). Conditions for hybridization, binding of a streptavidin–horseradish peroxidase complex to the hybridized probe and subsequent enzymatic detection with the chromogen diaminobenzidine tetrahydrochloride (DAB) in the presence of hydrogen peroxide were as outlined in Bio-Note 101 (Enzo Biochem., Inc.). Slides were viewed on a Nikon Optiphot microscrope and recorded on Polaroid type-55 Land film.

*UV cross-linking of DBP to RNA*
Mock- or Ad5-infected HeLa cell monolayer (100 mM plates) at 24 hr postinfection were washed twice with PBS containing 0.1 g/liter each of $CaCl_2$ and $MgCl_2 \cdot 6H_2O$, then immersed in 1–2 ml/plate of above buffer, before being UV-irradiated for 0 or 2 min (Dreyfuss et al. 1984). Cells were scraped from plates, pelleted, resuspended, and swelled in 1.25 ml/plate of 0°C high-salt RSB buffer (10 mM Tris [pH 7.4], 1 mM EDTA, 1.5 mM $MgCl_2$, 1 M NaCl) containing 10 µg/ml aprotinin, 1 µg/ml pepstatin A, and 1 µg/ml leupeptin for 5–10 min. Cells were solubilized by addition of Empigen BB (Albright and Wilson, Inc.) and NP-40 to a final concentration of 1% each for 5–10 min at 0°C. Viscosity was reduced by addition of DNase I (Worthington RNase-free DNAse I, further treated with iodoacetate according to Zimmerman and Sandeen [1966] to inactivate residual RNase) to 50 µg/ml, $MgCl_2$ to 10 mM, and vanadyl ribonucleoside complex (Bethesda Research Laboratories) to 10 mM with subsequent incubation at 37°C for 20 min. Finally, the extracts were passed through a 25-gauge needle four times. Polyadenylated RNA present in the extract was selected by two rounds of oligo(dT) (Type 3, Collaborative Research) chromatography as follows. To the extracts were added SDS to 0.5% (w/v), β-mercaptoethanol to 1% (v/v), and EDTA to 2 mM. Extracts were heated to 65°C for 7 min, then rapidly chilled. LiCl was added to 750 mM, and the extracts were then warmed to room temperature before gentle mixing with prewashed oligo(dT)-cellulose (0.02 g dry-weight oligo(dT) per 100-mm plate) for 15 min. The oligo(dT)-cellulose extract mixture was packed into a column and washed with 10 volumes of 10 mM Tris (pH 7.4), 1 mM EDTA, 0.5% SDS, 750 mM LiCl. The polyadenylated RNA was eluted from the column in 10 mM Tris (pH 7.4), 1 mM EDTA, 0.5% SDS, then heated, chilled, and had LiCl added to 750 mM before chromatography directed onto an oligo(dT)-cellulose column. After elution and concentration by ethanol precipitation, the RNA and associated material was treated with 25 µg/ml of boiled RNase A and 400 units/ml micrococcal nuclease (Sigma) in 10 mM Tris (pH 7.4), 1 mM $CaCl_2$, 10 µg/ml aprotinin, 1 µg/ml leupeptin, 1 µg/ml pepstatin A at 37°C for 60 min. The residual material was concentrated by acetone precipitation and subjected to SDS-PAGE on a 10% gel before immunoblot analysis as described by Klessig et al. (1984b) with the modification that nonfat dried milk was used in place of bovine serum albumin.

## Results

### Construction of DBP deletion mutants
Five Ad5 mutants with deletions in the DBP (E2A) gene were constructed (Rice and Klessig 1985). Deletions were first made in cloned copies of E2A and then the altered alleles were introduced into the intact viral genomes by in vivo recombination. Our strategy for introducing a cloned E2A gene into the viral chromosome is a variation of the genetic mapping procedure called marker rescue (Frost and Williams 1978; Stow et al. 1978). In this technique a subgenomic fragment of adenovirus DNA is transfected with intact genomic DNA into permissive cells. A small but significant fraction of the progeny are recombinants that have acquired some or all of the subgenomic fragment. To select for the desired E2A recombinants, our scheme made use of a parental virus that contains temperature-sensitive mutations in two late viral genes that flank E2A. Selecting for viruses that had lost the temperature-sensitive phenotype insured that a reasonable portion of the progeny had acquired the desired DBP allele, providing of course that the introduced DBP defect was complemented. This selection and complementation was achieved by propagation of the recombinant virus mixture on the DBP-producing cell line, gmDBP2, in the presence of dexamethasone at the nonpermissive temperature.

### Characterization of DBP deletion mutants
The relevant segments of the genome of two of these DBP mutants (Ad5*dl*801 and Ad5*dl*802) are shown in Figure 2. The E2A deletions extend very near to the amino terminus of DBP coding sequences (Rice and Klessig 1985). Since these deletions also cause frameshift mutations, both mutants are predicted to encode small proteins with very limited identity to DBP. This expectation was upheld for the mutant Ad5*dl*802, which makes no detectable DBP-related polypeptide and thus represents the first DBP-negative adenovirus mutant (data not shown).

Both mutants appear to be absolutely defective for growth in HeLa cells, thus verifying the expectation that DBP is an essential adenovirus protein (Table 1). The mutants can, as anticipated, be propagated in the DBP-producing cell lines. However, their growth in these cells is quite depressed compared with wild-type virus (Table 1), presumably reflecting the suboptimal level of DBP produced from the endogenous, chromosomal copy of E2A present in these lines.

As expected, given the well-defined role of DBP in viral DNA replication, the deletion mutants were incapable of DNA replication. Since DNA replication is a prerequisite for late viral gene expression (Thomas and Matthews 1980), it was not surprising that the mutants failed to express their late genes in HeLa cells (for details, see Rice and Klessig 1985).

### The role of DBP in adenovirus early gene regulation
To study the role of DBP in viral early gene regulation, steady-state levels of early mRNAs were monitored in

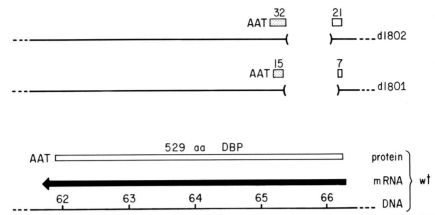

**Figure 2** Structure of the deleted E2A alleles. The lower portion of the diagram indicates the structure of the wild-type E2A gene and its RNA and protein. Adenovirus coordinates are indicated at the bottom. The large horizontal arrow indicates the coding exon of E2A mRNA, and the large open bar represents the open reading frame that encodes the 529-amino-acid DBP. The two lines above indicate the structures of the E2A genes in the deleted alleles. The parentheses denote the extent of the DNA sequences deleted, while the bars above each line represent the open reading frame predicted to be used by each mutant. Open bars correspond to translation in the normal DBP reading frame, and the stippled bars indicate that translation is in an alternate reading frame. The numbers above each bar correspond to the number of amino acids in each segment of the predicted polypeptide, while to the left of each bar is indicated the nonsense codon that should terminate translation.

cells infected with wild-type Ad5, Ad5ts125, and Ad5dl802. Since Ad5dl802 produced no DBP or detectable DBP fragment, this study allows analysis of adenovirus early gene expression in the complete absence of any DBP-mediated regulation. HeLa monolayers were infected and then incubated at the nonpermissive temperature. To insure proper comparison between wild-type Ad5 and the two DNA replication-defective mutants, the potent DNA replication inhibitor hydroxyurea was added to all the infections. At various times after infection, cytoplasmic DNA was prepared from the infected cells and equal amounts were subjected to dot-blot analysis using $^{32}$P-labeled DNAs specific for early regions 1A, 1B, 2A, 3, and 4 as the hybridization probes. Quantitation of the dot-blot data for two of the early regions, E1B and E2A, is illustrated in Figure 3.

Three general conclusions can be drawn from the results of these sets of experiments (for more details, see Rice and Klessig 1985). First, DBP does not appear to be a significant positive regulator of either E1B or E4

mRNA expression in infected HeLa cells since wild-type Ad5 and Ad5dl802 show similar patterns for both the induction and continued expression of these two early genes. Second, we have observed that E2A mRNA levels are reduced approximately fivefold in Ad5dl802 infections. This is consistent either with a positive role for DBP in the trans-acting regulation of its own gene or with a cis-acting defect caused by the Ad5dl802 deletion. Finally, DBP does not appear to negatively regulate expression of early regions 1A, 1B, 3, or 4 mRNAs when DNA replication is tightly blocked since Ad5dl802 shows little overexpression of these regions when compared with wild-type Ad5. In contrast, Ad5ts125-infected cells exhibited a reproducible tendency to overexpress early mRNAs at late times (12–20 hr p.i.) when compared with wild-type Ad5. The observed effect, however, was quite variable and never dramatic (maximally ~fourfold). More reproducible was the tendency for Ad5ts125-infected cells to express early viral genes sooner than wild-type Ad5 or Ad5dl802-infected cells and at higher levels at the earlier times. Since the DBP-negative mutant Ad5dl802 behaves much more similarly to wild-type Ad5 than to Ad5ts125, we suspect that DBP does not have a significant negative effect on the accumulation of early region 1A, 1B, 3, or 4 mRNAs in infected HeLa cell monolayers when DNA replication is tightly blocked.

### E2A deletion mutants transform rat cells at frequencies similar to wild-type Ad5

The temperature-sensitive adenovirus E2A mutants Ad5ts125 and Ad5ts107 transform rat cells at a threefold to eightfold enhanced rate compared with wild-type virus (Ginsberg et al. 1975; Williams et al. 1975). If wild-type DBP normally suppresses transformation in rat cells, as suggested by the above observation, then Ad5dl802 should express the high-transfor-

**Table 1** Yields of Virus in HeLa and gmDBP2a Cells

| Virus | Cell line[a] | pfu/cell[b] |
|---|---|---|
| Ad5 | HeLa | 2000 |
| Ad5dl801 | HeLa | 0.1 |
| Ad5 | gmDBP2a | 9000 |
| Ad5dl801 | gmDBP2a | 2 |
| Ad5dl801 | gmDBP2a (+dex)[c] | 90 |

[a]Confluent monolayers were infected at a multiplicity of 20 pfu/cell and incubated for 68 hr at 37°C.

[b]Progeny virus were harvested by freeze-thawing and sonication. The yields of Ad5 were determined by titration on HeLa cells. The yields of Ad5dl801 were determined by titration on HeLa cells using Ad5dl434 as a helper virus.

[c]The overlay media, including 0.6 μM dexamethasone.

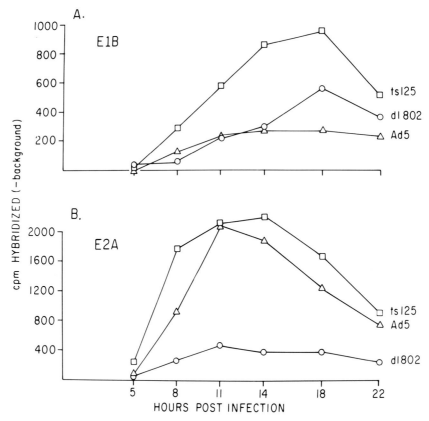

**Figure 3** Analysis of early adenovirus gene expression in DBP mutant–infected cells. Confluent monolayers of HeLa cells were infected with wild-type Ad5, Ad5dl802, or Ad5ts125 and incubated at 40°C in media containing 10 mM hydroxyurea. At various times after infection, cytoplasmic RNA was extracted and subjected to dot-blot analysis using probes for early region 1B (E1B) and early region 2A (E2A). For quantitation, individual dots were cut out, counted, and the average hybridized counts per minute minus background of the duplicate dots were plotted above.

mation phenotype like the temperature-sensitive mutants.

To test this, primary rat embryo cells were infected with wild-type Ad5 or various Ad5 mutants. As expected, at nonpermissive (38.5°C) and semipermissive (37°C) temperatures, Ad5ts125 transformed rat embryo cells at a frequency enhanced threefold to fivefold compared with wild-type Ad5. Furthermore, Ad5ts125-induced foci appeared 3–4 days earlier than other foci at these temperatures. In contrast, the Ad5dl802 and the other four E2A deletion mutants transformed rat cells at frequencies that were very similar to wild-type Ad5 (data not shown).

Since adenovirus-induced transformation of rat cells is somewhat dependent on the multiplicity of viral infection (Williams et al. 1975; Logan et al. 1981), the transformation frequencies of wild-type and mutant viruses were compared at 37°C as a function of input viral multiplicity (Fig. 4). Again, Ad5ts125 transformed with a threefold to fivefold higher frequency than wild-type Ad5, whereas Ad5dl802 transformed at the wild-type frequency. Therefore, it appears that the lack of a functional DBP does not, per se, cause an elevation in the frequency of rat cell transformation.

## Two nuclear subclasses of DBP

Although the genetic approach outlined above is a powerful and indispensible tool to explore and define the various functions performed by DBP, in itself it is insufficient to complete this difficult task. Biochemical studies, both in vivo and in vitro, as well as cell biological approaches will certainly be necessary. As an initial step in this direction, the intracellular location of DBP during the time course of wild-type Ad5 infection of HeLa cells was examined using immunofluorescence microscopy (Fig. 5). DBP was first readily detected at 10 hr p.i., at which time it was predominantly localized in the nucleus and exhibited a diffuse staining pattern (data not shown). As early as 12 hr p.i. a transition began to occur wherein some cell nuclei showed both diffuse staining DBP as well as more intensely staining, small globular concentrations of DBP. Increasing numbers of nuclei exhibited this sequestering process by 18–24 hr p.i., with the majority of DBP becoming localized in larger, randomly distributed globular structures that coexisted with a minor proportion of diffuse nuclear staining DBP (Fig. 5, top row). As the infectious cycle proceeded, the globular structures appeared to coalesce into large amorphous forms (data not shown). These results suggested

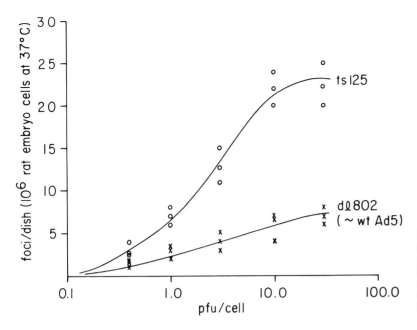

**Figure 4** Relationship between input multiplicity and transformation frequency in primary rat embryo cells infected by Ad5 E2A mutants. Rat embryo cells were infected at the above multiplicities indicated in plaque-forming units (pfu), incubated at 37°C, and split the next day into three replicate dishes at $10^6$ cells/dish.

that the nucleus contains at least two major populations of DBP.

To biochemically define these two populations, infected cells were sequentially extracted in situ with nonionic detergent and increasing salt concentrations. The residual cell structures were examined by immunofluorescence microscopy (Fig. 5) and the extracted proteins were analyzed by immunoblots (data not shown). Extraction with 1% NP-40, 150 mM NaCl removed the diffusely staining nuclear DBP, whereas the globular DBP structures remained intact within the residual cell structures (Fig. 5, middle row). Subsequent extraction with 2 M NaCl resulted in the loss of the globular structures. The nuclear matrix, which is resistant to this extraction procedure, exhibited only trace amounts of DBP (Fig. 5, bottom row). A minor fraction of DBP, which is located in the cytoplasm of cells at various times after infection, is also extractable with 1% NP-40, 150 mM NaCl (for details, see Voelkerding and Klessig 1986).

## Association of DBP with centers of active viral DNA replication

The DBP found in the globular structures seen in the nucleus by immunofluorescence microscopy and extractable with 2 M NaCl may represent a subclass of the protein associated with centers of viral DNA replication. For example, it is well known that the DBP not only binds single-stranded DNA (van der Vliet and Levin 1973) but is required for viral DNA replication both in vivo (Ensinger and Ginsberg 1972) and in vitro (Horwitz 1978). Moreover, the appearance and quantity of DBP in the globular structures parallels viral DNA replication activity.

We have achieved a more direct demonstration of this association by comparing size, number, and location of globular structures with the distribution of viral genomes during the infectious cycle using immunofluorescence

microscopy and in situ hybridization (Fig. 6), respectively. At 12 hr p.i., when DBP was first detected in globular structures, small focal centers of viral DNA were first evident. As infection proceeded, DBP was increasingly sequestered into the more prominent globular structures with a corresponding enlargement of the viral DNA centers.

If the globular structures represent a direct interaction of DBP with centers of active viral DNA replication, this association might be affected by perturbing viral DNA replication. To test this, viral DNA replication was manipulated in two ways (for details, see Voelkerding and Klessig 1986). When replication was blocked by hydroxyurea, the formation of globular structures was prevented and no centers of viral DNA could be detected by in situ hybridization. Moreover, when viral DNA synthesis was inhibited at the restrictive temperature in cells infected by a virus (Ad5ts149) carrying a temperature-sensitive mutation in the gene encoding the viral DNA polymerase, the globular structures did not form. When the block to viral DNA replication was removed by shifting to the permissive temperature, DBP globules rapidly appeared. Conversely, when Ad5ts149-infected cells were shifted from the permissive to the nonpermissive temperature late in the infectious cycle, preexisting globular structures were perturbed. These results suggest that this subclass of DBP is actively participating in viral DNA replication rather than only associating with sites of viral DNA accumulation.

## RNA-binding activity of DBP

Although DBP's ability to bind single-stranded DNA and its role in viral DNA replication are well-documented, evidence for DBP's participation in the regulation of gene expression through its interaction with RNA, although intriguing, is less compelling. The ability of either wild-type DBP to destabilize early viral mRNA (Babich

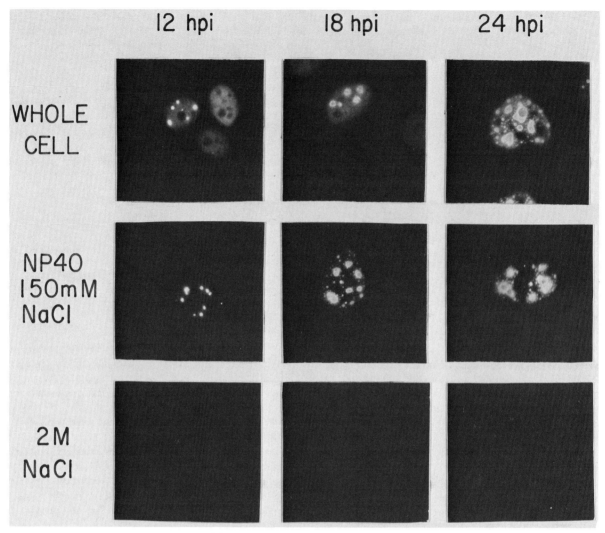

**Figure 5** Immunofluorescence analyses of DBP in Ad5-infected HeLa cells before and after in situ extraction. Ad5-infected HeLa cells (12, 18, and 24 hr p.i.) on parallel coverslips were analyzed by immunofluorescence directly (*top* row) or after extraction in situ with 1% NP-40, 150 mM NaCl (*middle* row) or 1% NP-40, 150 mM NaCl followed by 2 M NaCl (*bottom* row). Intact cells or residual cell structures remaining after extraction were fixed in −20°C methanol for 20 min and incubated with an anti-DBP monoclonal antibody followed by reaction with an FITC-conjugated secondary antibody.

and Nevins 1981) or temperature-sensitive DBP to stabilize them (Rice and Klessig 1985) implies a direct or indirect association of DBP with RNA. This association is also suggested by studies on the nature of the block to human adenovirus multiplication in monkey cells. This complex block, which involves altered late viral mRNA transcription, processing, and/or utilization, is overcome by mutations in the DBP gene (Klessig and Grodzicker 1979; Klessig and Chow 1980; Anderson and Klessig 1983, 1984).

We have directly demonstrated, both in vitro and in vivo, DBP's ability to bind to RNA. Purified DBP could be shown to bind to a variety of RNAs in a salt-dependent manner using a filter-binding assay (data not shown). Moreover, after in situ extraction with 1% NP-40, 150 mM NaCl, as described above, additional DBP could be released by RNase A treatment of residual cell structures (data not shown). But more importantly, DBP

was found associated with RNA in vivo. When HeLa monolayers (20–24 hr p.i. with Ad5) were subjected to UV irradiation for several minutes, a small portion of the total DBP was found covalently attached to polyadenylated RNA (Fig. 7). This cross-linking was specific to RNA since treatment with RNase, but not with DNase, prior to oligo(dT) chromatography eliminated any DBP association with the oligo(dT)-selected material (data not shown). This direct demonstration, together with the circumstantial evidence presented above, suggest that some of DBP's functions may be mediated by its RNA-binding activity.

**Discussion**

Many functions have been ascribed to the adenovirus-encoded DNA-binding protein. One of these includes a complex set of roles in the control of early gene ex-

**In situ hybrid.
Ad**

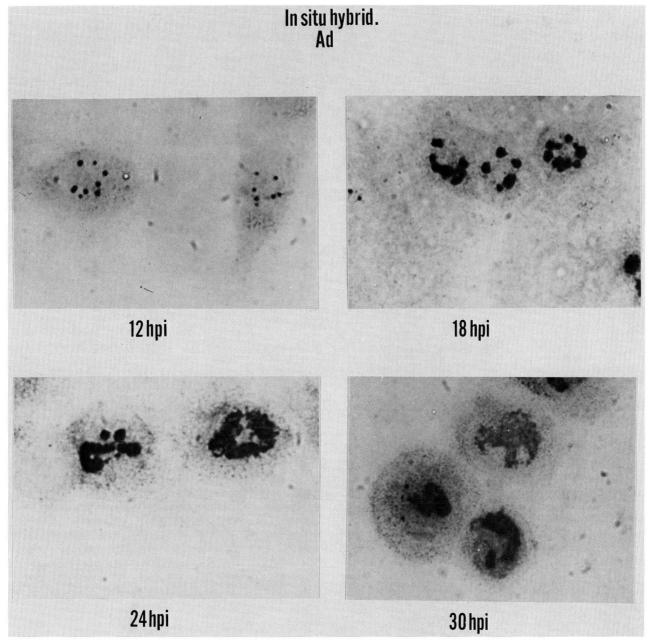

**Figure 6** Localization of viral DNA during Ad5 infection of HeLa cells. Infected cells grown on glass coverslips were analyzed at the indicated times postinfection (hpi) by in situ hybridization using a biotinylated Ad2 probe. Detection of the hybridized probe was accomplished by reacting first with streptavidin-linked horseradish peroxidase followed by incubation with the chromogen DAB in the presence of hydrogen peroxide. The reaction product, a dark brown to black precipitate, denotes the hybridization signal.

pression. Isolation of the first DBP-negative mutant Ad5*dl*802 has enabled us to analyze early gene expression in the complete absence of DBP-mediated regulation to test the validity of these purported functions.

Previous studies that involved the microinjection of purified adenovirus genes or mRNAs into mammalian cells suggested that DBP may be required for or have a stimulatory effect on the expression of E1B (Rossini 1983) and E4 (Richardson and Westphal 1981). In contrast, our work with Ad5*dl*802 failed to demonstrate

that DBP is necessary for normal accumulation of early mRNAs in infected HeLa cells. It should be noted, however, that since both microinjection studies assayed gene expression at the protein level, it is possible that DBP stimulated E1B and E4 gene expression at a post-transcriptional level.

Unexpectedly, Ad5*dl*802-infected cells expressed much-reduced steady-state levels of E2A mRNAs. This might indicate that DBP is a positive *trans*-acting regulator of its own gene. An alternative, and perhaps more

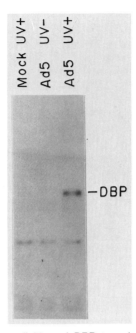

**Figure 7** UV cross-linking of DBP to polyadenylated RNA. Mock- or Ad5-infected HeLa cell monolayers were subjected to 0 (UV−) or 2 min (UV+) of UV light at 24 hr postinfection. The cell contents were solubilized with detergent and high salt, treated with DNase to reduce viscosity, and chromatographed on oligo(dT)-cellulose in the presence of the denaturant SDS. Polyadenylated RNA and covalently associated material were eluted from the column, digested with RNase A, and subjected to immunoblot analysis with anti-DBP serum.

likely possibility, is that the deletion within the DBP gene of Ad5*dl*802 causes a *cis*-acting defect that prevents efficient E2A expression. The sequences removed by the deletion may be required for efficient E2A transcription or RNA processing, or perhaps the altered Ad5*dl*802 E2A mRNA is less stable.

Studies with temperature-sensitive E2A mutants such as Ad5*ts*125 suggested that DBP negatively regulates expression from several early regions, since Ad5*ts*125-infected cells accumulate abnormally high levels of early mRNAs at intermediate and late times after infection (Carter and Blanton 1978a; Nevins and Jensen-Winkler 1980; Babich and Nevins 1981). If DBP was an important negative regulator of early mRNA levels, then the DBP-negative mutant Ad5*dl*802 should exhibit a phenotype that is as dramatic, if not more so, as that of Ad5*ts*125. This was never observed. Although Ad5*dl*802-infected HeLa cells sometimes show slightly enhanced levels of early mRNAs compared with wild type−infected cells, most often little or no difference was observed. In no case did the DBP-negative mutant accumulate higher amounts of early mRNAs than did Ad5*ts*125.

Taken together, our results suggest that DBP plays neither a significant positive nor a significant negative role in the regulation of E1A, E1B, E3, or E4 mRNA levels. Its role, if any, in controlling the expression of its own gene is yet to be defined.

Transformation studies with temperature-sensitive and deletion DBP mutants also gave disparate results. Our own studies, as well as those of others (Ginsberg et al. 1975; Williams et al. 1975; Logan et al. 1981), indicated that several DBP temperature-sensitive mutants transform rat cells at threefold to eightfold higher frequencies than wild-type adenovirus. Although this suggests that wild-type DBP normally suppresses transformation, this hypothesis is inconsistent with our observation that DBP-negative deletion mutants exhibit transformation phenotypes similar to that of wild-type Ad5.

Since the high transformation frequency exhibited by the temperature-sensitive DBP mutants cannot be due simply to inactivation of DBP, it is likely that the temperature-sensitive mutants encode a DBP with altered properties. The altered DBP presumably carries out an activity in infected rat cells that serves to enhance the frequency of transformation. Consistent with this idea is our observation that the transformation phenotypes of temperature-sensitive mutants are dominant over the wild-type phenotype in coinfections. Although it is not immediately apparent why an E2A temperature-sensitive mutant should have a phenotype not possessed by an E2A deletion mutant, it may reflect the multifunctional nature of the DBP. Perhaps the temperature-sensitive DBP acquires altered and even pathological activities at the nonpermissive temperature due to the differential inactivation of functional domains.

In addition to its purported roles in the modulation of early gene expression and the frequency of morphological transformation of rodent cells, DBP is involved in viral DNA replication and late gene expression. The performance of such diverse functions may require unique subclasses of the protein. Immunofluorescent localization of DBP, both in our studies and those of others (Ginsberg et al. 1977; Sugawara et al. 1977), reveals that DBP's cellular location changes during the infectious cycle. These distributional alterations may reflect changes in the interactions of the protein not only with other viral products but also with host factors and structures. Thus, throughout an infection, intracellular partitioning of functionally distinct subclasses of DBP may occur.

By utilizing an in situ cell fractionation procedure developed by Staufenbiel and Deppert (1984), we have defined two major nuclear subclasses of DBP that are present in HeLa cells during infection with Ad5. One nuclear subclass has a diffuse staining pattern and can be released during in situ fractionation with 1% NP-40, 150 mм NaCl. Initially seen during the early phase of infection, it subsequently becomes less prominent as infection proceeds to intermediate and late times. The function of this subpopulation is unclear at present.

The second nuclear subclass of DBP is sequestered into globular structures that first appear around 12 hr p.i. as small, brightly staining foci. As infection proceeds, the number and size of these globular structures increase. The globules are resistant to 1% NP-40, 150 mм NaCl but can be extracted with high salt (2 м NaCl). Sugiwara et al. (1977) and Ginsberg et al. (1977)

have suggested that the DBP globular structures may represent sites of adenovirus DNA replication. Our observations are consistent with this hypothesis. First, the temporal appearance of DBP globular structures seen by immunoflurescence microscopy correlated well with the intranuclear distribution of adenovirus genomes detected by in situ hybridization. Second, the formation and maintenance of DBP globules was dependent on the onset and continuation of viral DNA replication.

Though a large proportion of DBP associated with these globular structures may be involved in viral DNA replication, it may be premature to designate this as its sole function. Our preliminary studies with actinomycin D, an inhibitor primarily of RNA transcription, showed that the DBP globular structures underwent partial breakdown upon addition of this drug during the late phase of infection. Analysis of the blocks to late gene expression in wild-type adenovirus—infected monkey cells, which can be overcome by mutations in the DBP gene, also suggest that this protein may be directly or indirectly involved in viral RNA transcription and processing (Klessig and Grodzicker 1979; Klessig and Chow 1980; Anderson and Klessig 1984; Johnston et al. 1985). Moreover, we have directly demonstrated DBP's capacity to bind RNA not only in vitro but also in vivo via UV cross-linking.

The ability (1) to construct any type of DBP mutant using the complementing, DBP-producing cell lines, (2) to isolate at least two major nuclear subclasses of DBP, and (3) to analyze the chemical (e.g., phosphorylation) and biochemical (e.g., DNA binding, RNA binding, and DNA replication) properties of the protein should provide a rich environment in which to further characterize this multifunctional protein.

# References

Anderson, C.W., M.M. Hardy, J.J. Dunn, and D.F. Klessig. 1983. Independent, spontaneous mutants of adenovirus type 2—simian virus 40 hybrid Ad2$^+$ND3 that grow efficiently in monkey cells possess identical mutations in the adenovirus type 2 DNA-binding protein gene. *J. Virol* **48**: 31.

Anderson, K.P. and D.F. Klessig. 1983. Posttranscriptional block to synthesis of a human adenovirus capsid protein in abortively infected monkey cells. *J. Mol. Appl. Genet.* **2**: 31.

————. 1984. Altered mRNA splicing in monkey cells abortively infected with human adenovirus may be responsible for inefficient synthesis of the virion fiber polypeptide. *Proc. Natl. Acad. Sci.* **81**: 4023.

Ariga, H., H. Klein, A. Levine, and M. Horwitz. 1980. A cleavage product of the adenovirus DNA binding protein is active in DNA replication *in vitro*. *Virology* **101**: 307.

Axelrod, N. 1978. Phosphoproteins of adenovirus 2. *Virology* **87**: 366.

Babich, A. and J.R. Nevins. 1981. The stability of early adenovirus mRNA is controlled by the viral 72 kd DNA-binding protein. *Cell* **26**: 371.

Brough, D.E., S.A. Rice, S. Sell, and D.F. Klessig. 1985. Restricted changes in the adenovirus DNA binding protein that lead to extended host range or temperature-sensitive phenotypes. *J. Virol.* **55**: 206.

Carter, T.H. and R.A. Blanton. 1978a. Possible role of 72,000-dalton DNA-binding protein in regulation of adenovirus type 5 early gene expression. *J. Virol.* **25**: 664.

————. 1978b. Autoregulation of adenovirus type 5 early gene

expression. II. Effect of temperature-sensitive early mutations on virus RNA accumulation. *J. Virol.* **28**: 450.

Dreyfuss, G., S.A. Adam, and Y.D. Choi. 1984. Physical change in cytoplasmic messenger ribonucleoproteins in cells treated with inhibitors of mRNA transcription. *Mol. Cell. Biol.* **4**: 415.

Ensinger, M.J. and H.S. Ginsberg. 1972. Selection and preliminary characterization of temperature-sensitive mutants of type 5 adenovirus. *J. Virol.* **10**: 328.

Friefeld, B.R., M.D. Krevolin, and M.S. Horwitz. 1983. Effects of the adenovirus H5ts125 and H5ts107 DNA binding proteins on DNA replication *in vitro*. *Virology* **124**: 380.

Frost, E. and J. Williams. 1978. Mapping temperature-sensitive and host-range mutants of adenovirus type 5 by marker rescue. *Virology* **91**: 39.

Ginsberg, H.S., U. Lundholm, and T. Linne. 1977. Adenovirus DNA-binding protein in cells infected with wild-type 5 adenovirus and two DNA-minus, temperature-sensitive mutants, H5ts125 and H5ts149. *J. Virol.* **23**: 142.

Ginsberg, H.S., M.J. Ensinger, R.S. Kauffman, A.J. Mayer, and U. Lundholm. 1975. Cell transformation: A study of regulation with types 5 and 12 adenovirus temperature-sensitive mutants. *Cold Spring Harbor Symp. Quant. Biol.* **39**: 419.

Handa, H., R.E. Kingston, and P.A. Sharp. 1983. Inhibition of adenovirus early region IV transcription *in vivo* by a purified DNA binding protein. *Nature* **302**: 545.

Horwitz, M.S. 1978. Temperature-sensitive replication of H5ts125 adenovirus DNA *in vitro*. *Proc. Natl. Acad. Sci.* **75**: 4291.

Johnston, J.M., K.P. Anderson, and D.F. Klessig. 1985. Partial block to transcription of human adenovirus type 2 late genes in abortively injected monkey cells. *J. Virol.* **56**: 378.

Klein, H., W. Maltzman, and A.J. Levine. 1979. Structure-function relationships of the adenovirus DNA-binding protein. *J. Biol. Chem.* **254**: 11051.

Klessig, D.F. and C.W. Anderson. 1975. Block to multiplication of adenovirus serotype 2 in monkey cells. *J. Virol.* **16**: 1650.

Klessig, D.F. and L.T. Chow. 1980. Incomplete splicing and deficient accumulation of the fiber messenger RNA in monkey cells abortively infected by human adenovirus type 2. *J. Mol. Biol.* **139**: 221.

Klessig, D.F. and T. Grodzicker. 1979. Mutations that allow human Ad2 and Ad5 to express late genes in monkey cells map in the viral gene encoding the 72K DNA binding protein. *Cell* **17**: 957.

Klessig, D.F. and M.P. Quinlan. 1982. Genetic evidence for separate functional domains on the human adenovirus specified, 72 kd, DNA binding protein. *J. Mol. Appl. Genet.* **1**: 263.

Klessig, D.F., D.E. Brough, and V. Cleghon. 1984a. Introduction, stable integration, and controlled expression of a chimeric adenovirus gene whose product is toxic to the recipient human cell. *Mol. Cell. Biol.* **4**: 1354.

Klessig, D.F., T. Grodzicker, and V. Cleghon. 1984b. Construction of human cell lines which contain and express the adenovirus DNA binding protein gene by cotransformation with HSV-1 tk gene. *Virus Res.* **1**: 169.

Kruijer, W., F.M.A. van Schaik, and J.S. Sussenbach. 1981. Structure and organization of the gene coding for the DNA binding protein of adenovirus type 5. *Nucleic Acids Res.* **9**: 4439.

Levinson, A. and A.J. Levine. 1977. The isolation and identification of the adenovirus group C tumor antigens. *Virology* **76**: 1.

Linné, T., H. Jörnvall, and L. Philipson. 1977. Purification and characterization of the phosphorylated DNA-binding protein from adenovirus-type-2-infected cells. *Eur. J. Biochem.* **76**: 481.

Logan, J., J.C. Nicolas, W.C. Topp, M. Girard, T. Shenk, and A.J. Levine. 1981. Transformation by adenovirus early region 2A temperature-sensitive mutants and their revertants. *Virology* **115**: 419.

Maniatis, T., E.F. Fritsch, and J. Sambrook. 1982. *Molecular *cloning: A laboratory manual.* Cold Spring Harbor Laboratory, Cold Spring Harbor, New York.

Nagata, K., R.A. Guggenheimer, T. Enomoto, J.H. Lichy, and J. Hurwitz. 1982. Adenovirus DNA replication *in vitro*: Identification of a host factor that stimulates synthesis of the preterminal protein-dCMP complex. *Proc. Natl. Acad. Sci.* **79**: 6438.

Nevins, J.R. and J.J. Jensen-Winkler. 1980. Regulation of early adenovirus transcription: A protein product of early region 2 specifically represses region 4 transcription. *Proc. Natl. Acad. Sci.* **77**: 1893.

Nicolas, J.C., P. Sarnow, M. Girard, and A.J. Levine. 1983. Host range temperature—conditional mutants in the adenovirus DNA binding protein are defective in the assembly of infectious virus. *Virology* **126**: 228.

Rice, S.A. and D.F. Klessig. 1984. The function(s) provided by the adenovirus-specified, DNA-binding protein required for viral late gene expression is independent of the role of the protein in viral DNA replication. *J. Virol.* **49**: 35.

———. 1985. Isolation and analysis of adenovirus type 5 mutants containing deletions in the gene encoding the DNA binding protein. *J. Virol.* **56**: 767.

Richardson, W.D. and H. Westphal. 1981. A cascade of adenovirus early functions is required for expression of adeno-associated virus. *Cell* **27**: 133.

Rossini, M. 1983. The role of adenovirus early region 1A in the regulation of early regions 2A and 1B expression. *Virology* **131**: 49.

Schechter, N.M., W. Davies, and C.W. Anderson. 1980. Adenovirus coded deoxyribonucleic acid binding protein. Isolation, physical properties, and effects of proteolytic digestion. *Biochemistry* **19**: 2802.

Staufenbiel, M. and W. Deppert. 1984. Preparation of nuclear matrices from cultured cells: Subfractionation of nuclei *in situ*. *J. Cell Biol.* **98**: 1886.

Stow, N.D., J.H. Subak-Sharpe, and N.W. Wilkie. 1978. Physical mapping of herpes simplex virus 1 mutations by marker rescue. *J. Virol.* **28**: 182.

Sugawara, K., Z. Gilead, W.S.M. Wold, and M. Green. 1977. Immunofluorescence study of the adenovirus type 2 single-stranded DNA binding protein in infected and transformed cells. *J. Virol.* **22**: 527.

Thomas, G.P. and M.B. Matthews. 1980. DNA replication and the early to late transition in adenovirus infection. *Cell* **22**: 523.

van Bergen, B.G.M. and P.C. van der Vliet. 1983. Temperature-sensitive initiation and elongation of adenovirus DNA replication in vitro with nucelar extracts from H5ts36-, H5ts149-, and H5ts125-infected HeLa cells. *J. Virol.* **46**: 642.

van der Vliet, P.C. and A.J. Levine. 1973. DNA-binding proteins specific for cells infected by adenovirus. *Nat. New Biol.* **246**: 170.

van der Vliet, P. and J. Sussenbach. 1975. An adenovirus type 5 gene function required for initiation of viral DNA replication. *Virology* **67**: 415.

van der Vliet, P.C., A.J. Levine, M.S. Ensinger, and H.S. Ginsberg. 1975. Thermolabile DNA binding proteins from cells infected with a temperature-sensitive mutant of adenovirus defective in viral DNA synthesis. *J. Virol.* **15**: 348.

Voelkerding, K. and D.F. Klessig. 1986. Identification of two nuclear subclasses of the adenovirus type 5 encoded DNA binding protein. *J. Virol.* (in press).

Williams, J.F., C.S.H. Young, and P.E. Austin. 1975. Genetic analysis of human adenovirus type 5 in permissive and nonpermissive cells. *Cold Spring Harbor Symp. Quant. Biol.* **39**: 427.

Zimmerman, F.D. and G. Sandeen. 1966. The ribonuclease activity of crystallized pancreatic deoxyribonuclease. *Anal. Biochem.* **14**: 269.

# Cis-acting Signals Involved in the Replication and Packaging of Herpes Simplex Virus Type-1 DNA

## N.D. Stow, M.D. Murray, and E.C. Stow
Medical Research Council Virology Unit, Institute of Virology, Glasgow, G11 5JR United Kingdom

We have developed assays using cloned fragments of herpes simplex virus type-1 (HSV-1) DNA to analyze the cis-acting signals involved in the initiation of DNA replication and encapsidation of the viral genome. One sequence specifying an origin of DNA replication ($ori_S$) is present at two copies per genome and is located within the inverted repetitions flanking the $U_S$ region. A striking homology exists between the sequence of $ori_S$ and that recently reported for the other HSV-1 origin ($ori_L$), located in the $U_L$ region. Plasmids containing $ori_S$ appear to be replicated by a rolling-circle mechanism in the presence of an appropriate helper virus, but a second cis-acting signal, supplied by the approximately 400-bp terminal redundancy or "a" sequence, is necessary to allow encapsidation of the amplified plasmid DNA. The presence of these two signals enables foreign DNA sequences to be serially propagated as components of defective virus genomes, and this property is being exploited in the construction of helper-dependent HSV-1 vectors.

The genome of herpes simplex virus type-1 (HSV-1) is a linear DNA duplex of approximately 155 kb, the sequence arrangement of which is depicted in Figure 1a (for review, see Roizman 1979). Two covalently linked components (L and S) each comprise a region of unique sequences ($U_L$ and $U_S$) bounded by a set of inverted repetitions ($IR_L/TR_L$ and $IR_S/TR_S$). Common to the two sets of reiterations is a region of approximately 400 bp (the "a" sequence) that occurs as a direct repeat at the genomic termini and in inverted orientation at the junction between L and S (Fig. 1b; Davison and Wilkie 1981; Mocarski and Roizman 1981).

Although the precise mode of replication of the HSV-1 genome is not known, several of the principal events have been outlined. Input viral genomes appear initially to become circularized, possibly as a result of direct ligation of the termini (Poffenberger and Roizman 1985), and subsequently replicate in an "endless" form (Jongeneel and Bachenheimer 1981). Late in infection, rapidly sedimenting viral DNA molecules consisting of tandem head-to-tail repeats of the viral genome are produced, probably by a rolling-circle mechanism of replication (Jacob et al. 1979). Site-specific cleavage events must therefore occur to generate the unit-length molecules that are packaged into virus particles. The structures of the genomic termini suggest that cleavage may be directed by signals within the "a" sequence (Davison and Wilkie 1981; Mocarski and Roizman 1982), but it is not known whether additional sequences must be present to enable encapsidation. The observation that temperature-sensitive (ts) mutants that fail to package replicated viral DNA at the nonpermissive temperature also fail to cleave it indicates that these processes are tightly coupled (Ladin et al. 1980; C. Addison and V.G. Preston, in prep.).

Serial passage of HSV-1 at a high multiplicity of infection results in the generation of defective genomes in which small subsets of viral DNA sequences are tandemly repeated (for review, see Frenkel et al. 1980). That these molecules can be serially propagated demonstrates the presence of cis-acting signals necessary for their replication and encapsidation (Vlazny and Frenkel 1981). Analysis of the structures of defective genomes, supported by evidence obtained from electron microscopic examination of replicating HSV-1 DNA (Friedmann et al. 1977), thus suggested the presence of two origins of DNA replication on the viral genome; one possibly within the $TR_S/IR_S$ sequences ($ori_S$; Denniston et al. 1981; Vlazny and Frenkel 1981) and the other with $U_L$ ($ori_L$; Kaerner et al. 1979; Cuifo and Hayward 1981; Spaete and Frenkel 1982). Similarly, the ubiquitous presence in defective genomes of DNA from the terminus of the S component indicated the possible presence within this region of signals required for packaging.

In this communication we describe our attempts, using well-characterized fragments of HSV-1 DNA cloned into bacterial plasmid vectors, to define more precisely the cis-acting signals that facilitate viral DNA replication and packaging. A functional test for $ori_S$ activity was initially developed (Stow 1982), and further analysis enabled the minimum sequence requirements to be defined (Stow and McMonagle 1983). A striking homology is now apparent between these sequences and the more recently characterized sequences shown to be necessary for $ori_L$ function (Weller et al. 1985). We also demonstrated that the additional presence within a plasmid of the viral "a" sequence sufficed to enable the encapsidation of amplified vector DNA sequences, indicating that the cis-acting signals necessary for cleav-

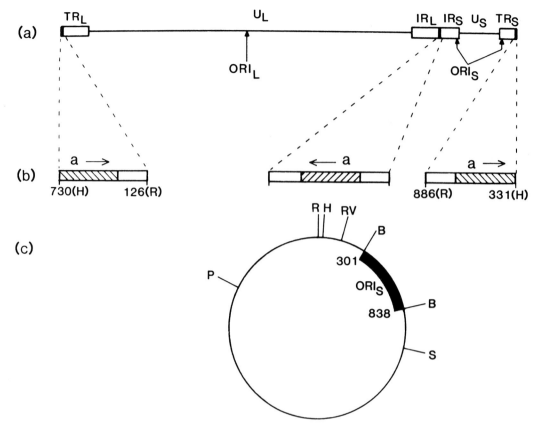

**Figure 1** (*a*) Structure and sequence arrangement of the HSV-1 genome, indicating the locations of *ori*$_L$ (ORI$_L$) and the two copies of *ori*$_S$ (ORI$_S$). (*b*) Orientation of the "a" sequence at the termini and joint between IR$_L$ and IR$_S$. The numbers underneath the terminal fragments refer to nucleotide positions in the published joint sequence of Davison and Wilkie (1981). These fragments were cloned between the *Eco*RI (R) and *Hin*dIII (H) sites of plasmid pS1 (Stow et al. 1983). (*c*) Structure of plasmid pS1, containing a 537-bp HSV-1 fragment specifying *ori*$_S$ inserted into the *Bam*HI (B) site of the vector pAT153 (Stow and McMonagle 1983). Other sites indicated are: *Eco*RI (R), *Hin*dIII (H), *Eco*RV (RV), *Sal*I (S), and *Pst*I (P).

age and packaging are contained within this region (Stow et al. 1983). Foreign DNA sequences linked to both *ori*$_S$ and the "a" sequence can consequently be serially propagated in virus stocks containing an appropriate helper virus, and here we describe the use of the bacterial chloramphenicol acetyltransferase (CAT) gene to demonstrate the potential of this system in the development of helper-dependent HSV-1 vectors for the expression of foreign genes.

## Materials and Methods

### Cells and virus
BHK-21 C13 cells were grown in Eagle's medium supplemented with 10% tryptose phosphate broth and 10% calf serum (Macpherson and Stoker 1962). Subconfluent cell monolayers ($2 \times 10^6$ cells per 50-mm plastic petri dish) were used for DNA transfections and for experiments with the derived virus stocks. Wild-type HSV-1 (Glasgow strain 17) or a temperature-sensitive mutant, *tsK*, generated from it (Marsden et al. 1976) were used for superinfections.

### Analysis of plasmid replication and packaging
Replication experiments were performed as previously described (Stow and McMonagle 1983). Briefly, mono-

layers of BHK cells were transfected with supercoiled plasmid DNA in the presence of calf thymus carrier DNA, using the calcium phosphate technique (Graham and van der Eb 1973), followed by a DMSO boost (Stow and Wilkie 1976). Monolayers received amounts of recombinant plasmids containing numbers of molecules equivalent to 0.24 μg of the vector (pAT153) DNA. Six hr after transfection the cells were superinfected with wild-type HSV-1 or *tsK* at a multiplicity of infection of 5 pfu/cell and incubated for a further 16 hr at 37°C or 31°C, respectively.

Virus stocks were produced by scraping the cells into the growth medium (4 ml), followed by extensive sonication. Samples (1.0 ml) of the stocks were used to infect fresh BHK cell monolayers in order to assay the ability of vector sequences to be serially propagated (Stow et al. 1983) or to measure the production of CAT activity.

Total cellular DNA was prepared from transfected monolayers using previously described methods (Stow et al. 1983). The preparation of encapsidated (DNase-resistant) DNA was also as previously described (Stow et al. 1983) except that on some occasions the sonication step was omitted and DNase digestion was performed in the presence of 0.5% NP-40.

DNA was digested with restriction endonucleases, and the products were separated by electrophoresis through agarose gels and transferred to nitrocellulose sheets using the Southern (1975) blotting technique. Hybridization with in vitro [32]P-labeled probes was as described previously (Stow and McMonagle 1983).

### Assay of CAT activity

Monolayers of BHK cells were infected with virus stocks as described above and incubated for 12 hr at 37°C (wild-type helper virus), 31°C (*tsK* helper at permissive temperature), or 38.5°C (*tsK* helper at nonpermissive temperature). Cell extracts were prepared and their ability to acetylate [14]C-labeled chloramphenicol was assayed as described by Mackett et al. (1984).

### Plasmids

Plasmid pS1 (Stow and McMonagle 1983) contains a 537-bp *Sau*3A-I fragment specifying a functional HSV-1 $ori_S$ inserted into the *Bam*HI site of the vector pAT153. Copies of the "a" sequence from the joint region between the L and S segments or from the genomic termini were inserted between the *Eco*RI and *Hin*dIII sites of pS1 (Stow et al. 1983). The *Eco*RV site of one such plasmid (pT091; Stow et al. 1983), containing an approximately 550-bp fragment from the L terminus, was converted to an *Xba*I site using oligonucleotide linkers. Fragments containing the bacterial CAT gene under the control of various HSV-1 promoters and attached to an HSV-2 transcription termination signal were inserted between the *Hin*dIII and *Xba*I sites of the modified plasmid. The CAT-coding sequences and HSV-2 transcription termination signal were obtained on a *Bam*HI plus *Xba*I fragment of a plasmid, pLW2, essentially similar to pTER5 (McLauchlan et al. 1985). The structures of the resulting plasmids and sources of HSV-1 promoters are shown in Figure 6. Plasmid DNAs were propagated and purified as described by Davison and Wilkie (1981).

## Results

### Replication of plasmids containing a cloned copy of $ori_S$

Analysis of the structures of HSV-1 defective genomes originally suggested that $ori_S$ was probably located within, or close to, the $TR_S$ and $IR_S$ regions. Direct evidence for its presence was provided by Vlazny and Frenkel (1981), who cotransfected cells with restriction endonuclease-generated monomer units of defective genomes containing sequences from $U_S$ and $TR_S$ and nondefective HSV-1 DNA as helper and showed that the individual units replicated and regenerated tandemly repeated genomic structures. To define accurately the sequences upon which replication may be initiated, we employed plasmids containing well-characterized fragments from these regions of nondefective HSV-1 DNA.

To assay for $ori_S$ activity, BHK cells were transfected with circular plasmid molecules and superinfected with wild-type HSV-1 to provide helper functions. Total cellular DNA was prepared and the plasmid vector content

was examined by hybridization. Amplification of the vector DNA was detected only with plasmids containing $ori_S$ and was dependent upon the provision of helper virus functions. The products of plasmid replication were high-molecular-weight DNA molecules consisting of tandem reiterations of the complete plasmid, suggesting that amplification took place by a rolling-circle mechanism (Stow 1982). A 537-bp fragment entirely from within the $TR_S/IR_S$ regions that contained all the *cis*-acting sequences necessary for origin function was identified, and the structure of a plasmid, pS1, containing this fragment inserted into the *Bam*HI site of the vector pAT153 is shown in Figure 1c (Stow and McMonagle 1983).

A time course for pS1 replication after superinfection with wild-type HSV-1 is shown in Figure 2. Samples of total cellular DNA, prepared at various times after infection, were digested with a combination of *Dpn*I and *Eco*RI. *Dpn*I cleaves its recognition sequence, GATC, only when the A residue is methylated (Lacks and Greenberg 1977), and plasmid DNA isolated from our normal host, the *dam*[+] *Escherichia coli* strain DH1 (Hanahan 1983), is therefore susceptible to digestion. In contrast, plasmid DNA replicated in eukaryotic cells is no longer methylated at this sequence and is consequently resistant to cleavage, allowing it to be easily discriminated from unreplicated input DNA. After digestion, the fragments were separated by agarose gel

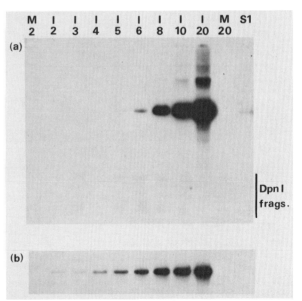

**Figure 2** Time course of plasmid pS1 replication. Replicate monolayers were transfected with pS1 and subsequently either mock-infected (lanes *M*) or infected with wild-type HSV-1 (lanes *I*). Total cellular DNA was prepared at the indicated number of hours after virus addition or mock-infection. After cleavage with *Eco*RI plus *Dpn*I, the resulting fragments were separated by agarose gel electrophoresis, transferred to nitrocellulose, and hybridized to nick-translated pAT153 DNA. (*a*) An autoradiograph of the filter; (*b*) an autoradiograph of a region of the filter in (*a*) hybridized to a probe containing sequences from the HSV-1 *Eco*RI-H fragment ($U_S$ region). Lane S1 contains linear pS1 (cleaved with *Eco*RI alone).

electrophoresis, transferred to nitrocellulose, and hybridized to $^{32}$P-labeled pAT153 DNA. Small fragments resulting from DpnI cleavage of unreplicated pS1 DNA are visible in each track. A fragment comigrating with EcoRI-digested pS1 is first observable 4 hr after virus addition. Because hemimethylated DNA is DpnI-resistant, the synthesis of the equivalent of at least one complete strand must have occurred. As infection proceeds, replicated pS1 DNA rapidly accumulates. Hybridization of the same samples to a probe containing HSV-1 DNA sequences indicates that pS1 replication initially lags behind that of the helper virus DNA but continues at a relatively higher rate late in infection. It is not yet clear whether this is indicative of the normal

kinetics of $ori_S$ activity or reflects slow uptake of the transfected plasmid.

## Comparison of $ori_S$ and $ori_L$

We have performed deletion analyses on $ori_S$-containing plasmids and localized the sequences required for origin activity to a 90-bp region that contains an almost perfect palindromic sequence of 45 bp (Stow and McMonagle 1983). Replacement of a portion of the palindrome with a synthetic oligonucleotide linker abolished origin function, demonstrating an essential role for sequences within the palindrome in the initiation of DNA synthesis (Fig. 3a; Stow 1985).

Studies on defective HSV-1 genomes suggested that

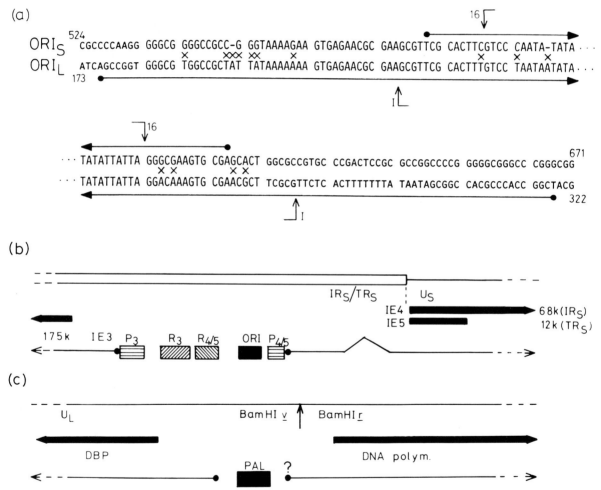

**Figure 3** (a) Homology between the sequences specifying $ori_S$ (Stow and McMonagle 1983) and $ori_L$ (Weller et al. 1985). The 92-bp regions exhibiting high homology are shown in larger capital letters, with nonidentical residues indicated by ×. The extent of the palindromes are indicated by horizontal arrows. Replacement of the sequences between the two arrows marked "16" with a synthetic 8-bp linker abolished $ori_S$ function (Stow 1985). Similarly, a plasmid lacking the seqences between the two arrows marked "I" did not specify a functional $ori_L$ (Quinn 1984; N. Stow, unpubl.). (b) Location of $ori_S$. The 90-bp region (ORI) required for $ori_S$ function (Stow and McMonagle 1983) is shown relative to the positions of the promoter (P) and upstream regulatory (R) regions of the immediate-early genes IE3 (3) and IE4/5 (4/5). The 5′ termini of the mRNAs are indicated (●), and the position of the splice in the IE4/5 mRNA is shown. The IE4 and IE5 mRNAs utilize the same upstream sequences and 5′ ends (located in IR$_S$ and TR$_S$, respectively) but encode different polypeptides (heavy bars) from the opposite sides of U$_S$ (see Preston et al. 1984 for further discussion and references). (c) Location of $ori_L$. The position of the long palindromic sequence (PAL) within the $ori_L$ region (Weller et al. 1985) is shown relative to the sequences encoding the major DNA-binding protein (DBP) and DNA polymerase, as deduced from sequencing studies (Quinn 1984). The 5′ end of the DBP mRNA was located by Quinn (1984), whereas that for the DNA polymerase has not been precisely defined.

another viral origin, $ori_L$, was located near the middle of $U_L$, but attempts to clone fragments from this region in bacterial plasmids invariably resulted in the deletion of a portion of the viral sequences (Spaete and Frenkel 1982; Weller et al. 1983). Such deleted plasmids were unable to replicate in assays of the type described above (Weller et al. 1985; N.D. Stow, unpubl.). Spaete and Frenkel (1982) showed, however, that when a deleted, cloned fragment from a defective genome containing $ori_L$ was cotransfected with helper HSV-1 DNA and the resulting virus was serially propagated, plasmid amplification did occur. This amplification was consistently associated with the restoration of the deleted sequences, presumably by recombination with the helper virus, suggesting an essential role for the deletion-prone sequences in $ori_L$ function. This hypothesis was recently proven by Weller et al. (1985), who succeeded in cloning DNA from this region in an undeleted form by using a yeast vector and directly demonstrated the presence of a functional viral origin.

The DNA sequence of the deletion-prone region comprising $ori_L$ has now been determined from both defective (Gray and Kaerner 1984) and nondefective (Quinn 1984; Weller et al. 1985) HSV-1 genomes. Examination of this sequence (Fig. 3a) reveals the presence of a 144-bp perfect inverted repeat, probably responsible for the failure to maintain these sequences in *E. coli* plasmids. A fragment cloned in pAT153 and lacking 60 bp from the center of the palindrome (Quinn 1984 and Fig. 3a) did not specify a functional $ori_L$ (N. Stow, unpubl.), in agreement with the results of Weller et al. (1985).

Figure 3a shows that the $ori_S$ and $ori_L$ sequences can be aligned to reveal a high degree of homology (85% identical residues) throughout the $ori_S$ palindrome and extending approximately 40 bp to one side of it. A further interesting similarity between the two origins is that $ori_S$ is flanked by divergently transcribed immediate early mRNAs and $ori_L$ is flanked by divergently transcribed early mRNAs (Fig. 3b).

### Encapsidation of amplified plasmid DNAs

To identify *cis*-acting sequences necessary for encapsidation, two assays were employed (Stow et al. 1983). Packaging of DNA into virions renders it resistant to digestion by exogenously added DNase (Gibson and Roizman 1972; Vlazny et al. 1982), and this property provided the basis for a direct test. An alternative, indirect assay was based upon the fact that in the absence of specific facilitating agents, unencapsidated viral DNA is noninfectious (Graham et al. 1973; Sheldrick et al. 1973; Farber 1976), and consequently serial passage of viral DNA in tissue-culture cells is dependent upon its packaging into virus particles.

Figure 4a shows that although plasmid pS1 was replicated in the presence of wild-type HSV-1 helper virus, the vector sequences were not maintained when a virus stock derived from such cells was used in a high-multiplicity infection of fresh BHK cells. Similarly, DNase-resistant DNA isolated from cells transfected with pS1 and subsequently superinfected with wild-type

HSV-1 did not hybridize to the labeled vector DNA probe (Fig. 4b). These results therefore indicate the absence of *cis*-acting encapsidation signals from pS1.

Because studies on defective genomes implicated DNA from close to the S terminus as being essential for the maturation of replicated viral DNA, additional fragments from both genomic termini were inserted into pS1. Plasmids pT071 and pT085 each contain an approximately 550-bp fragment from the S terminus, and pT091 contains a similarly sized fragment from the L terminus (Fig. 1b). When tested as described above, it was found that amplified DNA sequences of all three plasmids could be detected after passaging of virus progeny (Fig. 4a) and were also protected from the action of DNase (Fig. 4b). Only the "a" sequence is common to the three terminal fragments, suggesting that signals that enable encapsidation of HSV-1 DNA reside within this region.

The nucleotide sequences for the "a" sequence of several strains of HSV-1 have been determined (Davison and Wilkie 1981; Mocarski and Roizman 1981; Mocarski et al. 1985), and the structures of the genomic termini have been defined. At the joint between the L and S segments, the "a" sequence is flanked by a short direct repeat (DR1) of 17–21 bp. The structures of the genomic termini suggest that they result from staggered cleavage toward one side of DR1 that leaves a single-nucleotide 3' overhang (Mocarski and Roizman 1982). The position of this cleavage in HSV-1 strain 17 is shown in Figure 4c. In the case of terminal fragments "tailed" with dC residues prior to cloning, a complete 21-bp DR1 is regenerated at the L terminus, whereas only one nucleotide of DR1 is present at the S terminus (Davison and Rixon 1985). The observation that fragments from the S terminus enable packaging therefore indicates that two complete copies of DR1 are not essential for encapsidation.

The above conclusion raises the possibility that the single DR1 within a cloned S-terminal fragment might function as a cleavage site to generate both termini of encapsidated molecules. If this were the case, these molecules, in contrast to the standard HSV-1 genome, would contain an "a" sequence at only one terminus. The sizes of the terminal fragments produced after replication and encapsidation of plasmid pT011 DNA (which contains a cloned S-terminal fragment and is essentially similar to pT071) were therefore analyzed as illustrated in Figure 5a. It can be seen that both terminal fragments comigrate with marker fragments that include an "a" sequence (Fig. 5b). In addition, hybridization with a labeled probe containing only the HSV-1 "a" sequence verified the presence of DNA from the "a" sequence in both terminal fragments (Fig. 5c).

Further evidence suggestive of the presence of an "a" sequence at each terminus of the packaged molecules was provided by examining the sizes of DNase-resistant molecules accumulating in the nuclei. Vlazny et al. (1982) previously showed that although only capsids containing defective genomes approximately the size of standard virion DNA were present in the cytoplasm, defective DNA from nuclear capsids consisted of molecules comprising various numbers, from one up, of tandem repeats of the basic monomeric unit. Figure 5d, lane 8, demonstrates that DNase-resistant DNA from the nuclei of cells that had received pTO11 occurred in discrete size classes consisting of one or more copies of the input plasmid. Comparison of the size of the smallest species revealed that it was slightly larger than linearized pTO11 DNA, consistent with the presence of an "a" sequence at each terminus. In contrast, monomeric units generated from within packaged genomes comigrated with the linearized plasmid marker (lanes 6 and 7), indicating that the "a" sequence was not duplicated internally.

## Use of *ori*$_S$ and the "a" sequence in a helper-dependent viral vector

The experiments described above demonstrate that the combined presence of two *cis*-acting signals, namely *ori*$_S$ and an "a" sequence, enables serial propagation of linked foreign DNA sequences when helper functions are provided. Plasmids containing both signals are maintained in virus stocks as linear, defective genomes comprising tandem reiterations of the entire molecule, suggesting their use in the development of defective HSV vectors for the expression of foreign genes.

To examine this possibility, the bacterial CAT-coding sequences were inserted into a modified pTO91 either in the absence of a promoter or under the control of HSV-1

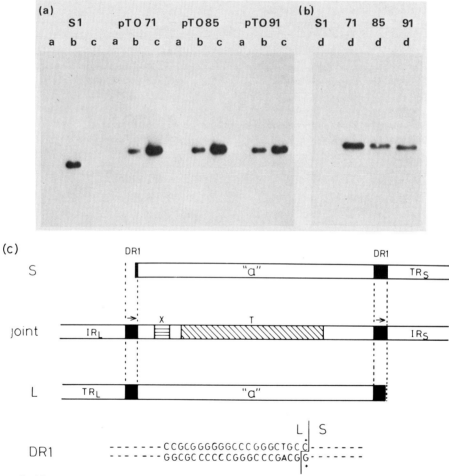

**Figure 4** HSV-1 terminal fragments contain encapsidation signals. (*a*) Serial propagation of DNAs. Total cellular DNA was cleaved with *Eco*RI and subjected to electrophoresis through an agarose gel. The gel was blotted and hybridized to nick-translated pAT153 DNA. The plasmids used are indicated: Lanes labeled *a* and *b* are DNA from transfected cells that were then mock-infected or superinfected, respectively; lanes labeled *c* are DNA from cells infected with virus stocks obtained from monolayers that had received both the indicated plasmid and HSV-1. (*b*) DNase-resistant DNA (lanes labeled *d*) was prepared from cells transfected with the plasmids indicated and superinfected with wild-type HSV-1. Samples were analyzed as described above. 71, 85, and 91 refer to the respective pTO plasmids. (*a* and *b* are adapted from Stow et al. 1983.) (*c*) Structure of the HSV-1 (strain 17) termini and joint (Davison and Wilkie 1981). DR1 is a direct repeat of 21 bp flanking the "a" sequence in the joint region. T represents a region comprising 18 copies of a 12-bp sequence, and × is the highly conserved sequence found near a terminus of several herpesvirus genomes. Cleavage of concatemers produces termini with a single-nucleotide 3' overhang such that direct ligation of the two ends generates precisely a single copy of DR1 (Mocarski and Roizman 1982). Terminal fragments used in the above experiments were tailed with dC residues, resulting in the regeneration of a complete DR1 at the L terminus and the presence of 1 bp of DR1 at the S terminus (Stow et al. 1983; Davison and Rixon 1985). The nucleotide pair marked + is therefore present at both cloned termini.

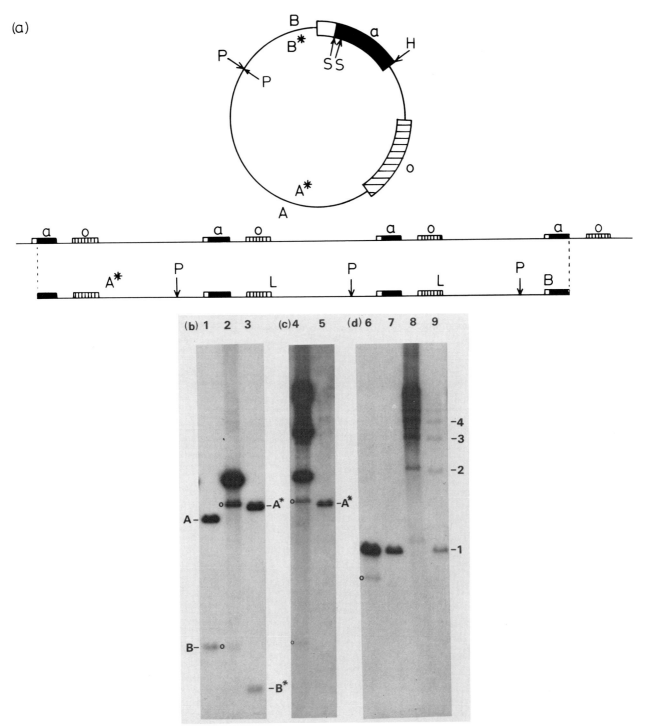

**Figure 5** Encapsidated molecules contain an "a" sequence at both termini. (a) Structure of pTO11 indicating the positions of *Pst*I (P), *Sst*II (S), *Hind*III (H) sites, and the *ori*$_S$-containing fragment (o). *Pst*I plus *Hind*III fragment A lacks an "a" sequence, whereas *Pst*I plus *Sst*II fragment A* contains an "a" sequence truncated by 100 bp. Similarly, fragments B and B*, respectively, contain or lack an entire "a" sequence. Replication by a rolling-circle mechanism generates concatemers of the structure shown, and cleavages resulting in the presence of an "a" sequence at each end of packaged molecules would produce *Pst*I terminal fragments, one 100 bp larger than A* and the other the same size as B. (b) (Lane *1*) pTO11 cleaved with *Pst*I and *Hind*III; (lane *2*) DNase-resistant DNA obtained from cells transfected with pTO11 and superinfected with HSV-1 cleaved with *Pst*I; (lane *3*) pTO11 cleaved with *Pst*I and *Sst*II. The fragments were analyzed by hybridization to pAT153 DNA and terminal fragments are indicated by o. (c) Lanes *4* and *5* contain the same samples as lanes *2* and *3*, but the pAT153 probe was removed in H$_2$O at 100°C and the filter was rehybridized with a probe containing only the "a" sequence. Note that this probe does not detect fragment B*. (d) (Lane *6*) The same as lane *2*; (lane *7*) pTO11 cleaved with *Eco*RI; (lane *8*) uncleaved DNase-resistant DNA from the nuclei of cells transfected with pTO11 and superinfected with HSV-1; (lane *9*) ladder of fragments produced by ligation of *Eco*RI-cleaved pTO11. Bands containing 1, 2, 3, and 4 copies of the plasmid are indicated. The samples were analyzed on a 0.5% gel, and the probe was pAT153.

promoters for immediate-early gene 3 (IE3), the viral thymidine kinase (TK) gene, or the major capsid protein (VP5) gene. In each instance a small HSV-2 fragment specifying transcription termination and polyadenylation was present downstream of the coding region (Fig. 6a).

Gene expression during HSV-1 infection is temporally regulated (for review, see Wagner 1985). Immediate-early genes are transcribed in the absence of viral protein synthesis, and their products are required for the expression of the remaining viral genes (Clements et al. 1977; Anderson et al. 1980). Temperature-sensitive mutants containing lesions in the IE3 gene (e.g., *tsk*) are blocked at the immediate-early stage of infection and overproduce the products of these genes at the nonpermissive temperature (Preston 1981). Early genes, which include the viral TK, are expressed at near maximum levels in the absence of viral DNA synthesis, whereas reduced or undetectable expression of "leaky" late (e.g., VP5) or true late genes occurs under these conditions (Wagner 1985).

Virus stocks containing defective genomes generated from the above plasmids were produced using wild-type HSV-1 or *tsK* as helper as described in Materials and Methods. Induction of CAT activity was measured after the infection of BHK cells with these stocks.

As expected, no acetylation of chloramphenicol occurred in the absence of a promoter fragment upstream of the CAT-coding region. With wild-type HSV-1 helper, CAT activity was detected with each of the three remaining plasmids. The levels obtained with the TK and VP5 promoters were substantially greater than with the IE3 gene promoter, reflecting the relative levels of accumulation of these gene products during HSV-1 infection. When stocks in which *tsK* was helper for the plasmid containing the IE3 gene promoter were assayed at the permissive and nonpermissive temperatures, greatly increased production of CAT activity was observed at the higher temperature (Fig. 6b).

These results therefore demonstrate that plasmid pTO91 can be used as a vector for the bacterial CAT gene in BHK cells and that regulation of the synthesis of the bacterial enzyme can be achieved through the use of different viral promoters and helper viruses.

(a)

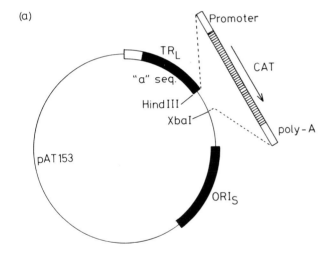

(b)

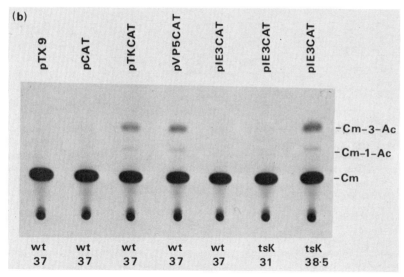

**Figure 6** Expression of CAT using defective HSV-1 vectors. (*a*) Structure of plasmid pTX9 in which the *Eco*RV site of pTO91 has been converted to *Xba*I. The position of insertion of CAT-containing fragments is shown. pCAT contains no promoter fragment. pTKCAT, pVP5CAT, and pIE3CAT contain, respectively, the *Bam*HI/*Bgl*II promoter fragment of the HSV-1 thymidine kinase gene (Shih et al. 1984), the *Bam*HI f' promoter fragment from the gene of the HSV-1 major capsid protein, VP5 (Costa et al. 1984), and the *Bam*HI/*Hind*III promoter fragment of the HSV-1 IE3 gene from plasmid pGX161 (Preston and Fisher 1984). (*b*) Virus stocks containing defective genomes generated using the indicated plasmids were assayed for their ability to induce the synthesis of CAT in infected cells. The helper virus was wild-type HSV-1 or *tsK* and extracts were prepared 12 hr p.i. from cells incubated at the temperatures indicated at bottom. The positions of chloramphenicol (Cm) and its monoacetylated products are shown. A longer autoradiographic exposure more clearly revealed the presence of low amounts of acetylated products in the pIE3CAT 37°C and 31°C lanes.

## Discussion

### HSV-1 origins of DNA replication

Although the sequences comprising HSV-1 $ori_S$ and $ori_L$ have now been characterized and localized on the viral genome, many questions remain to be answered concerning their activity. Defective genomes generated after serial, high-multiplicity passage of virus stocks are clearly capable of utilizing either $ori_S$ or $ori_L$ for their replication. The standard genome, however, contains two copies of $ori_S$ and one of $ori_L$, and it is not yet known whether the presence of all three origins is essential. Although the striking sequence homology between $ori_S$ and $ori_L$ is indicative of related function, the observation that the sequences are not identical suggests that there may be important operational differences. These might include the time during infection at which the two origins are active, the relative frequency at which initiation occurs, or the directionality of function. A further, intriguing observation is that in some virus strains the palindromic region of $ori_S$ or $ori_L$ may be duplicated, as has been reported for $ori_L$ of HSV-1 strain Angelloti (Gray and Kaerner 1984) and $ori_S$ of HSV-2 strain HG52 (Whitton and Clements 1984). It is not known whether this sequence arrangement has any effect on origin function.

It is interesting to note that $ori_S$ and $ori_L$ are flanked by divergently transcribed mRNAs, both belonging to the immediate-early and early gene classes, respectively (Fig. 3; Wagner 1985), and that $ori_S$ appears to be embedded between the promoter region and upstream regulatory sequences for the IE4 (and IE5) gene (Preston et al. 1984). These observations suggest there may be a relationship between gene expression and initiation of DNA synthesis at these loci that remains to be elucidated.

### HSV-1 encapsidation signals

The signals required for encapsidation of HSV-1 DNA reside within the approximately 400-bp terminal redundancy or "a" sequence. Also involved in packaging of viral DNA are the closely associated cleavage events that generate, from long concatemers, unit-length genomes containing an "a" sequence at each terminus. The precise region(s) of the "a" sequence that specifies the cleavages has not been defined. The observation that cleavage can occur at both ends of an "a" sequence flanked by only one complete copy of DR1 (Fig. 5 and Mocarski et al. 1985) suggests that the signal is probably not DR1 itself. Evidence has recently been presented that suggests that the genomic termini are generated by cleavage events that occur at fixed distances from signals located within and close to the opposite ends of an "a" sequence (Varmuza and Smiley 1985). It is therefore interesting to note the presence of a short, highly conserved sequence approximately 30 bp from one terminus of a number of different herpesvirus genomes (Davison and Rixon 1985; Spaete and Mocarski 1985).

Replication of plasmids like pTO71 and pTO11, prob-

ably by a rolling-circle mechanism, results in the generation of concatemers in which single copies of the "a" sequence are present as direct repeats at intervals corresponding to the length of the input plasmid. The cleavage of concatemers into molecules containing an "a" sequence at each terminus therefore poses a problem. One mechanism for regenerating an "a" sequence on either side of a cutting site would involve a staggered cleavage across the "a" sequence followed by repair synthesis. The alternative of a double-stranded cut would result in a free end lacking an "a" sequence (Fig. 5a), and cleavage and packaging of the concatemer might simply resume from the next available "a" sequence with the concomitant production of an unpackageable DNA fragment lacking terminal "a" sequences. An additional possibility to allow packaging to continue from these free ends might involve the restoration of the "a" sequence through recombination events. Current data do not allow discrimination between these models.

The presence in the nucleus of short, encapsidated, defective DNA molecules indicates that HSV-1 encapsidation does not take place by a simple "headful" mechanism and that cleavage events can be separated by a variety of distances if they occur between direct repeats of the "a" sequence (Vlazny et al. 1982; Fig. 5d). In the case of concatemers of the standard virus genome, it should be noted that the distance between direct repeats of the "a" sequence corresponds to the genomic unit length. Although encapsidation itself does not appear to operate by a "headful" mechanism, Vlazny et al. (1982) have demonstrated that only those capsids containing an approximate genome equivalent of DNA can become enveloped and translocated to the cytoplasm as mature virions.

### HSV-1 defective vectors

We have demonstrated that the HSV-1 $ori_S$ and encapsidation signals enable the propagation and expression of the bacterial CAT gene in the presence of an appropriate helper virus. Expression of CAT in the absence of helper virus replication was possible through the use of the IE3 gene promoter in combination with tsK helper virus at the nonpermissive temperature.

Plasmids such as pTX9, which contains approximately 1 kb of HSV-1 DNA specifying $ori_S$ and encapsidation functions, should provide useful starting points for the expression of foreign genes by helper-dependent vectors by enabling the initial constructs to be cloned and propagated in E. coli and subsequently introduced into tissue-culture cells by transfection. After superinfection with helper virus, plasmid sequences are maintained in the form of defective genomes, which accumulate during serial, high-multiplicity passage, facilitating the introduction of high copy numbers of cloned genes into cells. The overproduction of several virus polypeptides encoded by HSV-1 defective genomes (Frenkel et al. 1980) suggests that it should be possible to obtain high-level expression of foreign gene products using this approach.

Nondefective HSV-1 has also recently been successfully used as a vector for the expression of a foreign gene (hepatitis B virus surface antigen; Shih et al. 1984), and we are currently comparing the efficiency of expression of the bacterial CAT gene in defective and nondefective vectors. The large genome size and wide host range of HSV-1 make it an attractive system for the introduction of a variety of DNA sequences into different mammalian cell types. Nevertheless, the recent report by Kwong and Frenkel (1984) that constructed defective genomes with repeat units larger than approximately 15 kb suffered deletions when serially propagated illustrates that many aspects of the replication and packaging of HSV-1 DNA still remain to be unraveled, and hopefully the lessons learned from future experiments will prove useful in the design of improved vectors.

## Acknowledgments

We thank Professor J.H. Subak-Sharpe for his continued interest and support, Dr. A.J. Davison for helpful discussions and the gift of cloned terminal fragments, and Dr. J.L. Whitton for pLW2. M.D.M. is the recipient of a Medical Research Council Research Scholarship.

## References

Anderson, K.P., R.H. Costa, L.E. Holland, and E.K. Wagner. 1980. Characterization of herpes simplex virus type 1 RNA present in the absence of de novo protein synthesis. *J. Virol.* **34:** 9.

Clements, J.B., R.J. Watson, and N.M. Wilkie. 1977. Temporal regulation of herpes simplex virus type 1 transcription: Location of transcripts on the viral genome. *Cell* **12:** 275.

Costa, R.H., G. Cohen, R. Eisenberg, D. Long, and E. Wagner. 1984. Direct demonstration that the abundant 6-kilobase herpes simplex virus type 1 mRNA mapping between 0.23 and 0.27 map units encodes the major capsid protein VP5. *J. Virol.* **49:** 287.

Cuifo, D.M. and G.S. Hayward. 1981. Tandem repeat defective DNA from the L segment of the HSV genomes. In *Herpesvirus DNA* (ed. Y. Becker), p. 107. Martinus Nijhoff, The Netherlands.

Davison, A. and F. Rixon. 1985. Cloning of the DNA of alphaherpesvirinae. In *Recombinant DNA research and viruses* (ed. Y. Becker), p. 103. Martinus Nijhoff, Boston.

Davison, A.J. and N.M. Wilkie. 1981. Nucleotide sequences of the joint between the L and S segments of herpes simplex virus types 1 and 2. *J. Gen. Virol.* **55:** 315.

Denniston, K.J., M.J. Madden, L.W. Enquist, and G. Vande Woude. 1981. Characterization of coliphage lambda hybrids carrying DNA fragments from herpes simplex virus type 1 defective interfering particles. *Gene* **15:** 365.

Farber, F.E. 1976. Comparison of DNA facilitators in the uptake and intracellular fate of infectious herpes simplex virus type 2 DNA. *Biochim. Biophys. Acta* **454:** 410.

Frenkel, N., H. Locker, and D.A. Vlazny. 1980. Studies of defective herpes simplex viruses. *Ann. N.Y. Acad. Sci.* **354:** 347.

Friedmann, A., J. Shlomai, and Y. Becker. 1977. Electron microscopy of herpes simplex virus DNA molecules isolated from infected cells by centrifugation in CsCl gradients. *J. Gen. Virol.* **34:** 507.

Gibson, W. and B. Roizman. 1972. Proteins specified by herpes simplex virus. VIII. Characterization and composition of multiple capsid forms of subtypes 1 and 2. *J. Virol.* **10:** 1044.

Graham, F.L. and A.J. van der Eb. 1973. A new technique for the assay of infectivity of human adenovirus 5 DNA. *Virology* **52:** 456.

Graham, F.L., G. Veldhuisen, and N.M. Wilkie. 1973. Infectious herpesvirus DNA. *Nat. New Biol.* **245:** 265.

Gray, C.P. and H.C. Kaerner. 1984. Sequence of the putative origin of replication in the $U_L$ region of herpes simplex virus type 1 ANG DNA. *J. Gen. Virol.* **65:** 2109.

Hanahan, D. 1983. Studies on transformation of *Escherichia coli* with plasmids. *J. Mol. Biol.* **166:** 557.

Jacob, R.J., L.S. Morse, and B. Roizman. 1979. Anatomy of herpes simplex virus DNA. XII. Accumulation of head-to-tail concatemers in nuclei of infected cells and their role in the generation of the four isomeric arrangements of viral DNA. *J. Virol.* **29:** 448.

Jongeneel, C.V. and S.L. Bachenheimer. 1981. Structure of replicating herpes simplex virus DNA. *J. Virol.* **39:** 656.

Kaerner, H.C., I.B. Maichle, A. Ott, and C.H. Schröder. 1979. Origin of two different classes of defective HSV-1 Angelotti DNA. *Nucleic Acids Res.* **6:** 1467.

Kwong, A.D. and N. Frenkel. 1984. Herpes simplex virus amplicon: Effect of size on replication of constructed defective genomes containing eucaryotic DNA sequences. *J. Virol.* **51:** 595.

Lacks, S. and B. Greenberg. 1977. Complementary specificity of restriction endonucleases of *Diplococcus pneumoniae* with respect to DNA methylation. *J. Mol. Biol.* **114:** 153.

Ladin, B.F., M.L. Blankenship, and T. Ben-Porat. 1980. Replication of herpesvirus DNA. V. Maturation of concatemeric DNA of pseudorabies virus to genome length is related to capsid formation. *J. Virol.* **33:** 1151.

Mackett, M., G.L. Smith, and B. Moss. 1984. General method for production and selection of infectious vaccinia virus recombinants expressing foreign genes. *J. Virol.* **49:** 857.

Macpherson, I. and M. Stoker. 1962. Polyoma transformation of hamster cell clones—An investigation of genetic factors affecting cell competence. *Virology* **16:** 147.

Marsden, H.S., I.K. Crombie, and J.H. Subak-Sharpe. 1976. Control of protein synthesis in herpesvirus-infected cells: Analysis of the polypeptides induced by wild type and sixteen temperature-sensitive mutants of HSV strain 17. *J. Gen. Virol.* **31:** 347.

McLauchlan, J., D. Gaffney, J.L. Whitton, and J.B. Clements. 1985. The consensus sequence YGTGTTYY located downstream from the AATAAA signal is required for efficient formation of mRNA 3′ termini. *Nucleic Acids Res.* **13:** 1347.

Mocarski, E.S. and B. Roizman. 1981. Site-specific inversion sequence of the herpes simplex virus genome: Domain and structural features. *Proc. Natl. Acad. Sci.* **78:** 7047.

———. 1982. Structure and role of the herpes simplex virus DNA termini in inversion, circularization and generation of virion DNA. *Cell* **31:** 89.

Mocarski, E.S., L.P. Deiss, and N. Frenkel. 1985. Nucleotide sequence and structural features of a novel $U_S$-a junction present in a defective herpes simplex virus genome. *J. Virol.* **55:** 140.

Poffenberger, K.L. and B. Roizman. 1985. A noninverting genome of a viable herpes simplex virus 1: Presence of head-to-tail linkages in packaged genomes and requirements for circularization after infection. *J. Virol.* **53:** 587.

Preston, C.M., M.G. Cordingley, and N.D. Stow. 1984. Analysis of DNA sequences which regulate the transcription of a herpes simplex virus immediate early gene. *J. Virol.* **50:** 708.

Preston, V.G. 1981. Fine-structure mapping of herpes simplex virus type 1 temperature-sensitive mutations within the short repeat region of the genome. *J. Virol.* **39:** 150.

Preston, V.G. and F.B. Fisher. 1984. Identification of the

herpes simplex virus type 1 gene encoding the dUTPase. *Virology* **138**: 58.

Quinn, J.P. 1984. "Sequence of the DNA polymerase locus in the genome of herpes simplex virus type 1." Ph.D. thesis, University of Glasgow, Scotland.

Roizman, B. 1979. The structure and isomerization of herpes simplex virus genomes. *Cell* **16**: 481.

Sheldrick, P., M. Laithier, D. Lando, and M.L. Ryhiner. 1973. Infectious DNA from herpes simplex virus: Infectivity of double-stranded and single-stranded molecules. *Proc. Natl. Acad. Sci.* **70**: 3621.

Shih, M.-F., M. Arsenakis, P. Tiollais, and B. Roizman. 1984. Expression of hepatitis B virus S gene by herpes simplex virus type 1 vectors carrying alpha- and beta-regulated gene chimeras. *Proc. Natl. Acad. Sci.* **81**: 5867.

Southern, E.M. 1975. Detection of specific sequences among DNA fragments separated by gel electrophoresis. *J. Mol. Biol.* **98**: 503.

Spaete, R.R. and N. Frenkel. 1982. The herpes simplex virus amplicon: A new eucaryotic defective-virus cloning-amplifying vector. *Cell* **30**: 295.

Spaete, R.R. and E.S. Mocarski. 1985. The *a* sequence of the cytomegalovirus genome functions as a cleavage/packaging signal for herpes simplex defective genomes. *J. Virol.* **54**: 817.

Stow, N.D. 1982. Localization of an origin of DNA replication within the TR$_S$/IR$_S$ repeated region of the herpes simplex virus type 1 genome. *EMBO J.* **1**: 863.

———. 1985. Mutagenesis of a herpes simplex virus origin of DNA replication and its effect on viral interference. *J. Gen. Virol.* **66**: 31.

Stow, N.D. and E.C. McMonagle. 1983. Characterization of the TR$_S$/IR$_S$ origin of DNA replication of herpes simplex virus type 1. *Virology* **130**: 427.

Stow, N.D. and N.M. Wilkie. 1976. An improved technique for obtaining enhanced infectivity with herpes simplex virus type 1 DNA. *J. Gen. Virol.* **33**: 447.

Stow, N.D., E.C. McMonagle, and A.J. Davison. 1983. Fragments from both termini of the herpes simplex virus type 1 genome contain signals required for the encapsidation of viral DNA. *Nucleic Acids Res.* **11**: 8205.

Varmuza, S.L. and J.R. Smiley. 1985. Signals for site-specific cleavage of herpes simplex virus DNA: Maturation involves two separate cleavage events at sites distal to the recognition sequences. *Cell* **41**: 793.

Vlazny, D.A. and N. Frenkel. 1981. Replication of herpes simplex virus DNA: Localization of replication recognition signals within defective virus genomes. *Proc. Natl. Acad. Sci.* **78**: 742.

Vlazny, D.A., A. Kwong, and N. Frenkel. 1982. Site-specific cleavage/packaging of herpes simplex virus DNA and the selective maturation of nucleocapsids containing full-length viral DNA. *Proc. Natl. Acad. Sci.* **79**: 1423.

Wagner, E.K. 1985. Individual HSV transcripts: Characterization of specific genes. In *The herpesviruses* (ed. B. Roizman), vol. 3, p. 45. Plenum Press, New York.

Weller, S.K., K.J. Lee, D.J. Sabourin, and P.A. Schaffer. 1983. Genetic analysis of temperature-sensitive mutants which define the gene for the major herpes simplex virus type 1 DNA binding protein. *J. Virol.* **45**: 354.

Weller, S.K., A. Spadaro, J.E. Schaffer, A.W. Murray, A.M. Maxam, and P.A. Schaffer. 1985. Cloning, sequencing, and functional analysis of *ori*$_L$, a herpes simplex virus type 1 origin of DNA synthesis. *Mol. Cell. Biol.* **5**: 930.

Whitton, J.L. and J.B. Clements. 1984. Replication origins and a sequence involved in the coordinate induction of the immediate-early gene family are conserved in an intergenic region of herpes simplex virus. *Nucleic Acids Res.* **12**: 2061.

# Multiple Pathways for Simian Virus 40 Nonhomologous Recombination

E. Winocour, T. Chitlaru, K. Tsutsui, R. Ben-Levy, and Y. Shaul

Department of Virology, The Weizmann Institute of Science, Rehovot 76100, Israel

With the development of systems for monitoring nonhomologous recombination between the simian virus 40 (SV40) genome and defined substrates in monkey cells, it has become apparent that the types of recombinant products—and the recombination pathways from which they arise—are more varied than expected. We focus in this article on the data leading to the conclusion that SV40 recombination can proceed via at least two distinct pathways: one independent of DNA replication and one associated with a replication process. In addition, we present and discuss the nucleotide sequences across 16 recombinant junctions derived from the products of recombination between SV40 and $\phi$X174 RFI DNAs, SV40 and adeno-associated virus, and polyoma DNA and the DNA of various plasmid (SV40 origin-containing) constructs.

To gain a better understanding of the mechanisms of nonhomologous recombination in animal cells, we developed a protocol based upon the observation that cotransfection (DEAE-Dextran mediated) of simian virus 40 (SV40) and unrelated DNAs into monkey cells gives rise to SV40 hybrids detectable by infectious-center in situ plaque hybridization (Winocour and Keshet 1980; Dorsett et al. 1983). Using this assay to monitor recombination between circular SV40 and $\phi$X174 RFI DNAs in monkey cells, we have shown that the frequency of recombinant-producing cells (expressed as a fraction of the number of SV40-replicating cells) depends upon the concentration of SV40 DNA in the transfection mixture and that replication-defective SV40 mutants compete with the wild type for recombination with $\phi$X174 DNA (Dorsett et al. 1983). Both of these observations argue that a limiting stage in the recombination pathway pursued by circular SV40 DNA occurs before the onset of DNA replication. For example, if the limiting stage were to occur after the onset of SV40 DNA replication, the intracellular concentration of the replication-defective DNA would be insignificant compared with that of the replication-competent wild-type SV40 DNA, and effective competition between the mutant and wild-type DNA, for recombination with $\phi$X174 DNA, would be unlikely under such conditions. By similar reasoning, the dependency of the number of recombinant-producing cells (among the virion-producing cells) upon the initial input concentration of SV40 DNA would not be expected if recombination occurred after the onset of massive SV40 DNA replication. The results of the analyses of the SV40/$\phi$X174 recombinant structures, generated by cotransfection with circular DNAs, are consistent with the formation of recombinants by a nonreplicative process. In most cases, the recombinant structures were of a simple deletion/insertion type, in which a portion of the SV40 genome was replaced by a single, slightly shorter insert of randomly derived $\phi$X174 DNA. Other than the region of the substitution, the remaining sectors

of the recombinant genome were identical with those of wild-type SV40, as judged by heteroduplex mapping, restriction analysis, and complementation tests (Winocour et al. 1983). The recombinant junctions (see below) are not flanked by the target sequence duplication characteristic of prokaryotic transposition events (for review, see Kleckner 1981) or retrovirus insertions (for review, see Varmus 1983).

The evidence that recombination between SV40 DNA and $\phi$X174 DNA can also proceed via a different pathway, of a replicative nature, stems from the observation that linear origin-containing ($ori^+$) forms of SV40 DNA, when added to transfection mixtures containing circular SV40 and $\phi$X174 RFI DNAs, enhanced the recombination frequency threefold to fivefold (Dorsett et al. 1985). No such enhancement was observed when the transfection mixture was supplemented with an equivalent dose of circular SV40 DNA or linear $\phi$X174 RFI DNA. Several lines of evidence indicate that the $ori^+$ linear forms of SV40 DNA promote a recombination pathway that is associated with a replication process. In contrast to the simple deletion/insertion structures of SV40/$\phi$X174 recombinants arising from the transfection of circular DNA substrates alone, the structures that arose from transfections supplemented with the linear $ori^+$ SV40 DNA additions were highly complex. Their most striking feature was the interspersion of $\phi$X174 DNA segments within tandem head-to-tail repeats specifically derived from the $ori^+$-enhancing linear DNA fragment. The head-to-tail repeat units were not formed by homologous recombination. Although circular SV40 DNA normally replicates bidirectionally from a unique origin, yielding $\phi$-type intermediates (for review, see Acheson 1980), we have found that the recombination-enhancing linear $ori^+$ forms of SV40 DNA exhibit a strong preference for the rolling-circle mode of DNA replication (Deichaite et al. 1985). This replication mode gives rise to intermediates comprising a circular domain (the "rolling template") and a tail of newly synthesized, tandemly repeated DNA (Gilbert and Dressler 1969). We have proposed that φX174 RFI DNA recombines with the circular template portion of the rolling-circle intermediate such that continuation of the replication process generates tandem repeats containing both φX174 and SV40 *ori*⁺ DNA (Dorsett et al. 1985). We assume that the enhancement of recombination mediated by linear SV40 *ori*⁺ DNA arises because rolling-circle intermediates are highly recombinogenic.

The tendency of linear SV40 DNA to undergo amplification by a rolling-circle mode of replication may also explain some features of integrative chromosomal recombination. The structure of chromosomally integrated SV40 DNA (Botchan et al. 1976) and polyoma DNA (Della-Valle et al. 1981) frequently exhibits head-to-tail repeats. Linearization of SV40 DNA enhances the frequency of cell transformation in culture (Z. Grossman and E. Winocour, unpubl.), and linearization of polyoma DNA enhances its oncogenic potential in animals (Bouchard et al. 1984). These observations are consistent with the notion that linearization of papovavirus DNA promotes an active replicative recombination pathway associated with the formation of rolling-circle replication intermediates. Conceivably, the SV40 host-substituted variants (Lavi and Winocour 1972; Winocour et al. 1975) whose structure is characterized by tandem repeats of cellular and SV40 *ori*⁺ DNA (Oren et al. 1978) may arise by a similar mechanism.

Recombination studies between SV40 and adeno-associated virus (AAV) have also revealed that two distinct classes of recombinant products can arise (Grossman et al. 1984, 1985). In this system, the recombining substrates can be introduced into the monkey cell either by DNA cotransfection or by virus coinfection. In terms of the dosage-response data for the frequency of recombination, the results obtained by using both methods of DNA delivery were quite comparable. In one respect, however, the recombination process after DNA cotransfection differed from that after covirion infection. DNA cotransfection gave rise to simple insertion/deletion recombinant structures containing single inserts of AAV DNA (derived from any part of the AAV genome) similar to those arising from DNA cotransfection with circular SV40 and φX174 RFI DNA. In contrast, most of the recombinants arising from SV40 and AAV covirion infection exhibited a tandem repeat structure composed of the AAV terminal sequences and SV40 origin-region sequences. The AAV genome in the virion is a linear, single-stranded DNA molecule that contains an inverted terminal repetition of 145 bases, the first 125 of which are palindromic and capable of forming a hairpin structure (for review, see Berns and Hauswirth 1982). Nucleotide sequence analysis of the recombinant junctions in two SV40/AAV recombinants that arose after covirion infection has shown that the crossover points are closely allied with the palindromic parts of the AAV inverted terminal repeats. The joining with SV40 DNA occurred within a few hundred nucleotides of the SV40 replication origin (Grossman et al. 1985; see also Fig. 2). The AAV terminal sequences are known to be important for repli-

cation (Berns and Hauswirth 1982), integrative recombination (Cheung et al. 1980), and excision/replication from plasmid vectors (Samulski et al. 1983). Why the reiterated recombinant structures containing the AAV terminal repeats occur only after SV40 and AAV covirion infection (and not after DNA cotransfection) remains an open question. However, it should be noted that the DNA cotransfections, which gave rise to the simple insertion/deletion structures, were carried out with duplex AAV DNA formed by annealing in vitro the complementary strands isolated from a population of AAV virus particles (plus and minus AAV strands are encapsidated separately), whereas in the virion infection the only segment of the AAV genome capable of duplex formation is the terminal portion. It is possible that the AAV terminal repeat, in the conformation in which it is released from the virion, is predisposed to recombine with the SV40 origin region, triggering a rolling-circle mode of DNA replication that gives rise to the tandemly repeated structure characteristic of this class of recombinant products. Whatever the reason, it is clear that recombination between SV40 and AAV generates two distinctly different classes of products—depending upon the way substrates are delivered to the cell—which presumably arise from quite different recombinatorial processes.

The SV40/φX174 and SV40/AAV recombination studies reviewed above were carried out using a SV40 helper–dependent in situ plaque hybridization assay, which measures the number of recombinant-producing cells. This procedure scores only those recombinants that retain the SV40 origin of replication, are of a size commensurate with SV40 encapsidation restrictions, and are capable of undergoing the multiple cycles of helper-dependent infection required for plaque development. The restriction on size imposed by the SV40 packaging process is severe; recombinant genomes larger than 6000 bp cannot be detected. Furthermore, because the recombinant plaque contains a large excess of the wild-type SV40 helper genome, laborious cloning procedures are required to isolate the recombinants for structural analyses. To circumvent these restrictions, we have devised another type of assay in which the recombinant products, generated in monkey cells, are amplified and identified by bacterial transformation procedures (D. Vernade et al., in prep.). We describe below recent results, obtained with this assay, on plasmid recombination in monkey COS cells. In addition, we describe the nucleotide sequences across the recombinant junctions in a variety of different recombinant structures, isolated either by infectious-center in situ plaque hybridization or by bacterial transformation.

## Methods

### Plasmids

pSVK₁H is a pBR322-derived shuttle vector that contains the SV40 replication/regulatory region (nucleotides [nt] 5171–5243/0–296), an ampicillin-resistance (Apʳ) gene, and an *Escherichia coli* replicon

(Dorsett et al. 1985). The *Bacillus subtilis* plasmid pHV660, which was constructed and kindly provided by M. Michel and S.D. Ehrlich, is a derivative of pUB110 and contains two copies of the kanamycin-resistance (Km$^r$) gene (Michel and Ehrlich 1984). pHV660 does not replicate in *E. coli* cells.

*Recombinant isolates*
Recombinants were identified and isolated either by infectious-center in situ plaque hybridization (and cloned as described in Winocour et al. 1983; Grossman et al. 1985) or by the bacterial transformation procedures outlined in Figure 1, details of which are given elsewhere (D. Vernade et al., in prep.).

*Nucleotide sequencing*
The recombinant junction fragments were sequenced either according to Maxam and Gilbert (1980), or by the dideoxynucleotide-chain-termination method (Sanger et al. 1977) described by Smith et al. (1979), using synthetic oligonucleotide primers prepared by the unit for oligonucleotide synthesis at the Weismann Institute. In some cases, single-stranded DNA sequencing using the M13 system was employed (Hu and Messing 1982).

## Results and Discussion

### Recombination monitored by bacterial transformation

In one form of the assay (Fig. 1), monkey COS cells, which synthesize the SV40 large T antigen constitutively (Gluzman 1981), were cotransfected with (1) pBR322-derived SV40 shuttle vectors that replicate both in COS cells and in *E. coli* (conferring Ap$^r$) and (2) a plasmid of *B. subtilis* origin (carrying Km$^r$ genes) that replicates neither in COS cells or in *E. coli* and whose DNA does not hybridize with that of the shuttle vector. Extrachromosomal DNA was isolated from the cells 2–3 days posttransfection and plasmid recombinants were detected by transformation of *E. coli* DH1 (Hanahan 1983) to ampicillin resistance followed by colony hybridization. Alternatively, since the *B. subtilis* plasmid's replicon is not recognized by *E. coli*, recombinant plasmids that contain the Km$^r$ marker linked to the pBR322 replicon were detected by transformation of *E. coli* to kanamycin resistance (positive selection). The bacterial transformation assays differ from the infectious-center in situ plaque hybridization assay in that (1) they measure the yield of recombinant DNA molecules rather than the number of recombinant-producing monkey cells, (2)

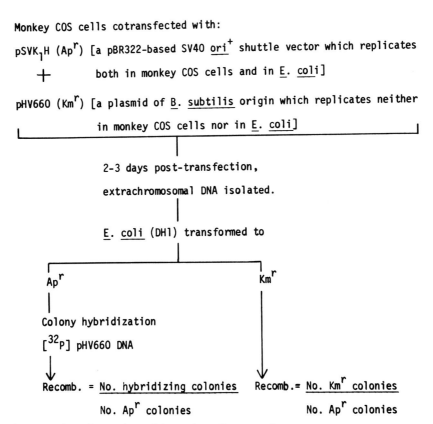

Figure 1 A recombination assay based upon bacterial transformation procedures.

they are independent of SV40 encapsidation restrictions (they depend, however, upon the retention of an *E. coli* replicon and an antibiotic-resistance marker in the recombinant), and (3) they rapidly clone out the recombinant products for structural analysis.

The data in Table 1 illustrate the use of the bacterial transformation assays to investigate the influence of DNA substrate conformation on recombination in monkey COS cells. The recombination frequencies were determined both by transformation to ampicillin resistance followed by colony hybridization ("CH" in Table 1) and by transformation to kanamycin resistance. The values obtained for the efficiency of recombination (normalized to the total number of $Ap^r$ colonies) were similar in both assays, indicating that the selective pressure for the $Ap^r$ marker was roughly equivalent to that for the $Km^r$ marker. When competent *E. coli* DH1 cells were exposed to $pSVK_1H$ and pHV660 DNAs that had not previously been passaged through monkey COS cells, the background level of recombination was less than 0.001%. The resistance of the recombinant plasmids in the 48-hr extracts of COS cells to *Dpn*I (Table 1, exp. 2), which cuts DNA replicated in bacterial cells but not DNA replicated in animal cells, is additional confirmation that the recombinants were made in the monkey cells rather than in the bacterial cells.

The data in Table 1 show the striking effect of substrate DNA conformation on the recombination frequency in monkey COS cells. Relative to the values obtained with circular DNA inputs, linearization of either plasmid DNA increased the recombination frequency. Significantly, however, linearization of $pSVK_1H$ DNA, which is capable of replication in COS cells, was much more effective (22-fold to 27-fold increase) than linearization of the pHV660 plasmid DNA incapable of replication (1.8-fold to 2.5-fold increase). Linearization of both transfected DNAs resulted in a synergistic, rather than an additive, response and increased the recombination frequency 107-fold to 121-fold higher than that obtained with circular plasmid DNAs. Linearization of $pSVK_1H$ DNA decreased the total number of $Ap^r$ colonies, presumably because the linear form of the SV40 origin-containing $pSVK_1H$ ($Ap^r$) DNA replicates more poorly

than the circular form in monkey COS cells, generating a lower yield of progeny DNA. Normalization of the recombination efficiency to the total number of $Ap^r$ colonies takes this factor into account (that linearization of the cotransfected substrate DNAs increases the absolute number of recombinants can also be deduced from the nonnormalized data).

That linearization of the recombination substrate capable of replication results in much higher recombination frequencies than linearization of the nonreplicating substrate is fully consistent with previous studies in which the recombination between SV40 DNA and φX174 RFI DNA was monitored by infectious-center in situ plaque hybridization (Dorsett et al. 1985). The results obtained by the two scoring systems differ, however, in one respect. Linearization of the nonreplicative φX174 RFI DNA produced no increase in the recombination frequency measured by in situ plaque hybridization (Dorsett et al. 1985), whereas linearization of the nonreplicative pHV660 DNA resulted in a small increase (1.8-fold to 2.5-fold; Table 1) in the bacterial transformation assay. It is possible that this apparent discrepancy arises because the two types of recombination assays measure different quantitative parameters; for example, linearization of nonreplicating DNA may increase the yield of recombinants per cell (the parameter scored by bacterial transformation) but not the number of recombinant-producing cells (the parameter scored by the in situ plaque hybridization assay). Alternatively, the bacterial transformation protocol may detect classes of recombinants that cannot be scored by the infectious-center in situ plaque hybridization procedure. Other than this apparent discrepancy, the results obtained by both scoring systems clearly establish that the linear form of a recombination substrate *capable of replication* is much more recombinogenic than the linear form of a nonreplicative substrate.

In the experiment described in Table 1, the DNA was extracted from the COS cells at 48 hr posttransfection. When extracts were made at different times posttransfection, a significant increase in the relative proportion of recombinants (normalized to the number of $Ap^r$ colonies) occurred between 24 and 48 hr posttransfection

**Table 1** Plasmid Recombination in Monkey COS Cells Monitored by Bacterial Transformation

| DNAs transfected into COS cells | No. colonies | | | Recombination (%) | |
| --- | --- | --- | --- | --- | --- |
| | $Ap^R$ | $CH^a$ | $Km^R$ | $CH/Ap^R$ | $KM^R/Ap^R$ |
| Circular $pSVK_1H$ + circular pHV660 | 38,900 | 25 | 22 | 0.06 (=1)$^b$ | 0.06 (=1) |
| Circular $pSVK_1H$ + linear pHV660 | 36,900 | 41 | 56 | 0.11 (1.8) | 0.15 (2.5) |
| Linear $pSVK_1H$ + circular pHV660 | 9,500 | 155 | 128 | 1.63 (27.2) | 1.35 (22.5) |
| Linear $pSVK_1H$ + linear pHV660 | 11,200 | 714 | 811 | 6.40 (107) | 7.24 (121) |
| Linear $pSVK_1H$ + linear pHV660 | 1,710 | 164 | 254 | 9.6 | 14.8 |
| Linear $pSVK_1H$ + linear pHV660 | 1,505 | 172 | 139 | 11.4 (+*Dpn*I)$^c$ | 9.2 (+*Dpn*I) |

$^a$Colony hybridization = no. of $Ap^R$ colonies that hybridize with [$^{32}$P]pHV660 DNA.
$^b$Figures in brackets are the fold increase in percentage recombination relative to the values obtained with circular DNAs (=1).
$^c$The transfected monkey cell extracts were treated with *Dpn*I before bacterial transformation under conditions that completely digest $pSVK_1H$ DNA that had not been passaged through COS cells.

(D. Vernade et al., in prep.). In considering the interpretation of this observation, it is necessary to distinguish between (1) amplification of the recombinant products (the resistance to *Dpn*I digestion indicates that most of the recombinants formed between the SV40 *ori*$^+$ pSVK$_1$H DNA and pHV660 DNA replicated in COS cells; Table 1, exp. 2), (2) amplification of the pSVK$_1$H parental DNA, which will increase the intracellular concentration of the substrate available for recombination, and (3) a dependency of the initial recombination events upon a DNA replication process. Because the recombination frequencies are normalized to the total number of Ap$^r$ colonies, the increase in the proportion of recombinants between 24 and 48 hr posttransfection cannot be attributed to an increase in the number of parental pSVK$_1$H (Ap$^r$) molecules available for recombination. Since there are no a priori reasons for assuming that the recombinants (most of which contain only a single SV40 origin of replication) replicate faster than the parental pSVK$_1$H DNA molecules, the increase in the proportion of recombinants is unlikely to be due to enrichment. Evidence in support of consideration number 3, namely, that a rate-limiting event in the initial recombination process is associated with the replication of pSVK$_1$H, has been obtained from inhibitor experiments. When COS cells cotransfected with linear pSVK$_1$H DNA and linear pHV660 (or polyoma DNA) were exposed to inhibitors of DNA replication at 24 hr posttransfection, the 48-hr extracts contained a lower relative proportion of recombinants compared with extracts of control cells (D. Vernade et al., in prep.).

The conformation of the DNA transfected into COS cells influences the structure of the recombinant products in addition to the frequency of recombination. In an investigation of the recombinant products formed between pSVK$_1$ (similar to pSVK$_1$H except that an intact copy of the SV40 genome is present; Dorsett et al. 1985) and pHV660 or polyoma DNA, two main classes of structures, whose abundance depends upon the input conformation of the pHV660 or polyoma DNA, have been discerned (D. Vernade et al., in prep.). In class-I structures, which arose primarily after transfection with circular DNAs, the pHV660 or polyoma sectors were flanked by large, 1–2-kb repeats of DNA derived specifically from the SV40 early region segment of pSVK$_1$. The flanking repeats were present in the same orientation at both junctions. Superficially, these structures resemble the cointegrate transposition products of bacteriophage Mu (Mizuuchi 1984). However, as shown in Figure 2 (recombinant 6) the nucleotide sequences at the recombinant junctions do not exhibit the short duplications characteristic of transposition products. Class-II structures, which arose mainly from the cotransfection of circular pSVK$_1$ DNA and linear pHV660 or polyoma DNA, exhibited large deletions at the junctions of the parental DNAs and no large duplicated sectors (the size of the class-II structures is 5–6-kb; that of class-I structures is 15–17-kb). We do not now understand the nature of the recombination events that give rise to the duplications in the cointegratelike class-I structures.

## The nucleotide sequence of the recombinant junctions

In Figure 2 we show the nucleotide sequences across 16 recombinant junctions. Recombinants 1–3 arose after cotransfection of monkey BSC-1 cells with circular SV40 and φX174 RFI DNAs and were isolated by in situ plaque hybridization. Their structure is of the simple deletion/insertion type (Winocour et al. 1983). Recombinants 4 and 5, which arose after covirion infection of monkey BSC-1 cells with SV40 and AAV, contain tandemly repeated segments of SV40 origin–region DNA linked to AAV palindromic terminal sequences; the nucleotide sequence data is taken from Grossman et al. (1985), which also describes the isolation and structure of these recombinants. Recombinants 6–10, which were generated by DNA cotransfection of monkey COS cells with circular polyoma (Py) DNA and various plasmid constructs, were isolated and cloned by D. Vernade using the bacterial transformation procedures described above. The cotransfected plasmid DNAs were circular pSVK$_1$ (recombinant 6, a class-I structure), linear pSVK$_1$H (recombinant 8), and circular pBR322-related constructs containing SV40 DNA and hepatitis B virus (HBV) DNA (recombinants 7, 9, and 10) (Laub et al. 1983). The plasmid sectors that recombined with polyoma DNA are indicated in each recombinant. In several cases, the exact crossover points are ambiguous (multiple arrows in Fig. 2) either because they occur at nucleotides common to both parents, or, in a few cases, because they contain sequences that are not identical with those of either parent (lower-case letters in Fig. 2). The sequence given for the parental polyoma DNA in recombinants 6–10 is that of the A2 strain (Griffin et al. 1980); however, the recombinants were generated with DNA from an unsequenced strain of the virus and this may account for a few of the unidentified nucleotides.

Short regions of parental DNA homology (underlined in Fig. 2) at or closely allied with the junction regions can be observed in each case. Common parental nucleotides at the crossover points occur at 13 of the 16 junctions; 8 of the 16 exhibit a common nucleotide, 2 of the 16 exhibit a common dinucleotide, and 3 of the 16 exhibit a common trinucleotide. Such nucleotide redundancy at the exact crossover points and the presence of short regions of parental DNA homology closely allied with the junctions have been frequently observed by others (e.g., see Gutai and Nathans 1978; Stringer 1982; Wilson et al. 1982; Bullock et al. 1984). These features are not, however, found in all SV40 nonhomologous recombinant products; in an analysis of six Ad2/SV40 recombinant junctions, Ling et al. (1982) observed abrupt transitions from Ad2 to SV40 DNA and essentially no homology between the joining regions of the parental genomes.

The significance of "patchy" parental homology in nonhomologous recombination is still a matter of conjecture. Some have suggested that it may introduce a degree of specificity by allowing the parental DNA strands to partially base-pair at the recombining regions (Gutai and Nathans 1978). However, both the statistical

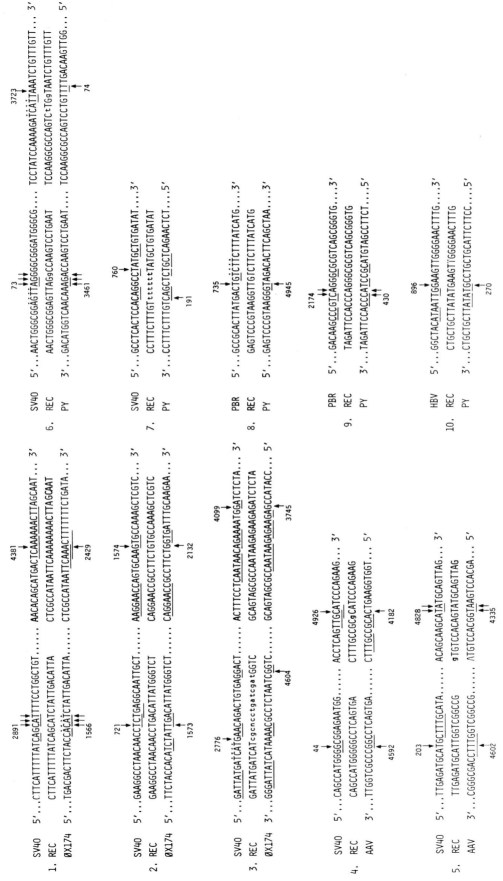

**Figure 2** Nucleotide sequences at the recombination junctions. The origin of recombinants 1–10 is described in the text. For each recombinant, the upper and bottom lines show the nucleotide sequence of the two parental DNAs; the middle line shows the sequence of the recombinant. The direction of the parental DNA sequence (5′ to 3′ or 3′ to 5′) is written to correspond with that of the recombinant so that the parental origin of the DNA in the recombinant can be identified. Recombinant DNA nucleotides not identical with those of either parent are designated by lower-case letters. The crossover points are indicated by arrows; short regions of parental DNA homology are underlined; dots above nucleotides designate potential cutting sites for eukaryotic topoisomerase I (see text). Only one of the two junctions in recombinants 7–10 has been sequenced so far.

significance of the distribution of short homologous regions around SV40 recombinant joints and the stability of the postulated base-pairing have been questioned (Savageau et al. 1983; Bullock et al. 1984). An alternative interpretation of "patchy" homology between the recombining regions of the parental DNAs and of the nucleotide redundancy at the exact crossover points is that the common nucleotides represent recognition sites for topoisomerases involved in the recombination process (Marvo et al. 1983). An association of SV40 recombinant junctions with topoisomerase I cleavage sites has been reported by Bullock et al. (1984, 1985). We have noted five junctions (in recombinants 1, 3, 6, and 8, Fig. 2) where the crossover points are adjacent to a string of 4 nt that conform to the consensus sequence for eukaryotic topoisomerase I (5′. . . [A or T]-[G or C]-[A or T]-T . . . 3′) derived by Been et al. (1984). In addition, 3-bp remnants of the consensus can be discerned at several other junctions. However, overall, only about 50% of the crossover points shown in Figure 2 occur at the nucleotide T (the 3′ T of the consensus sequence is considered to be the break site of topoisomerase I; Been et al. 1984). Furthermore, the consensus sequence for topoisomerase I occurs in SV40 DNA as frequently as once every 30–35 nt and is thus unlikely to be the sole factor controlling access of the enzyme (Been et al. 1984). It therefore seems likely that other factors, such as the conformation of the SV40 chromatin structure, play an equally crucial role in determining topoisomerase access.

## References

Acheson, N.H. 1980. Lytic cycle of SV40 and polyoma virus. In *Molecular biology of tumor viruses*, 2nd edition: *DNA tumor viruses* (ed. J. Tooze), p. 125. Cold Spring Harbor Laboratory, Cold Spring Harbor, New York.

Been, M.D., R.R. Burgess, and J.J. Champoux. 1984. Nucleotide sequence preference at rat liver and wheat germ type 1 DNA topoisomerase breakage sites in duplex DNA. *Nucleic Acids Res.* **12**: 3097.

Berns, K.I. and W.W. Hauswirth. 1982. Organization and replication of parvovirus DNA. In *Organization and replication of viral DNA* (ed. A. Kaplan), p. 3. Chemical Rubber Company, New York.

Botchan, M., W. Topp, and J. Sambrook. 1976. The arrangement of simian virus 40 sequences in the DNA of transformed cells. *Cell* **9**: 269.

Bouchard, L., C. Gelinas, C. Asselin, and M. Bastin. 1984. Tumorigenic activity of polyoma virus and SV40 DNAs in newborn rodents. *Virology* **135**: 53.

Bullock, P., J.J. Champoux, and M. Botchan. 1985. Association of crossover points with topoisomerase I cleavage sites: A model for non-homologous recombination. *Science* **230**: 954.

Bullock, P., W. Forrester, and M. Botchan. 1984. DNA sequence studies of simian virus 40 chromosomal excision and integration in rat cells. *J. Mol. Biol.* **174**: 55.

Cheung, A.K.M., M.D. Hoggan, W. Hauswirth, and K.I. Berns. 1980. Integration of the adeno-associated virus genome into cellular DNA in latently infected Detroit 6 cells. *J. Virol.* **33**: 739.

Deichaite, I., Z. Laver-Rudich, D. Dorsett, and E. Winocour. 1985. Linear simian virus 40 DNA fragments exhibit a propensity for rolling-circle replication. *Mol. Cell. Biol.* **5**: 1787.

Della-Valle, G., R.G. Fenton, and C. Basilico. 1981. Polyoma

large T antigen regulates the integration of viral DNA sequences into the genome of transformed cells. *Cell* **23**: 347.

Dorsett, D., I. Deichaite, and E. Winocour. 1985. Circular and linear simian virus 40 DNAs differ in recombination. *Mol. Cell. Biol.* **5**: 869.

Dorsett, D., I. Keshet, and E. Winocour. 1983. Quantitation of a simian virus 40 nonhomologous recombination pathway. *J. Virol.* **48**: 218.

Gilbert, W. and D. Dressler. 1969. DNA replication: The rolling circle. *Cold Spring Harbor Symp. Quant. Biol.* **33**: 473.

Gluzman, Y. 1981. SV40-transformed simian cells support the replication of early SV40 mutants. *Cell* **23**: 175.

Griffin, B.E., E. Soeda, B.G. Barrel, and R. Staden. 1980. Sequence analysis of polyoma virus DNA. In *Molecular biology of tumor viruses*, 2nd edition: *DNA tumor viruses* (ed. J. Tooze), p. 843. Cold Spring Harbor Laboratory, Cold Spring Harbor, New York.

Grossman, Z., K.I. Berns, and E. Winocour. 1985. Structure of simian virus 40/adeno-associated virus recombinant genomes. *J. Virol.* **56**: 457.

Grossman, Z., E. Winocour, and K.I. Berns. 1984. Recombination between simian virus 40 and adeno associated virus: Virion coinfection compared to DNA cotransfection. *Virology* **134**: 125.

Gutai, M.W. and D. Nathans. 1978. Evolutionary variants of simian virus 40: Cellular DNA sequences at the recombinant joints of substituted variants. *J. Mol. Biol.* **126**: 275.

Hanahan, D. 1983. Studies on transformation of *Escherichia coli* with plasmids. *J. Mol. Biol.* **166**: 557.

Hu, N. and J. Messing. 1982. The making of strand specific M13 probes. *Gene* **17**: 271.

Kleckner, N. 1981. Transposable elements in prokaryotes. *Annu. Rev. Genet.* **15**: 341.

Laub, O., L.B. Rall, M. Truett, Y. Shaul, D.N. Standring, P. Valenzuela, and W.J. Rutter. 1983. Synthesis of hepatitis B surface antigen in mammalian cells: Expression of the entire gene and the coding region. *J. Virol.* **48**: 271.

Lavi, S. and E. Winocour. 1972. Acquisition of sequences homologous to host DNA by closed circular simian virus 40 DNA. *J. Virol.* **9**: 309.

Ling, L.E., M.M. Manos, and Y. Gluzman. 1982. Sequence of the junction in adenovirus 2–SV40 hybrids: Examples of illegitimate recombination. *Nucleic Acids Res.* **10**: 8099.

Marvo, S.L., S.R. King, and S.R. Jaskunas. 1983. Role of short regions of homology in intermolecular illegitimate recombination events. *Proc. Natl. Acad. Sci.* **80**: 2452.

Maxam, A.M. and W. Gilbert. 1980. Sequencing end-labelled DNA with base-specific chemical cleavages. *Methods Enzymol.* **65**: 499.

Michel, B. and S.D. Ehrlich. 1984. Recombination efficiency is a quadratic function of the length of homology during plasmid transformation of *Bacillus subtilis* protoplasts and *Escherichia coli* competent cells. *EMBO J.* **3**: 2879.

Mizuuchi, K. 1984. Mechanism of transposition of bacteriophage Mu: Polarity of the strand transfer reaction at the initiation of transposition. *Cell* **39**: 395.

Oren, M., S. Lavi, and E. Winocour. 1978. The structure of a cloned substituted SV40 genome. *Virology* **85**: 404.

Samulski, R.J., A. Srivastava, K.I. Berns, and N. Muzyczka. 1983. Rescue of adeno-associated virus from recombinant plasmids. Gene correction within the terminal repeats of AAV. *Cell* **33**: 135.

Sanger, R., S. Nickleu, and A.R. Coulson. 1977. DNA sequencing with chain-terminating inhibitors. *Proc. Natl. Acad. Sci.* **74**: 5463.

Savageau, M.A., R. Metter, and W.W. Brockman. 1983. Statistical significance of partial base-pairing: Implications for recombination of SV40 DNA in eukaryotic cells. *Nucleic Acids Res.* **11**: 6559.

Smith, M., D.W. Leung, S. Gillaur, G.R. Asfell, D.L. Montgomery, and B.D. Hall. 1979. Sequencing of the gene for iso-1-cytochrome C in *Saccaromyces cerevisiae*. *Cell* **16**: 753.

Stringer, J.R. 1982. DNA sequence homology and chromosomal deletion at a site of SV40 DNA integration. *Nature* **296:** 363.

Varmus, H.E. 1983. Retroviruses. In *Mobile genetic elements* (ed. J.A. Shapiro), p. 411. Academic Press, New York.

Wilson, J.H., P.B. Berget, and J.M. Pipas. 1982. Somatic cells efficiently join unrelated DNA segments end-to-end. *Mol. Cell. Biol.* **2:** 1258.

Winocour, E. and I. Keshet. 1980. Indiscriminate recombination in SV40-infected monkey cells. *Proc. Natl. Acad. Sci.* **77:** 4861.

Winocour, E., V. Lavie, and I. Keshet. 1983. Structure of simian virus 40-φX174 recombinant genomes isolated from single cells. *J. Virol.* **48:** 229.

Winocour, E., N. Frenkel, S. Lavi, M. Osenholts, and S. Rozenblatt. 1975. Host substitution in SV40 and polyoma DNA. *Cold Spring Harbor Symp. Quant. Biol.* **39:** 101.

# A 2.7-kb Rearranged DNA Fragment from Epstein-Barr Virus Capable of Disruption of Latency

**J.K. Countryman,\* H. Jenson,[†] E. Grogan,[‡] and G. Miller\*[†‡]**

Yale University School of Medicine, Departments of \*Molecular Biophysics and Biochemistry, [†]Pediatrics, and [‡]Epidemiology and Public Health, New Haven, Connecticut 06510

Our experiments provide genetic clues to the mechanism of latency in an oncogenic herpesvirus system. A population of defective Epstein-Barr (EB) virions present in one cellular subclone of the P3HR-1 cell line (clone HH543-5) is able to activate expression of latent EBV in lymphoid cells (Miller et al. 1984). To identify EBV genes responsible for induction of viral replication, three novel BamHI fragments, present only in the defective genome, were transferred into D98/HR-1 cells, a somatic cell hybrid that contains latent EBV. One of the three fragments (BamHI-WZ het) caused activation of a large number of early and late viral replicative polypeptides detectable by immunofluorescence and immunoblotting (Countryman and Miller 1985).

The present studies concern the structure and biologic activity of the active BamHI-WZ het fragment. This 2.7-kb fragment represents a fusion and rearrangement of two regions (BamHI-W and -Z) not normally contiguous in the genome of standard EBV. The fragment contains a single, complete open reading frame (BZLF 1) and several partial open reading frames (BWRF 1, BZLF 2, BRLF 1). Compared with the available DNA sequence of BZLF 1 from the prototype EBV strain B95-8, which encodes a protein of 200 amino acids, BZLF 1 from the active "het" fragment has 15 scattered amino acid changes, an insertion of 9 amino acids, followed by a stretch of 9 different amino acids caused by a change in reading frame. Thus the activity of disruption of latency by WZ het could be due either to alterations in regulation of expression consequent to the genomic rearrangement or to differences in the polypeptide made by the het fragment or both.

Deletion mutagenesis suggests that the activity of the WZ het fragment is enhanced by adjacent sequences derived from BamHI-W. These sequences are evidently needed in cis.

Lymphoid cells containing latent EBV become virus producers when stably transformed by WZ het under selection with the antibiotic G418. Thus it is likely that the full viral replicative cycle can be activated by the product(s) encoded by the rearranged piece of DNA.

For many years we have been studying an Epstein-Barr virus (EBV) called P3J-HR-1 (HR-1), which is an immortalization-incompetent laboratory strain (Miller et al. 1975). Using this naturally occurring mutant, we have attempted to study the molecular mechanism of lymphocyte immortalization. The inability of HR-1 virus to immortalize lymphocytes is most likely due to a large genomic deletion in a region that encodes one of the EBV nuclear antigens (Bornkamm et al. 1982; Rabson et al. 1982; Hennessy and Kieff 1985). This laboratory strain has also provided us with unexpected clues to the mechanism of latency.

In addition to a standard (though deleted) EBV genome, the virions produced by HR-1 cells contain "heterogeneous" viral DNA (het DNA) that differs in organization from the standard EBV genome (Cho et al. 1984). Het DNA has the features of defective viral DNA; only portions of the standard genome are represented, and many genomic rearrangements have occurred.

Assignment of biologic traits to standard or defective HR-1 virus has been made possible by subcloning cell lines from the parental HR-1 line. The majority of cell subclones that have been isolated harbor only the standard HR-1 genome (Heston et al. 1982). A rare cell clone contains the standard genome plus het DNA (Rabson et al. 1983). The defective virus, containing het DNA, has not yet been propagated in the absence of standard virus. The defective virus may rely on helper functions of the standard genome.

The parental HR-1 line and the rare clone that contains het DNA both spontaneously enter the viral replicative cycle at a high frequency (about 10% or more of the cells). By comparison, cell clones that have lost het DNA (het⁻) have a low level of spontaneous virus production, usually less than 1%. However, virion synthesis can be induced in such het⁻ clones by chemicals such as phorbol ester and butyrate. Thus het DNA is not required for chemical induction.

Virion populations containing het DNA, but not those lacking such a sequence, are able to induce early antigens in Raji cells (Rabson et al. 1983). Similarly, virus stocks that contain standard and defective genomes activate complete expression of a latent immortalizing EBV present in a cell line designated X50-7 (Miller et al. 1984). Because the virus recovered is of the transforming phenotype and has restriction endonuclease polymorphisms of the endogeneous EBV genome, it is clear that the defective HR-1 virus activates EBV in *trans*.

In an effort to define which portion of the defective genome was responsible for activating latent virus, novel *Bam*HI fragments, contained only in the defective but not the standard genome, were cloned on recombinant plasmids and introduced into cells that harbored a latent EBV genome.

For our initial experiments, we used D98/HR-1 cells, a somatic cell hybrid that contains latent EBV and is readily transfected (Stoerker and Glaser 1983). All three *Bam*HI fragments of defective DNA encoded nuclear antigens in cells that lacked an EBV genome. However, only one fragment, *Bam*HI-WZ het, induced many more antigen-positive cells when it was introduced into cells already infected with EBV. These antigens, which were both nuclear and cytoplasmic in location, were shown to be viral replicative products, early and late antigens, which were not encoded on the WZ het fragment itself. The WZ het fragment encodes a polypeptide of about 36 kD, whereas a group of polypeptides with a range of sizes was induced in D98/HR-1 cells transfected by WZ het. Thus the WZ het fragment itself appears to be able to *trans*-activate expression of latent EBV. The two other defective EBV DNA fragments studied, *Bam*HI-S and *Bam*HI-MB' het, did not have the capacity to activate EBV expression (Countryman and Miller 1985).

The present studies, both structural and biologic in nature, are ultimately directed at understanding the mechanism by which the WZ het fragment promotes expression of an ordinarily quiescent EBV genome.

## Results

### Structure of the 2.7-kb *Bam*HI-WZ het fragment, which is capable of disrupting latency
Using a combination of Southern blot hybridization, partial digestion mapping of end-labeled fragments, and ultimately DNA sequencing, we have determined the structure of the rearranged DNA fragment that is capable of activating latency (Fig. 1). It consists exclusively of sequences from *Bam*HI-W (the first internal repeat) and *Bam*HI-Z. It is composed of 1133 bp of *Bam*HI-W and 1549 bp of *Bam*HI-Z. In the standard EBV genome, these sequences are normally separated by 52 kb. Within this 2.7-kb fragment there is only one intact open reading frame of 602 bp, which is designated BZLF 1 in the Cambridge terminology (Baer et al. 1984). There are portions of three other open reading frames, including BWRF 1, BRLF 1, and BZLF 2. If the sequences from *Bam*HI-W are represented in their orientation in the standard EBV genome at the left of the map, then the sequences from *Bam*HI-Z are reversed in orientation. Thus, for example, BZLF 1, normally a leftward open reading frame, now reads rightward. The het DNA was derived from the HR-1 strain, whereas the prototype EBV strain (B95-8), which has been sequenced, was used for comparison. There are certain differences in DNA sequence between B95-8 and HR-1 het DNA, both in the region of *Bam*HI-W and *Bam*HI-Z. These will be reported in detail later (H. Jenson et al., in prep.). For example, a *Hind*III site present in the BZLF 1 B95-8 strain is missing from the HR-1 het DNA but not from HR-1 DNA.

### Comparison of DNA sequences of BZLF 1's from EBV strains B95-8 and HR-1 WZ het
It is likely that the BZLF 1 reading frame wholly or partially encodes the product that is responsible for disruption of latency. If the WZ het fragment is first cut with *Hind*III, which transects BZLF 1, before it is transfected into cells, the fragment loses its ability to disrupt latency (Countryman and Miller 1985). Furthermore, WZ het cloned on plasmids containing the SV40 promoter and enhancer is more active than the same fragment cloned on plasmids lacking these sequences (Countryman and Miller 1985). Thus, transcription is probably

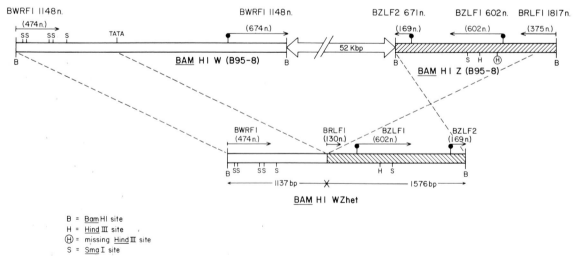

**Figure 1** A diagram of the structure of the *Bam*HI-WZ het fragment. Letters such as BWRF refer to open reading frames in the Cambridge terminology; e.g., BWRF is *Bam*HI-W right frame (Baer et al. 1984).

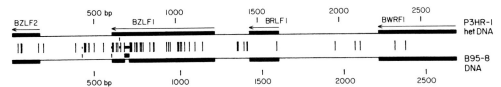

**Figure 2** Diagram comparing the nucleotide sequence of *Bam*HI-WZ het (upper horizontal line) with the analogous sequence of B95-8 EBV (lower horizontal line). Note that this rearrangement of sequences does not exist in B95-8. Solid bars indicate open reading frames. Arrows show the direction of transcription. Single-base-pair differences are marked by long vertical lines between the two sequences. The 28-bp gap in the B95-8 sequence and the 1-bp gap in the HR-1 sequence are indicated by short vertical lines.

required for induction. Therefore, we have compared the DNA sequence of BZLF 1 from the het fragment with that of B95-8, the prototype EBV strain that has already been sequenced. The results of this comparison are diagramed in Figure 2. BZLF 1 from B95-8 virus encodes a protein of 200 amino acids; the comparable reading frame from HR-1 *Bam*HI-WZ het encodes a protein of 209 amino acids due to a 28-bp insertion followed by a single-base-pair deletion near the carboxyl end of the protein. Immediately after the insertion, 9 additional amino acids are different between B95-8 and HR-1 WZ het due to a change in reading frame. Throughout the remainder of BZLF 1, there are 27 additional base pair changes between B95-8 and HR-1 WZ het. Fifteen of these changes result in an amino acid change, and 12 are silent since they fall in the wobble position. Further DNA sequencing is underway to compare BZLF 1 in WZ het with that in the standard HR-1 genome. The sequencing done so far indicates that the WZ het fragment could possibly become active as the result of altered protein structure as well as altered regulation of expression.

## Comparison of polypeptides encoded by WZ het and *Bam*HI-Z from strains HR-1 and FF41

The standard *Bam*HI-Z fragment from a transforming strain (FF41; Fig. 3), *Bam*HI-Z from a cell subclone of HR-1 lacking het DNA, and the WZ het fragment were all cloned on pSV2neo and introduced into COS-1 cells. Polypeptides were sought on immunoblots reacted with a polyvalent human antiserum. The most prominent polypeptide varied in size between 33 and 36 kD. The apparent mobility was different for each of the polypeptides; the larger polypeptide was encoded by the WZ het fragment. Another less-prominent polypeptide of about 19 kD, which also varied in size, was seen as well from each of the clones.

## Are the sequences from *Bam*HI-W required for disruption of latency?

To attempt to answer this question, we made a series of deletions from the 5' end of the active WZ het fragment. Plasmids containing these deleted fragments were assayed for their capacity to disrupt latency by inducing antigen-positive cells after transfection into D98/HR-1 cells. Those cells were scored as antigen-positive that expressed cytoplasmic products indicative of viral replication (Table 1). A mutant (no. 13) that removed only 79

bp was as active as the intact WZ het fragment. The other four mutants produced a diminished signal. One mutant (no. 28) that removed nearly 2000 bp and therefore invaded the BZLF 1 reading frame was inactive at inducing viral expression. Three other mutants (nos. 22, 81, and 80) caused induction of antigen at a frequency about 7-fold to 10-fold above background, but less than the intact WZ het fragment, which stimulated replication by 30-fold to 40-fold. These three mutants removed sequences from the *Bam*HI W portion of WZ het but left the BZLF 1 open reading frame intact. With the exception of mutant 28, which invades BZLF 1, all the other mutants produce a polypeptide in COS-1 cells of 36 kD,

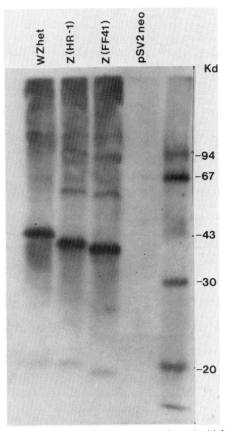

**Figure 3** Immunoblot of COS-1 cells transfected with WZ het, *Bam*HI-Z (HR-1), and *Bam*HI-Z (FF41) cloned on pSV2neo. The immunoblot reacted with a 1:200 dilution of human antiserum S.C.

**Table 1** Induction of Antigens in D98/HR-1 Cells by WZ het and Five Deletion Mutants from the 5' End

| Plasmid | No. of bp removed from 5' end | No. of antigen-positive cells[a] exp. 1 | No. of antigen-positive cells[a] exp. 2 |
|---|---|---|---|
| pSV2-WZ het | 0 | 214 | 120 |
| pSV2-Δ WZ-13 | 79 | 244 | NT |
| pSV2-Δ WZ-22 | 886 | 66 | 31 |
| pSV2-Δ WZ-81 | 984 | NT | 41 |
| pSV2-Δ WZ-80 | 1134 | NT | 28 |
| pSV2-Δ WZ-28 | 1986 | 7 | NT |
| pSV2neo | NA | 8 | 4 |

(NA) Not applicable; (NT) not tested.
[a]Indirect antiimmunoglobulin immunofluorescence with a 1:50 dilution of human serum S.C. Numbers are the no. of cells per coverslip with replicative antigens.

similar in size to the polypeptide made by WZ het (Fig. 3).

The data are consistent with the idea that sequences in *Bam*HI-W enhance the expression of the WZ product in D98/HR-1 cells.

### Are *Bam*HI-W sequences required in *cis* or *trans*?

Results of an experiment that attempts to answer this question are found in Table 2. In this experiment *Bam*HI-Z caused the appearance of antigen-positive cells at about four times the background level, whereas there was a 37-fold stimulation of antigen expression by WZ het. Plasmids were constructed that contained either the left or right *Bam*HI-to-*Hin*dIII pieces of WZ het. These were cotransfected into D98/HR-1 cells together with *Bam*HI-Z and *Bam*HI-WZ het. Neither subfragment of WZ het induced antigens above the background level. Nor did either subfragment of WZ het cause antigen induction to increase when it was cotransfected with *Bam*HI-Z. There seemed to be a decrease in antigen expression when either of the two subfragments of WZ het was transfected together with the intact WZ het fragment.

**Table 2** *Bam*HI-W Sequences Enhance Expression of *Bam*HI-Z Sequences in *Cis*

| Plasmid | No. of antigen-positive D98/HR-1 cells[a] |
|---|---|
| pSV2-Z | 138 |
| pSV2-WZ het | 1227 |
| pSVod 1.7[b] + Z | 68 |
| pSVod 1.7 + WZ het | 281 |
| pSVod 1.0[c] + Z | 82 |
| pSVod 1.0 + WZ het | 323 |
| pSVod 1.7 | 37 |
| pSVod 1.0 | 21 |
| pSV2neo | 33 |

[a]No. of cells per coverslip with replicative antigens detectable with a 1:50 dilution of human serum S.C.
[b]pSVod 1.7: The 1.7-kb *Bam*HI/*Hin*dIII left subfragment of *Bam*HI-WZ het was cloned in pSVod.
[c]pSVod 1.0: The 1.0-kb *Hin*dIII/*Bam*HI right subfragment of *Bam*HI-WZ het was cloned in pSVod (see Fig. 1).

These experiments suggest that supplying the *Bam*HI-W sequences in *cis* is required for enhanced activity of WZ het.

### Stable conversion of a lymphoblastoid cell line from latency to virus production by gene transfer with WZ het

In our previous experiments, the biologic activity of WZ het was tested in D98/HR-1 cells, which contain a latent EBV genome. These cells were not appropriate for answering the question of whether WZ het induced the *entire* viral replicative cycle with production of mature, biologically active virus. There is no bioassay for the EBV variant contained within D98/HR-1 cells. Furthermore, D98/HR-1 cells, a somatic cell hybrid between a lymphoid cell and an epithelial cell, might regulate the expression of EBV in an unusual way.

Therefore WZ het cloned on pSV2neo was transferred into X50-7 cells, a lymphoid line with a latent genome, using the electroporation technique. A series of X50-7 cell clones surviving selection in the antibiotic G418 was examined for viral replication. Of 25 clones studied (Table 3), 15 expressed viral replicative antigens, which were detected with a polyvalent anti-EBV human serum. The level of antigen-positive cells varied between 0.7% and 8%. Ten clones that were G418-resistant did not express antigen. Most of the clones that expressed replicative antigens were also reactive with a mouse monoclonal antibody, 72A1, directed against membrane antigen (Hoffman et al. 1980). The supernatants of 5 out of 12 antigen-positive clones released small amounts of biologically active EBV detectable by immortalization of human umbilical cord lymphocytes. None of the 7 clones tested that were antigen-negative released transforming virus.

Additional clones of neomycin-resistant X50-7 cells have been isolated that have been stably transformed by pSV2neo or pSV2neo–*Bam*HI-Z. None of 25 X50-7 cell clones containing pSV2neo make replicative antigens. One of 20 clones stably transformed by pSV2neo–*Bam*HI-Z expresses replicative antigens but does not release virus.

**Table 3** Replicative Expression of Epstein-Barr Virus in Cellular Subclones of X50-7 Cells Stably Transformed by pSV2neo-WZ het after Electroporation

| Cell clone | Antigen expression | | Release of transforming virus[c] |
|---|---|---|---|
| | polyvalent human ab (%)[a] | monoclonal mouse ab[b] | |
| 1 | + (0.7) | + | ND |
| 2 | + (0.8) | + | + (2/24) |
| 8 | + (0.7) | − | − |
| 9 | ++ (7.6) | + | ND |
| 11 | + (2.8) | − | + (1/24) |
| 13[d] | + (2.0) | + | + (3/16) |
| 15[d] | + (1.4) | + | − |
| 16 | ++ (3.5) | ++ | ND |
| 18[d] | + (8.0) | + | − |
| A8 | + (1.7) | + | − |
| B3 | + (2.6) | ND | − |
| B4 | ++ (7.5) | + | + (2/24) |
| C3 | ++ (7.0) | − | − |
| C12 | + (2.5) | ND | + (1/24) |
| D6 | + (3.1) | ND | − |
| 3[d] | − (<0.1) | − | − |
| 4[d] | − | − | − |
| 5[d] | − | − | − |
| 6 | − | − | − |
| 7 | − | − | ND |
| 10 | − | − | − |
| 12 | − | − | ND |
| 14 | − | − | − |
| 17 | − | − | ND |
| CI | − | − | − |

[a] Indirect antiimmunoglobulin immunofluorescence with a 1 : 50 dilution of human polyvalent antiserum S.C.

[b] Indirect antiimmunoglobulin immunofluorescence with a 1 : 10 dilution of mouse monoclonal antibody 72A1 directed against membrane antigen.

[c] Supernatants were obtained from cell clones grown in the absence of G418 for one week. They were assayed for immortalizing virus by a microwell assay using human umbilical cord lymphocytes. Numbers in parenthesis = no. of transformed wells per no. of wells inoculated.

[d] See immunoblot (Fig. 4).

To obtain a more biochemical description of the phenomenon of activation of latency in the X50-7 cells, six clones—three virus producers and three non-producers—after stable conversion by WZ het, were examined for viral proteins by immunoblotting, using as a source of antibody a human serum with high titers to the viral replicative products (Fig. 4). Three clones that were found to release transforming virus (Table 3) expressed a number of viral replicative polypeptides. The three nonproducer clones expressed only latent EBV products, that is, the EB nuclear antigens.

## Discussion

Present in the defective virions of the EBV HR-1 strain are rearranged EBV sequences that are capable of disrupting the latent state of the virus and initiating viral replication. We have shown, using a combination of biologic, immunologic, and biochemical assays that the sequences found on one specific piece of rearranged EBV DNA are capable of driving the entire EBV replicative cycle to completion with the production of mature virus.

How does this sequence bring about this remarkable biologic effect? There are two possibilities: Either the DNA sequences themselves titrate out a repressor present in the cell, or the DNA sequences encode a product that acts to inactivate a repressor or to activate the viral genome in a positive manner. The bulk of evidence, though not definitive, is against the idea that the DNA sequences themelves are able to titrate out a repressor. The magnitude of the effect of induction of viral expression we have observed is always much greater when the responsible fragment, WZ het, is cloned on a vector with an SV40 promoter and enhancer than one in which the SV40 enhancer is partially deleted or one without a eukaryotic promoter. Furthermore, when the active DNA fragment cloned on pSV2neo is cleaved with HindIII, which cuts in the middle of the single intact open reading frame, the inducing or disrupting effect is no longer evident.

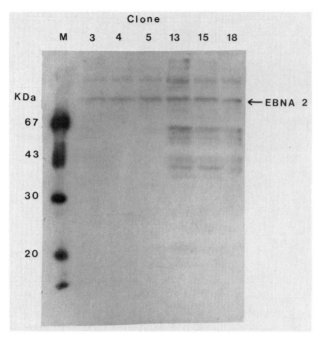

**Figure 4** Immunoblot of six clones of X50-7 cells stably resistant to G418 after transfection with pSV2neo WZ het. The immunoblot was reacted with a 1:200 dilution of human antiserum S.P.

## What is the role of the DNA rearrangement?

Our experiments indicate that the biologic effect of disruption of latency, as measured by induction of replicative antigens and polypeptides in D98/HR-1 cells, is increased through the juxtaposition of sequences (in *Bam*HI-W and *Bam*HI-Z) that are ordinarily far apart on the genome (see Table 1). Our present hypothesis is that disruption of latency is brought about by products within the *Bam*HI-Z fragment of EBV DNA whose expression is enhanced by sequences from *Bam*HI-W. We can now only guess at the function of the *Bam*HI-W sequences; much further work is needed. They may provide an origin for DNA replication. They may promote transcription of the *Bam*HI-Z open reading frame by acting as a promoter or enhancer. On the basis of results shown in Table 2, it is doubted that they *trans*-activate *Bam*HI-Z transcription. They may provide leader sequences to the mRNA that in turn might alter the efficiency of translation. Finally, we have not excluded the possibility that W sequences allow a fusion protein to be made. That the predominant polypeptide encoded by WZ het appears to be slightly larger than that encoded by *Bam*HI-Z (Fig. 3) is consistent with this hypothesis.

## What is the active product?

In one experiment (Table 2), transfection of D98/HR-1 cells with the *Bam*HI-Z fragment cloned on pSV2neo caused induction of replicative antigens at a level 3-fold to 4-fold higher than background, but 10-fold lower than that effected by WZ het. This suggests that the active product is encoded within *Bam*HI-Z. Since transfection of BZLF 1 by *Hin*dIII abolishes the activity, it is likely that this reading frame encodes at least part of the responsible product. The protein encoded by this reading frame would be 200–209 amino acids long, whereas the polypeptide seen in eukaryotic expression systems is about 33–36 kD. This larger-than-predicted size could be due to anomalous migration or modification. Alternatively, the 36-kD protein could be encoded by more than one exon. Deletion mutants that have removed as much as 984 bp from *Bam*HI-W still encode a protein that is the same size as that encoded by WZ het. Thus, if a fusion protein includes *Bam*HI-W sequences, they must come from the 150 bp found at the 3' end of the *Bam*HI-W portion.

## What is the target of the active product?

At this point we cannot even guess about the function of the active product. We can only list some alternative possibilities: It may interact with or modify a cellular product that acts to keep the virus latent. The cell product may be a normal constituent of B lymphocytes or its synthesis may in some way be activated by the EBV-encoded immortalization event. Alternatively, the virus itself may encode a repressor that is modified in some way by the WZ product. The other model is that the product of WZ is needed to activate transcription or translation of another EBV gene that is next in line in a linked cascade of functions. The model, of course, draws on the analogy with the cascade of viral gene expression in herpes simplex virus and the role of immediate-early genes in initiating this cascade (Honess and Roizman 1975).

These hypotheses do permit different kinds of experimental tests. If a "repressor" mechanism, either virally or cellularly encoded, is operative, then the WZ product might only be active in cells that have an already established latent infection. On the other hand, if the product acts to provide something that is essential for the next step in the viral replicative cycle, then WZ het might be shown to activate expression of specific viral genes that are cotransfected into cells.

## What is the control on expression of the Z product?

We believe that the activity of the rearranged fragment is a clue to a normal mechanism that controls latency and activates viral expression in a cell immortalized by an EBV without rearranged DNA. If this idea is correct, then the product from *Bam*HI-Z is normally negatively regulated in the immortalized cell. To activate the viral replicative cycle, this control on expression of the Z product must be released. Upon production of an active Z protein, the remainder of the viral replicative cycle goes on to completion.

## Acknowledgments

This work was supported by grants from the National Institutes of Health (CA 12055) and the American

Cancer Society (MV-173F). J.C. is supported by Training Grant GM 07223, and H.J. is supported by a Physician Scientist Award from the National Institutes of Health (AI 00651). We thank K. Papov for preparing the manuscript.

## References

Baer, R., A.T. Bankier, M.D. Biggin, P.L. Deininger, P.J. Farrell, T.J. Gibson, G. Hatfull, G.S. Hudson, S.C. Satchwell, C. Seguin, P.S. Tuffenell, and G. Barrell. 1984. DNA sequence and expression of the B95-8 Epstein-Barr virus genome. *Nature* **310:** 207.

Bornkamm, G.W., J. Hudewentz, W.K. Freese, and U. Zimber. 1982. Deletion of the non-transforming Epstein-Barr virus strain P3HR-1 causes fusion of the large internal repeat to the DSL region. *J. Virol.* **43:** 952.

Cho, M.-S., G.W. Bornkamm, and J. zur Hausen. 1984. Structure of defective DNA molecules in Epstein-Barr virus preparations from P3HR-1 cells. *J. Virol.* **51:** 199.

Countryman, J. and G. Miller. 1985. Activation of expression of latent Epstein-Barr herpesvirus after gene transfer with a small cloned subfragment of heterogeneous viral DNA. *Proc. Natl. Acad. Sci.* **82:** 4085.

Hennessy, K. and E. Kieff. 1985. A second nuclear protein is encoded by Epstein-Barr virus in latent infection. *Science* **227:** 1238.

Heston, L., M. Rabson, N. Brown, and G. Miller. 1982. New Epstein-Barr virus variants from cellular subclones of P3JHR-1 Burkitt lymphoma. *Nature* **295:** 160.

Hoffman, G.J., S.G. Lazarowitz, and S.D. Hayward. 1980. Monoclonal antibody against a 250,000 dalton glycoprotein of Epstein-Barr virus identified a membrane antigen and a neutralizing antigen. *Proc. Natl. Acad. Sci.* **77:** 2979.

Honess, R.W. and B. Roizman. 1975. Regulation of herpesvirus macromolecular synthesis: Sequential transition of polypeptide synthesis requires functional viral polypeptides. *Proc. Natl. Acad. Sci.* **72:** 1276.

Miller, G., M. Rabson, and L. Heston. 1984. Epstein-Barr virus with heterogeneous DNA disrupts latency. *J. Virol.* **50:** 174.

Miller, G., J. Robinson, and L. Heston. 1975. Immortalizing and non-immortalizing laboratory strains of EBV. *Cold Spring Harbor Symp. Quant. Biol.* **39:** 773.

Rabson, M., L. Heston, and G. Miller. 1983. Identification of a rare Epstein-Barr virus variant which enhances early antigen expression in Raji cells. *Proc. Natl. Acad. Sci.* **80:** 2762.

Rabson, M., L. Gradoville, L. Heston, and G. Miller. 1982. Non-immortalizing P3J-HR-1 Epstein-Barr virus: A deletion mutant of its transforming parent, Jijoye. *J. Virol.* **44:** 834.

Stoerker, J. and R. Glaser. 1983. Rescue of transforming Epstein-Barr virus (EBV) from EBV genome-positive epithelial hybrid cells transfected with subgenomic fragments of EBV DNA. *Proc. Natl. Acad. Sci.* **80:** 1726.

# Sequence-specific Interactions of Cellular Nuclear Factor I and Epstein-Barr Virus Nuclear Antigen with Herpesvirus DNAs

D.R. Rawlins,*[†] P.J. Rosenfeld,* T.J. Kelly, Jr.,* G.R. Milman,[‡] K.-T. Jeang,[§]** S.D. Hayward,[§] and G.S. Hayward[§]

*Department of Molecular Biology and Genetics and [§]Virology Laboratories, Department of Pharmacology, Johns Hopkins University School of Medicine, and [‡]Department of Biochemistry, Johns Hopkins School of Hygiene, Baltimore, Maryland 21205

We have been studying the potential roles of nuclear factor I (NFI) and Epstein-Barr virus nuclear antigen (EBNA-1) in the regulation of herpesvirus DNA replication and latency. NFI is a cellular protein that recognizes and binds with high affinity to the DNA sequence TGGA/C(N)$_5$GCCAA. Recognition sites for NFI occur near the replication origins in adenoviruses and papovaviruses and are located at several distinct sites in the genomes of herpes simplex virus (HSV) and cytomegalovirus (CMV). Interestingly, in simian CMV(Colburn), a cluster of 20 adjacent NFI-binding sites was found within a series of 30-bp tandem repeats upstream from the strong promoter-regulatory region for the immediate-early gene IE94. An additional set of 5 tandem sites was also detected in the first intron of this gene. In human CMV(Towne), a total of four NFI-binding sites were found in similar positions encompassing the promoter-enhancer region of the IE68 gene. The functional significance of this conservation and amplification of NFI-binding sites in CMV is not yet known.

EBNA-1 is known to be a *trans*-acting viral factor required for the establishment and maintenance of Epstein-Barr virus (EBV) DNA genomes in a multicopy plasmid state in transformed cells. Using a 28K carboxyterminal fragment of bacterially synthesized EBNA-1, we have demonstrated specific binding to two *cis*-acting DNA loci within the ori$_p$ plasmid maintenance region and also to a single additional distal locus within the EBV genome. DNase footprinting analyses revealed that one binding locus within ori$_p$ consists of a set of 20 adjacent copies of a 30-bp tandemly repeated element containing the 12-bp palindromic sequence TAGCATATGCTA. The second ori$_p$ locus contains a cluster of two pairs of overlapping binding sites with a similar core concensus sequence. These findings suggest that EBNA-1 may mediate its *trans*-acting functions in plasmid maintenance via direct protein-DNA interactions at multiple loci within the ori$_p$ DNA region.

Little is known about the viral and cellular factors that regulate the initiation of DNA replication or the establishment of latency after herpesvirus infection. Two partially homologous lytic DNA replication origins (ori$_S$ and ori$_L$) have been defined in the herpes simplex virus (HSV) genome (Ciufo and Hayward 1981; Stow and McMonagle 1983; Weller et al. 1985), and several viral proteins essential for replication, including the DNA polymerase and major DNA-binding protein, have been studied extensively. However, no HSV protein (similar to simian virus 40 (SV40) T antigen, for example) that is likely to be involved in specific initiation events has yet been identified by either genetic or biochemical proce-

dures. The locations of lytic DNA replication origins are unknown for Epstein-Barr virus (EBV) and cytomegalovirus (CMV), although the *cis*-acting DNA region shown to be necessary for maintenance of the plasmid state of EBV genomes presumably includes a replication origin (Yates et al. 1984, 1985).

The initiation of adenovirus DNA replication in vitro is enhanced by a cellular protein component of extracts from uninfected HeLa cells referred to as nuclear factor I (NFI) (Nagata et al. 1982). The NFI protein binds specifically to a DNA sequence within the adenovirus origin (Nagata et al. 1983; Rawlins et al. 1984), and mutational analyses of this sequence revealed a direct correlation between the binding of NFI to the origin and its ability to enhance the initiation of adenovirus replication in vitro (Guggenheimer et al. 1984; Rawlins et al. 1984). A consensus sequence for NFI, TGGA/C(N)$_5$GCCAA, was derived from comparative binding studies with the mutated adenovirus origins, other viral

*Present addresses*: [†]Department of Microbiology and Immunology, Emory University, School of Medicine, Atlanta, Georgia; **Laboratory of Molecular Virology, National Cancer Institute, Building 41, Suite 200, Bethesda, Maryland 20205.

DNAs (including those reported here), and cellular DNAs (Borgmeyer et al. 1984; Gronostajski et al. 1985; Hennighausen et al. 1985; Rosenfeld et al., this volume), and the protein has been purified by chromatography on a DNA affinity matrix containing multiple copies of this recognition sequence (Rosenfeld and Kelly 1986).

The conservation of a TGGA/C(N)$_5$GCCAA-binding activity in many animal species suggests that NFI may be a fundamental element of cellular biochemistry. However, the mechanism by which NFI enhances the initiation of adenovirus replication and the cellular role(s) of the protein remain mysteries. Thus, an analysis of the binding sites for NFI in viral DNAs has twofold potential. First, the physical localization of the binding sites in genetically defined backgrounds may indicate whether the protein has a common or dissimilar role in the metabolism of viruses other than adenovirus, and it may provide some idea as to the extent of cellular participation in the regulation of viral functions. Second, the cellular role(s) of NFI may be more readily derived from the use of viral models since the ease of genetically manipulating viral genomes makes them preferable to direct studies of randomly selected chromosomal NFI-binding sites.

The EBNA-1 protein is a component of the viral nuclear antigen that is expressed in the tumor tissues of Burkitt's lymphoma and nasopharyngeal carcinoma, in permanent B-lymphoblastoid cell lines derived from infectious mononucleosis patients, and in nearly all cells of lines established by in vitro immortalization of primary lymphocytes with EBV. EBNA-1 is a polypeptide with a molecular weight of approximately 80,000 (80K) encoded by the *Bam*HI-K fragment of the EBV genome (Summers et al. 1982; Hennessy et al. 1983; Baer et al. 1984), and it is one of only four known viral proteins that are expressed in EBV latently infected nonproducer cells. The correlation of EBNA-1 with the nonlytic persistence of the EBV genomes in B lymphoblasts suggests a potentially intimate role for this protein in the establishment or maintenance of EBV latency. In the latent state, the genome of EBV is maintained in a multicopy, extrachromosomal circular or plasmid form (Lindahl et al. 1976). Sugden and his associates have described experiments identifying two *cis*-acting DNA elements in the 1800-bp *ori*$_P$ region of EBV DNA that contain the minimal signals necessary for plasmid maintenance (Yates et al. 1984). In addition, they have shown that neomycin-resistant (Neo$^R$) pBR322-derived DNA containing *ori*$_P$ can be maintained in a circular, plasmid state in primate cell lines that have also been transfected with the EBV *Bam*HI-K fragment and express EBNA-1 (Yates et al. 1985). Thus EBNA-1 appears to be a *trans*-acting viral function required for plasmid maintenance. In addition to these in vivo demonstrations of a functional linkage between *ori*$_P$ and EBNA-1, partially purified EBNA-1 preparations from mammalian cells and an EBNA-1 fragment synthesized in *Escherichia coli* have been reported to have single-stranded and nonspecific, double-stranded DNA-binding properties (Sculley et al. 1983; Hearing and Levine 1985; Milman et al. 1985a).

## Materials and Methods

### Viral and plasmid DNAs

HSV-1(MPcl22) and HSV-2(333cl5) were grown in VERO cells, and cytoplasmic virions from clarified Dounce homogenates were banded in sucrose density gradients. Human (H) CMV(Towne) virions were prepared similarly from stocks with low multiplicity of infection (moi) grown in diploid human fibroblast cells. The purified virions were treated with DNase and lysed with 1% SDS, 0.1 M EDTA, 0.1 M Tris · HCl (pH 8.4), followed by phenol extraction. Purified virion DNA was recovered after equilibrium banding in CsCl density gradients, followed by ethanol precipitation and extensive dialysis.

Plasmids pKB67 and pKB67-88 contain either 1 or 88 copies, respectively, of the 67 terminal bp of the adenovirus type-2 (Ad2) genome (Challberg and Rawlins 1984; Rosenfeld and Kelly 1986). They were constructed by inserting an *Eco*RI/*Bam*HI fragment containing the Ad2 sequences into pKP45, a 2675-bp pBR322 derivative kindly provided by K. Peden. The plasmid pTJ148 contains the previously described IE94 gene and its 5'-sequences from simian (S) CMV(Colburn) (Jeang et al. 1984). The plasmid pKCMV is a *Bam*HI/*Hind*III subclone of pTJ148 that includes the first intron of the IE94 gene and approximately 3500 bp 5' to the transcriptional start site (in pKP45). pTJ280 is an IE94-CAT deletion construct that contains 260 bp from immediately 5' to the IE94 transcriptional start site. The *Xba*I-E fragment of HCMV(Towne) in pMSDT-E (provided by M. Stinksi) contains the complete IE68 gene and 5'-flanking sequences from the viral genome. Recombinant clone pGH12 contains a *Bam*HI to *Hind*III segment of HSV-1 *Bam*HI-M, including *ori*$_S$ and the upstream promoter-regulatory region of IE175. pGH1 contains a deleted fragment from the *ori*$_L$ and DNA polymerase promoter region of HSV-1 as part of the repeat unit from class-II defective DNA (Ciufo and Hayward 1981). Plasmid pSL93 and pSL87 are independent clones containing the EBV(B95-8) *Bam*HI-C fragment in pBR322. Plasmid pPL7 is a pBR322 subclone of pSL93 containing the *ori*$_P$ region between the *Eco*RI and *Hinc*II sites at bases 7315–9132 of the EBV(B95-8) genome (Baer et al. 1984). pHSV106 and pGR18 contain the intact HSV-1 (3.5-kb *Bam*HI-P) and HSV-2 TK genes (4.8-kb *Sal*I to *Hind*III), respectively.

Additional recombinant clones containing viral DNAs were kindly provided by several other investigators, including clones of JC(UD), BK(UD), and human papillomavirus (HPV-1a) DNAs (P. Mounts); polyomavirus (J. Corden); SV40 (J. Li); pHEBO-1, an EBV *ori*$_P$ and hygromycin-resistant clone from EBV(B95-8) (W. Sugden); pMC54, the *Bam*HI-C fragment from EBV(P3HR-1) defective DNA (M.-S. Cho); and pM765-10, containing the EBV(M-ABA) DNA region between *Sal*I at position 644 and *Bgl*II at position 13,944 (G. Bornkamm).

### NFI and 28K EBNA proteins

The NFI used for these studies was partially purified from uninfected HeLa cells as described previously

(Rawlins et al. 1984). The single-stranded DNA-cellulose fraction was routinely employed for the filter-binding and footprinting assays. The bacterially synthesized 28K carboxyterminal fragment of EBNA-1 was purified to near homogeneity by phosphocellulose and hydroxylapatite chromatography from uninduced bacterial cultures. Column fractions containing the protein were stored at $-70°C$ in 250 mM sodium phosphate (pH 7.5), 250 mM NaCl, 50 mM Tris·HCl (Waldman et al. 1983; Milman et al. 1985b).

*Nitrocellulose filter–binding assays*
After cleavage of the plasmid or viral DNAs with the appropriate restriction enzymes (see Figures), the DNA fragments generated were 3′-labeled at their termini by filling in 5′-overhanging ends with dNTPs (one or two of which were $^{32}$P-labeled) as described previously (Rawlins et al. 1984). Either *Micrococcus luteus* DNA polymerase or *E. coli* DNA polymerase I (Klenow fragment) was used in the labeling procedure. Protein-DNA binding was carried out at 4°C for 30 min in a reaction mixture (40–50 μl) containing either 150 mM NaCl (NFI) or 200 mM NaCl (EBNA). The reaction mixtures also routinely contained 50 mM HEPES (pH 7.5), 5 mM MgCl$_2$, 1 mM DTT, 250 μg/ml BSA, 125 μg/ml tRNA, 10–20 fmoles of $^{32}$P-labeled DNA, and binding protein. The reaction mixture was filtered through nitrocellulose, washed five times with 6–8 volumes of wash buffer (150 mM or 200 mM NaCl, 5 mM MgCl$_2$, 1 mM DTT, 10% glycerol). DNA bound to the filters was eluted with 25 mM HEPES (pH 7.5), 10 mM NaCl, 0.2% SDS. After gel electrophoresis, ethanol fixation, and drying, the DNA was visualized by autoradiography. For quantitation of binding in the competition assay, protein-DNA complexes were determined directly by scintillation counting of the dried nitrocellulose filters after filtration and washing (Rosenfeld and Kelly 1986).

*DNase footprinting*
DNA-binding protein–mediated protection of regions of the DNAs from cleavage by DNase was determined as described previously (Rawlins et al. 1984). DNA sequencing reactions were carried out by using a modified chemical sequencing procedure (Maxam and Gilbert 1977; Bencini et al. 1984).

*Altered mobility assay*
Each 5-μl sample contained 200 mM NaCl, 10 mM Tris·HCl (pH 7.6), 0.1 mM EDTA, 2 mM MgCl$_2$, 1 mM DTT, 50 μg/ml BSA, 0.15 μg DNA, and varying amounts of 28K EBNA protein. Incubations were carried out at 23°C for 20 min, followed by agarose gel electrophoresis and autoradiography.

**Results**

**NFI-binding sites in viral DNA**
NFI-binding sites have been described within the inverted terminal repetitions in adenovirus DNA and throughout mammalian genomic DNA. However, as shown by filter-binding assays in Figure 1, NFI-binding sites also

occur in other DNA viruses, and at least two additional sites were detected in the Ad5 genome. Although NFI-binding sites were absent from SV40 and polyomavirus DNA, human JC papovavirus DNA bound NFI at several loci, and BK virus DNA showed weak binding. Human papillomavirus (HPV) DNA also gave one strong binding fragment.

Similar filter-binding results were obtained with the HSV and CMV genomes. The pattern of NFI binding to a total *Hin*dIII digest of HCMV(Towne) DNA and to total *Bam*HI digests of HSV-1(MP) and HSV-1(333) are shown in Figure 1b. In each case approximately 5 of the more than 30 fragments present displayed relatively strong binding, and several weak-binding fragments were also detected. The use of other restriction enzyme digests of these same DNA samples in similar assays (not shown) has revealed that differences in the relative sizes of individual fragments used in the assay can lead to considerable distortions in their apparent relative affinity for the NFI protein. Positive DNA fragments of high molecular weight usually showed aberrantly weak binding, whereas cleavage of the same DNA fragments into smaller-sized subfragments frequently revealed the presence of loci with strong binding properties under otherwise identical experimental conditions. In HSV-1(MP) DNA the strongest binding sites occurred within fragments *Bam*HI-C (0.738–0.809) and -E (0.020–0.079), which contain portions of the L-segment inverted repeats; *Bam*HI-M (0.864–0.896) encompassing ori$_S$ plus unique-S-region sequences; and *Bam*HI-O (0.292–0.316) encompassing the TK gene. Among several weaker-binding fragments, we were also able to identify *Bam*HI-U (0.391–0.408), which contains ori$_L$, and *Bam*HI-X (0.951–0.965), encompassing the second copy of ori$_S$ within the inverted repeats on the opposite side of the S segment. Filter-binding assays using pBR322-derived clones from these regions confirmed the existence of weak binding sites within the vicinity of both ori$_L$ (0.392–0.408) and at least one copy of ori$_S$ (0.865–0.876), and strong binding 3′ to the TK coding region, but not on the promoter side of the HSV-1 TK gene (Fig. 2a). In comparison, the cloned HSV-2 TK gene coding region and 3′-flanking sequence did not bind NFI, although relatively weak binding was detected upstream from the HSV-2 TK promoter.

In the HCMV(Towne) genome, we were particularly intrigued by relatively strong binding to the *Hin*dIII-C and *Bam*HI-J fragments that encompass the major IE68 gene of that virus. Cloned fragments containing the HCMV(Towne) *Xba*I-E fragment (pMSDT-E) and the SCMV(Colburn) *Hin*dIII-H fragment (pTJ148) also both showed strong binding (Fig. 2a), thus confirming the existence of NFI sites in the vicinity of the major IE genes. Additional mapping with cleaved and deleted plasmid DNAs from both genes (not shown) revealed a relatively complex situation. In HCMV(Towne) we localized sites both within 1000 bp upstream of the IE68 mRNA start site and within 900 bp downstream of the start site in the major intron region. In SCMV(Colburn) strong NFI binding was again localized in the far upstream IE94 promoter-regulatory region, although on

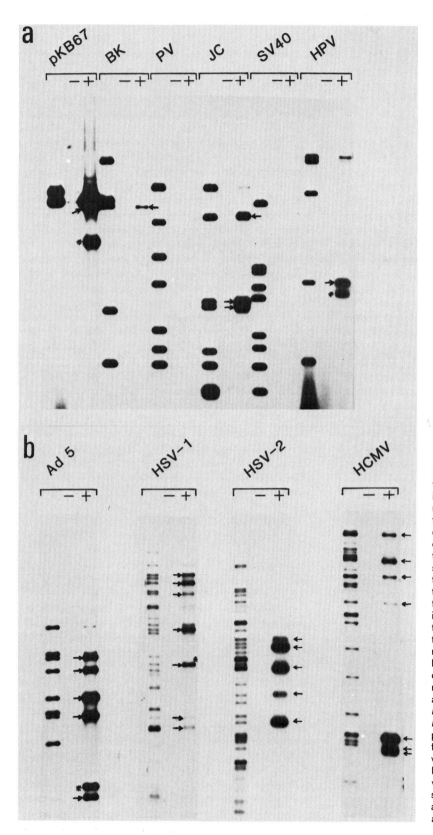

**Figure 1** Binding of HeLa cell NFI to DNA virus genomes. Binding reactions containing NFI and $^{32}$P-labeled restriction fragments of cloned or virion DNA were filtered through nitrocellulose. The DNAs retained on the filters were eluted and electrophoresed on a 1.4% agarose gel. In each set the first lane represents input DNA, the second lane shows the DNA retained on the filter in the absence of NFI (−), and the third lane shows the DNA retained on the filter in the presence of NFI (+). Arrows indicate all specifically bound fragments of the viral DNA, and asterisks indicate denatured forms of the specifically bound fragments. (*a*) Papovaviruses and human papillomavirus. All samples represent viral DNA cloned into pBR322 at the *Bam*HI site. The restriction enzyme digests used for this analysis were: *Bam*HI (pKB67); *Bam*HI/*Hin*dIII (BK, PV, JC, and SV40); *Bam*HI/*Eco*RI (HPV). pKB67 is a recombinant plasmid containing terminal nucleotides from Ad2 DNA. *Bam*HI-cleaved pKP45 vector DNA was included in the pKB67 lane as a nonspecific control. (*b*) Adenovirus and herpesviruses. Total virion DNA samples were cleaved with *Hin*dIII (Ad5 and HCMV[Towne]) or *Bam*HI (HSV-1[MP] and HSV-2[333]).

both sides of a *Nco*I site at position −1020. No binding occurred in the enhancer and proximal promoter region between −440 and +220, but an additional locus mapped in the downstream IE94 intron region between positions +220 and +1430.

## Competition binding assays with purified NFI

Although concensus NFI sites with various different binding affinities and of unknown or doubtful functional significance occur relatively frequently within the adenovirus and herpesvirus genomes, the pattern of binding within the CMV IE gene regions appeared to be highly unusual. First, both CMV IE loci bound more strongly than the known strong binding site shown to be essential for adenovirus DNA replication; and second, the binding activity was distributed widely over an approximately 2-kb region at the 5′ end of the gene. Therefore, we carried out quantitative measurements of the relative ability of the CMV loci to compete with single and multicopy forms of the Ad2 concensus NFI site for binding to a constant amount of the purified NFI protein (Fig. 2b). The control Ad2 DNA plasmids used in these studies were a single copy of the 67-bp terminus of Ad2 DNA (pKB67) and an engineered 88-tandem-repeat copy of the 67-bp sequence (pKB67-88). The amount of NFI protein used in the assay gave 50% filter binding of $^{32}$P-labeled pKB67 DNA in the presence of 0.2 nM unlabeled competitor pKB67 DNA. Control vector plasmids pKP45 and pSV2CAT that lack binding sites reduced binding to pKB67 by 50% at concentrations above 7nM, whereas (by extrapolation) the 88-copy plasmid reduced binding to 50% at concentrations of only 0.0025 nM. In comparison, concentrations of 0.04 nM and 0.0045 nM of the plasmids containing the complete IE68 gene (pMSDT-E) and the complete IE94 gene (pTJ148) were required for 50% competition with the single Ad2 site. Therefore, the 88-copy plasmid showed 80-fold greater affinity for the protein than did the single copy Ad2 site and the HCMV and SCMV IE regions behaved as though they contained 5 and 45 sites, respectively, with equivalent affinities to that of the single Ad2 site. The use of a 5′-deletion CAT plasmid containing only the first 260 bp of IE94 5′-promoter-region sequences (pTJ280) gave no significant binding, indicating that all of the upstream IE94 sites lie beyond position −260.

## A cluster of tandemly repeated NFI-binding sites upstream from the IE94 enhancer region

DNA sequencing analysis of the far-upstream region 5′ to the strong promoter-regulatory region of IE94 revealed a remarkable pattern of directly repeated sequences within the region from −600 to −1280 (relative to the IE94 mRNA start site), including a central, nearly perfect 11× 30-bp tandem repeat structure. Although individual repeats within this region were generally highly diverged in sequence, we recognized a total of 23 adjacent copies of a conserved consensus sequence element (T)TGGA/CN$_5$GCCAA (see Fig. 3c). To confirm that these repeat elements represented a cluster of tandemly repeated NFI-binding sites, we carried out

DNase I footprinting studies with partially purified NFI protein and isolated DNA fragments asymmetrically labeled at the *Nco*I site at position +1020. The results presented in Figure 3a show the pattern of binding sites both upstream and downstream from the *Nco*I site. A series of approximately 25-bp protected sites with interspersed 4–5-bp hypersensitive regions occur at approximately 30-bp intervals across the entire region, and in each case the center of the protected region corresponds closely to one of the 23 conserved concensus elements identified in the DNA sequence (Fig. 3b). Individual sites in the cluster varied considerably in their observed affinity for the NFI protein. In general, those sites with a perfect match to the concensus sequence TTGGA/CN$_5$GCCAA bound strongly, whereas those with a cytosine at position 1 (sites 9, 11, 13, 16, and 19) were variable, and those with an adenine at position 1 (sites 10, 12, and 14) were weak binding sites. The first three members of the series (sites 1, 2, and 3), which are spaced further apart and deviate from the concensus sequence at either the first or last positions, appeared not to bind to NFI at all in this assay. The most distal three concensus repeat elements (sites 21, 22, and 23), which lie in an area where the spacing between individual repeats has been reduced to only 27 and 21 bp, respectively, appeared in the autoradiograph as a single, large, protected region without hypersensitive spacers.

## Other binding sites in the CMV IE94 and IE68 promoter-regulatory regions

The overall pattern of NFI binding sites both upstream and downstream of the human and simian versions of the major IE promoter are summarized in the diagram in Figure 4. Known or predicted strong NFI-binding sites are indicated by solid squares, and known or predicted weak binding sites are denoted by solid circles. Consistent with our findings from the filter-binding assays, DNA sequence analysis revealed additional NFI consensus sequences upstream of the complex enhancer region in IE68 and also downstream within the first intron in both genes. IE68 contains three copies of the consensus sequences, all in reverse orientation, at −628, −684, and −729 (Boshart et al. 1985). The first copy of this element (TTGGAN$_5$GCCAA) would be predicted to be a strong binding site, but the other two (both ATGGA/CN$_5$GCCAA) correspond to weak binding sites only. In the first intron of IE94, a cluster of five TGGA/CN$_5$GCCAA concensus NFI sites occur between positions +468 and +557, although from the binding affinity observed in the tandem repeat cluster, only one site— that at +557 (TTGGAN$_5$GCCAA)—would be predicted to be a strong binding site, and the others with either cytosine or adenine at position 1 should all represent weak sites. Similarly, in the IE68 intron (Ackrigg et al. 1985) a single perfect concensus site (TTGGCN$_5$GCCAA) occurring at +334 would presumably represent a strong binding site, whereas a second, closely related but diverged sequence (ATGGCN$_5$GCCAC) at +318 would not be expected to produce detectable binding in a footprinting assay.

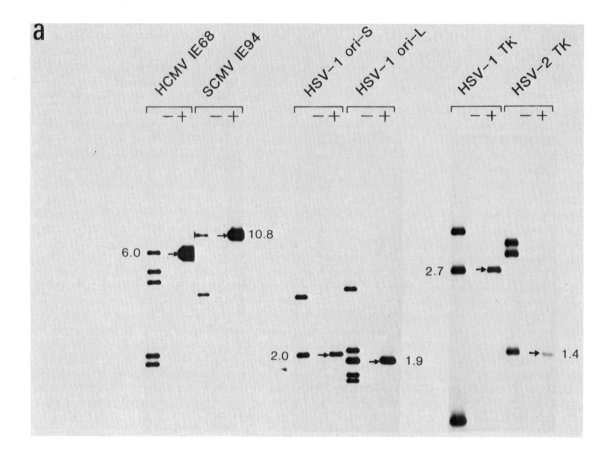

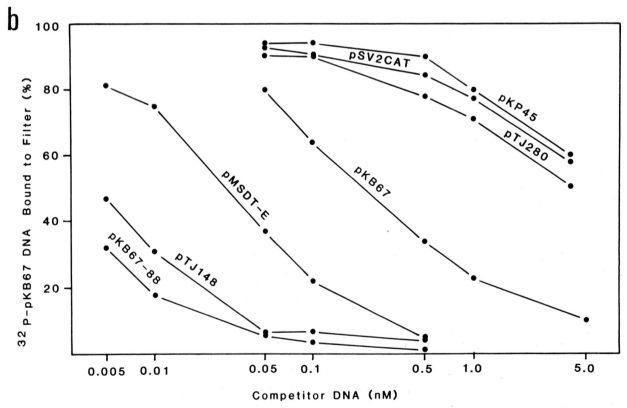

**Figure 2** *See facing page for legend.*

## Sequence-specific binding properties of EBNA-1

A 28K fusion polypeptide from the carboxyl terminus of EBNA-1 has been synthesized in an *E. coli* expression system vector (Waldman et al. 1983), and we have found it to have both double-stranded and sequence-specific DNA–binding properties. The polypeptide was purified by phosphocellulose and hydroxylapatite chromatography and used to prepare monospecific and monoclonal antibodies that specifically recognized the 80K EBNA protein in both EBV-positive Raji cells and in a *Bam*HI-K DNA–transfected mouse Ltk$^+$ cell line (Milman et al. 1985b). In initial experiments, the addition of the 28K EBNA protein was found to inhibit the mobility of both linear and supercoiled pBR322 DNA and of all *Eco*RI fragments of phage λ DNA in agarose gel electrophoresis assays (Milman et al. 1985a). However, subsequent studies with a plasmid containing the 1800-bp $ori_P$ region of EBV DNA revealed larger migration inhibition effects on the $ori_P$ DNA segment than on the pBR322 vector DNA (Fig. 5a). The specificity for $ori_P$ sequences was considerably enhanced when the same two DNA fragments were subjected to a filter-binding assay in the presence of an excess of nonspecific competitor phage λ DNA. The $ori_P$ fragment bound quantitatively in the presence of 5–10 μg/ml of 28K EBNA protein, whereas the vector DNA fragment failed to bind significantly even at 80 μg/ml of the protein (not shown). Thus, the 28K EBNA appeared to give a strong sequence-specific interaction with the plasmid maintenance region of the EBV genome. To compare the $ori_P$ sequences with other portions of the EBV genome, we also carried out a filter-binding assay with a set of cosmids containing a library of fragments from the EBV genome. This experiment revealed that only the *Bam*HI-C fragment (encompassing $ori_P$) and also the *Bam*HI-Q fragment of EBV(M-ABA) DNA bound to 28K EBNA (Rawlins et al. 1985).

## Two distinct clusters of EBNA-1–binding sites within the $ori_P$ region

The results of a filter-binding assay carried out with a number of recombinant pBR322-derived plasmid DNAs containing the entire *Bam*HI-C fragment from several different EBV isolates, M-ABA (pM765-10), B95-8 (pSL93), and P3HR-1 (pMC54), are shown in Figure 5b. In all three cases, we detected two EBNA-1–binding loci separated by internal *Nco*I cleavage sites. Clones pSL93 and pSL87 containing *Bam*HI-C were derived independently in our laboratory from EBV(B95-8) virion DNA, but their nucleotide sequence (obtained from the

subclone pPL7) differs significantly on the left-hand side of $ori_P$ from that in both the pHEBO-1 clone (received from Bill Sugden) and the total EBV (B95-8) DNA sequence presented by the Cambridge group (Baer et al. 1984). However, all four EBV(B95-8) clones tested retained two EBNA-binding loci separated by a 590-bp internal *Nco*I fragment. In contrast, the EBV(M-ABA) and EBV(P3HR-1) clones possessed only one *Nco*I site between the two binding loci and both differed further from each other and from the EBV(B95-8) clones in both the size and cleavage pattern of the region on the left-hand side of $ori_P$. The structure and map coordinates of the viral inserts in the two variants of the EBV(B95-8) $ori_P$ clones are presented in Figure 6a. The EBNA-1–binding locus on the left-hand side of $ori_P$ encompasses a region of 20 × 30-bp direct tandem repeats in pHEBO-1 (Fig. 6b). In contrast, pPL7 (and its parent pSL93) contain 7 fewer repeats in an inverted and rearranged organization that gives rise to 120-bp and 240-bp deleted forms in recA$^-$ *E. coli* (denoted by open triangles in Fig. 5b).

## DNase footprinting of tandemly repeated EBNA-1–binding sites in $ori_P$ region I

Evidence that the 30-bp tandem repeats on the left-hand side of $ori_P$ (region I) do indeed represent EBNA-1–binding sites was obtained by direct footprinting analysis of protected sequences in the presence of *E. coli* 28K EBNA. An example of the pattern obtained is shown in Figure 7a. The *Eco*RI-*Nco*I fragment of pPL7 and a *Sal*I-*Nco*I fragment of pHEBO-1 were both $^{32}$P-labeled at the *Nco*I site and treated to partial digestion with DNase I in the presence of increasing amounts of protein. A series of 25-bp protected regions were observed that (with few exceptions) were spaced exactly 30 bp apart with 3–4-bp hypersensitive sites between them. Unlike the situation with NFI sites in the CMV tandem repeats, all of these EBNA-1–binding sites appeared to have equal and strong affinity, and in each case the protected region centered approximately over a perfect or nearly perfect copy of an 18-bp palindromic concensus sequence of GGATAGCATATGCTACCC. A total of 20 adjacent binding sites were found in pHEBO-1 DNA and 13 in pPL7 DNA, and these all corresponded precisely in spacing and orientation to the patterns of the 30-bp tandem repeats present in each clone (see Fig. 6b). A detailed analysis of the protected and hypersensitive nucleotides over two adjacent concensus copies of the 30-bp tandem repeat sites from pHEBO-1 is given in Figure 6c.

**Figure 2** (*see facing page*) Strong NFI-binding loci near the major IE genes of CMV. (*a*) Additional filter-binding assays with pBR322-derived subclones containing selected regions of CMV and HSV DNA. Fragments exhibiting positive binding properties are identified by their approximate sizes in kilobase pairs. The following plasmid DNA samples and restriction enzymes were used: HCMV IE68 (pMSDT-E, *Bam*HI); SCMV IE94 (pTJ148, *Hind*III); HSV-1 $ori_S$ (pGH12, *Bam*HI/*Hind*III); HSV-1 $ori_L$ (pGH1, *Eco*RI/*Bam*HI), HSV-1 TK (pHSV106, *Bam*HI/*Bgl*II); HSV-2 TK (pGR18, *Bam*HI/*Sal*I/*Hind*III). (*b*) Competition filter-binding assay comparing relative NFI-binding affinities of single-copy (pKB67) and 88-copy (pKB67-88) Ad2 terminal fragment plasmids with those of plasmids containing the major IE68 gene of HCMV (pMSDT-E) and the major IE94 gene of SCMV (pTJ148). The assay measured the amount of $^{32}$P pKB67 DNA bound in the presence of a constant amount of NFI protein and increasing amounts of unlabeled competitor DNA.

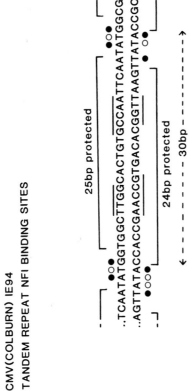

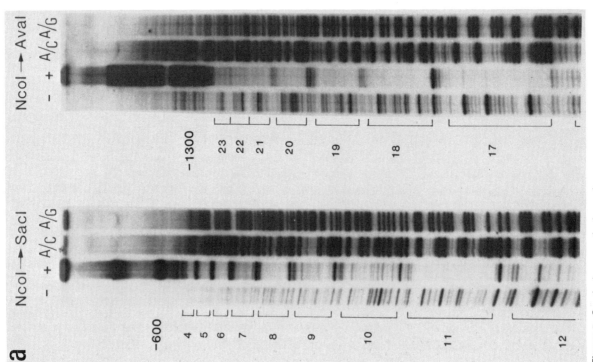

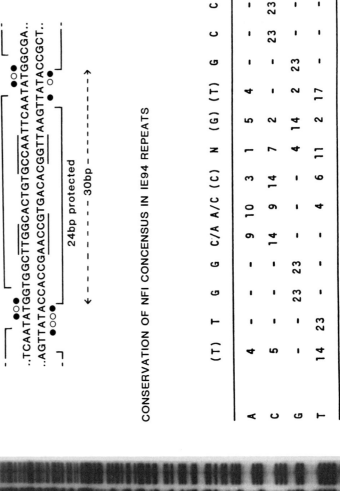

b

CMV(COLBURN) IE94
TANDEM REPEAT NFI BINDING SITES

```
                                              25bp protected
                                          ●○                    ●●
                                                                ●●
..TCAATATGGTGGCTTGGCACTGTGCCAATTCAATATGGCGA...
..AGTTATACCACCGAACCGTGACACGGTTAAGTTATACCGCT...
                ●●                       ○●
                ○○                       ●
                      24bp protected
              <------ - - - 30bp - - - - - ->
```

CONSERVATION OF NFI CONCENSUS IN IE94 REPEATS

|   | (T) | T | G | G | C/A | A/C | (C) | N | (G) | (T) | G | C | C | A | A |
|---|-----|---|---|---|-----|-----|-----|---|-----|-----|---|---|---|---|---|
| A | 4 | - | - | - | 9 | 10 | 3 | 1 | 5 | 4 | - | - | - | 23 | 21 |
| C | 5 | - | - | - | 14 | 9 | 14 | 7 | 2 | 2 | - | 23 | 23 | - | - |
| G | - | - | 23 | 23 | - | - | - | 4 | 14 | 2 | 23 | - | - | - | 1 |
| T | 14 | 23 | - | - | - | - | 6 | 11 | 2 | 17 | - | - | - | - | 1 |

**Figure 3** *See facing page for legend.*

**Figure 3** (*see facing page*) Tandemly repeated NFI-binding sites upstream of the CMV IE94 promoter-enhancer region. (*a*) DNase I footprint analysis on isolated 3′-end-labeled DNA fragments from pTJ148 in the presence (+) and absence (−) of NFI protein. Two reference lanes of Maxam and Gilbert sequencing reactions on the same DNA samples are also shown (A/C; A/G). The left-hand panel (*Nco*I → *Sac*I) represents the coding strand proceeding from the *Nco*I site at −990 in the middle of the cluster toward the promoter. The right-hand panel (*Nco*I → *Ava*I) represents the noncoding strand in the direction away from the promoter. Individual binding sites are indicated by brackets and identified relative to the numerical order of tandemly repeated NFI concensus elements in the DNA sequence. (*b*) (*Top*) Typical example of one of the 30-bp tandemly repeated NFI-binding sites showing the location and extent of DNase I protection on both DNA strands and the usual pattern of hypersensitive sites between adjacent protected regions. (*Bottom*) Table illustrating the extraordinary conservation of the concensus NFI-binding sequence (T)TGGA/CN$_5$GCCAA within the otherwise highly diverged DNA sequences of the 23 × 30-bp tandem repeats.

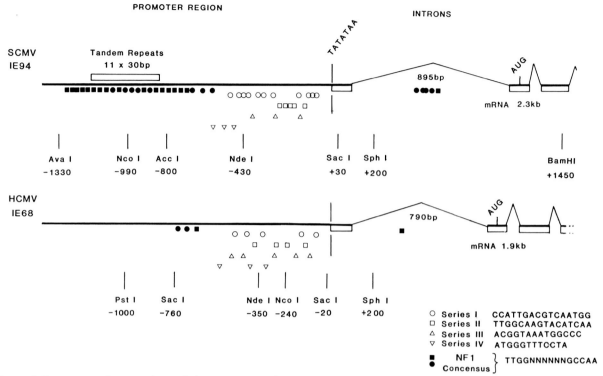

**Figure 4** Summary and comparison of the structure and arrangement of repeat elements and NFI-binding sites within the promoter-regulatory region and introns of IE94 and IE68. The 5′ and aminoterminal portions of the two genes are drawn oriented with transcription proceeding 5′ to 3′ toward the right (although their orientation in the standardized physical map of the CMV viral genome is leftward). The open symbols (○, △, △, and □) denote sets of interspersed repetitive elements within the "enhancer" region (−70 to −550). The solid symbols refer to known or predicted strong NFI-binding sites (■) and known or predicted weak NFI-binding sites or concensus sequences (●).

## A different pattern of EBNA-binding sites in the region II dyad symmetry locus of ori_P

A similar footprinting analysis of protected sites in the EBNA-binding locus on the right-hand side of $ori_P$ (region II) was carried out on a BamHI-NcoI fragment of pPL7 DNA labeled at the BamHI linker site. In this region the sequence of pPL7 DNA was identical with that given by Baer et al. (1984). Both short and long gel tracts from the same sample are shown in Figure 7b. The protection pattern displayed by region II again proved to contain multiple binding sites, but with different spacing and orientation than in the tandem repeats of region I. Two large 46–47-bp protected regions were found separated by an 8-bp hypersensitive spacer sequence. Each of the 46-bp sites appeared to represent two overlapping binding sites, and this interpretation was confirmed by inspection of the DNA sequence, which revealed a total of four related core binding-site sequences, each showing extensive homology with the concensus EBNA-1–binding-site sequence from region I (see Fig. 6d). The DNA sequence across region II can be drawn as a 114-bp imperfect stem-loop structure in which the top of the loop corresponds to the hypersensitive region observed between the two 46-bp protected sites. More-detailed analysis of the data indicated that each 46-bp site probably consists of two head-to-head binding sites centered 21 bp apart, in contrast to the

head-to-tail 30-bp spacing found in the tandem repeat cluster. Computer analysis revealed a similar homologous region in BamHI-Q with two core concensus sites spaced 23 bp apart and apparently in a head-to-head configuration (Rawlins et al. 1985).

## Discussion

During the transcriptional and replicative metabolic processes of cells, the cis-acting regulatory effects of specific DNA sequences are mediated, at least in part, by the interactions of these sequences with DNA-binding proteins. In the past few years, investigators have identified many cis-acting control elements of viral and cellular DNAs by using a variety of in vivo and in vitro assays to provide functional evidence of their roles. Included among these regulatory sequences are promoter elements, transcriptional enhancers, plasmid maintenance sequences, and origins of replication. We have studied two such interactions involving herpesvirus genomes, one with a cellular protein (NFI) and one with a virus-encoded protein (EBNA-1). Both proteins are strongly implicated, directly or indirectly, in DNA replication processes, and both may also potentially function in transcriptional regulation.

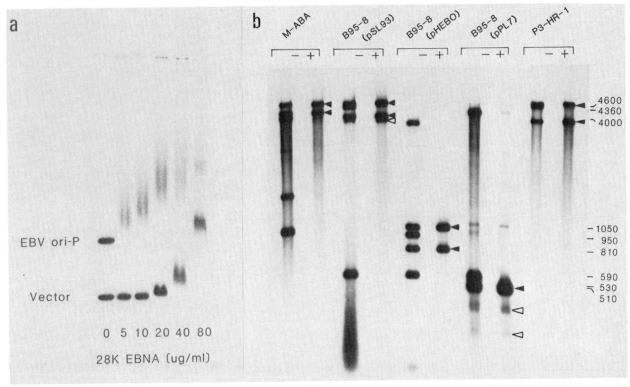

**Figure 5** Evidence for preferential EBNA-1 binding at two loci within EBV DNA clones containing *ori*P. (*a*) Mobility shift-up assay with *Eco*RI-cleaved $^{32}$P-labeled pHEBO-1 DNA in the presence of increasing amounts of purified *E. coli* 28K EBNA protein. (*b*) Filter-binding assay with cleaved *Bam*HI-C- or *ori*P-containing DNA clones from different EBV isolates: (M-ABA (p765-10, *Bam*HI/*Nco*I); B95-8 (pSL93, *Bam*HI/*Nco*I); B95-8 (pHEBO-1, *Sal*I/*Bam*HI/*Nco*I); B95-8 (pPL7, *Bam*HI/*Eco*RI/*Nco*I); P3-HR-1 (pMC54, *Bam*HI/*Nco*I). Solid arrowheads indicate positive bands; open arrowheads denote minor deleted forms.

## Occurrence and significance of NFI-binding sites in CMV and other DNA viruses

NFI-binding loci occur at a frequency of approximately one in every 100,000 bp in mammalian DNA and apparently at higher frequency in papovavirus, adenovirus, and herpesvirus DNAs. This number is close to the random frequency of occurrence of an average 9–10-bp sequence element. Therefore, questions arise as to whether all of these sites, even though they may bind to NFI in vitro, are likely to be functionally significant. We expect that most of the functionally important NFI-binding sites will have one or more of the following properties: (1) be found consistently at equivalent, significant regulatory loci (such as replication origins or in the upstream promoter regions) of related genes or genomes; (2) lie adjacent to other specific DNA-protein binding sites; and (3) occur in multiple copies or within short, tandemly repeated sequences. At present, the only proven functional role of NFI binding is in adenovirus DNA replication, where the binding sites lie adjacent to other conserved essential elements within 50 bp of the genomic termini and are themselves highly conserved among different adenovirus genomes. More importantly, the NFI protein itself has been shown to be essential for efficient and specific initiation in vitro of the adenovirus DNA replication process and to act synergistically with other protein components of the system. The

exact mechanism of NFI action is unknown, but it may be noteworthy that in some experiments a significant portion of the bound adenovirus terminal DNA fragment was recovered from the filter-binding assay in a single-stranded form (see asterisks in Fig. 1 for pKB67, HPV, and Ad5).

The pattern of tandemly repeated, multiple binding sites found adjacent to the strong enhancer region in the 5'-upstream region of the SCMV IE94 promoter clearly indicates a functionally significant interaction. The almost perfect conservation of the (T)TGGA/ CN$_5$GCCAA elements, among considerable sequence divergence within these tandem repeats, indicates not only that the original duplication events probably occurred a long time ago, but that there was considerable selective pressure to retain the NFI-binding interaction during their "evolution." However, the nature of this functional role is unclear at present. The location of replication origins interspersed with or adjacent to the "immediate-early" promoter-enhancer regions is a common theme found throughout papovaviruses and adenoviruses and also in HSV. The organization into two sets of multiple binding sites, one being a cluster of 30-bp tandem repeats, is also remarkably similar to the organization of the *ori*P region in EBV. However, there is no evidence for or against a replication role in CMV at present, and the absence of tandem repeats in the

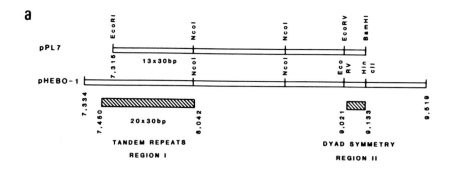

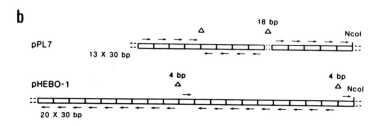

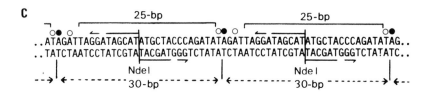

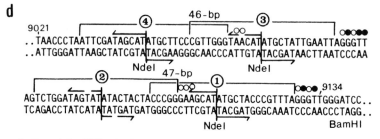

**Figure 6** Structural organization of the EBV *ori*ₚ plasmid maintenance region and nucleotide sequence of EBNA-1 binding sites. (*a*) Physical map of viral DNA inserts in two variants of the EBV(B95-8) *ori*ₚ region represented by the pPL7 and pHEBO-1 plasmids. Nucleotide coordinates refer to the Cambridge EBV DNA sequence (Baer et al. 1984). The boundaries of the two genetic elements found to be essential for plasmid maintenance functions are shown by hatched bars. (*b*) Diagram illustrating our interpretation of the number, arrangement, and orientation of the 30-bp tandem repeat elements (and EBNA-1-binding sites) within *ori*ₚ region I in the two EBV(B95-8) variants studied here. (△) The location of deletions in the standard 30-bp repeat pattern. (*c*) Location of DNase I–protected regions (bracketed) and strong (●) and weak (○) hypersensitive sites in the nucleotide sequence of the EBNA-1-binding region on the coding strand over two copies of the concensus 30-bp tandem repeats (region I). (*d*) Similar analysis as in *c* for dyad symmetry locus (region II).

HCMV IE68 gene (at least in the standard fibroblast-adapted laboratory strains examined) weakens the analogy.

In other studies we have found that the upstream clusters of NFI-binding sites do not contribute to the fourfold greater activity of IE94 promoter-regulatory region compared with the complete SV40 early promoter in transient chloramphenicol acetyltransferase expression assays in VERO cells (see O'Hare et al., this volume). Nor does the NFI region represent part of the strong enhancer region identified in the human IE68 promoter-regulatory region (Boshart et al. 1985). In both cases, a complex 550-bp region is involved that consists of four sets of unrelated, interspersed repeat elements lying immediately upstream from the mRNA start site. We have also found that a single-copy cloned, synthetic concensus NFI-binding site does not act as an enhancer of transcription from the minimal SV40 early promoter. Therefore, the NFI clusters appear to surround the CMV IE enhancer-promoter region on both sides but to have no direct transcriptional effects in vitro. Nevertheless, SCMV(Colburn) displays a much broader host range for expression of the IE protein than does HCMV (R.L. LaFemina and G.S. Hayward, in prep.). For example, IE68 mRNA and protein are synthesized in diploid human fibroblasts in culture and in human teratocarcinoma cells after differentiation by retinoic acid, but not at all after infection in the undifferentiated human teratocarcinoma stem cells, adenovirus E1A-expressing human cells (293), SV40-transformed HEK cells (NBE), or VERO cells. In contrast, IE94 is synthesized abundantly after infection of all of these cell types. Since input HCMV DNA enters the nucleus in all of the cell lines examined, we have speculated that the tandemly repeated NFI-binding sites in IE94 play a role in overcoming whatever cellular and/or viral factors prevent transcription of IE68 in the "transformed" lines.

In the case of the papovavirus and papillomavirus genomes, relatively strong NFI-binding loci were found in several human viruses (BK, JC, and HPV), but not in SV40 or polyoma. In the three positive examples, concensus NFI sequences occur in the close vicinity of the known DNA replication origins and seem likely to be of functional relevance. However, in HSV-1 DNA, although binding occurred within 1 to 2 kb of both $ori_S$ and $ori_L$, there are no NFI concensus elements in the DNA sequences within the immediate vicinity of the known boundaries of either site (Stow and McMonagle 1983; Weller et al. 1985). A binding site was also found in the long terminal repeat (LTR) region of murine sarcoma virus proviral DNA but not in adenovirus-associated virus DNA (not shown). Preliminary filter-binding analysis of a set of EBV cosmid clones revealed only weak binding relative to the other herpesvirus genomes, and indeed there were no good matches in the entire EBV DNA sequence to any of the preferred binding concensus sequences defined here from the CMV footprinting analysis. Cellular NFI-binding sites with good matches to our concensus sequence have also been detected in a mouse IgM gene intron region by footprinting analysis and 5' of the human c-*myc* gene by filter binding (Siebenlist et al. 1984; Hennighausen et al. 1985).

The results of a computer search of the Los Alamos GenBank DNA library for all mammalian viral and genomic DNA sequences that match our NFI concensus of $(T)TGGA/CN_5GCCAA$ is presented in Table 1, together with a summary of all of our observations concerning both positive and negative binding by the filter-binding assay and, where appropriate, the relative levels of affinity in footprinting assays. The list also includes other known examples of matches from CMV DNA, etc. that are not in the data bank and includes some selected examples of incomplete matches to the concensus. The

analysis indicates a far-above-random occurrence of potential NFI-binding sites both among small DNA viruses and in the LTR regions of several retroviruses. In CMV, the strongest and most consistent binding was observed with concensus sequences having an additional thymine in the first position and weaker and less consistent results were obtained when cytosine or adenine lies at that position. Very few examples of positive binding have been observed with variations at other positions, although this may represent a bias in the sequences of the samples tested rather than strict specificity of the NFI enzyme. For additional information from the results of mutational analysis of sequence requirements across the Ad2 binding site, see Rosenfeld et al. (this volume). Borgmeyer et al. (1984) have reported binding of "TGGCCA" protein(s) from chicken oviduct extracts to three sites in the far upstream regions of the chicken lysozyme gene that differ slightly from our concensus and also to a site in the left arm of phage λ DNA. However, our HeLa cell NFI preparation, which had been purified using an adenovirus DNA replication assay, did not bind to phage λ DNA, thus implying either that NFI preparations from different sources may have different specificities or perhaps that multiple NFI-like DNA-binding proteins with different but related specificities may exist.

### Specific-binding of EBNA-1 to the $ori_p$ region in EBV DNA

The 3' exon of the EBNA-1 gene that is contained within the *Bam*HI-K fragment of EBV DNA apparently includes the entire coding region for the protein and is expressed efficiently (presumably through a cryptic pBR322 promoter) in both long-term coselection and transient DNA-transfection systems (Fischer et al. 1984; Hearing et al. 1984). Using cultured fibroblast cell lines expressing *Bam*HI-K EBNA, Sugden and his colleagues showed that this gene product, acting in *trans* together with the *cis*-acting $ori_p$ DNA segment of the EBV genome, is both necessary and sufficient for the establishment and maintenance of a multicopy plasmid state with a selectable pBR322-Neo$^R$ DNA marker transfected into the cells (Yates et al. 1984, 1985). Our studies showing that $ori_p$ contains two distinct clusters of specific binding sites for EBNA-1 imply that the function of EBNA-1 in plasmid maintenance must be mediated by direct protein-DNA interactions.

The stable *E. coli*–synthesized 28K fusion protein used in our experiments consists of 33 amino acids from the amino terminus of the phage λ N protein linked to the carboxyterminal 191 amino acids of *Bam*HI-K EBNA. The specific binding and protection pattern observed within $ori_p$ with this protein fragment implies that the complete binding domain of EBNA-1 lies within the carboxyterminal one-third of the protein and that the binding interaction itself presumably does not require the phosphorylated modifications present in the in vivo form of the protein. One would expect that the intact EBNA-1 protein, like that of SV40 T antigen will have additional pleiotropic functions in vivo and that this part

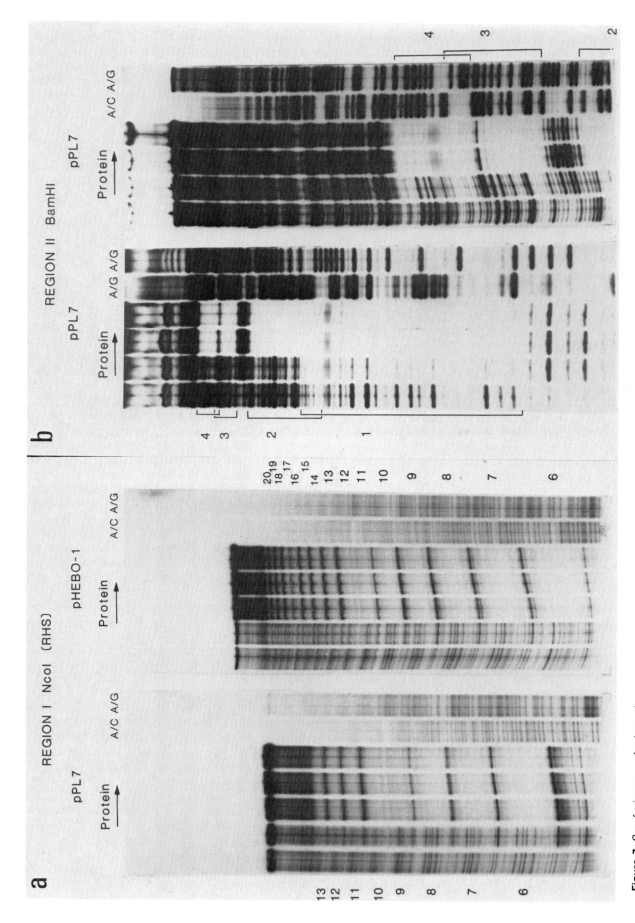

**Figure 7** *See facing page for legend.*

**Figure 7** (*see facing page*) Footprint analysis of EBNA-1 binding in the tandem repeat and dyad symmetry regions of EBV $ori_p$. The first lane in each panel shows a partial DNase I digestion in the absence of protein, and subsequent lanes show the patterns obtained in the presence of increasing amounts of 28K EBNA. Lanes A/C and A/G show parallel Maxam and Gilbert chemical sequencing tracts from the same DNA samples. Individual protected binding sites are numbered from right to left in the standard genome orientation (Baer et al. 1984). (a) Region I: DNase I protection of the 510-bp $NcoI \rightarrow EcoRI$ fragment of pPL7 and the 1000-bp $NcoI \rightarrow SalI$ fragment from pHEBO-1 DNA (both 3′-end-labeled at the $NcoI$ site) in the presence of increasing amounts of 28K EBNA protein. (b) Region II: Similar analysis of the isolated 530-bp $BamHI \rightarrow NcoI$ DNA fragment from pHEBO-1 labeled at the $BamHI$ site. The left- and right-hand panels show short and long fractionations of the same samples.

**Table 1** Correlation of Known Concensus Sites with Observed NFI-binding Properties

| Sequence | Virus or gene | Location | NFI-binding properties[a] |
|---|---|---|---|
| TTGGCN5GCCAA | Ad-2, Ad-5 | Ori | FP + + + |
| TTGGCN5GCCAA | HTLV-I | LTR | |
| TTGGCN5GCCAA (10×) | SCMV | IE 5' | FP 9X + + +, 1X + + |
| TTGGCN5GCCAA | HCMV | IE intron | FBA + VE |
| TTGGAN5GCCAA | LA cross virus | | |
| TTGGAN5GCCAA | Ad-FL | Ori | FBA + VE |
| TTGGAN5GCCAA | HCMV | IE 5' | FBA + VE |
| TTGGAN5GCCAA (2×) | SCMV | IE 5' | FP 2X + + + |
| TTGGAN5GCCAA | H-1 parvovirus | | |
| TTGGAN5GCCAA | human HLA DC-B | | |
| CTGGCN5GCCAA | HEP-B | | |
| CTGGCN5GCCAA | BPV | Ori | |
| CTGGCN5GCCAA (2×) | SCMV | IE 5' | FP + + +, + |
| CTGGCN5GCCAA | influenza nucleoprotein | | |
| CTGGCN5GCCAA | MLV, MSV | LTR | FBA + VE |
| CTGGCN5GCCAA | HSV-2 TK | coding | FBA − VE |
| CTGGAN5GCCAA | Ad-12(Ad-18,Ad-31) | Ori | FP + VE |
| CTGGAN5GCCAA (5×) | SCMV | IE 5' | FP + + +, +, 3X − |
| CTGGAN5GCCAA | SCMV | IE intron | FBA + VE |
| CTGGAN5GCCAA | human c-*myc* | 5'−1500 | FBA + VE, FP − |
| CTGGAN5GCCAA | human HLA DR | | |
| CTGGAN5GCCAA | murine HLA *H2-B* | | |
| CTGGAN5GCCAA | goat α-globin | | |
| ATGGCN5GCCAA | JC | Ori | FBA + VE |
| ATGGCN5GCCAA | human endogenous retrovirus | LTR | |
| ATGGCN5GCCAA | SCMV | IE 5' | FP + |
| ATGGCN5GCCAA (2×) | SCMV | IE intron | |
| ATGGCN5GCCAA (2×) | HCMV | IE 5' | |
| ATGGCN5GCCAA | human IgM constant | intron | FP + + |
| ATGGCN5GCCAA | human P21 *ras*(T-24) | | |
| ATGGCN5GCCAA | human α-tubulin | | |
| ATGGAN5GCCAA | Ad-7(Ad-3) | Ori | FBA + VE |
| ATGGAN5GCCAA | BK | Ori | FBA + VE |
| ATGGAN5GCCAA | HTLV-III(ARV-2) | | |
| ATGGAN5GCCAA (3×) | SCMV | IE 5' | FP +, +, − |
| GTGGCN5GCCAA | JC | | |
| GTGGCN5GCCAA | HSV-1 IE5 | | |
| GTGGCN5GCCAA | EBV 117,000 | | |
| GTGGCN5GCCAA | phage λ A gene | | *, FBA − VE |
| GTGGAN5GCCAA | EBV ribonucleotide reductase | | |
| GTGGAN5GCCAA | VZV US | | |
| GTGGAN5GCCAA | BK | | |
| TTGGCN5GCCCA | chicken lysozyme | 5' | * |
| TTGGAN5CCCAA | human kpn repeat | | |
| TTGGAN5ACCAA | MMTV | LTR | |
| TTGATN5GCCAA | SCMV | enhancer | FBA − VE |
| TTGATN5GCCAA (2×) | HCMV | enhancer | |
| TGGGCN5GCCAA | SA7 | Ori | |
| CGGGCN5GCCAA | equine Ad | Ori | FBA + VE |
| ATGGCN5CCCAA | human Alu repeat | | FBA − VE |
| ATGGCN5ACCAA | Ad-5 Hexon | | FBA − VE |
| GTGGCN5CCCAA | HSV GP-B, GP-D | | |
| GTGGCN5CCCAA | RAV-O, RSV | LTR | |
| CTGGCN5GTCAA | chicken lysozyme | 5' | * |
| CTGGCN5GCCAC | chicken lysozyme | 5' | * |
| TTGGCN6GCCAA | SCMV | enhancer | FBA − VE |

[a]Key: (FBA) filter-binding assay; (FP) DNase I footprinting; (+, + +, + + +) increasing relative affinity in footprinting analysis; (*) positive results with chicken oviduct "TGGCCA"-binding protein (Borgmeyer et al. 1984).

of the protein alone is insufficient to support maintenance of the plasmid state. The interaction of EBNA-1 with the *ori*~P~ or *Bam*HI-Q binding sites could be involved in one or more of the following functions: (1) controlling the rate of initiation of EBV plasmid DNA synthesis (i.e., once per DNA molecule per cell cycle); (2) controlling the copy number of the EBV plasmids; (3) ensuring uniform segregation of the EBV genomes among the daughter cells; (4) repression or prevention of lytic cycle functions; and (5) autoregulation of its own expression.

However, the immortalization functions of EBV appear to be associated with a different latency product, another nuclear antigen referred to as EBNA-2.

We have suggested elsewhere a model in which the region-I repeats might represent a high-affinity "sink" binding locus and the dyad symmetry a lower-affinity "origin" or initiation site for plasmid DNA replication (Rawlins et al. 1985). This could potentially also provide control over plasmid copy number but probably requires that very limiting amounts of EBNA-1 protein, or at least of an "activated" DNA-binding form of EBNA, be present in the cell. An alternative role for the binding interaction may involve segregation or active prevention of integration. For example, the propensity of the EBNA-1 protein to associate with metaphase chromosomes during cell division could have the advantageous effect of distributing the attached EBV genomes relatively uniformly among the host cell chromosomes, thus ensuring approximately equal segregation into the daughter cell nuclei.

Although the coding region for EBNA-1 in *Bam*HI-K lies over 100 kb away from *ori*$_P$, it represents only a 2.0-kb 3' exon in a larger 3.7-kb mRNA, and recent studies have indicated that latency mRNAs exist that contain multiple, short spliced portions of far-upstream sequences linked to the *Bam*HI-K exon (Bodescot et al. 1985; S. Speck, pers. comm.). In particular, splicing signals within the 12 × 3072-bp *Bam*HI-W repeats appear to be involved, which lie 60 to 90 kb upstream. Moreover, Reisman et al. (1985) have discovered that the 30-bp tandem repeats from region I of *ori*$_P$ act as EBNA-dependent transcriptional enhancers in transient expression assays. Therefore, it seems very likely that the EBNA-1 protein positively regulates its own transcription during latency by direct interactions at the tandem repeat binding sites.

## Acknowledgments

This work was funded by National Cancer Institute grants CA28473 and CA37314 awarded to G.S.H., grant CA30356 awarded to S.D.H., and grant CA16579 awarded to T.J.K., Jr. D.R. was supported by fellowship DRG-607 from the Damon Runyon–Walter Winchell Cancer Fund. K.-T.J. and P.R. were supported by the Medical Scientist Training Program at Johns Hopkins Medical School (NIH GM07309). G.S.H. was supported by a Faculty Research Award from the American Cancer Society (FRA # 247).

We thank Mabel Chiu for technical assistance, Keith Peden, Phoebe Mounts, Georg Bornkamm, Myung-Sam Cho, Barbera Sollner-Webb, and Mark Stinski for gifts of plasmids or DNA, and Judy DiStefano and Pamela Wright for assistance in preparation of the manuscript.

## References

Ackrigg, A., G.W.G. Wilkinson, and J.D. Oram. 1985. The structure of the major immediate early gene of human cytomegalovirus strain AD169. *Virus Res.* **2**: 107.

Baer, R., A.T. Bankier, M.D. Biggin, P.L. Deininger, P.J. Farrell, T.J. Gibson, G. Hatful, G.S. Hudson, S.C. Satchwell, C. Seguin, P.S. Tuffnell, and B.G. Barrell. 1984. Organization of the B95-8 Epstein-Barr virus genome. *Nature* **310**: 207.

Bencini, D.A., G.A. O'Donovan, and J.R. Wild. 1984. Rapid chemical degradation sequencing. *Biotechniques* **2**: 4.

Bodescot, M., B. Chambraud, P. Farrell, and M. Pericaudet. 1985. Spliced RNA from 1R1-U2 region of Epstein-Barr virus: Presence of an open reading frame for a repetitive polypeptide. *EMBO J.* **3**: 1913.

Borgmeyer, U., J. Nowock, and A.E. Sippel. 1984. The TGGCCA-binding protein: A eucaryotic nuclear protein recognizing a symmetrical sequence on double-stranded linear DNA. *Nucleic Acids Res.* **12**: 4295.

Boshart, M., F. Weber, G. Jahn, K. Dorsch-Hasler, B. Fleckenstein, and W. Schaffner. 1985. A very strong enhancer is located upstream of an immediate-early gene of human cytomegalovirus. *Cell* **41**: 521.

Challberg, M.D. and D.R. Rawlins. 1984. Template requirements for the initiation of adenovirus DNA replication. *Proc. Natl. Acad. Sci.* **81**: 100.

Ciufo, D.M. and G.S. Hayward. 1981. Tandem repeat defective DNA from the L-segment of herpes simplex virus genome. In *Herpesvirus DNA: Developments in molecular virology* (ed. Y. Becker), vol. 1, p. 107. Martinus-Nijhoff, Boston.

Fischer, D.K., M.F. Robert, D. Shedd, W.P. Summers, J.E. Robinson, J. Wolak, J.E. Stephano, and G. Miller. 1984. Identification of Epstein-Barr nuclear antigen polypeptide in mouse and monkey cells after gene transfer with cloned 2.9 kilobase pair subfragment of the genome. *Proc. Natl. Acad. Sci.* **81**: 41.

Gronostajski, R.M., S. Adhya K. Nagata, R.A. Guggenheimer, and J. Hurwitz. 1985. Site-specific DNA binding of nuclear factor I: Analysis of cellular binding sites. *Mol. Cell. Biol.* **5**: 964.

Guggenheimer, R.A., B.W. Stillman, K. Nagata, F. Tamanoi, and J. Hurwitz. 1984. DNA sequences required for the *in vitro* replication of adenovirus DNA. *Proc. Natl. Acad. Sci.* **81**: 3069.

Hearing, J.C. and A.J. Levine. 1985. The Epstein-Barr virus nuclear antigen (*Bam*HI K antigen) is a single-stranded DNA binding phosphoprotein. *Virology* **145**: 105.

Hearing, J.C., J.-C. Nicolas, and A.J. Levine. 1984. Identification of Epstein-Barr virus sequences that encode a nuclear antigen expressed in latently infected lymphocytes. *Proc. Natl. Acad. Sci.* **81**: 4373.

Hennessy, K., M. Heller, V. van Santen, and E. Kieff. 1983. Simple repeat array in Epstein-Barr virus DNA encodes part of the Epstein-Barr nuclear antigen. *Science* **220**: 1396.

Hennighausen, L., U. Siebenlist, D. Danner, P. Leder, D. Rawlins, P. Rosenfeld, and T.J. Kelly, Jr. 1985. High-affinity binding site for a specific nuclear protein in the human IgM gene. *Nature* **314**: 289.

Jeang, K.-T., M.-S. Cho, and G.S. Hayward. 1984. Abundant constitutive expression of the immediate-early 94K protein from cytomegalovirus (Colburn) in a DNA-transfected mouse cell line. *Mol. Cell. Biol.* **4**: 2214.

Lindahl, T., A. Adams, G. Bjursell, G.W. Bornkamm, C. Kaschka-Dierich, and U. Jehn. 1976. Covalently closed circular duplex DNA of EBV in a human lymphoid cell line. *J. Mol. Biol.* **102**: 511.

Maxam, A.M. and W. Gilbert. 1977. A new method for sequencing DNA. *Proc. Natl. Acad. Sci.* **74**: 560.

Milman, G., D.K. Ades, M.-S. Cho, S.C. Hartman, G.S. Hayward, A.L. Scott, and S.D. Hayward. 1985a. Bacterially synthesized EBNA as a reagent for enzyme linked immunosorbent assays. In *Epstein-Barr virus and associated diseases* (ed. P. Levine et al.), p. 426. Martinus-Nijhoff, Boston.

Milman, G., A.L. Scott, M.-S. Cho, S.C. Hartman, D.K. Ades, G.S. Hayward, P.-F. Ki, J.T. August, and S.D. Hayward. 1985b. Carboxyl-terminal domain of the Epstein-Barr virus nuclear antigen in highly immunogenic in man. *Proc. Natl. Acad. Sci.* **82**: 6300.

Nagata, K., R.A. Guggenheimer, and J. Hurwitz. 1983. Specific binding of a cellular DNA replication protein to the origin of replication of adenovirus DNA. *Proc. Natl. Acad. Sci.* **80:** 6177.

Nagata, K., R.A. Guggenheimer, T. Enomoto, J.H. Lichy, and J. Hurwitz. 1982. Adenovirus DNA replication *in vitro*: Identification of a host factor that stimulates synthesis of the preterminal protein-dCMP complex. *Proc. Natl. Acad. Sci.* **79:** 6438.

Rawlins, D.R., G. Milman, S.D. Hayward, and G.S. Hayward. 1985. Sequence specific DNA binding of the Epstein-Barr virus nuclear antigen (EBNA-1) to clustered sites in the plasmid maintenance region. *Cell* **42:** 859.

Rawlins, D.R., P.J. Rosenfeld, R.J. Wides, M.D. Challberg, and T.J. Kelly, Jr. 1984. Structure and function of the adenovirus origin of replication. *Cell* **37:** 309.

Reisman, D., J. Yates, and B. Sugden. 1985. A putative origin of replication of plasmids derived from Epstein-Barr virus is composed of two *cis*-acting components. *Mol. Cell. Biol.* **5:** 1822.

Rosenfeld, P. and T.J. Kelly. 1986. Purification of nuclear factor I by DNA recognition site affinity chromatography. *J. Biol. Chem.* **261:** 1398.

Siebenlist, U., L. Hennighausen, J. Bettey, and P. Leder. 1984. Chromatin structure and protein binding in the putative regulatory region of the c-*myc* gene in Burkitt lymphoma. *Cell* **37:** 381.

Sculley, T.B., T. Kreofsky, G.R. Pearson, and T.C. Spelsberg. 1983. Partial purification of the Epstein-Barr virus nuclear antigen(s). *J. Biol. Chem.* **258:** 3974.

Stow, N.D. and E.C. McMonagle. 1983. Characterization of the $TR_s/IR_s$ origin of DNA replication of herpes simplex virus type 1. *Virology* **130:** 427.

Summers, W.P., E.A. Grogan, D. Shedd, M. Robert, C.R. Liu, and G. Miller. 1982. Stable expression in mouse cells of nuclear neo-antigen after transfer of a 3.4 megadalton cloned fragment of Epstein-Barr virus DNA. *Proc. Natl. Acad. Sci.* **79:** 5688.

Waldman, A.S., E. Haeusslein, and G. Milman. 1983. Purification and characterization of herpes simplex virus (type 1) thymidine kinase produced in *Escherichia coli* by a high efficiency expression plasmid utilizing a (lambda) $P_L$ promoter and a $cl_{857}$ temperature-sensitive repressor. *J. Biol. Chem.* **258:** 11571.

Weller, S.K., A. Spadaro, J.E. Schaffer, A.W. Murray, A.M. Maxam, and P.A. Schaffer. 1985. Cloning, sequencing, and functional analysis of $ori_L$, a herpes simplex virus type 1 origin of DNA synthesis. *Mol. Cell. Biol.* **5:** 930.

Yates, J.L., N. Warren, and B. Sugden. 1985. Stable replication of plasmids derived from Epstein-Barr virus in various mammalian cells. *Nature* **313:** 812.

Yates, J.L., N. Warren, D. Reisman, and B. Sugden. 1984. A *cis*-acting element from the Epstein-Barr viral genome that permits stable replication of recombinant plasmids in latently infected cells. *Proc. Natl. Acad. Sci.* **81:** 3806.

# Characterization of the Genetic Signals Required for Epstein-Barr Virus Plasmid Maintenance

### S. Lupton and A.J. Levine

Department of Molecular Biology, Princeton University, Princeton, New Jersey 08544

The Epstein-Barr virus (EBV) genome becomes established as a multicopy plasmid in the nucleus of infected B lymphocytes. A *cis*-acting DNA sequence previously discovered within the *Bam*HI-C fragment of the EBV genome (Yates et al. 1984) allows stable extrachromosomal plasmid maintenance in latently infected cells, but not in EBV-negative cells. In agreement with Yates et al. (1984), a deletion analysis assigned this function to a 2208-bp region of the *Bam*HI-C fragment containing a striking repetitive sequence and a large dyad symmetry. A recombinant vector, p410$^+$, was constructed that carries the *Bam*HI-K fragment, encoding the EBV-associated nuclear antigen (EBNA-1), the *cis*-acting sequence from the *Bam*HI-C fragment, and a dominant, selectable marker gene. After transfection of HeLa cells, this plasmid conferred G418 resistance (G418$^r$) and persisted extrachromosomally. Mutations in the *Bam*HI-K–derived portion of p410$^+$, altering the carboxyterminal portion of EBNA-1, destroyed the ability of the plasmid to persist extrachromosomally in HeLa cells. These observations indicate that in combination with the *cis*-acting sequence located in the *Bam*HI-C fragment, EBNA-1 is in part responsible for EBV-derived extrachromosomal plasmid maintenance in HeLa cells. Deletion mapping demonstrated that a portion of the *Bam*HI-C fragment can functionally substitute for the SV40 enhancer and promoter. The enhanced frequency of G418$^r$ colonies formed by some EBV-derived plasmids when introduced into D98-Raji cells did not correlate with the ability to be maintained extrachromosomally. One interpretation of these data is that the *Bam*HI-C fragment contains a promoter element stimulated by an EBV- or lymphocyte-induced gene function.

Epstein-Barr virus (EBV) DNA persists as an autonomous circular plasmid in B-lymphoblastoid cells derived from African Burkitt's lymphoma and in epithelial cells derived from nasopharyngeal carcinomas in southern China (zur Hausen et al. 1970; Kaschka-Dierich et al. 1976). Infection of B cells in culture with EBV can produce immortalized or permanent B-cell lines that also contain multiple copies of an extrachromosomal, circular EBV plasmid DNA (Henle et al. 1967; Lindahl et al. 1976). The duplication of this viral episome and its efficient segregation into daughter cells requires virally encoded *cis*-acting and *trans*-acting functions in addition to several cellular gene products that regulate viral DNA replication in a cell-cycle-dependent fashion. The *Bam*HI-C DNA fragment of the EBV genome has been shown to contain two *cis*-acting sequences required for the stable persistence of plasmids in cells containing the EBV genome (and expressing *trans*-acting viral functions) (Sugden et al. 1984; Yates et al. 1984; Lupton and Levine 1985; Reisman et al. 1985). This *cis*-acting element fails to promote plasmid maintenance in cells not carrying the EBV genome (Yates et al. 1984, 1985; Lupton and Levine 1985). The virus-encoded *trans*-acting factor, important for long-term persistence of plasmids containing the EBV *Bam*HI-C *cis*-acting signals, was shown to be encoded by the gene for the EBV nuclear antigen 1 (EBNA-1) contained in the *Bam*HI-K

DNA fragment (Lupton and Levine 1985). Recombinant plasmids containing the *Bam*HI-C (*cis*-acting) and *Bam*HI-K (*trans*-acting) gene functions were able to persist as autonomous episomes in cells that did not contain the EBV genome (Lupton and Levine 1985; Yates et al. 1985). Deletion and frameshift mutations in the coding region of the EBNA-1 gene eliminated the ability of these plasmids to persist in EBV-negative (but not EBV-positive) cell lines. In addition, the EBNA-1 antigen has been shown to bind to the *cis*-acting sequences contained in the *Bam*HI-C DNA fragment (Rawlins et al. 1985 and this volume). Thus, both biochemical and genetic evidence implicate the interaction of the *Bam*HI-K EBNA-1 protein with the *cis*-acting regulatory signals in the *Bam*HI-C DNA fragment, resulting in the long-term persistence of EBV-derived plasmids in cells.

In this paper the experimental evidence supporting these arguments is presented and these *cis*- and *trans*-acting EBV signals and gene products are mapped and identified. During the course of these experiments two observations were made that suggested that the *Bam*HI-C DNA fragment contains regulatory signals for the transcription of associated genes. First, a positive selection for the persistence of EBV DNA-containing plasmids was employed to select cells containing these transfected DNAs. The G418-resistance (G418$^r$) gene placed under the regulatory control of the SV40 enhancer and promoter (in pSLneo) provided this positive selection. Deletion of this SV40 enhancer-promoter region eliminated the G418$^r$ phenotype of the plasmid. Interestingly, the *Bam*HI-C DNA fragment restored the functional enhancer-promoter activity in these plasmid constructions. Second, some of the deletion mutants in the EBV *Bam*HI-C DNA fragment enhanced the number of G418$^r$ colonies produced after transfection of the plasmid DNA, which was subsequently found in an integrated form rather than as an autonomous episome. This separation of the ability to enhance the number of G418$^r$ colonies produced from the ability to promote the extrachromosomal state of the episome is similar to the effect of an SV40 enhancer upon transfection frequencies. Sugden and his colleagues (Reisman et al. 1985 and pers. comm.) have recently shown that nucleotide sequences in the *Bam*HI-C fragment can act as an enhancer for a foreign gene construction, the gene for chloramphenicol acetyltransferase (CAT) under the regulatory control of the SV40 early promoter. That one of the *cis*-acting EBV regulatory signals required for the persistence of EBV-derived plasmids can act as an enhancer of gene expression is similar to the role of the polyomavirus enhancer in the replication of polyomavirus DNA (de Villiers et al. 1984; Veldman et al. 1985). EBNA-1 gene function, acting upon the enhancer, might then regulate replication or gene expression in EBV-transformed cells.

## Experimental Procedures

### Cell culture

HeLa cells were obtained from the American Type Culture Collection. D98-Raji cells (Glaser and Nonoyama 1974), a somatic hybrid line of adherent cells that harbor the EBV genome as a multicopy plasmid, were obtained from R. Glaser. Monolayer cells were grown in Dulbecco's modified Eagle's medium (GIBCO) supplemented with 100 μg/ml penicillin, 100 μg/ml streptomycin, and 10% heat-inactivated fetal bovine serum (Flow Labs).

### DNA transfer procedure

Recombinant plasmids were introduced into monolayer cells growing in 100-mm dishes by the calcium phosphate–DNA coprecipitation procedure (Graham and van der Eb 1973), with modifications as described previously (Lupton and Levine 1985).

### Selection and analysis of G418$^r$ monolayer cell lines

At approximately 48 hr posttransfection, the cell monolayer was transferred to four 150-mm dishes, and at about 60 hr after transfection, G418 (46% active, GIBCO) was added to the medium at a concentration of 500 μg/ml. Medium containing 500 μg/ml G418 was replaced every 4 days. G418$^r$ colonies were stained with Giemsa and counted or isolated by trypsinization in steel cylinders and individually expanded into cell lines at about 3 weeks after transfection. The resulting cell

lines were maintained in medium containing 200 μg/ml G418.

Low-molecular-weight DNA was isolated by selective salt precipitation (Hirt 1967). Portions of each extract were digested with restriction enzymes and analyzed by Southern blot hybridization as described previously (Southern 1975; Lupton and Levine 1985).

### Recombinant plasmids

The recombinant plasmids constructed for this study are described in more detail in Lupton and Levine (1985).

## Results

### Extrachromosomal plasmid maintenance in EBV-positive and -negative cells is promoted by EBV-derived sequences

To explore the viral functions that promote extrachromosomal maintenance of EBV-derived plasmids, we utilized a series of plasmids containing portions of the EBV *Bam*HI-C and *Bam*HI-K fragments (Dambaugh et al. 1980; Baer et al. 1984). The parental plasmid to these constructions was pSLneo, a vector constructed by joining the *Pvu*II-to-*Bam*HI fragment of pSV2neo (Southern and Berg 1982), encoding G418 resistance (G418$^r$) in animal cells, to an *Eco*RI-to-*Nru*I fragment of pML2 (Lusky and Botchan 1981) (Fig. 1). Synthetic oligonucleotides were used to create *Xho*I and *Bam*HI sites at the junctions, while preserving the *Eco*RI site of pML2. A 5097-bp *Eco*RI-to-*Pvu*II segment from the EBV *Bam*HI-C fragment (nucleotides [nt] 7315–12,413 of EBV strain B95-8) was inserted, producing a plasmid called p404 (Fig. 1). To construct a plasmid containing the viral EBNA-1 coding region (p429[K$^+$]), the EBV *Bam*HI-K fragment (nt 107,565–112,625 of strain B95-8) was inserted into pSLneo at the unique *Bam*HI site, such that the left end of the *Bam*HI-K fragment was closest to the 3' end of the G418$^r$ gene. Finally, the *Bam*HI-K fragment was inserted at the unique *Bam*HI site of p404 to give p410$^+$ (Fig. 1).

These plasmids were introduced into D98-Raji and HeLa cells by the calcium phosphate–DNA coprecipitation method, and selection for G418$^r$ was applied. The vector plasmids, pSLneo and pSV2neo (with no inserted EBV sequences) gave rise to small numbers of G418$^r$ colonies (Table 1). The number of G418$^r$ colonies obtained in D98-Raji cells (EBV-positive) was stimulated about 200-fold when the EBV *Bam*HI-C–derived nucleotide sequences were inserted into pSLneo (producing p404, Fig. 1), but little or no stimulation was seen when these plasmids were introduced into HeLa cells (EBV-negative) (Table 1). When the EBV *Bam*HI-K fragment was inserted into pSLneo, the resulting plasmid (p429[K$^+$]) gave rise to fewer G418$^r$ colonies than pSV2neo or pSLneo in either D98-Raji or HeLa cells (Table 1). However, when the *Bam*HI-K fragment was inserted into p404, the number of G418$^r$ colonies obtained (with p410$^+$) was elevated by about 200-fold in D98-Raji cells (EBV-positive) and about 10-fold to 15-

**Table 1**

| Plasmid | Avg. no. of G418$^R$ colonies obtained[a] | | Avg. no of episomal plasmids per cell[b] | |
|---|---|---|---|---|
| | D98-Raji | Hela | D98-Raji | Hela |
| p465 | 0 | 0 | – | – |
| pSV2sst | 0 | 0 | – | – |
| pSV2neo | 17 | 51 | ND | NT |
| pSLneo | 32 | 108 | NT | ND |
| p404Δ | 970 | 31 | 10–20 | ND |
| p404 | 4180 | 117 | 2–10 | ND |
| pKpn3'Δ | 4000 | 816 | 1 | NT |
| pSst3'Δ | 1390 | 144 | 1 | NT |
| pEco3'Δ | 1040 | 124 | ND | NT |
| pMlu3'Δ | 60 | 392 | ND | NT |
| pMlu5'Δ | 45 | 4 | ND | NT |
| pEco5'Δ | 54 | 74 | ND | NT |
| pKpn5'Δ | 124 | 258 | ND | NT |
| p429(K+) | 4 | 8 | ND | ND |
| p410+ | 4000 | 319 | 10 | 2 |
| pCK(S1) | 4000 | 19 | 10–20 | ND |
| pCK(S2) | 4000 | 41 | 2–10 | ND |
| pCK(S3 − 1) | 4000 | 649 | 10–20 | 1–2 |

[a] Number of G418$^R$ colonies obtained when 25 µg of plasmid DNA was transfected into cells without carrier DNA, as described in Lupton and Levine (1985).

[b] The average copy number was obtained by comparing the intensity of hybridization in experimental samples with that of copy-number reconstructions of the type illustrated in Figs. 2, 3A, and 5. (ND) Not detectable; (NT) not tested; (−) no colonies were obtained for analysis.

fold in HeLa cells (EBV-negative) over that obtained with the vector alone, pSLneo (Table 1).

These experiments demonstrate that DNA sequences located in the EcoRI-to-PvuII region of the EBV BamHI-C fragment (nt 7315–12,413 of strain B95-8) stimulate the production of G418$^r$ colonies by 200-fold in an EBV-positive cell line (D98-Raji) but not in a related, EBV-negative cell line (HeLa). The EBV BamHI-C–derived sequences and the EBV BamHI-K fragment (nt 107,565–112,625 of strain B95-8) together enhance the formation of G418$^r$ colonies in EBV-negative HeLa cells, but the BamHI-K fragment alone is not able to stimulate the formation of G418$^r$ colonies either in EBV-positive D98-Raji cells or in EBV-negative HeLa cells.

One explanation for the ability of the EBV-derived nucleotide sequences to stimulate the number of G418$^r$ colonies formed in these transfection experiments (Table 1) is that an enhanced number of colonies might result when the plasmid more readily becomes established as an extrachromosomal element rather than in an integrated form. To investigate this, low-molecular-weight DNA fractions were prepared from cell lines derived from individual G418$^r$ colonies obtained in the above experiments. These DNA preparations were screened for the presence of extrachromosomal plasmid sequences related to the sequences employed in the transfections. Southern blot hybridization determined that p404 was maintained in an extrachromosomal, closed circular form in D98-Raji (EBV-positive) cells.

Normal (not rearranged) intact p404 plasmid was also recovered from these low-molecular-weight fractions by transforming *Escherichia coli* to ampicillin resistance, thus confirming this result. All further analyses, however, were conducted using low-molecular-weight fractions derived from cells in culture that had been digested with restriction enzymes, because supercoiled DNA molecules derived from these cells were not efficiently retained on nitrocellulose filters. This eliminated considerable variations in hybridization efficiency, which could have caused an underestimation of the copy number of extrachromosomal plasmids.

Low-molecular-weight DNA fractions were digested with BamHI and XhoI, fractionated by gel electrophoresis, and analyzed by Southern blot hybridization employing pSLneo as the radioactive probe (Fig. 2). The approximate copy number of extrachromosomal plasmids was determined by including a reconstruction control in the analysis. Low-molecular-weight DNA extracted from D98-Raji cells (with no plasmid) was mixed with 1000 pg, 500 pg, 100 pg, or 50 pg of p404 plasmid DNA before digestion with BamHI and XhoI, so as to reconstruct 20, 10, 2, and 1 copy of p404 per cell, respectively. D98-Raji cells transfected with pSV2neo failed to maintain the plasmid in an extrachromosomal form (Fig. 2). The inclusion of the EcoRI-to-PvuII portion of the EBV BamHI-C fragment consistently allowed plasmid p404 to be maintained at between 2 and 20 copies in the low-molecular-weight fraction (Table 1; Fig. 2) in D98-Raji cells. In HeLa cells, only p410$^+$ (containing both EBV BamHI-C–derived sequences and the EBV BamHI-K fragment) was maintained extrachromosomally at 1–2 copies/cell (Fig. 2). Plasmid p429(K$^+$), containing the BamHI-K fragment alone, failed to become established extrachromosomally in either D98-Raji and HeLa cells, demonstrating that the BamHI-K fragment is by itself insufficient for plasmid maintenance. Low-molecular-weight DNA extracted from HeLa cells transfected with pSLneo or p404 (Fig. 2) contained undetectable levels of extrachromosomal plasmid DNA.

These studies demonstrate that the EcoRI-to-PvuII portion of the EBV BamHI-C fragment (nt 7315–12,413 of the B95-8 strain) allows plasmid sequences to be maintained extrachromosomally in D98-Raji (EBV-positive) cells and that both this region of the EBV BamHI-C fragment and the EBV BamHI-K fragment (nt 107,565–112,625 of the B95-8 strain) are required for efficient extrachromosomal persistence of plasmids in HeLa (EBV-negative) cells.

### Sequence requirements for stable extrachromosomal plasmid propagation in EBV-positive cells

Deletion mutations were introduced into the EBV BamHI-C–derived portion of p404 by dropping out restriction fragments from the 5′ or 3′ end of the insert (Fig. 1). To characterize in greater detail the region of the BamHI-C fragment, which is required for extrachromosomal plasmid maintenance in EBV-positive D98-

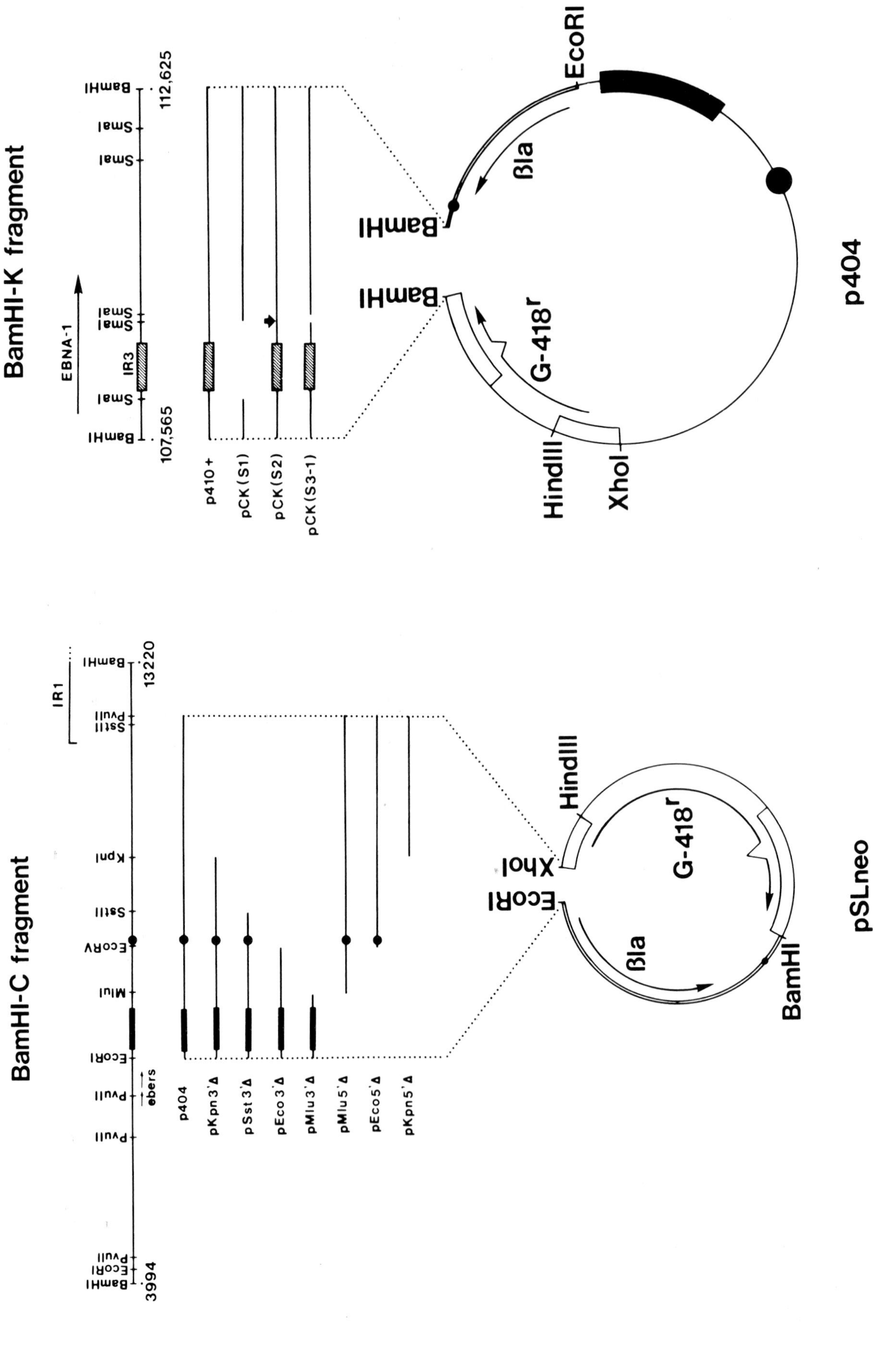

**Figure 1** *See facing page for legend.*

**Figure 1**(*see facing page*) Structure of recombinant plasmids. (*Left*) *Top*: Restriction map of the EBV B95-8 *Bam*HI-C fragment. Nucleotide sequence numbering is as in Baer et al. (1984). The region encoding two small nonpolyadenylated RNA polymerase III transcripts (EBV-encoded RNAs; "ebers"; "ebers'') and the position of the U1-IR1 junction are indicated. (■) Repeated element consisting of 20 tandem copies of a 30-bp sequence; (●) 65-bp dyad symmetry (these features are shown in detail in Fig. 6). Beneath the restriction map, the subfragments inserted between the *Eco*RI and *Xho*I sites of pSLneo and the resulting plasmids are shown. *Bottom*: Diagram of pSLneo. Double line, pML2-derived sequences encoding β-lactamase (βla) and containing the ColE1 replication origin (●); open rectangles represent SV40 sequences controlling expression of the transposon *Tn*5-derived G418ʳ gene. (*Right*) *Top*: Restriction map of the EBV B95-8 *Bam*HI-K fragment. Nucleotide sequence numbering is as in Baer et al. (1984). (Hatched rectangle) IR3; (arrow) region encoding EBNA-1. Below the restriction map are shown the wild-type and mutated forms of the *Bam*HI-K fragment inserted at the *Bam*HI site of p404 and the resulting plasmids. Breaks in the lines indicate the extent of deletion in pCK(S1) and pCK(S3-1); the position of the *Bgl*II linker insertion mutation in pCK(S2) is indicated by the arrow (these mutations are shown in detail in Fig. 4). *Bottom*: Diagram of p404. Symbols are as described for pSLneo.

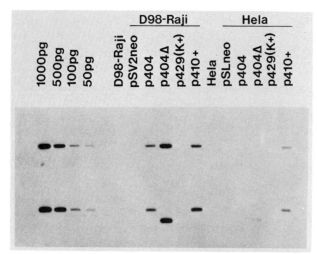

**Figure 2** Southern blot analysis of low-molecular-weight DNA extracted from D98-Raji and HeLa cell lines derived from transfection with recombinant plasmids. Markers containing reconstructions of 20, 10, 2, and 1 plasmid copy per cell, respectively, were obtained by mixing 1000, 500, 100, or 50 pg, respectively, of p404 plasmid DNA with low-molecular-weight DNA extracted from ~5 × 10⁶ D98-Raji cells (this number of cells was used for all extractions). DNA samples were digested with *Bam*HI and *Xho*I, fractionated by agarose gel electrophoresis, transferred to nitrocellulose, and probed with radiolabeled pSLneo. The *Bam*HI-K fragment released from p429(K⁺) and p410⁺ by *Bam*HI digestion is not seen because the *Bam*HI-K fragment was not included in the probe.

Raji cells, these deleted plasmids were transfected into D98-Raji cells and G418ʳ colonies were selected as described above (Table 1). Individual G418ʳ colonies were isolated and grown into large-scale cultures, from which low-molecular-weight DNA was prepared. These DNAs were then incubated with *Bam*HI and *Xho*I restriction enzymes. The digested DNA fragments were separated by agarose gel electrophoresis and analyzed by Southern blot hybridization using pSLneo as the radioactive probe to determine whether or not the deleted plasmids had persisted extrachromosomally in these cells (Fig. 3).

Deletions removing the 5′ end of the p404 insert eliminated the ability of the EBV sequences both to enhance the number of G418ʳ colonies formed in transfection experiments and to promote extrachromosomal plasmid maintenance in EBV-positive D98-Raji cells (Table 1; Fig. 3). DNA sequences located between the *Eco*RI and *Mlu*I sites of the EBV *Bam*HI-C fragment are required for both enhancement of colony formation in G418-containing medium and plasmid persistence in these cells.

Curiously, progressive deletion from the 3′ end of the p404 insert eliminated the plasmid maintenance function before eliminating the ability of the EBV *Bam*HI-C sequences to stimulate the numbers of G418ʳ colonies formed in these transfections (Table 1; Fig. 3). Deleting EBV *Bam*HI-C—derived sequences as far as the *Sst*II site (cf. p404 with pKpn3′Δ and pSst3′Δ) resulted in plasmids that gave rise to G418ʳ cell lines maintaining

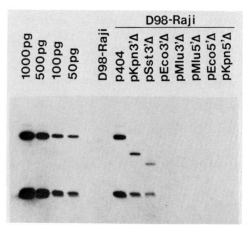

**Figure 3** Effect of deletions in EBV *Bam*HI-C—derived sequences on EBV-derived plasmid propagation in D98-Raji cells. Markers (left lanes) were obtained as described in the legend to Fig. 2. Low-molecular-weight DNA was extracted from ~5 × 10⁶ cells. DNA samples were digested with *Bam*HI and *Xho*I fractionated by agarose gel electrophoresis, transferred to nitrocellulose, and probed with radiolabeled pSLneo. D98-Raji cells transfected with p404 maintained between 2 and 10 copies of it per cell in an extrachromosomal form.

about 1 copy of the transfected plasmid per cell in an extrachromosomal form (Fig. 3). Upon further deletion, the *Bam*HI-C—derived sequences (in pEco3′Δ) failed to promote extrachromosomal plasmid maintenance but nonetheless gave rise to a substantially stimulated number of G418ʳ colonies in D98-Raji cells compared with the vector containing no EBV sequences (Table 1; Fig. 3).

These deletion mutations locate the 5′ boundary of the EBV *Bam*HI-C—derived sequences, which promote extrachromosomal plasmid maintenance in EBV-positive D98-Raji cells, between the *Eco*RI and *Mlu*I sites (Fig. 1). Similarly, the 3′ boundary can be placed between the *Eco*RV and *Sst*II sites (Fig. 1). Maximally, 2208 bp located between the *Eco*RI and *Sst*II sites (nt 7315–9517 of strain B95-8) are required for the extrachromosomal persistence of EBV-derived plasmids in D98-Raji cells. These results are consistent with and confirm those of Sugden and colleagues (Yates et al. 1984; Reisman et al. 1985), who utilized different selectable markers and plasmid constructions. In addition, the discordance between the numbers of G418ʳ colonies formed in transfection experiments and the ability to become established extrachromosomally was the first indication that the EBV *Bam*HI-C fragment may contain sequences capable of enhancing the expression of a linked marker gene, possibly in a manner similar to viral or cellular enhancers.

### Sequence requirements for stable extrachromosomal plasmid propagation in EBV-negative cells

The EBV *Bam*HI-K fragment, which acts in concert with the EBV *Bam*HI-C—derived sequences to promote extrachromosomal plasmid maintenance in EBV-negative

HeLa cells (Table 1; Fig. 2) (Lupton and Levine 1985; Yates et al. 1985), encodes EBNA-1 (Summers et al. 1982; Hennessy et al. 1983; Hennessy and Kieff 1983; Fischer et al. 1984; Hearing et al. 1984; Robert et al. 1984). Deletion and insertion mutations have been constructed in the EBV *Bam*HI-K fragment that lead to the production of altered forms of EBNA-1 (Hearing et al. 1984) (Figs. 1 and 4). p410⁺ contains the wild-type EBV *Bam*HI-K fragment, in addition to the EBV *Bam*HI-C—derived sequences and the G418ʳ gene (Fig. 1). pCK(S1) contained a 1122-bp deletion between two adjacent *Sma*I sites in the EBV *Bam*HI-K—derived portion of the plasmid, which maintains the reading frame encoding EBNA-1 but removes 374 amino acids from the middle of the protein, including the glycine-alanine repeat and the surrounding amino acids (Hearing et al. 1984) (Figs. 1 and 4). pCK(S2) carries a *Bgl*II linker insertion mutation in the region of the EBV *Bam*HI-K—derived portion of the plasmid encoding the carboxyter-

minal domain of EBNA-1. This mutation alters the reading frame at the site of insertion, resulting in premature termination and the production of a truncated EBNA-1—related protein that lacks the authentic carboxyterminal domain (Hearing et al. 1984) (Figs. 1 and 4). Finally, pCK(S3-1) carries a 59-bp deletion between two adjacent *Sma*I sites in the EBV *Bam*HI-K—derived portion of the plasmid, with a *Bgl*II linker inserted at the site of deletion. These alterations maintain the EBNA-1 reading frame, but replace 19 amino acids with 2 in the carboxy-terminal region of the protein (Hearing et al. 1984) (Figs. 1 and 4). The wild-type plasmid, p410⁺, and each of the mutant plasmids (Fig. 1) were introduced into D98-Raji and HeLa cells, and selection for G418ʳ was applied (Table 1). All four plasmids produced equally high numbers of G418ʳ colonies in D98-Raji cells, similar to the number obtained with p404 and elevated about 200-fold over the number obtained with the vector containing no EBV sequences. In HeLa cells, p410⁺ (wild-

# BamHI-K fragment

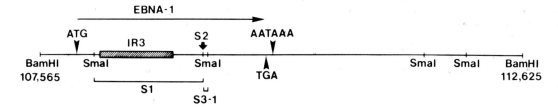

Mutation S1:

```
      P   G   A   P             G   A   I   E
      CCAGGAGCCCC ——— GGGAGCGATAGAG
              108,119   109,242
```

Deletion
( in - frame )

Mutation S2:

```
  D   V   P   P   R   S   G   E   R  Stop
  GACGTGCCCCC- CAGATCTG- GGGAGCGATAG
          109,241       109,242
```

Insertion
(frameshift)

Mutation S3-1:

```
  D   V   P   P   R   S   G   G   Q   G
  GACGTGCCCCC- CAGATCTG- GGGGTCAGGGT
          109,241       109,301
```

Deletion and insertion
( in - frame )

**Figure 4** Mutations in the *Bam*HI-K fragment. A restriction map of the EBV B95-8 *Bam*HI-K fragment is shown at the top, with nucleotide sequence numbering as in Baer et al. (1984). (Hatched rectangle) IR3; (arrow) position of the open reading frame encoding EBNA-1, with the probable translational start and stop codons and polyadenylation signal indicated. The regions deleted in mutants S1 and S3-1 are shown beneath the map, and the position of the *Bgl*II linker insertion mutation, S2, is shown by the arrow above the map. The precise nucleotide sequence changes and the impact of these alterations on EBNA-1 are shown below the map.

type EBV *Bam*HI-K fragment) and pCK(S3-1) (small in-frame deletion of the EBNA-1 coding region) gave rise to stimulated numbers of G418ʳ colonies compared with the vector plasmid. But pCK(S1) (large in-frame deletion of the EBNA-1 coding region) and pCK(S2) (premature termination of the EBNA-1 coding region) both failed to produce high numbers of G418ʳ colonies (Table 1).

G418ʳ colonies resulting from transfection of D98-Raji and HeLa cells with each plasmid were isolated and expanded individually into large-scale cultures, from which low-molecular-weight DNA was extracted. The purified DNA was digested with *Bam*HI and *Xho*I, separated by agarose gel electrophoresis, and examined by Southern blot hybridization to determine whether or not the plasmid had become established extrachromosomally (Fig. 5). All four plasmids persisted extrachromosomally in D98-Raji cells, as anticipated (because all four plasmids carried the EBV *Bam*HI-C–derived sequences that are sufficient for plasmid maintenance in EBV-infected cells). However, in HeLa (EBV-negative) cells, only the wild-type plasmid (p410⁺) and the derivative carrying mutation S3-1 (pCK[S3-1]) were able to persist extrachromosomally. Mutations S1 and S2 both destroyed the plasmid maintenance function. These experiments demonstrate that the *trans*-acting factor, which is supplied by D98-Raji cells but lacking in HeLa cells and which acts in concert with the EBV *Bam*HI-C sequences to promote extrachromosomal plasmid maintenance, is EBNA-1.

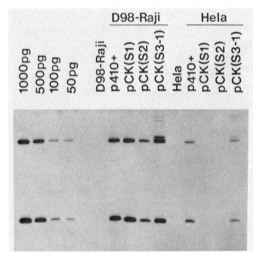

**Figure 5** Effect of mutations in the EBNA-1 coding region on EBV-derived plasmid propagation in D98-Raji and HeLa cells. Markers (left lanes) were obtained as described in the legend to Fig. 2. Low-molecular-weight DNA was extracted from ~5 × 10⁶ cells. DNA samples were digested with *Bam*HI and *Xho*I, fractionated by agarose gel electrophoresis, transferred to nitrocellulose, and probed with radiolabeled pSLneo. The *Bam*HI-K fragments released by *Bam*HI digestion are not seen because the *Bam*HI-K fragment was not included in the probe. Plasmids p410⁺, pCK(S1), pCK(S2), and pCK(S3-1) persist extrachromosomally at between 2 and 20 copies per cell in D98-Raji cells. Plasmids p410⁺ and pCK(S3-1) persist extrachromosomally at between 1 and 10 copies per cell in HeLa cells.

## A cell-specific regulatory element located in the EBV *Bam*HI-C fragment

The results presented above provide several examples in which plasmids that were not able to become established extrachromosomally nonetheless gave rise to stimulated numbers of G418ʳ colonies upon transfection. For example, in D98-Raji cells, pEco3′Δ produced about 1000 colonies but was not maintained extrachromosomally. In the same experiments, pSV2neo also failed to persist extrachromosomally and gave rise to only about 20 G418ʳ colonies. These results suggested that the EBV *Bam*HI-C sequences might contain a transcriptional regulatory element. To test this hypothesis, a plasmid was produced by deleting the small *Xho*I-to-*Hind*III fragment of p404 so as to remove the SV40 enhancer and promoter region and create a negative control plasmid. As anticipated, deleting the SV40 enhancer and promoter region from pSLneo destroyed the ability of the resulting plasmid, p465, to confer G418ʳ on D98-Raji and HeLa cells (Table 1). However, deleting this same region in the p404 plasmid to give p404Δ had only a modest effect upon the ability of the plasmid to confer G418ʳ. The number of G418ʳ colonies produced by p404Δ in D98-Raji cells remained about 50-fold greater than the number obtained with the vectors pSLneo or pSV2neo and was only reduced about four-fold compared with the parent p404, which contained the SV40 enhancer-promoter. Similarly, p404Δ produced, in HeLa cells, about one-fourth the number of G418ʳ colonies obtained with p404 (Table 1). When low-molecular-weight DNA was extracted from G418ʳ cell lines derived from G418ʳ colonies obtained by transfecting p404Δ into D98-Raji cells and examined for the presence of plasmid-related sequences, p404Δ was found to persist extrachromosomally (Fig. 2). Apparently, when the EBV *Bam*HI-C sequences are adjacent to the G418ʳ gene, the SV40 enhancer and promoter region is now dispensable. This observation was the second, and quite direct, indication that a sequence element that can functionally substitute for the SV40 enhancer and promoter region resides within the *Eco*RI-to-*Pvu*II segment of the *Bam*HI-C fragment.

## Discussion

The goal of these studies was to genetically map and identify the *cis*- and *trans*-acting functions of the EBV genome that are involved in the maintenance or persistence of an autonomously replicating episomal element. The experiments presented here and elsewhere (Lupton and Levine 1985) and those of others (Yates et al. 1984, 1985; Reisman et al. 1985) have described two *cis*-acting signals in the EBV *Bam*HI-C DNA fragment (in a 2208-bp region: nt 7315–9417 of the B95-8 strain [Baer et al. 1984]) that are required for the persistence of EBV-derived plasmids in cells expressing EBV gene functions in *trans* (e.g., D98-Raji cells). One of these signals is composed of a 30-bp motif repeated 20 times with some variation of a consensus sequence (Fig. 6).

This repeated set of sequences constructs a higher-order repeat reflecting duplications of multiple 30-bp repeat units. The basic 30-bp repeat contains an 18-bp dyad symmetry within it. The second element in the BamHI-C DNA fragment is located about 1000 bp downstream (3' to the first element). This cis-acting sequence is related to the 30-bp consensus sequence by being composed of four partial homologies to it. Two of these 30-bp homologs reside within a 65-bp dyad symmetry (Fig. 6). The pattern of A + T−rich sequences and repeated nucleotide sequences containing elements of dyad symmetries is similar to sequences that constitute the origins of DNA replication in other viruses (DePamphilis and Wassarman 1982; Stow and McMonagle 1983; Veldman et al. 1985). Although these cis-acting elements in the BamHI-C DNA fragments are clearly required for persistence of the extrachromosomal episome in D98-Raji cells, no formal evidence has been presented to show that this region is indeed an origin of DNA replication.

The trans-acting viral function required for efficient extrachromosomal maintenance of plasmids containing the BamHI-C repeated sequences has been identified as the EBNA-1 gene product. Deletion mutants and frameshift insertion mutants in this gene eliminate the ability of a plasmid containing the cis- and trans-acting functions to persist in HeLa (EBV-negative) cells. This genetic evidence suggests an interaction between the EBNA-1 gene product and the repeated cis-acting elements, and such an interaction via DNA binding has

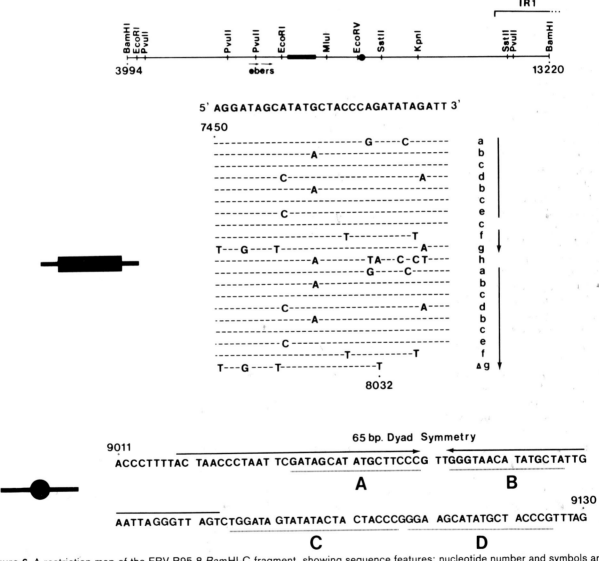

**Figure 6** A restriction map of the EBV B95-8 BamHI-C fragment, showing sequence features; nucleotide number and symbols are described in the legend to Fig. 1. Beneath the restriction map is a schematic representation of the 582-bp repeated element. The 30-bp consensus sequence is shown, and deviations from the consensus sequence in each repeat within the repeated element are indicated. Lower-case letters are assigned to each variation from the consensus sequence, and increasing order of repetition is represented by the vertical arrows. Below the 30-bp repeats, the sequence of the 65-bp sequence of dyad symmetry is presented. Four partial homologs of the 30-bp repeated motif are marked A, B, C, and D.

been shown directly by Hayward and his colleagues (Rawlins et al. 1985 and this volume). It is likely that additional elements of cellular origin (and possibly of viral origin) remain to be identified and are involved in efficient persistence of the EBV-derived plasmids in EBV-negative cell lines.

The first clue in these studies as to the possible biological activity or function of the *cis*-acting elements in the *Bam*HI-C DNA fragment came from the deletion-mapping experiments of these *cis*-acting elements. In D98-Raji cells the pEco3'Δ plasmid produced about 1000 G418$^r$ colonies, whereas the plasmid pSV2neo clone gave rise to about 20 G418$^r$ colonies. This stimulation or enhancement in the number of G418$^r$ colonies formed did not result from an enhanced copy number of the plasmid nor from the existence of extrachromosomal plasmid DNA. In both cases the plasmid was found to be associated with the high-molecular-weight DNA fraction (integrated DNA). This enhancement of the frequency of transfected G418$^r$ colonies produced is similar to the results obtained when a retrovirus enhancer is placed adjacent to a thymidine kinase gene (Luciw et al. 1983). Indeed, when the SV40 enhancer-promoter region, which regulates the G418$^r$ gene in this plasmid, was deleted, these signals could be functionally substituted for by sequences in the *Bam*HI-C DNA fragment (in p404Δ) (Table 1; Figs. 2 and 3). There was a 50-fold increase in the number of G418$^r$ colonies produced by p404Δ (controlled by the EBV repeated-DNA signals) when compared with pSV2neo. Furthermore, this increase was only observed in cells containing the EBV genome (D98-Raji) and not in HeLa cells (EBV-negative cells). These data suggest that the repeated elements in the *Bam*HI-C DNA fragments of the EBV genome could act as an enhancer (substitute for the SV40 enhancer) but are dependent upon EBV-encoded *trans*-acting gene products, such as EBNA-1. The 30-bp repeated-DNA-sequence motif, indeed, behaved as an orientation-independent element enhancing the activity of the CAT gene in cells containing the EBV genome but not in lymphoblastoid cells without the EBV genome (Reisman et al. 1985). Experiments by Sugden and his colleagues (D. Reisman et al., pers. comm.) have provided evidence that this enhancer activity is dependent upon the presence of EBNA-1. The DNA-binding studies of Hayward and his colleagues (Rawlins et al. 1985 and this volume) suggest that EBNA-1 elicits the enhancer activity of the 30-bp repeated sequence by interacting directly with these sequences. That one portion of the *cis*-acting element in EBV can act as an enhancer dependent upon a virally encoded function for activity (EBNA-1) begins to define possible events that lead to replication or persistence of EBV episomes in lymphoblastoid cells.

## Acknowledgments

We thank Furzana Chaudry for invaluable discussions. Terry Van Dyke, Lenny Kaplan, and Janet Hearing provided useful suggestions and recombinant plasmids, and Noel Mann helped to prepare the manuscript. This work was supported by grants M47E and M47F from the American Cancer Society.

## References

Baer, R., A.T. Bankier, M.D. Biggin, P.L. Deininger, P.J. Farrell, T.J. Gibson, G. Hatful, G.S. Hudson, S.C. Satchwell, C. Seguin, P.S. Tuffnell, and B.G. Barrell. 1984. DNA sequence and expression of the B95-8 Epstein-Barr virus genome. *Nature* **310**: 207.

Dambaugh, T., C. Beisel, M. Hummel, W. King, S. Fennewald, A. Cheung, M. Heller, N. Raab-Traub, and E. Kieff. 1980. Epstein-Barr virus (B95-8) DNA. VII. Molecular cloning and detailed mapping. *Proc. Natl. Acad. Sci.* **77**: 2999.

DePamphilis, M.L. and P.M. Wassarman. 1982. Organization and replication of papovavirus DNA. In *Organization and replication of viral DNA* (ed. A.S. Kaplan), p. 37. CRC Press, Inc., Boca Raton, Florida.

de Villiers, J., W. Schaffner, C. Tyndall, S. Lupton, and R. Kamen. 1984. Polyomavirus DNA replication requires an enhancer. *Nature* **312**: 242.

Fischer, D.K., M.F. Robert, D. Shedd, W.P. Summers, J.E. Robinson, J. Wolak, J.E. Stefano, and G. Miller. 1984. Identification of Epstein-Barr nuclear antigen polypeptide in mouse and monkey cells after gene transfer with a cloned 2.9-kilobase-pair subfragment of the genome. *Proc. Natl. Acad. Sci.* **81**: 43.

Glaser, R. and M. Nonoyama. 1974. Host cell regulation of induction of Epstein-Barr virus. *J. Virol.* **14**: 174.

Graham, F.L. and A.J. van der Eb. 1973. A new technique for the assay of infectivity of human adenovirus 5 DNA. *Virology* **52**: 456.

Hearing, J.C., J.-C. Nicolas, and A.J. Levine. 1984. Identification of Epstein-Barr virus sequences that encode a nuclear antigen expressed in latently infected lymphocytes. *Proc. Natl. Acad. Sci.* **81**: 4373.

Henle, W., V. Diehl, G. Kohn, H. zur Hausen, and G. Henle. 1967. Herpes-type virus and chromosome marker in normal leukocytes after growth with irradiated Burkitt's cells. *Science* **157**: 1064.

Hennessy, K. and E. Kieff. 1983. One of two Epstein-Barr virus nuclear antigens contains a glycine-alanine copolymer domain. *Proc. Natl. Acad. Sci.* **80**: 5665.

Hennessy, K., M. Heller, V. Van Santen, and E. Kieff. 1983. Simple repeat array in Epstein-Barr virus DNA encodes part of the Epstein-Barr nuclear antigen. *Science* **220**: 1396.

Hirt, B. 1967. Selective extraction of polyoma DNA from infected mouse cell cultures. *J. Mol. Biol.* **26**: 365.

Kaschka-Dierich, C., A. Adams, T. Lindahl, G.W. Bornkamm, G. Bjurnsell, G. Klein, B.C. Giovanella, and S. Singh. 1976. Intracellular forms of Epstein-Barr virus DNA in human tumor cells *in vivo*. *Nature* **260**: 303.

Lindahl, T., A. Adams, G. Bjursell, G.W. Bornkamm, C. Kaschka-Dierich, and U. Jehn. 1976. Covalently closed circular duplex DNA of Epstein-Barr virus in a human lymphoid cell line. *J. Mol. Biol.* **102**: 511.

Luciw, P.A., J.M. Bishop, H.E. Varmus, and M.R. Capecchi. 1983. Location and function of retroviral and SV40 sequences that enhance biochemical transformations after microinjection of DNA. *Cell* **33**: 705.

Lupton, S. and A.J. Levine. 1985. Mapping genetic elements of Epstein-Barr virus that facilitate extrachromosomal persistence of Epstein-Barr virus-derived plasmids in human cells. *Mol. Cell. Biol.* **5**: 2533.

Lusky, M. and M. Botchan. 1981. Inhibition of SV40 replication in simian cells by specific pBR322 DNA sequences. *Nature* **293**: 79.

Rawlins, D.R., G. Milman, S.D. Hayward, and G. Hayward. 1985. Sequence-specific DNA binding of the Epstein-Barr virus nuclear antigen (EBNA-1) to clustered sites in the plasmid maintenance region. *Cell* **42**: 859.

Reisman, D., J. Yates, and B. Sugden. 1985. A putative origin of replication of plasmids derived from Epstein-Barr virus is composed of two *cis*-acting components. *Mol. Cell. Biol.* **5:** 1822.

Robert, M.F., D. Shedd, R.J. Weigel, D.K. Fischer, and G. Miller. 1984. Expression in COS-1 cells of Epstein-Barr virus nuclear antigen from a complete gene and a deleted gene. *J. Virol.* **50:** 822.

Southern, E.M. 1975. Detection of specific sequences among DNA fragments separated by gel electrophoresis. *J. Mol. Biol.* **98:** 503.

Southern, P.J. and P. Berg. 1982. Transformation of mammalian cells to antibiotic resistance with a bacterial gene under control of the SV40 early region promoter. *J. Mol. Appl. Genet.* **1:** 327.

Stow, N.D. and E.C. McMonagle. 1983. Characterization of the TRs/IRs origin of DNA replication of herpes simplex virus type 1. *Virology* **130:** 427.

Sugden, B., K. Marsh, and J. Yates. 1984. A vector that replicates as a plasmid and can be efficiently selected in B-lymphocytes transformed by Epstein-Barr virus. *Mol. Cell. Biol.* **5:** 410.

Summers, W.P., E.A. Grogan, D. Shedd, M. Robert, C.-R. Liu, and G. Miller. 1982. Stable expression in mouse cells of nuclear neoantigen after transfer of a 3.4-megadalton cloned fragment of Epstein-Barr virus DNA. *Proc. Natl. Acad. Sci.* **79:** 5688.

Veldman, G.M., S. Lupton, and R. Kamen. 1985. Polyomavirus enhancer contains multiple redundant sequence elements that activate both DNA replication and gene expression. *Mol. Cell. Biol.* **5:** 649.

Yates, J.L., N. Warren, and B. Sugden. 1985. Stable replication of plasmids derived from Epstein-Barr virus in various mammalian cells. *Nature* **313:** 812.

Yates, J., N. Warren, D. Reisman, and B. Sugden. 1984. A *cis*-acting element from the Epstein-Barr viral genomes that permits stable replication of recombinant plasmids in latently infected cells. *Proc. Natl. Acad. Sci.* **81:** 3806.

zur Hausen, H., H. Schulte-Holthausen, G. Klein, W. Henle, G. Henle, P. Clifford, and L. Sanesson. 1970. EBV DNA in biopsies of Burkitt tumors and anaplastic carcinomas of the nasopharynx. *Nature* **228:** 1056.

# Copy Number Control of DNA Replication in SV40–BPV Hybrid Replicons

## J.M. Roberts and H. Weintraub

Department of Genetics, Fred Hutchinson Cancer Research Center, Seattle, Washington 98104

We have begun an analysis of DNA sequences that function in replication control. Simian virus 40 (SV40) demonstrates a replication pattern that is uncoupled from the host cell's regulatory mechanisms so that each viral genome replicates multiple times within each cell cycle. Bovine papillomavirus (BPV), however, replicates in synchrony with the host cell genome, thus displaying a regulated form of replication. To approach the mechanisms of replication control, we have designed a simple model system consisting of SV40 and BPV DNA sequences linked to create a hybrid replicon. We have found that in this configuration, the BPV mode of replication is dominant to that of SV40. This system has permitted us to define those sequences in BPV that are able to impose replication control onto SV40 runaway replication. The BPV replication control system involves at least three elements. Two cis-acting sequences required for replication control are closely associated with BPV replication origins. A third sequence encodes a trans-acting product. We intend to use this system to isolate similarly acting control sequences from the cellular genome.

Although DNA replication apparently is one of the most carefully regulated processes in eukaryotic cells, virtually nothing is understood about those mechanisms that insure that each DNA sequence within the cellular genome is copied once, but not twice, within each cell cycle. To begin to understand the components of replication control, we have constructed a simple model system consisting of simian virus 40 (SV40) and bovine papillomavirus (BPV) DNA sequences. BPV exists in transformed cells as a stable, extrachromosomal element at about 100 copies/cell (Moar et al. 1981). The viral genome contains a replication origin, at least two genes that encode proteins necessary for normal viral replication, and two sequences capable of supporting autonomous plasmid maintenance in BPV-infected cells (Lusky and Botchan 1984, 1985; Waldeck et al. 1984). One of the plasmid maintenance sequences seems to be identical with the viral replication origin. BPV exhibits rigid replication control in infected cells, each viral genome replicating just once per cell cycle (M. Botchan, pers. comm.). In contrast to BPV, lytic viruses such as SV40 have learned to escape cellular factors that control DNA replication. Viral molecules in infected cells will replicate many times within a single cell cycle, ultimately killing the host cell.

Here we have constructed plasmids designed to ask whether the copy number control exhibited by BPV is dominant or recessive to the runaway replication observed for SV40. To do this, we studied the replicative properties of hybrid molecules containing SV40 and BPV sequences. These experiments have defined cis- and trans-acting DNA sequences in BPV that are involved in conferring replication control onto the SV40 origin. In the presence of BPV-encoded trans-acting factor(s), BPV cis-acting sequences suppress runaway replication from the SV40 replication origin present on the same plasmid.

## Methods

COS cells (Gluzman 1981) were cultured in Dulbecco's modified Eagle's medium (DME) plus 10% fetal calf serum. Cells were transfected at a density of $5 \times 10^5$ cells per 100-mm plate by calcium phosphate precipitation of plasmid DNAs. Cells were exposed to the precipitate for 3 hr and were then shocked for 1 min with DME plus 15% glycerol. Cells were then washed three times with DME, and then fresh medium was added. Total cellular DNA was isolated at 16, 24, and 48 hr posttransfection by treatment of whole cells with 0.2% SDS plus 25 μg/ml proteinase K followed by phenol/chloroform extraction and ethanol precipitation.

## Results

A hybrid replicon containing SV40 and BPV DNA sequences was constructed (Fig. 1A). This plasmid, pSV-BPV, contains the SV40 replication origin and enhancer sequences on a 550-bp HindIII-HpaI fragment, linked directly to a 5.4-kb fragment of the BPV genome (HindIII-BamHI) that contains all information necessary for regulated BPV replication (Law et al. 1981). This information includes the BPV replication origin, at least three promoters, and the "early" coding region, which encodes all proteins involved in BPV replication. The replicative properties of this composite plasmid were quantitatively compared with a control plasmid, pSV, which contains only the SV40 replication origin (Fig. 1A).

Quantitative comparisons of replication kinetics are

**555**

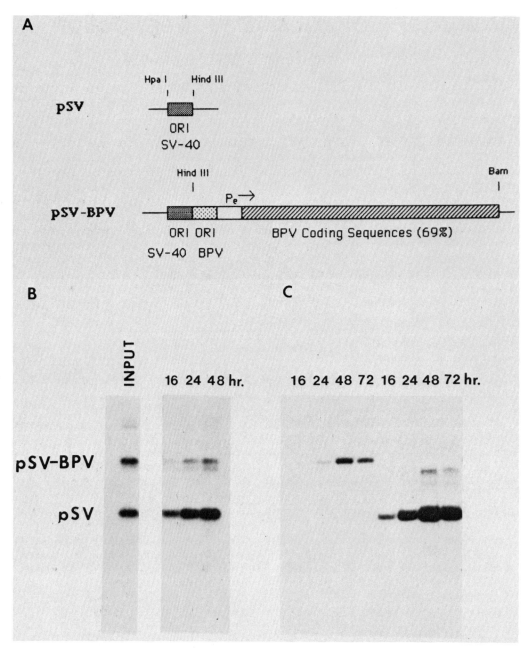

**Figure 1** Analysis of the relative replication rates of an SV40 plasmid and an SV-BPV hybrid replicon. (*A*) Plasmid pSV was constructed by cloning the *Hind*III-*Hpa*I fragment of the SV40 genome into the *Hind*III/*Eco*RI-cleaved pAT153. Plasmid pSV-BPV contains the 5.4-kb *Hind*III-*Bam*HI fragment of the BPV genome inserted into pSV. (*B*) Southern blot demonstrating the relative replication rates of pSV and pSV-BPV in COS cells. One hundred ng of pSV and 200 ng of pSV-BPV were cotransfected into COS cells with 2 μg of carrier DNA as described in Methods. The lane labeled INPUT shows the ratio of the two plasmids in the calcium phosphate precipitate and represents $10^{-3}$ of the DNA added per plate of cells. The blot was probed with nick-translated pBR322. (*C*) Southern blot demonstrating the replication of pSV and pSV-BPV in independent COS cell populations.

made by cotransfecting the experimental and control plasmids into COS cells, which supply SV40 T antigen in *trans*, and monitoring the accumulation of supercoiled DNA by Southern blot. Equal amounts of the pSV and pSV-BPV plasmids were cotransfected into COS cells and total cellular DNA was harvested at 16, 24, and 48 hr (Fig. 1B). In this experimental protocol, the pSV

replicon serves as an internal control for the replication of the hybrid SV-BPV replicon, allowing direct comparisons of replication rates within the same cell. In the blot shown (Fig. 1B), equal amounts of total cellular DNA were loaded per lane. Thus the accumulation of pSV with respect to total DNA demonstrates the unregulated "runaway" replication of this plasmid. The replication of the hybrid SV-BPV replicon is at least 10-fold suppressed relative to the control SV plasmid, and it accumulates only slightly with respect to total cellular DNA. Other plasmid constructs that separate the SV and BPV sequences by about 2.5 kb demonstrate up to 40-fold suppression of replication with respect to pSV and exhibit replication rates equivalent to the host cell genome. We believe, although have not yet directly demonstrated, that the suppression of replication observed in the hybrid replicon is a consequence of the imposition of BPV replication controls upon SV40 runaway replication.

We were initially concerned that this protocol might establish a situation in which the two plasmids were competing for a limiting amount of a factor necessary for DNA replication. Therefore we felt it necessary to show that similar replication properties are observed in the absence of any potential competition. Figure 1C shows a Southern blot after the accumulation of replicating SV and SV-BPV plasmids that have been transfected into separate cell populations. Again it is quite clear that the accumulation of the hybrid replicon is at least 10-fold lower than the pSV control. As a further control for any potential competition for limiting replication factors, we have shown that the amount of suppression observed is independent of the absolute amount of transfected DNA over at least a 10-fold range.

Although it is not obvious that size alone could account for the differences in replication observed, we have shown that plasmids of very different size can replicate with equal efficiency in this assay.

Our analysis relies on the differential accumulation of supercoiled DNA as an indicator of replication efficiencies; we have confirmed that supercoiled DNA is the replicating DNA form through the use of *Dpn* and *Mbo*I digestions. These enzymes assay the methylated state of the input DNA relative to replicated DNA and therefore provide an independent means of identifying the different replicated molecules.

### Two *cis*-acting sites are necessary for replication suppression

We initially suspected that sequences associated with the BPV replication origin would be one component of the replication suppression in the SV-BPV construct. This is directly demonstrated by the plasmid pSB-Ori⁻, which contains a 400-bp *Hind*III-*Mlu*I deletion that entirely removes the BPV replication origin. This plasmid replicates almost equivalently to the pSV control (Fig. 2). Although the BPV *ori* sequence is necessary for suppression, it is not sufficient. The plasmid pSB-d(Hpa) contains the BPV and SV *ori* sequences but is deleted for the BPV coding region. Its replication is equivalent to the pSV control. The runaway replication of this plasmid could be attributed either to the absence of BPV-encoded *trans*-acting factor(s) necessary for replication control and/or the absence of a second *cis*-acting replication control sequence. The replication of

pSB-d(Hpa) is not suppressed if BPV coding information is provided in *trans* (as pSV-BPV), suggesting the existence of a second *cis*-acting site necessary for replication suppression.

The presence of a second *cis*-acting sequence necessary for replication suppression is directly demonstrated by the plasmid pSB-(H/S)⁻. This construct contains the SV and BPV replication origins and most of the BPV coding region. This plasmid has a deletion that encompasses the E6 and E7 open reading frames of the BPV early coding region and the 5′ end of the E1 ORF. This plasmid is not self-suppressed in our replication assay; it replicates (within a factor of two) equivalently to the SV control. However, if BPV coding information is provided in *trans*, its replication is suppressed to the level of the parental SV-BPV construct (Fig. 3). Deletion mutagenesis indicates that this second *cis* site overlaps PMSII, a sequence previously demonstrated to support autonomous replication of plasmids in BPV-infected cells (Lusky and Botchan 1984). Note that both *cis*-acting sites are necessary to achieve replication suppression; the presence of either one alone has only a twofold effect on plasmid replication.

### Summary

We have found that linkage of BPV genomic sequences to the SV40 replication origin results in at least a 10-fold suppression of SV40 replication in COS cells. We believe that this suppression reflects the imposition of BPV replication control signals on SV40-driven replication. At least three elements from the BPV genome are necessary to achieve replication suppression. One *cis*-acting site is closely linked to, or identical with, the BPV replication origin. The second *cis* site overlaps but is not identical with the PMSII sequence. Both *cis*-acting sites must be present on the same molecule to observe suppression. The presence of either site alone has no effect. We do not know, at present, whether the two *cis* sites perform different functions. However, these sites can function independently of distance and orientation with respect to the SV40 replication origin. Finally, at least one *trans*-acting factor is involved in this control system, and this factor is probably encoded in the E6-E7 or E1 regions of the BPV genome. A major conclusion of these studies is that replication control functions are closely associated with DNA sequences that can serve as replication origins. It is our hope that this system will enable us to isolate similarly acting control sequences from the cellular genome.

These results also have implications regarding the mechanism by which SV40 has become an uncontrolled (''runaway'') replicon. SV40 apparently can manifest replication control if an SV40 replicon is provided with BPV replication control signals. Thus, SV40 does not seem to actively exclude or ignore these control functions. Rather, we suggest that wild-type SV40 is deficient for *cis*-acting sequences, which give cellular replication control factors access to the SV40 genome.

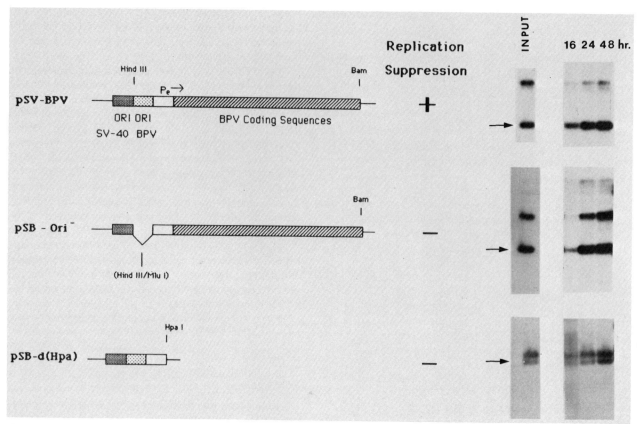

**Figure 2** The BPV replication origin is necessary for replication control. The plasmid pSB-Ori⁻ is a *Hind*III-*Mlu*I deletion of pSV-BPV. The plasmid pSB-d(Hpa) joins the *Hpa*I site of the BPV genome with the *Bal* site of pAT153, deleting all intervening BPV and pAT153 sequences. Beside each plasmid map is a Southern blot depicting the replication of each construct as compared with pSV (→) in COS cells.

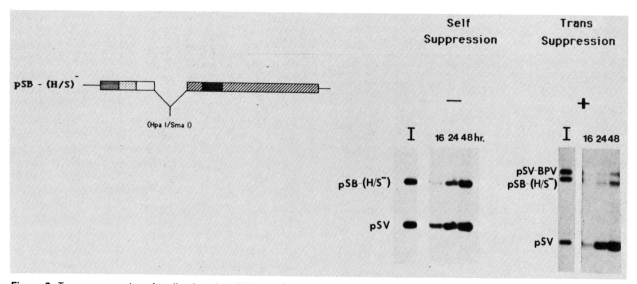

**Figure 3** *Trans* suppression of replication of an SV-BPV hybrid replicon. Plasmid pSB-(H/S)⁻ is derived from pSV-BPV by a deletion of BPV sequences between the *Hpa*I and *Sma*I sites. Its replication relative to pSV in COS cells is shown with and without cotransfection with an equimolar amount of pSV-BPV.

## Acknowledgments

We thank M. Lusky and M. Botchan for generously providing the E1$^-$ and E6/7$^-$ BPV genomes used in these studies. J.R. is a postdoctoral fellow of the Helen Hay Whitney Foundation.

## References

Gluzman Y. 1981. SV40-transformed simian cells support the replication of early SV40 mutants. *Cell* **23:** 175.

Law, M., D. Lowy, J. Dvoretzky, and P. Howley. 1981. Mouse cells transformed by bovine papillomavirus contain only extrachromosomal viral DNA sequences. *Proc. Natl. Acad. Sci.* **78:** 2727.

Lusky, M. and M. Botchan. 1984. Characterization of the bovine papillomavirus plasmid maintenance sequences. *Cell* **36:** 391.

———. 1985. Genetic analysis of bovine papillomavirus type 1 *trans*-acting replication factors. *J. Virol.* **53:** 955.

Moar, M., M. Campo, H. Laird, and W. Garrett. 1981. Persistence of non-integrated viral DNA in bovine cells transformed in vitro by bovine papillomavirus type 2. *Nature* **293:** 749.

Waldeck, W., F. Rosl, and H. Zentgraf. 1984. Origin of replication in episomal bovine papillomavirus type 1 DNA isolated from transformed cells. *EMBO J.* **3:** 2173.

# The Dual Role of the Polyomavirus Enhancer in Transcription and DNA Replication

## J.A. Hassell, W.J. Muller, and C.R. Mueller

Department of Microbiology and Immunology, McGill University, Montreal, Quebec, Canada H3A 2B4

We have used in vitro mutagenesis techniques to define the structure of the polyomavirus enhancer. The enhancer is made up of at least three sequence elements, which we have numbered 1, 2, and 3. Individual elements function poorly or not at all to augment transcription of marker genes, but pairwise combinations of elements enhance transcription nearly as well as all three elements together do. Enhancer elements 2 and 3 contain the auxiliary replication elements α and β within their borders. Either the α or β element must be juxtaposed next to another replication element, termed the core, to form a functional origin (ori) for DNA replication. Like the enhancer elements, the auxiliary elements α and β activate replication independent of their orientation relative to the core. But unlike enhancer elements, the auxiliary elements cannot activate DNA replication when placed at a distance from the core. From these observations, we propose that the initial reactions leading to transcription enhancement and replication activation occur by a common mechanism, but that different pathways are subsequently used to effect these processes.

The noncoding, 460 bp between the translational initiation signals for the polyomavirus early and late proteins contain numerous sequence elements important for transcription and DNA replication. These include the early and late promoters and the functional origin (ori) for DNA replication. We are interested in defining the sequences of these cis-acting regulatory elements and in identifying and purifying the viral and cellular proteins that interact with them to elucidate the mechanism of initiation of transcription and DNA replication. To this end we have begun to map the location of these various regulatory sequences within the polyomavirus noncoding region and to study the interaction of viral and cellular proteins with these sequences (Muller et al. 1983; Pomerantz et al. 1983; Mueller et al. 1984; Pomerantz and Hassell 1984).

The polyomavirus early promoter, like that of simian virus 40 (SV40), comprises multiple sequence elements, including a TATA box—cap site element, a functionally redundant upstream promoter element, and an enhancer (Jat et al. 1982; Mueller et al. 1984). The enhancer is contained within a 225-bp DNA fragment that maps between nucleotide (nt) 5039 and nt 5264 or between positions −403 and −178 relative to the major start sites for the early mRNAs (de Villiers and Schaffner 1981; Tyndall et al. 1981; Jat et al. 1982; Mueller et al. 1984). Interestingly, two functionally redundant replication elements, which we have named α and β, map within the borders of the enhancer (Muller et al. 1983). Either the α or β element in conjunction with another sequence, termed the core, is required to constitute a functional origin for polyomavirus DNA replication (Muller et al. 1983). To define more precisely the sequences required for transcriptional enhancement and those required for replication activation, we constructed li-

braries of deletion mutants of the enhancer region and separately tested their capacity to augment transcription and DNA replication. Analyses of the phenotype of the mutants lead us to suggest that the polyomavirus enhancer consists of multiple sequence elements. At least three elements have been identified and numbered 1, 2, and 3. Individual enhancer elements function poorly or not at all to augment transcription, but pairs of elements enhance transcription nearly as well as all three elements together do. The α and β replication elements map within the borders of enhancer elements 2 and 3. Like enhancers, the replication elements function to activate DNA replication independent of their orientation relative to the polyomavirus core. But by contrast to enhancers, these replication elements either individually or together fail to activate DNA replication when placed at a distance of 200 bp from the late border of the core or when placed near its early border. We propose a model for transcription enhancement and replication activation that takes these observations into account.

## Materials and Methods

### Cell culture and transfections

Mouse 3T3 cells were grown in Dulbecco's modification of Eagle's medium containing 10% calf serum at 37°C in a humidified, 7% $CO_2$ atmosphere. Transfections were performed according to Wigler et al. (1978) with the following modifications. The cells were trypsinized the day before transfection and seeded in fresh medium at a density of $1 \times 10^5$ to $3 \times 10^5$ cells per 100-mm dish, and the medium was not changed on the day of transfection. Duplicate plates of cells were generally transfected with 10 μg of plasmid DNA per plate. After incubation overnight, the cells were washed twice with PBS containing 1 mM EGTA and the medium was replenished. Cell lysates were prepared 48 hr after the addition of the DNA.

## CAT assays

After transfection, the cells were scraped into PBS and washed twice. The cell pellet was resuspended in 100 μl of 250 mM Tris · HCl (pH 7.5), the mixture was frozen and thawed three times, and the broken cells were then centrifuged for 15 min in a refrigerated microfuge. The supernatant was used as a source of chloramphenicol acetyltransferase (CAT) enzymatic activity. The assays to measure CAT activity were carried out essentially as described by Gorman et al. (1982). The reaction mixture for the CAT assay contained 0.2 μCi of (dichloroacetyl-1,2 $^{14}$C)chloramphenicol (New England Nuclear), 250 mM Tris · HCl (pH 7.5), 10 mM acetyl coenzyme A, and 5–50 μl of cell extract in a total volume of 160 μl. The reaction was carried out at 37°C for no longer than 1 hr. After extraction of the reaction mixture with ethyl acetate, the various species of chloramphenicol were separated by ascending thin-layer chromatography. The areas corresponding to the acetylated and nonacetylated forms were removed from the plate and counted using liquid scintillation fluid (OCS, Amersham). The amount of extract used in each assay and the period of incubation were varied to insure that comparisons between extracts were made during the linear phase of the reaction. Two or more independent preparations of wild-type and mutant DNA were tested by transfection of 3T3 cells on at least three separate occasions. The CAT enzymatic activity obtained after transfection of cells with the various mutant DNAs was expressed as a percentage of that obtained with an equivalent amount of extract from cells transfected with the parent plasmid, pdPyEcat, in the same experiment. In this way we could compare the capacity of the various DNAs to express the CAT gene both within an experiment and between different experiments.

## DNA replication assay

The replication in mammalian cells of recombinant plasmids bearing the polyomavirus origin was measured as described previously (Muller et al. 1983).

## Recombinant DNA methods

The manipulations used to construct, isolate, and sequence the various recombinant plasmids are described in Maniatis et al. (1982). The vector pTE1 was constructed by T. Edlund and was a gift from J. Drouin.

## Results

### Structure of the polyomavirus enhancer

To facilitate mapping the polyomavirus enhancer, we substituted the polyomavirus enhancer-promoter region for that of SV40 in the plasmid pSV2-cat (Gorman et al.

1982) to generate a new plasmid, named pdPyEcat (Fig. 1). pdPyEcat is made up of the polyomavirus early enhancer-promoter region (nt 4632–152, or −810−+2) joined to the CAT gene and an SV40−β-globin DNA segment that provides splicing, transcription termination, and polyadenylation signals.

To delineate the borders of the polyomavirus enhancer, we isolated a set of 5′ and 3′ unidirectional deletion mutants of the enhancer, measured their capacity to express the CAT gene after transfection of mouse 3T3 cells, and compared this activity with that obtained after transfection of these same cells with pdPyEcat. Analysis of eight 5′ unidirectional deletion mutants revealed that the 5′ border of the enhancer was between nt 5073 (−369) and nt 5113 (−329) (Fig. 2). Two mutants with deletions between these end points expressed the CAT gene at intermediate levels (~30% of wild type).

To map the 3′ border of the enhancer, we began mutagenesis at nt 90 (−60) and extended the deletion in the upstream direction. All of these deletion mutants retain the early promoter (TATA box−cap site element), which together with the enhancer is sufficient to effect expression of downstream genes (Mueller et al. 1984). Characterization of these mutants revealed that sequences up to nt 5130 (−312) could be removed without appreciably affected expression of the CAT gene (Fig. 2). However, deletion of sequences upstream of nt 5130 to nt 4947 (−495) abolished the capacity of the plasmid to express the CAT gene. Taken together, these results suggested that the polyomavirus enhancer was contained within a 57-bp fragment located between nt 5073 and nt 5130 (−369 to −312).

To determine whether this 57-bp fragment possessed enhancer activity on its own, we cloned it in its natural orientation before the polyomavirus early promoter in the CAT vector. Transfection of this plasmid into 3T3 cells yielded unexpectedly low levels of CAT activity (8−17% of the wild-type plasmid) (Fig. 3). This value is at best only twofold greater than that obtained after transfection of 3T3 cells with a recombinant plasmid that lacks this 57-bp DNA segment and retains only the early promoter (Fig. 2). This result and those described previously suggested to us that the enhancer might be composed of a central, 57-bp region and one of two flanking segments either upstream or downstream of the central segment. To test this possibility and to delineate the approximate borders of the flanking segments, we effectively restored sequences alternately downstream and upstream to the 57-bp fragment and measured the capacity of the resulting plasmids to express the CAT gene in 3T3 cells. In practice this was achieved by deleting sequences in the 3′-to-5′ direction from a CAT vector that maintained the 5′ border of the polyomavirus enhancer (nt 5073, or −369) to generate one set of mutants, and by deleting sequences in the 5′-to-3′ direction from a similar vector that maintained the previously mapped 3′ border of the enhancer (nt 5130, or −312) to obtain another set of mutants. Analysis of the phenotype of these two groups of mutants allowed us to

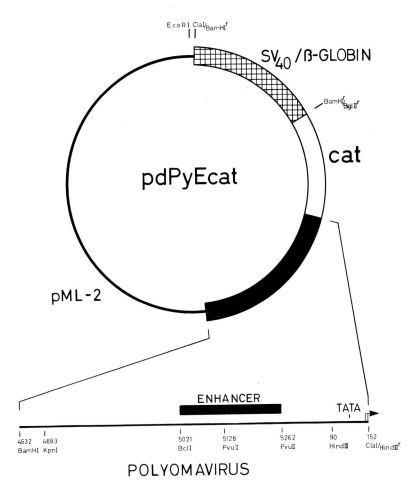

**Figure 1** Structure of pdPyEcat. Polyomavirus sequences from nt 4632 (a *Bam*HI site) to nt 152 (a *Cla*I site) were joined to the CAT gene and accompanying SV40 β-globin sequences that encode RNA processing and transcription termination signals. This hybrid transcription unit was cloned between the *Bam*HI and *Cla*I sites of the plasmid pML-2. The filled area on the circular map represents polyomavirus DNA, the open band represents CAT DNA, and the hatched area represents SV40 and β-globin DNA. The linear representation of polyomavirus DNA shows the position of the enhancer region, the early TATA box, and the major start sites for transcription (arrows) as well as relevant restriction endonuclease cleavage sites.

identify the locations of two partially overlapping enhancers (Fig. 3). One enhancer maps between nt 5057 (−385) and nt 5130 (−312), whereas the other is located between nt 5073 (−369) and nt 5229 (−213). These enhancers have in common those 57 bp located between nt 5073 (−369) and nt 5130 (−312). Noteworthy is the observation that neither minimal enhancer was as effective as the entire region (nt 5039−5229) in augmenting CAT gene expression (cf. Figs. 2 and 3).

Although the difference between the capacity of any one of the two minimal enhancer regions (nt 5039−5130 or nt 5073−5229) and the entire region (nt 5039−5229) to potentiate CAT gene expression was small, this difference was reproducible. Therefore, we could not rule out the possibility that the polyomavirus enhancer comprised three independent elements capable of functioning in pairwise combinations. To examine this, we deleted sequences in the region of overlap between the two minimal enhancers and tested the capacity of the resulting mutants to express the CAT gene in 3T3 cells. The mutant bearing the largest deletion, which spanned the region between nt 5092 (−350) and nt 5130 (−312), previously shown to be essential for the activity of each enhancer, expressed the CAT gene at levels equivalent to that of the two minimal enhancer-bearing plasmids (data not shown). This result suggested that

the polyomavirus enhancer is composed of at least three functionally equivalent elements. Apparently any pair of elements function nearly as well as all three elements together to enhance CAT expression. In the remaining text we refer to these DNA segments as enhancer elements and number them 1 (nt 5039−5073), 2 (nt 5073−5130), and 3 (nt 5131−5229).

To confirm this proposed structure for the polyomavirus enhancer, we cloned various combinations of enhancer elements in another CAT plasmid named pTE1 and tested the capacity of the resulting DNAs to express the CAT gene. pTE1 contains a hybrid transcription unit made up of the HSV-1 thymidine kinase (TK) promoter (−109 to +50) fused to the CAT gene and accompanying downstream signals from SV40 and rabbit β-globin to effect proper processing of its transcript (Fig. 4). Polyomavirus enhancer fragments were cloned at a site about 700 bp upstream from the start site for CAT transcription in both possible orientations (Fig. 4). A total of 14 different plasmids were constructed and assayed for their capacity to express the CAT gene after transfection of 3T3 cells by comparison with the parental plasmid, pTE1. The results of several experiments are summarized in Table 1. In general, single enhancer elements poorly enhanced CAT gene expression (from 3-fold to 4-fold). One notable exception was element 2,

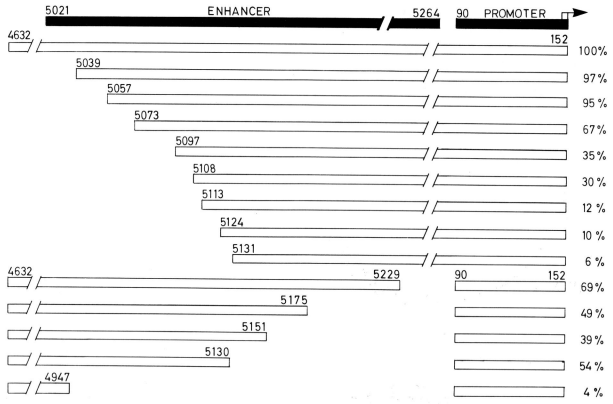

**Figure 2** Structure and efficiency of CAT expression of a set of 5′ and 3′ unidirectional deletion-mutant DNAs. The topmost, filled line shows the position of the polyomavirus enhancer and early promoter as established from previous work. The arrow denotes the direction and positions of the early start sites for transcription. The boxed, open areas represent the polyomavirus sequences of various deleted DNAs that drive expression of the CAT gene in the same sequence context as shown in Fig. 1. The sequenced end points of all the 5′ and 3′ unidirectional deletion mutants are numbered. A *Bam*HI linker abuts all the 5′ deletion-mutant DNAs, whereas a *Hin*dIII linker joins the 3′ enhancer deletions to the early polyomavirus promoter. The values cited as percentages at right refer to the efficiency with which the deleted DNAs were expressed compared with the parental, wild-type DNA, pdPyEcat. Each DNA was tested on at least three separate occasions in duplicate by transfection of 3T3 cells and subsequent assay of CAT enzymatic activity. The value shown represents the average of these individual measurements.

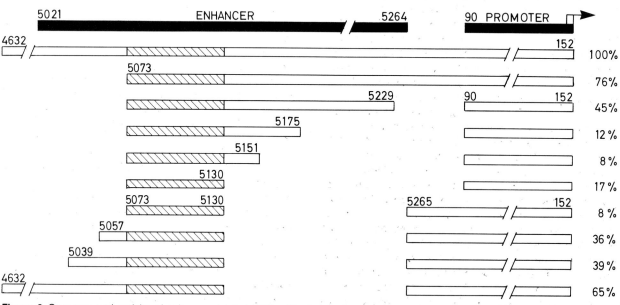

**Figure 3** Structure and activity of enhancer-deleted DNAs. The hatched areas denote those sequences that were defined by the analysis of 5′ and 3′ unidirectional deletion-mutant DNAs to represent the minimal enhancer. The open boxes represent those polyomavirus sequences that were joined to the CAT gene to promote its expression. A *Hin*dIII linker served to join the enhancer to the promoter in those plasmids where the joint occurs at nt 90, whereas those DNAs with the joint at nt 5265 contain no linker at this site.

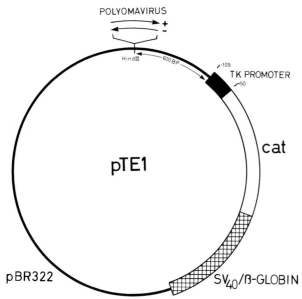

**Figure 4** Structure of pTE1. pTE1 is composed of sequences that include the HSV-TK promoter (filled box), the CAT gene (open box), portions of the SV40 early and β-globin genes (hatched box), and pBR322 DNA (dark line). A unique *Hind*III site, embedded within a polylinker, was used to clone various polyomavirus DNA fragments as described in the text.

which could enhance CAT gene expression 12-fold in one orientation (−), but only 4-fold in the other (+). Pairs of enhancer elements potentiated CAT gene expression several-fold better than the sum of each element, and this value was nearly as high as that of all three elements together.

To ensure that the enhanced levels of CAT activity were the result of increased transcription of the CAT gene, we measured by S1 analysis the abundance and position of the 5′ termini of the CAT mRNAs that resulted after transfection of 3T3 cells with the various plasmids. This analysis showed that increased levels of correctly initiated transcripts resulted after inclusion of

either all three or pairs of polyomavirus enhancer elements in pTE1 (data not shown).

### Polyomavirus replication activator elements

We also mapped the α and β replication elements using methods described previously (Muller et al. 1983; W.J. Muller and J.A. Hassell, in prep.). The α element maps within the borders of enhancer element 2, between nt 5097 and nt 5126, whereas the β element maps within the borders of enhancer element 3, between nt 5172 and nt 5202.

Because the replication activators and the enhancer elements appeared to be coincident, we were curious to learn whether the α and β elements, like enhancers, could activate replication independent of their orientation and position relative to the core element. Therefore, we measured the replication of a number of recombinant plasmids in which the orientation and/or position of the α and β elements were altered relative to the core. Replication was measured after transfecting the various recombinant plasmids into MOP-8 cells by the method of Peden et al. (1980). MOP-8 cells are permissive 3T3 cells that constitutively synthesize polyomavirus large, middle, and small T antigens (Muller et al. 1984). A recombinant plasmid that contained the β element at its native site, but inverted relative to its natural orientation, replicated as well as its wild-type counterpart (Fig. 5A). Similarly, a plasmid bearing an inverted α element in essentially its native position relative to the core also replicated, but poorly relative to the wild-type control (Fig. 5A). These results showed that the α and β elements were capable of activating replication independent of their orientation relative to the core.

To determine whether the replication elements could also act at a distance from the core, we placed them either individually or as a unit in both possible orientations at two sites in a recombinant plasmid bearing the core (Fig. 5B). None of the inserts containing either α or β or both elements together could activate replication from a position 200 bp removed from the late core border or from a site 50 bp distal to the early border of

**Table 1** Elevation of CAT Gene Expression by Polyomavirus Enhancer Elements

| Polyoma DNA segment | Enhancer element | Relative orientation | Enhancement factor |
|---|---|---|---|
| 5039–5092 | 1 | + | 1 |
| 5039–5092 | 1 | − | 1 |
| 5073–5130 | 2 | + | 4 |
| 5073–5130 | 2 | − | 12 |
| 5131–5264 | 3 | + | 3 |
| 5131–5264 | 3 | − | 3 |
| 5039–5130 | 1 + 2 | + | 7 |
| 5039–5130 | 1 + 2 | − | 83 |
| 5039–5092/ 5131–5264 | 1 + 3 | + | 31 |
| 5039–5092/ 5131–5264 | 1 + 3 | − | 26 |
| 5073–5229 | 2 + 3 | + | 43 |
| 5073–5229 | 2 + 3 | − | 61 |
| 5039–5264 | 1 + 2 + 3 | + | 83 |
| 5039–5264 | 1 + 2 + 3 | − | 101 |

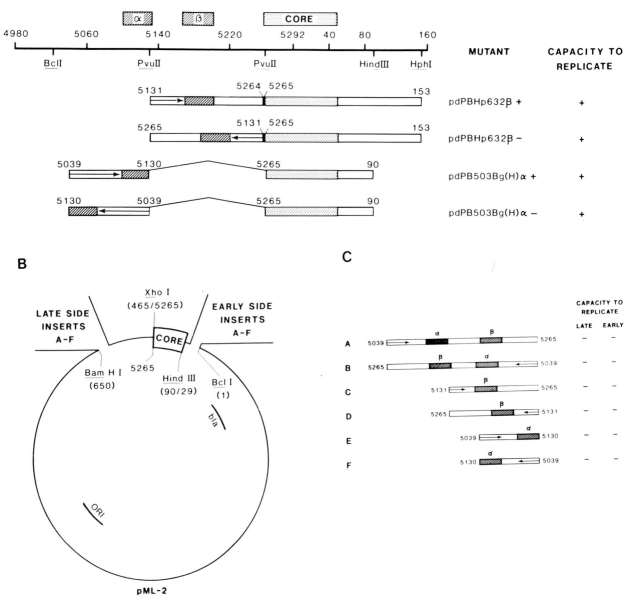

**Figure 5** The structure and replicational activity of various configurations of the polyomavirus origin for DNA replication. (*A*) The structure and activity of the polyomavirus origin containing inversions of the β (pdPBHp632β⁻) and α (pdPB503Bg[H]α⁻) replication activators. (*B*) The structure of the vector used to clone and assess the replicational activity of recombinant plasmids containing the replication activators positioned at a distance from the core. The DNA segments shown in *C* were cloned with *Bam*HI linkers in the vector at a unique *Bam*HI site (late-side inserts) or *Bcl*I site (early-side inserts) in both possible orientations to yield six recombinant plasmids. (*C*) Composition and orientation of the various DNA fragments that were cloned in the vector shown in *B*. The arrow within each segment refers to the orientation of the insert relative to the core. Arrows pointing to the right are meant to depict the activator in the same orientation as it naturally occurs in polyomavirus DNA relative to the core.

the core (Fig. 5C). This was true despite the fact that the fragment containing both replication activator elements (fragments A and B in Fig. 5C) carried all three enhancer elements.

## Discussion

The position of the enhancer and replication elements relative to other features in the noncoding region of the

polyomavirus genome is illustrated in Figure 6. The entire enhancer region occupies no more than 190 bp (nt 5039–5229). It can be divided into at least three, apparently functionally redundant segments that we have named enhancer elements 1, 2, and 3. The replication activators, α and β, are individually no larger than 30 bp, and they map respectively in enhancer elements 2 and 3. The borders of the enhancer elements have not been defined to the same extent as those of the replica-

# ENHANCER / REPLICATION ELEMENTS

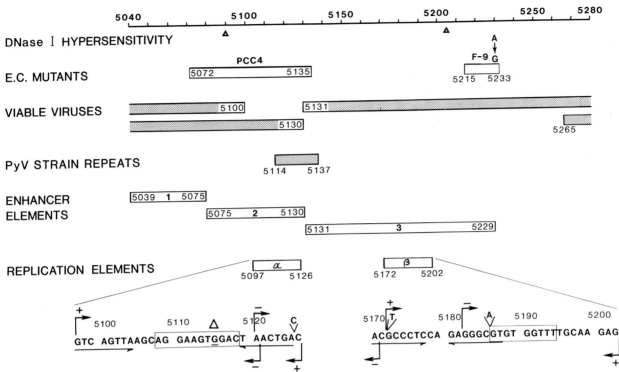

**Figure 6** Schematic of features within the polyomavirus enhancer region. The top line represents noncoding polyomavirus DNA numbered according to the scheme proposed by Soeda et al. (1979). The open triangles below the line depict sites in viral chromatin that are hypersensitive to DNase I cleavage (Herbomel et al. 1981). The borders of DNA that are commonly duplicated in the genome of polyomavirus mutants selected for their capacity to replicate in PCC4 or F-9 embryonal carcinoma (E.C.) cells are shown by the open boxes. One E.C. mutant, F441, capable of efficient replication in F-9 cells contains only a nucleotide substitution at nt 5230 (A → G transition). The stippled boxes represent those sequences that are present within the genomes of two viable deletion mutants of polyomavirus. The sequences between nt 5100 and nt 5131 and between nt 5130 and nt 5265 are missing from these viral genomes. The hatched box shows a region in polyomavirus DNA that is commonly duplicated in various wild-type strains of polyomavirus (Ruley and Fried 1983). The positions of the three DNA segments that are thought to be enhancer elements are shown as three numbered, open boxes. The α and β replication elements are shown together with their sequences. The arrows that underline sequences in α and β represent the inverted repeat motif, whereas the boxed sequences represent enhancer core sequences. The arrows above and below the sequence represent deletion end points on the late (left) or early (right) side of each element. The + and − symbols designate whether a mutant DNA with that end point does or does not replicate in MOP-8 cells. Note that the DNA sequence of our polyomavirus strain differs slightly from that reported by Soeda et al. (1979). For example, our DNA lacks the guanine at nt 5115 and contains several 1-bp insertions between nt positions 5125 and 5126, 5172 and 5173, and 5185 and 5186.

tion activators. Consequently, we do not know if the borders of the replication elements and enhancer elements are identical or not.

The organization of the polyomavirus enhancer and replication elements described here is consistent with the structure of viable viruses with deleted or rearranged genomes. For example, the virus dl2039L lacks sequences between positions 5100 and 5131 yet is viable (Cowie et al. 1981). This virus does not contain enhancer element 2 and the α replication element (Fig. 6). Similarly, another virus (dl1024) lacks enhancer element 3 and the β replication element (Luthman et al. 1982). There are also numerous strains and host-range mutants of polyomavirus whose genomes contain duplications of enhancer elements 2 and 3, or α and β, as well as mutations within these elements (Ruley and Fried 1983). Duplications may increase the strength of the

elements, whereas mutations may alter their specificity, thereby allowing for their interaction with proteins found only in embryonic cells or particular differentiated cells in the mouse. In this regard it is noteworthy that the α element fails to activate the polyomavirus core for replication in early mouse embryos whereas the β element does (Wirak et al. 1985).

Inspection of the primary structure of each replication activator reveals the presence of conserved sequence motifs that have been found in the enhancers of other viruses (Fig. 6). For example, there is a perfect match between a 10-bp stretch of DNA in α and a core sequence (5' $^A/_C$GGAAGTGA$^A/_C$ 3') in the adenovirus type-5 E1A enhancer (Hearing and Shenk 1983). Similarly, there is a match of 9 out of 10 nucleotides in the β element to a different core sequence (5' $^G/_C^G/_C^C/_T$GTGG$^A/_T^A/_T^A/_T^A/_T$ 3') first described in the SV40 enhancer (Weiher et al. 1983). The enhancer core sequences in the replication activators are closely flanked by inverted repeats (Fig. 6). Interestingly, the inverted repeats in each element share a common sequence, 5′ ACTG$^A$/$_C$ C 3′.

We have as yet not been able to individually assess the functional significance of each of these sequence motifs with respect to transcription enhancement or replication activation. However, we do know that the deletion of the inverted repeat motif (nt 5097–5108) to the left of the adenovirus enhancer core sequence does not impair either enhancer element 2 or α function (Fig. 6). By contrast, deletion of the inverse of this sequence (nt 5120–5126) to the right of the adenovirus enhancer core sequence abolishes both enhancer element 2 and α function. These results imply that the hairpin structure that could be generated between these two sequences is not important for the functioning of the elements and that the two sequences, despite their common primary structure, are not functionally equivalent. Perhaps it is the orientation or position of the conserved inverted repeat sequence relative to the adenovirus enhancer core that is important for the proper functioning of element 2 and α.

We do not know whether the adenovirus core sequence in element 2 or α plays a functional role or not. Deletion of the first 5 nt of this sequence (nt 5108–5113) causes a threefold reduction in the activity of enhancer element 2 (Fig. 2). However, an equivalent mutant has not been tested for α function. It may be noteworthy that only part of this 10-bp sequence, the last 5 bp (5′ TGACT 3′) is commonly duplicated in various wild-type strains of polyomavirus (Fig. 6). The commonly duplicated sequence spans nt 5114–5137 (Ruley and Fried 1983). Whatever the role of this sequence, it is clear that it cannot function by itself to endow either enhancer element 2 or α with activity. If the enhancer core sequence is important, then it must be coupled to the inverted repeat motif on its right to be active (Fig. 6).

The β replication element, which lies within the borders of enhancer element 3, contains the SV40 enhancer core sequence within its borders (Fig. 6). We have no data that bear on its role in the functioning of either the β replication activator or enhancer element 3. However, if this sequence is functionally important, it cannot act alone, because deletions that remove the inverted repeat motif but retain the SV40 enhancer core sequence inactivate the β replication element. We do not know whether the inverted repeat sequence or enhancer core sequence is important for enhancer element 3 to function, because we have not yet tested the mutants described above for their capacity to enhance transcription.

At this stage it is not clear to us whether the minimal sequences required to constitute an enhancer element and those required to form a replication activator element are the same or not. However, it is clear that each replication activator and enhancer element share sequences. They include a common sequence that is repeated in both replication activators on opposite DNA strands, the inverted repeat motif (5′ ACTG$^A$/$_C$ C 3′), as well as enhancer core sequences, whose primary structure differs in each element. From these results we infer that at least three different *trans*-acting factors must be required to act at these various sequences. One factor may interact with the inverted repeat motif, whereas two different factors may bind to the two enhancer core sequences. There is both indirect (Schöler and Gruss 1984) and direct (see Gluzman 1985) evidence to support the existence of such factors. In this regard the fact that the α replication element fails to activate the polyomavirus DNA replication in early mouse embryos, under conditions where the β element functions well, suggests that the factor that interacts with the adenovirus enhancer core sequence in α is absent or inactive in early mouse embryo cells.

We have been unable to activate polyomavirus DNA replication with enhancer element 1. This element is also completely incapable of enhancing CAT expression when placed alone in the pTE1 (Table 1). These observations suggest that element 1 is intrinsically weak, perhaps because it is unable to serve as an independent binding site for *trans*-acting factors. In this regard it is noteworthy that element 1 contains a poor match with the SV40 enhancer core sequence (an 8 out of 10 match; 5′ CGGGAGGAAA 3′ compared with the consensus sequence 5′ $^G$/$_C$$^G$/$_C$$^C$/$_T$GTGG$^A$/$_T$$^A$/$_T$$^A$/$_T$ 3′) and an equally poor match with the conserved inverted repeat motif that is present in enhancer elements 2 and 3 (5′ TACTGTG 3′ compared with the consensus sequence 5′ $^A$/$_C$ACTG$^A$/$_C$ C 3′).

We have attempted to devise a model for enhancer function that takes into account its capacity to enhance transcription as well as to activate DNA replication. The observation that enhancers can potentiate transcription but not replication from a distance suggests to us that although the initial reactions leading to these processes might be the same, subsequent pathways are likely to be different. We propose that one of the consequences of the binding of factors to the enhancer is to cause the separation of the DNA strands. Single-stranded DNA could provide an entry site for RNA polymerase II, which is known to prefer single-stranded over double-stranded DNA as a template for transcription in vitro (for review, see Roeder 1976). Once on the template, RNA polymerase may then move along it until it encounters transcription factor–DNA complexes that align it for initiation. Alternatively, the opening of the helix could facilitate the bending or collapse of the DNA to permit protein-protein interactions to occur between factors bound to the enhancer and other factors bound to the promoter (see Gluzman 1985). This complex of proteins bound to DNA may serve as a template for RNA polymerase to initiate transcription.

Replication activation may also require the separation of DNA strands, but this must occur at a specific site near the A·T–rich, late border of the core. The A·T stretch is an essential component of the polyomavirus core as well as the SV40 origin for DNA replication

(Muller et al. 1983). The A·T−rich region may facilitate local denaturation in the core. Local denaturation of the DNA may permit access of replication factors required to initiate DNA synthesis. Another possibility is that the separation of the DNA strands may promote folding of each strand into a structure required for initiation. The cruciform that the G·C-rich palindrome in the core could adopt may be this structure.

In this model the structural alteration to the DNA (viz., separation of its strands) must occur locally to effect DNA replication, whereas it can occur at a distance to effect transcription. We believe this reflects the idiosyncracies of the enzymes required to carry out these different processes. For example, RNA polymerase II may be able to move along DNA, whereas a replication factor such as primase may not; or primase may require single-stranded DNA to initiate the synthesis of RNA primers, whereas RNA polymerase may require a different template (e.g., a DNA-protein complex) to initiate mRNA synthesis. Whatever the mechanism, it seems clear that in vitro assays will have to be developed to permit the isolation of the various factors that participate in transcription enhancement and replication activation. Using isolated components, it should be possible to more readily elucidate the molecular mechanism of transcription enhancement and replication activation.

## Acknowledgments

We thank Monica Naujokas and Cathy Collins for providing technical assistance and Fiona Lees for typing the manuscript. This research was supported by funds from the Medical Research Council and the National Cancer Institute of Canada. W.J.M. and C.R.M. are supported by research fellowships from the Medical Research Council. J.A.H. is a research associate of the National Cancer Institute of Canada.

## References

Cowie, C., C. Tyndall, and R. Kamen. 1981. Sequences at the capped 5′-ends of polyomavirus late region mRNAs: An example of extreme terminal heterogeneity. *Nucleic Acids Res.* **9:** 6305.

de Villiers, J. and W. Schaffner. 1981. A small segment of polyomavirus DNA enhances the expression of a cloned β-globin gene over a distance of 1,400 base pairs. *Nucleic Acids Res.* **9:** 6251.

Gluzman, Y., ed. 1985. *Eukaryotic transcription.* Cold Spring Harbor Laboratory, Cold Spring Harbor, New York.

Gorman, C.M., L.F. Moffat, and B.H. Howard. 1982. Recombinant genomes which express chloramphenicol acetyltransferase in mammalian cells. *Mol. Cell. Biol.* **2:** 1044.

Hearing, P. and T. Shenk. 1983. The adenovirus type 5 E1A transcriptional control region contains a duplicated enhancer element. *Cell* **33:** 695.

Herbomel, P., S. Saragosti, D. Blangy, and M. Yaniv. 1981. Fine structure of origin-proximal DNase I−hypersensitive region in wild-type and E.C. mutant polyoma. *Cell* **25:** 651.

Jat, P., U. Novak, A. Cowie, C. Tyndall, and R. Kamen. 1982. DNA sequences required for specific and efficient initiation of transcription at the polyoma virus early promoter. *Mol. Cell. Biol.* **2:** 737.

Luthman, H., M.G. Nilsson, and S. Magnusson. 1982. Noncontiguous segments of the polyoma gene required in *cis* for DNA replication. *J. Mol. Biol.* **161:** 533.

Maniatis, T., E.F. Fritsch, and J. Sambrook. 1982. Analysis of recombinant DNA clones. In *Molecular cloning: A laboratory manual.* p. 363. Cold Spring Harbor Laboratory, Cold Spring Harbor, New York.

Mueller, C.R., A.-M. Mes-Masson, M. Bouvier, and J.A. Hassell. 1984. Location of sequences in polyomavirus DNA that are required for early gene expression *in vivo* and *in vitro*. *Mol. Cell. Biol.* **4:** 2594.

Muller, W.J., M.A. Naujokas, and J.A. Hassell. 1984. Isolation of large T antigen−producing mouse cell lines capable of supporting replication of polyomavirus-plasmid recombinants. *Mol. Cell. Biol.* **4:** 2406.

Muller, W.J., C.R. Mueller, A.-M. Mes, and J.A. Hassell. 1983. Polyomavirus origin for DNA replication comprises multiple genetic elements. *J. Virol.* **47:** 586.

Peden, D.W.C., J.M. Pipas, S. Pearson-White, and D. Nathans. 1980. Isolation of mutants of an animal virus in bacteria. *Science* **209:** 1392.

Pomerantz, B.J. and J.A. Hassell. 1984. Polyomavirus and simian virus 40 large T antigens bind to common DNA sequences. *J. Virol.* **49:** 925.

Pomerantz, B.J., C.R. Mueller, and J.A. Hassell. 1983. Polyomavirus large T antigen binds independently to multiple, unique regions on the viral genome. *J. Virol.* **47:** 600.

Roeder, R.G. 1976. Eukaryotic nuclear RNA polymerases. In *RNA polymerase* (ed. R. Losick and M. Chamberlin), p. 285. Cold Spring Harbor Laboratory, Cold Spring Harbor, New York.

Ruley, H.E. and M. Fried. 1983. Sequence repeats in a polyomavirus DNA region important for gene expression. *J. Virol.* **47:** 233.

Schöler, H.R. and P. Gruss. 1984. Specific interaction between enhancer-containing molecules and cellular components. *Cell* **36:** 403.

Soeda, E., J.R. Arrand, N. Smolar, and B.E. Griffin. 1979. Coding potential and regulatory signals of the polyomavirus genome. *Nature* **283:** 445.

Tyndall, C., G. Lamantia, C.M. Thacker, H. Favaloro, and R. Kamen. 1981. A region of the polyoma virus genome between the replication origin and late protein coding sequences is required in *cis* for both early gene exression and viral DNA replication. *Nucleic Acids Res.* **9:** 6231.

Weiher, H., M. König, and P. Gruss. 1983. Multiple point mutations affecting the simian virus 40 enhancer. *Science* **219:** 626.

Wigler, M., A. Pellicer, S. Silverstein, and R. Axel. 1978. Biochemical transfer of single-copy eukaryotic genes using total cellular DNA as donor. *Cell* **14:** 725.

Wirak, D.O., L.E. Chalifour, P.M. Wassarman, W.J. Muller, J.A. Hassell, and M.L. DePamphilis. 1985. Sequence-dependent DNA replication in preimplantation mouse embryos. *Mol. Cell. Biol.* **5:** 2924.

# Cellular Transformation by Bovine Papillomavirus

J.T. Schiller, E.J. Androphy, W.C. Vass, and D.R. Lowy

Laboratory of Cellular Oncology, National Cancer Institute, National Institutes of Health, Bethesda, Maryland 20205

We have genetically analyzed the transforming activity of the bovine papillomavirus type-1 (BPV-1) genome and determined that it contains two genes that can independently transform mouse C127 cells. One of these genes is encoded by the E6 open reading frame located at the 5' end of the early region. This reading frame was cloned into a bacterial expression vector that directed the synthesis of large amounts of an E6 fusion protein that was used to generate antibodies to E6. These antibodies specifically immunoprecipitated the predicted 15.5-kD protein from C127 cells transformed by the E6 gene. Analysis of the E6 protein in transformed cells indicated that it is not in disulfide linkage with other proteins and is distributed equally between the nucleus and the nonnuclear membranes. The second transforming gene has been localized to the 3' end of the early region and appears to lie entirely within the E5 open reading frame. The predicted E5 gene product is a 44-amino-acid hydrophobic peptide that remains to be identified in transformed cells.

Papillomaviruses induce benign tumors (warts or papillomas) of the cutaneous and mucosal epithelia. In some instances, a subset of these papillomas appear to undergo malignant degeneration (Orth et al. 1980; Ostrow et al. 1982). A close association between human papillomavirus (HPV) infection and certain epithelial malignancies has given added significance to the elucidation of papillomavirus genetics and biology (Durst et al. 1983; Boshart et al. 1984). Bovine papillomavirus type 1 (BPV-1), which induces fibropapillomas in its natural host, is the papillomavirus whose genetics has been studied in greatest detail. Since DNA sequence analysis suggests that all papillomavirus genomes share a similar genetic organization (Danos et al. 1984; Schwarz et al. 1983), it is probable that an understanding of BPV genetics will have direct relevance to other papillomaviruses, including HPVs.

Infection with BPV or its viral DNA genome can induce morphologic transformation of certain established cell lines in vitro, such as NIH-3T3 and C127 mouse cells (Dvoretzky et al. 1980; Lowy et al. 1980). Cells transformed by BPV display a fully transformed phenotype: They form foci of proliferating cells in monolayer culture, colonies in agar, and tumors in nude mice. In these cells the viral DNA genome replicates as an unintegrated, multicopy episome (Law et al. 1981). This extrachromosomal replication of the viral DNA is not required for morphologic transformation; after DNA-mediated gene transfer, some mutant viral genomes can transform the cells despite the viral genetic information integrating into the host chromosomal DNA (Nakabayashi et al. 1983; Lusky and Botchan 1985). Cultured cells that contain the papillomavirus genome do not release progeny virus particles, since the genes that encode structural viral proteins are not transcribed in these cells (Heilman et al. 1982).

Genetic analysis of BPV has begun to assign specific functions to defined regions within the viral DNA genome. The transforming and extrachromosomal maintenance functions have been mapped to a 5.4-kb segment of the 8-kb viral genome; this segment is called 69T because it represents 69% of the viral genome (Fig. 1) (Lowy et al. 1980). Since 69T lacks the open reading frames (ORFs, potential protein-coding segments) that encode the structural virion proteins (ORFs L1 and L2), it is clear that ORFs L1 and L2 are dispensable for these functions. As shown in Figure 1, DNA sequence analysis of BPV indicates the presence of at least eight ORFs (E1–E8) within this 5.4-kb transforming segment (Chen et al. 1982). Sequences from these ORFs are expressed as RNA in transformed cells (and in warts) (Amtmann and Sauer 1982; Heilman et al. 1982; Engel et al. 1983).

At present, the viral functions that control replication and expression of the viral genome are poorly understood. Since these functions may indirectly affect the induction of morphologic transformation by the virus, we have studied BPV-induced transformation principally by analyzing the transforming activity of subgenomic fragments that integrate into host DNA and are activated by a retroviral long terminal repeat (LTR) rather than by BPV controlling elements. Using this approach, two nonoverlapping transforming fragments have been isolated, and their respective transforming genes, E6 and E5, have been identified genetically.

To begin to understand the mechanisms by which BPV genes transform cells, we have initiated studies to identify their protein products in transformed cells. No protein encoded by a papillomavirus early ORF has previously been identified, presumably because of the low level of expression of these genes in infected cells. After expressing the E6 ORF in bacteria and making

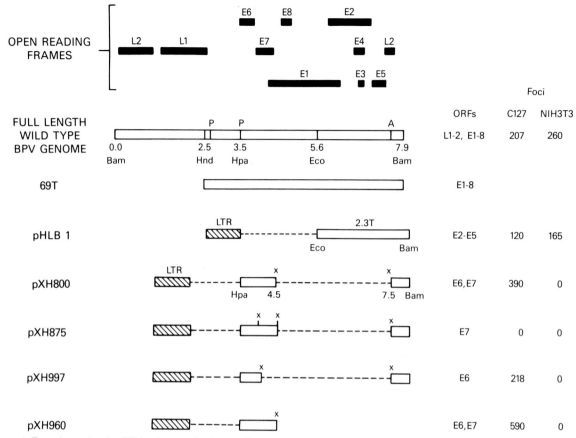

**Figure 1** Transformation by BPV subgenomic clones. The genome of BPV-1 linearized at the unique *Bam*HI site in which it was cloned into the pBR322 derivative pML2d is shown. The numbers below the BPV genome indicate the distance in kilobases from the *Bam*HI site at 0.0. The locations of putative promoters and polyadenylation signal are designated P and A, respectively. Open bars represent BPV sequences included in the clones, and dashed lines indicate deleted sequences. Sites of *Xho*I linker insertions are denoted by × . Recognition sites: (Bam) *Bam*HI; (Hnd) *Hind*III; (Hpa) *Hpa*I; (Eco) *Eco*RI.

antibodies to the resulting polypeptide encoded by the E6 ORF, we have identified an E6-encoded protein in C127 cells transformed by this gene and have begun its characterization. This manuscript summarizes several recent reports from our laboratory (Schiller et al. 1984; Androphy et al. 1985; Schiller et al. 1985, 1986).

## Methods

### Construction of mutants

*Xho*I linker-insertion mutations were introduced into the full-length BPV genome cloned in pML2d (Sarver et al. 1982) after random cleavage with DNase I (Heffron et al. 1978). Deletion mutants were generated by elimination of the sequences between pairs of *Xho*I linker inserts. pXH800 was constructed by the ligation of the 0.9-kb LTR containing the *Eco*RI-to-*Bal*I fragment of Moloney murine sarcoma virus (Dhar et al. 1980) to a BPV segment that included the sequences between the *Hpa*I site and an *Xho*I linker insertion at nucleotide (nt) 1019, which in turn was linked to a 0.4-kb *Xho*I-to-*Bam*HI fragment (nt 4040–4450) from the 3′ end of the BPV early region (Fig. 1). *Xho*I linker insertions were

introduced into pXH800 by replacing its *Hpa*I-to-*Xma*I fragment with the corresponding fragment from full-length BPV insertion mutants (Schiller et al. 1984). Clone pHLB1 contains the 2.3-kb *Eco*RI-to-*Bam*HI fragment from the 3′ end of the early region of BPV-1 and the 4.0-kb *Eco*RI-to-*Bam*HI fragment of pBR322 with the 0.6-kb permuted Harvey retroviral LTR inserted into the *Eco*RI site in a promoter-positive orientation relative to the BPV sequences (Nakabayashi et al. 1983). The linker-insertion mutants were generated by substituting the 2.05-kb *Bst*EII-to-*Bam*HI BPV fragment of each of the appropriate *Xho*I linker-insertion mutants of the full-length genome for the corresponding fragment of pHLB1. Deletion derivatives of pHLB1 were constructed by removing the sequences between pairs of *Xho*I linker-insertion mutants of pHLB1.

### Cell transformation assay

In the transfection assays, 0.5 μg of undigested DNA was precipitated with $CaCl_2$ (Graham and van der Eb 1973) and added in 0.2 ml to a 35-mm plate seeded on the previous day with $2.5 \times 10^5$ C127 or NIH-3T3 cells.

The cells were treated as described previously (Lowy et al. 1978), except that dimethyl sulfoxide was not used.

### Expression of E6 ORF in bacteria

The prokaryotic expression vector pCO-5, which is identical with a previously published vector (Lautenberger et al. 1983), was constructed by inserting a 736-bp fragment of pOG7 (which contains the phage λ $p_L$ promoter, ribosome-binding site, and amino terminus of the phage λ cII protein) into the ClaI site of pBR322. A 589-bp HpaII fragment of BPV was inserted; in-frame translation predicted a fusion of the aminoterminal 13 amino acids of cII to 140 amino acids of the E6 ORF. This DNA was introduced into bacteria N6405, an N⁺ Escherichia coli that contains a lysogenic, temperature-sensitive repressor (cI857ts) of $p_L$. Synthesis of the fusion protein was induced at 42°C, and rabbits were immunized with acrylamide gel slices containing this fusion protein.

### Immunoassays

Immunoprecipitations were performed using the rabbit antisera and protein A bound to Sepharose beads (Pharmacia) in RIPA buffer (50 mM Tris [pH 7.4], 1 mM EDTA, 150 mM NaCl, 1% deoxycholate, 0.1% SDS, 1% aprotinin), and run on 15% polyacrylamide gels. For blocking experiments, nonradiolabeled bacterial protein extracts were preincubated with antisera, washed, and then incubated with radiolabeled lysates from transformed cells. Subcellular fractionations were performed using minor modifications of two different procedures (Cooper and Hunter 1982; Hann and Eisenman 1984).

## Results

### BPV contains more than one transforming gene

To establish which gene(s) might be required for BPV transformation, we generated a series of single-site mutants of this genome by randomly inserting an 8-bp XhoI linker into the full-length BPV-1 genome. Forty single-site insertions were analyzed, and although every major ORF was interrupted by at least one of the linker-insertion mutations, each mutant retained at least some capacity to induce focal transformation in mouse C127 cells (data not shown). Since the majority of the linker insertions should have resulted in a frameshift mutation within the interrupted ORF, these results suggest that BPV-1 contained more than one transforming gene.

To confirm that BPV encoded multiple transforming genes, two nonoverlapping subgenomic fragments were identified that, when activated by a retroviral LTR, could induce morphologic transformation. One of the transforming clones, pXH800, contained the first 1.0 kb of the early region and the other, pHLB1, contained a 2.3-kb fragment from the 3′ end of the early region (Fig. 1). pHLB1 induces transformation in both C127 and NIH-3T3 cells. In contrast, pXH800 induces a fully transformed phenotype in C127 cells but is unable to transform NIH-3T3 cells (Schiller et al. 1984).

### E5 encodes the transforming gene of the 3′ transforming fragment

The 2.3-kb EcoRI-to-BamHI fragment of BPV (2.3T) in the transforming clone pHLB1 contains the complete E2, E3, E4, and E5 ORFs as well as the 3′ end of the E1 ORF (Fig. 2). As in the analysis described above, a series of XhoI linker insertions and deletions were introduced into the BPV sequences of pHLB1 in order to determine which ORFs were required for pHLB1-induced transformation. The transforming activities of the mutants were assayed in NIH-3T3 cells rather than C127 cells because, in our hands, the foci induced in NIH-3T3 cells by pHLB1 and its transformation-competent derivatives were easier to distinguish from the background of nontransformed cells than were the foci induced in the C127 cells.

The transforming activities in NIH-3T3 cells of a series of pHLB1 mutants are shown in Figure 2. Mutants with insertions of the linker into the E1, E2, E3, or E4 ORF (even those mutants with shifts in reading frame, such as pHLB793, pHLB709, and pHLB113) did not have significantly reduced transforming activities (Fig. 2A). In contrast, an insertion that resulted in a frameshift immediately downstream from the first ATG of the E5 ORF abolished the transforming activity of the pHLB1 clone (pHLB500). Another linker-insertion mutant, pHLB717, in which the first 25 bp from the 5′ end of the E5 ORF were deleted but the ATG and downstream sequences remained intact, did not have reduced transforming activity. Similarly, a mutant with an insertion 11 bp downstream from the end of the E5 ORF (pHLB133) was transformation-competent. These results indicated that the E5 ORF encodes sequences that are required for pHLB1-induced transformation but that expression of an intact E1, E2, E3, or E4 ORF is not required.

Due to the limited coding capacity of the E5 ORF, we considered the possibility that upstream coding segments might be spliced onto E5 sequences to generate the functional transforming protein. To test this hypothesis, a series of relatively short, adjacent deletion mutations were introduced into pHLB1 by deleting the sequences between pairs of inserted linkers in the mutants described above. Mutants in which segments of the E1, E2, E3, and/or E4 ORFs were deleted still transformed the NIH-3T3 cells (Fig. 2B). Only the mutant pHLB113-133, in which the E5 ORF was deleted, failed to transform the cells. Since any of the BPV sequences upstream of the E5 ORF in pHLB1 can be removed without abolishing the transforming activity of pHLB1, our working hypothesis is that the transforming gene is encoded by the sequences between the first AUG of the ORF and the end of the ORF, from which a short peptide of only 44 amino acids could be synthesized. However, since neither an E5-specific transcript nor protein product has been unequivocally identified, we cannot rule out the possibility that the putative E5-encoded transforming peptide is a functional domain of a larger BPV protein that, in our constructions, has transforming activity in a truncated form. We speculate that some of the deletion mutants have a reduced transforming activity because they alter the expression of the E5 ORF.

### E6 encodes the transforming gene of the 5′ transforming fragment

An analogous genetic analysis of the 5′ transforming fragment has also been carried out (Schiller et al. 1984). pXH800 contains the entire E6 and E7 ORF and the 5′ end of the E1 ORF as well as a 0.4-kb fragment from the 3′ terminus of the early region that contains the polyadenylation site (Fig. 1). To determine which of these elements were active in the transformation of C127 cells induced by pXH800, a series of XhoI linker-insertion and deletion mutants were introduced into the BPV sequences of this clone. Mutations that interrupted the E6 ORF rendered pXH800 nontransforming (e.g., pXH875; Fig. 1). In contrast, mutations that interrupted

the E7 ORF but left the E6 ORF intact had little if any effect on the transforming activity of the clone (e.g., pXH997; Fig. 1). A mutant that deletes the BPV segment containing the polyadenylation signal (pXH960) was also highly transforming. These results indicate that the E6 ORF encodes the transforming gene of this clone and that the E7 ORF and 3′ sequences are not required for transformation.

An E6-7 fusion gene product is predicted from the analysis of BPV-specific mRNA in C127 cells transformed by the full-length genome (Stenlund et al. 1985; Yang et al. 1985). The mutation in pXH875 is located in the intron of the E6-E7 message. The fact that this mutant is transformation-defective but can still synthesize the E6-7 message (M. Botchan, pers. comm.) suggests that the putative E6-7 spliced protein does not have transforming activity. Analysis of a cDNA clone of an E6-7 message has led to a similar conclusion (Yang et al. 1985).

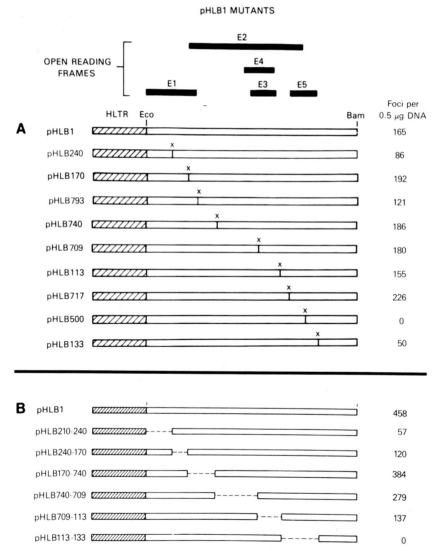

**Figure 2** Mutational analysis of the 3′ transforming region. Designations are the same as in Fig. 1. Only the carboxyl terminus of E1 is present in the 2.3-kb *Eco*RI-to-*Bam*HI fragment of pHLB1. DNAs were transfected onto NIH-3T3 cells in both experiments *A* and *B*.

## Generation of E6 antisera

Having demonstrated that E6 was a transforming gene, we used cells transformed by this gene to identify the E6 protein that presumably mediates this function. First we employed a prokaryotic expression vector in order to produce large quantities of an E6 polypeptide that could be used to generate antibodies specific to E6. The E6 ORF was inserted into the vector so that the plasmid encoded a fusion protein containing the aminoterminal 13 amino acids of the phage λ cII protein and 140 amino acids of the E6 ORF. The E6 sequences in the vector began three codons upstream from the first (translation initiation) AUG of E6 and extended down past the termination codon of the ORF. When placed in *E. coli*, this vector directed the synthesis of the predicted 17.5-kD E6 fusion protein (data not shown). This band was excised from SDS-polyacrylamide gels and rabbits were immunized with this protein preparation. The immunization protocol resulted in rabbit sera that reacted with the fusion protein under reducing and nonreducing conditions by immunoprecipitation and by immunoblotting (not shown).

## Identification of the E6 protein

The above experiments confirmed the presence of the E6 ORF, as predicted from DNA sequence analysis (Chen et al. 1982). However, they did not prove that this ORF actually encoded a polypeptide in mammalian cells that had been morphologically transformed by E6. In principle, it should be easier to detect a putative E6 protein in cells that expressed the gene at high levels. We therefore sought to identify the protein in C127 cells transformed by the LTR-activated clones pXH800 and pXH997 (Fig. 1), since E6 RNA levels were found to be at least 10 times greater in cells transformed by these constructions than in cells transformed by the full-length BPV genome.

When cells transformed by either of these LTR-activated clones were metabolically labeled with [$^{35}$S]cysteine, the sera from the rabbit that possessed antibodies to the bacterial fusion protein specifically immunoprecipitated a 15.5-kD protein (Fig. 3; pXH800-1, 800-2, and 997). This protein was not immunoprecipitated from control C127 cells transformed by the 3′ BPV transforming clone or by a *ras*$^H$ oncogene (Fig. 4; pHLB-1 and HamCl8). 15.5 kD corresponds to the molecular mass predicted from DNA sequence analysis for the E6 protein, if the protein is initiated from the first AUG within the ORF and is not subjected to posttranslational modification. Since much of E7 is deleted from clone pXH997 (Fig. 1), the presence of this protein in pXH997-transformed cells rules out the possibility that the 15.5-kD molecule represents an E7 or E6-7 fusion protein. When NIH-3T3 and C127 cells transformed by the full-length BPV genome were analyzed, they also synthesized the 15.5-kD protein, but at levels approximately 10-fold lower than those found in cells transformed by the LTR-driven E6 ORF (not shown).

Immunoprecipitation inhibition experiments confirmed that the protein found in the transformed mouse cells

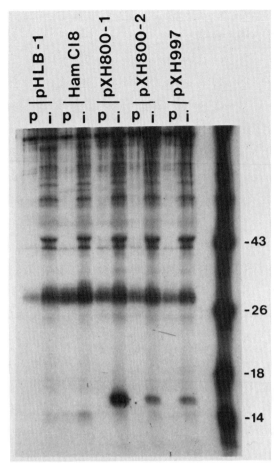

**Figure 3** Immunoprecipitation of E6 from transformed cell lines (13-day autoradiogram exposure). (p) Preimmune sera; (i) immune sera. 25 × 10$^6$ cpm of [$^{35}$S]cysteine-labeled cells were used per lane. Cell lines are designated by the plasmids (Figs. 1 and 2) that were used to generate them. HamCl8 is a *ras*$^H$-transformed C127 cell line.

was encoded by E6 (Fig. 4). Immunoprecipitation of the 15.5-kD band from E6-transformed C127 cells (lane 2) and the 17.5-kD E6 fusion protein from bacteria (lane 4) were specifically blocked only by a nonradiolabeled bacterial extract containing the E6 fusion protein. An extract from bacteria that did not contain E6 failed to block the immunoprecipitation of the 15.5-kD band from transformed cells (lane 1) or the fusion protein from bacteria (lane 3). As expected, neither extract blocked the immunoprecipitation of a *ras*$^H$ fusion protein by *ras* antibody.

## Subcellular localization of the E6 protein

After determining that the 15.5-kD protein is the product of E6, we sought to localize the protein in transformed cells by subcellular fractionation. Cells transformed by pXH800 were metabolically labeled with [$^{35}$S]cysteine and separated into nuclei, cytoplasm, and membranes (Fig. 5). Approximately equal quantities of the E6 protein were found in the nuclear and membrane fractions by two different methods of fractionation (Cooper and with [$^{14}$C]palmitate. Using conditions that clearly labeled p21, no labeling of E6 was detected (D. Lowy, unpubl.).

## E6 detected only as a monomer in transformed cells

When the immunoprecipitation of E6 from transformed cells was performed under nonreducing conditions, the 15.5-kD band was the only specific band observed. This result suggests that the protein is not in disulfide linkage with other E6 molecules or with other proteins. Since the association of the cellular protein p53 with SV40 large T antigen may contribute to the transforming activity of T antigen (Rotter and Wolf 1985), we decided to examine the relationship of E6 and p53 in E6-transformed cells. Antisera to E6 did not immunoprecipitate p53 under conditions that T antisera immunoprecipitate p53. Furthermore, p53 antisera did not immunoprecipitate E6, nor were p53 levels elevated in cells transformed by pXH800 (data not shown). Therefore, it is unlikely that an interaction between E6 and p53 is involved in E6-induced transformation of C127 cells.

## Discussion

We have identified two BPV-encoded genes that can independently induce focal transformation in cultured mouse cells. One is encoded by the E5 ORF, which is located at the 3′ end of the early region. If, as suggested from the genetic data, the first AUG of the ORF represents the translation initiation codon, then E5 could encode a 44-amino-acid hydrophobic peptide. It is therefore possible that the BPV E5 ORF encodes a low-molecular-weight, membrane-associated protein. In support of the hypothesis that translation of the E5 ORF is required for transformation, the analysis of a series of E5 mutants of the full-length genome has demonstrated the exact correspondence between the ability of the E5 ORF to be translated and the transforming activity of the mutants (D. DiMaio et al., unpubl.).

From genetic analysis of the mutants in the full-length genome (Sarver et al. 1984) and cDNA clones (Yang et al. 1985), it had previously been suggested that E2 might encode the primary 3′ transforming gene. Our results, coupled with the recent identification of the transcriptional activation activity of E2 (see Spalholz et al. 1985 and this volume), argue that the effect of E2 on transformation may be indirect, perhaps by positively regulating the expression of the E6 and/or E5 transforming genes.

Several papillomaviruses that exclusively induce epidermal proliferation in vivo (HPV-1 and cottontail rabbit papillomavirus [CRPV]) do not appear to contain an ORF homologous to the BPV E5 (although the epidermotropic HPV-6 does appear to have a homologous ORF), whereas the deer papillomavirus (DPV) and European elk papillomavirus (EEPV), which, like BPV, normally induce fibroblastic proliferation in vivo, apparently do encode a homologous E5 (Groff and Lancaster 1985; U. Pettersson, pers. comm.). This may indicate

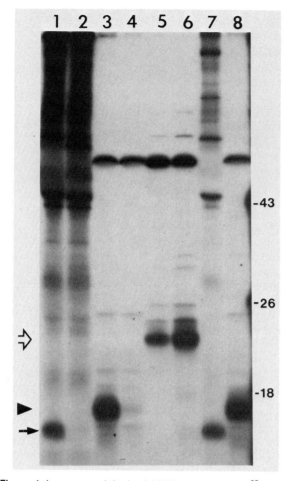

**Figure 4** Immunoprecipitation inhibition experiment. [$^{35}$S]Cysteine-labeled lysates from pXH800 cells (lanes *1*, *2*, and *7*), E6-producing bacteria (lanes *3*, *4*, and *8*), or *ras*$^{H}$-synthesizing bacteria (lanes *5* and *6*) were immunoprecipitated. Antisera were preincubated with control bacterial lysate (lanes *1*, *3*, and *5*) or E6 bacteria lysate (lanes *2*, *4*, and *6*). Lanes *1–4*, *7*, and *8* used rabbit antisera to E6, and lanes *5* and *6* used rabbit antisera to *ras*. In lanes *7* and *8*, the E6 antiserum was not blocked.

Hunter 1982; Hann and Eisenman 1984). As controls, C127 cells transformed by the unrelated *ras* oncogene (HamC18) did not contain the 15.5-kD band when fractionated by these methods. The completeness of the fractionation was confirmed by identification of c-*ras*, which is found in membranes, and c-*myc*, which is found in nuclei, in the appropriate fractions of the pXH800 lysate (Fig. 6), as well as by the use of other markers.

It is unclear how E6 becomes associated with the membrane fraction. It has no obviously hydrophobic domains that might interact directly with membranes, no aminoterminal signal sequence for synthesis in membrane-bound ribosomes, nor the proposed aminoterminal signal for myristylation found in the *src* transforming protein (Pellmann et al. 1985). Since lipid binding through a carboxyterminal cysteine is critical for the localization of *ras* (p21) to the plasma membrane (Willumsen et al. 1984), we labeled E6-transformed cells

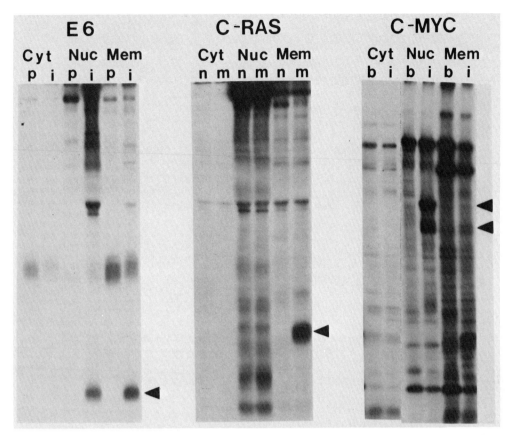

**Figure 5** Immunoprecipitation of BPV E6 from subcellular fractions of transformed cells. pXH800 cells were fractionated using a Dounce homogenizer with cells suspended in 0.25 M sucrose in the presence of MgCl$_2$ using a modification of the procedure of Hann and Eisenman (1984). HamCl8 is a Ha-*ras*–transformed C127 cell line. (p) Preimmune sera; (i) immune sera; (Nuc) nuclei; (cyt) cytoplasm; (mem) membranes.

that the E5 gene product is in part responsible for the fibromal aspects of BPV-, DPV-, and EEPV-induced fibropapillomas. It may also help to explain why the members of this latter group readily transform cultured fibroblasts in vitro.

The other transforming gene is encoded by the E6 ORF at the 3' end of the early region. Each papillomavirus genome that has been subjected to DNA sequence analysis has been found to have an ORF that is analogous to the BPV E6 ORF (Danos et al. 1983; Schwarz et al. 1983; W. Rowekamp and K. Seedorf, pers. comm.). There is some homology at the amino acid level among these different E6 ORFs (Fig. 6). All of these polypeptides are relatively rich in cysteines. Many of the cysteine residues are arranged in repeats of Cys-X-X-Cys, which has been suggested to be a characteristic feature of nucleic acid–binding proteins (NBPs) (Wain-Hobson et al. 1985). These E6 polypeptides are also relatively rich in basic amino acids, which is another feature of NBPs.

Recent data suggest that E6 may be functionally significant in malignant tumors. Carcinomas induced by CRPV as well as human cervical carcinomas that contain HPV DNA appear to selectively retain and express sequences from the E6 ORF (Danos et al. 1985;

Schwarz et al. 1985). It may be significant that clones expressing E6 can induce a fully transformed phenotype in C127 cells but are unable to morphologically transform NIH-3T3 cells. This observation raises the possibility that C127 cells might be responsive to a class of transforming genes that are negative in the NIH-3T3 transfection assay, which is the one used by most investigators to screen for transforming genes.

As predicted from genetic and sequence analysis, the BPV E6 transforming gene encodes a protein product in transformed cells. The 15.5-kD E6 protein appears to be initiated at the first AUG of the ORF and to be encoded entirely by the E6 ORF. Our results suggest further that the protein is not in disulfide linkage to other E6 molecules or to other proteins. Since some of the E6 protein is found in the nucleus, it may be a NBP. SV40 large T antigen has a nucleotide-binding region and, like E6, is found in the nucleus and in membranes (Staufenbiel and Deppert 1983). However, only small amounts of large T antigen are found on the plasma membrane, whereas the BPV E6 protein is distributed relatively equally between the nucleus and the nonnuclear membranes (the membrane fraction that contains E6 has not been identified). Furthermore, large T antigen can be extracted from the nucleoplasm (Staufenbiel and Dep-

```
                         *              *  *
BPV  1: MDLKPF..ARTNPFSG.....LDC.LWCR.......EPLTEVDAFR.CMVK.DF.HVVIREGCRY..
HPV  1: MAT.PIRTVRQLSES......L.C.IPY...I.DVLLPCNFCNYFLSNAEKLLFDHFDLHLVWRDNL
HPV  6: MESANASTSATTID......QL.CKT.FNLSMHTLQINCVFCKNALTTAE........IYSYA.YKH
HPV16: MHQKRTAMFQDPQERPRKLPQL.C.TELQTTIHDIILECVYCKQQLLRREVYDFAFRDLCIV..YRD
        ++  +  ++   ++      +X X  +      +++  X+XX+X+ +X  +X+ +X + +X  + +X+


                  *    *                                        *   *
..GACTI.CLGN.CLATERRLWQGVPV..........TGEEAELLHGK..TLDRLC....IRCCYCG
VFG.....CCQG.CARTVSLLEF.VLYYQES......Y.EVPEIE....GILDRPLLQIELRCVTCI
LKVLFRGGYPYAACAC....LEFHGKINQYRHFDYAGYATTVEEET.KQDILD.VL....IRCYLCH
..GNPYAVC.DN CLKF......YSKISEYRHYCYSLYGTTLEQQYNK.PLCD.LL....IRCINCQ
   X     X  + X+ +    X++ ++X ++++   +  X +++X +  X   XXX+X+    XXX+ X


                  *   *   *
....GKLTKN.GKHRHVLFN.EPFCKTRAN..IIR....GRCYDC...CRHGSRSKY....P
K....KLSVA.EKLEVVS.NGERVHRVR.N......RLKAKCSL....CRLYA.......I
KPLCHKPLCEVEKVKHIL....T..KAR..FIKLNCTWKGRCLHCWTTCMEDM.......LP
KPLC..PE.EKQKQRHLD....K..KQR..FHNIRGRWTGRCMSC...CRS.SRTRRETQL
X+++ X+  + +X++XX+ + +   X X ++  ++ +++XXX  X   XX  ++      ++
```

**Figure 6** Comparison of E6 proteins. The amino acid sequences of the E6s of BPV-1 and HPV-1, -6, and -16, based on their DNA sequences (Chen et al. 1982; Danos et al. 1983; Schwarz et al. 1983; W. Rowekamp and K. Seedorf, pers. comm.), have been aligned for comparison of their homologous peptides. (★) Conserved cysteine; (×) 3/4 or 4/4; (+) 2/4.

pert 1983), whereas the E6 protein is not present in this fraction (E. Androphy, unpubl.).

The approach we have used to identify the BPV E6 protein should be applicable to the analogous E6 ORF of other papillomaviruses as well as to other early papillomavirus gene products. Since most papillomavirus-associated dysplasias and carcinomas do not synthesize detectable levels of structural viral antigens, identification of nonstructural gene products such as E6 may eventually prove useful diagnostically. The identification of the E6 protein should permit analysis of the biochemical properties of the molecule, as has been carried out for other transforming proteins, and facilitate the study of the pathogenesis of warts and papillomavirus-associated malignancies.

## Acknowledgments

We thank Max Gottesman for plasmid pOG7 and *E. coli* strain N6405, Robert Eisenman for antibody to c-*myc*, Lionel Crawford for antibody to p53, and Nancy Hubbert for helpful discussions and technical assistance. J.T.S. is the recipient of National Institutes of Health Postdoctoral Fellowship (PSH grant 5-F32-CA-07237) from the National Cancer Institute (DHHS).

## References

Amtmann, E. and G. Sauer. 1982. Bovine papillomavirus transcription: Polyadenylated RNA species and assessment of the direction of transcription. *J. Virol.* **43:** 59.

Androphy, E.J., J.T. Schiller, and D.R. Lowy. 1985. Identification of the protein encoded by the E6 transforming gene. *Science* **230:** 442.

Boshart, M., L. Gissmann, H. Ikenberg, A. Kleinheinz, W. Scheurlen, and H. zur Hausen. 1984. A new type of papillomavirus DNA, its presence in genital cancer biopsies and in cell lines derived from cervical cancer. *EMBO J.* **3:** 1151.

Chen, E.Y., P.M. Howley, A.D. Levenson, and P.H. Seeburg. 1982. The primary structure and genetic organization of the bovine papillomavirus type 1 genome. *Nature* **299:** 529.

Cooper, J.A. and T. Hunter. 1982. Discrete primary locations of a tyrosine protein kinase and of three proteins that contain phosphotyrosine in virally transformed chick fibroblasts. *J. Cell Biol.* **94:** 287.

Danos, O., E. Georges, G. Orth, and M. Yaniv. 1985. Fine structure of the cottontail rabbit papillomavirus mRNAs expressed in the transplantable VX2 carcinoma. *J. Virol.* **53:** 735.

Danos, O., I. Giri, F. Thierry, and M. Yaniv. 1984. Papillomavirus genomes: Sequences and consequences. *J. Invest. Dermatol.* **83:** 8s.

Danos, O., L.W. Engel, E.Y. Chen, M. Yaniv, and P.M. Howley. 1983. Comparative analysis of the human type 1a and bovine type 1 papillomavirus genomes. *J. Virol.* **46:** 557.

Dhar, R., W.L. McClements, L.W. Enquist, and G.F. Vande Woude. 1980. Nucleotide sequences of integrated Moloney sarcoma provirus long terminal repeats and their host and viral junctions. *Proc. Natl. Acad. Sci.* **77:** 3937.

Durst, M., L. Gissmann, H. Ikenberg, and H. zur Hansen. 1983. A papillomavirus DNA from a cervical carcinoma and its prevalence in cancer biopsy samples from different geographic regions. *Proc. Natl. Acad. Sci.* **80:** 3812.

Dvoretzky, I., R. Shober, and D. Lowy. 1980. Focus assay in mouse cells for bovine papillomavirus. *Virology* **103:** 369.

Engel, L.W., C.A. Heilman, and P.M. Howley. 1983. Transcriptional organization of the bovine papillomavirus type 1. *J. Virol.* **47:** 516.

Graham, F.L. and A.J. van der Eb. 1973. A new technique for the assay of infectivity of human adenovirus 5 DNA. *Virology* **52**: 456.

Groff, D.E. and W.D. Lancaster. 1985. Molecular cloning and nucleotide sequence of deer papillomavirus. *J. Virol.* **56**: 85.

Hann, S.R. and R.N. Eisenman. 1984. Proteins encoded by the human c-*myc* oncogene: Differential expression in neoplastic cells. *Mol. Cell. Biol.* **4**: 2486.

Heffron, F., M. So, and B. McCarthy. 1978. *In vitro* mutagenesis of a circular DNA molecule by using synthetic restriction sites. *Proc. Natl. Acad. Sci.* **75**: 6012.

Heilman, C.A., L. Engel, D.R. Lowy, and P.M. Howley. 1982. Virus-specific transcription in bovine papillomavirus-transformed mouse cells. *Virology* **119**: 22.

Lautenberger, J.A., D. Court, and T.S. Papas. 1983. High-level expression in *Escherichia coli* of the carboxyterminal sequences of the avian myelocytomatosis virus (MC29) v-*myc* protein. *Gene* **23**: 75.

Law, M.-F., D.R. Lowy, I. Dvoretzky, and P.M. Howley. 1981. Mouse cells transformed by bovine papillomavirus contain only extrachromosomal viral DNA sequences. *Proc. Natl. Acad. Sci.* **78**: 2727.

Lowy, D.R., E. Rands, and E.M. Scolnick. 1978. Helper-independent transformation by unintegrated Harvey sarcoma virus DNA. *J. Virol.* **26**: 291.

Lowy, D.R., I. Dvoretsky, R. Shober, M.-F. Law, L. Engel, and P.M. Howley. 1980. *In vitro* transformation by a defined subgenomic fragment of bovine papillomavirus DNA. *Nature* **287**: 72.

Lusky, M. and M.R. Botchan. 1985. Genetic analysis of bovine papillomavirus type 1 *trans*-acting replication factors. *J. Virol.* **53**: 955.

Nakabayashi, Y., S.K. Chattopadhyay, and D.R. Lowy. 1983. The transforming functions of bovine papillomavirus DNA. *Proc. Natl. Acad. Sci.* **80**: 5832.

Orth, G., M. Favre, F. Breitburd, O. Croissant, S. Jablonska, S. Obalek, M. Jarzabek-Chorzelska, and G. Rzesa. 1980. Epidermodysplasia verruciformis: A model for the role of papilloma viruses in human cancer. *Cold Spring Harbor Conf. Cell Proliferation* **7**: 259.

Ostrow, R.S., M. Bender, M. Niimura, T. Seki, M. Kawashima, F. Pass, and A.J. Faras. 1982. Human papillomavirus DNA in cutaneous primary and metastasized squamous cell carcinomas from patients with epidermodysplasia verruciformis. *Proc. Natl. Acad. Sci.* **79**: 1634.

Pellman, D., E. Garber, F.R. Cross, and H. Hanafusa. 1985. An N-terminal peptide from p60$^{src}$ can direct myristylation and plasma membrane localization when fused to heterologous proteins. *Nature* **314**: 374.

Rotter, V. and D. Wolf. 1985. Biological and molecular analysis of p53 cellular-encoded tumor antigen. *Adv. Cancer Res.* **43**: 113.

Sarver, N., J.C. Byrne, and P.M. Howley. 1982. Transformation and replication in mouse cells of a bovine papillomavirus-pML2 plasmid vector that can be rescued in bacteria. *Proc. Natl. Acad. Sci.* **79**: 7147.

Sarver, N., M.S. Rabson, Y.-C. Yang, J.C. Byrne, and P.M. Howley. 1984. Localization and analysis of bovine papillomavirus type 1 transforming functions. *J. Virol.* **52**: 377.

Schiller, J.T., E.J. Androphy, and D.R. Lowy. 1985. The bovine papillomavirus E6 gene: Identification of its transforming function in protein products. *UCLA Symp. Mol. Cell. Biol.* **32**: 457.

Schiller, J.T., W.C. Vass, and D.R. Lowy. 1984. Identification of a second transforming region in bovine papillomavirus DNA. *Proc. Natl. Acad. Sci.* **81**: 7880.

Schiller, J.T., W.C. Vass, K.H. Vousden, and D.R. Lowy. 1986. E5 open reading frame of bovine papillomavirus type 1 encodes a transforming gene. *J. Virol.* **57**: 126.

Schwarz, E., U.K. Freese, L. Gissmann, W. Mayer, B. Roggenbuck, A. Stremlau, and H. zur Hausen. 1985. Structure and transcription of human papillomavirus sequences in cervical carcinoma cells. *Nature* **314**: 111.

Schwarz, E., M. Durst, C. Demankowski, O. Lattermann, R. Zech, E. Wolfsperger, S. Suhai, and H. zur Hausen. 1983. DNA sequence and genome organization of genital human papillomavirus type 6b. *EMBO J.* **2**: 2341.

Spalholz, B.A., Y.-C. Yang, and P.M. Howley. 1985. Transactivation of a bovine papillomavirus transcriptional regulatory element by the E2 gene product. *Cell* **42**: 183.

Staufenbiel, M. and W. Deppert. 1983. Differential structure systems of the nucleus are targets for SV40 large T antigen. *Cell* **33**: 173.

Stenlund, A., J. Zabielski, H. Ahola, H. Moreno-Lopez, and U. Pettersson. 1985. Messenger RNAs from the transforming region of bovine papillomavirus type I. *J. Mol. Biol.* **182**: 541.

Wain-Hobson, S., P. Sonigo, O. Danos, S. Cole, and M. Alizon. 1985. Nucleotide sequence of the AIDS virus, LAV. *Cell* **40**: 9.

Willumsen, B.M., A. Christensen, N.L. Hubbert, A.G. Papageorge, and D.R. Lowy. 1984. The p21 *ras* C-terminus is required for transformation and membrane association. *Nature* **310**: 583.

Yang, Y.-C., H. Okayama, and P.M. Howley. 1985. Bovine papillomavirus contains multiple transforming genes. *Proc. Natl. Acad. Sci.* **82**: 1030.

# Expression of Human Papillomavirus Type-18 DNA in Cervical Carcinoma Cell Lines

**E. Schwarz and A. Schneider-Gädicke**

Institut für Virusforschung, Deutsches Krebsforschungszentrum, D-6900 Heidelberg 1, Federal Republic of Germany

Cell lines established from human cervical carcinomas were found to contain DNA from human papillomavirus (HPV) types 18 and 16 integrated into the host genome, thus providing in vitro cell systems for the analysis of those HPV types associated with human genital cancer. We have studied transcription of HPV-18 DNA in the cell lines HeLa, C4-1, and SW756 by northern analysis and subsequently by cDNA cloning and sequencing. Very similar patterns of HPV-18 expression emerged for the three different lines, suggesting a functional role of early HPV-18 genes in these cells. From the cDNA sequences it can be deduced that the HPV-18–positive mRNAs most likely direct translation of virus-specific gene products encoded in the 5′ part of the early region (E6, E6*, and E7), although most of the sequenced cDNAs were found to be derived from hybrid transcripts in which 3′ cotranscribed host cell sequences were spliced to the 5′-proximal HPV-18 exons. Due to differential splicing, two proteins (E6 and E6* with 158 and 57 amino acids, respectively) seem to be encoded within the HPV-18 E6 segment that share 43 amino acids at their amino termini. Generation of a spliced E6* cistron is probably specific for HPV-18 and HPV-16 among genital papillomaviruses.

Evidence has rapidly accumulated in recent years supporting a role for specific types of human papillomaviruses (HPVs) in genital cancer (for review, see zur Hausen and Schneider 1985), thereby greatly enhancing interest in this former wallflower group of DNA tumor viruses (Bishop 1985). DNA sequences of HPV types 16 and 18 have been found in about 70% of all malignant genital tumors analyzed, whereas the most prevalent types in benign warts are HPV-6 and HPV-11 (for review, see Gissmann and Schneider 1986). Furthermore, several human cell lines all established from cervical carcinomas have been identified to contain HPV-16 or HPV-18 DNA (Boshart et al. 1984; Pater and Pater 1985; Schwarz et al. 1985; Yee et al. 1985), thus providing interesting, hitherto missing in vitro systems for analysis of the clearly cancer-associated HPV types. In benign warts and in premalignant precursor lesions, the viral genomes persist as extrachromosomal episomes. Malignant tumors and cervical cancer cell lines, however, maintain the papillomavirus DNA integrated into the host chromosomal DNA, and some of the tumors have additional multimeric HPV episomes (Dürst et al. 1985, 1986).

Mainly due to the lack of suitable in vitro systems for either propagation of or transformation by HPVs, the most extensively studied papillomavirus is the bovine papillomavirus type 1 (BPV-1), which is able to transform certain rodent cells in vitro. Functions involved in transcription, transformation, and replication have been assigned to the different coding and noncoding parts of the BPV-1 genome.

The transforming (early) region can be subdivided into a 5′ noncoding region with cis-acting elements for transcription and replication, followed 3′ by a coding segment that contains several open reading frames (ORFs) in the order E6, E7, E1, E2 (overlapping with E4), and E5 with regard to the 5′-to-3′ direction of transcription (see Fig. 1 for the similar arrangement of ORFs in HPV-6). ORFs E6 and E5 have been identified to encode functions involved in transformation (Nakabayashi et al. 1983; Sarver et al. 1984; Schiller et al. 1984; Yang et al. 1985; see also Schiller et al., this volume) in addition to E2, which was recently shown to code for a trans-activation factor (Spalholz et al. 1985; see also Howley et al., this volume). The BPV-1 genome is maintained as a multicopy plasmid in the transformed cells (Law et al. 1981). Autonomous replication requires a function encoded in ORF E1, whereas the plasmid copy number is affected by mutations located in ORF E7 (Sarver et al. 1984; Lusky and Botchan 1985; see also Botchan et al., this volume).

Complete sequence analysis of HPV-6, -11, and -16 DNA (Schwarz et al. 1983; Seedorf et al. 1985; Dartmann et al. 1986) has revealed a genome organization corresponding very well to that of other human and animal papillomaviruses, including BPV-1. Thus, the insights already gained into the functions of the different BPV-1 genomic segments in BPV-1–transformed cells should provide a useful basis for interpretation and comparison in the attempts to identify HPV-16 or -18 functions possibly involved in human genital cancer. We have used the three different cervical carcinoma cell lines HeLa, C4-1, and SW756 to characterize the organization and transcription of HPV-18 DNA (Schwarz et al. 1985). cDNA cloning and sequence analysis further allowed us to define the viral genes probably expressed in the carcinoma cells (A. Schneider-Gädicke and E. Schwarz, in prep.).

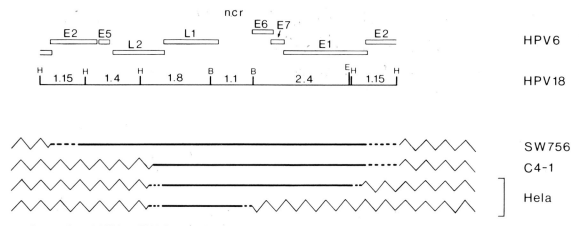

**Figure 1** Integration of HPV-18 DNA into the host genome in cervical carcinoma cell lines HeLa, C4-1, and SW756. At *bottom*, integrated HPV-18 DNA sequences are indicated by solid lines; virus-cell DNA junctions are given by dotted lines, unless determined by DNA sequence analysis, and zig-zag lines represent flanking cellular sequences. Aligned with the integrated HPV-18 sequences, a restriction map of the linearized prototype HPV-18 DNA (Boshart et al. 1984) is presented at *top*, indicating the cleavage sites for *Bam*HI (B), *Eco*RI (E), and *Hinc*II (H). Sizes of restriction fragments used as subgenomic hybridization probes are given in kilobases. The 1.15-kb *Hinc*II fragment is drawn at both sides of the linearized DNA. The arrangement of ORFs in HPV-6 DNA is shown aligned with the HPV-18 restriction map according to sequence homologies (Schwarz et al. 1985). (ncr) Noncoding region.

## Experimental Procedures

### DNA and RNA hybridization analyses

Analyses of DNA and RNA (either cytoplasmic or poly(A)$^+$ RNA) from cervical carcinoma cell lines HeLa, C4-1, and SW756 by Southern and northern blot hybridization were performed under conditions essentially as described (Schwarz et al. 1985). Restriction fragments of HPV-18 DNA, which were used as subgenomic hybridization probes, were prepared by digestion of cloned HPV-18 DNA with *Bam*HI, *Eco*RI, and *Hinc*II and fractionation of the cleavage products on 0.8% agarose gels.

### Construction of cDNA libraries

Cytoplasmic RNA was isolated from HeLa, C4-1, and SW756 cells as described (Freese et al. 1983) with subsequent purification of poly(A)$^+$ RNA by oligo(dT)-cellulose chromatography. Double-stranded cDNA of HeLa, C4-1, and SW756 poly(A)$^+$RNA, respectively, was prepared according to standard protocols (Land et al. 1981; Maniatis et al. 1982) as described by Schwarz-Sommer et al. (1985). Second-strand synthesis was followed by nuclease S1 treatment, and *Eco*RI linkers were then ligated to the flush ends of the double-stranded cDNA molecules. After digestion with *Eco*RI and size fractionation on a Sepharose 4B column, cDNA molecules with sizes greater than 600 bp were ligated into the *Eco*RI site of the vector λNM1149 (Murray 1983). Recombinants were selected by plating the phages on *Escherichia coli* POP-13b, which is a derivative of strain POP-101 (Murray 1983) and were screened by plaque hybridization using $^{32}$P-labeled, cloned HPV-18 DNA (Boshart et al. 1984) as hybridization probe. Twenty and 17 HPV-18–positive cDNA clones were isolated from the C4-1 and SW756 cDNA libraries, respectively. The HeLa cDNA library contained more than 200 HPV-18–positive clones, from which 20 were selected for further analysis.

### DNA sequence analysis

cDNA inserts selected for sequence analysis were cloned into vectors M13mp18 or M13mp19 (Messing 1983). Nucleotide sequences were determined by the dideoxy method (Sanger et al. 1977), using various synthetic oligonucleotides as sequencing primers. Oligonucleotides were synthesized using an automated DNA synthesizer (Applied Biosystems, Inc.). Sequence data were analyzed by a computer program developed at Deutsches Krebforschungszentrum. cDNA sequences were compared with the genomic sequences of HPV-18 (Schwarz et al. 1985; K. Seedorf and K. Dartmann, unpubl.), HPV-16 (Seedorf et al. 1986), HPV-11 (Dartmann et al. 1986), and HPV-6 (Schwarz et al. 1983).

## Results

### Analysis of HPV-18 DNA and RNA

The presence of HPV DNA in cell lines established from cervical carcinomas was first demonstrated in the cell lines HeLa, C4-1, and SW756, which all contained HPV-18 DNA (Boshart et al. 1984; Schwarz et al. 1985). In the meantime, additional cervical cancer lines were shown to harbor either HPV-18 or HPV-16 DNA (Pater and Pater 1985; Schwarz and Schneider-Gädicke 1985; Yang et al. 1985). In our comparative study of the organization and transcription of HPV-18 DNA in HeLa, C4-1, and SW756 cells, the initial results (which have been mainly obtained by Southern and northern blot analysis using HPV-18 DNA restriction fragments as radiolabeled subgenomic probes) can be summarized as follows (Schwarz et al. 1985; see Fig. 1): In all three cell lines, HPV-18 DNA is integrated into the host genome. The viral DNA is amplified about 10-fold to 50-fold together with the flanking cellular sequences in HeLa and SW756 cells, whereas C4-1 cells contain only a single copy of HPV-18 DNA per cell. Integration of HPV-18 DNA always causes a disruption of the early region within the segment covering ORFs E1 and E2. In HeLa and C4-1 cells, a portion of the HPV-18 genome extending from ORF E2 into ORF L2 is deleted. Additional deletions affecting the E6-E7-E1 part of the early region are observed in a subpopulation of integrated HPV-18 copies present in HeLa cells (Fig. 1; W. Mayer and E. Schwarz, unpubl.).

In all three cell lines, only sequences from the E6-E7-E1 part of the HPV-18 early region are transcribed into poly(A)$^+$ RNAs that range in size from about 1.2 to 6.5 kb (Schwarz et al. 1985). Each cell line contains two or three major HPV-18–positive mRNA species (HeLa: 1.6 and 3.5 kb; C-41: 1.5, 4.2, and 5.5 kb, SW756: 1.5 and 6.5 kb) and additional minor ones.

## Structures of HPV-18–positive cDNAs

To analyze and compare the structures and possible coding properties of the HPV-18–positive mRNAs present in the three cell lines, libraries of cloned cDNAs were constructed and HPV-18–positive cDNA clones

isolated. The complete or partial nucleotide sequences of cDNA inserts from 16 different clones (4, 5, and 7 clones from the HeLa, C4-1 and SW756 cDNA libraries, respectively; sizes of 1.2–3.0 kb) were determined and gave the following results (A. Schneider-Gädicke and E. Schwarz, in prep.).

The nucleotide sequences of the cDNA inserts that correspond to the 5′-terminal sequences of the mRNAs are all HPV-18–specific, starting either upstream from or within the 5′ portion of ORF E6 (with the exception of one HeLa cDNA clone; Fig. 2). Their positions within the HPV-18 sequence indicate that transcription is likely to initiate at HPV-18–specific promoters. The most probable candidates include two TATA-box sequences (5′ TATATAAAA 3′) located 82 bp and 34 bp, respectively, upstream of the putative translation-initiation codon of ORF E6 (Fig. 2). Since nuclease S1 treatment was employed during cDNA preparation, however, the 5′ ends of the cDNAs do not represent the true 5′ ends of the mRNAs. Therefore on the basis of some of the cDNA sequences, an ambiguity remains in predicting the 5′-proximal cistrons of the corresponding mRNAs that are the most likely candidates for being translated into proteins (Kozak 1981). This applies to the two groups of cDNAs in which the 5′ ends are located either 36–38 nucleotides (nt) or 91–120 nt downstream from the E6 start codon (Fig. 2).

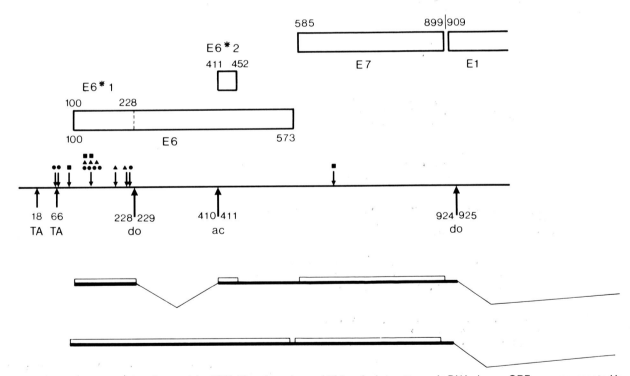

**Figure 2** Organization of the 5′ part of the HPV-18 early region and 5′-terminal structures of cDNA clones. ORFs are represented by open boxes. Numbers denote the putative AUG start codons, the last nucleotides preceding the stop codons and, for E6*1 and E6*2, the positions at the splice junctions. The 5′ ends of the 16 sequenced cDNA clones are indicated by small arrows: (■) HeLa; (●) SW756; (▲) C4-1. The positions of splice donor (do) and acceptor (ac) sites and of two TATA-box sequences (TA) are indicated. At *bottom*, the two principal 5′-terminal structures of cDNAs up to the E1 splice site are presented. Thin, slanted lines denote excised intron sequences. The possible 5′-proximal cistrons (spliced E6*, continuous E6 or E7, respectively, depending on the positions of the mRNA cap sites) are shown by the narrow, open boxes.

### Potential coding regions E6, E6*, and E7

The majority of sequenced cDNAs (9 of 16) turned out to be derived from mRNAs in which a 182-bp intron located in the E6 region was removed by splicing. This splicing resulted in fusion of the 5' part of ORF E6 (E6*1, 129 bp) with a second coding region (E6*2) that is terminated by a stop codon only 42 nt after the 3' splice junction. This 3' coding region completely overlaps with the 3' part of ORF E6 but uses a different reading frame (Fig. 2). From the cDNA sequences it can thus be deduced that the E6 region of HPV-18 may encode two proteins (E6 and E6*, 158 and 57 amino acid residues with calculated molecular weights of about 18,900 and 6600, respectively) that are identical in their aminoterminal 43 amino acids but differ both in number and sequence of the amino acid residues of their carboxyterminal portions. The E6 protein is encoded by the continuous ORF E6, whereas the E6* coding sequence is composed of two segments separated by a 182-bp intron. All three HeLa cDNAs with 5' ends located before or in E6 contain the spliced E6* version. The sequenced C4-1 and SW756 cDNAs show either the continuous E6 (C4-1: 1 of 5; SW756: 5 of 7) or the spliced E6* coding sequence, suggesting that the two proteins may be simultaneously present in the cells.

Comparison of the highly conserved E6 nucleotide sequences of genital papillomaviruses HPV-18, -16, -11, and -6 interestingly reveals that HPV-16 contains putative splice donor and acceptor sites at the exact corresponding positions (Fig. 3), whereas in the HPV-6 and HPV-11 sequences these splice sites are "inactivated" due to nucleotide exchanges just in the G-T and A-G dinucleotide sequences, which are rigorously conserved in splice donor and acceptor sequences, respectively (Mount 1982). The intriguing possibility thus arises that this splicing reaction is specific for HPV-18 and HPV-16 (i.e., for those HPV types associated with malignant genital tumors).

Splicing affecting only sequences from the E6 region has not been observed in mRNA and cDNA analyses of

```
                10          20          30          40          50
                .           .           .           .           .
HPV18   ATG--------------GCGCGCTTTGAGGATCCAACACGGCGACCCTA
        ***              **R  *** ****Y***  R R*******  R
HPV16   ATGCACCAAAAGAGAACTGCAATGTTTCAGGACCCACAGGAGCGACCCAG


        CAAGCTACCTGATCTGTGCACGGAACTGAACACTTCACTGCAAGACATAG
        ***Y****  *  Y*R*****R**R*** * **  *  *R** **Y***R
        AAAGTTACCACAGTTATGCACAGAGCTGCAAACAACTATACATGATATAA


                                                      ↓
        AAATAACCTGTGTATATTGCAAG--ACAG-TATTGGAACTTACAGAGGTA
        *  **R *****R**Y******  ****  **Y** R**  *  ******
        TATTAGAATGTGTGTACTGCAAGCAACAGTTACTGCGACGT---GAGGTA


        TTTGAATTTGCATTCAAAGATTTATTTGTGGTGTATAGAGACAGTATACC
        *  *** *****  **Y RR******* YR*R**R**********YR* *  **
        TATGACTTTGCTTTTCGGGATTTATGCATAGTATATAGAGATGGGAATCC


        GCATGCTGCATGCCATAAATGTATAGCTTTTTATTCTAGAATTAGAGAAT
        RY*******Y***Y ******** **R  ***********R****** **R*
        ATATGCTGTATGTGATAAATGTTTAAAGTTTTATTCTAAAATTAGTGAGT


        TTAGACATTATTCAGACTCTGTGTATGGAGACACATTGGAAAAACTAACT
        **********   *Y * ********R  *****R*** *R* *  Y
        ATAGACATTATTGTTATAGTTTGTATGGAACAACATTAGAACAGCAATAC


                          ↓
        AACACTGGGTTATACAATTTATTAATAAGGTGCCTGCGGTGCCAGAAACC
        ****     ***R*RYR****R***** *****Y *  R  **Y**R**R**
        AACAAACCGTTGTGTGATTTGTTAATTAGGTGTATTAACTGTCAAAAGCC


        GTTGAATCCAGCAGAAAAACTTAGACACCTTAATGAAAAACGACGATTTC
        RY** R*** * ******R* ******Y** R*YR****R*R* ****Y*
        ACTGTGTCCTGAAGAAAAGCAAAGACATCTGGACAAAAAGCAAAGATTCC


        ACAACATAGCTGGGCACTATAGAGGCCAGTGCCATTCGTGCTGCAACCGA
        *Y**Y**R  ** *R *R *  **Y*RR**Y  **  **Y****R  Y *
        ATAATATAAGGGGTCGGTGGACCGGTCGATGTATGTCTTGTTGCAGATCA


        GCACG-ACA-GGAAAGACTCCAACGACGCAGAGAAACACAAGTATAA
        ** * *** * *R***  ** * * Y*Y*R
        TCAAGAACACGTAGAGAAACCCA-GCTGTAA
```

**Figure 3** Comparison of E6 nucleotide sequences of HPV-18 and HPV-16 DNA. The alignment extends from the putative ATG start codons until the termination codons of ORF E6. (*) Identical nucleotides; (R and Y) purine and pyrimidine residues, respectively. Splice donor and acceptor sites are indicated by vertical arrows.

BPV-1 and CRPV (Nasseri and Wettstein 1984; Danos et al. 1985; Yang et al. 1985).

Furthermore, it should be added that from the four Cys-X-X-Cys tetrapeptide sequences present in a highly conserved spatial arrangement in the E6 sequences of all papillomaviruses analyzed so far, only the first amino-terminal one would be preserved in the putative E6* proteins of HPV-18 and HPV-16.

In addition to the E6 or E6* coding segment, all cDNAs (with the exception of one HeLa clone) contain the complete ORF E7 (Fig. 2). This would be the 5'-proximal cistron and thus could direct translation of an E7 gene product (Kozak 1981) in HPV-18–positive mRNAs initiated downstream from the E6 translation start codon. Experiments to map the 5' ends of the mRNAs are in progress and will allow more-thorough discussion of the possibility of translation of E7.

Some direct evidence for expression of ORF E7 has already been obtained in the HPV-16–positive CaSki cervical cancer cell line by the use of specific antisera (F.O. Wettstein, pers. comm.), which, however, are not yet available in the case of HPV-18.

### Presence of 3' cotranscribed host cell sequences in cDNAs

Most cDNA inserts sequenced so far are derived from virus-cell fusion transcripts in which the 5'-terminal HPV-18–specific nucleotide sequences are spliced to 3' cotranscribed cellular sequences by using a splice donor site (5' AAG GTACAG 3') that is located 16 nt downstream of the putative AUG start codon of ORF E1 (Fig. 2). The E1 splice donor site is well conserved among papillomaviruses and is known to be used in generating mature early transcripts of BPV and CRPV, in addition to a splice acceptor site located in the E2-E4 region and a polyadenylation signal at the 3' end of the early region (Danos et al. 1985; Yang et al. 1985). Generation of virus-cell fusion transcripts in the cervical cancer cells is a consequence of the integration mode of HPV-18 DNA since disruption of the early region within the E1-E2 segments removes the 3' early mRNA maturation signals from the upstream transcribed sequences.

The sizes of the excised (virus-cell hybrid) introns have not yet been determined. From the presence of different host cell sequences at the splice junctions of cDNA clones derived from the same cell line (either C4-1 or SW756), we can deduce that splicing involves different cellular splice acceptor sites. Sequence comparison further reveals that additional splicing may occur within the cotranscribed cellular sequences.

No sequence homology is observed when comparing the cotranscribed cellular sequences of cDNA clones from the three different cell lines. This is in agreement with the results of other experiments that gave no indication of HPV DNA integration into specific cellular sites in cervical carcinomas (Dürst et al. 1985, 1986).

In all cDNA clones examined, splicing of cellular sequences to 5' HPV-18 sequences does not create any significantly large hybrid ORF starting with the first 16 nt of the viral ORF E1, then continuing in-frame with cellu-

lar sequences (maximal coding capacity observed is 24 amino acids). The largest ORFs present in the cellular parts of the different cDNA clones would encode proteins of 16 to 87 amino acids. Regarding the small sizes of these ORFs together with their internal localization in the cDNAs (thus not representing 5'-proximal cistrons) and further taking into account the probably random distribution of cellular integration sites in cervical carcinomas, it seems quite unlikely that the HPV-18–cell hybrid mRNAs are synthesized in the cells in order to direct translation of either virus-cell fusion or cell-specific proteins. Another possibility that cotranscription of host cell sequences may stabilize the mRNA products has already been discussed (Schwarz et al. 1986).

Whereas the structures of all C4-1 and SW756 cDNAs sequenced so far reflect the involvement of the E1 splice donor site in a splicing reaction, this is not the case with two HeLa cDNA clones. The use of E1 "exon"-specific probes in northern analysis will clarify whether mRNA species with a continuous E1 sequence are present in all three cell lines.

### Discussion

Analysis of the three different HPV-18–positive cervical carcinoma cell lines HeLa, C4-1, and SW756 has revealed interesting common features in the integration and transcription of the papillomavirus DNA. Integration of the circular HPV-18 DNA into the host genome leads to a disruption of the early region, thereby joining the 5' part (ORFs E6, E7, and E1) to downstream cellular sequences. Transcription of HPV-18 DNA in the cells is restricted to sequences from the E6-E7-E1 part of the early region, which are transcribed in their sense-direction. From the structures of all 16 cDNA clones determined by DNA sequence analysis (A. Schneider-Gädicke and E. Schwarz, in prep.), it can be deduced that the HPV-18–positive mRNAs most likely direct translation of virus-specific gene products (E6, E6*, and E7), although most or all of them are virus-cell fusion transcripts in which 3' cotranscribed cellular sequences are spliced to the 5' HPV-18 exons. Taken together, these data may point to a functional role of (some of) these early papillomavirus genes in cells of cervical carcinoma cell lines. It is tempting to further speculate that expression of the same papillomavirus functions is also required in the development and/or maintenance of cervical carcinomas:

1. Transcription of papillomavirus DNA was demonstrated at least in some of the HPV-16 DNA–containing cervical cancer biopsies analyzed so far (Lehn et al. 1985; Schwarz et al. 1985).
2. Integration of HPV DNA by opening within the E1-E2 segment of the early region was observed in the majority of HPV-16–positive cervical carcinomas analyzed in detail and in the SIHA cell line (Dürst et al. 1985, 1986; Lehn et al. 1985).
3. ORF E6 of BPV-1 has been shown to encode a function that can independently transform mouse

cells, and ORFs E6 and E7 are expressed in BPV-1–transformed mouse cells and in CRPV-induced rabbit carcinomas (Nasseri and Wettstein 1984; Schiller et al. 1984; Danos et al. 1985; Yang et al. 1985).

By analysis of additional HPV-18– or HPV-16–positive cervical carcinoma cell lines, it will become apparent whether transcription of the HPV E6-E7 region is a regular phenomenon of these cells. As predicted from the cDNA sequences, initiation of translation at the putative E6 start codon could give rise to two different proteins E6 and E6*, respectively, depending on whether or not part of the E6 sequence has been spliced out from the primary transcript.

What are the functions of the viral genes E6 and E6* and perhaps also of E7 in the established carcinoma cells? Are they involved in growth regulation? One may tackle these questions now by analyzing (1) whether the growth properties of these cells can be influenced by inhibition of E6, E6*, or E7 expression or (2) what effects the expression of any of these genes may have after transfection into HPV-negative target cells.

According to the nucleotide sequence comparison, the putative E6* gene product is presumably specific for HPV-18 and HPV-16. It can be considered as the released aminoterminal part of E6 with 14 (HPV-18) or only 2 (HPV-16) carboxyterminal amino acids added after the splice junction. Further work is needed to analyze the various interesting questions arising from the sequence predictions. First of all, it is not yet known whether this small protein is actually synthesized in the cells. If yes, it will be interesting to analyze the occurrence of E6* in different cells. Is it present only in cell lines or also in cervical carcinomas or, in addition, in benign and precursor lesions? In other words, is it possible to correlate E6* expression with the state of the cell and with the state of the viral DNA in the cells? Furthermore, one may ask whether splicing and subsequent translation of E6* is required to activate a function residing in the aminoterminal part of E6 and whether any cooperation between E6* and E6 or other proteins can be observed.

Most of the questions—especially those concerning the correlation between the expression of a particular gene, the state of the viral DNA (integrated vs. episomal), and the phenotype of the cell (malignant vs. benign)—are as yet equally unsolved for the other coding regions of HPV-18 (or HPV-16), those transcribed (E6 and E7) as well as those not expressed (E2, E4, and E5) in cervical cancer cell lines. Further analysis of HPV-18 (and HPV-16) gene expression and regulation and of the possible interactions between viral and host cell factors will be of central interest in future work.

## Acknowledgments

We thank Zsuzsanna Schwarz-Sommer for advice in cDNA cloning, Wolfram Scheurlen and Tomas Kahn for synthesis of oligonucleotides, Klaus Dartmann for advice in dideoxy sequencing, and Sandor Suhai for help in computer analysis. We further thank Barbara Kürschner for technical assistance, F.O. Wettstein and K. Seedorf for communicating unpublished data, Lutz Gissmann and Harald zur Hausen for comments on the manuscript, and Martina Deschner for typing the manuscript. This work was supported by the Deutsche Forschungsgemeinschaft (SFB 31: Tumorentstehung und -entwicklung).

## References

Bishop, J.M. 1985. Viral oncogenes. *Cell* **42:** 23.

Boshart, M., L. Gissmann, H. Ikenberg, A. Kleinheinz, W. Scheurlen, and H. zur Hausen. 1984. A new type of papillomavirus DNA, its presence in genital cancer biopsies and in cell lines derived from cervical cancer. *EMBO J.* **3:** 1151.

Danos, O., E. Georges, G. Orth, and M. Yaniv. 1985. Fine structure of the cottontail rabbit papillomavirus mRNAs expressed in the transplantable VX2 carcinoma. *J. Virol.* **53:** 735.

Dartmann, K., E. Schwarz, L. Gissmann, and H. zur Hausen. 1986. The nucleotide sequence and genome organization of human papillomavirus type 11. *Virology* (in press).

Dürst, M., E. Schwarz, and L. Gissmann. 1986. Integration and persistence of human papillomavirus DNA in genital tumors. *Banbury Rep.* **21:** (in press).

Dürst, M., A. Kleinheinz, M. Hotz, and L. Gissmann. 1985. The physical state of human papillomavirus type 16 DNA in benign and malignant genital tumors. *J. Gen. Virol.* **66:** 1515.

Freese, U.-K., G. Laux, J. Hudewentz, E. Schwarz, and G.W. Bornkamm. 1983. Two distant clusters of partially homologous small repeats of Epstein-Barr virus are transcribed upon induction of an abortive or lytic cycle of the virus. *J. Virol.* **48:** 731.

Gissmann, L. and A. Schneider. 1986. Papillomavirus DNA in premalignant and malignant genital lesions. *Banbury Rep.* **21:** (in press).

Kozak, M. 1981. Mechanism of mRNA recognition by eukaryotic ribosomes during initiation of protein synthesis. *Curr. Top. Microbiol. Immunol.* **93:** 81.

Land, H., M. Grez, H. Hauser, W. Lindenmaier, and G. Schütz. 1981. 5'-terminal sequences of eukaryotic mRNA can be cloned with high efficiency. *Nucleic Acids Res.* **9:** 2251.

Law, M.-F., D.R. Lowy, J. Dvoretzky, and P.M. Howley. 1981. Mouse cells transformed by bovine papillomavirus contain only extrachromosomal viral DNA sequences. *Proc. Natl. Acad. Sci.* **78:** 2727.

Lehn, H., P. Krieg, and G. Sauer. 1985. Papillomavirus genomes in human cervical tumors: Analysis of their transcriptional activity. *Proc. Natl. Acad. Sci.* **82:** 5540.

Lusky, M. and M. Botchan. 1985. Genetic analysis of bovine papillomavirus type 1 *trans*-acting replication factors. *J. Virol.* **53:** 955.

Maniatis, T., E.F. Fritsch, and J. Sambrook. 1982. *Molecular cloning: A laboratory manual.* Cold Spring Harbor Laboratory, Cold Spring Harbor, New York.

Messing, J. 1983. New M13 vectors for cloning. *Methods Enzymol.* **101:** 20.

Mount, S.M. 1982. A catalogue of splice junction sequences. *Nucleic Acids Res.* **10:** 459.

Murray, N.E. 1983. Phage lambda and molecular cloning. In *Lambda II* (ed. R.W. Hendrix et al.), p. 395. Cold Spring Harbor Laboratory, Cold Spring Harbor, New York.

Nakabayashi, Y., S.K. Chattopadhyay, and D.R. Lowy. 1983. The transforming function of bovine papillomavirus DNA. *Proc. Natl. Acad. Sci.* **80:** 5832.

Nasseri, M. and F.O. Wettstein. 1984. Differences exist between viral transcripts in cottontail rabbit papillomavirus-induced benign and malignant tumors as well as non-virus producing and virus-producing tumors. *J. Virol.* **51:** 706.

Pater, M.M. and A. Pater. 1985. Human papillomavirus types 16 and 18 sequences in carcinoma cell lines of the cervix. *Virology* **145**: 313.

Sanger, F., S. Nicklen, and A.R. Coulson. 1977. DNA sequencing with chain-terminating inhibitors. *Proc. Natl. Acad. Sci.* **74**: 5463.

Sarver, N., M.S. Rabson, Y.-C. Yang, J.C. Byrne, and P.M. Howley. 1984. Localization and analysis of bovine papillomavirus type 1 transforming functions. *J. Virol.* **52**: 377.

Schiller, J.T., W.C. Vass, and D.R. Lowy. 1984. Identification of a second transforming region of bovine papillomavirus DNA. *Proc. Natl. Acad. Sci.* **81**: 7880.

Schwarz, E. and A. Schneider-Gädicke. 1985. Organization and expression of human papillomavirus DNA in cervical cancer cell lines. In *Herpes and papillomaviruses: Their role in the carcinogenesis of the lower genital tract* (ed. G. de Palo). Raven Press, New York. (In press.)

Schwarz, E., A. Schneider-Gädicke, B. Roggenbuck, W. Mayer, L. Gissmann and H. zur Hausen. 1986. Expression of human papillomavirus DNA in cervical carcinoma cell lines. *Banbury Rep.* **21**: (in press).

Schwarz, E., U.-K. Freese, L. Gissmann, W. Mayer, B. Roggenbuck, A. Stremlau, and H. zur Hausen. 1985. Structure and transcription of human papillomavirus sequences in cervical carcinoma cells. *Nature* **314**: 111.

Schwarz, E., M. Dürst, C. Demankowski, O. Lattermann, R. Zech, E. Wolfsperger, S. Suhai, and H. zur Hausen. 1983. DNA sequence and genomic organization of genital human papillomavirus type 6b. *EMBO J.* **2**: 2341.

Schwarz-Sommer, Z., A. Gierl, H. Cuypers, P.A. Peterson, and H. Saedler. 1985. Plant transposable elements generate the DNA sequence diversity needed in evolution. *EMBO J.* **4**: 591.

Seedorf, K., G. Krämmer, M. Dürst, S. Suhai, and W.G. Röwekamp. 1985. Human papillomavirus type 16 DNA sequence. *Virology* **145**: 181.

Spalholz, B.A., Y.-C. Yang, and P.M. Howley. 1985. Transactivation of a bovine papillomavirus transcriptional regulatory element by the E2 gene product. *Cell* **42**: 183.

Yang, Y.-C., H. Okayama, and P.M. Howley. 1985. Bovine papillomavirus contains multiple transforming genes. *Proc. Natl. Acad. Sci.* **82**: 1030.

Yee, C., I. Krishnan-Hewlett, C.C. Baker, R. Schlegel, and P.M. Howley. 1985. Presence and expression of human papillomavirus sequences in human cervical carcinoma cell lines. *Am. J. Pathol.* **119**: 361.

zur Hausen, H. and A. Schneider. 1985. The role of papillomaviruses in human anogenital cancer. In *The papovaviridae: The papillomaviruses* (ed. P.M. Howley and N. Salzman). Plenum Press, New York. (In press.)

# Human Papillomaviruses of the Genital Mucosa: Electron Microscopic Analyses of DNA Heteroduplexes Formed with HPV Types 6, 11, and 18

**T.R. Broker and L.T. Chow**
Biochemistry Department, University of Rochester, School of Medicine and Dentistry, Rochester, New York 14642

Pairwise DNA heteroduplexes of human papillomavirus (HPV) types 6b, 11, and 18 revealed that all were collinear and therefore had very similar genetic organization. HPV-6b and -11 were closely related and maintain extensive but regional pairing even at $T_m$ 20°C, indicating that certain genes diverged less than 23% and that virtually all regions exceeded 60% homology. HPV-6b × HPV-18 made interspersed heteroduplexes at $T_m$ 41°C, indicating these paired regions exceed 60% homology. In all cases, the regions most likely to differ corresponded to the long control region (LCR) and the E5 region located at the 3' end of the E region and the 5' end of the L region. The L1 open reading frame was the most consistently conserved region among these papillomaviruses, and much of the E1 (replication) open reading frame was conserved to a similar extent. Other regions show intermediate levels of divergence.

More than 40 types of human papillomaviruses (HPVs) have been cloned from various epithelial lesions. HPV types 6b, 11, 16, and 18 were among the first prototypes isolated from lesions of the genital or oral mucosa. HPV-6b was originally isolated from a benign genital condyloma (de Villiers et al. 1981), HPV-11 was from benign laryngeal papilloma (Gissmann et al. 1982), and HPV-16 (Durst et al. 1983) and HPV-18 (Boshart et al. 1984) were from high-grade dysplasias and carcinomas of the uterine cervix. With the possible exception of HPV-18, these viruses can infect both oral and genital tissues of both sexes. Similar lesions sometimes test negative for all prototype HPVs under stringent hybridization conditions (e.g., $T_m$ 25°C) but are positive for HPV-related sequences under nonstringent conditions ($T_m$ 4°C), suggesting the presence of new types of HPV. Indeed, several more types have been cloned out of genital or oral lesions, including HPV-31, HPV-35 (A.T. Lorincz and G.F. Temple, pers. comm.), and HPV-13 (Pfister et al. 1983).

The DNA genomes of quite a few papillomaviruses have been completely sequenced in the past few years, including HPV-1 (Danos et al. 1983), HPV-6b (Schwarz et al. 1983), HPV-11 (K. Dartmann et al., pers. comm.), HPV-16 (Seedorf et al. 1985), bovine papillomavirus type 1 (BPV-1) (Danos et al. 1983), cottontail rabbit papillomavirus (CRPV) (Giri et al. 1985), and deer papillomavirus (Groff and Lancaster 1985). Clear sequence homologies are evident in certain genetic regions. Moreover, the open reading frames (ORFs) of each virus are organized in very similar fashion and have comparable lengths, indicating that all the papillomaviruses originated from a common ancestral virus.

We have been interested in the evolution of these viruses for several reasons: (1) to determine whether some of the virus types might be genetic recombinants of other prototypes, a possible basis for the considerable plurality of the HPVs, and (2) to determine both the genetic locations and the degrees of sequence divergence between pairs of viruses, a step toward correlating sequence information with biological function and disease spectra. This is particularly important as a close relative of a virus that generally causes benign tumors can be found to be associated with malignant lesions. By studying many pairs of these viruses, we expect a picture of the key elements to emerge. Of course, eventual DNA sequence analyses of these elements and an experimental animal system for testing the properties of the whole viruses or isolated genes are essential to the final interpretation. Here we report our initial electron microscopic studies of the DNA:DNA heteroduplexes of HPV types 6b, 11, and 18, which are trophic for the genital and oral mucosa.

## Materials and Methods

### HPV DNAs

All cloned human papillomavirus DNAs studied were the generous gifts of Dr. Lutz Gissmann and colleagues. HPV-6b was received as two separate BamHI-EcoRI restriction fragments cloned into pBR322 (de Villiers et al. 1981). HPV-11 (Gissmann et al. 1982) was inserted at its BamHI site (map position, 89%) into a pBR322 derivative, pML-d, also containing the 69% transforming fragment of BPV-1 (Lowy et al. 1980). HPV-18 was cloned at its single EcoRI site (Boshart et al. 1984). The recombinant DNAs were linearized with a restriction endonuclease prior to heteroduplex formation. The restriction endonuclease was chosen according to two essential criteria. First, HPV sequences must be located asymmetrically in the resulting linear molecule. Second, the HPV locations in the vectors for the pair of clones to be studied must be different so that they can be distinguished in the heteroduplexes and the homology map can be aligned to the restriction maps as well as to the genetic maps all at once. The second criterion can also be fulfilled by recloning the HPVs into vectors of different lengths, as we have done with HPV-6b and -11.

When high degrees of homology are expected or observed, it is more advantageous to hybridize clones in which the HPV DNAs are in opposite orientations so that the vector sequences serve only as orientation and position markers and do not hybridize to one other. On the other hand, when little homology is experienced, clones with HPVs in the same orientations should be used so that the vector sequences serve to nucleate the heteroduplex and hold the HPV sequences in close proximity for a pseudo-first-order reaction. Therefore, it is useful to have on hand clones of one of the viruses to be studied in either orientation in the vector, especially when dealing with HPVs about which little is known.

*Heteroduplex formation*

After restriction digestion, the linearized recombinant DNAs were purified by phenol and ether extraction. The DNAs were mixed in stoichiometric amounts, denatured with 0.1–0.3 M NaOH for 5 min at room temperature, neutralized with 2 M Tris · Cl (pH 7.2), and mixed with spectro-grade formamide to give final concentrations of 30% formamide, 0.1 M Tris · Cl (pH 8.5), 0.01 M Na$_3$ EDTA, and 1.5 µg/ml for each DNA. These were incubated for 15–20 min at room temperature. Reannealing was terminated by cooling in ice. The samples were spread in cytochrome *c* monolayers at room temperature (22°C) under isodenaturation conditions (Davis and Hyman 1971), containing 30%, 50%, or 60% formamide in the hyperphase (corresponding to $T_m$ 41°, $T_m$ 28°, and $T_m$ 21°C), and were mounted onto Parlodion-coated, 200-mesh copper grids (Davis et al. 1972). They were examined in a Zeiss EM 10C transmission electron microscope, and well-displayed heteroduplexes were photographed at 10,000 × magnification.

*Data analysis*

Electron micrographs were projected from a photographic enlarger onto a white matte surface. The length of each section of the molecule was measured with a Numonics electronic planimeter. In most cases, the two single-stranded sides of a substitution bubble were averaged, unless it was apparent that one was anomalously distorted. In heteroduplexes where the HPVs are in opposite orientation with respect to the vector (e.g., HPV-6b × HPV-11), the HPV DNAs formed a topological circle and the vectors remained single-stranded to provide index coordinates for evaluating the positions of the homologous duplexes and heterologous bubbles. The lengths of appropriate sections were summed and segments were reported as percentages of the full genome;

absolute map positions were assigned on the basis of their positions relative to the cloning site and the known orientation of the HPV in the vector.

In the HPV-6 × HPV-18 heteroduplexes, the recombinants chosen for heteroduplex formation had their vectors in the same orientation. The vector sequences therefore formed part of the circular structures and had to be identified clearly and subtracted from the aggregate length when evaluating the map coodinates of the HPV sequences.

## Results

### Heteroduplexes of HPV-6b and HPV-11

The HPV-11 clone was linearized with *Xba*I restriction endonuclease, which we determined to be a single-cut enzyme in the clone received, and we mapped it at 85% of the linear map as presented in Gissmann et al. (1982). This was hybridized separately to both of the HPV-6b *Eco*RI-*Bam*HI fragments in pBR322 that were linearized with *Sal*I, at $T_m$ 41°C hybridization conditions. Electron microscopic grids were also prepared at $T_m$ 41°C.

The large HPV-6b fragment, when hybridized to HPV-11, formed a circular heteroduplex in which the HPV-6b sequence was entirely double-stranded. A portion of the circle was single-stranded HPV-11. Vector sequences were present as a single-stranded insertion loop (from the HPV-11 clone) or as single-stranded branches (from the HPV-6b clone) flanking the unhybridized HPV-11 sequences. This indicated that the cloning sites of the two viruses were circularly permuted to each other, and from the distance between the single-stranded vector loop of the HPV-11 clone and the shorter, single-stranded vector tail (*Sal*I to *Bam*HI) of the HPV-6b clone, we determined that the *Bam*HI sites of the two HPVs were about 30% apart from each other and that the large *Bam*HI-*Eco*RI fragment of HPV-6b was completely homologous to the comparable region of the HPV-11. The small *Bam*HI-*Eco*RI fragment formed a linear heteroduplex with the HPV-11 clone in which both HPV and pBR322 sequences hybridized. The heteroduplex showed one or frequent nonhomology loop. Using the landmarks of the HPV-11 and the vector sequences that are present in one clone but absent in the other (and vice versa), we determined the nonhomologous region to be near the E5 ORFs. Having aligned the two viral genomes, we reassembled the HPV-6b from the two *Eco*RI-*Bam*HI fragments into one piece and inserted the intact HPV-6b at its *Bam*HI site into a pBR322 derivative to facilitate analyses at different stringencies.

Heteroduplexes were formed between the two clones as before, except the HPV-6b clone was digested with *Sal*I, which cleaves the vector twice, on either side of the HPV-6b sequences, and results in short, single-stranded branches of vector sequences in the heteroduplexes.

The grids were prepared at $T_m$ 41°C (30% formamide), at $T_m$ 28°C (50% formamide), and at $T_m$ 21°C (Fig. 1). The homology map was aligned to the restric-

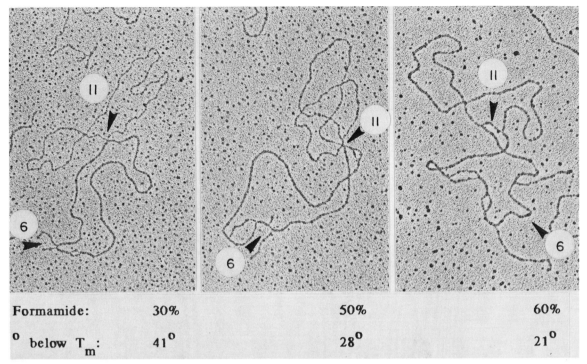

**Figure 1** Electron micrographs of heteroduplexes between HPV-6b and HPV-11 under different stringencies. The heteroduplexes were prepared at $T_m$ 41°C and mounted onto grids at $T_m$ 41°C, $T_m$ 28°C, and $T_m$ 21°C as described in Materials and Methods. Arrowheads point to the cloning sites. The single-stranded vector sequences for the HPV-6 and -11 are identified.

tion maps of both viruses and to the genetic map of HPV-6b (Schwarz et al. 1983). The results are summarized in Figure 2, which shows HPV-6b and HPV-11 were collinear and were highly homologous to each other. Only at $T_m$ 21°C, substantial denaturation can be seen in the E5, E1, L2, E2, and the long control regions. Using the formula of 1% formamide depresses $T_m$ by 0.65°C for DNAs of 50% AT base composition and 1% sequence mismatch depresses $T_m$ by about 0.95°C (D. Sporn, L. Chow, and T. Broker, unpubl.), we estimated the regions denatured at $T_m$ 41°C have no more than 57–61% homology; those that denatured between $T_m$ 41°C and 28°C had 61–74% homology; those denatured between $T_m$ 27°C and 21°C had 74–80% homology; and finally, those remaining paired at $T_m$ 21°C had at least 80–84% homology. Since the helix-coil transition is a dynamic, breathing process over an interval of about 8°C, even finer estimates can be made depending on the fraction of molecules in the population paired in a particular region. This high degree of sequence homology, unexpected on the basis of renaturation kinetics in liquid phase (Gissmann et al. 1982), has been confirmed by direct sequence comparison of these two viruses when the DNA sequence of HPV-11 was made available to us prior to publication (K. Dartmann et al., pers. comm.).

### Heteroduplexes of HPV-6b and HPV-18

Because very little homology was expected between these two viruses (Boshart et al. 1984), we adopted the strategy of taking advantage of vector sequences to nucleate the hybridization, as described in Materials and Methods.

First, we recloned HPV-6b at its *Bam*HI site in either orientation into the multiple cloning site of a vector called pSVO10 (described in Chow and Broker 1984), which is deleted for half of the pBR322 sequence, including the *tet* gene (R. Myers and R. Tjian, pers. comm.). Both HPV-6b clones were then linearized at the *Cla*I site in the multiple cloning site and hybridized separately to the *Sal*I-linearized HPV-18 DNA clone. The heteroduplexes were formed at $T_m$ 41°C (in 30% formamide) and grids were also prepared at $T_m$ 41°C. Only the preparations in which HPV-6 and HPV-18 were found to be in the same orientation relative to the vector sequence were scanned, for these indeed had more HPV heterduplexes for the reason discussed. The reference points in these heteroduplexes were the portions of the pBR322 sequences that were present in both clones and hybridized to each other, as well as the rest of the pBR322 sequence, which was only present in the HPV-18 clone and thus remained single-stranded. Knowing the orientation of the HPVs in the clones used, we were able to align the HPV-6b × HPV-18 homology map to their respective restriction maps and to the genetic map of HPV-6b, as shown in Figure 3. Our result showed that the two DNAs were collinear and of approximately the same length. The homology was not extensive even at $T_m$ 41°C. The most homologous regions were in L1, encoding the major capsid protein, and in E1, specifying functions required for episomal replication.

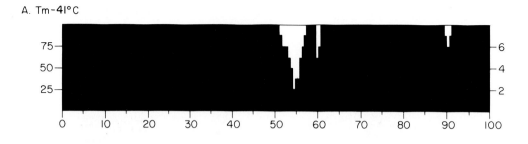

A. Tm-41°C

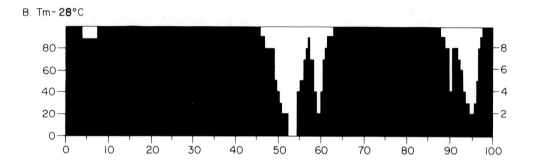

B. Tm-28°C

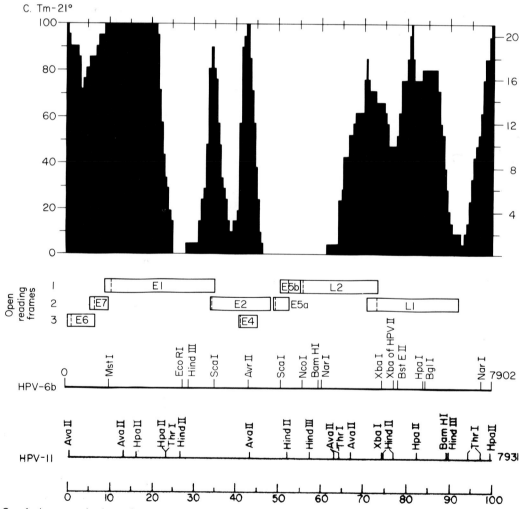

C. Tm-21°

**Figure 2** *See facing page for legend.*

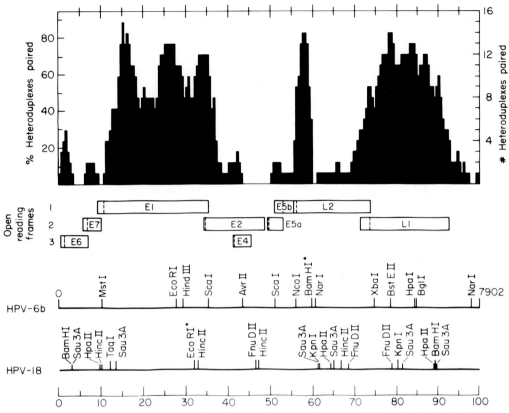

**Figure 3** Histogram of heteroduplexes between HPV-6b and -18 under nonstringent conditions. The heteroduplexes were prepared at $T_m$ 41°C. The regions of homology (black) and nonhomology (white) were aligned and plotted against the restriction maps of both and the genetic map of HPV-6b. The percent and number of heteroduplexes paired are represented as the left and right ordinates, respectively.

## Discussion

We have compared the sequence relationship between HPV-6b and HPV-11 and between HPV-6b and HPV-18. All three viral genomes were collinear and were similar in length. We found that, in addition, HPV-11 was highly homologous to HPV-6b in spite of the original reassociation kinetics indicating the contrary. On the other hand, HPV-18 did not show a high degree of homology to HPV-6b even under conditions equivalent to 41°C below the melting temperature for homoduplexes. The different degrees of homology among these three viruses are reflected in the associated disease spectra. HPV-6b and -11 are primarily associated with benign condylomata and low-grade dysplasias, whereas HPV-18 is associated with high-grade dysplasias and malignancy. With all three viruses, the most homologous regions include L1 ORF, encoding the major capsid antigen (Engel et al. 1983), and the E1 ORF, which is required for episomal replication (Sarver et al. 1984; Lusky and Botchan 1985). The least homologous regions were E5, a transformation gene in BPV-1 (DiMaio et al. 1985; D.

Schiller et al., pers. comm.), and the long control region (formerly called the noncoding region). This pattern was also observed in other HPVs we have compared (our unpublished results). This is rather different from other viruses, where the major capsid antigens tend to have diverged more than other genes, presumably due to the selection pressure of host immune surveillence. The considerable conservation in the virion structural protein of the papillomavirus group suggests that these virus particles are not particularly subject to the scrutiny of the host immune system and that capsid antigens may not be a good choice with which to generate potential vaccines.

Subsequent to the heteroduplex mapping and with the recent availability of the HPV-11 DNA sequence, we have aligned the two DNA sequences using the computer programs available through BIONET (D. Sporn, L. Chow, and T. Broker, unpubl.). The degrees and locations of homology between HPV-6 and -11 based on our electron microscopic heteroduplex were in excellent agreement with the DNA sequence analyses (Schwarz

**Figure 2** (*See facing page*) Histograms of heteroduplexes between HPV-6b and -11 under different stringencies. Regions of homology (black) and nonhomology (white) visualized under $T_m$ 41°C, $T_m$ 28°C, and $T_m$ 21°C were aligned and plotted along the linearized restriction maps of both and the genetic map of HPV-6b. The percent and number of heteroduplexes paired are represented as the left and right ordinates, respectively.

et al. 1983; K. Dartmann et al., pers. comm.). The DNA sequence for HPV-18 is not determined yet. The only minor discrepancies between the predicted and observed homologies occur in the vicinity of junctions between HPV and vector sequences, where single-stranded branches or loops apparently facilitate some denaturation of heteroduplex sequences when conditions are approaching their melting temperature. Therefore, we believe the use of DNA heteroduplex analysis is rapid and accurate and produces concurrent data on both the degree and the location of homology.

## Acknowledgments

A portion of this research was conducted while we were at Cold Spring Harbor Laboratory, with support from National Cancer Institute Program Project Grant CA13106. Additional aspects have been supported by grant MV-144 from the American Cancer Society and grant 1587 from the Council for Tobacco Research–USA. We are most grateful to Drs. Lutz Gissmann and Harold zur Hausen and their colleagues for sharing with us their recombinant DNA clones of HPB-6b, -11, and -18 and for making available the HPV-11 DNA sequence prior to publication.

## References

Boshart, M., L. Gissmann, H. Ikenberg, A Kleinheinz, W. Scheurlen, and H. zur Hausen. 1984. A new type of papillomavirus DNA and its presence in gential cancer biopsies and in cell lines derived from cervical cancer. *EMBO J.* **3:** 1151.

Chow, L.T. and T.R. Broker. 1984. Human papillomavirus type 1 RNA transcription and processing in COS-1 cells. *Prog. Cancer Res. Ther.* **30:** 125.

Danos, O., L.W. Engel, E.Y. Chen, M. Yaniv, and P.M. Howley. 1983. A comparative analysis of the human type 1a and bovine type 1 papillomavirus genomes. *J. Virol.* **46:** 557.

Davis, R.W. and R.W. Hyman. 1971. A study in evolution: The DNA base sequence homology between coliphages T7 and T3. *J. Mol. Biol.* **62:** 287.

Davis, R.W., M.N. Simon, and N. Davidson. 1972. Electron microscrope heteroduplex methods for mapping regions of base sequence homology in nucleic acids. *Methods Enzymol.* **21:** 413.

de Villiers, E.M., L. Gissmann, and H. zur Hausen. 1981. Molecular cloning of viral DNA from human genital warts. *J. Virol.* **40:** 932.

DiMaio, D., J. Metherall, K. Neary, and D. Guralski. 1985. Genetic analysis of cell transformation by bovine papillomavirus. *UCLA Symp. Mol. Cell. Biol. New Ser.* **32:** 437.

Durst, M., L. Gissmann, H. Ikenberg, and H. zur Hausen. 1983. A papillomavirus DNA from a cervical carcinoma and its prevalence in cancer biopsy samples from different geographical regions. *Proc. Natl. Acad. Sci.* **80:** 3812.

Engel, L.W., C.A. Heilman, and P.M. Howley. 1983. Transcriptional organization of the bovine papillomavirus type 1. *J. Virol.* **47:** 516.

Giri, I., O. Danos, and M. Yaniv. 1985. Genomic structure of the cottontail rabbit (Shope) papillomavirus. *Proc. Natl. Acad. Sci.* **82:** 1580.

Gissmann, L., V. Diehl, H.J. Schultz-Coulon, and H. zur Hausen. 1982. Molecular cloning and characterization of human papillomavirus DNA derived from a laryngeal papilloma. *J. Virol.* **44:** 393.

Groff, D.E. and W.D. Lancaster. 1985. Molecular cloning and nucleotide sequence of deer papillomavirus. *J. Virol.* **56:** 85.

Lowy, D.R., I. Dvoretzky, R. Shober, M.F. Law, L. Engel, and P.M. Howley. 1980. In vitro tumorigenic transformation by a defined sub-genomic fragment of bovine papillomavirus DNA. *Nature* **287:** 72.

Lusky, M. and M.R. Botchan. 1985. Genetic analysis of bovine papillomavirus type 1 *trans*-acting replication factors. *J. Virol.* **53:** 955.

Pfister, H., I. Hettich, U. Runne, L. Gissmann, and G.N. Chilf. 1983. Characterization of human papillomavirus type 13 from focal epithelial hyperplasia Heck lesions. *J. Virol.* **47:** 363.

Sarver, N., M.S. Rabson, Y.C. Byrne, and P.M. Howley. 1984. Localization and analysis of bovine papillomavirus type 1 transforming functions. *J. Virol.* **52:** 377.

Schwarz, E., M. Durst, C. Demankowski, O. Lattermann, R. Zech, E. Wolfsperger, S. Suhai, and H. zur Hausen. 1983. DNA sequence and genome organization of genital human papillomavirus type 6b. *EMBO J.* **2:** 2341.

Seedorf, K., G. Krammer, M. Durst, S. Suhai, and W. Rowekamp. 1985. Human papillomavirus type 16 DNA sequence. *Virology* **145:** 181.

# Anatomy and Expression of the Transforming Region of Bovine Papillomavirus Type 1

U. Pettersson,* P. Bergman,* H. Ahola,*[†] M. Ustav,[‡] A. Stenlund,[§] J. Zabielski,* B. Vennstrom,** and J. Moreno-Lopez[†]

*Department of Medical Genetics, University of Uppsala, Biomedical Center, S-751 23 Uppsala, Sweden; [†]Department of Veterinary Microbiology (Virology), Swedish University of Agricultural Sciences, Biomedical Center, S-751 23 Uppsala, Sweden; [‡] Laboratory of Molecular Biology, Tartu University, 202 400 Tartu, Estonia, USSR; ** European Molecular Biology Laboratory, 6900 Heidelberg, Federal Republic of Germany

The viral mRNAs that are present in C127 cells transformed by bovine papillomavirus type 1 (BPV-1) have been characterized by a variety of methods. The results revealed a very complex mRNA pattern, comprising at least five types of spliced mRNAs in addition to several unspliced cytoplasmic RNA species. Both unspliced and partially processed nuclear RNAs have also been identified.

The transcriptional start sites were studied by S1 nuclease analysis and primer extension. The results revealed several promoters within the transforming region of BPV-1. The major promoter region, located between coordinates 0 and 1, directs the synthesis of at least four differentially spliced mRNAs. An additional promoter region, located at coordinate 31, appears to direct the synthesis of an abundant 1-kb-long mRNA. A third putative promoter was mapped at coordinate 39, apparently generating a less abundant colinear RNA species. mRNAs transcribed from all three promoters terminate at a common polyadenylation site, located at coordinate 53.

A functional analysis of the transforming region of BPV-1 has been initiated, using retrovirus vectors. We report here that the physical state of the viral genome in the transformed cell is influenced by the transcriptional activity of a retroviral long terminal repeat. The effect is most likely caused by interference from antisense transcripts.

Although the papillomaviruses were already discovered in the beginning of this century (Ciuffo 1907; Shope 1933), studies of their molecular properties have progressed slowly until the last decade. Today the papillomavirus field is one of the most rapidly progressing areas of virology. Gene technology has probably had a greater impact on this field than almost any other branch of virology. Also, the discovery of efficient in vitro systems for cellular transformation allows certain biological properties of the viral genome to be studied (for review, see Pfister 1984). Complete nucleotide sequences are known for six human papillomaviruses and three animal papillomaviruses. Although the human and animal papillomavirus genomes generally exhibit a low degree of sequence homology (Law et al. 1979; Heilman et al. 1980), they have many properties in common, particularly regarding the organization of the open translational reading frames (ORFs). General conclusions can thus be drawn from experiments with the papillomavirus that is most amenable to experimental studies, bovine papillomavirus type 1 (BPV-1). The BPV-1 genome consists

of 7945 bp (Chen et al. 1982; Danos and Yaniv 1983; revised by Stenlund et al. 1985), and the transforming segment is located between the cleavage sites for endonucleases HindIII (coordinate 87) and BamHI (coordinate 56). Eight ORFs of a significant size have been detected in this segment of the BPV-1 genome (Fig. 1) and they are designated E1 to E8 (Danos and Yaniv 1983). Five of these, namely E1, E2, E4, E6, and E7, have equivalent counterparts in all papillomavirus genomes sequenced so far and are thus likely to encode functional polypeptides. The L region, which covers approximately 3000 bp in the BPV-1 genome, is located between coordinates 53 and 89 (Chen et al. 1982). Two major ORFs, designated L1 and L2, have been identified in this region. Corresponding ORFs are found in the other sequenced papillomavirus genomes, and they code for proteins that are present in the papillomavirus capsid.

The approximately 1-kb-long region in the BPV-1 genome that is located between the end of the L1 ORF and the beginning of the E6 ORF appears to be noncoding. In this communication we discuss the transcriptional organization and expression of the transforming region of BPV-1.

[§]*Present address*: Department of Molecular Biology, University of California, Berkeley, California 94720.

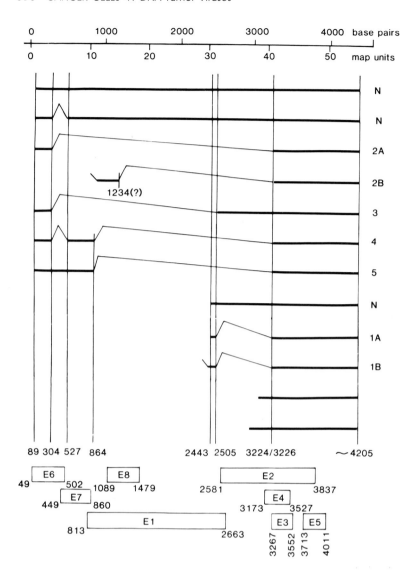

**Figure 1** Structure of mRNAs from the transforming region of BPV-1. The exons are indicated with thick lines, and the mRNAs are designated according to Stenlund et al. (1985). The figure is based on information published by Stenlund et al. (1985) and Yang et al. (1985). Three nuclear RNA species (N), one of which is spliced, are also shown. The positions of the different splice acceptor and donor sites are indicated, as well as the positions of the major ORFs in the BPV-1 genome.

## Materials and Methods

### Cells
A Swedish isolate of BPV-1 (Ahola et al. 1983) was used to transform C127 mouse cells. The transformed cells were cloned in agar prior to being used.

### RNA preparation
Cytoplasmic RNA was prepared according to the methods of Brawerman et al. (1972) or Maniatis et al. (1982). Total RNA was prepared as described by Vennström and Bishop (1982).

### Sequence analysis
The method of Maxam and Gilbert (1980) was used.

### S1 nuclease analysis
The protocol of Favaloro et al. (1980) was followed. Sequencing ladders were separated on the same gel as the S1-resistant material (Weaver and Weissman 1979).

### Northern blot analysis
Formaldehyde/formamide-denatured poly(A)-containing RNA was fractionated in 1% or 1.4% (w/v) agarose gels containing 2.2 M formaldehyde (Lehrach et al. 1977). The RNA was subsequently blotted onto a nitrocellulose filter (Thomas 1980), and the filter was hybridized in 50% (v/v) formamide, 50 mM HEPES (pH 7.4), 5× Denhardt's solution, 250 µg/ml yeast RNA, 3× SSC (SSC = 0.15 M NaCl, 0.015 M Na citrate, pH 7.0), 100 µg/ml single-stranded calf thymus DNA, and 0.1% (w/v) SDS at 42°C for 36 hr with nick-translated $^{32}$P-labeled fragments as probes (Rigby et al. 1977). After hybridization, the filter was washed three times for 40 min in 0.1× SSC, 0.1% SDS at 50°C before autoradiography. The filters to be probed with RNA probes were hybridized and washed according to DeLeon et al. (1983).

### Electron microscopy
The procedure was recently described in detail by Stenlund et al. (1985).

*Primer extension*
The protocol of Virtanen and Pettersson (1985) was followed. Oligomers 20 nucleotides (nt) long served as primers. Avian myeloblastosis virus (AMV) polymerase was used for extension as described above.

*Construction of the pML and pMR recombinants*
The pMV5 vector which contains two Harvey murine sarcoma virus (Na-MSV) long terminal repeats (LTRs), was a kind gift from Dr. P. Luciw. The vector contains two cleavage sites for endonuclease *Hind*III and one cleavage site for *Bam*HI between the LTRs (Fig. 2). These were used for insertion of the 69% transforming fragment from the BPV-1 genome (69T), and the design of the constructs is shown in Figure 2. The pMR recombinant was constructed in the following way: The pBH1 plasmid (Ahola et al. 1983), which contains the entire BPV-1 genome, was cleaved with *Bam*HI and the excised fragment was inserted into the corresponding cleavage site in the pMV5 vector. The pML recombinant was constructed in the following way: The pMV5 vector was cleaved to completion with *Bam*HI and partially with *Hind*III. The 69T fragment was then excised from the pB69 plasmid (containing the 69T fragment cloned in pBR322) by using a mixture of *Bam*HI and *Hind*III and subsequently ligated into the pMV5 vector, precleaved as described above. The structure of the recombinant genomes was verified by restriction enzyme cleavage.

## Results

### Transcriptional organization of the BPV-1 genome
Studies on papillomavirus transcription have been hampered by the absence of tissue-culture systems for viral replication and the low abundance of viral mRNAs in papillomavirus-transformed cells. Between 0.01% and 0.2% of the mRNA in a BPV-1−transformed cell appears to be of viral origin, and some mRNA species represent

a very small fraction of the total viral mRNA (Heilman et al. 1982; Yang et al. 1985). It has consequently been very difficult to establish a complete transcriptional map for any papillomavirus genome, and several mRNA species that can be predicted to exist from the DNA sequence remain yet to be discovered.

*Spliced mRNAs expressed in BPV-1−transformed cells*
Northern blot and S1 nuclease analyses show that BPV-1−transformed cells contain several overlapping mRNA species, ranging in size between approximately 1.1 and 4.1 kb. Most major mRNA species in BPV-1−transformed cells consist of two or more exons. A major cap site for early BPV-1 mRNAs is located at nt 89 in the BPV-1 sequence. This cap site defines one transcription unit, extending between nt 89 and nt 4203, where the common polyadenylation site for all E-region mRNAs is located (Stenlund et al. 1985; Yang et al. 1985). Nuclear mRNA precursors have been identified that extend between these positions in the BPV-1 genome, and a variety of spliced mRNAs appear to be generated from these nuclear transcripts by differential splicing (Fig. 1).

The positions of splice acceptor and donor sites within this transcription unit have been determined by the S1 nuclease protection technique and by sequencing of cDNA clones (Stenlund et al. 1985; Yang et al. 1985). A set of mRNAs consisting of different exon combinations is generated by combining the splice donor and acceptor sites in different fashions. From electron microscopic heteroduplex analysis (Stenlund et al. 1985), the spliced mRNAs from the E region of BPV-1 have been divided into five categories, designated types 1 to 5 (Fig. 1).

The type-1 mRNA is the most abundant viral mRNA species present in BPV-1−transformed C127 cells. It has a length of 1.0 kb and consists of two exons (Fig. 1). Two different type-1 mRNAs appear to exist that differ with regard to the structure of their 5′ exons. One type

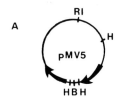

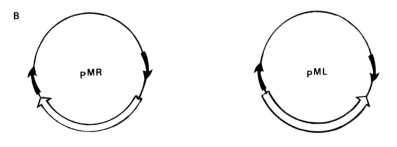

**Figure 2** Structure of the retrovirus recombinants containing the 69T fragment of BPV-1. The black and white arrows denote LTRs and the 69T fragment, respectively. Endonuclease cleavage sites are: (RI) *Eco*RI; (H) *Hind*III; (B) *Bam*HI. (*A*) The HaMSV−based retrovirus plasmid vector, pMV5. Cleavage sites used for cloning are shown between the LTRs. (*B*) The pMV5-based vectors pMR (BPV transcription on the same strand as the LTRs) and pML (BPV transcription on the opposite strand as the LTRs).

(1A) has the structure outlined above, whereas the second type (1B) contains a longer 5' exon. The latter is also located at coordinate 31, although its precise position remains to be determined. It is possible that the 5' exon in the type-1B mRNA is linked to yet another, very short leader since heteroduplexes formed between the type-1B mRNAs and BPV-1 DNA often contain a short 5' tail of unhybridized RNA (Stenlund et al. 1985).

The 3' exon of the type-1A and -1B mRNAs covers the entire E5 ORF as well as the 3' portions of the E2, E3, and E4 ORFs. The latter ORFs, however, lack ATG triplets beyond the splice acceptor site. The leader that is present in the type-1A mRNAs contains an ATG that is followed by a translational reading frame that is uninterrupted until the splice junction. However, a protein that is only 16 amino acids long would be translated under these conditions.

The type-2 mRNAs that, disregarding the poly(A) tail, have a length of 1.2 kb resemble the type-1 mRNAs in that they also contain the common 3' exon, although it is linked to a different leader, which maps much further upstream (Fig. 1). Electron microscopic heteroduplex analysis has shown that the type-2 mRNAs are heterogeneous with regard to their 5' exons, and at least two subclasses have been identified, designated types 2A and 2B (Fig. 1). The type-2B mRNA contains a leader that maps between coordinates 10 and 15 (Stenlund et al. 1985), although its precise structure has not yet been determined. A candidate splice donor site has been mapped at nucleotide 1234.

From its structure, it can be predicted that the type-2A mRNAs encode a polypeptide that consists of sequences derived from both the E6 and E4 ORFs. Its predicted molecular weight would be 19,000, provided that translation initiates at the 5'-proximal ATG in the E6 ORF. Of course, the coding capacity of the type-2B mRNA cannot be predicted since its precise structure is unknown.

The type-3 mRNAs have an estimated length of approximately 1.8 kb, excluding the poly(A) tail. They differ from the other spliced BPV-1 mRNAs in containing a 3' exon that is longer and extends between coordinate 32 and the common polyadenylation site at coordinate 53. The type-3 mRNAs are apparently very rare, since exceptionally few molecules of this kind were detected in an electron microscopic study of the mRNA pool that is present in BPV-1-transformed cells. An interesting feature of the type-3 mRNA is that it covers the E2 ORF.

The type-4 mRNA is unique in being composed of three exons (Fig. 1). Its total length is 1.5 kb, excluding the poly(A) tail, and it can be predicted from the structure that it encodes a polypeptide with a molecular weight of 21,000. This polypeptide consists of sequences both from ORF E6 and ORF E7 since the removal of the proximal intron results in an in-phase fusion of these two ORFs.

The type-5 mRNAs are 1.9 kb long and are composed of two exons. This mRNA thus has a structure similar to that of the type-4 mRNAs, except that the intervening sequence between the 5' and middle exons is maintained. The E6 and E7 ORFs will consequently not be fused in the type-5 mRNAs, and the expected translation product is the E6 protein product, having a molecular weight of 16,000.

In addition to the spliced cytoplasmic mRNAs, a prominent spliced nuclear RNA species was described by Stenlund et al. (1985) (Fig. 1). An intron is removed from this RNA that has an approximate length of 3.9 kb. It is present in approximately the same quantities as the unspliced RNA precursor, which extends from the cap site at nt 89 to the polyadenylation site at coordinate 53. The existence of this spliced nuclear mRNA species suggests that splicing may be retarded at an intermediate stage and, due to its structure, it is likely to be a precursor of the type-4 mRNAs.

*Colinear RNA species, expressed in BPV-1–transformed cells*

In addition to these spliced mRNA species, several colinear RNAs appear to be present in BPV-1–transformed cells. One set of colinear RNAs extends from the polyadenylation site at coordinate 53 to either coordinates 39 (nt 3100), 38 (nt 3000), 32 (nt 2530), or 31 (nt 2430). The RNA species whose 5' end maps at coordinate 31 may be an unspliced precursor for the type-1 mRNA since it is rather abundant in nuclear RNA and extends from the position of the 5' end of the type-1A mRNA to the polyadenylation site. The colinear RNA species that starts at coordinate 39 has been studied by S1 nuclease analysis and primer extension. It is heterologous, with a major 5' end located at nt 3070. An interesting property of the latter RNA is that it covers the entire E4 ORF and it might therefore encode the E4 protein. The fact that ATG triplets are located near the 3' end of the E4 ORF of all the sequenced papillomavirus genomes predicts that mRNAs indeed should exist, having their capped 5' ends near the start of the E4 ORF.

The RNAs whose 5' ends map at coordinates 31 and 32 have not yet been mapped at the nucleotide level. They cover the entire E2 ORF.

*Promoters in the E region of the BPV-1 genome*

The 5' ends of mRNAs that are expressed from the transforming region of BPV-1 have been mapped both by S1 nuclease analysis and primer extension. The primers that were used are shown in Table 1.

One major cap site maps at nt 89. This cap site is shared by the mRNAs of types 2A, 3, 4, and 5 (Fig. 1). It is preceded by a TATA motif starting at nt 58. The promoter region is furthermore juxtaposed to an enhancer-like sequence element that is *trans*-activated by another BPV function, most likely the product of the E2 ORF (see Rabson et al., this volume).

The type-1A mRNAs appear to have different 5' ends than the other major E-region mRNAs. It has been established by S1 nuclease analysis that the leader sequence that is attached to the type-1A mRNA maps between nt 2443 and nt 2505. An important question is whether the 5' end of the leader is a cap site or,

**Table 1** Synthetic Oligonucleotides Used for Primer Extension to Map Exon/Intron Junctions and 5' Ends of mRNAs from the Transforming Region of BPV-1

|  | Coordinate | Nucleotides |
|---|---|---|
| Primer 1 | 7 | 563–582 |
| Primer 2 | 31 | 2484–2503 |
| Primer 3 | 41 | 3271–3290 |

alternatively, a splice acceptor site that is joined to a very short upstream exon. To answer this question, primer extension experiments were carried out, using synthetic oligonucleotides as primers. Two separate oligonucleotides were used for the analysis, derived from sequences located at coordinates 31 (primer 2 in Table 1) and 42 (primer 3 in Table 1), respectively. The primers were 5'-end-labeled and then extended by AMV DNA polymerase using RNA from BPV-1–transformed C127 cells as template. The extension products were fractionated by polyacrylamide gel electrophoresis. Extension products obtained with RNA from untransformed C127 cells were separated in parallel to determine which products were virus-specific. A heterogenous array of 5' ends was observed, all of which belonged to type-1A mRNAs. These mapped between nt 2440 and nt 2450 in the BPV-1 sequence. The extension products were also subjected to sequence analysis by the Maxam and Gilbert (1980) procedure. It was, however, impossible to read the sequence accurately all the way to the 5' ends. This was due to the fact that several closely spaced extension products failed to separate from each other. The results indicated nevertheless that the sequence of the extension products was colinear with the DNA sequence around coordinate 31, suggesting that the type-1A mRNAs indeed are initiated in this region. An examination of the nucleotide sequence in this region reveals the sequence TAATAT between nt 2414 and nt 2419. This is the expected position of a TATA box if the cap site is located around nt 2440.

Two sets of RNAs appear to be transcribed from this putative promoter, one being represented by the type-1A mRNAs. A second set consists of colinear RNAs extending from coordinate 31 to the polyadenylation site at coordinate 53. RNA species with this structure have been observed by S1 nuclease analysis. It is, however, unclear whether these represent the unspliced precursor for the type-1A mRNAs or whether they also represent a rare mRNA species. If the latter is true, they are likely to encode the product of the E2 ORF since they cover this ORF in its entirety.

A third promoter appears to be located around coordinate 39. Using the S1 nuclease protection method, a set of heterogenous colinear RNAs have been identified, whose 5' ends map in this region. Primer extension analysis, using primer 3 (Table 1) showed that a major 5' end maps at nt 3070. No sequence typical for a splice acceptor site is found in the region where the 5' ends map, further supporting the notion that they represent

capped 5' ends. No characteristic TATA box is, however, present around coordinate 39. It is noteworthy that the E4 ORF begins at nt 3173. These colinear mRNAs may thus encode the E4 protein product.

## Functional analysis of the transforming region of BPV-1, using retrovirus vectors

Retrovirus vectors are convenient tools for introducing foreign genes into cells. The retrovirus genome facilitates integration into the cellular genome, and a novel type of retrovirus vector is also capable of producing RNAs that can be packaged into viral envelopes (for a review, see Mulligan 1983). These packaged RNAs can then be introduced into animal cells with a high efficiency. We have constructed several recombinant genomes in which the transforming region of BPV-1 is inserted into different retrovirus vectors. We describe below a retrovirus–BPV-1 construct that has some interesting consequences for the BPV-1 replicon.

### Construction of recombinant genomes with different orientations

The 69% fragment of the BPV-1 genome (69T), which is located between the cleavage sites for *Bam*HI and *Hind*III and is capable of transformation as well as episomal replication, was cloned between two LTRs originating from Ha-MSV. Two constructs were made, one having the 69T fragment in the same transcriptional orientation as the LTRs (construct pMR), and the other one containing the BPV-1 fragment in the opposite transcription orientation relative to that of the LTRs (construct pML). The procedure for construction of the recombinant genomes is described in Experimental Procedures and their structures are illustrated in Figure 2. Both constructs were used to transform mouse C127 cells. Transformed foci appeared 10–14 days after transfection, irrespective of which construct was used, and the transformation frequency was approximately the same in the two cases, 10–20 foci/μg. The transformation frequency was lower than usually achieved with BPV-1 DNA, presumably due to the presence of plasmid sequences in the constructs. Dot-blot analysis of DNA from individual clones revealed that pML- and pMR-transformed cells both contained BPV-1 genomes in high copy numbers. From an analysis of serially diluted DNA, we estimate that the transformed cells contained 100–500 genome equivalents per cell and there was no significant difference between pML- and pMR-transformed cells.

### State of the viral genome in pML- and pMR-transformed cells

Individual foci of pML- and pMR-transformed cells were cloned in soft agar and the total cellular DNA was extracted from three pML-transformed clones and three pMR-transformed clones when the appropriate cell density was reached. The DNA was then analyzed by Southern blotting, either without cleavage or after cleavage with different restriction enzymes. The 69T fragment of BPV-1 DNA was used as the hybridization probe. The results show that the viral DNA in all analyzed pML-transformed clones was integrated (data not shown). This conclusion is based on the finding that the viral DNA comigrated with the cellular DNA in absence of cleavage. Cleavage with *Hin*dIII excised the 69T fragment together with a short segment from the multilinker in the vector (data not shown). Cleavage with endonucleases *Eco*RI or *Cla*I, which cut within the 69T fragment, generated multiple bands of different sizes. They presumably contain BPV sequences linked to flanking cellular sequences. The latter result suggests that the BPV-1 sequences are integrated at several different sites.

DNA extracted from pMR-transformed clones gave a dramatically different picture. Analysis of uncleaved DNA revealed the presence of multiple copies of episomal DNA. It is not possible to exclude from the results that a few copies were integrated, although the bulk of the viral DNA is clearly present as monomeric plasmids. The size of the plasmids varied between different clones, most likely depending on the presence in the constructs of bacterial plasmids that are known to cause deletions (Sarver et al. 1982).

*Analysis of BPV-1–specific mRNA in pML- and pMR-transformed cells*
Cytoplasmic poly(A)–containing RNA was extracted and analyzed by northern blotting, using the 69T fragment as a probe. pMR-transformed cells revealed several prominent mRNA species that ranged in size between approximately 1 and 5 kb. The mRNA pattern in these cells was similar to that observed in cells transformed with the complete BPV-1 genome (Stenlund et al. 1985). RNA extracted from cells transformed by pML showed a different pattern; the concentration of BPV-1–specific RNA was 5-fold to 10-fold lower and the only clearly detectable BPV-1 mRNA species was the type-1 mRNA. A few, very large RNA species were detected in addition ( > 5 kb); they apparently represent transcripts initiated in the LTRs (see below).

*Cells transformed by pML contain abundant antisense transcripts*
It was suspected that the difference between the state of the viral genome in pML- and pMR-transformed cells was due to the transcriptional activity of the LTRs. One possibility that was considered was that the LTR in the pML construct gave rise to antisense transcripts that interfered with BPV-1 gene expression. To examine this possibility, an RNA probe was used to study transcripts coming from the antisense strand of the recombinant by northern blot analysis. The probe was prepared in the following way: A *Pst*I fragment located between nt 1861 and nt 2775 was cloned in the pSP64 vector (Melton et al. 1984) in an orientation that yielded RNA that would detect BPV-1 transcripts of the antisense polarity. The results of the analysis are shown in Figure 3. A clear-cut difference was observed when the two types of transformed cells were analyzed; pML-transformed cells contained easily detectable antisense transcripts that were

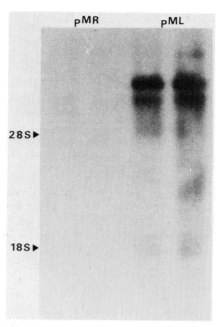

**Figure 3** Detection of antisense transcripts in pML-transformed cells. Total poly(A) RNA was analyzed by northern blotting. An SP6 RNA polymerase–generated $^{32}$P-labeled probe (described in the text) was used to detect large transcripts originating from the noncoding strand of BPV-1.

absent from pMR-transformed cells. The most likely explanation for the difference in the physical state of the viral DNA in pML- and pMR-transformed cells is that the antisense transcripts interfere with those mRNAs that are required for episomal replication, most likely the mRNA for the E1 protein. The mRNA that encodes the transforming function is apparently less sensitive to interference by antisense RNA, since both constructs transform C127 cells with approximately the same frequency. The explanation is perhaps that this mRNA is present in a greater abundance. Since the type-1 mRNA is the only clearly detectable BPV-1 mRNA in pML-transformed cells, the results imply that this mRNA plays a crucial role in transformation.

**Discussion**

The structures of the mRNA species that are present in BPV-1–transformed BPV-1 cells are outlined in Figure 1. This figure is a compilation of results obtained by electron microscopic heteroduplex mapping, S1 nuclease analysis, cDNA cloning, and primer extension (Stenlund et al. 1985; Yang et al. 1985).

Although a very complex mRNA pattern has emerged, it seems likely that the picture is still incomplete. An important question remaining to be answered concerns the coding capacity of the most abundant mRNA species present in transformed cells, the type-1A mRNA. This mRNA covers the entire E5 ORF as well as the 3' portions of the E2 and E4 ORFs. The portions of the E2 and E4 ORFs that are included in the type-1A mRNA lack ATG triplets. A possibility is that translation initiates in an upstream exon. The leader sequence present in the type-1A mRNAs does indeed contain an ATG. However, if translation is initiated at this ATG, a protein with a predicted length of 16 amino acids would be synthesized, which seems to be an unlikely size for a functional product. Thus, E5 remains; this ORF is a strong candidate for playing a major role in transformation (see Schiller et al., this volume). If the type-1A mRNA encodes the E5 protein product, the 5′ noncoding region is exceptionally long (~700 nt), and it is furthermore puzzling that the first methionine is located in the middle of the ORF.

Additional mRNAs to those depicted in Figure 1 can be predicted to exist based on the sequence organization of the BPV-1 genome. No mRNA has, for instance, been characterized that has the capacity to encode the E1 protein. The E1 sequence is spliced out from all hitherto characterized BPV-1 mRNAs except for two nuclear RNA species that, however, are more likely to be RNA precursors rather than mRNAs (Fig. 1). Since the E1 ORF is well conserved between different papillomaviruses, it is likely to be expressed as a protein product, and the most probable explanation is that the true E1 mRNA has escaped detection due to its low concentration. It seems likely that future studies will unravel a more complex picture regarding BPV-1 gene expression, although the genome at first glance looks rather uncomplicated.

The functional analysis of the BPV-1 genome has just begun. Regions of the genome that are required for DNA replication and transformation have been identified by constructing deletion mutants (see Schiller et al.; Botchan et al.; both this volume). We have started to use retrovirus vectors to facilitate the functional analysis of the BPV-1 genome. The use of such vectors offers many advantages, one being that the retrovirus packaging system could be utilized to facilitate the introduction of BPV-1 genomes into a wide range of cells. While making recombinants between retrovirus vectors of the 69T fragment of BPV-1, we have noticed that the retroviral sequences have a profound effect on the BPV replication functions. When the 69T fragment is in the same transcriptional orientation as the LTRs, the recombinant replicates as a multicopy plasmid as opposed to constructs having the LTRs and the 69T fragments in opposite orientations, which consistently integrate. From an analysis of the BPV-1–specific RNA in cells transformed by the two types of recombinants, it appears that the integration is caused by antisense transcripts that presumably interfere with the E1 mRNA, believed to be present at an extremely low level. The transforming functions, however do not seem to be affected by the antisense transcripts. It has been noticed before that the BPV-1 genomes sometimes integrate when linked to foreign genes (for a review, see Di Maio et al. 1984). This has in many cases restricted the use of the BPV-1 genome as a cloning vector. When constructing BPV-1 recombinants, it is obviously important to keep in mind that the transcriptional activity of the cloned fragment may interfere with the BPV-1 functions.

## Acknowledgments

The authors are grateful to Elisabeth Sandberg and Jeanette Backman for typing the manuscript. We thank Hans Hultberg and Staffan Josephson (KabiGen AB, Stockholm) for oligonucleotides. This investigation was supported by grants from the Swedish Cancer Society, the Swedish National Board for Technical Development, and the Swedish Council for Forestry and Agricultural Research.

## References

Ahola, H., A. Stenlund, J. Moreno-Lopez, and U. Pettersson. 1983. Sequences of bovine papillomavirus type 1 DNA—Functional and evolutionary implications. *Nucleic Acids Res.* **11:** 2639.

Brawerman, G., J. Mendecki, and S.Y. Lee. 1972. A procedure for the isolation of mammalian messenger ribonucleic acid. *Biochemistry* **11:** 637.

Chen, E.Y., P.M. Howley, A.D. Levinson, and P.H. Seeburg. 1982. The primary structure and genetic organization of the bovine papillomavirus type 1 genome. *Nature* **299:** 529.

Ciuffo, G. 1907. Innesto positivo con filtrado di verrucae volgare. *G. Ital. Mal. Venereol.* **48:** 12.

Danos, O. and M. Yaniv. 1983. Structure and function of papillomavirus genomes. *Viral Oncol.* **3:** 59.

DeLeon, D.V., K.H. Cox, L.M. Angerer, and R.C. Angerer. 1983. Most early-variant histone mRNA is contained in the pronucleus of sea urchin eggs. *Dev. Biol.* **100:** 197.

Di Maio, D., V. Corbin, E. Sibley, and T. Maniatis. 1984. High level expression of a cloned HLA heavy chain gene introduced into mouse cells on a bovine papillomavirus vector. *Mol. Cell. Biol.* **4:** 340.

Favaloro, J., R. Treisman, and R. Kamen. 1980. Transcription maps of polyoma virus–specific RNA: Analysis by two-dimensional nuclease S1 gel mapping. *Methods Enzymol.* **65:** 718.

Heilman, C., L. Engel, D.R. Lowy, and P.M. Howley. 1982. Virus-specific transcription in bovine papillomavirus–transformed mouse cells. *Virology* **119:** 22.

Heilman, C.A., M.-F. Law, M.A. Israel, and P.M. Howley. 1980. Cloning of human papilloma virus genomic DNAs and analysis of homologous polynucleotide sequences. *J. Virol.* **36:** 395.

Law, M.-F., W.D. Lancaster, and P.M. Howley. 1979. Conserved polynucleotide sequences among the genomes of papilloma viruses. *J. Virol.* **32:** 199.

Lehrach, H., D. Diamond, J.M. Wozney, and H. Boedtker. 1977. RNA molecular weight determinations by gel electrophoresis under denaturing conditions, a critical reexamination. *Biochemistry* **16:** 4743.

Maniatis, T., E.F. Fritsch, and J. Sambrook. 1982. *Molecular cloning: A laboratory manual.* Cold Spring Harbor Laboratory, Cold Spring Harbor, New York.

Maxam, A.M. and W. Gilbert. 1980. Sequencing end-labeled DNA with base-specific chemical cleavages. *Methods Enzymol.* **65:** 499.

Melton, D.A., P.A. Krieg, M.R. Rebagliati, T. Maniatis, K. Zinn, and M.R. Green. 1984. Efficient *in vitro* synthesis of biologically active RNA and RNA hybridization probes from plasmids containing a bacteriophage SP6 promoter. *Nucleic Acids Res.* **12:** 7035.

Mulligan, R.C. 1983. Construction of highly transmissible mammalian cloning vehicles derived from murine reretroviruses. In *Experimental manipulation of gene expression* (ed. M. Inouye), p. 155. Academic Press, New York.

Pfister, H. 1984. Biology and biochemistry of papillomaviruses. *Rev. Physiol. Biochem. Pharmacol* **98:** 111.

Rigby, P.W.J., M. Dieckmann, C. Rhodes, and P. Berg. 1977. Labeling deoxyribonucleic acid to high specific activity *in vitro* by nick translation with DNA polymerase I. *J. Mol. Biol.* **113:** 237.

Sarver, N., J.C. Byrne, and P.M. Howley. 1982. Transformation and replication in mouse cells of a bovine papillomavirus–pML2 plasmid vector that can be rescued in bacteria. *Proc. Natl. Acad. Sci.* **79:** 7147.

Shope, R.E. 1933. Infectious papillomatosis of rabbits; with a note on the histopathology. *J. Exp. Med.* **58:** 607.

Southern, E.M. 1975. Detection of specific sequences among DNA fragments separated by gel electrophoresis. *J. Mol. Biol.* **98:** 503.

Stenlund, A., J. Zabielski, H. Ahola, J. Moreno-Lopez, and U. Pettersson. 1985. The messenger RNAs from the transforming region of bovine papillomavirus type 1. *J. Mol. Biol.* **182:** 541.

Thomas, P.S. 1980. Hybridization of denatured RNA and small DNA fragments transferred to nitrocellulose. *Proc. Natl. Acad. Sci.* **77:** 5201.

Vennström, B. and J.M. Bishop. 1982. Isolation and characterization of chicken DNA, homologous to the two putative oncogenes of avian erythroblastosis virus. *Cell* **28:** 135.

Virtanen, A. and U. Pettersson. 1985. Organization of early region 1B of human adenovirus type 2: Identification of four differentially spliced mRNAs. *J. Virol.* **54:** 383.

Weaver, R.F. and C. Weissman. 1979. Mapping of RNA by a modification of the Berk-Sharp procedure: The 5′ termini of 15S β-globin mRNA have identical map coordinates. *Nucleic Acids Res.* **7:** 1175.

Yang, Y.C., H. Okayama, and P.M. Howley. 1985. Bovine papillomavirus contains multiple transforming genes. *Proc. Natl. Acad. Sci.* **82:** 1034.

# Transcription of Human Papillomavirus Types 1 and 6

**L.T. Chow, A.J. Pelletier,\* R. Galli,[†] U. Brinckmann, M. Chin, D. Arvan,
D. Campanelli,[‡] S. Cheng, and T.R. Broker**
Biochemistry Department, University of Rochester School of Medicine, Rochester, New York 14642

Unlike other eukaryotic genes, the transcriptional promoters of cloned human papillomavirus (HPV) are not active when transiently transfected into monkey COS cells, whether or not the SV40 origin of DNA replication and/or the SV40 enhancer are present in the same recombinant DNA. To obtain HPV mRNAs for structural analysis, we have attempted to establish permanent cell lines that contain HPV DNA, based upon either morphological transformation or cotransfection with selectable drug-resistance genes. We have also constructed recombinant DNA clones containing HPV-1 or -6 coding regions placed downstream of three different surrogate promoters. Our results indicate that HPV-1, -6, and -11 do not cause morphological transformation in rodent, human, or monkey cells in culture. Furthermore, drug-resistant colonies containing HPV DNA do not express detectable amounts of viral RNA. However, viral mRNAs were successfully obtained from the expression vectors. These mRNAs have identical splice patterns irrespective of which of the promoters is used. The ratio of the various mRNA species changed with the strength of the promoters. mRNAs spanning the early region alone and the early and late regions together were observed with transfected HPV-1 clones. HPV-6 does not have a conventional polyadenylation signal at the end of the early region. We found that an alternative signal was used efficiently, for all mRNAs were polyadenylated at the end of the early region.

The human papillomaviruses (HPVs) induce hyperproliferation or dysplasia of various kind of epithelia in a tissue-specific manner. The virion contains a double-stranded, circular DNA chromosome of about 7900 bp that generally is maintained, in benign lesions, as a multicopy episome in the nucleus of an infected cell. The DNAs of over 42 viral types have been cloned and characterized on the basis of significant polynucleotide sequence heterology (see Table 1 in Broker and Botchan, this volume). Virus production appears to be coupled to tissue differentiation, since only some cells of the upper layers of the stratified epithelium in benign lesions contain the viral capsid antigen. Several types of HPVs are associated with cervical carcinomas. In those cases, virus is no longer produced and, instead, integration of viral DNA into host chromosomes is usually detected.

HPV-1 causes plantar and palmar warts, in which virus production is usually ample. The closely related pair of viruses HPV-6b and -11 cause genital and laryngeal condylomata and mild dysplasias, but virion synthesis is often negligible. The genomes of these and several other human and animal papillomaviruses have been completely sequenced (Chen et al. 1982; Danos et al. 1983; Schwarz et al. 1983; Giri et al. 1985; Groff and Lancaster 1985; Seedorf et al. 1985). Despite

*Present addresses*: \*Department of Molecular, Cellular, and Developmental Biology, University of Colorado, Boulder, Colorado 80304; [†]Department of Otolaryngology, Long Island Jewish-Hillside Medical Center, New Hyde Park, New York 10042; [‡]The Rockefeller University, New York, New York 10021.

considerable DNA sequence heterology among papillomaviruses, all exhibit a highly conserved genetic organization of potential protein-coding open reading frames (ORFs) (see Figs. 1 and 2 in Broker and Botchan, this volume). At present, there is no cell culture system available that allows the lytic propagation of any papillomavirus. This lack of a productive, in vitro system has greatly hampered studies of these viruses. However, bovine and cottontail rabbit papillomaviruses (BPV-1 and CRPV) can transform rodent cells in culture (Dvoretsky et al. 1980; Watts et al. 1983), and the viral DNA remains episomal, as in vivo (Lancaster 1981; Law et al. 1981; Groff et al. 1983; Watts et al. 1983). The mRNAs of BPV-1 and CRPV from transformed mouse cells, from lesions, and from transplantable tumors have recently been characterized by electron microscopy of R-loops (Georges et al. 1984; Stenlund et al. 1985), by S1 nuclease mapping (Heilman et al. 1982; Engel et al. 1983; Nasseri and Wettstein 1984a,b; Phelps et al. 1985), by primer extension (Danos et al. 1985; Stenlund et al. 1985), and by cDNA cloning and sequencing (Yang et al. 1985a). Furthermore, with the availability of the in vitro transformation system, it has been possible to dissect the functions of the various ORFs of BPV-1 using deletion analysis and site-directed mutagenesis (Lusky and Botchan 1984, 1985; Sarver et al. 1984; Schiller et al. 1984; DiMaio et al. 1985) and by expression of cDNAs or selected ORFs from foreign promoters (Androphy et al. 1985; Schiller et al. 1985; Spalholz et al. 1985; Yang et al. 1985a,b).

In contrast to the rapid progress in exploring the

molecular genetics of BPV-1, studies of HPVs have, for the most part, remained clinical. Human tissue specimens of sufficient size for isolation of viral gene products are rarely available, and not all of them are suitable for detailed molecular analyses. Unlike BPV or CRPV, HPVs do not readily transform rodent cells in culture, as described in this report. Consequently, there is relatively little information on the genetic expression of HPVs. In this paper, we describe our systematic investigation of the transcriptional activity of HPV-1 in different cell systems, the inability of several wild-type or genetically engineered HPVs to transform rodent cells in culture, and the successful expression of HPV-1 and HPV-6 RNAs from surrogate promoters in transient transfection experiments.

We reported previously (Chow and Broker 1984) that low levels of HPV-1 RNA transcripts can be produced in monkey COS cells (Gluzman 1981) transiently transfected with HPV-1 viral DNA cloned into shuttle vectors. These vectors contained the simian virus 40 (SV40) origin of replication (nucleotide [nt] numbers 5171–5243/1–160), but no SV40 promoter activity was expected for lack of enhancer sequences and major cap sites for early or late SV40 mRNAs. These vectors were chosen so the normal promoter activity of HPV-1 would not be obscured. The extremely low levels of HPV-1 RNAs that we detected in these experiments ( < 0.001% of total mRNA) in fact all originated from the nominally disarmed SV40 early or late promoters, as judged from the positions of the 5′ ends of the RNAs. Only one DNA strand of HPV-1 served as template for stable transcripts, and these were processed into two spliced species. When the DNA sequence of HPV-1 became available (Danos et al. 1983), we were able to align the transcripts to the ORFs and to putative regulatory signals such as consensus splice sites and the polyadenylation signals located at the 3′ ends of the E and L regions. The main body of the predominant HPV-1 RNA corresponds to that of a major species observed in mouse cells transformed by BPV-1 (Heilman et al. 1982); it spans the overlapping E4 and E2 ORFs and extends through the E5 region. The main body of the HPV-1 minor species corresponds to the L1 mRNA

encoding the major capsid antigen from bovine fibro-papillomas (Engel et al. 1983).

Having recognized that there was no detectable transcriptional activity from HPV-1 promoters in these recombinants, we set out to examine three other approaches to generating HPV RNA: (1) the production of HPV RNAs from their own promoter in transient transfection assays using other shuttle vectors; (2) the establishment of stably transformed cells lines containing HPV DNA that might be transcribed; and (3) the production of HPV RNAs from surrogate promoters in transient transfection experiments. We anticipated that one or more of these methods would generate adequate amounts of HPV RNAs to allow accurate mapping and eventual cDNA cloning so that the ORFs fused by mRNA splicing could be ascertained.

## Results and Discussion

### Additional shuttle vectors lacking SV40 promoters

We have cloned HPV-1 opened at the EcoRI site at nt 7778 (in the noncoding or long control region [LCR] situated upstream from all the ORFs), in both orientations, into shuttle vectors that have the SV40 enhancer (the 72-bp repetitions) with or without the SV40 origin of DNA replication. The SV40 21-bp repetitions known to be essential for early transcription were deleted from this class of vectors (Fig. 1A). These plasmids generated no more HPV-1 RNAs than our original constructions when transfected into COS or HeLa cells. These results demonstrate that the SV40 enhancer was inactive on the promoters of HPV-1 in these cells. This inactivity was not entirely surprising, for in vivo, active papillomavirus RNA transcription and DNA replication occur only in more-differentiated human epithelial cells.

### Cell lines containing HPV DNAs

It could not be ruled our that our inability to detect HPV promoter activity was due to an inherently low activity of HPV promoters compounded by the inefficiency of the transfection procedure, in which only a few percent of the cultured cells took up the recombinant DNA. We therefore attempted to establish cell lines that stably

**Figure 1** (*see facing page*) Transcriptionally passive shuttle vectors for HPV-1. (*A*) pSVΔpH-HPV-1:I24. HPV-1 DNA from purified viral particles recovered from plantar warts was linearized at the EcoRI site in the LCR (nt 7778) by partial digestion and inserted into a vector that was modified from pSVΔpH (Treisman et al. 1982) to include the following features: The BamHI site of pBR322 was changed to EcoRI for HPV-1 insertion in either orientation (only one is shown here); it contained two functional SV40 large-T-antigen-binding sites. The third site and the 21-bp repetitions were deleted together with a portion of one of the 72 repetitions. This modification in the SV40 control region still allowed the plasmid to replicate in COS cells, albeit to a lesser extent, but rendered the SV40 early promoter transcriptionally inactive (M. Green, pers. comm.). This plasmid still has one intact copy of the SV40 72-bp enhancer. The pBR322 "poison" sequences that inhibit replication in eukaryotic cells (Lusky and Botchan 1981) were deleted as modified by Mellon et al. (1981). In this and subsequent figures, the restriction site in parentheses indicates that it has been changed to a new site. (*B*) pBR-TK-HPV-1:N46. HPV-1 DNA from purified virus particles was cloned at the EcoRI site at nt 7778 into a vector containing a truncated herpes simplex virus thymidine kinase gene (p*tk* 2.4 of Zipser et al. 1981) in pBR322. An extra EcoRI fragment (nt 7779–967) was inserted in tandem during ligation. A clone that had HPV-1 in the opposite orientation but did not have the duplication was also obtained (not shown). In addition, HPV-1 DNA was cloned in either orientation into pBR322 in which the TK gene was in the configuration shown in *C*. (*C*) pSVΔpH-TK-HPV-1:W38. HPV-1 DNA was cloned at the EcoRI site at nt 7778 in either orientation into the vector illustrated in *A* except that it also contained a truncated TK gene (only one orientation is shown). In addition, viral DNA was cloned at the same EcoRI site in either orientation in the same basic vector except that the TK gene was in the configuration shown in *B*.

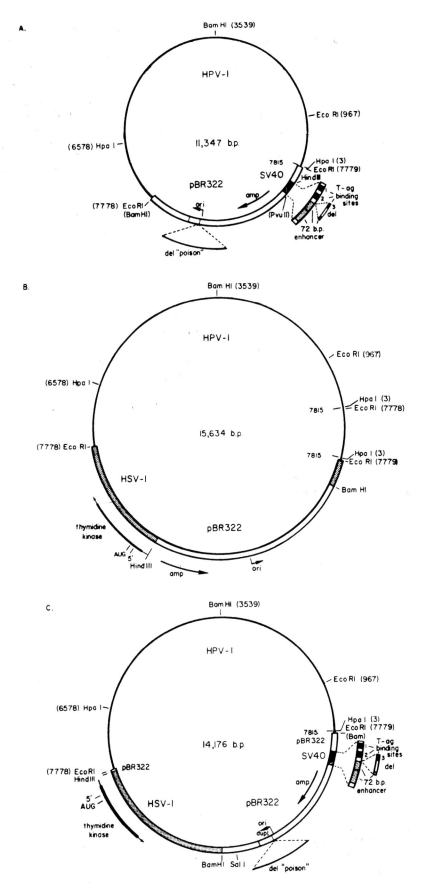

**Figure 1** *See facing page for legend.*

contain HPV DNAs, as was achieved with BPV-1 and CRPV DNAs. Several different approaches were taken, as described below.

*Morphological transformation of rodent cells in culture*
Cloned BPV-1 and CRPV DNA cause morphological transformation of mouse C127 or NIH-3T3 cells, and the DNA remains episomal (Dvoretsky et al. 1980; Law et al. 1981; Watts et al. 1983). Several of the recombinant HPV-1 DNAs in which the viral sequences were linearized and cloned at different sites located all around the chromosome were used for transfection into mouse C127 cells and into primary cultures of baby Fischer rat kidney cells following the protocols of Wigler et al. (1979) and DiMaio et al. (1982). The restriction sites in HPV-1 selected for cloning included the *Bam*HI site, both of the *Eco*RI sites, and all of the *Bgl*II sites except the one at nt 1667. Another clone contained the SV40 early promoter linked to nt 75 of HPV-1 (see below). All the vectors lacked the "poison" sequence of pBR322 (Lusky and Botchan 1981), which reduces the efficiency of replication and transformation in rodent cells, but they contained the SV40 enhancer and/or SV40 origin of replication. Rarely, we did detect small foci, usually only after more than 4 weeks of culturing. When these apparent foci were expanded and the DNA was analyzed by Southern blots, they proved to be false positives or abortive transformants, since they contained no HPV-1 DNA. These experiments were repeated several times. Occasionally we did find foci that appeared to harbor HPV-1 DNA at one or two copies per cell in an integrated state, but they produced no detectable HPV-1 mRNA. In parallel control experiments, BPV-1 readily transformed C127 cells. In addition, we also cotransfected HPV-1 and BPV-1 DNA into C127 cells. In the morphological transformants, only BPV-1 DNA stayed episomal whereas HPV-1 DNA became integrated. We therefore concluded that transformation of C127 or primary rat kidney cells with HPV-1 occurs rarely, if at all, or is irreproducible, at least under the conditions used.

*Transformation of cell lines in culture using selectable markers*
The selectable markers utilized were the thymidine kinase (TK) gene of the herpes simplex virus type 1 (HSV-1) and the neomycin (Neo)-resistance gene (*neo*) from *Escherichia coli* transposon Tn5. We assembled several types of TK-gene-containing recombinants into which HPV-1 was cloned in either orientation. One type of vector was pBR322 (Fig. 1B). The second type was a pBR322 derivative that contained the SV40 origin of replication and enhancer but had no early promoter activity (due to a deletion in the 21-bp repetitive sequences) nor the "poison" sequences of pBR322 (Fig. 1C). Viral HPV-1 was linearized by partial restriction digestion and cloned at the *Eco*RI site at nt 7778, upstream from all the major ORFs. In these recombinant DNAs, the TK gene was placed at different locations and orientations relative to the HPV-1 DNA. The HSV TK gene used was truncated at its 5' end (to nt −9). It has

been observed that such a partially deleted HSV TK gene (Zipser et al. 1981; Roberts and Axel 1982) must be amplified to hundreds of copies per cell to assure cell survival in the presence of HAT-selective medium. Apparently, a foreshortened TK protein, lacking its normal aminoterminal region, can be generated from an abbreviated message originating from a cryptic promoter located within the 5'-coding portion of the gene, and this residual TK protein has reduced specific activity, insufficient to confer HAT resistance unless present in abundance (Zipser et al. 1981; Roberts and Axel 1982). Usually, DNA sequences linked to the TK gene are coamplified. Accordingly, we anticipated that, in the TK$^+$ transformants, there would be an amplified copy number of HPV-1 DNA and perhaps increased transcription, which would facilitate mapping the viral RNA. Such HPV-1–TK recombinant plasmids were transfected into mouse L TK$^-$ and human 143 TK$^-$ cells, and cells were then selected for HAT resistance according to the procedure of Wigler et al. (1979). TK$^+$ transformants were obtained from both cell lines with all HPV-1 recombinants used. However, none were morphologically transformed. Very few TK$^+$ colonies were recovered when the plasmid contained the wild-type pBR322. The transformation efficiency was increased by at least two orders of magnitude (to about 3000 transformants/μg DNA) when the plasmids contained an SV40 enhancer and lacked the pBR322 "poison" sequences. In general, the transformation efficiency of human 143 TK$^-$ cells was about 1/10 to 1/20 of that for mouse L TK$^-$ cells. A number of TK$^+$ colonies derived from transformation by each of the constructions were expanded and the DNA was analyzed by Southern blotting. All lines contained the HPV-1 sequences in an integrated state (not shown). But surprisingly, none of the lines contained more than 10 copies of the plasmid, whether it had an SV40 enhancer or not. In some lines the DNA appeared to be integrated in tandem arrays, as there were very few junction bands. In others, the band patterns appeared more complicated, indicative of multiple integration events or rearrangement of the plasmid before or after integration. In any case, the lack of gene amplification in the presence or in the absence of an SV40 enhancer suggested that HPV-1 DNA might contain an enhancer of its own that could up-regulate the transcription of the truncated TK genes in the transformants. Enhancer elements from BPV-1 have been identified that stimulate transcription of HSV TK or chloramphenicol acetyltransferase (CAT) genes in *cis* (Lusky et al. 1983; Spalholz et al. 1985).

In addition to HPV-TK recombinants, we also constructed two recombinant plasmids that contained HPV type-6b DNA in either orientation relative to the expression cassette (obtained from D. Hanahan) of the bacterial *neo* gene from the *E. coli* transposon Tn5 (Beck et al. 1982). First, full-length HPV-6b was reassembled from the two *Eco*RI-*Bam*HI restriction fragments originally received from Dr. Lutz Gissmann. The reconstructed HPV-6 was then cloned at the *Bam*HI site (located in its L2 ORF) into the vector that was deleted for the pBR322 "poison" sequence (Fig. 2). Recombinants were transfected into C127 cells. In addition HPV-11, which is highly homologous to HPV-6, was also cloned at its *Bam*HI site (located in the L1 region) into a modified pBR322 vector lacking the "poison" sequences. This was cotransfected with the Neo-resistance plasmid into Rat-4 cells. Two clones in which the SV40 promoters were placed at nt 75 of HPV-1, upstream of all the ORFs (cf. Figs. 4 and 5A), were also transfected into monkey CV-1 cells together with the Neo-expression plasmid. Clones of HPV-1 driven by the SV40 early promoter were also cotransfected with the Neo-resistance plasmid into the BPV-1–transformed C127 cell line called ID14 (Dvoretsky et al. 1980). In each case, cells were selected with G418, and a number of surviving colonies were obtained. None was morphologically transformed (except for ID14). They were picked at 3 weeks posttransfection and expanded, and total DNA was harvested and analyzed by Southern blotting. All contained integrated HPV-1, -6, or -11 DNA at low copy numbers (data not shown).

Several TK$^+$ or G418-resistant colonies were harvested for cytoplasmic RNA. HPV-specific RNA was not detected by dot-blots or by electron microscopic searches for RNA:DNA heteroduplexes, which would reveal HPV mRNA if present in greater than 0.001% of mRNA.

In summary, it appears that, in our hands, neither HPV-1, HPV-6, nor HPV-11 could transform the rodent, monkey, or human cells that we tested, whether the HPV DNA was cloned in the LCR or in the L region. The inability to transform these cells could be due to inadequate gene expression and/or to the incompatibility of the transforming proteins with the cells. The latter is most likely the case for CV-1 cells since the clone containing the SV40 promoter did not transform CV-1

cells even though it appeared to generate properly spliced mRNAs when transfected into COS cells (see below). Watts et al. (1984) have reported morphological transformation of mouse C127 cells by a *Bam*HI clone of HPV-1 in pBR322, as well as by HPV-5 cloned in pBR322. We have used a similar HPV-1 clone, in addition to others, without success. At this point, we do not know the basis for this discrepancy. Potentially, it could be due to a difference in the HPV-1 isolates or to subtle variations in the cell lines or procedures of transfection.

Our results with the selectable markers are also in contrast to observations made with BPV-1, in which BPV-recombinant DNAs remain episomal (Law et al. 1983; Lusky et al. 1983; Matthias et al. 1983). We believe these differences are due to the compatibility of the host replication and transcription factors with BPV-1, but not with the HPVs studied. BPV-1, which infects both epithelial cells and fibroblasts of either bovine or equine origin, has a wider tissue and host specificity than HPVs, which infect only certain types of human epithelium. The integration of HPV-1–TK recombinant plasmids and the lack of morphological transformation of TK$^+$ transformants have also been reported recently by Burnett and Gallimore (1985), who used rat fibroblasts as recipient cells.

*Expression of HPV-1 RNA from surrogate promoters*
Because of the inability of HPV-1 to generate RNA from its own promoter to any detectable level (<0.001% of total mRNA) in rodent, monkey, or human cells, we next chose to insert strong foreign promoters into the HPV DNA. We reasoned that active mRNA synthesis would depend on a promoter matched to the host cell transcription factors. The placement of the foreign promoter close to the position of a natural HPV promoter should yield a nearly normal primary transcript, and post-

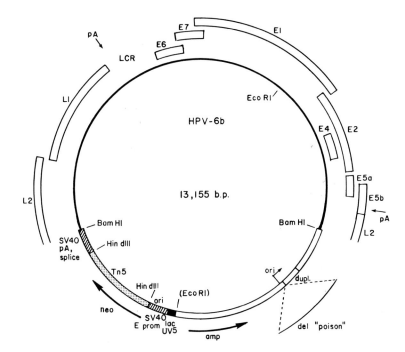

**Figure 2** HPV-6b clone in a vector containing a general selectable marker (pKO-*neo*-M-HPV-6b:S). The basic vector pKO-*neo* (obtained from Dr. D. Hanahan) was modified to be pBR322 "poison"–minus as described in Fig. 1. The modified vector contains the SV40 origin of DNA replication and the early promoter, enhancer, and SV40 small-T-antigen splice site and polyadenylation signal for the expression of the neomycin-resistance gene from *E. coli* transposon Tn*5* (Beck et al. 1982). The two HPV-6b *Bam*HI-*Eco*RI fragments (received from L. Gissmann) were reassembled, then cloned at the *Bam*HI site of the vector in either orientation (only one is shown here).

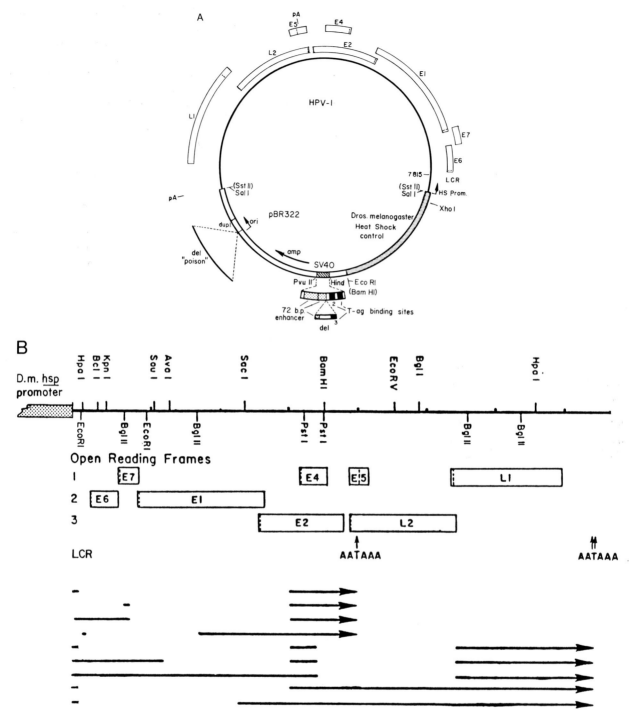

**Figure 3** HPV-1 mRNA generated from a heat-shock promoter. (*A*) Structure of plasmid pF1-I24-HPV-1. The *Sst*II site (at nt 7632) of a *Bam*HI clone of HPV-1 was changed to be a *Sal*I site. The two *Bam*HI-*Sal*I restriction fragments of HPV-1 were then excised out of the vector, religated at the *Bam*HI site, and inserted at the *Sal*I site of the vector pF1-I24 just downstream of the *Drosophila melanogaster* hsp70 promoter (at nt −10 for RNA initiation). Transcription would initiate in HPV sequences. The vector pF1 (Pelham 1982) was modified to contain pBR322 and SV40 sequences as described in Fig. 1A. (*B*) mRNAs derived from pF1-I24-HPV-1. The recombinant DNA was transfected into COS-1 cells and allowed to undergo replicative amplification. Forty hours after transfection, they were heat-shocked at 43°C for 5 hrs. Recovered cytoplasmic RNAs were mapped by electron microscopic heteroduplex analysis after hybridization to a linearized form of the same HPV-1 DNA recombinant used for transfection. The ORFs of HPV-1 are indicated beneath the restriction map. The RNA species observed are represented by interrupted arrows, where gaps between each segment of any one line represent intervening sequences spliced out of the RNA. The 5′ ends of the RNAs are to the left and the 3′ ends are indicated by arrowheads. The major transcripts are the first and fifth species shown. Some E-region species apparently arose from an HPV promoter located near the 3′ end of the E7 ORF. Some of the species that were not seen in mRNA preparations derived from SV40 early and late promoters were most likely RNA processing intermediates that accumulated during the long heat-shock treatment and leaked into the cytoplasm.

transcriptional processing such as 3' cleavage, poly-adenylation, and RNA splicing should follow their natural course as dictated by the signals in the RNA sequence. We cloned HPV-1 into several eukaryotic expression vectors; each had the SV40 origin of DNA replication and the SV40 transcriptional enhancer but lacked the "poison" sequences present in pBR322. One expression vector incorporated the promoter for the 70-kD *Drosophila melanogaster* heat-shock protein (hsp70), which has been shown to function in COS cells (Pelham 1982) (Fig. 3A). A second utilized the SV40 early promoter (nt 272–1/5243–5171) (Fig. 4). A third employed the SV40 late promoter (nt 2533–5243/1–346) (Fig. 5) (Tooze 1981). The cloning site for HPV-1 was at nt 75 for placement of all three promoters, immediately upstream from the E6 ORF, or at nt 7632 (for the hsp70 promoter), or at nt 7778 (for the SV40 early promoter) in the LCR. When SV40 promoters were used, RNA initiation was expected to be within SV40 sequences. When the *Drosophila* hsp70 promoter was used, the anticipated 5' ends would be within HPV-1 sequences, since the cloning site for the promoter segment was at nt −10, relative to the hsp70 mRNA cap site.

Each of the HPV-1 clones was introduced into COS cells by the calcium phosphate precipitate procedure (Mellon et al. 1981) and allowed to undergo replicative amplification using the SV40 origin of replication in the presence of the SV40 large T antigen. Cells receiving the HPV-1 plasmid with the hsp70 promoter were heat-shocked at 43°C for the 5 hr prior to harvesting, as described by Pelham (1982). To help stabilize and accumulate mRNAs, the translational inhibitor aniso-mycin was added 3 hr before harvesting, and cytoplasmic RNA was isolated at different times over the interval

from 24 to 74 hr posttransfection. The recovered RNAs were hybridized to the linearized HPV-1 DNA clone that was used for each transfection, and the resulting R-loops were examined in the electron microscope. All of the HPV-1 RNAs were spliced. Most of the RNAs indeed originated from the surrogate promoters, as judged from the positions of the 5' ends of the mRNAs. Occasionally, mRNAs appeared to be derived from an HPV-1 promoter, for their 5' ends were in the overlapping E7/E1 ORF (cf. Figs. 3B, 5B, and 6). This would suggest that active transcription of the viral DNA from the foreign promoter could facilitate initiation from otherwise inactive promoters.

When the SV40 early promoter or the *Drosophila* hsp70 promoter was used to drive transcription, HPV-1–specific RNAs ranged from 0.01% to 0.05% of the total cytoplasmic mRNAs from the cultures (in which only a few percent of the cells took up DNA). Ninety percent of the HPV-1 RNAs were polyadenylated at the E region poly(A) site near nt 4000. The remaining 10% contained sequences derived from both the E and L regions and were polyadenylated at the L-region poly(A) site near nt 7400 (Fig. 3B). These were essentially the same HPV species we reported earlier (Chow and Broker 1984). Specifically, the RNA had a short 5'-leader segment of SV40 and HPV-1 sequences (when derived from the SV40 promoter) or HPV-1 sequences (from the hsp70 promoter). In the major species, the leader was spliced to the E2/E4 ORF overlapping region, and the main body of the RNA was polyadenylated at the end of the E region. In the minor species, the leader was spliced to a shorter exon from the E2/E4 overlapping region and then spliced again to the 5' end of the L1 ORF, with polyadenylation at the end of the L region. The RNA

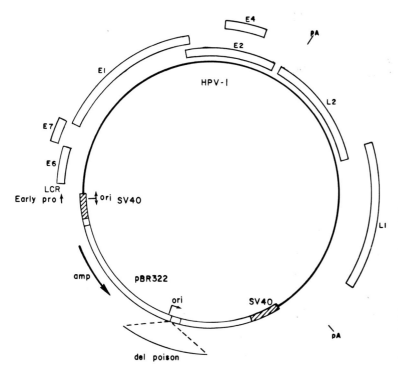

**Figure 4** HPV-1 cloned in expression vectors with an SV40 early promoter. The SV40 promoter was placed either at the *Eco*RI site of HPV-1 at nt 7778 or, alternatively, at nt 75 via an *Xho*I linker after BAL-31 nuclease deletion between nt 7633 (*Sst*II site) (through nt 7815) to nt 75, immediately upstream of the E6 ORF. The SV40 control region included nt 5171 through the origin to nt 272. The "poison" sequences of pBR322 were absent. mRNAs species recovered from transfected COS cells were virtually identical with those shown in Fig. 5, except that the major mRNA from the early promoter is the fourth species depicted.

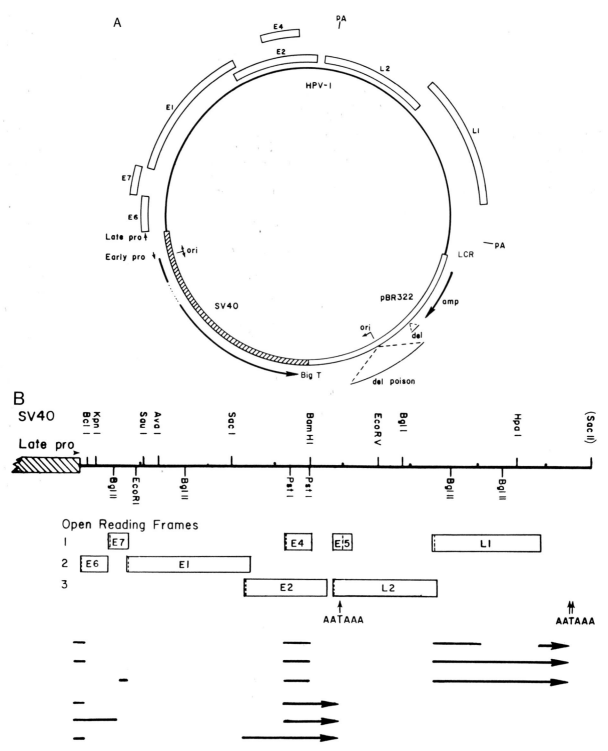

**Figure 5** Transcriptional map of HPV-1 RNAs from an SV40 late promoter. (*A*) Structure of pPX$_L$-HPV-1 plasmid. HPV-1 was cloned downstream of the SV40 late promoter in a vector called pPX$_L$ (obtained from J. Sambrook and M.J. Gething). The cloning site for HPV-1 was at nt 75, immediately upstream from the E6 ORF. Vector pPX$_L$ contains the SV40 sequence between its *Bam*HI and *Hpa*II restriction sites. It has the early region encoding the T antigen and the origin of DNA replication and is carried in a modified pBR322, pXf$_3$ (Maniatis et al. 1982), lacking the ''poison'' sequences. (*B*) mRNAs derived from pPX$_L$-HPV-1 plasmids. The cloned HPV-1 DNA was introduced into COS cells by the calcium phosphate precipitation procedure. Cytoplasmic RNA was extracted 24–72 hr posttransfection and hybridized to pPX$_L$ linearized with *Sal*I (immediately downstream of the SV40 T antigen gene). The RNA-DNA heteroduplexes (or R-loops) were then examined by electron microscopy. Most of the RNAs originated from the SV40 late promoter and had a short SV40 leader. The major transcript is the second species depicted. Minor species started from within the HPV-1 sequences at the E7/E1 overlapping region. Essentially all the species were also seen in RNA preparations derived from the SV40 early promoter or the *Drosophila melanogaster* hsp70 promoter (see Fig. 3B).

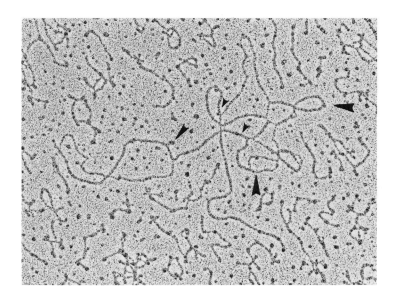

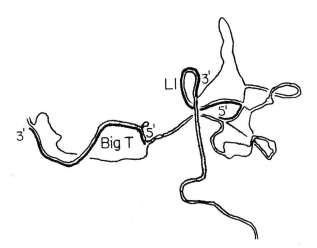

**Figure 6** Electron micrograph of an R-loop with an L1 mRNA recovered from COS cells transfected with a pPX_L-HPV-1 plasmid HPV-1 clone containing an SV40 late promoter (see Fig. 2). R-loop preparations were made as described in Fig. 5B. The L1 mRNA was derived from an HPV-1 promoter located in the E7 ORF; it had a short leader from the E7/E1 overlapping region spliced to a segment in the E2/E4 ORF overlap region and then to the L1 ORF. The 5′ and 3′ ends of the RNA (small arrowheads) are indicated in the tracing. The DNA loops corresponding to intron sequences are indicated on the micrograph with large arrowheads. A schematic representation of this molecule is presented Fig. 4, species 3. An SV40 large-T-antigen mRNA can also be seen hybridized near one end of the DNA, and its splice loop is marked with a medium arrowhead.

containing both E- and L-region sequences is most unusual, and comparable structures have not been reported before in other viral systems. The increased abundance of HPV-specific RNA generated from these vectors led to the detection of new RNA species. These include examples that contained a much longer 5′ exon covering the entire E6-E7 region, which was then spliced to the E2/E4 region. Another had a short 5′ leader spliced to the beginning of the E2 ORF (Fig. 5B). Both forms were polyadenylated solely near nt 4000.

These RNA species were also seen in preparations generated from the SV40 late promoter (Fig. 5B). In this case, the HPV-1 mRNAs were highly abundant and were estimated to be around 0.1–0.5% of the total mRNA pool and therefore must have been dominant species within the small percentage of successfully transfected cells. Curiously, the relative amounts of the various HPV E- and L-region RNA species were reversed relative to RNAs derived from the SV40 early promoter: More than 90% of the viral mRNA was that form presumed to encode the L1 capsid protein and contained both E and L segments (Figs. 5B and 6). There are several possible

explanations for this change of ratio between L and E RNAs. One is that a strong promoter attracts many more RNA polymerases than a relatively weak promoter. These polymerases pile up on the template and allow the readthrough of the upstream polyadenylation signals much more frequently than when there are relatively few polymerases on the template. The second is that one of the early transcripts encodes a viral product that modulates the cleavage and polyadenylation at the upstream E-region site and allows the virus to progress into a simulated late phase of gene expression. In RNAs derived from all three types of expression vectors, practically no transcripts of E1 or L2 were seen, although the primary transcripts span one or both regions.

*Expression of HPV-6b from surrogate promoters*

The SV40 early promoter and the *D. melanogaster* hsp70 promoter were used to produce HPV-6 mRNA after transfection of COS cells, with expression vectors comparable to those made with HPV-1, as described in Figures 4 and 7 (see also Fig. 8). All of the mRNAs were apparently polyadenylated at the end of the E region

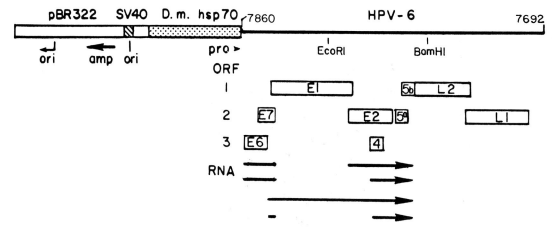

**Figure 7** HPV-6b expression vector and mRNAs. Plasmid pF1-I24-HPV-6b:I3 was constructed as follows: HPV-6b was placed downstream of the promoter for the *Drosophila melanogaster* hsp70 (as described in the legend to Fig. 2) at nt 7859 via an *Sal*I linker after BAL-31 deletion of nt 7693–7859. Cytoplasmic RNAs were recovered 48 hr posttransfection of COS cells after heat-shock at 43°C and were mapped by electron microscopic heteroduplex analysis. The third and fourth species shown seemed to have arisen from an HPV-6 promoter. Determination of the 3′ ends of the mRNAs apparently was in response to an unconventional signal at the end of the E region, since no consensus AAUAAA sequence is present in the nucleotide sequence. An SV40 early promoter placed at nt 378 generated only the species with E6/E7 spliced to E4 (similar to the second species in the diagram). No mRNA covering the L region was seen with either expression vector.

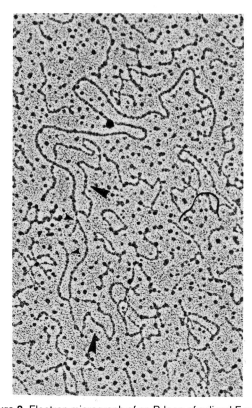

**Figure 8** Electron micrograph of an R-loop of spliced E-region RNA from COS cells transfected with pF1-I24-HPV-6:I3. R-loops were prepared with cytoplasmic RNA obtained after heat-shock as described in the legend to Fig. 7, using pF1-I24-HPV-6b:I3 DNA linearized with *Xho*I, which cuts 180 bp upstream of the cloning site near E6 (cf. Fig. 3A). The mRNA had the entire E6-E7 region spliced to E4 and it terminated around nt 4400. Small arrowheads indicate the 5′ (near one end of the heteroduplex molecule) and 3′ ends of the RNA. Large arrowheads point to the intron and the displaced single-stranded DNA.

since an unhybridized tail was always seen in the heteroduplexes. Interestingly, there is no conventional AAUAAA signal, nor variants that are known to function in other transcription units, in the vicinity.

No mRNA from the L region was found at all, suggesting that the 3′ determination signal recognized in COS cells was "stronger" than the consensus sequence used by virtually all eukaryotic mRNAs. No RNA was seen to end at either of the two AAUAAA sequences found in the HPV-6b genome (one in E6, one in E1), suggesting they strictly serve other (protein-coding) roles.

## Conclusion

One might question the validity of our approach of using surrogate promoters in a heterologous cell system. Several lines of evidence suggest that the RNA patterns we observed in vitro are authentic for the viruses (except, of course, at the promoter-proximal 5′ end of the RNA). First, the structures of mRNAs correlate well with the ORFs and the processing signals (polyadenylation, consensus splicing sequences). Second, the mRNAs covering the E region for HPV-1 and -6b are very similar to those seen in BPV-1–transformed C127 cells (Yang et al. 1985a). BPV-1 is organized into similar patterns of ORFs, but it does not share a great deal of sequence homology with HPV-1 or -6. Third, in a collaboration with S. Reilly and L. Taichman, we have recently examined HPV-1 RNA recovered from human plantar warts and from primary cultures of human foreskin keratinocytes infected with HPV-1 viruses (L. Chow et al., unpubl.). Identical RNA splicing patterns were observed. In the latter cases, the 5′ ends of the HPV-1 mRNAs were either in E7 where the minor species of RNA initiated in transfected cells or were very close to the cloning sites used in our expression vectors. Furthermore, our recent studies of mRNAs and cDNA derived from an HPV-11–associated genital condyloma also gave quite similar mRNA species (M. Nasseri et al., unpubl.).

These results strongly suggest that our expression vectors mimic natural transcription and will be able to generate moderate amounts of mRNAs in COS cells. The studies of HPV mRNAs will no longer be limited by the availability, quality, and quantity of human tissue specimens. Production of derivative reagents such as cDNAs, proteins expressed from cDNAs, and antibodies generated against expressed proteins should become possible, leading to probes for the study of HPV gene expression in vivo.

## Acknowledgments

This project was initiated while some of us were at Cold Spring Harbor Laboratory. The work was supported by research grants from the National Cancer Institute (CA-36200), the American Cancer Society (MV-744), and the Council for Tobacco Research-USA (No. 7589) to L.T.C. and T.R.B., an NCI Program Project grant (CA-13106) to Cold Spring Harbor Laboratory, and a fellowship (DGF Br863/1-1) to U.B. M.C. is a recipient of an MSTP grant (T32-GM07356) from the National Institutes of Health to the University of Rochester School of Medicine and Dentistry.

We thank our colleagues Drs. Michael Green, David Zipser, Douglas Hanahan, Joe Sambrook, and Mary Jane Gething for sharing vectors and Drs. Lutz Gissmann and Harold zur Hausen for providing cloned HPV-6b and -11 DNAs.

## References

Androphy, E.J., J.T. Schiller, and D.R. Lowy. 1985. Identification of the protein encoded by the E6 transforming gene of bovine papillomavirus. *Science* **230**: 442.

Beck, E., G. Ludwig, E.A. Auerswald, B. Reiss, and H. Schaller. 1982. Nucleotide sequence and exact localization of the neomycin phosphotransferase gene from transposon Tn5. *Gene* **19**: 327.

Burnett, T.S. and P.H. Gallimore. 1985. Introduction of cloned human papillomavirus 1a DNA into rat fibroblasts: Integration, de novo methylation and absence of cellular morphological transformation. *J. Gen. Virol.* **66**: 1063.

Chen, E.Y., P.M. Howley, A.D. Levinson, and P.H. Seeburg. 1982. The primary structure and genetic organization of the bovine papillomavirus type 1 genome. *Nature* **299**: 529.

Chow, L.T. and T.R. Broker. 1984. Human papilloma virus type 1 RNA transcription and processing in COS-1 cells. *Prog. Cancer Res. Ther.* **30**: 125.

Danos, O., E. Georges, G. Orth, and M. Yaniv. 1985. Fine structure of the cottontail rabbit papillomavirus mRNAs expressed in the transplantable Vx2 carcinoma. *J. Virol.* **53**: 735.

Danos, O., L.W. Engel, E.Y. Chen, M. Yaniv, and P.M. Howley. 1983. A comparative analysis of the human type 1a and bovine type 1 papillomavirus genomes. *J. Virol.* **46**: 557.

DiMaio, D., R. Treisman, and T. Maniatis. 1982. Bovine papillomavirus vector that propagates as a plasmid in both mouse and bacterial cells. *Proc. Natl. Acad. Sci.* **79**: 4030.

DiMaio, D., J. Metherall, K. Neary, and D. Guralski. 1985. Genetic analysis of cell transformation by bovine papillomavirus. *UCLA Symp. Mol. Cell. Biol. New Ser.* **32**: 437.

Dvoretzky, I., R. Shober, S.K. Chattopadhyay, and D.R. Lowy. 1980. A quantitative in vitro focus assay for bovine papilloma virus. *Virology* **103**: 369.

Engel, L.W., C.A. Heilman, and P.M. Howley. 1983. Transcriptional organization of the bovine papillomavirus type 1. *J. Virol.* **47**: 516.

Georges, E., O. Croissant, N. Bonneaud, and G. Orth. 1984. Physical state and transcription of the genome of the cottontail rabbit papillomavirus in the warts and in the transplantable Vx2 and Vx7 carcinomas of the domestic rabbit. *J. Virol.* **51**: 530.

Giri, I., O. Danos, and M. Yaniv. 1985. Genomic structure of the cottontail rabbit (Shope) papillomavirus. *Proc. Natl. Acad. Sci.* **82**: 1580.

Gluzman, Y. 1981. SV40-transformed simian cells support the replication of early SV40 mutants. *Cell* **23**: 175.

Groff, D.E. and W.D. Lancaster. 1985. Molecular cloning and nucleotide sequence of deer papillomavirus. *J. Virol.* **56**: 85.

Groff, D.E., J.P. Sundberg, and W.D. Lancaster. 1983. Extrachromosomal deer fibromavirus DNA in deer fibromas and virus-transformed mouse cells. *Virology* **131**: 546.

Heilman, C.A., L. Engel, D.R. Lowy, and P.M. Howley. 1982. Virus-specific transcription in bovine papillomavirus-transformed mouse cells. *Virology* **119**: 22.

Lancaster, W.D. 1981. Apparent lack of integration of bovine papillomavirus DNA in virus-induced equine and bovine tumor cells and virus-transformed mouse cells. *Virology* **108**: 251.

Law, M.F., J.C. Byrne, and P.M. Howley. 1983. A stable bovine papillomavirus hybrid plasmid that expresses a dominant selective trait. *Mol. Cell. Biol.* **3**: 2110.

Law, M.F., D.R. Lowy, I. Dvoretzky, and P.M. Howley. 1981. Mouse cells transformed by bovine papillomavirus contain only extrachromosomal viral DNA sequences. *Proc. Natl. Acad. Sci.* **78**: 2727.

Lusky, M. and M. Botchan. 1981. Inhibition of SV40 replication of simian cells by specific pBR322 DNA sequences. *Nature* **293**: 79.

———. 1984. Characterization of the bovine papilloma virus plasmid maintenance sequences. *Cell* **36**: 391.

———. 1985. Genetic analysis of bovine papillomavirus type 1 *trans*-acting replication factors. *J. Virol.* **53**: 955.

Lusky, M., L. Berg, H. Weiher, and M. Botchan. 1983. Bovine papilloma virus contains an activator of gene expression at the distal end of the early transcription unit. *Mol. Cell. Biol.* **3**: 1108.

Maniatis, T., E.F. Fritsch, and J. Sambrook. 1982. *Molecular cloning: A laboratory manual.* Cold Spring Harbor Laboratory, Cold Spring Harbor, New York.

Matthias, P.D., H.U. Bernard, A. Scott, G. Brady, T. Hashimoto-Gotoh, and G. Schotz. 1983. A bovine papilloma virus vector with a dominant resistance marker replicates extrachromosomally in mouse and *E. coli* cells. *EMBO J.* **2**: 1487.

Mellon, P., V. Parker, Y. Gluzman, and T. Maniatis. 1981. Identification of DNA sequences required for transcription of the human α1-globin gene in a new SV40 host-vector system. *Cell* **27**: 279.

Nasseri, M. and F. Wettstein. 1984a. Differences exist between viral transcripts in cottontail rabbit papillomavirus-induced benign and malignant tumors as well as non-virus-producing and virus-producing tumors. *J. Virol.* **51**: 706.

———. 1984b. Cottontail rabbit papillomavirus-specific transcripts in transplantable tumors with integrated DNA. *Virology* **138**: 362.

Pelham, H.R.B. 1982. A regulatory upstream promoter element in the *Drosophila hsp*70 heat-shock gene. *Cell* **30**: 517.

Phelps, W.C., S.L. Leary, and A.J. Faras. 1985. Shope papillomavirus transcription in benign and malignant rabbit tumors. *Virology* **146**: 120.

Roberts, J.M. and R. Axel. 1982. Amplification and correction of transformed genes. In *Gene amplification* (ed. R.T. Schimke), p. 251. Cold Spring Harbor Laboratory, Cold Spring Harbor, New York.

Sarver, N., M.S. Rabson, Y.C. Yang, J.C. Byrne, and P.M. Howley. 1984. Localization and analysis of bovine papillomavirus type 1 transforming functions. *J. Virol.* **52:** 377.

Schiller, J.T., W.C. Vass, and D.R. Lowy. 1984. Identification of a second transforming region in bovine papillomavirus DNA. *Proc. Natl. Acad. Sci.* **24:** 7880.

Schiller, J.T., E.J. Androphy, W.C. Vass, and D.R. Lowy. 1985. The bovine papillomavirus E6 virus: Identification of its transforming function and protein product. *UCLA Symp. Mol. Cell. Biol. New Ser.* **32:** 457.

Schwarz, E., M. Durst, C. Demankowski, O. Lattermann, R. Zech, E. Wolfsperger, S. Suhai, and H. zur Hausen. 1983. DNA sequence and genome organization of genital human papillomavirus type 6b. *EMBO J.* **2:** 2341.

Seedorf, K., G. Krammer, M. Durst, S. Suhai, and W. Rowekamp. 1985. Human papillomavirus type 16 DNA sequence. *Virology* **145:** 181.

Spalholz, B.A., Y.C Yang, and P.M. Howley. 1985. *Trans*-activation of a bovine papilloma virus transcriptional regulatory element by the E2 gene product. *Cell* **42:** 183.

Stenlund, A., J. Zabielski, H. Ahola, J. Moreno-Lopez, and U. Pettersson. 1985. Messenger RNAs from the transforming region of bovine papilloma virus type 1. *J. Mol. Biol.* **182:** 541.

Tooze, J., ed. 1981. *Molecular biology of tumor viruses*, 2nd edition, revised: *DNA tumor viruses*. Cold Spring Harbor Laboratory, Cold Spring Harbor, New York.

Treisman, R., B. Seed, P. Little, M. Green, N. Proudfoot, and T. Maniatis. 1982. An approach to the analysis of the structure and expression of mutant globin genes. In *Eukaryotic viral vectors* (ed. Y. Gluzman), p. 63. Cold Spring Harbor Laboratory, Cold Spring Harbor, New York.

Watts, S.L., R.S. Ostrow, W.C. Phelps, J.T. Prince, and A.J. Faras. 1983. Free cottontail rabbit papillomavirus DNA persists in warts and carcinomas of infected rabbits and in cells in culture transformed with virus or viral DNA. *Virology* **125:** 127.

Watts, S.L., W.C. Phelps, R.S. Ostrow, K.R. Zachow, and A.J. Faras. 1984. Cellular transformation by human papillomavirus DNA *in vitro*. *Science* **225:** 634.

Wigler, M., A. Pellicer, S. Silverstein, R. Axel, G. Urlaub, and L. Chasin. 1979. DNA-mediated transfer of the adenine phosphoribosyltransferase locus into mammalian cells. *Proc. Natl. Acad. Sci.* **76:** 1373.

Yang, Y., H. Okayama, and P.M. Howley. 1985a. Bovine papillomavirus contains multiple transforming genes. *Proc. Natl. Acad. Sci.* **82:** 1030.

Yang, Y.C., B.A. Spalholz, M.S. Rabson, and P.M. Howley. 1985b. Dissociation of transforming and *trans*-activation functions for bovine papillomavirus type 1. *Nature* **318:** 575.

Zipser, D., L. Lipsich, and J. Kwoh. 1981. Mapping functional domains in the promoter region of the herpes thymidine kinase gene. *Proc. Natl. Acad. Sci.* **78:** 6276.

# Author Index

# Subject Index